ENCYCLOPEDIA OF

# CONTROLLED DRUG DELIVERY

VOLUME 1

# ENCYCLOPEDIA OF CONTROLLED DRUG DELIVERY

# ENCYCLOPEDIA OF

# CONTROLLED DRUG DELIVERY

## VOLUME 1

**Edith Mathiowitz**
Brown University
Providence, Rhode Island

A Wiley-Interscience Publication
**John Wiley & Sons, Inc.**
New York / Chichester / Weinheim / Brisbane / Singapore / Toronto

For ordering and customer service, call 1-800-CALL-WILEY.

*Library of Congress Cataloging-in-Publication Data:*

Mathiowitz, Edith, 1952–
    Encyclopedia of controlled drug delivery / Edith Mathiowitz.
      p. cm.
    Includes index.
    ISBN 0-471-14828-8 (set : cloth : alk. paper).—ISBN
0-471-16662-6 (vol 1 : alk. paper).—ISBN 0-471-16663-4 (vol 2 :
alk. paper)
    1. Drugs—Controlled release Encyclopedias.  I. Title.
    [DNLM: 1. Drug Delivery Systems Encyclopedias—English.  2. Drug
Carriers Encyclopedias—English.  QV 13 M431e 1999]
RS201.C64M38 1999
615.7—dc21
DNLM/DLC
for Library of Congress                    .                    99-24907
                                                                CIP

Printed in the United States of America.

10 9 8 7 6 5 4 3 2 1

To my loving husband, George
To my dear children, Daphne and Ariel

# CONTRIBUTORS

**Patrick Aebischer,** *University of Lausanne Medical School, Lausanne, Switzerland,* Immunoisolated Cell Therapy

**V. Baeyens,** *University of Geneva, Geneva, Switzerland,* Mucosal Drug Delivery, Ocular

**Roland Bodmeier,** *Freie Universität Berlin, Berlin, Germany,* Nondegradable Polymers for Drug Delivery

**Nicholas Bodor,** *University of Florida, Gainesville, Florida,* Chemical Approaches to Drug Delivery

**Clarissa Bonnano,** *Health Advances, Inc., Wellesley, Massachusetts,* Economic Aspects of Controlled Drug Delivery

**Lisa Brannon-Peppas,** *Biogel Technology, Indianapolis, Indiana,* Microencapsulation

**Henry Brem,** *Johns Hopkins University School of Medicine, Baltimore, Maryland,* Central Nervous System, Drug Delivery to Treat

**Steve Brocchini,** *University of London, London, United Kingdom,* Pendent Drugs, Release from Polymers

**P. Buri,** *University of Geneva, Geneva, Switzerland,* Mucosal Drug Delivery, Ocular

**Martin Burke,** *Johns Hopkins University School of Medicine, Baltimore, Maryland,* Central Nervous System, Drug Delivery to Treat

**Paul A. Burke,** *Amgen, Inc., Thousand Oaks, California,* Characterization of Delivery Systems, Magnetic Resonance Techniques

**Gerardo P. Carino,** *Brown University, Providence, Rhode Island,* Vaccine Delivery

**F.J. Lou Carmichael,** *Hemosol, Inc., Toronto, Canada,* Blood Substitutes: A Review of Clinical Trials

**Pravin R. Chaturvedi,** *Vertex Pharmaceuticals, Cambridge, Massachusetts,* Pharmacokinetics

**Donald Chickering,** *Brown University, Providence, Rhode Island,* Bioadhesive Drug Delivery Systems

**Kadriye Ciftci,** *Temple University, Philadelphia, Pennsylvania,* Veterinary Applications

**Paolo Colombo,** *University of Parma, Parma, Italy,* Mucosal Drug Delivery, Nasal

**Nora E. Cutcliffe,** *Hemosol, Inc., Toronto, Canada,* Blood Substitutes: A Review of Clinical Trials

**Wenbin Dang,** *Guilford Pharmaceuticals, Baltimore, Maryland,* Biodegradable Polymers: Poly(phosphoester)s; Fabrication of Controlled-Delivery Devices

**M.C. Davies,** *University of Nottingham, Nottingham, United Kingdom,* Characterization of Delivery Systems, XPS, SIMS and AFM Analysis

**Fanny De Jaeghere,** *University of Geneva, Geneva, Switzerland,* Nanoparticles

**Arati A. Deshpande,** *University of Geneva, Geneva, Switzerland,* Mucosal Drug Delivery, Intravitreal

**Eric Doelker,** *University of Geneva, Geneva, Switzerland,* Nanoparticles

**Ruth Duncan,** *University of London, London, United Kingdom,* Pendent Drugs, Release from Polymers

**Elazer R. Edelman,** *Massachusetts Institute of Technology, Cambridge, Massachusetts,* Cardiovascular Drug Delivery Systems

**David A. Edwards,** *Advanced Inhalation Research (AIR), Cambridge, Massachusetts,* Respiratory System Delivery

**Jonathan D. Eichman,** *University of Michigan Medical School, Ann Arbor, Michigan,* Mucosal Drug Delivery, Vaginal Drug Delivery and Treatment Modalities

**Suzanne Einmahl,** *University of Geneva, Geneva, Switzerland,* Mucosal Drug Delivery, Intravitreal

**James P. English,** *Absorbable Polymer Technologies, Pelham, Alabama,* Fabrication of Controlled-Delivery Devices

**O. Felt,** *University of Geneva, Geneva, Switzerland,* Mucosal Drug Delivery, Ocular

**Gregory T. Fieldson,** *ALZA Corporation, Palo Alto, California,* Characterization of Delivery Systems, Spectroscopy

**Joseph Fix,** *Yamanouchi Shaklee Pharmaceutical Research Center, Palo Alto, California,* Oral Drug Delivery, Small Intestine & Colon

**Virginia Fleming,** *ALZA Corporation, Palo Alto, California,* Pumps/Osmotic—ALZET® System

**Achim Göpferich,** *University of Regensburg, Regensburg, Germany,* Biodegradable Polymers: Polyanhydrides

**Robert Gale,** *ALZA Corporation, Palo Alto, California,* Transdermal Drug Delivery, Passive

**A. Gerson Greenburg,** *The Miriam Hospital and Brown University, Providence, Rhode Island,* Blood Substitutes: A Review of Clinical Trials

**Gregory Gregoriadis,** *University of London School of Pharmacy, London, United Kingdom,* Liposomes

**M.J. Groves,** *University of Illinois at Chicago, Chicago, Illinois,* Parenteral Drug Delivery Systems

**Suneel K. Gupta,** *ALZA Corporation, Palo Alto, California,* In Vitro–In Vivo Correlation

**Robert Gurny,** *University of Geneva, Geneva, Switzerland,* Mucosal Drug Delivery, Intravitreal; Mucosal Drug Delivery, Ocular; Nanoparticles; Poly(ortho esters)

**Jorge Heller,** *Advanced Polymer Systems, Redwood City, California,* Poly(ortho esters)

**Benjamin A. Hertzog,** *Brown University, Providence, Rhode Island,* Cardiovascular Drug Delivery Systems

**Mohammad Ashlaf Hossain,** *Nagasaki University School of Medicine, Nagasaki, Japan,* Infectious Disease, Drug Delivery to Treat

**Eric Kai Huang,** *Brown University, Providence, Rhode Island,* Microencapsulation for Gene Delivery

**James Hunt,** *ALZA Corporation, Palo Alto, California,* Transdermal Drug Delivery, Passive

**Stephen S. Hwang,** *ALZA Corporation, Palo Alto, California,* In Vitro–In Vivo Correlation

**Jules S. Jacob,** *Brown University, Providence, Rhode Island,* Bioadhesive Drug Delivery Systems; Characterization of Delivery Systems, Microscopy

**OluFunmi L. Johnson,** *Alkermes Incorporated, Cambridge, Massachusetts,* Peptide and Protein Drug Delivery

**Yong Shik Jong,** *Brown University, Providence, Rhode Island,* Microencapsulation for Gene Delivery

**Irina Kadiyala,** *The Johns Hopkins University School of Medicine, Baltimore, Maryland,* Biodegradable Polymers: Poly(phosphoester)s

**James J. Kaminski,** *Schering-Plough Research Institute, Kenilworth, New Jersey,* Chemical Approaches to Drug Delivery

**Hyun D. Kim,** *Genetics Institute, Andover, Massachusetts,* Protein Therapeutics for Skeletal Tissue Repair

**Christopher J. Kirby,** *Cortecs Research Laboratory, London, United Kingdom,* Liposomes

**Shigeru Kohno,** *Nagasaki University School of Medicine, Nagasaki, Japan,* Infectious Disease, Drug Delivery to Treat

**Joseph Kost,** *Ben-Gurion University, Beer-Sheva, Israel,* Intelligent Drug Delivery Systems

**Mark R. Kreitz,** *Brown University, Providence, Rhode Island,* Microencapsulation; Synthetic Vascular Grafts and Controlled-Release Technology

**Connie Kwok,** *University of Washington Engineered Biomaterials (UWEB), Seattle, Washington,* Characterization of Delivery Systems, Surface Analysis and Controlled Release Systems

**Robert Langer,** *Johns Hopkins University School of Medicine, Baltimore, Maryland,* Central Nervous System, Drug Delivery to Treat

**Kathleen J. Leach,** *University of Washington, Seattle, Washington,* Cancer, Drug Delivery to Treat—Local & Systemic

**Kam W. Leong,** *The Johns Hopkins University School of Medicine, Baltimore, Maryland,* Biodegradable Polymers: Poly(phosphoester)s

**Robert J. Levy,** *Children's Hospital of Philadelphia and the University of Pennsylvania, Philadelphia, Pennsylvania,* Calcification, Drug Delivery to Prevent

**Suming Li,** *Centre de Recherche sur les Biopolymères Artificiels, Montpellier, France,* Biodegradable Polymers: Polyesters

**Anthony M. Lowman,** *Drexel University, Philadelphia, Pennsylvania,* Hydrogels

**Michael J. Lysaght,** *Brown University, Providence, Rhode Island,* Immunoisolated Cell Therapy

**Fiona C. MacLaughlin,** *Valentis, Inc., The Woodlands, Texas,* Polymeric Systems for Gene Delivery, Chitosan and PINC Systems

**Shigefumi Maesaki,** *Nagasaki University School of Medicine, Nagasaki, Japan,* Infectious Disease, Drug Delivery to Treat

**Judy Magruder,** *ALZA Corporation, Palo Alto, California,* Pumps/Osmotic—VITS Veterinary Implant

**Henry J. Malinowski,** *U.S. Food and Drug Administration, Rockville, Maryland,* Food and Drug Administration Requirements for Controlled Release Products

**Surya K. Mallapragada,** *Iowa State University, Ames, Iowa,* Release Kinetics, Data Interpretation

**Hai-Quan Mao,** *The Johns Hopkins University School of Medicine, Baltimore, Maryland,* Biodegradable Polymers: Poly(phosphoester)s

**Patrick J. Marroum,** *U.S. Food and Drug Administration, Rockville, Maryland,* Food and Drug Administration Requirements for Controlled Release Products

**Edith Mathiowitz,** *Brown University, Providence, Rhode Island,* Bioadhesive Drug Delivery Systems; Microencapsulation

**C. Russell Middaugh,** *University of Kansas, Lawrence, Kansas,* Oligonucleotide Delivery

**Geoffrey Moodie,** *Brown University, Providence, Rhode Island,* Tissue–Implant Interface, Biological Response to Artificial Materials with Surface-Immobilized Small Peptides

**Balaji Narasimhan,** *Rutgers University, Piscataway, New Jersey,* Release Kinetics, Data Interpretation

**Padma Narayan,** *Drexel University, Philadelphia, Pennsylvania,* Diagnostic Use of Microspheres

**Patrea Pabst,** *Arnall Golden & Gregory, LLP, Atlanta, Georgia,* Patents and Other Intellectual Property Rights in Drug Delivery

**Clarisa Peer,** *ALZA Corporation, Palo Alto, California,* Pumps/Osmotic—ALZET® System

**Nicholas A. Peppas,** *Purdue University, West Lafayette, Indiana,* Hydrogels; Release Kinetics, Data Interpretation

**Lorri Perkins,** *ALZA Corporation, Palo Alto, California,* Pumps/Osmotic—ALZET® System

**S.C. Porter,** *Colorcon, West Point, Pennsylvania,* Coatings

**Mary E. Prevo,** *ALZA Corporation, Palo Alto, California,* Transdermal Drug Delivery, Passive

**Venkatesh Raman,** *Massachusetts Institute of Technology, Cambridge, Massachusetts,* Cardiovascular Drug Delivery Systems

**Suneel K. Rastogi,** *University of Minnesota, Minneapolis, Minnesota,* Characterization of Delivery Systems, X-Ray Powder Diffractometry

**Michael J. Rathbone,** *InterAg, Hamilton, New Zealand,* Mucosal Drug Delivery, Vaginal Drug Delivery and Treatment Modalities; Veterinary Applications

**Buddy D. Ratner,** *University of Washington Engineered Biomaterials (UWEB), Seattle, Washington,* Characterization of Delivery Systems, Surface Analysis and Controlled Release Systems

**C.T. Rhodes,** *PharmaCon, Inc., West Kingston, Rhode Island,* Coatings

**C.J. Roberts,** *University of Nottingham, Nottingham, United Kingdom,* Characterization of Delivery Systems, XPS, SIMS and AFM Analysis

**Joseph R. Robinson,** *University of Wisconsin, Madison, Wisconsin,* Mucosal Drug Delivery, Vaginal Drug Delivery and Treatment Modalities

**Alain P. Rolland,** *Valentis, Inc., The Woodlands, Texas,* Polymeric Systems for Gene Delivery, Chitosan and PINC Systems

**G.J. Russell-Jones,** *Biotech Australia Pty Ltd., Roseville, NSW, Australia,* Carrier-Mediated Transport, Oral Drug Delivery

**J. Howard Rytting,** *University of Kansas, Lawrence, Kansas,* Characterization of Delivery Systems, Differential Scanning Calorimetry

**Maryellen Sandor,** *Brown University, Providence, Rhode Island,* Cardiovascular Drug Delivery Systems

**Camilla A. Santos,** *Brown University, Providence, Rhode Island,* Bioadhesive Drug Delivery Systems; Characterization of Delivery Systems, Gel Permeation Chromatography; Fertility Control

**K.M. Shakesheff,** *University of Nottingham, Nottingham, United Kingdom,* Characterization of Delivery Systems, XPS, SIMS and AFM Analysis

**Robert G.L. Shorr,** *United Therapeutics, Washington, D.C.,* Cancer, Drug Delivery to Treat—Prodrugs

**Jürgen Siepmann,** *Freie Universität Berlin, Berlin, Germany,* Nondegradable Polymers for Drug Delivery

**Pierre Souillac,** *University of Kansas, Lawrence, Kansas,* Characterization of Delivery Systems, Differential Scanning Calorimetry

**Mark Speers,** *Health Advances, Inc., Wellesley, Massachusetts,* Economic Aspects of Controlled Drug Delivery

**Cynthia L. Stevenson,** *ALZA Corporation, Palo Alto, California,* Pumps/Osmotic—Introduction;    Pumps/Osmotic—DUROS® Osmotic Implant for Humans

**Gregory R. Stewart,** *ALZA Corporation, Palo Alto, California,* Pumps/Osmotic—DUROS® Osmotic Implant for Humans

**Raj Suryanarayanan,** *University of Minnesota, Minneapolis, Minnesota,* Characterization of Delivery Systems, X-Ray Powder Diffractometry

**Robert M. Swift,** *Brown University and the VA Medical Center, Providence, Rhode Island,* Alcoholism and Drug Dependence, Drug Delivery to Treat

**Cyrus Tabatabay,** *University of Geneva, Geneva, Switzerland,* Mucosal Drug Delivery, Intravitreal

**Kanji Takada,** *Kyoto Pharmaceutical University, Kyoto, Japan,* Oral Drug Delivery, Traditional

**Janet Tamada,** *Cygnus, Inc., Redwood City, California,* Transdermal Drug Delivery, Electrical

**S.J.B. Tendler,** *University of Nottingham, Nottingham, United Kingdom,* Characterization of Delivery Systems, XPS, SIMS and AFM Analysis

**Chris Thanos,** *Brown University, Providence, Rhode Island,* Cardiovascular Drug Delivery Systems

**Mark A. Tracy,** *Alkermes Inc., Cambridge, Massachusetts,* Characterization of Delivery Systems, Particle Sizing Techniques; Peptide and Protein Drug Delivery

**Robert F. Valentini,** *Brown University, Providence, Rhode Island,* Protein Therapeutics for Skeletal Tissue Repair; Tissue–Implant Interface, Biological Response to Artificial Materials with Surface-Immobilized Small Peptides

**Michel Vert,** *Centre de Recherche sur les Biopolymères Artificiels, Montpellier, France,* Biodegradable Polymers: Polyesters

**Narendra R. Vyavahare,** *Children's Hospital of Philadelphia and the University of Pennsylvania, Philadelphia, Pennsylvania,* Calcification, Drug Delivery to Prevent

**K.L. Ward,** *Alkermes, Inc., Cambridge, Massachusetts,* Characterization of Delivery Systems, Particle Sizing Techniques

**Wendy Webber,** *Brown University, Providence, Rhode Island,* Mucosal Drug Delivery, Buccal

**Margaret A. Wheatley,** *Drexel University, Philadelphia, Pennsylvania,* Diagnostic Use of Microspheres

**Leonore Witchey-Lakshmanan,** *Schering-Plough Research Institute, Kenilworth, New Jersey,* Veterinary Applications

**Jeremy C. Wright,** *ALZA Corporation, Palo Alto, California,* Pumps/Osmotic—Introduction; Pumps/Osmotic—DUROS™ Osmotic Implant for Humans; Pumps/Osmotic—Ruminal Osmotic Bolus

**Hiroshi Yoshikawa,** *Toyama Medical and Pharmaceutical University, Toyama, Japan,* Oral Drug Delivery, Traditional

**Zhong Zhao,** *Guilford Pharmaceuticals, Inc., Baltimore, Maryland,* Biodegradable Polymers: Poly(phosphoester)s; Fabrication of Controlled-Delivery Devices

**M. Zignani,** *University of Geneva, Geneva, Switzerland,* Mucosal Drug Delivery, Ocular

# FOREWORD

Drug delivery systems that can precisely control drug release rates or target drugs to a specific body site, although a relatively recent technology, have had an enormous medical and economic impact. New drug delivery systems impact nearly every branch of medicine and annual sales of these systems are far in excess of 10 billion dollars. However, to intelligently create new delivery systems or to understand how to evaluate existing ones, much knowledge is needed. Dr. Edith Mathiowitz of Brown University has, in this encyclopedia, successfully put together a remarkable amount of information to achieve that goal of knowledge.

The *Encyclopedia of Controlled Drug Delivery* provides an up-to-date analysis of critical areas in this promising field. New approaches in treating diseases such as alcoholism, cancer, heart disease, and infectious diseases are examined. Delivery of vaccines, contraceptive agents, anticalcification agents, orthopedic agents, and veterinary agents is discussed. Novel polymeric materials including polyanhydrides, chitosan polyesters, polyphosphates, polyphosphazenes, hydrogels, bioadhesive materials, and poly(ortho esters) are evaluated. Extensive characterization approaches including differential scanning calorimetry, gel permeation chromatography, spectroscopy, X-ray photoelectron spectroscopy, X-ray powder diffraction, and surface characterization are explored.

New areas related to drug delivery such as gene therapy, blood substitutes, food ingredients, and tissue engineering are discussed. An analysis of various routes of administration including parenteral, intravitreal, oral, rectal, ocular, nasal, buccal, vaginal, and the central nervous system is provided. Different controlled release designs such as osmotic pumps, pendent-chain systems, membrane systems, nanoparticles, and liposomes are examined. Finally, patents, regulatory issues, manufacturing approaches, economics, in vitro–in vivo correlations, pharmacokinetics, release kinetics, assays, diagnostics, and related issues are considered.

This encyclopedia is a very complete compendium of the state-of-the-art of this burgeoning field and should be of considerable value for those who wish to enter it.

ROBERT LANGER
Massachusetts Institute of Technology

# PREFACE

The two-volume *Encyclopedia of Controlled Drug Delivery* will provide extensive, yet easily accessible, A–Z coverage of state-of-the-art topics in drug delivery systems. The encyclopedia can be used by research departments in industry, research institutes, universities, libraries, and consultants. The readers may range from undergraduate and graduate students to professional engineers, biologists, chemists, and medical researchers. In addition, research managers, as well as business venturers, will find this encyclopedia a very useful source of information. Scientists unfamiliar with the field of drug delivery will find a good introduction to the field, while experts will find the book to be a good, current source of information. The contents of these two volumes provide coverage of many aspects of drug delivery systems, including:

- The history and development of the field from 1975 to date
- Advantages and disadvantages of controlled release technology as compared to conventional delivery systems (pharmaceutical as well as veterinary applications)
- Detailed descriptions of the various systems for achieving controlled drug release, including nonerodible reservoir and matrix devices, bioerodible polymers, pendent drug substitutes, and osmotic pumps
- Pharmaceutical applications of drug delivery systems, approaches to achieve zero-release kinetics, development of pulsatile delivery systems including approaches to develop self-regulated systems by using responsive hydrogels or encapsulated live cells
- Stabilization and release characterization of proteins
- Characterization of specific delivery systems such as oral, nasal, ocular, and other routes of administration
- Methods to fabricate controlled delivery systems, including microencapsulation, liposome preparation, film casting, and membrane formation
- Factors affecting regulatory considerations
- Economic aspects of controlled drug delivery devices
- Patents and other intellectual property rights in drug delivery
- Oligonucleotides and gene delivery

- Specific topics in polymer technology which are peculiar to drug delivery systems, including polymer synthesis, structure, morphology, amorphous polymers, glassy and rubbery states, polymer networks, and a variety of methods to characterize polymers and drug delivery systems

The encyclopedia can easily be used as an advanced text or reference book for a drug delivery system course. It is written by some of the greatest experts and most diligent educators in the field. These volumes should constitute an important research reference tool, a desktop information resource, and supplementary reading for teaching professionals and their students.

## ACKNOWLEDGMENT

The idea to start this two-volume book came from Hannah Ben-Zvi, of John Wiley and Sons, and caught me by surprise. I had been thinking for many years that there was no encyclopedia for the field of drug delivery systems, which has emerged since the 1970s. The encouragement to start this project came from my students. It was with the help of Kathleen Leach (Pekarek), Wendy Webber, and Camilla Santos that this project was born. I am also extremely thankful to the editorial board for the time and effort they contributed to this endeavor. I would like particularly to mention Howard Bernstein, Robert Gurny, and Nicholas Peppas, who contributed above and beyond to the success of this project. I would also like to thank Pierre Galletti, whose untimely death prevented him from seeing the completion of the encyclopedia. Special thanks to Jules Jacob, Jong Yong, Ben Hertzog, Mark Kreitz, Gerardo Carino, Chris Thanos, MaryEllen Sandor, and Don Chickering. I am also grateful to Glenn Collins of John Wiley, who was always there to help and comfort, to Susan Hirsch, who always kept me on track, and perhaps most importantly to the team of authors who put an enormous amount of time into writing this book. To my husband, who helped with graphics, I am eternally grateful for always being patient, even during the days when he hardly saw me.

EDITH MATHIOWITZ
Brown University

# CONVERSION FACTORS, ABBREVIATIONS, AND UNIT SYMBOLS

## SI Units (Adopted 1960)

The International System of Units (abbreviated SI), is being implemented throughout the world. This measurement system is a modernized version of the MKSA (meter, kilogram, second, ampere) system, and its details are published and controlled by an international treaty organization (The International Bureau of Weights and Measures) (1).

SI units are divided into three classes:

<table>
<tr><td colspan="2" align="center">BASE UNITS</td><td colspan="2" align="center">SUPPLEMENTARY UNITS</td></tr>
<tr><td>length</td><td>meter† (m)</td><td>plane angle</td><td>radian (rad)</td></tr>
<tr><td>mass</td><td>kilogram (kg)</td><td>solid angle</td><td>steradian (sr)</td></tr>
<tr><td>time</td><td>second (s)</td><td></td><td></td></tr>
<tr><td>electric current</td><td>ampere (A)</td><td></td><td></td></tr>
<tr><td>thermodynamic temperature‡</td><td>kelvin (K)</td><td></td><td></td></tr>
<tr><td>amount of substance</td><td>mole (mol)</td><td></td><td></td></tr>
<tr><td>luminous intensity</td><td>candela (cd)</td><td></td><td></td></tr>
</table>

## DERIVED UNITS AND OTHER ACCEPTABLE UNITS

These units are formed by combining base units, supplementary units, and other derived units (2–4). Those derived units having special names and symbols are marked with an asterisk in the list below.

| Quantity | Unit | Symbol | Acceptable equivalent |
|---|---|---|---|
| *absorbed dose | gray | Gy | J/kg |
| acceleration | meter per second squared | $m/s^2$ | |
| *activity (of a radionuclide) | becquerel | Bq | 1/s |
| area | square kilometer | $km^2$ | |
| | square hectometer | $hm^2$ | ha (hectare) |
| | square meter | $m^2$ | |
| concentration (of amount of substance) | mole per cubic meter | $mol/m^3$ | |
| current density | ampere per square meter | $A//m^2$ | |
| density, mass density | kilogram per cubic meter | $kg/m^3$ | g/L; $mg/cm^3$ |
| dipole moment (quantity) | coulomb meter | $C \cdot m$ | |
| *dose equivalent | sievert | Sv | J/kg |
| *electric capacitance | farad | F | C/V |
| *electric charge, quantity of electricity | coulomb | C | $A \cdot s$ |
| electric charge density | coulomb per cubic meter | $C/m^3$ | |
| *electric conductance | siemens | S | A/V |
| electric field strength | volt per meter | V/m | |
| electric flux density | coulomb per square meter | $C/m^2$ | |
| *electric potential, potential difference, electromotive force | volt | V | W/A |
| *electric resistance | ohm | $\Omega$ | V/A |

---

†The spellings "metre" and "litre" are preferred by ASTM; however, "-er" is used in the encyclopedia.

‡Wide use is made of Celsius temperature $(t)$ defined by

$$t = T - T_0$$

where $T$ is the thermodynamic temperature, expressed in kelvin, and $T_0 = 273.15$ K by definition. A temperature interval may be expressed in degrees Celsius as well as in kelvin.

| Quantity | Unit | Symbol | Acceptable equivalent |
|---|---|---|---|
| *energy, work, quantity of heat | megajoule | MJ | |
| | kilojoule | kJ | |
| | joule | J | $N \cdot m$ |
| | electronvolt† | eV† | |
| | kilowatt-hour† | $kW \cdot h$† | |
| energy density | joule per cubic meter | $J/m^3$ | |
| *force | kilonewton | kN | |
| | newton | N | $kg \cdot m/s^2$ |
| *frequency | megahertz | MHz | |
| | hertz | Hz | 1/s |
| heat capacity, entropy | joule per kelvin | J/K | |
| heat capacity (specific), specific entropy | joule per kilogram kelvin | $J/(kg \cdot K)$ | |
| heat-transfer coefficient | watt per square meter kelvin | $W/(m^2 \cdot K)$ | |
| *illuminance | lux | lx | $lm/m^2$ |
| *inductance | henry | H | Wb/A |
| linear density | kilogram per meter | kg/m | |
| luminance | candela per square meter | $cd/m^2$ | |
| *luminous flux | lumen | lm | $cd \cdot sr$ |
| magnetic field strength | ampere per meter | A/m | |
| *magnetic flux | weber | Wb | $V \cdot s$ |
| *magnetic flux density | tesla | T | $Wb/m^2$ |
| molar energy | joule per mole | J/mol | |
| molar entropy, molar heat capacity | joule per mole kelvin | $J/(mol \cdot K)$ | |
| moment of force, torque | newton meter | $N \cdot m$ | |
| momentum | kilogram meter per second | $kg \cdot m/s$ | |
| permeability | henry per meter | H/m | |
| permittivity | farad per meter | F/m | |
| *power, heat flow rate, radiant flux | kilowatt | kW | |
| | watt | W | J/s |
| power density, heat flux density, irradiance | watt per square meter | $W/m^2$ | |
| *pressure, stress | megapascal | MPa | |
| | kilopascal | kPa | |
| | pascal | Pa | $N/m^2$ |
| sound level | decibel | dB | |
| specific energy | joule per kilogram | J/kg | |
| specific volume | cubic meter per kilogram | $m^3/kg$ | |
| surface tension | newton per meter | N/m | |
| thermal conductivity | watt per meter kelvin | $W/(m \cdot K)$ | |
| velocity | meter per second | m/s | |
| | kilometer per hour | km/h | |
| viscosity, dynamic | pascal second | $Pa \cdot s$ | |
| | millipascal second | $mPa \cdot s$ | |
| viscosity, kinematic | square meter per second | $m^2/s$ | |
| | square millimeter per second | $mm^2/s$ | |
| volume | cubic meter | $m^3$ | |
| | cubic diameter | $dm^3$ | L (liter) (5) |
| | cubic centimeter | $cm^3$ | mL |
| wave number | 1 per meter | $m^{-1}$ | |
| | 1 per centimeter | $cm^{-1}$ | |

---

†This non-SI unit is recognized by the CIPM as having to be retained because of practical importance or use in specialized fields (1).

In addition, there are 16 prefixes used to indicate order of magnitude, as follows:

| Multiplication factor | Prefix | Symbol |
|---|---|---|
| $10^{18}$ | exa | E |
| $10^{15}$ | peta | P |
| $10^{12}$ | tera | T |
| $10^{9}$ | giga | G |
| $10^{6}$ | mega | M |
| $10^{3}$ | kilo | k |
| $10^{2}$ | hecto | $h^{a}$ |
| 10 | deka | $da^{a}$ |
| $10^{-1}$ | deci | $d^{a}$ |
| $10^{-2}$ | centi | $c^{a}$ |
| $10^{-3}$ | milli | m |
| $10^{-6}$ | micro | $\mu$ |
| $10^{-9}$ | nano | n |
| $10^{-12}$ | pico | p |
| $10^{-15}$ | femto | f |
| $10^{-18}$ | atto | a |

[a]Although hecto, deka, deci, and centi are SI prefixes, their use should be avoided except for SI unit-multiples for area and volume and nontechnical use of centimeter, as for body and clothing measurement.

For a complete description of SI and its use the reader is referred to ASTM E380 (4).

A representative list of conversion factors from non-SI to SI units is presented herewith. Factors are given to four significant figures. Exact relationships are followed by a dagger. A more complete list is given in the latest editions of ASTM E380 (4) and ANSI Z210.1 (6).

## CONVERSION FACTORS TO SI UNITS

| To convert from | To | Multiply by |
|---|---|---|
| acre | square meter (m²) | $4.047 \times 10^{3}$ |
| angstrom | meter (m) | $1.0 \times 10^{-10}$† |
| are | square meter (m²) | $1.0 \times 10^{2}$† |
| astronomical unit | meter (m) | $1.496 \times 10^{11}$ |
| atmosphere, standard | pascal (Pa) | $1.013 \times 10^{5}$ |
| bar | pascal (Pa) | $1.0 \times 10^{5}$† |
| barn | square meter (m²) | $1.0 \times 10^{-28}$† |
| barrel (42 U.S. liquid gallons) | cubic meter (m³) | 0.1590 |
| Bohr magneton ($\mu_{B}$) | J/T | $9.274 \times 10^{-24}$ |
| Btu (International Table) | joule (J) | $1.055 \times 10^{3}$ |
| Btu (mean) | joule (J) | $1.056 \times 10^{3}$ |
| Btu (thermochemical) | joule (J) | $1.054 \times 10^{3}$ |
| bushel | cubic meter (m³) | $3.524 \times 10^{-2}$ |
| calorie (International Table) | joule (J) | 4.187 |
| calorie (mean) | joule (J) | 4.190 |
| calorie (thermochemical) | joule (J) | 4.184† |
| centipoise | pascal second (Pa·s) | $1.0 \times 10^{-3}$† |
| centistokes | square millimeter per second (mm²/s) | 1.0† |
| cfm (cubic foot per minute) | cubic meter per second (m³/s) | $4.72 \times 10^{-4}$ |
| cubic inch | cubic meter (m³) | $1.639 \times 10^{-5}$ |
| cubic foot | cubic meter (m³) | $2.832 \times 10^{-2}$ |
| cubic yard | cubic meter (m³) | 0.7646 |
| curie | becquerel (Bq) | $3.70 \times 10^{10}$† |
| debye | coulomb meter (C·m) | $3.336 \times 10^{-30}$ |
| degree (angle) | radian (rad) | $1.745 \times 10^{-2}$ |

†Exact.

## CONVERSION FACTORS TO SI UNITS

| To convert from | To | Multiply by |
| --- | --- | --- |
| denier (international) | kilogram per meter (kg/m) | $1.111 \times 10^{-7}$ |
|  | tex‡ | 0.1111 |
| dram (apothecaries') | kilogram (kg) | $3.888 \times 10^{-3}$ |
| dram (avoirdupois) | kilogram (kg) | $1.772 \times 10^{-3}$ |
| dram (U.S. fluid) | cubic meter (m³) | $3.697 \times 10^{-6}$ |
| dyne | newton (N) | $1.0 \times 10^{-5}$† |
| dyne/cm | newton per meter (N/m) | $1.0 \times 10^{-3}$† |
| electronvolt | joule (J) | $1.602 \times 10^{-19}$ |
| erg | joule (J) | $1.0 \times 10^{-7}$† |
| fathom | meter (m) | 1.829 |
| fluid ounce (U.S.) | cubic meter (m³) | $2.957 \times 10^{-5}$ |
| foot | meter (m) | 0.3048† |
| footcandle | lux (lx) | 10.76 |
| furlong | meter (m) | $2.012 \times 10^{-2}$ |
| gal | meter per second squared (m/s²) | $1.0 \times 10^{-2}$† |
| gallon (U.S. dry) | cubic meter (m³) | $4.405 \times 10^{-3}$ |
| gallon (U.S. liquid) | cubic meter (m³) | $3.785 \times 10^{-3}$ |
| gallon per minute (gpm) | cubic meter per second (m³/s) | $6.309 \times 10^{-5}$ |
|  | cubic meter per hour (m³/h) | 0.2271 |
| gauss | tesla (T) | $1.0 \times 10^{-4}$ |
| gilbert | ampere (A) | 0.7958 |
| gill (U.S.) | cubic meter (m³) | $1.183 \times 10^{-4}$ |
| grade | radian | $1.571 \times 10^{-2}$ |
| grain | kilogram (kg) | $6.480 \times 10^{-5}$ |
| gram force per denier | newton per tex (N/tex) | $8.826 \times 10^{-2}$ |
| hectare | square meter (m²) | $1.0 \times 10^{4}$† |
| horsepower (550 ft·lbf/s) | watt (W) | $7.457 \times 10^{2}$ |
| horsepower (boiler) | watt (W) | $9.810 \times 10^{3}$ |
| horsepower (electric) | watt (W) | $7.46 \times 10^{2}$† |
| hundredweight (long) | kilogram (kg) | 50.80 |
| hundredweight (short) | kilogram (kg) | 45.36 |
| inch | meter (m) | $2.54 \times 10^{-2}$† |
| inch of mercury (32°F) | pascal (Pa) | $3.386 \times 10^{3}$ |
| inch of water (39.2°F) | pascal (Pa) | $2.491 \times 10^{2}$ |
| kilogram-force | newton (N) | 9.807 |
| kilowatt hour | megajoule (MJ) | 3.6† |
| kip | newton (N) | $4.448 \times 10^{3}$ |
| knot (international) | meter per second (m/S) | 0.5144 |
| lambert | candela per square meter (cd/m³) | $3.183 \times 10^{3}$ |
| league (British nautical) | meter (m) | $5.559 \times 10^{3}$ |
| league (statute) | meter (m) | $4.828 \times 10^{3}$ |
| light year | meter (m) | $9.461 \times 10^{15}$ |
| liter (for fluids only) | cubic meter (m³) | $1.0 \times 10^{-3}$† |
| maxwell | weber (Wb) | $1.0 \times 10^{-8}$† |
| micron | meter (m) | $1.0 \times 10^{-6}$† |
| mil | meter (m) | $2.54 \times 10^{-5}$† |
| mile (statute) | meter (m) | $1.609 \times 10^{3}$ |
| mile (U.S. nautical) | meter (m) | $1.852 \times 10^{3}$† |
| mile per hour | meter per second (m/s) | 0.4470 |
| millibar | pascal (Pa) | $1.0 \times 10^{2}$ |
| millimeter of mercury (0°C) | pascal (Pa) | $1.333 \times 10^{2}$† |
| minute (angular) | radian | $2.909 \times 10^{-4}$ |
| myriagram | kilogram (kg) | 10 |

†Exact.
‡See footnote on p. xvi.

## CONVERSION FACTORS TO SI UNITS

| To convert from | To | Multiply by |
|---|---|---|
| myriameter | kilometer (km) | 10 |
| oersted | ampere per meter (A/m) | 79.58 |
| ounce (avoirdupois) | kilogram (kg) | $2.835 \times 10^{-2}$ |
| ounce (troy) | kilogram (kg) | $3.110 \times 10^{-2}$ |
| ounce (U.S. fluid) | cubic meter (m³) | $2.957 \times 10^{-5}$ |
| ounce-force | newton (N) | 0.2780 |
| peck (U.S.) | cubic meter (m³) | $8.810 \times 10^{-3}$ |
| pennyweight | kilogram (kg) | $1.555 \times 10^{-3}$ |
| pint (U.S. dry) | cubic meter (m³) | $5.506 \times 10^{-4}$ |
| pint (U.S. liquid) | cubic meter (m³) | $4.732 \times 10^{-4}$ |
| poise (absolute viscosity) | pascal second (Pa·s) | 0.10† |
| pound (avoirdupois) | kilogram (kg) | 0.4536 |
| pound (troy) | kilogram (kg) | 0.3732 |
| poundal | newton (N) | 0.1383 |
| pound-force | newton (N) | 4.448 |
| pound force per square inch (psi) | pascal (Pa) | $6.895 \times 10^{3}$ |
| quart (U.S. dry) | cubic meter (m³) | $1.101 \times 10^{-3}$ |
| quart (U.S. liquid) | cubic meter (m³) | $9.464 \times 10^{-4}$ |
| quintal | kilogram (kg) | $1.0 \times 10^{2}$† |
| rad | gray (Gy) | $1.0 \times 10^{-2}$† |
| rod | meter (m) | 5.029 |
| roentgen | coulomb per kilogram (C/kg) | $2.58 \times 10^{-4}$ |
| second (angle) | radian (rad) | $4.848 \times 10^{-6}$† |
| section | square meter (m²) | $2.590 \times 10^{6}$ |
| slug | kilogram (kg) | 14.59 |
| spherical candle power | lumen (lm) | 12.57 |
| square inch | square meter (m²) | $6.452 \times 10^{-4}$ |
| square foot | square meter (m²) | $9.290 \times 10^{-2}$ |
| square mile | square meter (m²) | $2.590 \times 10^{6}$ |
| square yard | square meter (m²) | 0.8361 |
| stere | cubic meter (m³) | 1.0† |
| stokes (kinematic viscosity) | square meter per second (m²/s) | $1.0 \times 10^{-4}$† |
| tex | kilogram per meter (kg/m) | $1.0 \times 10^{-6}$† |
| ton (long, 2240 pounds) | kilogram (kg) | $1.016 \times 10^{3}$ |
| ton (metric) (tonnee) | kilogram (kg) | $1.0 \times 10^{3}$† |
| ton (short, 2000 pounds) | kilogram (kg) | $9.072 \times 10^{2}$ |
| torr | pascal (Pa) | $1.333 \times 10^{2}$ |
| unit pole | weber (Wb) | $1.257 \times 10^{-7}$ |
| yard | meter (m) | 0.9144† |

†Exact.

## BIBLIOGRAPHY

1. The International Bureau of Weights and Measures, BIPM (Parc de Saint-Cloud, France) is described in Appendix X2 of Ref. 4. This bureau operates under the exclusive supervision of the International Committee for Weights and Measures (CIPM).
2. *Metric Editorial Guide (ANMC-78-1)*, latest ed., American National Metric Council, 5410 Grosvenor Lane, Bethesda, Md. 20814, 1981.
3. *SI Units and Recommendations for the Use of Their Multiples and of Certain Other Units (ISO 1000-1981)*, American National Standards Institute, 1430 Broadway, New York, 10018, 1981.
4. Based on *ASTM E380-89a (Standard Practice for Use of the International System of Units (SI))*, American Society for Testing and Materials, 1916 Race Street, Philadelphia, Pa. 19103, 1989.
5. *Fed. Reg.*, Dec. 10, 1976 (41 FR 36414).
6. For ANSI address, see Ref. 3.

R. P. Lukens
ASTM Committee E-43 on SI Practice

# A

## ALCOHOLISM AND DRUG DEPENDENCE, DRUG DELIVERY TO TREAT

ROBERT M. SWIFT
Brown University and the VA Medical Center
Providence, Rhode Island

**KEY WORDS**

Addictive disorders

Alcoholism

Buprenorphine

Clonidine

Disulfiram

Drug abuse

Naltrexone

Nicotine

Opioids

**OUTLINE**

## ADDICTIVE DISORDERS AND THEIR TREATMENT

Addictive disorders, including alcoholism, drug dependence, and nicotine dependence, afflict over 30% of Americans (1) and are associated with considerable morbidity, mortality, social problems, and health care costs (2,3). Addiction is characterized by impaired control over drinking or drug use, increased tolerance to the effects of alcohol and drugs, preoccupation with alcohol and drugs, and use despite adverse consequences (4).

One way to reduce the impact of addictive disorders is through effective drug and alcohol treatment. Treatment consists of medical, psychological, and social interventions to reduce or eliminate the harmful effects of substances on the individual, his or her family and associates, and others in society. The treatment consists of two components: detoxification and rehabilitation. Detoxification refers to the removal of the drug from the body and the treatment of physiological withdrawal signs and symptoms that may occur with drug discontinuation. Rehabilitation provides the patient with strategies and techniques to avoid psychoactive substances, to develop better methods of coping with stress and distress, and to improve self-esteem and self-efficacy.

Medications are frequently used as a component of both detoxification treatment and rehabilitation treatment, along with psychosocial therapies, such as counseling and self-help groups (e.g., Alcoholics Anonymous). Pharmacotherapies can treat alcohol and drug dependence through several mechanisms that may reduce some of the impetus for drug use. These mechanisms and the medications that may operate through these mechanisms are depicted in Table 1.

Controlled drug delivery systems are particularly applicable to the treatment of addictive disorders. Several advantages of controlled drug delivery systems are as follows:

- They mimic the pharmacokinetics of the abused drug, including the rise and fall of plasma drug concentrations.
- They facilitate the attainment of constant plasma concentrations of drug to prevent intoxication symptoms caused by high plasma concentrations and to prevent the development of withdrawal caused by low plasma drug concentrations.
- They improve therapeutic medication bioavailability.
- They improve therapeutic medication compliance in drug and alcohol treatment.

For a medication used as a substitution treatment, a controlled delivery system can be used to mimic the pharmacokinetics of the abused drug, without the dangers as-

**Table 1. Medications Used for Addiction Treatment**

| Medication property | Example |
| --- | --- |
| Substitution treatment with a cross-tolerant medication | Methadone maintenance treatment for opioid dependence; transdermal nicotine or nicotine gum for smoking cessation |
| Administration of agents to block the signs and symptoms of withdrawal | Transdermal clonidine in opioid detoxification and withdrawal |
| Administration of a medication to block drug intoxication | Depot naltrexone for opioid dependence |
| Aversive therapy | Depot disulfiram treatment in alcoholism |
| Administration of a medication to suppress craving | Depot naltrexone in alcohol dependence |

sociated with the usual mode of administration. For example, several methods of controlled administration of nicotine can keep plasma nicotine levels high enough to prevent nicotine withdrawal, without the inherent dangers of smoking. Two of these nicotine delivery methods can actually induce the rise and fall of plasma nicotine levels that occurs with smoking. Patients with addictive disorders are frequently noncompliant with medications and may either overuse or underuse therapeutic medications. Controlled drug delivery systems offer methods for controlling drug use and plasma concentrations to improve compliance. This chapter discusses use of controlled drug delivery in the treatment of addictive disorders, the putative mechanisms of action of the medications, and the evidence for their efficacy.

## TREATMENT FOR NICOTINE DEPENDENCE (SMOKING)

The treatment of nicotine dependence provides several examples of the therapeutic use of controlled medication delivery systems in treatment. Nicotine is an alkaloid drug present in the leaves of the tobacco plant, *Nicotiana tabacum*, used for centuries by Native Americans in rituals and folk medicine. Today, nicotine has become one of the most commonly used psychoactive drugs. Over 50 million persons in the U.S. are daily users of cigarettes (one-third of adults), and 10 million use another form of tobacco (5). Although the overall number of Americans who smoke has declined, the numbers of young women who smoke, and the use of other tobacco products such as smokeless tobacco, has increased. Tobacco use is increasing in developing countries.

The medical consequences of nicotine use are common and constitute a significant public health problem. These include coronary artery disease, vascular disease, respiratory disease, and cancer, particularly of the lung, oral cavity, and pharynx. Many deleterious effects of tobacco are not due to nicotine, but are due to other toxic and carcinogenic compounds present in tobacco extract or smoke.

To maximize the absorption of nicotine, tobacco products are usually smoked in pipes, cigars, or cigarettes, or instilled intranasally or orally as snuff or smokeless tobacco. Following absorption from the lungs or buccal mucosa, nicotine levels peak rapidly and then decline with a half-life of 30 to 60 minutes.

Nicotine has several effects on the peripheral autonomic and central nervous systems. It is an agonist at nicotinic cholinergic receptors in parasympathetic and sympathetic autonomic ganglia. Nicotine produces salivation, increases gastric motility and acid secretion, and releases catecholamines, resulting in cardiac stimulation and peripheral vasoconstriction. Nicotine is a central-nervous-system stimulant, producing increased alertness, increased attention and concentration, and appetite suppression. The fact that tobacco use can prevent weight gain makes the drug especially attractive to young women.

Repeated use of nicotine produces tolerance and dependence. The degree of dependence is considerable, as over 70% of dependent individuals relapse within one year of stopping use. Cessation of nicotine use in dependent individuals is followed by a withdrawal syndrome characterized by increased irritability, decreased attention and concentration, an intense craving for and preoccupation with nicotine, anxiety, and depression (6,7). Withdrawal symptoms begin within several hours of cessation of use or reduction in dosage, and typically last about a week. Increased appetite with weight gain occurs in the weeks and months following cessation of chronic nicotine use.

The treatment of nicotine dependence consists of reducing or stopping use of tobacco use and minimizing nicotine withdrawal symptoms. Brief education and advice on smoking cessation provided by physicians has been shown to be effective in helping patients stop smoking, and it is now recommended that all physicians provide their patients with smoking cessation tools (8,9). The most successful treatment programs use cognitive-behavioral techniques to educate patients about the health hazards of tobacco and provide the patient with behavioral methods of coping with urges. Such programs can achieve 25–45% abstinence rates at 6 to 12 months. Although some programs use gradual reduction in tobacco use over days to weeks (nicotine fading) for detoxification, others suggest abrupt discontinuation (cold turkey).

### Nicotine Replacement

Pharmacologic therapy with nicotine replacement is increasingly popular in the treatment of nicotine dependence. The principle of nicotine replacement therapy is to provide the nicotine-dependent patient with nicotine in a form not associated with the carcinogenic and irritant elements in tobacco products. The substitution of tobacco with alternative nicotine delivery systems allows the patient to address behavioral aspects of the habit without having to experience nicotine withdrawal. At a later time, plasma nicotine levels can be reduced and eventually discontinued in a slow and controlled fashion. Several systems of controlled nicotine delivery have been developed and introduced into clinical practice.

**Transdermal Nicotine.** Thin film impregnated with various doses of nicotine (7 mg, 14 mg, and 21 mg) are made into adhesive patches for transdermal administration. Nicotine patches deliver a predictable amount of nicotine and achieve steady-state plasma nicotine levels in the ranges achieved by smoking 10 to 30 cigarettes per day. The labeled dose refers to the amount of nicotine delivered rather than the amount of nicotine present in the patch. Transdermal nicotine is well absorbed, but the peak plasma concentrations are delayed by up to 10 hours after patch application.

Placebo-controlled clinical trials with transdermal nicotine show efficacy in smoking cessation treatment. Nicotine replacement therapy with transdermal nicotine significantly reduces nicotine withdrawal symptoms and increases the likelihood of successful smoking cessation (10,11). A recent meta-analysis of 17 studies involving over 5,000 patients indicated that transdermal nicotine patches produced quit rates of 27.1% at end of treatment and 21.8% at 6 month follow-up, compared to 13.1% and 8.4% for placebo groups, respectively.

Transdermal nicotine is used as follows. After stopping tobacco use, one patch is applied to uncovered skin each 24-hour period and the previous patch discarded. Some patches are labeled for use during the daytime only and are not worn while the patient is asleep. Typical transdermal nicotine treatment involves applying a high-dose patch to skin for 2 to 4 weeks, an intermediate dose patch for 2 to 4 weeks, and then the smallest dose patch for 2 to 4 weeks. Side effects include irritation from the patch and nicotine effects (nausea, cardiac effects, etc.). It is important that patients not use tobacco products or other nicotine replacement methods while using the patch, as toxic nicotine blood levels may occur. It is also important that patients receive behavior-oriented treatment while using the patch.

**Nicotine Gum.** This gum is a sweet, flavored polacrilex resin containing 2 or 4 mg of nicotine that is released slowly when the resin is chewed. Up to 90% of the nicotine in the resin is released into the saliva, although the amount of release and the rate of release depends upon the rate and duration of chewing. Nicotine is absorbed across the buccal mucosa, with a time to peak concentration of 15 to 30 minutes after start of chewing. The gum is marketed as Nicorette® and has recently been made available as an over-the-counter medication, as well as being available by prescription.

Patients must be instructed in the proper use of the gum, which is chewed slowly and intermittently whenever the individual feels the need for tobacco. When tingling of the mouth or tongue is perceived, chewing should cease for a short time, while still holding the gum in the mouth. Rapid chewing releases excess nicotine and may cause nausea and other side effects of nicotine toxicity. The gum is chewed for 20 to 30 minutes and then discarded. This method reduces tobacco craving and withdrawal discomfort with nicotine blood levels that rise and fall, mimicking smoking (12). Nicotine gum has been shown to be more effective than placebo in most clinical trials of its use in smoking cessation (13). Recent studies suggest that the 4 mg nicotine gum is more effective than the 2-mg gum in high nicotine-dependent smokers (14). Side effects of gum use include symptoms of nicotine toxicity and occlusive injuries of the teeth and dental appliances.

Most patients achieve stable gum use within a few days if they can stop smoking. A schedule for tapering gum use is planned after the daily maintenance dose is established. For example, a patient using 15 pieces of gum each day during the first week of treatment gradually reduces the number to 10 per day by the end of the first month and 5 per day by the end of the second month. Patients are typically able to discontinue nicotine gum after 3 to 6 months of treatment.

**Nicotine Nasal Spray.** Nicotine has recently become available in a solution for intranasal administration, to be used as source of nicotine replacement during smoking cessation treatment. The medication is administered in a concentration of 0.5 mg nicotine per 50 $\mu$L of metered spray vehicle. The recommended nicotine nasal dose is 1–2 mg (2–4 sprays) per hour, with a maximum recommended daily dose of 40 mg. When nicotine is administered intranasally, the time to peak plasma concentration is 4 to 15 minutes. The bioavailability of nicotine nasal spray solution is approximately 53%, although peak plasma concentrations achieved vary considerably due to individual differences in absorption and variations in usage. Rhinitis or other nasal abnormalities may reduce absorption, reduce peak plasma nicotine concentrations, and increase the time to peak plasma concentrations. Side effects of the spray include irritation of the nasal and pharyngeal cavities, in addition to the physiological side effects of nicotine. It is recommended that tapering of nasal nicotine doses begin after 2–4 weeks and that use not exceed 8 weeks to minimize the chances of developing dependence on the nicotine spray.

**Nicotine Inhaler.** Nicotine replacement using a nicotine inhaler (sometimes called a smokeless cigarette) to deliver nicotine orally by inhalation through a plastic tube has recently become available and is marketed as the Nicotrol® Inhaler. An active, disposable nicotine cartridge consisting of a porous plug impregnated with 10 mg of nicotine is placed in the plastic tube, and the patient inhales through a plastic mouthpiece as if smoking a cigarette. The device delivers a dose of 13 $\mu$g nicotine per puff, for up to 400 puffs. The delivered nicotine is primarily absorbed through the buccal mucosa. This method most mimics smoking, as it involves bringing the device to the mouth and inhaling to obtain nicotine dosing. In a study of 247 smokers who had previously failed other nicotine replacement therapy, continuous abstinence rates were 28% with the active drug, compared to 18% with the placebo inhaler (15). Common side effects include dyspepsia, transient coughing, and mouth irritation.

### Other Medications for Smoking Cessation

**Clonidine: Oral and Patch.** Clonidine has been found to be effective in the treatment of smoking cessation and the amelioration of nicotine withdrawal symptoms, reducing cigarette craving and other symptoms of nicotine withdrawal in dependent cigarette smokers who stopped cold turkey (16). Several other investigators have confirmed the efficacy of clonidine in reducing nicotine withdrawal and improving quitting rates. However, in other studies on the treatment of nicotine withdrawal, the effects of clonidine have been more equivocal.

A transdermal clonidine therapeutic system, marketed as Catapres-TTS®, was developed for the treatment of hypertension and is approved by the U.S. FDA for that indication. The medication is incorporated into a transdermal delivery device designed to adhere to the skin (a patch) and to provide stable therapeutic levels of drug for period of one week. A microporous membrane controls the rate of clonidine delivery to the skin surface, whereupon the drug diffuses into the skin. The dose received is proportional to the patch surface area. Bioavailability and efficacy studies demonstrate comparability to oral clonidine preparations (17). However, due to cutaneous compartmental pharmacokinetics, there is a delay of 48 to 72 hours before the therapeutic blood levels are achieved and a similar persistence in clonidine levels after the patch is removed. A large multicenter clinical trial used transdermal clonidine combined with behavioral treatment for smoking cessation (18). The clonidine doses used in this study were 0.1 to 0.2 mg per day. Although there was some decrease in nicotine withdrawal symptoms with clonidine, there was no significant increase in quit rates, compared with placebo. Nevertheless, the reduction in nicotine withdrawal symptoms with clonidine may have benefits in selected patients (19). Some investigators have found an improved response in female smokers.

Side effects of clonidine include hypotension and sedation. Clonidine is usually prescribed for a period of 3 to 4 weeks, with the dose gradually reduced over the detoxification period. As with other forms of treatment for nicotine dependence, clonidine should be used in conjunction with a behavioral recovery program.

## TREATMENTS FOR OPIOID DEPENDENCE

Opioid abuse and dependence are significant social and medical problems in the U.S., with an estimated opioid addict population of greater than 500,000. These patients are frequent users of medical and surgical services because of the multiple medical sequelae of intravenous drug use, including infections (especially human immunodifficiency virus) and overdose. The crime and violence associated with the addict lifestyle engenders serious injuries to the addicts and to others.

Opiate drugs affect organ systems due to stimulation of receptors for endogenous hormones, enkephalins, endorphins, and dynorphins. There exist at least three distinct opioid receptors, which are designated by the Greek letters $\mu$, $\kappa$, and $\delta$ (20). Drugs that act primarily through $\mu$-receptor effects include heroin, morphine, and methadone; such drugs produce analgesia, euphoria, and respiratory depression. Drugs mediated through the $\kappa$-receptor include the so-called mixed agonist-antagonists, buprenorphine, butorphanol, and pentazocine, which produce analgesia, but less respiratory depression. The $\delta$-receptor appears to bind endogenous opioid peptides.

The treatment of opioid dependence includes detoxification, followed by long-term rehabilitation. Several medications that use the principles of controlled drug delivery systems are used clinically for detoxification and rehabilitation of the opioid-dependent patient.

### Opioid Detoxification: Clonidine

Opiate withdrawal, although rarely life threatening, is subjectively distressing. It is marked by increased sympathetic activity, intestinal hyperactivity, hypersensitivity to pain, and an intense craving to use more opiates. The peak period of acute withdrawal depends on the opiate used: for short-acting opiates such as morphine, heroin, or meperidine, the peak withdrawal is 1–3 days and duration is 5–7 days. For longer-acting opiates, such as methadone, the peak is 3–5 days and the duration 10–14 days.

Clonidine hydrochloride is an imidazoline derivative originally approved as an antihypertensive medication. Clonidine is an agonist at presynaptic $\alpha$-2 adrenergic re-ceptors and blocks the release of central and peripheral norepinephrine. Noradrenergic neurons in the locus ceruleus of the brain show increased neuronal activity during opiate withdrawal and this effect can be blocked by $\alpha$-2 adrenergic agonists such as clonidine (21). On the basis of this observation, clonidine was tested clinically as a blocker of opioid withdrawal signs and symptoms (22,23). Subsequent double-blind clinical trials confirmed that clonidine was more effective than placebo and slightly less effective then a slow methadone taper in reducing signs and symptoms of opiate withdrawal in both inpatients and outpatients. Clonidine was found to be effective in patients withdrawing from either short-acting opioids, such as heroin, or long-acting opioids, such as methadone. However, clonidine was never formally approved by the FDA for the treatment of withdrawal in opioid-dependent patients.

Although most studies have been performed using oral clonidine, clonidine is available as a transdermal delivery system (described earlier), and this modality has been used successfully to treat opioid withdrawal (24). Advantages of the clonidine transdermal system include attainment of more constant plasma levels of clonidine (i.e., the avoidance of peaks and troughs) and the psychological benefit of an addicted patient not taking pills to relieve discomfort. A disadvantage of transdermal clonidine is a lag time of up to 72 hours for medication effect after applying the patch to the skin. Because of the lag time application of the clonidine transdermal patch and the attainment of new steady-state blood levels, supplementation of the patch with the more rapidly absorbed oral clonidine may be required to treat emergent withdrawal symptoms. Likewise, hypotension from overmedication with transdermal clonidine will take several hours to resolve after dose reduction.

When used clinically to treat opioid withdrawal, clonidine suppresses approximately 75% of opioid withdrawal signs and symptoms, especially autonomic hyperactivity (tremor, piloerection, tachycardia), anxiety, and gastrointestinal symptoms (cramps and diarrhea). Clonidine is administered in increasing doses such that opioid withdrawal signs and symptoms are decreased but blood pressure is maintained. A typical schedule for transdermal clonidine uses 0.1 mg on day 1 following discontinuation of opiates, 0.2 mg on day 2, 0.3 mg on day 3, and 0.4 mg on day 4. This maximal dose is continued for 5–7 additional days for short-acting opioids and 10–14 additional days for long-acting opioids.

Opioid withdrawal symptoms not significantly ameliorated by clonidine include drug craving, insomnia, and arthralgias and myalgias. Insomnia is best treated with a short-acting hypnotic such as chloral hydrate, and pain may respond to nonnarcotic analgesics such as a nonsteroidal antiinflammatory medication or acetaminophen. Side effects of clonidine include dry mouth, sedation, and orthostatic hypotension. Clonidine should be used cautiously in hypotensive patients and in those receiving other antihypertensive, antidepressants, stimulants, or antipsychotics.

### Opioid Maintenance

The most widely used pharmacological treatments for opioid-dependent individuals include pharmacological maintenance treatments with the opiate agonists methadone and L-α-acetylmethodol (LAAM), maintenance with the partial opiate agonist buprenorphine, and opiate antagonist therapy with naltrexone. All of these medications are best used in the setting of a structured, maintenance treatment program, which includes monitored medication administration; periodic, random urine toxicological screening to assess compliance; and intensive psychological, medical, and vocational services. Maintenance treatments reduce use of illicit opiates by increasing drug tolerance, thereby decreasing the subjective effects of illicitly administered opiates, and by stabilizing mood, thereby decreasing self-medication. Maintenance treatments also provide an incentive for treatment so that they may be exposed to other therapies.

**Methadone Maintenance.** Methadone is a synthetic opiate, which is orally active, possesses a long duration of action, produces minimal sedation or high, and has few side effects at therapeutic doses. A single daily dose of methadone will prevent the onset of withdrawal for at least 24 hours. Since its introduction in 1965, methadone maintenance has become a major modality of long-term treatment of opioid abuse and dependence (25). Currently, over 100,000 individuals are maintained on methadone in the U.S.

Although orally administered, an elaborate medication delivery system has evolved to ensure the controlled administration of this medication. Methadone is dissolved in small aliquots of a sweetened, flavored liquid vehicle and stored in single-dose plastic bottles. Each dose of methadone is dispensed daily and must be consumed under the direct observation of the dispensing nurse or pharmacist to ensure compliance. Frequent urine samples are obtained at randomly determined intervals for toxicological screening to confirm the presence of methadone and the lack of other illicit drugs. In addition, patients receive counseling, medical, and social services to assist them in achieving a drug-free lifestyle. Long-standing program participants are allowed contingency take-home doses of methadone, which patients may self-administer. However, these are withdrawn for missing appointments or for evidence of illicit drug use. Doses of methadone usually range from 20 mg per day to over 100 mg per day. Higher doses are shown to be generally associated with better retention in treatment.

Many studies have shown the efficacy of methadone maintenance in the treatment of addicts who are dependent on heroin and other opiates. Methadone-treated patients show increased treatment retention, improved physical health, decreased criminal activity, increased employment, and decreased chance of becoming HIV positive (26). Methadone is most effective in the context of a program that provides intensive psychosocial and medical services, and adequate methadone dosing. The use of methadone for maintenance is highly regulated by government agencies. Senay (27) provides an excellent recent review of the theory and practice of methadone maintenance.

**L-α-Acetylmethodol Acetate (LAAM).** A long-acting, orally active opiate, the pharmacological properties of LAAM are similar to methadone. Studies on LAAM have shown it to be equal or superior to methadone maintenance in reducing IV drug use, when used in the context of a structured maintenance treatment program (28). The advantages of LAAM include a slower onset of effects and a longer duration of action than methadone. This allows LAAM to be administered only 3 times per week, reducing the cost of preparation and monitoring of medication and reducing the use of take-home medications that may be diverted to illicit use. Patients treated with LAAM should be started on 20 mg administered 3 times weekly, with the dose increased weekly in 10 mg increments as necessary. Doses up to 80 mg 3 times weekly are safe and effective.

**Sublingual Buprenorphine.** Buprenorphine is a partial agonist opiate medication (mixed agonist-antagonist), originally used medically as an analgesic. It has high affinity for the μ opioid receptor and the κ receptor. Buprenorphine possesses both agonist and antagonist properties—agonist properties predominate at lower doses and antagonist properties predominate at higher doses. Cessation in buprenorphine in dependent individuals results in a withdrawal syndrome that is much milder than that observed with pure opioid agonists. These properties of the drug suggested its use as a maintenance medication in the treatment of chronic opioid dependence. However, oral buprenorphine has poor bioavailability and only parenteral preparations of the medication are used.

Sublingual (SL) administration is an effective way of administering medication that may not be orally active. SL administration of buprenorphine results in effective medication plasma levels and was therefore tested in the treatment of opioid dependence. In the setting of a structured treatment program, daily dosing of SL buprenorphine was found to effective in the maintenance treatment of narcotics addicts, reducing illicit drug use (29–31). Buprenorphine may also reduce concomitant cocaine use in opiate addicts (32).

Buprenorphine doses usually range from 4 mg per day to up to 16 mg per day, administered sublingually. Advantages of buprenorphine include a milder withdrawal syndrome upon discontinuation and less potential for abuse, as agonist effects diminish at higher doses. Opioid-dependent patients may be started on 2 to 4 mg buprenorphine immediately after opiates are discontinued, and the dose of buprenorphine titrated to 8 to 16 mg over several days (33).

**Naltrexone.** Opioid antagonist therapy reduces the use of illicit drugs by blocking the effect of the drugs at neurotransmitter receptors, leading to decreased use. There is some evidence that opiate antagonists may block craving for opiates as well. Naltrexone (Trexan®) is a long-acting, orally active opioid antagonist, which when taken regularly, entirely blocks μ opioid receptors and thus blocks the euphoric, analgesic, and sedative properties of opioids (34). Oral naltrexone is administered either daily to detoxified opioid users at a dose of 50 mg, or 3 times weekly at doses of 100 mg, 100 mg, and 150 mg. Although naltrexone is quite effective when taken as prescribed, most studies have demonstrated poor medication compliance among subjects. The drug appears to be most effective in motivated individuals with good social support and appears less helpful for heroin addicts or less motivated individuals.

Because of the low compliance with oral naltrexone formulations, there has been interest in the development of alternative drug delivery systems to improve medication compliance. Two parenteral depot formulations of naltrexone have been developed; injectable naltrexone microspheres coated with poly(DL-lactic acid) (35) and an implantable biodegradable copolymer polylactic/glycolic matrix delivery system (36,37). The injectable microspheres have been tested on animals and humans and found to result in plasma naltrexone levels that would effectively block exogenous opioids and that are stable for at least 30 days. Problems with residual organic solvent used in the microsphere preparation have delayed FDA approval of this method; however, new preparation techniques avoid the problem with organic solvents, and the injectable microspheres are being tested again. Both of these methodologies remain experimental but are undergoing continued testing.

## TREATMENTS FOR ALCOHOL DEPENDENCE

### Disulfiram

An irreversible inhibitor of the enzyme acetaldehyde dehydrogenase, disulfiram (Antabuse®) is used as an adjunctive treatment in selected alcoholic patients (38,39). If alcohol is consumed in the presence of disulfiram, the toxic metabolite acetaldehyde accumulates in the body, producing tachycardia, skin flushing, diaphoresis, dyspnea, nausea, and vomiting. Hypotension and death may occur if large amounts of alcohol are consumed. This unpleasant and potentially dangerous reaction provides a strong deterrent to the consumption of alcohol. Patients using disulfiram must be able to understand its benefits and risks. Alcohol present in foods, shaving lotion, mouthwashes, or over-the-counter medications may also produce a disulfiram reaction and must be avoided.

As increased disulfiram compliance improves treatment success, there has been interest in the development of long-lasting depot formulations of disulfiram to improve medication compliance (40). Several methods for depot disulfiram administration have been developed, including the implantation of disulfiram surrounded by a semipermeable membrane to retard absorption, direct tablet implantation, and the subcutaneous injection of disulfiram in vehicles such as methylcellulose or polysorbate 80. Unfortunately, most clinical studies have shown an inability of implanted forms of disulfiram to induce a significant disulfiram–alcohol reaction when subjects are challenged with alcohol (41).

### Naltrexone

In clinical trials with recently abstinent human alcoholics, subjects treated with the opioid antagonist naltrexone had lower rates of relapse to heavy drinking, and more total abstinence, than did a placebo group (42,43). Subjects receiving naltrexone also report decreased craving and decreased high from alcohol. Naltrexone is thought to act by blocking the alcohol-induced release of dopamine in the nucleus accumbens and other brain areas that control the reinforcing properties of drugs and alcohol.

One factor that appears important for the efficacy of naltrexone is medication compliance. Two placebo-controlled clinical trials with oral naltrexone demonstrated significant efficacy in reducing drinking only in subjects that showed high compliance with medication ingestion (44,45). Thus, there is interest in the development of alternative drug delivery systems for naltrexone that will enhance compliance and optimize medication effects to improve treatment.

A recent report comparing oral naltrexone with injectable sustained-release naltrexone microspheres (35) in 20 patients found comparable effects of the two different preparations in reducing alcohol consumption (46). Side effects of the oral and sustained-release preparations were also similar. More research needs to be conducted on the sustained-release forms of naltrexone; however, the initial results are encouraging.

## SUMMARY

Controlled drug delivery systems have been applied to the treatment of several addictive disorders, including nicotine dependence, opioid dependence, and alcohol dependence. In the case of nicotine dependence, several commercial products are available, giving clinicians and patients considerable flexibility in drug dosing. Controlled drug delivery systems are currently being studied in the maintenance treatment of alcoholism and opioid dependence.

## BIBLIOGRAPHY

1. D.A. Regier, M.E. Farmer, and D.S. Rae, *J. Am. Med. Assoc.* **264**, 2511–2518 (1990).
2. D.P. Rice, *Alcohol Health Res. World* **17**, 10–11 (1990).
3. J.M. McGinnis and W.H. Foege, *J. Am. Med. Assoc.* **270**, 2201–2212 (1993).
4. American Psychological Association, *Diagnostic and Statistical Manual*, 4th ed., APA Press, Washington, D.C., 1994.
5. Centers for Disease Control, *Morbid. Mortal. Wkly. Rep.* **43**, 342–346 (1994).
6. J.R. Hughes and D.K. Hatsukami, *Arch. Gen. Psychiatry* **43**, 289–294 (1986).
7. N. Breslau, M. Kilbey, and P. Andreski, *Am. J. Psychiatry* **149**, 464–469 (1992).
8. H.L. Greene, R. Goldberg, and J.K. Ockene, *J. Gen. Intern. Med.* **3**, 75–87 (1988).
9. Agency for Health Care Policy and Research, *Clin. Pract. Guideline* **18B**, 1–10 (1996).
10. J.L. Tang, M. Law, and N. Wald, *Br. Med. J.* **308**, 21–26 (1994).
11. Transdermal Nicotine Study Group, *J. Am. Med. Assoc.* **266**, 3133–3138 (1991).
12. N.G. Schneider, *Addict. Behav.* **9**, 149–156 (1984).
13. M.G. Goldstein and R.S. Niaura, in J.A. Cocores, ed., *Clinical Management of Nicotine Dependence*, Springer-Verlag, New York, 1991, pp. 181–195.
14. N. Herrera et al., *Chest* **108**, 447–451 (1995).

15. A. Hjalmarson, F. Nilsson, L. Sjostrom, and O. Wiklund, *Arch. Intern. Med.* **157**, 1721–1728 (1997).

16. A.H. Glassman, F. Stetner, B.T. Walsh, and P.S. Raizman, *J. Am. Med. Assoc.* **259**, 2863–2866 (1988).

17. D.T. Lowenthal et al., *Am. Heart J.* **112**, 893–900 (1986).

18. A.V. Prochazka et al., *Arch. Intern. Med.* **152**, 2065–2069 (1992).

19. J.J. Green and D.H. Cordes, *West. J. Med.* **151**, 79–80 (1989).

20. T. Reisine and G. Pasternack, in J.G. Hardman and L.E. Limbird, eds., *Goodman and Gilman's The Pharmacological Basis of Therapeutics*, 9th ed., McGraw-Hill, New York, 1996, pp. 521–555.

21. G. Aghajanian, *Nature (London)* **276**, 186–187 (1978).

22. D.S. Charney et al., *Arch. Gen. Psychiatry* **38**, 1273–1278 (1981).

23. M.S. Gold, A.C. Pottash, D.R. Sweeney, and H.D. Kleber, *J. Am. Med. Assoc.* **234**, 343–344 (1979).

24. L. Spencer and M. Gregory, *J. Substance Abuse Treat.* **6**, 113–117 (1989).

25. V.P. Dole and M. Nyswander, *J. Am. Med. Assoc.* **93**, 646–650 (1965).

26. J.C. Ball, W.R. Lange, C.P. Myers, and S.R. Friedman, *J. Health Soc. Behav.* **29**, 214–226 (1988).

27. E. Senay, *Int. J. Addict.* **20**, 803–821 (1985).

28. W. Ling, R.A. Rawson, and P.A. Compton, *J. Psychoact. Drugs* **26**, 119–128 (1994).

29. P.A. Compton, D.R. Wesson, V.C. Charuvastra, and W. Ling, *Am. J. Addict.* **5**, 220–230 (1996).

30. R.E. Johnson, J.J. Jaffe, and P.J. Fudala, *J. Am. Med. Assoc.* **267**, 2750–2755 (1992).

31. T.R. Kosten, C. Morgan, and H.D. Kleber, *Am. J. Drug Alcohol Abuse* **17**, 119–128 (1991).

32. T.R. Kosten, H.D. Kleber, and C. Morgan, *Biol Psychiatry* **26**, 637–639 (1989).

33. T.R. Kosten, C. Morgan, and H.D. Kleber, *NIDA Res. Monogr.* **121**, 101–119 (1992).

34. R.B. Resnick, E. Schuyten-Resnick, and A.M. Washton, *Annu. Rev. Pharmacol. Toxicol.* **20**, 463–470 (1980).

35. E.S. Nuwayser, D.J. DeRoo, P.D. Balskovich, and A.G. Tsuk, *NIDA Res. Monogr.* **105**, 532–533 (1991).

36. R.H. Reuning et al., *J. Pharmacokinet. Biopharm.* **11**, 369–387 (1983).

37. A.C. Sharon and D.L. Wise, *NIDA Res. Monogr.* **28**, 194–213 (1981).

38. C. Brewer, *Alcohol Alcohol.* **28**, 383–395 (1993).

39. J. Chick, K. Gough, and W. Falkowski, *Br. J. Psychiatry* **161**, 84–89 (1992).

40. M.D. Faiman, K.E. Thompson, and K.L. Smith, in C.A. Naranjo and E.M. Sellers, eds., *Novel Pharmacological Interventions for Alcoholism*, Springer-Verlag, New York, 1992, pp. 267–272.

41. J.C. Hughes and C.C. Cook, *Addiction* **92**, 381–395 (1997).

42. S.S. O'Malley et al., *Arch. Gen. Psychiatry* **49**, 881–887 (1992).

43. J.R. Volpicelli, A.I. Alterman, M. Hayashida, and C.P. O'Brien, *Arch. Gen. Psychiatry* **49**, 876–880 (1992).

44. J. Chick, *10th World Psychiatry Conf.*, Madrid, Spain, Aug. 24, 1996.

45. J.R. Volpicelli et al., *Arch. Gen. Psychiatry*, **54**, 737–742 (1997).

46. H.R. Kranzler, V. Modesto-Lowe, and E.S. Nuwayser, *Alcohol Clin. Exp. Res.* **22**, 1074–1079 (1998).

# B

## BIOADHESIVE DRUG DELIVERY SYSTEMS

Edith Mathiowitz
Donald Chickering
Jules S. Jacob
Camilla Santos
Brown University
Providence, Rhode Island

**KEY WORDS**

Adsorption theory
Bioadhesion
Bioadhesive drug delivery systems
Bioadhesive polymers
Electronic theory
Fracture theory
Mucus
Oral delivery
Wetting theory

**OUTLINE**

The goals of this article are to give a general introduction to the field of bioadhesion, to discuss some specific theories of bioadhesion as well as methods involved in its quantification, and end with an overview of general applications of bioadhesion in the medical field. The current literature contains an enormous amount of material on bioadhesion, preventing us from including all the information available in the field, so we advise interested readers desiring further information to consult general books on bioadhesion (1–3).

Our approach to discussing the current trends in bioadhesion in this article is a bit unusual; we first cover the general concepts of bioadhesion and discuss current applications of bioadhesive polymers. Then we deviate from the norm and discuss what we know best: our work in bioadhesion over the past six years. We thought it more worthwhile to introduce, as a case study, the target of research in our laboratory. Thus, in the following section we discuss the general concept of bioadhesion, and in the section after that, "Development of Bioadhesive Microspheres: A Case Study," we describe in more detail several aspects of our research, including where appropriate the work of others as well. Our research covers five main focal points: (*1*) the development of an accurate and reproducible method to analyze the bioadhesive interaction between individual polymer microspheres and soft tissues in a controlled environment closely mimicking in vivo conditions; (*2*) the quantification of the bioadhesive properties of hydrogel and thermoplastic microspheres, including the characterization of a series of polymers suitable for bioadhesive drug delivery systems; (*3*) the use of several in vitro characterization techniques to gain insight into the mechanism and forces responsible for bioadhesion; (*4*) the determination of the effects bioadhesive polymers have on in vivo drug delivery by measuring gastrointestinal (GI) transit time; and (*5*) the evaluation of the bioavailability of encapsulated drugs such as dicumarol, insulin, and plasmid DNA.

## BACKGROUND OF BIOADHESION

### Introduction

Over the past 20 years, interaction among the fields of polymer and material science and the pharmaceutical industry has resulted in the development of what are known as drug delivery systems (DDSs), or controlled-release systems (4–7). The advantages of using polymer-based devices over traditional dosage forms include (*1*) the ability to optimize the therapeutic effects of a drug by controlling its release into the body; (*2*) lower and more efficient doses; (*3*) less frequent dosing; (*4*) better patient compliance; (*5*) flexibility in physical state, shape, size, and surface; (*6*) the ability to stabilize drugs and protect against hydrolytic or enzymatic degradation; and (*7*) the ability to mask unpleasant taste or odor (4–7).

DDSs are already in widespread use, and their applications seem limited only by the imaginations of their inventors. They have been designed in the forms of reservoirs, homogeneous matrices, emulsions, capsules, rods, tablets, patches, and pumps (7–13). Systems have been engineered to release in response to pH, ultrasonic, magnetic, photochemical, or thermal stimulation (7,14–16). Administration routes, selected to target specific absorption pathways, have included oral, rectal, transdermal, subcutaneous, inhalatory, intrauterine, intravaginal, and intravenous (4).

Owing to the fact that intimate contact between a delivery device and the absorbing cell layer will improve both effectiveness and efficiency of the product, many researchers have recently focused on developing bioadhesive drug delivery systems (BDDSs) (17). The term *bioadhesion* refers to either adhesion between two biological materials or adhesion between some biological material (including cells, cellular secretions, mucus, extracellular matrix, and so on) and an artificial substrate (metals, ceramics, polymers, etc.). Bioadhesive materials have found numerous applications in the medical field. For example, cyanoacrylates (commonly known as "super glues") have frequently been used for both orthopedic and dental applications. In terms of the pharmaceutical industry, bioadhesion generally refers to adhesion between a polymer-based delivery system and soft tissue in the presence of water (18). Possible means of administration for BDDSs include the occular, respiratory, buccal, nasal, GI, rectal, urethral, and vaginal routes (17–21). GI bioadhesive devices that can be administered orally are of considerable interest owing to the ease of administration and targeted contact with the absorbing intestinal epithelium (1,4,22,23).

Systemic drug delivery via absorption into the bloodstream through the GI epithelium can be limited by drug degradation during the first pass through the liver; however, the GI mucosa offers several advantages as an administration site over other mucus membranes. These advantages include the following: (1) the oral administration route is familiar, convenient, and an accepted means of dosing for most people; (2) the GI epithelium offers a large surface area for absorption; and (3) the GI epithelium provides a close connection with a vast blood supply. The development of efficient orally delivered BDDSs could enable the following four important effects: (1) enhanced bioavailability and effectiveness of drug due to targeted delivery to a specific region of the GI tract, (2) maximized absorption rate due to intimate contact with the absorbing membrane and decreased diffusion barriers, (3) improved drug protection by polymer encapsulation and direct contact with absorbing cell layers, and (4) longer gut transit time resulting in ≅ extended periods for absorption (23–25).

Many researchers have developed various techniques for determining and evaluating bioadhesion. Each of these systems has typically been tailored to the needs of a particular experiment (see "Methods to Evaluate Bioadhesive Interactions"). For this reason, the experimental conditions and resulting data have varied greatly between studies. It has been very difficult to compare one set of experimental findings with another. Also, owing to variations in mucus, tissue, tissue preparation techniques, polymers (their molecular weight, degree of cross-linking, degree of hydration, etc.), and polymer geometry, many of the experiments have lacked reproducibility and shown statistically high variations in data (26,27). Surprisingly, our lab has been the only group to attempt to mimic both physiological conditions and final delivery device geometries in a single in vitro experiment (28).

In developing orally administered BDDSs, it is important to realize that the targeted GI tissue is coated with a continuous layer of protective secretions known as mucus. With this in mind, BDDSs must be designed to either penetrate this boundary layer and bind to the underlying epithelium or adhere directly to the mucus (29). In either event, because mucus and epithelial cells are continuously sloughed off and replaced, it is unrealistic to expect a delivery device to adhere permanently to the luminal surface of the digestive system (30,31). Instead, a more feasible goal for oral bioadhesive systems should be delayed transit through the gut, during which time intimate contact is achieved between bioadhesive and target tissue. Once adhesion has occurred, the device can deliver its bioactive contents directly to the absorbing cell layer, minimizing losses to the luminal environment.

## Mucus and the Mucosal Layer

The epithelium of the gut is protected by a continuous coating of mucus, which is secreted by a number of different cells: (1) mucus neck cells, from the necks of the gastric glands in the stomach, produce soluble mucus; (2) surface epithelial goblet cells produce visible mucus in the stomach as well as in the small and large intestines; (3) cells collectively known as Brunner's glands produce mucus in the proximal duodenum; and (4) goblet cells lining the crypts of Lieberkühn produce mucus in both the large and small intestines (32,33). Once produced, mucus is stored in large granules and then released either by exocytosis or exfoliation of the entire cell (32). Mucus in the GI tract functions mainly as a lubricant, protecting against abrasive, mechanical damage from food. It also creates a barrier against destruction of the epithelial cell layer by harsh gastric pH conditions and digestive enzymes (34–36). Mucus secretion is directly stimulated by even the smallest mechanical or chemical irritation of the mucosa (32).

Both the composition and thickness of the mucus layer vary with location, degree of GI activity, sex, and state of health (37,38). In general, it can be as thick as 1 mm in humans and consists mainly of water (up to 95%), electrolytes, proteins, lipids, and glycoproteins (29,39,40).

The glycoprotein mucins (molecular weight of approximately 2 million) give mucus its unique gel-like characteristics. These glycoproteins consist of a protein core (≅18.6–25.6% by weight) with covalently attached carbohydrate side chains (≅81.4–74.4% by weight) (39,41–43). Although specific composition varies with location (44), in general there are about four times as many amino acid groups in the protein core as there are side chains (Fig. 1) (29,42). Each of the side chains is anywhere from 2 to 20 sugars in length and terminates with either L-fucose or sialic acid. Mucus glycoproteins in solution are anionic polyelectrolytes at pH greater than 2.6. Disulfide, electrostatic, and hydrophobic interactions help to entangle mucin chains to produce the gel-like properties characteristic of mucus (18,43,45).

A pH gradient exists in the mucus layer from the cell surface to about 750 $\mu$m into the center of the lumen that varies with location along the gut (46). The pH at the epithelial surface is maintained around 7 in all areas of the

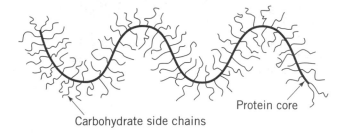

**Figure 1.** Schematic diagram of a glycoprotein mucin, showing protein core with carbohydrate side chains.

GI tract owing to alkaline fluids that are secreted along with mucus, while luminal pH may range anywhere from 2.0 (stomach) to 7.6 (duodenum) (32,36).

Below the mucus layer of the intestine are the cellular layers of the gut wall, consisting of epithelium, lamina propria, and muscularis mucosa. These tissues and mucus layers are collectively known as mucosa. The epithelium is a single layer of columnar cells that are arranged into fingerlike projections called villi. Each epithelial cell is blanketed with a coating of microvilli on its apical surface, known as the brush border. Within the lamina propria are numerous capillaries and lymphatic vessels or lacteals as well as mucus-secreting glands (32,47). Drugs and other molecules must first pass the barriers of the mucus layer, brush border, and epithelium before reaching the capillaries of the lamina propria. The muscularis mucosa is a thin, smooth muscle layer that produces folds in the epithelium known as valves or folds of Kerckring. The folds help to increase the absorbing surface area of the gut (33,47). In fact, the total absorptive area of the small intestine is approximately 250 m$^2$, which is about the size of a tennis court (33).

## Bioadhesion and Mucoadhesion

Although the term *bioadhesion* can refer to any bond formed between two biological surfaces or a bond between a biological and synthetic surface, in this article it will specifically be used to describe the adhesion between polymer samples, either synthetic or natural, and soft tissue (GI mucosa). Although the target of orally administered BDDSs may be the epithelial cell layer, the actual adhesive bond may form with either the cell layer, the continuous mucus layer, or a combination of the two. In the instances when the bond involves the mucus coating and the polymer device, many authors use the term *mucoadhesion* (48,49). In this article, *bioadhesion* is meant to include all three possible adhesive conditions, and *mucoadhesion* is used only when describing a bond between polymer and mucus.

## Mechanisms of Bioadhesion

The mechanisms involved in the formation of a bioadhesive bond are not completely clear. To develop ideal BDDSs, it is important to try to describe and understand the forces that are responsible for adhesive bond formation. Most research has focused on analyzing the bioadhesive bond between polymer hydrogels and soft tissue.

The process involved in the formation of such bioadhesive bonds has been described in three steps: (*1*) wetting and swelling of polymer to permit intimate contact with biological tissue, (*2*) interpenetration of bioadhesive polymer (BP) chains and entanglement of polymer and mucin chains, and (*3*) formation of weak chemical bonds (17,19). In the case of hydrogels, it has been determined that several polymer characteristics are required to obtain adhesion: (*1*) sufficient quantities of hydrogen-bonding chemical groups (-OH and -COOH), (*2*) anionic surface charges, (*3*) high molecular weight, (*4*) high chain flexibility, and (*5*) surface tensions that will induce spreading into the mucus layer (49). These characteristics favor the formation of bonds that are either mechanical or chemical in nature.

**Mechanical or Physical Bonds.** Mechanical bonds can be thought of as a physical connection of polymer and tissue, similar to interlocking puzzle pieces. On a macroscopic level, they can be caused by the inclusion of one substance into the cracks or crevices of another (50). On a microscopic scale, they involve the physical entanglement of mucin strands with flexible polymer chains and the interpenetration of mucin strands into the porous structure of a polymer substrate. The rate of penetration of polymer strands into the mucin layer is dependent on chain flexibility and diffusion coefficients of each. The strength of the adhesive bond is directly proportional to the depth of penetration of the polymer chains. Other factors that influence bond strength include the presence of water, the time of contact between the materials, and the length and flexibility of the polymer chains (18).

**Chemical Bonds.** Chemical bonds can include strong primary bonds (i.e., covalent bonds) as well as weaker secondary forces such as ionic bonds, Van der Waals' interactions, and hydrogen bonds. Although it is possible to develop polymers with reactive functional groups that could theoretically form permanent covalent bonds with proteins on the surface of the epithelial cells, there are several reasons why investigators have not focused on such systems. First, the mucus barrier may inhibit direct contact of polymer and tissue. Second, permanent chemical bonds with the epithelium may not produce a permanently retained delivery device because the epithelial cells are exfoliated every three to four days. Third, biocompatibility of such binding has not been thoroughly investigated and could pose significant problems (18). For these reasons, most researchers have focused on developing hydrogel, mucoadhesive systems that bond through either Van der Waals' interactions or hydrogen bonds. Although these forces are very weak, strong adhesion can be produced through numerous interaction sites. Therefore, polymers with high molecular weights and greater concentrations of reactive polar groups (such as -COOH and -OH) tend to develop more intense mucoadhesive bonds (26,51).

## Theories on Bioadhesion

Several theories have been developed to describe the processes involved in the formation of bioadhesive bonds. These theories have been used as guidelines in engineering possible BDDSs. Some are based on the formation of mechanical bonds, while others focus on chemical interactions.

### The Electronic Theory.

The electronic theory is based on an assumption that the bioadhesive material and the glycoprotein mucin network have different electronic structures. On this assumption, when the two materials come in contact with each other, electron transfer will occur in an attempt to balance Fermi levels, causing the formation of a double layer of electrical charge at the interface. The bioadhesive force is believed to be due to attractive forces across this electrical double layer. This system is analogous to a capacitor, where the system is charged when the adhesive and substrate are in contact and discharged when they are separated (52). The electronic theory has produced some controversy regarding whether the electrostatic forces are an important *cause* or the *result* of the contact between the bioadhesive and the biological tissue (53).

### The Adsorption Theory.

This theory states that the bioadhesive bond formed between an adhesive substrate and intestinal mucosa is due to Van der Waals' interactions, hydrogen bonds, and related forces (54,55). The adsorption theory is the most widely accepted theory of adhesion and has been studied in depth by both Kinloch and Huntsberger (50,56,57).

### The Wetting Theory.

The ability of bioadhesive polymers or mucus to spread and develop intimate contact with their corresponding substrate is one important factor for bond formation. The wetting theory, which has been used predominantly in regards to liquid adhesives, uses interfacial tensions to predict spreading and, in turn, adhesion (49,58–60).

Figure 2 schematically represents a BP spreading over soft tissue. The contact angle ($\phi$), which should be zero or near zero for proper spreading, is related to interfacial tensions ($\gamma$) through Young's equation:

$$\gamma_{tg} = \gamma_{bt} + \gamma_{bg} \cos\phi \qquad (1)$$

where the subscripts t, g, and b stand for tissue, gastrointestinal contents, and bioadhesive polymer, respectively. For spontaneous wetting to occur, $\phi$ must equal zero and, therefore, the following must apply (49):

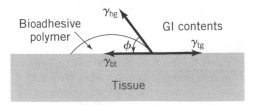

**Figure 2.** Schematic diagram showing the interfacial tensions involved in spreading.

$$\gamma_{tg} \geq \gamma_{bt} + \gamma_{bg} \qquad (2)$$

The spreading coefficient, $S_{b/t}$, of a bioadhesive over biological tissue in vivo can be used to predict bioadhesion and can be determined as follows:

$$S_{b/t} = \gamma_{tg} - \gamma_{bt} - \gamma_{bg} \qquad (3)$$

For the bioadhesive to displace GI luminal contents and make intimate contact with the biological tissue (i.e. spreading), the spreading coefficient must be positive. Therefore, it is advantageous to maximize the interfacial tension at the tissue/GI contents interface ($\gamma_{tg}$) while minimizing the surface tensions at the other two interfaces ($\gamma_{bt}$ and $\gamma_{bg}$) (49).

It is theoretically possible to determine each of the parameters that make up the spreading coefficient. The interfacial tension of the tissue/GI contents interface ($\gamma_{tg}$) can be determined in vitro using classical Zisman analysis (61,62), although it has not been done. The interfacial tension at the BP/GI contents interface ($\gamma_{bg}$) can be experimentally determined using traditional, surface tension–measuring techniques such as the Wilhelmy plate method. Lastly, it has been shown that the BP/tissue interfacial tension ($\gamma_{bt}$) can be calculated as follows (63,64):

$$\gamma_{bt} = \gamma_b + \gamma_t - 2F(\gamma_b\gamma_t)^{1/2} \qquad (4)$$

where values of the interaction parameter ($F$) can be found in previously published papers (65,66). Extensive studies have been conducted to determine the surface tension parameters for several biological tissues ($\gamma_t$) and many commonly used biomaterials ($\gamma_b$) (67).

The BP/tissue interfacial tension ($\gamma_{bt}$) has been shown to be proportional to the square root of the polymer–polymer Flory interaction parameter ($c$):

$$\gamma_{bt} \sim c^{1/2} \qquad (5)$$

When $c$ is small, the bioadhesive and biological components are similar structurally. This results in increased spreading and, therefore, greater adhesive bond strength (68,69).

Besides the spreading coefficient, another important parameter that may indicate the strength of an adhesive bond is the specific work of adhesion ($W_{bt}$). According to the Dupré equation, this is equal to the sum of the surface tensions of the tissue and bioadhesive, minus the interfacial tension (70):

$$W_{bt} = \gamma_b + \gamma_t - \gamma_{bt} \qquad (6)$$

Thus, using the wetting theory, it is possible to calculate spreading coefficients for various bioadhesives over biological tissues and predict the intensity of the bioadhesive bond. By measuring surface and interfacial tensions, it is possible to calculate work done in forming an adhesive bond. Both spreading coefficients and bioadhesive work directly influence the nature of the bioadhesive bond and therefore provide essential information for development of BDDSs.

**The Diffusion Theory.** The diffusion theory suggests that interpenetration and entanglement of BP chains and mucus polymer chains produce semipermanent adhesive bonds (Fig. 3), and bond strength is believed to increase with the depth of penetration of the polymer chains (71).

Penetration of BP chains into the mucus network, and vice versa, is dependent on concentration gradients and diffusion coefficients. Obviously, any cross-linking of either component will tend to hinder interpenetration, but small chains and chain ends may still become entangled. It has not been determined exactly how much interpenetration is required to produce an effective bioadhesive bond, but it is believed to be in the range of 0.2–0.5 $\mu$m. It is possible to estimate penetration depth ($l$) with the following equation:

$$l = (tD_b)^{1/2} \tag{7}$$

where $t$ is time of contact and $D_b$ is the diffusion coefficient of the biomaterial in mucus. The maximum achievable bioadhesive bond for a given polymer is believed to occur when the depth of penetration is approximately equal to the end-to-end distance of the polymer chains (72,73).

For diffusion to occur, it is important to have good solubility of one component in the other; the BP and mucus should be of similar chemical structure. Therefore, the strongest bioadhesive bonds should form between those biomaterials whose solubility parameters ($\delta_b$) are similar to those of mucus glycoproteins ($\delta_g$) (50). Thus, the diffusion theory states that interpenetration and entanglement of polymer chains are responsible for bioadhesion. The more structurally similar a bioadhesive is to mucus, the greater the mucoadhesive bond will be.

**The Fracture Theory.** The most useful theory for studying bioadhesion through tensile experiments has been the fracture theory, which analyzes the forces required to separate two surfaces after adhesion. The maximum tensile stress ($\sigma_\mu$) produced during detachment can be determined by dividing the maximum force of detachment, $F_m$, by the total surface area ($A_o$) involved in the adhesive interaction:

$$\sigma_m = F_m/A_o \tag{8}$$

In a uniform single-component system, fracture strength ($\sigma_f$), which is equal to the maximum stress of detachment ($\sigma_m$), is proportional to fracture energy ($\gamma_c$), Young's modulus of elasticity ($E$), and the critical crack

length ($c$) of the fracture site, as described in the following relationship (74):

$$\sigma_f \sim (\gamma_c E/c)^{1/2} \tag{9}$$

Fracture energy ($\gamma_c$) can be obtained from the sum of the reversible work of adhesion, $W_r$ (i.e., the energy required to produce new fracture surfaces) and the irreversible work of adhesion, $W_i$ (i.e., the work of plastic deformation at the tip of the growing crack), where both values are expressed per unit area of the fracture surface ($A_f$):

$$\gamma_c = W_r + W_i \tag{10}$$

The elastic modulus of the system ($E$) is related to stress ($\sigma$) and strain ($\epsilon$) through Hooke's law:

$$E = \left[\frac{\sigma}{\epsilon}\right]_{\epsilon \to 0} = \left[\frac{F/A_0}{\Delta l/l_0}\right]_{\Delta l \to 0} \tag{11}$$

In this equation, stress is equal to the changing force ($F$) divided by the area ($A_0$), and strain is equal to the change in thickness ($\Delta l$) of the system divided by the original thickness ($l_0$) (75).

One critical assumption in equation 11 is that the system being investigated is of known physical dimensions and composed of a single uniform-bulk material. Considering this, equations 9 and 11 cannot be applied to analyze the fracture site of a multicomponent bioadhesive bond between a polymer microsphere and either mucus or mucosal tissue. For such analysis, the equations must be expanded to accommodate dimensions and elastic moduli of each component (76). Furthermore, to determine fracture properties of an adhesive union from separation experiments, failure of the adhesive bond must be assumed to occur at the bioadhesive interface (73). However, it has been demonstrated that fracture rarely, if ever, occurs at the interface but instead occurs close to it (75,77).

Although these limitations exist, because the fracture theory deals only with analyzing the adhesive force required for separation, it does not assume or require entanglement, diffusion, or interpenetration of polymer chains. Therefore, it is appropriate to use equation 8 to calculate fracture strengths of adhesive bonds involving hard, bioadhesive materials in which the polymer chains may not penetrate the mucus layer (though one must be aware that the measurement may be the fracture strength of the cohesive properties of the mucus instead of the fracture strength of the adhesive bond). Other theories that could be applicable to such a system are the wetting, electronic, and adsorption theories, so long as it is assumed that Van der Waals' interactions and hydrogen bonds can form between flexible mucin chains and presumably rigid polymer chains of hard, bioerodible surfaces.

## Methods to Evaluate Bioadhesive Interactions

Most researchers in the field of bioadhesion have developed their own techniques, which suit their particular systems and interests, for measuring and evaluating the interactions between BPs and biological substrates.

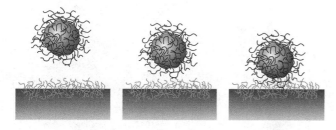

**Figure 3.** Interpenetration of bioadhesive and mucus polymer chains.

However, no attempt has been made to develop a standardized method of evaluation. Each technique has its own set of experimental conditions, and, therefore, it has been difficult to compare experimental findings among investigators. The following sections highlight some of the more significant methods that have been used in the past.

**In Vitro Techniques.** In vitro techniques have involved testing BPs against synthetic mucus, natural mucus, frozen tissue samples, or freshly excised tissue samples. For mucoadhesion studies, there have been two common techniques: the Wilhelmy plate technique and a shear test (26).

The Wilhelmy plate technique has traditionally been used for dynamic contact-angle measurement and involves a microbalance or tensiometer. A glass slide is coated with the polymer of interest and then dipped into a beaker of synthetic or natural mucus (Fig. 4a). The surface tension, contact angle, and adhesive force can be automatically measured using available software (78,79).

The shear test measures the force required to separate two polymer-coated glass slides joined by a thin film of natural or synthetic mucus (Fig. 4b) (80). The results of this technique often correlate well with in vivo test results (26).

A majority of the in vitro experiments have been variations of a simple tensile test that use either large tensile machines, modified tensiometers, or electrobalances (Fig. 5). Either a slab of polymer or a polymer-coated stopper is brought in contact with a section of biological tissue (either fresh or previously frozen). The samples are left in contact for a certain adhesion time, and the force required to break the adhesive bond is measured. These types of studies have been conducted with both hydrogels and thermoplastics, in air and various physiological buffers, and with varying temperatures and pH (2,24,48,51,81–85). The major flaw with most of these techniques is that they fail to incorporate both physiological conditions and delivery device geometries in a single, controlled experiment.

Some attempts have been made to mimic in vivo conditions in an in vitro environment. One such system consists of a unique flow chamber (Fig. 6) (73,86,87) in which a polymer microsphere is placed on the surface of a layer of natural mucus. Fluid, moving at physiologic rate, is introduced to the chamber, and the movement of the microsphere is monitored using video equipment. By measuring the size and speed of the microsphere, it was possible to calculate the bioadhesive force.

**In Vivo Techniques.** Most in vivo measurements of GI bioadhesive performance involve administering BPs to laboratory animals and monitoring their transit rate through the gut. Experiments have varied in both administration routes and tracking techniques. Bioadhesives have been orally force-fed by gavage (88–90), surgically implanted in the stomach (24) and infused with a perfusion pump through an in situ loop of the small intestine (48). Investigators have monitored transit using radiopaque markers (90,91), radioactive elements (92,93), and fluorescent labeling techniques (24).

### Bioadhesive Polymers

From current scientific literature, two classes of polymers appear to be of interest for bioadhesion: hydrophilic polymers and hydrogels. Recent studies have suggested that in the large class of hydrophilic polymers, those containing carboxyl groups exhibit the best bioadhesive properties (75,94). Therefore, tremendous effort has been made to develop polyacrylic acid–based BDDSs (27,80,82,83,95–99). In other studies (51,78,82,100,101), promising bioadhesive polymers have included sodium alginate, methylcellulose, carboxymethylcellulose, hydroxymethylcellulose, and cationic hydrogels such as chitosan. In general, hydrogels have most often been used for bioadhesive drug delivery because of the belief that polymer–mucin chain entangle-

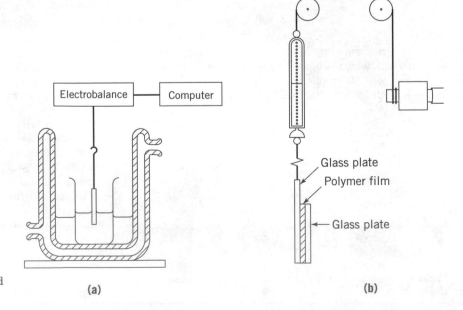

**Figure 4.** The Wilhelmy plate method (**a**) and the shear test for mucoadhesion studies (**b**).

(a)

(b)

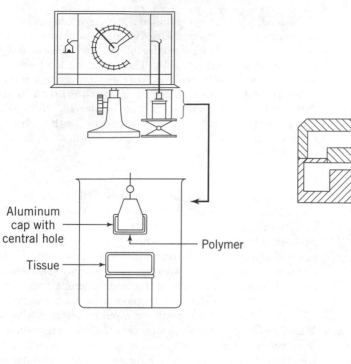

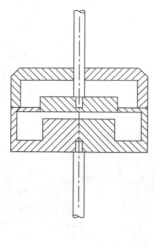

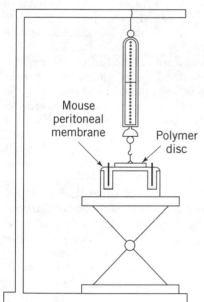

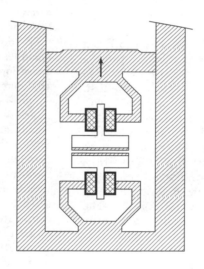

**Figure 5.** Tensile experiments for bioadhesive force measurements.

ment is an essential component in bioadhesive bond formation. However, other factors, such as surface energy, surface texture, electrical charge, and hydrophilic functional groups, may be equally important. It has recently been shown that nonhydrogel polymers that are high in hydrophilic functional groups can also produce intense bioadhesive interactions (25,28,102) and can be utilized to improve bioavailability of orally administered compounds (91). The case studies detailed in the following sections will describe this avenue of research in further detail.

### Applications of Bioadhesive Polymers

The bioadhesive polymers discussed in the previous section have been used in various forms as drug delivery ve-

hicles. It has been hypothesized that bioadhesive contact with an absorptive cell layer could improve drug transport, thereby increasing delivery efficiency. Furthermore, adhesives provide a means of securing DDSs to specific sites and localizing administration to targeted tissues. Common BDDS shapes and forms used in current research efforts have included tablets, powders, gels, patches, liposomes, and microspheres (1,21,23,73,103,104).

Bioadhesive tablets have been proposed for buccal, gingival, and vaginal delivery of such drugs as fluoride, miconazole, lidocaine, metronidazole, and morphine (17,99,105–116). These systems are usually produced by compressing a dry mixture of drug and polymer (most often polyacrylic acid). The tablets are administered by pressing

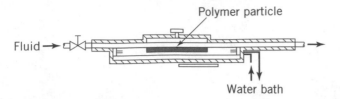

**Figure 6.** Fluid flow chamber for bioadhesive microsphere studies.

against a dried area in the mouth, and as saliva swells the polymer hydrogel, drug is released. Sustained release and adhesion are typically maintained for 8–10 h.

One group has investigated the possibility of delivering drug-containing bioadhesive powders to the mouth (117). The systems, designed to form an adhesive gel coating in situ, were shown to improve buccal retention and prolong delivery compared with systems delivered initially as a gel.

Bioadhesive gels have been investigated as drug carriers for delivery to the eye (12,95,118), mouth (2,12,117,119,120), rectum (121), vagina (13,110,122,123), and open wounds (124). As with bioadhesive tablets, polyacrylic acid (namely polycarbophil and Carbopol 934) has been the most commonly used polymer, most likely owing to its biocompatibility, strong adhesive characteristics, and U.S. FDA approval.

Other forms of bioadhesive systems have included patches, liposomes, and microspheres. Adhesive patches being developed by 3M, are intended for transdermal as well as buccal drug delivery (125,126). Bioadhesive liposomes, based on phosphatidylcholine, collagen, and hyaluronic acid, have been investigated as a means to enhance topical drug delivery (127–129). Microspheres, although widely studied as DDSs and often used as model systems for bioadhesive analysis (3,18,25,28,48,87,91,130–132), have thus far found fewer applications than other forms of BDDSs, with investigations focusing mainly on nasal (133,134) and GI administration (10,48,82,135–137). The following sections describe in depth our ongoing research efforts to utilize bioadhesive polymer microspheres as DDSs to the GI tract.

## DEVELOPMENT OF BIOADHESIVE MICROSPHERES: A CASE STUDY

### A Novel Electrobalance-Based Tensiometer

Although bioadhesive microspheres have been investigated as DDSs (18,82,87,131,133,138,139), only two methods have been reported for the measurement of bioadhesive interactions between individual polymer microspheres and soft tissues. One technique took advantage of the relationships between directional contact angles and physical forces to estimate strength of mucoadhesive bonds (131), whereas the other utilized a flow channel and basic fluid mechanic principles to calculate adhesive forces between microspheres and mucus (87,138). Neither of these systems was capable of directly measuring bioadhesive forces, and neither mimicked physiologic conditions by using viable tissue.

By modifying the operation of a sensitive microbalance, originally designed to measure dynamic contact angles, we developed a simple and reproducible method for measuring bioadhesion of microspheres. The system allows microspheres to be tested with freshly excised tissue while maintaining specific physiologic conditions, such as pH and temperature. Eleven parameters can be derived from the load-versus-deformation curves generated with each experiment, including fracture strength, deformation to failure, and tensile work.

**Experimental Design and Methods** *Microbalance.* A Cahn Dynamic Contact Angle Analyzer (Model DCA-322; CAHN Instruments, Inc.; Cerritos, CA) was modified to perform adhesive measurements (28,140). Although this piece of equipment is designed for measuring contact angles and surface tensions using the Wilhelmy plate technique, it is also an extremely accurate microbalance. The DCA-322 system includes a microbalance stand assembly, a Cahn DACS IBM-compatible computer, and an Okidata Microline 320 dot matrix printer (Fig. 7). The microbalance unit consists of stationary sample and tare loops and a z-translation stage powered by a stepper motor. The balance can be operated with samples weighing up to 3.0 g and has a sensitivity rated at $1 \times 10^{-5}$ mN. The stage speed can be varied from 20 to 264 $\mu$m/s.

To develop an automated, reproducible method for bioadhesion measurements, it was necessary to modify the operation of the balance and stage. The standard DACS IBM-compatible computer system was replaced with an Apple Macintosh II computer. The computer–microbalance interface was through the modem port, with an $-10$ V signal supplied by an external power supply (DC Power Supply 1630; BK Precision, Chicago, IL), hardwired through the RS-232 connection. Labview II software was used to write a user-friendly, menu-driven package to automatically run tensile experiments, with easily adjustable settings for stage speed, applied load, and time of adhesion. After each run, graphs of load versus stage position and load versus time were plotted, and 11 parameters were automatically calculated: (*1*) compressive deformation, (*2*) peak compressive load, (*3*) compressive work, (*4*) yield point, (*5*) deformation to yield, (*6*) returned work, (*7*) peak tensile load, (*8*) deformation to peak tensile load, (*9*) fracture strength, (*10*) deformation to failure, and (*11*) tensile work (28,140).

*Tissue Chamber.* To maintain physiological temperature and pH throughout experiments, a temperature-controlled tissue chamber was constructed (Fig. 8). The chamber was fabricated of plexiglass and consisted of a 3-mL tissue cell jacketed by a circulating water bath connected to a Fisher Scientific Isotemp refrigerated circulator (model 9000). Two stainless-steel clamps with thumbscrews were used to secure tissue samples to the bottom of the tissue cell.

*Mounting Microspheres.* To attach microspheres to the microbalance, it was necessary to first mount microspheres on rigid support wires. Thermoplastic microspheres were melt-mounted by piercing with red-hot, 280-$\mu$m-diameter, iron wires (Leeds & Northrup Company, Philadelphia, PA) cut to $\approx$2 cm in length (although this technique could alter

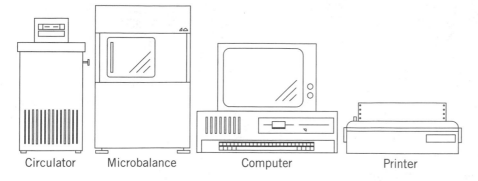

**Figure 7.** Cahn DCA 322 dynamic contact angle analyzer.

Circulator    Microbalance    Computer    Printer

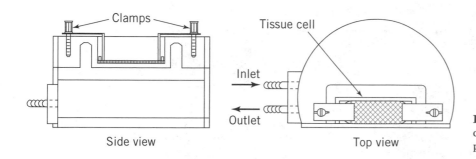

Side view    Top view

**Figure 8.** Tissue chamber showing tissue clamps and inlet and outlet flow ports for temperature control.

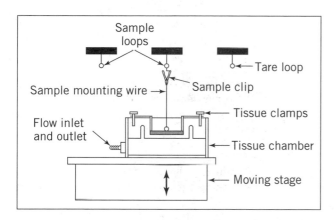

**Figure 9.** Microbalance enclosure with tissue chamber and polymer sample. The mobile stage can be raised and lowered at controlled rates while monitoring the load on the balance and the displacement of the stage. *Source:* Reproduced with permission from *Journal of Controlled Release* (Ref. 28).

the surface morphology of some materials, SEM analysis, not reported here, showed that the structure of the microspheres remained unchanged on the hemisphere intended for intestinal contact). Hydrogel microspheres were attached to the same wires using cyanoacrylate glue (Super Glue Corp., Hollis, NY). After mounting, wires were suspended from a sample clip in the microbalance enclosure (Fig. 9).

*Microbalance Operation.* Operation of the Cahn system is simple. The circulating water bath is adjusted to physiologic temperature (37°C), freshly excised tissue is placed in the chamber and submerged in phosphate-buffered saline (PBS), and the microsphere is suspended from the sample loop. The computer program is started, and the

stage rises until the applied load between microsphere and tissue reaches the preprogrammed set point. At this position, motion stops for the duration of adhesion time, and then the stage is lowered. Once the stage returns to its initial position, parameters are calculated and graphs are plotted.

**Results and Discussion** *Interpretation of the Microbalance Recordings.* Figure 10 shows a typical load-versus-deformation graph obtained using the Cahn microbalance where the stage position and the mass exerted on the microbalance are represented by the horizontal and vertical axes, respectively. Letters have been added to Figure 10 to

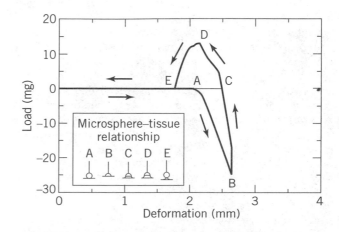

**Figure 10.** Typical load-versus-deformation curve: arrows indicate direction of tracing, and inset depicts relationship between tissue and microspheres at each lettered point. *Source:* Reproduced with permission from *Journal of Controlled Release* (Ref. 28).

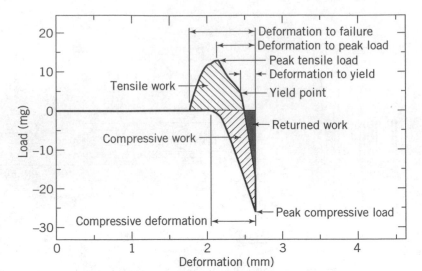

**Figure 11.** Typical load-versus-deformation curve indicating parameters of interest for bioadhesive analysis. *Source:* Reproduced with permission from *Journal of Controlled Release* (Ref. 28).

aid in description, and arrows have been added to indicate the order in which the curve is drawn over time. Experimental measurements have been indicated on Figure 11. The curve is a plot of the recorded forces (presented as a load) detected by the sample loop holding the microsphere as the stage holding the tissue is raised to contact the microsphere. Deformation is the distance the stage has moved. The initial state of the experiment begins with zero applied load and zero deformation. The stage then rises (increase deformation) to point A, the initial point of contact between the microsphere and tissue. The stage continues to rise until the set applied load ($-25$ mg) is recorded, point B, peak compressive load. The region from A to B represents the compressive deformation of the system. The stage position is held at point B for the set time of the experiment (usually seven minutes), during which the tissue often relaxes and the force decreases. This is seen as a completely vertical line in the plot just above point B. During this time the microsphere and mucus form an adhesive bond. After seven minutes, the stage is slowly stepped down, and the experiment enters the tensile region. The compressive load is released as the stage begins to move away from the microsphere. The recorded forces pass through point C, indicating zero applied force on the tissue. The stage continues to pull the tissue away from the microsphere, and a maximum force of adhesion is recorded, point D. In between points C and D, however, is the yield point of the system, which is identified by a change in slope of the curve from C to D. The initial slope of the curve above point C is linear and is indicative of the stiffness of the bioadhesive bond. The force increases during this time, and we assume that the contact area remains the same and that the earliest disruption of the bioadhesive bond between the microsphere and tissue is at the yield point. The distance from B to the yield point is the deformation to yield. As the stage continues to move, the adhesive bond is further disrupted, and the maximum adhesive force is measured, the peak tensile load (point D). The deformation to peak load is the distance from point B to point D. As the stage continues to move away, the recorded forces decrease as the

adhesive contact between microsphere and tissue is completely disrupted and no force reading can be detected, point E. Point E is considered to be the point of failure. During the stage movement from point D to point E, we assume the adhesive contact area is decreasing until final separation is achieved. In some microsphere–tissue interactions, this region of the data plot is elongated and marked by several peaks and valleys, indicating existence of several individual adhesive bonds. Some bonds are stretched until the breaking point, causing the "peak" to appear, and then the curve drops momentarily as the load is transferred to other bonds (forming the "valley" on the data plot). Figure 12 schematically depicts the disruption of the adhesive union where several focal attachments may be responsible for the adhesive bond strength. In Figure 10, the distance from point B to point E is known as the deformation to failure. A large deformation to failure indicates the adhesive contact is extremely plastic and compliant and less susceptible to outside disruptions. The stage continues to move downward until its home position is reached (the point at which there is zero deformation on the $y$ axis).

Three different work values can be obtained from these curves. Compressive work is the work done in joining the two surfaces and is calculated as the area between the compressive region of the curve and the zero load line (upward cross-hatched area in Fig. 11). Returned work is the work done by the tissue on the microsphere when the compressive load is released. This component is the area bounded by the negative region of the tensile curve and the

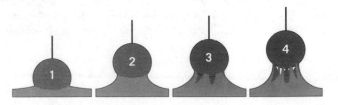

**Figure 12.** Schematic diagram showing the progression of bioadhesive fracture.

zero load line (dotted area in Fig. 11). Tensile work is the work done in separating the two surfaces and is the area between the positive portion of the tensile curve and the zero load line (downward cross-hatched area in Fig. 11).

To compute fracture strength ($F_m/A_o$) of a bioadhesive bond, the contact area has to be determined. The area of contact between the microsphere and mucosa was estimated to be the projected surface area of the spherical cap defined by the depth of penetration of the microsphere below the surface level of the tissue (Fig. 13). [Area $= A_o = \pi R^2 - \pi(R - a)^2$, where $R$ is the microsphere radius and $a$ is the depth of penetration. The depth of penetration, $a$, was visually estimated to be approximately equal to the microsphere radius for microspheres between 710 and 850 $\mu$m. Thus, $A_o = \pi R^2$.] Fracture strengths were obtained by normalizing peak load by the projected area of this cap.

***Limitations of the System.*** The microbalance-based system has several limitations: (*1*) microspheres smaller than 300 $\mu$m are difficult to mount, (*2*) overshoot of the applied load must be accounted for, (*3*) very small and very large applied loads are difficult to control, and (*4*) changes in applied load over time due to tissue relaxation or contraction are not accounted for once the stage has stopped moving.

First, because of the mounting method, it is difficult to attach microspheres smaller than 300 $\mu$m to the microbalance. A more sophisticated method using micropipettes and a micropositioner (141) could possibly be employed to mount microspheres as small as 10 $\mu$m. However, estimations in contact area for fracture strength calculations would be difficult.

Second, since the sampling rate of the microbalance is 1 Hz, stage motion commands can be sent only every second. Therefore, once the microbalance reaches the applied load set point, there is a 1-s delay between when the computer commands the stage to stop moving and when the motion actually ceases. This results in overshoot of the applied load in the range of 5–50%. For moderate applied loads (i.e., 10–50 mg), the overshoot is easily adjusted for and is rarely greater than 20%.

Third, very small and very large applied loads are difficult to control due to response time of the microbalance and the nature of the motor controlling the stage movement. The stepper motor supplied with the DCA 322 moves in incremental jumps (as opposed to smooth motion). When an attempt is made to apply very small loads, a small

amount of overshoot can be greater than 100% of the intended value. When very large loads are attempted, the slope of the load-versus-deformation curve is so steep that the overshoot is always very large. Therefore, moderate forces (10–50 mg) are most easily attained.

Fourth, in many cases the applied load changes while the stage is not in motion. This occurs due to relaxation or contraction of the smooth muscle in the tissue. Because of unsuitable response time and relatively large incremental movements of the stepper motor, accurate adjustments can not be made in the stage position to account for these small dynamic effects.

In summary, the microtensiometer described in this section offers many advantages over previous bioadhesion measurement devices. The setup enables determination of bioadhesive forces between a single microsphere and intestinal mucosa. This technique allows measurement of bioadhesive properties of a candidate material in the exact geometry of the proposed microsphere delivery device. The use of a physiological tissue chamber mimics temperature and pH conditions of a living animal and maintains viable tissue. Because experiments are conducted in an aqueous environment, problems in distinguishing surface tension forces at the air/liquid interface from forces at the microsphere/mucus interface are eliminated.

Eleven parameters relating to the bioadhesive event can be computed from every experiment. Although some are independent of the material studied, we believe that three may be direct predictors of bioadhesive potential: fracture strength, deformation to failure, and tensile work (28). The following section details a thorough analysis of adhesion data collected with the Cahn machine on various bioadhesive polymer microspheres.

### Bioadhesive Measurements

Traditionally, two classes of polymers have been of interest for use as bioadhesives: hydrogels (21,101,142,143) and other hydrophilic polymers containing carboxylic groups [e.g. poly[acrylic acid]) (23,48,75,94)]. The formation of bioadhesive bonds between soft tissue and these materials has been described in three steps: (*1*) wetting and swelling of the polymer to permit intimate contact with the biological tissue, (*2*) interpenetration of bioadhesive polymer chains and entanglement of polymer and mucin chains, and (*3*) formation of weak chemical bonds (17,19). Because

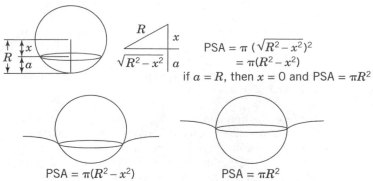

**Figure 13.** Calculation of projected surface area.

water-insoluble, bioerodible materials are often hydrophobic and typically do not exhibit highly flexible chains, swelling and chain entanglement appear unlikely to occur. Hence, investigation of these materials for bioadhesive applications has been neglected.

However, realizing that hydrogen bonding between surface carboxyl groups and mucus or epithelial cell glycoproteins could be a key contributor to nonspecific bioadhesion as dictated by the adsorption theory of adhesion, we hypothesized that some bioerodible thermoplastic materials might exhibit bioadhesive interactions. When polymers such as polyesters and polyanhydrides hydrolytically degrade, carboxylic acid groups are formed at the transected polymer chain ends. Rapidly degrading materials may quickly form surfaces rich in carboxylic acid ideally suited for bioadhesion. To test this concept, we used the Cahn microbalance (described in the previous section) to compare the bioadhesive properties of one hydrogel and five bioerodible thermoplastics.

**Polymer Samples.** Polymers investigated in this study were selected based on properties believed to favor bioadhesion (i.e., carboxylic acid) (23,26,51) and can be separated into two distinct categories: a natural hydrogel (alginic acid); and synthetic, degradable, thermoplastic polymers (polyesters, polyanhydrides). Figure 14 shows the chemical structures of the polymers. Powdered, medium-viscosity alginate (Kelgin® MV) was obtained from Kelco, Division of Merck & Co., Inc. (Chicago, IL). Low-molecular-weight (2K) polylactic acid (PLA) was purchased through Polysciences (Warrington, PA). The polyanhydrides studied were a copolymer of 1,3-bis(p-carboxyphenoxy)-propane (CPP) and sebacic acid (SA) (molar ratio of 20:80), and copolymers of fumaric acid (FA) and SA (molar ratios of 20:80 and 50:50). P(CPP:SA) 20:80, P(FA:SA) 50:50, and some P(FA:SA) 20:80 were obtained as a gift from Nova Pharmaceutical Corporation (Baltimore, MD). Other P(FA:SA) 20:80 was synthesized in our laboratory.

**P(FA:SA) 20:80 Synthesis.** P(FA:SA) 20:80 was synthesized according to the methods established by Domb and Langer (144,145). FA and SA monomers (Sigma Chemical Company, St. Louis, MO) (Fig. 15) were purified by recrystallization from 95% ethanol or 100% ethanol, respectively, and drying with vacuum in a calcium chloride desiccator.

Prepolymers (PP) were produced by refluxing purified monomer in acetic anhydride (FA:172 mmol/250 mL, 1 h; SA: 148 mmol/100 mL, 30 min), evaporating 80–90% at 50–60°C (Buchi RE-111 Rotavapor and 461 Water Bath system), and recrystallizing at −20°C. Precipitates were filtered, dissolved in toluene (100 mL or 30 mL, for FAPP or SAPP, respectively) with heat, precipitated overnight at room temperature, and then further recrystallized in an ice bath for 10 h. Precipitates were filtered, washed (in 200 mL petroleum ether or 1:1 volume ratio of ethyl/petroleum ether, for FAPP or SAPP, respectively), and dried under vacuum.

For polymerization, prepolymers (8 mmol SAPP, 2.288 g, and 2 mmol FAPP, 0.4 g) were melted under reduced pressure and periodically purged with nitrogen gas to re-

move acetic anhydride. After 30 min, the resulting viscous polymer was cooled, dissolved in methylene chloride, filtered, precipitated in hexane, dried, and stored at −20°C. Molecular weight was determined with gel permeation chromatography (GPC) to be between 5,000 and 15,000, depending on the batch.

**Microsphere Preparation** *Hydrogel Microspheres.* Microspheres made of alginate were produced through a standard ionic gelation technique (146). The polymer was first dissolved in an aqueous solution (2% w/v) and then extruded through a microdroplet-forming device, which employed a flow of nitrogen gas to break off the droplet. A slowly stirred (≈100–170 rpm) calcium chloride bath (1.3% w/v) was positioned below the extruding device to catch the forming microdroplets. Microspheres were left to incubate in the bath for 20–30 min to allow sufficient time for gelation to occur. Microsphere particle size was controlled by using various size extruders or by varying either the nitrogen gas or polymer solution flow rates.

*Bioerodible Thermoplastic Microspheres.* Several methods for preparation of bioerodible microspheres exist (147–152). Certain methods are more appropriate for particular polymers. The two methods used in this study were hot-melt microencapsulation and solvent evaporation microencapsulation.

In hot-melt microencapsulation (147), 2 g of polymer (either PLA, P[CPP:SA] 20:80, P[FA:SA] 20:80, or P[FA:SA] 50:50) were melted in a crucible on a hot plate (melting points ranged from 60–125°C). The molten polymer was added to 250 mL of rapidly stirred silicon oil (200 Fluid; Dow Corning, Midland, MI), heated to ≈10°C above the melting point of the polymer, to form an emulsion. The emulsion was allowed to stabilize and was then rapidly cooled with an ice jacket until polymer particles solidified. The resulting microspheres were washed with petroleum ether to produce free-flowing beads. Microspheres with diameters between 10 and 1,000 $\mu$m were obtained with this method. External surfaces of spheres prepared with this technique ranged from very smooth to somewhat pitted, depending on the polymer used (148,153).

Solvent evaporation was used to manufacture PLA microspheres (154,155). Polymer solutions were produced by dissolving PLA in methylene chloride. A 20% (w/v) solution of PLA was suspended as an emulsion in 300 mL of 1% (w/v) poly(vinyl alcohol) (Sigma Chemical Co., St. Louis, MO). The solution was vigorously stirred at a rate of 1,200 rpm using a Caframo (Wiarton, Ontario) overhead stirrer. One drop of octanol (Sigma Chemical Company, St. Louis, MO) was added to the solution to reduce foaming. After 4 h of stirring to allow for evaporation of the organic solvent, resulting microspheres were collected, washed with water, and lyophilized (Lyph-Lock 6; Labconco, Kansas City, MO). Microspheres obtained with this technique ranged in size from 10 to 1,000 $\mu$m.

**Tissue Collection and Handling.** Unfasted, male, VAF, CD® rats (250–350 g, Charles River Labs, Wilmington, MA) were anesthetized by ip injection of nembutal (sodium pentobarbital, 55 mg/kg) and perfused transcardially with ice cold, 0.9% NaCl. Approximately 10–15 cm of jejunum

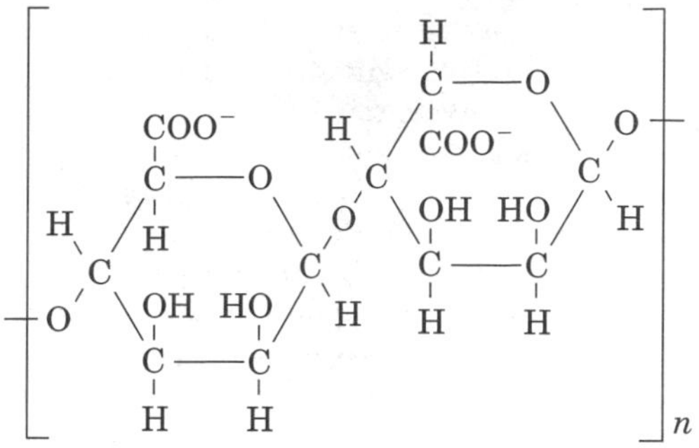

Alginate

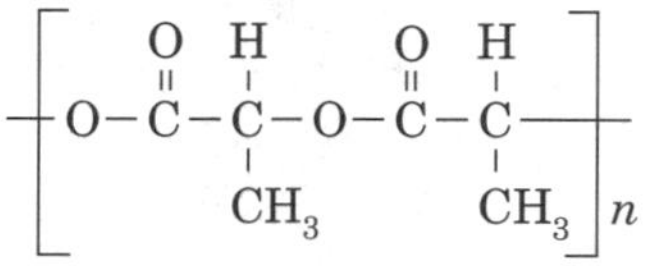

Polylactic acid

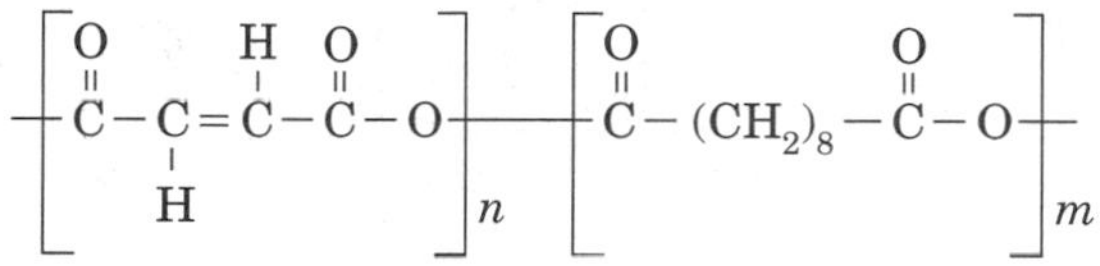

P(FA:SA)

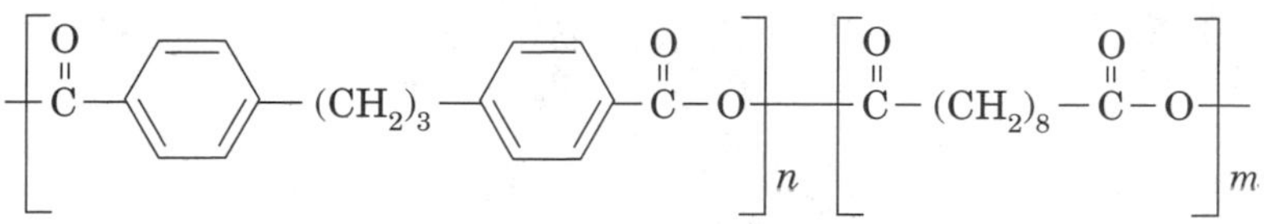

P(CPP:SA)

**Figure 14.** Chemical structures of polymers.

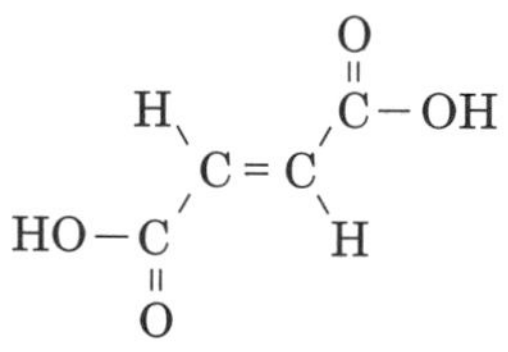

Fumaric acid 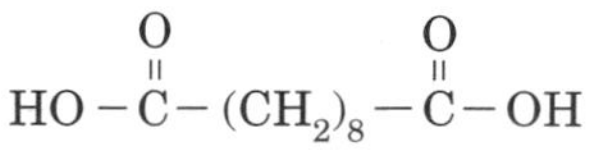                     Sebacic acid

**Figure 15.** Chemical structures of monomers.

were excised, flushed with Dulbecco's phosphate-buffered saline containing 100 mg/dL of glucose (DPBSG), and stored in DPBSG at 4°C for no more than 3 h.

**Bioadhesive Measurements.** Adhesive forces between polymer microspheres and segments of rat jejunum were measured using the previously described microbalance. Polymers were evaluated for their bioadhesive properties using microspheres that were sieved to diameters ranging

from 710 to 850 $\mu$m. Sections of rat jejunum ($\approx$2 cm in length) were cut longitudinally (to expose the lumen), submerged at the bottom of a 9-mm layer ($\approx$3 mL) of DPBSG, and clamped at their edges inside the tissue chamber. The chamber was subsequently placed on the moving stage below a suspended microsphere. The stage was raised manually until microspheres were completely submerged in buffer, but still 1–2 mm above the surface of the mucosa. At this point the microbalance was tared and data collection began. The stage was raised at a rate of 50.2 $\mu$m/s until contact was made between microsphere and tissue. Microspheres were left in contact with the tissue for 7 min with an applied force of $\sim$0.25 mN and then pulled vertically away from the tissue sample (at 50.2 $\mu$m/s) while recording the required force of detachment. For each experiment, buffer was replaced with fresh DPBSG, and a new microsphere was brought in contact with a different site along the tissue. Tissue sections were used for three experiments and then replaced with fresh segments.

**Data Analysis.** As described previously, five theories of adhesion have been adapted and developed to analyze bioadhesive interactions: electronic (52,53), adsorption (50,54–56), wetting (49,68,69), diffusion (49,50,68,69, 71,72), and fracture (74,75). Though these theories involve independent mechanisms, they are not mutually exclusive of one another. For analyzing the results of the tensile experiments we have applied the fracture theory; in the discussion that follows, we will assert that the adsorption and wetting theories of adhesion are most appropriate for explaining the observed bioadhesive phenomena.

**Load-versus-Deformation Curve Morphology.** Careful study of the load-versus-deformation curves produced from the microbalance tensile experiments can reveal many important phenomena. If the interaction between two materials is completely nonadhesive and both materials behave elastically, the curve would retrace itself and no adhesive force would be measured. There would be no hysteresis, and the compressive work imparted on the system would be returned completely as the stage was lowered to its original position (i.e., the interaction would be totally reversible). All materials investigated in this study showed positive adhesive forces, and therefore some form of adhesion occurred between the mucosa and the polymers. Representative load-versus-deformation curves for the six microsphere types studied are displayed in Figure 16. Although the recordings from the tensile experiments all have a similar general shape, three common trends were observed for the different materials investigated.

Some materials, such as PLA, exhibited adhesions that fractured in what appeared to be a brittle manner. In these instances, the resultant load-versus-deformation curves rose quickly during the descent of the stage but then returned sharply to zero as separation occurred (Fig. 16).

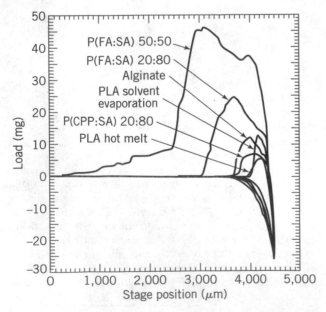

**Figure 16.** Characteristic load-versus-deformation curves generated with the microbalance for each of the six microsphere types studied. All curves have been scaled identically to emphasize the differences between them. *Source:* Reproduced with permission from *Journal of Controlled Release* (Ref. 28).

Failure took place at nearly the same stage position as the point of initial contact (i.e., deformation to failure = compressive deformation). The fracture strengths were low to moderate, depending on the microsphere fabrication technique, but the interaction between the microsphere and mucus seemed weak. We believe that the forces produced were mainly due to either mechanical engulfment of the microspheres by the mucus or by mechanical penetration of the mucus into crevices of the microspheres (in the solvent evaporation case). When fracture occurred, the microspheres almost "popped" out of the mucus layer as if the mucin was not actually bound to the microspheres.

Other materials, such as P(CPP:SA) 20:80, produced load curves that were more rounded in shape (Fig. 16). Again, no difference was observed between the deformation to failure and the compressive deformation. In these cases, we believe that some bioadhesive interaction occurred between the microsphere and the mucosal layer; however, the fracture strengths produced by these materials were only moderate at best. The bonds produced with the mucus appeared to be weaker than the adhesion of the mucus to the epithelial cell layer. The eventual fracture was witnessed to occur at the mucus/microsphere interface, and SEM examination revealed no mucus adherent to the microsphere surfaces.

In the final group of curves, both the peak load and deformation to failure were large. Copolymers of FA and SA (both P[FA:SA] 20:80 and 50:50) exhibited such load-versus-deformation curves (Fig. 16). In these cases, the bond between the microsphere and the mucosal layer was observed to be very strong. The deformation to failure was larger than the deformation due to the compressive load. In other words, the intensity of the bond was great enough to deform the adhesive site and elongate the mucus/mucosal surface beyond its original position. Prior to fracture, strands of mucin were observed bridging the gap between the microsphere and the mucosa. Fracture after peak load almost always occurred in a brittle manner, but the load rarely returned to zero. The fact that the load did not return to zero suggests that some mucin or cellular material was transferred from the tissue's surface to the microsphere and that the site of fracture occurred either within the mucus layer or at the mucus/epithelium interface instead of at the mucus/microsphere interface. Therefore, we believe that the weak link in the adhesive system is due to the properties of the biological substrate and not the bioadhesive interaction.

While overlapping of the compressive region of the six curves helped to support the repeatability of the experiment, size and shape of the tensile regions of the curves indicated there were differences between polymers. The most striking differences were found between both of the P(FA:SA) copolymers and all of the other materials; however, further statistical analysis was necessary to confirm these differences.

***Comparison of Bioadhesive Parameters.*** Of the 11 parameters measured from the load-versus-deformation curves, 9 were tabulated (Tables 1 and 2) and compared: compressive deformation, deformation to yield, deformation to peak load, deformation to failure, yield force, fracture

**Table 1. Mean Deformations (±SEM) Calculated from Load-versus-Deformation Plots Generated during the Electrobalance Tensile Experiments**

| Polymer | Compressive deformation (mm) | Deformation to yield (mm) | Deformation to peak load (mm) | Deformation to failure (mm) |
|---|---|---|---|---|
| PLA hot melt ($n = 9$) | $0.656 \pm 0.058$ | $0.155 \pm 0.022$ | $0.335 \pm 0.034$ | $0.583 \pm 0.054$ |
| P(CPP:SA)20:80 ($n = 9$) | $0.647 \pm 0.071$ | $0.238 \pm 0.030$ | $0.459 \pm 0.053$ | $0.647 \pm 0.063$ |
| PLA solvent evaporation ($n = 6$) | $0.516 \pm 0.087$ | $0.202 \pm 0.043$ | $0.381 \pm 0.067$ | $0.625 \pm 0.140$ |
| Alginate ($n = 10$) | $0.957 \pm 0.134$ | $0.256 \pm 0.052$ | $0.649 \pm 0.144$ | $0.835 \pm 0.189$ |
| P(FA:SA)20:80 ($n = 23$) | $0.635 \pm 0.037$ | $0.256 \pm 0.039$ | $1.219 \pm 0.185$ | $1.859 \pm 0.173$ |
| P(FA:SA)50:50 ($n = 7$) | $0.866 \pm 0.176$ | $0.376 \pm 0.137$ | $1.367 \pm 0.264$ | $2.020 \pm 0.198$ |

*Source:* Ref. 140.

**Table 2. Mean Yield Force, Fracture Strength, and Work Values (±SEM) Calculated from Load versus Deformation Plots Generated during the Electrobalance Tensile Experiments**

| Polymer | Yield force ($\mu$N) | Fracture strength (mN/cm$^2$) | Compressive work (nJ) | Returned work (nJ) | Tensile work (nJ) |
|---|---|---|---|---|---|
| PLA hot melt ($n = 9$) | $40.77 \pm 4.41$ | $11.39 \pm 0.97$ | $46.01 \pm 5.77$ | $2.25 \pm 0.78$ | $17.37 \pm 2.80$ |
| P(CPP:SA)20:80 ($n = 9$) | $64.09 \pm 4.31$ | $17.89 \pm 1.09$ | $70.14 \pm 11.34$ | $10.99 \pm 4.69$ | $25.86 \pm 4.39$ |
| PLA solvent evaporation ($n = 6$) | $70.66 \pm 18.03$ | $24.30 \pm 3.13$ | $47.62 \pm 5.74$ | $5.43 \pm 1.86$ | $34.88 \pm 12.65$ |
| Alginate ($n = 10$) | $90.23 \pm 15.23$ | $30.16 \pm 4.18$ | $78.67 \pm 21.82$ | $6.21 \pm 2.37$ | $62.81 \pm 14.50$ |
| P(FA:SA)20:80 ($n = 23$) | $138.57 \pm 9.51$ | $51.63 \pm 3.38$ | $64.01 \pm 5.21$ | $7.45 \pm 1.70$ | $255.45 \pm 33.04$ |
| P(FA:SA)50:50 ($n = 7$) | $196.78 \pm 71.44$ | $76.85 \pm 13.53$ | $53.91 \pm 5.50$ | $3.18 \pm 1.05$ | $366.01 \pm 64.07$ |

*Source:* Ref. 140.

strength, compressive work, returned work, and tensile work.

The outcome of a detailed statistical analysis is discussed thoroughly in previous reports (28,132), and demonstrates that fracture strength, deformation to peak load, deformation to failure, and tensile work are the most important factors in characterizing bioadhesive materials.

*Deformations.* Table 1 compares four deformation parameters for the six varieties of microspheres investigated. Because the microbalance measures the interaction of the microsphere–tissue composite, the deformations are functions of the material properties of both the tissue and the microspheres. The system cannot distinguish between the elastic moduli of the composite's components. Microspheres with higher elastic moduli (i.e., hard, bioerodible materials) will undergo insignificant deformations under the loads used in this experiment. Therefore, we hypothesized that compressive deformations measured for these materials would depend on the mechanical properties of the tissue alone, while tensile deformations (i.e., deformation to yield, deformation to peak load, and deformation to failure) would depend on mechanical properties of the tissue and the adhesive bond. This would result in similar compressive deformations for all thermoplastics but different tensile deformations. In contrast, we believed that the low modulus of elasticity of alginate would contribute to both tensile and compressive deformations.

As hypothesized, compressive deformations of bioerodible microspheres were not statistically different, indicating that for those materials with considerably high moduli, this parameter is not a function of polymer or bioadhesive interaction but instead is a function of mucus and tissue moduli. However, alginate, which can be expected to have a low modulus compared to the bioerodible

materials, produced a compressive deformation statistically different from P(FA:SA) 20:80 ($P \leq .005$). In this instance, the measured value for compressive deformation consists of the deformation of the tissue plus the deformation of the microsphere.

Deformation to yield represents the distance the stage travels before the tensile portion of the curve changes slope and indicates the point at which the adhered surfaces initially start to yield. Because nearly all of deformations to yield were statistically identical (excluding PLA hot melt), we concluded that the change in slope was caused by yielding of the biological component, not yielding of the microsphere or the adhesive union. Thus, we believe that within the microsphere/adhesive-bond/mucus/tissue composite, the component with the lowest modulus is most likely the mucus. In the case of PLA hot melt, however, deformation to yield was significantly less than other materials tested. In this instance, we believe the adhesive interaction may have been the first component to give way. The explanation for this effect may be found in the absence of surface texture of PLA hot-melt microspheres, which limits the degree of mechanical bonding.

Analysis of deformation to peak load and deformation to failure showed that copolymers of FA and SA were statistically different from all other materials (deformation to peak load: $P \leq .05$, deformation to failure: $P \leq .005$). These results clearly demonstrated that interactions formed between P(FA:SA) polymers and intestinal tissue were strong and unlike any other materials investigated. The adhesive bonds were resistant enough to elongate the bioadhesive/tissue composite to a length greater than twice that of any other materials studied.

*Fracture Strengths.* Fracture strengths of the microspheres are compared in Table 2 and were obtained by dividing the peak load by the projected contact area of adhesion, which for microspheres between 710 and 850 $\mu$m was estimated to be approximately 0.00487 cm$^2$. Polymers with the greatest bioadhesive fracture strengths were P(FA:SA) copolymers and alginate. All fracture strengths were statistically different from one another except for PLA solvent evaporation compared to alginate (Table 2). P(FA:SA) copolymers were different from nearly all other materials, with a confidence level of $P \leq .0005$. These results support the hypothesis that polymers with high concentrations of carboxylic acid groups, such as alginate and polyanhydrides, produce strong bioadhesive bonds. The extremely high fracture strengths obtained for P(FA:SA) 20:80 and P(FA:SA) 50:50 (51.63 mN/cm$^2$ and 76.85 mN/cm$^2$, respectively) suggest that bioerodible polymers could be promising bioadhesive delivery systems.

In analyzing fracture strengths, we also observed that fabrication technique influenced the bioadhesive properties of a material. PLA microspheres made by solvent evaporation produced greater fracture strengths than PLA microspheres made by hot-melt microencapsulation (24.30 $\pm$ 3.13 mN/cm$^2$ versus 11.39 $\pm$ 0.97 mN/cm$^2$, $P \leq .0005$). This effect could be attributed to increased surface texture of solvent evaporation microspheres (see "Characterization of Potential Bioadhesive Polymers"). Greater texture promotes mechanical bond formation and enlarges effective surface area, increasing sites for chemical bonding.

*Works.* The three work values analyzed in this study were compressive work, returned work, and tensile work (Table 2). Compressive work represents the work done in joining the microsphere and tissue together. This component is a function of the applied load and compressive deformation. Because an attempt was made to exert a consistent applied load of 25 mg and compressive deformation for bioerodible microspheres was shown to be a function of tissue property, compressive work was expected to be identical for all bioerodible materials. Statistical analysis proved that compressive works measured for all materials in this study were not significantly different [excluding PLA hot melt compared to P(FA:SA) 20:80, which showed weak statistical difference: $P \leq .05$].

During the initial tensile region of the curve (i.e., when compressive force is first released), elastic properties of the tissue respond by exerting a force upward on the microsphere. Work done by the tissue on the microsphere is what we have termed returned work. If the tissue and microsphere are completely elastic, returned work would equal compressive work. However, in our experiments energy was lost through plastic deformation, most likely as heat, and returned work was considerably less than compressive work for all materials (Table 2). Analysis of variance on returned work showed no statistical difference among the means (.10 $\leq P \leq$ .25), indicating that returned work was dependent on tissue properties and not on microspheres.

Conversely, polymer type did affect tensile work (Table 2). As expected, tensile work was very large for those polymers that displayed high fracture strengths and large deformations to failure. P(FA:SA) 20:80 and 50:50 produced mean tensile works of 255.45 nJ and 366.01 nJ, respectively. These values were 4 to 20 times higher than any other material tested and statistically different.

Two explanations can be given for these results. One theory is that interfacial energies favor spreading of the mucus layer over the surface of the polymer microspheres, and the adhesive interaction is a function of polymer wettability by mucus, as suggested by Lehr (156,157). The other is that numerous surface carboxyl groups on the P(FA:SA) copolymers form hydrogen bonds with the mucin or epithelial cell layer during the compressive portion of the experiment. The intensity of the interaction would be dependent on the concentration of carboxylic acid. As anhydride bonds are degraded through hydrolysis, more carboxylic acid is produced, thus enhancing the bonding capacity of the microspheres. Both of these explanations probably contribute to the unique bioadhesion seen with these materials.

**Summary.** Using the microbalance as a microtensiometer, we were able to evaluate bioadhesive performance of both hydrogel and bioerodible microspheres. Although hard, bioerodible materials have not previously been studied for bioadhesive applications, we identified one group of polyanhydrides [P(FA:SA)] that exhibits high fracture strengths (>50 mN/cm$^2$), long deformations to failure (>1.7 mm), and large tensile works (>250 nJ). These findings strongly support the existence of bioadhesion between soft tissue and hard, bioerodible, thermoplastic polymers. In spite of the fact that in these experiments every attempt was made to use freshly excised, viable tissue under physiologic conditions, we should note that this method is an in vitro technique and therefore results may not exactly reflect bioadhesive properties of these materials in vivo. Correlation between these findings and in vivo performance is investigated later in this article.

Although hydrogels and hydrophilic polymers have been traditionally thought of as ideal BDDSs, new results suggest that novel materials with greater bioadhesive properties may exist (156). The findings in this study show that certain bioerodible polyanhydrides produce strong bioadhesive interactions, in vitro, compared to other thermoplastic and hydrogel materials. In bioerodible polymers lacking chain flexibility, bioadhesion could largely be a result of surface energy effects and/or hydrogen bonding between mucin and carboxyl groups. Thus, as suggested by Lehr et al. (83,156,157), chain flexibility and interpenetration of polymer and mucin chains should not be considered prerequisites for bioadhesion.

### Characterization of Potential Bioadhesive Polymers

To gain a better understanding of the mechanisms responsible for bioadhesive phenomena observed with polyanhydride microspheres, three common polymer characterization techniques were employed. First, surface morphology and morphological changes produced through polymer degradation were investigated and documented using scanning electron microscopy (SEM). Second, bulk chemical analysis was conducted with Fourier transform infrared spectroscopy (FTIR). Our intent was to determine whether a correlation could be observed between high carboxylic acid content and high bioadhesive properties. Also, we were interested in determining if chemical composition was affected by degradation and how these changes could alter a material's bioadhesive performance. Third, because interfacial forces are believed to play a large role in bioadhesion (58–60,64,67,156,157), both water and mucus contact angles were measured with a goniometer. Correlation between tensile experiment measurements and mucus contact angles was investigated (140).

**Morphological Characterization Using SEM.** To assess the effects of surface morphology on bioadhesive properties, microsphere samples were lyophilized and analyzed under SEM at both 150× and 1,000×. We analyzed both thermoplastics and hydrogels in the same manner even though we realize that alginate microspheres in the lyophilized state may appear quite different than those in the hydrated state. Although free of pores, surfaces of the alginate microspheres were marked with distinct arrays of raised stripes (Figs. 17a and 17b). The origin of these structures is unknown, but we believe that they may be a mechanical stretching artifact of the freeze-drying process. At 150× both P(CPP:SA) 20:80 and PLA microspheres, appeared fairly smooth in appearance (Figs. 17c and 17e), but closer examination (1,000×) revealed some differences. High-magnification analysis of P(CPP:SA) 20:80 microspheres demonstrated a granular texture as well as a few scattered, shallow pockmarks (Fig. 17d). PLA hot-melt microspheres were mostly smooth and uniform even at high power (Fig. 17f). Though some dispersed particulate matter could be found, surfaces were completely free of any pits or indentations. This smooth texture could account for the weak bioadhesive performance of these microspheres including the significantly low deformation to yield.

Among bioerodible microspheres, four were manufactured using a hot-melt technique: PLA, P(CPP:SA) 20:80, P(FA:SA) 20:80, and P(FA:SA) 50:50. Owing to the nature of this fabrication procedure, we expected the resulting microspheres to be generally smooth and dense in appearance. However, depending on the polymer employed, surfaces ranged from very smooth to highly pitted. SEMs of the two P(FA:SA) copolymers (Fig. 18a–d) demonstrated different surface morphologies. P(FA:SA) 20:80 copolymer microspheres possessed a granular topography and were riddled with deep pits approximately 10 μm in diameter and approximately 30–50 μm apart (Fig. 18b). We believe these structures developed through growth of spherulites during solidification of the microspheres, although fine lamellar morphologies could not be seen, and that the pits are actually crevices between adjacent spherulites. These structures were found on all P(FA:SA) 20:80 microspheres analyzed. However, P(FA:SA) 50:50 microspheres showed no sign of spherulite formation or pock marks but instead displayed a fine granular texture very similar to P(CPP:SA) 20:80 (compare Figs. 17d and 18d). This difference in spherulite development could be attributed to different cooling rates during the fabrication procedures and/or different glass transition temperatures.

PLA microspheres were also manufactured using a solvent evaporation technique. Because this process involves the removal of a solvent from the bulk of the polymer, we expected to find more porous microspheres than those created with the hot-melt technique. Examination of the surfaces did indeed reveal a highly porous morphology (Fig. 18e). At 1,000×, spherulites could be clearly seen with distinct lamellar structures (Fig. 18f). The irregular pores and channels, created as the solvent evaporated from the system, were approximately 3–10 μm in diameter and spaced 6–10 μm apart. This structure was completely different from the smooth surface observed on the hot-melt PLA microspheres (compare Figs. 17f and 18f). We believe this difference in surface morphology is the main explanation for the difference in bioadhesive properties measured using the microbalance (as discussed in "Bioadhesive Measurements".). Coarser surface texture improved adhesion of solvent evaporation microspheres through stronger mechanical interactions.

**Morphological Characterization of Short-Term Degradation Using SEM.** Because the polyanhydrides are known for their rapid hydrolytic degradation and our in vitro bioadhesion analysis is performed in an aqueous bath, it was important to study the morphological surface changes occurring on the polymers when incubated in buffer. All three polyanhydrides were degraded for 0, 7, 15, 30, and 60 min in PBS and analyzed by SEM (Figs. 19 and 20). The most obvious result was the change in surface texture of the P(FA:SA) 50:50 microspheres after only 7 min. The entire surface of the microspheres became porous and flaky (Fig. 19a) as compared to the nondegraded samples (Fig. 18d). After 15 min, the surfaces had further changed, looking more uniform but displaying small cracks and fissures (Fig. 19b). Thus, as degradation proceeded, the original flakes and scurf were swept away, and surface cracks developed. After 30 min, the cracks enlarged, and by 60 min more pieces of polymer had crumbled away from the surface (Figs. 19c and 19d). Cross sections of these microspheres showed a solid and relatively intact core, with only the outer 50 microns showing signs of degradation (158,159).

Analysis of the other two polyanhydrides showed less significant changes (Fig. 20). After 7 min we observed an increase in the size and number of pores or pits on the surface of P(FA:SA) 20:80 microspheres (Fig. 20a). These pits continued to grow in number and dimensions as degradation proceeded, but no cracks developed as was observed in the 50:50 copolymer (Fig. 20b). These microspheres may have degraded more slowly than the P(FA:SA) 50:50 microspheres owing to higher crystallinity, as indicated by pockmarks in the nondegraded samples. P(CPP:SA) 20:80 exhibited even fewer changes than the P(FA:SA) 20:80. At 1,000× there was no discernable difference between the microspheres at time zero and the microspheres after 7 min degradation. After 30 min, scattered pores developed across the surface, and granulation that was observed in nondegraded samples either degraded or washed away (Fig. 20c). By 60 min, very small cracks appeared (Fig. 20d). However, these changes were minimal compared to those shown by copolymers of P(FA:SA).

The results of SEM analysis may indicate one possible explanation for the tremendous adhesive interactions measured for some polyanhydride materials. The rapid surface degradation of copolymers of FA and SA may im-

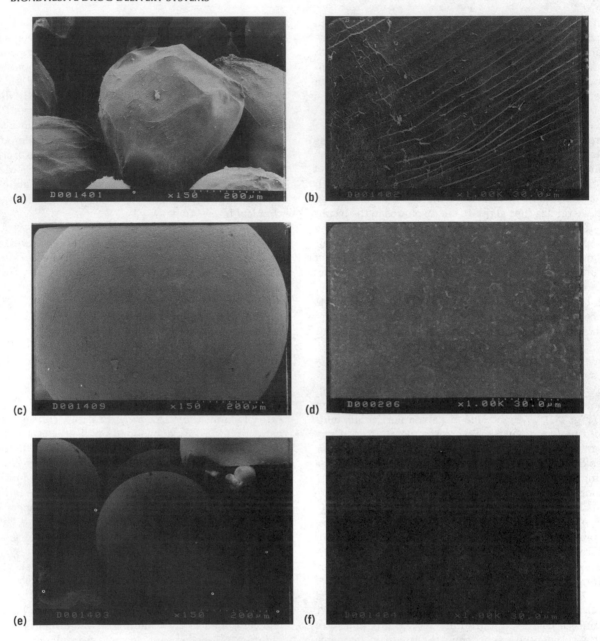

**Figure 17.** Scanning electron micrographs of three polymer microspheres. (**a**) Lyophilized alginate microsphere (150×). The nonspherical shape is an artifact of the lyophilization process; (**b**) higher magnification (1,000×) of (**a**) showing artifactual striations generated during the freeze-drying process; (**c**) P(CPP:SA) 20:80 microsphere made by hot melt (150×); (**d**) higher magnification (1,000×) of the microsphere in (**c**) showing a dense structure with a slightly granular surface. (**e**) PLA microsphere made by the hot-melt fabrication technique (150×); (**f**) higher magnification (1,000×) of the microsphere in (**e**) showing a very dense surface free of any pores. *Source:* Reproduced with permission from *Reactive Polymers* (Ref. 140).

prove their bioadhesive properties. As each anhydride bond is cleaved, two carboxylic acid groups are produced. These hydrophilic functional groups are then able to form hydrogen bonds with surrounding mucin strands that can in turn penetrate newly created surface cracks. The result is the formation of both chemical and mechanical bonds between the biological tissue and microsphere. Further degradation renews the surface carboxyl concentration, allowing for even greater adhesion.

**FTIR Analysis.** FTIR transmission spectra of polymer microspheres were taken to verify the presence of functional carboxyl groups believed to play a role in the formation of hydrogen bonds with biological tissues. Figure 21 shows transmission data from 2,000 cm$^{-1}$ to 1,500 cm$^{-1}$ for the polymers investigated in this study.

Peaks present in the spectrum for alginate were very broad, indicating an amorphous structure, but make it difficult to pinpoint specific chemical compo-

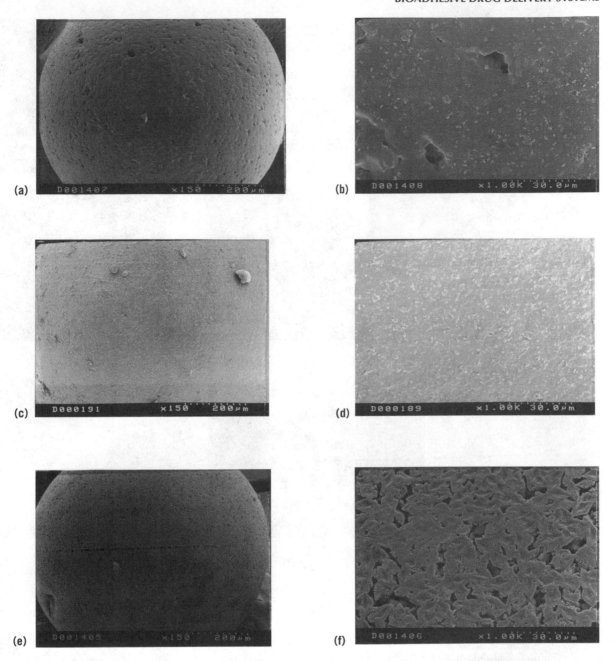

**Figure 18.** Scanning electron micrographs of three polymer microspheres. (**a**) P(FA:SA) 20:80 microsphere made by hot melt (150×); (**b**) higher magnification (1,000×) of the microsphere in (**a**) showing several large pits, which most likely represent spaces between adjacent spherulites; (**c**) P(FA:SA) 50:50 microsphere made by hot melt (150×); (**d**) higher magnification (1,000×) of the microsphere in (**c**); note that the external surface of this microsphere bears a closer resemblance to the P(CPP:SA) 20:80 (Fig. 17d) than it does to the P(FA:SA) 20:80 in (**b**); (**e**) PLA microsphere made by the solvent evaporation technique (150×); (**f**) higher magnification (1,000×) of the microsphere in (**e**) revealing the highly porous and corallike structure of the PLA. *Source:* Reproduced with permission from *Reactive Polymers* (Ref. 140).

nents. The broad band at 1630 cm$^{-1}$ is typical of carboxyl groups.

Both PLA preparations showed distinct peaks around 1,760 cm$^{-1}$ characteristic of the ester linkage as well as subtle bands at 1,625 cm$^{-1}$ that we attribute to the presence of carboxylic acid. The occurrence of carboxylic acid

in PLA is due to end-group contribution. Each hydrolytically degraded ester linkage results in the formation of one carboxyl and one hydroxyl end group (Fig. 22). The PLA used in this study was a low-molecular-weight material ($M_r = 2K$), and therefore the concentration of end groups appears as a significant fraction of the entire polymer sam-

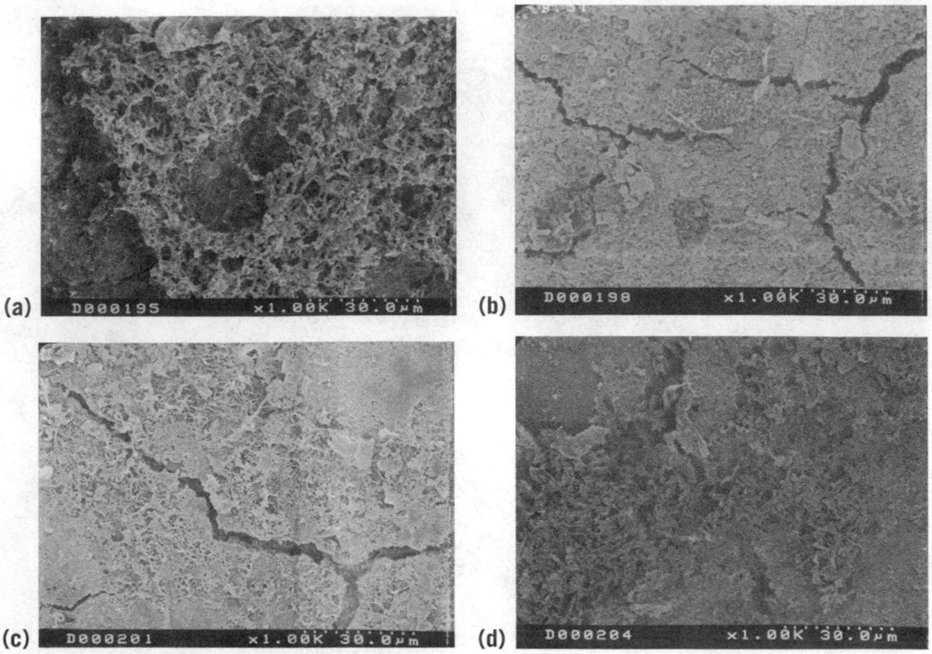

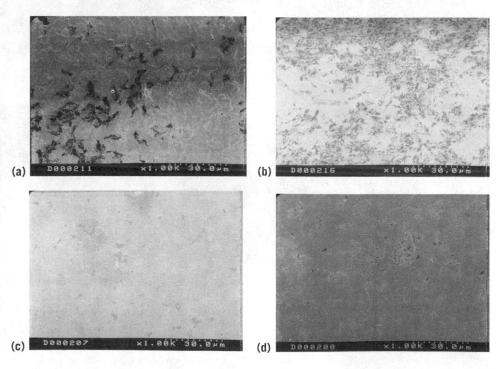

**Figure 19.** Scanning electron micrographs (1,000×) of the external surfaces of the poly(fumaric-co-sebacic anhydride) 50:50 microspheres after incubation in 37°C PBS (pH 7.4). (**a**) 7-min incubation, (**b**) 15-min incubation (note the crack formation), (**c**) 30-min incubation, (**d**) 60-min incubation. *Source:* Reproduced with permission from *Reactive Polymers* (Ref. 140).

**Figure 20.** Scanning electron micrographs (1,000×) of the external surfaces of the poly(fumaric-co-sebacic anhydride) 20:80 and poly[1,3-bis(p-carboxyphenoxy)-propane-co-sebacic anhydride] 20:80 microspheres after incubation in 37°C PBS (pH 7.4). (**a**) P(FA:SA) 20:80 microsphere after a 7-min incubation, (**b**) P(FA:SA) 20:80 microsphere after a 60-min incubation. (**c**) P(CPP:SA) 20:80 microsphere after a 30-min incubation, (**d**) P(CPP:SA) 20:80 microsphere after 60-min incubation. *Source:* Reproduced with permission from *Reactive Polymers* (Ref. 140).

ple. In high-molecular-weight PLA, decrease in the relative number of end groups may make carboxylic acid undetectable by FTIR and reduce its bioadhesive properties.

All three polyanhydrides (P[CPP:SA] 20:80, P[FA:SA] 20:80, and P[FA:SA] 50:50) displayed multiple peaks from 1,840 cm$^{-1}$ to 1,680 cm$^{-1}$, which are typical of anhydride linkages (159). Particular peaks correspond to either SA–SA, SA–CPP, CPP–CPP, FA–SA, or FA–FA diads or a combination of these. The small peak around 1,710 cm$^{-1}$ is due to the presence of carboxylic acid (161) and is believed to increase as anhydride bonds degrade. Figure 23

depicts the the reaction scheme for hydrolytic degradation of polyanhydrides.

Spectra generated after short-term degradation experiments of P(FA:SA) 20:80 (Fig. 24) helped confirm the hypothesis that the peak at 1,710 cm$^{-1}$ becomes more pronounced with degradation. The rapid hydrolytic degradation characteristic of polyanhydrides is believed to predominantly be a surface rather than bulk phenomenon. In fact, the degradation is often characterized by an erosion zone, made of high-molecular-weight, crystalline polymer containing carboxylic acid end groups (159). The hy-

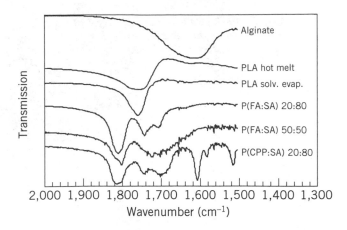

**Figure 21.** FTIR transmission spectra, from 2,000 to 1,500 cm$^{-1}$, of the polymers investigated in this study. *Source:* Reproduced with permission from *Reactive Polymers* (Ref. 140).

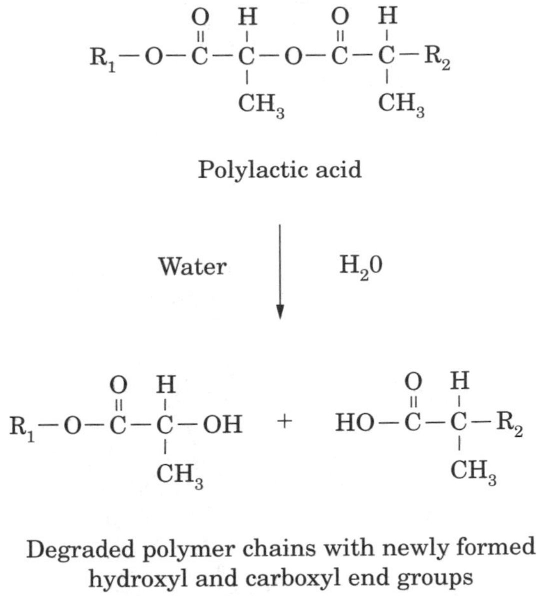

$$R_1-O-\overset{\overset{\displaystyle O}{\|}}{C}-\overset{\overset{\displaystyle H}{|}}{\underset{\underset{\displaystyle CH_3}{|}}{C}}-O-\overset{\overset{\displaystyle O}{\|}}{C}-\overset{\overset{\displaystyle H}{|}}{\underset{\underset{\displaystyle CH_3}{|}}{C}}-R_2$$

Polylactic acid

Water $\quad\Big\downarrow\quad$ H$_2$0

$$R_1-O-\overset{\overset{\displaystyle O}{\|}}{C}-\overset{\overset{\displaystyle H}{|}}{\underset{\underset{\displaystyle CH_3}{|}}{C}}-OH \quad + \quad HO-\overset{\overset{\displaystyle O}{\|}}{C}-\overset{\overset{\displaystyle H}{|}}{\underset{\underset{\displaystyle CH_3}{|}}{C}}-R_2$$

Degraded polymer chains with newly formed
hydroxyl and carboxyl end groups

**Figure 22.** Degradation of polylactic acid.

drolysis of each anhydride bond results in the formation of two carboxylic acid groups (Fig. 23). Figure 24 shows transmission spectra from 2000 to 1,200 cm$^{-1}$ collected for P(FA:SA) 20:80 at four degradation times. In all three polyanhydrides (P[FA:SA] 20:80, P[FA:SA] 50:50 and P[CPP:SA] 20:80), the peak at 1,710 cm$^{-1}$ increased in size proportional to other anhydride peaks at 1,800 and 1,734 cm$^{-1}$ suggesting an increase in carboxyl end-group concentration as anhydride bonds are hydrolyzed (Fig. 24).

Another peak of interest was found at 1,260 cm$^{-1}$. Single-bond stretching and bending deformation associated with carboxylic acid are known to produce strong absorptions at this wavelength (160). Although this peak was nearly undetectable in the nondegraded sample of P(FA:SA) 20:80, after 7 min in buffer a distinct peak emerged that persisted throughout the degradation study (Fig. 24). Because this peak could not be found in either of the pure monomers or pure polymers of FA or SA (158), we suspect that it may have arisen as the ratio of ordered to

nonordered regions increased in the degrading polymer. As amorphous areas were rapidly degraded, crystalline regions were uncovered. These erosion-resistant zones may contain high-molecular-weight polymer chains as well as smaller chains, such as trimers, tetramers, pentamers, and so on, all with carboxylic end groups (158,159).

We believe that rapid degradation of P(FA:SA) copolymers resulting in carboxylic acid production plays an important role in the formation of the intense bioadhesion exhibited during the in vitro studies. For a material to form strong bioadhesive interactions through carboxyl group hydrogen bonding, carboxylic acid must be present as polymer end groups and not free monomer (which could not provide the necessary mechanical strength). The narrow

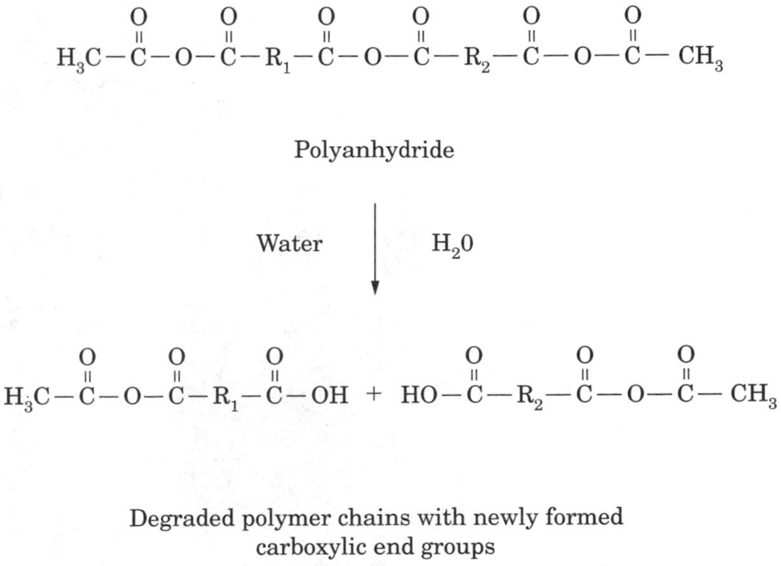

$$H_3C-\overset{\overset{\displaystyle O}{\|}}{C}-O-\overset{\overset{\displaystyle O}{\|}}{C}-R_1-\overset{\overset{\displaystyle O}{\|}}{C}-O-\overset{\overset{\displaystyle O}{\|}}{C}-R_2-\overset{\overset{\displaystyle O}{\|}}{C}-O-\overset{\overset{\displaystyle O}{\|}}{C}-CH_3$$

Polyanhydride

Water $\quad\Big\downarrow\quad$ H$_2$0

$$H_3C-\overset{\overset{\displaystyle O}{\|}}{C}-O-\overset{\overset{\displaystyle O}{\|}}{C}-R_1-\overset{\overset{\displaystyle O}{\|}}{C}-OH \quad + \quad HO-\overset{\overset{\displaystyle O}{\|}}{C}-R_2-\overset{\overset{\displaystyle O}{\|}}{C}-O-\overset{\overset{\displaystyle O}{\|}}{C}-CH_3$$

Degraded polymer chains with newly formed
carboxylic end groups

**Figure 23.** Hydrolytic degradation of polyanhydrides.

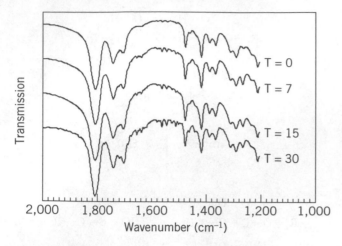

**Figure 24.** FTIR transmission spectra of P(FA:SA) 20:80 microspheres during the course of hydrolytic degradation (time $T_0$ to 30 min). Entire microspheres were ground with KBr and formed into pellets. Note the increasing prominence of peaks at 1,710 and 1,260 cm$^{-1}$. *Source:* Reproduced with permission from *Reactive Polymers* (Ref. 140).

shape of the peak at 1,710 cm$^{-1}$ is a strong indication of carboxylic end groups because free acids typically appear as a broad band (158).

**The Role of FA in Polyanhydride Copolymers.** We suggest that bioadhesion of these hard, bioerodible materials is not due to chain entanglement, as required by the diffusion theory of bioadhesion, but due to numerous hydrogen bonds generated between hydrophilic functional groups (-COOH) and mucus glycoproteins. In addition, we found a strong correlation between the amount of fumaric acid and the adhesive forces, summarized in Table 3. Larger amounts of FA in the polymer result in larger tensile work measurements as well as higher fracture strength measurements. All experiments were performed as described in "Bioadhesive Measurements" using rat jejunum. Statistical analysis using the unpaired t-test is summarized in Table 3. As seen, the formulation made of P(FA:SA) 70:30 is statistically different from any formulation composed of 10 or 20% FA but not different from those containing higher concentrations of FA. The same trend is found for P(FA:SA) 60:40 and 50:50 (Tables 4a and 4b).

**Everted Sac Experiments.** We initiated a new study aimed at mapping the exact amount of microspheres adhering to the GI tract using the everted sac bioassay, which provides important information about the relative bioadhesion of different formulations at potentially any location along the length of the gut. The everted sac experiments were performed using viable segments of rat jejunum. Unfasted rats (300–400 g, male, CD strain) were sacrificed, and the tissue was excised and flushed with 10 mL of ice-cold pBS, pH 7.2, containing 200 mg/dL glucose (PBSG). Six-centimeter segments of jejunum were everted using a stainless steel rod and lightly washed with PBSG to remove the contents. Ligatures were placed at both ends of the segment, and the sac was filled with 1.0–1.5 mL of

**Table 3. Tensile Work and Fracture Strength of Microspheres As Determined by Cahn Force Measurements for Different P(FA:SA) Copolymer Ratios**

| Polymer | Diam. ($\mu$m) | Tensile wk. (nJ) | Fract. str. (mN/cm$^2$) |
|---|---|---|---|
| *P(FA:SA)10:90 (n = 11)* | | | |
| Average | 784.45 | 82.99 | 31.58 |
| stdev | 43.14 | 40.35 | 17.61 |
| *P(FA:SA)20:80 (n = 15)* | | | |
| Average | 775.67 | 32.95 | 17.46 |
| stdev | 48.02 | 20.27 | 6.64 |
| *P(FA:SA)30:70 (n = 10)* | | | |
| Average | 788.90 | 66.89 | 24.26 |
| stdev | 44.42 | 60.10 | 15.29 |
| *P(FA:SA)50:50 (n = 23)* | | | |
| Average | 737.26 | 315.29 | 58.04 |
| stdev | 68.91 | 208.93 | 22.11 |
| *P(FA:SA)60:40 (n = 21)* | | | |
| Average | 763.10 | 319.53 | 67.44 |
| stdev | 49.63 | 209.96 | 31.92 |
| *P(FA:SA)70:30 (n = 4)* | | | |
| Average | 781.75 | 453.23 | 99.71 |
| stdev | 66.77 | 82.67 | 30.06 |

**Table 4a. Fracture Strength Statistics Using the Unpaired T-Test (Alternate Welch T-test); Two-Tailed $P$ Value Reported**

| P(FA:SA) | 10:90 | 20:80 | 30:70 | 50:50 | 60:40 | 70:30 |
|---|---|---|---|---|---|---|
| 10:90 | n.s. | * | n.s. | *** | *** | * |
| 20:80 | — | — | n.s. | *** | *** | * |
| 30:70 | — | — | — | *** | *** | * |
| 50:50 | — | — | — | — | n.s. | n.s. |
| 60:40 | — | — | — | — | — | n.s. |
| 70:30 | — | — | — | — | — | — |

* = significant, $P < .05$; ** = very significant, $P < .01$; *** = extremely significant, $P < .001$; n.s. = not significant.

**Table 4b. Tensile Work Statistics Using the Unpaired T-Test (Alternate Welch T-Test); Two-Tailed $P$ Value Reported**

| P(FA:SA) | 10:90 | 20:80 | 30:70 | 50:50 | 60:40 | 70:30 |
|---|---|---|---|---|---|---|
| 10:90 | — | ** | n.s. | *** | *** | ** |
| 20:80 | — | — | n.s. | *** | *** | ** |
| 30:70 | — | — | — | *** | *** | ** |
| 50:50 | — | — | — | — | * | n.s. |
| 60:40 | — | — | — | — | — | * |
| 70:30 | — | — | — | — | — | — |

* = significant, $P < .05$; ** = very significant, $P < .01$; *** = extremely significant, $P < .001$; n.s. = not significant.

PBSG. Tissue was maintained at 4°C prior to incubation. The sacs were incubated with 60 mg of spheres and 5 mL of PBSG at 37°C. The incubated sacs were agitated end over end. After 30 min, the sacs were removed, and the solution of PBSG and unbound microspheres was centrifuged for 30 min. The supernatant fluid was discarded, the remaining spheres were washed with 5.0 mL distilled water, centrifuged, the water discarded, and the microspheres frozen and dried via lyophilization for 24–48 h. The weight of the bound spheres was determined by subtraction of the tared weight of the tube and lyophilized beads from the initial tare weight and is reported as percent binding.

One of the most interesting results was that the amount of binding is strongly related to the amount of FA in the polymer. Table 5 demonstrates this relationship for hot melt microspheres.

We found that spray-dried polyanhydride microspheres [P(FA:SA)20:80] showed tremendous bioadhesion to segments of small and large intestine in vitro. Spray-dried microspheres also display a much rougher surface and tend to be smaller in diameter than hot-melt microspheres. As much as 65% of the initial dose (100 mg microspheres/g wet weight of tissue) adhered to the mucosa after a 30-min incubation (data not shown). Additionally, the microspheres were found to adhere to mechanically damaged sites on the intestine (161).

**Cahn and Everted Sac Correlation.** To validate the results of our unique Cahn system to determine bioadhesion between individual microspheres and intestinal tissue in vitro, we correlated the results of these measurements to the everted sac bioassay for several polymers. Microspheres were prepared by the hot-melt method, collected by filtration, washed with petroleum ether, and sieved into two size ranges, 600 $\mu$m and below for everted sac measurements and 600–850 $\mu$m for Cahn measurements.

As seen in Figure 25, polymers that exhibit a high binding percentage in the everted sac assay also demonstrate high fracture strengths as determined by the Cahn microbalance. Specifically, [P(FA:SA)50:50]:FAPP (poly[fumaric-co-sebacic anhydride] blended with FA prepolymer) PCL:FAPP (a blend of polycaprolactone and FA prepolymer), and SAPP:FAPP (SA prepolymer and FA prepolymer) all show high bioadhesion in each method. In the everted sac assay [P(FA:SA) 50:50]:FAPP shows 71 $\pm$ 9% (mean $\pm$ SEM) binding, PCL:FAPP shows 82 $\pm$ 15% binding, and SAPP:FAPP shows 64 $\pm$ 4% binding. In the Cahn assay, the fracture strengths for each of these microspheres is 185 $\pm$ 45, 255 $\pm$ 37, and 333 $\pm$ 72 mN/cm$^2$, respectively. Interestingly, polymers such as PCL that show low percentage binding from everted sac experiments also demonstrate low fracture strengths in Cahn measurements. Everted sac experiments with PCL result in 2 $\pm$ 3% binding, and the Cahn assay shows a fracture strength of 15 $\pm$ 2 mN/cm$^2$, both of which are relatively low compared to the other polymers tested. These studies demonstrate for the first time a good correlation between the two methods.

**Contact Angle Measurements with Water and Mucus.** Static contact measurements between polymer films and rat intestinal mucus (Pel Freeze, Rogers, AR) were made using a goniometer. Interestingly, water contact angle for the series of polymers studied gave similar hydrophobic readings (between 97° to 68°), while the mucus reading showed very low contact angle, which is a strong indication of high adhesive forces as dictated by the wetting theory of adhesion. Moreover, the series indicate that higher FA concentrations within the copolymer result in lower contact angles (Table 6). These results strongly correlate with the Cahn data described previously.

Static contact-angle measurements were measured for several polymers (Table 7). As expected, water contact-angle measurements for the bioerodible materials all gave similar hydrophobic readings (~64°), while the hydrated alginate sample showed zero (0°) contact angle (i.e., complete wetting).

Mucus measurements provided the most notable results. Though no correlation was seen between water contact angles and mucus contact angles, an interesting trend was observed between mucus contact angle and bioadhesive force. Both alginate and P(FA:SA) copolymers produced very low contact angles with mucus and produced strong bioadhesive interactions (see "Bioadhesive Measurements"), whereas P(CPP:SA) 20:80 produced a contact angle of 35° and weaker bioadhesive interactions. This supports the wetting theory of adhesion, which states that strong adhesive forces result from sufficient wetting of the two components, and these results further advocate the notion of bioadhesion between hard, bioerodible materials and soft tissue.

It is possible that surface energetics could promote bioadhesion in thermoplastic materials. All thermoplastics studied were rather hydrophobic ($\geq$60°), suggesting a driving force away from water. Hydrogels, on the other hand, are naturally hydrophilic. Although both P(FA:SA) 50:50 and alginate produced similar mucus contact angles owing to the hydrophobic property of P(FA:SA), it may experience a greater driving force towards mucus. Figure 26 depicts a schematic representation of two microspheres in the GI lumen. Assuming the center of the lumen to be most like pure water and the edge of the lumen to be most like pure mucus, vectors are indicated representing the driving forces produced by interfacial energies. Alginate is both hydrophilic and mucophilic and thus experiences opposing driving forces, whereas P(FA:SA) 50:50 is hydrophobic and mucophilic and is thus driven towards the mucosal surface. Although this pure water/pure mucus condition does not exist in vivo, mucus concentration gradients are found in the GI lumen, and therefore similar driving forces may in-

**Table 5. Percent Binding of Hot-Melt Microspheres As Determined by Everted Sac Experiments**

| Polymer | $M_r$ (weight average) | % binding $\pm$ SEM | No. of sacs |
|---|---|---|---|
| P(FA:SA)20:80 | 10,000 | 19.9 $\pm$ 8.3 | 6 |
| P(FA:SA)50:50 | 3,000 | 10.8 $\pm$ 7.4 | 6 |
| P(FA:SA)70:30 | 3,000 | 61.3 $\pm$ 17.1 | 5 |

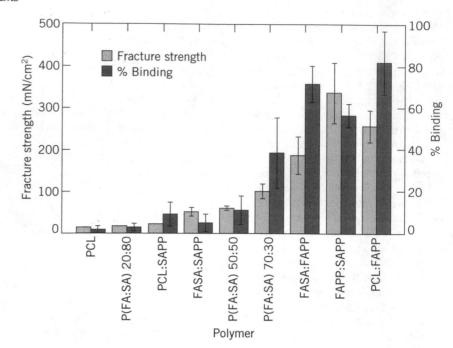

**Figure 25.** Correlation between Cahn and everted sac measurements. *Note:* FASA:SAPP and FASA:FAPP denote [P(FA:SA)50:50]:SAPP and [P(FA:SA)50:50]:FAPP.

**Table 6. Contact Angle Measurements on Melt-Cast Films Made of P(FA:SA) Copolymers**

| P(FA:SA) | 20:80 | 30:70 | 40:60 | 50:50 | 60:40 | 70:30 |
|---|---|---|---|---|---|---|
| Water | 97 | 89 | 81 | 88 | 71.5 | 68.5 |
| Mucus | 34 | 28 | 19 | 20 | 20 | 29 |

**Table 7. Water and Mucus Static Contact Angles Measured on Polymer Films Cast on Glass Slides**

| Polymer | Water contact angle (°) | Mucus contact angle (°) |
|---|---|---|
| P(CPP:SA)20:80 | 64 | 35 |
| PLA (solvent evaporation) | 63 | 20 |
| Alginate | 0 | 14 |

duce contact of P(FA:SA) microspheres with the mucosal layer.

In papers by Lehr et al. (156,157) and Baszkin et al. (60), a more sophisticated captive bubble technique was employed to measure contact angles and interfacial energies of hydrated bioadhesives. Lehr et al. found a strong correlation between mucus wetting of bioadhesives and bioadhesive bond strength. Furthermore, they suggested that interpenetration and chain entanglement was not a necessity for bioadhesion. In light of their analysis and the results of our study, we agree that mucus contact angle and surface energy analysis can be strong predictors of bioadhesive potential, and we believe that mechanisms other than chain entanglement are responsible for the bioadhesive properties of copolymers of P(FA:SA).

To summarize this section, we have shown that fracture theory analysis of the microbalance experiments clearly showed that copolymers of FA and SA produced bioadhesive bonds with rat intestine in vitro. Both the fracture strengths and tensile works calculated for these bioerodible microspheres indicated stronger adhesive interactions than those observed for any other material in this study, including alginate microspheres. We believe interactions between P(FA:SA) copolymers and intestinal tissue rely on more than simple physical entrapments. Furthermore, because these materials lack flexible polymer chains, it seems unlikely that polymer–mucin chain entanglement plays a significant role in bioadhesion. The results of our studies suggest that adhesion of these materials to mucus is due to surface energy effects and secondary bonding phenomena, but other mechanisms, yet to be determined, may also play a role.

SEMs of P(FA:SA) microspheres confirmed their ability to erode rapidly in the presence of water. Visible surface changes were observed after even very short incubation

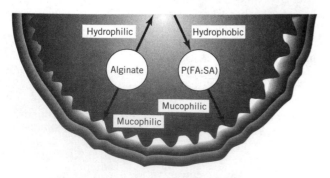

**Figure 26.** Proposed surface energy theory.

periods (7 min). Degradation of melt-cast polyanhydrides has been shown to be a surface rather than bulk phenomena, characterized by a unique erosion zone (158,159). The remaining polymer in the eroding surface layer, although porous and fragile in appearance, is highly crystalline and contains specific spherulitic structures (158,159). We believe that rapid degradation of these materials may enhance their bioadhesive properties by not only creating a rougher and more porous surface but also by increasing the quantity of surface carboxylic acid.

An interesting trend was noted between bioadhesive fracture strengths and mucus contact angles. Those materials with low mucus contact angles (i.e., alginate and P[FA:SA] 50:50) produced strong adhesive interactions. These results support the wetting theory of adhesion, which states that strong bioadhesive bond formation is dependent on favorable spreading and wetting of the biological component on the polymer or vice versa. It should also be noted here that water contact angle seemed to have no correlation with mucoadhesive properties.

### GI Transit and Bioavailability of Low-Molecular-Weight Drugs

Having established P(FA:SA) as a strong BP and realizing that polyanhydrides are well suited for drug delivery, we set out to determine if its bioadhesive properties could positively effect GI transit and bioavailability of two model drugs. Alginate was selected as a comparison polymer owing to its moderate adhesive performance, ease of microsphere fabrication, low cost, and the fact that it is a hydrogel.

To determine effects on transit through the gut, radiopaque barium sulfate was encapsulated in both polymers. After administration to rats, serial feces collections were made, feces were x-rayed, and microspheres were counted to determine GI residence time.

For bioavailability studies, two drugs were investigated: salicylic acid (an analgesic and nonsteroidal antiinflammatory drug [NSAID] rapidly absorbed by the stomach and duodenum) and dicumarol (an anticoagulant that is poorly absorbed in the small intestine) (Fig. 27). To determine bioavailability, serum concentration-time curves were generated and maximum serum concentration ($C_{max}$), time to maximum serum concentration ($T_{max}$), and area under the serum-concentration-time curves (AUC) were compared to unencapsulated controls.

**Material Selection and Animal Models.** Bioadhesive polymers investigated in this study were alginate and

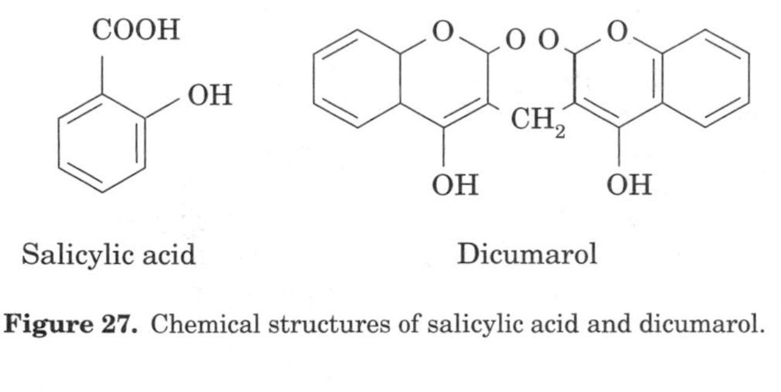

**Figure 27.** Chemical structures of salicylic acid and dicumarol.

P(FA:SA) 20:80 and were chosen based on in vitro results. A commercially available formulation of 95% w/w barium sulfate (Tonopaque) was chosen as a radiopaque marker for GI transit studies. Two model drugs [salicylic acid and 3,3'-methylene-bis(4-hydroxy-coumarin) (dicumarol)] were selected for bioavailability studies based on absorption characteristics and assessability using simple spectrophotometric techniques. Polyvinylphenol (PVP; Polysciences, Warrington, PA) was used to coat salicylic acid, as a barrier to dissolution, prior to encapsulation in P(FA:SA).

**Preparation of Drug-Loaded Microspheres** *Drug Preparation.* To produce proper DDSs and controls for those systems, processing of the stock drugs was required. A LabPlant SD-4 spray-drier (Virtis, Gardiner, NY) equipped with a 1.0-mm nozzle was used to micronize dicumarol particles to improve dissolution and to coat salicylic acid particles with PVP to inhibit dissolution. Dicumarol was dissolved in methylene chloride (0.5% w/v) and sprayed with an inlet temperature of 90°C. Resulting particles were less than 5 $\mu$m in diameter, porous, and very fragile. Equal amounts of salicylic acid and PVP were dissolved in acetone (10% w/v for each component) to give a 1:1 ratio and sprayed at 65°C. Resulting particles were less than 5 $\mu$m in diameter, spherical, and dense.

*Microsphere Fabrication and Ionic Gelation.* Alginate microspheres were fabricated according to a modified method of Lim (146,162). Powdered drug (4, 25, or 50%, w/v of dicumarol, salicylic acid, or Tonopaque, respectively) was added to a 2% alginate solution. Each mixture was vigorously blended with a Virtis 23 microhomogenizer (Virtis, Gardiner, NY) for 3 min. The alginate suspension was loaded into a 10-cc syringe and pumped through the inner lumen (18 gauge) of a double-lumen spinneret at a rate of 0.8 mL/min using a syringe pump (model 22; Harvard Apparatus, South Natick, MA). Droplet size was adjusted by controlling the flow of nitrogen gas (30 psi) through the spinneret's outer lumen. Droplets were collected and allowed to gel for 20 min in a slowly stirred calcium chloride bath (1.3% w/v). Resulting microspheres were wet-sieved and collected into two size ranges (drug loaded: $d < 212$ $\mu$m and Tonopaque loaded: $600 \ \mu m < d < 850 \ \mu m$). Microspheres were stored damp in sealed vials. Final drug concentrations were assayed to be 16 and 25% weight/wet weight microspheres for dicumarol and salicylic acid, respectively.

*Hot-Melt Microencapsulation.* Hot-melt microencapsulation (147), was used to form large, radiopaque microspheres of P(FA:SA) for transit studies. P(FA:SA) was melted in a crucible over low heat (<100°C), and powdered Tonopaque was mixed with the polymer (50% w/w). The mixture was poured into a beaker containing 200 mL of silicon oil (200 Fluid; Dow Corning Inc., Midland, MI) at 80°C and stirred to produce droplets of molten P(FA:SA) containing barium sulfate. The oil system was cooled, causing droplets to solidify into microspheres. Microspheres were collected, washed with petroleum ether, and seived to a range of 600–850 $\mu$m.

*Phase Inversion.* To produce drug-loaded polyanhydride microspheres, a phase-inversion microencapsulation technique was developed. Briefly, the polymer was dissolved in a good solvent (5% w/v in methylene chloride); drug (40 and 16% w drug/w polymer for dicumarol and spray-dried salicylic acid/PVP, respectively) was added to the solution; and the mixture was poured into a strong nonsolvent (petroleum ether) that was highly miscible with the solvent (solvent to nonsolvent ratio was 1:20). The result was spontaneous production of microparticles ($0.5 \mu m < d < 5 \mu m$) through phase inversion. The microparticles were filtered, briefly washed with petroleum ether, and dried with air. The advantages of this technique over methods that could produce similar-sized particles (spray-drying or solvent evaporation) are very low polymer loss, low drug loss (for dicumarol), simplicity, and speed.

***In Vitro Dissolution and Release Studies.*** Drug and bioadhesive drug formulations (40 and 5 mg drug for salicylic acid and dicumarol, respectively) were incubated in 10 mL PBS in polypropylene centrifuge tubes at 23°C for 10 h. Periodically, 100-$\mu$L samples of the aqueous solution were removed from each tube and centrifuged through a 0.2-$\mu$m filter to eliminate particulates. Samples were analyzed to determine dissolved drug concentrations. After each sampling, tubes were shaken to resuspend drug and microspheres.

**Animal Studies** *GI Transit of Radiopaque Microspheres.* Prior to administration, $\approx$200-mg microspheres (either barium-loaded P[FA:SA] 20:80 made by hot-melt or barium-loaded alginate) were suspended in 1 mL 0.9% sodium chloride. Male, VAF, CD® rats (250–350 g, Charles River Labs, Wilmington, MA) that had been fasted overnight, were anesthetized with methoxyfurane, and a silicone feeding tube (2-mm o.d.) was used to deliver microsphere suspensions directly into their stomachs. Following administration, rats were allowed to recover from anesthesia.

Animals were reanesthetized and x-rayed using a Hitachi model DGC-1010 mobile X-ray unit (Hitachi Co., Tokyo, Japan) and Kodak Ready Pak II Type M X-ray film (Kodak, Rochester, NY), thereby verifying the presence of microspheres in the stomach lumen. From these X rays it was possible to determine the number of microspheres initially administered to each animal.

After x-raying to determine initial dose, animals were placed in metabolic cages positioned above a custom-designed, automatic, feces-collecting machine. Feces were separated from the urine and collected at 2 h–50 min intervals over a three-day period. During this time the animals were provided with food and water ad libitum. Feces were x-rayed, and microspheres were counted to determine their rate of passage from stomach to anus as well as the percentage of microspheres remaining in the animal at each time point.

Excel 5.0 (Microsoft, Redmond, WA) was used to perform one-tailed student t-tests (unpaired, equal variance) on raw data at each collection point to assess statistical differences between the groups.

***Bioavailability of Drug and Encapsulated Drug Formulations.*** Catheters were required for repeated blood sampling during the bioavailability study. Male, VAF, CD® rats (250–350 g, Charles River Labs, Wilmington, MA) were anesthetized with an ip injection of sodium pentobarbital (60 mg/kg), and with the animals supine, a midline incision was made from the top of the neck to the clavicle. Blunt dissection was used to locate and isolate the external jugular vein. Three ligatures were placed around the vessel. One ligature was used to tie off the cranial end of the isolated segment, a small incision was made in the side of the vessel, and a silicone catheter (Bio-Sil medical grade silicone tubing (500-$\mu$m i.d., 940-$\mu$m o.d., 53-cm length); Sil-Med Corporation, Taunton, MA) filled with heparin (666 U/mL) was inserted into the vessel lumen and fed towards the heart. The other two ligatures were tied off around the vessel and catheter, caudal to the point of insertion, while being careful not to deform and restrict the catheter.

Next, animals were turned prone, and a second incision was made from the base of the skull to the midpoint between the scapulae. Forceps were used to tunnel subcutaneously to the opening of the ventral incision. The free end of the catheter was pulled beneath the skin to emerge from the dorsal incision at the base of the neck. The ventral incision was sutured. The free end of the catheter was fed through a 30.5-cm-long stainless steel spring tether with 22-gauge swivel (Harvard Apparatus, South Natick, MA). The button-end of the tether was sutured to the fascial covering of the muscles in the back of the neck, and a purse-string stitch was used to close the dorsal incision.

Tall cages were constructed to allow for untangled movement of the tether. To produce an enclosure 33 cm high, two rat cages (36 cm $\times$ 30 cm $\times$ 16 cm) were joined, one inverted on top of the other, with 4 delrin spacers (1 cm thick) to allow adequate air flow. Additional air holes were drilled, and a delrin collar was glued into the center of the top cage. The swivel-end of the tether was passed through the top of the upper cage and secured in the collar with a set screw. A three-way stopcock was attached to the swivel on the outside of the cage with a small section of silicone tubing ($\approx$8 cm) to facilitate syringe changes without breaking the closed-catheter system. This setup allowed blood samples to be drawn from outside the cages without handling or anesthetizing the animals. Additionally, animals have near normal mobility.

On the day following surgery, the animals were anesthetized with methoxyfurane and a flexible silicone feeding tube was introduced via the esophagus to the stomach. Either microspheres or drug (25 or 200 mg drug weight/kg rat body weight for dicumarol or salicylic acid, respectively) were suspended in 1.5 mL of maple syrup (Camp Inc., Plessisville, Canada) and administered to the animals via the silicon tube. A positive displacement plunger was passed through the tube to ensure proper dosage.

Following drug administration, blood samples were drawn from the animals approximately 1 h for the first 4 h and approximately every 4 h after that point for the next 68 h. Samples were centrifuged, 100 $\mu$L of serum was removed, and blood cells were resuspended in 100 $\mu$L of saline (0.9% NaCl) and returned to the animals. Catheters were flushed with saline and plugged with heparin (1000 U/ml) after returning blood cells.

Salicylic acid concentrations were measured using a colorimetric assay (163). One hundred microliters of serum was diluted with 600 $\mu$L of distilled water and reacted with 1 mL 1.0% ferric nitrate in 0.07 N nitric acid. The presence of salicylate was revealed by a change in color from faint yellow to deep purple. The samples were measured spectrophotometrically at 530 nm using a Beckman model DU65 spectrophotometer (Beckman Instruments, Inc.; Fullerton, CA).

Dicumarol concentrations were assayed spectophotometrically in the UV range (164). One hundred microliters of serum was acidified with 20 $\mu$L of 3 M HCl. Dicumarol was extracted from the aqueous phase into 700 $\mu$L of $n$-heptane. Six hundred microliters of the heptane phase was mixed with 900 $\mu$L of 2.5 M NaOH to back-extract the dicumarol. Dicumarol concentration in NaOH was measured spectrophotometrically (315 nm) in quartz cuvettes.

Readings from individual time points were averaged to produce mean concentrations with standard errors for graphical analysis. Pharmacokinetic software (R-Strip; Micromath Scientific Software, Salt Lake City, UT) was used to perform least-squares regression on the average values to produce exponential curve fits. Raw data points were analyzed with a students $t$-test (one-tailed, unpaired, equal variance) to determine statistical significance at each sampling time. $C_{max}$, $T_{max}$, and AUC were compared for statistical differences using a students $t$-test (two-tailed, unpaired, equal variance) on raw data for $T_{max}$ and on logarithmically transformed values for $C_{max}$ and AUC.

**Results and Discussion.** The effects of bioadhesive polymers on delaying microsphere transit and on bioavailability of encapsulated drugs have been studied. Two polymers were chosen for this study based on our in vitro bioadhesive strength measurements (see "Bioadhesive Measurements" through "GI Transit and Bioavailability of Low-Molecular-Weight Drugs"): alginate and poly(fumaric-co-sebacic anhydride) [P(FA:SA)]; calcium alginate microspheres were shown to have weak-to-moderate bioadhesive properties, whereas P(FA:SA) microspheres were shown to be very strong bioadhesives.

*Radiopaque Microspheres.* To provide a means of tracking microsphere transit through the GI tract, radiopaque barium sulfate was encapsulated in the polymers. Encapsulation was performed using ionic gelation (alginate) and hot-melt [P(FA:SA)20:80) methods. Because barium sulfate is a heavy powder that is insoluble in aqueous solutions and oil, encapsulation by these techniques was straightforward and nearly 100% efficient.

*Drug Encapsulation.* The hot-melt technique, which typically produces very dense microspheres larger than 100 $\mu$m in diameter (147) could have been used to encapsulate salicylic acid and dicumarol in P(FA:SA) 20:80, but preliminary investigations revealed that microspheres of this size and morphology produced inappropriate delivery profiles (lasting several days to several weeks, depending on size and loading). To speed release kinetics, a phase-inversion microencapsulation technique was used to produce very small, quick-releasing microparticles (Fig. 28). This procedure worked well for encapsulating dicumarol (initial loading: 40% w/w, final loading: 38% w/w), because dicumarol is soluble in methylene chloride and only partially soluble in petroleum ether. Also, because dicumarol is poorly soluble in neutral to acid aqueous solutions, the small particle size may help dissolution in vivo or enhance uptake.

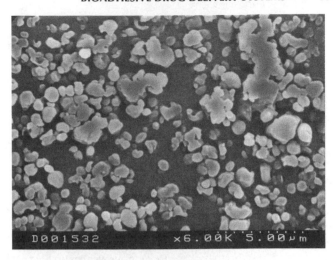

**Figure 28.** SEM of P(FA:SA) particles formed by the phase inversion method. Microparticles are very small (0.5 $\mu$m $< d < 5$ $\mu$m) and dense.

On the other hand, salicylic acid is quite soluble in petroleum ether and aqueous solutions. These two properties resulted in poor encapsulation (<1% w/w final loading) and extremely rapid release. Therefore, a double encapsulation technique was employed; salicylic acid was first spray-dried with PVP (from a 10% w/v of solution of both components in acetone), and then the spray-dried particles were encapsulated in P(FA:SA) using the phase-inversion technique. PVP was chosen because it has a common solvent with salicylic acid (acetone) and because it is nonsoluble in both methylene chloride and petroleum ether. The result of the double encapsulation was a much higher loaded final product (9% w/w) with a longer release profile.

*Drug Controls.* Appropriate controls were needed to assess the effect of bioadhesive polymers on bioavailability. It was essential to provide controls that had similar solubility and absorptivity as each bioadhesive system. No special controls were required for either of the salicylic acid–containing devices, because stock salicylic acid is known to be freely absorbed in the stomach.

Dicumarol, as received from Sigma, was in the form of particulate crystals (ranging in size from 10 to 200 $\mu$m) and was poorly soluble in neutral-to-acid aqueous solutions (165). Because drug particle size was unchanged by alginate encapsulation, stock dicumarol was determined to be an appropriate control for that system. However, to obtain a suitable control for the 0.5–5 $\mu$m, P(FA:SA)-encapsulated dicumarol-loaded microspheres produced through phase inversion, we felt it was necessary to reduce particle size and thereby improve dissolution. Dicumarol was micronized by dissolving in methylene chloride (0.5% w/v) and spray-drying. The resulting particles were less than 5 $\mu$m in diameter, porous, and very fragile.

*In Vitro Dissolution.* In vitro dissolution and release experiments were performed on all drugs and polymer-encapsulated drugs to investigate the effects of processing techniques and particle size on drug solubility and release. Figures 29 and 30 show dissolution and release profiles for salicylic acid and dicumarol formulations, respectively. In

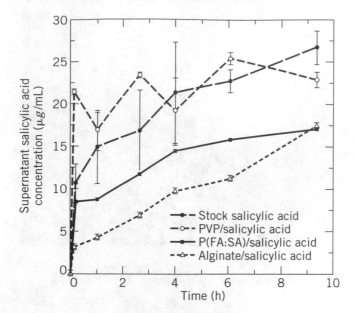

**Figure 29.** In vitro dissolution of stock salicylic acid and release curves for spray-dried salicylic acid/PVP, salicylic acid encapsulated in alginate, and salicylic acid/PVP encapsulated in P(FA:SA). Stock salicylate was chosen as the in vivo control for both alginate and P(FA:SA)-encapsulated forms.

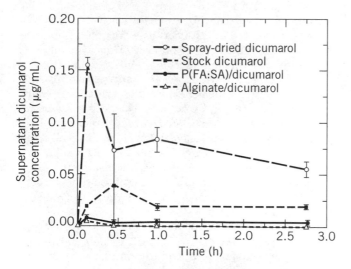

**Figure 30.** In vitro dissolution of stock dicumarol, spray-dried dicumarol, dicumarol encapsulated in alginate, and dicumarol encapsulated in P(FA:SA). Stock dicumarol was used as control for the alginate-encapsulated system, while spray-dried dicumarol served as control for the P(FA:SA) encapsulated system.

both graphs, controls (unencapsulated forms) result in equal or faster dissolution than polymer encapsulated formulations.

Of the dissolution and release curves for salicylic acid, stock salicylic acid and spray-dried PVP–salicylic acid produced the fastest release rates. The spray-dried material was released quickly owing to its very small particle size. However, release was significantly decreased by encapsu-

lating PVP–salicylic acid in P(FA:SA) using the phase-inversion technique.

Dicumarol is normally insoluble in neutral-pH solutions. Dissolution and release curves for stock dicumarol, dicumarol encapsulated in alginate, and dicumarol encapsulated in P(FA:SA) supported this fact (Fig. 30). However, an interesting dissolution profile was observed with spray-dried dicumarol. After 5 min incubation, supernatant concentrations rose to 11 $\mu$g/mL, corresponding to approximately 2.2% dissolution. As the incubation continued, more drug became insoluble; the supernatant concentration dropped in half after 3 h. A possible explanation is that spray-drying not only reduces particle size and increases surface area but also creates an amorphous soluble structure. Although initially soluble with agitation, dicumarol quickly crystallizes as it precipitates around scattered drug particles acting as nucleation sites. We believe that in the living animal, the soluble material may be absorbed into the circulation prior to having the opportunity to recrystallizing. These results and beliefs are in agreement with others (165).

Alginate encapsulation appears to limit drug release for both salicylic acid and dicumarol even though alginate is a loose hydrogel network offering a poor diffusion barrier. However, the curves are misleading. During the ionic gelation fabrication process, most of the freely soluble drug is lost to the supernatant. Drug that remains encapsulated is of larger particle size and mainly crystalline. Therefore, dissolution of this material is slower than stock samples, and release rates of these devices tend to be less than expected.

***In Vivo Transit.*** A simple and straightforward procedure using radiopaque markers, feces collection, and X rays was designed to monitor GI passage in vivo. By encapsulating 50% Tonopaque in polymer microspheres (using hot melt for P[FA:SA] and ionic gelation for alginate) and sieving the spheres to a fairly large size (600 > d ≥ 850), we were able to produce polymer devices that were easily traced with X rays. Not only could microspheres be identified in feces, but immediately following oral administration they could also be clearly observed and counted in the lumen of the GI tract. Feces collection and X ray inspection provided us with a noninvasive method of monitoring total GI residence time without affecting normal GI motility. Designing and constructing an automated feces-collection machine (Figs. 31 and 32) simplified retrieval of the passed microspheres.

Twelve animals were used in this study. Each animal was used twice (once for alginate and once for P(FA:SA) 20:80), with greater than 1 week rest period between experiments. The mean percentages of microspheres retained in the animals at each time point with standard errors are shown in Figure 33.

Distinct differences were observed between the two polymers. Alginate spheres were eliminated quickly, appearing after 8.5 h; P(FA:SA) spheres never appeared before 11 h 20 min. By 17 h, only 27.41 ± 7.19% of the alginate microspheres remained in the animals, whereas 53.93 ± 11.14% remained for the P(FA:SA) 20:80. Although percentages beyond 27 h were not statistically different, the time required for 90% passage was about 20 h

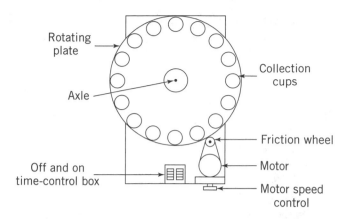

Figure 31. Automatic feces-collecting machine. Top view: shows arrangement of collection cups. *Source:* Reproduced with permission from *Journal of Controlled Release* (Ref. 91).

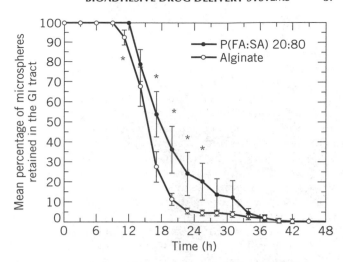

**Figure 33.** Percentage of microspheres retained over time for radiopaque alginate and P(FA:SA) 20:80. Time points marked by asterisks (*) indicate points that were statistically different in the two curves ($P \leq .05$). *Source:* Reproduced with permission from *Journal of Controlled Release* (Ref. 91).

for alginate and nearly 34 h for P(FA:SA) 20:80. These results clearly demonstrate the ability of bioadhesive P(FA:SA) 20:80 to slow GI transit in the rat.

***Bioavailability Studies.*** Both plain drugs and BDDS microspheres were administered orally as either solutions or suspensions in maple syrup. A thick fluid, such as maple syrup, was required to prevent the microspheres from packing in the feeding tube during administration. Table 8 outlines the delivery systems and controls used in this study.

Catheters were implanted in the jugular veins to facilitate remote blood sampling without anesthetizing the animals and to provide a means of reinjecting sampled blood. To protect the fragile catheters from the animals, armored-steel spring tethers were used, requiring the fabrication of

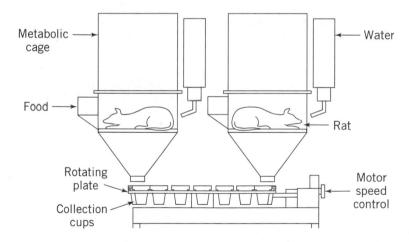

**Figure 32.** Automatic feces-collecting machine. Side view: shows position of collection cups, rotating wheel, and metabolic cages. *Source:* Reproduced with permission from *Journal of Controlled Release* (Ref. 91).

**Table 8. Bioavailability Experiment Layout Showing Controls for Each Experimental Group**

| Drug | Experiment | Control |
|---|---|---|
| Salicylic acid (200 mg/kg) | Stock salicylic acid encapsulated in alginate suspended in maple syrup (25%, $d \leq 212$ μm) | Stock salicylic acid suspended in maple syrup |
| | Stock salicylic acid spray-dried with PVP encapsulated in P(FA:SA) 20:80 suspended in maple syrup (9%, $d = 1$–15 μm) | Stock salicylic acid suspended in maple syrup |
| Dicumarol (25 mg/kg) | Stock dicumarol encapsulated in alginate suspended in maple syrup (16%, $d \leq 212$ μn) | Stock dicumarol suspended in maple syrup |
| | Spray-dried dicumarol encapsulated in P(FA:SA) 20:80 suspended in maple syrup (38%, $d = 0.5$–5 μm) | Spray-dried dicumarol suspended in maple syrup |

*Note:* Drug-loading per percentages and microsphere diameters are given in parentheses.

unusually tall cages to ensure untangled movement of the animals (Fig. 34). Blood samples were taken at regular intervals (every 3–5 h) not only to obtain data but also to reduce the chance of irreversible clot formation. After reinjecting resuspended cells, the catheters were flushed with saline and plugged with a heparin solution (1,000 U/mL) to minimize intraluminal coagulation. However, several catheters did clot, and, therefore, data for those animals were collected only up until that point.

*Salicylic Acid.* Salicylic acid is rapidly absorbed following protonation in the acid environment of the stomach. When orally administered in an unencapsulated form, most absorption occurs in the first 20–30 min, prior to gastric emptying. Once the drug passes through the duodenum into the higher pH of the lower segments of the small intestine, very little absorption takes place (166,167).

Figure 35 shows the serum-concentration-time curves for the three salicylic acid formulations. Both alginate- and P(FA:SA)-encapsulated formulations showed statistically

lower AUC than the control ($P \le .002$) and P(FA:SA) showed statistically lower $C_{max}$ than the control ($P \le .05$). No statistical differences were observed for $T_{max}$. Statistical differences for individual time points are displayed on the graph (Fig. 35).

In both encapsulated formulations we observed less salicylate entering the systemic circulation than in controls. This could be explained by the following: If the encapsulated formulations slow the release of salicylic acid, as observed by in vitro dissolution and release profiles, and gastric emptying occurs prior to the majority of release, then very little drug is available for absorption. The result is lower systemic concentrations. Thus, DDSs known to adhere to the small intestine have little effect on improving bioavailability of drugs that are rapidly absorbed in the stomach and duodenum, such as salicylates.

*Dicumarol.* Dicumarol is widely used clinically as an anticoagulant. Its mode of action is attributed to competitive inhibition of vitamin K, thus limiting the production of vitamin K-dependent coagulation factors VII, IX, X, and II (168). As received from Sigma Chemicals, it was a white, crystalline powder that was poorly soluble in all but very basic aqueous solutions. When given orally, dicumarol is taken up mainly in the small intestine, and its absorption has been characterized as poor and erratic (165,168,169). It was our hypothesis that a bioadhesive delivery system designed to adhere to the intestinal lining and deliver its contents over 4- to 8-h would improve absorption of dicumarol.

Bioavailability plots of stock dicumarol and alginate encapsulated dicumarol are compared in Figure 36. None of the concentrations were statistically different at any of the sampling points, and no statistical differences were observed for the $C_{max}$, $T_{max}$, and AUC. The two formulations produced identical results. The reason for this may be twofold: In vivo dissolution and release profiles of the formulations may not be significantly different, and physical properties of alginate (e.g., unfavorable surface energy for

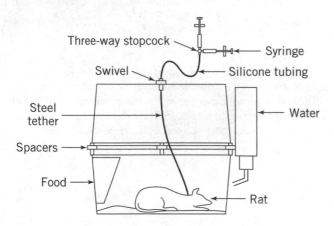

**Figure 34.** Diagram of catheterized rats. Tall cages were constructed to allow for untangled movement of animals. This system allowed blood to be sampled and returned to the animals without need for handling or anesthetizing. *Source:* Reproduced with permission from *Journal of Controlled Release* (Ref. 91).

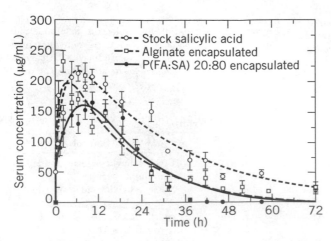

**Figure 35.** Mean salicylic acid serum concentrations (±SEM) versus time. Neither encapsulated form resulted in improved bioavailability.

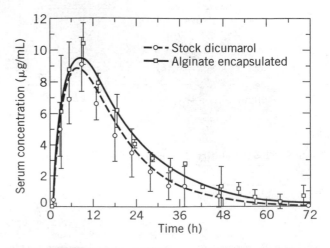

**Figure 36.** Mean dicumarol serum concentrations (±SEM) versus time for stock dicumarol and dicumarol encapsulated in alginate. No effect was observed on bioavailability. *Source:* Reproduced with permission from *Journal of Controlled Release* (Ref. 91).

spontaneous mucoadhesion) may produce bioadhesive interactions that are too weak to improve drug delivery and bioavailability.

Dicumarol encapsulated in P(FA:SA) 20:80 produced strong support for bioadhesion (Fig. 37). Although no significant increase was observed in $C_{max}$ and $T_{max}$, the AUC for P(FA:SA)-encapsulated dicumarol was statistically higher than controls ($P \le .05$). In fact, a 56.6% increase in mean AUC was observed for the bioadhesive device (363.6 and 232.1 $\mu$g h/mL for experimental and controls, respectively). Furthermore, beyond the first 12 h, 9 out of 13 time points showed statistically higher serum concentrations compared to spray-dried controls. Detectable serum dicumarol levels were maintained for at least 72 h in animals given bioadhesive formulations, while those given unencapsulated controls exhibited undetectable serum levels after 48 h.

Observation that controls exhibit more rapid dissolution profiles than encapsulated forms (Fig. 30) eliminates the possibility that improved bioavailability of bioadhesive dosage forms is a result of faster solubilization of the drug. If encapsulated dosages show slower dissolution rates, then improvements in bioavailability must be due to other mechanisms such as increased absorption rate through intimate contact with absorbing tissues or increased residence time at the site of absorption, both of which can be attributed to bioadhesion.

These results indicate that bioadhesive P(FA:SA) improves bioavailability by either enhancing absorption through intimate contact, minimizing drug degradation prior to absorption, or increasing residence time in the absorbing portion of the GI tract. Although statistical differences were measured, degradation, dissolution, and release characteristics of P(FA:SA) DDSs could possibly be tailored to produce even stronger effects.

***Overview of Bioadhesive Performance.*** We have utilized four methods to assess bioadhesion: in vitro mechanical

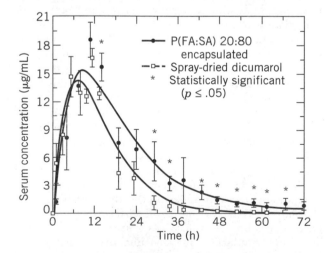

**Figure 37.** Mean dicumarol serum concentrations ($\pm$ SEM) versus time for spray-dried and encapsulated dicumarol. Both formulations were administered orally via a silicone feeding tube. Encapsulation resulted in a 56% improvement in AUC compared to controls. *Source:* Reproduced with permission from *Biotechnology & Bioengineering* (Ref. 25).

testing, polymer characterization techniques, measurement of in vivo transit, and assessment of in vivo bioavailability of encapsulated drugs. In all evaluations we observed better performance from P(FA:SA) 20:80 microspheres compared to calcium alginate microspheres. Table 9 outlines some relevant measurements. P(FA:SA) produced significantly greater deformation to failure, fracture strength, and tensile work than alginate as measured by the Cahn technique; significantly longer GI retention times compared to alginate; and increases in blood serum $C_{max}$, $T_{max}$, and AUC as compared to controls. P(FA:SA) polymers appear to offer promise for oral BDDSs.

***Conclusions.*** The retention times obtained using noninvasive technique (the oral feeding of radioopaque microspheres) correlate well with in vitro results obtained previously using a Cahn microbalance to measure bioadhesion between microspheres and rat intestinal epithelium and with the properties measured using polymer characterization techniques. P(FA:SA) 20:80 showed greater fracture strength and tensile work than alginate, indicating stronger bioadhesive properties. This pattern was repeated in the in vivo transit study, where a significant delay in GI transit was observed for P(FA:SA) 20:80 microspheres when compared to alginate microspheres of a similar size. The results support the hypothesis that bioadhesive materials can prolong GI residence.

Even though no improvement was observed in salicylic acid absorption, dicumarol delivery was improved by encapsulating in P(FA:SA) 20:80. In light of these findings and based on known absorption properties of salicylic acid and dicumarol, we could assume a scenario where bioadhesive polyanhydride DDSs retained significant quantities of drug while in the stomach, delayed intestinal transit, and released most of their contents while adhered to the intestinal mucosa. We believe adhesion and the resulting increased residence time was most likely a result of hydrogen bond formation between surface carboxylic acid and mucosal glycoproteins. Through bioadhesion, intimate contact with absorptive epithelium may have reduced diffusion distances and drug loss, resulting in improved delivery.

Our findings show that bioadhesive, controlled-release systems are a potential solution for improving bioavailability of orally delivered drugs. To bring bioadhesive systems to practical use, each particular bioactive agent will require a well-engineered vehicle providing proper release kinetics as well as appropriate adhesive characteristics.

**Table 9. Comparison of Bioadhesive Properties ($\pm$ SEM) of Alginate and P(FA:SA) 20:80**

| Parameter | Alginate | P(FA:SA)20:80 |
|---|---|---|
| Deformation to failure (mm) | $0.649 \pm 0.144$ | $1.219 \pm 0.185$ |
| Fracture strength (mN/cm$^2$) | $30.16 \pm 4.18$ | $51.63 \pm 3.38$ |
| Tensile work (nJ) | $62.18 \pm 14.50$ | $255.45 \pm 33.04$ |
| % remaining @ 11 h 20 min | $90.68 \pm 3.83$ | $100.0 \pm 0.00$ |
| % remaining @ 17 h | $27.41 \pm 7.19$ | $53.93 \pm 11.14$ |
| % remaining @ 25 h 30 min | $4.82 \pm 1.72$ | $20.59 \pm 9.44$ |
| %$\Delta$ in $C_{max}$ (dicumarol) | | $+38$ |
| %$\Delta$ in $T_{max}$ (dicumarol) | $-26.9$ | $+12.7$ |
| %$\Delta$ in AUC (dicumarol) | $+23.5$ | $+56.6$ |

*Source:* Ref. 91.

Site-specific targeting of the devices will be paramount to improving delivery efficiency. Overall, the findings of this study show that materials that have been proven to exhibit strong bioadhesive interactions in vitro may perform well as bioadhesive delivery systems in vivo. By fine-tuning the release kinetics of the vehicles, it should be possible to further improve bioavailability. We believe that bioadhesive polyanhydrides, such as P(FA:SA), will eventually find clinical use for the oral delivery of bioactive agents.

## Bioadhesive Microspheres for Oral Delivery of Proteins and Genes

Previous investigators overlooked the possibility that hard thermoplastic materials may be suitable bioadhesives, based on the notion that diffusion and chain entanglement were essential to bioadhesion. However, our work proves that other factors, such as hydrogen bonding, surface texture, and surface energy, can play equally important roles in the formation of bioadhesive interactions and that thermoplastic, bioadhesive, controlled-release systems can be used to improve bioavailability of orally delivered drugs. We have found that polyanhydride copolymers of FA and SA [P(FA:SA)] demonstrated the highest bioadhesive properties (140) using two in vitro techniques—the Cahn microbalance and the everted sac technique (161). In vivo radiographic studies on P(FA:SA) microspheres showed longer residence times in the stomach as compared to that of other polymers (36 versus 18 h) (25). All these results suggest that strong bioadhesive interactions delay microspheres passage through the gastrointestinal system. Intrigued by this data, we then wondered what would happen if the bioadhesive microspheres were manufactured at a very small size. Others have demonstrated that small microspheres of polystyrene (PS), poly-lactide-co-glycolide (PLGA), and polycyanoacrylate could be taken up by lymphoid tissue associated with the Peyer's patches (PP) of the GI tract (170–178).

To explore the potential of uptake of our formulations in rats, we used optical microscopy to follow the fate of orally administered P(FA:SA) 20:80 microspheres (0.1–10 $\mu$m) fabricated using the phase-inversion (PIN) method. As early as 1 h postfeeding of a single dose, microspheres were observed to traverse both the mucosal epithelium (through and between individual cells) and the follicle-associated epithelium (FAE) covering the lymphatic elements of the PP. Microspheres were also observed within goblet cells. In addition, spheres were seen crossing through FAE and into the PP. After 3 and 6 h, an intense uptake was observed in absorptive cells and PP. The same observations were made 12 to 24 h postfeeding. Both spleen and liver tissue samples, with apparently normal-looking hepatocytes, showed a large number of microspheres. Similar patterns were never observed in sections of tissue from untreated control animals.

Transmission electron microscopy (TEM) experiments using an electron-opaque, 5-nm colloidal gold tracer that had been PIN-encapsulated with P(FA:SA) 20:80 were also conducted (137). The range of particle sizes observed in the cytoplasm of cells was 40–120 nm, well below the resolution of normal light optics. Colloidal gold-labeled nano-

spheres were visualized in the cytoplasm, inside Golgi apparatus and in secretory vesicles predominantly in the supranuclear (apical) portion of the absorptive cells. The mechanism of entry was not elucidated as relatively few pinocytotic vesicles were observed to account for the uptake. Microspheres were often found between microvilli, in vesicles near the lateral borders of the cell, in the intercellular spaces and in the basal aspects of the cells, in lymphatics, and also in Goblet cells (137). Unequivocal identification of the particles was difficult because the microspheres had different appearances depending upon their location in the cell or the intercellular spaces. This may have resulted from degradation or processing of the polymer in the various environments. However, in each case the particles were identified on the basis of the colloidal gold tag. Microspheres were never observed in control tissues from untreated animals (137). These findings suggest that translocation of polymer particles via the transcellular route as well as via paracellular transport occurred. Similar work, conducted with colloidal gold-labeled PS showed high uptake.

**Oral Insulin Delivery.** To evaluate the possibility of orally delivered insulin, three groups of fasted rats were injected subcutaneously with an initial glucose load (179) and then fed either (1) a suspension of P(FAO-PLGA) microspheres containing 20 IU zinc–insulin with micronized FeO included as an electron dense transmission electron microscopy (TEM) tracer in saline ($n = 6$), (2) 24 IU of soluble insulin in saline ($n = 8$), or (3) saline without insulin as controls ($n = 7$). (Note: FAO is fumaric anhydride oligomer, also known as fumaric acid prepolymer, FAPP.) The control animals showed the expected response to the glucose load (137). In saline-fed animals, blood glucose levels (BGL) rose by 46 mg/dL after 3 h and then slowly started to return toward baseline. Rats fed soluble insulin reached a maximum BGL of 36 mg/dL above baseline at 1.5 h postfeeding. At all timepoints, the BGL of animals receiving the insulin solution were not significantly different from those of control animals receiving saline only. In contrast, the BGL of animals fed the PIN-encapsulated insulin formulation never increased above the fasting levels and were statistically lower than both control groups at all four time points (137). Clearly, animals fed the P(FAO-PLGA)-encapsulated insulin preparation were better able to regulate the glucose load than the controls, suggesting that the insulin crossed the intestinal barrier and was released from the microspheres in a bioactive form.

Formulations of blends of PLA and FAO, up to ratios of 1:3 (w/w) and loaded with the same amount of insulin (with similar drug release kinetics, bioactivity and particle size), did not show any effect in the glucose tolerance test. This observation correlated well with the low level of PLA particle uptake. The differences observed between the two polymer formulations may be due to either the lesser bioadhesive properties or slower degradation rate of the blends of P(FAO-PLA) polymers.

To study the bioavailability of the insulin formulation, two normal rats, weighing 300 g each, were fed 20 IU of insulin in the P(FAO-PLGA) blend containing 1.6% insulin. Both rats had critically low BGLs (<60 mg/dL) at 4 h and then began to return to normal at about 6 h. When lower doses were administered, we could not see the same effect. The results can be used to calculate the oral bioavailability by the following equation:

$$f = \frac{AUC_{0-6 \text{ oral}}}{AUC_{0-6 \text{ ip}}} \times \frac{(weight \ dose)_{\text{oral}}}{(weight / dose)_{\text{ip}}}$$

$$= \frac{118\%}{257\%} \times \frac{300_g/20_{IU}}{270_g/4_{IU}} = 10.2\% \qquad (12)$$

Information on the ip serum levels was obtained from Touitou (180). Thus, for the rats fed the microspheres orally, the calculated bioavailability is 10.2%.

These are only preliminary results, but further study indicated that in that particular formulation 25% of the microspheres were smaller than 4 $\mu$m and 10% were smaller than 1.6 $\mu$m (size distribution was determined by volume). If we take those numbers into consideration, the estimated bioavailability numbers rise to 40% for the 4 $\mu$m cutoff and to 100% for the 1.6 $\mu$m cutoff. These numbers seem extremely high; however, they indicate that transmucosal delivery of only a small population of insulin-loaded nanoparticles is necessary to reduce BGL. Furthermore, we believe that if we can shift the microsphere size range below 2 $\mu$m, then we can dramatically improve the oral bioavailability of insulin to levels above 50%.

Now if we remember that the oral bioavailability of dicumarol in the microspheres was estimated as 19% and the microspheres were loaded with 40% (w/w) dicumarol, up to 47% of the microparticles were estimated to be absorbed in the rats in the study. Again, the results between the oral insulin and the oral dicumarol are very intriguing, and more work will be done to further evaluate those systems.

**Oral Gene Delivery.** Perhaps the most surprising results were found when we applied the system to the delivery of plasmid DNA for in vivo gene transfer (137). Because "naked" plasmid DNA transfection efficiencies are generally low owing to inefficient uptake as well as rapid degradation of the DNA, encapsulation of plasmid DNA might serve to protect DNA and promote cellular uptake following oral administration. A bioadhesive polymer delivery system composed of P(FA:SA) 20:80 was used to encapsulate pCMV/$\beta$-gal (0.1% w/w), a commercially available reporter plasmid with bacterial $\beta$-galactosidase gene activity. The migration pattern of plasmid DNA extracted from the microspheres was identical to stock unencapsulated DNA, indicating that the fabrication process did not alter its structural integrity.

Five days following a single oral dose of either plasmid-loaded PIN microspheres or unencapsulated pCMV/$\beta$-gal, $\beta$-galactosidase activity was quantified in the stomach, small intestine, and liver (137). Animals fed encapsulated pCMV/$\beta$-gal showed increased levels of $\beta$-galactosidase activity in both the small intestine and the liver compared with those receiving unencapsulated material as well as unfed animals. The reporter gene activity measured in intestinal tissue of animals that received the encapsulated pCMV/$\beta$-gal was highest (>54 mU versus 24 mU for the unencapsulated plasmid and 18 mU for the background levels in untreated control animals; however, nsd for the three groups). These same animals averaged 11 mU of activity in the liver versus < 1 mU for plain CMV-fed ($P <$ .045) or untreated control animals ($P <$ .025). Reporter gene expression in stomach homogenates was generally low (<11 mU) in all groups.

Visual localization of transfected cells following oral administration was performed using X-gal histochemical staining techniques on whole tissue and frozen sections (181). Because epithelial cells on villi have endogenous lactase activity that can mask the bacterial $\beta$-galactosidase (182), we focused on the PP, which do not contain background activities.

The serosal surface of small intestine from encapsulated pCMV/$\beta$-gal-fed rats showed intensely stained cells in areas containing PP (137). Frozen sections of PP revealed that although there were a few $\beta$-galactosidase positive cells within the central lymphoid tissue mass (137), the majority of transfected cells were located in the muscularis mucosae and adventitia below the PP (137). This distribution of staining was consistent with previous studies that showed retention of microspheres in the PP. Neither the animals receiving unencapsulated pCMV/$\beta$-gal nor unfed normal rats showed any false-positive $\beta$-galactosidase staining in the PP region. Histological examination showed no evidence of mucosal damage or inflammation. This study confirmed that plasmid DNA can be delivered by the oral route using PIN formulations, with the encapsulated DNA incorporated into cells lining the small intestine and hepatocytes and expressing functional gene products at levels detectable using common histological and luminometric techniques.

In conclusion, when using the PIN system in combination with bioerodible bioadhesive polymers, the enhanced bioavailability of dicumarol might be explained by the delay in the dissolution of the drug caused by the encapsulation process (25), the increased absorption rate resulting from more intimate contact with the intestinal epithelium, or an increased residence time at the local absorption site—all of which can be attributed to bioadhesion. However, the increased bioactivity of insulin and plasmid DNA might also be due to the uptake of the microspheres into the circulation by the cells lining the GI tract. The results with insulin, a 6-kDa protein susceptible to proteolytic degradation by GI enzymes and having limited mucosal uptake, support this hypothesis. Clearly, the same uptake pathways can be used as a platform for a variety of therapeutic agents with poor oral absorption including nucleic acids and plasmids.

## Concluding Remarks

Traditionally, bioadhesive researchers have stated that bioadhesion in hydrogels is mainly a function of chain entanglement (75,78,183). Recent findings, however, have indicated that other components, such as surface energies, may play major roles in the formation of bioadhesive interactions (156,157). Thus far, the study of bioadhesion in bioerodible polymers has been overlooked owing to the fact that these materials lack sufficient chain flexibility for interpenetration. The results of our studies suggest that bioadhesion of bioerodible polymers lacking chain flexibility could largely be a result of secondary bond formation, such as hydrogen bonding between mucin and carboxyl groups. Rapid degradation of P(FA:SA) polymers may enhance their bioadhesive nature through production of carboxylic acid and increases in surface roughness. In light of our findings, we believe that bioadhesive analysis using the diffusion theory of chain entanglement may be inappropriate for hard, bioerodible materials. Presently more work is needed to clarify the exact mechanisms of bioadhesion in thermoplastics, but the results here suggest that bioerodible, bioadhesive polymers may be excellent candidates for the development of orally administered DDSs. Selection of the appropriate delivery system, with proper release kinetics as well as appropriate adhesive characteristics, is essential to producing desired effects. Not every drug will benefit from bioadhesive delivery, but if targeted bioadhesive systems can be developed many pharmaceutical agents may be administered more efficiently than by conventional means. This field is still in its infancy, but with decades of research ahead, it is sure to have an impact on future drug delivery systems.

## ACKNOWLEDGMENTS

We would like to thank the National Institutes of Health and NSF for supporting this research.

## BIBLIOGRAPHY

1. V. Lenaerts and R. Gurny, *Bioadhesive Drug Delivery Systems*, CRC Press, Boca Raton, Fla., 1990.

2. R. Gurny, J.M. Meyer, and N.A. Peppas, *Biomaterials* **5**, 336–340 (1984).

3. E. Mathiowitz, D.E. Chickering, and C.M. Lehr, *Bioadhesive Drug Delivery Systems: Fundamentals, Novel Approaches, and Development*, Dekker, New York, 1999.

4. W.J. Passl, *Prog. Polym. Sci.* **14**, 629–677 (1989).

5. R. Langer, *Methods Enzymol.* **73**, 57–75 (1981).

6. R. Langer and N. Peppas, *Biomaterials* **2**, 201–214 (1981).

7. R. Langer, *Science* **249**, 1527–1532 (1990).

8. S.S. Davis and L. Illum, in F.H. Roerdink and A.M. Kroon, eds., *Drug Carrier Systems*, Wiley, Chichester, England, 1989, pp. 131–153.

9. R. Langer, L.G. Cima, J.A. Tamada, and E. Wintermantel, *Biomaterials* **11**, 738–745 (1990).

10. G. Ponchel, J.M. Irache, C. Durrer, and D. Duchêne, *Proc. Int. Symp. Controlled Release Bioact. Mater.* **21**, 31–32 (1994).

11. W.D. Rhine, D.S.T. Hsieh, and R. Langer, *J. Pharm. Sci.* **69**, 265–268 (1980).

12. D.H. Robinson and J.W. Mauger, *Am. J. Hosp. Pharm.* **48**, S14–S23 (1991).

13. J.R. Robinson, *Proc. Int. Symp. Controlled Release Bioact. Mater.* **18**, 75–76 (1991).

14. J. Kost and S. Shefer, *Biomaterials* **11**, 695–698 (1990).

15. J. Kost, K. Leong, and R. Langer, *Proc. Nat. Acad. Sci. U.S.A.* **86**, 7663–7666 (1989).

16. E.R. Edelman, J. Kost, H. Bobeck, and R. Langer, *J. Biomed. Mater. Res.*, **19**, 67–83 (1985).

17. G. Ponchel and D. Duchêne, in M. Szycher, ed., *High Performance Biomaterials*, Technomic Publishing, Lancaster, Pa., 1991, pp. 231–242.

18. K.V. Ranga Rao and P. Buri, in M. Szycher, ed., *High Performance Biomaterials*, Technomic Publishing, Lancaster, Pa., 1988, pp. 259–268.

19. D. Duchêne, F. Touchard, and N.A. Peppas, *Drug Dev. Ind. Pharm.* **14**, 283–318 (1988).

20. D. Duchêne and G. Ponchel, *Biomaterials* **13**, 709–715 (1992).

21. J.D. Smart, *Adv. Drug Delivery Rev.* **11**, 253–270 (1993).

22. V. Lenaerts, P. Couvreur, L. Grislain, and P. Maincent, in V. Lenaerts and R. Gurny, eds., *Bioadhesive Drug Delivery Systems*, CRC Press, Boca Raton, Fla., 1990, pp. 93–104.

23. J.M. Gu, J.R. Robinson, and S.-H.S. Leung, *CRC Crit. Rev. Ther. Drug Carrier Syst.* **5**, 21–67 (1988).

24. H.S. Ch'ng, H. Park, P. Kelly, and J.R. Robinson, *J. Pharm. Sci.* **74**, 399–405 (1985).

25. D.E. Chickering, J. Jacob, and E. Mathiowitz, *Biotechnol. Bioeng.* **52**, 96–101 (1996).

26. J.L. Chen and G.N. Cyr, in R.S. Manly, ed., *Adhesion in Biological Systems*, Academic Press, New York, 1970, pp. 163–181.

27. C.-M. Lehr et al., *J. Controlled Release* **18**, 249–260 (1992).

28. D.E. Chickering and E. Mathiowitz, *J. Controlled Release* **34**, 251–262 (1995).

29. C. Marriot, *Proc. Int. Symp. Controlled Release Bioact. Mater.* **18**, 67–68 (1991).

30. C.-M. Lehr, F.G.J. Poelma, H.E. Junginger, and J.J. Tukker, *Int. J. Pharm.* **70**, 235–240 (1991).

31. A. Rubinstein and B. Tirosh, *Pharm. Res.* **11**, 794–799 (1994).

32. R.M. Berne and M.N. Levey, *Physiology*, Mosby, St. Louis, Mo., 1988.

33. A.C. Guyton, *Textbook of Medical Physiology*, Saunders, Philadelphia, 1986.

34. K.R. Bhaskar et al., *Nature (London)* **360**, 458–461 (1992).

35. A. Allen, L.A. Sellers, and M.K. Bennet, *Scand. J. Gastroenterol., Suppl.* **128**, 6–13 (1987).

36. A. Garner, G. Flemstrom, and A. Allen, *Scand. J. Gastroenterol.* **19**, 78–80 (1984).

37. B. Sandzén, H. Blom, and S. Dahlgren, *Scand. J. Gastroenterol.* **23**, 1160–1164 (1988).

38. L. Szentkuti et al., *Histochem. J.* **22**, 491–497 (1990).

39. R.G. Spiro, *Annu. Rev. Biochem.* **39**, 599–638 (1970).

40. J. Labat-Robert and C. Decaeus, *Pathol. Biol.* **24**, 241 (1979).

41. M. Scawen and A. Allen, *Biochem. J.* **163**, 363–368 (1977).

42. M.I. Horowitz and W. Pigman, *The Glycoconjugates*, Academic Press, New York, 1977.

43. W. Pigman and A. Gottschalk, in A. Gottschalk, ed., *Glycoproteins: Their Composition, Structure and Function*, Elsevier, Amsterdam, 1966, pp. 434–445.

44. S. Ohara, K. Ishihara, and K. Hotta, *Comp. Biochem. Physiol.* **106B**, 147–152 (1993).

45. P. Johnson and K.D. Rainsford, *Biochem. Biophys. Acta* **286**, 72–78 (1972).

46. G. Flemstrom and E. Kivilaakso, *Gastroenterology* **84**, 787–794 (1983).

47. P.R. Wheater, G.H. Burkitt, and V.G. Daniels, *Functional Histology*, Churchill-Livingstone, Edinburgh, 1987.

48. C.M. Lehr, J.A. Bouwstra, J.J. Tukker, and H.E. Junginger, *J. Controlled Release* **13**, 51–62 (1990).

49. N.A. Peppas and P.A. Buri, *J. Controlled Release* **2**, 257–275 (1985).

50. A.J. Kinloch, *J. Mater. Sci.* **15**, 2141–2166 (1980).

51. H. Park and J.R. Robinson, *J. Controlled Release* **2**, 47–57 (1985).

52. B.V. Derjaguin and V.P. Smilga, *Adhes.: Fundam. Pract.* **1**, 152–163 (1966).

53. B.V. Derjaguin, Y.P. Toporov, V.M. Muller, and I.N. Aleinikova, *J. Colloid Interface Sci.* **58**, 528–533 (1977).

54. R.J. Good, *J. Colloid Interface Sci.* **59**, 398–419 (1977).

55. D. Tabor, *J. Colloid Interface Sci.* **58**, 2–13 (1977).

56. J.R. Huntsberger, in R.L. Patrick, ed., *Treatise on Adhesion and Adhesives*, Dekker, New York, 1966, pp. 119–149.

57. J.R. Huntsberger, *J. Paint Technol.* **39**, 199–211 (1967).

58. A.G. Mikos and N.A. Peppas, *Int. J. Pharm.* **53**, 1–5 (1989).

59. R.T. Spychal, J.M. Marrero, S.H. Saverymuttu, and T.C. Northfield, *Gastroenterology* **97**, 104–111 (1989).

60. A. Baszkin, J.E. Proust, P. Monsenego, and M.M. Boissonnade, *Biorheology* **27**, 503–514 (1990).

61. C.T. Reinhart and N.A. Peppas, *J. Memb. Sci.* **18**, 227–239 (1984).

62. R.P. Campion, *J. Adhes.* **7**, 1–23 (1974).

63. L.A. Girifalco and R.J. Good, *J. Phys. Chem.* **61**, 904–909 (1957).

64. B.O. Bateup, *Int. J. Adhes. Adhes.*, 233–239 (July 1981).

65. R.J. Good, *J. Colloid Interface Sci.* **52**, 308–313 (1975).

66. S. Wu, *J. Phys. Chem.* **74**, 632–638 (1970).

67. D.H. Kaelble and J. Moacanin, *Polymer* **18**, 475–482 (1977).

68. E. Helfand and Y. Tagami, *J. Chem. Phys.* **57**, 1812–1813 (1972).

69. E. Helfand and Y. Tagami, *J. Chem. Phys.* **56**, 3592–3601 (1972).

70. P.C. Hiemenz, *Principles of Colloid and Surface Chemistry*, Dekker, New York, 1986.

71. S.S. Voyutskii, *Autohesion and Adhesion of High Polymers*, Wiley, New York, 1963.

72. A.G. Mikos and N.A. Peppas, *Proc. Int. Symp. Controlled Release Bioact. Mater.* **13**, 97–98 (1986).

73. A.G. Mikos and N.A. Peppas, *S.T.P. Pharmacol.* **2**, 705–716 (1986).

74. H.W. Kammer, *Acta Polym.* **34**, 112 (1983).

75. G. Ponchel, F. Touchard, D. Duchêne, and N. Peppas, *J. Controlled Release* **5**, 129–141 (1987).

76. D.H. Kaelble, *J. Appl. Polym. Sci.* **18**, 1869–1889 (1974).

77. J.J. Bikerman, *The Science of Adhesive Joints*, Academic Press, New York, 1968.

78. J.D. Smart, I.W. Kellaway, and H.E.C. Worthington, *J. Pharm. Pharmacol.* **36**, 295–299 (1984).

79. A.P. Sam, J.T.M. van der Heuij, and J.J. Tukker, *Int. J. Pharm.* **79**, 97–105 (1992).

80. M. Ishida, N. Nambu, and T. Nagai, *Chem. Pharm. Bull.* **31**, 1010–1014 (1983).

81. M. Ishida, Y. Machida, N. Nambu, and T. Nagai, *Chem. Pharm. Bull.* **29**, 810–816 (1981).

82. C.-M. Lehr and V.H.L. Lee, *Proc. Int. Symp. Controlled Release Bioact. Mater.* **19**, 94–95 (1992).

83. C.-M. Lehr, J.A. Bowstra, H.E. Boddé, and H.E. Junginger, *Pharm. Res.* **9**, 70–75 (1992).

84. M.F. Saettone, S. Burgalassi, L. Panichi, and B. Giannaccini, *Proc. Int. Symp. Controlled Release Bioact. Mater.* **21**, 559–560 (1994).

85. F.J. Otero-Espinar et al., *Proc. Int. Symp. Controlled Release Bioact. Mater.* **21**, 700–701 (1994).

86. N.A. Peppas, L. Achar, and N. Wisniewski, *Proc. Int. Symp. Controlled Release Bioact. Mater.* **21**, 581–582 (1994).

87. L. Achar and N.A. Peppas, *J. Controlled Release* **31**, 271–276 (1994).

88. D.E. Chickering, J.S. Jacob, and E. Mathiowitz, *Proc. Int. Symp. Controlled Release Bioact. Mater.* **20**, 244–245 (1993).

89. D.E. Chickering et al., *Proc. Mater. Res. Soc. Fall Meet. Biomater. Drug Cell Delivery*, Boston, March 31, 1993, pp. 67–71.

90. E. Mathiowitz et al., *Proc. Int. Symp. Controlled Release Bioact. Mater.* **21**, 27–26 (1994).

91. D.E. Chickering et al., *J. Controlled Release* **48**, 35–46 (1997).

92. S.S. Davis, J.G. Hardy, and J.W. Fara, *Gut* **27**, 886–892 (1986).

93. S.S. Davis, F. Norring-Christensen, R. Khosla, and L.C. Feely, *J. Pharm. Pharmacol.* **40**, 205–207 (1988).

94. N.A. Peppas, ed., *Hydrogels in Medicine and Pharmacy*, CRC Press, Boca Raton, Fla., 1987.

95. C.-M. Lehr, Y.-H. Lee, and V.H.L. Lee, *Invest. Ophthalmol. Visual Sci.* **35**, 2809–2814 (1994).

96. D.Q.M. Craig, S. Tamburic, G. Buckton, and J.M. Newton, *J. Controlled Release* **30**, 213–223 (1994).

97. D.R. Mack, P.L. Blain-Nelson, and J.W. Mauger, *J. Biomed. Mater. Res.* **27**, 1579–1583 (1993).

98. S. Anlar, Y. Capan, and A.A. Hincal, *Pharmazie* **48**, 285–287 (1993).

99. M. Ishida, N. Nambu, and T. Nagai, *Chem. Pharm. Bull.* **30**, 980–984 (1982).

100. H. Park and J.R. Robinson, *Pharm. Res.* **4**, 457–464 (1987).

101. K. Park, S.L. Cooper, and J.R. Robinson, in N.A. Peppas, ed., *Hydrogels in Medicine and Pharmacy*, CRC Press, Boca Raton, Fla., 1987, pp. 151–175.

102. D.E. Chickering, J.S. Jacob, and E. Mathiowitz, *Proc. Int. Symp. Controlled Release Bioact. Mater.* **21**, 776–777 (1994).

103. M.R. Jimenez-Castellanos, H. Zia, and C.T. Rhodes, *Drug Dev. Ind. Pharm.* **19**, 143–194 (1993).

104. M. Helliwell, *Adv. Drug Delivery Rev.* **11**, 221–251 (1993).

105. K. Satoh et al., *Chem. Pharm. Bull.* **37**, 1366–1368 (1989).

106. P. Bottenberg et al., *Proc. Int. Symp. Controlled Release Bioact. Mater.* **18**, 631–632 (1991).

107. P. Bottenberg et al., *J. Pharm. Pharmacol.* **44**, 684–686 (1992).

108. S. Bouckaert et al., *J. Pharm. Pharmacol.* **44**, 684–686 (1992).

109. S. Bouckaert, M. Temmerman, M. Dhont, and J.P. Remon, *Proc. Int. Symp. Controlled Release Bioact. Mater.* **21**, 585–586 (1994).

110. S. Bouckaert, R. Van WeissenBrauch, H. Nelis, and J.P. Remon, *Proc. Int. Symp. Controlled Release Bioact. Mater.* **21**, 543–544 (1994).

111. S. Miyazaki et al., *Biol. Pharm. Bull.* **17**, 745–747 (1994).

112. T. Nagai and R. Konishi, *J. Controlled Release* **6**, 353–360 (1987).

113. H.E. Boddé, M.E. de Vries, and H.E. Junginger, *J. Controlled Release* **13**, 225–231 (1990).

114. J. Voorspoels and J.P. Remon, *Proc. Int. Symp. Controlled Release Bioact. Mater.* **21**, 539–540 (1994).

115. E. Beyssac et al., *Proc. Int. Symp. Controlled Release Bioact. Mater.* **21**, 553–554 (1994).

116. E. Beyssac et al., *Proc. Int. Symp. Controlled Release Bioact. Mater.* **21**, 891–892 (1994).

117. C. Charrueau et al., *J. Controlled Release* **32**, 9–15 (1994).

118. F. Thermes, A. Rozier, B. Plazonnet, and J. Grove, *Proc. Symp. Controlled Release Bioact. Mater.* **18**, 627–628 (1991).

119. M.E. de Vries, H.E. Boddé, H.J. Busscher, and H.E. Junginger, *J. Biomed. Mater. Res.* **22**, 1023–1032 (1988).

120. H.L. Luessen et al., *J. Controlled Release* **29**, 329–338 (1994).

121. H. Umejima et al., *J. Pharm. Sci.* **82**, 195–199 (1993).

122. L.E. Nachtigall, *Fertil. Steril.* **61**, 178–180 (1994).

123. J.R. Robinson and W.J. Bologna, *J. Controlled Release* **28**, 87–94 (1994).

124. N. Celebi, N. Erden, B. Gönül, and M. Koz, *J. Pharm. Pharmacol.* **46**, 386–387 (1994).

125. J.-H. Guo, *Proc. Int. Symp. Controlled Release Bioact. Mater.* **21**, 545–546 (1994).

126. J.-H. Guo, *Proc. Int. Symp. Controlled Release Bioact. Mater.* **21**, 547–548 (1994).

127. N. Yerushalmi and R. Margalit, *Biochim. Biophys. Acta* **1189**, 13–20 (1994).

128. N. Yerushalmi, A. Arad, and R. Margalit, *Arch. Biochem. Biophys.* **313**, 267–273 (1994).

129. R. Margalit, M. Okon, N. Yerushalmi, and E. Avidor, *J. Controlled Release* **19**, 275–288 (1992).

130. A.G. Mikos, and N.A. Peppas, *J. Controlled Release* **12**, 31–37 (1990).

131. A.G. Mikos, E. Mathiowitz, N.A. Peppas, and R. Langer, *Proc. Int. Symp. Controlled Release Bioact. Mater.* **18**, 109–110 (1991).

132. D.E. Chickering, W.P. Harris, and E. Mathiowitz, *13th Ann. Meet. Expo. American Association of Medical Instrumentation*, Anaheim, Calif., 1995.

133. N.F. Farraj, B.R. Johansen, S.S. Davis, and L. Illum, *J. Controlled Release* **13**, 253–261 (1990).

134. P. Vidgren et al., *Proc. Int. Symp. Controlled Release Bioact. Mater.* **18**, 638–639 (1991).

135. C.-M. Lehr, W. Kok, and J. Koninkx, *Proc. Int. Symp. Controlled Release Bioact. Mater.* **21**, 344–345 (1994).

136. N. Hussain, P.U. Jani, and A.T. Florence, *Proc. Int. Symp. Controlled Release Bioact. Mater.* **21**, 29–30 (1994).

137. E. Mathiowitz et al., *Nature (London)* **386**, 410–416 (1997).

138. A.G. Mikos, E. Mathiowitz, R. Langer, and N.A. Peppas, *J. Colloid Interface Sci.* **143**, 367–373 (1991).

139. C. Pimienta et al., *Pharm. Res.* **7**, 49–53 (1990).

140. D.E. Chickering, J.S. Jacob, and E. Mathiowitz, *React. Polym.* **25**, 189–206 (1995).

141. W.H. Guilford and R.W. Gore, *Am. J. Physiol.* **263**, C700–C707 (1992).

142. C.M. Lehr et al., *J. Pharm. Pharmacol.* **44**, 402–407 (1992).

143. J.D. Smart, B. Carpenter, and S.A. Mortazavi, *Proc. Int. Symp. Controlled Release Bioact. Mater.* **18**, 629–630 (1991).

144. A.J. Domb, F.C. Gallardo, and R. Langer, *Macromolecules* **22**, 3200–3204 (1989).

145. A. Domb and R. Langer, *J. Polym. Sci.* **25**, 3373–3386 (1987).

146. F. Lim and D. Moss, *J. Pharm. Sci.* **70**, 351–354 (1981).

147. E. Mathiowitz and R. Langer, *J. Controlled Release* **5**, 13–22 (1987).

148. E. Mathiowitz, M.D. Cohen, and R. Langer, *React. Polym.* **6**, 275–283 (1987).

149. E. Mathiowitz et al., *J. Appl. Polym. Sci.* **35**, 755–774 (1988).

150. E. Mathiowitz, D. Kline, and R. Langer, *Scanning Microsc.* **4**, 329–340 (1990).

151. E. Mathiowitz, P. Dor, C. Amato, and R. Langer, *Polymer* **31**, 547–555 (1990).

152. E. Mathiowitz et al., *J. Appl. Polym. Sci.* **45**, 125–134 (1992).

153. E. Mathiowitz, E. Ron, G. Mathiowitz, and R. Langer, *Macromolecules* **23**, 3212–3218 (1990).

154. L.R. Beck et al., *Fertil. Steril.* **31**, 545–551 (1979).

155. L.R. Beck et al., *Am. J. Obstet. Gynecol.* **135**, 419–426 (1979).

156. C.-M. Lehr, H.E. Boddé, J.A. Bowstra, and H.E. Junginger, *Eur. J. Pharm. Sci.* **1**, 19–30 (1993).

157. C.-M. Lehr, *J. Pharm. Pharmacol.* **44**, 402–407 (1992).

158. E. Mathiowitz, J. Jacob, D. Chickering, and K. Pekarek, *Macromolecules* **26**, 6756–6765 (1993).

159. E. Mathiowitz, M. Kreitz, and K. Pekarek, *Macromolecules* **26**, 6749–6755 (1993).

160. D.J. Pasto and C.R. Johnson, *Organic Structure Determination*, Prentice-Hall, Englewood Cliffs, N.J., 1969.

161. J. Jacob et al., *Proc. Int. Symp. Controlled Release Bioact. Mater.* **22**, 312–313 (1995).

162. N.N. Salib, M.A. El-Menshawy, and A.A. Ismail, *Pharm. Ind.* **40** (11A), 1230 (1978).

163. P. Trinder, *Biochem. J.* **57**, 301–303 (1954).

164. J. Axelrod, J.R. Cooper, and B.B. Brodie, *Proc. Soc. Biol. Med.* **70**, 693–695 (1949).

165. H. Sekikawi et al., *Chem. Pharm. Bull.* **31**, 1350–1356 (1983).

166. S. Goodman et al., *J. Appl. Biomater.* **6**, 161–165 (1995).

167. R.J.L. Davidson, *J. Clin. Pathol.* **24**, 537–541 (1971).

168. E.S. Vesell and J.G. Page, *J. Clin. Invest.* **47**, 2657–2663 (1968).

169. L.S. Goodman and A. Gilman, eds., *The Pharmacological Basis of Therapeutics*, Macmillan, New York, 1975.

170. E. Sanders and C.T. Ashworth, *Exp. Cell Res.* **22**, 137–145 (1961).

171. T.H. Ermak et al., *Cell Tissue Res.* **279**, 433–436 (1995).

172. A. Florence and P.U. Jani, *Drug. Saf.* **10**, 233–266 (1994).

173. A.T. Florence, A.M. Hillery, N. Hussain, and P.U. Jani, *J. Controlled Release* **36**, 39–46 (1995).

174. M.E. LeFevre, R. Olivo, J.W. Vanderhoff, and D.D. Joel, *Proc. Soc. Exp. Biol. Med.* **159**, 298–302 (1978).

175. M.E. Lefevre, D.D. Joel, and G. Schidlovsky, *Proc. Soc. Exp. Biol. Med.* **179**, 522–528 (1985).

176. P. Jani, G.W. Halbert, J. Langridge, and A.T. Florence, *J. Pharm. Pharmacol.* **42**, 821–826 (1990).

177. A.M. Hillery, P.U. Jani, and A.T. Florence, *J. Drug Target.* **2**, 151–156 (1994).

178. D.E. Bockman and M.D. Cooper, *Am. J. Anat.* **136**, 455–478 (1973).

179. G.F. Tutwiler, T. Kirsch, and G. Bridi, *Diabetes* **27**, 856–867 (1978).

180. E. Touitou and A. Rubinstein, *Int. J. Pharm.* 95–99 (1986).

181. J.A. Wolff et al., *BioTechniques* **11**, 474–85 (1991).

182. C. Lau et al., *Hum. Gene Ther.* **6**, 1145–1151 (1995).

183. S.-H.S. Leung and J.R. Robinson, *J. Controlled Release* **12**, 187–194 (1990).

See also MUCOSAL DRUG DELIVERY, BUCCAL; MUCOSAL DRUG DELIVERY, INTRAVITREAL; MUCOSAL DRUG DELIVERY, NASAL; MUCOSAL DRUG DELIVERY, OCULAR; MUCOSAL DRUG DELIVERY, VAGINAL DRUG DELIVERY AND TREATMENT MODALITIES.

# BIODEGRADABLE POLYMERS: POLY(ORTHO ESTERS). See POLY(ORTHO ESTERS).

# BIODEGRADABLE POLYMERS: POLY(PHOSPHOESTER)S

HAI-QUAN MAO
IRINA KADIYALA
KAM W. LEONG
The Johns Hopkins University School of Medicine
Baltimore, Maryland

ZHONG ZHAO
WENBIN DANG
Guilford Pharmaceuticals, Inc.
Baltimore, Maryland

**KEY WORDS**

Biocompatibility

Biodegradable polymers

Controlled delivery

Degradation

Drug and protein delivery

Microspheres

Poly(phosphoesters)

Synthesis

Viscous liquid

**OUTLINE**

In controlled drug delivery, a biodegradable drug carrier may offer features difficult to attain with nonbiodegradable systems (1,2). Other than obviating the need to remove the drug-depleted devices, a biodegradable system is also applicable to a wider range of drugs. For nonbiodegradable matrices, drug release in most cases is controlled by diffusion through the polymer phase (3,4). Biomacromolecules with low permeability through polymers are released only through pores and channels created by the dissolved drug phase. Proteins might sometimes aggregate and precipitate in the matrix, clogging the channels for diffusion. A matrix-degradation–controlled drug release mechanism can leach out the entire drug content provided that there is no strong interaction between the drug and the degrading matrix. A diffusion-controlled release mechanism would yield a drug release rate that exponentially decays with time; a matrix-degradation–controlled mechanism in many cases would produce a more steady release.

Several classes of synthetic polymers, including polyesters (5–7), poly(amino acid)s (8,9), polyamides (10), polyurethanes (11), poly(orthoester)s (12), poly(anhydride)s (13,14), poly(carbonate)s (15,16), poly(iminocarbonate)s (17,18), and poly(phosphazene)s (19–23), have been proposed for controlled drug delivery; the poly(lactide-*co*-glycolide) copolymers still dominate the field (5,24). There is justification for continuing to develop new biodegradable drug carriers, however, because carriers may be required for more than just passive delivery. The widening scope of applications may require the carrier to assume different configurations, and the carrier may need to serve additional functions. For instance, active targeting would involve conjugation of ligands to the carriers, requiring the polymeric carrier to contain functional groups for derivatization. Applying the controlled-release device as more than just a monolithic matrix—for example, as a coating material for a drug-eluting stent—may require the polymer to have elastomeric properties. In the new and exciting field of tissue engineering, where local and sustained delivery of growth factors may influence the course of tissue development, the drug carrier may also need to perform the double duty of providing structural support or scaffolding functions. With such a broad use for these biodegradable drug carriers, no one single material can be expected to satisfy all of the requirements of different applications. In recognizing that there is still room for new biodegradable drug carriers with distinct characteristics, in the past decade we have been studying polymers with a phosphoester linkage in the backbone. In this article we describe our initial effort to assess the structure–property relationship of these polymers with a bisphenol A backbone (25–28), and we describe our most recent effort in developing these polymers with three different types of backbones, affording drug carriers with a wide range of physicochemical properties.

## BACKGROUND OF POLY(PHOSPHOESTER)S

The general chemical structure of poly(phosphoester)s (PPEs) is shown in this section. Depending on the nature of the side chain connected to the phosphorus, these polymers are conventionally called polyphosphates (P-O-C), polyphosphonates (P-C), or polyphosphites. Interests in phosphorus-containing polymers in the past have primarily been centered on their flame-retardant property (29). Because of a combination of factors—their high cost of synthesis as compared with the carbon analogs and their perceived hydrolytic instability—research interest has faded since the 1960s. Our optimism on these polymers stems from the following considerations:

1. *Adjustable properties*. The versatility of these polymers comes from the versatility of the phosphorus atom, which is known for its multiplicity of reactions. Its bonding can involve the 3p orbitals or various 3s–3p hydrids; spd hybrids are also possible because of the accessible d orbitals. Examining the general structure of the PPEs (Scheme 1), one can see that their physicochemical properties can be readily altered by varying either the R or R' group. The biodegradability of the polymer is due to the physiologically labile phosphoester bond in the backbone. By manipulating the backbone or the side chain, a wide range of biodegradation rates should be attainable.

2. *Favorable physicochemical properties*. The speculation that these polymers would have favorable physical properties is based on the fact that high molecular weights, which should yield good mechanical strength, have been reported for these polymers. Average molecular weights ($M_r$) of over 100,000 have been reported for a polyphosphate obtained by a ring-opening polymerization. Considering the high reactivity of the phosphoryl chloride, there is reason to believe that through optimization of the polymerization process, these polymers can be obtained in high molecular weights. In addition, the P-O-C group in the backbone is known to provide a plasticizing effect. This would lower the glass transition temperature of the polymer and most importantly confer the polymer solubility in common organic solvents (30,31). Although not immediately apparent, this is highly desirable so that the polymers can be easily characterized and processed.

3. *Biocompatibility*. The issue of biocompatibility is less amenable to prediction. However, in principle, the ultimate hydrolytic breakdown products of a polyphosphate are phosphates, alcohols, and diols, all of which have the potential to be nontoxic (Scheme 1). Naturally, the intermediate oligomeric products of the hydrolysis might have different properties, and the toxicology of a biodegradable polymer built from even innocuous monomeric structures can be determined only by careful in vivo studies.

An additional feature of these polymers is the availability of functional side groups. This is a unique advantage of the phosphorus atom, which can be pentavalent, allowing the chemical linkage of drug molecules or other agents to the polymer (32). For instance, drugs with hydroxyl groups may be coupled to phosphorus via an ester bond, which is hydrolyzable. There are only a handful of pendent delivery systems based on a carbon–carbon backbone and even fewer with a heteroatom backbone. These PPEs are natural candidates for a completely biodegradable pendent delivery system.

## PROPERTIES OF PPES WITH A BISPHENOL A BACKBONE

To perform a systematic study, we initially focused our attention on a series of bisphenol A–based PPEs shown in Scheme 2, the poly(oxyphosphoryl-oxy-1,4-phenylene-isopropylidene-1,4-phenylenes) (25,33): poly(bisphenol A-ethyl phosphate) (BPA-EOP), poly(bisphenol A-phenyl phosphate) (BPA-POP), poly(bisphenol A-phenyl phosphonate) (BPA-PP), and poly(bisphenol A-ethyl phosphonate) (BPA-EP). Bisphenol A (BPA), is chosen for this study because of the high reactivity of its phenoxide group and its relative hydrolytic stability, allowing the possibility of obtaining reasonable molecular weight with interfacial polymerization.

The polymers are obtained by interfacial polycondensation in methylene chloride–water, with potassium hydroxide as the acid receptor and cetyltrimethyl ammonium bromide as the phase transfer catalyst. With the parameters of the interfacial polycondensation optimized, the polymers have a $M_r$ in the range of 30,000–50,000 (33). Degradation is observed for the four polymers studied under both in vitro and in vivo conditions and is affected by polymer side-chain structure. The ethyl side-chain polymer (BPA-EOP) degrades faster than its phenyl counterparts. Weight loss of this polymer in the intramuscular space of rabbits reaches more than 80% in 70 weeks. The most hydrophobic polymer in the series, BPA-PP, shows approximately 12% loss of mass in the same period, but the mass does not change significantly after week 5, suggesting that after the low molecular weight fragments are leached out, the polymers become quite stable. Tissue response to the PPEs in rabbits is characterized by slight or no lymphocyte, giant cell, or macrophage activity (25).

The swelling behavior of the polymers follows the relative hydrophobicity of the side chains, in the order of EOP > EP > POP > PP (25,27). The BPA-EOP swells up to the dimensions of the container and exhibits large voids and pores observable upon gross examination. For the other polymers, further water uptake ceases after 250 days in the absence of drug but continues beyond that when loaded with *p*-nitroaniline. Relative in vitro mass loss of the four polymers also matches the swelling trend. BPA-EOP degrades the most and BPA-PP the least, most likely because increased water uptake by the more hydrophilic polymers leads to increased hydrolytic cleavage of the polymer backbone and leaching of degradation products. Blank sample of BPA-EOP shows 10% mass loss after 150 days, while BPA-EP, BPA-POP, and BPA-PP degrade by 5, 2, and 1%, respectively (Fig. 1). The values increase to 22, 7, 3, and 2%, respectively, after 222 days. Cleavage of polymer chains is confirmed by gel permeation chromatography (GPC) in all cases.

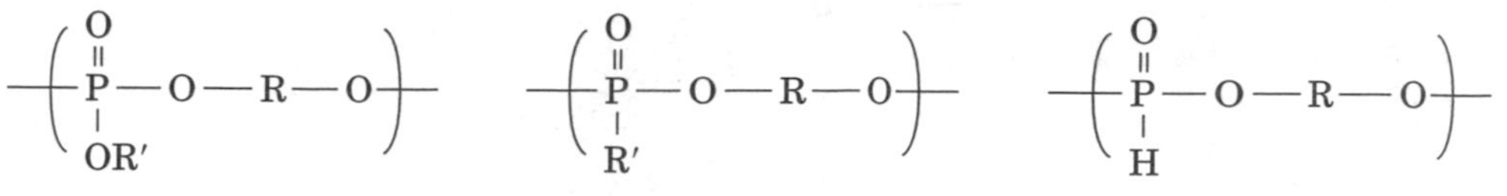

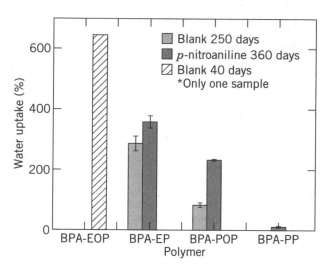

Scheme 1. General structure of PPEs.

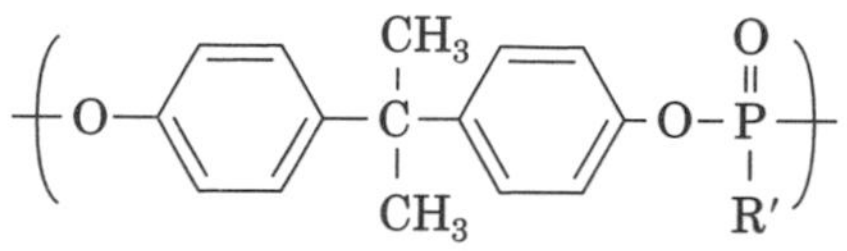

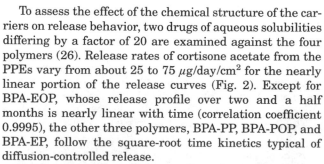

$R' = -OC_2H_5$ (BPA-EOP), $-OC_6H_5$ (BPA-POP),
$-C_2H_5$ (BPA-EP), $-C_6H_5$ (BPA-PP)

Scheme 2. Structures of Bisphenol A–based PPEs.

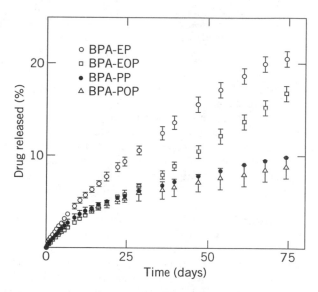

Figure 1. In vitro swelling of blank and p-nitroaniline–loaded PPE discs in phosphate-buffered saline at 37°C. Assay was carried out according to ASTM-D570 method.

Figure 2. In vitro release of cortisone acetate from BPA-based PPE matrices at 37°C and pH 7.4.

To assess the effect of the chemical structure of the carriers on release behavior, two drugs of aqueous solubilities differing by a factor of 20 are examined against the four polymers (26). Release rates of cortisone acetate from the PPEs vary from about 25 to 75 μg/day/cm² for the nearly linear portion of the release curves (Fig. 2). Except for BPA-EOP, whose release profile over two and a half months is nearly linear with time (correlation coefficient 0.9995), the other three polymers, BPA-PP, BPA-POP, and BPA-EP, follow the square-root time kinetics typical of diffusion-controlled release.

Release of p-nitroaniline, a compound more hydrophilic than cortisone acetate and 20 times more soluble in phosphate buffer, is far more rapid except in the case of BPA-PP (Fig. 3). Release of p-nitroaniline from BPA-EOP and BPA-EP follows root-time kinetics for the first 60% of release and shows complete release after 20 and 360 days, respectively. Faster drug transport is likely due to higher water uptake and, to a lesser extent, greater degradation of these aliphatic side-chain polymers. A mass balance on

BPA-EP shows an 11% degradation of the matrix by the end of drug release. It appears that in this case, release is predominantly controlled by the swelling of the matrices.

This initial structure–property relationship study confirms the influence of the side-chain structure on the swelling and degradation behavior of the polymers, which in turn regulates the drug release kinetics. It also suggests the polyphosphate might be the most promising to study because of its complete biodegradability. Subsequent studies therefore focused on the polyphosphates and with backbone structures that are potentially more likely to be innocuous for biocompatibility considerations.

## DESIGN, SYNTHESIS, AND PHYSICOCHEMICAL PROPERTIES OF PPES

### PPEs Based on Poly(Ethylene Terephthalate)

Poly(ethylene terephthalate) (PET) is one of the most commonly used biomedical polymers, as in vascular graft ap-

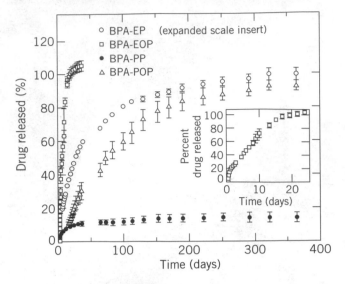

**Figure 3.** In vitro release of *p*-nitroaniline from BPA-based PPE matrices at 37°C and pH 7.4.

plications (34,35). Its appeal partly lies in its superb mechanical properties, which are derived from the liquid crystalline characteristics of the ethylene terephthalate structure. Hypothesizing that a biodegradable polymer with good mechanical properties can be built on this structure, we introduce phosphate units into PET (36). The general structure of poly(terephthalate-*co*-phosphate) is shown in Scheme 3 and Tables 1a and 1b. This series of PPEs can be synthesized by a two-step polycondensation. A diol such as 1,4-bis(hydroxyethyl)-terephthalate (BHET) is first reacted with ethyl phosphorodichloridate (EOP) to yield a hydroxyl-terminated prepolymer, which is further polymerized with the more energetic terephthaloyl chloride (TC). A typical synthetic procedure for P(BHET-EOP/TC, 80/20) is given in Example 1 (the numbers refer to the molar ratio of EOP to TC). The addition sequence and the ratio of reactive chlorides (EOP and TC) are crucial to obtain a polymer with high molecular weight and good solubility in common organic solvents. Adding the TC first or adding the two reactive dichlorides at the same time yields a polymer ($T_m$ = 184°C) that is only slightly soluble in dimethylformamide (DMF) and dimethyl sulfoxide (DMSO) but not in chloroform or methylene chloride. As the chain extension is achieved more efficiently in the second step by the reactive TC, the molecular weight is increased but the solubility in chloroform or methylene chloride decreases because of the lower phosphate content. When the terephthalate unit is equimolar with the ethyl phosphoester unit, the polymer P(BHET-EOP/TC, 50/50) becomes insoluble in chloroform.

The composition of poly(terephthalate-*co*-phosphate) (contents of TC and phosphoester units) as calculated from the nuclear magnetic resonance (NMR) spectrum is consistent with the charging ratio. The relative molecular weights (using polystyrene as standards) of poly(terephthalate-*co*-phosphate)s increase with the TC content, which is also confirmed by the intrinsic viscosity data. This confirms that the reaction of the extended diol with terephthaloyl chloride is more energetic than that with ethyl phosphorodichloridate.

### EXAMPLE 1

**Synthesis of P(BHET-EOP/TC, 80/20) (Scheme 4).**

Under an argon stream, a two-neck round-bottom flask (1 L) fitted with a condenser and addition funnel was charged with a solution of 50.85 g of 1,4-bis(hydroxyethyl) terephthalate (BHET, 0.2 mol). 48.9 g of 4-dimethylaminopyridine (DMAP, 0.4 mol) in 400 mL of methylene chloride. The mixture in the flask was cooled down to −40°C in a dry ice–acetone bath. A solution of 26.7 g of EOP (0.164 mol, 2.4% excess) in 50 mL of methylene chloride was added dropwise to the flask through the funnel. Following the addition of the EOP, the mixture was refluxed for 4 h, and then a solution of 8.12 g of TC (0.04 mol) in 20 mL of methylene chloride was added to the flask through the addition funnel. The temperature was gradually brought up to 50°C and allowed to react overnight. The mixture was cooled down to room temperature, washed three times with 0.1 N HCl solution saturated with NaCl (300 mL each time) and once with saturated NaCl solution (300 mL). The organic layer was dried over anhydrous sodium sulfate and filtered. The filtrate was concentrated and quenched with ether. The precipitate was collected and reprecipitated twice by methylene chloride–ether quenching. The precipitate was dried under vacuum at 60°C overnight to give a polymer (51 g, 73%) as a white, tough, chunky solid, with $M_r$ = 14,910 and $M_n$ = 5810 (by GPC measured in chloroform at 40°C using polystyrene as standards). It was found that a 2–5% excess of phosphorodichloridate yields a higher molecular weight polymer. This might be due to the high susceptibility of ethyl phosphorodichloridate to moisture.

Dacron (PET) is a highly crystallized polymer ($T_m$ = 265°C, $T_g$ = 69°C) (37). Introduction of an ethyl phosphate bond into the backbone significantly changes the polymer's thermal properties. In the P(BHET-EOP/TC) series, when the EOP/TC ratio is higher than 1, the PPEs are amorphous with no melting point ($T_m$) observable from differential scanning calorimetric (DSC) analysis; the polymers begin to soften at around 150°C. When the EOP/TC charging ratio is reduced to 1, crystalline phase begins to form in the polymer ($T_m$ = 201°C for P[BHET-EOP/TC, 50/50]). As expected, the glass-transition temperature ($T_g$) increases as the charging ratio of EOP/TC decreases (Fig. 4). To evaluate the effect of a more sterically hindered diol on

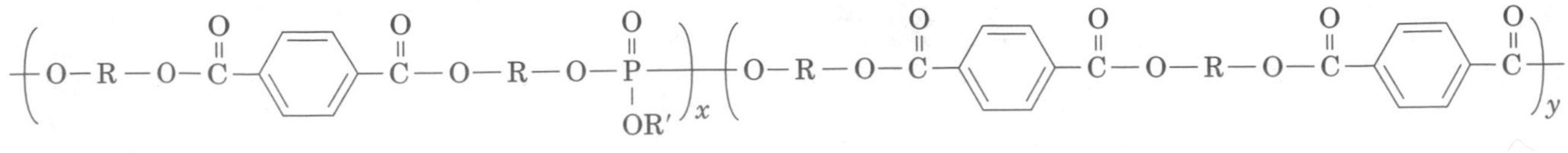

**Scheme 3.** Structures of poly(terephthalate-*co*-phosphate)s.

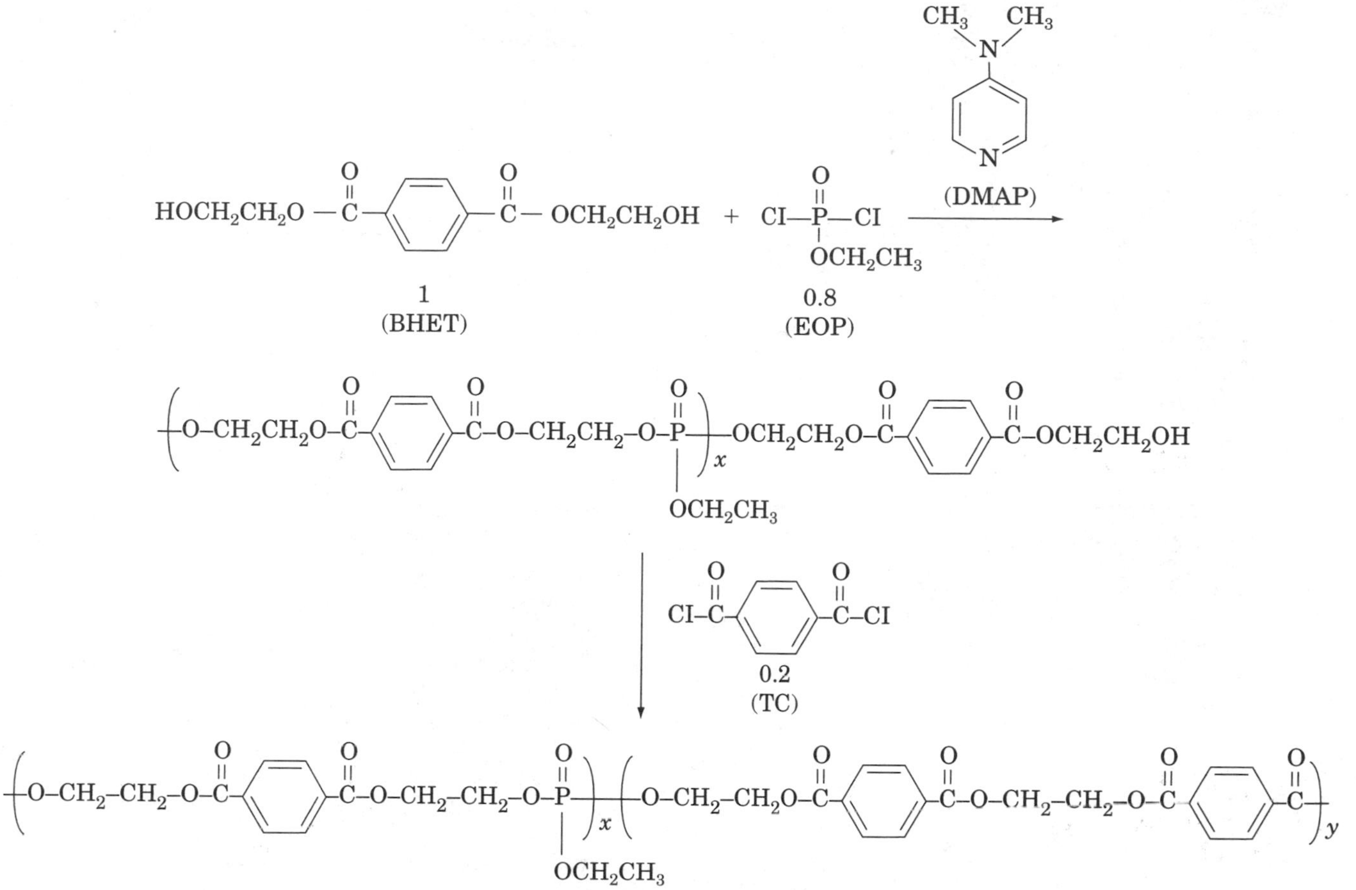

**Scheme 4.** Synthesis of P(BHET-EOP/TC, 80/20).

**Table 1a. Structural Variation of Poly(terephthalate-*co*-phosphate)s: Diol in the Backbone**

| R (R' = CH$_3$CH$_2$-) | | Polymer[a] |
|---|---|---|
| -CH$_2$-CH$_2$- | EG | P(BHET-EOP/TC) |
| -CH$_2$-CH$_2$-CH$_2$- | PD | P(BHPT-EOP/TC) |
| -CH$_2$-CH-CH$_2$-<br>    &vert;<br>    CH$_3$ | MPD | P(BHMPT-EOP/TC) |
|     CH$_3$<br>    &vert;<br>-CH$_2$-C-CH$_2$-<br>    &vert;<br>    CH$_3$ | DMPD | P(BHDPT-EOP/TC) |

[a]Numbers in the parentheses appearing in the text refer to the molar ratio of the two chloride monomers, ethyl phosphorodichloridate to terephthaloyl chloride.

**Table 1b. Structural Variation of Poly(terephthalate-*co*-phosphate)Side Chains**

| R' (R = -CH$_2$-CH$_2$- | Ratio[b] | Polymer |
|---|---|---|
| CH$_3$-CH$_2$- | 80 | P(BHET-EOP/TC, 80/20) |
| CH$_3$CH$_2$- and CH$_3$- | 60/20 | P(BHET-EMOP/TC, 60/20/20) |
| CH$_3$CH$_2$- and CH$_3$- | 40/40 | P(BHET-EMOP/TC, 40/40/20) |
| CH$_3$CH$_2$- and Na | 60/20 | P(BHET-ENaOP/TC, 60/20/20) |
| CH$_3$CH$_2$- and Na | 40/40 | P(BHET-ENaOP/TC, 40/40/20) |

[b]Molar ratio of EOP to MOP.

the polymer biodegradation and physical properties, we have also synthesized a series of copolymers by substituting ethylene glycol (EG) with propylene diol (PD), 2-methylpropylene diol (MPD), or 2,2-dimethylpropylene diol (DMPD), respectively. Comparison of the aliphatic diol component indicates that the polymer chain rigidity decreases in the order of EG > DMPD > MPD > PD.

The relative hydrophobicity of the P(BHET-EOP/TC) series has been assessed by measuring the water-in-air contact angle. The contact angle decreases dramatically from 65° for P(BHET-EOP/TC, 80/20) to 8° for P(BHET-EOP). While the charging molar ratio of EOP/TC is maintained at 4 (or 80/20), substituting the ethoxy side chain with methoxy side chain (MOP) would also increase the hydrophilicity (Table 2). At the ratio of EOP/MOP/TC of 40/40/20, the contact angle decreases to a value close to that of P(BHET-EOP). Furthermore, converting the meth-

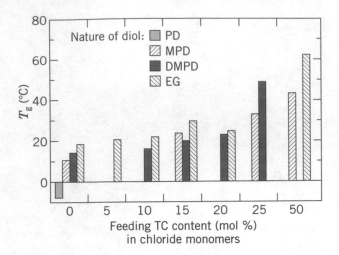

**Figure 4.** Glass-transition temperatures of poly(terephthalate-*co*-phosphate)s with different structures. Poly(terephthalate-*co*-phosphate)s were synthesized using propylene diol (PD), 2-methylpropylene diol (MPD), 2,2-dimethylpropylene diol (DMPD), or ethylene glycol (EG) as the initial diol (Scheme 1). The EOP/TC content is consistent with the charging ratio, calculated according to their NMR spectra.

oxy groups into sodium salt renders the polymer insoluble in chloroform and methylene chloride. The negatively charged P(BHET-ENaOP/TC)s become more hydrophilic by gross observation. As expected, the methoxy group substituted P(BHET-EOP/TC) and the negatively charged poly(terephthalate-*co*-phosphate)s have a much higher $T_g$ than their parent polymers (Table 2). P(BHET-NaOP/TC, 80/20) has a clear sharp melting peak at 180°C as determined by DSC.

The poly(terephthalate-*co*-phosphate) shows typical tensile properties for plastics with low degrees of crystallinity. At compositions with an EOP/TC charging ratio of 80/20 to 85/15, the P(BHET-EOP/TC)s could form thin elastomeric films from solvent casting. At an EOP/TC ratio higher than 90/10, the film becomes brittle. Substituting EOP with MOP (P[BHET-EMOP/TC]s, see Table 2) does not affect the film formation. However, the film becomes more brittle and less ductile. Fibers could be drawn from polymer melts at 160–180°C from all the polymers with a 20% molar ratio of TC, even for the negatively charged P(BHET-ENaOP/TC)s (Table 3).

The in vitro degradation rates of solvent-cast films of P(BHET-EOP/TC, 80/20) and P(BHET-EOP/TC, 85/15), with relative molecular weights ($M_r$ polystyrene as a standard) in the range of 5,000 to 7,000, lose 21 and 43%, respectively, of their original weight in 18 days when incubated in 0.1 M PBS at 37°C (Fig. 5). Higher molecular

weight PPEs with the same composition degrades at a slower rate. P(BHET-EOP/TC, 80/20) with a $M_r$ of 14,300 shows only a 5% weight loss in 14 days of incubation under the same condition. Increasing the TC content in the polymer would also retard the degradation rate.

As discussed above, in the P(BHET-EOP/TC, 80/20) series, substituting the ethoxy with a methoxy group yields polymers with higher hydrophilicity, which in turn accelerates the in vitro degradation as demonstrated in Figure 6. Furthermore, converting the methoxy group to a sodium salt dramatically increases the degradation rate.

## PPEs Based on Chain Extension of Oligomeric Lactides by Phosphates

Poly(lactide-*co*-glycolide)s (PLGAs) remain the most popular and well-characterized biodegradable polymeric biomaterials. Their regulatory approval and extensive database of human use render them an obvious choice in medical applications that range from controlled drug delivery to tissue engineering (38). This series of PPEs contains phosphate bonds distributed between oligomeric blocks of lactides in the backbone. The rationale for designing such a structure is to explore the possibility of extending the physicochemical properties of the most commonly used PLGA polymers. The inclusion of the phosphate linkage in the backbone offers an extra degree of freedom to fine-tune the properties of the polymers (Scheme 5) (39). The degradation rate of a poly(lactide-*co*-phosphate) thus constructed is mainly controlled by the percentage of phosphate component introduced into the backbone. This has the effect of eliminating the biphasic degradation behavior typically exhibited by the crystalline polylactic acid (PLA). The higher the phosphate content in the backbone, the faster the degradation rate of the polymers. An additional factor that plays an important role in

**Table 2. Glass-Transition Temperatures and Water-in-Air Contact Angles of Some Poly(terephthalate-*co*-phosphate)s**

| Poly(terephthalate-*co*-phosphate) | $T_g$ (°C) | Water-in-air contact angle |
|---|---|---|
| P(BHET-EOP/TC, 80/20) | 25 | 65° |
| P(BHET-EOP/TC, 85/15) | 28 | 54° |
| P(BHET-EOP/TC, 95/5) | 21 | 10° |
| P(BHET-EOP/TC, 100/0) | 19 | 8° |
| P(BHET-EMOP/TC, 60/60/20) | 32 | 14° |
| P(BHET-EMOP/TC, 40/40/20) | 39 | 9° |
| P(BHET-NaOP/TC, 80/20) | 58 ($T_m = 180$) | N.D. |
| P(BHET-ENaOP/TC, 60/20/20) | 52 | N.D. |
| P(BHET-ENaOP/TC, 40/40/20) | 63 | N.D. |

**Table 3. Tensile Properties of Poly(terephthalate-*co*-phosphate)s**

| Poly(terephthalate-*co*-phosphate) | Tensile stress at yield (psi) | Elongation at break (%) | Modulus of elasticity (MPa) |
|---|---|---|---|
| P(BHET-EOP/TC, 80/20) | 230.89 ± 1.44 | 113.5 ± 42.3 | 13.35 ± 0.32 |
| P(BHET-EMOP/TC, 60/20/20) | 370.27 ± 67.95 | 74.6 ± 8.7 | 26.77 ± 2.19 |
| P(BHET-EMOP/TC, 40/40/20) | 404.22 ± 145.87 | 48.9 ± 1.6 | 27.59 ± 9.36 |

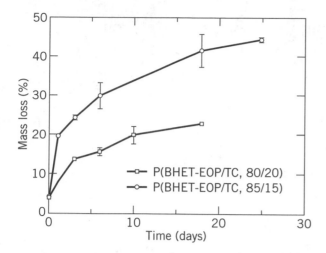

**Figure 5.** In vitro degradation of P(BHET-EOP/TC) discs in 0.1 M PBS at 37°C. The numbers in parentheses refer to the molar ratio of EOP to TC.

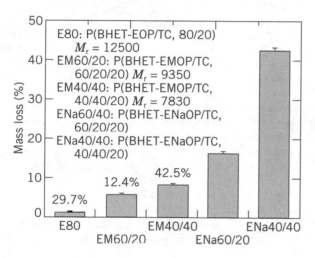

**Figure 6.** In vitro mass loss of P(BHET-EMOP/TC) discs in two weeks in 0.1 M PBS at 37°C. Numbers on top of the bars represent the average $M_r$ decreases.

determining the overall degradation rate is the stereoisomerism of the lactide, with the D,L-lactide (DLLA) producing polymers with lower crystallinity, lower $T_g$, and faster degradation rate than polymers synthesized from L-lactide (LLA).

The synthesis of poly(lactide-*co*-phosphate)s starts with the ring opening of lactide with an aliphatic diol (Scheme

6). This step is accomplished by thermal ring opening either in the absence or presence of a tin catalyst. The resulting macromonomer diol is further chain extended by phosphorodichloridate carried out in melt. An example of the synthetic protocol is given in Example 2. Propylene glyol (PG) has also been used as the initiator of the ring-opening reaction because it has a better safety profile than EG. L-lactide can be substituted by D,L-lactide to afford a more rapid degradation profile of the final polymer. Small amount of stannous octoate (200 ppm) can be added to the prepolymerization to shorten the reaction time and result in a more uniform molecular weight distribution of the prepolymer. We have explored different molar ratios between DLLA and PG such as 5:1, 10:1, and 20:1, etc. to create prepolymers of different molecular weights and subsequently different degradation rates of the final polymers. This section will focus on two polymers in this series (Scheme 6), P(LAEG-EOP) and P(DAPG-EOP).

---

**EXAMPLE 2**

**Synthesis of P(LAEG-EOP) Through Bulk Polymerization (Scheme 6).**

To a 250-mL round-bottom flask flushed with dried argon, 20 g (0.139 mol) of (3S)-*cis*-3,6-dimethyl-1,4-dioane-2,5-dione (L-lactide) and 0.432 g (6.94 mmol) of EG were added. The flask was closed under vacuum, placed in a 140°C oven, and kept in the oven for 2 to 4 days with occasional shaking. Then it was filled with dried argon and placed in a 135°C oil bath. Under argon stream, 1.13 g of EOP was added through a funnel with stirring. A low vacuum (~20 Torr) was applied to the system after 1 h of stirring and let stand overnight, and a high vacuum (0.05 Torr) was applied for at least 2 h. The polymer was cooled down and dissolved in 250 mL of chloroform. The solution was quenched into 1 L of ether. The dissolving–quenching step was repeated twice. The polymer was dried under vacuum to yield a white or slightly yellowish powder, yield 72–80%. Parameters tested: $M_r = 33,000, M_n = 4,800$. $[\eta] = 0.315$ (chloroform, 40°C); $T_g = 52°C$.

---

The poly(lactide-*co*-phosphate)s have excellent solubility in a broad range of organic solvents including ethyl acetate and acetonitrile, which are poor solvents for PLGA. Changing the initiation diol does not significantly change the physicochemical properties. As expected, P(DAPG-EOP) with a D,L-lactide block has an even higher solubility in common organic solvents.

The in vitro degradation has been studied in phosphate-buffered saline (PBS) at 37°C using microspheres prepared by the solvent evaporation method (40). The degradation rate is quite sensitive to the composition and molecular weights of the polymers. The feeding ratio of lactide to

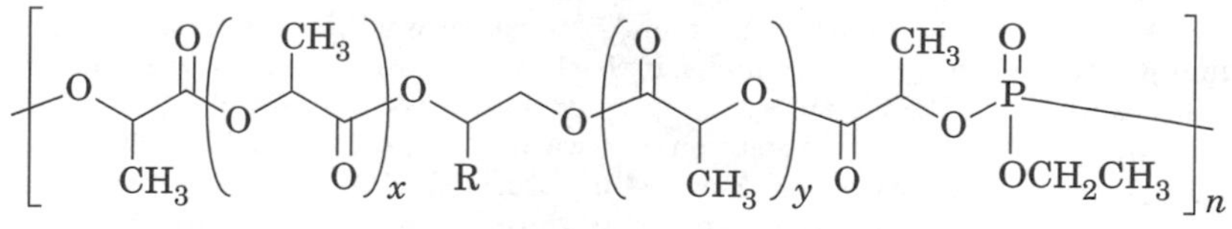

P(LAEG-EOP); R = H, all L-lactide
P(DAPG-EOP); R = CH₃, D,L-lactide

**Scheme 5.** Structures of poly(lactide-*co*-phosphate)s.

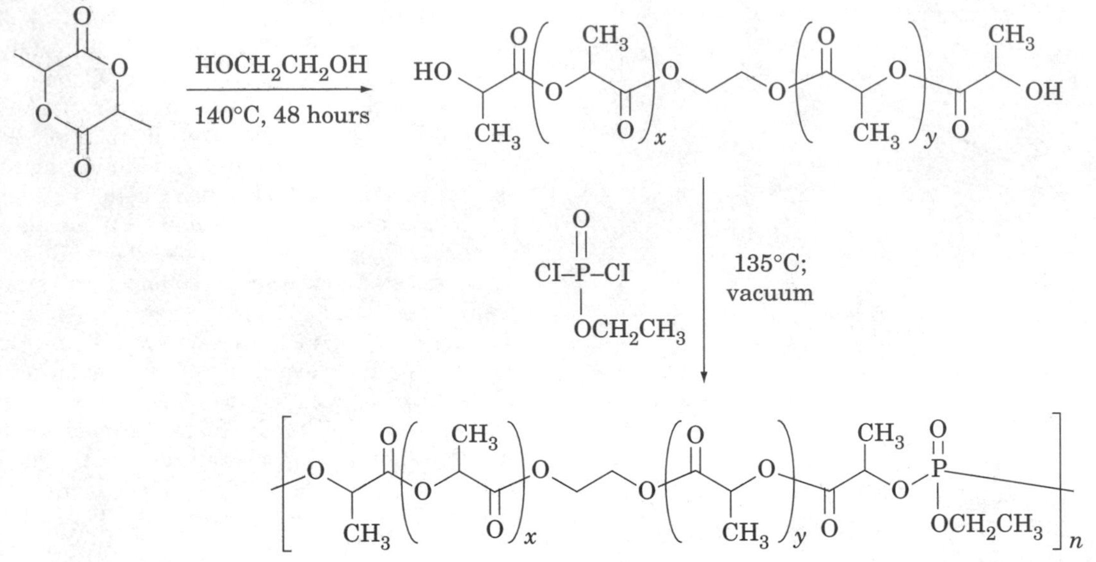

**Scheme 6.** Synthesis of P(LAEG-EOP).

initiation diol or phosphorodichloridate determines the length of the oligomeric lactide block. Lowering the molecular weight and increasing the phosphate content of the PPEs will increase the degradation rate of microspheres, owing to the higher hydrophilicity of the polymers. Polymers with D,L-lactide oligomers as the repeating unit degrade faster than those prepared from the L-lactide oligomers. Molecular weight also plays an important role in the degradation. At the same DLA/PG feeding ratio of 10:1, polymer with $M_r$ 7,900 degrades much faster than that with $M_r$ 12,000 by about 20% mass loss in two months (Fig. 7). Polymers with a faster degradation rate could in principle be custom designed by increasing the phosphate content and lowering the molecular weight.

### PPEs Based on Cyclohexane-1,4-Dimethyl Phosphate Backbone

The third series of PPEs are obtained from the polycondensation of aliphatic diols and aliphatic phosphorodichloridates, one of the simplest structures possible for PPEs. In contrast to the other PPEs described earlier, these polymers contain neither aromatic rings nor crystallizable blocks, and owing to a combination of the flexible phosphate and alkane structures, they have flowing points significantly below room temperature. The rationale of developing this series is to examine if these polymers, which exist as viscous liquids, would be useful as drug carriers, particularly for protein delivery, and may be applied with minimal fabrication processing. In principle, drugs can be blended into these polymers at a wide range of loading levels at room temperature and can be applied topically or injected into anatomical sites such as corneas and joints. In the case of proteins, the absence of any contact with organic solvents, low temperature, and possibility of inclusion of protein stabilizers in a simple loading process might be favorable for preserving the bioactivity of the proteins.

We focus on a cyclohexane dimethanol (*cis* or *trans*, or a mixture of both) backbone with different aliphatic side

chains (39,41) (Scheme 7). A typical protocol of preparing P(*trans*-CHDM-HOP) is described in Example 3. These aliphatic polyphosphates are obtained as clear colorless or pale yellow liquids with $M_n$ ranging from 3,000 to 10,000. Polymers can be obtained without a catalyst if the polymerization is conducted at high temperatures, but the molecular weights are generally lower. Side reactions such as transesterification involving the side chain may also lead to cross-linking. The viscosity profile of the P(*trans*-CHDM-HOP) as a function of temperature and frequency sweep is shown in Figure 8. Although viscous at room temperature with a viscosity of 327 Pas, the liquid can be injected with a gauge needle of 20 and below. The viscosity of the polymer is decreased if the backbone is composed of a mixture of *cis* and *trans* instead of the pure *trans* cyclic diol.

---

### EXAMPLE 3

#### Synthesis of P(*trans*-CHDM-HOP) By Solution Polymerization (Scheme 8).

Under an argon stream, 10 g of *trans*-1,4-cyclohexane dimethanol (CHDM), 0.847 g of 4-dimethylaminopyridine (DMAP), 15.25 mL (14.03 g) of *N*-methyl morpholine (NMM) and 50 mL of methylene chloride were transferred into a 250 mL flask equipped with a funnel. The solution in the flask was cooled down to −40°C with stirring. A solution of 15.19 g of hexyl phosphorodichloridate (HOP) in 20 mL of methylene chloride was added through the funnel (10 mL of methylene chloride was used to flush through the funnel), and then the mixture was brought to boiling temperature gradually. It was kept refluxing for 4 h and was filtered. The filtrate was evaporated and redissolved in 100 mL of chloroform. This solution was washed twice with 0.5 M of HCl NaCl solution and once with saturated NaCl solution, dried over Na$_2$SO$_4$, and quenched into an ether–petroleum (1:5) mixture. The precipitate (oily) was collected and dried under vacuum. A clear, pale yellow, viscous liquid was obtained.

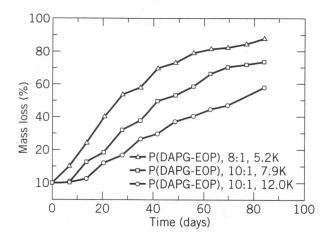

**Figure 7** . In vitro degradation of microspheres made from P(DAPG-EOP) with different $M_r$ and compositions. Ratios refer to DLLA/PG, followed by $M_r$ of the polymers.

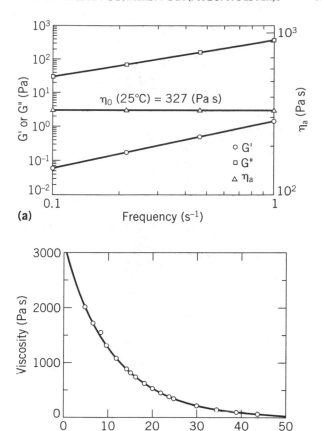

**Figure 8.** (**a**) Apparent viscosity ($\eta_a$), storage modules (G′) and loss modules (G″) of P(*trans*-CHDM-HOP) as a function of frequency; (**b**) temperature dependency of apparent viscosity ($\eta_a$) of P(*trans*-CHDM-HOP) at 25°C at a frequency of 1 s$^{-1}$.

The polymers degrade in vitro as a function of the alkyl chain length in the side chain (39). Following the trend of the lipophilicity of the side chains, the degradation rate of the polymers decreases in the order of P(CHDM-EOP) > P(CHDM-BOP) > P(CHDM-HOP) (Figure 9). Higher degradation rates can be obtained by substituting CHDM with the linear aliphatic 1,6-hexane diol. The degradation rate would be even further enhanced if 1,4-butane diol is used. Poly(ethylene glycol) (PEG) of different molecular weights have also been studied as the backbone. These linear aliphatic PPEs become water soluble with low molecular weights of PEG and waxy with PEG of molecular weights above 8,000. In the series of PPEs with hexyl side chains, the viscosity of these linear aliphatic PPEs is in general lower than that of P(CHDM-HOP).

## STORAGE STABILITY AND STERILIZATION STABILITY

Designed to be biodegradable, the PPEs are naturally susceptible to hydrolysis even in air. When stored in air at room temperature, their molecular weights decline slightly (42). However, all three series of polymers are reasonably stable in a desiccator at room temperature, without the need for storage under inert gas, as monitored by GPC and intrinsic viscosity. For example, the molecular weight of a P(LAEG-EOP) stored in a dessicator remains unchanged throughout a 3-month period (Fig. 10), but in contrast, the molecular weight drops about 70% when the same sample is placed in air.

Sterilizability of a biomedical polymer is an important parameter (43). These PPEs have been placed in an aluminum foil pouch, heat sealed, packed with dry ice, and $\gamma$-irradiated at 18.3–25.0 KGy. Both the structure and molecular weight are unaffected, as monitored by GPC, intrinsic viscosity measurement, and Fourier transform infrared spectroscopy (FTIR) (36,44).

The effect of $\gamma$-irradiation on the stability of a PPE/paclitaxel microsphere formulation has also been investigated by sterilizing the microspheres containing 10% Paclitaxel under the same condition (45). Neither the polymer molecular weights nor the microsphere characteristics are affected by the irradiation process, and paclitaxel in PPE microspheres remains stable during $\gamma$-irradiation. Furthermore, paclitaxel release profiles are comparable before and after $\gamma$-irradiation, suggesting that it can be used to sterilize paclitaxel/PPE microspheres.

## IN VIVO DEGRADATION, BIOCOMPATIBILITY, AND CYTOTOXICITY

The biodegradation and biocompatibility of these PPEs have been evaluated in an SPF Sprage-Dawley rat model.

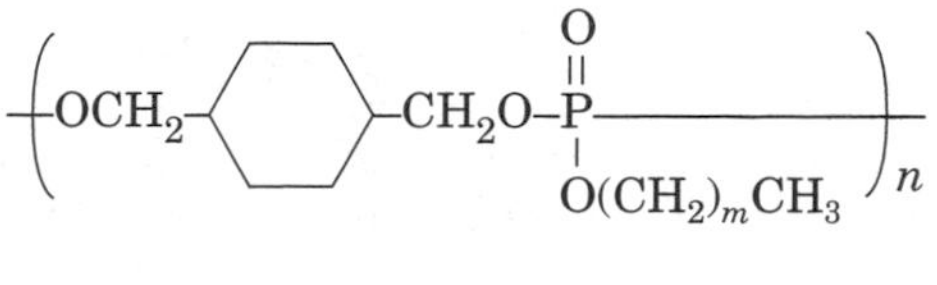

$$\left(OCH_2-\!\!\!\bigcirc\!\!\!-CH_2O-\overset{\displaystyle O}{\underset{\displaystyle O(CH_2)_mCH_3}{\overset{\|}{P}}}\right)_n$$

$m = 5$, P(CHDM-HOP)
$m = 3$, P(CHDM-BOP)
$m = 1$, P(CHDM-EOP)

**Scheme 7.** Structures of CHDM-based PPEs.

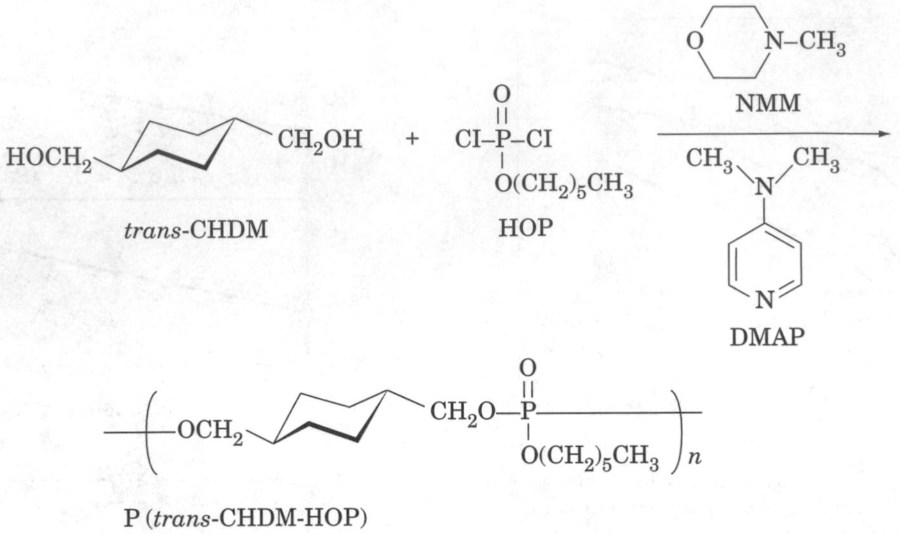

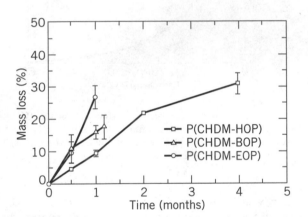

**Scheme 8.** Synthesis of P(CHDM-HOP).

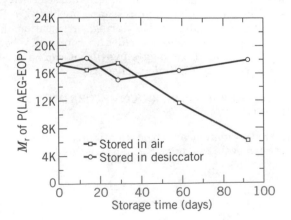

**Figure 9.** In vitro degradation of PPEs with different side chains in PBS at 37°C. Refer to Scheme 7 for the symbols.

**Figure 10.** Changes in $M_r$ of P(LAEG-EOP) samples stored at room temperature with or without desiccation.

P(BHET-EOP/TC, 80/20), P(LAEG-EOP), and P(CHDM-HOP) have been investigated as the representatives for each class of PPEs, compared with PLGA (75:25, RG755). PPE discs are fabricated by compression molding at 200 MPa or by a film-casting method and sterilized by γ-irradiation at 25 KGy on dry ice. The discs and viscous liquid are implanted into the muscles at the right hind limb of the adult rat. The reaction surrounding the implantation site is characterized as slight to mild for up to 4 months for all four polymers. At the 4- and 6-month time points, all tissue reactions to the four groups are characterized as only slightly irritating (Table 4). Hematology data, differential leukocyte counts, cellular morphology findings, and clinical chemistry values are overall unremarkable and comparable among the four groups at each interval. No histomorphological change in all major organs at the 6-month time point is observed (42).

The analysis of polymer discs retrieved at different time points shows a continuous mass loss of the P(LAEG-EOP) discs with time, compared with the typical biphasic degradation behavior of PLGA (Fig. 11). P(BHET-EOP/TC, 80/20), on the other hand, shows approximately 20% mass loss in 2 weeks and remains unchanged for up to 4 months (42). As discussed earlier for the in vitro degradation rate of the P(BHET-EOP/TC)s, it is expected that their in vivo degradation rates would be accelerated by increasing the phosphate content in the backbone.

The cytotoxicity of P(BHET-EOP/TC, 80/20) has been assessed by culturing human embryonic kidney cells (HEK293) on a polymer-coated cover slip. Cells exhibit a normal morphology and proliferate at a rate comparable to those cultured on tissue culture polystyrene (TCPS) surface. The proliferation assay on human gastric carcinoma cells (GT3TKB) showed no inhibition of cell growth for either P(BHET-EOP/TC, 80/20) or P(LAEG-EOP) microspheres at up to at least 0.5 mg/mL and 1.0 mg/mL concentration, respectively. The Ames test and International

**Table 4. Tissue Reaction Score for PPE Discs and Viscous Liquid Implanted between the Muscle Layers of the Hind Limb of Male Sprague-Dawley Rats**

| PPE | Time after the implantation[a] (d = days, m = months) | | | | | | |
|-----|------|------|------|------|------|------|------|
|     | 3 d | 7 d | 14 d | 1 m | 2 m | 4 m | 6 m |
| P(LAEG-EOP) | SI (130) | SI (123) | SI (180) | SI (198) | SI (106) | SI (99) | SI (30) |
| P(BHET-EOP/TC, 80/20) | SI (151) | SI (116) | SI (163) | SI (98) | SI (60) | SI (35) | SI (52) |
| P(CHDM-HOP) | MI (226) | SI (197) | SI (196) | MI (225) | MI (207) | MI (122) | SI (40) |
| PLGA (RG755) | SI (148) | SI (98) | SI (137) | SI (105) | SI (94) | SI (43) | SI (44) |

*Note:* The extent of inflammatory response at the implantation sites has been assessed by an independent pathologist. The number of monocytes and macrophages under 400× microscopic field are counted. The arithmetic mean of four different fields is used to grade the tissue reaction according to the following system: NI, no irritation (0); SI, slight irritation (1–200); MI, mild irritation (201–400); MOI, moderate irritation (401–600); SEI, severe irritation (>601).
[a]Numbers in parentheses represent average monocyte/macrophage count/fields at 400×.

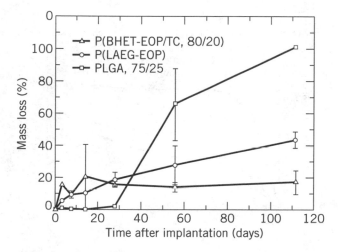

**Figure 11.** Degradation profiles of PPEs in rat muscle compared with PLGA (75/25).

Organization for Standardization (ISO) agarose overlay tests suggest that all three classes of polymers are neither mutagenic nor cytotoxic (36,39).

## DRUG RELEASE PROPERTIES

Controlled release of several low-molecular-weight drugs, proteins, and plasmid DNA with different physicochemical properties have been studied in the forms of microspheres, thin films, or viscous liquids as described in this section.

### Controlled Release through PPE Microspheres

**Controlled Release of Paclitaxel, Cisplatin and Lidocaine.** Most of our effort of delivering low-molecular weight drugs using PPE microspheres has been concentrated on two types of drugs, chemotherapeutic agents (Paclitaxel and cisplatin) and analgesics (lidocaine). All microspheres have been prepared by a solvent evaporation method.

The effect of side-chain length on drug release has been investigated using two poly(terephthalate-*co*-phosphate)s as a model system (45). Lidocaine encapsulated into P(BHDPT-EOP/TC, 50/50) or P(BHDPT-HOP/TC, 50/50) at a loading level of 5 wt % yields microspheres with a size range of 20 to 100 $\mu$m. PPE with a longer side chain

P(BHDPT-HOP/TC, 50/50) causes a slower release kinetics for lidocaine (Fig. 12), consistent with the relative degradation rates of the polymers.

The correlation between degradation of and drug release from the PPE microspheres has been demonstrated in a P(DAPG-EOP) microsphere system (40). High encapsulation efficiency can be achieved greater than 96% with an average drug-loading level of 10 wt %. Paclitaxel release rate is the fastest for microspheres with 5:1 lactide-to-PG ratio, followed by polymer microspheres with 8:1 and 10:1 ratios. A continuous decrease in molecular weight and weight loss for all the PPE microspheres are observed (Figs. 7 and 13). Paclitaxel release from microspheres has shown similar kinetics.

Microspheres containing cisplatin or lidocaine have also been studied (43,46). The sizes of microspheres are generally between 3 and 50 $\mu$m with a loading level of 2.4 wt % for cisplatin or 5 wt % for lidocaine. A burst release of cisplatin (45%) is observed, followed by slower release for 3 days (Fig. 14). A near-zero-order release is found for lidocaine from the microspheres for the first 7 days, and the release reaches about 81% after 13 days.

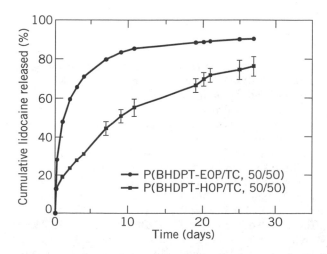

**Figure 12.** In vitro release of lidocaine from P(BHDPT-EOP/TC, 50/50) and P(BHDPT-HOP/TC, 50/50) microspheres in PBS at 37°C. Lidocaine was microencapsulated at 5 wt % loading in the microspheres. Both polymers were synthesized using DMPD as the initial diol, and the EOP/TC ratio was 50/50.

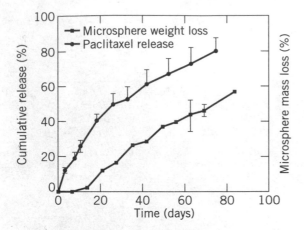

**Figure 13.** In vitro release of Paclitaxel from P(DAPG-EOP) microspheres (10 wt % loading) and mass-loss profile of these drug-loaded microspheres in PBS at 37°C. P(DAPG-EOP) used in this study had $M_r$ 12,000 and a D,L-lactide–EOP ratio of 10:1.

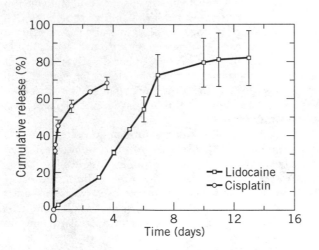

**Figure 14.** In vitro release of lidocaine (5 wt % loading) and cisplatin (2.4 wt % loading) from P(LAEG-EOP) microspheres in PBS at 37°C.

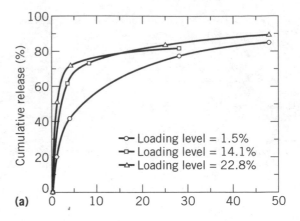

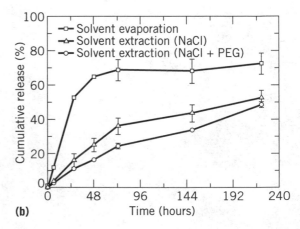

**Figure 15.** (a) Effect of loading level of FITC-BSA in P(BHET-EOP/TC, 80/20) microspheres on its in vitro release. (b) In vitro release of FITC-BSA from P(LAEG-EOP) microspheres prepared by different protocols. Microspheres were prepared by a solvent evaporation method (open square), a solvent extraction method using 0.3% poly(vinyl alcohol) (PVA) solution containing 5% sodium chloride as the second aqueous phase (open triangle), or 0.3% PVA–5% NaCl–1% PEG$_{8,000}$ solution (open circle).

**Controlled Release of Fluorescence-Labeled Bovine Serum Albumin (FITC-BSA) As a Model Protein Drug.** FITC-BSA has been used as a model protein drug to study the controlled release properties of these PPEs (36,43). Microspheres are prepared from either poly(terephthalate-*co*-phosphate) or poly(lactide-*co*-phosphate) by the double-emulsion technique. A wide range of protein-loading levels (1.5–22.8 wt %) could be achieved with an encapsulation efficiency of 70–90%. The average size of PPE microspheres prepared by this method ranges from 2 to 20 μm with a smooth surface morphology.

The in vitro release of FITC-BSA from P(BHET-EOP/TC, 80/20) microspheres is relatively fast. As expected, higher loading levels result in increased release rates. More than 80% of the protein is released within the first 24 h for microspheres with 14.1 and 22.8 wt % loading levels, whereas nearly 75% is released from 1.5 wt % protein-loaded microspheres (Fig. 15a).

The formulation method significantly affects the protein-release kinetics. The second aqueous phase in the double-emulsion method is of the major factor. Using a 5% sodium chloride solution containing 1% PEG$_{8,000}$ as the second aqueous phase yields microspheres the most linear release profile for FITC-BSA (Figure 15b).

**Controlled Release of Plasmid DNA.** Genetic immunization using naked DNA has spawned intense interest because of the promising immune responses that can be generated against different antigens in various animal models. It has also been argued that in gene therapy, the exogenous gene should be increasingly viewed as a drug for practical considerations (47). As such, many of the pharmaceutical issues confronting the successful application of delicate proteins are also relevant for nonviral gene delivery. Conceivable advantages of the controlled-release approach may include improving the bioavailability of the DNA to the target tissue and prolonging the expression of the gene product in the transfected cells. We have therefore evaluated the feasibility of delivering plasmid DNAs using these PPEs in both the microsphere and the viscous liquid configurations.

Microspheres containing plasmid DNA have been prepared using a double-emulsion method (48,49). Two polymers that have been studied are P(DAPG-EOP) ($M_r$ = 13.6 K), with the D,L-lactide oligomer as the building block, and P(LAEG-EOP) ($M_r$ = 33 K), with the L-lactide oligomer as the macromonomer. With a typical double-emulsion protocol, microspheres containing 0.9–2% plasmid DNA are obtained with an encapsulation efficiency of 88–95%. A typical batch of microspheres has an average size of 8–10 $\mu$m and average volume of 42–57 $\mu$m. At such a low loading level, the release mechanism is almost purely matrix degradation controlled, which is quite slow. Bovine or mouse serum albumin (BSA or MSA, 8.5–10% w/w) has therefore been coencapsulated with the DNA to increase the DNA release rate. Confocal microscopic images of microspheres containing plasmid with a fluorescent dye suggest that the majority of DNA is concentrated in the core.

The in vitro release shows that microspheres prepared from P(DAPG-EOP) containing both DNA and MSA have an average DNA release rate of 380 ng/day/mg of microspheres, compared with 125 ng/day/mg of microspheres released from microspheres containing DNA alone (Fig. 16). Both rates are much higher than the microspheres made from higher-molecular-weight P(LAEG-EOP), which releases DNA at a rate of 20–29 ng/day/mg of microspheres. DNA released from these microspheres is intact as revealed by electrophoretic mobility analysis (Fig. 17).

## Controlled Release through Poly(Terephthalate-*co*-Phosphate) Film

Paclitaxel (10 wt %) has been embedded in P(BHET-EOP/TC, 80/20) during solvent casting (50). The film is flexible with a smooth surface. DSC and X-ray analyses fail to show any crystallinity of the drug in the polymer. This formulation can deliver paclitaxel over a period of several

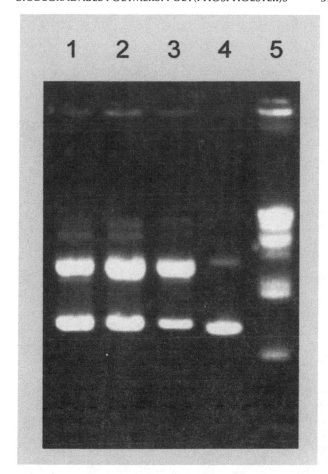

**Figure 17.** Electrophoretic mobility of DNA released from P(DAPG-EOP) microspheres. Supernatant of a microsphere–PBS suspension collected after 1 h (lane 1), 2 days (lane 2), or 6 days of incubation at 37°C (lane 3) was run on a 0.8% agarose gel, as compared with the original plasmid (lane 4). Lane 5 is the molecular-weight marker (λDNA HindIII digested).

months (Fig. 18). Less than 30% mass loss of the polymer is observed in this period. The degradation profile approximates that of the release profile of paclitaxel, indicating degradation controlled release mechanism. The unreleased paclitaxel remains stable in the film during the degradation of the polymer.

## Controlled Release through P(*trans*-CHDM-HOP) Viscous Liquid

**Release of Lidocaine and Doxorubicin.** Lidocaine (25 wt %) or doxorubicin (10 wt %) has been incorporated into the PPE viscous liquid by blending the drugs with P(*trans*-CHDM-HOP) viscous liquid until homogeneity is achieved. As shown in Figure 19, lidocaine is almost completely released in 1 week. The rapid release is expected of such a low molecular weight compound. Doxorubicin, which is more hydrophobic, unexpectedly produces a near linear release rate over 18 days (39). Initially it was thought that such a viscous liquid would not be able to provide a prolonged release for low-molecular-weight drugs. Apparently the P(*trans*-CHDM-HOP) is hydrophobic enough to retard

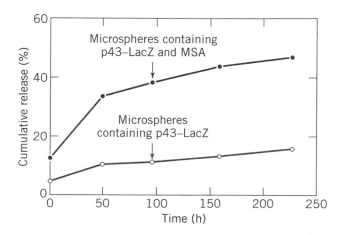

**Figure 16.** In vitro release of plasmid DNA from P(DAPG-EOP) microspheres containing 1.84–2% p43–LacZ plasmid coencapsulated with 8.9% MSA (filled circle) or without MSA (open circle) in PBS at 37°C.

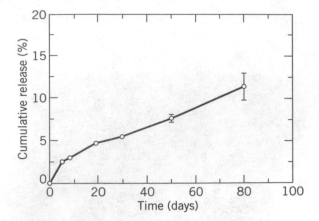

**Figure 18.** In vitro release of Paclitaxel from P(BHET-EOP/TC, 80/20) film (10 wt % Paclitaxel loading) in PBS at 37°C.

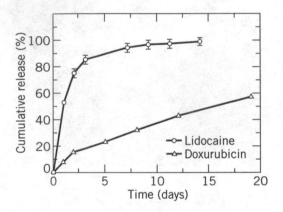

**Figure 19.** In vitro release of lidocaine (25 wt % loading) and doxorubicin (10 wt % loading) from P(*trans*-CHDM-HOP) in PBS at 37°C.

rapid penetration of water, and when a lipophilic drug is embedded, sustained release over several weeks is possible.

**Release of FITC-BSA and Interleukin 2 (IL-2).** The effect of loading level of FITC-BSA on the release kinetics is shown in Figure 20. With a 30% loading, the hydrophilic protein would strongly attract water absorption into the sample. Not surprisingly, over 80% of the protein is released in 1 week (39). In protein delivery by the nonbiodegradable poly(ethylene-vinyl acetate) (EVAc), a loading level above 30% is required to effect a complete release (51). Protein powder exposed to the surface leads to pores and channels for subsequently dissolved protein molecules to diffuse out. It is not clear if a similar mechanism is at work here because the P(*trans*-CHDM-HOP) is so much more fluid that any channel should collapse after the dissolution of the protein. Protein diffusion through its matrix is evident by the release of the 1% loaded sample. The release mechanism for such a low loading level may be dominantly matrix degradation controlled. However, because the matrix-degradation rate is low and lagged behind the release rate significantly, it suggests that the pro-

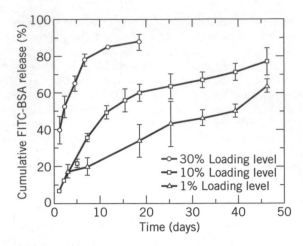

**Figure 20.** Effect of loading levels of FITC-BSA on its in vitro release from P(*trans*-CHDM-HOP) in PBS at 37°C.

tein can also permeate through the polymer phase. The effect of the side chain of the P(*trans*-CHDM-phosphate) series on the protein release kinetics is shown in Figure 21. When the loading level is kept constant at 10 wt %, the relative release rate, more clearly so at the early time points, decreases with the lipophilicity of the side chain (39).

To assess whether the bioactivity of the released protein is preserved, IL-2 release from P(*trans*-CHDM-HOP) has been studied (Fig. 22). The bioactive concentration of IL-2 is determined by a cellular assay where the NSF-60 cells proliferate in a dose-dependent manner in the presence of IL-2. Correlation of this biological concentration with the total IL-2 concentration measured by ELISA yields the retention of the bioactivity. The bioactive portion ranges from 16 to 35%. Because there is a finite time interval, ranging from hours to days, before a released IL-2 molecule is measured, the degradation of the protein while sitting in the PBS at 37°C may be the main source of loss of bioactivity.

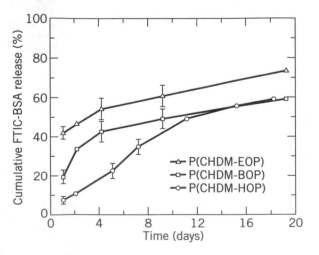

**Figure 21.** Effect of side-chain structure on the in vitro release kinetics of FITC-BSA from PPE in PBS at 37°C. Refer to Scheme 7 for the symbols.

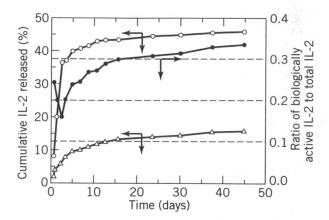

**Figure 22.** In vitro release of IL-2 from P(*trans*-CHDM-HOP) in RPMI-1640 medium at 37°C. Bioactive IL-2 (open triangle) was determined by NSF-60 cell proliferation assay, whereas total IL-2 released (open circle) was measured by ELISA. Bioactivity retention (filled circle) was defined as the ratio of the two.

The values presented are therefore the lower limit that can be expected of IL-2 delivery by such a system.

**Release of Plasmid DNA.** A viscous liquid containing plasmid DNA has been prepared by physically blending the lyophilized DNA–mannitol powder (1:10 or 1:50 weight ratio) with P(CHDM-HOP) at room temperature until the sample reaches homogeneity. The in vitro release of DNA has been studied on two batches of gels with a 1% DNA loading level. Mannitol can be used as an excipient (10 and 33 wt % loading in the gel, respectively) to help disperse the DNA throughout the gel and increase the release rate. As shown in Figure 23, 70 and 90% of the DNA is released within 2 weeks from PPE viscous liquids coloaded with 10 and 33 wt % mannitol, respectively.

## SUMMARY

This article suggests that PPEs may be added to the list of biodegradable polymers for controlled drug delivery. The versatility of the polyphosphates should prove advantageous in different therapeutic applications. Matrix properties can range from viscous to crystalline to elastomeric. Device configurations of viscous liquids, wafers, microspheres, and flexible coatings are possible. Controlled release of bioactive agents spanning a wide range of molecular weights from doxurubicin and IL-2 to plasmid DNA has been demonstrated. The potential of this new class of drug carriers needs to be defined in animal and eventually clinical studies, and work is proceeding toward that end.

## BIBLIOGRAPHY

1. R. Langer, *Science* **249**, 1527–1533 (1990).
2. K.W. Leong, in P. Tarcha, ed., *Synthetic Polymeric Drug Delivery Systems*, CRC Press, Boca Raton, Fla., 1991.
3. W.M. Saltzman and M.L. Radomsky, *Chem. Eng. Sci.* **46**(10), 2429–2444 (1991).
4. K.L. Leong and R. Langer, *Adv. Drug Delivery Rev.* **1**, 199 (1987).
5. R. Jain, N.H. Shah, A.W. Malick, and C.T. Rhodes, *Drug Dev. Ind. Pharm.* **24**(8), 703–727 (1998).
6. B. Buntner et al., *J. Controlled Release* **56**(1–3), 159–167 (1998).
7. T.W. Atkins and S.J. Peacock, *J. Biomater. Sci., Polym. Ed.* **7**, 1065–1073 (1996).
8. Y. Ogawa, *J. Biomater. Sci., Polym. Ed.* **8**, 391–409 (1997).
9. J. Kohn and R. Langer, *J. Am. Chem. Soc.* **109**, 817–820 (1987).
10. I. Gachard, S. Bechaouch, B. Coutin, and H. Sekiguchi, *Polym. Bull.* **38**(4), 427–431 (1997).
11. B.I. Dahiyat, E. Hostin, E.M. Posadas, and K.W. Leong, *J. Biomater. Sci., Polym. Ed.* **4**(5), 529–543 (1993).
12. J. Heller, R.V. Sparer, and G.M. Zentner, in M. Chasin and R. Langer, eds., *Biodegradable Polymers as Drug Delivery Systems*, Dekker, New York, 1990, pp. 121–162.
13. K.W. Leong et al., in J.I. Kroschwitz, ed., *Encyclopedia of Polymer Science and Engineering*, Wiley, New York, 1989.
14. M. Chasin et al., in M. Chasin and R. Langer, eds. *Biodegradable Polymers as Drug Delivery Systems*, Dekker, New York, 1990, pp. 43–70.
15. T. Kojima, M. Nakano, and K. Juni, *Chem. Pharm. Bull.* **33**, 5119–5125 (1985).
16. S.J. Shieh, M.C. Zimmerman, and J.R. Parsons, *J. Biomed. Mater. Res.* **24**, 789–808 (1990).
17. S. Pulapura, C. Li, and J. Kohn, *Biomaterials* **11**, 666–678 (1990).
18. S. Pulapura and J. Kohn, *Biopolymers* **32**, 411–417 (1992).
19. S.M. Ibim et al., *Pharm. Dev. Technol.* **3**(1), 55–62 (1998).
20. C.T. Laurencin et al., *J. Biomed. Mater. Res.* **30**(2), 133–138 (1996).
21. F. Langone et al., *Biomaterials* **16**(5), 347–353 (1995).
22. H.R. Allcock, in M. Chasin and R. Langer, eds., *Bioerodible Polymers as Drug Delivery Systems*, Dekker, New York, 1990, pp. 163–194.
23. H.R. Allcock, S.R. Pucher, and A.G. Scopelianos, *Biomaterials* **15**(8), 563–569 (1994).
24. C.G. Pitt et al., *J. Appl. Polym. Sci.* **26**, 3779–3787 (1981).
25. M. Richards et al., *J. Biomed. Mater. Res.* **25**, 1151–1167 (1991).

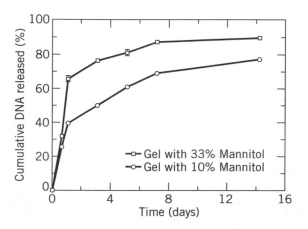

**Figure 23.** In vitro release of p43-LacZ (1 wt % loading) from P(*trans*-CHDM-HOP) in PBS at 37°C. Mannitol was coloaded with DNA at a loading level of 10 wt % (open circle) or 33 wt % (open square).

26. B.I. Dahiyat, M. Richards, and K.W. Leong, *J. Controlled Release* **33**(1), 13–21 (1995).

27. S. Kadiyala, H. Lo, M.S. Ponticiello, and K.W. Leong, in J.O. Hollinger, ed., *Biomedical Application of Synthetic Biodegradable Polymers*, CRS Press, 1995, pp. 33–57.

28. S. Kadiyala, P. Axtel, J.D. Michelson, and K.W. Leong, *Annu. Meet. Soc. Biomater.* Birmingham, Ala., 1993, p. 321.

29. H.W. Coover, R. McConnel, and M. McCall, *Ind. Eng. Chem.* **52**, 409 (1960).

30. M. Sander and E. Steininger, *J. Macromol. Sci., Rev. Macromol. Chem.* **C1**(1), 91 (1967).

31. F. Ignatious, A. Sein, I. Cabasso, and J. Smid, *J. Polym. Sci., Part A: Polym. Chem.* **31**(1), 239–247 (1993).

32. J.C. Brosse, D. Derouet, L. Fontaine, and S. Chairatanathavorn, *Makromol. Chem.* **190**, 2339–2345 (1989).

33. M. Richards et al., *J. Polym. Sci., Part A: Polym. Chem.* **29**, 1157–1165 (1991).

34. A. Tunstall, R.C. Eberhart, and M.D. Prager, *J. Biomed. Mater. Res.* **29**(10), 1193–1199 (1995).

35. M.S. Aronoff, *J. Biomater. Appl.* **9**(3), 205–261 (1995).

36. H.-Q. Mao et al., *Proc. Am. Inst. Chem. Eng. Top. Conf. Biomater., Carriers Drug Delivery Scaffolds Tissue Eng.*, Los Angeles, Nov. 17–19, 1997, p. 141.

37. J. Brandrup and E.H. Immergut, eds., *Polymer Handbook*, 2nd ed., Wiley-Interscience, New York, 1975.

38. M. Chasin and R. Langer, eds., *Biodegradable Polymers as Drug Delivery Systems*, Dekker, New York, 1990.

39. H.-Q. Mao et al., *Pharm. Res.* **14**(11), Suppl., 601 (1997).

40. H. Wang et al., *Proc. Am. Assoc. Pharm. Sci. Annu. Meet.*, San Francisco, Nov. 15–19, 1998, p. 140.

41. I. Shipanova-Kadiyala, H.-Q. Mao, and K.W. Leong, *Proc. Am. Inst. Chem. Eng. Top. Conf. Biomater. Carriers Drug Delivery Scaffolds Tissue Eng.*, Los Angeles, Nov. 17–19, 1997, p. 40.

42. Z. Zhao et al., *Pharm. Res.* **14**(11), Suppl., 293 (1997).

43. W. Dang et al., *Proc. Am. Assoc. Pharm. Sci. Annu. Meet.*, San Francisco, Nov. 15–19, 1998, p. 419.

44. H.-Q. Mao et al., *Proc. Am. Inst. Chem. Eng. Top. Conf. Biomater., Carriers Drug Delivery Scaffolds Tissue Eng.*, Los Angeles, Nov. 17–19, 1997, p. 193.

45. A. Kader et al., *Proc. Am. Assoc. Pharm. Sci. Annu. Meet.*, San Francisco, Nov. 15–19, 1998, p. 413.

46. W. Dang et al., *Pharm. Res.* **14**(11), Suppl., 287–288 (1997).

47. R.G. Crystal, *Nat. Med.* **1**, 15–17 (1995).

48. H.-Q. Mao, B.S. Hendriks, K.Y. Lin, and K.W. Leong, *Proc. Int. Symp. Controlled Release Bioact. Mater.* **25**, 203 (1998).

49. H.-Q. Mao et al., *Proc. Int. Symp. Controlled Release Bioact. Mater.* **26** (1999).

50. A. Kader et al., *Proc. Am. Assoc. Pharm. Sci. Annu. Meet.*, San Francisco, Nov. 15–19, 1998, p. 413.

51. W.M. Saltzman and R. Langer, *Biophys. J.* **55**(1), 163–172 (1989).

See also BIODEGRADABLE POLYMERS: POLYANHYDRIDES;
BIODEGRADABLE POLYMERS: POLYESTERS; POLY(ORTHO ESTERS).

# BIODEGRADABLE POLYMERS: POLYANHYDRIDES

ACHIM GÖPFERICH
University of Regensburg
Regensburg, Germany

## KEY WORDS

Bulk erosion

Degradation

Drug stability

Erosion

Erosion-controlled release

Erosion zone

Modeling

Polyanhydride

Polycondensation

Porosity

Pulsatile drug release

Surface erosion

## OUTLINE

Historical Development and Significance of Polyanhydrides as Biodegradable Polymers

Survey of the Various Types of Polyanhydrides

Monomers

Aliphatic Polyanhydrides

Aromatic Polyanhydrides

Cross-linked and Branched Polyanhydrides

Polyanhydride Synthesis and Characterization

Synthesis

Physicochemical Characterization of Polyanhydrides

Biocompatibility

Polyanhydride Degradation and Erosion

Kinetics of Degradation

Polyanhydride Erosion

Theoretical Description of Polyanhydride Erosion

Drug Delivery Systems Made of Polyanhydrides

Considerations Regarding the Kinetics of Drug Release

Drug Stability

Drug Delivery Systems Made of Polyanhydrides

Drug Release Kinetics

Summary and Outlook

Bibliography

## HISTORICAL DEVELOPMENT AND SIGNIFICANCE OF POLYANHYDRIDES AS BIODEGRADABLE POLYMERS

When polymers are used for the parenteral administration of drugs, the most important advantage is the circumvention of the postapplication removal of the material. Degradation and erosion processes the polymers into oligomers and monomers that can be metabolized and excreted. One of the major goals in the past decades of research on degradable polymers in medicine and pharmacy has been to obtain materials that allow controlled drug release to be controlled by polymer erosion. The development of polyanhydrides and poly(ortho esters) paved the way to understand under which circumstances this goal might be reached. Concomitantly, the tremendous efforts to synthesize new polymers in both classes led to a better understanding of polymer erosion in general and finally to new drug delivery systems on the market.

All degradable polymers consist of monomers that are connected to one another by functional groups that break down during the degradation process. Hydrolysis is the major cause for degradation (1) and can be investigated by following the loss of molecular weight. Degradation induces the subsequent erosion of the material, which is defined as the mass loss of the material. The degradation velocity depends amongst other factors on the type of hydrolyzable functional group that the polymer is built from and determines how a polymer erodes (2). Slowly degrading polymers such as poly($\alpha$-hydroxy esters) have been reported to be bulk eroding, while fast-degrading polymers have been reported to be surface eroding (3) (see "Polyanhydride Erosion"). Polyanhydrides have been reported to be surface eroding (1), which is not surprising since carboxylic acid anhydrides are among the functional groups that hydrolyze the most rapidly. The correlation between erosion and drug release that exists for some drugs allows the use of polyanhydrides for a number of very sophisticated drug delivery applications (4,5). Historically, polyanhydrides were first made for other reasons than obtaining a biomaterial: the manufacture of fibers. The work originated in 1909 with the synthesis of the first polyanhydrides (6). Extensive efforts were made in the 1930s and the 1950s to synthesize aliphatic and aromatic polyanhydrides with enhanced chemical stability (7–9). However, due to their rapid hydrolysis, these turned out to be too unstable to be used as a raw material for the manufacture of fibers. In the 1980s polyanhydrides were finally discovered for the purpose of drug delivery (10). Since then, a number of monomers were identified that can be used to synthesize a variety of biocompatible polyanhydrides. The general structure of polyanhydrides is shown in Figure 1.

## SURVEY OF THE VARIOUS TYPES OF POLYANHYDRIDES

### Monomers

It is obvious that the monomers used for the synthesis of polyanhydrides are bifunctional with at least two carbox-

ylic acid groups per molecule. Figure 2 lists some of some molecules that have been used for the manufacture of polyanhydrides. The IUPAC names of some polymers made of these monomers would have rather long names. Therefore, the abbreviations given in parenthesis in Figure 2 are used in the literature to shorten polymer names for convenience. A copolymer made of 1,3-bis-($p$-carboxyphenoxy)propane (CPP) and sebacic acid (SA), for example, is abbreviated p(CPP-SA). The monomer ratio is given by the figures after the abbreviation. p(CPP-SA) 20:80, for example, contains 20% (w/w) CPP and 80% (w/w) SA. A homopolymer such as poly(fumaric acid) is abbreviated p(FA).

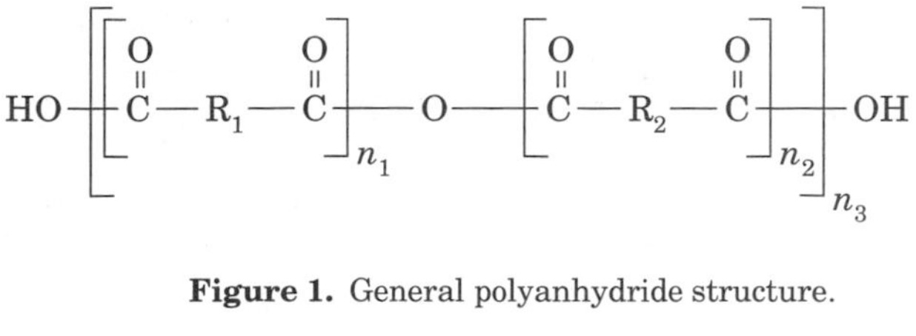

A glance at Figures 1 and 2 illustrates that there is a substantial variability regarding the design of polyanhydrides. They can be manufactured as aliphatic or aromatic homopolymers and copolymers as well as cross-linked or branched polymers. As some of the homopolymers have poor mechanical properties and an undesired stability or instability against degradation, copolymers have been used for most medical and pharmaceutical applications (11).

### Aliphatic Polyanhydrides

A group of polyanhydrides that were synthesized for drug delivery purposes are aliphatic polyanhydrides (12,13). One class of aliphatic polyanhydrides that proved to be useful for drug delivery purposes is p(FAD-SA) (14). Many other aliphatic polyanhydrides, however, have properties that are not of advantage for the manufacture of drug delivery systems. p(SA), for example, is highly crystalline and very brittle. p(FAD), on the other hand, is a liquid and not well suited for the manufacture of solid drug delivery systems. Aliphatic polyanhydrides hydrolyze on average much faster than aromatic ones due to the better accessibility of the bonds to water.

### Aromatic Polyanhydrides

Aromatic polyanhydrides erode slower than aliphatic ones (15), which is due to their increased hydrophobicity and the hindered approach of water to the anhydride bond (16). The erosion rate can be increased by copolymerization with aliphatic monomers. p(CPP-SA) can serve as a good example (Fig. 3). Copolymerization allows the adjustment of erosion rates and, therefore, the duration of drug release in drug delivery applications. Depending on the composition, p(CPP-SA) erodes within weeks or months, whereas p(CPP) homopolymer has been reported to be stable for years (15). Aromatic polyanhydrides that have been under investigation for drug delivery applications are p(CPP) and p(CPP-IPA). In the form of homopolymers, however, some of these polymers cannot be processed at all. p(CPP) for example cannot be melt-processed as it has a high melting point at which it also begins thermal degradation. Furthermore, the solubility in common solvents is very low. This illustrates why tremendous efforts have been undertaken to improve the properties of polyanhydrides by copolymerization.

**Figure 1.** General polyanhydride structure.

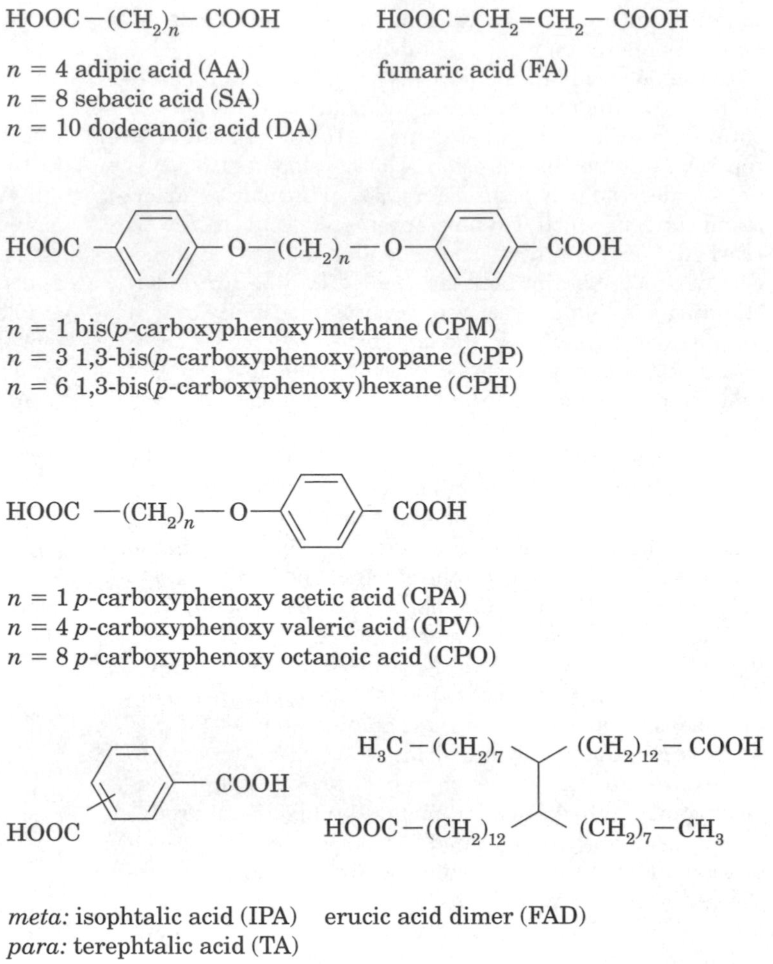

$$\text{HOOC}-(\text{CH}_2)_n-\text{COOH} \qquad\qquad \text{HOOC}-\text{CH}_2=\text{CH}_2-\text{COOH}$$

$n = 4$ adipic acid (AA)          fumaric acid (FA)
$n = 8$ sebacic acid (SA)
$n = 10$ dodecanoic acid (DA)

$$\text{HOOC}-\text{C}_6\text{H}_4-\text{O}-(\text{CH}_2)_n-\text{O}-\text{C}_6\text{H}_4-\text{COOH}$$

$n = 1$ bis(p-carboxyphenoxy)methane (CPM)
$n = 3$ 1,3-bis(p-carboxyphenoxy)propane (CPP)
$n = 6$ 1,3-bis(p-carboxyphenoxy)hexane (CPH)

$$\text{HOOC}-(\text{CH}_2)_n-\text{O}-\text{C}_6\text{H}_4-\text{COOH}$$

$n = 1$ p-carboxyphenoxy acetic acid (CPA)
$n = 4$ p-carboxyphenoxy valeric acid (CPV)
$n = 8$ p-carboxyphenoxy octanoic acid (CPO)

$$\text{HOOC}-\text{C}_6\text{H}_4-\text{COOH}$$

$$\text{H}_3\text{C}-(\text{CH}_2)_7\diagdown(\text{CH}_2)_{12}-\text{COOH}$$
$$\text{HOOC}-(\text{CH}_2)_{12}\diagup(\text{CH}_2)_7-\text{CH}_3$$

**Figure 2.** Monomers that have been used for the manufacture of polyanhydrides.

*meta:* isophtalic acid (IPA)   erucic acid dimer (FAD)
*para:* terephtalic acid (TA)

## Cross-linked and Branched Polyanhydrides

The intention to manufacture cross-linked polyanhydrides was to increase the mechanical stability of polyanhydrides. This can be important for their use as load-bearing biomaterials in orthopedic applications. Cross-linked polyanhydrides can be obtained after introducing double bonds into the polymer backbone. A monomer that has been used for that purpose is FA in combination with SA.

The goal that was pursued with the manufacture of branched polyanhydrides was to improve the mechanical and film-forming properties, especially for the crystalline polyanhydrides. When p(SA) was compared with branched p(SA), an impact on drug release was noticeable but there was little change in physical and mechanical properties (17). By increasing the amount of branching agent benzenetricarboxylic acid from 0 to 2% it was possible to reduce the release of morphine from approximately 70% to approximately 40% within 8 days.

## POLYANHYDRIDE SYNTHESIS AND CHARACTERIZATION

### Synthesis

There are numerous ways to synthesize polyanhydrides from carboxylic acid monomers (18). The most frequently used technique for the manufacture of linear polyanhydrides is melt polycondensation. A common method of initiation of polycondensation is the activation of the carboxylic acids using acetic acid anhydride (19,20). For the manufacture of copolymers all individual monomers are activated separately. The resulting mixed anhydrides, i.e., the so-called prepolymers, are first purified and isolated. They usually consist of a few monomers that are connected to one another via anhydride bonds and form a mixed carboxylic anhydride group with acetic acid at each end of the molecule. For the actual polymerization these prepolymers are heated to 180°C under vacuum. For the synthesis of copolymers two types of prepolymer are mixed prior to the polycondensation. The oligomers polymerize under acetic anhydride formation, which is removed by distillation during the reaction. The advantage of polycondensation is the high molecular weight that can be obtained. A disadvantage of the method is the thermal stress to which the monomers are subjected. An alternative method for the synthesis of polyanhydrides is the use of phosgene or trichloromethyl chloroformate, which is a liquid diphosgene derivative (21). Both reagents activate the carboxylic acid group, which is reacted with nonactivated monomer under anhydride formation. This method of synthesis has the advantage that it allows the synthesis of polyanhydrides under mild conditions. However, the molecular weights that were reported for polyanhydrides synthesized by this method are only approximately 14,000 (21) compared to over 100,000 when prepared by melt polycondensation (20).

The manufacture of cross-linked polyanhydrides requires the manufacture of linear polyanhydrides that contain unsaturated carbon–carbon bonds in their backbone. For reasons of biocompatibility fumaric acid is one of the preferred compounds but also other unsaturated bicarboxylic acids such as such as 1,4'-stilbendicarboxylic acid have been used as model compounds for the synthesis of unsaturated polyanhydrides (22). These unsaturated polyanhydrides are cross-linkable by a radical mechanism. p(FA-SA) 50:50, for example, was cross-linked either in CH$_2$Cl$_2$ or in bulk by adding 2% (w/v) comonomer, 2% (w/v) catalyst, and 0.2% (w/v) accelerator. Styrene and methylmethacrylate were used as comonomers, benzoyl peroxide or 2-butanon peroxide as catalyst and dimethyltoluidine or cobalt naphtanoate as accelerator (22). As the cross-linked products are insoluble in most organic solvents, their use for pharmaceutical applications and especially drug delivery applications is limited. It is, however, possible that they might be used as biomaterials with enhanced mechanical stability for a limited period of time.

Branched polyanhydrides have been synthesized using either multifunctional carboxylic acids such as 1,3,5-benzenetricaboxylic acid or poly(acrylic acid) (17). Branched polymers made of these compounds and sebacic acid have also been synthesized by melt polycondensation. The components were again activated using acetic anhydride, which yields the necessary prepolymers. The molecular weights that were obtained using such multifunctional components in combination with sebacic acid were 250,000 compared to 80,000 for pure p(SA) (17).

## Physicochemical Characterization of Polyanhydrides

The careful characterization of a biodegradable polymer is essential for the successful investigation of its erosion

mechanism. The understanding of the latter is a must for the successful application of these materials in drug delivery applications. For that reason polyanhydrides have been characterized extensively (23).

One of the most important properties is crystallinity. The crystallinity of polyanhydrides was investigated by wide angle X-ray diffraction (WAXD). Some of the homopolymers such as p(SA), p(CPP), and p(FA) were found to be partially crystalline (24); others such as p(FAD) were found to be amorphous (25). Crystallinities as high as 60% have been recorded. The crystallinity of copolymers was shown to depend on the monomer ratio. As shown in Figure 4(a), the lowest degree of crystallinity is reached at a copolymer composition of 1:1 for many polyanhydrides. Polyanhydrides derived from monomers such as FAD or CPH in combination with SA are an exemption. As p(CPH) is almost amorphous the crystallinity of p(CPH-SA) increases only with increasing SA content. The same can be observed for p(FAD-SA) (Fig. 4(b)). When copolymers are made of one crystallizable type of monomer such as SA and one that does not form crystallites such as FAD, its crys-

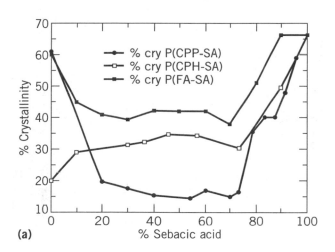

(a)

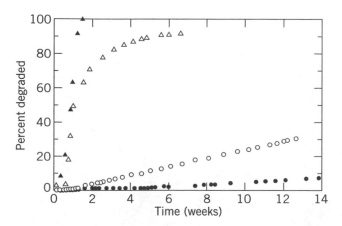

**Figure 3.** Degradation profiles of compression molded poly[bis(*p*-carboxyphenoxy)propane anhydride] and its copolymers with sebacic acid in 0.1 M pH 7.4 phosphate buffer at 37°C. ●, p(CPP); ○, p(CPP-SA) 85:15; △, p(CPP-SA) 45:55; ▲, p(CPP-SA) 21:79. *Source:* Reproduced with permission from Ref. 15.

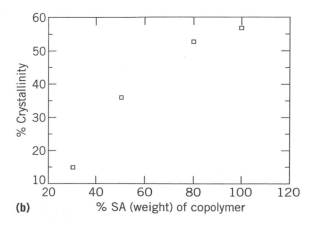

(b)

**Figure 4.** (a) Crystallinity of p(CPP-SA), p(CPH-SA), and p(FA-SA) as a function of the SA content. *Source:* Reproduced with permission from Ref. 24. (b) Crystallinity of p(FAD-SA) as a function of the SA content. *Source:* Reproduced with permission from Ref. 25.

tallinity has also been calculated from its heat of fusion data measured by differential scanning calorimetry (DSC). In addition to the melting enthalpy of the copolymer, the melting enthalpy of the homopolymer made of the crystallizable monomer has to be known (25).

Besides the melting point and the heat of fusion, which are related to the crystalline phase of a polymer, the glass-transition temperature, $T_g$, of the amorphous phase also can be determined by DSC. The melting points were found to be as high as 246°C for p(FA), 240°C for p(CPP), and 143°C for p(CPH). The melting points drop substantially after copolymerization. For all homopolymers made of SA, FA, CPP, and CPH and all copolymers made of SA in combination with FA, CPP, and CPH, $T_g$ values ranged from 2°C to 60°C. Only p(CPP) is the exception with a $T_g$ of 90°C. The lowest values were obtained for the copolymers with equal molar composition (24). More detailed data is shown for some polyanhydrides in Tables 1 and 2.

The distribution of monomers inside copolymer chains was investigated by NMR. The results differentiate between randomly distributed monomers in the polymer backbone and a more block-like structure. For copolymers

### Table 1. Melting Point, Glass-Transition Temperature, and Heat of Fusion

| Polymer | $T_m$ (°C) | $T_g$ (°C) | Heat of fusion (cal/g) |
|---|---|---|---|
| p(SA) | 86.0 | 60.1 | 36.6 |
| p(CPH-SA) 10:90 | 74.4 | 58.3 | 17.8 |
| p(CPH-SA) 18:82 | 66.4 | | 13.1 |
| p(CPH-SA) 27:73 | 57.2 | 45.0 | 9.3 |
| p(CPH-SA) 44:56 | 52.7 | 6.1 | 7.2 |
| p(CPH-SA) 55:45 | 49.5 | 11.5 | 3.2 |
| p(CPH-SA) 64:36 | 43.3 | 11.8 | 2.5 |
| p(CPH-SA) 70:30 | 110.5 | 14.5 | 3.0 |
| p(CPH-SA) 10:90 | 133.1 | 34.8 | 7.3 |
| p(CPH-SA) 80:20 | 136.2 | 26.0 | 18.4 |
| p(CPH-SA) 90:10 | 143.1 | 47.0 | 1.7 |
| p(FA) | 246.2 | 41.2 | 16.0 |
| p(FA-SA) 90:10 | 213.0 | 56.0 | 13.8 |
| p(FA-SA) 80:20 | 185.7 | 58.0 | 22.8 |
| p(FA-SA) 70:30 | 106.0 | 46.0 | 5.7 |
| p(FA-SA) 60:40 | 94.7 | — | 4.7 |
| p(FA-SA) 50:50 | 69.0 | — | 12.5 |
| p(FA-SA) 40:60 | 68.0 | 47.0 | 9.0 |
| p(FA-SA) 30:70 | 67.0 | 46.0 | 7.9 |
| p(FA-SA) 20:80 | 73.9 | — | 21.9 |
| p(FA-SA) 10:90 | 83.0 | 55.0 | 25.8 |
| p(CPP) | 240.0 | 96.0 | 26.5 |
| p(CPP-SA) 4:96 | 76.0 | 41.7 | 24.9 |
| p(CPP-SA) 9:91 | 78.0 | — | 25.7 |
| p(CPP-SA) 13:87 | 75.0 | 47.0 | 20.7 |
| p(CPP-SA) 17:83 | 72.0 | 47.0 | 19.3 |
| p(CPP-SA) 22:78 | 66.0 | 47.0 | 15.3 |
| p(CPP-SA) 27:73 | 66.0 | 44.0 | 10.2 |
| p(CPP-SA) 31:69 | 66.0 | 40.0 | 5.1 |
| p(CPP-SA) 41:59 | 178.0 | 4.2 | 2.0 |
| p(CPP-SA) 46:54 | 185.0 | 1.8 | 3.1 |
| p(CPP-SA) 60:40 | 200.0 | 0.2 | 6.0 |
| p(CPP-SA) 80:20 | 205.0 | 15.0 | 8.2 |

*Source:* From Ref. 24.

### Table 2. Molecular Weight and Viscosity of p(FAD-SA)

| Polymer | $M_r$ | $M_n$ | $T_m$ (°C) | Viscosity ($\eta$) |
|---|---|---|---|---|
| p(SA) | 133,000 | 21,600 | 80–82 | 0.88 |
| p(FAD-SA) 10:90 | 110,000 | 17,500 | 76–78 | 0.90 |
| p(FAD-SA) 20:80 | 92,000 | 22,000 | 72–77 | 0.85 |
| p(FAD-SA) 30:70 | 175,200 | 16,900 | 70–76 | 1.10 |
| p(FAD-SA) 40:60 | 54,044 | 14,300 | 64–68 | 0.42 |
| p(FAD-SA) 50:50 | 235,000 | 25,600 | 62–66 | 1.12 |
| p(FAD-SA) 60:40 | 34,800 | 12,800 | 45–52 | 0.38 |
| p(FAD-SA) 70:30 | 22,500 | 7,100 | 35–42 | 0.35 |
| p(FAD-SA) 80:20 | 21,500 | 7,000 | — | 0.33 |
| p(FAD-SA) 90:10 | 18,910 | 5,900 | — | 0.29 |
| p(FAD) | 37,000 | 12,500 | — | 0.25 |

*Source:* From Ref. 14.

made of SA in a combination with CPP or CPH, it was found that the monomers are mainly randomly distributed when the content of both monomers was equal (26) (Fig. 5). The extent of randomness in the distribution is important with respect to erosion. A block-like arrangement of the monomers inside the polymer chain might lead to the discontinuous erosion of the material when the two blocks exhibit different resistance against degradation and erosion.

### Biocompatibility

The biocompatibility of polyanhydrides was investigated extensively. Early studies assessed the biocompatibility of p(CPP), p(TA), p(CPP-SA), and p(PTA-SA) as well as the toxicity of their monomers (27). The monomers were tested nonmutagenic, were nontoxic, and were found to have a low teratogenic potential in vitro. The polymers did not lead to inflammatory responses after 6 weeks of implantation into rabbit cornea and showed no signs of inflammation after subcutaneous implantation into rats. The encapsulation by fibrous tissue that was observed is not unusual for an implanted biomaterial (28). The biocompatibility of p(CPP-SA) 20:80 was tested subcutaneously

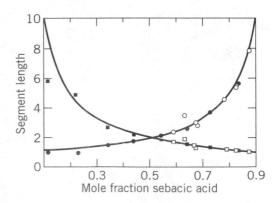

**Figure 5.** Comparison of block length as measured by NMR to that predicted by a random copolymer distribution for p(CPH-SA) and p(CPP-SA). For p(CPH-SA): ●, SA-SA; ■, CPH-CPH. For p(CPP-SA): ○, SA-SA; □, CPP-CPP. *Source:* Reproduced with permission from Ref. 16.

in rats. The polymer showed excellent biocompatibility in doses to rats of 2,400 mg/kg (29). The biocompatibility of p(CPP-SA) 20:80 in the brain was first assessed in rodent models (30,31). There was a slight transient inflammatory response to the polymer but it was comparable to the response provoked by Surgicel, an oxidized regenerated cellulose and established hemostatic agent used routinely in neurosurgery. Similar results were obtained for p(CPP-SA) 50:50 in rabbits. The polymer showed again no signs of toxicity and a tissue reaction comparable to Gelfoam, a resorbable gelatin sponge. The brain compatibility of p(FAD-SA) was also assessed in rats. It showed an acute inflammatory response after 3–6 days comparable to p(CPP-SA) and Surgicel, but was nontoxic as well (32). Finally the brain biocompatibility was verified in a monkey model (33). In the same animal model carmustine-loaded implants made of p(CPP-SA) 20:80 were found to be a safe dosage form.

## POLYANHYDRIDE DEGRADATION AND EROSION

One of the most important characteristics of a biodegradable polymer is its degradation and erosion behavior. Degradation, which is the process of chain cleavage, can be investigated by following the molecular weight change of a substance. Erosion is the sum of all processes leading to the loss of mass from a polymer matrix. It should be kept in mind that degradation is not mandatory for a polymer matrix to erode. If the polymer is at least partially soluble inside the erosion medium, for example, dissolution processes might contribute to erosion as well. Conversely, if the polymer has degraded completely, it does not necessarily erode. In the case of polyanhydrides the polymers are not water-soluble and must, therefore, degrade at least to water-soluble oligomers prior to erosion.

### Kinetics of Degradation

Polymer degradation is usually followed by investigating the loss of molecular weight. This was also done in the case of poly(anhydrides). When p(CPP-SA) 20:80 matrix discs with 8 mm diameter and 1.6 mm height were degraded in phosphate buffer solution pH 7.4 at 37°C it was found that the molecular weight drops exponentially during the first 24 hours (34). Such investigations reveal the time scale on which degradation occurs and, therefore, yields precious information on the expected time over which drugs may be released. However, the result of investigating large matrix discs does not allow the assessment of the degradation properties of a material unequivocally. With increasing dimensions, the result depends on other processes in addition to degradation, such as the diffusion of water into the polymer bulk. If water diffusion is slow, the degradation of the polymer matrix disc is affected tremendously because the lack of water prevents the degradation inside the polymer matrix. From the resulting molecular weight changes one would, however, draw the conclusion that polyanhydrides degrade across their entire cross section for geometries of reasonable size. Other aspects that have to be considered are autocatalytic effects that stem from the free monomers created during degradation. More recently, NMR investigations were performed to monitor the deg-

radation of individual bonds in polyanhydride copolymers (35). Studies on p(CPP-SA) confirmed that bonds in which SA is involved are cleaved faster than bonds between CPP molecules.

The fast degradation of polyanhydrides is their strength and concomitantly their weakness. This is illustrated by experiments with polyanhydrides in solution. Even when dissolved in anhydrous chloroform they have been reported to lose molecular weight. p(CPV) $M_r$ 18,500 and p(CPO) $M_r$ 25,950 lost 50% of their molecular weight within approximately 1.5 hours (36). These examples illustrate that polyanhydrides have to be stored under anhydrous conditions.

### Polyanhydride Erosion

Besides its biocompatibility the erosion of a degradable polymer is perhaps the most crucial property with respect to its performance as a carrier material for drug delivery. When research on degradable polymers for drug delivery intensified, a basic classification was proposed for degradable polymers. Bulk-eroding or homogeneously eroding polymers were distinguished from surface-eroding or heterogeneously eroding ones (37). The difference is illustrated in Figure 6. While a bulk-eroding polymer degrades and erodes over its entire cross section, a surface-eroding polymer erodes mainly from its surface. An essential condition for a water-insoluble polymer to undergo surface erosion is the fast degradation of its polymer backbone (38). It is not surprising that polyanhydrides and poly(ortho esters) are among the few polymer groups that have been reported to be surface eroding, since they are assembled from fast-hydrolyzing functional groups. However, one must bear in mind that surface erosion and bulk erosion are ideal cases. For most polymers, erosion has features of both mechanisms, which is also the case for most polyanhydrides. As a general rule one can assume that surface-eroding polymers erode faster than bulk-eroding ones. The differences in erosion between polyanhydrides and crystalline poly(lactides) might serve as an example:

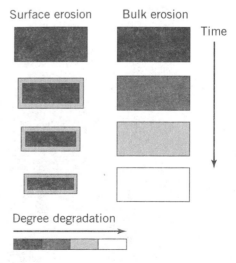

**Figure 6.** Schematic illustration of surface erosion and bulk erosion of polymers.

erosion is a matter of weeks for 2-mm thick p(CPP-SA) 20:80 and p(FAD-SA) 20:80 wafers. For poly(L-lactic acid), which is a partially crystalline bulk-eroding polymer, it is a matter of several months. However, some polyanhydrides such as those containing aromatic monomers such as p(CPP) have been reported to be extremely erosion-resistant as well.

The erosion of most of the clinically relevant polyanhydrides has been investigated extensively and revealed useful information for the manufacture of drug delivery systems. Originally erosion was followed by the determination of monomer release from the polymers (23) and the mass loss of polymer matrices during erosion (39). Figures 7 and 8 illustrate that such data can be very confusing. Although the mass loss of p(CPP-SA) matrix discs is almost linear (Fig. 7) as one would expect from a surface-eroding polymer, the release profiles of the monomers do not follow the same kinetics (Fig. 8). The two monomers CPP and SA are not only released in a nonlinear way, but also with marked differences regarding their overall kinetics. To solve this paradox, one must take into account other factors with an impact on erosion besides degradation. The microstructure of a polymer is one of them. The microstructure of a polymer and its changes due to erosion have to be assessed using physicochemical techniques. This information is essential to understand how polymers erode. For p(CPP-SA) the crystallinity changes were investigated using DSC, WAXD, scanning electron microscopy (SEM) (39,40), and solid-state NMR (D.L. McCann, F. Heatley, and A. D'Emanuelle, personal communication). It was found that these polymers do not erode according to a perfect surface erosion mechanism. The amorphous polymer parts were found to erode substantially faster than the crystalline ones (19, 39, D.L. McCann, F. Heatley, and A. D'Emanuelle, personal communication). As a consequence erosion zones form in which the amorphous polymer disappears first and is replaced by a network of pores that stretch through the crystalline areas of noneroded polymer. The foremost line of eroded polymer, the erosion

front, moves from the surface of the polymer matrix into its center (39). Figure 9(a) illustrates schematically the changes that p(CPP-SA) polymers undergo during erosion. This erosion mechanism, which creates porous erosion zones, has a marked impact on the release of substances. A good example are the two monomers that are released during erosion.

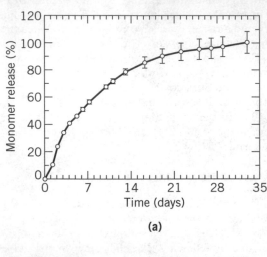

(a)

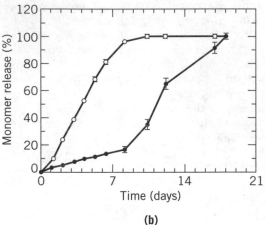

(b)

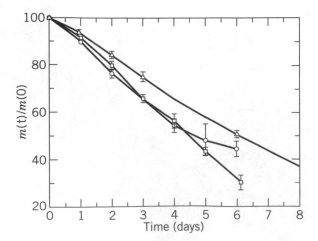

**Figure 7.** Mass loss of polyanhydride matrix discs (diameter 7 mm, height approximately 2 mm) during erosion ○, p(SA); □, p(CPP-SA) 20:80; △, p(CPP-SA) 50:50. *Source:* Reproduced with permission from Ref. 39.

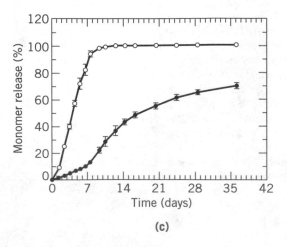

(c)

**Figure 8.** SA (○) and CPP (●) release polyanhydride matrix discs during erosion: (**a**) p(SA), (**b**) p(CPP-SA) 20:80, (**c**) p(CPP-SA) 50:50. *Source:* Reproduced with permission from Ref. 39.

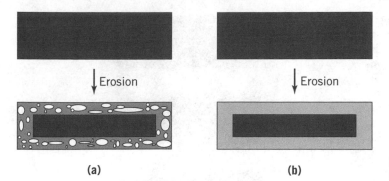

(a)                                      (b)

**Figure 9.** Schematic illustration of (a) p(CPP-SA) erosion and (b) p(FAD-SA) erosion.

Other investigations of the erosion mechanism of polyanhydrides focused on the characterization of the erosion zones and the chemical conditions that prevail within them. Investigations by confocal fluorescence microscopy using pH-sensitive fluorescent probes revealed that the pH on the surface of p(CPP-SA) 20:80 matrices is one unit lower than in the surrounding pH 7.4 buffer (39). From these results in combination with findings by DSC and WAXD, which indicate that both monomers crystallize inside the erosion zone, it was concluded that the pH inside the erosion zone is about 5 (39). These assumptions were confirmed more recently by spectral spatial electron paramagnetic resonance imaging using pH-sensitive spin probes (41). The effect that this pH microclimate has on the release of substances can again be seen from the release profiles of the monomers. The profile for SA is rapid concave downward as one would expect for a monolithic device. The profile of CPP, in contrast, exhibits a slow leakage before its rapid release. This can be explained with the solubility of the two compounds. In the pH range of interest, the solubility of SA was found to be 10 times higher than that of CPP (39). Given that both monomers are created at the same velocity by degradation, and that they have similar pKa values one can assume that SA determines the pH and keeps the solubility of CPP even lower while it is created in the erosion zone. Once SA is released completely, the solubility of CPP increases and so does its release velocity. These results confirm that even though the mass loss kinetics appear to be simple, the individual processes of erosion can become quite complicated.

That the erosion kinetics are an individual characteristic of a polymer group can be seen when comparing these results obtained for p(CPP-SA) with the erosion mechanism of p(FAD-SA) (25) (Fig. 9(b)). Although p(FAD-SA) polymers are also partially crystalline, their erosion mechanism is different from that of p(CPP-SA) (25,42). One major reason is the nature of FAD, a water-insoluble liquid. During the erosion of p(FAD-SA), FAD precipitates on the surface of polymer matrix discs while SA is released (25). The erosion zones are a semisolid layer of FAD and its salt and are less well defined as in the case of p(CPP-SA).

The erosion zones that are created during polyanhydride erosion may have some effect on the release of low molecular weight substances besides their impact on monomer release. p(CPP-SA) and p(FAD-SA) can serve again as a good example. Comparing the release of SA from both polymers the release is slightly faster from p(CPP-SA)

compared to p(FAD-SA). Most likely the different nature of the erosion zones account for this effect. Whereas in one case SA can diffuse through a network of pores, it has to pass through an amorphous lipid layer in the other. The impact of the erosion zones on drug or monomer release from p(CPP-SA) has also been illustrated by applying diffusion theory (43). Under the assumption that a saturated solution of SA, which is in equilibrium with suspended SA, exists at the erosion front, a diffusion model simulates the release of SA through the porous and tortuous erosion zone. Assuming further that SA controls the solubility of CPP, the sigmoid release profile for CPP was confirmed by this modeling approach. The poor solubility of the monomers is also reflected by the fast release of drugs from p(CPP-SA) matrix discs. Indometacin, for example, was found to be released faster than SA (44).

### Theoretical Description of Polyanhydride Erosion

There have been a number of approaches to describe the erosion of biodegradable polymers in the past. For bulk-eroding polymers this is very complicated, because their erosion mechanism is still not completely understood (45). For surface-eroding polymers early approaches assume that erosion fronts move at linear speed leaving no erosion zones behind (46,47). Such assumptions work well if erosion zones have no impact on the erosion kinetics, which is, however, not the case for most polyanhydrides. In order to take into account the effect of such zones on erosion, two-dimensional models were developed (48) similar to the ones used previously for modeling erosion-controlled drug release (49). For the simulation of erosion the matrices are first represented using two-dimensional rectangular grids such as the one shown schematically in Figure 10(a). The

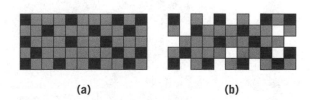

(a)                           (b)

**Figure 10.** Theoretical representation of a cylindric polymer matrix cross-section by a two-dimensional grid (black pixels represent amorphous polymer; white pixels, amorphous polymer): (a) prior to erosion, (b) during erosion.

grid covers the cross-section through a cylindrical poly-anhydride matrix disc and consists of a multitude of small polymer parts (pixels). The partial crystallinity of the polymer is taken into account by assigning some of the pixels the quality of being crystalline and others the quality of being amorphous. In the easiest case this can be done at random by simply taking the overall crystallinity of the material into account. To simulate erosion it is then necessary to define an erosion algorithm. In the case of p(CPP-SA) this required two basic assumptions:

1.  A polymer pixel can only erode after it has contacted the erosion medium (at time 0 these are only the pixels located on the "surface" of the grid). After this contact is established, the lifetime of a pixel is calculated.
2.  The lifetimes of all polymer pixels are distributed according to a first order Erlang distribution (equation 1). Lifetimes are determined at random from this distribution and once the lifetime expired the pixels are taken from the grid and are considered eroded (Fig. 10(b)). Amorphous polymer pixels have higher chances to erode at early times than crystalline ones. This is accounted for by a higher erosion rate constant, $\lambda$.

$$e(t) = \lambda \cdot e^{-\lambda \cdot t} \qquad (1)$$

where e(t) is the probability that a pixel has the lifetime t. When this algorithm is applied to a large grid, simulations such as the one shown in Figure 11 are obtained. The appearance of these cross-sections is typical for eroding p(CPP-SA) and matches exactly the appearance of cross-sections of this material when observed by SEM (39). There is an erosion front that moves from the surface towards the center of the matrix and separates a non-eroded polymer core from a highly porous erosion zone. The erosion zone consists mainly of crystalline pixels, which is in agreement with experimental findings. Such models have two advantages. First they provide a better understanding of how polymer surface erosion can be envisioned to proceed, and second, they can be used to predict certain parameters such as the porosity of the polymers during erosion and under certain conditions of drug release.

## DRUG DELIVERY SYSTEMS MADE OF POLYANHYDRIDES

### Considerations Regarding the Kinetics of Drug Release

The fast erosion of polyanhydrides makes them attractive candidates for the purpose of drug delivery. There are three major mechanisms by which drugs can be released from polymers. Release can be controlled by diffusion, swelling, or erosion. In the case of a biodegradable polymer the three mechanisms compete against one another,

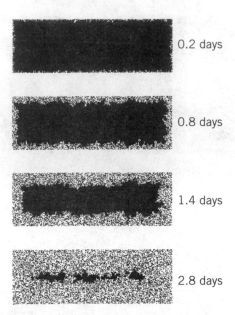

0.2 days

0.8 days

1.4 days

2.8 days

**Figure 11.** Simulation of the erosion of a p(CPP-SA) 20:80 matrix using discrete two-dimensional models. (Black pixels represent noneroded polymer; white pixels, eroded polymer.) *Source:* Reproduced with permission from Ref. 50.

and it will be the slowest of these processes that will control the release of drugs. If the three processes proceed at similar speed, drug release will be controlled by all three simultaneously. In order to have optimal control over drug release from degradable polymers, it is desirable that it be mainly erosion controlled. The only way that this can be achieved is by using fast-eroding polymers, which is the case for polyanhydrides. In some cases polyanhydrides have been reported to release drugs by perfect erosion control (15,51). An example is shown in Figure 12.

### Drug Stability

The fast erosion of polyanhydrides can be attributed to the fast hydrolysis of the carboxylic anhydride bonds in the polymer chain. Water is, however, not the only nucleophile that might react with carboxylic anhydrides. Primary amines might react as well under irreversible amide formation. Although such reactions pose a potential threat to the stability of incorporated drugs, systematic investigations have shown that amide formation is strongly pH dependent (52). Whereas the interaction was reported to be strong at pH 7.4, it was most effectively suppressed at pH 5. From these results one can expect that the stability of drug-carrying amines could, on the other hand, be minimized, as the pH inside the erosion zones of polyanhydrides was found to be below 5 (39,41). The impact of polyanhydrides on the stability may not be too severe as evidenced by studies of encapsulated enzymes. It was found that polyanhydrides can even protect such systems from inactivation (53).

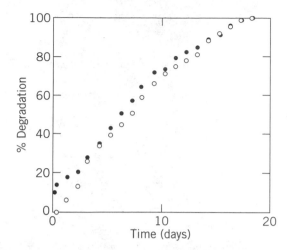

**Figure 12.** Release of $p$-nitroaniline (10% loading) from injection-molded poly[bis($p$-carboxyphenoxy)propane anhydride] in 0.1 M pH 7.4 phosphate buffer at 37°C. (○) polymer degradation; (●) drug release. *Source:* Reproduced with permission from Ref. 51.

## Drug Delivery Systems Made of Polyanhydrides

Polyanhydrides have been used for the manufacture of a number of challenging drug delivery applications involving the manufacture of new drug delivery systems (54) as well as new types of therapy (4,55). It is beyond the scope of this chapter to give a complete survey on all drug delivery applications for which polyanhydrides have been used. It is instead the intention to give some examples that illustrate how one can take advantage of the specific properties of polyanhydrides. More information on polyanhydride drug delivery systems can be found in chapters on CENTRAL NERVOUS SYSTEM, DRUG DELIVERY TO TREAT and MICROENCAPSULATION.

Although many of the applications already mentioned require the use of polyanhydrides for their quick erosion, which coincides with the goal of short-term drug delivery, some applications depend on the exact performance of these polymers. An example is the development of pulsatile drug delivery systems. The simulation shown in Figure 10 illustrates that polyanhydrides are surface eroding. If a polyanhydride matrix disc was made of several layers of polymer carrying different doses of the same drug, the drugs should be released one after another. This principle could be used for a number of drug delivery applications such as vaccination or local tumor therapy. Early investigations showed that such release behavior can indeed be achieved. In the simplest case, a composite matrix, such as the one shown in Figure 13(a), could be used to release drugs one after another. The drug release profile for two model compounds is shown in Figure 14(a). The drug release behavior can completely be understood on the basis of the erosion mechanism of polyanhydrides. When the erosion front moves into the core of the polymer matrix discs, the drug in the areas that are located inside the erosion zone are liberated by the process. As the erosion profiles move parallel to the matrix surface, first the drug from the perimeter is liberated and then from the core of the composite. Figure 14(a) also shows, however, that the possi-

(a)                                    (b)

**Figure 13.** (**a**) Cross-section through a composite polyanhydride matrix cylinder, (**b**) Cross-section through a complex composite matrix cylinder made of polyanhydrides and poly(lactic acid). ■, drug-carrying polyanhydride perimeter; ▨, drug-carrying polyanhydride core; □, drug-free polyanhydride layer; ▨, drug-free bulk-eroding polymer (e.g., PLA).

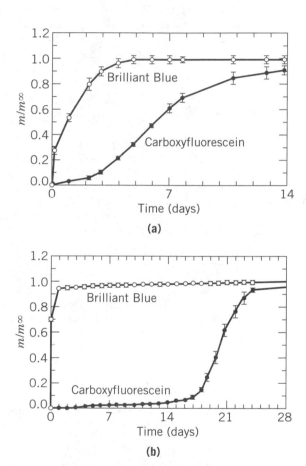

(a)

(b)

**Figure 14.** (○) Brilliant blue (perimeter) and (●) carboxyfluorescein (core) release from composite polymer matrix discs. (**a**) composite polyanhydride cylinder (see Fig. 13a). (**b**) composite polyanhydride cylinder in combination with poly(lactic acid) (see Fig. 13b). *Source:* Reproduced with permission from Ref. 50.

bilities to postpone the release of the second dose are limited. The lag period between the start of erosion and the release of the second dose depend on the geometry of the device. To postpone the release of the second drug for 10 days, the perimeter would have to be unreasonably thick. Therefore, the combination of polyanhydrides with slow-eroding polymers such as poly(lactic acid) was shown to be a useful alternative as illustrated in Figure 13(b). When drugs are released from such a composite device, release profiles such as the one shown in Figure 14(b) can be obtained (50).

## DRUG RELEASE KINETICS

The release kinetics depend on several factors. One of them is certainly the nature of the polyanhydride that is used as a drug carrier. The slower the erosion of the material, the slower the release of drug. Another aspect is the nature of the incorporated drug. Lipophilic drugs tend to be released more slowly from the polymers than hydrophilic ones (56). As many drugs possess some acid or base functionality the microclimate inside the eroding polymer can affect the property of a drug tremendously. The solubility of carboxylic acids, for example, will be decreased during the erosion of polyanhydrides, because the pH values inside these polymers are low.

## SUMMARY AND OUTLOOK

Polyanhydrides are a very versatile and interesting group of polymer. The most important characteristic is their fast degradation, which is followed by the rapid erosion of the material. Polyanhydrides are ideal candidates for drug delivery applications that last from days to weeks. The clean-cut erosion mechanism makes polyanhydrides a family of polymer the erosion behavior of which can readily be understood. Therefore, polyanhydrides have been used for a number of drug delivery applications: the encapsulation of proteins and peptides, the treatment of osteomyelitis, and the treatment of brain tumors. They have been shown to be effective and safe, and one can assume that polyanhydrides will have their place in the family of biodegradable polymers for parenteral applications in the future.

## BIBLIOGRAPHY

1. K. Park, W.S.W. Shalaby, and H. Park, *Biodegradable Hydrogels for Drug Delivery*, Technomic Publishing, Lancaster, Pa., 1993.
2. A. Göpferich, *Biomaterials* 17, 103–114 (1996).
3. A. Göpferich and R. Langer, *AIChE J.* 41, 2292–2299 (1995).
4. C.T. Laurencin et al., *J. Orthop. Res.* 11, 252–262 (1993).
5. H. Brem, K.A. Walter, and R. Langer, *Eur. J. Pharm. Biopharm.* 39, 2–7 (1993).
6. J.E. Bucher and W.C. Slade, *J. Am. Chem. Soc.* 31, 1319–1321 (1909).
7. J.W. Hill, *J. Am. Chem. Soc.* 52, 4110–4115 (1930).
8. J.W. Hill and W.H. Carothers, *J. Am. Chem. Soc.* 54, 1569–1579 (1932).
9. A. Conix, *J. Polym. Sci.* 29, 343–353 (1958).
10. H.B. Rosen et al., *Biomaterials* 4, 131–133 (1983).
11. A. Domb, E. Ron, and R. Langer, in H.F. Mark et al., eds., *Encyclopedia of Polymer Science and Engineering*, 2nd ed., suppl. vol., Wiley, New York, 1989, pp. 648–665.
12. A.J. Domb and R. Nudelman, *Biomaterials* 16 319–323 (1995).
13. M. Davies et al., *J. Appl. Polym. Sci.* 42, 1597–1605 (1991).
14. A.J. Domb and M. Maniar, *J. Polym. Sci., Part A* 31, 1275–1285 (1993).
15. K.W. Leong, B.C. Brott, and R. Langer, *J. Biomed. Mater. Res.* 19, 941–955 (1985).
16. J. Tamada and R. Langer, *J. Biomater. Sci., Polym. Ed.* 3, 315–353 (1992).
17. M. Maniar, X. Xie, and A.J. Domb, *Biomaterials* 11, 690–694 (1990).
18. K.W. Leong, V. Simonte, and R. Langer, *Macromolecules* 20, 705–712 (1987).
19. A.-C. Albertsson and S. Lundmark, *Br. Polym. J.* 23, 205–212 (1990).
20. A.J. Domb and R. Langer, *J. Polym. Sci., Part A* 25, 3373–3386 (1987).
21. A.J. Domb, E. Ron, and R. Langer, *Macromolecules* 21, 1925–1929 (1988).
22. A.J. Domb et al., *J. Polym. Sci., Part A* 29, 571–579 (1991).
23. J. Tamada and R. Langer, *Proc. Natl. Acad. Sci. U.S.A.* 90, 552–556 (1993).
24. E. Mathiowitz et al., *Macromolecules* 23, 3212–3218 (1990).
25. L. Shieh et al., *J. Biomed. Mater. Res.* 28, 1465–1475 (1994).
26. E. Ron et al., *Macromolecules* 24, 2278–2282 (1991).
27. K.W. Leong, P.D'Amore, M. Marletta, and R. Langer, *J. Biomed. Mater. Res.* 20, 51–64 (1986).
28. D. Bakker, C.A. van-Blitterswijk, S.C. Hesseling, and J.J. Grote, *Biomaterials* 9, 14–23 (1988).
29. C. Laurencin et al., *J. Biomed. Mater. Res.* 24, 1463–1481 (1990).
30. R. Tamargo et al., *J. Biomed. Mater. Res.* 23, 253–266 (1989).
31. H. Brem et al., *Select. Cancer Ther.* 5, 55–65 (1989).
32. H. Brem et al., *J. Controlled Release* 19, 325–330 (1992).
33. H. Brem, *Biomaterials* 11, 699–701 (1990).
34. A. D'Emanuelle et al., *Pharm. Res.* 9, 1279–1283 (1992).
35. F. Heatley, M. Humadi, R.V. Law, and A. D'Emanuele, *Macromolecules* 31, 3832–3838 (1998).
36. A.J. Domb and R. Langer, *Makromol. Chem. Macromol. Symp.* 19, 189–200 (1988).
37. R. Langer and N. Peppas, *J. Macromol. Sci., Rev. Macromol. Chem. Phys.* C23, 61–126 (1983).
38. A. Göpferich, *Eur. J. Pharm. Biopharm.* 42, 1–11 (1996).
39. A. Göpferich and R. Langer, *J. Polym. Sci., Part A:* 31, 2445–2458 (1993).
40. E. Mathiowitz, J. Jacob, K. Pekarek, and D. Chickering, III, *Macromolecules* 26, 6756–6765 (1993).
41. K. Mäder et al., *Polymer* 38, 4785–4794 (1997).
42. A. Göpferich, L. Schedl, and R. Langer, *Polymer* 37, 3861–3869 (1996).
43. A. Göpferich and R. Langer, *J. Controlled Release* 33, 55–69 (1995).
44. A. Göpferich, D. Karydas, and R. Langer, *Eur. J. Pharm. Biopharm.* 41, 81–87 (1995).
45. A. Göpferich, *Macromolecules* 30, 2598–2604 (1997).
46. D.O. Cooney, *AIChE J.* 18, 446–449 (1972).
47. H.B. Hopfenberg, *ACS Symp. Ser.* 33, 26–32 (1976).
48. K. Zygourakis, *Chem. Eng. Sci.* 45, 2359–2366 (1990).
49. A. Göpferich and R. Langer, *Macromolecules* 26, 4105–4112 (1993).

50. A. Göpferich, *J. Controlled Release* **44**, 271–281 (1997).
51. A.J. Domb, C.F. Gallardo, and R. Langer, *Macromolecules* **22**, 3200–3204 (1989).
52. A.J. Domb, L. Turovsky, and R. Nudelman, *Pharm. Res.* **11**, 865–868 (1994).
53. Y. Tabata, S. Gutta, and R. Langer, *Pharm. Res.* **10**, 487–496 (1993).
54. K.J. Pekarek, J.S. Jacob, and E. Mathiowitz, *Nature (London)* **367**, 258–260 (1994).
55. H. Brem et al., *Lancet* **345**, 1008–1012 (1995).
56. L. Shieh et al., *J. Controlled Release* **29**, 73–82 (1994).

See also BIODEGRADABLE POLYMERS: POLY(PHOSPHOESTER)S; BIODEGRADABLE POLYMERS: POLYESTERS; POLY(ORTHO ESTERS).

# BIODEGRADABLE POLYMERS: POLYESTERS

SUMING LI
MICHEL VERT
Centre de Recherche sur les Biopolymères Artificiels
Montpellier, France

## KEY WORDS

Aliphatic polymer

Biodegradation

Biomaterial

Crystallinity

Degradation

Drug delivery

Hydrolysis

Lactic acid polymer

Lactic acid–glycolic acid copolymer

Poly($\epsilon$-caprolactone)

Polyglycolide

Polylactide

Stereocopolymer

## OUTLINE

## INTRODUCTION

During the past 50 years, synthetic polymers have changed the everyday life of humans due to the possibility of covering a wide range of properties by modification of macromolecular structures and introduction of additives (fillers, plasticizers, etc.). In the meantime, surgeons and pharmacists tried to use these materials as biomaterials (1,2). About 30 years ago, distinction was made between permanent and temporary therapeutic uses. The former requires biostable polymeric materials, and the main problem is resistance to degradation in the body. In contrast, the latter needs a material only for a limited healing time. In this regard, degradable polymers became of great interest in surgery as well as in pharmacology. The first degradable synthetic polymer was poly(glycolic acid) (PGA), which appeared in 1954 (3). This polymer was first discarded because of its poor thermal and hydrolytic stabilities, which precluded any permanent application. Later on, people realized that one could take advantage of the hydrolytic sensitivity of PGA to make polymeric devices that can degrade in a humid environment and, thus, in a human body. This led to the first bioabsorbable suture material made of a synthetic polymer (4,5). It is worth noting that terminology is one of the sources of confusion in the field. Nowadays, people tend to use the word *degradable* as a general term and reserve *biodegradable* for polymers that are biogically degraded by enzymes introduced in vitro or generated by surrounding living cells. The possibility for a polymer to degrade and to have its degradation by-products assimilated or excreted by a living system is thus designated as *bioresorbable* (6).

Most of the degradable and biodegradable polymers identified during the past 20 years have hydrolyzable linkages, namely ester, orthoester, anhydride, carbonate, amide, urea, and urethane in their backbone (7,8). The ester bond–containing aliphatic polyesters are the most attractive because of their outstanding biocompatibility and versatility regarding physical, chemical, and biological properties (9–11). The main members of the aliphatic polyester family are listed in Table 1. Only a few have reached the stage of clinical experimentations as bioresorbable devices in drug delivery. This is primarily due to the fact that being degradable or biodegradable is not sufficient. Many other prerequisites must be fulfilled for clinical use and commercialization (Table 2) (10).

The delivery of drugs to a human body can be achieved through oral, transdermal, topical, and parenteral admin-

**Table 1. Main Members of the Aliphatic Polyester Family**

| Polymer | Structure |
|---|---|
| Poly(glycolic acid) | $-(-O-CO-CH_2-)_n-$ |
| Poly(lactic acid) | $-(-O-CO-CH-)_n-$<br>$\vert$<br>$CH_3$ |
| Poly($\epsilon$-caprolactone) | $-(-O-CO-(CH_2)_5-)_n-$ |
| Poly($para$-dioxanone) | $-(-O-CO-(CH_2)_2-O-CH_2)_n-$ |
| Poly(hydroxybutyrate) | $-(-O-CO-CH-CH_2-)_n-$<br>$\vert$<br>$CH_3$ |
| Poly($\beta$-malic acid) | $-(-O-CO-CH-CH_2)_n-$<br>$\vert$<br>$COOH$ |

**Table 2. Prerequisites for Biomedical Applications of Degradable Polymers**

| Prerequisite | Polymer properties |
|---|---|
| Biocompatibility | Leachables (e.g., residual oligomers and monomers, degradation products) |
| | Shape |
| | Surface |
| Biofunctionality | Physical |
| | Mechanical |
| | Biological |
| Stability | Processing |
| | Sterilization |
| | Storage |
| Bioresorbability | Degradability |
| | Resorption of degradation products |

istrations (12). A great deal of work has been done during the past two decades to develop degradable, controlled drug delivery systems adapted to these various routes. Large-size implants require surgery, whereas needle-like implants can be injected subcutaneously (s.c.) or intramuscularly (i.m.) using a trochar. Microparticles can also be injected s.c. and i.m. and behave as tiny implants. Intravenous (i.v.) injection is possible with microparticles, the size of which must be below 7 $\mu$m to avoid lung capillary embolization. However, microparticles can be taken up very rapidly by the macrophages of the reticuloendothelial system to finally end up in Kupffer cells in the liver. Although nanoparticles have been proposed to overcome the size limitation imposed by capillary beds; they can also be taken up by macrophages. Stealth nanoparticles with a surface covered by a brush of poly(ethylene oxide) (PEO) have been proposed to avoid macrophage uptake (13). In any event, nanoparticles, as well as microparticles, can hardly leave the vascular compartment. Recently, colloidal particles have been considered in the form of macromolecular micelles of amphiphilic biblock copolymers (14,15) or of aggregates of hydrophilic polymers bearing hydrophobic side chains (16). These systems can serve as a drug carrier via physical entrapment of a lipophilic drug within the hydrophobic microdomains formed by the core of micelles or aggregates. Macromolecular prodrugs, where a drug mol-

ecule is temporarily attached to a polymeric carrier, have also been proposed (17). Last but not least, the next century might see the development of polymeric drugs, because any synthetic polymer can advantageously interact with elements of living systems, such as molecules, cell membranes, viruses, and tissues (18).

So far, only drug delivery devices based on polymers and copolymers deriving from lactic acid (LA) enantiomers, glycolic acid, and $\epsilon$-caprolactone (abbreviated as PLA, PGA, and PCL, respectively) have been commercialized. The prospective applications include devices to treat cancer, drug addiction, and infection, as well as drugs for contraception, vaccination, and tissue regeneration. A number of products are commercially available such as Decapeptyl®, Lupron Depot®, Zoladex®, Adriamycin®, and Capronor® (12).

High molar mass PLA, PGA, PCL, and their copolymers are obtained by ring-opening polymerization of cyclic esters, i.e., lactides, glycolide and $\epsilon$-caprolactone, respectively (19,20). The direct polycondensation of corresponding hydroxy acids leads to low molar mass oligomers only (21–23). In the case of LA-containing polymer chains, chirality of LA units provides a worthwhile means to adjust bioresorption rates as well as physical and mechanical characteristics (24–26). The use of the PGA homopolymer is limited to suture material because of its high crystallinity and of the absence of practical solvent. In contrast, PLAGA copolymers have been largely used to make implants, microparticles, and nanoparticles.

For the sake of simplicity, PLA stereocopolymers and PLAGA copolymers are identified in this paper by the acronyms $PLA_X$ or $PLA_XGA_Y$, where X is the percentage of L-LA units present in the monomer feed, Y is that of glycolic acid (GA) units, the rest being the percentage of D-LA units (Table 3). Similarly, copolymers of lactides and $\epsilon$-caprolactone (CL) can be denoted as $PLA_XCL_Y$ where X and Y represent the percentages of L-LA and CL units, respectively. This nomenclature presents the advantage of reflecting clearly the chemical and configurational compositions of the polymers, the average polymer chain composition being generally close to that of the feed (27). Nevertheless, it is far from being standardized. In fact, many other acronyms are used in the literature.

The properties of degradable polyesters depend on many factors, including those related to polymer synthesis and device processing (11). Therefore, the first part of this chapter recalls the various routes available to synthesize polymers and to fabricate drug delivery devices. In the second part, the various mechanisms by which an aliphatic polyester can degrade in a living environment are discussed. Thereafter, the various factors that can affect the degradation and release characteristics of the devices are examined. The discussion is largely based on recent advances in the field. Convergences and discrepancies are outlined when it is reasonably possible.

## SYNTHESIS

From the practical aspect of drug delivery, users rarely prepare polymers themselves. This is particularly true nowadays because various PLA, PLAGA, and PCL compounds are marketed as raw materials.

**Table 3. Homo- and Copolymers Deriving From Lactides and Glycolide**

| Polymer and acronym | Structure |
|---|---|
| | *Poly(glycolic acid)* |
| PGA | -(-O-CO-CH$_2$-)$_n$- |
| | *Poly(L-lactic acid)* |
| PLA$_{100}$ | $\begin{array}{c} H \\ \vert \\ \text{-(-O-CO-C-)}_n\text{-} \\ \vert \\ CH_3 \end{array}$ |
| | L-LA/D-LA stereocopolymers |
| PLA$_X$ {X = 100$n$/($n + p$)} | $\begin{array}{cc} H & CH_3 \\ \vert & \vert \\ \text{-(-O-CO-C-}\vert_n\text{-O-CO-C-)}_p\text{-} \\ \vert & \vert \\ CH_3 & H \end{array}$ |
| | L-LA/D-LA/GA terpolymers |
| PLA$_X$GA$_Y$ {X = 100$n$/($n + p + q$)} {Y = 100$q$/($n + p + q$)} | $\begin{array}{ccc} H & CH_3 & \\ \vert & \vert & \\ \text{-(-O-CO-C-}\vert_n\text{-O-CO-C-}\vert_p\text{-O-CO-CH}_2\text{-)}_q\text{-} \\ \vert & \vert & \\ CH_3 & H & \end{array}$ |

There are two main routes to synthesize these aliphatic polyesters: polycondensation of bifunctional hydroxy acids and ring opening polymerization of cyclic ester monomers.

Lactic acids, glycolic acid, or hydroxycaproic acid can be condensed at low pressure and high temperature, according to a step-growth mechanism. This route generally leads to low molar mass chains ($M_r < 5,000$ daltons) terminated by equimolar amounts of OH and COOH endgroups.

PLA compounds with rather high molar masses issued from step-growth polycondensation are mentioned in literature (28,29). Hiltunen et al. reported obtaining PLA with $\bar{M}_n$ (NMR) up to 30,000 by polycondensation of L-lactic acid in the presence of different catalysts at high temperatures (180–220°C). The best catalyst was found to be sulfuric acid, which yielded the highest molar masses (28). Compounds obtained by the postcondensation of PLA oligomers using a coupling reagent such as dicyclohexyl carbodiimide (DCC) were also reported (29).

The main route to high molar mass PLA, PLAGA, and PCL is the ring opening polymerization of heterocyclic monomers, namely lactide, glycolide, or ε-caprolactone (19,20).

**Monomer Synthesis**

In practice, the synthesis of lactides and glycolide consists of two steps as shown in Figure 1. Lactic acid or glycolic acid is first polycondensed to yield low molar mass oligomers. Then, the oligomers are thermally depolymerized to form the corresponding cyclic diester, which is recovered by distillation at low pressure (30). Catalysts such as zinc

metal or zinc oxide are used in the second step to improve the yield (31).

Glycolide is an achiral molecule, whereas the lactide cycle bears two asymmetric carbon atoms. Therefore, there exist three diastereoisomeric forms of lactides, namely L-lactide (L,L cyclic dimer), D-lactide (D,D cyclic dimer) and meso-lactide (D,L cyclic dimer) as shown in Figure 2.

The equimolar mixture of L- and D-lactides, namely racemic- or DL-lactide, is also commercially available and largely used. Both lactides and glycolide need to be purified prior to polymerization if one wants to obtain well-defined, high molar mass polymers. Purification is generally achieved by sublimation or by recrystallization from acetone or ethyl acetate solutions.

ε-Caprolactone is generally manufactured by oxidation of cyclohexanone with peracetic acid in an efficient continuous process (32), as shown in Figure 3. It needs to be dried by distillation in the presence of either diphenylmethane 4,4'-diisocyanate (33), calcium hydride (34–36), or sodium metal (36) prior to polymerization.

**Polymer Synthesis**

A great deal of work has been done to investigate the ring opening polymerization of lactides and lactones. Carothers et al. were the first to polymerize lactides by heating at high temperature (250–270°C) (37). Only low molar mass compounds were obtained. In 1954, high molar mass PGA was obtained from the polymerization of glycolide in the presence of zinc chloride as initiator. The procedure included purification of monomer by recrystallization and degassing of the reaction mixture prior to bulk polymerization under vacuum (38,39). The same procedure was patented for lactides.

The conversion of cyclic monomers to polymer chains requires the use of initiators or catalysts. Many initiation systems have been reported in literature during the past two decades. Among them, two compounds are used industrially, namely tin(II) 2-ethyl hexanoate (stannous octoate or SnOct$_2$) and Zn metal. Stannous octoate has been approved by the U.S. FDA for surgical and pharmacological applications, although it is very unstable and usually contains impurities. This compound is the most widely used initiator. It provides high reaction rate, high conversion ratio, and high molar mass even under relatively mild conditions. However, the use of tin derivatives in the biomedical field remains dubious. In fact, it has been shown that stannous octoate is slightly cytotoxic (40). On the other hand, the presence of residues has been suspected and indirectly shown, such as octanoyl moieties combined to some of the alcoholic chain ends, octanoic acid and hydroxy tin octoate (41). Zn metal was retained as an alternative because Zn is an oligoelement with daily allowance for the metabolism of mammalian bodies. Zn ions are also regarded as bacteriostatic compounds. The use of Zn leads to slower polymerization as compared with stannous octoate. Nevertheless, no modification of chain ends has been detected so far. This feature is a source of difference of polymer properties.

From the structural viewpoint, the ring opening polymerization of cyclic dimers such as lactide and glycolide

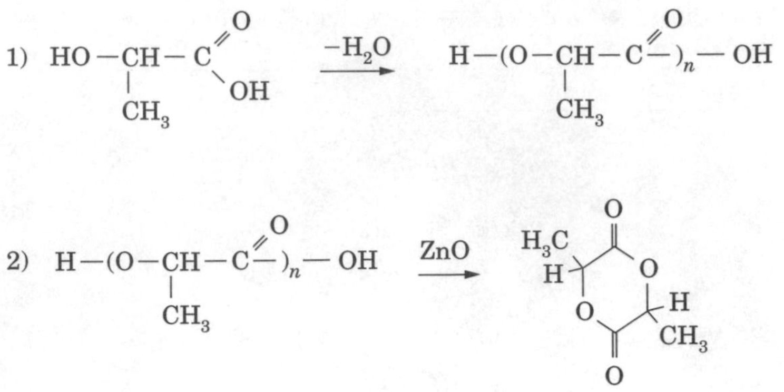

1) $HO-CH-C{\displaystyle \mathop{}_{OH}^{O}} \xrightarrow{-H_2O} H-(O-CH-C{\overset{O}{\diagup}})_n-OH$
   $\quad\quad\ \ CH_3$

**Figure 1.** Synthesis of lactide by polycondensation of lactic acid and cyclization.

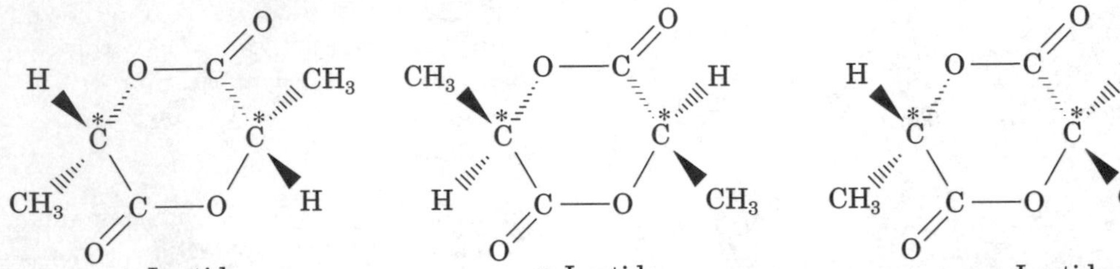

2) $H-(O-CH-C{\overset{O}{\diagup}})_n-OH \xrightarrow{ZnO}$
   $\quad\quad\ \ CH_3$

L-Lactide        D-Lactide        *meso*-Lactide

**Figure 2.** Diastereoisomeric forms of lactide: L-lactide, D-lactide, and *meso*-lactide.

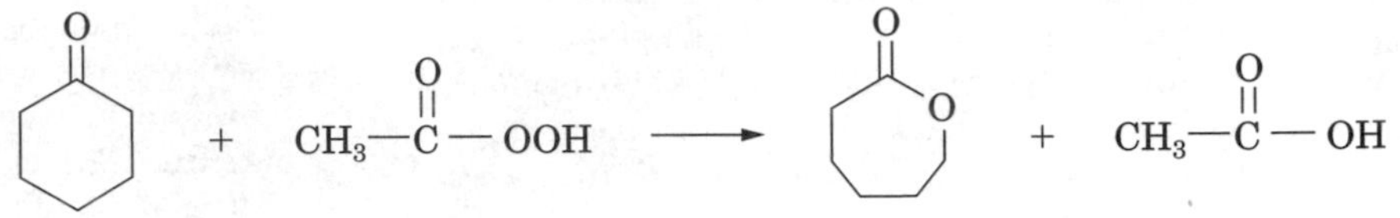

$+ \ CH_3-C-OOH \longrightarrow \quad + \ CH_3-C-OH$

**Figure 3.** Synthesis of ε-caprolactone by oxidation of cyclohexanone with peracetic acid. *Source:* From Ref. 76.

proceeds via pair addition of repeat units. Thus, nonrandom copolymer and stereocopolymer chains are obtained, depending on the composition of the feed (27). The case of ε-caprolactone is simpler since it contains only one repeat unit in the cycle. Transesterification reactions can be promoted by certain initiators at high temperatures. These reactions tend to randomize the distribution of repeat units issued from the pair addition mechanism (42).

The polymerization of lactones can proceed through anionic, cationic, or coordination routes. A great deal of work has been devoted to the understanding of these mechanisms because they determine the chain end structures, and to some extent, the degradability. However, many unknowns still remain in the literature due to the complexity of the lactone polymerization chemistry.

**Anionic Polymerization.** Anionic polymerization is generally conducted in solution under mild reaction conditions as compared with other methods. Anionic-type initiators such as calcium carbonate and lithium hydride had been used early in the 50s and 60s for the polymerization of

lactides and glycolide (31,43). Later on, many other initiators were reported, mainly in the form of acetate, carbonate, or octoate salts of calcium, sodium, magnesium, potassium, and lithium metals (44–48).

The active species for the chain propagation have been identified. They are alcoholate or carboxylate groups resulting from the opening of lactone rings after nucleophilic substitution. In the case of strong nucleophilic attacks of basic alcoholate-type on the carbonyl, the ring is opened by cleavage of O-acyl bonds. Retention of configuration is thus observed during lactide polymerization. In contrast, initiation by weak nucleophiles of the carboxylate-type seems to lead to O-alkyl scission after attack on the methylated carbon, leading to racemization (49). The propagation step depends on the nature and size of the counterion, a pair of ions being formed at the active chain end. The stronger the interaction, the slower the chain propagation. The propagation kinetics are also influenced by the solvation phenomena affecting both monomer and polymer.

Anionic polymerization is considered as a living process, chain growth continuing until the monomer is exhausted.

The living nature of the reaction is of interest since it allows functionalization and the obtainment of various block copolymers (50–52). From the structural viewpoint, anionic polymerization leads to polymers with various chain ends depending on initiation and termination mechanisms. Transesterification reactions, which are usually very limited due to the mild reaction conditions, can contribute to chain end modifications too.

**Cationic Polymerization.** Various initiators have been used for the cationic polymerization of lactones: strong acids (53–55), Lewis acids (56–58), acylating or alkylating agents (59,60). Some of them can promote rapid degradation of the polymers. However, cationic polymerization is a means to yield otherwise inaccessible copolymers.

During the initiation step, active species of the oxonium or carbocation-type are formed. The oxonium cycles which are stabilized by inductive effects have been observed by NMR (61). In contrast, carbocation cycles are generally instable. In the absence of other strong nucleophiles (e.g., water, alcohols), the propagation step consists in a nucleophilic attack of the active species by the endocyclic oxygen of another monomer molecule. For the lactide oxonium active species, the attack occurs on the carbonyl rather than on the methine due to the presence of the bulky methyl group. The attack is of $S_N2$-type, with cleavage of $O$-acyl bonds and retention of configuration. If the attack happens on the methine, racemization can be observed to various extents. For the carbenium one, the attack is considered to be of $S_N1$-type on the methine, leading to $O$-alkyl bond cleavage and racemization. Termination reactions are generally caused by an external agent or by the intra- or interchain nucleophilic attack of the active species by endochain oxygen atoms. Transesterification reactions have been reported (61).

Most of the mechanisms described in literature are based on coinitiation by impurities such as water, lactic acid, or lactoylactyl acid. Introduction of an alcohol is a means to insure reproductibility, but the molar mass is consequently decreased. Anyhow, one must keep in mind that the coinitiated polymerizations lead to polymers with different chain ends depending on the coinitiator.

**Insertion-Coordination Polymerization.** The insertion–coordination polymerization mechanism have been proposed for a class of initiators deriving from transition metals such as Zn, Al, Ti, Zr, Sn, Y, and lanthanides (62–68). This is regarded as the most versatile and efficient method to prepare PLA, PGA, PCL, and various copolymers. High molar mass and high conversion ratio can be easily achieved. As in the case of anionic polymerization, the product is a living polymer. In some cases, it has been shown that trace amounts of water or other nucleophiles are the initiator, the coordination compounds serving as catalyst.

The coordination–insertion mechanism can be considered as an intermediate between anionic and cationic ones. A coordination between lactide and the metal with vacant p or d orbitals occurs via the carbonyl oxygen, followed by a concerted attack by an endocyclic oxygen, which leads to the insertion of lactide into the initiator. Nevertheless, this mechanism has been demonstrated for few initiators only, and was derived from chain end modifications. Moreover, distinction from ionic mechanisms is not clear, especially in the case of stannous octoate.

No racemization has been observed with these initiators because the ring opening occurs by $O$-acyl cleavage. Transesterification reactions can be detected to variable extents depending on the polymerization conditions and on the nature of the initiator (69).

Recently, Schwach et al. carried out a systemic study on the ring opening polymerization of DL-lactide using stannous octoate or Zn metal as initiators (41,70–72). Data collected for low monomer–initiator ratios showed that the opening of lactide ring led to octoate-terminated chains. None of the experimental findings agreed with the coordination–insertion mechanism. A cationic-type mechanism coinitiated by 2-ethyl hexanoic acid was proposed with $O$-acyl cleavage and retention of lactyl configuration. This mechanism involves the formation of esterified alcohol chain ends in the form of octoate and the formation of hydroxy tin(II) octoate (41). In the presence of lactic acid, hydroxy tin(II) octoate gives rise to a side-product, hydroxy tin(II) lactate, which also initiates the lactide polymerization. This double initiation mechanism well accounted for the poor correlation between obtained molar masses and theoretical ones. In the case of Zn metal, it has been shown that the active species was Zn lactate. The initiation was also of cationic-type, with coinitiation by lactic acid. Use of Zn lactate was suggested instead of Zn metal since the former yielded higher conversion ratios in shorter times (69,70).

The initiation mechanisms by tin octoate (or hydroxy tin(II) lactate) and Zn metal (or Zn lactate) led to polymers with different chain ends. The presence of purification-resistant C8 hydrophobic derivatives was detected in the case of tin initiation. These hydrophobic compounds seemed to dramatically affect the general behavior of the PLA matrices (71,72). In contrast, Zn-based initiators led only to hydroxylated chain ends without hydrophobic residues.

Therefore, the variety of synthetic routes provides a means to modify the molecular characteristics of polymers such as molar mass, molar mass distribution, endgroup nature, configuration of lactyl units, and chain structure. These characteristics greatly determine the physicochemical properties, especially the degradability. On the other hand, the manufacturing methods used to make drug delivery devices can also be the source of structural modifications.

## MANUFACTURING METHODOLOGY

Various PLA, PLAGA, and PCL polymers have been used to prepare controlled drug delivery devices. These devices present a number of advantages such as variable release rates and profiles, targeting of recipients, protection of immediate environment, increasing stability, separation of incompatible components, conversion of liquids to free-flowing solids, masking of odor, taste, and activity (73).

The various drug delivery devices can be divided in two types: (*1*) monolithic, the therapeutic agents being dissolved or dispersed in a polymeric matrix, and (2) reservoir, the active agents forming a core surrounded by a polymeric barrier (73–78). Macromolecular prodrugs where the active agent is covalently bound to a macromolecule constitute a more sophisticated type of drug delivery system. So far, no such system derived from PLA, PLAGA, and PCL has been reported because of the lack of pendent functional groups to attach drug molecules.

A wide variety of monolithic and reservoir drug delivery systems exist in the form of microspheres, nanospheres, fibers, films, discs, pellets, cylinders, and gels. These devices can be classified into two categories: implantable and injectable devices.

### Implantable Drug Delivery Devices

Implantable large devices such cylinders, pellets, slabs, discs, and films thicker than 0.1 mm are usually prepared by compression molding an intimate polymer-drug mixture (79–82). However, there is risk of thermal degradation. Tubings and needle-like implants can be obtained by extrusion (76,83,84). The temperature of compression molding or extrusion depends on the morphological characteristics of the polymer. For crystalline polymers, the processing temperature has to be above the melting temperature ($T_m$). In the case of amorphous polymers, temperatures above the glass transition ($T_g$) are usually sufficient. Prior to processing, the polymer should be thoroughly dried to prevent thermal and/or hydrolytic degradation.

Thin films can be prepared by casting a polymer-drug solution (85–90). Hollow fibers of highly crystalline $PLA_{100}$ can be prepared by using a "dry-wet" coagulation spinning process (91,92). The use of different spinning systems (i.e., different solvent–nonsolvent pairs and with or without additive) leads to hollow fibers with varying asymmetric membrane structures. PCL fibers containing tetracycline hydrochloride, with an outer diameter of 0.5 mm, have been prepared by melt spinning at 161°C (93). When organic solvents are used, the elimination of residual solvents is of major importance because they can generate toxicity regardless of the polymer matrix. Implantable mesh sheets were also reported (94).

Implantable systems provide various advantages such as prolonged release of drugs, reproductibility of drug release profiles, and ease of fabrication. However, implantation of such systems requires surgery with risk of infection.

### Injectable Drug Delivery Devices

Injectable systems have been largely investigated in recent years because they circumvent the need for a surgical incision. Shah et al. used PLAGA copolymers in a formulation that forms a gel matrix immediately on contact with aqueous fluids (95). Prolonged release of drugs for up to months was achieved. Miyamoto and Takaoka developed composite delivery systems for bone morphogenetic protein (BMP) by mixing semipurified BMP and low molar mass PLA-poly(ethylene glycol) copolymer (96).

Among the various injectable drug delivery systems, microparticles are the most widely investigated. There are basically three methods to manufacture monolithic microparticles: grinding, phase separation, and solvent evaporation. Grinding can be applied to solid matrices issued from fusion of polymer-drug blends or from drying of polymer-drug solutions (97). The presence of residual solvents depends on drying and grinding temperatures. The thermal unstability of some drugs may render solvent removal difficult. This method is suitable for the preparation of monolithic microparticles containing a water-soluble drug because the drug is totally entrapped in the matrix.

The phase separation method consists in emulsifying an aqueous drug solution (dispersion phase) in an organic solvent solution of polymer (continuous phase). A second organic solvent, which is a nonsolvent of the polymer but miscible with the first organic solvent, is then added under vigorous stirring. The resulting particles precipitate, inserting the droplets of the aqueous drug solution (phase separation). Thereafter, the nonsolvent is extracted with an appropriate solvent to harden the spherical precipitate, and then the organic and the appropriate solvents are removed, leaving the microspheres (78). The first application of this technique was to produce microspheres that released drugs such as enzymes, hormones, and vaccines over a prolonged period (98). A great deal of effort has been devoted to the encapsulation of highly water-soluble drugs because of the total entrapment. Typical nonsolvents are silicone oil and alkanes such as n-hexanes and n-heptanes. Tice et al. (99) and Sanders et al. (100) successfully used this method for the microencapsulation of highly water-soluble peptides into PLAGA copolymers. They used methylene chloride as the solvent of the polymer, silicone oil as the nonsolvent, and a large volume of n-heptane or n-hexane as the solvent to extract the silicone oil. The major disadvantage of this method is the large amounts of n-alkanes that have to be used and cannot be totally removed from the matrix. The microencapsulation of hydrophobic or lipophilic drugs by this technique has also been described (101,102).

The solvent evaporation method has been largely used to encapsulate lipophilic drugs (103–106). The preparation of microspheres by this method consists of two steps. First, the drug and the polymer are dissolved or dispersed in a water-immiscible solvent, and then the solution is poured into an aqueous medium while being stirred to form small polymer droplets containing the drug. In the second step, the droplets are hardened to yield microspheres. The hardening process is achieved by solvent evaporation under normal or reduced pressure. Hardening of droplets can also be achieved by solvent extraction using a third solvent that is a precipitant for the polymer and miscible with both water and the first organic solvent. The latter is usually methylene chloride, chloroform, or some other relatively volatile solvent that is a good solvent for PLA, PLAGA, and PCL polymers but difficult to remove completely. Solvent removal by extraction is generally much faster than evaporation (<1 h versus 4–20 h, respectively). Furthermore, solvent extraction can lead to microspheres that are more porous than can solvent evaporation. A combination of both methods is often preferred. Emulsification usually requires the use of a surfactant in rather large amounts. The most commonly used surfactants are poly(vinyl alcohol), poly(vinylpyrrolidone), alginates, and gelatins. The surfactant molecules attached to the surface of microspheres are very difficult to eliminate and thus remain as an uncontrolled component in the formulation. Carrio et al. reported preparation of surfactant-free microspheres based on a PLA$_{37.5}$GA$_{25}$ copolymer, amphiphilic PLA$_{50}$ oligomers being used for emulsification (107).

In the case of more or less water soluble drugs, the oil-in-water (O/W) emulsion system leads to poor entrapment. To solve this problem, the solvent evaporation method has been adapted to the cases of oil-in-oil (O/O) and of water-in-oil-in-water (W/O/W)–types emulsions. Tsai et al. were the first to describe an O/O emulsion technique to entrap mitomycin-C (108). A polar organic solvent dissolving a water-soluble drug and the polymer constitutes the dispersed phase, whereas a nonsolvent oil such as liquid paraffin dissolving a low hydrophilic-lipophilic-balanced value (HLB) emulsifier constitutes the continuous phase. Subsequently, the polar solvent is evaporated to yield a suspension of microspheres in the nonsolvent. After filtration, the microspheres are washed with an appropriate solvent to eliminate the residual oil. Tsai et al. (108) and Jalil and Nixon (109) used acetonitrile as the polar solvent, liquid paraffin as the nonsolvent, and Span 65 or 41 as the emulsifier. Ikada et al. used dimethylformamide as the polar solvent, caster oil or liquid paraffin as the nonsolvent, and lecithin or Span 80 as the emulsifier, to prepare PLA microspheres containing cisplatin by the O/O emulsion technique (110,111). These methods can also introduce pollutants, especially the nonsolvent as it is slightly soluble in the polar organic solvent. The presence of these pollutants must be taken into account as they may affect the properties of microspheres.

An alternative to these methods is spray-drying (112,113). This technique is very convenient for the incorporation of highly water-soluble drugs. Large scale production under mild conditions is possible as it is rapid and sequential. Nevertheless, agglomerates can be formed during the spray process, which require the use of surfactants. Another disadvantage is the possible contamination by minute particles of alien substances.

When one deals with degradable drug delivery devices, the first question is to know whether the manufacturing process affected the structural characteristics of polymers and whether solvents and surfactants were present. As mentioned earlier, the properties of the polymeric matrix can be dramatically changed during the manufacturing process.

## DEGRADATION MECHANISMS

According to literature, the degradation of polymeric materials in a living environment can result from either enzymatically mediated or chemically mediated cleavages. The two mechanisms can act separately or simultaneously. Although in vivo environments are different from outdoor ones, there is no fundamental difference between the biodegradation of a polymer by animal cells and that by microorganisms. Both involve water, enzymes, metabolites, and ions that interact with the material (114). Under these conditions, one can distinguish enzymatic, hydrolytic, and microbial degradations.

### Enzymatic Degradation

It is now well identified that biopolymers such as proteins, polysaccharides, polynucleotides, and even bacterial poly($\beta$-alkanoates) degrade enzymatically (115–118), in agreement with the two main characteristics of living systems (i.e., degradation and recycling). The situation is totally different in the case of synthetic polymers. In fact, there has been much debate about the involvement of enzymes in the in vivo degradation of PLA, PGA, and PCL homo- and copolymers. Some authors argued in favor of a substantial enzymatic degradation (117,119,120), whereas most relegated the enzymatic involvement to a second role (116,121–124). The secondary differences between parenteral and outdoor conditions further increased the confusion. In some cases, enzymatic degradation was shown from differences between the behaviors of samples in the presence and in the absence of living organisms. In the case of in vitro studies, comparison was generally made between data in the presence and in the absence of enzymes. However, the observation of such differences is not conclusive because there are many other factors that may interfere with polymer degradation when experimental conditions are not similar (10). Therefore, the demonstration of enzymatic degradation (biodegradation) must be based on the concordance of data issued from different analytical methods and from a careful monitoring of the generation and the fate of the degradation products.

Various enzymes have been investigated in attempts to clarify their effect on the degradation of PLAGA polymers. Among them, bacterial proteinase K was shown to be able to strongly accelerate the degradation rate of PLA stereocopolymers (21,125–128), L-LA units being preferentially degraded as compared to D-LA (127,128). Enzymes such as tissue esterases, pronase, and bromelain also affect PLA degradation (120,125). In contrast, many other enzymes seem to be inactive. Pitt et al. investigated the in vivo degradation of a series of elastomeric homo- and copolymers of PCL and poly(valerolactone) cross-linked with biscaprolactone. These compounds were subject to bioerosion involving immediate attack at the surface (129,130). This finding was inferred by the rubbery nature of the materials, polymer chains having enough freedom to take on chain conformations convenient for enzymatic attack. The authors also noted that for degradable polymers in the glassy state, this conformation could hardly be achieved and thus small (if any) enzymatic degradation could occur. Mochizuki et al. examined the enzymatic degradation of PCL fibres by Lipase (131). Scanning electron microscopy (SEM) photographs showed that the enzyme preferentially attacked amorphous regions rather than crystalline ones. As enzymatic degradation proceeded, the diameter of the fibers became gradually slimmer.

It should be noted that enzymes can be inactive on high molar mass material and become active at the later stages of degradation when the chain fragments become small and soluble in surrounding fluids. Once formed during degradation, tiny crystalline particles can be phagocytosed and undergo intracellular degradation (132). Microspheres can also be easily phagocytosed (133).

In conclusion, it is now generally admitted that in the case of glassy aliphatic polyesters such as PLA, PGA, and their copolymers, enzyme involvement is unlikely at the early stages of degradation in vivo or under outdoor conditions. However, enzymes contribute at the later stages, especially when soluble by-products are released. In contrast, for rubbery polymers like cross-linked PCL, enzymes seem to be active from the very beginning via surface erosion phenomena (129,130). Actually, enzymatic degradation of aliphatic polyesters should not be claimed unless mass loss and dimensional changes without molar mass decrease are shown, and nonenzymatic degradation is excluded. So far, one can hardly consider that the biodegradation of linear PCL in the presence of enzymes has been clearly depicted.

## Hydrolytic Degradation

Basically, the hydrolytic degradation of aliphatic polyesters proceeds through ester bond cleavage according to the reaction:

$$R{\sim}COO{\sim}R' + H_2O \longrightarrow R{\sim}COOH + HO{\sim}R'$$

From the macroscopic viewpoint, degradation of aliphatic polyesters has been regarded as homogeneous, although surface erosion was claimed occasionally. Ginde and Gupta investigated the in vitro degradation of PGA pellets and fibers in aqueous media at different pH values (134). The authors found that pellets showed considerable surface degradation, whereas fibers showed little surface changes. Singh et al. studied a drug delivery system based on $PLA_{50}$ microcapsules and concluded erosion-based degradation, the hydrolytic cleavage of ester bonds in the polymer backbone at the surface of microcapsules leading to the formation of lactic acid monomers (135). Kimura et al. investigated the in vitro and in vivo degradations of fibers deriving from a copoly(ester-ether) composed of $PLA_{100}$ and polyoxypropylene blocks (136). Surface erosion was observed in both cases and the authors concluded that hydrolysis was limited to the surface.

In contrast, many authors argued either explicitly or implicitly in favor of autocatalyzed bulk degradation. Hutchinson investigated the in vitro and in vivo release of polypeptides ($633 < \bar{M}_n < 22,000$ Da) from PLAGA copolymer matrices containing from 25% to 100% DL-lactic acid. The degradation process of the polymer matrix was considered as homogeneous (137). Sanders et al. studied a $PLA_{22}GA_{56}$ microsphere-based delivery system and observed a homogeneous (bulk) rather than heterogeneous (surface) degradation (138). This conclusion was derived from the biological response and from changes in microsphere aspects in vivo. Kenley et al. examined a series of PLAGA copolymers representing a range of monomer ratios and molar masses, with the goal of studying polymer degradation kinetics in vivo and in vitro (139). Hydrolysis was supposed to proceed throughout the bulk of the polymer structure because the onset of mass loss lagged behind molar mass decrease. Schakenraad et al. (140) and Helder et al. (141) investigated a series of glycine/DL-lactic acid copolymers in vivo and in vitro. Bulk hydrolysis was described as the degradation mechanism in both cases, molar mass decreasing continuously with the aging time. Cohen et al. used a PLAGA (75/25) copolymer for long-term delivery of high molar mass water-soluble proteins (142). At all times, molar mass distribution displayed a unimodal pattern, suggesting homogeneous degradation. St. Pierre and Chiellini reviewed the degradability of synthetic polymers for pharmaceutical and medical uses. These authors concluded that hydrolysis of PLAGA polymers was a bulk process with random cleavage of ester functions (143). In another review concerning the controlled release of bioactive agents from PLAGA polymers, Lewis concluded that degradation of aliphatic polyesters occurred in the bulk (144).

Pitt et al. studied the in vivo degradation of films of $PLA_{50}$, PCL, and corresponding copolymers. The authors suggested that the first stage of degradation was confined to a molar mass decrease due to random hydrolytic ester cleavage autocatalyzed by the carboxyl endgroups, the second stage being characterized by the onset of mass loss and a decrease in the rate of chain scission (82,145).

The kinetics of the autocatalyzed hydrolytic degradation proposed by Pitt et al. were derived according to the following equations:

$$d[E]/dt = -d[COOH]/dt = -k[COOH] \cdot [H_2O] \cdot [E] \tag{1}$$

where [COOH], [$H_2O$], and [E] represent, respectively, carboxyl endgroup, water, and ester concentrations in the polymer matrix. By using the following relationships:

$$[COOH] = M/(\bar{M}_n \cdot V) = \rho/\bar{M}_n$$

$$[COOH] = [E]/(\overline{DP}_n - 1)$$

$$\bar{M}_n = m \cdot \overline{DP}_n$$

where M is the polymer matrix mass, V its volume, $\rho$ its mass volume unit, $\bar{M}_n$ the number average molar mass, $\overline{DP}_n$ the number average degree of polymerization and m the repeat unit mass, one obtains equation 2:

$$d(1/\overline{DP}_n)/dt = k(\rho/m)[H_2O](\overline{DP}_n - 1)\overline{DP}_n^{-2} \tag{2}$$

Integration of equation 2 leads to equation 3:

$$\ln\{(1 - \overline{DP}_n)/(1 - \overline{DP}_{n0})\} = k't \tag{3}$$

where $k' = k(\rho/m)[H_2O]$, and $\overline{DP}_{n0}$ is the $\overline{DP}_n$ at time zero. This kinetic expression is valid before the onset of mass loss. If $\overline{DP}_n \gg 1$, equation 3 can be simplified to equation 4:

$$\ln(\overline{DP}_n/\overline{DP}_{n0}) = \ln(\bar{M}_n/\bar{M}_{n0}) = -k't \tag{4}$$

According to this relationship, semilog plots of $\overline{DP}_n$ or of $\bar{M}_n$ versus hydrolysis time should be linear prior to the onset of mass loss, a feature that was observed experimentally (145).

A few years ago, the discovery of a faster degradation inside large-size PLA$_{50}$ specimens greatly changed the understanding of the hydrolytic degradation of PLAGA polymers (146). The heterogeneous degradation was assigned to diffusion-reaction phenomena. Typically, the polymer matrix is initially homogeneous in the sense that the average molar mass is the same throughout the matrix. Once placed in an aqueous medium, water penetrates into the specimen leading to hydrolytic cleavage of ester bonds. Each ester bond cleavage forms a new carboxyl endgroup that, according to autocatalysis, accelerates the hydrolytic reaction of the remaining ester bonds (147). At the very beginning, degradation occurs in the bulk and is macroscopically homogeneous. However, the situation becomes totally different when soluble oligomeric compounds are generated in the matrix. The soluble oligomers that are close to the surface can escape from the matrix before total degradation, whereas those located inside the matrix can hardly diffuse out of the matrix. This difference results in a higher acidity inside than at the surface. When the aqueous medium is buffered at neutral, as is the case in vivo, the neutralization of carboxyl endgroups present at the surface can also contribute to decrease the surface acidity. Therefore, autocatalysis is larger in the bulk than at the surface, thus leading to a surface-interior differentiation. As the degradation proceeds, more and more carboxyl endgroups are formed inside to accelerate the internal degradation and enhance the surface-interior differentiation. Bimodal molar mass distributions are observable due to the presence of two populations of macromolecules degrading at different rates (Fig 4). Finally, hollow structures are formed when the internal material, which is totally transformed to soluble oligmers, dissolves in the aqueous medium (Fig. 5). Similar features were observed for other amorphous polymers like PLA$_{62.5}$, PLA$_{75}$, and PLA$_{37.5}$GA$_{25}$ (148–153). In contrast, in the case of crystallizable polyesters like PLA$_{87.5}$, PLA$_{96}$, PLA$_{100}$, PLA$_{75}$GA$_{25}$, and PLA$_{85}$GA$_{15}$, no hollow structures were obtained due to the crystallization of degradation products (Fig. 6), although degradation was faster inside than outside (148–153).

The faster internal degradation of PLA and PLAGA polymers is now regarded as a general phenomenon. It was confirmed by many authors (154–160), and for PLACL co-

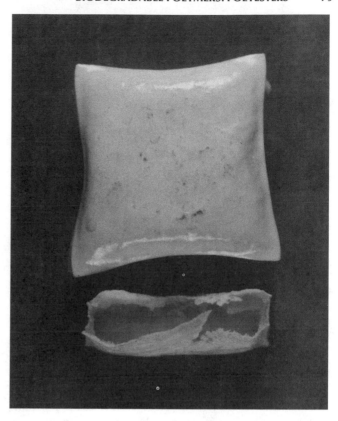

**Figure 5.** Hollow structure and cross section of a PLA$_{50}$ plate after 17 weeks in a pH 7.4 phosphate buffer at 37°C. *Source:* From Ref. 182.

**Figure 6.** Cross section of a PLA$_{96}$ plate after 40 weeks in a pH 7.4 phosphate buffer at 37°C. *Source:* From Ref. 150.

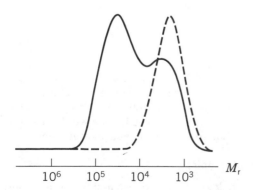

**Figure 4.** SEC chromatograms of PLA$_{50}$ plates after 7 weeks in saline at 37°C: ——— (surface), --- (interior). *Source:* From Ref. 146.

polymers (161), lactide-glycine copolymers (140,141), lactide-1,5-dioxepan-2-one copolymers (162), poly(trimethylene carbonate) (163), and poly(*para*-dioxanone) as well (164). According to diffusion-reaction phenomena, small devices should degrade more slowly (*i.e.*, at the surface rate) than large ones since thinness favors the elimination of soluble oligomeric compounds and minimizes the autocatalytic effect. This trend was confirmed by comparing the degradation rates shown by submillimetric microparticles and films and millimeric plates made from the same PLA$_{50}$ (165). As expected, the plates degraded faster than the microparticles and films.

In the case of PCL, hydrolytic degradation is very slow. However, the degradation mechanism is the same as that of PLAGA polymers. Typically, degradation of PCL begins with random hydrolytic cleavage of the ester linkages autocatalyzed by the carboxyl endgroups. The molar mass decreases continuously, but there is no weight loss during this first stage of degradation. The second phase is characterized by the onset of weight loss due to the formation of low molar mass fragments small enough to diffuse out of the bulk (82,161,166). No faster internal degradation has been reported for the PCL homopolymer so far, probably because of the combination of high crystallinity, hydrophobicity, and low degradation rate.

Therefore, the hydrolytic degradation of aliphatic polyesters is a complex process involving four main phenomena, namely water absorption, ester cleavage, diffusion of soluble oligomers, and solubilization of fragments (167). These phenomena depend on many factors such as matrix morphology, chemical composition and configurational structure, molar mass, size, distribution of chemically reactive compounds within the matrix, and nature of the degradation media (11). Considering the interdependence of degradation and drug release in controlled drug delivery systems, all the factors listed here can affect the characteristics of drug release, as will be shown later.

### Microbial Degradation

More and more attention is paid to aliphatic polyesters of the PLA, PGA, and PCL types with respect to the problems of solid waste accumulation, delivery of chemicals to plants, and seed protection. Whether these polymers can biodegrade in a wild environment or whether it is the byproducts generated by abiotic hydrolysis that can be bioassimilated by fungi or bacteria are two currently investigated questions, particularly in the area of composting.

Torres et al. studied the ability of some microorganisms to use lactic acids, PLA oligomers and polymers, and PLAGA copolymers as sole carbon and energy sources under controlled or natural conditions (168–170). Among the 14 filamentous fungal strains tested, two strains of *Fusarium moniliforme* and one strain of *Penicillium roqueforti* were identified as able to totally assimilate L- and DL-lactic acids as well as $PLA_{50}$ oligomers (168,169). In contrast, $PLA_{100}$ oligomers appeared biostable because of the crystallinity. A synergistic effect was observed when both microorganisms were present in the same culture medium containing $PLA_{50}$ oligomers.

High molar mass polymers were also considered via degradation tests in soil and in selected culture media. *Fusarium moniliforme* filaments were found to grow on the surface of a $PLA_{37.5}GA_{25}$ copolymer and through the bulk. On the other hand, five strains of different filamentous fungi were isolated from the soil as capable of bioassimilating $PLA_{50}$ soluble oligomers in mixed cultures. Based on the results obtained by SEM, size exclusion chromatography (SEC), pH, and water absorption measurements, the authors proposed a mechanism including abiotic hydrolysis of macromolecules followed by bioassimilation of soluble oligomers (170). Therefore, when a PLA material is placed in a degradation medium in the presence of microorganisms, only abiotic hydrolytic degradation occurs unless the few enzymes mentioned already as capable of degrading high molar mass PLA are present. Whether the

degradation is homogeneous or heterogeneous depends on the size of the material and on the physical state of the medium (liquid or moisture). In a liquid medium, degradation is heterogeneous due to larger internal autocatalysis for large-size devices, whereas in the case of a humid environment, degradation has to be homogeneous since the soluble oligomers cannot diffuse out of the matrix. Anyhow, sooner or later, low molar mass assimilable oligomers are formed throughout the polymer matrix. The bioassimilation process then takes place. At this time, fungal filaments penetrate the partially degraded mass to take advantage of the internal oligomers, thus turning the abiotic degradation to a biotic one. Under wild conditions, this should happen after a long lag period. However, PLA polymers are definitely degraded and assimilated because the ultimate degradation products (L- and D-lactic acids) have been shown to be metabolized by some common microorganisms (170). In particular, it is now well known that PLA polymers degrade completely and rather rapidly in compost where the temperature is usually between 50 and 60°C.

Jarrett et al. investigated the microbial degradation of cross-linked PCL films and of PCL single crystals in the presence of a yeast, *Cryptococcus laurentii*, and a fungus, *Fusarium*. Data were compared with those obtained for the chemical degradation in a 40% methyl amine aqueous solution (171). These microorganisms produced both endo- and exoenzymes and a cofactor (surfactant), which worked together in the degradation of the polymer. In the absence of the cofactor, the access of enzyme to the hydrophobic polymer was limited. Lefebvre et al. examined the biodegradation of hydroxy- or methoxy-terminated PCL samples by mixed cultures of microorganisms issued from a suspension of compost, and also by a pure culture of an actinomycete isolated from compost (172). The authors suggested that the initiation of degradation took place in the vicinity of the chain ends. Methylation of the chain ends did not affect the biodegradability. Akahori and Osawa investigated the biodegradability of PCL-paper composites by outdoor exposure and outdoor soil burial (173). All the samples crumbled within 6 months. Molar mass of residual PCL was found to decrease slowly, suggesting the contribution of hydrolytic degradation.

## FACTORS INFLUENCING POLYMER DEGRADATION AND DRUG DELIVERY

In the preceding paragraphs, we have underlined some of the main structural, physical, chemical, and biological characteristics of PLAGA and PCL polymers. Let us now consider the various factors that can affect the degradation of aliphatic polyesters as well as the drug delivery profiles. For the sake of clarity, these factors will be discussed separately, although most of them are interdependent.

### Polymer Morphology

The morphology of a polymeric material (i.e., amorphousness or semicrystallinity) plays a critical role in both the enzymatic and nonenzymatic degradation processes. It is now known that degradation of semicrystalline polyesters in aqueous media occurs in two stages. The first stage consists of water diffusion into the amorphous regions with random hydrolytic scission of ester bonds. The second stage starts when most of the amorphous regions are degraded. The hydrolytic attack then progresses from the edge towards the center of crystallites (174–176). This phenomenon was first observed by Fischer et al. who investigated the structure of solution-grown crystals of PLA stereocopolymers by means of chemical reactions (174). Disordered regions of single crystals were found to be selectively degraded by methanolic sodium hydroxide. In the case of a $PLA_{92.5}$ stereocopolymer, the authors obtained a trimodal molar mass distribution with three distinct peaks corresponding respectively to onefold, twofold, and threefold the crystalline lamellae thickness. Later on, many authors reported a preferential degradation of amorphous areas in the case of semicrystalline Vicryl® pellets or sutures (175–177), Dexon® sutures (177,178), and $PLA_{100}$ plates and rods (179,180). An increase of crystallinity was often detected (175–180). On the other hand, Miller et al. observed that fast-cured PGA degraded more rapidly than slow-cured one (26), the in vivo half-lives being 0.85 and 5 months, respectively. The difference was related to the higher crystallinity of the latter. Gutwald et al. also observed a faster resorption of amorphous $PLA_{100}$ implants as compared to crystalline ones (181). Last but not least, the degradation rate of PGA pellets was found to be much faster than that of fibers with similar crystallinity (134). This feature was assigned to the presence of long range order in fibers. Therefore, chain orientation could also play an important role in the hydrolytic degradation.

Recent systemic investigation on the in vitro degradation of PLAGA polymers well agreed with the preferential degradation in amorphous zones of semicrystalline $PLA_{100}$. Bimodal molar mass distributions were detected at the later stages of degradation, the two peaks corresponding to onefold and twofold the thickness of crystalline lamellae, respectively (149). On the other hand, crystallization was observed in the case of initially amorphous $PLA_{100}$ and $PLA_{96}$ obtained by quenching (149,150), as shown in Figure 7. Crystallization mainly resulted from low molar mass chains whose $T_g$ was lower than that of longer chains. With the plasticizing effect of absorbed water, which further decreased the $T_g$, these short chains were mobile enough to crystallize under the degradation conditions. The resulting crystallites were more resistant to further degradation. As a consequence, a very narrow peak corresponding to low molar mass crystalline zones was detected by SEC (Fig. 8).

Among the various $PLA_x$ stereocopolymers, the polymer is intrinsically amorphous if X is in the 10 to 90 range (11). For $PLA_XGA_Y$ copolymers, the polymer is intrinsically amorphous when Y is in the 10 to 70 range (for L-LA) or in the 0 to 70 range (for DL-LA) (19). However, one must be careful with these divisions because of the variable extents of transesterification reactions, which tend to randomize the comonomer distribution along the chains. The fate of intrinsically amorphous polymers during degradation appeared more complex depending on chain structures. Two classes can be distinguished. The first class consists of polymers that can never crystallize. The second one groups together polymers that can crystallize during deg-

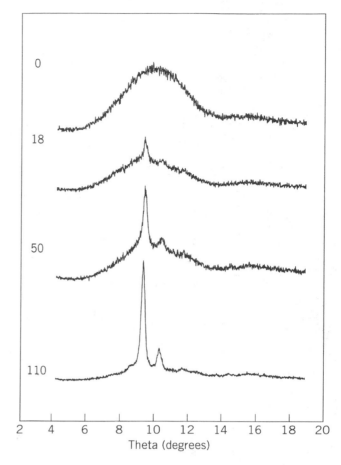

**Figure 7.** X-ray diffraction spectra of initially amorphous $PLA_{100}$ plates after 0, 18, 50, and 110 weeks in a pH 7.4 phosphate buffer at 37°C. *Source:* From Ref. 149.

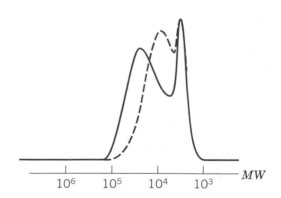

**Figure 8.** SEC chromatograms of initially amorphous $PLA_{100}$ plates after 70 weeks in a pH 7.4 phosphate buffer at 37°C: ——— (surface), ---- (interior). *Source:* From Ref. 149.

radation. $PLA_{37.5}GA_{25}$ is a good example of the first class, whereas $PLA_{50}$, $PLA_{62.5}$, $PLA_{75}$, $PLA_{87.5}$, $PLA_{75}GA_{25}$, and $PLA_{85}GA_{15}$ belong to the second class. $PLA_{37.5}GA_{25}$ cannot crystallize even after degradation due to the high irregularity of its chain structure. The others are crystallizable because of the presence of isotactic blocks which, once released by degradation, can crystallize. For example, a crystalline oligomeric stereocomplex composed of L-LA and D-LA–rich segments was detected at the end of PLA$_{50}$ degradation (182). The stereoregular segments were initially present within the PLA$_{50}$ matrix, which is known to have a predominantly isotactic structure resulting from the pair-addition mechanism of DL-lactide polymerization (27). PLA$_{62.5}$ leads also to slightly crystalline residues of the stereocomplex type (153). In contrast, PLA$_{75}$ and PLA$_{87.5}$ degradation residues have the same crystalline structure as PLA$_{100}$ (153). Insofar as PLA$_{75}$GA$_{25}$ and PLA$_{85}$GA$_{15}$ are concerned, GA units are more hydrophilic than LA ones and, as a consequence, are more inclined to hydrolysis. The preferential degradation of GA units results in L-LA-enriched segments which can crystallize under the degradation conditions, as shown by the appearance of a melting peak on differential scanning calorimetry (DSC) thermograms (Fig. 9). However, the structure of the crystalline residues is slightly different from that of PLA$_{100}$ due to the inclusion of GA units in the crystalline domains (148,150).

A schematic representation is given in Figure 10. The triangle can be divided in two types of zones: C zones are composed of intrinsically crystalline polymers and A zones of intrinsically amorphous polymers. For polymers in C zones, if they are prepared in an amorphous state by quenching, degradation is characterized by crystallization

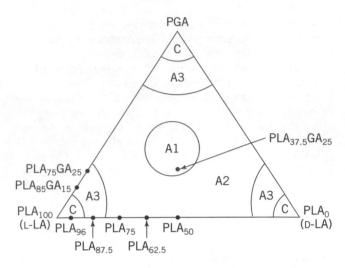

**Figure 10.** Schematic presentation of the morphological changes of PLAGA polymers with degradation. *Source:* From Ref. 11.

of degradation by-products. If they are prepared in a semi-crystalline state by annealing, degradation is characterized by selective degradation of amorphous regions. In both cases, no hollow structures are obtained for large-size devices in spite of the heterogeneous degradation. In the case of amorphous polymers, three subzones can be distinguished: A1, A2, and A3. For A1 polymers, degradation leads to hollow structures that remain amorphous due to the high irregularity of polymer chain structures. For A2 polymers, degradation also leads to hollow structures that partially crystallize, the crystalline structure depending on the initial composition. For A3 polymers, no hollow structures can be obtained, both the surface and interior crystallize. It is noteworthy that degradation of small-size amorphous polymers does not lead to hollow structures, but morphological changes are comparable to large-size ones (183).

All these examples and comments demonstrate that polymer morphology and morphological changes have to be taken into account to control the hydrolytic degradation of PLAGA polymers.

PCL and its copolymers with PLA have also been largely investigated. Pitt et al. observed a steady increase in crystallinity of PCL films from 45% to nearly 80% after 120 weeks' implantation, which was attributed to the crystallization of tie segments after chain cleavage in the amorphous phase. The low glass-transition temperature of PCL ($T_g = -60°C$) facilitated the recrystallization (166). In a detailed study on the microbial and enzymatic degradation of PCL films, Jarrett et al. showed that amorphous material on the top and bottom faces of the crystallites degraded most rapidly, leading to multimodal molar mass distributions (171). Li et al. observed a crystallinity increase during the in vitro degradation of PCL: from initial 47% to 75% after 200 weeks (161). Water absorption was very limited (less than 2%) during the first 63 weeks due to the high crystallinity and hydrophobicity of PCL. Thereafter, water absorption increased more rapidly. It reached 8.1% at 133 weeks and 12.5% at 200 weeks, which can be related

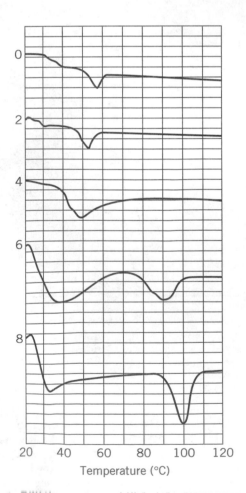

**Figure 9.** DSC thermograms of PLA$_{75}$GA$_{25}$ plates after 0, 2, 4, 6, and 8 weeks in vivo degradation. *Source:* From Ref. 151.

to molar mass decrease and formation of more carboxyl endgroups. On the other hand, weight loss reached 3.4% at 63 weeks, 10.7% at 133 weeks and 14.2% at 200 weeks. Insofar as molar mass changes are concerned, it decreased steadily from initial 58,700 to 7,000 after 200 weeks. After 133 weeks, the molar mass distribution became trimodal due to the selective degradation of amorphous zones and of the crystallite edges, the molar masses of the three peaks being 2,600, 5,200, and 8,800, respectively (Fig. 11). In the case of a Zn metal–initiated $PLA_{60}CL_{40}$ copolymer, the degradation rate was very much enhanced as shown by the increased water absorption and weight loss rates compared with the parent homopolymers PCL and $PLA_{100}$. Crystallinity increased from initial 14% to 54% at 133 weeks. The crystalline structure was of the $PLA_{100}$ type, showing a phase separation between the two components (161).

From the results presented here, it can be predicted that the morphology of the polymeric matrix should greatly influence the drug release profiles, because for both diffusion and degradation controlled mechanisms, the matrix morphology is a determining factor. Izumikawa et al. investigated $PLA_{100}$-progesterone microspheres of different morphologies obtained by changing the pressure in the solvent evaporation process (184). Microspheres made of amorphous $PLA_{100}$ were obtained by removing the solvent under reduced pressure. Data showed that the microspheres of crystalline polymeric matrices had rough surfaces with large surface areas and exhibited a rapid drug release. In contrast, the microspheres of amorphous polymeric matrices had smooth surfaces with smaller surface areas and provided a slower drug release. Similarly, Jalil and Nixon investigated the release of phenobarbitone from $PLA_{100}$ microspheres (185). The authors observed that the release profile was highly dependent on the polymer molar mass. The porous nature of the microspheres prepared from high molar mass polymer allowed a very fast release and even the smooth microsphere surface obtained from low molar mass $PLA_{100}$ failed to retain the drug for any extended time. This was attributed to the semicrystallinity of $PLA_{100}$. Fong et al. evaluated microspheres of thioridazine and ketotifen with a series of PLAGA polymers (186). Drug release was frequently slower with microspheres made from the amorphous $PLA_{50}$ and faster with those from the crystalline $PLA_{100}$. Recently, Mauduit et al. prepared gentamycin-PLA microparticles by grinding and sieving the solid mass resulting from mixing various amounts of drug and polymer in an acetone solution. The aim was to design a delivery system for local antibiotic therapy administered peroperatively. For $PLA_{50}$ systems, the release profile was biphasic with an initial burst followed by a sustained release lasting for more than two months (97). In contrast, the total load was released within the first 6 hours in the case of $PLA_{100}$ microparticles. In both cases, the drug release profile was correlated to the morphology of the polymeric matrix rather than to its degradability.

Therefore, it can be concluded that amorphous polymers are preferable for sustained drug delivery systems because they usually yield a smooth surface and an uniform structure, which retains the drug for long periods of time. In contrast, semicrystalline polymers generally lead to rough surfaces and porous structures, which are not suitable for sustained delivery of drugs.

### Chemical Composition and Configurational Structure

It is now well known that the composition of polymer chains (i.e., the contents in L-LA, D-LA, and/or GA units) greatly determines the degradation rate of PLAGA polymers (9–11,24). $PLA_X$ stereocopolymers have variable degradation rates. $PLA_{50}$ degrades the most rapidly, whereas $PLA_{100}$ is the most stable member of the series (24). The copolymerization of LA with GA enhances considerably the degradation rate as compared with the parent homopolymers (26,187–192).

Cutright et al. carried out a systemic in vivo degradation study on a series of PLAGA homo- and copolymers. The degradation rate of these polymers was found to increase as follows: PGA < $PLA_{100}$ < $PLA_{75}GA_{25}$ < $PLA_{50}GA_{50}$ < $PLA_{25}GA_{75}$ (187), showing that degradation rates can be adjusted by varying the proportion of LA to GA. Later, Miller et al. found a slightly different and more logical order: $PLA_{100}$ < PGA < $PLA_{75}GA_{25}$ < $PLA_{25}GA_{75}$ < $PLA_{50}GA_{50}$ (26). The half-life of these polymers decreased from 5 months for PGA to 1 week for $PLA_{50}GA_{50}$ and rapidly increased to 6.1 months for $PLA_{100}$. Reed and Gilding investigated comparatively the in vitro degradation of $PLA_{100}$, PGA, and $PLA_{50}GA_{50}$ (188). The authors observed that PGA sutures were totally degraded within

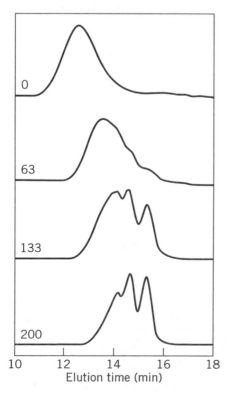

**Figure 11.** SEC chromatograms of PCL plates after 0, 63, 133, and 200 weeks in a pH 7.4 phosphate buffer at 37°C. *Source:* From Ref. 161.

10 to 12 weeks, $PLA_{50}GA_{50}$ discs within 2 to 4 weeks, whereas $PLA_{100}$ discs lost only 10–15% of mass over a period of 16 weeks. Kaetsu et al. developed implantable luteinizing hormone releasing hormone (LHRH) agonist–PLAGA polymer composites and observed that in vivo degradation and release rates could be altered over a wide range by changing molar mass, stereoisomerism, crystallinity, and composition of copolymers (84). $PLA_{50}$ was found to degrade faster than $PLA_{100}$ and, more surprisingly, $PLA_{100}$ degraded faster than $PLA_0$ (84). Zhu et al. carried out an in vivo degradation study of several PLAGA homo- and copolymers and found that all corresponding threads were completely absorbed by 60 days (189). Ogawa et al. studied the in vivo degradation of $1 \times 10 \times 10$ mm plates made of PLAGA polymers. Both the lag time and the half-life decreased with the increase in GA content (190). Amarpreet and Hubbell prepared a series of 66 terpolymers of DL-lactide, glycolide, and $\epsilon$-caprolactone (191). The half-lives of these polymers were found to vary from a few weeks to several months. Thomasin et al. examined the in vitro degradation of a series of $PLA_{50}$ and PLAGA microspheres (192). With similar molar masses, $PLA_{25}GA_{50}$ degraded faster than $PLA_{37.5}GA_{25}$, $PLA_{50}$ exhibiting the slowest degradation rate.

In the early 80s, Vert et al. pointed out the discrepancies in the literature concerning the degradation rates of various PLAGA polymers and attributed them to the fact that polymeric materials identified by the same name can be very different compounds due to secondary factors related to the origins of the compounds (24). Their analysis led to the conclusion that the time period required for complete resorption of these polymers should be substantially longer than previously supposed. Eitenmüller et al. showed similar trends (193).

A systemic investigation of the degradation characteristics of various large-size PLAGA plates under comparable conditions provided a more consistent pattern. The half-life in terms of mass loss of initially amorphous $PLA_{100}$ was found to be about 110 weeks and that of crystallized $PLA_{100}$ was even longer (149). The introduction of 4% of D-LA units to L-LA chains considerably enhanced the degradation rate since $PLA_{96}$ had a half-life of about 90 weeks (150). $PLA_{50}$ degraded much more rapidly than $PLA_{100}$ and $PLA_{96}$ with a half-life of 10 weeks (146). In the case of the copolymers, $PLA_{37.5}GA_{25}$ had a half-life of 3 weeks only (148). All these data are recalled in Table 4.

**Table 4. In Vitro Degradation Rates of PLAGA Polymers**

| Polymer | $\overline{M}_w$ (E−3) | $\overline{M}_w/\overline{M}_n$ | Half-life (weeks) |
|---|---|---|---|
| $PLA_{100}$ | 130.0 | 1.8 | 110 |
| $PLA_{96}$ | 100.0 | 1.7 | 90 |
| $PLA_{87.5}$ | 190.4 | 1.9 | 80 |
| $PLA_{75}$ | 48.6 | 1.6 | 22 |
| $PLA_{62.5}$ | 67.6 | 1.9 | 23 |
| $PLA_{50}$ | 65.0 | 1.6 | 10 |
| $PLA_{85}GA_{15}$ | 112.0 | 1.8 | 20 |
| $PLA_{75}GA_{25}$ | 111.0 | 1.8 | 10 |
| $PLA_{37.5}GA_{25}$ | 51.0 | 1.6 | 3 |

It is noteworthy that the degradation rate of a polymer is closely related to the rate of water absorption or swelling. Intrinsically semicrystalline polymers are generally less inclined to water diffusion and swelling, in contrast to amorphous ones. For example, $PLA_{100}$ and $PLA_{96}$ absorbed respectively 7% and 12% after 10 weeks in phosphate buffer, whereas $PLA_{50}$ absorbed 150% and largely swelled (146,149,150). In the cases of PLAGA copolymers, $PLA_{75}GA_{25}$ absorbed only 3% after 3 weeks, whereas $PLA_{37.5}GA_{25}$ absorbed 23% in the meantime (148). Again, the rate and ratio of water absorption depend on many factors such as initial morphology, chemical composition and configurational structure, molar mass and molar mass distribution, and polymerization conditions.

Attention has scarcely been paid to chemical composition changes of the copolymers in the field of drug delivery. Hutchinson and Furr reported that the LA/GA ratio increased with degradation of PLAGA copolymers as GA blocks degraded preferentially (137). In contrast, Fredericks et al. claimed that there was no preferential attack on either the LA or GA units of Vicryl® polymer chains despite the difference in hydrophilicity (176). Degradation-related composition changes were observed in the case of $PLA_{75}GA_{25}$ and $PLA_{85}GA_{15}$. The L-LA-enriched segments crystallized under in vitro degradation conditions to form crystalline zones of relatively low molar masses (148,150). The crystallization limited in return the preferential degradation of GA units included in crystalline zones, as confirmed recently by Grijmpa et al. (194). In the case of $PLA_{37.5}GA_{25}$, no significant preferential degradation of GA units was observed probably due to the fact that $PLA_{37.5}GA_{25}$ degraded too fast (148).

Copolymers of PLA or PGA with PCL have also been considered. Pitt et al. investigated the in vivo degradation of copolymers of $\epsilon$-caprolactone with DL-lactide in rabbits (82). Copolymers with 11, 23, 47, and 90 mol % DL-lactide degraded much more rapidly than corresponding homopolymers. Song et al. studied the degradation of random and block copolymers of $\epsilon$-caprolactone with DL-lactide and found that the random copolymers degraded faster than the parent homopolymers, block copolymers degrading at intermediate rates (195). Li and Feng investigated the degradation behaviors of ABA, ABC, and ACB triblock copolymers of $\epsilon$-caprolactone (A), DL-lactide (B), and glycolide (C) in water at 37°C and 50°C (196). The order of degradation rates of the different phases was found to be C > B ≫ A, which permitted various degradation rates by changing copolymer composition. During the degradation process, the crystallinity of the A phase increased. The composition changed too, with an increase in A content and a decrease in B and C. Fukuzaki et al. synthesized a series of low molar mass L-LA-CL copolymers by direct condensation, and studied the in vivo degradation by subcutaneous implantation in the back of rats (22). Pasty copolymers (30–70 mol % CL) degraded faster than solid (0–15 mol % CL) and waxy copolymers (85–100 mol % CL).

The chain structure of the copolymers can be characterized by NMR. In Figure 12, the $^1H$ NMR spectra of a $PLA_{60}CL_{40}$ copolymer and of the homopolymers PCL and $PLA_{100}$ are shown (161). Comparative analysis of these spectra allowed a complete assignment of all the signals

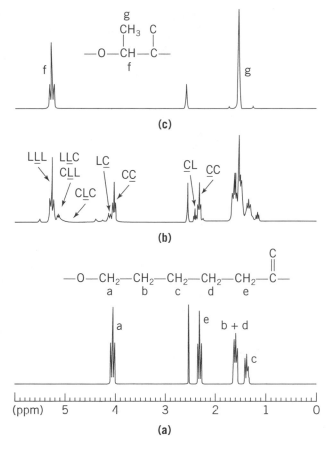

**Figure 12.** $^1$H NMR spectra of (**a**) PCL (**b**), PLA$_{60}$CL$_{40}$, and (**c**) PLA$_{100}$ in DMSO-d$_6$ solutions. Spectra (**a**) and (**c**) were obtained at 80°C due to their insolubility at 30°C. *Source:* From Ref. 161.

observed for each type of protons. Average sequence lengths $\bar{L}_{PLA}$ and $\bar{L}_{PCL}$ were determined by using the following equations:

$$\bar{L}_{PLA} = ([LLL] + [LLC] + [CLL] + [CLC])$$
$$\div \{[CLC] + 1/2([LLC] + [CLL])\} \quad (5)$$

$$\bar{L}_{PCL} = ([CC] + [CL])/[CL] \quad (6)$$

Compositional changes with degradation were also followed by $^1$H NMR. Appearance of lactic acid and of lactyl end units was detected on the $^1$H NMR spectrum. On the other hand, the signal of CC dyads progressively diminished as compared to that of LC ones, which was related to the crystallization of lactyl blocks (161).

Espartero et al. carried out a detailed analysis on the $^1$H and $^{13}$C NMR spectra of L- and DL-lactic acid oligomers (197). The use of DMSO-d$_6$ as a solvent led to a higher spectral resolution as compared to chlorinated solvents. Systematic comparison of oligomer spectra led to the identification of a linearly additive neighboring effect. Assignment of all the $^{13}$C NMR chemical shifts belonging to the different constitutive units was realized up to the octamer. Besides, the well-identified positions of the resonances of chain end units in the $^1$H NMR spectra permitted deter-

mination of the absolute average degree of polymerization of oligomers issued from polycondensation. These findings are of great interest for the study of the ultimate degradation stages of PLA and various copolymers (197).

The drug delivery profiles are also dependent on chemical composition and configurational structure. Ogawa et al. studied the in vivo release of leuprolide acetate from microspheres based on PLAGA copolymers. The drug release profile well agreed with that of the polymer degradation (190). Heya et al. investigated the release of thyrotropin-releasing hormone (TRH) from DL-LA/GA copolymer microspheres of rather low molar masses with LA/GA ratio varying from 100/0 to 50/50. The authors found that after the initial burst, the release rate was dominated mainly by the degradation of the copolymer (198). The microspheres prepared with a higher GA content exhibited a faster release rate.

Pitt et al. developed progesterone release systems (films and microspheres) of homo- and copolymers deriving from ε-caprolactone, DL-lactide, and glycolide (87). The release from PCL homopolymer and PLACL copolymer was diffusion controlled. This was assigned to the high permeability of PCL to drug molecules. The release from PLA$_{50}$ was very slow when diffusion controlled, in agreement with the low solubility and diffusion coefficients of progesterone in this polymer. The release from PLAGA copolymer matrices was associated with polymer degradation. Fong et al. evaluated various PLA, PLAGA, and PLACL microsphere systems containing thioridazine, ketotifen, or hydrocortisone acetate (186). A rather complex pattern of drug release was observed. The release of drugs from semicrystalline PLA$_{100}$ was more rapid than from amorphous PLA$_{100}$. This feature can be explained by the presence of channels and pores within the PLA$_{100}$ matrix, as also observed by Mauduit et al. for a gentamycin-releasing system (97). Both types of copolymers, namely PLAGA (PLA$_{90}$GA$_{10}$, PLA$_{75}$GA$_{25}$, PLA$_{37.5}$GA$_{25}$) and PLACL (PLA$_{80}$CL$_{20}$ and PLA$_{45}$CL$_{10}$), led to faster release than PLA$_{50}$. In the case of the copolymers, it was difficult to correlate drug release to specific parameters. The authors suggested that the influence of polymer composition on release rates may be due to a number of polymer properties such as crystallinity, degradation rate, comonomer ratio, permeability, glass transition, and solubility of the drug in the polymer. When the polymer has a slow degradation rate like PLA$_{100}$, the predominant mode of drug release is theoretically diffusion through the matrix. With fast-degrading polymers like PLA$_{37.5}$GA$_{25}$, the drug release can occur through diffusion and concomitant release of drug due to the matrix degradation (186). Song et al. investigated long-acting release of Norgestrel (MCN) from PLACL block copolymers (195). The block copolymer showed a double release mechanism: diffusion release due to PCL segment and degradation-caused release due to PLA segment. The effects of these two pathways can be balanced by tailoring the ratio of LA/CL to such an extent that the degradation-caused release could just compensate for the declined amount resulted from diffusion release and a whole zero-order kinetics can be actually achieved.

## Molar Mass and Molar Mass Distribution

Molar mass and molar mass distribution are important factors in the polymer degradation and drug release process because of the autocatalytic character of aliphatic polyester hydrolysis. Pitt et al. found that $PLA_{50}$ films with an initial $\overline{M}_n$ of 14,000 were absorbed by week 28, against 60 weeks for those with an initial $\overline{M}_n$ of 49,000 (82). Chawla and Chang investigated the in vivo degradation of four $PLA_{50}$ samples having different high molar masses. At the end of a 48-week implantation period, samples with lower molar mass degraded faster (199). Kaetsu et al. studied the in vivo degradation of four $PLA_{50}$ needle samples with relatively low molar mass (1,100 to 2,200 determined by terminal group titration) (84). In these cases, the higher the initial molar mass, the slower the degradation rate. Ogawa et al. observed that both the lag time and the half-life of $PLA_{50}$ plates increased with molar mass during in vivo degradation (190). Fukuzaki et al. studied the degradation of different $PLA_{50}$ with relatively low molar masses (21). The authors proposed three types of degradation profile according to molar mass, namely parabolic, linear, and S types. The parabolic type observed for $\overline{M}_n$ = 1,500 Da can be explained by the release of the soluble fraction present within the initial oligomers, whereas the S type ($\overline{M}_n$ = 3,500 daltons) reflects a lag time due to higher molar mass.

So far, the effect of molar mass distribution, or polydispersity ($\overline{M}_w/\overline{M}_n$), on PLAGA polymer degradation has been investigated only qualitatively. Nevertheless, the presence of low molar mass species and/or monomers does accelerate the degradation. The elimination of low molar mass and residual monomers is thus very important if one wants a relatively stable polymer. Leeslag et al. showed that the purification of $PLA_{100}$ led to a material more resistant to hydrolytic degradation (179). Nakamura et al. observed an important increase of degradation rate of PLA stereocopolymers in the presence of DL-lactide and glycolide (25). Recently, Mauduit et al. showed that oligomers accelerated the rate of degradation of $PLA_{50}$ films (86).

From these published data, one can conclude that the lower the molar mass, the faster the degradation rate, in agreement with the presence of more carboxylic acid catalyzing groups. Drug release profiles are also affected. In the case of TRH/PLAGA microspheres, Heya et al. observed that the lower the molar mass (in the 6,000–11,000 range), the faster the drug release. Moreover, the initial burst increased with decreasing molar mass (198). Ogawa et al. also observed a more rapid release of drug from lower molar mass PLA or PLAGA matrices (190). Cohen et al. investigated the release of high molar mass, water-soluble proteins from PLAGA microspheres. The authors found that the burst decreased with increasing PLAGA molar mass from 5,000 to 10,000 and 14,000 daltons (142), in agreement with the data of Heya et al. Wada et al. examined low molar mass $PLA_{100}$ and $PLA_{50}$ microspheres containing an anticancer agent (aclacinomycin or ACR) for selective lymphatic delivery. The release rate was reduced as the molar mass was increased from 3,600 to 10,000 for $PLA_{100}$ or from 3,300 to 9,600 for $PLA_{50}$. Polymers with molar mass higher than 10,000 were regarded as not use-

ful (200,201). Similar observations were made by Juni et al. for the release of local anesthetics from PLA microspheres (202). On the other hand, Bodmeier et al. obtained an acceleration of drug release from $PLA_{50}$ films and microspheres containing low molar mass oligomers (85). In the case of PCL-based drug delivery systems, Huatan et al. found that release of bovine serum albumin (BSA) was slower from high molar mass PCL films than from high-low molar masses PCL blends (88).

## Size and Porosity

The size of polymer samples has been recently recognized as an important factor for the degradation of aliphatic polyesters. Kwong et al. observed a rapid disintegration of insulin-containing $PLA_{100}$ pellets 28 days after implantation as compared with microbeads, which remained intact. The faster degradation of pellets was attributed to porosity (90). Ginde and Gupta examined the influence of macroscopic dimensions on the chemical degradation of PGA fibers and pellets (134). The degradation rate was found to be much faster in pellets than in fibers. This feature was assigned to the lack of long range order in pellets. Visscher et al. investigated the effect of particle size on the in vitro and in vivo degradations of three $PLA_{25}GA_{50}$ microsphere samples whose average diameters $\overline{X}$ were respectively less than 45–75 ($\overline{X}$ = 30), 75–106 ($\overline{X}$ = 79) and 106–177 $\mu$m ($\overline{X}$ = 130). It was concluded that over the examined size ranges, there were minimal differences in the degradation properties of polymeric matrices and thus of microcapsules (203). Törmälä et al. evaluated the strength retention of self-reinforced PGA rods with different diameters (1.5, 2.0, 3.2, and 4.5 mm) in a phosphate buffer (204). Small rods were found to lose strength more rapidly than large ones. Differences were related to surface area/mass ratios, which controlled water diffusion and thus hydrolysis. Mauduit et al. investigated the degradation of low and high molar mass $PLA_{50}$ particles and films in vitro (97). In contrast to large-size devices, corresponding SEC data remained monomodal during the whole degradation period. Zhu et al. observed the same phenomenon for microspheres of PLA and two PLAGA copolymers (205). Grizzi et al. investigated the influence of size on the degradation of $PLA_{50}$. Films, powder, and microspheres degraded much less rapidly as compared to large-size specimens (165), showing conclusively that the smaller the polymer size, the slower the degradation rate. This unexpected behavior was well explained by the mechanism of autocatalytic degradation described earlier. In fact, if the size of the polymer matrix is very small, there is no surface-center differentiation. It is of interest to keep in mind that the literature is often misleading in comparing degradation rates of microspheres, fibers, films, pellets, and other massive implants (90,134). Insofar as the drug delivery profiles are concerned, Mauduit et al. observed a faster release of gentamycin from $PLA_{50}$ particles in the size range 0.125–0.25 mm than from larger ones (0.25–0.5 mm and 0.5–1.0 mm). This was assigned to the larger surface area for smaller particles and not to degradation phenomena (97).

The porosity of the polymer matrix is also an important factor. Lam et al. evaluated the in vitro and in vivo degradation of $PLA_{100}$ films. The authors observed a faster degradation of nonporous films as compared with porous ones (206). Smith and Hunneyball prepared 1–10 $\mu$m $PLA_{50}$ microspheres by solvent evaporation and microparticles by grinding the glassy solid formed by the cooling of molten PLA, both containing various amounts of prednisolone (207). The microparticles exhibited a more prolonged drug release profile, indicating that the fusion process may have substantial advantages for thermostable drugs requiring long-term release. Eenink et al. developed hollow fibers for the delivery of hormones (tritium labeled–levonorgestrel). The release was found to be dependent on the membrane structure of the hollow fiber wall. A zero-order release was achieved both in vitro and in vivo for as long as 6 months (91).

### Drug Load and Drug/Polymer Interactions

Many authors reported that drug release rate increased with increasing the drug load (79,108,198,200,201, 208,209). This can be assigned to the morphology and distribution of drugs within the polymer matrix. With small loads, the release is slow because drug molecules are dissolved or molecularly dispersed within the matrix. In contrast, with high loads, drug crystals can be formed. The initial burst generally leads to channels and pores that facilitate diffusion later on (105,210). Kwong et al. examined the in vitro and in vivo releases of insulin from $PLA_{100}$ microbeads and pellets (90). The microbeads exhibited a burst of 50% during the first hour. In contrast, the pellets showed a relatively small burst (10%) and an almost constant release for more than 48 h. The authors suggested that the release from microbeads consists of a rapid dissolution of surface insulin crystals followed by a pore-release mechanism, whereas the release from the pellets is governed by the pore-release model only. It is of interest to note that the two systems were prepared by different methods, namely solvent casting with total entrapment of insulin for the pellets, and solvent evaporation for the microbeads.

In some systems prepared by solvent evaporation, drug molecules can form a metastable molecular dispersion even for high loads. For example, Benoît et al. detected the absence of crystalline progesterone domains within $PLA_{50}$ matrix with a load of 23 wt % (211). Fitzgerald and Corrigan found that levamisole (14.3 wt %) exists in an amorphous form within $PLA_{25}GA_{50}$ microspheres (89).

Matrix-drug interactions can be critical for both matrix degradation and drug delivery profiles. For acidic drugs, one can expect faster hydrolysis of ester bonds because of acid catalysis. In contrast, in the case of basic drugs, two phenomena can be expected: base catalysis of ester bond cleavage and neutralization of carboxyl endgroups of polymer chains, which minimizes or eliminates the autocatalytic effect. Thus, the degradation can be accelerated or slowed down by a base depending on the relative importance of the two effects.

Maulding et al. reported an acceleration of the degradation of $PLA_{50}$ in the presence of a basic drug, thioridazine, which is a tertiary amine compound. Catalysis was attributed to the nucleophilic nature of the amino group

(212,213). Kishida et al. obtained similar results (214). The authors found that the degradation rates of $PLA_{100}$ matrix, which normally hydrolyzes slowly in aqueous media, could be increased by incorporating basic compounds, such as cinnarizine, thioridazine, indenolol, and clonidine, the catalytic effect increasing with the load. The degradation rate depended also on the apparent $pK_a$ of the conjugate acid of each basic compound. The higher the $pK_a$, the faster the degradation according to the following order: indenorol > clonidine > thioridazine > cinnarizine (214). Cha and Pitt examined the effect of four tertiary amines: methadone, naltrexone, promethazine, and meperidine on the hydrolytic rate of $PLA_{100}$ and of $PLA_{80}GA_{20}$ (215). It was found that the catalytic effect of the amines increases with the load, but its order (meperidine > methadone > promethazine ≫ naltrexone) did not correlate with the $pK_a$ nor with $\log(P_{oct})$ (octanol-water partition coefficient) of the amines, in contrast to data reported by Kishida et al. The authors suggested that the steric accessibility of the unsolvated nitrogen might be the determining factor. Moreover, drug release were degradation controlled, faster degradation corresponding to shorter induction time prior to drug release (215). The same authors also examined the case of PLACL copolymer (75–85 mol % L-LA units). In contrast to $PLA_{100}$ and of $PLA_{80}GA_{20}$, Fickian diffusion was responsible for the drug release kinetics (216). Fitzgerald and Corrigan investigated the mechanism governing the release of levamisole from $PLA_{25}GA_{50}$ microspheres and discs (89). Drug release profiles were sigmoidal and fit a degradation-controlled release model. On the other hand, polymer degradation was dramatically accelerated with increasing the load.

Other authors observed a decrease of the degradation rate or of the drug release rate due to interactions between polymer chain ends and drugs. Bodmeier and Chen evaluated degradable PLA pellets prepared by pressing without heat or solvent (79). They detected a lag time in drug release in the case of low molar mass $PLA_{50}$ pellets. This was assigned to interactions between drugs (quinidine sulfate or propranolol hydrochloride) and carboxyl endgroups. Drug-polymer interactions were absent in the case of $PLA_{100}$ pellets. The same authors suspected similar interactions in the case of microspheres and films made of high and low molar masses $PLA_{50}$ blends containing quinidine (85). Mauduit et al. investigated various gentamycin $PLA_{50}$ blends (86,97,217). It was found that the base form of the drug was able to neutralize polymer chain ends on blending in acetone, whereas the sulfate form required an aqueous environment. In the case of gentamycin sulfate–low molar mass $PLA_{50}$ systems, interactions between drug and chain ends stabilized the matrix, the free drug in excess being rapidly released (217). Both phenomena acted against base catalysis.

The effects of sparingly soluble additives were also investigated. Verheyen et al. examined the physicochemical properties of hydroxyapatite-$PLA_{100}$ composites in solution tests (218). Although the authors did not mention it, data showed that the higher the hydroxyapatite content, the slower the decrease of molar mass. Li and Vert investigated the hydrolytic behavior of $PLA_{50}$-coral blends (219). It was found that coral, which is composed of calcium carbonate, significantly slowed down the degradation of PLA$_{50}$ matrix and suppressed the faster internal degradation. Zhang et al. observed a decrease of degradation rate by 1.7 to 3.0 times with incorporation of salts such as Mg(OH)$_2$, MgCO$_3$, CaCO$_3$, and ZnCO$_3$ into PLA$_{25}$GA$_{50}$ films (220). The stabilizing effect of hydroxyapatite, coral, and other metal salts on the hydrolysis of polymer chains can be assigned to the neutralization of carboxyl end groups by the additives and/or by the degradation medium that penetrated the matrix due to the presence of the polymer-additive interfaces.

Li et al. investigated the mechanism of hydrolytic degradation of PLA$_{50}$ matrix in the presence of a tertiary amine, namely caffeine, in order to elucidate the influence of this basic compound on the hydrolytic cleavage of polyester chains (183). Caffeine was incorporated into PLA$_{50}$ in various contents (0% to 20%) by blending in acetone followed by solvent evaporation. The resulting blends were processed to 1.5-mm-thick plates and 0.3-mm-thick films by compression molding. Degradation was carried out in isoosmolar pH = 7.4 phosphate buffer at 37°C. The effects of caffeine on degradation characteristics were rather complex and largely depended on the load (Fig. 13). For low contents ($\leq$2%), caffeine was molecularly dispersed in matrices and accelerated considerably the degradation with respect to caffeine-free devices. However, the increase of degradation rate was not proportional to the caffeine load due to the combined effects of base–carboxyl endgroup interaction, crystallization, and matrix-controlled or channeling-controlled diffusion of caffeine. In the early stages of degradation, the overall catalytic effect was larger for devices with low caffeine contents than for highly loaded ones where caffeine was in crystallized state and thus less available for basic catalysis. At the later stages, however, the neutralization of carboxyl endgroups became predominant and governed the degradation in the case of highly loaded devices (Fig. 13). From a general viewpoint, plates degraded slightly faster than films.

Therefore, the degradation of PLA polymers in the presence of basic drugs appeared rather complex because of the contribution of a number of parameters (i.e., base catalysis, neutralization of carboxyl endgroups, porosity, dimensions of devices, drug load, and drug morphology). They all contribute to make the drug itself an important and specific factor for the control of both drug release and matrix degradation. None of these factors can be considered separately if one wants to understand the effect of basic drugs on the properties of drug delivery systems.

### γ Irradiation

Several authors examined the influence of $\gamma$ irradiation on the degradation of PLAGA biomedical devices. Data are of interest in the field of drug delivery too. Gupta and Deshmukh claimed that irradiation in air or in nitrogen atmosphere can generate chain scission and cross-linking simultaneously (221). Chu et al. considered chain scission as predominant on the basis of the reduction in mechanical properties of Dexon® sutures upon $\gamma$ irradiation (222). The scission process was speculated to depend on free radical chemistry. In addition, the authors observed formation of surface cracks during the hydrolysis of Dexon® sutures whose number and regularity increased with increasing irradiation dosages. In another study, Chu showed that $\gamma$ irradiation of Dexon® and Vicryl® fibers resulted in an earlier appearance of pH fall in the degradation medium (177). All these observations are in agreement with a molar mass decrease on irradiation although no molar mass data are available.

Tsai et al. carried out effective sterilization of PLA–mitomycin C microcapsules for parenteral use by $^{60}$Co $\gamma$ irradiation (108). The authors found that the sterilization did not affect microcapsule structure, release rate, and drug stability. However, no molar mass measurements were performed. Spenlehauer et al. found that $\gamma$ sterilization dramatically decreased the molar masses of PLA$_{50}$, PLA$_{45}$GA$_{10}$, and PLA$_{37.5}$GA$_{25}$ microspheres, and this degradation continued on storage for GA-containing compounds (223). $\gamma$ sterilization also modified the release pattern of cisplatin-loaded microspheres. Birkinshaw et al. studied $\gamma$ irradiation of compression-molded PLA$_{50}$ samples in air with doses as high as 10 Mrad (224). Substantial embrittlement occurred at higher dose levels and the irradiated material absorbed water at a slightly slower rate than the unirradiated one. The authors concluded that the primary effect of irradiation on hydrolytic degradation was associated with the initial reduction in molar mass, the degradation mechanism remaining the same. Recently, Volland et al. investigated the influence of $\gamma$ sterilization on captopril-containing PLA$_{25}$GA$_{50}$ microspheres (225). The molar mass of the polymer decreased with increasing irradiation dose, but the polydispersity remained unchanged, suggesting a random chain cleavage rather than an unzipping process.

### pH and Ionic Strength

Chu examined the effect of pH on the degradation of Dexon® and Vicryl® sutures by using three different buffer solutions with pH = 5.25, 7.44, and 10.06, respectively (226,227). For Vicryl® sutures, maximum retention of tensile properties occurred around pH 7.0. For Dexon® su-

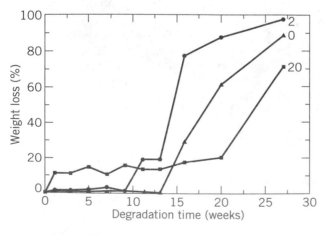

**Figure 13.** Weight loss of PLA$_{50}$ plates containing 0%, 2%, and 20% of caffeine base with degradation in a pH 7.4 phosphate buffer at 37°C. *Source:* From Ref. 183.

tures, no major difference was observed between physiological and acidic media. In contrast, the alkaline buffer had a dramatic effect on Dexon® degradation (226,227). Makino et al. observed that PLA$_{100}$ degraded rapidly in a strongly alkaline solution or in solutions of high ionic strength. The effects of pH and ionic strength were interpreted in terms of electric potential distribution at the polymer–solution interface. Degradation was also found affected by salt concentration in buffer solutions, suggesting that the cleavage of polymer ester bonds was accelerated by conversion of the acidic degradation products into neutral salts (228).

For Reed and Gilding, pH changes had no effect on the in vitro degradation of PLAGA polymers in citrate-phosphate buffers at pH 5 and pH 7 and boric acid–borax buffer at pH 9. Any sensitivity of PLAGA polymers to pH was actually dominated by combined effects of crystallinity and hydrophobicity, according the authors (188). Kenley et al. studied the in vitro degradation of PLA$_{25}$GA$_{50}$ cylindrical samples in aqueous buffers with pH varying from 4.5 to 7.4 at 37°C (139). They obtained superimposable total mass and molar mass profiles for all samples, suggesting pH independence for the in vitro hydrolysis. Ginde and Gupta studied the in vitro degradation of PGA pellets and fibers in four different buffer solutions with pH 4.7, 7.0, 9.2, and 10.6, respectively (134). No major difference was found between the slightly acidic and neutral media. However, the two alkaline solutions accelerated considerably the degradation of PGA samples.

Li et al. comparatively investigated the in vitro degradation of large-size PLAGA polymers in three different media: distilled water, iso-osmolar phosphate buffer, and saline (146,148,149). There was no significant difference between the two iso-osmolar media despite the pH fall observed in saline at the later stages of degradation. In contrast, the absence of ionic strength in distilled water promoted water absorption and thus the surface-center differentiation in the early stages (Fig. 14). Furthermore, the phosphate buffer enhanced solubilization of degradation products in the later stages (Fig. 15), the carboxylate form RCOO$^-$ of organic acids being more hydrophilic than the carboxylic form RCOOH (148).

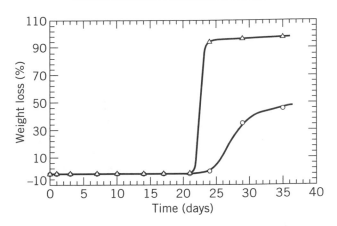

**Figure 15.** Weight loss of PLA$_{37.5}$GA$_{25}$ plates with degradation at 37°C in a pH 7.4 phosphate buffer ($\triangle$) and in water ($\bigcirc$). *Source: From Ref. 148.*

Therefore, one can reasonably conclude that alkaline and strong acidic media accelerate polymer degradation. The difference between slightly acidic and physiological media, however, is much less pronounced due to autocatalysis by carboxyl endgroups. The effects of pH and ionic strength on drug release profiles are rarely discussed in the literature. Nevertheless, one can assume that for degradation-controlled release, drug release should be enhanced if degradation is enhanced. In the case of diffusion-controlled release, one should take into account the solubility of the drug in the external medium.

## Comparison Between in Vitro and in Vivo Behaviors

Many authors studied comparatively the matrix degradation behaviors in vitro and in vivo. Chegini et al. carried out a comparative SEM study on degradation of Lactomer® (PLA$_{70}$GA$_{30}$) ligating clips in vivo by implantation in the incision of rabbits and in vitro by incubation in pH 7.3 phosphate-buffered saline (229). Lactomer® clips were found to show a greater change in proportion of breakdown comparing in vivo to in vitro. Törmälä et al. observed a faster loss of mechanical strength of self-reinforced PGA rods and PLA$_{100}$ screws and plates in vivo than in vitro (204,230). This difference in rates of strength loss was assigned to the effect of cellular enzymes or other biological or biochemical factors as well as the mechanical stresses caused to the implants by the rabbits' movements. Albertsson and Karlsson also suggested that the higher in vivo degradation rate resulted probably from the dynamics in in vivo systems (231). Thérin et al. (151) and Spenlehauer et al. (223) also found a faster degradation in vivo than in vitro. Differences were assigned to mechanical stresses due to muscular movements of the rabbits (151), in agreement with data reported by Suuronen et al. (230) and Albertsson and Karlsson (231).

Leeslag et al. investigated the in vivo degradation of high molar PLA$_{100}$ devices used for fixation of mandibular fractures in sheep and dogs or subcutaneously implanted in rats. They also considered the in vitro degradation in

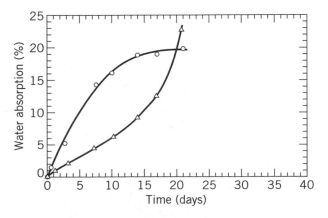

**Figure 14.** Water absorption profiles of PLA$_{37.5}$GA$_{25}$ plates with degradation at 37°C in a pH 7.4 phosphate buffer ($\triangle$) and in water ($\bigcirc$). *Source: From Ref. 148.*

pH = 6.9 phosphate-buffered saline (179). The authors showed that except for dynamically loaded bone plates, there were no significant differences between in vivo and in vitro degradation, thus excluding the additional effect of enzymes. Kenley et al. also found no major difference between $PLA_{25}GA_{50}$ copolymer degradation kinetics in vitro in buffer solutions and in vivo after subcutaneous implantation in rats (139). In contrast, Visscher et al. observed a much faster degradation of $PLA_{25}GA_{50}$ in vitro than in vivo and attributed it to the totally aqueous environment and constant physical movement that the microspheres were subjected to in vitro (203).

Authors also compared the drug release behaviors in vitro and in vivo. Ikada et al. investigated degradation and release behaviors of a dideoxykanamycin B (DKB)–containing composite material prepared from hydroxyapatite and low molar mass $PLA_{50}$ in vitro in pH 8 phosphate buffered saline and in vivo by implantation into the fenestrated tibias of rats (232). Both the amount and the period of retention of DKB were much greater in vivo than in vitro, indicating a faster degradation in vitro. In contrast, Smith and Hunneyball observed a faster release of prednisolone from $PLA_{50}$ microspheres in vivo than in vitro, which was assigned to the faster penetration by biological fluids and/or faster degradation of PLA in vivo (207). On the other hand, Kwong et al. (90), Sanders et al. (138), and Heya et al. (198) obtained a good correlation between in vivo and in vitro drug release profiles.

Therefore, the literature shows that in most cases, drug delivery systems based on PLAGA polymers exhibited comparable in vitro and in vivo matrix degradation and drug release behaviors. Whenever differences are observed, effects of physical factors (temperature, stirring, pH) or physiological ones related to implantation sites (subcutaneous, intramuscular, or in bony tissues) should be considered prior to any reference to enzymatic activities.

## CONCLUSION

Degradable aliphatic polyesters, in particular those deriving from lactides, glycolide, and $\epsilon$-caprolactone, have been most widely investigated for applications in the field of sustained drug delivery during the past two decades. Factors that can affect the hydrolytic degradation of polymers and the release profiles of drugs have been gradually identified. This chapter has tried to present the state-of-the-art on the basis of recent advances in this domain. One of the first conclusions is that polymers aimed at making drug delivery systems can bear the same name and be very different for many reasons. Sources of differences can be found at any stage of the history of a drug delivery system: polymer synthesis, device processing, drug-polymer blending, sterilization, release conditions, and implantation sites. Most of these sources are sometimes neglected, ignored, or not taken into account in scientific papers. It is only by considering all of them simultaneously that one can expect success in a real control of the drug delivery profiles from polymeric matrices.

## BIBLIOGRAPHY

1. A.S. Hoffman, *ACS Symp. Ser.* **256**, 13–29 (1984).
2. D.V. Rosato, in M. Szycher, ed., *Biocompatible Polymers, Metals and Composites*, Technomic Publishing, Lancaster, Pa., 1983, pp. 1019–1067.
3. U.S. Pat. 2,668,162 (Feb. 2, 1954), E.L. Charles and N.Y. Buffalo (to E.I. DuPont de Nemours).
4. U.S. Pat. 3,297,033 (Jan. 10, 1967), E.E. Schmitt and R.A. Polistina (to Am. Cyanamid Co.).
5. E.J. Frazza and E.E. Schmitt, *J. Biomed. Mater. Res. Symp.* **1**, 43–58 (1971).
6. S. Li and M. Vert, in G. Scott and D. Gilead, eds., *Degradable Polymers: Principles and Applications*, Chapman & Hall, London, 1995, pp. 43–87.
7. S.J. Huang et al., *Am. Chem. Soc. Stab. Degradation Polym.* **17**, 209–214 (1978).
8. R.L. Kronenthal, in R.L. Kronenthal, Z. User, and E. Martin, eds., *Polymers in Medicine and Surgery*, Plenum, New York, 1974, pp. 119–137.
9. S.J. Holland, B.J. Tighe, and P.L. Gould, *J. Controlled Release* **4**, 155–180 (1986).
10. M. Vert, S. Li, G. Spenlehauer, and P. Guérin, *J. Mater. Sci., Mater. Med.* **3**, 432–446 (1992).
11. S. Li, *J. Biomed. Mater. Res., Appl. Biomater.* **48**, 342–353 (1999).
12. R.L. Dunn, in J.O. Hollinger, ed., *Biomedical Applications of Synthetic Biodegradable Polymers*, CRC Press, Boca Raton, Fla., 1995, pp. 17–31.
13. R.H. Müller and K.H. Wallis, *Int. J. Pharm.* **89**, 25–31 (1993).
14. C. Braud and M. Vert, in S.W. Shalaby, A.S. Hoffman, B.D. Ratner, and T.A. Horbett eds., *Polymers as Biomaterials*, Plenum, New York, 1984, pp. 1–16.
15. M. Yokoyama et al., *Makromol. Chem., Rapid Commun.* **8**, 431–435 (1987).
16. S. Gautier, M. Boustta, and M. Vert, *J. Bioact. Compat. Polym.* **12**, 77–98 (1997).
17. H. Ringsdorf, *J. Polym. Sci., Polym. Symp.* **51**, 135–153 (1975).
18. M. Vert, *CRC Crit. Rev. Ther. Drug Carrier Syst.* **2**, 291–327 (1985).
19. A.M. Reed and D.K. Gilding, *Polymer* **20**, 1459–1464 (1979).
20. M. Spinu, C. Jackson, M.Y. Keating, and K.H. Gardner, *J. Macromol. Sci. Pure Appl. Chem.* **A33**, 1497–1530 (1996).
21. H. Fukuzaki, M. Yoshida, M. Asano, and M. Kumakura, *Eur. Polym. J.* **25**, 1019–1026 (1989).
22. H. Fukuzaki, M. Yoshida, M. Asano, and M. Kumakura, *Polymer* **31**, 2006–2014 (1990).
23. N. Wang et al., *Proc. Am. Chem. Soc.* **76**, 373–374 (1997).
24. M. Vert, P. Christel, F. Chabot, and J. Leray, in G.W. Hastings and P. Ducheyne, eds., *Macromolecular Biomaterials*, CRC Press, Boca Raton, Fla., 1984, pp. 119–142.
25. T. Nakamura et al., *J. Biomed. Mater. Res.* **23**, 1115–1130 (1989).
26. R.A. Miller, J.M. Brady, and D.E. Cutright, *J. Biomed. Mater. Res.* **11**, 711–719 (1977).
27. F. Chabot, M. Vert, S. Chapelle, and P. Granger, *Polymer* **24**, 53–60 (1983).
28. K. Hiltunen, J.V. Soppälä, and M. Härkönen, *Macromolecules* **30**, 373–379 (1997).

29. European Pat. 0,443,542 A2 (Feb. 20, 1991), B. Buchholz (to Boehringer Ingelheim International GmbH).

30. S. Bezzi, L. Riccoboni, and C. Sullam, *Mem. R. Accad. Ital. Cl. Sci. Fis. Mat. Nat.* **8**, 201–211 (1937).

31. J. Kleine and H. Kleine, *Makromol. Chem.* **30**, 23–38 (1959).

32. V.F. Jenkins, *Polym. Paint Colour J.* **167**, 626–627 (1977).

33. E.F. Cox, F. Hostettler, and R.R. Kiser, *Macromol. Synth.* **3**, 111–113 (1969).

34. M. Morton and M. Wu, *ACS Symp. Ser.* **286**, 175–182 (1985).

35. R.S. Velishkova, V.D. Toncheva, and I.M. Panayotov, *J. Polym. Sci., Part A* **25**, 3283–3292 (1987).

36. P. Dreyfuss, T. Adaway, and J.P. Kennedy, *Appl. Polym. Symp.* **30**, 183–192 (1977).

37. W.H. Carothers, G.L. Dorough, and F.J. Van Natta, *J. Am. Chem. Soc.* **54**, 761–772 (1932).

38. R.K. Kulkarni, K.C. Pani, C. Neuman, and F. Leonard, *Arch. Surg.* **93**, 839–843 (1966).

39. U.K. Pat. 779,291 (Jan. 28, 1955), J. Kleine.

40. M.C. Tanzi et al., *J. Mater. Sci., Mater. Med.* **5**, 393–396 (1994).

41. G. Schwach, J. Coudane, R. Engel, and M. Vert, *J. Polym. Sci., Part A* **35**, 3431–3440 (1997).

42. E. Lillie and R.C. Schulz, *Makromol. Chem.* **176**, 1901–1906 (1975).

43. K. Chujo et al., *Makromol. Chem.* **100**, 262–266 (1967).

44. H.R. Kricheldorf and A. Serra, *Polym. Bull.* **14**, 497–502 (1985).

45. R. Dunsing and H.R. Kricheldorf, *Polym. Bull.* **14**, 491–496 (1985).

46. H.R. Kricheldorf and I. Kreiser-Saunders, *Makromol. Chem.* **191**, 1057–1066 (1990).

47. Z. Jedlinski and W. Walach, *Makromol. Chem.* **192**, 2051–2057 (1991).

48. L. Sipos, M. Zsuga, and T. Kelen, *Polym. Bull.* **27**, 495–502 (1992).

49. N. Spassky, *Makromol. Chem., Macromol. Symp.* **42/43**, 15–49 (1991).

50. P. Kurcok, J. Penczck, J. Franek, and Z. Jedlinski, *Macromolecules* **25**, 2285–2289 (1992).

51. M. Bero, G. Adamus, J. Kasperczyk, and H. Janeczek, *Polym. Bull.* **31**, 9–14 (1993).

52. Z. Jedlinski et al., *Makromol. Chem.* **194**, 1681–1689 (1993).

53. H.R. Kricheldorf and R. Dunsing, *Makromol. Chem.* **187**, 1611–1625 (1986).

54. H.R. Kricheldorf and I. Kreiser-Saunders, *Makromol. Chem.* **188**, 1861–1873 (1987).

55. H.R. Kricheldorf and I. Kreiser-Saunders, *J. Macromol. Sci., Chem.* **A24**, 1345–1356 (1987).

56. F.E. Kohn, J.G. Van Ommen, and J. Feijen, *Eur. Polym. J.* **19**, 1081–1088 (1983).

57. A.J. Nijenhuis, D.W. Grijmpa, and A.J. Pennings, *Macromolecules* **25**, 6419–6424 (1992).

58. M. Bero, J. Kasperczyk, and Z. Jedlinski, *Makromol. Chem.* **191**, 2287–2296 (1990).

59. A.M. Jonte, R. Dunsing, and H.R. Kricheldorf, *J. Macromol. Sci., Chem.* **A23**, 495–514 (1986).

60. R. Dunsing and H.R. Kricheldorf, *Eur. Polym. J.* **24**, 145–150 (1988).

61. E.J. Goethals, *Makromol. Chem., Macromol. Symp.* **42/43**, 51–68 (1991).

62. M. Bero, J. Kasperczyk, and G. Adamus, *Makromol. Chem.* **194**, 907–912 (1993).

63. H.R. Kricheldorf, J.M. Jonte, and M. Berl, *Makromol. Chem., Suppl.* **12**, 25–38 (1985).

64. X. Zhang, D.A. MacDonald, M.F.A. Goosen, and K.B. Mcauley, *J. Polym. Sci., Part A* **32**, 2965–2970 (1994).

65. U.S. Pat. 5,028,667 (July 2, 1991), S.J. McLain and N.E. Drysdale (to E.I. DuPont de Nemours).

66. A. Le Borgne, V. Vincens, M. Jouglard, and N. Spassky, *Makromol. Chem., Macromol. Symp.* **73**, 37–46 (1993).

67. N. Spassky, M. Wisniewski, C. Pluta, and A. Le Borgne, *Macromol. Chem. Phys.* **197**, 2627–2637 (1996).

68. P. Dubois, C. Jacobs, R. Jérôme, and P. Teyssié, *Macromolecules* **24**, 2266–2270 (1991).

69. H.R. Kricheldorf, I. Kreiser-Saunders, and C. Boettcher, *Polymer* **36**, 1253–1259 (1995).

70. G. Schwach, J. Coudane, R. Engel, and M. Vert, *Polym. Bull.* **37**, 771–776 (1996).

71. G. Schwach, J. Coudane, R. Engel, and M. Vert, *Polym. Int.* **46**, 177–182 (1998).

72. M. Vert, G. Schwach, R. Engel, and J. Coudane, *J. Controlled Release* **53**, 85–92 (1998).

73. R. Arshady, *J. Bioact. Compat. Polym.* **5**, 315–342 (1990).

74. D.A. Wood, *Int. J. Pharm.* **7**, 1–18 (1980).

75. A. Schindler et al., *Contem. Top. Polym. Sci.* **2**, 251–289 (1977).

76. C.G. Pitt, *Drugs Pharm. Sci.* **45**, 71–119 (1990).

77. S.Y. Jeong and S.W. Kim, *Arch. Pharm. Res.* **9**, 63–73 (1986).

78. Y. Ogawa, *J. Biomater. Sci., Polym. Ed.* **8**, 391–409 (1997).

79. R. Bodmeier and H. Chen, *J. Pharm. Sci.* **78**, 819–822 (1989).

80. D.L. Wise, H. Rosenkrantz, J.B. Gregory, and H.J. Esber, *J. Pharm. Pharmacol.* **32**, 399–403 (1980).

81. M.C. Meikle et al., *Biomaterials* **14**, 177–183 (1993).

82. C.G. Pitt et al., *Biomaterials* **2**, 215–220 (1981).

83. M. Yoshida et al., *Polym. J.* **14**, 941–950 (1982).

84. I. Kaetsu et al., *J. Controlled Release* **6**, 249–263 (1987).

85. R. Bodmeier, K.H. Oh, and H. Chen, *Int. J. Pharm.* **51**, 1–8 (1989).

86. J. Mauduit, N. Bukh, and M. Vert, *J. Controlled Release* **25**, 43–49 (1993).

87. C.G. Pitt et al., *J. Pharm. Sci.* **68**, 1534–1538 (1979).

88. H. Huatan, J.H. Collett, D. Attwood, and C. Booth, *Biomaterials* **16**, 1297–1303 (1995).

89. J.F. Fitzgerald and O.I. Corrigan, *J. Controlled Release* **42**, 125–132 (1996).

90. A.K. Kwong et al., *J. Controlled Release* **4**, 47–62 (1986).

91. M.J.D. Eenink et al., *J. Controlled Release* **6**, 225–247 (1987).

92. J.M. Schakenraad et al., *Biomaterials* **9**, 116–120 (1988).

93. J.M. Goodson et al., *J. Periodontol.* **54**, 575–579 (1983).

94. Y. Kinoshita et al., *Biomaterials* **14**, 729–736 (1993).

95. N.H. Shah et al., *J. Controlled Release* **27**, 139–147 (1993).

96. S. Miyamoto and K. Takaoka, *Annal. Chir. Gynaecol.* **82**, 69–76 (1993).

97. J. Mauduit, N. Bukh, and M. Vert, *J. Controlled Release* **23**, 221–230 (1993).

98. T.M.S. Chang, *J. Bioeng.* **1**, 25–32 (1976).

99. R.H. Asch et al., *J. Androl.* **6**, 83–88 (1985).

100. L.M. Sanders et al., *J. Pharm. Sci.* **73**, 1294–1297 (1984).

101. V. Vidmar, S. Pepeljnjak, and I. Jalsenjak, *J. Microencapsul.* **2**, 289–292 (1985).

102. M. Nakano et al., *Int. J. Pharm.* **4**, 291–298 (1980).

103. L.R. Beck, V.Z. Pope, T.R. Tice, and R.M. Gilley, *Adv. Contracept.* **1**, 119–129 (1985).

104. K. Susuki and J.C. Price, *J. Pharm. Sci.* **74**, 21–24 (1985).

105. G. Spenlehauer, M. Verllard, and J.P. Benoit, *J. Pharm. Sci.* **75**, 750–755 (1986).

106. R. Jalil and J.R. Nixon, *J. Microencapsul.* **6**, 473–484 (1989).

107. A. Carrio, G. Schwach, J. Coudane, and M. Vert, *J. Controlled Release* **37**, 113–121 (1995).

108. D.C. Tsai et al., *J. Microencapsul.* **3**, 181–193 (1986).

109. R. Jalil and J.R. Nixon, *J. Microencapsul.* **7**, 375–383 (1990).

110. O. Ike et al., *Biomaterials* **13**, 230–234 (1992).

111. M. Kyo, S.H. Hyon, and Y. Ikada, *J. Controlled Release* **35**, 73–77 (1995).

112. R. Bodmeier and H. Chen, *J. Pharm. Pharmacol.* **40**, 754–757 (1988).

113. S. Takada, Y. Uda, H. Toguchi, and Y. Ogawa, *J. Pharm. Sci. Technol.* **49**, 180–184 (1995).

114. M. Vert et al., in Y. Doi and K. Fukuda, eds., *Biodegradable Plastics and Polymers*, Elsevier, Amsterdam, 1994, pp. 11–23.

115. D.K. Gilding, *Biocompat. Clin. Implant. Mater.* **2**, 209–232 (1981).

116. R.D. Gilbert, V. Stannett, C.G. Pitt, and A. Schindler, in N. Grassie, ed., *Development in Polymer Degradation*, vol. 4, Applied Science Publishers, London, 1982, pp. 259–293.

117. D.F. Williams, *J. Bioeng.* **1**, 279–294 (1977).

118. J.E. Kemnitzer, R. Gross, and S.P. McCarthy, *Macromolecules* **25**, 5227–5234 (1992).

119. A.M. Reed, Ph.D. Thesis, University of Liverpool, 1978.

120. J.B. Herrmann, R.J. Kelly, and G.A. Higgins, *Arch. Surg. (Chicago)* **100**, 486–490 (1970).

121. T.N. Salthouse and B.F. Matlaga, *Surg., Gynecol. Obstet.* **142**, 544–550 (1975).

122. J.M. Schakenraad et al., *J. Biomed. Mater. Res.* **24**, 529–545 (1990).

123. H. Younes et al., *Biomater. Artif. Cells, Artif. Organs* **16**, 705–719 (1988).

124. G.E. Zaikov, *J. Macromol. Sci., Rev. Macromol. Chem. Phys.* **C25**, 551–597 (1985).

125. D.F. Williams, *Eng. Med.* **10**, 5–7 (1981).

126. S.L. Ashley and J.W. McGinity, *Congr. Int. Technol. Pharm.* **5**, 195–204 (1989).

127. M.S. Reeve, S.P. McCarthy, M.J. Downey, and R.A. Gross, *Macromolecules* **27**, 825–831 (1994).

128. R.T. MacDonald, S.P. McCarthy, and R.A. Gross, *Macromolecules* **29**, 7356–7361 (1996).

129. A. Schindler and C.G. Pitt, *Polym. Prepr., Am. Chem. Soc., Div. Polym. Chem.* **23**, 111–112 (1982).

130. C.G. Pitt, R.W. Hendren, A. Schindler, and S.C. Woodward, *J. Controlled Release* **1**, 3–14 (1984).

131. M. Mochizuki et al., *J. Appl. Polym. Sci.* **55**, 289–296 (1995).

132. S.C. Woodward et al., *J. Biomed. Mater. Res.* **19**, 437–444 (1985).

133. P.A. Kramer and T. Burnstein, *Life Sci.* **19**, 515–520 (1976).

134. R.M. Ginde and R.K. Gupta, *J. Appl. Polym. Sci.* **33**, 2411–2429 (1987).

135. M. Singh, A. Singh, and G.P. Talwar, *Pharm. Res.* **8**, 958–961 (1991).

136. Y. Kimura, Y. Matsuzaki, H. Yamane, and T. Kitao, *Polymer* **30**, 1342–1349 (1989).

137. F.G. Hutchinson and B.J.A. Furr, *Biochem. Soc. Trans.* **13**, 520–523 (1985).

138. L.M. Sanders, G.I. McRae, K.M. Vitale, and B.A. Kell, *J. Controlled Release* **2**, 187–195 (1985).

139. R.A. Kenley, M.O. Lee, T.R. Mahoney, II, and L.M. Sanders, *Macromolecules* **20**, 2398–2403 (1987).

140. J.M. Schakenraad et al., *J. Biomed. Mater. Res.* **23**, 1271–1288 (1989).

141. J. Helder, P.J. Dijkstra, and J. Feijen, *J. Biomed. Mater. Res.* **24**, 1005–1020 (1990).

142. S. Cohen et al., *Pharm. Res.* **8**, 713–720 (1991).

143. T. St. Pierre and E. Chiellini, *J. Bioact. Compat. Polym.* **2**, 4–30 (1987).

144. D.H. Lewis, *Drugs Pharm. Sci.* **45**, 1–41 (1990).

145. C.G. Pitt, in M. Vert et al., eds., *Biodegradable Polymers and Plastics*, Royal Society of Chemistry, London, 1992, pp. 7–19.

146. S. Li, H. Garreau, and M. Vert, *J. Mater. Sci., Mater. Med.* **1**, 123–130 (1990).

147. K.R. Huffman and D.J. Casey, *J. Polym. Sci., Polym. Chem. Ed.* **23**, 1939–1954 (1985).

148. S. Li, H. Garreau, and M. Vert, *J. Mater. Sci., Mater. Med.* **1**, 131–139 (1990).

149. S. Li, H. Garreau, and M. Vert, *J. Mater. Sci., Mater. Med.* **1**, 198–206 (1990).

150. M. Vert, S. Li, and H. Garreau, *J. Controlled Release* **16**, 15–26 (1991).

151. M. Thérin et al., *Biomaterials* **13**, 594–600 (1992).

152. M. Vert, S. Li, and H. Garreau, *Clin. Mater.* **10**, 3–8 (1992).

153. S. Li and M. Vert, *Macromolecules* **27**, 3107–3110 (1994).

154. E.A. Schmitt, D.R. Flanagan, and R.J. Linhardt, *Macromolecules* **27**, 743–748 (1994).

155. S.A.M. Ali, P.J. Doherty, and D.F. Williams, *J. Biomed. Mater. Res.* **27**, 1409–1418 (1993).

156. X. Zhang, U.P. Wyss, D. Pichora, and M.F.A. Goosen, *J. Bioact. Compat. Polym.* **9**, 80–100 (1994).

157. H. Pistner et al., *Biomaterials* **14**, 671–677 (1993).

158. H. Pistner et al., *Biomaterials* **15**, 439–450 (1994).

159. H. Pistner, D.R. Bendix, J. Müling, and J. Reuther, *Biomaterials* **14**, 291–298 (1993).

160. T.G. Park, *Biomaterials* **16**, 1123–1130 (1995).

161. S. Li, J.L. Espartero, P. Foch, and M. Vert, *J. Biomater. Sci., Polym. Ed.* **8**, 165–187 (1996).

162. A. Löfgren and A. Albertsson, *J. Appl. Polym. Sci.* **52**, 1327–1338 (1994).

163. A.C. Albertsson and M. Eklund, *J. Appl. Polym. Sci.* **57**, 87–103 (1995).

164. H. Schliephake, D. Klosa, and M. Rahlff, *J. Biomed. Mater. Res.* **27**, 991–998 (1993).

165. I. Grizzi, H. Garreau, S. Li, and M. Vert, *Biomaterials* **16**, 305–311 (1995).

166. C.G. Pitt et al., *J. Appl. Polym. Sci.* **26**, 3779–3787 (1981).

167. M. Vert, in G. Walenkamp, ed., *Biomaterials in Surgery*, G.T. Verlag, Stuttgart, 1998, pp. 97–101.

168. A. Torres, S. Roussos, S.M. Li, and M. Vert, *Appl. Environ. Microbiol.* **62**, 2393–2397 (1996).

169. A. Torres, S. Li, S. Roussos, and M. Vert, *J. Environ. Polym. Degradation* **4**, 213–223 (1996).

170. A. Torres, S. Li, S. Roussos, and M. Vert, *J. Appl. Polym. Sci.* **62**, 2295–2302 (1996).

171. P. Jarrett et al., in S.W. Shalaby, A.S. Hoffman, B.D. Ratner, and T.A. Horbett, eds., *Polymers as Biomaterials*, Plenum, New York, 1985, pp. 181–192.

172. F. Lefebvre, C. David, and C. Vander Wauven, *Polym. Degradation Stab.* **45**, 347–353 (1994).

173. S.I. Akahori and Z. Osawa, *Polym. Degradation Stab.* **45**, 261–265 (1994).

174. E.W. Fischer, H.J. Sterzel, and G. Wegner, *Kolloid-Z. Z. Polym.* **251**, 980–990 (1973).

175. B.K. Carter and G.L. Wilkes, in S.W. Shalaby, A.S. Hoffman, B.D. Ratner, and T.A. Horbert, eds., *Polymers as Biomaterials*, Plenum, New York, 1984, pp. 67–92.

176. R.J. Fredericks, A.J. Melveger, and L.J. Dolegiewtz, *J. Polym. Sci., Polym. Phys. Ed.* **22**, 57–66 (1984).

177. C.C. Chu, *Polymer* **26**, 591–594 (1981).

178. C. Chu, *J. Appl. Polym. Sci.* **26**, 1727–1734 (1981).

179. J.W. Leeslag et al., *Biomaterials* **8**, 311–314 (1987).

180. T. Pohjonen and P. Törmälä, in K.-S. Leung, L.-K. Hung, and P.C. Leung, eds., *Biodegradable Implants in Fracture Fixation*, Chinese University of Hong Kong, Hong Kong, 1993, pp. 75–88.

181. R. Gutwald, H. Pistner, J. Reuther, and J. Mühling, *J. Mater. Sci., Mater. Med.* **5**, 485–490 (1994).

182. S. Li and M. Vert, *Polym. Inter.* **33**, 37–41 (1994).

183. S. Li, S. Girod-Holland, and M. Vert, *J. Controlled Release* **40**, 41–53 (1996).

184. S. Izumikawa, S. Yoshioka, Y. Aso, and Y. Takeda, *J. Controlled Release* **15**, 133–140 (1991).

185. R. Jalil and J.R. Nixon, *Proc. 5th Inter. Conf. Pharm. Technol.*, Paris, *1989*, vol. V, pp. 94–103.

186. J.W. Fong, J.P. Nazareno, J.E. Pearson, and H.V. Maulding, *J. Controlled Release* **3**, 119–130 (1986).

187. D.E. Cutright et al., *Oral Surg. Oral Med. Oral Pathol.* **37**, 142–152 (1971).

188. A.M. Reed and D.K. Gilding, *Polymer* **22**, 494–498 (1981).

189. J.M. Zhu, Y.M. Shao, S.Z. Zhang, and W.M. Sui, *J. China Tex. Univ. (Engl. Ed.)* **8**, 57–61 (1991).

190. Y. Ogawa, H. Okada, M. Yamamoto, and T. Shimamoto, *Chem. Pharm. Bull.* **36**, 2576–2581 (1988).

191. S.S. Amarpreet and J.A. Hubbell, *J. Biomed. Mater. Res.* **24**, 1937–1411 (1990).

192. C. Thomasin et al., *J. Controlled Release* **41**, 131–145 (1996).

193. J. Eitenmüller, G. Muhr, K.L. Gerlach, and T. Schmickal, *J. Bioact. Compat. Polym.* **4**, 215–241 (1989).

194. D.W. Grijpma, A.J. Nijenhuis, and A.J. Pennings, *Polymer* **31**, 2201–2206 (1990).

195. C.X. Song, H.F. Sun, and X.D. Feng, *Polym. J.* **19**, 485–491 (1987).

196. Y.X. Li and X.D. Feng, *Makromol. Chem., Macromol. Symp.* **33**, 253–264 (1990).

197. J.L. Espartero et al., *Macromolecules* **29**, 3535–3539 (1996).

198. T. Heya, H. Okada, Y. Ogawa, and H. Toguchi, *Int. J. Pharm.* **72**, 199–205 (1991).

199. A.S. Chawla and T.M.S. Chang, *Biomater., Med. Devices, Artif. Organs* **13**, 153–162 (1985–1986).

200. R. Wada et al., *J. Bioact. Compat. Polym.* **3**, 126–136 (1988).

201. R. Wada, Y. Tabana, S.-H. Hyon, and Y. Ikada, *Bull. Inst. Chem. Res., Kyoto Univ.* **66**, 241–250 (1988).

202. N. Wakiyama, K. Juni, and M. Nakano, *Chem. Pharm. Bull.* **30**, 2621–2628 (1982).

203. G.E. Visscher et al., *J. Biomed. Mater. Res.* **22**, 733–746 (1988).

204. P. Törmälä et al., *Angew. Makromol. Chem.* **185/186**, 293–302 (1991).

205. J.H. Zhu, Z.R. Shen, L.T. Wu, and S.L. Yang, *J. Appl. Polym. Sci.* **43**, 2099–2106 (1991).

206. K.H. Lam et al., *J. Mater. Sci., Mater. Med.* **5**, 181–189 (1994).

207. A. Smith and I.M. Hunneyball, *Int. J. Pharm.* **30**, 215–220 (1986).

208. K. Juni et al., *Chem. Pharm. Bull.* **33**, 311–318 (1985).

209. Z. Ramtoola, O.I. Corrigan, and C.J. Barrett, *J. Microencapsul.* **9**, 415–423 (1992).

210. G. Spenlehauer et al., *J. Controlled Release* **7**, 217–229 (1988).

211. J.P. Benoit, F. Courteille, and C. Thies, *Int. J. Pharm.* **29**, 95–102 (1986).

212. H.V. Maulding et al., *J. Controlled Release* **3**, 103–117 (1986).

213. H.V. Maulding, *J. Controlled Release* **6**, 167–176 (1987).

214. A. Kishida, S. Yoshioka, Y. Takeda, and M. Uchiyama, *Chem. Pharm. Bull.* **37**, 1954–1956 (1989).

215. Y. Cha and C.C. Pitt, *J. Controlled Release* **8**, 259–265 (1989).

216. Y. Cha and C.C. Pitt, *J. Controlled Release* **7**, 69–78 (1988).

217. J. Mauduit, N. Bukh, and M. Vert, *J. Controlled Release* **23**, 209–220 (1993).

218. C.C.P.M. Verheyen et al., *J. Mater. Sci., Mater. Med.* **4**, 58–65 (1993).

219. S. Li and M. Vert, *J. Biomater. Sci., Polym. Ed.* **7**, 817–827 (1996).

220. Y. Zhang, S. Zale, L. Sawyer, and H. Bernstein, *J. Biomed. Mater. Res.* **34**, 531–538 (1997).

221. M.C. Gupta and V.G. Deshmukh, *Polymer* **24**, 827–830 (1983).

222. C.C. Chu, *J. Biomed. Mater. Res.* **16**, 417–430 (1982).

223. G. Spenlehauer, M. Vert, J.P. Benoit, and A. Boddaert, *Biomaterials* **10**, 557–563 (1989).

224. C. Birkinshaw, M. Buggy, G.G. Henn, and E. Jones, *Polym. Degradation Stab.* **38**, 249–253 (1992).

225. C. Volland, M. Wolff, and T. Kissel, *J. Controlled Release* **31**, 293–305 (1992).

226. C.C. Chu, *J. Biomed. Mater. Res.* **15**, 19–27 (1981).

227. C.C. Chu, *J. Biomed. Mater. Res.* **15**, 795–804 (1981).

228. K. Makino, H. Ohshima, and T. Kondo, *J. Microencapsul.* **3**, 203–212 (1986).

229. N. Chegini, D.L. Hay, J.A. von Fraunhofer, and B.J. Masterson, *J. Biomed. Mater. Res.* **22**, 71–79 (1988).

230. R. Suuronen et al., *J. Mater. Sci., Mater. Med.* **3**, 426–431 (1992).

231. A.C. Albertsson and S. Karlsson, in S.A. Barenberg, J.L. Brash, R. Narayan, and A.E. Redpath, eds., *Degradable Materials: Perspectives, Issues and Opportunities*, CRC Press, Boca Raton, Fla., 1990, pp. 263–286.

232. Y. Ikada et al., *J. Controlled Release* **2**, 179–186 (1985).

See also BIODEGRADABLE POLYMERS:
POLY(PHOSPHOESTER)S; BIODEGRADABLE POLYMERS:
POLYANHYDRIDES; POLY(ORTHO ESTERS).

# BLOOD SUBSTITUTES: A REVIEW OF CLINICAL TRIALS

NORA E. CUTCLIFFE
F.J. LOU CARMICHAEL
Hemosol, Inc.
Toronto, Canada

A. GERSON GREENBURG
The Miriam Hospital and Brown University
Providence, Rhode Island

## KEY WORDS

Artificial blood
Clinical studies
Clinical trials
Elective surgery
HBOC
Healthy volunteers
Hemoglobin
Oxygen carrier
Oxygen delivery
Oxygen therapeutic
Product characteristics
Red cell substitute
Review
Safety studies
Trauma

## OUTLINE

## INTRODUCTION

For nearly a century the search for a blood substitute has been an ongoing evolutionary process. Indeed, the recognition of the essential nature of blood in ensuring survival preceded the establishment of the very elements that comprise the essential functions of blood. These functions include oxygen delivery, maintenance of vascular volume, provision of clotting factors, delivery and removal of nutrients and waste products, and a communications route for signaling between various organs. It has long been rec-ognized that humans and other mammals are obligate oxygen-dependent organisms. Any decrease in the ability to provide adequate tissue perfusion via efficient oxygen delivery results in harmful metabolic consequences affecting organ function and ultimately survival.

The adverse events associated with transfusion of allogeneic blood have been recognized for some time. In recent years the concern with the transmission of infectious disease has captured the spotlight and driven the search for alternatives to traditional allogeneic transfusion practices. In addition, newer issues concerning the immunologic consequences of the transfusion of blood and blood products with respect to infection and recurrence of malignancy invokes a heightened sense of concern on the part of clinicians (1–3). Newer strategies for the use of red cell transfusion are evolving and the potential for use of a red cell substitute, i.e., an oxygen carrying solution, is deemed both possible and reasonable in many clinical situations (4–6). Several recent reviews have been published, encompassing a variety of blood substitute topics (7–10).

## HISTORICAL PERSPECTIVE

There are a variety of approaches that can be used to define a red cell substitute, assuming the primary objective is the delivery of oxygen to tissues. Behind each approach is a unique chemistry and rationale for investigation. It must be recalled that the modern era of hemoglobin-based oxygen carriers (HBOCs) began in the late 1960s (11). Problems with these solutions were considered to be (1) altered oxygen delivery due to an increased oxyhemoglobin affinity resulting from the loss of 2,3-diphosphoglycerate (DPG), the allosteric modifier of hemoglobin, when the red cell membrane is removed, and (2) a circulating half-life too short to be useful in the then-perceived primary application, trauma resuscitation. To address these issues, chemical modification of stripped or stroma-free hemoglobin to alter these properties was undertaken by many investigators. Many different concepts were explored and many modifiers defined and evaluated (10,12). The details of the modifications are perhaps less critical than the overall concepts as they may apply to the eventual acellular hemoglobin (Hb) solution.

Chemically modified acellular Hb may be stabilized as a 64 kD molecule (cross-linked), stabilized and polymerized or oligomerized, conjugated (molecules added), or stabilized and conjugated. All of these conceptual approaches address the issues of improving oxygen off-loading by decreasing oxyhemoglobin affinity and increasing intravascular persistence by decreasing Hb dimer formation or otherwise preventing its renal excretion. The site of chemical modification may vary (e.g., $\alpha$-$\alpha$ or the DPG pocket) and the degree of chemical specificity of the modification varies considerably. The agents used to polymerize the stabilized Hb tetramers vary as well. The source of the acellular Hb is another variable accounting for differences in solutions. Indeed, the source of Hb may have an eventual role in clinical utility from a pharmacoeconomic analysis vantage; an assessment of which solution is better may in the long run be determined by the economics of production. Current source material for acellular Hb solutions includes processing of outdated human red cells, bovine Hb, and recombinant technologies using *E. coli* as an expression system (10).

## THE NEED FOR RED CELL SUBSTITUTES

Fueled by reports of transfusion-related transmission of infectious disease and other adverse consequences, the medical community as well as the general public are increasingly concerned that the transfusion of allogeneic red blood cells (RBCs) and other blood products is risky. Hence, one overriding principle of modern transfusion therapy is transfusion avoidance; unwanted outcomes are avoided if the offending agent is not given (2,13). To this end alternative strategies, including cell salvage, hemodilution, and presurgery deposit have been developed. In addition, application of the ever-increasing knowledge of the relationship of oxygen delivery to outcome in terms of transfusion triggers (14,15) has helped to limit or define the indications for transfusion (2,4,5,10,13,14). In the context of these concerns various investigators have approached the design of clinical Phase II and III trials to address the issue of avoidance of allogeneic transfusion as one significant end point.

Of the many functions of blood, the substitutes, as now perceived, will primarily serve the roles of oxygen delivery and vascular volume expansion. With enhanced knowledge other roles will be identified based on detailed explanations of the physiology related, in large part, to the effects of replacing or enhancing tissue perfusion in a variety of clinical settings. In that sense red cell substitutes will become a delivery system for oxygen, the *essential drug of survival*.

## PURPOSE, SCOPE, AND METHODS

This chapter is not intended to be an all-inclusive review of the red cell substitutes. Rather, it is intended to review the current state of the field with respect to published clinical results of present generation Hb-based products. One major goal of this analysis is to place into perspective the various HBOC solutions in current use as well as the models being used for their clinical evaluation. In addition, it is aimed at establishing a framework or reference for ongoing developments in this field. In the absence of published results, clinical trials announced to be *in progress* at the time of submitting this manuscript are not referenced in this review.

Although perfluorocarbon-based red cell substitutes are now well advanced in humans testing (16–18), this literature was not included. For the key results presented in the next section, supportive animal/preclinical studies that have been reported elsewhere (19–23), as well as reports on liposome encapsulated hemoglobins that have not yet entered clinical trials (24), were also excluded. Publications were further narrowed to those reporting results in healthy volunteers and patients undergoing elective surgical and trauma treatment. The reader is referred to other papers for results of HBOC clinical trials in other more novel indications, such as septic shock (25–28), renal dialysis (29), and sickle cell anemia (30,31).

Publications included in this review were compiled mainly through Medline, *Current Contents*, and manual literature searches, with December 1997 as the cutoff date. Material presented at recent North American and European meetings on blood substitutes was also included, where published conference proceedings were available. A small number of nonrefereed publications was also used to supplement primary literature sources. Collectively, these available reports have permitted analysis of the various directions taken by investigators in this field.

## RESULTS

The overall approach taken in this section was to describe the products currently in clinical trials insofar as possible from published information (Table 1) and to then describe the various trials in terms of safety and efficacy, two key elements in moving forward to eventual clinical use. Specifically, Tables 2, 3, and 4 correspond to HBOC studies in healthy volunteers, elective surgical indications, and traumatic shock, respectively. In these tables, all product doses are presented in grams Hb. Where mg/kg doses are reported in the literature without total milligram or gram doses, it is assumed that the average patient weighs 70 kg. Results presented in Tables 2, 3, and 4 refer to treated volunteers or patients in comparison with control subjects. These detailed charts are accompanied by an analytical discussion of key results.

### HBOC Product Characteristics

It should be noted that direct comparisons of acellular Hb solutions in clinical trials involving acute blood loss, the focus of this chapter, is complicated by the enormous number of variables by which these solutions can be described. For purposes of this work a few have been selected, as presented in Table 1, since they are the logical ones that can be related to clinically useful parameters where the physiologic implications are known. These variables have been detailed elsewhere (10). The fact that there are differences not only in products, but also in models for establishment of safety and efficacy parameters allows a critical comparison of the approaches with reference to known physiology.

### HBOC Studies in Healthy Volunteers

Table 2 presents an outline of all Phase I HBOC safety trials conducted in healthy volunteers, as published prior to December 1997. These trials were conducted in North America (U.S. and Canada), and were completed between 1993 and 1995. For each trial, the table specifies the product tested (see also Table 1), the trial location, the number of treated and control subjects, and the dose range of Hb administered. A brief description of protocol design is presented, and key results from each study are summarized. These *initial safety studies* were conducted predominantly in adult males. The majority of these trials were controlled, with the exception of one study (54), in which all subjects were treated with HBOC product. For the remainder of these studies, control articles included crystalloid solutions, human serum albumin, and, in one study (42), autologous blood. Note that several of the publications associated with these studies correspond to subcomponents of larger trials. In addition, some were published in abstract form as initial segments of the studies were completed.

**Table 1. HBOC Product Characteristics**

| Company and product name(s) | Hb source | Chemical modification | Hb (g/dL) | Molecular weight distribution | $P_{50}$ at 37°C (mm Hg) | Viscosity at 37°C (cp) | Colloid osmotic pressure (mm Hg) | Half-life (hs) | MetHb (%) | pH (37°C) |
|---|---|---|---|---|---|---|---|---|---|---|
| Baxter Healthcare Corp. DCLHb™ (HemAssist™) | Human red cells | Hb is $\alpha\alpha$ cross-linked with bis(3,5-dibromosalicyl) furmarate (29,32) | 9.5–10.5 (29,32,33,34) | 64.5 kD (96–98%) 130 kD (2–3%) (32) | 32–33 (32,33,35,36) | 1 (35) | 42–44 (32) | 2.1–4.3 (25–100 mg/kg) 10.5 (658–1500 mg/kg) (29,33,37) | <5 at release (33,38) 14.6 near expiry (32) | 7.3–7.5 (29,32,33) |
| Biopure Corp. HBOC-201 (Hemopure®) | Bovine red cells | Glutaraldehyde polymerized Hb (39) | 13 (39,40,41,42,43,44) | 68–500 kD (43) Predominantly 130–500 kD (39,44) | 34–38 (39,40,41,42,43,44) | 1.3 (39,40,43,44) | 17 (39,40,41,43) | 8.5 (400 mg/kg) (45) 20–24 (600–800 mg/kg) (39,40,41,42,44) | <15 (39,41,42,43,44) | 7.6–7.9 (39,40,41,43,44) |
| Enzon Inc. PEG-Hb | Bovine red cells | Hb is conjugated to polyethylene glycol (PEG) (46) | 6 (46) | 64 kD species, conjugated to 9–14 PEG molecules (46) | 9–16 (46,47) | 3.39 (47) | 118 (47) | ~48 (27,48) | <10 (46) | 7.5 (46) |
| Hemosol Inc. Hemolink™ | Human red cells | Hb is $\beta\beta$ cross-linked & polymerized with o-raffinose (49) | 10 (49) | 64 kD (~33%) 126–600 kD (~63%) >600 kD (≤3%) (49) | 34 (49) | 1 (49) | 24 (49) | At 500 mg/kg: 14 (total Hb), 18.4 (oligomeric Hb) 7.4 (cross-linked tetrameric Hb) (49) | <10 (49) | 7.5 (49) |
| Northfield Labs PolyHeme™ (Poly SFH-P) | Human red cells | Pyridoxylated, glutaraldehyde polymerized Hb (50) | 8–10 (50,51,52,53,54,55) | Predominantly polymer, 64 kD (<1%) (50,54,55) | 28–30 (50,51,52,53,54,55) | 1.9–2.2[a] (53) | 20–25[a] (53,54) | 24 (50,55) | <3 (50,52,53,54,55) | Not available |
| Somatogen Inc. rHb1.1 (Optro™) | E. coli expression system | Genetically fused $\alpha\alpha$ Hb, with mutant $\beta$ chains (56) | 5–8 (56,57,58) | 64 kD (100%) (56) | 33 (56,58) | 0.8 (59) | 12 (58) | 2.8–12 at plasma [Hb] = 0.5–5 mg/ml (58) | Not available | Not available |

[a]For *polymerized pyridoxylated hemoglobin*, not necessarily the polymerized pyridoxylated stroma-free hemoglobin solution used in clinical trials.

Since the vasoactivity of cell-free Hb has been previously reported (23,60), safety studies with HBOCs have focused on product-induced hemodynamic effects. In general however, there is a wide range of reported cardiovascular responses to HBOC products. Some HBOCs have been found to have very little or no vasoactivity in human volunteers (27,54), whereas others have produced a variable dose-dependent increase in blood pressure with a maximum response reached at Hb doses of 0.05 to 0.1 g/kg (33). In studies where blood pressure increased, there was a reciprocal reduction in heart rate that would appear to be a normal physiological response to the rise in blood pressure (33,41,49,58,69). Where studied, cardiac outputs were found to be reduced following the administration of the HBOC in those cases where the blood pressures were elevated. Again, this is most likely an appropriate autonomic nervous system response to increased blood pressure.

With regard to specific product-induced effects, Przybelski et al. (33) studied 24 healthy volunteers receiving DCLHb™ compared to lactated Ringer's (LR) solution in a crossover study design. They reported a dose-related increase in mean arterial pressure (MAP; to a maximum of 20–25% at a the highest dose of 0.1 g/kg) that was mainly the result of increased diastolic pressure. There was a concurrent 16% reduction in heart rate at the 0.1 g/kg (~8.5 g) dose level. Electrocardiogram recordings were found to be normal throughout the study using Holter monitoring or telemetry. Similarly, Viele et al. (58) administered up to 25 g recombinant Hb, rHb1.1, to 34 volunteers and found an increase in MAP of 10–15 mmHg, while heart rate decreased by 15–20 beats per minute. Hughes et al. (62) reported on 9 volunteers receiving up to 28 g of Hemopure® who exhibited a 16% increase in MAP and increased systemic vascular resistance (SVR), along with a reduced heart rate. Cardiac index, as measured by impedance methods, was reduced by 1.3 L/m/m² at the maximum dose level. Later, Hughes et al. (41) reported on a Phase I trial in 18 subjects also receiving the bovine product (also referred to as HBOC-201), in doses of 16 to 45 g, following phlebotomy of 15% of the estimated blood volume and a 3:1 replacement with LR. In this follow-up Phase Ib study, they also reported an increase in MAP up to 10% above baseline and a 6% decrease in heart rate, with a 15% reduction in cardiac index. More recently, Adamson et al. (49) reported results of a Phase I safety trial in 42 volunteers of which 33 received Hemolink™ in doses ranging from ~2 to 43 g Hb. Following a similar trend as for other HBOC products, a 10–12% increase in MAP was observed, along with a reduction in heart rate.

It is noteworthy that Gould et al. (52,54) briefly reported on a Phase I safety trial in 30 subjects receiving up to 63 g of PolyHeme™ in which only a small (5%) increase in MAP and a 5% reduction in heart rate was seen. Also, in contrast to results described already for other safety trials, no increase in MAP was reported in a Phase I study in which up to ~38 g of Hb conjugated with polyethylene glycol (PEG-Hb) was administered to human volunteers (27).

The range of hemodynamic responses induced by these HBOC products underscores the fact that each product has a different impact on physiological function, likely due to underlying differences in chemical modification. Preclinical studies have suggested that scavenging of nitric oxide (NO), as well as other potential mediators, could play a major role in the vasopressor effects of these agents (23). This hypothesis would suggest that some HBOC products do not interact with NO, although preclinical studies to confirm the presence or absence of NO scavenging activity has not been reported for all products. The significance of vasopressor activity in the clinical setting varies greatly, depending on how these products are used. For example, in trauma and high blood loss surgical procedures, the vasoactivity *may be* of benefit if it does not compromise blood flow to vital end organs. Maintenance of vascular volume is a common benefit of all HBOC products; specific formulations (e.g., with high colloid osmotic pressure) may facilitate resuscitation.

Where reported (33,40,56,62), respiratory function was not adversely affected by the HBOC products. Arterial saturation was found to be maintained, and a good correlation between pulse oximetry and arterial oxygen saturation has been noted (40). Pulmonary function tests were normal where assessed by spirometry and flow volume studies. Both arterial oxygen partial pressures and carbon dioxide partial pressures were unaffected by the HBOCs. In one particularly interesting study (41), the diffusion capacity for carbon monoxide was found to be increased by 20% by HBOC-201. These data support the hypothesis that the cell-free Hb permits a more efficient transfer of gases between the alveolus and the circulation.

Historically, there has been considerable concern with respect to the effects of cell-free Hb on renal function. In this light, it is of interest that no evidence of renal toxicity has been noted in any safety study with current HBOC products. There were no reported adverse effects on serum creatinine or creatinine clearance. Likewise, the tubular cell marker, $N$-acetyl-$\beta$-glucosaminidase, was not affected (58). In general, investigators have not found Hb in the urine of the volunteers (41,42,56). These results strongly suggest a lack of HBOC-induced effects on renal function in the test subjects.

Most investigators have reported dose-dependent plasma retention times for the various HBOC products. Although Table 2 shows that relatively low Hb doses were administered in safety studies (e.g., up to 63 g Hb, approximately equivalent to one unit of packed red cells), circulation times may well be longer at higher Hb doses. Although the mechanism of this dose-dependent clearance has not yet been determined, it is likely that different mechanisms are recruited and possibly become saturated as the plasma levels of stabilized cell-free Hb rise. It appears that the tetrameric stabilized products have relatively shorter half-lives, of less than 12 hours, at the doses presented in Table 1 (33,58). In contrast, PEG-Hb at up to 38 g Hb has a half-life of about 48 hours (27). As reported in Phase I safety studies, polymerized HBOCs have intermediate half-lives of ~18–24 hours (39,49,50), again depending on the Hb dose.

There are a number of reports of altered clinical chemistry analytes, although at the doses studied in Phase I HBOC safety trials, none were clinically significant.

Table 2. HBOC Safety Studies in Healthy Volunteers

| Product, trial phase, and location | Subjects, [treated + controls], [male (M) + female (F)], age range | Dose range (Hb) | Control | Protocol design | Year of completion | Results | Refs. |
|---|---|---|---|---|---|---|---|
| DCLHb® Phase I U.S. | n = 24, [of which n = 22 completed crossover], [23 M + 1 F], age 20–49 yrs | ~2–8.5 g (25–100 mg/kg) | Lactated Ringer's (LR) | Randomized, controlled, double blind, crossover study: DCLHb® (or LR) was given on day 1; the alternate was given on day 6 | 1993 | *Hemodynamic.* Dose-related increase in mean arterial pressure (MAP), with a greater effect on diastolic blood pressure (DBP) than systolic blood pressure (SBP). No evidence of vasoconstriction. Heart rate (HR) was significantly lower for subjects given 100 mg/kg. Normal electrocardiograms. *Respiratory/oxygenation.* No changes in laser Doppler, pulse oximetry, or toe temperature measurements. *Clinical chemistry.* Dose-related increased in lactate dehydrogenase (LDH)-5 isoenzyme (n = 4, after control; n = 12, after treatment), with no increase in circulating concentrations of total LDH, alanine aminotransferase (ALT), aspartate aminotransferase (AST), alkaline phosphatase (ALP). Total serum creatine kinase (CK) increased at 100 mg/kg; the isoenzyme CK-myocardial band (CK-MB) did not increase. *Pharmacokinetics.* Half-life of 2.5 h at 25 and 50 mg/kg; 3.3 h at 100 mg/kg. *Adverse Events (AEs).* Mild to moderate abdominal pain (n = 3, after control; n = 6 after treatment). No serious adverse events, no evidence of organ dysfunction or toxicity. | 32,33,60,61 |
| Poly SFH-P Phase I U.S. | n = 30 [30 + 0], [? M + ? F], adults | Maximum = 63 g | None used | Uncontrolled study | 1993 | *Hemodynamics.* No evidence of vasoconstriction. *Clinical Chemistry.* Normal *Hematology.* Normal *Renal effects.* No evidence of renal toxicity. *AEs.* None reported. No gastrointestinal (GI) effect or other adverse events reported. | 50,52–54 |
| Hemopure® Phase I U.S. | n = 11 [9 + 2], [11 M], adults | 14–28 g | Normal saline | Randomized, controlled, single-blind, single-dose study | 1993 | *Hemodynamics.* Transient increases in SBP, DBP, MAP, and total peripheral resistance (TPR). Decreased cardiac index (CI) and pulse. *Respiratory/oxygenation.* Respiration and oximetry unchanged. *AEs.* None reported. | 62 |
| rHb1.1 Phase I U.S. | n = 93, [76 + 17], [93 M], age unknown[a] | ~1–25 g (15–320 mg/kg) | Human serum albumin (HSA) | Design unspecified | 1993 | *Hemodynamics.* No cardiovascular toxicities. *Respiratory/oxygenation.* No pulmonary toxicities. *Clinical Chemistry.* Hepatic function enzymes normal. *Hematology/coagulation.* Hematological measurements and coagulation parameters normal. *Renal.* No evidence of renal impairment or appreciable renal excretion of rHb1.1. *AEs.* No serious adverse events, but some subjects developed gastrointestinal discomfort, which was managed effectively with pharmacologic intervention. | 56 |

| Product/Phase/Country | n | Dose | Solution | Study design | Year | Notes | Ref. |
|---|---|---|---|---|---|---|---|
| rHb1.1 Phase I U.S. | n = 48 [34 (of which 33 completed the infusion) + 14], [48 M], age 18–35 yrs[a] | ~1–25 g (15–320 mg/kg) | HSA | Randomized, controlled, single-blind study | 1993 | *Hemodynamics.* Non-dose-related increases in MAP, 1 and 2 h after infusion; HR decreased during and for 2 h after infusion. *Clinical chemistry.* No increase in N-acetyl-β-glucosaminidase, creatinine clearance increased significantly 1 day after rHb1.1 infusion. Of the 23 subjects who experienced GI effects, 6 had elevated lipase, and 4 of these also had elevated amylase (both effects observed 2–4 h after infusion). *Renal.* No evidence of rHb1.1-mediated nephrotoxicity. *Pharmacokinetics.* Half-life of 2.8 h at plasma concentration of 0.5 mg/mL, and 12 h at 5 mg/mL. *AEs.* GI upset (n = 23/34), fever, chills, headaches, and backache in some subjects. | 57,58 |
| rHb1.1 Phase I? U.S. | n = 12[a] [9 + 3], [12 M], age 21–32 yrs[a] | ~8–11 g (110–150 mg/kg) | HSA (5%) | Controlled, single-blind study | 1994 | *AEs.* Increased velocities of peristaltic contractions in 6/9 treated subjects. Increased amplitude and duration of contractile waves in the esophagus. No consistent effect on the resting tone of lower esophageal sphincter (LES), but LES relaxation was inhibited. Spontaneous, simultaneous high-pressure contractions in 8/9 subjects, lower retrosternal chest pain during swallowing in 4 subjects. | 63 |
| HBOC-201 Phase Ib U.S. | n = 24, [18 + 6], [12 M + 12 F], age 18–45 yrs[b] | Maximum = 45 g (600–800 mg/kg) | LR | Randomized, controlled, single-blind, single-dose study[c] | 1994 | *Hematology/coagulation.* Serum iron levels paralleled HBOC-201 concentration, peaking at 8 h (up to 220 μg/dL). Ferritin levels peaked at 48 h (up to 180 ng/mL). Serum erythropoietin increased by 2–6-fold over baseline levels at 24 h. No significant changes in white blood cell or platelet counts, or in prothrombin time (PT) or activated partial thromboplastin time (PTT). *Pharmacokinetics.* Plasma half-life of HBOC-201 was ~20 h. | 39 |
| HBOC-201 Phase Ib U.S. | n = 24, [18 + 6], [12 M + 12 F], age 18–45 yrs[b] | Maximum = 45 g (600–800 mg/kg) | LR | Randomized, controlled, single-blind study[c] | 1994 | *AEs.* None reported. *Respiratory/oxygenation.* Percent oxygen saturation of Hb (Spo₂) can be reliably measured by pulse oximetry or arterial blood gas analysis in the presence of HBOC-201 over the normal range of Spo₂. *Pharmacokinetics.* Plasma half-life of HBOC-201 was ~24 h. | 40,64 |
| HBOC-201 Phase Ib U.S. | n = 41, [18 + 23], [41 M], age 18–45 yrs | 16.5–45 g (250–600 mg/kg) | LR | Randomized, controlled, single-blind, dose escalation study[c] | 1994 | *AEs.* None reported. *Hemodynamics.* Small but transient decreases in CI and HR, concomitant with increased BP during the first 8 h after the beginning of infusion. *Respiratory/oxygenation.* Diffusion capacity of oxygen was increased up to 20% above baseline levels in the 38.0 and 45.0 g groups in comparison with controls (~14% below baseline) at 2–4 h after infusion. Other pulmonary function tests and arterial blood gas measurements were unremarkable. Arterial oxygen content and oxygen delivery tended to be greater in active groups than in controls. *Pharmacokinetics.* Elimination of the product was a linear, first-order process, with no renal excretion. Peak plasma | 41,44,65 |

**Table 2. HBOC Safety Studies in Healthy Volunteers** (*continued*)

| Product, trial phase, and location | Subjects, [treated + controls], [male (M) + female (F)], age range | Dose range (Hb) | Control | Protocol design | Year of completion | Results | Refs. |
|---|---|---|---|---|---|---|---|
| HBOC-201 Phase Ib U.S. | n = 6, [3 + 3], [6 M], age 25–45 yrs | 45 g (600 mg/kg) | autologous blood | Randomized, controlled, single-blind, two-way crossover study: on day 1, HBOC-201 or autologous blood was given, with the opposite given on day 8[c] | 1994 | [Hb] was 1–2 g/dL and plasma half-life approached 20 h at the highest dose. No detectable Hb in the urine. Dosing of HBOC-201 to a target plasma Hb concentration can be achieved using pharmacokinetic principles with measurable effects on oxygen physiology. *AEs.* Minor transient gastrointestinal events such as "gas" were seen in some subjects in both active and control groups, but required no treatment. *Respiratory/oxygenation.* Subjects had similar exercise and diffusion capacity but lower lactate levels (for up to 24 h) during HBOC-201 than during autologous transfusion periods. $O_2$ use (uptake) and $CO_2$ production at rest were greater during HBOC-201 than during autologous transfusion periods. Under the study conditions, the physiological effects of 1 g HBOC-201 were similar to 3 g Hb from autologous transfusion. *Pharmacokinetics.* Half-life was ~23 h. No Hb detected in the urine. *AEs.* None reported | 42,44,66 |
| Hemolink® Phase I Canada | n = 42 [33 + 9], [42 M], age 18–40 yrs | ~2–43 g (25–600 mg/kg) | LR | Randomized, controlled, double-blind, dose escalation study | 1995 | *Hemodynamics.* MAP increased to a maximum of 10–12%, but was not clinically significant. HR decreased for up to 24 h postinfusion. *Clinical chemistry.* At the highest dose, clinically insignificant and transient changes in liver enzymes (AST, ALT, gamma-glutamyltransferase [GGTP] and pancreatic enzymes [amylase and lipase] were observed, but did not appear to reflect clinically significant events. *Hematology/coagulation.* No coagulation abnormalities, as determined by platelets counts, PT, and PTT. *Renal.* The product did not impair normal renal function. *Pharmacokinetics.* Plasma half-life for Hemolink® increased with dose level, ranging from ~1.6 h at a dose of 25 mg/kg to ~15.6 h at 500 mg/kg. There was also a direct relationship between increasing half-life and increasing molecular weight for the various (oligomeric, tetrameric, and non-cross-linked dissociable) Hb fractions of Hemolink®. *AEs.* At the highest dose, subjects had moderate to severe GI discomfort, which transiently interefered with normal everyday activities. Airway function, body temperature, and respiration rate remained unchanged. | 49,67 |
| PEG-Hb Phase I U.S. | n = 38, [32 + 6], [38 M], age unknown | ~4–38 g (0.83–8.33 ml/kg) | LR | Controlled, unblinded, single infusion study | 1995 | *Hemodynamics.* No changes in blood pressure. *Renal.* No renal or other organ toxicities. *Pharmacokinetics.* PEG-Hb remains in circulation long enough to be consistent with weekly dosing and current fractionated radiation therapy. *AEs.* Subjects given 5.83 and 8.33 ml/kg experienced transient episodes of GI pain. | 48,68,69 |

[a] The patient population (*n* = 93) described by Gerber et al. (56) may include the patients enrolled in studies published by Viele et al. (58) and Murray et al. (63).

[b] The patient population described by Hughes et al. (39) may be the same population as that presented in Hughes et al. (40).

[c] In studies reported by Hughes et al. (39–42), 15% of blood volume was withdrawn, and LR was given in a ratio of 3:1 prior to infusion of HBOC-201 or control solution.

Increases in lactate dehydrogenase (LDH) and creatine kinase (CK) above baseline have been reported (33,49), with the cardiac subfractions being negative, suggesting a skeletal muscle source and not a cardiac source for these elevations. In general, analytes measured as markers of liver function fell within normal ranges in all studies. Slightly increased lipase and amylase activity have also been reported (49), in contrast to another study in which serum lipase was elevated to 10 times the upper limit of normal (ULN) and amylase was elevated to 3–4 times the ULN (58). These reported elevations returned to normal levels within 24 to 48 hours, with no clinical evidence of pancreatitis. Detailed iron kinetics were measured in only one safety study (39). Other authors reported no change in the already mentioned clinical chemistry parameters or simply did not comment. It is also noteworthy that, where reported, no antibodies were detected in response to HBOC products at the dose levels administered in safety studies (33,49).

Although there were some striking findings in Phase I safety studies, there were no serious adverse events reported in any of these trials. The most consistent adverse events induced by HBOCs have been gastrointestinal (GI) effects. Most investigators have reported GI symptomatology in awake participants in the Phase I safety studies. The symptoms appear to be dose dependent and include abdominal pain, flatulence, and dysphagia (27,33,41,48,49,56,58). In an elegant study these effects were shown to be due at least in part to an interruption of normal esophageal motility mechanisms (63). Preclinical studies identified NO scavenging by the hemoglobins as a likely mechanism for adverse GI effects, and these effects appear to be mimicked by the inhibition of NO synthase (70). Interestingly, the recombinant Hb product was also reported to produce fever, chills, headache, and dermatological abnormalities following administration (56). The potential role of residual endotoxin from *E. coli* may be called into question by this finding.

## HBOC Studies in Elective Surgery

A large number of abstracts, along with a small number of full text articles, is now available regarding HBOC trials in surgical patients. These publications, which correspond to Phase I, Ib, I/II, II, and III trials with HBOC products, are referenced in Table 3. The areas of surgery that have been studied in these trials were classified here as orthopedic, cardiovascular, and general surgery (i.e., trials in which one or more of the following surgical groups were included: urological, gynecological, hepatic, abdominal, and orthopedic surgery, or where no particular surgical population was specified). Within each of these classifications, trials are organized in chronological order of completion, between 1994 and 1997.

These studies were typically randomized and controlled, with either single or double blinding. Control articles used in these studies varied more widely than for safety trials in volunteers. That is, control solutions included crystalloids (saline and LR), colloids (hydroxyethyl starch), as well as both allogeneic RBCs and autologous blood. Trials involving acute normovolemic hemodilution

(ANH) procedures are identified in the column titled "Protocol Details." Dose ranges (up to ~100 g Hb) were generally greater than those administered in the volunteer studies described already. Collectively, these studies in surgical patients addressed both safety and efficacy issues, with key results reported in Table 3. Pharmacoeconomic results (e.g., duration of hospital stay) are also presented, if reported in the published literature.

All studies reported safety with hemodynamic stability and varying degrees of hemodynamic activity. In vascular and cardiac surgery, an increase in MAP and SVR has been observed. A small increase in pulmonary vascular resistance (PVR) has been reported by two groups (34,76), whereas one found no effect on PVR (43). Where reported, cardiac output has been depressed, an appropriate response to increased MAP. In orthopedic and general surgery, hemodynamics were found to be relatively stable following administration of the specific HBOC products used. Due to the variability of the surgical settings and anesthesia, it is not possible to directly compare the hemodynamic effects induced by individual HBOC products.

Global oxygen delivery has been noted to be reduced, commensurate with the reduction in cardiac output (34,43,81,82). Oxygen consumption was depressed or maintained, while the oxygen extraction ratio generally increased. These reported effects may be due in part be due to the right-shifted nature of the products studied (i.e., they have right-shifted oxygen-Hb dissociation curves, associated with $P_{50}$ values higher than that of whole blood). HBOC-induced effects on respiratory function in these trials have not been described in the literature.

Consistent with the safety studies in awake individuals discussed earlier, and consistent with preclinical studies in various animal species (49,88), renal function has not been adversely affected by HBOCs in any of the trials in surgical patients reported to date. However, investigators have reported hemoglobinuria in patients following administration of 75 g DCLHb™, and jaundice has been reported in the same surgical trials (involving cardiac, orthopedic, and abdominal patients) without evidence of hepatic failure (36,85).

Pharmacokinetic data for HBOCs obtained in the surgical setting are similar to the findings in the awake volunteers. Clinical chemistry values generally remain within the normal range or are not significantly different from control patient values (77,81). In some surgical studies, increased serum amylase and/or lipase has been noted (36,72,78,85); consistent with the observations in earlier safety trials, no acute pancreatitis has been reported.

Only a small number of publications have reported on the immunological effects of HBOC administration. One study reported negative DCLHb™ antibody titres up to 6 weeks following administration of a low hemoglobin dose (~3.5 g) in patients undergoing abdominal aortic aneurysm repair (76). Interestingly, patients with hepatic surgery did not have IgE antibodies to HBOC-201 but developed very low levels of IgG to HBOC-201 14 days after administration of ~28 g of the product (45). In a novel, double-dose study design (87), HBOC-201 was administered both intraoperatively and on postoperative day 1 for a total dose of ~100 g Hb. There was no explicit mention

**Table 3. HBOC Studies in Elective Surgery**

| Indication/patient population | Product, trial phase, [treated + control], and location | Subjects, [M + F], age range | Dose range (Hb) | Protocol details | Completion date | Primary endpoint(s) | Results (safety, efficacy, pharmacoeconomic) | Refs. |
|---|---|---|---|---|---|---|---|---|
| *Orthopedic* | | | | | | | | |
| Surgery | rHb1.1 Phase I/II U.S. | n = ? [? + ?], [?M + ?F], age unknown | ≤25 g | Controlled, single-blinded, dose escalation study. *rHb1.1* or *saline* given perioperatively. | Mid 1994 | Safety | *Safety.* Hemodynamics. No evidence of hemodynamic impairment. Renal. No renal impairment. Hematology/coagulation. No evidence of any toxicity or impairment of function. AEs. No serious adverse events, and GI symptoms seen in Phase I were not encountered. *Efficacy.* None reported. *Pharmacoeconomic.* None reported. | 57 |
| Total hip replacement | DCLHb® Phase II Europe | n = 80, [? + ?], [?M + ?F], age unknown | ~2–18 g (25–200 mg/kg) | Double-blind, controlled, randomized study. *DCLHb®* or *placebo* given preoperatively, prior to anesthesia and surgery. | Early 1995 | Safety | *Safety.* Hemodynamics. Mean increase in MAP = 8–16 mm Hg. AEs. No toxicities reported. *Efficacy.* None reported. *Pharmacoeconomic.* None reported. | 35,71 |
| Total hip and knee replacement | rHb1.1 Phase II U.S. | n = 10, [7 + 3], [?M + ?F], age unknown | 12.5–50 g | Controlled dose escalation study. 250–800 ml of blood withdrawn and replaced with saline in a ratio of 2:1, then *rHb1.1* or *saline* given preoperatively, prior to anesthetic and surgery. [acute normovolemic hemodilution (ANH) protocol] | Late 1995 | Safety | *Safety.* Hemodynamics. One patient given 12.5 g experienced an increase in systolic BP 20% above baseline; no other patients demonstrated significant systolic hypertension. No intraoperative myocardial ischemia or infarctions were noted. Clinical chemistry. Increased amylase and lipase at 2 h in two patients, but not suggestive of pancreatitis. AEs. No treatment-related GI symptomatology was reported by any patient during or postoperatively. No fever or hypersensitivity effects. *Efficacy.* None reported. *Pharmacoeconomic.* None reported. | 57,72,73 |
| Surgery | DCLHb® Phase II? Europe? | n = ?, [12 + ?], [?M + ?F], age unknown | ≤75 g | Randomized, controlled study. *DCLHb®* or *allogeneic RBCs* given postoperatively, within 24 h of surgery. | Early 1997 | Reduced Allogeneic RBC Transfusion | *Safety.* Well tolerated. *Efficacy.* 33% of patients receiving DCLHb avoided blood transfusion over 7 days. *Pharmacoeconomic.* None reported. | 74 |

*Cardiovascular*

| Indication | Product / Phase / Region | n / demographics | Dose | Study design | Date | Objective | Results | Ref. |
|---|---|---|---|---|---|---|---|---|
| Abdominal aortic repair | DCLHb® Phase II Europe | $n = 71$, [35 + 36] [?M + ?F], age unknown | 3.5–14 g (50–200 mg/kg) | Randomized, controlled, single-blind study. *DCLHb®* or *LR* given preoperatively, after anesthetic, prior to surgery. | Late 1995 | Safety | *Safety.*<br>*Hemodynamics.* Maximum increase in MAP ~20 torr. Duration, but not magnitude, of this pressor response is dose dependent. Fewer hypotensive episodes in the perioperative period. CI and oxygen delivery ($DO_2$) maintained or decreased, oxygen consumption ($VO_2$) maintained.<br>*Efficacy.* None reported.<br>*Pharmacoeconomic.* None reported. | 60,75 |
| Abdominal aortic surgery | HBOC-201 Phase II Europe | $n = 13$, [7 + 6], [10M + 3F], age 18–70 yrs | ~33 g (3 ml/kg @ 13% Hb) | Controlled, randomized, single-dose study. Following anesthesia, 1 liter of blood withdrawn and replaced with LR, then *HBOC-201* or *hydroxyethyl starch (HES)* (6%, mean molecular weight ($M_r$) = 70,000 (substitution ratio ~ 0.5) given preoperatively, prior to surgery. [ANH protocol] | Mid 1996 | Safety | *Safety.*<br>*Hemodynamics.* 30 min after HBOC-201 infusion, MAP, SVR and CI were 149%, 169%, and 75% of preinfusion values, respectively. No change in HR or pulmonary vascular resistance (PVR).<br>*Respiratory/oxygenation* $O_2$ delivery index ($Do_2I$) and $O_2$ consumption index ($Vo_2I$) 30 min after infusion were 79% and 76% of preinfusion values, respectively, whereas $O_2$ content in arterial blood ($CaO_2$) and $O_2$ extraction ratio ($O_2ER$) remained unaffected.<br>*Efficacy.* None reported.<br>*Pharmacoeconomic.* No change in intensive are unit (ICU) or postoperative hospital length to stay (LOS); 11 ± 2 days total stay for both groups. | 43 |
| Abdominal aortic aneurysm repair | DCLHb® Phase I/II? U.S. | $n = 10$, [5 + 5], [?M + ?F], age unknown | ~3.5 g (50 mg/kg) | Randomized, controlled study, *DCLHb®* or *Hespan* given postoperatively. | Late 1996 | Safety | *Safety.*<br>*Hemodynamics.* DCLHb-treated patients had higher systemic, pulmonary vascular and arterial pressures, as well as higher SVR and PVR. These effects were transient, with no differences between groups at 2 h after administration.<br>*Immunological.* DCLHb antibody titres were negative in all patients at 6 weeks.<br>*AEs.* No patients experienced severe adverse reactions directly attributable to DCLHb. One treated subject suffered a myocardial infarction 36 h after infusion, but it was not deemed attributable to DCLHb.<br>*Efficacy.* None reported.<br>*Pharmacoeconomic.* None reported. | 76 |
| After bypass surgery | DCLHb® Phase III Europe | $n = 209$ [104 + 105], [?M + ?F], age unknown | 75 g | Randomized, controlled study. *DCLHb®* or *allogeneic RBCs* given postoperatively, in the first 24 h postbypass. | Early 1997 | Safety and reduced allogeneic RBC transfusion | *Safety.*<br>*Hemodynamics.* Following the first infusion, $O_2ER$ increased significantly (+7%) in the DCLHb group, compared with a 3% fall in the control group. DCLHb increased SVR, most significantly after the first infusion. CK-MB, $LDH_1$ and troponin I levels were comparable or lower following DCLHb infusion at 1, 3, and 7 days postsurgery.<br>*Mortality.* Comparable for both groups. | 34,36,74 |

**Table 3. HBOC Studies in Elective Surgery** (continued)

| Indication/ patient population | Product, trial phase, and location | Subjects, [treated + control], [M + F], age range | Dose range (Hb) | Protocol details | Completion date | Primary endpoint(s) | Results (safety, efficacy, pharmacoeconomic) | Refs. |
|---|---|---|---|---|---|---|---|---|
| | | | | | | | *AEs.* DCLHb-related events (in some patients) included: yellow skin discoloration; hematuria/hemoglobinuria; hypertension; increases in AST, amylase, and ALT; hyperbilirubinemia; and abnormal hepatic function. *Efficacy.* At 24 h postsurgery, 39% avoided allogeneic RBCs; at 7 days postsurgery (or hospital discharge), 19% avoided allogeneic RBCs. The total number of RBC units given over the 7-day period was similar in both groups. *Pharmacoeconomic.* None reported. | |
| *General surgery* | | | | | | | | |
| Urological (radical prostatectomy) | HBOC-201 Phase I U.S. | n = 20 [12 + 8], [20M + 0F], age unknown[a] | ~34–45 g (400–600 mg/kg) | Controlled, single-blind, dose escalation study. HBOC-201 or LR was given postoperatively, on the first postoperative day. | Late 1995 | Safety | *Safety.* *Hemodynamics.* A slight increase in mean BP and slight decrease in HR observed. *Clinical chemistry.* No clinically significant changes in liver function (AST, ALT, GGTP, bilirubin). *Renal.* No clinically significant changes in renal function (creatinine). *Hematology/coagulation.* No clinically significant changes in coagulation profile (PT, PTT, platelet count, fibrinogen). *AEs.* No serious AEs or mortality occurred. *Efficacy.* None reported. *Pharmacoeconomic.* No change in hospital LOS. | 77 |
| No surgical population specified | rHb1.1 Phase II U.S. | n = 23, [16 + 7], [?M + ?F], age 18–75 years | 25–100 g | Controlled, randomized single-blind study. Optro® or control therapy (allogeneic RBCs or autologous blood) given to treat blood loss intraoperatively, during surgery. | Late 1995 | Safety | *Safety.* *Hemodynamics.* Electrocardiogram (ECG) and HR were stable during transfusion period. No consistent changes in BP were observed. *Oxygenation/pulmonary.* Oxygen saturation was stable. *Clinical chemistry.* No significant changes in serum chemistries, with no evidence of renal function abnormality. Elevated amylase and lipase were observed in some patients, but was not suggestive of acute pancreatitis. *Hematology/coagulation.* No significant changes in hematology, coagulation tests. *AEs.* No drug-related serious adverse events; no signs of GI dysmotility. *Efficacy.* None reported. *Pharmacoeconomic.* None reported. | 57,78 |

| Indication | Product/Phase/Location | n / demographics | Dose | Study design | Date | Category | Findings | Ref. |
|---|---|---|---|---|---|---|---|---|
| Orthopedic, gynecological, and urological | HBOC-201 Phase Ib? U.S. | $n \sim 54$ (3 groups of $\sim$18), [ratio = 2:1], [?M + ?F], age unknown[a] | ~28–42 g (400–600 mg/kg) | Randomized, placebo-controlled, single-blind, parallel group studies. HBOC-201 or LR given postoperatively (when Hb < 11 g/dl). | Late 1995 | Erythropoietic effects | Safety.<br>Hematology/coagulation. HBOC-201 accelerated erythropoiesis compared to controls as measured by corrected absolute reticulocyte count (CARC) and hematocrit (Hct), equivalent to endogenous generation of ≥1/2 unit of blood in less than one week. Blood loss and allogeneic and/or autologous transfusion and volume replacement were similar in active and control groups.<br>Efficacy. See hematological effects.<br>Pharmacoeconomic. None reported. | 79 |
| No surgical population specified | HBOC-201 Phase II? U.S. | $n = 58$, [37 + 21], [?M + ?F], age unknown | ~42–84 g (600–1200 mg/kg) | Randomized, placebo-controlled, single-blind, dose escalation, study. Following anesthesia and blood loss of ≥500 ml, HBOC-201 or LR was given intraoperatively. | Mid 1996 | Hemodynamic effects | Safety. Hemodynamics. No significant differences in vital signs (SBP, DBP, HR) when changes from baseline were compared between patients receiving LR or HBOC-201.<br>Efficacy. None reported.<br>Pharmacoeconomic. None reported. | 80 |
| Hepatic (liver resection) | HBOC-201 Phase II Europe | $n = 12$, [6 + 6?], [6M + 6F], avg = 59 yrs | ~28 g (400 mg/kg) | Randomized, controlled, single-dose, prospective study. Following anesthesia, 1 L of blood withdrawn and replaced with 1 litre LR, then HBOC-201 or HES (6%, mean MW = 70,000, substitution ratio 0.5) given preoperatively, prior to surgery. [ANH protocol] | Late 1996 | Safety | Safety.<br>Hemodynamics. MAP increased by 18%, and SVR increased by 42%.<br>Respiratory/oxygenation. CO, mixed venous oxygen content, and oxygen delivery decreased. Oxygen extraction ratio increased, even in the ICU.<br>Clinical chemistry. No changes in amylase.<br>Hematology/coagulation. Treated patients developed a more pronounced increase in leukocytes, reticulocytes, and arterial methemoglobin during postoperative days 2–3 compared to controls. No changes in coagulation profiles.<br>Renal. No Hb detected in urine. No changes in creatinine.<br>Pharmacokinetics. Mean half-life ~8.5 h.<br>Immunological. No IgE antibodies to HBOC-201 were observed, but mean concentration of IgG to HBOC-201 increased after day 7 to a maximum of ~10 ng/mL on day 14.<br>AEs. No allergic or adverse reactions.<br>Efficacy. None reported.<br>Pharmacoeconomic. No change in hospital LOS. | 45,81–83 |

**Table 3. HBOC Studies in Elective Surgery** (continued)

| Indication/ patient population | Product, trial phase, and location | Subjects, [treated + control], [M + F], age range | Dose range (Hb) | Protocol details | Completion date | Primary endpoint(s) | Results (safety, efficacy, pharmacoeconomic) | Refs. |
|---|---|---|---|---|---|---|---|---|
| Orthopedic and abdominal | DCLHb® Phase II? U.S. | n = 36?, [12 + 24]?, [?M + ?F], age unknown[b] | 75 g | Randomized, controlled, unblinded, prospective study. Following anesthesia, DCLHb® or allogeneic RBCs was given intraoperatively, within 12 h after surgery start. | Late 1997 | Effects on methemoglobin (MetHb), organ function, coagulation, pharmacokinetics, and allogeneic RBC transfusion | *Safety.* *Hemodynamics.* No cardiac complications. *Clinical chemistry.* No hepatic complications. Amylase was elevated to 3 times the upper limit of normal (ULN) at POD 2, returning to normal by POD 4. *Hematology/coagulation.* Significantly higher metHb levels were observed in the treated group at several time points, but did not affect patient outcome. No adverse effects on hematology or coagulation tests. *Renal.* No renal complications. *Pharmacokinetics.* Harmonic mean half-life was 10.5 h. *AEs.* Transient mild to moderate GI effects, skin discoloration, and hemoglobinuria. *Efficacy.* In combination with autologous predonation, DCLHb may have potential to reduce RBC transfusion and to avoid allogeneic RBCs. *Pharmacoeconomics.* None reported. | 37,38,84–86 |
| No surgical population specified | HBOC-201 Phase II? U.S. | n = 33, [19 + 14], [?M + ?F], age unknown | ~70–105 g [Intra-op + post-op] (600 + 400 mg/kg to 900 + 600 mg/kg) | Randomized, controlled, single-blind, dose escalation study. Following anesthesia, and blood loss ≥ 500 mL, HBOC-201 or LR was given intraoperatively in OR, followed by a postoperative dose of HBOC-201 or LR, respectively, on postoperative day (POD) 1, at 24 h after the OR dose. | Late 1997 | Kinetics and hemodynamic effects | *Safety.* *Hemodynamics.* No significant hemodynamic changes (HR, mean BP) in patients infused during anesthesia and postoperatively. *Pharmacokinetics.* A dose response in plasma Hb levels (<2 g/dL) was observed with both initial and second infusion of HBOC-201; both were safe and well tolerated. *Efficacy.* None reported. *Pharmacoeconomics.* No change in hospital LOS. | 87 |

[a]The general surgical patient population (n ~ 54) described by Hughes et al. (79) may overlap with the urological patient population (n = 20) enrolled in the study published by Monk et al. (77).
[b]It is assumed that the abstracts by Schubert et al. (85,86) and O'Hara et al. (37,38,84) refer to the same patient population (n = 36?). Some results are reported for subsets (n = 12) of this population.

of any immunological change. Although single and double HBOC doses appear to be well tolerated, it remains unclear if repeat administration of HBOCs at any dose will have immunological consequences.

Several HBOC trials have reported no clinically significant changes in hematology tests or coagulation profiles at doses up to 100 g Hb (77,78,81,84) consistent with results from earlier safety studies. In one study, HBOC-201 was shown to accelerate erythropoiesis (compared to LR treatment), as determined by corrected absolute reticulocyte counts and hematocrit measurements (79). These data support the hypothesis that, in addition to their role in boosting oxygen-carrying capacity in the immediate term, HBOC products might be useful in treating patients with various anemias (39,47) by potentially supplementing endogenous iron sources and/or stimulating erythropoiesis, thereby enhancing red cell mass.

In the surgical setting, no serious adverse events have been reported in response to HBOC administration. Moreover, there is an almost complete lack of GI complaints in these patients. The apparent absence of GI effects in the surgical setting may be multifactorial. The degree of postoperative pain may be more marked than the HBOC-induced GI discomfort. The effect may also be blunted by concomitant analgesic administration or the use of other agents. Although GI events do not appear to be of great concern in anesthetized surgical patients, further investigation of relevant mechanisms and clinical outcomes will be required in future studies.

Only two studies have reported on product efficacy in terms of transfusion avoidance. In one small study, the combination of autologous predonation and DCLHb™ was reported to have *the potential* to reduce exposure to allogeneic red cells (86). In a larger Phase III trial, 39% of postbypass patients avoided allogeneic blood in the first 24 hours after surgery; this was reduced to 19% by 7 days postsurgery or hospital discharge (34,36). The total number of red cell units transfused was similar in the control and treatment arms, despite the fact that approximately one-fifth of treated patients avoided blood completely.

Some pharmacoeconomic parameters have also been reported by groups conducting HBOC trials in surgical patients. The most common measure is hospital length of stay (LOS), which has been found to be unchanged in HBOC-treated versus control patients in several groups, including hepatic (45), abdominal aortic (43), urological (77), and general surgery (87). The use of LOS as an appropriate evaluation parameter in these settings can be debated.

### HBOC Studies in Traumatic Shock

In comparison with HBOC trials in elective surgery, only a small number of trauma studies have been completed and published (55,91). This situation may be explained by the general belief that it is best to accumulate data in the controlled surgical setting before embarking upon trials in the less controlled trauma arena. Not all HBOC manufacturers have initiated trauma studies, although the majority have expressed an intent to do so.

Table 4 presents the protocol details of published trauma trials, including specific products and trial locations, as well as the number of treated and control subjects, Hb dose ranges, and key results. One of the three published trials was not a controlled study, with all patients receiving the HBOC product, at doses up to 300 g Hb, once the clinical decision to initiate transfusion was made (54,55). In a follow-up, controlled trial in which the same maximum dose was achieved (indeed the highest Hb dose for all clinical trials reported to date), allogeneic red cells were given to patients in the control arm (93). In the third study, in which low doses, up to 20 g Hb, were administered, saline was used as the control solution (91).

Reported hemodynamic parameters showed little if any change. In response to the specific products and dose levels used in these trials, MAP levels were maintained or increased slightly (55,89). Renal function was reported to be normal (55,89). Results of clinical chemistry analyses are presented for only one trial; patients with normal preinfusion values were claimed to exhibit no changes in creatinine, aspartate aminotransferase (AST), alanine aminotransferase (ALT), bilirubin, or amylase levels (55). However, even after excluding patients with abnormal preinfusion values, levels of both AST ($n = 26$) and amylase ($n = 19$) on day 1 were approximately double baseline values, with slight reductions but still elevated levels on day 3. In trauma studies, the trend towards elevated levels of amylase appears consistent with results of other trials in healthy volunteers and elective surgical patients. Further studies will be required to determine the significance and mechanism for these amylase elevations. To date, little if any information has been reported on HBOC-induced effects on immunology, hematology, or coagulation profiles in published trauma studies. One study reported the absence of HBOC-related GI distress (50).

It should be emphasized that at least two different clinical approaches and endpoints have been adopted in the HBOC trauma trials conducted to date. The low-dose trauma trial was primarily conducted as a safety study to support subsequent, larger scale trials in the trauma setting, in which DCLHb™ would be used as a low-volume, pressor, perfusion agent to reduce mortality and morbidity (32,90). In contrast, the high-dose studies have aimed to reduce allogeneic red cell transfusion as the primary endpoint, by using PolyHeme™ as a blood substitute in the traditional sense (55,93).

With regard to efficacy, HBOCs have been shown to reduce both mortality and allogeneic blood exposure, the two key targeted endpoints. In one study protocol, administration of DCLHb™ reduced death and serious adverse events by 25% and 15%, respectively, compared to saline-treated controls, in the first 50 patients who were treated with up to 10 g Hb (91). However, for the entire group of 139 patients, mortality and adverse event rates were not statistically different between treated and control groups (89,90). A larger efficacy trial now underway in DCLHb™-treated trauma patients aims to reduce 28-day mortality by 25% (i.e., from 40% to 30%) (10,90,95). Using different protocol designs, transfusion avoidance has also been demonstrated to some degree in other trauma trials. In an uncontrolled study with PolyHeme™, 59% of patients avoided allogeneic red cell transfusion within the first 24 hours after blood loss, although transfusion requirements during

**Table 4. HBOC Studies in Traumatic Shock**

| Indication/ patient population | Product, trial phase, and location | Subjects, [treated + control], [M + F], age range | Dose range (Hb) | Protocol details | Completion date | Primary endpoint | Results (safety, efficacy, pharmacoeconomic) | Refs. |
|---|---|---|---|---|---|---|---|---|
| Hemorrhagic hypovolemic shock (class II–IV) (74% of shock was trauma-related) | DCLHb® Phase I/II U.S. and Europe | n = 139 [71 + 68], [97M + 42F], 17–91 years | 5–20 g (50–200 mL) | Prospective, controlled, randomized, single-blind, dose escalation study. *HemAssist®* or *saline* given perioperatively, within 4 h of shock episode. | Mid 1995 | Safety | *Safety.* *Hemodynamics.* 10-g and 20-g doses induced a slight increase in MAP. *Renal.* No increase in renal insufficiency and failure. *AEs.* No increase in rate of complications, adverse events, or overall mortality. No allergic or transfusion reactions. *Efficacy.* None reported *Pharmacoeconomic.* None reported. | 89–91 |
| Trauma (blunt and penetrating trauma or non-trauma-related surgery) | Poly SFH-P Phase I/II U.S. | n = 39 [39 + 0], [28M + 11F], 19–83 years | 50–300 g | Prospective, uncontrolled, nonrandomized, unblinded study. *PolyHeme®* given pre-, intra- and postoperatively to both awake and anesthetized patients. | Late 1995 | Reduced allogeneic RBC transfusion | *Safety.* *Hemodynamics.* No changes in cardiovascular function (including MAP and HR). *Oxygenation/pulmonary.* Utilization of $O_2$ (extraction ratio) was $27 \pm 16\%$ from RBCs vs. $37 \pm 13\%$ from Poly SFH-P. *Clinical chemistry.* No changes in creatinine, AST, ALT, bilirubin, or amylase in patients with normal preinfusion values. *Hematology/coagulation.* Plasma Hb concentration after infusion of 300 g of Poly SFH-P was $4.8 \pm 0.8$ g/dL. Poly SFH-P maintained total Hb despite the marked fall in red cell Hb due to blood loss. *Renal.* No kidney dysfunction. *AEs.* No fever, GI distress, or other significant adverse events related to product infusion. *Efficacy.* 23 patients (59%) avoided allogeneic transfusion in the first 24 h after blood loss. *Pharmacoeconomics.* None reported. | 50,53–55,92 |
| Trauma and emergent surgery | PolyHeme® Phase II U.S. | n = 44 [21 + 23], [33M + 11F], age 19–83 years | ≤300 g | Prospective, controlled, randomized study. *PolyHeme®* or *allogeneic RBCs* (un-cross-matched?) given perioperatively. | Late 1996 | Reduced allogeneic RBC transfusion | *Safety.* *AEs.* No adverse clinical events attributable to PolyHeme®. *Efficacy.* Treated patients received (an average of) 4.4 units of PolyHeme® plus 7.8 units of allogeneic RBCs, whereas control patients received 11.3 units allogeneic RBCs, prior to hospital discharge. PolyHeme® maintained total Hb and $O_2$ transport despite the marked fall in RBC Hb due to acute blood loss. *Pharmacoeconomic.* None reported. | 50,93,94 |

the entire hospital stay were not disclosed (55). More convincing results from the controlled, follow-up study showed that requirements for allogeneic RBCs were reduced by an average of 3–4 units per patient during the entire hospital stay (94). In these published trauma trials, actual duration of hospital stay, a key pharmacoeconomic parameter, was not reported.

## DISCUSSION

### What We Have Learned

It is clear from the text and tables herein that the clinical testing of Hb-based red cell substitutes has progressed significantly since the beginning of this decade. At least six HBOC products of various composition (10) have been administered to over 600 subjects in Phase I, II, or III clinical trials, supporting their use in acute blood loss indications. Although many of the cited reports have presented limited data, it appears that, most importantly, these products are safe and well tolerated in the human volunteers and various patient groups evaluated to date.

Earlier concerns regarding the potential for nephrotoxicity and influences on the coagulation mechanism have not been realized. Likewise, pulmonary function does not appear to be affected by HBOCs, nor does the central nervous system. There is no clear data from any reports to date, either written or oral, to validate any concerns in these areas. However, it is also apparent that some adverse events, according to the true definition of the term, have been observed. In awake, healthy volunteers, unexpected GI disturbance and/or esophageal dysfunction have been reported for most of the products tested. Interestingly, the solution of human Hb not containing any 64 kD Hb species is the only product for which these effects have not been reported in either Phase I or II testing (53,55). It is noted that when general anesthesia is a part of the experimental protocol, as it has been with many of the Phase II tests in various patient groups, the GI symptoms are absent or certainly minimized (57,72,78,85). No other significant adverse events or unexpected findings have been reported from the infusions of this diverse group of products that are either universal or systematically recorded.

It has also been observed that there is some degree of vasoactivity associated with the infusion of many of these Hb solutions. Certain investigators consider this a potentially beneficial effect, perceiving an increase in systemic pressure as a good in the resuscitation models. Others view it as potentially harmful, particularly if there is redistribution of flow resulting from the vasoconstriction. The concept that these solutions are more than oxygen carriers is now acknowledged, yet their acceptance as pharmacologically active agents requires further inquiry into the nature of their effects and the potential for harm in both normal and compromised host situations. For example, vasoconstriction is generally believed to be an active physiologic process requiring expenditure of energy and hence consumption of oxygen. If the consumption of additional oxygen is sufficient to affect the total amount of oxygen available for delivery to tissues, then the beneficial net increase in oxygen delivery will be decreased.

With the potential availability of an oxygen carrying cell-free solution we have also learned that the medical community as a whole, and surgery and anesthesia in particular, are willing to alter their existing concepts of transfusion triggers. That is, once a solution with acceptable characteristics is available, the need for alternatives to allogeneic transfusion will change. Indeed, other options will also be exercised in an attempt to minimize exposure to allogeneic blood. The availability of autologous blood through predeposit or intraoperative hemodilution, the latter without compromise of oxygen-carrying capacity, in combination with an HBOC product will likely be a significant factor in altering transfusion decisions in the surgical area. In many of the Phase II (72,81,86) and some of the Phase III (36) protocols, these concepts are now being exercised with the goal of demonstrating transfusion avoidance without an increase in ischemic events, the obvious outcome to be avoided.

Cell-free Hb solutions were once considered to be an ideal replacement for blood or packed RBCs in trauma situations. The brass ring prize was to develop a commercial product for such use. With time, a great deal of knowledge has been acquired in the area of shock resuscitation physiology, making this an even more attractive transfusion alternative. Furthermore, the equivalence of acellular Hb solutions to whole blood has now been demonstrated in various animal and clinical models. These observations are both exciting and encouraging. Specifically, a solution of recombinant Hb (at 5 g/dl) was shown to possess oxygen-carrying capability superior to autologous rat blood, which contains ~8–9 g/dl Hb (57). In a separate study, which evaluated exercise capacity in humans using a bicycle exercise stress test, the physiological effects of 1 unit of HBOC-201 were shown to be equivalent to 3 units of autologous whole blood with respect to exercise function (42,44). The acellular property of these solutions is believed to be a significant factor in enhancing their relative efficaciousness in terms of oxygen delivery.

### Future Research Directions

Although a great deal has been learned from these early clinical trials, which have demonstrated safety and to some extent efficacy, there are still gaps in the knowledge that need to be filled to give a richer and more comprehensive picture. Is it necessary to define all of the physiologic nuances before proceeding with new generations of solutions? How will the composition of the new solutions differ in terms of basic characteristics and traits? What new features will be added to address the various problems perceived as the list of indications is expanded? Should these solutions be designed to address reperfusion phenomena along with the free radical interactions? Should they address NO issues along with oxygen delivery concepts? It must be recalled that the generation of NO is also an oxygen-consuming process, such that some of the increased oxygen-carrying capacity will not be used for nutrient tissue perfusion—one initial objective of using these solutions.

At present, little is known of the effects of these products in the presence of certain diseases. This is especially true of patients with hepatic or renal disease, for the liver and kidney are the primary sites of excretion and metabolism along with the reticuloendothelial system. Physiological effects induced by HBOC products in patients with compromised function in these organs requires further exploration, although there is currently no evidence to suggest additional adverse events will occur.

There is also a need to develop an understanding of the mechanism(s) of HBOC-induced smooth muscle contraction both in the vasculature and the GI tract. Understanding the underlying mechanism will allow agents to be added to existing products to counter these effects, or will permit the design of new Hb solutions to obviate such effects. There are difficulties of course, particularly from a regulatory perspective, when the composition of the solution is altered. The combinatorics of evaluating a product with many biological and physiological effects in animal and clinical models imply an enormous challenge.

It must be noted that most of the solutions presently being evaluated are truly first generation products, relying on some degree of stabilization of the Hb tetramer and then, in some cases, using a polymerization process to enlarge the molecular weight. The former provides an improvement in oxygen delivery and off-loading by decreasing oxyhemoglobin affinity, whereas the latter aids in promoting intravascular persistence, both desirable traits. Whether newer solutions will address these basic issues or add new complexity is not known. Over time, we will need to know the details of models to evaluate the effects of the newer and potentially more complex solutions (24).

On a cautionary note, we will need to understand the pharmacokinetics and the pharmacodynamics of Hb solutions to deal with their complete assessment in the clinical arena. Hence, it will be necessary to understand the details of product metabolism, including the effects of any directly or indirectly derived by-product(s). It is possible that some of the adverse events being observed are not directly induced by HBOC products, but are the effects of molecular species produced by normal metabolic degradation of the hemoglobin material. Due to the magnitude of anticipated clinical doses, these mechanisms could become overloaded or stimulated well beyond normal levels and thereby exert unexpected and untoward up-regulation effects on usually balanced physiological systems. Indeed, since off-label use of HBOCs could open questions as yet unanswered, it will be necessary to develop a better appreciation of these issues before any of the existing solutions reach the market.

As a greater appreciation of oxygen-dependent physiology evolves, as the indications for transfusion are refined, and as alternatives to traditional transfusions continue to be developed, the role of red cell substitutes remains an open question. Certainly, we will need to carefully address the effectiveness of these solutions in reaching universally accepted efficacy endpoints. For now, HBOC products are viewed as part of a continuum of options, to be used in many situations as part of an overall practice pattern where blood conservation and avoidance of allogeneic transfusion are primary objectives. Seen in that light, we have gained enough knowledge now to say they have a bright and useful future, probably well beyond their originally conceived use as resuscitation fluids in the field (5,8,10).

## ACKNOWLEDGMENTS

Since submission of this paper, Baxter Healthcare Corporation acquired Somatogen, Inc., in May, 1998. By July, 1998, Baxter halted all clinical trials with HemAssist® (DCLHb®). The company then announced its decision to end the clinical development of this product as of September 16, 1998. At that time, Baxter also stated that it would focus its research and development efforts on second generation products based on recombinant technology, as acquired from Somatogen. Although Baxter halted clinical development of Optro® (rHb1.1) during 1998, it is now developing a new second-generation recombinant hemoglobin product that is targeted for Phase I clinical trials after the year 2000.

## BIBLIOGRAPHY

1. A.G. Greenburg, World J. Surg. 20, 1189–1193 (1996).
2. American Society of Anesthesiologists Task Force on Blood Component Therapy (A.G. Greenburg, M.D., Ph.D., member), Anesthesiology 84, 732–747 (1996).
3. H.G. Klein, Am. J. Surg. 170, 21S–26S (1995).
4. A.G. Greenburg, Am. J. Surg. 173, 49–52 (1997).
5. A.G. Greenburg and H.W. Kim, Artif. Cells, Blood Substitutes Immob. Biotechnol. 25(1 and 2), 25–29 (1997).
6. H.W. Kim and A.G. Greenburg, Int. Symp. Korean Sci. Eng., Seoul, Korea, June 24–28, 1996, pp. 1300–1307.
7. N.M. Dietz, M.J. Joyner, and M.A. Warner, Anesth. Analg. 82, 390–405 (1996).
8. R.M. Winslow, Expert Opin. Invest. Drugs 5(11), 1443–1452 (1996).
9. J.E. Ogden and S.L. MacDonald, Vox Sang. 69, 302–308 (1995).
10. A.G. Greenburg, Adv. Surg. 31, 149–165 (1998).
11. G.W. Peskin, K. O'Brien, and S.F. Rabiner, Surgery 66, 185 (1969).
12. A.G. Greenburg, Crit Care Med. 14, 325–351 (1992).
13. R.K. Spence, A.G. Greenburg, and Members of the Blood Management Practice Guidelines Conference, Am. J. Surg. 170(6A), Suppl., 3S–15S (1995).
14. A.G. Greenburg, Am. J. Surg. 170(6A), 44S–48S (1995).
15. J.L. Carson, Am. J. Surg. 170, 32S–36S (1995).
16. S.F. Flaim, in A.S. Rudolph, R. Rabinovici, and G.Z. Feuerstein, Red Blood Cell Substitutes: Basic Principles and Clinical Applications, Dekker, New York, 1997, pp. 437–464.
17. P.E. Keipert, Artif. Cells, Blood Substitutes Immob. Biotechnol. 23(3), 381–394 (1995).
18. P.E. Keipert and M.G. Conlan, Artif. Cells, Blood Substitutes Immob. Biotechnol. 24(4), 359 (Abstr.) (1996).
19. T.M.S. Chang and R.P. Geyer, eds., Blood Substitutes, Dekker, New York, 1989.
20. T.M.S. Chang, ed., Blood Substitutes and Oxygen Carriers, Dekker, New York, 1993.
21. R.M. Winslow, ed., Hemoglobin-Based Red Cell Substitutes, John Hopkins University Press, Baltimore, Md., 1992.
22. R.M. Winslow, K.D. Vandegriff, and M. Intaglietta, eds., Blood Substitutes, Physiological Basis of Efficacy, Birkhaeuser, Boston, 1995.
23. R.M. Winslow, K.D. Vandegriff, and M. Intaglietta, eds., Blood Substitutes, New Challenges, Birkhaeuser, Boston, 1996.

24. R. Rabinovici, W.T. Phillips, G.Z. Feuerstein, and A.S. Rudolph, in A.S. Rudolph, R. Rabinovici, and G.Z. Feuerstein, eds., *Red Blood Cell Substitutes: Basic Principles and Clinical Applications* Dekker, New York, 1997, pp. 263–286.

25. G. Rhea et al., *Crit. Care Med.* **24**(1), Suppl., A39 (Abstr. 3) (1996).

26. G. Reah et al., *Crit. Care Med.* **25**(9), 1480–1488 (1997).

27. P. Bassett, *Drug Market Dev.* **7**(7), 164–168 (1996).

28. R.G. Kilbourn, J. DeAngelo, and J. Bonaventura, in J.-L. Vincent, ed., *Yearbook of Intensive Care and Emergency Medicine*, Springer-Verlag, Berlin, 1997, pp. 230–239.

29. S.K. Swan et al., *Am. J. Kidney Dis.* **26**(6), 918–923 (1995).

30. P. Gonzalez et al., *Blood* **84**(10), Suppl., 413a (Abstr. 1639) (1994).

31. P. Gonzalez et al., *J. Invest. Med.* **45**(5), 258–264 (1997).

32. D.J. Nelson, in A.S. Rudolph, R. Rabinovici, G.Z. Feuerstein, eds., *Red Cell Substitutes: Basic Principles and Clinical Applications*, Dekker, New York, 1997, pp. 353–400.

33. R.J. Przybelski et al., *Crit. Care Med.* **24**(12), 1993–2000 (1996).

34. J.F. Baron et al., *Anesthesiology* **87**(3A), Abstr. A217 (1997).

35. M.A. Garrioch, E.K. Daily, and R.J. Przybelski, *Abstr., Proc. 2nd Congr. Jpn. Soc. Blood Substitutes*, June 29, 1995.

36. J.F. Baron et al., *Poster, Am. Soc. Anesthesiol. Ann. Meet.*, San Diego, Calif., October 18–22, 1997.

37. J.F. O'Hara et al., *Anesthesiology* **87**(3A), Abstr. A344 (1997).

38. J.F. O'Hara et al., *Anesthesiology* **87**(3A), Abstr. A205 (1997).

39. G. Hughes et al., *J. Lab. Clin. Med.* **126**(5), 444–451 (1995).

40. G. Hughes et al., *Ann. Emerg. Med.* **27**(2), 164–169 (1996).

41. G. Hughes et al., *Crit. Care Med.* **24**(5), 756–764 (1996).

42. G. Hughes et al., *Clin. Pharmacol. Ther.* **58**(4), 434–443 (1995).

43. S.M. Kasper et al., *Anesth. Analg.* **83**, 921–927 (1996).

44. W.R. Light et al., in A.S. Rudolph, R. Rabinovici, and G.Z. Feuerstein, eds., *Red Blood Cell Substitutes: Basic Principles and Clinical Applications*, Dekker, New York, 1997, pp. 421–436.

45. T. Standl et al., *Crit. Care* **1** (Suppl. 1), 51 (Abstr. P92) (1997).

46. K.L. Shum et al., *Artif. Cells, Blood Substitutes Immob. Biotechnol.* **24**(6), 655–683 (1996).

47. P. Bassett, *Drug Market Dev.* **8**(6), 126–130 (1997).

48. R. Shorr, *Proc. Int. Bus. Commun., 4th Annu. Blood Substitutes Conf.*, San Diego, Calif., November 21, 1996.

49. J.G. Adamson et al., in A.S. Rudolph, R. Rabinovici, and G.Z. Feuerstein, eds., *Red Blood Cell Substitutes: Basic Principles and Clinical Applications*, Dekker, New York, 1997, pp. 335–352.

50. S.A. Gould, L.R. Sehgal, G.S. Moss, and H.L. Sehgal, in A.S. Rudolph, R. Rabinovici, and G.Z. Feuerstein, eds., *Red Blood Cell Substitutes: Basic Principles and Clinical Applications*, Dekker, New York, 1997, pp. 401–420.

51. S.A. Gould, L.R. Sehgal, H.L. Sehgal, and G.S. Moss, *Proc. 5th Int. Soc. Blood Substitutes Meet.*, San Diego, Calif., March 19, 1993, Abstr. H13.

52. S. Gould et al., *Transfusion* **33**(Suppl.), 60S (Abstr. S231) (1993).

53. S.A. Gould, L.R. Sehgal, H.L. Sehgal, and G.S. Moss, *Transfusion Sci.* **16**(1), 5–17 (1995).

54. S.A. Gould and G.S. Moss, *World J. Surg.* **20**, 1200–1207 (1996).

55. S.A. Gould et al., *J. Trauma, Injury, Infect. Crit. Care* **43**(2), 325–332 (1997).

56. M.J. Gerber, G.L. Stetler, and D. Templeton, in E. Tsuchida, ed., *Artificial Red Cells*, Wiley, New York, 1995, pp. 187–197.

57. J.W. Freytag and D. Templeton, in A.S. Rudolph, R. Rabinovici, and G.Z. Feuerstein, eds., *Red Blood Cell Substitutes: Basic Principles and Clinical Applications*, Dekker, New York, 1997, pp. 325–333.

58. M.K. Viele, R.B. Weiskopf, and D. Fisher, *Anesthesiology* **86**, 848–858 (1997).

59. M.N. Stetter, G.M. Baerlocher, H.J. Meiselman, and W.H. Reinhart, *Transfusion* **37**, 1149–1155 (1997).

60. R.J. Przybelski, E.K. Daily, and M.L. Birnbaum, in R.M. Winslow, K.D. Vandegriff, and M. Intaglietta, eds., *Advances in Blood Substitutes: Industrial Opportunities and Medical Challenges*, Birkhaeuser, Boston, 1997, pp. 71–90.

61. R. Przybelski et al., *Crit. Care Med.* **22**(1), Suppl. (Abstr. A231) (1994).

62. G.S. Hughes and E. Jacobs, *5th Int. Soc. Blood Substitutes Meet.*, San Diego, Calif., March 19, 1993, Abstr. H15.

63. J.A. Murray et al., *Gastroenterology* **109**, 1241–1248 (1995).

64. G. Hughes, E. Jacobs, S. Francom, and J. Riggin, *Crit. Care Med.* **23**(1), Suppl. (Abstr.) (1995).

65. G. Hughes et al., *Crit. Care Med.* **23**(1), Suppl. (Abstr.) (1995).

66. G. Hughes et al., *Crit. Care Med.* **23**(1), Suppl. (Abstr.) (1995).

67. D. Wicks, S. Nakao, P. Champagne, and C. Mihas, *Artif. Cells, Blood Substitutes Immob. Technol.* **24**(4), 460 (Abstr.) (1996).

68. R. Shorr, A. Viau, and A. Abuchowski, *Artif. Cells, Blood Substitutes Immob. Biotechnol.* **24**(4), 425 (Abstr.) (1996).

69. J.D. Bristow et al., *Anesthesiology* **31**, 422–428 (1969).

70. L.T. Wong et al., *Artif. Cells, Blood Substitutes Immob. Biotechnol.* **26**, 529–548 (1998).

71. P. Bassett, *Drug Market Dev. Rep.* **907**, 5-86–5-94 (1995).

72. B.J. Leone et al., *Artif. Cells, Blood Substitutes Immob. Biotechnol.* **24**(4), 379 (Abstr.) (1996).

73. K. Berman, *Int. Blood Plasma News*, April 1996, p. 124.

74. M. Lamy, *Proc. Eur. Soc. Anesthesiol. Meet.*, Lausanne, Switzerland, May 5, 1997, Abstr.

75. M. Garrioch et al., *Crit. Care Med.* **24**(1), Suppl., A39 (Abstr. 4) (1996).

76. E.L. Bloomfield et al., *Anesthesiology* **85**(3A), Suppl. (Abstr. A220) (1996).

77. T. Monk, L. Goodnough, G. Hughes, and E. Jacobs, *Anesthesiology* 83(3A) (Abstr. A285) (1995).

78. R. Lessen et al., *Artif. Cells, Blood Substitutes Immob. Biotechnol.* **24**(4), 380 (Abstr.) (1996).

79. G. Hughes, E. Jacobs, S. Francom, and the Hemopure Surgical Study Group, *Crit. Care Med.* **24**(1), Suppl., A36 (Abstr. 2) (1996).

80. J.A. Wahr et al., *Anesthesiology* **85**(3A) (Abstr. A347) (1996).

81. T. Standl, *Acta Haematol.* **98**, Suppl. 1, 123 (Abstr. 484) (1997).

82. T. Standl et al., *Crit. Care* **1**, Suppl. 1, 52–53 (Abstr. P94) (1997).

83. T. Standl et al., *Anesthesiology* **87**(3A) (Abstr. A65) (1997).

84. J.F. O'Hara et al., *Anesthesiology* **87**(3A) (Abstr. A230) (1997).

85. A. Schubert et al., *Anesthesiology* **87**(3A) (Abstr. A220) (1997).

86. A. Schubert et al., *Anesthesiology* **87**(3A) (Abstr. A218) (1997).

87. T.G. Monk et al., *Anesthesiology* **87**(3A) (Abstr. A214) (1997).

88. W. Lieberthal, in A.S. Rudolph, R. Rabinovici, and G.Z. Feuerstein, eds., *Red Blood Cell Substitutes: Basic Principles and Clinical Applications*, Dekker, New York, 1997, pp. 189–217.

89. T. Estep, *Proc. Int. Bus. Commun. 3rd Annu. Blood Substitutes Conf.* Bethesda, Md., October 30, 1995.

90. E.P. Sloan, *Proc. Int. Bus. Commun. 4th Annu. Blood Substitutes Conf.* San Diego, Calif., November 21, 1996.

91. E.P. Sloan et al., *Acad. Emerg. Med.* **2**(5), 365 (Abstr. 078) (1995).

92. S.A. Gould et al., *J. Trauma, Injury, Infect. Crit. Care* **39**(1), 157 (Abstr.) (1995).

93. E.E. Moore, S.A. Gould, D.B. Hoyt, and J.M. Haenel, *Shock* **7**(Suppl.), 145 (Abstr. 576) (1997).

94. K.C. Tang and P.M. Cheng, *Life Sci. Biotechnol. Rep.*, 1–8 (1997).

95. E.P. Sloan, *Proc. Int. Bus. Commun. 5th Annu. Blood Substitutes Conf.*, Cambridge, Mass., November 21, 1997.

# C

## CALCIFICATION, DRUG DELIVERY TO PREVENT

Narendra R. Vyavahare
Robert J. Levy
Children's Hospital of Philadelphia and the University of
    Pennsylvania
Philadelphia, Pennsylvania

### KEY WORDS

Aluminum chloride
Bioprostheses
Bisphosphonates
Bovine pericardium
Calcium phosphates
Ferric chloride
Glutaraldehyde
Heterograft
Polymeric drug delivery
Porcine aortic valve

### OUTLINE

### BACKGROUND

#### Definitions: Cardiovascular Calcification

Calcification occurs in a wide variety of cardiovascular disease processes including arteriosclerosis, valvular degeneration, endocarditis, and myocardial infarction. Calcification is also associated with implanted medical devices such as bioprosthetic heart valves (BPHV) derived from glutaraldehyde cross-linked porcine aortic valves or bovine pericardium, vascular grafts, and blood pumps (1–3). In particular, calcification is the leading cause of failure of heart valve bioprostheses (4,5). We will focus on BPHV cal-cification as a model for cardiovascular calcification. Accumulation of crystalline calcium phosphate mineral is most often pathologic calcification in cardiovascular diseases. This pathologic calcification can either be metastatic or dystrophic. The calcification is termed metastatic when it occurs in hypercalcemic hosts with otherwise normal tissues, whereas calcification is termed dystrophic when it occurs in necrotic or damaged tissues in normocalcemic subjects. In contrast to the orderly and regulated pattern of normal skeletal mineralization, dystrophic calcification, although progressive, is haphazard in its localization and poorly regulated. In case of medical implants, the location of mineral nucleation may be intrinsic (i.e., within the boundaries of the tissue or biomaterial involving its original constituents) or extrinsic (i.e., associated with elements or tissue not initially implanted, such as within thrombus, vegetations, or pseudointima). Due to the localized nature of this disease process, site-specific drug delivery systems releasing anticalcification drugs locally would be an ideal approach.

#### Scope and Importance of the Clinical Problem of Bioprostheses

Thousands of patients undergo heart valve replacement surgeries annually (6). Lifetime anticoagulant therapy is needed for patients with mechanical prostheses. Bioprosthetic heart valves are excellent alternatives when long-term anticoagulation is contraindicated or undesirable, such as in patients with diabetes, women of childbearing age, and individuals in nonindustrialized geographic regions. However, long-term use of BPHV is limited due to cuspal tearing and valvular degeneration related to calcification.

Bioprosthetic heart valve calcification has the following determinants:

- Host factors (e.g., immature subjects calcify bioprosthetic tissue more rapidly than adults)
- Fixation conditions (e.g., glutaraldehyde cross-linking aggravates bioprosthetic calcification) (7)
- Mechanical factors (deposition of calcium phosphates occurs more readily at regions of stress concentrations in bioprosthetic leaflets) (8–10)

Calcific nodules interfere with leaflet function, and calcification has also been shown to occur in the aortic wall region of BPHV, which may compromise the lumen of stentless valve designs (11,12). Animal models for bioprosthetic calcification have been developed and characterized using noncirculatory subdermal implants (rat, rabbit, mouse) and circulatory valve replacements in sheep or calves. These animal models result in pathology comparable to that seen in clinical calcific bioprosthesis failure (13–19). Figure 1 shows the various steps involved in clinical failure of the valves. The treatments that may alter one

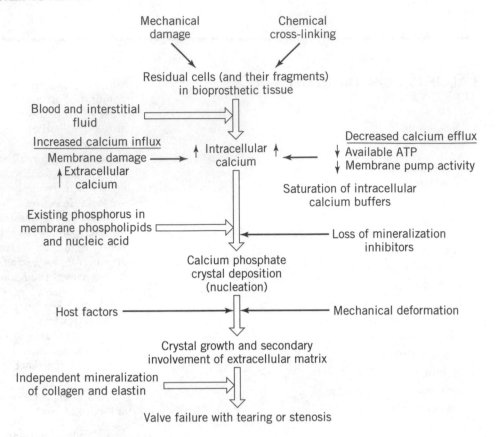

**Figure 1.** Extended hypothetical model for the calcification of bioprosthetic tissue. This model considers host factors, implant factors, and mechanical damage and relates initial sites of mineral nucleation to increased intracellular calcium in residual cells and cell fragments in bioprosthetic tissue. The ultimate result of calcification is valve failure, with tearing or stenosis. *Source:* Reproduced with permission from Ref. 20.

or more steps are hypothesized to reduce calcification of bioprosthetic valves.

The pathogenesis of BPHV calcification involves an interaction of host and implant factors. Initial calcification has been noted in the devitalized connective tissue cells intrinsic to the bioprosthetic leaflet material. Structural protein calcification, most prominently collagen in leaflets and elastin in aortic wall, then ensues (13). Tissue fixation protocols devitalize cells, thus disrupting their calcium regulation mechanisms. In living cells the intracellular calcium level is approximately $10^{-7}$ M, whereas extracellular calcium is $10^{-3}$ M. This 10,000-fold gradient across the plasma membrane is maintained by energy-requiring metabolic pumps, which extrude this ion as well as intracellular buffering mechanisms. In cells modified by glutaraldehyde cross-linking, necrosis, or mechanical injury, the mechanisms for calcium extrusion are no longer functional, thus the calcium influx occurs unimpeded. Plasma membranes and membrane bound organelles such as mitochondria are rich in phospholipids and provide phosphorus for appetite formation (13,14,21). Although calcification has been found to initiate within devitalized cells of glutaraldehyde-fixed BPHVs, extracellular matrix such as collagen (in BPHV cusps) and elastin (in aortic wall) has been shown to independently calcify in animal models (22,23). The role of immunologic factors such as nonspecific inflammation or antigenic-specific immunologic response in mediating calcification is controversial. Valve tissue implanted in congenitally athymic, T-cell deficient (nude)

mice calcifies to the same extent as implants in immunologically competent hosts (16).

A growing body of evidence indicates that mineral matrix proteins that function in physiologic calcification of bone and tooth are also associated with pathologic cardiovascular calcification (24–30). Levy et al. have demonstrated the role of exogenous alkaline phosphatase in bioprosthetic mineralization (26). The noncollagenous bone matrix proteins such as osteocalcin and osteopontin have been found to be associated with pathologic calcification of bioprostheses as well as in atherosclerosis (24,25). In particular, bone matrix proteins such as osteopontin, bone acidic glycoprotein (BAG 75) have been shown to be associated with pathologic calcification of BPHV in rat subdermal implants (31). These proteins have been extensively studied in physiologic bone mineralization, although their exact functions are not completely understood. Once their exact role in cardiovascular calcification is understood, site-specific, gene-based therapies for preventing calcification can be envisioned.

### Rationale for Controlled Drug Delivery

There are no conventional approaches to prevent cardiovascular calcification. Any systemic use of drugs to prevent this disease process would ultimately have unwanted toxic effects on bone and teeth development and calcium metabolism. This is particularly true for children and young adults where pathologic calcification of BPHV occurs rapidly. Bisphosphonates, a class of drugs approved for preventing calcification, represent this type of therapeutic dilemma. Bisphosphonates are poorly absorbed orally and must be given parenterally (32). The affinity of bisphosphonates for bone mineral leads to their accumulation in this organ system with parenteral administration. Osteomalacia and overall diminution in somatic growth are classic side effects of this drug therapy (32). Due to the localized nature of the BPHV calcification, the site-specific delivery of anticalcification agents such as bisphosphonates is ideal. Such site-specific therapy would require less drug, increase its therapeutic effectiveness, and reduce or eliminate its systemic side effects.

## TREATMENT STRATEGIES

Two classes of anticalcification agents, namely bisphosphonates and metallic salts such as iron chloride ($FeCl_3$) or aluminum nitrate ($Al(NO_3)_3$), have been studied in animal models in the form of controlled release drug delivery systems (33–45). Thus, drug delivery devices were placed in a close proximity of the heart valve tissue implant and their anticalcification effects were studied. Ethane hydroxybisphosphonate (EHBP) is one of the best known anticalcification agents. It has been approved by the FDA for human use for inhibition of pathologic calcification and for treating hypercalcemia of malignancy (46). The EHBP mechanism is based on a specific interaction with calcium and calcium-phosphate crystal formation and growth. EHBP also inhibits alkaline phosphatase activity (47). Alkaline phosphatase is an important enzyme in physiologic mineralization, and it may also have a role in cardiovascular calcification, because this enzyme has been shown to be present intrinsically in bioprosthetic tissue despite the glutaraldehyde fixation, and it is also adsorbed extrinsically from the host following implantation (48).

The mechanisms of action of $Fe^{3+}$ and $Al^{3+}$ in preventing calcification are incompletely understood but are thought to be due to the retardation of hydroxyapatite crystal growth by a crystal poisoning effect and reduction in the alkaline phosphatase activity (48). Their mechanism of action may also be related to their interactions with phosphorus-rich loci within membranes of devitalized cells and other phosphorus-rich organelles present in BPHV (49); these sites are thought to be the initial donors of phosphorus in the nucleation of calcium phosphate mineral (13,14).

### Controlled Release Drug Delivery Formulations for Preventing Cusp Calcification

Table 1 summarizes the different controlled release drug delivery systems studied so far for preventing cardiovascular calcification. The bisphosphonate compounds, in particular EHBP, have been used in the past as anticalcification agents. However, due to their systemic toxicity, conventional drug therapy results in many side effects in bone and calcium metabolism (54). Thus, controlled release delivery systems using polymers were formulated and investigated. The principle of this approach was to maintain the drug levels only at the sites (i.e., valve im-

plants) where it is needed and to prevent systemic side effects through local administration. Thus drug-releasing matrices were implanted in close proximity to BPHV tissue implant. Many studies incorporating EHBP in nondegradable polymers such as ethylene-vinyl acetate (EVA), polydimethylsiloxane (silicone), silastic, and polyurethanes have shown the effectiveness of this strategy in preventing BPHV calcification in a variety of animal models (33–37,39,40,43,44,55,56).

In the rat subdermal model, the controlled release of EHBP from EVA matrices was shown to prevent the calcification of porcine aortic valve cusps for up to 84 days (33,34). In this study, cusps in close proximity to control EVA films without any drug calcified with calcium levels comparable to those observed with clinical failure of heart valves (Fig. 2). The kinetics of drug release from matrices was controlled from hours to years by modifying various parameters such as percent drug loading, incorporation of the less soluble calcium salt of the EHBP, addition of an inert filler such as insulin, and fabrication of membrane coatings (35–37). Based on the rat subdermal studies, the optimal controlled release system with two salts of EHBP, namely sodium and calcium (1,000-fold less soluble than sodium salt) salts equally mixed (1:1) in silastic matrices (30% drug loading), was found to increase the extrapolated duration of EHBP-controlled release to several decades (38). These drug-loaded matrices were then tested for their efficacy in a sheep mitral valve replacement model (39). The matrices were fabricated as rings and were placed around the stent post of the bioprosthesis directly under the sewing cushion as shown in Figure 3. These polymeric implants were shown to be effective in inhibiting calcification in sheep tricuspid valve replacements with porcine aortic bioprostheses (39). However, the same controlled release implants were ineffective in preventing calcification of glutaraldehyde-pretreated bovine pericardium in mitral valve replacements in sheep (150 days) (58) (Table 2). The reasons for this lack of efficacy are not clear, but may be attributed to the fact that mitral valve calcification is more rapidly progressive and drug levels attained by controlled release systems may be insufficient due to washout effects in the circulation. Furthermore, Golomb et al. have prepared chitosan microspheres with encapsulated bisphosphonate (53) and have shown effective inhibition of bioprosthetic tissue calcification by injected microspheres near the BPHV cusp implants in the rat subdermal model.

Metal ion salts such as ferric chloride and aluminum nitrate were studied for their anticalcification efficacy when released from polymer matrices (43,45,50–53). Thus, $FeCl_3$ and $Al(NO_3)_3$ were incorporated into silicon polymer (silastic 6605-41) and polyether urethane (Biomer®) (10 weight% loading) and were implanted in close proximity to glutaraldehyde-fixed bovine pericardium in a rat subdermal model (50). Depending upon their release kinetics both the agents significantly inhibited calcification of bovine pericardium (Table 3). The release of metal ions from Biomer® was rapid with a large initial burst phase (up to 50% within first few hours); thus, less drug was available for prolong action, and the anticalcification effect was only partial. The release of metal ions ($Fe^{3+}$, $Al^{3+}$) from silastic polymer, on the other hand, was sustained for the implant

**Table 1. Summary of Controlled Release Drug Delivery Systems Studied to Prevent Cardiovascular Calcification**

| Drug | Matrix/geometry | Tissue | Model | Reference |
|---|---|---|---|---|
| $Na_2EHBP$ | EVA/hemisphere | Porcine aortic valve | Rat subdermal implant | 33,34 |
| $Na_2EHBP/Ca_2$ EHBP | Silicone-rubber, silastic/slab | Bovine pericardium | Rat subdermal implant | 35 |
| EHBP | EVA/hemisphere | Bovine pericardium | Rat subdermal implant | 36 |
| $Na_2EHBP/Ca_2$ EHBP | Polydimethylsiloxane/membrane-coated slabs | Bovine pericardium | Rat subdermal implant | 37 |
| $Na_2EHBP/Ca_2$ EHBP | Polydimethylsiloxane/membrane-coated slab and ring | Bovine pericardium/porcine aortic valve | Rat subdermal implant/sheep tricuspid valve replacement | 39 |
| $Fe^{3+}/Al^{3+}$ | Polydimethylsiloxane/slab | Bovine pericardium | Rat subdermal implant | 50 |
| Protamine sulfate | Polydimethylsiloxane/slab | Bovine pericardium | Rat subdermal implant | 50 |
| Levamisole | Polydimethylsiloxane/slab | Bovine pericardium | Rat subdermal implant | 50 |
| $Na_2EHBP/Ca_2$ EHBP | Polyurethane (mitrathane)/reservoir | Bovine pericardium | Rat subdermal implant | 40 |
| Sodium dodecyl sulfate (SDS) | Polydimethylsiloxane/slab | Porcine aortic valve | Rat subdermal implant | 42 |
| $EHBP/Fe^{3+}$ synergism | Polydimethylsiloxane/slab | Bovine pericardium | Rat subdermal implant | 43 |
| $EHBP/Fe^{3+}$ synergism | Polydimethylsiloxane/EVA/slab | Rat aortic wall | Rat aortic wall allograft model | 44,45 |
| $Fe^{3+}$ | Chitosan/slab | Bovine pericardium/polyurethane | Rat subdermal implant | 51,52 |
| EHBP | Chitosan/microparticles | Bovine pericardium | Rat subdermal implant | 53 |

period (21 days), and thus the anticalcification effect was substantial as compared to controls. Other researchers (51,52) have also shown the release of $Fe^{3+}$ ions from chitosan matrices prevents bovine pericardium calcification and polyurethane calcification in the rat subdermal model.

Other agents such as levamisole (alkaline phosphatase inhibitor) or protamine sulfate (tissue charge modifier) when released from polymer matrices near the bovine pericardium implants were ineffective in preventing bovine pericardium calcification in the rat subdermal model (50). In other studies, controlled release of sodium dodecyl sulfate (SDS) near the bovine pericardium implant was also shown to be completely ineffective in preventing bovine pericardium calcification (42). SDS and protamine sulfate were shown by others to be highly effective in preventing BPHV calcification when the BPHV cusps were incubated in these agents in solution prior to implantation (42,59,60). These results suggest that SDS and protamine sulfate may be modifying the bioprosthetic tissue itself in such a way that it makes it resistant to calcification. However, these agents have little role in preventing calcification related to progressive host–implant material interactions.

### Formulations for Preventing Aortic Root Calcification

Recent interests in stentless valve bioprostheses have raised concerns about valve failure due to aortic wall calcification, although there are no clinical results as yet to suggest this (11,12,61). It has been observed in the past that the mechanism of aortic wall calcification is different than cusp calcification due to differences in tissue morphology. Many agents that prevent cusp calcification do not prevent aortic wall calcification (61,62). Recently, a microsurgical rat aortic circulatory allograft model has been developed in our laboratory to study the calcification of rat

aorta (44). This model was used to study the effect of polymeric (EVA) implant releasing two drugs, EHBP and $FeCl_3$, which act synergistically in preventing aortic wall calcification (45). The results demonstrated that $FeCl_3$ alone was not effective and EHBP releasing from polymer was partially effective, whereas both the drugs released simultaneously from the same implant synergistically inhibited aortic wall calcification (Table 4) (45). No side effects of this drug therapy were found on overall weight gain, serum calcium levels, and bone growth of the animals.

### Limitations of Site Specific Drug Delivery

Although the controlled drug delivery systems as described before have shown promising results in preventing BPHV calcification in animal models, a great many difficulties need to be overcome in order to use them in clinical applications. For example, drug release (in effective concentrations) from matrix systems in the circulation may depend upon various factors such as inflammatory response causing capsule formation around the implant, degradation and cracking of implant, osmotic pressures caused by hydrophilic drugs. The matrix type devices (with dispersed drug) have only a limited lifetime and can be depleted of drug before the useful life of a heart valve (usually 15–20 years). Although the feasibility of using reservoir devices (a drug solution within a polymer membrane) has been shown in preventing BPHV calcification in the rat subdermal model, these devices have risks of dose dumping due to breakage of the membrane. The biocompatibility and toxicity of a drug delivery implant itself can become an issue in heart valve replacement surgery. Thus, a better approach could be to incorporate the drug in polymeric microspheres and then inject the microspheres near the im-

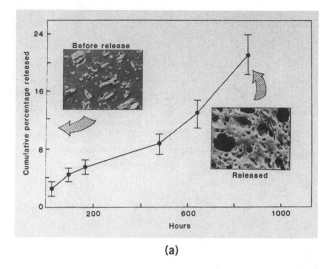

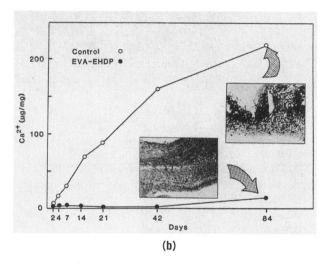

(b)

**Figure 2.** (a) In vitro cumulative percentage EHBP released from an EVA-EHBP matrix (20 weight%). Insets: scanning electron microscopy for drug delivery implants before and after 42 days of release in vitro. (b) Inhibition of BPHV calcification by site-specific drug delivery of EHBP from EVA-EHBP matrix (20 weight%) in a rat subdermal implantation model. Insets: light microscopy of BPHV tissue with von kossa staining (calciumphosphate-black) showed extensive intrinsic calcification at 84 days in a control specimen but no visible deposits in the BPHV tissue coimplanted with EVA-EHBP drug delivery matrix. *Source:* Reproduced with permission from Ref. 33.

plant site via catheter. The polymer could then slowly degrade and release the drug near the heart valve implant. Once the dose is completely released, another batch of microspheres could then be injected.

## FUTURE DIRECTIONS

Tissue-engineered heart valves are perceived by many as providing a solution to cardiac valve disease and cardiac valve regeneration (63). Tissue-engineered valves have been investigated most extensively by Mayer and his col-

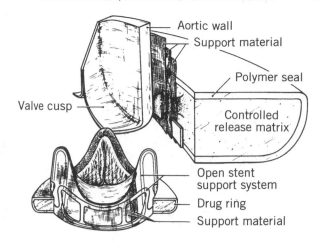

**Figure 3.** Site-specific delivery of drugs from controlled release matrix positioned at the circumferential support of a porcine aortic bioprosthesis. *Source:* Reproduced with permission from Ref. 57.

**Table 2. Inhibition of Calcification by Controlled Release Implants: Sheep Results (150 days)**

| Position | Type of implant | $n$ | Calcium (mg/mg of tissue ± SEM) |
|---|---|---|---|
| Tricuspid | Controlled release-EHBP | 5 | 2.7 ± 1.0 |
| Triscupid | Nondrug-control | 6 | 41.3 ± 14.9 |
| Mitral | Controlled release-EHBP | 10 | 39.9 ± 7.7 |
| Mitral | Nondrug control | 10 | 44.5 ± 7.5 |

*Source:* Based partly on Ref. 58.

**Table 3. Controlled Release of Metal Ions To Prevent Calcification of Glutaraldehyde Cross-linked Bovine Pericardium in a Rat Subdermal Implant Study (21 Day)**

| Group | Polymer | $n$ | Calcium levels (µg/mg tissue) | Percent calcification |
|---|---|---|---|---|
| $FeCl_3$ | Silastic | 10 | 14.25 ± 2.91 | 19.26 |
| $FeCl_3$ | Biomer | 10 | 44.67 ± 8.04 | 60.38 |
| $Al(NO_3)_3$ | Silastic | 10 | 14.28 ± 5.78 | 17.87 |
| $Al(NO_3)_3$ | Biomer | 10 | 31.04 ± 5.48 | 38.85 |
| No drug | Silastic | 30 | 52.27 ± 9.35 | 74.70 |
| No drug | Biomer | 20 | 64.20 ± 10.24 | 91.16 |
| Control BP | No polymer | 40 | 69.97 ± 10.76 | 100.0 |

*Source:* Based partly on Ref. 50.

leagues. This group has designed a polymer scaffold that can be seeded with autologous cells. In a series of sheep studies, they were able to carry out successful pulmonary valve cusp replacements (63). However, a number of major challenges face this approach, including biomechanical and design considerations, matrix biosynthesis following implantation, cell integration with the host, and potential problems with degeneration including calcification as well as primary material failure. Nevertheless, tissue-

**Table 4. Anticalfication Efficacy of Controlled Release Implants Releasing EHBP and FeCl₃ on 30-Day Rat Allograft Calcification**

| Group | Calcium (µg/mg of tissue) ± SEM | Phosphorus (µg/mg of tissue) ± SEM |
|---|---|---|
| Control (no polymer) | 168.68 ± 22.54 | 104.12 ± 14.44 |
| Control (only polymer) | 160.83 ± 8.66 | 107.22 ± 11.39 |
| CaEHBP release alone | 71.87 ± 17.31 | 65.93 ± 13.22 |
| FeCl₃ release alone | 151.13 ± 18.54 | 118.07 ± 11.46 |
| CaEHBP and FeCl₃ release together | 28.20 ± 7.46 | 26.60 ± 3.22 |

*Source:* Reproduced in part from Ref. 45.

engineered approaches have captured the imagination of the cardiovascular device industry. A number of efforts are underway to improve upon this first generation of tissue-engineered cardiac valve prosthesis. Tissue-engineered cardiac valve prostheses will likely be affected by a number of pathologic processes. In fact, it is predictable that they will be more likely to become diseased than native valves due to the lack of endogenous defense mechanisms. Thus, site-specific controlled release therapy may be relevant for tissue-engineered valves.

Gene-based therapies are becoming a reality both for genetic disease as well as tissue regeneration and local therapeutic purposes. Thus, it is conceivable that controlled release drug delivery for cardiovascular disease and cardiovascular calcification may use gene delivery. In terms of mineralization, antimineralization proteins could be expressed. These might include proteins such as statherin, an anticalcification protein normally found in saliva. Other gene therapy approaches could include antisense oligonucleotide delivery to counteract the local expression of bone-specific proteins that occur in valve calcification, such as bone morphogenic protein or osteopontin (24,25,31). Gene therapy could also take the approach of using genetically modified cells, implanted on diseased cardiac valves, that could resorb calcium deposits.

Synergistic approaches have already been used by our group in our bisphosphonate–ferric chloride studies (45). Synergistic approaches in the future could include combinations of not only pharmaceuticals but gene-based therapies as well as tissue engineering. Ideally, minimally invasive approaches should be the goal for therapy to prevent and reverse cardiovascular calcification. Ultimately this could be a targeted gene delivery through peripheral intravenous administration. Fibroblast growth factor (FGF)–targeted gene therapy has already been demonstrated to be feasible (64), and this represents the first example of peripheral targeting of DNA delivery. At the present time, the most reasonable goal in the near term would be DNA delivery through a site-specific cardiac catheter delivery implant.

## CONCLUSIONS

Calcification occurs in variety of cardiovascular diseases and with implanted biomaterials. Although calcification leads to failure of medical devices such as bioprostheses, no current therapy is available to prevent this disease process. Due to localized nature of the calcification, a variety of controlled release systems releasing anticalcification drugs have been investigated successfully in animal models for drugs such as bisphosphonates and metallic ions ($Fe^{3+}$ or $Al^{3+}$). Matrix and reservoir device configurations have limitations in terms of limited lifetime, maintaining sustained release, degradation, and fear of dose dumping for their clinical applications to prevent calcification. An ideal approach would be a catheter-based delivery implant such as microparticles containing either anticalcification drugs or antisense oligonucleotides to counteract local expression of mineral specific proteins, or gene-based therapies such as DNA delivery and tissue engineering.

## BIBLIOGRAPHY

1. F.J. Schoen, *Trans.—Am. Soc. Artif. Intern. Organs* **33**, 8–18 (1987).

2. H.C. Anderson, *Arch. Pathol. Lab. Med.* **107**, 341–348 (1983).

3. A. Milano et al., *Am. J. Cardiol.* **53**, 1066–1070 (1984).

4. F. J. Schoen, R. J. Levy, and H. R. Piehler, *Cardiovasc. Pathol.* **1**, 29–52 (1992).

5. G.L. Grunkemeier, W. R. E. Jamieson, D.C. Miller, and A. Starr, *J. Thorac. Cardiovasc. Surg.* **108**, 709–718 (1994).

6. L.H. Edmunds et al., *Eur. J. Cardiothorac. Surg.* **10**, 812–816 (1996).

7. G. Golomb et al., *Am. J. Pathol.* **127**, 122–130 (1987).

8. H.N. Sabbah, M.S. Hamid, and P.D. Stein, *Ann. Thorac. Surg.* **42**, 93–96 (1986).

9. G.M. Bernacca et al., *J. Biomed. Mater. Res.* **26**, 959–966 (1992).

10. H.E. Jorge et al., *J. Biomed. Mater. Res.* **30**, 411–415 (1996).

11. L.M. Biedrzycki, E. Lerner, R.J. Levy, and F.J. Schoen, *J. Biomed. Mater. Res.* **34**, 411–415 (1997).

12. T.E. David, J. Bos, and H. Rakowski, *J. Heart Valve Dis.* **1**, 244–248 (1992).

13. F.J. Schoen et al., *Lab. Invest.* **52**, 523–532 (1985).

14. F.J. Schoen, J.W. Tsao, and R.J. Levy, *Am. J. Pathol.* **123**, 134–145 (1986).

15. R. Levy et al., *Am. J. Pathol.* **113**, 143–155 (1983).

16. R.J. Levy, F.J. Schoen, and S.L. Howard, *Am. J. Cardiol.* **52**, 629–631 (1983).

17. M. Jones et al., *J. Cardiovasc. Surg.* **4**, 69–73 (1989).

18. M.C. Fishbein et al., *J. Thorac. Cardiovasc. Surg.* **83**, 602–609 (1982).

19. G.R. Barnhart et al., *Circulation* I150–I153 (1982).

20. F.J. Schoen, ed., *Interventional and Surgical Cardiovascular Pathology, Clinical Correlations and Basic Principles*, Saunders, Philadelphia, 1989.

21. F.J. Schoen and R.J. Levy, *Eur. J. Cardiothorac. Surg.* **6**(Suppl. 1), S91–S94 (1992).

22. R.J. Levy et al., *Am. J. Pathol.* **122**, 71–82 (1986).

23. C. Webb, N. Nguyen, F. Schoen, and R. Levy, *Am. J. Pathol.* **141**, 487–496 (1992).

24. L. Fitzpatrick, A. Severson, W. Edwards, and R. Ingram, *J. Clin. Invest.* **94**, 1597–604 (1994).

25. S. Srivatsa et al., *J. Clin. Invest.* **99**, 996–1009 (1997).

26. R.J. Levy, C. Gundberg, and C. Scheinman, *Atherosclerosis* **46**, 49–56 (1983).

27. S. Hirota et al., *Am. J. Pathol.* **143**, 1003–1008 (1993).

28. C.M. Giachelli et al., *J. Clin. Invest.* **92**, 1686–1896 (1993).

29. K. Bostrom, et al., *J. Clin. Invest.* **91**, 1800–1809 (1993).

30. C.M. Shanahan, N.R.B. Cary, J.C. Metcalfe, and P.L. Weissberg, *J. Clin. Invest.* **93**, 2393–2402 (1994).

31. T.A. Gura, K.L. Wright, A. Veis, and C.L. Webb, *J. Biomed. Mater. Res.* **35**, 483–495 (1997).

32. R.G.G. Russell and R. Smith, *J. Bone Joint Surg.* **55**, 66–86 (1973).

33. R.J. Levy et al., *Science* **228**, 190–192 (1985).

34. R.J. Levy et al., *Trans. Am. Soc. Artif. Intern. Organs* **31**, 459–463 (1985).

35. G. Golomb et al., *Trans. Am. Soc. Artif. Intern. Organs* **32**, 587–590 (1986).

36. G. Golomb et al., *J. Controlled Release* **4**, 181–194 (1986).

37. G. Golomb et al., *J. Pharm. Sci.* **76**, 271–276 (1987).

38. T.P. Johnston et al., *Trans. Am. Soc. Artif. Intern. Organs* **34**, 835–838 (1988).

39. T.P. Johnston et al., *Int. J. Pharm.* **52**, 139–148 (1989).

40. T.P. Johnston et al., *Int. J. Pharm.* **59**, 95–104 (1990).

41. T.P. Johnston, C.L. Webb, F.J. Schoen, and R.J. Levy, *J. Controlled Release* **25**, 227–240 (1993).

42. D. Hirsch et al., *J. Biomed. Mater. Res.* **27**, 1477–1484 (1993).

43. D. Hirsch et al., *Biomaterials* **14**, 705–711 (1993).

44. R.J. Levy et al., *J. Biomed. Mater. Res.* **29**, 217–226 (1995).

45. N.R. Vyavahare et al., *J. Controlled Release* **34**, 97–108 (1995).

46. H. Fleisch, in G.R. Mundy and T.J. Martin, eds., *Physiology and Pharmacology of Bone*, Springer-Verlag, New York, 1993, pp. 377–418.

47. D. Hirsch, F.J. Schoen, and R.J. Levy, *Biomaterials* **14**, 371–377 (1992).

48. R.J. Levy, F.J. Schoen, W.B. Flowers, and S.T. Staelin, *J. Biomed. Mater. Res.* **25**, 905–935 (1991).

49. C.L. Webb et al., *Am. J. Pathol.* **138**, 971–981 (1991).

50. Y. Pathak, J. Boyd, and R.J. Levy, *Biomaterials* **11**, 718–723 (1990).

51. T. Chandy et al., *Biomaterials* **17**, 577–585 (1996).

52. T. Chandy and C.P. Sharma, *Biomaterials* **17**, 61–66 (1996).

53. S. Patashnik, L. Rabinovich, and G. Golomb, *J. Drug Target.* **4**, 371–380 (1997).

54. R.J. Levy, F.J. Schoen, S.A. Lund, and M.S. Smith, *J. Thorac. Cardiovasc. Surg.* **94**, 551–557 (1987).

55. G. Golomb, M. Levi, G. Van, and M. Joel, *J. Appl. Biomater.* **3**, 23–28 (1992).

56. R.J. Levy, T.P. Johnston, A. Sintov, and G. Golomb, *J. Controlled Release* **11**, 245–254 (1990).

57. E. Bodner and R. Frater, eds., *Replacement Cardiac Valves*, Pergamon, Elmsford, N.Y., 1988.

58. R.J. Levy et al., *J. Controlled Release* **36**, 137–147 (1995).

59. G. Golomb and R.J. Levy, *Trans. Soc. Biomater.* **10**, 179 (1987).

60. M. Jones et al., *Trans. Am. Soc. Artif. Intern. Organs* **34**, 1027–1030 (1988).

61. W. Chen, F. J. Schoen, and R. J. Levy, *Circulation* **90**, 323–329 (1994).

62. M.N. Girardot, M. Torrianni, and J. M. D. Girardot, *Int. J. Artif. Organs* **17**, 76–82 (1994).

63. T. Shinoka et al., *Ann. Thorac. Surg.* **60**(6), Suppl., S513–S516 (1995).

64. B.A. Sosnowski et al., *J. Biol. Chem.* **271**(52), 33647–33653 (1996).

# CANCER, DRUG DELIVERY TO TREAT— LOCAL & SYSTEMIC

KATHLEEN J. LEACH
University of Washington
Seattle, Washington

## KEY WORDS

Cancer therapy
Chemoembolization
Depot systems
Immunoliposomes
Immunotherapy
Immunotoxins
Implantable pumps
Liposomes
Local delivery
Magnetic targeting
Microspheres
Nanoparticles
Stealth liposomes
Targeted delivery

## OUTLINE

Introduction
Local Delivery
    Implantable Pumps or Reservoirs
    Chemoembolization
    Internal Radiation Therapy Using Microspheres
    Locally Implanted Controlled-Release Systems
Systemic Delivery
    Depot Systems
    Targeted Delivery
Cancer Immunotherapy
Summary
Bibliography

## INTRODUCTION

In normal cells, growth is carefully regulated through an intricate web of physical and chemical signals. In adults, the rate of cell birth is maintained to be equal to the rate of cell death, whereas in children, slightly higher cell birth rates (or lower cell death rates) are necessary to allow for increases in body size during development. In either case, cell growth is carefully controlled. Cancer is the result of cell growth regulation gone awry. Any cell in the body has the potential to become a cancerous if it receives a series of genetic mutations that result in growth deregulation.

These mutations can result from spontaneous replication errors, chemical carcinogens, radiation, or even viral infection. The damage may be lethal to the cell, or it may free the cell from normal checks and balances on growth control, allowing it to proliferate when it should be quiescent. The mutated cell(s) will continue to divide, forming what is called the primary tumor. The tumor may be benign, or it may begin to invade surrounding tissues, breaking through basal lamina that define the natural boundaries of a particular tissue. Invasive tumors are described as malignant. Certain cells from these tumors may also mutate such that they gain the ability to leave the site of the primary tumor, enter the bloodstream, be carried to another tissue or organ, and begin forming a new tumor. These secondary tumors are called metastases and are often the cause of cancer deaths. Throughout the process of tumor development, many tumor cells lose their well-differentiated phenotypes and become more like immature or embryonic cells. They lose many of the unique features of the cell type from which they originated. Because tumor cells come from our own cells, they are very difficult to specifically target without cross-reactivity to other normal cells. As the tumor grows, more mutations occur, making the rapidly dividing cells within each tumor very heterogeneous in their cell surface markings and susceptibilities to various drugs and radiation damage. Tumor cells are also very adept at escaping from the constant patrol of the immune system, which should be able to recognize them as aberrant. For example, tumor cells often shed surface antigens that might identify them as abnormal. It is the combination of all these aspects that makes cancer such a difficult disease to cure and even to treat. In 1998, the American Cancer Society estimated that over half a million Americans are expected to die from cancer. Cancer is the second leading cause of death in the United States, with one in every four deaths attributed to the disease.

Conventional cancer therapies include surgery, chemotherapy, and radiation therapy. Each has its own limitations to providing a complete cure. Surgical resection is limited by the ability to expose and remove the tumor and can only remove those tumors detectable by current imaging techniques. Any cells that are not removed by the surgeon have the ability to proliferate, causing a recurrence. Surgery is also not effective against micrometastases that may have migrated from the site of the primary tumor. The efficacy of surgery is often limited by the surgeon's ability to remove enough healthy tissue at the margins of the tumor to ensure removal of all cancerous cells. Depending on the location of the tumor, this may or may not be feasible. For example, if the tumor is in the brain, the surgeon would probably not be as aggressive in removing healthy tissue at the tumor margins than if it were in the liver, which has regenerative abilities. This increases the risk of leaving behind malignant cells and subsequent tumor recurrence. Often, surgery is used in conjunction with chemotherapy or radiation therapy for improved efficacy. For example, the bulk of the tumor may be removed by surgery, and systemic chemotherapy may be prescribed after surgery if the tumor was found to have metastasized to local lymph nodes. Alternatively, radiation can be used after surgery to kill any parts of the tumor that were impossible to remove without destroying healthy tissue.

Chemotherapy, whether given systemically or by regional perfusion of a particular organ or tissue, is impeded by the lack of specificity of the drugs for cancer cells. Therefore, therapy is often limited by systemic toxicity before truly therapeutic drug levels in the tumor can be achieved. Drug concentrations in the tumor must also be sustained for prolonged periods of time for maximum efficacy so as to catch all the cancer cells during the portion of the cell cycle when they are susceptible to the drug. Chemotherapeutic drugs usually act on rapidly dividing cells, so cells of the intestinal lining and bone marrow can be extensively damaged during treatment. Depending on the health of the patient, toxic drug levels may not be achievable at the site of the tumor, because dosages must often be lowered or therapy halted altogether owing to this nonspecific action on healthy cells. Blood distribution to solid tumors may not be uniform throughout the mass, and therefore some regions of the tumor may not receive therapeutic doses of the drug. Tumor cell heterogeneity may also result in cancer cells that are resistant to a particular chemotherapeutic agent or even to multiple drugs. By treating the patient with a single-drug regimen, cells resistant to that drug are selected for survival; susceptible cells are killed, while the resistant cells are left behind to regrow a tumor composed entirely of resistant cells, making it impossible to effectively treat the tumor with that same drug a second time. By using several drugs in series or in combination, the problem of single-drug resistance can be significantly reduced. Multiple-drug resistance (MDR), where the tumor cells have gained the ability to protect themselves against a range of drugs by upregulating the expression of an efflux pump mechanism ($P$-glycoprotein) that ejects drug molecules from the interior of the cells, is more difficult to circumvent. Even drugs that are chemically dissimilar may be transported out of MDR tumor cells, so switching drugs may not successfully kill the cells. MDR is not as easy to work around than single-drug resistance, but many of the controlled drug delivery systems discussed later in this article have reportedly shown promise in circumventing MDR.

Radiation therapy can be specifically directed to the site of the tumor, but is also limited by the potential for damage to noncancerous tissue. The use of radiation to kill tumor cells is based on the idea that noncancerous cells divide more slowly and have a better chance to repair DNA damage caused by radiation before replicating. The cancer cells hopefully will not be able to make the necessary repairs, and the damage to them will be lethal. Radiation therapy, like surgical resection, is limited by the ability of imaging techniques to identify tumor sites for treatment. If the tumor is too small to be imaged, it will not be identified target for therapy and will be left to grow and possibly metastasize.

None of these traditional therapies, alone or in combination, have achieved complete cures for all cancer types in all patients. Therefore, many researchers have been exploring controlled-release or targeted-delivery options for the treatment of this disease. Although drug delivery systems cannot improve the specificity of the cytotoxic agents, the delivery vehicle itself can act to target the therapy to the tumor cells to focus the action of a nonspecific drug to the tumor or at least limit its effects to the tissue in closest proximity. Drug delivery systems act by locally concentrating the drugs for sustained periods of time, thereby reducing systemic toxicity.

The field of controlled release has made inroads into many forms of cancer therapy. Drug delivery systems have been proposed in three categories: (1) delivery systems that are implanted or injected in or near the tumor to act locally against the neoplasm; (2) delivery systems implanted or injected to act simply as a depot for the sustained release of the drug, which acts systemically; and (3) delivery systems administered systemically but that contain a targeting mechanism. Table 1 provides examples of controlled-release approaches to cancer therapy discussed in this article. The first type of system can be effective against the primary tumor or against metastases that occur near the primary tumor or are localized to a particular organ. The other two types of drug delivery system allow for the attack of the primary tumor as well as disseminated metastases, with targeted therapy having the ease of systemic administration and the advantage of affecting tumor cells without harming normal tissue. This article discusses examples of each of these types of delivery systems. The topic of immunotherapy for cancer is also covered briefly as it pertains to controlled-release systems. The information provided in this entry is not designed to exhaustively cover all cancer therapies that have been explored or are used clinically but focuses instead on the use of polymers for the sustained release or presence of the therapeutic agent. For reference, Table 2 summarizes the chemical structures, typical dosages, and mechanisms of action for the various chemotherapeutic drugs used in examples throughout this article.

## LOCAL DELIVERY

Damage to normal, healthy cells and tissue is usually what limits the aggressiveness of any given cancer therapy. Controlled-release systems implanted or injected at the tumor site can act to reduce systemic toxicity by delivering the drug locally in a sustained manner. This provides high concentrations of the drug at the site of the tumor, with the drug either being degraded before reaching the systemic circulation or diluted in the total blood volume of the patient, thus reducing side effects. The sustained-release aspect of the technology allows the patient more freedom during treatment while maintaining drug levels in the tumor at a constant, therapeutic level for an extended period of time. Conventional intravenous treatments are usually given in bolus form with periods of rest in between or as continuous infusions through percutaneous intravenous catheters. Patients are either restricted by an percutaneous catheter or are given intermittent treatment, which may not provide sustained drug concentrations necessary for maximum efficacy. Local delivery can also bypass physiologic barriers such as the blood–brain barrier, which can limit therapeutic efficacy of drugs delivered systemically. The polymer also acts to protect the drugs, which often have extremely short half-lives in vivo, from degradation until they are released from the device.

The following sections discuss the use of controlled-release systems for the localized treatment of tumors. These include systems implanted or injected in or near the tumor itself or infused into the local blood supply to a tumor or diseased organ. Local drug delivery therapies discussed include the infusion of drugs into the tumor vasculature via implantable pumps, the intratumoral infusion of polymer microspheres that may contain drugs or radioisotopes, or the injection or implantation directly into the tumor of polymer matrices loaded with drugs.

## Implantable Pumps or Reservoirs

Infusion systems are one of the simplest drug delivery systems that have been applied to cancer therapy. There are two types: (1) controllers, which use gravity as the driving force for fluid delivery; and (2) infusion pumps, which use positive pressure to deliver the drug solution. Both types use a drug reservoir that can be exchanged or refilled as needed. For controllers, the height of the reservoir and the resulting pressure head are critical in maintaining the desired level of flow. Infusion pumps are capable of overcoming the back pressure created by arterial infusion catheters, the pumping of viscous liquids, and minor occlusions of the vascular access line by the application of positive pressure, and they can accurately provide low infusion rates. Pumps may be external to the body with the drug administered via a percutaneous catheter, or the pumps may be totally implantable. External pumps are simple devices used extensively to deliver drugs to patients in hospital settings over prolonged periods of time. They may be connected directly to an intravenous (IV) or intraarterial (IA) catheter, or they may access the vascular system via a needle that percutaneously punctures the septum of a surgically implanted port. External pumps free medical personnel from having to slowly and/or repeatedly inject drug solutions using a syringe, but the patient is tethered to the pump during treatments. With an implantable pump, the patient can be freed from the hospital during therapy, a small but important factor in his or her quality of life. Only implantable pumps will be discussed further in this entry, with the focus on those used for the local infusion of drugs into tumor vasculature.

**Table 1. Examples of Drug-Delivery Systems Used to Treat Cancers**

| Administration | Action | Examples |
|---|---|---|
| Local | Local | Intraarterial infusion, chemoembolization, intratumoral injections, implants of sustained-release systems |
| Systemic | Systemic | Depot systems such as Zoladex® and Lupron Depot® |
| Systemic | Tumor cells | Drug–antibody conjugates, drug–polymer conjugates, liposomes, nanoparticles, magnetic localization |

**Table 2. List of Chemotherapeutic Agents Used in This Article**

| Drug name(s) | Chemical structure | Typical IV dosage | Mechanism of action |
|---|---|---|---|
| Adriamycin (ADR) Doxorubicin (DOX) | *(chemical structure)* · HCl | 60–75 mg/m² as bolus or continuous infusion over 2–4 days repeated every 3–4 weeks | DNA intercalation, preribosomal DNA and RNA inhibition, alteration of cell membranes, free radical formation |
| Bleomycin | *(chemical structure)* R = terminal amine | 10–20 U/m² weekly or twice weekly, 15–20 U/m²/day continuous infusion over 3–7 days | DNA strand scission by free radicals |
| Carmustine (BCNU) | *(chemical structure)* | 150–200 mg/m² every 6 weeks as a single dose or divided over 2 days | DNA cross-linking, DNA polymerase repair, RNA synthesis inhibition |
| Cisplatin, cisdiamine-dichloroplatinum (II) (CDDP) | *(chemical structure)* | 20–40 mg/m²/day for 35 days every 3–4 weeks; 20–120 mg/m² as a single dose every 3–4 weeks | DNA cross-linking, intercalation, DNA precursor inhibition, alteration of cell membranes |
| Daunorubicin | *(chemical structure)* | 45 mg/m²/day for 3 days | DNA intercalation, preribosomal DNA and RNA inhibition, alteration of cell membranes, free radical formation |
| Dexamethasone | *(chemical structure)* | 40 mg/day on days 1–4, 9–12, and 17–20, repeated every 4 weeks | Binds to specific proteins in cell, forming a steroid–receptor complex. Binding of this complex with nuclear chromatin alters mRNA and protein synthesis. |
| Doxorubicin (DOX) Adriamycin (ADR) | *(chemical structure)* · HCl | 60–75 mg/m² as bolus or continuous infusion over 2–4 days repeated every 3–4 weeks | DNA intercalation, preribosomal DNA and RNA inhibition, alteration of cell membranes, free radical formation |

**Table 2. List of Chemotherapeutic Agents Used in This Article** (*continued*)

| Drug name(s) | Chemical structure | Typical IV dosage | Mechanism of action |
|---|---|---|---|
| 5-Floxuridine (FUDR, 5-FUDR) | | 0.1–0.6 mg/kg/day (intrahepatic) or up to 60 mg/kg/week IV | Inhibition of thymidylate synthesis |
| 5-Fluorouracil (5-FU) | | 300–450 mg/m$^2$/day for 5 days (IV); 600–750 mg/m$^2$ given weekly (IV); 1,000 mg/m$^2$ for 4–5 days (IV over 24 h) | Inhibition of thymidylate synthase |
| Leucovorin | | 10–25 mg/m$^2$ every 6 h for 6–8 doses | Rescue agent for MTX |
| Methotrexate (MTX) | | 20–40 mg/m$^2$ every 1–2 weeks (solid tumors); 200–500 mg/m$^2$ every 2–4 weeks for leukemia and lymphoma | DHFR inhibition (stops thymidylate and purine synthesis) |
| Mitomycin, mitomycin-C (MMC) | | 10–20 mg/m$^2$ every 6–8 weeks | DNA cross-linking and depolymerization, free radical formation |
| Taxol, paclitaxel | | 135–175 mg/m$^2$ every 3 weeks | Promotes microtubule assembly and stabilizes tubulin polymers, resulting in formation of nonfunctional microtubules. |
| Vincristine (VRC) | R = CHO | 0.5–1.4 mg/m$^2$ every 1–4 weeks | Tubulin binding (microtubule assembly inhibition and dissolution of mitotic spindle structure) |

**Table 2. List of Chemotherapeutic Agents Used in This Article** (*continued*)

| Drug name(s) | Chemical structure | Typical IV dosage | Mechanism of action |
|---|---|---|---|
| Vinblastine | | 6–10 mg/m$^2$ every 2–4 weeks | Tubulin binding (microtubule assembly inhibition and dissolution of mitotic spindle structure) |

*Source:* Data taken from Ref. 117.

Implantable pumps have been shown to result in fewer treatment interruptions and better patient compliance than external pumps (1). Fordy and coworkers compared the use of a totally implantable pump (Infusaid model 400, Infusaid, Inc.) to the use of an implanted port (Infusaport, Infusaid, Inc.) in conjunction with an external pump in 95 patients with liver metastases of colorectal cancer (CRC). The drug was infused into the hepatic artery in all cases. They found a 3-fold lower incidence of treatment interruptions (due to line occlusions, infection, hematoma, etc.) and a 30-fold lower incidence of catheter blockage with the implanted pumps. Implanted pumps showed a lower incidence of line occlusion (0.3% of treatments) compared to 9% for implanted ports. When the IA line becomes occluded, treatment must be temporarily halted to reopen the vascular access. This is often achieved with flushes of heparin or other enzymes, but the flushing is frequently unsuccessful, and the occluded line must be replaced surgically. Because of the reduction in line occlusions with the use of an implanted pump, patients required fewer reoperations, procedures that can be risky in critically ill patients. Implantable pumps were more likely to result in complications due to the surgical implantation of the device (e.g. infection, hematoma, etc.), but overall, implantable pump treatments showed a lower rate of treatment problems (1.7%) than ports (5.7%). Costwise, the price of the implanted pump was 10 times higher than the price of the implanted port option, but the lower incidence of complications has the potential to improve therapy dramatically. The implantable pumps allowed for more extended therapies, which increased the efficacy of the anticancer drugs by ensuring their presence in therapeutic concentrations at critical points in the cell cycle. The Fordy study clearly showed the benefits of implantable pumps over external pumps (1).

In the past decade, two review articles have been published that describe designs of implantable pumps in great detail and discuss a few of their current applications, including the treatment of cancer and cancer-related pain (2,3). The clinical use of implantable pumps has now progressed to the point that most of the current literature does not describe clinical trials to test the efficacy of the pumps but rather the efficacies of various drug combinations, all delivered via these implantable pumps. The remainder of this section describes the features of the implantable pumps that have been approved by the U.S. FDA for the controlled delivery of various chemotherapy drugs and then discusses a few examples in the current literature in which these pumps were used to successfully treat cancer patients. Finally, some of the problems with current pump technology are addressed, with reference to current research being done to improve this type of therapy.

**FDA-Approved Devices.** Implantable infusion pumps were introduced for continuous drug delivery in 1969. Several implantable pumps have been approved by the FDA, and many more are being tested and used clinically around the world. The Infusaid Models 100, 100, and 550 (Pfizer Infusaid, Inc., Norwood, MA) were FDA approved starting in 1982 for the delivery of various cancer therapeutics, including floxuridine (FUDR), fluorouracil (5-FU), methotrexate sodium (MTX), and cisplatin (CDDP) (2). They have also been approved for the delivery of morphine sulfate for the treatment of pain resulting from incurable cancer. The Infusaid pumps consist of a flexible bellows containing the drug solution that is surrounded by a rigid chamber filled with a charging fluid (usually a volatile chlorofluorocarbon). Vapor pressure from the charging fluid expels the drug solution by compressing the internal drug reservoir. The release of the drug solution is regulated by a capillary flow restrictor or a valve/accumulator combination. The reservoir can be refilled by percutaneous puncture of a needle, which reexpands the reservoir and increases the pressure in the adjacent charging fluid chamber, causing the fluid to recondense. The pump is then ready for its next infusion cycle.

Medtronic, Inc. (Minneapolis, MN) received FDA approval for its SychroMed Infusion System in 1988 for the delivery of CDDP, FUDR, doxorubicin hydrochloride (DOX), and MTX and received FDA approval in 1996 for the intrathecal delivery of morphine sulfate for the treatment of pain (2). The SynchroMed pump delivers a drug solution from a percutaneously refillable reservoir via a peristaltic pump mechanism that is driven by a lithium battery with a life of 1 to 3 years (3). The pump can be programmed with an external computer and a magnetic field telemetry link, allowing for more complex delivery regimens, including those based on circadian rhythms. The doctor has the option of tailoring the drug dosage, flow rate, and dosage schedule to best fit the needs of the patient, while the patient is able to receive treatments untethered.

**Applications to Cancer Therapy.** The majority of the literature describing cancer therapy using implantable pumps centers on the treatment of primary and secondary tumors in the liver, especially liver metastases of CRC, with hepatic arterial infusion of chemotherapeutic agents. Only a few examples of the many studies published in the last decade are discussed here. Patt et al. report the results of a study that treated 48 patients with liver metastases of CRC, 19 of whom were treated via Infusaid implantable pumps (4). Therapy consisted of alternating delivery of recombinant human interferon $\alpha$-2b (6 h) and 5-FU (18 h) into the hepatic artery. This cycle was repeated daily for 5 days, with the cycle repeated every 4–7 weeks thereafter. They found a complete plus partial response rate of 33.3%, but they did not separate the results based on delivery via implantable pumps or external pumps. However, they report that 6 of the patients treated during the trial with implantable pumps elected to continue treatment past the end of the study, whereas none of the patients elected to continue treatments via percutaneous access past the completion of the study. Patient acceptance of the implantable pump was very high.

A 1994 paper by Kemeny et al. presented the results of a study involving 62 patients with liver metastases of CRC treated with hepatic arterial infusion of FUDR, leucovorin (LV), and dexamethasone (DEC) via Infusaid pumps (5). Treatment consisted of a 2-week infusion of FUDR (0.3 mg/kg/day), LV (15 mg/m$^2$/day), and DEC (20 mg total dose) alternated with a 2-week infusion of saline to keep the delivery catheter patent between treatment periods. Thirty-three of the patients had not previously received systemic chemotherapy for their disease, and this group saw a complete plus partial response rate of 78% (median survival 24.8 months). Patients who had previously received systemic therapy showed a complete plus partial response rate of 52% (median survival 13.5 months). Davidson et al. report that by alternating hepatic arterial infusion of FUDR (0.1 mg/kg/day; given as 7-day continuous infusion) and 5-FU (15 mg/kg; given as a weekly bolus for 3 weeks) by implantable pump, biliary toxicity could also be reduced (6). Out of 57 patients treated for hepatic metastases of CRC, 79% showed a reduction in tumor size or stabilization of disease, with a median survival of 19 months. More than 95% of the patients were free from all or at least irreversible hepatobiliary toxicity, a common side effect of hepatic arterial infusion of drugs.

A second group of cancers often treated by intraarterial infusion of chemotherapeutic agents via implantable pumps is cancers of the head and neck, including squamous cell carcinoma. Baker, Wheeler, and colleagues report the results of 37 patients treated over a 6-year period for advanced cancers of the head and neck using implantable pumps (7,8). The pumps (Infusaid model 400) were equipped with one or two silicone rubber catheters that allowed for singly treating the tumor via one external carotid artery or concurrently treating both sides of the head and neck. Forty-three percent of the 26 patients treated with a combination of cisplatin and FUDR for squamous cell carcinoma showed a partial or complete response to the treatment. The pumps were safe, effective, and had high patient compliance. Median infusion duration was 328 days, with only a small percentage of therapy-limiting complications: infection (5.4%), emboli-to-skin overlying tumor (5.4%), catheter occlusion by drug precipitation (8.1%), and infusion into the brain (21.6%).

Although not directly treating the malignancy itself, implantable pumps have been used extensively for the intrathecal delivery of pain medication for patients with incurable cancer. In 1990, Hassenbusch and colleagues showed great efficacy of pain treatment via the Infusaid model 400 pump to deliver morphine epidurally to patients with inoperable cancer (9). Before implanting the pumps, they used a temporary percutaneous catheter to epidurally deliver the morphine for 2 to 4 days to see if the patients' pain was responsive to morphine. Only those patients whose pain was relieved by the morphine had pumps implanted. Out of the 69 patients admitted to the study, 41 received implantable pumps, and 37 of these patients were evaluable. Mean survival time was 7.1 months, with more than 80% of the patients showing satisfactory pain relief (defined by a 30% decrease in pain analog scale value) at 1, 3, and 6 months. Hoekstra reported the results of a study on the delivery of morphine hydrochloride to 50 patients with terminal cancer via an implantable pump with an intrathecal or epidural catheter (10). They used a Secor® pump system (Cordis S.A., France) that did not provide continuous infusion of the drug but rather was activated by the patient to deliver a bolus dose of the drug when he or she percutaneously pressed two buttons on the pump in the correct sequence. This is a relatively cheap, totally implantable pump option (compared to continuous or programmable pumps) that is patient activated and has the capability of lasting up to 5 years in vivo. Half of the patients in this study achieved a pain-free state using the pumps, with a median implant duration of 3.3 months.

**New Directions.** Constant delivery of drugs may not always be the most effective or least toxic dosage schedule. A popular modification of traditional chemotherapy regimens is to infuse the drugs in cooperation with the body's natural circadian rhythm. In 1990, Damascelli et al. used a totally implantable, programmable Medtronic Sychromed pump to deliver drugs intravenously to 42 patients with metastatic renal cell carcinoma (11). The patients received 14 days of treatment (FUDR) alternated with 14 days of heparinized saline infusion to maintain catheter patency. Treatment was delivered as follows: 68% of the daily infusion was given between 3 P.M. and 9 P.M., 15% between 9 P.M. and 3 A.M., 2% between 3 A.M. and 9 A.M., and 15% between 9 A.M. and 3 P.M. The starting dose was 0.15 mg/kg/day and was increased by 0.025 mg/kg with each successive cycle. The study was continued for 2 years, with a median implant duration of 210 days. Median survival was 1 year after the start of therapy, with 55% of the patients alive after 1 year. The efficacy of treatment was equivalent to other standard treatments but was associated with a normal quality of life for the patients.

The future holds much for the application of these implantable pumps to cancer therapy, especially with the expanded use of programmable pumps that allow for increasingly complex dosage regimens to be delivered to ambulatory patients. The pumps can be used for local or systemic treatment, depending on the placement of the delivery catheter, and can be used to deliver any drug that can remain stable for the necessary storage times in the pump chamber. Pumps are the most versatile device for drug delivery, as the choice of drug, the infusion rate, and infusion schedule can be easily adapted to personalize the therapy for each individual patient.

## Chemoembolization

Other means of enhancing the local cytotoxicity of drugs have also been explored for use as stand-alone, alternative therapies or for use in conjunction with local infusion. Locally induced hypoxia (oxygen deprivation) is one such therapy. Hypoxia is most easily achieved for tumors by ligation or occlusion of the blood vessels that feed the tumor. Historically, ligation was the method of choice, although more recent research has focused on the development of occlusive techniques. Embolization, the use of particulates to effectively shut off blood flow to an organ or tumor so as to cause cell death by hypoxia, is an example of an occlusive technique. In addition to the direct effects of oxygen deprivation on tumor cells, embolization also has several indirect means of enhancing the effects of drugs administered with the emboli. The indirect means include (1) increasing the cytotoxic properties of the drugs themselves, (2) lowering the total tumor burden on which the drug needs to act, or (3) increasing the permeability of the tumor vasculature for better uptake of the drugs. An extension of embolization that uses drug-loaded polymer microspheres to cause local hypoxia in combination with local, sustained-delivery of cytotoxic drugs is called chemoembolization; it is a single-step procedure that does not require persistent percutaneous catheters to deliver the drug and is thus a significant improvement over past treatment strategies. The microspheres function both to induce local hypoxia and to serve as depots for the sustained release of the drug.

Most of the current research on chemoembolization has focused on primary or metastatic liver cancer. The anatomy of the liver is unique because it receives blood from two sources: the hepatic artery and the portal vein. The hepatic artery supplies the liver with fully oxygenated blood, whereas the portal vein drains a portion of gastrointestinal tract and then passes the absorbed nutrients to the liver for detoxification and storage. It was determined in the 1950s that most hepatic tumors derived their supply of oxygen from the hepatic artery and that normal liver tissue was able to survive on portal blood supply alone (12). This separation in blood supply offers a distinct advantage for targeted attack on liver tumors with minimal effects on the healthy liver parenchyma. The bulk of research has thus been focused against primary and secondary hepatic tumors, and cancers of the liver are the focus for the next sections.

**Historical Background.** Before discussing the use of drug delivery systems in the form of microspheres loaded with

a drug as embolizing agents, a brief history describing early research on oxygen deprivation of tumors is informative. As early as 1933, Graham published a paper describing multiple cases of accidental hepatic artery ligation that proved not to be fatal, as once believed, and even resulted in localized necrosis of malignant liver tissue (13). This opened the door for researchers to develop various techniques for clinically stopping blood flow to tumors by permanent or temporary ligation or occlusion of the hepatic artery (14–16). These techniques were used alone or in combination with arterial or local infusion of chemotherapeutic agents for added benefits. In a randomized, controlled study, permanent ligations of the hepatic artery or complete dearterializations of the liver increased survival times in patients with hepatocellular carcinoma (HCC), but the increases were not significant in comparison with the controls who received only palliative treatment (17). It was believed that this lack of therapeutic improvement was due to the rapid formation of collateral circulation, which was shown to come from as many as 26 different sources and had the ability to resupply hypoxic liver tissue with blood in less than 4 days (18). Thus, although the hepatic artery was permanently closed, the hypoxic condition of the tumor was only temporary. Hepatic artery occlusion using the percutaneous infusion of autologous or nondegradable, synthetic materials was less surgically invasive but did not circumvent the problems of collateral vessel formation. Thus, these permanent occlusion methods were not entirely successful in treating hepatic tumors, even in combination with arterial or portal venous infusion of chemotherapeutic drugs (19,20).

Because permanent occlusion resulted in the recruitment of collateral vessels, researchers began to investigate temporarily occlusive techniques that could provide the beneficial hypoxia within the tumor while avoiding the formation of alternative blood supplies. Research on temporary occlusion of the hepatic arterial blood supply in rats showed that if the circulation was halted for less than 4 hours at a time, collateral blood vessels would not form (21). Two generations of temporary occluding devices showed some success by causing localized hypoxia while avoiding collateral vessel formation; however, these treatments still involved surgical implantation and removal of the device (22,23). A nonsurgical alternative to temporary dearterialization is the percutaneous infusion of biodegradable polymer microspheres that could block flow temporarily until their hydrolytic or enzymatic degradation allowed blood flow to resume. By altering the properties of the microspheres, occlusion times could be optimized to allow for maximum tumoricidal effect without the formation of collateral circulation. Cytotoxic drugs could be infused with or loaded into the microspheres for a more effective combination approach. These treatments have been reviewed by Willmott (24) and are discussed in more detail in the following sections.

**Embolization Using Blank Microspheres.** The use of degradable polymer microspheres for the infarction of tumors has been reviewed extensively in the literature (24–27). The majority of the studies utilized degradable starch microspheres. The biodegradation (28) and patterns of blood flow (29) after starch microsphere embolization have been well characterized in animals and showed that blood flow to an organ can be halted after arterial administration of the microspheres until the particles had degraded sufficiently to pass through the capillary bed, generally over a period of hours. The effect of starch microsphere embolization alone as a treatment for primary and metastatic liver cancer showed some success in humans, although the results were not significantly better than standard chemotherapies (30).

The combination of starch microsphere embolization with the infusion of drugs was found to be more effective at tumor reduction against both primary (31–34) and secondary (30–37) liver cancer than embolization alone. The microspheres act to retain the drug in the tumor, slowing its release into the systemic circulation. The benefits of the combination therapy have been elucidated by pharmacokinetic studies done in animals showing that tumor uptake of drugs is higher after microsphere embolization (27,38,39).

In addition to the infusion of cytostatic drugs, vasoconstrictors such as angiotensin II have been included in the regimen (40,41) based on research showing that tumor blood vessels are less able to respond to vasoconstrictors than normal blood vessels (42–44). It has been suggested that tumor vessels lack the supportive smooth muscle cells that supply the main forces in the constriction of mature blood vessels. The injection of such vasoconstrictors would cause the blood vessels supplying normal tissue to constrict, reducing the flow of the chemotherapeutic agents to normal tissues whereby the vessels supplying the tumor would remain open, resulting in more selective tumor cell damage. A schematic of this shunting process is shown in Figure 1.

**Drug-Loaded Microspheres.** The utility of polymer microspheres is not limited to simply plugging a tumor's arterial blood supply. Microspheres can be loaded with drugs to provide sustained, local delivery of cytotoxic agents in the tumor vasculature. Similarly to blank microsphere embolization in conjunction with locally infused drugs, the hypoxia caused by the embolization increases vascular permeability so more drug can be absorbed by the tumor. In addition, some drugs, such as mitomycin C (MMC), show enhanced cytotoxicity against hypoxic cells (45,46). By loading MMC or other drugs into or onto the microspheres, the effects of local, sustained release of the drug may be synergistic with the effects of localized hypoxia. Researchers have loaded drugs into microspheres made of both nondegradable materials such as polystyrene-*co*-divinyl benzene (ion exchange resin) (47); ethylcellulose (48,49); and degradable materials such as collagen (50), poly(lactic acid) (PLA) or its copolymer, poly(lactide-*co*-glycolide) (PLGA) (51–53), and albumin (54).

Kato et al. published a series of papers in the late 1970s and early 1980s describing work with nondegradable, ethylcellulose microspheres loaded with MMC (48,49,55–57). This pioneering work introduced the term *chemoembolization*, and the microspheres were used to treat many types of cancer in clinical trials in Japan (49). One clinical trial showed significant tumor reduction in 65% of the tu-

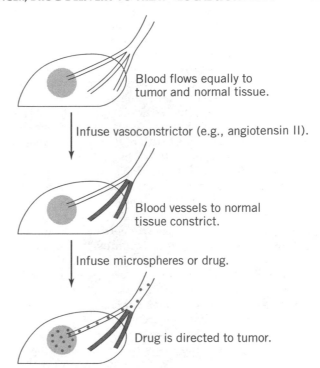

**Figure 1.** Schematic showing selective constriction of normal vasculature that directs flow of the chemotherapeutic agent (black circles) to the tumor vasculature, which is not able to respond. Shaded region represents tumor.

mors, pain relief in 80%, and hemostasis in 100% of patients treated with intraarterial infusions of MMC-loaded ethylcellulose microspheres (49). The trial involved 60 patients with a range of primary and secondary carcinomas. Patients were treated with single or multiple infusions. Codde and coworkers loaded nondegradable, ion exchange resin microspheres with doxorubicin and injected them into the hepatic artery of rats with colorectal adenocarcinoma transplanted in their livers (47); they found significantly reduced tumor growth as compared with treatment with free doxorubicin or blank microspheres alone.

As reports were published stating that the effectiveness of permanent occlusion of the vessels is often negated by collateral vessel formation, chemoembolization research switched focus to the development of methods to provide temporary occlusion in combination with the sustained delivery of cytotoxic agents. Degradable polymer microspheres that could be loaded with drugs provided an attractive solution. The majority of research on drug-loaded, degradable microspheres has focused on albumin microspheres, and this topic has been extensively reviewed by Morimoto and Fujimoto (54). The in vitro release characteristics (58) and biodegradation of albumin microspheres in both animals (59) and humans (60) have been studied. In rats, Willmott and colleagues showed that the time for 50% of the microspheres to disappear from the target organ for albumin microspheres with diameters of 10 to 30 $\mu$m was 3.6 days in the liver and 2.0 days in the lungs (59). Goldberg et al. showed that in seven patients with colorectal liver metastases, the median biological halftime of

[131]I-labeled albumin microspheres administered into the liver via the hepatic artery was 2.4 days with a range of 1.5 to 11.7 days (60). Kerr et al. studied the organ distribution and pharmacokinetics of doxorubicin-loaded albumin microspheres after renal artery administration and found that 97% of injected microspheres were retained in the treated kidney and that peak plasma levels in venous blood were significantly lowered from 135 ng/mL when the drug was administered in solution to 16 ng/mL when encapsulated in microspheres (61). Histologically the microspheres were shown to be trapped in small renal arterioles and capillaries. Fujimoto et al. infused MMC-loaded albumin microspheres into 19 patients with inoperable hepatic cancer and found objective tumor responses in 13 out of 19 patients treated with microspheres compared with 6 out of 13 treated by conventional infusion therapy with MMC and 5-FU (62).

These albumin microspheres were hampered by low drug loadings and release profiles characterized by an initial burst before leveling off for sustained release. Cremers et al. describes a method of increasing the amount of drug that can be loaded by combining heparin, a negatively charged mucopolysaccharide, with the albumin (63). Nishioka et al. studied the effects of adding chitin to the albumin microspheres to increase the loading and reduce the burst of cisplatin (64). Although the mechanism is still unclear, the chitin promoted degradation of the microspheres and the release of the cisplatin, increasing the efficacy of embolization therapy.

Degradable polymers other than albumin have also been studied. Daniels et al. loaded collagen microspheres with cisplatin and studied the pharmacokinetics and tissue tolerance in the rat liver and rabbit kidney (50). They found that arterially administered cisplatin-loaded collagen microspheres increased renal platinum concentrations to 250 times the levels in the contralateral, nontreated kidney. Platinum levels in the rat liver were doubled when the drug was delivered from microspheres as compared with when the drug solution was administered systemically (50). Spenlehauer et al. investigated the preparation and in vitro release characteristics of cisplatin-loaded PLA and PLGA microspheres (51,65). In the case of PLA microspheres, they found the optimal cisplatin concentration to be 45% (w/w) and that drug release was proportional to drug loading (51). For both PLA and PLGA, they determined that factors influencing the distribution of drug in or on the surface of the microspheres were more important in controlling the release kinetics from the PLGA microspheres than the degradation rate or molecular weight of the polymer (65). Verrijk et al. also studied cisplatin-loaded PLGA or PLGA-coated albumin microspheres for chemoembolization and found significant cytotoxicity in vitro (52). When the CDDP-loaded PLGA microspheres were administered to the liver via the portal vein, the liver-to-blood platinum ratio was 38 times higher than when free drug was injected by the same route (53).

The progression of the use of local hypoxia as a means to treat tumors from permanent arterial ligation to drug-loaded, degradable microspheres occurred over a period of 60 years. With each new development, the promise and efficacy of such treatments grew. They will always rely on direct access to the tumor blood supply, but advances in percutaneous, imaging-guided catheterizations will make this step as noninvasive as possible. Surgical exposure to blood vessels is no longer a necessary part of organ catheterization. Time will only tell if any of these controlled-release systems will make it into clinical practice as standard cancer therapies for tumors with an isolatable blood supply.

## Internal Radiation Therapy Using Microspheres

As an alternative to chemotherapeutic agents, radioactive isotopes such as yttrium-90 (Y-90) have also been loaded into microspheres and infused into the arterial blood supply of a tumor for localized radiation therapy. The microspheres are trapped in the vasculature of the tumor and are able to concentrate the effect of the radioisotopes on the tumor, thus reducing systemic side effects. The first attempts for internal radiotherapy by Grady and coworkers in 1963 used a particulate form of Y-90 injected intravenously for the treatment of lung cancer and injected into the celiac or hepatic artery for liver cancer (66,67). The procedure was deemed simple and nontoxic, and over half of the treated patients showed subjective or objective improvement (67). Y-90 was also adsorbed to ceramic microspheres by Kim et al. (68) and Ariel (69) and delivered intraarterially to humans. The ceramic spheres were shown to be physiologically inert, resisted leaching in body fluids, and were appropriate for the adsorption of any cationic isotope (68). Both researchers saw objective improvement in some of the patients treated by this method, but both agree that more research needs to be done to improve the selective delivery to tumor tissue (68,69).

Y-90 was also bound to ion exchange resin microspheres made from cross-linked poly(styrene-co-divinyl benzene) (70–74). Ariel and Padula administered Y-90 adsorbed to resin microspheres into the hepatic artery of 65 patients with symptomatic CRC liver metastases in combination with hepatic arterial infusion of drug (70). They found that 24 patients showed objective improvement and 40 showed subjective improvement. In patients with asymptomatic colorectal metastases in the liver, the average lifespan was increased an average of 28 months by this treatment (71). Gray and colleagues infused angiotensin II prior to infusion of the Y-90 resin microspheres to direct the spheres away from normal liver parenchyma and toward the tumor vasculature (72–74). Angiotensin II constricted only the fully developed, non-tumor-associated vasculature, leaving tumor vasculature open for the infusion of the microspheres (42). They documented an average ratio of tumor radiation dose to normal tissue dose of 6:1 (72) and found that normal liver tissue was tolerant to radiation delivered in this fashion (73). They also reported the regression of liver metastases in 18 of 22 patients treated with this therapy (74).

## LOCALLY IMPLANTED CONTROLLED-RELEASE SYSTEMS

Implantable pumps, chemoembolization, and internal radiation therapy provide local treatment of cancer by infusion directly into the blood vessels of the tumor. This direct access to tumor vasculature is not always easy to achieve and requires skilled medical professionals. Because of this, researchers have devoted much time and resources to developing delivery systems that can be injected or implanted directly into the tumor or left in the cavity created by surgical resection of the tumor. Many of these systems are biodegradable and therefore do not require removal at the end of therapy, unlike implantable pumps, which require a second surgery for device removal. Drug is distributed through the tumor by passive diffusion through the intracellular fluid that bathes the tumor cells. Controlled-release systems placed at the site of the tumor provide sustained delivery of the therapeutic agents in high concentrations locally without systemic side effects. The sustained-delivery property of such systems improves the therapy by providing prolonged exposure to drugs whose effects are generally cell cycle dependent. With delivery to the site of action, the drug is able to work at therapeutic concentrations and is degraded before reaching the circulation or diluted so as to not cause significant systemic side effects. A great deal of research at the preclinical level has been done on the local delivery of chemotherapeutic drugs from polymers, and a few illustrative examples are discussed here. A special section on drug delivery to the brain is included here to show how local delivery can be effective at circumventing anatomic barriers.

Lin et al. encapsulated DOX (ADR) into conical needle devices made of poly(ethylene-co-vinyl acetate) (EVAc) or PLA (75). The EVAc needle devices exhibited zero-order release in vitro and showed the greatest in vivo tumoricidal activity in mice bearing subcutaneously transplanted mammary carcinoma. The PLA devices exhibited rapid initial release of the drug in vitro and were not as effective in vivo because therapeutic tumor drug levels could not be sustained. Interestingly, another group found that PLA by itself can inhibit the growth of carcinoma cells in vitro (76). HTB-43 cells, a human hypopharyngeal squamous cell carcinoma cell line, were grown in the presence of extracts from stainless steel, titanium, and PLA that had been incubated in fetal bovine serum at 37°C for 24 hours. The extracts were filtered before addition to the cell cultures, so any inhibition seen would be due to a soluble factor. Cell growth was inhibited by extracts from PLA incubated at a 5% (v/v) concentration in cell culture media but not by extracts of the other materials. Presumably, the growth inhibition was due to products liberated by the hydrolytic degradation of the polymer.

Shikani et al. studied the treatment of nude mice with subcutaneously transplanted human floor-of-mouth squamous cell carcinoma with cisplatin encapsulated in a biodegradable polyanhydride made up of fatty acid dimer and sebacic acid, P(FAD:SA) (77). The polymer wafers contained 7% (w/w) cisplatin and were implanted into the tumors. Mean tumor size 70 days postimplantation for the P(FAD:SA) wafer group was an average 38.1% of the tumor size in the groups that had been treated by intraperitoneal (IP) injection of saline or cisplatin or intratumoral injection of free cisplatin. The mean number of days for the tumor to triple in size was approximately 20 days for the saline-treated controls, 22 days for those treated with blank rods, 24 days for those treated with IP injection of cisplatin solution, and 70 days for those treated with the cisplatin-loaded rods. The sustained release formulation was able to significantly slow the growth rate of the tumor.

Begg et al. encapsulated cisplatin in a rod of a polyether hydrogel and compared its antitumor activity to that of IT or IP injected drug solution (78). The rods were 1.5 mm in diameter and 5 mm in length and contained 2.5–8.0% (w/w) cisplatin. The drug-loaded rods were implanted subcutaneously in the center of RIF1 murine fibrosarcoma grown subcutaneously in mice and resulted in tumor growth rates equivalent to those treated with IT injection of free drug and slower tumor growth rates than IP injection of free drug. However, when combined with radiation therapy where the mice were given 5 days of intermittent radiation, the rod-encapsulated cisplatin was the most effective against tumor growth compared with both IT and IP injection of free drug. The combination of sustained drug release and radiation therapy proved to be beneficial.

Dordunoo et al. have been developing a biodegradable paste containing paclitaxel, an antineoplastic agent from the bark of *Taxus brevifolia* (79). The paste was designed to be spread onto the surface of the pocket left after tumor resection. The paste was made from poly(ε-caprolactone) (PCL), a low-melting, biocompatible, and biodegradable polyester that melts at temperatures below 60°C. The paste could therefore be applied to the tumor bed in molten form. Dordunoo and colleagues have evaluated various additives to the system, including gelatin, to modulate the release rate of the paclitaxel in vitro (79). The paste containing 20% (w/w) of gelatin provided the most desirable release rate in vitro and was tested for its antitumor activity in mice that had been subcutaneously inoculated with MDAY-D2 tumor cells. Six days after inoculating the mice with the tumor cells, the tumor site was surgically exposed, and 150 mg of the paste was extruded onto the tumor. As a control, some of the animals received a paste that contained only gelatin (20% w/w) and PCL. After 16 days, the taxol-gelatin-PCL paste–treated animals showed a tumor mass reduction of 63 ± 27% compared with the untreated controls. Taxol–PCl pastes without the gelatin did not show any antitumor effect, probably due to the low levels of release. The gelatin in the formulation increased the swelling of the paste and accelerated the drug release, allowing for tumoricidal levels of the drug to be achieved at the site of the tumor.

**Drug Delivery to the Brain.** The brain is a site which notably benefits from local treatment. Some examples of the delivery of chemotherapeutic agents to the brain are discussed here. For more information, see the article CENTRAL NERVOUS SYSTEM, DRUG DELIVERY TO TREAT. The efficacy of systemic chemotherapy against intracranial tumors is limited by the exclusion of chemotherapeutic agents by the blood–brain barrier (BBB). Anatomically, the BBB is the system of brain capillary endothelial cells that do not allow for significant nonspecific transport of molecules from the systemic circulation into the brain tissue. The function of the BBB is to protect the brain from systemic insults, but this mechanism also limits the efficacy of systemically administered drugs for the treatment of brain cancer, as only small fractions of the drug are able to reach the tumor. Local delivery directly into brain tumors would thus be very beneficial. Henry Brem and coworkers have actively been studying the treatment of recurrent malignant gliomas with 1,3-bis(2-chloroethyl)-1-nitrosourea (BCNU or carmustine) encapsulated in a biodegradable polyanhydride wafer made of poly[1,3-bis(p-carboxyphenoxy) propane-co-sebacic anhydride] [P(CPP:SA)20:80] (82–85). Malignant gliomas are brain cancers that progress rapidly, and the standard treatment of surgical resection plus external beam irradiation provides only median survival times of less than one year after surgery (80). Slight improvement in the number of long-term survivors resulted from the addition of systemic chemotherapy with BCNU to the therapy, but its use was limited by the systemic toxicity of the drug (80). BCNU also has a half-life in serum of only 15 minutes, further limiting its usefulness when delivered systemically (81). By protecting the drug from degradation and delivering it directly to the tumor site, Brem et al. showed in Phase I–II trials involving 21 patients with recurrent glioma that mean survival time was 48 weeks past reoperation and 94 weeks from the original surgical resection of the tumor after diagnosis (82). Hematology, blood chemistry, and urinalysis tests did not show any systemic effects in any of the patients. This product, called Gliadel® (Rhône-Poulenc Rorer Pharmaceuticals, Inc.) is now FDA approved for use as an adjunct to surgery to prolong survival in patients with glioblastoma multiforme (GBM) for whom surgical resection is also indicated. The disks (1.45 cm diameter and 1 mm thick) are placed into the cavity left behind when the tumor is removed. The usual number of disks per patient is seven to eight. As the polymer degrades hydrolytically into monomers over a period of weeks, the drug is released into the surrounding tissue and fluid. Each disk contains 192.3 mg of P(CPP:SA)20:80 and 7.7 mg of carmustine (BCNU). The chemical structures of the components are shown in Figure 2. After 3 weeks, more than 70% of the polymer has degraded, although monomers have been found at the site of the tumor 2–3 months after implantation (82,83). The CPP monomer is excreted by the kidney, and the SA is metabolized in the liver and expired as carbon dioxide. In a random, double-blind, placebo-controlled clinical trial, 56% of the patients with GBM treated with Gliadel® were still alive 6 months postsurgery, whereas only 36% of the patients treated with the blank wafers (placebo) were alive at that time. Median survival was increased from 20 weeks (placebo group) to 28 weeks with Gliadel® (82). Patients with malignant gliomas other than GBM showed no benefits from this therapy.

While the Gliadel® system was going through clinical trials, Brem's group continued to search for even more effective polymer/drug combinations for the treatment of brain cancers. They published results using a similar delivery system loaded with a different chemotherapeutic agent, 4-hydroperoxycyclophosphamide (4HC) against rat gliomas (84). In this study, animals with brain tumors treated with 4HC encapsulated in poly(dimer erucic acid-co-sebacic acid)1:1, another biodegradable polyanhydride, had a median survival time of 77 days compared with 21 days for animals treated by P(CPP:SA)20:80-encapsulated BCNU. Long-term survival (>80 days) was 40% for 4HC-treated rats versus 30% in the BCNU-treated animals. The 4HC appeared to be more effective than BCNU. In another study, minocycline, a semisynthetic tetracycline that had been shown to inhibit tumor-induced angiogenesis, was encapsulated in an ethylene vinyl acetate copolymer (EVAc) delivery system (85). Antiangiogenic agents have been explored for use in cancer treatment, with the hypothesis that by keeping the tumor from creating new blood vessels, the tumor will not be able to grow. Without the means to provide oxygen and nutrients to new cancer cells, they will die. The sustained-release formulation was tested by itself or in combination with systemic BCNU chemotherapy. Local treatment with minocycline-loaded EVAc extended median survival time by 530% compared with treatment with the blank polymer (85). Systemically administered minocycline did not affect survival time. In combination with systemically administered BCNU, locally released minocycline showed a 93% extension of median survival time compared to BCNU treatment alone and once again, systemic minocycline showed no benefit, even in this combination treatment.

For easier implantation, Boisdron-Celle et al. investigated a microsphere delivery system for the treatment of brain tumors (86). The microspheres were made of poly(lactide-co-glycolide)50:50 (PLGA 50:50, Resomer RG506) and were loaded with therapeutic doses of 5-FU. The spheres were small enough to be injected stereotaxically into the brain, and thus the surgeon would not be required to expose the tumor. The microspheres were shown to be biocompatible and degrade within 2 months

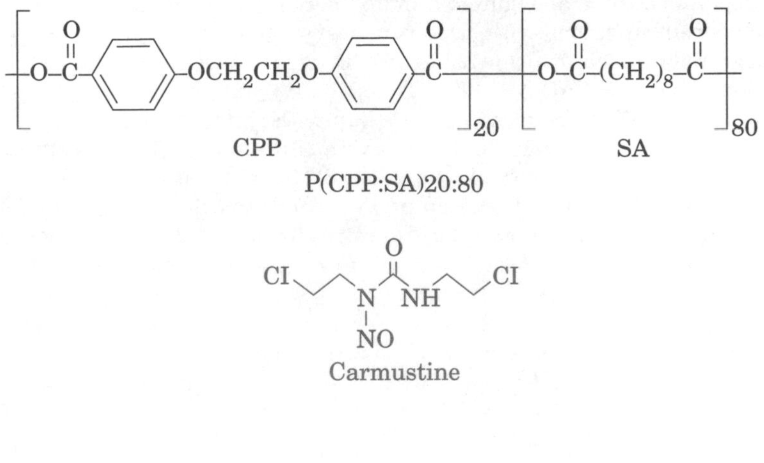

P(CPP:SA)20:80

Carmustine

**Figure 2.** Components of Gliadel® (Rhone-Poulenc Rorer Pharmaceuticals, Inc.). *Source:* From Ref. 92.

in the brain (87). To test the release kinetics and efficacy of this type of delivery system in vivo, Menei et al. (88) stereotactically implanted C6 malignant glioma cells into the brains of rats. One week later, the rats were treated with intratumoral (IT) injection of (1) 5-FU solution, (2) blank PLGA microspheres, (3) fast-releasing 5-FU–loaded microspheres (complete release of drug by 72 hours in vitro), or (4) slow-releasing 5-FU–loaded microspheres (complete release by 18 days in vitro). In vivo, the fast-releasing formulation showed entrapped 5-FU crystals until postimplantation day 12; in the slow-releasing formulation, until day 20. Release rates in vivo were slower than in vitro. This disparity in release rates was also reported by Wu et al. (83) for the Gliadel® system described previously. Only the slow-releasing formulation of the 5-FU microspheres showed any impact on the mortality of the animals. Half of the untreated control mice were dead 20 days after the tumor cells were implanted, whereas 50% of the mice treated with the slow-releasing formulation survived past 40 days (88). The mortality in the other two treatment groups (IT injection of 5-FU solution and the fast-releasing formulation) was not statistically different than the untreated controls.

**Polymer Implants Used in Developing Experimental Tumor Models.** Polymers and controlled-release systems have not just been used to treat cancer but have also been used to develop tumor models in animals. Shors et al. loaded a silicone polymer with benzo(a)pyrene, a known chemical carcinogen, to determine the carcinogenic potential of this chemical when delivered in small doses over an extended period of time (89). They surgically attached the implants to the tracheobronchial mucosa of hamsters and dogs and this resulted in bronchogenic cancers in 93% of the hamsters after 100 days, and squamous metaplasia was regularly found in dogs after 150 days. The authors suggest this technique as a means of producing a model of lung cancer in which neoplasia is induced at selected sites in a well-controlled manner. The models can then be used in the testing of cancer therapies.

Clinicians have known for a long time that certain tumors metastasize preferentially to particular organs. For example, CRC usually metastasizes to the liver. To study what aspects of particular organs or tissue are attractive to various tumor cells, Stackpole et al., devised a method to encapsulate extracts of various organs or defined cocktails of various factors or single agents into polymer devices. These were then implanted subcutaneously and monitored for their ability to attract metastatic cancer cells. In one report, they used polymer implants to influence the location of secondary metastases of B16 melanoma in mice (90). Cellulose disks with a central core of EVAc loaded with various angiogenic agents were implanted subcutaneously in the flank of the mice (90). The implant environment was shown to influence the rate of metastases to the implant site. EVAc loaded with tumor cell mitogenic agents (prepared from a crude extract of mouse lungs) was more efficient at localizing metastases than the implants that released heparin or endothelial cell growth factor.

## SYSTEMIC DELIVERY

Although local delivery of agents seems the most logical approach for achieving high concentrations of drug at the tumor site, there is often no easy access to the affected organ or tissue, or the tumor has metastasized and the patient requires systemic therapy. Systemic delivery systems range from simple, sustained-release depot formulations to drug carriers that can target specific organs or tissues. Depot systems are the easiest to develop but do not enhance the specificity of the drug. They simply allow the patient the freedom to receive treatment without the use of a percutaneous intravenous line. Depot systems are able to treat both the primary tumor and metastases equally, but side effects may not be reduced. On the other hand, targeted systems can also be designed to focus their effects on a particular organ, tissue, or even on a particular cell type through the use of antibodies or other targeting moieties. These systems can be designed to have minimal side effects, despite the systemic administration. Examples of targeted systems include nanoparticles, liposomes, and conjugates of drugs with polymers or antibodies. Both depot systems and targeted-delivery systems will be discussed in more detail in the following sections.

### Depot Systems

Currently, two controlled-release systems have been approved by the FDA to treat cancer. Both act as depots for the sustained release of anticancer peptides. The first, Zoladex® (Zeneca Pharmaceuticals), contains goserelin acetate, a synthetic decapeptide analogue of luteinizing hormone–releasing hormone (LH–RH) dispersed in a poly(lactide-co-glycolide) rod. Zoladex® is indicated for the palliative treatment of advanced prostate cancer and breast cancer. One formulation contains 3.6 mg of goserelin acetate and is designed to last 4 weeks; the second formulation contains 10.8 mg of the peptide and is designed as a 3-month implant. The 4-week rods are 1 mm in diameter, and the 3-month rods are slightly larger, with a diameter of 1.5 mm. The 3-month implant is indicated for prostate cancer patients only. The rods are supplied in a special syringe that is used to implant the delivery system subcutaneously in the upper abdominal wall. In a randomized, prospective clinical trial comparing radiation therapy alone to radiation therapy combined with Zoladex® implants, survival of patients with locally advanced prostate cancer was improved with the addition of Zoladex® to the treatment (91). Both groups of patients received 50 Gy of radiation to the pelvis over 5 weeks and an additional 20 Gy over the following two weeks. Patients in the combined-therapy group received the 4-week implants (Zoladex® 3.6 mg) starting on the first day of radiotherapy and again every 28 days for 3 years. In the combined therapy group, 85% of the patients were disease free at 5 years compared with only 48% of the radiotherapy-only group.

Lupron Depot® (TAP Pharmaceuticals, Inc.) has also been FDA approved for the treatment of prostate cancer. The formulations are supplied as lyophilized microspheres that are resuspended in a diluent for intramuscular injections every 1, 3, or 4 months. The composition of the three formulations is summarized in Table 3. Leuprolide acetate is a synthetic nonapeptide of LH–RH. The 1-month formulation is based on a poly(lactide-*co*-glycolide) copolymer, whereas the longer-acting systems use the homopolymer of lactic acid. The 1-month formulation also includes gelatin, which acts to modulate the release rate of the leuprolide acetate. In clinical trials, serum testosterone levels were suppressed to castrate level within 30 days for 95% of the patients (92). During the initial 24 weeks of treatment, 85% of the patients' disease did not progress. These depot formulations appear to be effective against cancer, but they do not necessarily avoid systemic side effects caused by nonspecific activity of the drugs, as can be achieved by the various targeted therapies discussed in the following sections.

## Targeted Delivery

As systemic delivery of encapsulated drugs does not circumvent the dangers of nonspecific toxicity, much research has been focused on targeting drugs to the tumor site without the requirement for surgical implantation of the device or exposure of tumor blood vessels. The topic of targeted therapy for cancer has been reviewed by Gupta (93). Targeting can improve the efficacy of cancer therapy by preferential distribution of the drug to the cancer cells or by maximizing the amount of drug that specifically acts against cancer cells. Preferential distribution can still result in adverse effects on normal cells in the vicinity of the tumor cells, whereas drug specificity should affect only the targeted cells. Therapies such as chemoembolization or arterial infusion could be included in the preferential distribution category, but in this article they are discussed under the heading of local therapy due to their specialized administration. The targeted therapies discussed here involve only systems that can be delivered systemically but that focus their effects on specific cells or organs. Targeting has been achieved by (*1*) antibodies conjugated to drugs; (*2*) antibodies conjugated to polymers loaded with drugs; (*3*) liposomes or nanoparticles, with or without conjugation to targeting moieties such as cell receptor ligands or antibodies; and (*4*) the use of external magnets to localize magnetized polymer microspheres. Each of these therapies will be discussed separately in the following sections.

**Antibody–Drug Conjugates.** More than 15 years of research has gone into the development of antibody–drug conjugates (immunotoxins) for cancer therapy. The hypothesis behind immunotoxin therapy is that the antibody will impart the necessary specificity to the cytotoxic drug (or

radionuclide), and the cytotoxic agent will thus affect only the targeted cells. Although many would argue that these conjugates should not be considered controlled-release systems, they do fit the description of controlled drug delivery and therefore will be discussed briefly in this article. For more information on this topic, please consult review articles that have been published during the past decade (94–97). Many immunotoxins are currently being tested in clinical trials and have been shown to have different side effects than the free drug, including hepatotoxicity, reversible vascular leak syndrome, and myalgias (95). Optimal regimens have yet to be determined, but most agree that plasma concentrations of <1 µg/mL will be necessary for therapeutic effect (95). The principle behind immunotoxin therapy has been proven, and now more subtle improvements must be made to enhance their specificity, increase cytotoxicity, and decrease immunogenicity so that the immunotoxins can circulate for longer periods of time to reach their target cells. Attempts are currently being made to use toxins that are less immunogenic, antibodies that are from humans or at least contain a human constant (Fc) region, and the coupling of poly(ethylene glycol) (PEG) to the immunotoxin so as to increase its circulation time (95). Targeting mechanisms other than antibodies have also been explored for targeted delivery of drugs (96). For example, many highly proliferative, malignant cells have high requirements for iron and overexpress the transferrin receptor on their surfaces (96). Transferrin therefore makes a good targeting moiety to direct cytotoxins to such tumor cells. As an alternative to the targeting of antineoplastic drugs to tumor cells, radioisotopes have also been conjugated to antibodies for cancer therapy. Jurcic recently reviewed the literature in the field and included an update on various clinical trials on such conjugates (98). Many different radioisotopes have been used, including $\beta$-particle emitters such as iodine-131 and yttrium-90 and $\alpha$-emitters such as bismuth-212 and astatine-211.

Sherwood et al. reported the development of controlled-release systems for antibodies that could be adapted for the sustained release of immunotoxins (99). These could allow for prolonged immunotoxin therapy without the conventionally required repeated intravenous injections. Sherwood et al. incorporated the antibodies into poly-(ethylene-*co*-vinyl acetate) (EVAc) and a biodegradable anhydride copolymer of stearic acid dimer and sebacic acid [P(SAD:SA)]. The release was shown to last for up to 30 days in vitro. The antibodies retained their specific binding after release from the matrices. The next step will be to test whether the systems are capable of releasing antibodies in vivo. To be effective, the antibodies or immunotoxins

**Table 3. Summary of Composition of Lupron Depot® Formulations**

|  | 7.5 mg (monthly) | 3-month 22.5 mg | 4-month 30 mg |
|---|---|---|---|
| Leuprolide acetate | 7.5 mg | 22.5 mg | 30 mg |
| Purified gelatin | 1.3 mg | – | – |
| Poly (DL-lactide-*co*-glycolide) | 66.2 mg | – | – |
| Poly(lactic acid) | – | 198.6 mg | 264.8 mg |
| *D*-Mannitol | 13.2 mg | 38.9 mg | 51.9 mg |

*Source:* From Ref. 92.

would have to either be released directly into the tumor or be absorbed into the bloodstream from the interstitial space after being released from the matrices.

For immunotoxins to be effective as anticancer agents, there are many obstacles to overcome. Some of these obstacles concern the properties and the design of immunotoxin itself, and others pertain to the constraints of the human anatomy and physiology. The problems with the immunotoxin itself, including antigenicity of the components (either the antibody or the toxin) and the ability of the toxin to kill cells, are much easier to study, and much research has gone on toward this end. However, there are other obstacles created by the tumor and the body itself that may be even more critical and even more difficult to overcome. As mentioned in the introduction to this article, tumor cells have an uncanny ability to make themselves blend in with normal cells. They can rapidly change the antigens they present on their surfaces, making it difficult to target an entire population of tumor cells with a single antibody. The cell population within each tumor can also be very heterogeneous with respect to susceptibility to various drugs. Some cells will be susceptible to a particular drug or toxin, and others will be resistant to it. A single drug or toxin may not be effective against all of the cells in the tumor. In this manner, treatment with a single antibody–toxin conjugate will just select for the cells that are not affected by the treatment, and then these cells can proliferate and replace the cells that were killed by the immunotoxin. The final barrier to overcome, and one that is critical for all of the systemically administered, targeted therapies, is the vascular endothelium of the tumor. To be effective, the targeted therapy must be able to exit the bloodstream and penetrate the tumor. This involves passing through the capillary endothelium and basement membrane. In some tumors, the vasculature has been shown to be "leaky," but this has not been proven to be the case for all tumors. And the degree of leakiness might be critical. There will always be a maximum molecular or particulate size that can effectively penetrate the tumor, and the delivery system must always be smaller than that critical size. These problems are difficult ones, and they will come up repeatedly in the discussion of the various types of targeted delivery systems. The targeting moiety must be able to come into contact with its target for the treatment to be effective.

**Polymer–Drug Conjugates.** See CANCER, DRUG DELIVERY TO TREAT—PRODRUGS.

**Liposomes.** The main component of liposomes are lipids, which contain two regions: (1) a polar, hydrophilic head group (usually a phosphate group) and (2) a nonpolar, hydrophobic tail that usually consists of a hydrocarbon chain containing 14–18 carbon atoms. These molecules have an interesting ability to spontaneously self-assemble in water so as to sequester the hydrophobic tail regions from the water. The simplest structure formed by lipids in an aqueous media is a flat bilayer, but this structure is thermodynamically unstable because of the exposed tail groups at the sheet edges. Therefore, the bilayers usually wrap around on themselves, creating vesicles with entrapped

water called liposomes. If a drug was included the aqueous phase, it would become entrapped with the water. In addition, liposomes can carry lipid-soluble drugs in the bilayer itself. The applications of liposomes to cancer therapy are discussed in this article; however, information on how to prepare and load the various types of liposomes can be found in LIPOSOMES. The applications of liposomes in fields other than cancer therapy (e.g., gene delivery or vaccines) can be found in those articles.

The placement of liposomes under the rubric of targeted delivery is somewhat controversial because they do not specifically target tumor cells but rather are selectively removed from the circulation by phagocytic cells of the reticuloendothelial system (RES). This makes liposomes good delivery systems for macrophage or tumors of the liver and spleen, but they are not broadly considered targeted systems unless surface modified by the attachment of antibodies to tumor-associated antigens (immunoliposomes) or other targeting molecules. Owing to the rapid removal of liposomes from the circulation by the RES, researchers have focused a great deal of effort into developing techniques to mask them from the immune system. By coating the liposomes so that they are not recognized as foreign by the immune system, circulation times have been increased to greater than 1 day; these liposomes are generally called "stealth liposomes." Standard liposomal formulations as well as the modified liposomes are discussed here.

A vast amount of research has been devoted to the topic of liposomes for cancer therapy. In fact, there are two liposomal formulations of chemotherapeutic agents approved by the FDA to treat AIDS-related Kaposi's sarcoma: Doxil® (Sequus Pharmaceuticals, Inc.) and DaunoXome® (NeXstar Pharmaceuticals, Inc.). Doxil® is doxorubicin hydrochloride encapsulated in STEALTH® liposomes. This product is discussed further under "Stealth Liposomes" in this article. DaunoXome® is an aqueous solution of the citrate salt of daunorubicin (DAU) encapsulated in the aqueous core of lipid vesicles composed of a lipid bilayer of distearoylphosphatidylcholine (DSPC) and cholesterol in a 2:1 molar ratio. The molar ratio of DSPC:cholesterol:DAU is 10:5:1. Figure 3 shows the chemical structures of the components. The liposome helps to protect the drug from chemical and enzymatic degradation, minimizes protein binding, and decreases uptake by normal (non-RES) tissues. The liposomes are 50–80 nm and are believed to penetrate tumor vasculature owing to the increased permeability of tumor blood vessels. Plasma clearance for DaunoXome® is reported as $17.3 \pm 6.1$ mL/min compared to $236 \pm 181$ mL/min for the unencapsulated DAU (92). The liposomal formulation is not cleared by the kidneys as rapidly as the free drug. A randomized, controlled clinical study compared treatment of HIV-related Kaposi's sarcoma by DaunoXome® ($40$ mg/m$^2$) to a conventional three-drug regimen including DOX ($10$ mg/m$^2$), bleomycin (15 U), and vincristine (1 mg) delivered intravenously every 2 weeks (100). Overall response rates (complete plus partial responses) were 25% for DaunoXome® and 28% for the three-drug regimen (a difference that is not statistically significant). Median survival times and median times to progression were also similar for the two therapies. DaunoXone® was thus shown to be safe and effective ther-

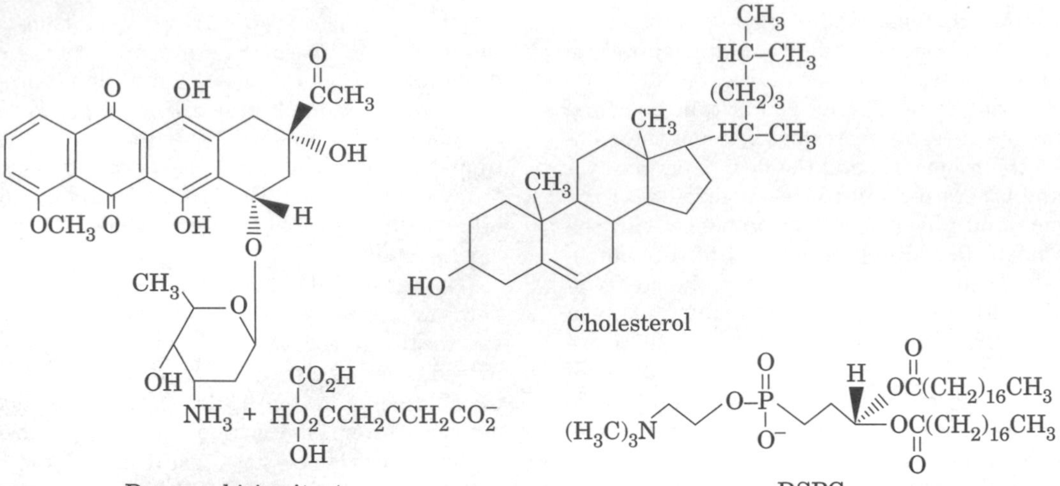

**Figure 3.** Chemical structures of the three components of DaunoXome®. *Source:* From Ref. 92.

apy for this disease. More importantly, the dose of the drug was able to be reduced with the liposomal encapsulation.

Other cancer drugs have also been investigated in liposomal formulations. Recently, Sharma et al. has reported on the antitumor activity of liposome-encapsulated paclitaxel (101). Paclitaxel has poor aqueous solubility and is currently given as a suspension in Cremophor and ethanol (Taxol®, Bristol-Myers Squibb), additives that cause significant side effects. Sharma and colleagues encapsulated the paclitaxel in two types of liposomes: (1) ETL, which contains a single lipid (egg phosphatidylcholine; PC); and (2) TTL, which contains a combination of dielaidoyl PC, dimyristoyl PC and 1-stearoyl 2-caproyl PC (molar ratio 1:1:0.9). Both ETL and TTL had drug:lipid molar ratios of 1:33, and both formulations showed a reduction in acute reactions normally seen after IV or IP injection of Taxol®. Mice were able to tolerate an IV dose of 50 mg/kg of the liposomal formulations. An equivalent dose of Taxol® was 100% lethal. At equivalent dosages and schedules, both liposomal formulations delayed tumor growth similar to Taxol® in mice containing subcutaneous xenografts of human ovarian A121 tumors. So although the antitumor efficacy was not improved by liposome encapsulation, the reduction in side effects would certainly be beneficial for the patient.

Mayer and colleagues encapsulated vincristine (VCR) in 120-nm diameter distearoylphosphatidylcholine/cholesterol (55:45 molar ratio) liposomes and investigated the pharmacokinetics, tumor uptake, and antitumor capabilities of this formulation (102). Plasma levels of VCR were found to be at least 100 times greater over the first 24 hours for the liposomal formulation compared to free drug injected intravenously at a dose of 2.0 mg/kg of VCR into mice with L1210 ascites tumors. The area under the tumor VCR concentration-time curve for the first 72 hours after administration ($AUC_{72}$) was 13 times higher for the liposomal formulation than that for the free drug. In a similar model using a solid tumor (B16/B16 melanoma), the $AUC_{72}$ for the liposomal formulation was 4 times

higher than that of the free drug. Drug levels found in noncancerous muscle tissue were similar for both formulations, showing that the liposomal formulations specifically enhanced drug delivery to the tumor, not to all tissues equally.

Liposome-encapsulated DOX has also been shown to overcome MDR. MDR is often associated with high levels of expression of P-glycoprotein, which acts as an ATP-driven pump to transport toxic drugs out of the cell. This mechanism is not drug specific but acts to expel many different drugs. Rahman et al. examined the use of liposomal-encapsulated DOX to overcome MDR in two MDR leukemia cell types: (1) HL-60/VCR, which expresses P-glycoprotein; and (2) HL-60/ADR, which has no P-glycoprotein (103). HL-60/VCR cells were shown to be 5-fold more sensitive to liposomal DOX than free DOX and accumulated up to 3-fold more drug from the liposomal formulation than from treatment with the free drug. The liposomal formulation had no beneficial effect on the HL-60/ADR cells. This suggested that the liposome specifically interacts with the P-glycoprotein, interfering with its function. Rahman et al. support this hypothesis by showing that blank or drug-loaded liposomes compete with azidopine (a known binding partner of P-glycoprotein) for binding with P-glycoprotein. It is not entirely surprising that lipids interfere with this efflux pump mechanism as lipophilic agents, such as cyclosporin A, have previously been shown to interfere with the function of P-glycoprotein, thereby reversing MDR.

**Targeted Liposomes.** To improve the tumor cell specificity of liposomes, targeting moieties such as antibodies have been conjugated to the surfaces of liposomes. Antibody-targeted liposomes are generally referred to as immunoliposomes. Ohta et al. conjugated the sulfhydryl groups of reduced IgM molecules to maleimide groups on the surface of a liposome (104). The maleimide groups resulted from the incorporation of N-(m-malemidobenzoyl)dipalmitoyl-phosphatidylethanolamine (2.5 μmol) with dipalmitoyl-phosphatidylcholine (25 μmol) and cholesterol (17.5 μmol) to form the liposomes. DOX was then loaded into the liposomes by the use of a $Na^+/K^+$ gradient. The liposomes had an average size of 86 nm and each contained approximately 400 molecules of DOX. The immunoliposomes showed specificity in vitro for the cell types corresponding to the antibody, but not all of the formulations retained the cytotoxicity of the free drug. Some antigen–antibody interactions seemed to facilitate the necessary uptake of the drug necessary for the cells to be killed, whereas others resulted in simple binding events without uptake. Preliminary in vivo experiments showed that the liposomes were immediately taken up by the RES. Other researchers have faced the same problem and have worked to include a PEG coating in addition to the antibodies (105–110). Examples of stealth immunoliposomes will be discussed under the heading of stealth liposomes.

Longman et al. developed a creative, two-step approach to targeting liposomes to tumor cells (111). They used the strong binding affinity between streptavidin (SA) and biotin to link the tumor cells to the DOX-loaded liposome in vitro and in vivo. First, they made liposomes of distearoylphosphatidylcholine, cholesterol, and $N$-[4-($p$-maleimidophenyl)butryl]phosphatidylethanolamine (MPB-PE) (molar ratio 54:45:1). Then they reacted these liposomes with SA, which had been made maleimide reactive by $N$-succinimidyl 3-(2-pyridyldithio) propionic acid (SDDP). This resulted in liposomes covalently coated with SA (SA-liposomes). DOX was then loaded into the liposomes using a pH gradient. The two-step targeting method involves the IV injection of a biotinylated antibody to the tumor cells, followed by the IV injection of the SA-liposomes. The antibody should find the tumor cells and bind to their surface, effectively biotinylating the cell surface, and then when the SA-liposomes are infused, the liposomes will stick to the biotinylated cells. For the in vitro study, P388 cells (murine lymphocytic leukemia cell line) were grown in culture and treated sequentially with the biotinylated antibody to the Thy 1.2 antigen (which is highly expressed on P388 cells) and the SA-liposomes. They demonstrated a 30-fold increase in cell associated lipid and a 20-fold increase in cell-associated DOX using the targeted liposomes as compared with the same two-step procedure using nontargeted liposomes. To test the in vivo tumor cell–binding efficacy of this technique, Longman and colleagues established intraperitoneal (IP) P388 tumors in mice that were injected with the biotinylated antibody, followed 24 hours later by the IP or IV injection of the SA-liposomes or control liposomes. The animals treated with both the biotinylated antibody and the SA liposomes showed statistically significant increases in cell-associated lipid than those controls using nontargeted liposomes. This two-step approach, although it may seem cumbersome at first glance, may allow for the ease of switching the antibody specificity of the treatment without have to reengineer the drug delivery vehicle. Only the various antibodies need be biotinylated for the specific application and used in combination with standard drug-loaded SA-liposomes.

Although antibodies are the most common means of targeting liposomes, other targeting moieties have also been used. To exploit the fact that epithelial cancer cells often overexpress receptors for the vitamin folic acid, Lee and

Low incorporated lipids conjugated to folate via a polyethylene glycol spacer (folate-polyethyleneglycol-distearoylphosphatidylethanolamine, or folate-PEG-DSPE) into liposomes of distearoylphosphatidylcholine (DSPC) and cholesterol (112). They studied the in vitro cellular uptake of DOX that had been loaded into these liposomes by KB cells (human nasopharyngeal epidermal carcinoma cell line) and HeLa cells (human cervical carcinoma cell line). For a formulation that included DSPC/cholesterol/folate-PEG-DSPE (molar ratio 56:40:0.1), the cellular uptake by KB cells was increased 45 times over that of nontargeted liposomes containing only the DSPC and cholesterol. Uptake by HeLa cells for the same formulation was 29 times higher for the folate-conjugated liposomes than for the nontargeted liposomes (112). The folate was able to mediate an increase in uptake of the drug because of close association of the liposome with the folic acid receptors on the cells' surfaces. However, this technique still needs to be tested in vivo.

**Stealth Liposomes.**  One of the main problems with standard and even targeted liposomes is their rapid clearance from circulation. Much research has focused on trying to increase circulation times and in vivo stability of liposomes for the delivery of chemotherapeutic agents. The topic has been reviewed by Lasic (113,114). The most common method of sterically stabilizing liposomes is by including a PEG-modified, phosphatidylethanolamine-based lipid into the liposome formulation. This results in liposomes with a "halo" of PEG (PEG-liposomes). Lasic et al. hypothesize that this modification is able to achieve significantly increased circulation times because of the high local concentration of hydrated groups at the surface of the PEG-liposomes that act to sterically inhibit both electrostatic and hydrophobic interactions of blood components (109). Lasic et al. hypothesize that it is the adsorption of blood components such as denatured proteins and opsonins onto the surfaces of liposomes that mark them for clearance by phagocytic cells in the body. Therefore, if these blood components are not able to adsorb or bind to the surface of the liposomes, the liposomes should be able to circulate for extended periods of time, increasing the likelihood that they reach the target.

Regardless of the mechanism of the origin of the extended circulation times, stealth liposomes are used successfully in the clinic. Doxil® (Sequus Pharmaceuticals, Inc.) is a formulation of doxorubicin hydrochloride (DOX-HCl) encapsulated in STEALTH® liposomes that has made it through FDA approval for the treatment of AIDS-related Kaposi's sarcoma. These liposomes are composed of $N$-(carbamoyl-methoxypolyethylene glycol 2000)-1,2-distearoyl-$sn$-glycero-3-phosphoethanolamine sodium salt (MPEG-DSPE), fully hydrogenated soy phosphatidylcholine (HSPC), and cholesterol. The weight ratio of MPEG-DSPE:HSPC:cholesterol in the liposomes is 1:3:1. The chemical structures of the components are shown in Figure 4. In a clinical trial, 53 patients who had experienced disease progression or intolerable toxicities while receiving standard intravenous doxorubicin/bleomycin/vincristine or bleomycin/vincristine treatment were given 20 mg/m$^2$ Doxil® every three weeks. The liposomal formulation increased the therapeutic effect of the DOX, with 49%

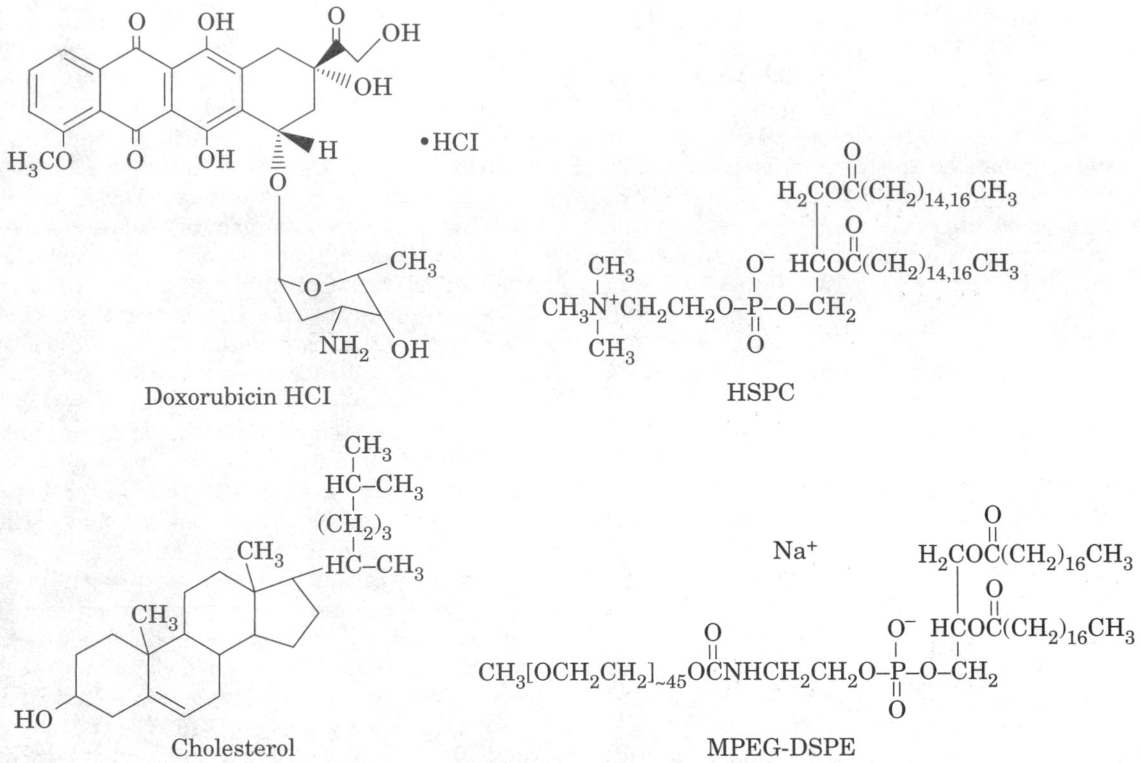

**Figure 4.** Chemical structures of the three components of Doxil®. *Source:* From Ref. 92.

of the patients experiencing more than one clinical benefit (115). The circulation half-life of the Doxil® formulation is approximately 2 days (116), compared to unencapsulated DOX with a half-life of 18 to 30 h (117).

Allen et al. investigated the in vivo cytotoxicity of VCR encapsulated in stealth liposomes (SL-VCR) against IP and IV tumor models in mice (107). When P388 cells were injected IP, SL-VCR treatment significantly increased mean survival times from 8.5 days (no treatment) or approximately 2 weeks (free VCR) to over 3 weeks. When the P388 cells were injected IV, no effect on mortality was seen for any of the treatments. The experiment was repeated with a different cell line (L1210 leukemia cells), and the liposomal formulation was effective only against tumors that had been started IP. The SL-VCR therapy appears to increase the efficacy only for tumors in the peritoneal cavity. It didn't seem to matter whether the SL-VCR therapy was given IP or IV. As long as the tumor was located in the peritoneal cavity, the SL-VCR therapy was effective in prolonging survival.

As mentioned previously, many researchers have combined the use of stealth liposomes with antibody targeting (105–110). A few examples are discussed here. Emanuel et al. encapsulated DOX in stealth liposomes with surface-linked antibodies (specific to tumor cells or nonspecific controls) and in nontargeted stealth liposomes (108). They showed that DOX accumulation in lung metastases of a polyoma virus-induced fibrosarcoma was higher in animals treated with the tumor-targeted stealth immunoliposomes than the nontargeted liposomes and the nonspecific stealth immunoliposomes. The highest tumor DOX

concentrations were achieved through the combination of steric stabilization and antibody targeting. They also showed that the addition of the antibodies to the surface did not effect the circulation time of the stealth liposomes. More recently, Harding et al. stress that the immunogenicity of the antibody can influence the circulation time of the immunoliposomes (110). If the antibody is immunogenic, the body will clear the liposomes quickly, despite the stealth coating. Care must be taken to match the antibody to the model species.

Ahmad et al. reported a study performed on mice injected IV with KLN-205 (murine lung squamous carcinoma) cells (106). A biotinylated monoclonal antibody to the KLN-205 cells (174H.64) was attached via a streptavidin linker to hydrogenated soy phosphatidylcholine (HSPC)/cholesterol/PEG-DSPE (molar ratio 2:1:0.1) liposomes to form the sterically stabilized immunoliposomes. The IV injection of the sterically stabilized immunoliposomes loaded with DOX resulted in 50% of the animals surviving more than 6 months. Animals treated with the untargeted sterically stabilized liposomes showed a mean survival time of approximately 66 days. By combining the benefits of steric stabilization by PEG and the tumor-specific targeting of a monoclonal antibody, the anticancer efficacy of the drug was improved.

**Nanoparticles.** To improve on the stability of liposomes as controlled-delivery devices for antineoplastic drugs, many researchers turned to polymeric nanoparticles. Nanoparticles are generally defined as particles with diameter less than 1 $\mu$m, small enough to be injectable IV.

The drugs may be dissolved in the polymer, entrapped within it, or adsorbed to the surface of the particles. Higher concentrations of hydrophilic drugs can be loaded into nanoparticles as there is no limitation of aqueous solubility as is inherent for loading drugs into liposomes. Leroux et al. have recently reviewed the topic of nanoparticles for the delivery of cancer therapies (118). Nanoparticles are considered targeted delivery systems, as the distribution and pharmacokinetics of anticancer drugs can be modified by encapsulation in polymers. This altered distribution is postulated to lower systemic toxicities and to provide better supply of drug to the tumor, assuming that the tumor location and the drug distribution match. After intravenous administration, hydrophobic nanoparticles are found predominantly in the reticuloendothelial system (RES) (118). The drug levels in the liver and spleen are accordingly very high. It has been hypothesized that intravenously administered nanoparticles can be passively targeted to tumor cells by means of enhanced permeation through tumor neovasculature, although this has not yet been clearly proven (118–120). So although the nanoparticles have enhanced stability and higher drug-loading capabilities than liposomes, they share the same problem of not being able to reach tumor cells efficiently. Similar means of enhancing nanoparticle delivery to tumors have been explored as have been investigated for liposomes, namely antibody targeting (119,121) and PEG coatings (122,123). These two topics are addressed in the following paragraphs.

To study the use of monoclonal antibodies against tumor cell antigens as targeting mechanisms for nanoparticles, Illum et al. coated poly(hexyl 2-cyanoacrylate) nanoparticles with a monoclonal antibody against osteogenic sarcoma cells. Although the antibodies were shown to retain their binding specificity in vitro, they found no enhanced localization of the particles to the tumor in vivo (121). This is probably due to the inability of the nanoparticles to access the tumor cells before they were taken up by the RES.

Recent research on nanoparticles has focused on creating long-circulating, or stealth, nanoparticles. The most common method for reducing RES uptake is by the covalent attachment of PEG to the surface of the particles (122,123). The density of the PEG is critical, and particles that showed highest PEG densities showed increased circulation times (123). It is hoped that such particles could avoid phagocytosis by the RES and could therefore have a better chance at accumulating at the tumor site. Similar results have been shown for the surface adsorption of Poloxamine 908 (124), a block copolymer of poly(ethylene oxide) (PEO) and poly(propylene oxide) (PPO) to hydrophobic polymer nanoparticles. The PPO region of these surface-active polymers adsorbs well to hydrophobic materials and makes the surface hydrophilic with the PEO tails. However, the possibility for desorption in vivo was a concern. In addition, the coating of the nanoparticles with a hydrophilic layer may make it even less likely for the nanoparticles to leave the bloodstream and accumulate in the tumor, so the enhanced circulation time may not be the complete solution.

The overall efficacy of nanoparticles for the delivery of cytotoxic agents has not been adequately proven as yet.

Leroux et al. reviewed the in vivo evaluation of drug-loaded nanoparticles (118). A wide range of polymers and drugs have been combined and tested in various animals tumor models. However, despite the large bulk of research, no clear consensus of efficacy was shown. Some formulations showed increased efficacy or decreased systemic toxicities over unencapsulated drug, whereas other formulations showed no benefit. A recent example of a positive result used poly(butylcyanoacrylate) (PBCA) nanoparticles, both uncoated and coated with poloxamine 1508 (125). The drug, mitoxantrone, was loaded in two different ways: adsorbed to the surface after the polymerization of the nanoparticles or incorporated during the polymerization. To test the in vivo cytotoxicity of the four types of nanoparticles, they were administered to separate groups of mice that had received intramuscular inoculations of B16 melanoma cells. Mitoxantrone-loaded liposomes and free drug were included for comparison. Of the four sets of nanoparticles, all but the uncoated/incorporated set showed a statistically significant decrease in tumor volume compared to the liposome formulation and the free mitoxantrone. But for every positive result showing that nanoparticle formulations are effective against tumors in vivo, others show no improvement over the free drug. Thus, room still exists for improvements on nanoparticulate systems for cancer therapy, including better antibody-conjugation techniques so as to maximize specific bindings and decrease undesirable antigenicity of the complex. Combining such antibody-targeted nanoparticles with PEG coatings to prevent RES clearance has the possibility of increasing circulation time and tumor cell specificity, although the problem of extravasation from tumor blood vessels may overwhelm all of these improvements.

**Magnetic Localization.** Research on targeting polymers to tumors for drug delivery using externally placed magnets has been studied since the early 1960s. Mosso and Rand mixed carbonyl iron microspheres into a silicone polymer that could be cured in vivo using a stannous oxide catalyst (126). The mixture could be infused into the tumor vasculature and held in place with an external magnet until it had cured. A double-lumen catheter was used for the procedure so that angiograms of the blood supply to the tumor could be obtained before and after treatment. The superconducting electromagent was then placed at the animal's flank for kidney infusions or next to the skin for hindlimb infusions. A bolus of 2 to 3 mL of the magnetized silicone/catalyst mixture was infused into the arteries of six dogs. The remaining uninfused silicone was used to measure vulcanization time (approximately 30 minutes). When the uninfused silicone was solidified, the external magnet was removed and the inner catheter was retracted. A second angiography was performed using the remaining guide catheter. The second angiogram confirmed that this technique occluded blood flow to the region. Generally the animals remained healthy, showing some transient ischemia in the region of the occluded vessels. At autopsy, silicone was found to have occluded the arteries, arterial branches, and even the capillary beds. No evidence of retrograde embolization into the aorta was found, and neither was any silicone found at necropsy in the lungs or venous system. There was some perivascular granulation and fibrosis in tissue surrounding the infarcted vessel, presumably in response to the iron, which may have helped form a direct cohesion between the vessel and the nonadherent silicone rubber.

Research switched focus to methods of temporary occlusion to solve the same problem of collateral blood vessel formation that plagued arterial ligation and other permanent arterial occlusions discussed earlier in this article. Researchers tried both magnetic degradable starch microspheres with covalently bound model drugs and proteins (127) and magnetic degradable albumin microspheres loaded with chemotherapeutic drugs (128–132). Mossbach and Schröder made starch microspheres with diameters less than 10 $\mu$m and loaded them with magnetite and covalently bound radio-labeled model proteins and drugs to the surface (127). The microspheres were suspended in NaCl solution and infused IV into one ear of a rabbit through a catheter. An external magnet was placed on the distal portion of the opposite ear ("magnetic ear") in attempts to localize the microspheres. After 10 min, the rabbits were euthanized, and the distal portions of both ears were removed to measure radioactivity from the proteins and drugs bound to the microspheres. Although the majority of the microspheres were retained in the capillary beds of the lungs (80%), the magnetic ear contained 4–8 times the radioactivity of the nonmagnetic ear.

Researchers also studied the use of magnetized albumin microspheres loaded with DOX and their distribution in rats after arterial (128,130,131) or venous (129,131) administration. Widder and Senyei and their colleagues prepared microspheres with an average diameter of 1 $\mu$m of human serum albumin (labeled with a small amount of $^{125}$I-bovine serum albumin) and loaded with 28.8% (w/w) of $Fe_3O_4$ particles and 5.8% (w/w) of DOX for infusion into the ventral caudal artery at the base of the tail. The tail was then demarcated into four equal segments approximately 3.5 cm in length. An external magnet was placed over the third segment from the base of the tail to localize the microspheres and was held in place for 30 minutes. Some of the rats were then sacrificed at that time, and the level of radioactivity in each of the four tail segments was measured. The concentration of DOX in each segment was also measured. Results showed that at the highest magnetic field (8,000 oersteds), 37–65% of the infused microspheres remained localized in the third tail segment, with 30–48% being found in the liver. In animals sacrificed 24 h after the removal of the magnet, approximately 50% of the injected microspheres remained localized in segment 3. Histological studies showed that some microspheres had been internalized by endothelial cells and others were trapped between the plasma membranes of adjacent endothelial cells. DOX concentrations in the third tail segment were 100-fold higher than levels achieved after injection of free drug. In a later study, the group showed that arterial administration of magnetic albumin microspheres loaded with DOX resulted in an average of 45% of the microspheres being localized in the third tail segment, when the magnetic field of 8,000 oe was applied for as little as 5 min and as much as 60 min (130).

Morimoto and colleagues developed magnetic albumin microspheres to treat cancer (131,133). The microspheres were made of bovine serum albumin with a small amount of $^{125}$I-human serum albumin for detectability, and magnetic fluid (suspension of $Fe_3O_4$ and colloidal magnetite) cross-linked a heat denaturation process. Microspheres were fractionated by centrifugation, and the <2-$\mu$m and 2- 4-$\mu$m fractions were saved for use in the in vivo studies. In an animal model to simulate treatment of localized lung cancer, they placed external magnets (3,000 gauss) ventrally and dorsally surrounding the chest cavity of mice and infused 1 mg of the microspheres into the tail vein. The mice were then sacrificed after 10 or 60 min, with the external magnet held in place for the duration of the experiment. Various tissues were isolated for detection of radioactivity to calculate the number of particles trapped in the various tissues. After 10 min, only 10.7% of the administered dose of microspheres 2–4 $\mu$m in diameter was found in the lungs without the use of magnets, and magnetic localization doubled this value. After 1 h, 17.0% of the administered dose of the 2- 4-$\mu$m spheres was found in the lung without localization, and this was increased to 28.2% with the external magnets. If the magnet was removed after the first 10 min, only 21% of the initial dose was found in the lungs after 1 h. Morimoto et al. also infused microspheres less than 2 $\mu$m into the left renal artery with two magnets placed on both sides of the left kidney. Magnetic localization of the microspheres increased 2.5 times over the nonlocalized control to 56.4% of the administered dose after 10 min and remained stable out to 1 h. If the magnet was removed after the first 10 min, only approximately 27% of the administered dose remained localized in the kidney.

Morimoto's group continued their research on the magnetic localization of albumin microspheres, showing the antitumor effect of DOX delivered from such microspheres (129). They administered microspheres (2–4 $\mu$m in diameter) containing 6% DOX and 47% magnetite encapsulated in bovine serum albumin into the tail vein of rats with AH 7974 (ascites hepatoma) cell lung tumors. This served as a model for the treatment of lung metastases. An external magnet was placed at the back and breast of the rat at the level of the chest cavity for the first 10 min after administration of the microspheres. The percent of the original microsphere dose localized to the lungs was doubled to 19.7% with the application of the magnet over control animals (no magnet). The levels of DOX in the magnet-localized group increased eightfold over IV injection of free DOX. The antitumor efficacy of this treatment was measured by changes in lung weight (higher weights signify tumor growth). After three treatments with the magnetically localized microspheres containing DOX at 1-week intervals, the average weight of the lungs was 2.4 g compared to 4.3 g for untreated rats and 1.1 g for rats without tumors. The magnetically localized microsphere treatment slowed the growth of the tumor. The antitumor effect was also monitored by animal survival, with median survival of 16.4 days for untreated rats with tumors and 23.6 days for animals with tumors treated by the magnetically localized microspheres containing DOX. Rats treated with microspheres without localization by external magnets showed a median survival of 17.4 days.

In 1983, Widder et al. reported the selective targeting of magnetic albumin microspheres containing DOX in rats with Yoshida sarcomas (132). In this study, magnetic microspheres were prepared by emulsion polymerization in which the aqueous phase consisted of human serum albumin, DOX, and magnetic iron oxide. Rats were subcutaneously inoculated with Yoshida tumor cells into the lateral aspect of the tail. Tumor size averaged 200 mm$^2$. Six to eight days after tumor cell inoculation, microspheres suspended in saline containing 0.1% Tween 80 were infused into the ventral caudal artery at a distance of 2 cm proximal to the tumor. A bipolar magnet with a field strength of 5,500 oe was placed with one of its faces in contact with the tumor and held in place for 30 minutes after the administration of the microspheres. Animals were kept for 30 days posttreatment (or until death) and monitored for weight change, tumor size, and metastases. Rats treated with magnetically localized microspheres containing DOX showed an approximately 90% decrease in tumor size, compared with a 120% increase in tumor size for untreated controls and 114% increase for those treated with nonlocalized DOX-loaded microspheres. Intraarterially administered free DOX at equivalent doses showed tumor size increases of 50 to 70%. All of the rats treated with magnetically localized microspheres showed tumor regression, with 75–80% showing total remissions for the DOX-loaded microspheres. All rats in the magnetically localized group survived the full 30 days, and no metastases were found upon autopsy. In the untreated controls, those treated by free DOX, those treated with blank microspheres (magnetically localized but containing no DOX), and those treated with nonlocalized DOX-loaded microspheres, 90–100% of the animals died before the 30-day limit, and no regressions or remissions were seen. Metastases were found in all animals. These dramatic results show the great promise of this form of treatment.

A different family of degradable polyalkylcyanoacrylates has also been developed for preparation of magnetic nanoparticles capable of adsorbing a wide variety of drugs (134). The length of the alkyl chain controls the rate of degradation so that a particular polymer can be chosen to match the specific requirements of an application. Ibrahim et al. made microspheres less than 0.3 $\mu$m in diameter and loaded with magnetite by emulsion polymerization of radioactive monomer in the presence of ultrafine magnetic particles. Dactinomycin was adsorbed to the surface of the spheres, and they were administered intravenously into mice. After 10 min the animals were sacrificed, and various tissue samples were homogenized, and their radioactivities was determined. They found that the placement of an external magnet over the kidney increased concentration of the drug in the kidney threefold and decreased liver concentrations of the drug by approximately 30%, showing the ability of the magnetite-loaded microspheres to be localized to a specific organ. Mortality of the animals was not increased with the addition of magnetite to the formulation, showing the magnetic microspheres to be as safe as the polymer itself.

Magnetic localization has also been applied to the area of internal radiotherapy. Häfeli et al. reported the use of poly(lactic acid) (PLA) microspheres with the ionic form of Y-90 loaded into the matrix (135). Magnetite (Fe$_3$O$_4$) was also incorporated to make the microspheres magnetically responsive. These microspheres could then be locally infused into a tumor and held there by the magnet or infused IV or into a body cavity and then attracted to a magnet held near the tumor. When the microspheres were injected into the peritoneal cavity with a magnet placed directly over a subcutaneous tumor, $73 \pm 32\%$ of the injected dose of radioactivity was found in the tumor 24 h after the injection of the microspheres. Without the magnet, radioactivity in the tumor was $6 \pm 4\%$.

The application of magnetically localized cancer therapies in humans has been relatively slow to gain universal popularity because of the need for magnets powerful enough to localize magnetic particles in humans and the controversy surrounding the biologic effects of high-intensity magnetic fields. As these problems are solved, magnetic localization could provide a simpler alternative to the more biological targeting approaches, such as antibody localization, discussed in previous sections. Targeted therapies in general are still actively being investigated, despite all of the problems still left to overcome. The desire for a cancer therapy that does not harm noncancerous tissues is very strong.

## CANCER IMMUNOTHERAPY

If the immune system functioned perfectly, neoplastic cells would be destroyed before they had the chance to proliferate and form tumors. However, some cancer cells do manage to slip through the surveillance of the immune system and proliferate uncontrolled. The field of cancer immunotherapy is based on investigating means of enhancing the ability of the immune systems to recognize and kill tumor cells, thereby achieving a cure. It uses the body's own natural immune mechanisms, enhanced by externally administered signals. The immune cells of the body generally rely on soluble signals, such as cytokines and growth factors, to stimulate and direct their activities against abnormal cells. If these signals could be applied artificially in the correct spatial and temporal fashion through controlled release systems, the functioning of the immune system could be enhanced without the use of artificial drug molecules. In a sense, if this sort of therapy were applied prophylactically, it would serve as a cancer vaccine, protecting the patient from ever developing a tumor. Thus, cancer immunotherapy and vaccines are often discussed interchangeably in the literature. The formulation of the therapy would be the same; it is just the time of administration that differs. Immunotherapy is given after cancer has been diagnosed, and vaccines are given before the patient has shown any signs of a tumor.

Various approaches to cancer immunotherapy have been studied. Much of the current research focuses on gene therapy where genes for various cytokines are added to the DNA of tumor cells removed from a patients tumor; then the cells are irradiated such that the cells cannot replicate and are reinfused into the patient (136,137). Other researchers are directly performing the gene transfer in vivo, which is often a less efficient procedure (138). The premise is that the immune system will see the cancer cell antigens and the immunostimulatory cytokines simultaneously and will be best activated to kill not only these reinfused, transfected cells but also any nontransfected tumor cells that remain in the patient. The downfall of this type of therapy is that it takes a certain amount of time to complete the transfection if done ex vivo, and transfection rates are low when performed in vivo. So other researchers have been developing other means of delivering the cytokines, including liposome and biodegradable and/or bioerodible microspheres.

By far, the most research has been on liposomal formulations of cytokines (139–145). The most common cytokine explored for use in tumor immunotherapy is interleukin 2 (IL-2). IL-2 controls the proliferation and cytotoxicity of natural killer (NK) cells and T lymphocytes and the proliferation of B lymphocytes (146). T lymphocytes play a role in both cellular and humoral immune responses. The cellular immune response acts against cells that appear abnormal in some way (e.g., expressing foreign antigens on their surface), and the humoral response involves antibodies that act by binding to foreign epitopes of proteins, marking them for engulfment and degradation by phagocytic cells. By enhancing the capabilities of both types of immune response, IL-2 is a good candidate for immunotherapy. However, IL-2 has a great number of side effects, including fever, malaise, fluid retention, and hypotension. The dose-limiting toxicity in humans results from vascular leak syndrome and cellular infiltration into the alveoli, which causes pulmonary edema (146). By altering the route and formulation of IL-2, therapies could be made more effective and less toxic.

Liposome-encapsulated IL-2 was shown by Anderson et al. to have improved biodistribution and depot characteristics more than free IL-2 (143). IL-2 was encapsulated in dimyristoylphosphatidylcholine (DMPC) liposomes and administered via various routes: intrathoracic, intraperitoneal (IP), IV, and subcutaneously (SC). Free IL-2 was used as a control. Two hours after IV injection, only about 7% of the free IL-2 dose was still circulating in the blood, 90% having been cleared from the animal. Approximately 37% of the liposomal IL-2 dose was still circulating in the blood after 2 hours, with >70% of the original dose remaining in the animal. The other routes of administration for the liposomal formulation showed similar results. Anderson and colleagues have studied the encapsulation of other cytokines, including IL-1, IL-6, granulocyte and macrophage colony stimulating factor (GM-CSF), and interferon γ (142). More recently, Khanna et al., in conjunction with Anderson, have been conducting research on inhalational therapy using liposomal IL-2 to treat lung cancer (140,145). Initial studies showed that the formulations were relatively stable through the nebulization process and that the liposome aerosols could be distributed evenly through the lungs of anesthetized dogs (145). To test the in vivo efficacy of the formulation against lung cancer, they treated nine dogs with primary or secondary lung tumors with the liposomal IL-2. Two of the dogs had primary lung

carcinoma, and the others suffered from lung metastases of various other forms of cancer (140). Three of the animals showed a complete response (regression), and one animal showed stable disease out to 8 months posttreatment.

Liposomal formulations have also been shown to improve the cytotoxicity of IL-1α and tumor necrosis factor α (TNF-α) (139). When these two cytokines were coencapsulated in phosphatidylcholine/phosphatidylserine (molar ratio 7:3) liposomes and administered IV to mice bearing B16F10 melanoma, it was shown that this combination was very effective at increasing survival of the animals. The combination was more effective than the individually encapsulated cytokines. The combination of free drugs at the same dose as that administered in the liposomes induced severe side effects, and therapy was discontinued. The liposomal formulation reduced side effects, thereby increasing the level of drug that can be safely administered for therapy.

Various types of polymeric microspheres have also been developed for the controlled release of cytokines (147–149). Golumbek et al. encapsulated GM-CSF within chondroitan sulfate microspheres and injected them subcutaneously in combination with irradiated tumor cells (149). The tumor cells used were B16/F10 murine melanoma cells, which have been shown to be poorly immunogenic. By themselves, the irradiated cells have not been sufficient to protect mice against subsequent challenges with live tumor cells. However, if these cells were transfected with the gene for GM-CSF prior to use as a vaccine, they were able generate potent systemic immunity against challenge. The simple combination of the GM-CSF-loaded microspheres and the irradiated tumor cells was examined as a replacement for the more complex gene transfer protocol. With no prior vaccination the mice would all die less than 40 days after challenge with live tumor cells. When the vaccine that included irradiated tumor cells plus the GM-CSF microspheres was administered before challenge, 70–80% of the mice were tumor free as long as 90 or 100 days postchallenge. Only 20–40% of the mice vaccinated with the irradiated tumor cells and blank microspheres prior to challenge with live cells were tumor free after 100 days only 20% of those vaccinated with irradiated tumor cells and free GM-CSF were tumor free 90 days post-challenge. The GM-CSF only seemed effective if encapsulated and administered in conjunction with the irradiated tumor cells.

Copolymers of lactic acid and glycolic acid (PLGA) were also investigated as controlled release systems for IL-1α for tumor immunotherapy (148). PLGA 75:25, with a molecular weight of 10,000 g/mol and loaded with IL-1α, was shown to have the most desirable release kinetics in vitro and was then tested in an in vivo tumor model in mice. Mice were inoculated with tumor cells in a foot pad and treated in the same footpad with an injection of the IL-1α microspheres 7 or 12 days after tumor cell inoculation. For controls, blank microspheres or the equivalent dose of soluble IL-1α were given by the same schedule. Untreated animals died after approximately 1 month, as did those treated with the empty microspheres or the soluble cytokine. Only the microsphere-encapsulated IL-1α produced a significant increase in survival, with 60% of the animals still alive after 43 days. Again, the sustained release of the cytokine was necessary for efficacy of the therapy.

Recently, Egilmez et al. have reported the use of PLA microspheres loaded with IL-2 for cytokine immunotherapy (150). They showed that coinjection of IL-2–loaded PLA microspheres with human squamous cell lung carcinoma cells (2E9) into a subcutaneous site resulted in complete suppression of tumor engraftment in 80% of the SCID mice treated. BSA-loaded microspheres and free IL-2 were not effective when coinjected with the tumor cells. The sustained release of the IL-2 from the PLA microspheres was the key to tumor rejection.

## SUMMARY

A large number of different therapies utilizing controlled-delivery technologies have been investigated over the past 60 years. Some of them have made it through the FDA approval process (implantable pumps for local infusion of drugs, Lupron Depot® and Zoladex® for the systemic delivery of peptide drugs, Gliadel® for the local treatment of malignant GBM, and liposomal formulations of doxorubicin and daunorubicin), but most are still at the clinical trial or research level. It is the hope of patients around the world that more of these therapies, including those that target the tumor cells selectively, will make it to the market in the coming years. Many of the therapies discussed in this article have the potential for improving cancer therapy and maybe even for curing additional types of cancers. But until cancer is completely eradicated, research into controlled-delivery applications to cancer therapy will continue.

## BIBLIOGRAPHY

1. C. Fordy et al., *Br. J. Cancer* **72**, 1023–1025 (1995).
2. H. Buchwald and T.D. Rohde, *ASAIO Trans. J.* **38**, 772–778 (1992).
3. J.W. Kwan, *Am. J. Hosp. Pharm.* **46**, 320–335 (1989).
4. Y.Z. Patt et al., *J. Clin. Oncol.* **15**, 1432–1438 (1997).
5. N. Kemeny et al., *J. Clin. Oncol.* **12**, 2288–2295 (1994).
6. B.S. Davidson et al., *Am. J. Surg.* **172**, 244–247 (1996).
7. S.R. Baker, R.H. Wheeler, W.D. Ensminger, and J.E. Niederhuber, *Head Neck Surg.* **4**, 118–124 (1981).
8. S.R. Baker, A.A. Forastiere, R. Wheeler, and B. Medvec, *Arch. Otolaryngol. Head Neck Surg.* **113**, 1183–1190 (1987).
9. S.J. Hassenbusch et al., *J. Neurosurg.* **73**, 405–409 (1990).
10. A. Hoekstra, *Int. J. Artif. Organs* **17**, 151–154 (1994).
11. B. Damascelli et al., *Cancer (Philadelphia)* **66**, 237–241 (1990).
12. C. Breedis and G. Young, *Am. J. Pathol.* **30**, 969–985 (1954).
13. R.R. Graham and D. Cannell, *Br. J. Surg.* **20**, 566–579 (1933).
14. S. Bengmark and B. Jeppsson, *Surg. Clin. North Am.* **69**, 411–418 (1989).
15. S. Bengmark and B. Jeppsson, *Surg. Annu.* **20**, 159–177 (1988).
16. B. Ahren and S. Bengmark, *HPB Surg.* **1**, 3–14 (1988).
17. E.C.S. Lai et al., *World J. Surg.* **10**, 501–509 (1986).
18. R.E. Koehler, M. Korobkin, and F. Lewis, *Radiology* **117**, 49–54 (1975).
19. R. Yamada et al., *Radiology* **148**, 397–401 (1983).
20. H. Stridbeck, L.E. Lorelius, and S.R. Reuter, *Invest. Radiol.* **19**, 179–183 (1984).
21. L.-Q. Wang, B.G. Persson, and S. Bengmark, *HPB Surg.* **6**, 105–113 (1992).
22. B. Jeppsson et al., *Eur. Surg. Res.* **11**, 243–253 (1979).
23. B.G. Persson et al., *World J. Surg.* **11**, 672–677 (1987).
24. N. Wilmott, *Cancer Treat. Rev.* **14**, 143–156 (1987).
25. L.H. Blumgart and D.J. Allison, *World J. Surg.* **6**, 32–45 (1982).
26. S. Wallace et al., *Curr. Probl. Cancer* **8**, 1–62 (1984).
27. S. Wallace et al., *Curr. Probl. Cancer* **8**, 1–76 (1984).
28. T. Laakso, P. Edman, and U. Brunk, *J. Pharm. Sci.* **77**, 138–144 (1988).
29. J.O. Forsberg, *Acta Chir. Scand.* **144**, 275–281 (1978).
30. D. Civalleri et al., *Br. J. Surg.* **81**, 1338–1341 (1994).
31. W.D. Ensminger, J.W. Gyves, P. Stetson, and S. Walker-Andrews, *Cancer Res.* **45**, 4464–4467 (1985).
32. J.W. Gyves et al., *Clin. Pharmacol. Ther.* **34**, 259–265 (1983).
33. C.E. Pfeifle, S.B. Howell, and J.J. Bookstein, *Cancer Drug Delivery* **2**, 305–311 (1985).
34. M. Andersson et al., *Acta Oncol.* **28**, 219–222 (1989).
35. T.M. Hunt et al., *Br. J. Surg.* **77**, 779–782 (1990).
36. K.F. Aronsen et al., *Eur. Surg. Res.* **11**, 99–106 (1979).
37. S. Dakhil et al., *Cancer (Philadelphia)* **50**, 631–635 (1982).
38. E.R. Sigurdson, J.A. Ridge, and J.M. Daly, *Arch. Surg. (Chicago)* **121**, 1277–1281 (1986).
39. B. Lindell, K.-F. Aronsen, B. Nosslin, and U. Rothman, *Ann. Surg.* **187**, 95–99 (1978).
40. J.A. Goldberg et al., *Nucl. Med. Commun.* **8**, 1025–1032 (1987).
41. J.A. Goldberg et al., *Br. J. Surg.* **77**, 1238–1240 (1990).
42. M.A. Burton et al., *Cancer Res.* **45**, 5390–5393 (1985).
43. L. Hafstrom, A. Nobin, B. Persson, and K. Sundqvist, *Cancer Res.* **40**, 481–485 (1980).
44. Y. Sasaki et al., 1985, *Cancer* **53**, 311–316 (1985).
45. K.A. Kennedy, S. Rockwell, and A.C. Sartorelli, *Cancer Res.* **40**, 2356–2360 (1980).
46. C.A. Pritsos and A.C. Sartorelli, *Cancer Res.* **46**, 3528–3532 (1986).
47. J.P. Codde et al., *Anticancer Res.* **13**, 539–544 (1993).
48. T. Kato et al., *Cancer (Philadelphia)* **48**, 674–680 (1981).
49. T. Kato et al., *JAMA, J. Am. Med. Assoc.* **245**, 1123–1127 (1981).
50. J.R. Daniels, M. Sternlicht, and A. Daniels, *Cancer Res.* **48**, 2446–2450 (1988).
51. G. Spenlehauer, M. Veillard, and J.-P. Benoit, *J. Pharm. Sci.* **75**, 750–755 (1986).
52. R. Verrijk, I.J.H. Smolders, J.G. McVie, and A.C. Begg, *Cancer Chemother. Pharmacol.* **29**, 117–121 (1991).
53. R. Verrijk, I.J.H. Smolders, N. Bosnie, and A.C. Begg, *Cancer Res.* **52**, 6653–6656 (1992).
54. Y. Morimoto and S. Fujimoto, *Crit. Rev. Ther. Drug Carrier Syst.* **2**, 19–63 (1985).
55. T. Kato, R. Nemoto, and T. Nishimoto, *Tohoku J. Exp. Med.* **127**, 99–100 (1979).
56. T. Kato, R. Nemoto, H. Mori, and I. Kumagai, *Cancer (Philadelphia)* **46**, 14–21 (1980).

57. R. Nemoto et al., *Urology* **17**, 315–319 (1981).

58. S. Fujimoto et al., *Cancer (Philadelphia)* **55**, 522–526 (1985).

59. N. Willmott et al., *J. Pharm. Pharmacol.* **41**, 433–438 (1989).

60. J.A. Goldberg et al., *Nucl. Med. Commun.* **12**, 57–63 (1991).

61. D.J. Kerr et al., *Cancer (Philadelphia)* **62**, 878–883 (1988).

62. S. Fujimoto et al., *Cancer (Philadelphia)* **56**, 2404–2410 (1985).

63. H.F.M. Cremers et al., *J. Controlled Release* **11**, 167–179 (1990).

64. Y. Nishioka et al., *Biol. Chem. Bull.* **16**, 1136–1139 (1993).

65. G. Spenlehauer et al., *J. Controlled Release* **7**, 217–229 (1988).

66. E.D. Grady, W.T. Sale, and L.C. Rollins, *Ann. Surg.* **157**, 97–114 (1963).

67. T.R. Nolan and E.D. Grady, *Am. Surg.* **35**, 181–188 (1969).

68. Y.S. Kim, J.W. LaFave, and L.D. MacLean, *Surgery* **52**, 220–231 (1962).

69. I.M. Ariel, *Ann. Surg.* **162**, 267–278 (1964).

70. I.M. Ariel and G. Padula, *J. Surg. Oncol.* **10**, 327–336 (1978).

71. I.M. Ariel and G. Padula, *J. Surg. Oncol.* **20**, 151–156 (1982).

72. M.A. Burton et al., *Eur. J. Cancer Clin. Oncol.* **25**, 1487–1491 (1989).

73. B.N. Gray et al., *Inter. J. Radiat. Oncol. Biol. Phys.* **18**, 619–623 (1990).

74. B.N. Gray et al., *Austr. N. Z. J. Surg.* **62**, 105–110 (1992).

75. S.Y. Lin et al., *Biomater. Artif. Cells. Artif. Organs* **17**, 189–203 (1989).

76. J.H. Campbell, L. Edsberg, and A.E. Meyer, *J. Oral Maxillofacial Surg.* **52**, 49–51 (1994).

77. A. Shikani, D.W. Eisele, and A. Domb, *Arch. Otolaryngol. Head Neck Surg.* **120**, 1242–1247 (1994).

78. A.C. Begg, M.J.M. Deurloo, W. Kop, and H. Bartelink, *Radiother. Oncol.* **31**, 129–137 (1994).

79. S.K. Dordunoo et al., *J. Controlled Release* **44**, 87–94 (1997).

80. P.I. Kornblith and M. Walker, *J. Neurosurg.* **68**, 1–17 (1988).

81. T.L. Loo, R.L. Dion, R.L. Dixon, and D.P. Rall, *J. Pharm. Sci.* **55**, 492–497 (1966).

82. H. Brem et al., *J. Neurosurg.* **74**, 441–446 (1991).

83. M.P. Wu, J.A. Tamada, H. Brem, and R. Langer, *J. Biomed. Mater. Res.* **28**, 387–395 (1994).

84. K.D. Judy et al., *J. Neurosurg.* **82**, 481–486 (1995).

85. J.D. Weingert, E.P. Sipos, and H. Brem, *J. Neurosurg.* **82**, 635–640 (1995).

86. M. Boisdron-Celle, P. Menei, and J.P. Benoit, *J. Pharm. Pharmacol.* **47**, 108–114 (1995).

87. P. Menei et al., *Biomaterials* **14**, 470–478 (1993).

88. P. Menei et al., *Neurosurgery* **39**, 117–124 (1996).

89. E.C. Shors et al., *Cancer Res.* **40**, 2288–2294 (1980).

90. C.W. Stackpole, E.F. Valle, and A.L. Alterman, *Cancer Res.* **51**, 2444–2450 (1991).

91. M. Bolla et al., *N. Engl. J. Med.* **337**, 295–300 (1997).

92. R. Arky, *Physicians' Desk Reference*, 52nd ed., Medical Economics Company, Montvale, N.J., 1998.

93. P.K. Gupta, *J. Pharm. Sci.* **79**, 949–962 (1990).

94. A.E. Frankel, *Oncology (Huntington)* **7**, 69–78 (1993).

95. M.-A. Ghetie and E.S. Vitetta, *Curr. Opin. Immunol.* **6**, 707–714 (1994).

96. W.A. Hall, *Neurosurg. Clin. North Am.* **7**, 537–545 (1996).

97. C.B. Siegall, *Cancer (Philadelphia)* **74**, 1006–1012 (1994).

98. J.G. Jurcic and D.A. Scheinberg, *Curr. Opin. Immunol.* **6**, 715–721 (1994).

99. J.K. Sherwood, R.B. Dause, and W.M. Saltzman, *Bio/Technology* **10**, 1446–1449 (1992).

100. P.S. Gill et al., *J. Clin. Oncol.* **14**, 2353–2364 (1996).

101. A. Sharma et al., *Int. J. Cancer* **71**, 103–107 (1997).

102. L.D. Mayer et al., *Br. J. Cancer* **71**, 482–488 (1995).

103. A. Rahman et al., *J. Natl. Cancer Inst.* **84**, 1909–1915 (1992).

104. S. Ohta et al., *Anticancer Res.* **13**, 331–336 (1993).

105. T.M. Allen et al., *Biochem. Soc. Trans.* **23**, 1073–1079 (1995).

106. I. Ahmad, M. Longenecker, J. Samuel, and T.M. Allen, *Cancer Res.* **53**, 1484–1488 (1993).

107. T.M. Allen et al., *Int. J. Cancer* **62**, 199–204 (1995).

108. N. Emanuel et al., *Pharm. Res.* **13**, 861–868 (1996).

109. D.D. Lasic et al., *Biochim. Biophys. Acta* **1070**, 187–192 (1991).

110. J.A. Harding et al., *Biochim. Biophys. Acta* **1327**, 282–292 (1997).

111. S.A. Longman et al., *Cancer Chemother. Pharmacol.* **36**, 91–101 (1995).

112. R.J. Lee and P.S. Low, *Biochim. Biophys. Acta* **1233**, 134–144 (1995).

113. D.D. Lasic, *Am. Sci.* **80**, 20–31 (1992).

114. D.D. Lasic, in S. Benita, ed., *Microencapsulation: Methods and Industrial Applications*, Dekker, New York, 1996, pp. 297–328.

115. D.W. Northfelt et al., *J. Clin. Oncol.* **15**, 653–659 (1997).

116. A. Gabizon et al., *Acta Oncol.* **33**, 779–786 (1994).

117. D.S. Fischer, M.T. Knobf, and H.J. Durivage, eds., *The Cancer Chemotherapy Handbook*, 5th ed., Mosby, St. Louis, Mo., 1997.

118. J.-C. Leroux, E. Doelker, and R. Gurney, in S. Benita, ed., *Microencapsulation: Methods and Industrial Applications*, Dekker, New York, 1996, pp. 535–575.

119. S.J. Douglas, S.S. Davis, and L. Illum, *Crit. Rev. Ther. Drug Carrier Syst.* **3**, 233–261 (1987).

120. V. Lenaerts et al., *J. Pharm. Sci.* **73**, 980–982 (1984).

121. L. Illum, P.D. Jones, R.W. Baldwin, and S.S. Davis, *J. Pharmacol. Exp. Ther.* **230**, 733–736 (1984).

122. M.T. Peracchia, C. Vauthier, F. Puisieux, and P. Couvreur, *J. Biomed. Mater. Res.* **34**, 317–326 (1997).

123. S.E. Dunn et al., *Pharm. Res.* **11**, 1016–1022 (1994).

124. L. Illum et al., *Life Sci.* **40**, 367–374 (1987).

125. R. Reszka et al., *J. Pharmacol. Exp. Ther.* **280**, 232–237 (1997).

126. J.A. Mosso and R.W. Rand, *Ann. Surg.* **178**, 663–668 (1973).

127. K. Mosbach and U. Schröder, *FEBS Lett.* **102**, 112–116 (1979).

128. K.J. Widder, A.E. Senyei, and D.G. Scarpelli, *Proc. Soc. Exp. Biol. Med.* **58**, 141–146 (1978).

129. K. Sugibayashi, M. Okamura, and Y. Morimoto, *Biomaterials* **3**, 181–186 (1982).

130. A.E. Senyei, S.D. Reich, C. Gonczy, and K.J. Widder, *J. Pharm. Sci.* **70**, 389–391 (1981).

131. Y. Morimoto, M. Okamura, K. Sugibayashi, and Y. Kato, *J. Pharmacobiodyn.* **4**, 624–631 (1981).

132. K.J. Widder et al., *Eur. J. Cancer Clin. Oncol.* **19**, 135–139 (1983).

133. Y. Morimoto, K. Sugibayashi, M. Okamura, and Y. Kato, *J. Pharmacobiodyn.* **3**, 264–267 (1980).

134. A. Ibrahim, P. Couvreur, M. Roland, and P. Speiser, *J. Pharm. Pharmacol.* **35**, 59–61 (1983).

135. U.O. Häfeli et al., *Nucl. Med. Biol.* **22**, 147–155 (1995).

136. G. Dranoff et al., *Proc. Natl. Acad. Sci. U.S.A.* **90**, 3539–3543 (1993).

137. J.W. Simons et al., *Cancer Res.* **57**, 1537–1546 (1997).

138. W.H. Sun et al., *Proc. Natl. Acad. Sci. U.S.A.* **92**, 2889–2893 (1995).

139. M. Saito, D. Fan, and L.B. Lachman, *Clin. Exp. Metastases* **13**, 249–259 (1995).

140. C. Khanna et al., *Cancer (Philadelphia)* **79**, 1409–1421 (1997).

141. S.M. Sugarman and R. Perez-Soler, *Crit. Rev. Oncol. /Hematol.* **12**, 231–242 (1992).

142. P.M. Anderson et al., *Cytokine* **6**, 92–101 (1994).

143. P.M. Anderson et al., *J. Immunother.* **12**, 19–31 (1992).

144. E. Kedar et al., *J. Immunother.* **16**, 47–59 (1994).

145. C. Khanna et al., *J. Pharm. Pharmacol.* **49**, 960–971 (1997).

146. P.M. Anderson and M.A. Sorenson, *Clin. Pharmacokinet.* **27**, 19–31 (1994).

147. L.-S. Liu et al., *J. Controlled Release* **43**, 65–74 (1997).

148. L. Chen, R.N. Apte, and S. Cohen, *J. Controlled Release* **43**, 261–272 (1997).

149. P.T. Golumbek et al., *Cancer Res.* **53**, 5841–5844 (1993).

150. N.K. Egilmez et al., *Cancer Immunol. Immunother.* **46**, 21–24 (1998).

See also CANCER, DRUG DELIVERY TO TREAT—PRODRUGS; CENTRAL NERVOUS SYSTEM, DRUG DELIVERY TO TREAT.

# CANCER, DRUG DELIVERY TO TREAT—PRODRUGS

ROBERT G.L. SHORR
United Therapeutics
Washington, D.C.

## KEY WORDS

Anticancer drugs

Camptothecin

Novel cancer therapeutics

Paclitaxel

Passive targeting

Polyethylene glycol

Polymer conjugates

Tumor accumulation

## OUTLINE

## INTRODUCTION

Approximately 550,000 deaths, or nearly one-quarter of disease-associated mortality, in America are cancer related. One in four Americans has been projected to experience cancer at some point during his or her life time. Cancer is the number two disease killer of Americans and is the leading cause of death from disease in children ages 1–14. Nearly 90% of cases are solid tumors with metastases distal to the primary tumor site. Cancers causing the most deaths in the United States for both sexes are lung and colorectal, followed by breast and uterine cancer in women and prostate cancer in men. The most frequently diagnosed cancer is cancer of the skin (over 800,000 cases); 40,000 of these are cases of advanced malignant melanoma. Management of metastatic disease remains one of the most challenging aspects of oncology care to date.

The principle treatment modalities for cancer are surgery, radiation therapy, and the use of chemotherapeutic agents. Of these, surgery, if performed at an early enough stage of disease progression, offers the greatest chance for cure. Radiotherapy does not discriminate between normal and healthy tissue and must rely on external methods of targeting radiation beams to the desired location. Collateral damage to healthy organs and tissues can be high and have devastating side effects. Chemotherapeutic agents, although capable of "killing" tumor cells also kill healthy cells, making mop up after radiation treatment or surgery and treatment of metastatic disease a balancing act between aggressive therapy and control of its side effects.

Whereas some progress has been made in the treatment of certain tumor types using chemotherapeutic agents, particularly liquid tumors, nearly half of all patients will fail to respond or relapse to metastatic disease. Despite the appearance of nearly 70 approved chemotherapeutic agents for the treatment of cancer, overall results have been disappointing, particularly for solid tumors (1).

## THE PROBLEM

A primary problem in the targeted treatment of cancer is that in most instances the differences between the normal cells of origin and malignant cells are subtle and difficult to quantify and translate to beneficial therapy. Shared properties between normal and tumor cells have confounded the development of tumor-specific agents.

### Properties of Tumors and Malignant Cells

Normal tissues may be considered to be those that are continuously renewing their cell populations (bone marrow, intestine), those that proliferate slowly but may regenerate in response to damage (liver, lung), or those that are relatively static (muscle and nerve).

Cancer can be defined as the emergence of cellular clusters that result from continuous production of abnormal cells that invade and destroy normal tissues.

Human tumors often first arise in renewing tissues. The tumor cells are able to divide endlessly, without differentiating into a mature state and without regard to the function of the tissue they are in. As they begin to multiply, angiogenesis factors, growth factors, and cytokines that stimulate the formation of collateral blood vessels and other "support" tissues are released. Some tumors may increase in malignancy with time and shed clonogenic cells that will give rise to metastatic nodules and eventually additional tumor masses. Nowell (2) has suggested that tumor cells are genetically unstable and that a high degree of random mutation during clonal expansion gives rise to lethal or disadvantageous mutations as well as greater autonomy and growth advantages to those cells that will go on to become tumor producers.

According to this model it is not surprising that tumors are heterogeneous in their origins and properties, extending to almost any property measured, including surface markers, karyotype, morphology, metastatic potential, and sensitivity to therapeutic agents (3). Further influencing heterogeneity, even as a tumor expands from a single cell, are differentiation-related diversity, nutritional heterogeneity, and the emergence of new subclones. Differentiation-related diversity may result in the production of different cellular products and membrane markers. Nutritional heterogeneity reflects the immature and often slower-growing vasculature associated with tumors and the appearance of nutritional and oxygen gradients as distance from the vascular tree increases. A dramatic influence on metabolic function and proliferation can be expected across cell layers. In some instances regions of tumor necrosis may emerge. The influence of tumor hypoxia associated with oxygen gradients and radiation therapy outcome is well characterized (see the section "PEG-Hemoglobin"). The development of subclones may be a key underlying mechanism for tumor progression and the emergence of resistance. Subclones with varying degrees of resistance may also contribute to a sense of "trying to hit a moving target" for the oncologist.

Indeed, it is the spread of cancerous cells and the appearance of metastases and resistance that represents a most difficult challenge to the oncologist. The routes and sites of metastases vary with different primary tumors. In general as cancers break through the surface of the organ in which they originated, cells may shed, lodge, and begin to grow on the surface of adjacent organs. Tumor cells may also migrate into the lymphatic system and be carried to the lymph nodes and blood vessels. As malignant cells migrate through the blood stream, many die. Others become trapped in vessels too small to let them pass. GI tract–originating tumor cells may be stopped in the liver or, eventually, lung. Other tumor cells originating in other parts of the body may be stopped first in the microvasculature of the lung. Smaller clusters of tumor cells, present as micrometasteses, may remain relatively dormant and escape detection.

Four stages for cancer progression have been defined with increasing severity of disease and seriousness:

*Stage 1.* Tumors are small and localized and, depending on location, may be removed surgically.

*Stages 2 and 3.* Tumors are larger and may have broken through the initial organ to attach to additional surrounding organs and tissues. Lymph node involvement is likely.

*Stage 4.* Tumors have metastasized to other parts of the body, usually with nodal involvement.

Human tumors vary widely in their properties. They may arise in different parts of the body, vary in retention of normal differentiation patterns or grade, and vary in extent of expansion into surrounding normal tissue and metastases or stage. It is likely that this variability reflects the different or individual initial genetic alterations at the root of the cancer emergence and the cells that ultimately will constitute the tumor mass. The collective biological properties of the tumor cells will determine the overall tumor phenotype. Cancer cells, even when spread throughout the body, are likely to retain at least some physical and biological characteristics of the tissue in which they originated. Tumors arising from endocrine tissue, for example may continue to produce and release hormones. The more de-differentiated a tumor, the more aggressive its growth and ability to generate metastases.

Although the cause of cancer is not completely understood, a combination of hereditary factors, virus or environmental exposures, and lifestyle has been implicated. Linkage between cigarette smoking and lung cancer, for example, is widely accepted. This year more than 140,000 people will be newly diagnosed with lung cancer. Epidemiologists predict that even with smoking cessation programs such numbers are likely to be seen for at least the next three decades.

Recently specific genetic events have been implicated in the emergence of cancer and the control of cell growth. Use of agents to specifically turn off or on these genes has become the focus of intense preclinical and clinical research. It is likely that these agents will find use in conjunction with known cytotoxic agents and newer formulations as part of a "combination" or "cocktail" approach to cancer treatment.

Three major cancer subtypes have been described:

*Sarcomas.* Arising from connective and supportive tissues (i.e., bone, cartilage, nerve, blood vessels, muscle, and fat)

*Carcinomas.* Arising from epithelial tissue (i.e., skin, body cavity and organ linings, glandular tissue, breast, and prostate); squamus tumors resemble skin; adenocarcinomas resemble glandular tissue

*Leukemias and lymphomas.* Arising from blood-cell-forming tissue and involving lymph nodes, spleen, and bone marrow, with overproduction of lymphocytes.

## THE SOLUTION: TUMOR TARGETING STRATEGIES

Given the similarities between normal and tumor tissue, the first challenge to the oncologist attempting a tumor targeting strategy is the localization of tumor sites, both primary and metastatic. In this regard magnetic resonance imaging techniques using metabolic markers, as well as the use of antibodies recognizing tumor-associated antigens and carrying radioisotopes, have been described. Once detected, direct injection of drug into a tumor (if reachable), placement of hydrogels containing drug, or infusion of an antitumor agent into the afferent blood supply feeding a tumor may be considered. This approach to tumor targeting, however, does not take into account the need to target and effectively treat metastatic and micrometastatic disease.

Beyond direct injection into a tumor or otherwise localized "physical delivering" of a therapeutic agent, the ideal tumor targeting strategy is one that allows for highly selective recognition of tumor masses as well as metastatic nodules and miocrometasteses even on systemic administration of drug. Little or no corecognition of normal or related healthy cells and tissue is especially desirable. Because the targeting event may or may not be of therapeutic benefit directly, drugs to be delivered via the targeting agent also need to be carefully considered. For example many drugs will need to be taken up into cells to be effective or further targeted to a specific organelle. For prodrugs that circulate in the blood as inactive or weakly active agents, tumor activation or potentiation of activity is essential.

## Active Targeting

Although a complete review is beyond the scope of this article, most tumor targeting efforts have been made using "active" agents with selective affinities for tumor markers. Briefly, potential sites for "active" tumor targeting may be at the cellular level and be structural membrane proteins, receptors, glycoproteins, or lipoproteins associated with malignant cells. They may be associated with supportive tissue such as the vasculature penetrating a tumor mass and stroma. Or they may be directed to necrotic regions of a tumor that expose regions of DNA and chromatin not normally exposed in healthy tissue. The uniqueness of these targets relative to tumor tissue will be a primary driver in the selectivity of any drug to be delivered via this mechanism.

Following are examples of active targeting agents:

Antibodies
Growth factor and hormone receptor ligands (including peptides and small synthetic organic molecules)
Nutrient transporters, vitamins, and virus

There are a number of anticancer agents that might be delivered:

Cytotoxic molecules (small synthetic organic molecules or peptides)
Ribosome-inhibiting proteins and toxins
Antisense nucleotides or genes
Hormones and differentiation promoters
Radioisotopes

Modifiers of resistance
Adjuvants for radiation therapy or conventional chemotherapeutic use

Kabanov and Alakhov (4) have summarized the following requirements for active "magic bullets":

A targeting moiety that is selective for tumor cells independent of their location anywhere in the body, with little or no recognition of normal tissue
An ability to recognize and bind to or be taken up by the targeted cells
An ability carry a "payload" without disintegrating or losing activity or potential activity along the way
An ability to deliver the payload while avoiding nonspecific drug or drug metabolite interactions with nontarget cells
An ability to be cleared from the body rapidly, with preferred accumulation at sites of disease only

Much research has been done concerning targeting moieties that carry a radioisotope for imaging or therapy, but in considering cytotoxic drugs and their analogues for delivery to tumors, the ideal molecule should be "inactive" except upon delivery to the targeted cell. In this regard the concept of a prodrug has emerged; target-site-specific modification can unleash the payload and destroy the tumor cell.

Antibody binding directed against specific antigens is among the most selective interactions known to science. It is not surprising then that in the search for "magic bullets" against cancer, antibodies that specifically recognize tumor-associated antigens have been sought.

The major forms of antibodies that have been used for tumor targeting are the complete intact antibody (typically IgG), or the F(ab′)2, Fab, or Fab′ fragments. Intact antibodies have been described as Y-shaped molecules composed of two identical heavy chains and two identical light chains. Antigen binding sites are at the tips of the Y. The F(ab′)2 fragment can be released from the intact antibody by proteolytic cleavage and contains both antigen-binding tips of the Y. The F(ab′)2 fragment is held together by a disulfide bridge that, on reduction, releases the Fab′ fragments with one binding site per Fab. The molecular weight of an intact IgG is 150,000; that of an F(ab′)2 and Fab are 100,000 and 50,000, respectively. Smaller fragments bind to antigens, often with both lower affinity and avidity.

Use of monoclonal antibodies for the treatment of leukemia and lymphoma has offered considerable success (5,6), but results with solid tumors have been disappointing (7,8). It has been suggested that the main reason for poor efficacy is resistance to uniform tumor penetration by antibody conjugates (7–9). In studies with radiolabeled antibodies only 0.001–0.01% of an administered antibody dose was found to be localized to each gram of solid tumors in humans, with much of it trapped in tumor perivascular space (10,11). Barriers to uniform antibody penetration have been suggested to arise from the endothelial layer and dense fibrous stroma associated with tumors, as well as the tight packing of tumor cells and the absence of lymphaticdrainage, which contribute to high interstitial pressure and opposition to influx of molecules into the tumor core (12) or nonspecific binding and entrapment (13).

One approach to increase tumor penetration has been to make antibody fragments as small as possible. Much of the rationale for preparation of F(ab′)2 and Fab fragments has come from the belief that smaller molecules enter and penetrate tumors more quickly and uniformly than intact antibodies, while being more rapidly cleared from the rest of the body. For imaging and diagnostic application, rapid tumor penetration and body clearance are desirable properties. Still, with the lower affinity and avidity associated with the smaller fragments, biodistribution studies indicate that the larger species are distributed more slowly and are retained for longer periods than the smaller fragments. The suggestion being that, for therapeutic use, higher molecular weights may be more desirable.

In searching for tumor-specific antigens it is necessary to identify targets with the following characteristics:

Ideally, expressed only by tumor cells and similarly by all cells of the tumor

Not shed into the circulation where they can "sponge up" drug directed to the tumor before it reaches the desired tumor mass

Allow for the drug conjugate to be taken up into the cell if required for activity, or to remain associated with the tumor cell surface long enough to be active

In the past much of the screening for tumor specific antigens and antibodies was based upon the immunization of mice and the screening for monoclonals in cultured pools of antibody producing cells. The development of display technology has now allowed for the creation of antigen-binding molecules in vitro (14,15). It is now no longer necessary to immunize animals to raise antibodies. Further, combinatorial approaches coupled with high-throughput screening capabilities can allow for antibody engineering to produce desired affinities, multivalencies, single-chain antibody fragments, and recombinant fusion molecules (16,17). Such engineered proteins can also be humanized by replacing murine framework regions with human framework regions or made entirely human, which eliminates the rejection of murine antibodies characterized as HAMA (human anti–mouse antibody)—a problem associated with early monoclonal antibody trials and products. Three advances have enabled the construction of antibodies and antigen-driven selection in vitro. These are (1) the ability to express the antigen-binding domains of the heavy and light chains of antibodies in *Escherichia coli*, either as Fab fragments or as single-chain antigen-binding molecules, (2) the large and diverse libraries of Fab and single-chain antigen-binding genes that can be generated by the polymerase chain reaction using heavy- and light-chain genes obtained before or after immunization and harvest of B lymphocytes, and (3) the ability to express the antibody fragments on the surface of bacteriophage to facilitate screening and selection.

Numerous tumor antigens have been identified and antibodies or their fragments characterized in animal or human clinical studies. A complete review of this effort is beyond the scope of this work. In general, although a particular marker may be overexpressed by some cells of a tumor, there is often a population that expresses substantially less, or normal healthy cells and tissue that express similar antigens. Thus tumor antigens have appeared to be more tumor selective than tumor specific. In the clinic, or in animal experiments where the tumor antigens are not uniformly available, initial treatment may appear quite promising only to result in the emergence of resistant clones that do not express the targeted antigen. Where the antibody is conjugated to a radioisotope for therapy, this problem is somewhat lessened due to the ability of radiation to penetrate more than one cell layer.

Immunotoxin constructs have been prepared using ribosome-inhibiting proteins and their subunits as chemical conjugates to antibodies or as fusion proteins produced by recombinant means. Ribosome-inhibiting proteins are extremely potent and function by inhibiting protein synthesis. To be effective they must be taken up into cells. Depending on the nature of the antibody–toxin conjugate, processing of the complex may or may not be necessary to release active inhibitor. Binding to normal healthy tissue can contribute to toxicity and side effects.

Despite the fact that a truly specific tumor antigen has not emerged, antibodies continue to be explored for diagnostic and therapeutic uses in cancer in a variety of formats. One interesting approach worthy of mention is the targeting of tumor vasculature with antibody drug and prodrug conjugates so as to cut off the "feeding" of a tumor mass. Attacking the endothelial cells instead of the tumor itself was first proposed by Denekamp (18). Killing of vascular endothelial cells within the tumor was reasoned to be likely to denude the endothelial lining and lead to platelet adhesion, activation of coagulation factors, and sealing off of the vasculature tree via occlusive thrombus formation (19). As each microvessel becomes blocked, tumor cells dependent on that vessel for nutrition and oxygen become unable to survive. It was reasoned that targeting and destruction of only a few relatively accessible endothelial cells should trigger a chain reaction.

Thorpe and Derbyshire (20) have prepared an IgM murine monoclonal antibody, TEC-11, by immunizing BALB/c mice with human umbilical vein endothelial cells in order to test this hypothesis. Hybridoma supernatants were prepared and screened for reactivity against proliferating human umbilical vein endothelial cells, lack of reactivity against quiescent frozen sections, and reactivity against sections of human malignant but not normal tissue. Characterization of TEC-11 indicated that the antibody recognized human endoglin, an essential component of the TGF-$\beta$ receptor on human endothelial tissue. Endoglin is a dimeric glycoprotein with two 95,000 MW subunits linked by a disulfide bridge (21) that binds TGF-$\beta$1 and TGF-$\beta$3 with high affinity. Endoglin is expressed on human endothelial cells, fetal syncytiotrophoblast, some macrophage, immature erythroid cells, and dermal endothelium under some conditions of inflammation (22).

Immunotoxins prepared by linking the A-chain of the ribosomal inhibiting protein ricin were tested as conjugates with tumor vasculature targeting antibodies in mice bearing solid tumors of greater than 1 cm diameter. In these studies, dose-dependent antitumor effects were observed. In following the time course of antitumor activity, the first loss of endothelial cells was observed 2 h postinjection of the antibody–toxin conjugate. By 6 h many blood vessels already showed signs of occlusion, which extended to nearly all vessels by 24 h, followed by physical collapse of the tumor mass. Other tumor-vasculature-selective antibodies have been identified that recognize additional endothelial cell antigens, basement membrane, or tumor stroma and are being explored as immunotoxin conjugates or as conjugates with other cytotoxic agents (23–28).

Although targeting macromolecular drug carrier conjugates to tumor vasculature using antibodies has attracted considerable attention, attacking tumor vasculature formation by blocking angiogenesis is also being explored. Naked antibodies as well as inhibitory molecules and adhesion molecule regulators have shown provocative results in experimental models (29). Targeting stroma using antibodies carrying radioisotope that bind to dead or dying tumor cells at a tumor core have entered phase I clinical trials for the treatment of malignant melanoma. In addition to radioisotope it is believed that such antibodies can be used to target other drugs. This approach has been described as attacking tumors from the inside out (30,31).

### Passive Targeting

In contrast to active tumor targeting, observations of hyperpermeability of vasculature and immaturity of the lymph drainage systems of tumor tissue have led to proposals of passive tumor targeting (32–39). It is the nature of the tumor vasculature that is believed to enable such an approach.

**Tumor Microvasculature.** Tumor microvasculature comprises both vessels from the preexisting organ or tissue in which the tumor has arisen, and new vessels stimulated to grow by the release of angiogenic factors (40,41). Distribution of nutrients as well as drugs through the tumor mass will be influenced by the diameter, length, number, and geometry of the available blood vessels, as well as blood flow (42–45). Jain and others (43–45) have suggested that the nature of the tumor vasculature will vary with location, tumor type, and growth rate. Red blood cell flow velocity may be slower in tumors than in normal tissue (46,47) because of viscous resistance in tumor vessels (48–51).

As tumors grow avascular regions of necrosis, seminecrotic and stable vascularized regions appear (52). As the tumor expands, additional cell layers are added and a metabolic gradient emerges (53). The precise nature of this gradient and its role in triggering metastases as well as chemotherapeutic responsiveness remains an area of intense investigation (54–56).

Interestingly, in addition to the differences in spatial geometry of tumor vasculature, differences over normal vasculature in both macroscopic and microscopic morphology have been reported (57). These features combine to make distribution of drugs to tumors both heterogeneous and challenging.

Generally vascular permeability and hydraulic conductivity of tumors is higher than for normal tissue (58–63). It has been suggested that molecules with positive charges have a higher permeability (64) and that tumor vascular is more porous than "leaky" (65). It has been hypothesized (66) that pore size and permeability vary not only between tumor types but within an individual tumor as well, in part reflecting the local cytokine environment (67).

In addition to the permeability differences just noted, tumors have been characterized as having high interstitial fluid pressures (32,68–79). Once extravasated molecular movement is likely to be governed by diffusion and convection, tumor pressures play an influential role in molecular entry as well as retention (80–82). Larger tumors may be expected to have reduced transvascular exchange over smaller tumors, and convective transport at a tumor's center might be less than at the tumor's periphery.

It has been suggested that the anatomical features and pressure gradients associated with tumors allow higher molecular weight molecules to accumulate in tumors with considerably longer retention times than expected for normal tissue. It has already been demonstrated that a correlation between molecular size and tumor retention exists (83,84). For "passive" tumor accumulation the "passively targeted" molecule should have the following characteristics:

Sufficient circulating lifetime so as to allow for sufficient tumor accumulation to achieve desired local concentrations of drug with little or no accumulation in normal tissue

An ability to be taken up by the targeted cells

Not lose activity or potential activity while circulating

An ability to deliver the anticancer activity while avoiding nonspecific drug or drug metabolite interactions with nontarget cells

To achieve the desired molecular weight relative to passive tumor accumulation, use of polymers as a conjugate to lower molecular weight cytotoxic agents has been proposed. These conjugates do not contain or require a tumor-directed binding moiety.

**Polymeric Drug Carriers.** There has long been the belief that the performance of many small-molecule drugs could be improved by their chemical transformation into polymer conjugates or macromolecular prodrugs. From the polymer perspective, dextran, inulin, polysaccharide B, and a variety of block copolymers have been explored (37,85–101).

Of these, perhaps the most advanced is SMANCS. SMANCS (molecular weight 16,000) is a conjugate of polystyrene-co-maleic acid half-*n*-butylate (SMA) with neocarzinostatin (NCS), developed by Maeda et al. (102,103) (Fig. 1). SMANCS has been approved in Japan for the treatment of primary hepatoma. Hepatic arterial administration has been found to be curative in a number of cases. In patients with little or mild concomitant liver cirrhosis and single-segment-confined tumor, 5-year survival rates are 90%. Overall 5-year survival for all patient

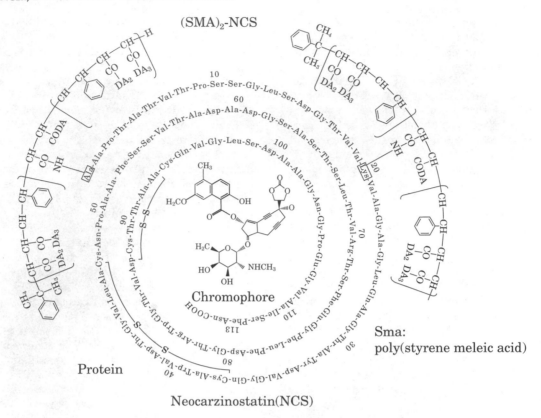

**Figure 1.** Illustration of the chemical structure of SMANCS.

types is nearly 35%. Additional clinical trials evaluating SMANCS for pleural and ascitic carcinomatosis via the intracavitary route are under way (104).

Little immunosuppression or marrow toxicity with SMANCS has been observed. The active component of SMANCS, neocarzinostatin, is an extremely potent cytotoxic agent effecting DNA cleavage and synthesis. The polymeric portion of SMANCS has been suggested by Maeda to promote tumor accumulation and prolonged tumor retention (105,106). Maeda has described the impact of tumor vasculature and flow on macromolecules as the "enhanced permeability and retention" effect. Lower molecular weight anticancer agents do not show the same degree of accumulation as smaller molecules that can more freely traverse interstitial space. Maeda has suggested that a molecular weight greater than 50,000 is required to begin seeing preferred tumor accumulation.

Other polymeric materials, liposomes, and drug carriers are being actively researched for tumor targeting. Daunorubicin when conjugated to polymers of carboxymethyl dextran, alginic acid, polyglutamate, or carboxymethyl cellulose irreversibly loses its cytotoxic properties. When linked by a hydrolyzable bond however, hydrolysis leads to recovery of the cytotoxic effects (107). Phase I clinical trials of a hydrolyzable dextran conjugate have been initiated (108). Conjugation of doxorubicin to *N*-(2-hydroxypropyl)methacrylamide (HPMA) via degradable tetrapeptide linkers has also been reported (109). Cleavage of the linker is believed to occur after cellular uptake of the drug-polymer by endocytosis. Phase I clinical trials of this conjugate have also been initiated (110).

*Polyethylene Glycol as a Drug Carrier.* The use of synthetic or natural polymers as direct chemical conjugates to provide for increased circulating lifetimes of proteins and peptides, as well as protection from different solutes, undesirable degradation, immune responses, and other interactions has been well accepted. Of the different protein polymer conjugates studied, polyethylene glycol (PEG) is the most advanced in terms of clinical usage. Already PEG-conjugated adenosine deaminase for the treatment of adenosine deaminase deficiency–mediated immunodeficiency, and PEG-asparaginase for the treatment of acute lymphoblastic leukemia (111,112) have been approved by the U.S. FDA. Other PEG-conjugated proteins such as PEG-hemoglobin (APEX, Enzon), PEG–interleukin 2 (Chiron), PEG-alpha-interferon (Schering Plough and Hoffman La Roche), and several PEG–colony stimulating factors (Amgen), and enzymes and other proteins (Knoll Pharmaceuticals) are in various stages of preclinical and clinical trial investigations. Examples of PEG activation chemistries used for conjugation to proteins are shown as Figure 2. While major pharmaceutical companies have internal PEG programs, contract CGMP PEGylation services using a variety of activation chemistries and linker groups are also available through Shearwater Polymers, Inc. (Huntsville, Alabama), and Polymasc, Inc. (London, England).

Although the use of higher molecular weight polymers to prolong the circulating life of small-molecule drugs has been proposed in the literature and several cytotoxic agent polymer conjugates have been tested in the clinic, less is known about the performance of such conjugates with PEG.

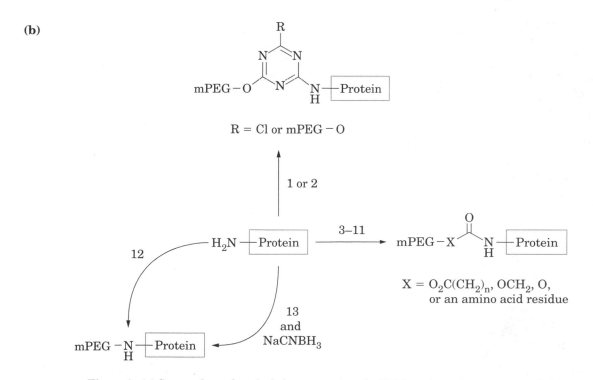

**Figure 2.** (**a**) Commonly used methods for preparation of mPEG-based modifying reagents. (**b**) Use of mPEG reagents for preparation of conjugates. mPEG represents $CH_3(OCH_2CH_2)n$. Su, succinimidyl; TCP, trichlorophenyl; pNP, *p*-nitrophenyl; AA, amino acid. *Source:* From Ref. 113.

In general, there are three ways in which PEG might be associated with a drug for conjugation. These are (*1*) nonreversible covalent attachment, (*2*) reversible covalent attachment, (*3*) association or complexation via ionic or hydrophobic interactions. Thus, in some cases the conjugated drug may be attached via a stable linker, with retention of its biological activity, or be linked via degradable bonds so that the long-circulating drug and carrier complex may serve as a depot for slow release. In other cases the polymer itself may be made to be degradable or be composed of degradable smaller fragments to enhance clearance.

The mechanism by which PEG mediates protection and prolongs circulation is not entirely understood, but it is generally agreed that much of the benefit is due to steric hindrance of recognition and uptake events. Thus, at the molecular level, interactions with degrading enzymes or receptors may be reduced, while at the cellular level, capture by the reticuloendothelial system and opsonization may be blocked (114,115).

Torchilin has suggested that an important feature of PEG contributing to the benefits observed is polymer flexibility (free rotation of individual polymer units around interunit linkages). As shown in Figure 3, each PEG chain is capable of creating a "protective" conformational cloud over the point at which it is attached (116). The number of PEGs and their position contribute greatly to the degree of protection and enhanced circulating life. Although it is possible that a covalent nonreversible conjugate might maintain activity, the likelihood of steric interactions and inhibition of small-molecule activity renders the use of reversible PEG conjugates as prodrugs the more likely strategy to succeed.

Already-long-circulating polymeric drug conjugates or macromolecular aggregates have been shown to slowly accumulate in pathological sites, such as tumors with affected vasculature. Long-circulating liposomes stabilized with PEG and containing doxorubicin have been shown to accumulate preferentially in tumors, with enhanced efficacy and decreased toxicity. Ikada et al. (117) have sought to examine the relationship between PEG molecular weight and tumor accumulation. In these studies, bishydroxyl PEGs of 31,000, 50,000, 136,000, 171,000, 215,000, and 277,000 MW were used. For quantitative studies the various PEGs were radiolabeled with $^{125}$I. For visualization of accumulation into tumors, the PEGs were bound to the dye RB19.

Mice bearing Meth A fibrosarcoma tumors were prepared by inoculation (0.05 mL) into the back subcutis or right footpad of female CDF mice (age 6 weeks) at $3 \times 10^6$ cells per mouse. When tumors had reached 5 mm average diameter (0.3 g) experimental studies were initiated.

Mice with or without tumors received i.v. injections of radiolabeled PEG; at various time intervals aliquots of blood were withdrawn via retroorbital plexus bleed, and radioactivity was measured. Figure 4 shows results obtained for the 31,000, 50,000, 215,000 and 277,000 MW PEGs. A clear correlation between blood retention time and molecular weight was observed. The authors reported no significant difference between tumor-bearing and non-tumor-bearing mice.

To measure tumor accumulation, tumor-bearing and normal mice were injected i.v. with radiolabeled PEGs. Blood samples as well as tissue samples (heart, lung, thymus, liver, spleen, kidney, GI tract, thyroid, and residual carcass) were taken at various time intervals, along with urine and feces. Blood was removed from tissues by washing twice with buffered saline prior to counting for radioactivity. For mice with tumor on their footpad, the foot with the tumor mass and the tumor-free foot were taken, and radioactivity levels were determined. Radioactivity from the normal foot was subtracted from that of the tumor-bearing foot to estimate tumor-selective accumulation.

For mice bearing back subcutis tumors, skins ($1.5 \times 1.5$ cm$^2$) with and without tumor mass were taken and counted for radioactivity. Radioactivity levels from non-tumor-bearing skins were subtracted from those of tumor-bearing skins to determine tumor-selective accumulation. Skin and thigh muscle of non-tumor-bearing mice were used as controls.

In some cases tumors of different sizes were utilized to examine the effect of tumor mass on passive accumulation.

As shown in Table 1, body distribution profiles of radiolabeled PEG with different molecular weights following i.v.

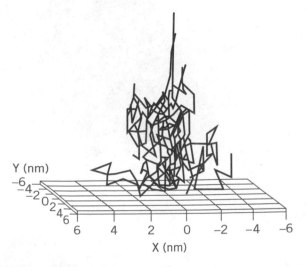

**Figure 3.** Computer simulation of a "conformational" cloud formation by a surface-attached polymer. Polymer molecule is conditionally assumed to consist of 20 segments, 1 nm each. This simulation was produced by random flight simulation. Unrestricted segment motion was assumed. Space restriction: $Z > 0$. Superposition of 11 random conformations. *Source:* From Ref. 116.

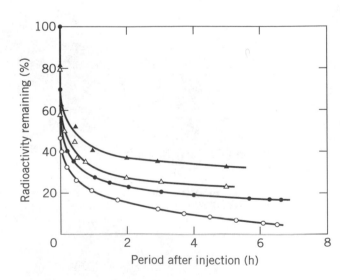

**Figure 4.** Time course of $^{125}$I-labeled PEG concentration in the blood circulation after intravenous injection into mice with a tumor mass at their footpad. Molecular weights of PEG are (○), 31,000; (●), 50,000; (△), 215,000; (▲), 277,000. *Source:* From Ref. 117.

**Table 1. Body Distribution of PEG with Different Molecular Weights 3 h after Intravenous Injection into Normal and Tumor Bearing Mice (Footpad)**

| | | Percent radioactivity | | | | | |
|---|---|---|---|---|---|---|---|
| | | Normal mouse | | | Tumor-bearing mouse[a] | | |
| Organ | MW of PEG = | 50,000 | 215,000 | 277,000 | 50,000 | 215,000 | 277,000 |
| Blood | | 35.6 ± 2.83 | 49.0 ± 1.08 | 52.3 ± 2.41 | 40.4 ± 5.10 | 46.2 ± 2.11 | 54.4 ± 0.70 |
| Heart | | 0.14 ± 0.05 | 0.03 ± 0.03 | 0.02 ± 0.02 | 0.24 ± 0.03 | 0.23 ± 0.04 | 0.24 ± 0.04 |
| Lung | | 0.15 ± 0.01 | 0.11 ± 0.10 | 0.17 ± 0.01 | 0.69 ± 0.14 | 0.62 ± 0.10 | 0.87 ± 0.43 |
| Thymus | | 0.05 ± 0.01 | 0.08 ± 0.01 | 0.08 ± 0.01 | 0.08 ± 0.01 | 0.11 ± 0.01 | 0.08 ± 0.03 |
| Liver | | 3.50 ± 0.11 | 6.49 ± 1.54 | 7.11 ± 1.46 | 3.78 ± 0.66 | 4.00 ± 0.69 | 6.54 ± 0.33 |
| Spleen | | 0.19 ± 0.03 | 0.15 ± 0.04 | 0.27 ± 0.05 | 0.15 ± 0.02 | 0.15 ± 0.03 | 0.28 ± 0.09 |
| Kidney | | 0.57 ± 0.19 | 0.64 ± 0.33 | 0.70 ± 0.00 | 1.16 ± 0.30 | 1.32 ± 0.05 | 1.23 ± 0.26 |
| GI tract | | 10.82 ± 2.48 | 15.6 ± 5.22 | 13.6 ± 2.43 | 8.29 ± 1.15 | 6.51 ± 0.55 | 8.50 ± 0.17 |
| Thyroid gland | | 0.71 ± 0.24 | 0.87 ± 0.24 | 0.23 ± 0.50 | 0.71 ± 0.20 | 0.13 ± 0.05 | 0.42 ± 0.27 |
| Carcass | | 10.8 ± 9.78 | 14.7 ± 9.87 | 8.21 ± 0.90 | 10.8 ± 1.20 | 14.5 ± 0.67 | 14.87 ± 0.36 |
| Excrement | | 40.1 ± 3.42 | 36.2 ± 1.22 | 11.21 ± 2.10 | 40.2 ± 3.12 | 28.0 ± 2.20 | 10.71 ± 0.88 |
| Tumor | | — | — | — | 0.23 ± 0.03 | 0.76 ± 0.03 | 0.53 ± 0.10 |

[a]Mice were loaded with a tumor mass at the footpad.

injection were not significantly different for tumor-bearing or tumor-free mice. Independent of PEG molecular weight, no specific affinity for any particular organ was found, with the exception that the longer the circulating life, the less accumulation in urine or feces at any given time point.

Figure 5 shows the results observed on examination of tumor accumulation as a function of PEG molecular weight. PEG was accumulated in the tumor tissue to a significantly higher extent than normal tissue, somewhat independently of molecular weight, albeit with the greatest degree of accumulation observed for the 200,000 MW PEG. Similar qualitative results were obtained for both footpad

and back subcutis models. Figure 6 shows the time course of accumulation. Clearly, up to 6 h, tumor accumulation increased, followed by a gradual decrease. As shown in Figure 5, the lower molecular weight PEG accumulated to a somewhat lesser degree than the larger polymers.

Light microscopic examination of tumors after i.v. injection of dye-labeled PEGs showed tumor accumulation within the tumor vascularized areas and surrounding tissue, although not in a homogenous manner. Differences as a result of tumor size beyond 0.3 gram were not pronounced because the larger tumors contained significant levels of necrotic tissue, which did not accumulate PEG.

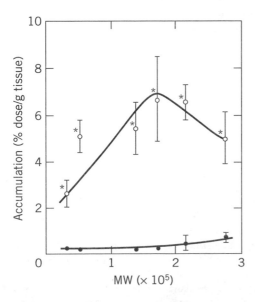

**Figure 5.** Effect of PEG molecular weight on PEG accumulation at the tumor tissue and the normal muscle after intravenous injection into mice with tumor mass at their footpad (3 h after injection): (○), tumor tissue; (●), normal muscle. $p > 0.05$, significant compared with corresponding accumulation at the normal muscle. *Source:* From Ref. 117.

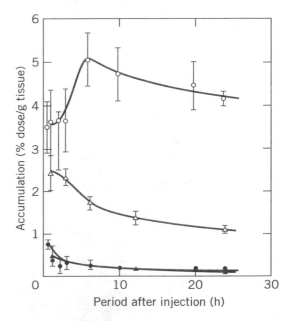

**Figure 6.** Time course of PEG accumulation at the tumor tissue (open symbols) and the normal muscle (closed symbols) after intravenous injection into mice with a tumor mass at their footpad. Molecular weights of PEG are (○,●), 215,000; (△,▲), 31,000. *Source:* From Ref. 117.

The authors also reported some quantitative differences between the levels and rates of tumor accumulation for the footpad and back subcutis models. The relevance of these differences to predicting outcomes of human disease is not known.

In general the authors concluded that smaller molecular weight PEGs accumulate more rapidly in tumors but are also retained for shorter periods. Higher molecular weight PEGs both accumulate more slowly and are retained for longer periods. A schematic representation of the author's view of the impact of these observations on the design of drug delivery systems is shown in Figure 7.

For those designing prodrugs, where release of active molecules from polymer conjugates occurs via nonselective release chemistries, consideration of the kinetics of accumulation and release are of paramount importance.

*PEG-Paclitaxel-2′-Glycinate.* Many of the cytotoxic compounds used as cancer chemotherapeutics are poorly soluble in aqueous conditions. In addition to the preferred tumor accumulation properties of PEG already noted, in-

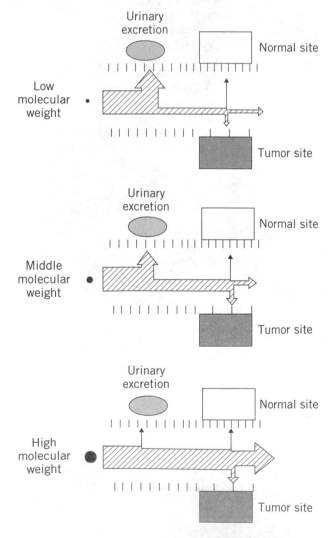

**Figure 7.** Schematic representation of models for PEG accumulation of low, middle, and higher molecular weights at tumor and normal tissues after intravenous injection. *Source*: From Ref. 117.

creased solubility of cytotoxic drug conjugates can also be achieved.

Paclitaxel isolated from the bark of the western yew tree (*Taxus brevifolia*) as well as other taxanes isolated from the needle leaves of the same tree have been developed as potent antitumor agents for the treatment of ovarian and breast cancer (118). The antitumor activity of the taxanes stems from their ability to promote tubulin polymerization in rapidly dividing cells such that stable microtubules are produced, cell division is blocked, and cell viability is compromised.

Due to limitations of solubility, taxanes are administered in vehicles containing Cremophor EL, ethanol, or detergents such as Tween 80. Corticosteroids and $H_1$ and $H_2$ histamine receptor antagonists may be coprescribed in order to limit allergic reactions to the formulation vehicle (119).

Considerable effort has focused on the development of more soluble analogues of paclitaxel and other taxanes. Non-PEG–amino acid paclitaxel prodrugs have been prepared and described in the literature (120). Although good aqueous solubility was reported (10 mg/mL), an in vitro half-life of <30 min in buffer at pH 7.4, and less than 5 min in human plasma were also observed.

Greenwald et al. (121) have sought to combine the benefits of PEG on drug solubility and long circulation life with enhanced tumor accumulation, by preparing PEG-paclitaxel analogues. In an attempt to maximize safety and efficacy, a prodrug strategy was also utilized. PEG of molecular weight 40,000 was used in these studies. A schematic of the synthesis of PEG-paclitaxel-2′-glycinate is shown in Figure 8.

For cytotoxicity assays and $IC_{50}$ determination or for syngeneic tumor burden studies a P388/0 cell line was used as described in Ref. 122. For solid tumor (human colon adenocarcinoma HT-29, human ovarian adenocarcinoma SKOV-3, and human lung A549) studies, female nu/nu mice (18–24 g) were inoculated subcutaneously at the left flank with $1 \times 10^6$ tumor cells. Animals bearing 100-$mm^3$ tumors were treated as shown in Table 3, and tumor growth inhibition was determined as in Ref. 121. Treatment and control groups were measured when control tumor median volume reached 800–1,100 $mm^3$.

In vitro cytotoxicity and hydrolysis rates for the 40,000 MW PEG conjugates made are shown in Table 2. The authors report that for the P388/0-inoculated animals, in comparison with paclitaxel controls, an increased life span of 21.8 days was observed for PEG-paclitaxel-2′glycinate, as opposed to 19 days for native paclitaxel. Further, at a total dose of 100 mg/kg many of the animals in the paclitaxel control group died due to drug toxicity, whereas in the PEG-paclitaxel-2′-glycinate group, life spans were increased by 82%, and some animals were cured of tumor burden.

Results of treatment of the different human solid tumor xenografts are shown in Table 3. For all three models the authors report an increase in tumor growth inhibition of 10%.

*PEG-Camptothecin.* Like paclitaxel, camptothecin is a natural product extracted from trees. In this case *Camptotheca acuminata*. Activity against a number of human

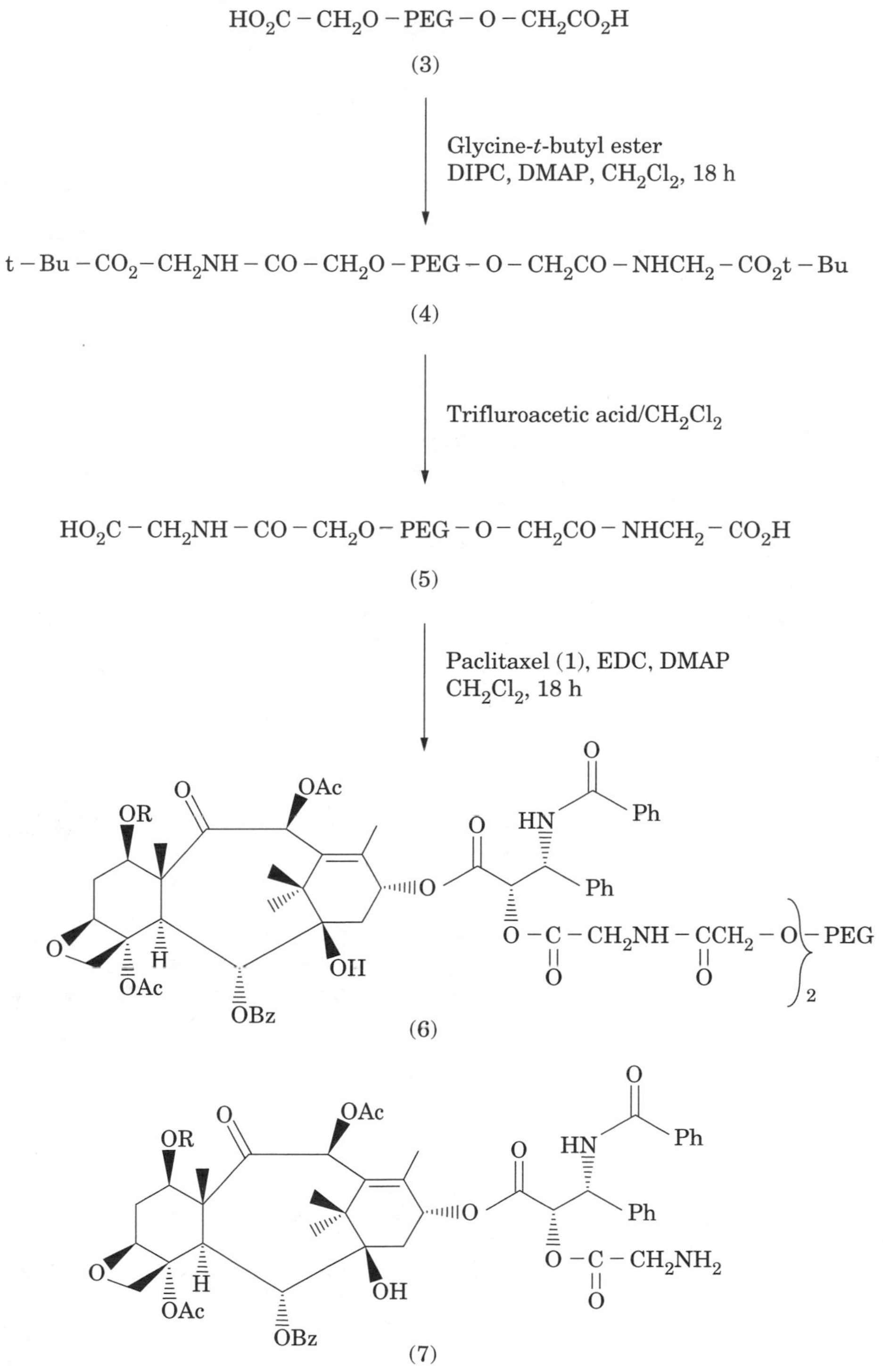

$$HO_2C-CH_2O-PEG-O-CH_2CO_2H$$

(3)

Glycine-$t$-butyl ester
DIPC, DMAP, CH$_2$Cl$_2$, 18 h

$$t-Bu-CO_2-CH_2NH-CO-CH_2O-PEG-O-CH_2CO-NHCH_2-CO_2t-Bu$$

(4)

Trifluroacetic acid/CH$_2$Cl$_2$

$$HO_2C-CH_2NH-CO-CH_2O-PEG-O-CH_2CO-NHCH_2-CO_2H$$

(5)

Paclitaxel (1), EDC, DMAP
CH$_2$Cl$_2$, 18 h

(6)

(7)

**Figure 8.** Schematic of PEG-gly-paclitaxel synthesis as described by Pendri et al. (121).

**Table 2. In Vivo Cytotoxicity and Rates of Hydrolysis for PEG-40kDa-Conjugated Paclitaxel Derivatives**

| Compound | Linker | IC50 (nM)$^a$ P388/0 | $t_{1/2}$ (h)$^a$ | |
|---|---|---|---|---|
| | | | PBS (pH 7.4) | Rat plasma |
| Paclitaxel (1) | – | 6 | – | – |
| PEG-paclitaxel (2) | -O-CO-CH$_2$-O-PEG | 10 | 5.5 | 0.4 |
| PEG-gly-paclitaxel (6) | -O-CO-CH$_2$-NHCO-CH$_2$OPEG | 14 | 7.0 | 0.4 |

*Source:* From Ref. 121.
$^a$All experiments were done in duplicate. Standard deviation of measurements: ±10%.

**Table 3. Antitumor Activity of PEG-Conjugated Gly-Paclitaxel against Subcutaneous Human Tumor Xenografts in Nude Mice**

| Tumor type | Treatment schedule[a] | Total Dose (mg/kg)[b] | % T/C[c] |
|---|---|---|---|
| HT-29 (colon) | Three cycles of daily × five at intervals of 14 days, i.p. | 225 | 11.3* |
| A549 (lung) | Daily × 5 for 2 weeks, i.p. | 150 | 9.5* |
| SKOV3 (ovarian) | 2 × week for 4 weeks, i.v. | 200 | 7.4* |

*Source:* From Ref. 121.

[a]Treatments initiated when tumor volumes reached approximately 100 mm$^3$.

[b]Based on paclitaxel content.

[c]Treatment and control groups were measured when the control group's median tumor volume reached approximately 800–1,100 mm$^3$.

*p < 0.01.

tumor xenograft models has been shown including lung, ovarian, breast, stomach, and pancreatic cancers (122). Also like paclitaxel, camptothecin is poorly soluble in water. Unlike paclitaxel, camptothecin functions as a topoisomerase inhibitor. Topoisomerase is a DNA repair enzyme necessary for cell replication and growth. When inhibited, topoisomerase-induced single-strand breaks cannot be religated, and eventual double-strand DNA breaks occur, triggering apoptosis.

Considerable efforts have focused on the preparation of more soluble analogues as well as prodrug forms of camptothecin. Several prodrug forms are inactive until processed by plasma enzymes or tumor-associated metabolism events. Camptothecin analogues topotecan and ironotecan have been FDA approved for the treatment of some solid tumors.

As described earlier for paclitaxel, Greenwald and coworkers have sought to improve camptothecin performance by conjugation to PEG (C.D. Conover et al., personal communication). Figure 9 illustrates an example of a disubstituted camptothecin conjugated to a 40,000 MW PEG prepared by the authors.

In attempting to prepare PEG conjugates with predictable rates of release, the linkers shown in Table 4 were prepared and characterized. Release kinetic "half-life" ranged from 2 to >24 h for the linkers evaluated. The authors suggest that the ability to predict release, as drug accumulates in tumors, is likely to be more effective than requiring uptake into cells prior to activation of a prodrug by a metabolic event.

Still, as shown in Table 5 in a P388/0 syngeneic-tumor-bearing mouse test, the PEG camptothecins were similar to or slightly less effective than native camptothecin. The authors report the camptothecin-di-20-O-ester of

PEG40000 glycine as having been selected for further development.

To evaluate accumulation of PEG camptothecin in solid tumor models, female nu/nu mice were inoculated with human colon (HT-29) tumor cells, and the tumors were allowed to grow to 200 mm$^3$ in size. Mice then received radiolabeled PEG camptothecin or radiolabeled camptothecin. All injections were i.v. via the tail vein. At various time points animals were bled and tissues removed for analysis. Lung, heart, spleen, kidney, muscle, tumor, and blood were collected, weighed, homogenized, and counted for radioactivity. Results were calculated as percent injected dose (%ID) per gram of tissue after correction for blood pool (C.D. Conover et al., personal communication) and are shown in Table 6. Clearly, as predicted by Ikada et al. (117) the PEG-conjugated camptothecin showed a greater degree of tumor accumulation than did its non-PEGylated counterpart.

A treatment comparison of PEG camptothecin to non-PEGylated camptothecin and the camptothecin analogue Topotecan as well as 5-fluorouracil (5-FU) is shown in Figure 10. The authors report that in these studies, at equivalent doses, PEG-camptothecin was more effective in delaying tumor growth than any of the other molecules tested. What is proposed is that the prolonged circulating lifetime of the PEG conjugate allows for a greater degree of tumor accumulation, followed by release of active cytotoxic drug and achievement of higher local concentrations.

***PEG-5-Fluorouracil.*** Nichifor et al. (123) have described polymeric conjugates of 5-FU linked via oligopeptide chains and a glycine ethyl ester bond. PEG, dextran, and poly(5N-(2-hydroxyethyl)-L-glutamine polymers were used. Oligopeptide bonds were designed to be sensitive to tumor associated enzymes such as collagenase, cathepsin

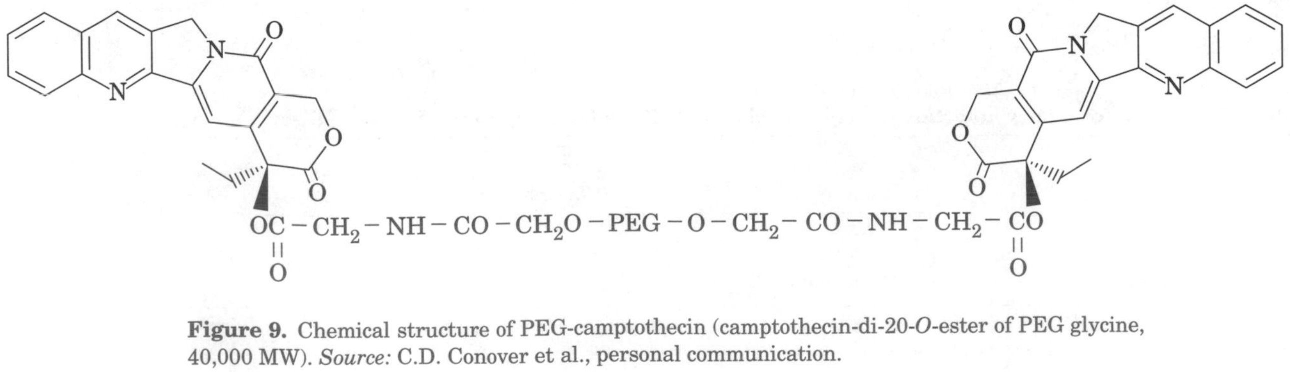

**Figure 9.** Chemical structure of PEG-camptothecin (camptothecin-di-20-O-ester of PEG glycine, 40,000 MW). *Source:* C.D. Conover et al., personal communication.

**Table 4. Rates of Hydrolysis and IC-50 Values for Various PEG-Camptothecin Derivatives as Described by Conover et al. (124)**

| Compound | Linker | IC50 (nM)[a] | $t_{1/2}$ (h)[b] PBS (pH 7.4) | $t_{1/2}$ (h)[b] Rat plasma |
|---|---|---|---|---|
| 1 | — | 7 | — | — |
| 3 | -O-CO-CH$_2$-O-PEG | 15 | 27 | 2 |
| 10 | -O-CO-CH$_2$-O-CH$_2$-CO-NH-PEG | 16 | 5.5 | 0.8 |
| 11 | -O-CO-CH$_2$-O-CH$_2$-CO-N(CH$_3$)-PEG | 21 | 27 | 3 |
| 16 | -O-CO-CH$_2$-O-CO-NH-PEG | 7 | 0.2 | ND |
| 17 | -O-CO-CH$_2$-O-CO-N(CH$_3$)-PEG | 18 | 28 | 5 |
| 24 | -O-CO-CH$_2$-NH-CO-CH$_2$-O-PEG | 12 | 40 | 6 |
| 25 | -O-CO-CH$_2$-N(CH$_3$)-CO-CH$_2$-O-PEG | 15 | 97 | 10 |
| 28 | -O-CO-CH$_2$-NH-PEG | 24 | 12 | 3 |
| 29 | -O-CO-CH$_2$-N(CH$_3$)-PEG | 42 | 102 | >24 |

*Note:* ND, not determined.
[a]All experiments were done in duplicate. Standard deviation of measurements: ±10%.
[b]These results more appropriately represent the half-lives by disappearance of the transport form.

**Table 5. Activity of Various PEG-Camptothecin Analogues Measured against P388 Murine Leukemia In Vivo**

| Test compound | Linker | Total dose[a] (mg/kg) | Mean time to death (days)[b] | % ILS[c] | Survivors on day 40 |
|---|---|---|---|---|---|
| Control | — | — | 13.0 | — | 0/10 |
| 1 | — | 16 | 38.0* | 192% | 7/10 |
| 3 | -O-CO-CH$_2$-O-PEG | 16 | 38.0* | 192% | 9/10 |
| 10 | -O-CO-CH$_2$-O-CH$_2$-CO-NH-PEG | 16 | 17.4* | 34% | 4/10 |
| 11 | -O-CO-CH$_2$-O-CH$_2$-CO-N(CH$_3$)-PEG | 16 | 31.6*,** | 143% | 6/10 |
| 17 | -O-CO-CH$_2$-O-CO-N(CH$_3$)-PEG | 16 | 23.4 | 80% | 0/10 |
| 24 | -O-CO-CH$_2$-NH-CO-PEG | 16 | 35.0* | 169% | 8/10 |
| 25 | -O-CO-CH$_2$-N(CH$_3$)-CO-PEG | 16 | 19.3*,** | 48% | 0/10 |
| 28 | -O-CO-CH$_2$-NH-PEG | 16 | 30.6* | 135% | 0/10 |
| 29 | -O-CO-CH$_2$-N(CH$_3$)-PEG | 16 | 21.4*,** | 65% | 0/10 |

*Source:* C.D. Conover et al., personal communication.
[a]Equivalent dose of camptothecin, mice dosed days 1–5.
[b]Kaplan–Meier estimates with survivors censored.
[c]Increased life span (ILS) is (T/C − 1) × 100.
*Significant, $p < 0.001$, compared to control (untreated).
**Significant, $p < 0.001$, compared to compound t.
In vivo efficacy study of the water soluble camptothecin derivatives using the P388/0 murine leukemia model. Compound 1 or prodrug derivatives were given daily [intraperitoneal(ip) × 5], 24 h following an injection of P388/0 cells into the abdominal cavity with survival monitored for 40 days.

**Table 6. Percent Injected Dose per Gram of PEG-Camptothecin and Camptothecin (CPT) at the Times Shown after Intravenous Injection in Athymic Mice Bearing Human Colon Adenocarcinoma Xenografts**

| Specimen | PEG-$\beta$-CPT 0.8 h | 2 h | 6 h | 24 h | 48 h | 72 h | CPT 0.8 h | 2 h | 6 h | 24 h | 48 h | 72 h |
|---|---|---|---|---|---|---|---|---|---|---|---|---|
| Tumor | 0.47 | 3.34 | 3.34 | 3.70 | 2.35 | 1.63 | 0.33 | 0.11 | 0.11 | 0.10 | 0.05 | 0.05 |
| Blood | 27.90 | 19.17 | 10.91 | 4.41 | 1.94 | 0.73 | 1.21 | 0.27 | 0.08 | 0.09 | 0.07 | 0.07 |
| Liver | 1.62 | 2.02 | 2.00 | 2.32 | 1.89 | 0.79 | 4.32 | 0.54 | 0.29 | 0.14 | 0.07 | 0.07 |
| Kidney | 0.01 | 0.02 | 0.04 | 0.16 | 0.21 | 0.22 | 10.32 | 0.32 | 0.10 | 0.07 | 0.07 | 0.03 |
| Spleen | 0.01 | 0.03 | 0.03 | 0.07 | 0.31 | 0.49 | 2.73 | 0.15 | 0.26 | 0.14 | 0.03 | 0.03 |
| Lung | 3.68 | 2.33 | 4.00 | 1.95 | 0.24 | 0.33 | 3.09 | 0.64 | 0.19 | 0.12 | 0.04 | 0.05 |
| Heart | 2.13 | 1.87 | 1.74 | 1.13 | 0.88 | 0.41 | 1.32 | 0.17 | 0.12 | 0.09 | 0.09 | 0.03 |
| Muscle | 0.58 | 1.11 | 1.50 | 0.91 | 0.96 | 0.44 | 0.71 | 0.12 | 0.09 | 0.09 | 0.03 | 0.03 |

*Source:* C.D. Conover et al., personal communication.
*Note:* Each value is the average from four mice.

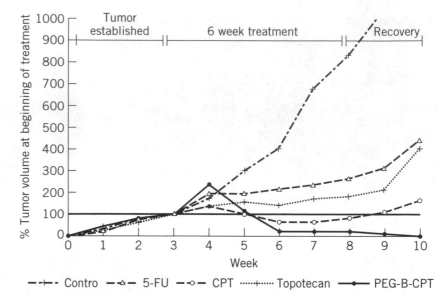

**Figure 10.** Growth curve of human colon adenocarcinoma (HT-29) tumor xenografts treated with 5-FU, camptothecin, topotecan, or PEG-camptothecin. All compounds were given over a 5-week period, with camptothecin, topotecan, and PEG camptothecin given five times per week (2.5 mg/kg per day) and 5-FU twice per week (80 mg/kg per day). PEG-camptothecin doses were based upon absolute equivalents of camptothecin. *Source:* C.D. Conover et al., personal communication.

B, and cathepsin D. The rationale of the authors was that polymer-promoted accumulated drug would be released under conditions selective for the tumor microenvironment. Preliminary in vitro studies suggest that, of the different polymers tested, rates of release were greatest for the PEG formulations.

*PEG-Hemoglobin.* Since the 1950s it has been known that as tumors grow, areas of hypoxia appear in direct correlation to distance from the tumor vascular tree. Across these gradients of hypoxia are also metabolic gradients and areas of necrotic tissue and tumor stroma. For many of the more common cytotoxic agents, and certainly for the use of radiation treatments, sufficient tumor oxygenation is a prerequisite for efficacious activity. Indeed, hypoxic tumor cells are known to be substantially more resistant to radiation treatment than their oxygenated counter parts (124,125).

Of the newly diagnosed patients each year in the United States nearly 60% will undergo radiation therapy. Tumor types already implicated or shown to contain substantial hypoxic regions and their approximated incidence are shown in Table 7. The extent of tumor hypoxia varies according to histological type as well as tumor size and stage of disease.

Attempts to sensitize tumors to radiation treatment have included the use of sensitizing molecules that mimic oxygen as well as the use of calcium channel blockers and other molecules or conditions such as hyperthermia designed to increase tumor blood flow, or the ability of hemoglobin to off-load oxygen. Attempts to improve tumor oxygenation directly have included use of hyperbaric oxygen chambers and perfluorochemical (PFC) emulsions. PFCs have been characterized as having a high oxygen carrying capacity in solution. While the initial purpose of PFC development was for the replacement of red blood cells in transfusion medicine, Teicher et al. (126–128) have shown that these agents can be used to deliver oxygen to a variety of hypoxic tumors, with resultant sensitization to radiation treatment.

Although it is clear that PFCs can indeed be used to sensitize tumors in experimental rodent models, clinical development has been hampered by the need for inspiration of 95% oxygen 5% carbon dioxide, as well as a variety of adverse reactions including lower back pain and cytopenia. It is possible that newer formulations of PFC may reemerge for further study.

Hemoglobin solutions, like PFC, have also been explored as replacements for red blood cells in transfusion or trauma medicine. An advantage of these solutions over PFCs is that they do not require inspiration of 95% oxygen 5% carbon dioxide; room air suffices. Side effects with some preparations (Baxter Healthcare), however, have been severe enough to slow clinical progress; with others (Northfield Laboratories, Hemosol, Enzon), esophageal spasm and other GI upsets or mild elevations in blood pressure only have been reported (129).

PEG-hemoglobin has been prepared by the conjugation of activated PEG to the amino groups of hemoglobin. Both APEX Biosciences and Enzon (129) have reported progress in this area. APEX has based its technology upon the use of human purified hemoglobin (hHb), but Enzon has used

**Table 7. Estimated Annual U.S. Incidence of Solid Tumor Types That May Contain Hypoxic and Therefore Radiation-Therapy-Resistant Regions**

| Tumor type | U.S. incidence |
|---|---|
| Breast | 183,000 |
| Lung | 172,000 |
| Colorectal | 149,000 |
| Prostate | 200,000 |
| Head and neck | 51,000 |
| Bladder | 51,000 |
| Brain metastases | 30,000 |
| Brain | 17,000 |
| Cervical, invasive | 15,000 |
| Fibrous histiocytoma | 8,000 |

purified preparations of bovine hemoglobin (bHb) for its solutions. An advantage of bovine hemoglobin over human is that the bovine hemoglobin tetramer is inherently more stable than the human and does not require covalent cross-linking to prevent dissociation into higher-oxygen-affinity dimers. Both companies use PEG strands of 5,000 molecular weight, although the Enzon product is PEGylated to a greater extent than the APEX product (9–15 PEGs versus 3–6 PEGs, respectively).

In a variety of animal models PEG-bHb has been shown to deliver oxygen in rodents and pigs from which 80–85% of the total blood volume had been replaced with hemoglobin solution. Using a 6 g% solution of PEG-bHb and measuring tissue oxygenation using phosphorescence quenching methods or following physiological markers, full oxygenation of liver, kidney, spleen, muscle, intestine, and brain could be shown, even under extreme conditions of transfusion or trauma (130).

To explore use of PEG-bHb for oxygenation of tumors, human or syngeneic tumors were implanted into rats and allowed to grow, and tumor oxygenation was followed by phosphorescence quenching methods before and after PEG-bHb i.v. administration via tail vein infusion. As shown in Table 8, oxygenation of hypoxic tumors increased 100–300% after administration of 6–15 mL/kg of a 6 g% solution of PEG-bHb (131). The authors reported a time course to maximum oxygenation levels of 2–4 h. It is possible that the delay in maximal oxygenation may reflect tumor accumulation of PEG-bHb and a locally increased concentration of the oxygen carrier.

When rats bearing osteogenic sarcoma tumor implants were treated with 6 mL/kg or 15 mL/kg PEG-bHb (6 g%) and low dose radiation, a greater than 80% complete response rate was observed (131) as compared with lactated Ringer's plus radiation or lactated Ringer's only controls (Fig. 11).

For human colon carcinoma implants, infusion with 15 mL/kg PEG-bHb (6 g%) and treatment with radiation, substantial growth delay was observed over no treatment or radiation treatment only controls. PEG-bHb as a sensitizer to radiation therapy in conjunction with breathing of 100% oxygen is now in phase Ib clinical trials (131). In these studies PEG-bHb is given weekly for up to three infusions, concomitant with daily external radiation. No dose-limiting toxicities have been reported.

In addition to resistance to radiation, hypoxic tumor cells have also been credited with resistance to a number of chemotherapeutic treatments. Teicher et al. (132) have examined the ability of PEG-bHb to reverse tumor-associated hypoxia, with sensitization to a variety of che-

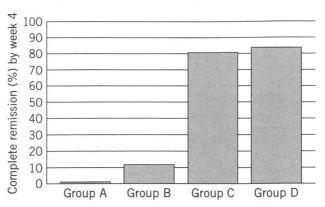

Group: Ringer's lactate (15 mL/kg), no radiation
Group B: Ringer's lactate (15 mL/kg), radiation (4 Gy)
Group C: PEG-Hb (6 mL/kg), radiation (4 Gy)
Group D: PEG-Hb (15 mL/kg), radiation (4 Gy)

**Figure 11.** Rats bearing syngeneic osteogenic sarcoma tumors (approximately 1 cm$^3$) were treated with PEG-bHb, at the doses shown, and a single whole-body dose of radiation (4 Gy). Alternatively rats received lactated Ringer's, at the doses shown, with or without radiation treatment.

motherapeutic agents. It had been demonstrated earlier that use of PFCs could enhance the cytotoxic activity of anticancer alkylating agents such as *cis*-diamminedichloroplatinum(II) (CDDP) (133), carboplatin, cyclophosphamide (134) *N,N',N''* triethylenethiophosphoramide, 1,3-bis(2-chloroethyl)-1-nitrosourea, and L-phenylalanine mustard (135–138).

In the PEG-bHb studies, rat mammary adenosarcoma 13672 tumors were used. These carcinogen induced tumor cells can metastasize to the lungs and abdominal organs. The tumor is composed of epithelial cells and acini and can grow to 100 mm$^3$, in size in approximately 2 weeks postimplantation. EMT-6 murine mammary carcinoma cell and tumors (132) were also studied. In both cases 2 × 10$^6$ cells were inoculated into the hindlimbs of rats or mice, respectively. To measure tumor oxygenation, a pO$_2$ histograph (Eppendorf, Inc., Hamburg, Germany) was used. The needle microelectrode was calibrated in aqueous solution saturated with air or 100% nitrogen.

The test animals were anesthetized and placed upon a heating pad, tumor sizes were estimated with calipers, and a small area of the skin covering the tumor mass was removed. A 20-gauge needle was used to perforate the tumor, and the electrode was positioned in the perforation for measurements as described in Ref. 132.

**Table 8. Measurement of Tumor Oxygen Tension before and after Administration of 15 mL/kg 6 g% PEG-Hemoglobin**

| Tumors examined | Pretreatment O$_2$ tension (mm Hg) | Posttreatment O$_2$ tension (mm Hg) | % increase |
|---|---|---|---|
| Rat osteogenic sarcoma | 3.7 ± 0.2 | 15.4 ± 0.4 | 316% |
| Rat glioma | 4.7 ± 0.6 | 11.2 ± 2.7 | 138% |
| Human prostate | 5.2 ± 1.3 | 10.4 ± 1.8 | 100% |
| Human ovarian | 15.0 ± 1.5 | 41.2 ± 5.6 | 175% |

*Note:* Oxygen measurements were by phosphoresence quenching methods 2–4 h after Intravenous Infusion.

To measure tumor growth delay, $2 \times 10^6$ EMT-6 cells were inoculated into the hindlimbs of mice. At 100 mm$^3$ tumor size (7 days postimplantation) cytotoxic therapy was initiated. Cyclophosphamide (150 mg/kg i.p.) or BCNU (15 mg/kg i.p.) on days 7, 9, and 11, 5-FU (30 mg/kg i.p.) and adriamycin (1.75 mg/kg i.p.) were given days 7 through 11. Taxol (24, 36, or 48 mg/kg i.v.) was given days 7 through 11. Animals received drug alone or in conjunction with PEG-bHb (6 g%, 6 mL/kg) and breathed room air, 28%, or 95% oxygen for 6 h after each drug injection. Tumor volumes were monitored and growth delay calculated as time to reach 500 mm$^3$ relative to untreated controls (132).

To measure tumor cell survival, 13,762 tumor-bearing rats were treated with a single dose of cyclophosphamide (250 mg/kg i.p.), melphalan (20 mg/kg i.p.), taxol (10mg/kg i.v.), or CDDP (8 mg/kg i.p.). EMT-6-bearing mice were treated with cyclophosphamide (100, 300, or 500 mg/kg i.p.), BCNU (50, 100, or 200 mg/kg i.p.), adriamycin (10, 25, or 50 mg/kg i.p.), or taxol (25, 50, or 75 mg/kg i.v.) alone or in conjunction with PEG-bHb (6 g%, 6 mL/kg i.v. just before chemotherapy). Twenty four hours later animals were killed and tumors excised and minced under sterile conditions in a laminar flow hood. A pool of four tumors for each treatment group was washed and treated with collagenase and DNAse prior to filtration and plating for colony-forming assay as described in Ref. 132.

To examine bone marrow toxicity, a pool of femur bone marrow from mice treated as just described was obtained by flushing with a 23-gauge needle. Colony-forming unit assays were carried out as described in Ref. 134.

To examine metastases, animals treated as just described were necropsied on day 20 after tumor implantation, and external metastatic nodules were counted manually and scored as greater than or equal to 3 mm in diameter (132).

As shown in Tables 9 and 10, substantial growth delays were also observed for the combination of PEG-bHb with chemotherapeutic agents over drug alone. For cyclophosphamide a significant decrease in the numbers of lung metastases was also noted. In Figure 12 the authors describe increased tumor cytotoxic effects as having occurred with little or no increase in bone marrow toxicity.

***The Case for Polyethylene Glycol.*** Liposomes stabilized with PEG and containing doxorubicin have gone far in proving the utility of passively targeted anticancer agents. However not all chemotherapeutic anticancer agents lend themselves to fascile formulation in liposomes, stabilized or otherwise.

Although conjugation of cytotoxic agents to SMANCS, HPMA, and other carriers has proceeded to clinical investigation and approvals, little is known about the perfor-

**Table 9. Growth Delay of the Murine-6 Mammary Carcinoma and Number of Lung Metastases on Day 20 Produced by Anticancer Chemotherapeutic Alone or in Combination with Administration of PEG-bHb**

| Treatment group | Tumor growth delay (days)[a] + PEG-Hb (6 mL/kg) | | | Number of lung metastases + PEG-Hb (6 mL/kg) | | |
|---|---|---|---|---|---|---|
| | Alone | Air | 95% $O_2$ | Alone | Air | 95% $O_2$ |
| Control | | | 16 ± 3 | | | |
| Cyclophosphamide (150 mg/kg) days 7, 9, 11 | 6.2 ± 0.3 | 7.8 ± 0.5 | 11.9 ± 0.8 | 7 ± 3 | 4 ± 1 | 2 ± 1 |
| Adriamycin (1.75 mg/kg) days 7–11 | 4.5 ± 0.3 | 8.3 ± 0.6 | 14.9 ± 1.1 | 13 ± 2 | 10 ± 1 | 9 ± 2 |
| 5-Fluorouracil (30 mg/kg) days 7, 9, 11 | 3.1 ± 0.3 | 7.5 ± 0.5 | 11.1 ± 0.9 | 12 ± 2 | 10 ± 3 | 9 ± 1 |
| BCNU (15 mg/kg) 5.1 ± 0.3 days 7, 9, 11 | 6.2 ± 0.4 | 11.7 ± 0.8 | 14 ± 2 | 12 ± 2 | 9 ± 2 | |

*Source:* From Ref. 132.

[a]Mean days ± S.E.M. for treated tumors to reach 500 mm$^3$ compared with untreated controls. Control tumors reached 500 mm$^3$ in 124 ± 1.2 days. The anticancer drugs were administered by intraperitoneal injection. PEG-hemoglobin (6 ml/kg) was administered by intravenous injection just prior to each drug injection. 95% oxygen (carbogen) breathing was maintained for 6 h after each drug injection. Each treatment group had five animals, and the experiment was repeated three times, therefore the number of animals per point was 15 ($n = 15$).

**Table 10. Growth Delay of the Murine-6 Mammary Carcinoma and Number of Lung Metastases on Day 20 Produced by Taxol Alone or in Combination with Administration of PEG-bHb**

| Treatment group | Tumor growth delay (days)[a] + PEG-Hb (6 ml/kg) | | | | Number of lung metastases + PEG-Hb (6 ml/kg) | | | |
|---|---|---|---|---|---|---|---|---|
| | Alone | Air | 28% $O_2$ | 95% $O_2$ | Alone | Air | 28% $O_2$ | 95% $O_2$ |
| Control | | | | 16 ± 3 | | | | |
| Taxol (24 mg/kg) days 7–11 | 2.9 ± 0.3 | 3.9 ± 0.3 | 4.4 ± 0.3 | 6.7 ± 0.5 | 14 ± 2 | 10 ± 2 | 10 ± 2 | 9.5 ± 2 |
| Taxol (36 mg/kg) days 7–11 | 4.9 ± 0.4 | 5.4 ± 0.5 | 6.1 ± 0.4 | 6.6 ± 0.5 | 13 ± 2 | 10 ± 2 | 9 ± 2 | 7.5 ± 2 |
| Taxol (48 mg/kg) days 7–11 | 5.4 ± 0.5 | 7.0 ± 0.5 | 8.0 ± 0.6 | 9.7 ± 0.6 | 12 ± 2 | 8.5 ± 2 | 6.5 ± 2 | 4.5 ± 1 |

*Source:* From Ref. 132.

[a]Mean days ± S.E.M. for treated tumors to reach 500 mm3 compared with untreated controls. Control tumors reached 500 mm3 in 12.4 ± 1.2 days. The taxol was administered by intravenous injection. PEG-hemoglobin (6 ml/kg) was administered by intravenous injection just prior to each taxol injection. 28% oxygen or 95% oxygen (carbogen) breathing was maintained for 6 h after each drug injection. Each treatment group had five animals and the experiment was repeated three times, therefore the number of animals per point was 15 ($n = 15$).

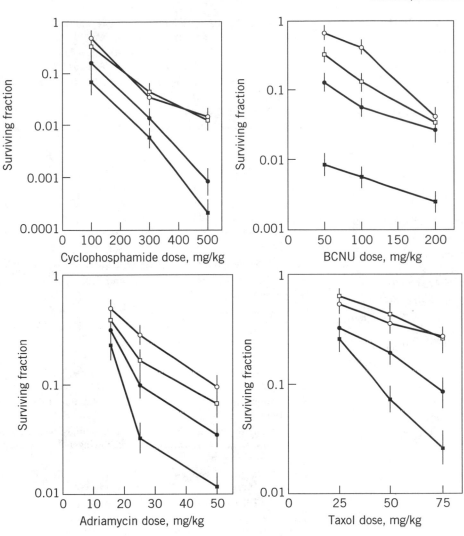

**Figure 12.** Survival of EMT-6 tumor cells (solid symbols) from tumors treated in vivo, and bone marrow CFU-GM (open symbols) from the same animals after treatment with the drugs and dosages shown alone (○, ●) or with PEG bHb (6 mL/kg, □, ■). Results are the means of three independent determinations, bars are S.E.M.

mance characteristics of small molecules directly linked to PEG, despite considerable experience with proteins.

As described earlier, conjugation to PEG of small-molecule anticancer chemotherapeutics has now been achieved. In addition to promotion of tumor accumulation, as predicted on the basis of molecular weight and tumor flow dynamics, increases in drug solubility have been reported. For compounds such as paclitaxel, increases in solubility are sufficient to abrogate the need for excipients such as Cremophor EL or ethanol.

PEG has been shown to be linkable by covalent reversible bonds and nonreversible covalent bonds to proteins and small-molecule anticancer agents. In some instances demonstration of tumor-selective enzymatic cleavage and release has been proposed and demonstrated. In others less-selective release via blood esterase has been shown.

The further use of PEG for exploration of passive tumor targeting is warranted by the long history of PEG usage as a formulation agent for pharmaceuticals and more than a decade of experience in the clinic with PEG-conjugated proteins for chronic treatment. The facility of PEG chemistry together with available pharmaceutical grade polymer also contributes greatly to the desirability of further exploration of this approach.

## BIBLIOGRAPHY

1. T. Beardsley, *Sci. Am.* **270**, 130–138 (1994).
2. P.C. Nowell, *Science* **194**, 23–28 (1976).
3. G.H. Heppner, *Cancer Res.* **44**, 2259–2265 (1984).
4. A.V. Kabanov and V.Y. Alakhov, *J. Controlled Release* **28**, 15–35 (1994).
5. J.N. Lowder et al., *Blood* **69**, 199–210 (1987).
6. E.S. Vitetta et al., *Cancer Res.* **15**, 4052–4058 (1991).
7. H.F. Dvorak, J.A. Nagy, and A.M. Dvorak, *Cancer Cells* **3**, 77–85 (1991).
8. L.T. Baxter and R.K. Jain, *Microvasc. Res.* **41**, 5–23 (1991).
9. F.J. Burrows, Y. Watanabe, and P.E. Thorpe, *Cancer Res.* **52**, 5954–5962 (1992).
10. H. Sands, *Antibody Immunoconjugates Immunoradiopharm.* **1**, 213–226 (1988).
11. A.A. Epenetos et al., *Cancer Res.* **46**, 3183–3191 (1986).
12. R.K. Jain, *Sci. Am.* **271**, 58–65 (1994).
13. M. Juweid et al., *Cancer Res.* **52**, 5144–5153 (1992).
14. C. Marks and J.D. Marks, *N. Engl. J. Med.* **335**, 730–733 (1996).
15. J.D. Marks et al., *J. Mol. Biol.* **222**, 581–597 (1991).

16. X. Cai and A. Garen, *Proc. Natl. Acad. Sci. U.S.A.* **93**, 6280–6285 (1996).

17. R. Schier et al., *J. Mol. Biol.* **255**, 28–43 (1996).

18. J. Denekamp, *Prog. Appl. Microcirc.* **4**, 28–38 (1984).

19. E.A. Jaffe, *Biology of Endothelial Cells*, Martinus Nijhoff, Boston, 1984.

20. P.E. Thorpe and E.J. Derbyshire, *J. Controlled Release* **48**, 277–288 (1997).

21. A. Gougos and M. Latarte, *J. Biol. Chem.* **265**, 8361–8364 (1990).

22. J.R. Westphal et al., *J. Invest. Dermatol.* **100**, 27–34 (1993).

23. A. Gougos and M. Letarte, *J. Immunol.* **141**, 1925–1933 (1988).

24. P.J. O'Connell et al., *Clin. Exp. Immunol.* **90**, 154–159 (1992).

25. H.J. Buhring et al., *Leukemia* **5**, 841–847 (1991).

26. W.J. Reftig et al., *Proc. Natl. Acad. Sci. U.S.A.* **89**, 10832–10836 (1992).

27. J.M. Wang et al., *Int. J. Cancer* **54**, 363–370 (1993).

28. B. Carnemolla et al., *J. Cell Biol.* **108**, 1139–1148 (1989).

29. G. Molema, L.F.M. De Leij, and D.K.F. Meijer, *Pharm. Res.* **14**, 2–10 (1997).

30. A.L. Epstein et al., in V.P. Torchilin, ed., *Handbook of Targeted Delivery of Imaging Agents*, CRC Press, Boca Raton, Fla., 1995, pp. 260–288.

31. L.A. Khawli and A.L. Epstein, *O. J. Nucl. Med.* **41**, 25–35 (1997).

32. R.H. Jain, *Cancer Res.* **47**, 3039–3051 (1987).

33. H.F. Dvorak, J.A. Nagy, J.T. Dvorak, and A.M. Dvorak, *Am. J. Pathol.* **133**, 95–109 (1988).

34. A. Gabizin and D. Papahadjopoulos, *Proc. Natl. Acad. Sci. U.S.A.* **85**, 6949–6953 (1988).

35. D. Liu, A. Mori, and L. Huang, *Biochim. Biophys. Acta* **1104**, 95–101 (1992).

36. H. Maeda and Y. Matsumura, *Crit. Rev. Ther. Drug Carrier Syst.* **6**, 193–210 (1989).

37. L.W. Seymour, *Crit. Rev. Ther. Drug Carrier Syst.* **9**, 135–187 (1992).

38. C. Sung, R.J. Youle, and R.L. Dedrick, *Cancer Res.* **50**, 7382–7392 (1990).

39. K. Uchiyama et al., *Int. J. Pharm.* **121**, 195–203 (1995).

40. J. Folkman, in P.M. Mendelsohn, M.A.P. Howley, and L.A. Liotta, eds., *The Molecular Basis of Cancer*, Saunders, Philadelphia, 1995.

41. R.K. Jain, *Cancer Res.* **48**, 2641–2658 (1988).

42. L.T. Baxter and R.K. Jain, *Microvasc. Res.* **40**, 246–263 (1990).

43. J.W. Baish et al., *Microvasc. Res.* **51**, 327–346 (1996).

44. Y. Gazit et al., *Phys. Rev. Lett.* **75**, 2428–2431 (1995).

45. J.R. Less, T.C. Skalak, E.M. Sevick, and R.K. Jain, *Cancer Res.* **51**, 265–273 (1991).

46. M. Leunig et al., *Cancer Res.* **52**, 6553–6560 (1992).

47. F. Yuan et al., *Cancer Res.* **54**, 4564–4568 (1994).

48. E.M. Sevick and R.K. Jain, *Cancer Res.* **49**, 3506–3512 (1989).

49. P.A. Netti et al., *Microvasc. Res.* **52**, 27–46 (1996).

50. E.M. Sevick and R.K. Jain, *Cancer Res.* **49**, 3513–3519 (1989).

51. E.M. Sevick and R.K. Jain, *Cancer Res.* **51**, 2727–2730 (1991).

52. B. Endrich, H.S. Reinhold, J.F. Gross, and M. Intaglietta, *JNCI, J. Natl. Cancer Inst.* **62**, 387–395 (1979).

53. C.J. Eskey, A.P. Koretsky, M.M. Domach, and R.K. Jain, *Proc. Natl. Acad. Sci. U.S.A.* **90**, 2646–2650 (1993).

54. R.K. Jain, S.A. Shah, and P.L. Finney, *JNCI, J. Natl. Cancer Inst.* **73**, 429–436 (1984).

55. M. Nozue et al., *J. Surg. Oncol.* **63**, 109–114 (1996).

56. K.A. Ward and R.K. Jain, *J. Hyperthermia* **4**, 233–250 (1988).

57. P. Vaupel and R.K. Jain, eds., *Tumor Blood Supply and Metabolic Microenvironment: Characterization and Therapeutic Implications*, Fischer, Stuttgart, 1991.

58. R.K. Jain, *Cancer Metastasis Rev.* **6**, 559–593 (1987).

59. H.F. Dvorak, L.F. Brown, M. Dermar, and A.M. Dvorak, *Am. J. Pathol.* **146**, 1029–1039 (1995).

60. L.E. Gerlowski and R.K. Jain, *Microvasc. Res.* **31**, 288–305 (1986).

61. E.M. Sevik and R.K. Jain, *Cancer Res.* **51**, 1352–1355 (1991).

62. F. Yuan, M. Leunig, D.A. Berk, and R.K. Jain, *Microvasc. Res.* **45**, 269–289 (1993).

63. F. Yuan et al., *Cancer Res.* **54**, 3352–3356 (1994).

64. P.A. Netti et al., *Proc. Natl. Acad. Sci.* **96**, 3137–3142 (1999).

65. S.K. Hobbs et al., *Proc. Natl. Acad. Sci.* **95**, 4607–4612 (1998).

66. W.G. Roberts and G. Palade, *Cancer Res.* **57**, 1207–1211 (1997).

67. F. Yuan et al., *Proc. Natl. Acad. Sci. U.S.A.* **93**, 14675–14770 (1996).

68. E. Arbit, J. Lee, and G. Diresta, in H. Nagai, K. Kamiya, and S. Ishi, eds., *Intracranial Pressure*, Springer-Verlag, Tokyo, 1994, pp. 609–614.

69. Y. Boucher, L.T. Baxter, and R.K. Jain, *Cancer Res.* **50**, 4478–4484 (1990).

70. Y. Boucher and R.K. Jain, *Cancer Res.* **52**, 5110–5114 (1992).

71. Y. Boucher, I. Lee, and R.K. Jain, *Microvasc. Res.* **50**, 175–182 (1995).

72. Y. Boucher et al., *Cancer Res.* **51**, 6691–6694 (1991).

73. Y. Boucher et al., *Br. J. Cancer* **75**, 829–836 (1997).

74. B.D. Curti et al., *Cancer Res.* **53**, 2204–2207 (1993).

75. R. Guttman et al., *Cancer Res.* **52**, 1993–1995 (1992).

76. J.R. Less et al., *Cancer Res.* **52**, 6371–6374 (1992).

77. S.D. Nathanson and L. Nelson, *Ann. Surg. Oncol.* **1**, 333–338 (1994).

78. H.D. Roh et al., *Cancer Res.* **51**, 6695–6698 (1991).

79. C.A. Znati et al., *Cancer Res.* **56**, 964–968 (1996).

80. H.C. Lichtenbeld, F. Yuan, C.C. Michel, and R.K. Jain, *Microcirculation* **3**, 349–357 (1996).

81. L.T. Baxter and R.K. Jain, *Microvasc. Res.* **37**, 77–104 (1989).

82. R.K. Jain and L.T. Baxter, *Cancer Res.* **48**, 7022–7032 (1988).

83. V.P. Torchilin, *Crit. Rev. Ther. Drug Carrier Syst.* **2**, 65–115 (1985).

84. G. Gregoriadis, ed., *Liposome Technology*, vols. 1–3, CRC Press, (Boca Raton, Fla., 1993).

85. T. Fujita et al., *J. Pharmacol. Exp. Ther.* **263**(3), 971–978 (1992).

86. R.A. Srivastava, *Enzyme Microb. Technol.* **13**, 164–170 (1991).

87. J. Suh and B.K. Hwang, *Bioorg. Chem.* **20**, 232–235 (1992).

88. J. Suh, Y. Cho, and G. Kwag, *Bioorg. Chem.* **20**, 236–244 (1992).

89. L. Gros, H. Ringsdorf, and H. Schlupp, *Angew Chem., Int. Ed. Engl.* **20**, 305–325 (1981).

90. J. Kopecek, *Biomaterials* **5**, 19–25 (1984).

91. C.R. Gardner, *Biomaterials* **6**, 153–160 (1985).

92. J. Kopecek and R. Duncan, *J. Controlled Release* **6**, 315–327 (1987).

93. D.R. Friend and S. Pangburn, *Med. Res. Rev.* **1**, 53–106 (1987).

94. J. Drobnik, *J. Adv. Drug Delivery Rev.* **3**, 229–245 (1989).

95. H. Sezaki, Y. Takakura, and M. Hashida, *Med. Res. Rev.* **3**, 247–266 (1989).

96. H. Maeda, *Med. Res. Rev.* **6**, 181–202 (1991).

97. R. Duncan, *Anticancer Drugs* **3**, 175–210 (1992).

98. R. Duncan, *Clin. Pharmacokinet.* **27**, 290–306 (1994).

99. Y. Takakura and M. Hashida, *Crit. Rev. Oncol. / Hematol.* **18**, 207–231 (1995).

100. S.E. Mathews, C.W. Pouten, and M.D. Threadgill, *Adv. Drug Delivery Rev.* **18**, 219–267 (1996).

101. R. Duncan, S. Dimitrijevic, and E.G. Evagorou, *S.T.P. Pharm. Sci.* **6**, 237–263 (1996).

102. H. Maeda, M. Ueda, T. Morinaga, and T. Matsumoto, *J. Med. Chem.* **28**, 455 (1985).

103. H. Maeda, L.W. Seymour, and Y. Miyamato, *Bioconjugate Chem.* **3**, 351–362 (1992).

104. M. Kimura et al., *Anticancer Res.* **13**, 1287–1292 (1993).

105. K. Iwai, H. Maeda, and T. Konno, *Cancer Res.* **44**, 2115–2121 (1984).

106. K. Iwai et al., *Anticancer Res.* **7**, 321–327 (1987).

107. E. Hurwitz, M. Wilchek, and J. Pitha, *J. Appl. Biochem.* **2**, 25–35 (1980).

108. S. Dunhauser-Reidi et al., *Invest. New Drugs* **11**, 187–195 (1993).

109. R. Duncan et al., *J. Makromol. Chem.* **184**, 197–200 (1984).

110. P.A. Vasey et al., *Clin. Cancer Res.* **5**, 83–94 (1999).

111. M.S. Hershfield et al., *N. Engl. J. Med.* **316**, 589–596 (1987).

112. B.L. Asselin et al., *J. Clin. Oncol.* **11**, 1780–1786 (1993).

113. S. Zalipsky, *Adv. Drug Delivery Rev.* **16**, 157–182 (1995).

114. J. Senior et al., *Biochim. Biophys. Acta* **1062**, 77–82 (1991).

115. V.P. Torchillin, *J. Liposome Res.* **6**, 99–116 (1996).

116. V.P. Torchillin and M.I. Papisov, *J. Liposome Res.* **4**, 725–739 (1994).

117. Y. Murakami, Y. Tabata, and Y. Ikada, *Drug Delivery* **4**, 23–31 (1997).

118. M.C. Wani et al., *J. Am. Chem. Soc.* **93**, 2325–2327 (1971).

119. R. Weiss et al., *J. Clin. Oncol.* **8**, 1263–1268 (1990).

120. A.E. Mathew et al., *J. Med. Chem.* **5**, 145–151 (1992).

121. A. Pendri, C.D. Conover, and R. Greenwald, *Anticancer Drug Des.* **13**, 87–395 (1998).

122. B. Giovanella et al., *Cancer Res.* **51**, 3052–3055 (1991).

123. M. Nichifor, E.H. Schacht, and L.W. Seymour, *J. Controlled Release* **39**, 79–92 (1996).

124. P. Vaupel, F. Kallinowski, and P. Okunieff, *Cancer Res.* **49**, 6449–6465 (1989).

125. R.A. Gatenby et al., *Int. J. Radiat. Oncol. Biol. Phys.* **14**, 831–838 (1988).

126. B.A. Teicher, J.M. Crawford, S.A. Holden, and K.N.S. Cathcart, *Cancer Res.* **47**, 5036–5041 (1987).

127. B.A. Teicher et al., *J. Cancer Res. Clin. Oncol.* **120**(10), 593–598 (1994).

128. B.A. Teicher et al., *Cancer J. Sci. Am.* **1**(1), 43–48 (1995).

129. R.G. Shorr, *J. Expert Opin., Int. Bus. Commun. Substitute Conf. Rev.* in press.

130. D. Song et al., *Transfusion* **35**, 552–555 (1995).

131. R. Linberg, C.D. Conover, K.L. Shum, and R.G. Shorr, *In Vivo* **12**, 167–174 (1998).

132. B.A. Teicher et al., *In Vivo* **11**, 301–312 (1997).

133. B.A. Teicher, S.A. Holden, A. Al-Achi, and T.S. Herman, *Cancer Res.* **50**, 3339–3344 (1990).

134. B.A. Teicher and S.A. Holden, *Cancer Treat. Rep.* **71**, 173–177 (1987).

135. B.A. Teicher et al., *Cancer Res.* **49**, 4996–5001 (1989).

136. B.A. Teicher, S.A. Holden, and C.M. Rose, *Cancer Res.* **38**, 285–288 (1986).

137. B.A. Teicher, S.A. Holden, and C.M. Rose, *Int. J. Cancer* **36**, 585–589 (1985).

138. B.A. Teicher, S.A. Holden, and J.L. Jacobs, *Cancer Res.* **47**, 5036–5041 (1987).

See also CANCER, DRUG DELIVERY TO TREAT—LOCAL & SYSTEMIC; CENTRAL NERVOUS SYSTEM, DRUG DELIVERY TO TREAT.

# CARDIOVASCULAR DRUG DELIVERY SYSTEMS

BENJAMIN A. HERTZOG
CHRIS THANOS
MARYELLEN SANDOR
Brown University
Providence, Rhode Island

VENKATESH RAMAN
ELAZER R. EDELMAN
Massachusetts Institute of Technology
Cambridge, Massachusetts

## KEY WORDS

Angioplasty
Atherosclerosis
Catheter delivery
Channel catheter
Double-balloon catheter
Endoluminal paving
Hydrogel-coated balloon
Intramyocardial delivery
Microparticle delivery
Neointimal hyperplasia
Pericardial delivery
Perivascular delivery
Porous balloon
Restenosis
Vascular stents

## OUTLINE

## INTRODUCTION

Because controlled drug delivery can release of precise amounts of drug with determinable and definable release kinetics, there is the potential for its use in both the study and treatment of disease. At the same time, the more the technology of controlled delivery is applied to different fields and indications, the more it is advanced. This symbiotic relationship among technological development, scientific investigation, and therapeutic innovation is evident in the field of cardiovascular medicine. Controlled drug delivery enabled the rigorous evaluation of the biological and physiological processes that govern cardiovascular disease and, in doing so, paved the way for whole new avenues of therapies. This chapter reviews the role of controlled drug delivery systems in the field of cardiovascular medicine.

## CLASSIFICATIONS

It is possible to classify drug delivery systems in many ways: on the basis of the drug to be applied, the delivery technology, and the target tissue. Classification of cardiovascular release systems, however, presents some limitations. An attempt to divide release systems based on the class of compounds is virtually impossible as the drugs available are so vast and overlapping in effect, physicochemical properties, and means of synthesis or derivation. Categorization might examine whether drugs are released into the lumen of the blood vessel or heart (endovascular or endocardial delivery), into the wall of the tissue (intravascular or intramyocardial) or external to the tissue (perivascular, epicardial, or pericardial). Alternatively, as the technologies employed span many sites of delivery and can be used for multiple compounds, we chose this means of defining this vast body of work. The chapter starts with a look at catheter delivery systems designed and used to deliver drugs locally. The next section explores the many attempts to use vascular stents as drug delivery systems. The final section discusses the cardiovascular applications of polymer-based drug delivery systems with special attention being paid to two promising therapies, endoluminal paving and microparticle delivery.

## CATHETER-BASED SYSTEMS

The technical power and therapeutic success of percutaneous transluminal angioplasty (PTA) often overshadow the dramatic conceptual changes that this procedure has provided. Cardiovascular diseases, particularly atherosclerosis, are processes that involved the entire organism, and systemic therapy was therefore de rigeur. Transluminal angioplasty of isolated vascular segments enabled the focal treatment of such lesions and heralded local therapy that might treat lesions without systemic dosing or toxicity. Angioplasty not only provided direct access to target tissue with the potential for site-specific drug delivery but also the conceptual basis for proceeding locally as well. The need for adjunctive therapy and motivation for pursuing local pharmacologic therapy also was driven by this procedure. From 35 to 50% of all patients undergoing angioplasty suffer some accelerated vascular biological response to the intervention to such an extent as to need to return in 3–6 months for repeat attempt at revascularization. This failure phenomenon, termed *restenosis* (1), remains a major health care problem by virtue of the number of patients involved and because no systemic pharmacologic approach to date has reduced its impact (2).

It has been suggested that one of the main reasons for the failure of systemic drug delivery is that submaximal doses were used in an effort to minimize the unwanted side effects of systemic administration (3). Local and site-specific catheter delivery systems were developed to circumvent these problems. Diverse technologies have been employed represented by devices that include double balloons, porous balloons, hydrogel-coated catheters, injection-needle-studded balloons, and iontophoresis catheters. Although these devices may vary drastically in design and function, they all locally deliver controlled quantities of pharmacological agents to the vessel wall. To date, most local delivery catheters are limited by one or more of a few specific problems, including (1) expulsion of contents down arterial side branches, (2) arterial wall injury, and (3) downstream loss of infused agents (1). Each new generation of local delivery catheters attempts to address some of these faults, although in many cases, the new catheters introduce entirely new limitations of their own.

### The Double-Balloon Catheter

One of the first vascular site-specific delivery catheters was the double-balloon catheter (4–8). These double-lumen catheters have two small polyethylene, latex, or poly(ethylene terephthalate) balloons attached to the periphery. One lumen (for balloon inflation) is ported into each of the two balloons, whereas the other serves as the drug infusion channel and is ported to the exterior of the catheter between the two balloons. Typical interballoon spacing is around 1.7 cm for dogs and 1.2 cm for humans (8). The diameter of the balloons is designed so that, when inflated, they expand gently against the endoluminal wall and form a seal sufficient to prevent blood flow. Typically, the catheter is threaded through the vessel until the desired site is reached. The pressure on the two balloons is slowly increased over several seconds until the vessel is occluded at both the proximal and distal balloons. The blood filling the interballoon space can be drained and replaced with pharmacological agents through the infusion port. In an alternative design the proximal balloon is inflated, the lumen is flushed of blood, and only then is the distal balloon inflated. The drugs in solution between the inflated balloons actually reach the isolated vessel wall segment by direct diffusion or as a result of an applied pressure. Goldman et al. demonstrated that the depth of penetration of an agent is directly related to the distending pressure (8). In their studies, it was found that a distending pressure of 300 mmHg applied for 45 seconds was sufficient to cause horseradish peroxidase (HRP, 40,000 daltons) to penetrate the entire media of a normal dog muscular artery or nearly the entire plaque thickness of mildly and moderately atherosclerotic human coronary arteries. In earlier studies, the same group demonstrated that relatively acute pressure elevations resulted in the entry of labeled albumin (60,000 daltons) into the normal dog aorta. Pressures of 100 mmHg were ineffective, whereas 240 mmHg resulted in increased permeability. The increase in permeability was thought to arise from stretching of the vessel wall, an indirect result of the increased pressure (9). Feasibility studies with HRP (8) have since given way to use of a variety of agents: 14C-labeled fluorouracil (5-FU) to dogs (4), r-hirudin and heparin to the carotid arteries of pigs (5), intact endothelial cells to the iliac artery in rabbits (6), and enzymatic plaque dissolving agents to the aortas of rabbits, femoral arteries of dogs, and carotid arteries of humans (7). Other studies have shown that 100 mmHg is sufficient for uptake of DNA plasmids into the arterial wall, and pressures greater than 500 mmHg exhibit a tendency toward additional vascular injury (10).

Although the double-balloon catheter is still used in a number of experimental settings, its use clinically has been limited by problems, including leakage of large volumes of the infused solution through side branches of the vessel. Additionally, the pressure required to form a seal against the inner wall of the vessel, coupled with the long incubation times of this technique (15–30 min), often result in additional vascular injury at the balloon contact sites (8).

## Porous Balloon Catheters

The porous balloon catheter attempted to avoid the balloon pressure-induced injury. In its first embodiment Wolinski et al. perforated a poly(ethylene terephthalate) angioplasty balloon with 300 holes of approximately 25 $\mu$m in size. The balloon was 2.0 cm in length, and the perforations were cut with a laser in a radial pattern of 6 rows with 50 holes per row. The porous balloon catheter uses the pressure of the perfusing fluid inside the lumen of the balloon to gently expand the balloon against the walls of the vessel. Consequently, the pressure used to distend the balloon also forces the perfusate out of the balloon via the 25 $\mu$m per-

forations and into the wall of the artery. In their first study, Wolinski and Thung demonstrated that HRP and heparin delivered at ~5 atm for 1 min could traverse the entire media of the normal canine artery (11). This device has been used for the delivery of various substances, including angiopeptin, antineoplastic agents such as methotrexate and doxorubicin, antimitotic agents such as colchicine (12), as well as for gene transfer (13,14). The main problem with the porous balloon catheter in its original, and what has now been termed "macroporous" form, is repeatable tissue damage caused by jetting of the high pressure fluids at the balloon pores. Fluid jetting out of the perforations resulted in medial necrosis by 48 hours, even when normal saline was used as the perfusate. When compared to standard angioplasty, simultaneous angioplasty and perfusion of saline resulted in similar lumen dilation as normal angioplasty but greater increase in intimal area at late time points (15).

In response to the limitations of the early porous balloon systems, a second generation of porous balloon catheters was developed, using new surface treatments or nonpressurized openings to deliver materials. Two unique strategies were employed to prevent fluid jetting from these catheters. The microporous balloon catheter consists of an inner porous balloon with an array of 25 $\mu$m holes surrounded by a permeable microporous membrane. The catheter described by Lambert et al. used a thin polycarbonate membrane with pores of approximately 0.8 $\mu$m in diameter (16). By using a much larger number of smaller pores, microporous catheters can deliver comparable fluid quantities at a given pressure with much lower fluid velocities at each pore. The resistance of this type of balloon to inflation is quite high. Consequently, the high pressure required for inflation causes the infusate to weep from the pores on the surface of the membrane (3).

Other devices were designed with separate inflation and infusion compartments (16–21). These catheters afford simultaneous high pressure angioplasty and low pressure infusion. The *Transport* coronary angioplasty catheter has an inner angioplasty balloon and an outer porous balloon with separate inflation port, which can be used to deliver agents before, during, or after angioplasty with the inner balloon (19). A similar design, often referred to as a *channel catheter*, typically consists of a standard (nonporous) angioplasty balloon covered by a network of perforated channels that can be perfused independently of the main balloon via an additional port (16,17,20,21). The *channeled balloon* catheter consists of 24 exterior channels, each with a single 100 $\mu$m drug-infusion pore, and is designed to perform simultaneous high-pressure angioplasty and low-pressure drug infusion. The pores are arranged in a spiral pattern along the entire length of the balloon. Therapeutic agents are infused at low pressure causing them to weep, rather than jetting from the exterior pores. The central lumen can be inflated at high pressure through a separate port, independent of the 24 drug delivery channels (17,20,21). The prototype balloons used in the 1993 study were made from poly(ethylene terephthalate), and they were 20 mm in length with a diameter of 2.75 mm. Drugs including tracer compounds like methylene blue, $^3$H heparin, and HRP were delivered with variable efficiency (25–50%) and generally rapid mural washout. Fluorescent and radiolabeled microspheres (15.5 $\mu$m Ce$^{141}$) can also be delivered in this fashion and with excellent vascular penetration (21).

A number of other novel catheters have been designed that use some modification of this technology or porous membranes to deliver pharmacological agents. The Local-med Infusion Sleeve is a multilumen array of perforated channels that can be employed in conjunction with a standard angioplasty balloon (18). The sleeve is loaded onto a standard dilation catheter prior to the procedure. After dilation of the vessel, the infusion sleeve is tracked over the catheter until the perforations are aligned with the balloon and site of injury. The balloon is reinflated, bringing the perforated channels into contact with the arterial wall. A catheter has been developed with a mechanical stent-like device inside the porous balloon (22). The mechanical infrastructure is used to distend and position the balloon against the wall of the vessel while fluids can be infused through the porous balloon at low pressures. This device was used to deliver 10 $\mu$m latex microspheres to atherosclerotic rabbit femoral arteries following angioplasty. The new device exhibited a delivery efficiency of 99.9% of the microspheres to the vascular and perivascular region as compared to 85.5% efficiency for a normal porous balloon.

The many experimental evaluations of porous catheter delivery systems have made one thing clear. The biggest challenge with the use of these devices is to minimize damage to the vessel wall while simultaneously producing efficient delivery of pharmacological agents. The damage-delivery relationship must be balanced in such a way that the benefits of local delivery are not overshadowed by the potentially negative consequences of the site specific delivery procedure (12).

## Hydrogel-Coated Balloon Catheters

Hydrogel-coated balloon catheters consist of a standard polyethylene balloon coated with a 5–10 $\mu$m (dry) layer of cross-linked polyacrylic acid chains. The coating can be made from high molecular weight, medical grade, soluble poly(acrylic acid) polymer covalently bound to the polyethylene balloon (23). The hydrogel coating swells to between 2 and 5 times its original thickness upon exposure to aqueous environments. Water and poly(acrylic acid) form a stable matrix into which any pharmacological or other agents can be dissolved for incorporation (24,25). Intramural delivery of the incorporated pharmacological agents is accomplished when the hydrogel coating comes into contact with the vessel wall upon balloon inflation. Loading of the hydrogel layer with drugs is achieved by one of two methods: either the inflated hydrogel-covered balloon can be immersed in an aqueous solution containing the agents or the balloon surface can be painted with known aliquots of the drug (24,26). Both the surface area of the hydrogel and the thickness of the coating are factors that limit the amount of water and drugs that can be absorbed by the device. Washout of drugs like heparin occurs within seconds from the standard hydrogel-coated balloon device (27,28), and a protective sleeve has been added to cover the hydrogel catheter and enable substantial (almost 75%) re-

tention on the order of minutes (29). In this way 2–5% of the loaded heparin has been delivered to the deep arterial wall during angioplasty (30), with small amounts of intramural heparin detected for over 48 hours (27,28). In addition to passive diffusion, hydrogel compression during balloon inflation results in "pressurized diffusion" of soluble agents from the hydrogel matrix directly into the arterial intimal surface (24). Thus, balloon pressure and the duration of inflation, as well as drug size and concentration, determine depth of penetration (29).

The hydrogel-coated balloon catheter has been used to deliver a vast array of pharmacological and biological agents to the arterial wall, including HRP (29), heparin (27–30), urokinase (26,31), D-phe-L-pro-L-arginyl chloromethyl ketone (PPACK) (23,25), and genes in the form of adenoviral vectors (32). Urokinase-coated balloons successfully reversed abrupt closure and dissolution of intracoronary thrombus in all 11 human patients in one trial (31).

Although the hydrogel balloon permits site-specific intravascular drug delivery to be performed directly during angioplasty with a standard balloon (3), with no additional tissue damage being observed (33), the system is limited by the rapid washout of loaded drugs upon exposure to the bloodstream. This necessitates limited dwell time in situ before deployment, and lack of applicability when large amounts of agent need to be deposited (3).

## Other Local Delivery Catheters

A number of other, novel, local and site-specific drug delivery catheters have been developed. A majority of them have been designed to overcome inherent problems in the more traditional balloon or porous balloon devices. One of the primary limitations of balloon-type catheter delivery systems is that the actual drug delivery time must be kept short because the balloon occludes the vessel and blocks distal blood flow. The Dispatch™ catheter (Scimed Life Systems, Maple Grove, MN) is specifically designed to overcome this limitation. The working component of the catheter consists of a POC-6 helical inflation coil that supports an inner urethane sheath. The sheath creates a large-diameter central lumen that maintains distal blood flow while the catheter is in place. In the inflated state, the helical coil and urethane sheath create channels, isolated from the bloodstream, into which a therapeutic fluid can be infused directly against the arterial wall (34–36).

The efficacy of the Dispatch™ catheter was assessed in vivo using $^3$H heparin delivered to 12 normal porcine coronary arteries. Following a simulated angioplasty procedure the Dispatch™ catheter was inflated to ~6 atmospheres at the angioplasty site, and $^3$H heparin was infused at 0.5 mL/min for either 60 seconds or 15 minutes. The study showed that, for the vessels infused for 15 minutes, 1.3% of the delivered heparin was still present in the arterial wall 5 minutes after the procedure, but that 85% of that heparin was gone 90 minutes after delivery (34). In another study, the Dispatch™ catheter was used to treat 6 human patients with evidence of intracoronary thrombus. Each of the patients successfully received 150,000 units of urokinase over a 30-minute period prior to, or following conventional angioplasty and/or directional atherectomy. In all 6 patients the local delivery of urokinase using the Dispatch™ catheter resulted in reduction of the stenosis and/or dissolution of the angiographic thrombus. The authors did note that one patient suffered from limited local ischemia resulting from a coronary side branch occlusion due to one of the catheter's coils (35).

The Dispatch™ catheter and hydrogel-coated balloons have been reported to provide a small intramural deposit of heparin and urokinase, respectively (26,34). It is thought that, at least in the case of urokinase, this creates a local reservoir of drug that may diffuse over time throughout the angioplasty site, resulting in a sustained local thrombolytic effect (35,36). In vitro experiments with human plasma have shown that a minimum urokinase concentration of 100 units/mL is required for clot lysis (37). Assuming that the spaces between the coils of the Dispatch™ catheter are filled entirely with the infusion fluid, the local urokinase concentrations is the same as that of the infusion solution. Indeed, 150,000 units of urokinase was delivered over approximately 30 minutes when reservoir solutions of approximately 10,000 units/mL were used (35,36).

Another family of local delivery catheters uses small needles to inject the therapeutic agent. The Infiltrator™ angioplasty balloon catheter is a type of microneedle injection catheter (38). In this particular design, three longitudinal metallic strips with six small injection nipples are mounted on the surface of a conventional balloon catheter. The nipples are connected to a central drug delivery port that is separated from the balloon lumen so that both high-pressure angioplasty and low-pressure drug infusion may be performed. Delivery of 0.4 mL of rhodamine B to porcine coronary arteries via a normal-sized balloon resulted in even distribution of the tracer within the wall of the vessel 10 min after infiltration. Barath et al. reported localized interruption of the internal elastic lamina at the site of nipple penetration and a mild thickening and edema of the media without dissection or separation of the vessel wall layers. The use of this catheter will expand into imaging, diagnosis and research as well as therapeutics. A second needle-injection catheter employs six, 250-$\mu$m (31-G) needles positioned symmetrically at the proximal tip of the catheter (39,40). The 5F polyethylene catheter was designed with a central lumen that allows it to be positioned under fluoroscopy using the guidewire from the previous angioplasty procedure. Once in place, the six preshaped needles can be extended laterally into the vessel wall, at which time drug can be injected into the perivascular tissue. It is hoped that the adventitia will act as a local reservoir, with subsequent drug delivery being achieved via diffusion and through the vasa vasorum. Relatively small doses of the fluorescent indicator (~5–7% of the systemically applied dose) resulted in a 500% increase in local tissue concentrations with a notable percentage of the applied marker still present at 14 days (40).

Iontophoresis, a form of electric current–facilitated transport, has been used, through various devices, to enhance the penetration of drugs through skin, directly to myocardium, and into blood vessel walls (12,41–45). A balloon catheter has recently been developed that uses iontophoresis through a semipermeable membrane to drive medications into the wall of a vessel. It has been reported that local tissue concentrations of hirudin were 50 times greater after iontophoretic therapy versus passive diffusion (45). Unfortunately, concentrations of the delivered agent, within the vessel wall, quickly decreased.

## Summary

Many of these catheter systems are currently being evaluated in human trials. Exciting preliminary reports have demonstrated reduced poststent restenosis with the transport-catheter delivery of Enoxparin low-molecular heparin. It is intriguing that efficacy requires heparin to be delivered at the time of deployment, not afterwards. Additional research will refine the technology and deployment strategies. Future developments will no doubt expand their potential in the therapeutic and diagnostic realm. The ability, for example, of a catheter to deliver contrast enhancing agents might be employed with sensitive noninvasive imaging modalities. The local delivery of radiotherapeutic sensitizers might further enhance the specificity of brachytherapy.

## DRUG DELIVERY FROM STENTS

### Introduction

Since its clinical introduction more than a decade ago, endoluminal stenting to relieve obstruction and restore coronary perfusion has become a mainstay in the repertoire of the interventional cardiologist. It was expected in 1998 that more than 500,000 percutaneous interventions involving coronary stents would be performed worldwide that year (46). Several trials have shown the superiority of stenting compared to balloon angioplasty in immediate and intermediate-term results (47,48). This has been largely due to a decrease in the need for repeat target lesion revascularization.

Early experience with stenting was notable for the frequent occurrence of acute and subacute thrombosis: up to 20% in a multicenter European registry from the early 1990s (49). Since then, two factors have contributed to the marked reduction in clinical thrombosis to less than 1%: (1) optimization of stent deployment techniques to include high pressure, in-stent angioplasty and (2) intensification of adjunct antiplatelet/anticoagulant regimens with the addition of ticlopidine and platelet glycoprotein IIb/IIIa receptor antagonists to conventional aspirin and heparin schedules. Attendant with this reduction in thrombosis has come an increase in hemorrhagic complications as well as an increase in other risks of systemic exposure to these medications, including neutropenia, thrombocytopenia, electrolyte and fluid shifts, and allergic reactions.

The efficacy of coronary stents has also been limited by intermediate-term restenosis from biologic events that primarily include neointimal hyperplasia. Unlike balloon angioplasty, coronary stenting induces more severe and chronic vessel injury that engenders a protracted response (50). Despite increased acute luminal gains compared with balloon angioplasty, however, the exuberant reactive and reparative responses to stent-induced injury still lead to clinical restenosis in 20–40% of stented arteries (46,47,49,51–54). Although a number of agents have been shown in experimental models of arterial injury to reduce the smooth muscle cell hyperplasia responsible for restenosis (55–58), all of these pharmacologic adjuvants have failed to show any benefit in clinical trials (59–64). Furthermore, weight-adjusted drug dosages in these animal studies were prohibitively high for use in clinical trials (65). Though this dichotomy might imply a lack of relevance of animal models to the pathogenesis of clinical restenosis, it is also likely that suboptimal drug delivery plays a role (66).

## Local Drug Delivery: Rationale, Experimental Basis and Biologic Targets

Given the limitations and risks of systemically administered drugs, increasing effort has been devoted to the evaluation of local means of drug delivery to the vasculature. As with the catheter systems, the motivation is the theoretical expectation of maximizing drug concentration and biological effect in local target tissues while maintaining low systemic levels. Achievement of these goals should produce desired therapeutic effects and avoid deleterious side effects. Experimental systems have indeed shown the effectiveness of local delivery. In one early study, rats subject to denuding carotid arterial injury were treated with heparin injected subcutaneously on a periodic basis, delivered continuously by intravenous pump, or released continuously from a drug-embedded matrix (ethylene–vinyl acetate copolymer [EVAc]) placed directly over the injured blood vessel or placed subcutaneously at a distant site (66). Heparin delivered periadventitially by local matrix was as effective as continuous intravenous heparin in reducing neointimal hyperplasia. This inhibition of luminal occlusion was achieved without a measurable increase in the activated partial thromboplastin time, verifying that local delivery can produce therapeutic effects without undesirable systemic effects and that the antiproliferative effects of heparin may well be divorced from the drug's antithrombotic effects. Perivascular release has now been extended to treatment of models of vascular injury using a variety of compounds, including antisense oligonucleotides (67), steroids (68), antimitotics (69), and antithrombotics (70).

In contrast to the acute and limited injury from balloon angioplasty and endothelial denudation, stent placement causes more severe vessel wall damage and a prolonged stimulus for neointimal hyperplasia (71). Cox et al. reported on the failure of intraluminally administered heparin or methotrexate to inhibit neointimal hyperplasia after stent-induced injury (72). To address whether this was a consequence of drug delivery or rather a lack heparin efficacy in stent-related complications, we evaluated several routes of heparin delivery on thrombosis and neointimal hyperplasia in a rabbit model of iliac artery injury (73). Arteries were subject to either balloon denudation alone or denudation followed by stent placement. Heparin was delivered intravenously or to the injured vessel from either the perivascular or endoluminal aspect of the artery. Perivascular administration was again achieved with EVAc matrices. Matrices were synthesized to provide either first-order release kinetics, with most drug released within three days, or release kinetics approaching zero-order, allowing prolonged steady release over two weeks. Heparin ionically bound to the stainless steel stents was rapidly released at the luminal aspect of the artery, with most drug eluted after a few hours. All forms of heparin delivery markedly reduced or eliminated thrombosis in stented arteries, although only prolonged perivascular administration reduced neointimal hyperplasia to the same degree as systemic heparin administration. Early perivascular release resulted in a less dramatic reduction in neointimal hyperplasia, and endoluminal delivery via heparin coating was no different compared to uncoated stents. These results confirmed that local delivery could treat stent-related sequelae in experimental models, but that drug efficacy is related to the site and duration of administration. The differential response of the variety of factors examined underscored the multifaceted nature of the vascular response to injury, suggesting the need to target multiple axes for successful treatment.

Our evolving understanding of the temporal and spatial characteristics of thrombosis and restenosis may help to explain the variability of drug effect with site and duration of delivery. Within the first few days of stent implantation, a marked deposition of fibrin, platelets, and erythrocytes sets the stage for subacute thrombotic occlusion, which typically occurs at 3–10 days after stent placement (74). This process is most pronounced at the stent struts where denudation and vessel injury is more severe. Drugs to inhibit this response may only need to be active at the stent–lumen interface and present for a short period of time. The process of neointimal hyperplasia leading to restenosis, however, appears to follow a different time course. The deep injury imposed by stent struts leads to an early inflammatory reaction involving mononuclear leukocytes adherent to the injured surface. The disappearance of these cells from the lumen by one week after injury coincides with the appearance of tissue-infiltrating macrophages in the nascent neointima. Subsequent smooth muscle migration from the media and proliferation in the neointima lead to vessel restenosis (50). This phase of the vascular response is quite prolonged in experimental models (75), suggesting that local therapy to modulate restenosis will likely require extended release of drug into the deeper vessel layers. Thus, treatment directed at thrombosis and restenosis may need to be targeted to different tissue elements at different points in time, reflecting the underlying pathobiology of these two processes.

## Coated Stents

Although perivascular matrices are useful in experimental models, this mode of drug delivery is not available currently for clinical percutaneous coronary interventions. A number of catheter-based systems have been developed for this purpose. Their utility is limited by low delivery efficiency (<1%) (76,77) and rapid drug washout (34). Delivery by catheter of several compounds including heparin (78), methotrexate (72), and colchicine (69) has not been shown to significantly reduce neointimal hyperplasia. Endovascular stents, in addition to providing a mechanical scaffolding, may be more attractive platforms from which to implement local therapy. This can be accomplished by two broad strategies: (1) immobilization of drug to the stent surface and (2) continuous release of drug from the stent.

**Drug Immobilization.** The interaction among blood-borne elements, the injured vessel surface, and the stent itself appear to be of the earliest events in the vascular response to device implantation. It is appealing to attach a drug to the stent capable of favorably modulating this interaction. Heparin, the systemic anticoagulant of choice, would seem to be an ideal candidate. There has been long-standing interest in heparin coating for a variety of catheters and indwelling devices (79). Several studies have evaluated heparin-coated stents in animal models. Chronos et al. showed a marked reduction in platelet attachment to such devices in a baboon arteriovenous shunt ex vivo system, and others demonstrated diminished occlusive thrombosis in rabbit and porcine arterial injury models (80,81). These studies in animal models have focused on acute indices of thrombosis, such as platelet deposition, or surrogate markers, such as animal survival.

Whereas experimental models frequently encounter acute thrombosis within the first 24 hours, clinical subacute thrombosis occurs between 3 and 10 days after stent implantation. To assess the effects of stent surface on thrombus generation within this time frame, our laboratory used histologic evaluation of mural thrombus 3 days after implantation of heparin-coated stents in rabbit iliac arteries (C. Rogers, M. Kjelsberg, P. Seifert, and E. Edelman, unpublished data). Denuded vessels received four types of stents: uncoated stainless steel, steel coated with the inert polymer poly(dichloroparaxylylene), steel with surface-bound heparin (via the binding agent, alkylbenzyldimethylammonium chloride), or steel coated with polymer and heparin. Histologic evaluation at 3 days showed a two-fold reduction in mural thrombus by stent-bound heparin (with or without polymer) compared to uncoated and polymer-coated stents. Yet, this reduction in thrombus deposition was not accompanied by a reduction in inflammatory cell number or in subsequent neointimal hyperplasia. Such data not only predict limited utility in combating the more chronic aspects of the vascular response to injury, but they also shed light on the basic biology of vascular repair. Despite our greatest hopes, the major portion of the thrombotic response appears as an event independent from the inflammation and proliferation that follow.

The sum total of these preclinical experiments was verified in the Benestent-II® Pilot Study (C. Rogers, M. Kjelsberg, P. Seifert, and E. Edelman, unpublished data), which evaluated the clinical safety and efficacy of heparin-coated stents in conjunction with alterations in adjuvant anticoagulant therapy. A total of 203 patients with stable angina, subdivided into four groups, successfully underwent implantation of heparin-coated stents for treatment of de novo lesions. Three groups received heparin and coumadin with progressively delayed institution of therapy, and the fourth group was treated with aspirin, adding ticlopidine as an adjunct. Although no subacute thrombotic events were observed, study investigators point out the probable multifactorial nature of the favorable outcomes. Although promising, these results await confirmation by larger randomized trials.

**Drug Release.** Whereas the events culminating in thrombosis take place principally early and at the lumen–vessel boundary, the processes leading to neointimal hyperplasia and restenosis appear to be protracted in duration and occur deeper within the arterial wall. Consequently, continuous and prolonged local drug therapy may be required to inhibit restenosis. This may be accomplished by directly coating the stent with drug or by coating the stent with a drug-eluting polymer matrix. In general, drug release from stent surfaces is rapid and on this basis may have a limited impact on the more prolonged phases of the vascular response to injury. The versatility of polymer matrices as drug reservoirs may expand local stent-based drug release. Although polymer materials have been used in drug delivery for some time, their use as vehicles for percutaneous coronary therapy is in its infancy. Polymeric bioerodible stents have been proposed as both drug reservoir and temporary scaffolding while reparative processes occur in the vascular wall. Despite the theoretical advantage of avoiding permanent placement of a foreign body, stents constructed solely from polymer materials have been limited thus far by significant thrombotic and inflammatory sequelae (82,83). Composite stents offer the mechanical strength of metal combined with the drug-carrying and eluting properties of polymer materials. Lambert et al. showed the feasibility of this type of system, using nitinol stents coated with a polyurethane matrix to locally deliver forskolin to rabbit carotid arteries (84). Most of the drug was released within 24 hours of stent deployment. Another group used a tantalum wire coil stent coated with high molecular weight polymer of poly(L-lactic acid) to deliver dexamethasone to porcine coronary arteries (85). Although this intervention had no effect on neointimal hyperplasia, the study did show sustained release evidenced by detectable serum dexamethasone levels at 28 days and polymer material biocompatibility supported by a lack of difference in tissue response to polymer-coated versus uncoated stents.

**Local Pharmacology.** Stent-based drug delivery is an appealing but complex potential therapeutic modality, dependent upon device and drug properties. The amount of drug that can be immobilized to or released from a stent is principally determined by the mode of drug attachment and stent configuration. The processes of coating and drug incorporation can modify the biological activity of the compound of interest. All of these issues must be optimized before we can expect a beneficial response from this means of drug delivery.

**Coating Process.** Specific processes of drug attachment may limit the availability of a drug's active sites or may expose drugs to denaturing or degrading stimuli. Polymer coatings or drug binding agents may themselves alter the activity of the drug, reduce the carrying capacity of the stent or induce a detrimental vascular response independent of the drug. The evolution of heparin coatings serves as a primary example of these phenomena. The early use of quaternary ammonium compounds to ionically bind heparin was notable for rapid drug leaching from the stent vehicle. Attempts at bonding heparin covalently to stents were initially confounded by loss of anticoagulant activity, presumably due either to masking of the active binding site or undesirable conformational changes that abrogated heparin's effect (79). More recent methods for attachment of heparin at inactive sites or at its end point (83) have maximized retention of biologic activity.

Many of these same factors influence drug delivery with release systems. As discussed earlier, a variety of polymers have been used in composite stents. Issues of biocompatibility may limit the utility of some of these polymers. With poly(glycolic acid), for instance, the resultant inflammatory response may be a function of the total mass of polymer in the stent (86), thereby limiting matrix drug-carrying capacity. In addition, the conditions required for matrix synthesis may alter drug activity, as we learned with the delivery of basic fibroblast growth factor (FGF-2) (87). Although incorporation into and release from EVAc matrices followed desired kinetics, over 99% of the growth factor's biological activity was lost. Basic FGF is intensely sensitive to extreme environmental conditions, and only very specific and gentle matrix production techniques enabled stabilization of the FGF-2 and retention of its effects.

### Stent Geometry.

The biologic activity of a drug is a function of its dose and distribution within tissue. Stents generally cover less than 10% of targeted vessel wall segments, limiting the surface area available for drug loading and, as a result, the maximum dose of drug available. The amount of drug deposited in distant tissues depends upon the solubility and diffusive properties of the drug as well as the individual design of the device. A careful balance exists between the release of the drug from each strut and the ability of the drug to diffuse through and be distributed within the surrounding tissue. As drug elution is centered at the stent struts a concentration gradient forms with highest drug levels just adjacent to stent struts and lower levels in the region between struts. There can therefore be a gradient of effect within a given tissue over a spectrum of activity ranges from no effect to the desired therapeutic effect to frank toxicity. Struts spaced too far apart leave drug-poor areas, whereas struts too close together may produce areas of overlap. A highly hydrophobic drug such as paclitaxel might be expected to accumulate and persist at potentially toxic levels in close proximity to stent struts.

### Drug Pharmacokinetics.

Release kinetics from the drug-eluting polymer coatings are determined by a number of factors, including the physicochemical characteristics of the drug and the material properties of the coatings and delivery vehicles. For example, insoluble compounds diffuse slowly from the coated stents into the intercellular milieu. Hydrophilic materials swell as they absorb water, thereby altering material characteristics, and potentially release kinetics as well. Porous polymer materials release drugs far more rapidly than less porous materials, further illustrating the profound influence of polymer-drug properties on drug delivery.

Although polymeric drug delivery systems have been well characterized and sophisticated control of kinetics can achieve almost any pattern of drug release, much less is known about the fate of a drug once it leaves the delivery device. Systemic pharmacokinetics treats an organism as a limited number of homogeneous compartments between which drugs flow with definable rate constants. Target tissues for local drug delivery, however, must be viewed as a continuum with the establishment of a concentration gradient from delivery site to target tissue. Factors influencing local drug levels include the ratio of bound versus soluble drug, volume of distribution, local metabolism and clearance, and specific versus nonspecific binding (12).

### Tissue Pharmacokinetics.

Recognition of the importance of all of these issues should caution against the assumption that local drug delivery necessarily translates into local drug deposition and a consequent focal therapeutic effect. A biologic or therapeutic agent in this aqueous intercellular milieu remains where released only if it is hydrophobic, largely protein bound, or partitions into the bound or internalized tissue fraction. Paclitaxel, a potent antitumor agent, possesses many of these characteristics. It is a complex natural product with bulky hydrophobic substituents, poor aqueous solubility, extensive protein binding in blood, and an intracellular site of action, where it promotes the abnormal polymerization of microtubules (88). Initial work with this compound and vascular tissue revealed its antiproliferative effect on rat smooth muscle cells in vitro and in vivo. Subsequently, Axel and colleagues locally delivered a single dose of paclitaxel via a microporous balloon system to the injured segment of rabbit carotid arteries and saw an approximate 20% reduction in stenosis in treated versus control animals (89). The efficacy of single dose paclitaxel may be due to its hydrophobicity, intracellular site of action, and its rapid and prolonged biological effect.

The results with paclitaxel-like compounds can be contrasted with the use of heparin. When heparin was rapidly released from stents to rabbit iliac arteries no effect was observed on neointimal hyperplasia (73). Other work from our laboratory showed minimal deposition of radiolabeled-heparin delivered by stents with ionically bound heparin (90). This was thought to arise from the marked solubility of heparin and its consequent absorption and dilution into the bloodstream, resulting in a markedly larger volume of distribution than anticipated for the amount of drug delivered. Computational models give us insight into the basis of experimental results. We have previously described a series of rigorous pharmacokinetic models to simulate vascular heparin deposition (71). When applied to a cross section of artery uniformly loaded with heparin, these models predicted low heparin concentrations in the artery after 1 hour. This supported rapid clearance of heparin and suggested that it and similarly soluble drugs would require continuous release to achieve persistently therapeutic local levels of drug. These types of simulations allow far greater spatial resolution of local forces and events than would be feasible otherwise. For example, delineation of drug partitioning within vessel layers as well as into soluble, bound, or internalized fractions within a section of tissue would be impossible with current analytical techniques. As valuable as these models are, it is impossible to rigorously define all of the parameters influencing local drug distribution within target tissues.

In addition to these factors, the state of the vessel and nonspecific binding play important roles in drug deposition and effect. We compared vessel wall deposition of basic fibroblast growth factor (FGF-2) in native versus injured rat carotid arteries (91). There was over a two-fold increase in FGF-2 deposition within injured arteries showing intimal hyperplasia versus control arteries, providing evidence that the vessel itself affects local pharmacology. Nonspecific binding, although removing drug from the "active" pool, may also sequester and alter biological efficacy. This phenomenon is most apparent in the interaction of growth factors with extracellular matrix. We looked at the response of cultured endothelial and smooth muscle cells to FGF-2 and transforming growth factor-beta 1 (TGF-$\beta$1) administered by bolus, continuous release, or following exposure of the growth factors to extracellular matrix (97). Continuous FGF-2 release was more potent than bolus administration in inducing cellular proliferation. With TGF-$\beta$1, however, bolus dosing proved more effective in inhibiting cellular growth. Importantly, both growth factors bound to extracellular matrix, but only FGF-2 was released from the matrix in a controlled fashion that significantly augmented its biological effect.

**Endothelial Cell Seeding of Stents.** Other attempts at limiting intimal proliferation with implantable devices include seeding of stents with genetically engineered endothelial cells (85,93,94). It has previously been postulated that the presence of endothelial cells alone can reduce the amount of thrombosis and intimal proliferation associated with stenting. Dichek et al. (93) coated stents with fibronectin and grew endothelial cells engineered to express either bacterial $\beta$-galactosidase or human tissue-type plasminogen activator. Expression of these proteins was monitored in vitro, and cells remained seeded after balloon expansion. Scott et al. (95) seeded human dermal microvascular endothelial cells with the simian virus 40 large T-antigen gene for 2 weeks in culture. This group was able to seed these cells directly onto the stents without coupling to fibronectin as well as use them in a simulated clinical setting. The same stents were seeded cells, frozen for 4 months and then implanted in vivo. Although they did find that many of the cells dislodged during balloon expansion, they were able to form another monolayer after 3 days in culture. Flugelman et al. (94) applied pulsatile forces to these endothelial cell-seeded stents, and found that substantial amounts of cells on the lateral surfaces of the stent remained intact, with fewer cells remaining on the luminal and abluminal surfaces. Only future technical innovations and experimental validation will determine whether there is promise to this methodology.

**Conclusion**

The success and prevalence of coronary stenting is limited by early thrombosis and late restenosis. Marked reduction in thrombotic sequelae have been achieved through the use of aggressive antiplatelet and anticoagulant regimens, which, nonetheless, place patients at risk for hemorrhage and other adverse effects. On the other hand, restenosis has not been appreciably affected by a variety of systemic therapeutics. Experimental studies have suggested a pharmacologic rather than biologic failure of these interventions. Local drug delivery is emerging as a technology to circumvent this phenomenon and deliver therapeutic doses of drug to target tissues while avoiding significant systemic exposure and consequent side effects.

Stents may be the ideal vehicles both to provide mechanical stability and to act as reservoirs for pharmacotherapy. Suitable drug candidates may be immobilized to the stent surface or may be released via drug-eluting polymer coatings. The clinical promise of this technology is evident from the Benestent-II™ Pilot Study, which used heparin-coated stents and reported no subacute thrombotic occlusions in a limited number of patients. Local drug therapy involves complex processes that affect delivery to and distribution within tissues. Determinants of drug loading onto stents include stent surface area, polymer and binding agent coating, mode of drug attachment, matrix fabrication, and specifics of matrix design. Once a drug is released, factors including local clearance, bound versus soluble fraction, nonspecific binding, and volume of distribution determine the fate and biologic effect of the agent. Our understanding of local pharmacology continues to evolve through complementary experimental and modeling systems with the goal of providing more rational design of delivery devices and more optimal choice of therapeutic agents.

## DELIVERY FROM POLYMER DEVICES

Polymer-based delivery of therapeutic agents has been employed in a wide range of applications from contraceptives to antianginal drugs. Its application to the cardiovascular system has been long investigated. The predictable and prolonged nature of release and the malleability of the technology enables delivery from a number of sites. A polymeric matrix or microsphere, encapsulating or impregnated with drug, first begins to release its agent through the pores and channels inherent in its structure. As the polymer degrades, drug can then also diffuse out between polymer chains. Transdermal adhesive drug patches have been used with antianginals and antihypertensives such as nitroglycerin, clonidine, the beta-blockers timolol and propanolol and, more recently, nitrendipine (96–98). More experimental systems now enable subcutaneous, extravascular, and pericardial administration as well.

The use of these systems has not only provided potential new therapeutic modalities but has also revealed much about the nature of biological disease. EVAc matrices, for example, have been used to release both standard, anticoagulant, and modified, nonanticoagulant heparin (99,100). Release kinetics were tailored to provide rapid first-order, or more prolonged near zero-order, release by alterations in the size, shape, and properties of the drug; geometry of the controlled release device; and application of a coating around the device. The ability of heparin released in this manner to inhibit smooth muscle cell proliferation and thrombosis was evaluated in three different models of vascular injury: after balloon catheter arterial de-endothelialization, after implantation of an expandable stainless steel stent, and after venous–arterial interposition grafting. In all instances the perivascular release of heparin limited intimal hyperplasia from proliferating smooth muscle cells to an identical extent as heparin systemic infusion and with one-fifth the intravenous dose, and without systemic effect. These results demonstrated that controlled release of a potent smooth muscle cell inhibitor, with profound systemic side effects, could deliver a dose low enough to avoid systemic effects but high enough to provide local protection against vascular disease. More importantly they showed that this accelerated form of vascular disease could be treated locally and paved the way for many other forms of local therapy. Since then a large range of agents have been administered from the perivascular space, including antithrombotics, antimitotics, genes and proteins, antisense oligonucleotides (101–104), growth factors and their antagonists, angiogenesis factors (105).

The work with heparin-binding growth factors is a wonderful example of how local release systems might mimic natural phenomena. Cells in culture respond optimally to the controlled release, rather than bolus administration, of FGF-2, and with good reason. Because this compound lacks a signal sequence, it relies on endogenous forms of controlled release for biological effect. Controlled release may be far more physiologic than bolus release of large doses. When FGF-2 controlled release devices were implanted into the perivascular space of injured rat carotid arteries intimal mass, cell proliferation and perivascular angiogenesis were increased in a dose dependent fashion (106). Thus, the perivascular release of this potent mitogen retained not only its biologic activity but may have reproduced the physiologic means by which the growth factor is released and metabolized in vivo.

### Intramyocardial, Pericardial Delivery

Drugs can additionally be released into or around the myocardium. Pacemaker leads have long served as delivery depots. Fibrosis and subsequent conduction block were markedly reduced when leads were coated with porous polyurethane or silicon impregnated with dexamethasone (107). Attempts were also made to release colchicine, hirulog, aspirin, and steroids (108,109). Polymer devices have also been made which can be implanted directly proximal to the ventricular myocardium. Polyurethane devices loaded with antiarrhythmic agents markedly increased the threshold for induction of ventricular tachycardia by rapid ventricular pacing, while intravenous or bolus administration exhibited no effect (110–112).

Drugs can also be delivered directly to the heart via synthetic heart valves. Porcine aortic valve replacements are highly subject to calcification within a span of five years from implantation. Ethanehydroxydiphosphonate and other anticalcification drugs released from polymeric valve ring collars inhibited calcification of valve cuspal implants in rats (113–115).

Drug-loaded polymer delivery devices have proven to be quite successful in their intended capacity as local convey-

ors of therapeutic agent. These devices have been fabricated in numerous forms from a vast array of polymers and for various indications and sites of application. In addition to the aforementioned polymer devices, two extensively studied methods of polymeric drug delivery include endoluminal paving and microparticle delivery.

### Endoluminal Paving

First described in 1988, endoluminal paving is a therapeutic approach to reduce restenosis and loss of patency providing vessel wall strength and support as well as a physical barrier to thrombogenic factors and a possibility for sustained drug release (116). In addition to favorable tissue reaction, endoluminal paving provides a temporary device that is biodegradable, a structural support for damaged tissue, and a method of preventing wall recoil (116). It also presents a barrier between the endoluminal surface and inflammation-inducing elements of the blood (116), is assumed to prevent endoluminal thickening following device degradation (117), and can be a method of physically targeting drugs, an option not available with current systemic drugs (116).

In all, three main types of paving have been developed. The first is solid paving, which involves the use of thin sheets or tubes of a biodegradable polymer. Polymer material surrounds an intraluminally placed thermoforming catheter. The catheter heats the polymer to just above its $T_m$, and propels the polymer to coat the endoluminal tissue as its balloon is inflated (118). The polymer coating is then allowed to cool to body temperature leaving a solid polymer coating on the endoluminal surface. The second type of endoluminal paving, gel paving, is performed in the same manner as solid paving but uses a polymeric hydrogel or colloidal system instead of a solid polymer film (116). Gel paving is less stable, lasting only days to weeks as the hydrogel biodegrades through bond scission or bioerodes through solvation and physical thinning. Moreover, paving focuses to a far lesser extent on mechanical support and far more on providing a barrier to blood-borne thrombogenic factors and other high molecular weight compounds. The final type of endoluminal paving is liquid paving, which makes use of "flowable" polymeric, macromeric, or prepolymeric solutions. With heating, these solutions adhere and interact with the endoluminal surface.

Whether using the thermal or photothermal procedure for paving, the reaction of the polymer with the underlying tissue is favorable. Vessel media has been shown to remain intact and without further damage following application of polymers as shown by histological cross section (118). When melted, the polymer is capable of molding to the exact shape of the vessel, inclusive of crevices and fissures in the surface, as the polymer binds to the tissue by physical interlocking (117). When the catheter balloon is removed, a smooth surface is left endoluminally despite the previous irregular and thrombogenic surface of the vessel (116). Liquid paving can reportedly interact with the vessel by altering the tissue surface charge, porosity, lubricity, and drug delivery capabilities (116).

Both in vitro and in vivo studies have been described using gel paving technology. Thrombi and platelet deposition have been reduced in hydrogel-paced segments of isolated perfused rat carotid (119), bovine coronary, and canine carotid arteries. In vivo studies yielded similarly promising results. Patency was determined in a rat carotid arterial crush models (117), a New Zealand White rabbit balloon injury model (120), and in canine carotid arteries (118). Tests still need to be performed to definitively determine paving efficacy and safety.

## Microparticle Delivery of Drugs to the Cardiovascular System

Inefficient delivery of drug to tissue, diffusion of drug away from infusion site (76), and injury to tissue caused by infusion pressure (121) all diminish the therapeutic value of catheter-based drug delivery systems. Superior drug vascular deposition might be achieved with polymeric microparticles as delivery vehicles (76). Injected microparticles adhere to and remain with the injected tissue sites far longer than free drug. The particles might become lodged in vasa vasorum as well and from these sites serve as local diffusion depots (122–124). Rates of release can be modulated by modifying the chemical composition or the molecular weight of the polymer (122). Other modifications to microsphere fabrication methods enable it to embody the requirements necessary to function as a successful delivery device. Prerequisites for polymeric drug carriers include a small enough size for infusion through contemporary catheter designs, the capability of carrying a high concentration of drug and releasing that drug in a controlled, sustained manner, and the properties of being both biodegradable and biocompatible (122).

Dexamethasone-encapsulating poly(lactide-co-glycolide) spheres, with prolonged release kinetics providing drug for over a month, were infused via porous balloon catheters into balloon denuded rat carotid arteries. Tissue levels of drug were much higher in the locally treated segment of artery than in adjacent segments or in contralateral control arteries, further substantiating the absence of distant drug distribution. Microspheres that persisted in the artery for at least seven days eliminated neointimal formation, but did not produce any significant inflammatory responses. Conversely, the systemic administration of dexamethasone given through intraperitoneal injection did not reduce neointimal growth (122). Radioactive and HRP-laden microspheres were retained for as long as 24 hours and were found mainly in the dissection planes of the artery and the adventitia (125). As studies are carried out for longer timepoints the potential of this therapy will become evident.

## BIBLIOGRAPHY

1. R. Wilensky et al., *Trends Cardiovasc. Med.* **3**, 163–170 (1993).
2. R.S. Schwartz, D.R. Holmes, and E.J. Topol, *J. Am. Coll. Cardiol.* **20**, 1284–1293 (1992).
3. R. Riessen and J.M. Isner, *J. Am. Coll. Cardiol.* **25**(5), 1234–1244 (1994).
4. J.H. Anderson, C. Gianturco, and S. Wallace, *Invest. Radiol.* **16**(6), 496–500 (1981).
5. B.J. Meyer et al., *Circulation* **90**(5), 2474–2480 (1994).
6. M.M. Thompson et al., *Eur. J. Vasc. Surg.* **8**(4), 423–434 (1994).
7. T. Kerenyi et al., *Exp. Mol. Pathol.* **49**, 330–338 (1988).
8. B. Goldman, H. Blanke, and H. Wolinsky, *Atherosclerosis* **65**, 215–225 (1987).
9. L.E. Duncan, K. Buck, and A. Lynch, *J. Atheroscler. Res.* **5**, 69–79 (1965).
10. E.G. Nabel et al., *J. Clin. Invest.* **91**, 1822–1829 (1993).
11. H. Wolinski and S.N. Thung, *J. Am. Coll. Cardiol.* **15**, 475–481 (1990).
12. A. Nathan and E.R. Edelman, in E.R. Edelman, ed., *Molecular Interventions and Local Drug Delivery*, Saunders, Philadelphia, 1995, pp. 29–52.
13. M.Y. Flugelman et al., *Circulation* **85**, 1110–1117 (1992).
14. G.D. Chapman et al., *Circ. Res.* **71**, 27–33 (1992).
15. M.L. Stadius, C. Collins, and R. Kernoff, *Am. Heart J.* **126**, 47–56 (1993).
16. C.R. Lambert, J.E. Leone, and S.M. Rowland, *Coronary Artery Dis.* **4**, 469–475 (1993).
17. M.K. Hong et al., *Circulation* **86**(Suppl. 1), I-380 (1992).
18. A.V. Kaplan et al., *J. Am. Coll. Cardiol.* **23**, 187A (1994).
19. D.C. Cumberland et al., *J. Am. Coll. Cardiol.* **23**, 186A (1994).
20. M.K. Hong et al., *Coronary Artery Dis.* **4**, 1023–1027 (1993).
21. M.K. Hong et al., *Catheter. Cardiovasc. Diagn.* **34**(3), 263–271 (1995).
22. R.L. Wilensky et al., *J. Am. Coll. Cardiol.* **21**, 185A (1993).
23. G.L. Nunes et al., *J. Am. Coll. Cardiol.* **23**(7), 1578–1583 (1994).
24. M. Azrin, *Proc. 1st Ann. Int. Symp. Local Cardiovasc. Drug Delivery*, Cambridge, Mass., Sept. 28–29, 1995.
25. G.L. Nunes et al., *Circulation* **86**, I-380 (1992).
26. J.F. Mitchel et al., *Circulation* **90**, 1979–1988 (1994).
27. M.A. Azrin et al., *Circulation* **88**, I-310 (1993).
28. M.A. Azrin et al., *Circulation* **90**, 433–441 (1994).
29. D.B. Fram et al., *J. Am. Coll. Cardiol.* **23**, 1570–1577 (1994).
30. D.B. Fram et al., *J. Am. Coll. Cardiol.* **21**, 118A (1993).
31. J.F. Mitchel et al., *Circulation* **88**, I-660 (1993).
32. P.G. Steg et al., *Circulation* **88**, I-660 (1993).
33. R. Riessen et al., *Hum. Gene Ther.* **4**, 749–758 (1993).
34. D.B. Fram et al., *J. Am. Coll. Cardiol.* **23**(1A–484A), 186A (1994).
35. R.G. McKay et al., *Catheter. Cardiovasc. Diagn.* **33**(2), 181–188 (1994).
36. J.F. Mitchel and R.G. McKay, *Catheter. Cardiovasc. Diagn.* **24**(2), 149–154 (1995).
37. C. Zamarron et al., *Thromb. Haemostasis* **52**, 19–23 (1984).
38. P. Barath and A. Popov, *Proc. 1st Ann. Int. Symp. Local Cardiovasc. Drug Delivery*, Cambridge, Mass., Sept. 28–29, 1995.
39. P. Gonschior et al., *J. Am. Coll. Cardiol.* **23**, 188A (1994).
40. P. Gonschior et al., *Coronary Artery Dis.* **6**(4), 329–334 (1995).
41. R.J. Levy and V. Labhasetwar, *Proc. 1st Ann. Int. Symp. Local Cardiovasc. Drug Delivery*, Cambridge, Mass., Sept. 28–29, 1995.

42. R.S. Schwartz, *Proc. 1st Ann. Int. Symp. Local Cardiovasc. Drug Delivery*, Cambridge, Mass., Sept. 28–29, 1995.

43. I. Soria et al., *Circulation* **88**, I-660 (1993).

44. B. Avitall et al., *Circulation* **85**, 1582–1593 (1992).

45. A. Fernandez-Ortiz et al., *Circulation* **89**(4), 1518–1522 (1994).

46. The Epistent Investigators, *Lancet* **352**, 87–92 (1998).

47. P. Serruys, P. De Jaegere, and F. Kiemeneij, *N. Engl. J. Med.* **331**, 489–495 (1994).

48. D. Fischman, M. Leon, and D. Baim, *N. Engl. J. Med.* **331**, 196–501 (1994).

49. P. Serruys, B. Straus, and K. Beat, *N. Engl. J. Med.* **324**, 13–17 (1991).

50. E. Edelman and C. Rogers, *Am. J. Cardiol.* **81**(7A), 4E–6E (1998).

51. J. Carrozza, R. Kuntz, and M. Levine, *J. Am. Coll. Cardiol.* **20**, 328–337 (1992).

52. M. Levine, B. Leonard, and J. Burke, *J. Am. Coll. Cardiol.* **16**, 332–339 (1990).

53. S. Ellis, M. Savage, and D. Fischman, *Circulation* **86**, 1836–1844 (1992).

54. P. Gordon, C. Gibson, and D. Cohen, *J. Am. Coll. Cardiol.* **21**, 1166–1174 (1993).

55. G. Ferns, M. Reidy, and R. Ross, *Am. J. Pathol.* **137**, 403–413 (1990).

56. M. Liu, G. Roubin, and K. Robinson, *Circulation* **81**, 1089–1093 (1990).

57. D. Muller, E. Topol, and G. Abrams, *Circulation* **82**, III-429 (Abstr.) (1990).

58. D. Handley et al., *Am. J. Pathol.* **124**, 88–93 (1986).

59. E. Hoberg, F. Schwarz, and A. Schomig, *Circulation* **82**, III-428 (1990).

60. K. Lehmann, R. Doria, and J. Feuer, *J. Am. Coll. Cardiol.* **17**, 181A (Abstr.) (1991).

61. H. Whitworth, G. Roubin, and J. Hollman, *J. Am. Coll. Cardiol.* **8**, 1271–1276 (1986).

62. D. Faxon, T. Spiro, and S. Minor, *J. Am. Coll. Cardiol.* **19**, 258A (1992).

63. The Mercator Investigators, *Circulation* **86**, 100–110 (1992).

64. C. Pepine, J. Hirshfeld, and R. Macdonald, *Circulation* **81**, 1753–1761 (1990).

65. A. Lincoff, E. Topol, and S. Ellis, *Circulation* **90**, 2070–2084 (1994).

66. E. Edelman, D. Adams, and M. Karnovsky, *Proc. Natl. Acad. Sci. U.S.A.* **87**, 3773–3777 (1990).

67. M. Simons, E. Edelman, and J. DeKeyser, *Nature (London)* **359**, 67–70 (1992).

68. A. Villa, L. Guzman, and W. Chen, *J. Clin. Invest.* **93**, 1243–1249 (1994).

69. R. Wilensky, I. Gradus-Pizlo, and K. March, *Circulation* **86**, I-52 (1992).

70. D. Muller, G. Golomb, and D. Gordon, *Circulation* **86**, I-381 (Abstr.) (1992).

71. C. Rogers et al., *Circulation* **90**, I-508 (Abstr.) (1994).

72. D. Cox, P. Anderson, and G. Roubin, *Circulation* **78**, II-71 (Abstr.) (1988).

73. C. Rogers, M.J. Karnovsky, and E.R. Edelman, *Circulation* **88**(3), 1215–1221 (1993).

74. J. Garasic, C. Rogers, and E. Edelman, in K. Beyar et al., eds., *Frontiers in Interventional Cardiology*, Martin Dunitz, UK, 1997, pp. 95–100.

75. H. Hanke, J. Kamenz, and S. Hassenstein, *Eur. Heart J.* **16**, 785–793 (1995).

76. R. Wilensky, K. March, and I. Gradus-Pizlo, *Am. Heart J.* **129**, 852–859 (1995).

77. C. Thomas, K. Robinson, and G. Cipolla, *J. Am. Coll. Cardiol.* **23**, 187A (Abstr.) (1994).

78. L. Gimple, R. Owen, and V. Lodge, *Circulation* **82**, III-338 (Abstr.) (1992).

79. A. Lunn, in U. Sigwart, ed., *Endoluminal Stenting* Saunders, Philadelphia, 1996 pp. 80–83.

80. S. Bailey et al., *Circulation* **86**, I-186 (1992).

81. P. Hardhammar, H. van Beusekom, and H. Emanuelsson, *Circulation* **93**, 423–430 (1996).

82. J. Murphy, R. Schwartz, and W. Edwards, *Circulation* **86**, 1596–1604 (1992).

83. The Benestent-II Investigators, *Circulation* **93**, 412–422 (1996).

84. T. Lambert, V. Dev, and R. Eldad, *Circulation* **90**, 1003–1111 (1994).

85. D. Holmes, A. Camrud, and M. Jorgenson, *J. Am. Coll. Cardiol.* **24**, 525–531 (1994).

86. A. Lincoff et al., *J. Am. Coll. Cardiol.* **23**, 18A (Abstr.) (1994).

87. M. Staab, D. Holmes, and R. Schwartz, in S.U, Editor. *Endoluminal Stenting*, Saunders, Philadelphia, 1996, pp. 34–44.

88. E. Edelman and M. Lovich, *Nat. Biotechnol.* **16**, 136–137 (1998).

89. S.J. Sollot, L. Cheng, and R. Pauly, *J. Clin. Invest.* **95**, 1869–1876 (1995).

90. D. Axel, W. Kunert, and C. Goggelmann, *Circulation* **96**, 636–645 (1997).

91. M. Lovich and E. Edelman, *Am. J. Physiol.* **271**, H2014–H2024 (1996).

92. E.R. Edelman, M.A. Nugent, and M.J. Karnovsky, *Proc. Natl. Acad. Sci. U.S.A.* **90**, 1513–1517 (1993).

93. D.A. Dichek et al., *Circulation* **80**, 1347–1353 (1989).

94. M.Y. Flugelman et al., *Circ. Res.* **70**, 348–354 (1992).

95. N. Scott, F. Candal, K. Robinson, and E. Ades, *Am. Heart J.* **129**, 860–865 (1995).

96. L. Ruan et al., *J. Controlled Release* **20**, 231–236 (1992).

97. P.H. Vlasses et al., *J. Cardiovasc. Pharmacol.* **7**(2), 245–250 (1985).

98. T. Nagai et al., *J. Controlled Release* **1**(3), 239–246 (1985).

99. R. Langer and J.B. Murray, *Appl. Biochem. Biotechnol.* **8**, 9–24 (1983).

100. E.R. Edelman et al., *Biomaterials* **12**, 619–626 (1991).

101. E.R. Edelman et al., *Circ. Res.* **76**(2), 176–182 (1995).

102. M.R. Bennett et al., *J. Clin. Invest.* **93**, 820–828 (1994).

103. M. Simons, E.R. Edelman, and R.D. Rosenburg, *J. Clin. Invest.* **93**, 2351–2356 (1994).

104. J. Abe et al., *Biochem. Biophys. Res. Commun.* **198**, 16–24 (1994).

105. F.W. Sellke et al., *Am. J. Physiol.* **267**, H1303–H1311 (1994).

106. E.R. Edelman et al., *J. Clin. Invest.* **89**, 465–471 (1992).

107. D.H. King et al., *Am. Heart J.* **106**, 1438–1440 (1983).

108. R.J. Levy et al., in C.G. Gebelein, ed., *Biotechnology and Bioactive Polymers*, Lionfire, Edgewater, Fla., 1994, pp. 259–268.

109. H.G. Mond and K.B. Stokes, *Pace Pacing Clin. Electrophysiol.* **15**, 95–107 (1992).

110. A. Sintov et al., *J. Cardiovasc. Pharmacol.* **16**, 812–817 (1990).

111. R. Siden et al., *J. Cardiovasc. Pharmacol.* **19**, 798–809 (1992).

112. V. Labhasetmar et al., *J. Pharm. Sci.* **83**, 156–164 (1994).

113. R.J. Levy et al., *Science* **228**, 190–192 (1985).

114. G. Golomb et al., *J. Pharm. Sci.* **76**, 271–276 (1987).

115. Y.V. Pathak et al., *Biomaterials* **11**, 718–723 (1990).

116. M.J. Slepian, *Proc. 1st Ann. Int. Symp. Local Cardiovasc. Drug Delivery*, Cambridge, Mass., Sept. 28–29, 1995.

117. J.L. Hill-West et al., *Proc. Natl. Acad. Sci. U.S.A.* **91**, 59–67 (1994).

118. M.J. Slepian, *Cardiol. Clin.* **12**, 715–737 (1994).

119. M.J. Slepian et al., *Circulation* **88**(4), I-319 (1993).

120. M.J. Slepian et al., *Circulation* **4** (P. 2), I-660 (1993).

121. K.A. Robinson et al., *J. Am. Coll. Cardiol.* **21**(2), 118A (1993).

122. L.A. Guzman et al., *Circulation* **94**(6), 1441–1448 (1996).

123. R.L. Wilensky, K.L. March, and D.R. Hathaway, *Am. Heart J.* **122**(4), 1136–1140 (1991).

124. J.J. Rome et al., *Arterioscler. Thromb.* **14**(1), 148–161 (1994).

125. A. Farb, G.C. Carlson, and R. Virami, *Circulation* **88**(Suppl. 1), I-310 (1993).

# CARRIER-MEDIATED TRANSPORT, ORAL DRUG DELIVERY

G.J. Russell-Jones
Biotech Australia Pty Ltd.
Roseville, NSW, Australia

## KEY WORDS

Cyanocobalamin
Drug delivery
Folate
Invasins
Oral
Peptides
Proteins

## OUTLINE

## INTRODUCTION

The intestinal absorption of low-molecular-weight vitamins, sugars, and other nutrients is highly efficient. In contrast, intestinally administered macromolecules are poorly absorbed and have very low oral bioavailability. Prior to absorption, most orally administered substances must first be digested to simple constituent molecules before being absorbed. Thus, oligosaccharides are hydrolyzed to monosaccharides that are in turn absorbed by active transport mechanisms, while dietary proteins are initially degraded by pepsin in the stomach before being further degraded by trypsin, chymotrypsin, and carboxypeptidases in the small intestine to become small peptides, individual amino acids, and dipeptides. Small peptides are hydrolyzed to free amino acids either during or soon after transport. The resultant material that appears in the portal blood is primarily, though not completely, free amino acids (1). Separate uptake mechanisms occur for neutral, basic, and acidic amino acids. Fats are first broken down to long-chain triglycerides in the stomach and upper small intestine and then complexed to and absorbed in association with bile salts. The absorption of fat-soluble vitamins probably occurs as a similar complex. In contrast to the highly efficient absorption of small molecules, the intraduodenal infusion of intact proteins such as $^3$H-BSA results in only 2% of the dose being transmitted in a macromolecular form to the lymph and blood (2). The absorption of orally administered macromolecules is even lower and ranges between 0.1–2.0% of the administered dose (3). Thus, the absorption of most peptides and proteins following oral administration is so low as to be economically unfeasible.

Despite the generally poor absorption of peptides and proteins, the oral route of delivery of peptide and protein still remains the most desirable method of chronic drug treatment. With the development of an increasing number of peptide and protein pharmaceuticals, a clear need exists for the development of a delivery system other than the current parenteral ones. Thus, molecules such as insulin,

LH–RH analogs, erythropoietin (EPO), granulocyte colony stimulating factor (G-CSF), thrombopoietin (TPO), heparin, vasopressin, and oxytocin, must still be administered parenterally. Perceived benefits for the development of such alternative delivery systems include increased patient comfort and compliance, ease of administration, reduced medical costs, and increased market. It must be noted, however, that additional costs may arise through the use of alternative carriers such as those described in this article owing to costs associated with the carrier, conjugation and scale up, and so on. The development of such alternative drug delivery systems also has the added benefit of "bringing new life to old drugs," particularly as many drugs come off patent.

In an attempt to overcome the problems of poor bioavailability of peptides and proteins, a number of carrier systems are currently being developed. This article describes the various specific carrier-mediated systems in the intestine that may be harnessed for oral drug delivery. While each of the carrier systems described is vastly different, they share the properties of initial binding to the apical surface of the intestinal epithelial cell (enterocyte), transfer across the cell membrane or entire cell, and release from the basal surface of the intestinal epithelial cell into the circulation.

## CARRIER-MEDIATED TRANSPORT SYSTEMS

### The Intestinal Peptide Transporter

Evidence from a number of sources has now shown that dipeptides are actively transported from the intestine into the circulation by a specific carrier system that differs from those involved in the transport of individual amino acids (4,5). This carrier system has been termed the intestinal peptide transporter or dipeptide transporter and has been implicated as a normal method of nutrient acquisition (6). This mechanism has been shown to be $Na^+$ independent (1,7) and to utilize an inward $H^+$ gradient as its driving force. The intestinal peptide transporter has been shown to be responsible for the transport of aminocephalosporin antibiotics such as cephalexin (7–9). The dipeptide transporter is separate from the brush border peptidases, and at least one report has suggested that a membrane protein of $M_r$ 127,000 is responsible for uptake (10). Several reports point to the possibility that two separate saturable transport systems for dipeptides exist in the small intestine. The Type I system prefers neutral pH and has a low affinity and a high capacity, while the Type II system prefers an acidic pH and has a high affinity and a low capacity. Thus Inui and coworkers (8) found that at neutral pH, cephradine was transported by the Type I system, while at an acidic pH it was transported via the Type II system. It may be possible that a single carrier, rather than two carriers, may exist that changes its conformation and affinity as the pH changes.

Although it was initially thought that the peptide transporter "required" an $N$-terminal nitrogen atom, the studies of Hu and Amidon (11) on the angiotensin converting enzyme inhibitor, Captopril, suggest this not to be the case. Thus, the peptide transporter shows broad substrate spec-

ificity (12,13) (Table 1). For optimal interaction with the transporter, dipeptides require a free carboxyl group, as it has been shown that reduction of this group to an alcohol reduces affinity for the transporter, as does cyclization (3). The transporter is stereospecific, showing higher transport of LL-dipeptides. Molecules and dipeptides shown to be transported by these transporters include glycylsarcosine, glycyl-L-proline, L-glutamyl-L-glutamate (8), amino $\beta$-lactam antibiotics (14,15), penicillins (amoxicillin, ampicillin, cyclacillin) (12,16), angiotensin-converting enzyme inhibitors (Captopril, Benazepril, Enalapril, Lisinopril, Quinapril) (17), cephalosporins (Cefaclor, Cefadroxil, Cefatrizine, Cefdinir, Cefixime, Ceftibuten) (12,18), and renin inhibitors (3) among others (Table 1). The maximal size of peptide transported via this mechanism appears to be a tripeptide (19), with transport of tetrapeptides occurring to only a limited extent, if at all (12). Intestinal cell models, such as the HT-29 and Caco-2 human carcinoma cell lines, have been shown to express the peptide transporter (12,20).

The identification of the peptide transport process has meant that many orally active molecules such as antibiotics (i.e., penicillins and cephalosporin), which were formerly thought to be taken up by passive diffusion, are in fact absorbed owing to their ability to interact with the normal uptake mechanism for dipeptides and tripeptides. This transport system has the advantage of a high capacity, but is limited to molecules of a small molecular weight (less than 600 Da) and to peptides or molecules having structural homologies with di- and tripeptides. A highly comprehensive treatment of the subject can be found in the review by Walter and co-workers (12).

### The Bile Acid Transport System

During the normal course of breakdown and absorption of fats, bile acids, synthesized from cholesterol in the liver, are secreted into bile and are then emptied into the small intestine where they aid the digestion and absorption of fats and fat-soluble vitamins. These bile acids are subsequently absorbed from the intestine and gain entry into the liver via the hepatic portal system. Bile salts taken up by hepatocytes may be resecreted into bile. Thus bile acids may circulate 6–10 times per day, representing a total bile acid turnover of 20–30 g bile salts per day in humans (21). The process of bile absorption is remarkably efficient as only 2–5% of bile acids are not absorbed and are excreted in the feces (21). Bile acids that have been recirculated back to the liver cause a feedback inhibition of cholesterol being converted to bile salts by modulating cholesterol-7$\alpha$-hydrolase activity. The possibility exists, therefore, that it might be possible to lower cholesterol through the interruption of the enterohepatic circulation of bile salts.

Bile acid absorption from the intestine is a $Na^+$-dependent process that relies on a specific bile acid transporter located in the ileum. Kramer and coworkers, using photoaffinity labeling, have identified the transporter in rabbits to be a 90–93-kDa protein (22,23), while in rats the putative transporter is a 99-kDa protein (24).

The bile acid transport system in the ileum shows a high degree of substrate specificity such that chenodeoxy-

**Table 1. Structural Homologies Between Di- and Tripeptides and Some Common Peptide Drugs**

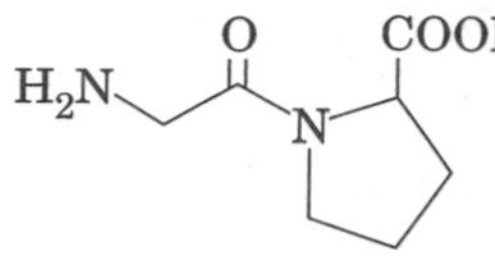

Gly-pro

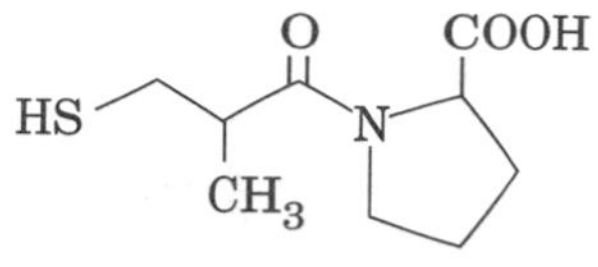

Captopril

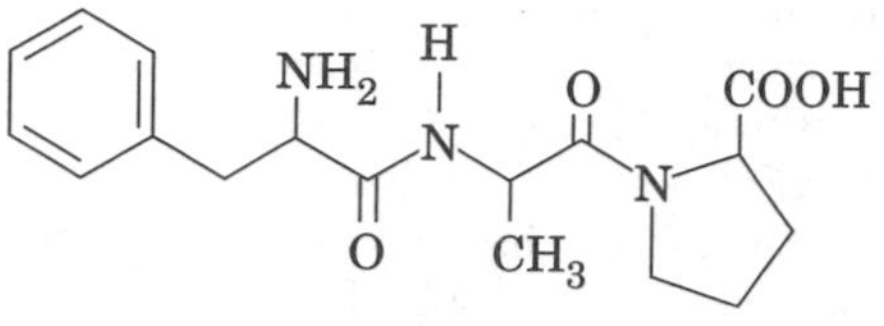

Phe-ala-pro

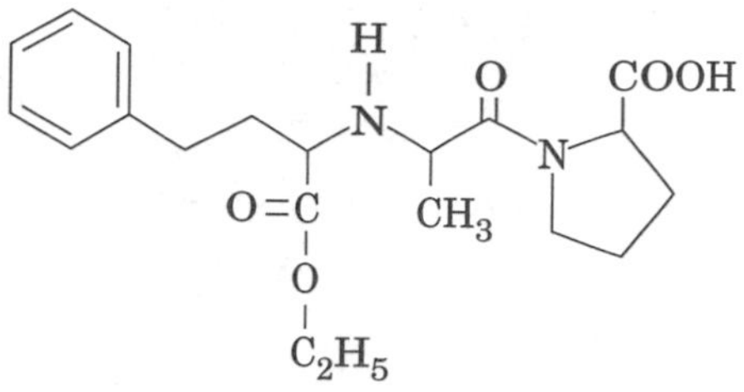

Enalapril

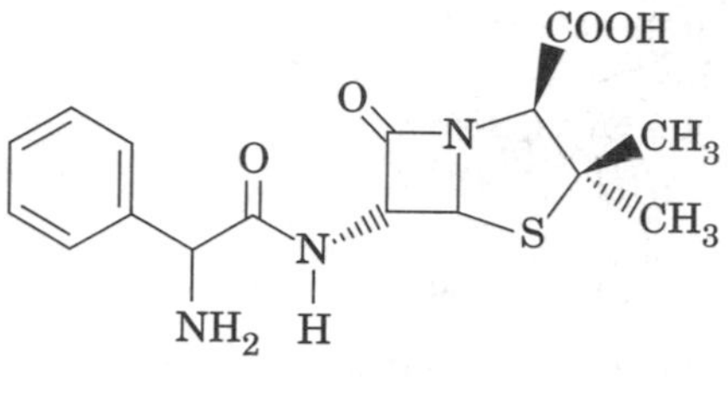

Ampicillin

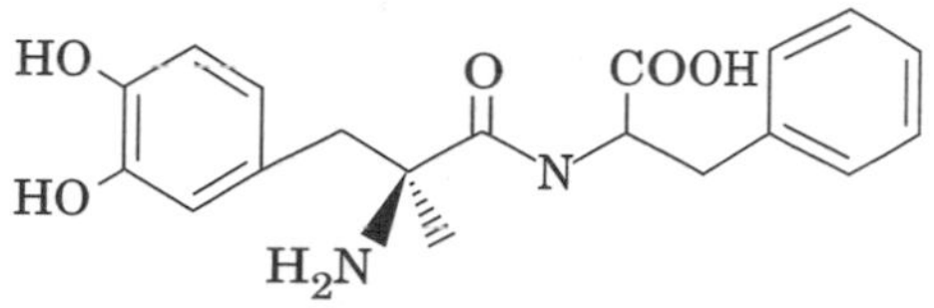

D-α-methyldopa-Phe

cholate uptake > deoxycholate > taurolithocholate > tauroursodeoxycholate > cholate ≫ lithocholate (22,23). The specificity for different bile acids for uptake by the small intestine is different than that of hepatic bile acid transport (22,23). Structural studies by Kramer and coworkers have shown that bile acid binding to the transporter depends on the presence of an essential cysteine residue at the binding site of the transporter. Binding is also inhibited following amino modification but not following carboxyl or hydroxyl modification of the transporter (22,23) Tables 2 and 3). Structural recognition of the bile salts by the transporter requires a negative charge on the bile acid molecule, and at least one hydroxyl group at positions 3, 7, or 12 of the steroid is also an essential feature. Similarly, for maximal binding the natural methylbutanoic acid or its taurine or glycine derivatives at position 21 should remain unchanged. Transport is greatly reduced for modified bile acids that contain dianionic, zwitterionic, or uncharged side chains (21,25). Further definition of structural requirements of the transporter may be obtained from studies using the recently cloned bile acid transporter (26).

The high uptake capacity of the intestine for bile salts (>20 per day), plus the subsequent targeting of bile salts to the liver, has led many researchers to attempt to use the bile acid transport system to improve intestinal absorption of small, poorly absorbed molecules and to specifically tar-

get drugs to the liver (27,28). In studies on model compounds, Kramer and coworkers found that the maximal size of a peptide that could be transported via the bile acid transporter was four amino acids, or around 600 Da (28). Several classes of compounds shown to have an affinity for the bile acid transporter include steroid analogs (ouabain, fusidic acid), cyclic peptides (antamanide), cyclic somatostatin derivatives (octreotide), and renin inhibitors (21,29). Liver-specific targeting with bile salts has been shown for conjugates to chlorambucil, phalloidin, and various oxaprolylpeptides (28).

Successful systemic delivery following oral administration of a bile acid drug conjugate would require the presence of a biodegradable linkage or prodrug approach for the drug to remain in the circulation rather than be delivered to the liver. Although the bile acid transport system appears to offer the possibility of a high-capacity oral uptake system for small peptides, to date the feasibility of this approach for systemic drug delivery remains to be demonstrated.

### Immunoglobulin Transport

**Neonatal IgG Transport.** Many neonatal animals have luminal receptors for the transport of IgG. These receptors specifically bind the Fc portion of IgG molecules but do not bind to other immunoglobulin classes; generally located in

**Table 2. Structure of Modified Bile Acids for Peptide Conjugation**

Bile acid—taurocholate

Peptide-modified bile acid

the proximal intestine, they show interesting binding characteristics in that they bind at pH 6.0 but not at pH 7.4. This makes them ideal for binding to intestinal IgG (pH < 6.5) and release in serosal plasma (pH 7.4) (30). IgG receptors have been shown to be responsible for the accumulation of maternal antibody from milk into the serum of newborn animals in many species, such as the rat (30–34), goat, sheep, pig, horse, cat, dog, and mouse (35,36). Jakoi and coworkers (34) found the IgG receptor to be present in neonatal rats (up to 21 days old) and mice (35,37) but lost following weaning.

During the identification of the IgG transport pathway, molecules such as ferritin have been linked to the IgG molecules to follow uptake and transcytosis. Thus, it would appear possible for large molecules such as ferritin to be carried into the neonate following conjugation to IgG. Similarly, as uptake and transcytosis depends on the Fc region of the IgG molecule, specific IgG antibody could conceivably be used to transport molecules bound to the antigen-binding site of the IgG across the neonatal small intestine into the circulation. Peppard and coworkers (38) have been able to demonstrate the transfer of immune complexes from the lumen of the small intestine of suckling rats to the bloodstream. The levels of transport were lower, however, than would have been expected from normal IgG transport. While little evidence exists for intestinal IgG transport in humans, this particular transport system has potential use in neonatal dogs, goats, sheep, pigs, cats and horses.

**The Polymeric Immunoglobulin Receptor.** The polymeric immunoglobulin receptor (pIgR) is responsible for the receptor-mediated transport of polymeric IgA and IgM across many types of epithelial surfaces. Transport occurs via binding of the Fc portion of these two molecules to the pIgR on the basal surface of the epithelial cell followed by transport of the molecules from the basolateral to apical surface. In this fashion, material bound to the IgA or IgM may be transported out of the body onto the epithelial surface (39–42). Transport occurs not only across epithelial cells but also via secretion into the bile and has been observed in several cell lines, including MDCK cells (40,41). Transport via this process is unidirectional and rapid, with a half time of less than 30 minutes (33).

Although the IgA has normally been implicated in protection of mucosal surfaces against infection, it also has been postulated that the IgA transport system may have a role in the removal of IgA-containing immune complexes by transport across epithelial surfaces (33). Other studies have shown that during the transport of IgA through the enterocyte, the IgA may also be able to neutralize virions within the cells (43). The possibility exists therefore to utilize this "reverse-transport" process for the removal of toxins, allergens, viruses, and bacteria from the circulation and into the intestine.

### Adhesins, Hemagglutinins, Toxins, and Lectins

A number of structures on the surface of bacteria and viruses are capable of binding specifically to the intestinal epithelium and can in many cases cause internalization of the bacterium or virus within the intestinal epithelial cells. Many plant lectins and toxins also share this property. The binding subunits of these molecules have the potential to be used as both intestinal epithelial targeting molecules for delaying the transit of pharmaceuticals down the intestine and as molecules for eliciting the uptake and transcytosis of the targeting molecule and attached pharmaceutical across the intestinal epithelial cell.

**Bacterial Adhesins.** The surface of many gram-positive bacteria is often covered with many proteins with potential roles in binding to surfaces and eliciting uptake. These include proteins isolated from various *Streptococcus* species such as the IgA-binding proteins ARP2, ARP4, and bac (44); the IgA-binding protein from Group B streptococci (45); fibrinogen-binding proteins (Mrp4, Sfb, PrtF, FnbA, FnbB, FnBP, FnBp [43]); and the collagen-binding factor (Cna) and clumping factor (ClfA) from *Streptococcus aureus*.

One of the first structures shown to be responsible for adhesion of bacteria to epithelial surfaces is the filamentous surface adhesion or pilus. These adhesins include the K88, K99 (46), F41, and 987P pili found on *Escherichia coli* inhabiting neonatal calves and piglets; the CFA1 and CFAII pili found on *E. coli* strains causing diarrhea in humans; and the *Pseudomonas aeruginosa* PAK pilus (47,48). There is also the type P pili isolated from *E. coli* strains associated with human pyelonephritis (49). Type I and Type 2 fimbriae found on *Aeromonas viscosis* and *Aeromonas naeslundii* (respectively) also have potential roles

**Table 3. Structure of Drugs with an Affinity for the Bile Acid Transporter**

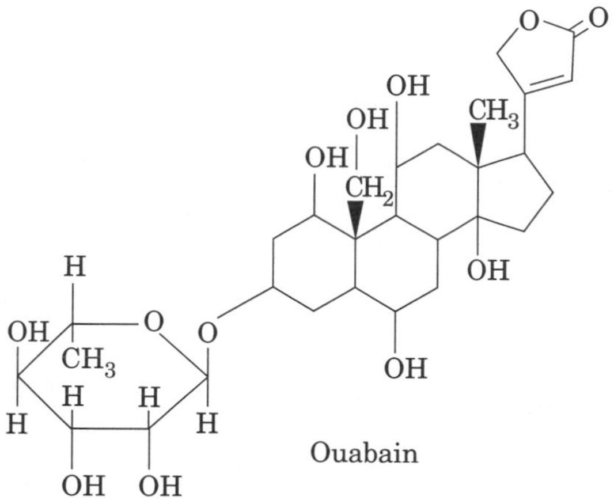

Ouabain

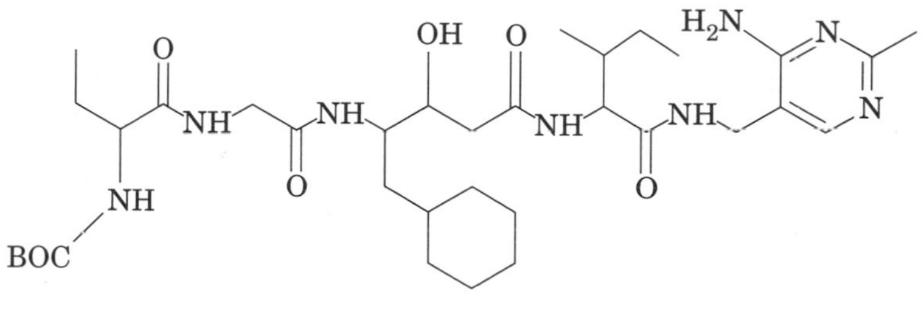

Renin Inhibitor

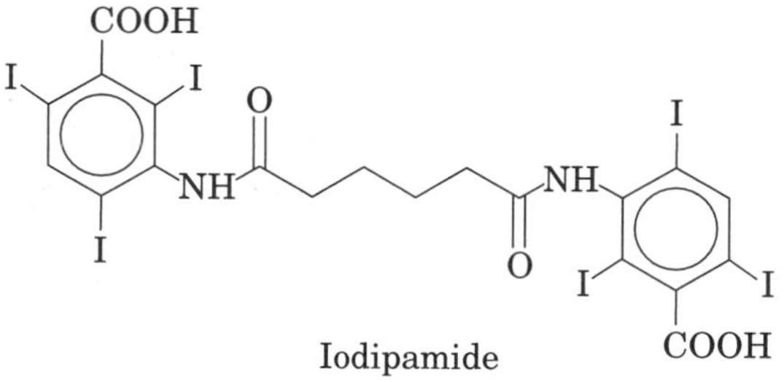

Iodipamide

in adherence and subsequent internalization of these bacteria (44) (Table 4). Similarly, a 36-kDa protein on the surface of *Neisseria gonorrhoeae* has been implicated in the binding of these organisms to surface lactosylceramides on human epithelial cells and may be responsible of uptake of these organisms by the epithelial cells (50).

Internalization by any of the above mechanisms is thought to be microfilament dependent and to be similar to the mechanism of uptake of *Campylobacter jejuni* and *Citrobacter freundii*, which have been shown to be able to initiate microtubule-dependent endocytosis with subsequent uptake into the endothelial endosome (51).

Binding of bacteria to epithelial cells results either in colonization of the epithelial cell surface or internalization of the bacteria within the target cell. While it is not entirely clear why binding sometimes results in internalization, it does depend on the type of cell, the type of cell ligand, and the nature of the interaction (49).

Russell-Jones and coworkers have found that it is possible to use the binding ability of these adhesins to cotransport other molecules into or across the epithelial cells. Thus, when haptens and proteins were covalently linked to 987P and K99 pili and fed to mice, the mice produced a serum antibody response that was significantly higher than that produced when the mice were fed the antigens alone (52) (Table 5).

**Bacterial Invasins.** Many bacteria possess surface structures apart from pili that have been shown to be responsible for the epithelial invasion of these bacteria. Inter-

**Table 4. Specificity of Binding of Various Pili**

| Pili | Specificity | Species |
|---|---|---|
| K99 | Gal, GM2 | Calves |
| F41 | | |
| Type 1 | D-mannose, methyl-$\alpha$-D-mannopyranoside | Many |
| P pili – Pap | Gal$\alpha$1 → 4Gal | Human urogenital epithelia |
| *Pseudomonas aeruginosa* PAK | | Human buccal epithelia |
| K88ab, ac | | Calves, pig |
| 987P | | Pig |

*Source:* From Refs. 44–49.

**Table 5. Anti-DNP Response to Orally Administered DNP-Carriers**

| | | Immune response | | | |
|---|---|---|---|---|---|
| | | Anti-DNP | | Anticarrier | |
| Immunogen | Dose ($\mu$g) | Serum IgG | Sec IgA | Serum IgG | Sec IgA |
| K99 | 100 | <4 | <4 | 1352 ± 128 | 16.7 ± 2.3 |
| DNP$_2$K99 | 500 | 1176 ± 164 | 28 ± 14.4 | 3565 ± 192 | 88 ± 21 |
| 987P | 100 | <4 | <4 | 1024 ± 89 | 88 ± 22.4 |
| DNP$_{26}$987P | 500 | 1024 ± 244 | 14 ± 3.1 | 111 ± 34.1 | 68 ± 19.2 |
| LTB | 20 | <4 | <4 | 1351 ± 211 | 12.2 ± 4.4 |
| DNP$_2$LTB | 20 | 24.3 ± 5.6 | <4 | 445 ± 35 | <4 |
| Concanavlin A | 20 | 666 ± 84 | <4 | nd | |
| PW mitogen | 20 | 641 ± 119 | <4 | nd | |
| *Phaseolus vulgaris* | 20 | 1378 ± 110 | 4.8 ± 2.3 | nd | |
| *Glycine max* | 20 | 1529 ± 65 | 3.1 ± 6.9 | nd | |

*Note:* Each value represents the mean of 5 mice ± 1 standard. Animals were fed on day 0 and 14. Serum and intestinal washouts were obtained on day 21 (52).

nalin, a surface protein encoded by the *inlA* gene of *Listeria monocytogenes* is responsible for internalization of *Listeria* within intestinal epithelial cells (53–57). Internalization of *Listeria* is triggered following binding of internalin to E-cadherin on epithelial cells (54,55). A protein with similar function to internalin is found on the surface *Yersinia pseudotuberculosis*; this protein, invasin, is a 986–amino acid protein located in the *Y. pseudotuberculosis* outer membrane (54,58). Expression of this protein on the surface of other gram-negative bacteria such as *E. coli* K-12 enables these cells to efficiently attach to and become internalized by epithelial cells (54,58). Similarly, latex particles coated with internalin are internalized by cultured mammalian cells (49). A second protein from *Y. pseudotuberculosis*, the *ail* gene product, is responsible for binding of these organisms to many eukaryotic cells but only promotes uptake in a few cell types (58).

**Viral Hemagglutinins.** Many viruses are known to gain access to the intestinal or respiratory epithelium following binding of surface structures on the virus. Normally such an event leads to the virus causing diarrhea in the infected subject. Specific binding proteins have been shown on the surface of rotaviruses (VP7), adenoviruses, and Norwalk virus (59,60). Similarly, surface hemagglutinins have also been implicated in inducing acid-induced fusion of viruses to membranes following initial binding and uptake of influenza virus (61).

Other hemagglutinin molecules exist on the surface of viruses such as rotavirus and aid in the binding, internalization, and transcytosis of these viruses across intestinal epithelial cells (62). Evidence suggests, however, that following binding of some viruses to the endocyte, the surface hemagglutin may be cleaved to yield a fusogenic protein that in turn enables the virus to enter the cell by direct cell membrane penetration rather than by endocytosis (59,63). Such molecules may have some utility in targeting pharmaceuticals to intestinal epithelial cells but may not be suitable for delivery of the pharmaceutical to the circulation as the endocytosed material may not be transcytosed.

**Toxins of Bacteria and Plants and Lectins**

Apart from the mucosal uptake and transport of adhesins such as pili and viral hemagglutinins, it has also been found that molecules can be covalently linked to the binding subunits of various toxin and plant lectins and to elicit uptake of the linked molecules following oral administration. Toxins of most relevance to initiate receptor-mediated uptake generally consist of an A–B$_n$ subunit structure in which the A subunit is responsible for toxicity while the B subunit functions as the specific binding unit of the toxin. These toxins include the cholera toxin–like molecules such as *E. coli* heat-labile toxin, cholera toxin, *C. jejuni* heat-labile toxin, *Clostridium botulinum* C2 toxin, tetanus toxin, diphtheria toxin, *Pseudomonas* exotoxin A (64–66), verotoxin, and Shiga toxin (66). All of these toxins share a common property in that they must first bind to and then cross membranes to exert their toxic activity. It is the ability of these toxins to bind to and be internalized by epithelial cells that makes them suitable vehicles for transporting molecules into and across epithelial cells.

A number of plant toxins of the general structure A–B$_n$ are also active orally and as such have the ability to bind to and be internalized by enterocytes. Toxins such as ricin, abrin, viscumin, modeccin, and volkensin all bind to D-galactose and are highly toxic following oral administration to rodents and humans (67) (Table 6); volkensin, for instance, is toxic at doses of as little as 60 ng/kg in rats. Other A–B toxins that could potentially bind to intestinal epithelial cells and cause uptake include botulinum toxin (from *C. botulinum*) (68).

Although there are few examples of oral delivery of pharmaceuticals using toxin-binding subunits or lectins, it has been shown that oral administration of these molecules elicits high titers of serum antibody to both the lectins/toxins and to haptens and proteins covalently linked to them (Table 6), suggesting that these molecules could also be used as vehicles for oral drug delivery. The repeated usefulness of these molecules for oral drug delivery is questionable, as all molecules studied so far are highly immunogenic and hence would elicit an antibody response to the carrier. These antibodies could block the binding of the lectins/toxins. It has also been found that some individuals are highly allergic to some food lectins and may suffer systemic anaphylaxis as a result of ingesting even small quantities of them.

## RECEPTOR-MEDIATED ENDOCYTOSIS OF TRANSFERRIN AND THE UPTAKE OF FE

The means by which iron (III) is taken up from the intestine is still a subject of controversy (69,70). Evidence is accumulating, however, to suggest that iron uptake is mediated by lactotransferrin and transferrin. Iron bound to transferrin is internalized by intestinal epithelial cells once the transferrin has been bound by a high-affinity transferrin receptor located in the brush border membrane (71,72). Once inside the cell, the lower pH of the endosomal vesicle (pH 5.5) causes dissociation of iron from the transferrin, thus delivering iron to the cell (73,74). The small size of the iron molecule most likely precludes this molecule from being used as an oral delivery system, although it is possible that molecules could be linked to transferrin and be taken up with it. A disadvantage of this approach would be the possibility of raising autoantibodies to the transferrin molecule itself. In a similar fashion, antibodies to the transferrin receptor could be used to target material to this receptor-mediated uptake system. Use of this strategy has already been demonstrated for overcoming the blood–brain barrier (75).

## USE OF THE VITAMIN B-12–MEDIATED TRANSPORT SYSTEM FOR UPTAKE OF PHARMACEUTICALS

Apart from the various uptake pathways described in the previous section, one natural uptake system offers itself as a unique opportunity for use in the oral uptake of pharmaceuticals. In this regard, the vitamin B-12 (VB$_{12}$) uptake system is unique in that absorption of this vitamin does not occur by facilitated diffusion or active transport but relies upon the receptor-mediated uptake facilitated by a VB$_{12}$-binding protein, intrinsic factor (IF). During this process, the [VB$_{12}$:IF] complex binds to an IF receptor located on the epithelium of the small intestine (predominantly in the ileum). Binding of the complex to the receptor triggers internalization of the complex via a yet-to-be-identified subepithelial vacuole (presumably an endosome), whereupon the IF is degraded by cathepsins, and the released VB$_{12}$ binds to intracellular transcobalamin II with subsequent transcytosis of the [VB$_{12}$:transcobalamin] complex into the circulation. VB$_{12}$ has many advantages as a carrier for oral drug delivery. Its molecular weight is small (1,356 Da), and as such it is poorly immunogenic. It is also relatively inexpensive and can easily be modified to

**Table 6. Specificity of Binding and Site of Action of Various Lectins and Toxins**

| Toxin | Specificity | Site of Action |
|---|---|---|
| Diphtheria toxin | | Elongation factor EF-2 |
| *Pseudomonas aeruginosa* exotoxin | | Elongation factor EF-2 |
| Cholera toxin | GM1-ganglioside | |
| *Escherichia coli* heat-labile toxin | GM1-ganglioside, galactose | |
| Shigella shigae toxin | N-acetyl-D-glucosamine | 60S ribosome |
| *Clostridium-perfringens* | | |
| Abrin | Galactose | 60S ribosome |
| Ricin | D-galactose | 60S ribosome |
| Visumin | D-galactose | 60S ribosome |
| Modeccin | D-galactose | 60S ribosome |
| Volkensin | D-galactose | 60S ribosome |
| Botulinum toxin | | |
| Tetanus toxin | | |

*Note:* Binding specificity from Refs. 67–68.

provide suitable functional groups for conjugation to peptide or protein pharmaceuticals. The extended period for uptake of $VB_{12}$, which lasts for 6–8 hours, results in it being a natural sustained-delivery system that is highly suitable for oral peptide/protein delivery (Fig. 1).

## Characteristics of the Uptake System

Uptake of $VB_{12}$ from the gut is a relatively slow process that takes place over several hours. After a single feed of $^{57}CoVB_{12}$ serum, $^{57}CoVB_{12}$ levels do not peak for 2–3 h and uptake continues for up to 18 h (76) (Figure 1). For many pharmaceuticals, delivery by such a slow trickle would be an advantage as it leads to a prolonged pharmaceutical window of drug delivery.

The capacity of the $VB_{12}$ uptake system is comparatively low. In humans only 1–2 nmol of $VB_{12}$ is taken up per oral dose (76). However, it is possible to repeat such oral dosing every 20–30 min, which means that the potential daily uptake using one-to-one conjugates of $VB_{12}$ to pharmaceutical is around 50–100 nmol per day. This level of delivery would be sufficient to deliver pharmaceutically relevant doses of peptides such as calcitonin analogues, vasopressin, oxytocin, LH–RH agonists, G-CSF, and EPO.

## Conjugation of Pharmaceuticals to Vitamin B-12

For pharmaceuticals to be cotransported across the intestinal epithelial cell layer and into the circulation, they must first be covalently linked to the carrier. In the case of $VB_{12}$-mediated transport, the $VB_{12}$ must first be modified to provide a suitable functional group for conjugation. This is readily achieved by the acid hydrolysis of one of the propionimide side chains of the corrin ring to form a monocarboxylic acid derivative (76). The isolated monocarboxylic acid derivative can then be conjugated directly to amino groups of proteins or peptides using a suitable carbodiimide (76,77). The direct linkage of $VB_{12}$ to small peptides may in some cases greatly reduce the bioactivity of the conjugated peptide. In such cases it may be necessary to use a spacer between the $VB_{12}$ and the peptide. Suitable spacers include diaminohexane, diaminododecane, adipyl hydrazine, or long-chain spacers formed between amino-

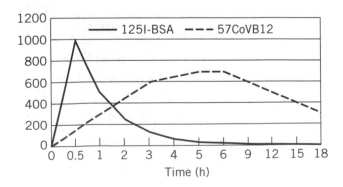

**Figure 1.** Comparison of uptake of $^{57}CoVB_{12}$ and $^{125}I$-BSA. Rats received either $^{57}CoVB_{12}$ or $^{125}I$-BSA by oral administration. At various time points the rats were bled and the radioactivity measured in serum.

derivatives of $VB_{12}$ and commercially available cross-linkers such as disuccinimyl suberate (DSS, Pierce), ethylene glycol bis(succinimidyl succinate) (EGS) (76,77). Alternatively, it may be necessary to form a biodegradable bond between the carrier and the drug. Thus Russell-Jones and coworkers (78) found it necessary to link $VB_{12}$ to D-Lys$^6$-ANTIDE (an LH–RH antagonist) using an iminothiolane derivative of the latter molecule with a succinimidyl 3-(2-pyridyldithio)propionate (SPDP; Pierce) derivative of aminoethyl-e$VB_{12}$ (Table 7). The resultant thiol-cleavable conjugate had a systemic bioactivity similar to that of native ANTIDE. It is also possible to use dithiopyridyl-derivatives of $VB_{12}$ to link directly to free thiols present on the peptide to be joined to $VB_{12}$. In this fashion it has been shown that forming a thiol-cleavable linkage to a buried thiol in G-CSF is possible (77,79).

## Cell Models of Receptor-Mediated Transport

A number of cell lines exist that polarize upon culture and show receptor-mediated apical-to-basal transport of carrier molecules in an analogous fashion to intestinal epithelial cells. These include two intestinally derived cell lines, the human colon carcinoma cell lines Caco-2 and HT29. Additionally, there are two kidney carcinoma derived lines, the OK cell line, derived from opossum kidney, and the LLCPK1 line, derived from porcine kidney (74,80). These cell lines all exhibit polarized transport when cultured on semipermeable membranes in two-chambered tissue culture wells. The intestinally derived cell lines differ from the kidney cell lines in that they are more heterogeneous and are microviliated, whereas the kidney lines show good homogeneity with little variation from passage to passage.

**Vitamin B-12 Transport in Polarized Epithelial Cell Lines.** Each of the four cell lines described in the previous section have also been shown to bind, internalize, and transcytose $VB_{12}$ in an IF-dependent fashion. Transport has been shown to be dependent upon polarization of the cells, which occurs when the cells are grown in two-chambered tissue culture wells, with each chamber separated by a permeable filter membrane (81,82). Caco-2 cells have been found to be highly variable in $VB_{12}$ transport with levels of transport ranging from 10–40 fmol/24 h (81,82) to 250 fmol/4 h (80). Caco-2 cells also have been found to transport $VB_{12}$ in an IF-independent fashion, possibly due to the expression of transcobalamin II receptors on their apical surface (83).

**Vitamin B-12–Mediated Transport of G-CSF by Caco-2 Cells.** Habberfield and coworkers (79) have examined the potential for $VB_{12}$ to act as a carrier to transport EPO and G-CSF across Caco-2 cells grown on semipermeable membranes. When G-CSF was added to the apical chamber of Caco-2 cell cultures, small but detectable levels of G-CSF (14.13 ± 4.4 fmol, or 0.27% of the added dose) were found to cross the monolayer and appear in the basal chamber (Fig. 2). The level of transported material increased dramatically when $VB_{12}$-G-CSF conjugates were added to the apical chamber. Thus, 398 ± 112, 187 ± 89, and 200 ±

**Table 7. Relative Bioactivity of Various Vitamin B-12–ANTIDE Conjugates**

| Conjugate | Spacer | IC50 | Mean LH (ng/ml; ± Std) |
|---|---|---|---|
| Control | – | – | 6.4 ± 1.9 |
| ANTIDE | – | 4.9 ± 2.8 | 0.3 ± 0.1 |
| D-Lys⁶-ANTIDE | – | 2.5 ± 0.5 | 6.4 ± 1.0 |
| VB₁₂-D-Lys⁶-ANTIDE | 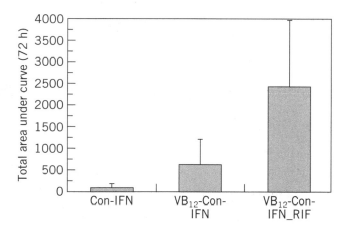 | 88 ± 47 | 6.4 ± 1.0 |
| VB₁₂-EGS-D-Lys⁶-ANTIDE | | 8.9 ± 3.5 | 8.4 ± 3.0 |
| VB₁₂-SS-D-Lys⁶-ANTIDE | 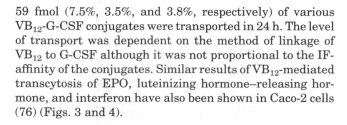 | 11.0 ± 4.0 | 0.3 ± 0.1 |

*Source:* Russell-Jones et al. (76).

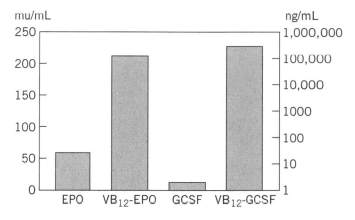

**Figure 2.** Vitamin B-12–mediated transport of consensus interferon across the rat intestine. Consensus interferon (Con-IFN) was administered to rats by intraduodenal pump infusion, and the quantity of interferon that reached the circulation was assessed. Data represents the area under the curve for 72 h of consensus interferon administered alone, conjugated to VB₁₂, or conjugated to VB₁₂ and administered in the presence of rat IP (85).

**Figure 3.** Serum levels of EPO and G-CSF in rats receiving VB₁₂-EPO and VB₁₂-G-CSF intraduodenally. Rats received EPO, VB₁₂-EPO, G-CSF, or VB₁₂-G-CSF by 24-hour intraduodenal pump infusion. The serum levels of EPO (mu/mL) and G-CSF (ng/mL) were calculated by an appropriate ELISA. Data is presented as mu/mL EPO or ng/mL G-CSF.

59 fmol (7.5%, 3.5%, and 3.8%, respectively) of various VB₁₂-G-CSF conjugates were transported in 24 h. The level of transport was dependent on the method of linkage of VB₁₂ to G-CSF although it was not proportional to the IF-affinity of the conjugates. Similar results of VB₁₂-mediated transcytosis of EPO, luteinizing hormone–releasing hormone, and interferon have also been shown in Caco-2 cells (76) (Figs. 3 and 4).

## Oral Delivery of Peptides and Proteins Using the Vitamin B-12 Transport System

**Oral Delivery of LH–RH Analogs.** Luteinizing hormone–releasing hormone (LH–RH) is a decapeptide released

from the hypothalamus, which stimulates the release of luteinizing hormone (LH) and follicle-stimulating hormone (FSH) from the anterior pituitary. This in turn stimulates ovulation in female animals and spermatogenesis in male animals. Over 1,800 analogs of LH–RH have been synthesized in the past 20 years; some are more potent than LH–RH in stimulating LH and FSH secretion (agonists), while some analogs are able to inhibit the action of LH–RH and are termed antagonists. Russell-Jones and co-workers (77,84) performed ovulation studies in mice using a direct conjugate between eVB₁₂ monocarboxylic acid and a D-Lys⁶ analogue of LH–RH. D-Lys⁶-LH–RH was highly active in stimulating ovulation in pregnant mare serum gonadotropin-primed (PMSG-primed) mice when injected

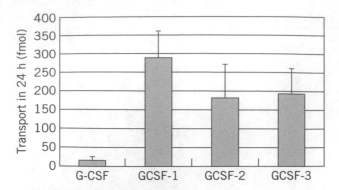

**Figure 4.** Vitamin B-12–mediated transport of G-CSF by Caco-2 cell cultures. Caco-2 cell cultures were grown on multicell transport chambers. G-CSF or $VB_{12}$-GCSF conjugates (GCSF-1, GCSF-2, or GCSF-3) were placed in the apical wells, and the quantity of G-CSF reaching the basal chamber after 24 hours was assessed by ELISA (79).

systematically; however, it was much less active when given orally. In contrast, $VB_{12}$-$\epsilon$-D-Lys[6]-LH–RH was highly active in stimulating ovulation when given either systematically or orally. The oral activity of this conjugate was completely abolished when the $VB_{12}$-$\epsilon$-D-Lys[6]-LH–RH was given in the presence of excess $VB_{12}$. In separate studies performed with Dr. J. Sandow (J. Sandow, personal communication, 1992) using $VB_{12}$ linked to D-Lys[6]-LH–RH-ethylamide ($VB_{12}$-D-$\epsilon$-Lys[6]-LH–RH-EA) using various spacers, oral bioavailabilities of between 10 and 45%, depending on the spacer, were obtained in rats. Dr. J. Alsenz and Dr. C. de Schmidt have reported similar oral bioavailabilities for $VB_{12}$-$\epsilon$-D-Lys[6]-LH–RH-EA conjugates orally administered to rats (J. Alsenz, personal communication, 1996).

**Oral Delivery of EPO, G-CSF, and Consensus Interferon via the Vitamin B-12 Transport System.** Habberfield and co-workers (79) have recently shown that it is possible to utilize the $VB_{12}$ transport system to promote the intestinal uptake of G-CSF (18,800 Da) and EPO (29,543 Da). In their studies, they implanted miniosmotic pumps into the peritoneal cavity of rats such that their contents drained into the duodenum of the rats. In this way they circumvented the need for multiple feeding of the animals. The pumps were designed to deliver their contents for 24 h and then stop. Rats received either buffer, unconjugated protein, or conjugates of $VB_{12}$ and protein.

Similarly, Habberfield and coworkers (85) found that conjugation of $VB_{12}$ to consensus interferon greatly increased the level of uptake of consensus interferon in rats receiving the $VB_{12}$-interferon complex duodenally (Fig. 4).

### Amplification of the Vitamin B-12 Uptake System Using Nanoparticles

Formulating the pharmaceuticals to be delivered within nanoparticles to which vitamin $B_{12}$ is linked could result in a 1,000- to 10,000-fold increase in the amount of material that could be delivered via the $VB_{12}$ uptake system. This delivery method would also protect the pharmaceutical during the passage down the intestine from enzymes

and other degradative processes. Furthermore, incorporating pharmaceuticals within nanoparticles has the added advantage that the peptide or protein to be delivered would not need to be chemically linked to $VB_{12}$ for uptake to occur. This would be particularly beneficial with peptides such as vasopressin, which has not been linked to $VB_{12}$ without loss of functional activity. For effective delivery, such nanoparticulate systems would need to be designed so that they remain intact within the intestine but would release their contents once they reached the circulation.

It has recently been shown that it is possible to initiate intestinal uptake of fluorescent nanoparticles in dogs, pigs, and rats by linking the nanoparticles either to $VB_{12}$, the binding subunit of the *E. coli* heat-labile toxin (LTB), or Concanavilin A lectin from *Canavalia eusiformis* (ConA). Histological examination of intestinal loops containing injected nanoparticles reveals that they can first be observed bound to the intestinal epithelial layer. Within 1 to 2 hours, the nanoparticles have crossed the epithelial cell layer and can be found below the mucosal cell layer lying in the central lacteal lymph vessel (84). Studies in pigs and dogs have shown that the nanoparticles can then be found in sections of the mesenteric lymph nodes, some 10–20 cm from the intestinal loops.

In separate studies on uptake of nanoparticles into Caco-2 cells, it has been shown that the quantity of uptake of nanoparticles is inversely proportional to size, with 100-nm particles showing 15 to 250-fold higher uptake than larger (500 nm, 1 $\mu$m) particles (86).

## FUTURE DIRECTIONS

### The Use of Pathogenic Methods of Invasion for Carrier-Mediated Transport

The normal method of invasion of epithelial cells by pathogens represents an untapped mechanism by which intestinal uptake of peptide and protein pharmaceuticals may be initiated: Intestinal epithelial cells internalize large nanoparticulate structures such as viruses or bacteria, and in some cases these structures are transported across the cell and into the subepithelial layer of the intestine. Uptake and transport has in many cases been found to depend on the presence of a surface protein on the pathogens that initiated internalization. Further study of these mechanisms and molecules that initiate uptake will inevitably lead to new and exciting methods for oral drug delivery; some of these are Internalin (*L. monocytogenes*). Invasin (*Y. pseudotuberculosis*), and other molecules with similar functional properties from other pathogens. The utility of such systems will ultimately lie in their ability to be used repeatedly without the induction of an immune response to the internalizing molecule.

### Carrier-Mediated Transport of Nanoparticles

In an extension to the transport systems described in this article, it should also be possible in the future to use many of the carriers described previously (LTB, toxin-binding subunits, lectins, $VB_{12}$, etc) to stimulate the uptake of drug-loaded nanoparticles. The carrier molecules would be covalently linked to the surface of the biodegradable nanoparticles and would be cotransported across the cells during the uptake of the carriers. Nanoparticles used for this purpose would need to be stable in the intestinal environment and to release their contents once they have reached the circulation. While it is possible that many of the carriers described in this review could be used to coat nanoparticles, the low immunogenicity of the $VB_{12}$ molecule may mean that it will ultimately be the delivery system of choice for repeated oral drug delivery.

## CONCLUSIONS

The drug delivery specialist would do well to study the mechanisms of pathogenesis of viral and bacterial pathogens, including their method of invasion and transcytosis of cells in a highly specific fashion. It is through such a study that new methods of oral drug delivery will be discovered and developed. Presently, however, the vitamin $VB_{12}$ uptake system shows particular promise in being able to cause the uptake and transport of peptides and proteins linked to the $VB_{12}$ molecule from the intestine into the circulation. In like fashion, the oral uptake of nanoparticles containing entrapped peptide or protein pharmaceuticals may also be possible.

## BIBLIOGRAPHY

1. V. Ganapathy and F.H. Leibach, *Am. J. Physiol.* **12**, G153–G160 (1985).
2. A.L. Warshaw, W.A. Walker, and K.J. Isselbacher, *Gastroenterology* **66**, 987–992 (1974).
3. G.M. Pauletti et al., *J. Controlled Release* **41**, 3–17 (1996).
4. D.M. Matthews and S.A. Adibi, *Gastroenterology* **71**, 151–161 (1976).
5. D.M. Matthews, *Physiol. Rev.* **55**, 537–608 (1975).
6. F. Navab and A.M. Asatoor, *Gut* **11**, 373–379 (1970).
7. V. Ganapathy and F.H. Leibach, *Life Sci.* **30**, 2137–2146 (1982).
8. K.I. Inui et al., *J. Pharmacol. Exp. Ther.* **247**, 235–241 (1988).
9. I.J. Hidalgo, F.M. Ryan, G.J. Marks, and P.L. Smith, *Int. J. Pharm.* **98**, 83–92 (1993).
10. W. Kramer et al., *Biochim. Biophys. Acta* **1030**, 41–49 (1990).
11. M. Hu and G.L. Amidon, *J. Pharm. Sci.* **77**, 1007–1011 (1988).
12. E. Walter, T. Kissel, and G.L. Amidon, *Adv. Drug Delivery Rev.* **20**, 33–58 (1996).
13. P.W. Swaan and J.J. Tukker, *J. Pharmacol. Exp. Ther.* **27**, 242–247 (1995).
14. T. Kimura et al., *J. Pharmacobio-Dyn.* **6**, 246–253 (1983).
15. T. Hori et al., *J. Pharm. Pharmacol.* **40**, 646–647 (1988).
16. K. Iseki et al., *J. Pharm. Pharmacol.* **41**, 628–632 (1989).
17. P.W. Swaan, M.C. Stehouwer, and J.J. Tukker, *Biochim. Biophys. Acta* **1236**, 31–38 (1995).
18. W. Kramer, F. Girbig, I. Leipe, and E. Petzoldt, *Biochem. Pharmacol.* **37**, 2447–2435 (1988).
19. S.A. Adibi and Y.S. Kim, in L.R. Johnson, ed., *Physiology of the Gastrointestinal Tract*, Raven Press, New York, 1981, pp. 1073–1096.
20. A.H. Dantzig and L. Bergin, *Biochem. Biophys. Res. Commun.* **155**, 1082–1087 (1988).
21. P.W. Swaan, F.C. Szoka, and S. Øie, *Adv. Drug Delivery Rev.* **20**, 59–82 (1996).
22. W. Kramer et al., *J. Biol. Chem.* **267**, 18598–18604 (1992).
23. W. Kramer et al., *Biochim. Biophys. Acta* **1111**, 93–102 (1992).
24. W. Kramer et al., *J. Biol. Chem.* **268**, 18035–18046 (1993).
25. M. Kågedahl et al., *Pharm. Res.* **14**, 176–180 (1997).
26. W.H. Wong, P. Oelkens, and P.A. Dawson, *J. Biol. Chem.* **270**, 27228–27234 (1995).
27. W. Kramer, G. Burckhardt, F.A. Wilson, and G. Kurz, *J. Biol. Chem.* **258**, 3623–3627 (1983).
28. W. Kramer et al., *J. Biol. Chem.* **269**, 10621–10627 (1994).
29. P.W. Swaan, K.M. Hillgren, F.C. Szoka, and S. Øie, *Bioconjugate Chem.* **8**, 520–525 (1997).
30. R. Rodewald, *J. Cell Biol.* **85**, 18–32 (1980).
31. R. Rodewald, *J. Cell Biol.* **71**, 666–670 (1976).
32. D.R. Abrahamson and R. Rodewald, *J. Cell Biol.* **91**, 270–280 (1981).
33. K.E. Mostov, *Annu. Rev. Immunol.* **12**, 63–84 (1994).
34. E.R. Jakoi, J. Cambier, and S. Saslow, *J. Immunol.* **135**, 3360–3364 (1985).
35. J.P. Kraehenbuhl and M.A. Campiche, *J. Cell Biol.* **42**, 345–365 (1969).
36. D.M. Neville and T.-M. Chang, *Curr. Top. Membr. Transp.* **10**, 65–150 (1978).
37. N. Mackenzie, *Immunol. Today* **5**, 364–366 (1984).
38. J.V. Peppard, L.E. Jackson, J.G. Hall, and D. Robertson, *Immunology* **53**, 385–393 (1984).
39. K.E. Mostov, M. Friedlander, and G. Blobel, *Nature (London)* **308**, 37–43 (1984).
40. R.P. Hirt et al., *Cell (Cambridge, Mass.)* **74**, 245–255 (1993).
41. G. Apodaca, L.A. Katz, and K.E. Mostov, *J. Cell Biol.* **125**, 67–86 (1994).
42. K.E. Mostov, *Cell (Cambridge, Mass.)* **43**, 389–390 (1985).
43. M.B. Mazanec et al., *Proc. Natl. Acad. Sci. U.S.A.* **89**, 6901–6905 (1992).
44. V.A. Fischetti, *ASM News* **62**, 405–410 (1996).
45. G.J. Russell-Jones, E.C. Gotschlich, and M.S. Blake, *J. Exp. Med.* **160**, 1467–1475 (1984).
46. M. Mouricout, J.M. Petit, J.R. Carias, and R. Julien, *Infect. Immun.* **58**, 98–106 (1990).
47. R.T. Irvin et al., *Infect. Immun.* **57**, 3720–3726 (1989).
48. P. Doig et al., *Infect. Immun.* **58**, 124–130 (1990).
49. R.R. Isberg, *Science* **252**, 934–938 (1991).
50. D.K. Paruchuri et al., *Proc. Natl. Acad. Sci. U.S.A.* **87**, 333–337 (1990).
51. T.A. Oelschlaeger, P. Guerry, and D.J. Kopecko, *Proc. Natl. Acad. Sci. U.S.A.* **90**, 6884–6888 (1993).
52. H.J. de Aizpurua and G.J. Russell-Jones, *J. Exp. Med.* **167**, 440–451 (1987).
53. P. Cossart, *J. Cell. Biochem. Suppl.* 18 Part A, 36 (1994).
54. S. Falkow, *Cell (Cambridge, Mass.)* **65**, 1099–1102 (1991).
55. J. Mengaud et al., *Cell (Cambridge, Mass.)* **84**, 923–932 (1996).
56. A. Lingnau et al., *Infect. Immun.* **63**, 3896–3903 (1995).
57. D.A. Drevets, R.T. Sawyer, T.A. Potter, and P.A. Campbell, *Infect. Immun.* **63**, 4268–4276 (1995).
58. R.R. Isberg and S. Falkow, *Nature (Lond.)* **317**, 262–264 (1985).
59. N. Fukuhara, O. Yoshie, S. Kitaoka, and T. Konno, *J. Virol.* **62**, 2209–2218 (1988).
60. R.H. Yolken, *Infect. Dis. Antimicrob. Agents* **6**, 273–291 (1985).
61. S.A. Wharton et al., *J. Biol. Chem.* **263**, 4474–4480 (1988).

62. D.J. Keljo and A.K. Smith, *J. Pediatr. Gastroenterol. Nutr.* **7**, 249–256 (1988).

63. K.T. Kaljot, R.D. Shaw, D.H. Rubin, and H.B. Greenberg, *J. Virol.* **62**, 1136–1144 (1988).

64. C. Sears and J.B. Kaper, *Microbiol. Rev.* **60**, 167–215 (1996).

65. A. Waksman et al., *Biochim. Biophys. Acta* **604**, 249–296 (1980).

66. L. Okerman, *Vet. Microbiol.* **14**, 33–46 (1987).

67. F. Stirpe and L. Barbieri, *FEBS Lett.* **195**, 1–8 (1986).

68. R.O. Blaustein, W.J. Germann, A. Finkelstein, and B.R. DasGupta, *FEBS Lett.* **226**, 115–120 (1987).

69. R.J. Simpson et al., *Biochim. Biophys. Acta* **884**, 166–171 (1986).

70. M.E. Conrad, R.T. Parmley, and K. Osterloch, *J. Lab. Clin. Med.* **110**, 418–426 (1987).

71. J. Mazurier, J. Montreuil, and G. Spik, *Biochim. Biophys. Acta* **821**, 453–460 (1985).

72. G. Johnson, P. Jacobs, and L.R. Purves, *J. Clin. Invest.* **71**, 1467–1476 (1983).

73. A. Dautry-Varsat, A. Ciechanover, and H.F. Lodfish, *Proc. Natl. Acad. Sci. U.S.A.* **80**, 2258–2262 (1983).

74. R.D. Klausner et al., *Proc. Natl. Acad. Sci. U.S.A.* **80**, 2263–2266 (1983).

75. R. Cecchelli et al., *Proc. Int. Symp. Controlled Release Bioact. Mater.* **24**, 221–222 (1997).

76. G.J. Russell-Jones, in M.D. Taylor and G.L. Amidon, eds., "Peptide-Based Drug Design: Controlling Transport and Metabolism," American Chemical Society, Washington, D.C., 1995, pp. 181–198.

77. G.J. Russell-Jones, S.W. Westwood, and A.D. Habberfield, *Bioconjugate Chem.* **6**, 459–465 (1996).

78. G.J. Russell-Jones et al., *Bioconjugate Chem.* **6**, 34–42 (1994).

79. A. Habberfield et al., *Int. J. Pharm.* **145**, 1–8 (1996).

80. K.S. Ramanujam, S. Seetharam, M. Ramasamy, and B. Seetharam, *Am. J. Physiol.* **260**, 6416–6422 (1991).

81. D.J. Dix et al., *Gastroenterology* **88**, 1272–1279 (1990).

82. G. Wilson et al., *J. Controlled Release* **11**, 25–40 (1990).

83. S. Bose, S. Seetharam, N.M. Dahm, and B. Seetharam, *J. Biol. Chem.* **272**, 3538–3543 (1997).

84. G.J. Russell-Jones, S.W. Westwood, and A. Habberfield, *Proc. Int. Symp. Controlled Release Bioact. Mater.* **23**, 49–50 (1996).

85. U.S. Pat. 5,574,018 (February 15, 1996), A.D. Habberfield, O.B. Kinstler, and C.G. Pitt.

86. M.P. Desai, V. Labhasetwar, G.L. Amidon, and R.J. Levy, *Pharm. Res.* **13**, 1838–1845 (1996).

See also BIOADHESIVE DRUG DELIVERY SYSTEMS; VACCINE DELIVERY.

# CENTRAL NERVOUS SYSTEM, DRUG DELIVERY TO TREAT

MARTIN BURKE
ROBERT LANGER
HENRY BREM
Johns Hopkins University School of Medicine
Baltimore, Maryland

## KEY WORDS

BCNU
Blood–brain barrier
Blood–brain barrier disruption
Blood–tumor barrier
Brain tumors
Carrier-mediated drug delivery
Central nervous system
Gene therapy
Immunotherapy
Microspheres
pCPP:SA
Polyanhydride polymers
Prodrugs
Redox chemical delivery system
Vector-mediated drug delivery

## OUTLINE

## INTRODUCTION

Billions of dollars in research and development spending aimed at curing human diseases has yielded a vast arsenal of powerful therapeutic agents. However, while many of these drugs demonstrate extraordinary promise in the laboratory, they are often far less successful when applied to patients in clinical trials. In no area of medicine has this recurrent inconsistency been more pronounced, or more tragic, than in the treatment of diseases of the central nervous system (CNS). Despite aggressive research, patients suffering from fatal and/or debilitating CNS diseases, such as brain tumors, HIV encephalopathy, epilepsy, cerebrovascular disease, and neurodegenerative disorders, far outnumber those dying of all types of systemic cancer or heart disease (1).

The clinical failure of many potentially effective therapeutics is often not due to a lack of drug potency but rather to shortcomings in the method by which the drug is delivered. Treating CNS diseases is particularly challenging because a variety of formidable obstacles often impede drug delivery to the brain and spinal cord. In response to the insufficiencies in conventional delivery mechanisms, aggressive research efforts have recently focused on the development of new strategies to more effectively deliver drug molecules to the CNS. This work is crucial to the development of successful treatments for CNS diseases, for the therapeutic potential of any drug can only be realized if the drug reaches its intracranial target. The first half of this article focuses on the specific barriers that inhibit drug delivery to the CNS and various strategies designed to overcome them. The latter half reviews our work and that of other groups involving the use of biocompatible polymer systems to achieve sustained drug delivery directly to the brain interstitium for the treatment of intracranial tumors as well as several nonneoplastic CNS diseases.

## BARRIERS TO CNS DRUG DELIVERY

The most conventional methods to deliver drug molecules for the treatment of human illness utilize the cardiovascular system. Typically, drugs are delivered to the cardiovascular system by oral ingestion, intravenous injection or infusion, or injection into tissue such as skin or muscle and subsequently distributed throughout the body. As blood-borne drug molecules pass through capillary networks, they are forced by diffusion and convection from the plasma into the surrounding interstitium where they enter target cells and elicit a desired biological response. This method of drug delivery is often highly effective for the treatment of a broad range of systemic ailments. However, when a pathological condition plagues the CNS, the cardiovascular route of drug delivery is often tragically unsuccessful. A variety of serious CNS diseases are insensitive to systemically delivered drugs, and therefore many neurological patients are often left with few therapeutic options. The failure of systemically delivered drugs to effectively treat many CNS diseases can be rationalized by considering a number of barriers that inhibit drug delivery to the CNS.

### Blood–Brain Barrier

When a drug is administered systemically to treat a CNS disease, its delivery is first inhibited by a physiological impediment known as the blood–brain barrier (BBB). An early teleological development, the BBB serves a homeostatic purpose. Through selective permeability to blood-borne molecules, the BBB maintains a constant milieu in the brain parenchyma. This is important for the proper function of neurons, which are highly sensitive to changes in concentrations of both extracellular ions and signaling molecules such as neurotransmitters. However, this same selectivity also limits the penetration of drug molecules into the brain parenchyma. The mechanisms underlying

BBB selectivity have therefore been the subject of aggressive research for the past 30 years. Today, the term BBB is often used in a general sense to describe the collection of physiological, metabolic, and biochemical characteristics that distinguish the cerebral capillary endothelium from the endothelium of systemic organ systems.

The first distinguishing physiological characteristic is that cerebral capillary endothelial cells are fused together by pentalaminar tight junctions, named for their five-layered appearance in the electron microscope: the cytoplasmic lipid layer of the first cell, a lucent line, the fused outer lipid layers of the first and second cell, another lucent line, and the cytoplasmic lipid layer of the second cell. This five-layered configuration virtually fuses neighboring endothelial cells in the cerebral microvasculature (2). In contrast, the endothelial cells in systemic capillaries are more loosely joined through a series of focal connections between the two outer lipid layers of adjacent cells. Furthermore, while endothelial cells in most systemic capillaries are fenestrated with gaps and pores that allow for the exchange of ions, large molecules, and even cells between the blood and the surrounding interstitium, these gaps and pores are lacking in the brain endothelium (3–5). As a result, in comparison to the systemic microvasculature, cerebral capillaries are much less permeable to ions and small molecules and are virtually impermeable to most peptides and macromolecules. For example, the electrical resistance, a measurement of resistance to ionic current flow, across most systemic capillary endothelium focal connections is $5–10 \ \Omega \ \text{cm}^2$. In contrast, the electrical resistance in the microvasculature of the CNS is typically $1,000 \ \Omega \ \text{cm}^2$ or more (6).

Additionally, in comparison to their systemic counterparts, cerebral endothelial cells have a marked deficiency in pinocytic vesicles (2,3). In systemic endothelium, pinocytic vesicles result from the invagination of the cell membrane to engulf an extracellular material (endocytosis). After traversing the cell cytoplasm, the vesicles can fuse with the opposite membrane and release their contents to the surrounding interstitium (exocytosis) (2). This process, known as transcytosis, is important for the extravasation of molecules that are otherwise unable to traverse tightly fused endothelial sheets. The apparent down-regulation of this process in the CNS provides an even greater selectivity to the cerebral capillary endothelium.

Unfortunately, the pentalaminar tight junctions and paucity of pinocytic vesicles that empower the cerebral capillary endothelium to maintain homeostasis in the brain parenchyma also inhibit the intracranial delivery of most drugs via the cardiovascular system. Because the tightly fused endothelial membranes essentially form a continuous layer of lipids, a drug's lipophilicity (defined by the octanol:water partition coefficient) correlates strongly with cerebrovascular permeability (Fig. 1). Because many drugs are functionalized hydrophilic organic compounds, they are unable to traverse the highly hydrophobic lipid barrier and are therefore ineffective in the treatment of CNS diseases.

The selectivity of the cerebral microvasculature is further enhanced by the so-called metabolic blood–brain barrier (MBBB) (7). The presence of a MBBB was first predicted after the discovery that endothelial cells in the

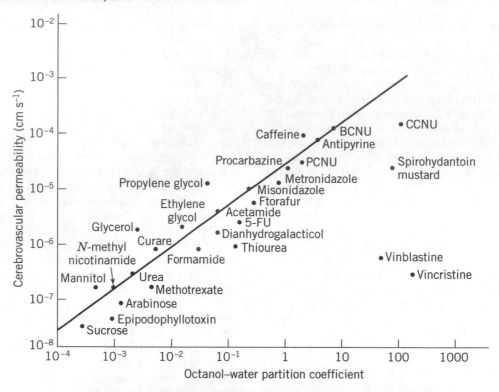

**Figure 1.** Relationship between the cerebrovascular permeability and the octanol–water partition coefficient of selected chemicals and drugs. Vinblastine and vincristine are clear "outliers," falling well below the line and hence penetrating much less than is expected from their lipophilicity. These drugs are excluded from the brain parenchyma by the active-drug-efflux-transporter protein in the luminal membrane of the cerebral capillary endothelium. *Source:* Reproduced with permission from Ref. 1.

cerebromicrovasculature have four times the mitochondrial content of systemic endothelial cells (2,3). Some molecules that freely enter brain endothelial cells through the luminal membrane undergo rapid metabolic chemical transformations that inhibit them from crossing the antiluminal membrane and reaching the surrounding brain interstitium (7). For example, L-dopa, an amino acid precursor of the neurotransmitters dopamine and norepinephrine, readily traverses the luminal membrane of brain endothelial cells. However, rapid metabolic conversion into dopamine by intracellular enzymes precludes L-dopa from crossing the antiluminal membrane. This metabolic conversion limits the therapeutic potential of exogenous L-dopa for the treatment of neurodegenerative disorders such as Alzheimer's disease.

Finally, the BBB is further reinforced by a high concentration of the *P*-glycoprotein (Pgp) active-drug-efflux-transporter protein in the luminal membrane of the cerebral capillary endothelium (1). This efflux transporter actively removes a broad range of drug molecules from the endothelial cell cytoplasm before they can cross into the brain parenchyma. For example, the octanol:water partition coefficient allows the antineoplastic agents vincristine and vinblastine to readily traverse endothelial membranes (Fig. 1). However, because they are actively pumped from the endothelial cytoplasm into the blood by the Pgp efflux system, these drugs are excluded from the brain parenchyma.

### Blood–Cerebrospinal Fluid Barrier

The second barrier that a systemically administered drug encounters before entering the CNS is known as the blood–cerebrospinal fluid barrier (BCB). The ventricles within the brain and the subarachnoid space that surrounds its outer surface are filled with an ultrafiltrate of plasma secreted by the choroid plexus called cerebrospinal fluid (CSF) (2). Because the CSF can exchange molecules with the interstitial fluid of the brain parenchyma, the passage of blood-borne molecules into the CSF is also carefully regulated by the BCB.

Physiologically, the BCB is found in the epithelium of the choroid plexus. The capillary endothelial sheets of the choroid plexus are fused not by pentalaminar fusions but by fenestrated tight junctions and are therefore somewhat more permeable than the capillary networks of the brain. However, these fenestrated capillaries are encased by a sheet of choroid epithelial cells bound by tight junctions, which restricts the rate of extravasation from the choroid capillary networks. This arrangement limits the passage of molecules and cells into the CSF, although, because the choroid endothelial sheets lack pentalaminar fusions, the BCB is somewhat more permeable to ions than the BBB (2).

In addition, the BCB is fortified by an active organic acid transporter system in the choroid plexus capable of driving CSF-borne organic acids into the blood. As a result, a variety of therapeutic organic acids such as the antibiotic penicillin, the antineoplastic agent methotrexate, and the antiviral agent zidovudine (AZT) are actively removed from the CSF and therefore inhibited from diffusing into the brain parenchyma (1).

Furthermore, substantial inconsistencies often exist between the composition of the CSF and interstitial fluid of the brain parenchyma, suggesting the presence of what is sometimes called the CSF–brain barrier (2,8). This barrier is attributed to the insurmountable diffusion distances required for equilibration between the CSF and the brain interstitial fluid. Therefore, entry into the CSF does not guarantee a drug's penetration into the brain.

### Blood–Tumor Barrier

The obstacles to intracranial drug delivery are even more challenging when the target is a CNS tumor. Solid tumors, including those of the brain, support their own growth by stimulating the intratumoral formation of new blood vessels in a process known as angiogenesis (9,10). The resultant intratumoral blood supply delivers oxygen and nutrients to internal neoplastic cells and removes metabolic wastes. Therefore, the cardiovascular system could theoretically be used to deliver drugs to the interstitium of solid tumors. However, research by Jain and others has shown that a variety of obstacles, collectively referred to as the blood–tumor barrier (BTB), often impede this mechanism of drug delivery to solid tumors, especially to solid tumors of the CNS (11–13).

In the CNS, the first of these obstacles is the BBB. The BBB is functional, albeit variably, in the microvasculature of primary and metastatic brain tumors (6,14). Low-grade primary brain tumors and small metastatic foci appear to have a relatively intact BBB, and in larger tumors the functional capacity of the BBB is regionally variable. Toward the center of a solid CNS tumor, the BBB is often severely compromised, but in the tumor periphery, especially at the outermost infiltrative regions of a primary brain tumor, the BBB is often highly functional. Between these two extremes the BBB is variably functional in the tumor microvasculature.

The presence of a BBB in the microvasculature of CNS tumors has clinical consequences. For example, even when primary and secondary systemic tumors respond to chemotherapy delivered via the cardiovascular system, intracranial metastases often continue to grow (6,15). In recent clinical trials for the treatment of systemic malignancies, patients receiving immunotherapy (16) and new chemotherapeutic treatment protocols (17–19) remained disease free systemically while succumbing to CNS metastases. Furthermore, the most effective chemotherapeutic drugs in the treatment of brain tumor patients are those that most efficiently cross the BBB in normal brain capillaries, such as the alkylating agent 1,3-bis-chloroethylnitrosourea (BCNU) (Fig. 1) (20,21).

Even for CNS malignancies in which the BBB is significantly compromised, a variety of physiological barriers common to all solid tumors inhibit drug delivery via the cardiovascular system. First of all, drug delivery to neoplastic cells in a solid tumor is compromised by a hetero-

geneous distribution of microvasculature throughout the tumor interstitium, which leads to spatially inconsistent drug delivery. Furthermore, as a tumor grows larger, the vascular surface area decreases, leading to a reduction in transvascular exchange of blood-borne molecules. At the same time, the intercapillary distance increases, leading to a greater diffusional requirement for drug delivery to neoplastic cells (11). Moreover, poorly vascularized regions of the tumor receiving insufficient nutrients for growth can house neoplastic cells resting in $G_0$, rendering them insensitive to cell cycle–dependent antineoplastic agents (6).

Another physiological characteristic of both systemic and CNS solid tumors that restricts drug delivery is an elevation of interstitial pressure inside the tumor (6,11). As the leakiness of the tumor vasculature increases, so does the rate of fluid extravasation leading to a buildup of interstitial fluid. Because the tumor does not possess a sufficient lymphatic system to remove the excess fluid, the intratumoral interstitial pressure increases (12). The net result is a pressure gradient: interstitial pressure is highest at the center of the tumor, while the pressure at the outermost regions approaches that of the surrounding tissue (11). As the elevated interstitial pressure approaches the pressure inside the tumor capillary networks, the rate of extravasation of fluid and macromolecules into the tumor is retarded. Additionally, the high pressure at the tumor's center results in a net flow of fluid from the center into the periphery and the surrounding interstitium, known as peritumoral edema. This continuous fluid flow creates a constant, radially directed convective force diametrically opposed to intratumoral drug penetration (11).

In the CNS, these inherent physiological barriers to solid-tumor drug delivery and the variably intact BBB may work synergistically to restrict the delivery of antineoplastic drugs to malignant cells. For example, the combination of a variably intact BBB and inconsistently dispersed tumor microvasculature can lead to an intratumoral drug distribution that is exceedingly heterogeneous, with some areas left completely unexposed. In the most highly vascularized and therefore potentially accessible peripheral regions of a CNS tumor, the BBB is in fact the most functional and most restricting of transvascular drug exchange. Infiltrative cells at the periphery can also escape into the surrounding interstitium where the BBB is fully functional (20). Furthermore, the high interstitial tumor pressure and associated peritumoral edema leads to increases in hydrostatic pressure in the normal brain parenchyma adjacent to the tumor. As a result, the cerebral microvasculature in these tumor-adjacent regions of normal brain may be even less permeable to drugs than normal brain endothelium, leading to exceptionally low extratumoral interstitial drug concentrations (20). This surrounding region can therefore act as a diffusion sink and further limit intratumoral drug penetration (6).

In conclusion, the delivery of drugs to the CNS via the cardiovascular system is often precluded by a variety of formidable barriers including the BBB, the BCB, and the BTB. In the last 30 years, aggressive research has elucidated the physiological, metabolic, and biochemical mechanisms underlying these barriers. The next section will explore how this knowledge has been used to rationally design therapeutic strategies for enhanced drug delivery to the CNS.

## STRATEGIES FOR ENHANCED CNS DRUG DELIVERY

To circumvent the multitude of barriers inhibiting CNS penetration by potentially therapeutic agents, numerous drug delivery strategies have been developed. These strategies generally fall into one or more of the following three categories: manipulating drugs, disrupting the BBB, and finding alternative routes for drug delivery.

### Enhancing BBB Permeability by Drug Manipulation

**Lipophilic Analogs.** A variety of pharmacological parameters influence a drug's cerebrovascular permeability. Specifically, CNS penetration is favored by low molecular weight, lack of ionization at physiological pH, and lipophilicity (2). Because a drug's lipophilicity correlates so strongly with cerebrovascular permeability (Fig. 1), hydrophobic analogues of small hydrophilic drugs ought to more readily penetrate the BBB. This strategy has been frequently employed by a number of researchers, but the results have often been disappointing.

The best example of such attempts are the series of lipophilic analogues of the nitrosoureas. Nitrosoureas, such as BCNU, elicit antineoplastic activity by alkylating purine and pyrimidine bases in DNA. The modest yet significant clinical success of BCNU against malignant brain tumors has been attributed to the compound's relatively high lipophilicity ($\log P$ 1.53) (2). It was therefore hypothesized that further latentiation of BCNU would enhance its antineoplastic activity. To this end, more than 20 BCNU analogs including 1-(2-chloroethyl)-3-cyclohexyl-1-nitrosourea (CCNU, $\log P$ 2.83) and methyl-CCNU ($\log P$ 3.3) were synthesized.

Quantitative structure activity relationship studies with this series of analogs indicated a surprising result (6). As was later confirmed in clinical trials, these studies demonstrated that the antineoplastic activity of the nitrosourea analogs was inversely proportional to their lipophilicity (2,5). These observations were rationalized as follows. First, the more lipophilic analogs demonstrate diminished alkylating activity and increased dose-limiting toxicity when compared to BCNU. In addition, the more lipophilic analogs are less soluble in the aqueous plasma and bind more readily to plasma proteins, leading to lower concentrations of drug available for diffusion into the CNS. Finally, CCNU and methyl-CCNU are less soluble in the extracellular/intracellular fluid of the brain, limiting their activity against tumor cells. The experience with the nitrosoureas indicates that if a drug is to be efficacious when delivered via the circulatory system for the treatment of a CNS disease, a delicate balance between cerebrovascular permeability and plasma solubility is required. Specifically, the optimal octanol:water partition coefficient is approximately $\log P$ 1.5 to 2.5 (22).

A second strategy for increasing the lipophilicity of a hydrophilic therapeutic agent is to surround the hydrophilic molecule with a sphere of lipids in the form of a liposome. Theoretically, encapsulated drugs may diffuse through the walls of the cerebromicrovasculature owing to the high affinity of their lipid coating for the cell membranes of capillary endothelial cells. Once inside the brain, encapsulated drugs may be released into the interstitium and act on target cells. Unfortunately, this strategy has not yet lived up to its theoretical potential. Because liposomes are preferentially taken up by fixed macrophages (3), most enter the reticuloendothelial system of the liver and spleen and never reach the brain (2). Additionally, because of their large size, liposomes that do reach the CNS are unable to traverse the BBB (3).

**Prodrugs.** Another strategy for drug modification to enhance CNS penetration is the design of prodrugs. A prodrug consists of a drug covalently attached to an unrelated chemical moiety that improves the drug's pharmacokinetic properties. The prodrug itself is not active but becomes active when the attached moiety is cleaved in vivo by enzymatic or hydrolytic processes. To enhance a drug's penetration into the CNS, prodrugs are often designed by attaching chemical moieties that increase the drug's lipophilicity. The best example of this approach is the series of analogs of the pain reliever morphine. As shown in Figure 2, the alkaloid morphine, which possesses two polar hydroxyl groups and a $\log P$ value of $-1.66$, does not readily cross the BBB. Methylation of one hydroxyl group, however, yields the more powerful alkaloid codeine, which has enhanced brain uptake. Further latentiation via acetyla-

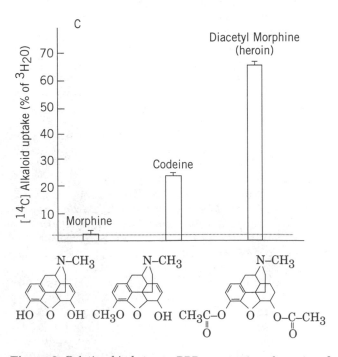

**Figure 2.** Relationship between BBB penetration of a series of [$^{14}$C] alkaloids (as a percent of penetration of $^3$H$_2$O) and their chemical structure. Morphine (left) penetrates the BBB poorly because it contains two highly polar hydroxyl groups. Addition of methyl (codeine) and acetyl (heroin) groups increases lipophilicity and facilitates brain uptake. *Source:* Reproduced with permission from Ref. 1.

tion of both hydroxyl groups yields the hallucinogenic heroin, which demonstrates a dramatic increase in activity over the parent compound. Heroin readily traverses the BBB, and subsequent hydrolytic cleavage of the acetyl groups yields high concentrations of morphine trapped in the brain due to its hydrophilicity (2).

**Redox Chemical Delivery Systems.** Bodor et al. have developed a creative approach to exploit the permeability characteristics of the BBB to achieve site specific and/or sustained release of drugs to the brain (23–25). The method involves linking a drug to a lipophilic dihydropyridine carrier, creating a complex that, after systemic administration, readily traverses the BBB because of its lipophilicity. Once inside the brain parenchyma, the dihydropyridine moiety is enzymatically oxidized to the ionic pyridinium salt. The acquisition of charge has the dual effect of accelerating the rate of systemic elimination by the kidney and bile and trapping the drug–pyridinium salt complex inside the brain. Subsequent cleavage of the drug from the pyridinium carrier leads to sustained drug delivery in the brain parenchyma (24). This methodology increases intracranial concentrations of a variety of drugs including neurotransmitters, antibiotics, and antineoplastic agents. Recently this methodology has been extended to deliver neuroactive peptides such as enkephalin to the brain. Protecting the peptide–dihydroxypyridine complex with a biolabile lipophilic steroid derivative shields the peptide from enzymatic degradation prior to traversing the BBB (26). This methodology has demonstrated promise in laboratory models, and evaluation of clinical efficacy in neurological patients is awaited with interest.

**Carrier-Mediated Drug Delivery.** An alternative strategy for enhanced CNS penetration is carrier-mediated drug delivery, which takes advantage of the facilitative transport systems present in the brain endothelium. The cerebrovascular membranes are densely populated with facilitative carrier systems for both glucose and large neutral amino acids (LNAAs). The carrier system for glucose, the primary mechanism whereby the brain receives its only source of energy, is highly specific for monosaccharides, rendering it useless for drug delivery (1). The LNAA carrier system, on the other hand, is capable of transporting numerous endogenous as well as exogenous LNAAs with great structural variety (1,2,6). This characteristic has made exploitation of the LNAA carrier system an attractive strategy for CNS drug delivery.

For example, neurons of the substantia nigra in patients with Parkinson's disease are markedly deficient in the neurotransmitter dopamine (22). Systemic administration of dopamine has no therapeutic benefit because dopamine does not cross the BBB to an appreciable extent (2). The LNAA carrier system has been used to deliver L-3,4-dihydroxyphenalanine (levodopa), an endogenous precursor to dopamine, to the brain. Unlike dopamine, levodopa has a high affinity for the LNAA carrier system. After traversing the antiluminal membrane of the cerebral endothelium, levodopa is decarboxylated to yield dopamine (2). The success of this therapeutic strategy is hindered by the peripheral conversion of levodopa to dopamine leading

to systemic toxicity. However, concurrent administration of the carboxylase inhibitor carbidopa ameliorates unwanted side effects and makes the systemic administration of levodopa an effective therapy for Parkinson's disease (2).

The LNAA carrier system could theoretically be exploited to also deliver the antineoplastic agent melphalin, a nitrogen–mustard derivative of phenylalanine, to the brain. However, under physiological conditions, melphalin's affinity for the LNAA carrier protein is lower than that of endogenous LNAAs, and its transport is therefore severely limited (6). A newly synthesized analog of melphalin, D,L-NAM, demonstrates enhanced affinity for the LNAA carrier (27). The addition of hydrophobic side chains seems to allow D,L-NAM to interact more strongly with a hydrophobic binding site in the LNAA carrier protein (1). This strategy may lead to enhanced penetration via the LNAA carrier system.

**Receptor/Vector-Mediated Drug Delivery.** Other CNS drug delivery strategies focus on manipulating the cerebral endothelial systems for receptor-mediated transcytosis. The BBB possesses several specific transcytosis systems for the extravasation of important nutrients and signaling molecules that cannot diffuse through the cerebromicrovasculature. These include systems for the transport of insulin, transferrin, and insulinlike growth factor. Pardridge et al. propose as a mechanism for intracranial drug delivery the covalent coupling of a therapeutic protein normally excluded from the brain to a protein that is readily transcytosed across the cerebral capillary endothelium (4). For example, the nontransportable protein β-endorphin was linked to the transportable protein-cationized albumin via a disulfide linkage. This "chimeric peptide" was successfully transcytosed into the brain and enzymatically cleaved in the parenchyma by thiol reductase.

This chimeric peptide methodology has been further developed by Friden et al., who proposed the covalent coupling of drug molecules with antibodies to the transferrin receptor (28–30). Based on the observation that cerebral capillary endothelial cells express an abundance of transferrin receptors relative to systemic capillaries (31), it was proposed that the use of antitransferrin receptor monoclonal antibodies may lead to brain-selective targeting of drug molecules. This hypothesis was tested by systemically administering radiolabeled antirat transferrin-receptor antibody OX-26, separating the brain vasculature from the parenchyma via centrifugation through dextran, and quantitating OX-26 content in the two fractions by autoradiography (28). These experiments demonstrated that the OX-26 antibody was indeed transcytosed into the brain parenchyma, reaching peak levels of 0.4–0.5% of the injected dose per gram of brain tissue at 24 hours postadministration. Similar results were obtained when the OX-26 antibody was covalently linked to radiolabeled nerve growth factor (NGF) (28). The systemic delivery of this NGF–OX-26 conjugate also significantly enhanced the survival of cholinergic neurons in a murine model when compared to systemic NGF alone, indicating the therapeutic potential of this approach in the treatment of Alzheimer's disease.

Pardridge has extended this methodology to achieve the intracranial delivery of polyamide ("peptide") nucleic acids (PNAs) (32). PNAs are a form of antisense oligodeoxynucleotides in which a polyamide backbone is substituted for the phosphate backbone. These molecules possess great therapeutic potential in the treatment of neurological disorders, yet due to their hydrophilicity and large size, they are unable to cross the BBB. Encouraging results have been obtained with a model PNA, antisense to the genome of the human immunodeficiency virus type 1. This PNA can successfully be delivered to the murine brain parenchyma at 0.1% of the initial dose per gram of tissue when covalently coupled to the OX-26 monoclonal antibody.

More recently, this technique has been expanded to successfully deliver drug-loaded immunoliposomes to the brain (33). Immunoliposomes consist of liposomes coated with the inert polymer polyethylene glycol conjugated to the OX-26 antitransferrin-receptor monoclonal antibody. In a recent application, maximum uptake of only 0.03% of the initial dose per gram of brain tissue was achieved for immunoliposomes loaded with the antineoplastic agent daunomycin. However, because a single immunoliposome may carry up to 10,000 drug molecules, this methodology might greatly enhance the capacity of the OX-26 antibody for intracranial drug delivery. Although these techniques hold great promise and warrant further study, the intracranial drug concentrations that may be achieved via the receptor-mediated drug delivery strategy may be limited by saturation of the target receptors and by competitive inhibition from endogenous substrates.

### Disrupting the BBB

In spite of recent developments in drug modification for enhanced CNS penetration, the BBB remains a formidable obstacle that compromises the successful treatment of many neurological disorders. The second general class of strategies for enhanced CNS drug delivery involves the systemic administration of drugs in conjunction with transient BBB disruption (BBBD). Theoretically, with the BBB weakened, systemically administered drugs can undergo enhanced extravasation rates in the cerebral endothelium, leading to increased parenchymal drug concentrations. A variety of techniques that transiently disrupt the BBB have been investigated; however, albeit physiologically interesting, many are unacceptably toxic and therefore not clinically useful. These include the infusion of solvents such as dimethyl sulfoxide or ethanol and metals such as aluminum; X-irradiation; and the induction of pathological conditions including hypertension, hypercapnia, hypoxia, or ischemia (3). The mechanisms responsible for BBBD with some of these techniques are not well understood. A somewhat safer technique involves the systemic delivery of the convulsant Metrazol (pentylenetetrazol), which transiently increases the BBB permeability while causing seizures. Concurrent administration of the anticonvulsant pentobarbital blocks seizing while allowing BBBD to persist (26). The BBB can also be compromised by the systemic administration of several antineoplastic agents including VP-16, cisplatin, hydroxyurea, fluorouracil, and etoposide (3). In animal models, the intracarotid administration of etoposide results in a 3-fold increase in cerebrovascular permeability that persists for 4–5 days. This technique yielded a statistically significant increase in the CNS uptake of indium-labeled liposomes in a rat model. However, the potential toxicity associated with long term BBBD is a major concern.

**Osmotic BBBD.** The most frequently applied clinical technique for achieving BBBD is the intraarterial infusion of a hyperosmolar solution of mannitol (34,35). By infusing the solution directly into an artery that feeds the target area of the brain, it is somewhat possible to achieve localized BBBD and/or drug delivery (36). As the hyperosmolar solution flows through the cerebral capillaries, acute dehydration of endothelial cells results in cell shrinkage, which in turn widens the tight junctions connecting adjacent membranes (36). Subsequent rehydration in the presence of normal plasma leads to complete restoration of the BBB about 4 h following treatment. The hyperosmolar technique of BBBD has been studied extensively in animal models (37–39). These studies demonstrate that osmotic BBBD can be safe and is able to increase concentrations of chemotherapeutic agents in the brain parenchyma 10–90-fold (40). This technique has been applied clinically for more than 15 years, most extensively by Neuwelt et al. (34,35).

The clinical protocol for BBBD in neurooncological patients generally consists of the following. A 25% mannitol solution is infused at 5–10 mL/s for 30 seconds into the carotid or vertebral artery that feeds the area of the brain containing an intracranial tumor. Very soon before or after BBBD, the patient is administered a chemotherapeutic regimen via the circulatory system. In a recent clinical study, Neuwelt et al. examined the efficacy of the coadministration of the antineoplastic agents carboplatin and etoposide in conjunction with BBBD in 34 patients with intracranial tumors (40). In 4 out of 4 patients with primitive neuroectodermal tumors and 2 out of 4 patients with CNS lymphomas, the authors report a complete response (a complete response was defined as "the resolution of enhancing tumor as shown on computed tomograms or magnetic resonance images, no need for steroids, stable or improved neurological function, and negative results on CSF tests.") No response, however, was seen in patients with glioblastoma multiforme, mixed astrocytoma-oligodendroglioma, or metastatic carcinoma.

Unfortunately, the treatment regimen was associated not only with expected reversible myelosuppression but also with unexpected, irreversible high-frequency hearing loss. The latter is believed to be associated with ototoxic effects of carboplatin. It has since been reported that the administration of the chemoprotective agent sodium thiosulfate (which converts carboplatin into a noncytotoxic compound) 1 h after carboplatin administration in conjunction with osmotic BBBD may provide a mechanism to rescue from carboplatin toxicity (41). The authors conclude that this treatment protocol warrants further investigation, especially for the treatment of highly responsive tumors. This study exemplifies both the therapeutic and toxic potential of BBBD in conjunction with systemic drug administration.

**Biochemical BBBD.** Recently, new and potentially safer biochemical techniques have been developed to disrupt the BBB. One reason for the unfavorable toxic/therapeutic ratio often observed with hyperosmotic BBBD is that this methodology results in only a 25% increase in the permeability of the tumor microvasculature, in contrast to a 10-fold increase in the permeability of normal brain endothelium (42). Therefore, the normal brain is left vulnerable to potentially neurotoxic chemotherapeutic agents. Black et al. recently developed biochemical techniques to selectively disrupt the intratumoral BBB while minimally altering the normal BBB (42). For example, in animal studies the intracarotid administration of the vasoactive agent leukotriene C4 (LTC4) selectively increases the permeability in brain tumor capillaries while leaving the normal brain unaffected. The selectivity for the tumor microvasculature is attributed to the secretion by endothelial cells in the normal brain of the enzyme $\gamma$-glutamyl transpeptidase, which inactivates LTC4. However, because this enzyme requires a glial inductive influence for expression, it is downregulated in tumor endothelial cells, leaving the tumor microvasculature vulnerable to LTC4-mediated BBBD.

Recently, this group has shown that the administration of another potent vasoactive agent, the synthetic bradykinin analogue RMP7, results in even greater selective increase in permeability in brain tumor capillaries. Mechanistic studies suggest that the effects of RMP7 on tumor endothelial cells are mediated by nitric oxide released by the enzyme nitric oxide synthase, which is expressed at higher levels in RG2 rat gliomas than in normal rat brain (43). In efficacy studies with the RG2 model, animals treated with a combination of carboplatin in conjunction with RMP7 exhibited a significantly higher survival rate at 31 days when compared to controls and carboplatin alone. Carboplatin penetration into brain tumors was increased 2.7-fold when the drug was administered in conjunction with RMP7-induced BBBD, with no enhanced uptake into the normal brain parenchyma (44). In a recent clinical study involving nine patients with recurrent malignant glioma, the intracarotid infusion of RMP7 demonstrated a selective increase in the penetration of the radioactive tracer [68]GaEDTA into brain tumors without increasing its transport into normal brain (45). Although the efficacy of this strategy has yet to be demonstrated clinically, the results are encouraging and warrant further study.

## Alternative Routes to CNS Drug Delivery

Despite advances in rational CNS drug design and BBBD, many potentially efficacious drug molecules still cannot penetrate into the brain parenchyma at therapeutic concentrations. Most of the aforementioned techniques aim to enhance the CNS penetration of drugs delivered via the circulatory system. This method results in delivery of the injected agent throughout the entire body, frequently causing unwanted systemic side effects. Additionally, systemically administered agents must penetrate the BBB to enter the brain, which has proven to be a formidable task. A third class of strategies aimed at enhancing CNS penetration of drug molecules is composed of delivery methodologies that do not rely on the cardiovascular system. These alternative routes for controlled CNS drug delivery obviate the need for drug manipulation to enhance BBB permeability and/or BBBD by circumventing the BBB altogether.

**Intraventricular/Intrathecal Route.** One strategy for bypassing the BBB that has been studied extensively both in the laboratory and in clinical trials is the intralumbar injection or intraventricular infusion of drugs directly into the CSF. Drugs can be infused intraventricularly using an Ommaya reservoir, a plastic reservoir implanted subcutaneously in the scalp and connected to the ventricles within the brain via an outlet catheter. Drug solutions can be subcutaneously injected into the implanted reservoir and delivered to the ventricles by manual compression of the reservoir through the scalp.

When compared to vascular drug delivery, intra-CSF drug administration theoretically has several advantages. Intra-CSF administration bypasses the BBB and results in immediate high CSF drug concentrations. Because the drug is somewhat contained within the CNS, a smaller dose can be used, potentially minimizing systemic toxicity (6). Furthermore, drugs in the CSF encounter minimized protein binding and decreased enzymatic activity relative to drugs in plasma, leading to longer drug half-life in the CSF (46). Finally, because the CSF freely exchanges molecules with the extracellular fluid of the brain parenchyma, delivering drugs into the CSF could theoretically result in therapeutic CNS drug concentrations.

However, this delivery method has not lived up to its theoretical potential for several reasons. These include a slow rate of drug distribution within the CSF space; increases in intracranial pressure associated with fluid injection or infusion into small ventricular volumes; and a high clinical incidence of hemorrhage, CSF leaks, neurotoxicity, and CNS infections (4). Most importantly, the success of this approach is limited by the CSF-brain barrier, composed of barriers to diffusion into the brain parenchyma. Because the extracellular fluid space of the brain is extremely tortuous, drug diffusion through the brain parenchyma is very slow and inversely proportional to the molecular weight of the drug. For example, 1 h after intraventricular administration, the methotrexate concentration 3.2 mm into the brain has been calculated to be only 1% of that in the CSF. For macromolecules, such as proteins, brain parenchymal concentrations following intra-CSF administration are undetectable (6). For these reasons, intra-CSF chemotherapy in the treatment of intraparenchymal CNS tumors has not proven to be effective. The greatest utility of this delivery methodology has been in cases where high drug concentrations in the CSF and/or the immediately adjacent parenchyma are desired, such as in the treatment of carcinomatous meningitis or for spinal anesthesia/analgesia (46).

**Olfactory Pathway.** An alternative CNS drug delivery strategy that has received relatively little attention is the intranasal route. Drugs delivered intranasally are transported along olfactory sensory neurons to yield significant concentrations in the CSF and olfactory bulb. In recent studies, intranasal administration of wheat germ agglutinin–horseradish peroxidase resulted in a mean olfactory bulb concentration in the nanomolar range. In theory, this delivery strategy could be effective in the delivery of therapeutic proteins such as brain-derived neurotrophic factor (BDNF) to the olfactory bulb as a treatment for Alzheimer's disease (47). An obvious advantage of this method is that it is noninvasive relative to other strategies. In practice, however, further study is required to determine if therapeutic drug concentrations can be achieved following intranasal delivery.

**Interstitial Delivery.** The most direct way of circumventing the BBB is to deliver drugs directly to the brain interstitium. By directing agents uniquely to an intracranial target, interstitial drug delivery can theoretically yield high CNS drug concentrations with minimal systemic exposure and toxicity. Furthermore, with this strategy, intracranial drug concentrations can be sustained, which is crucial in the treatment of many neurodegenerative disorders and for the antitumor efficacy of many chemotherapeutic agents.

*Injections, Catheters, and Pumps.* Several techniques have been developed for delivering drugs directly to the brain interstitium. Until recently, the most widely used method has been the interstitial injection or infusion of drugs using an Ommaya reservoir or implantable pump. The use of this methodology has recently been extensively reviewed by Walter et al. (48) and will only be summarized here. The adaptation of the Ommaya reservoir to achieve interstitial drug delivery simply involves placing the outlet catheter directly in the intracranial target area. This technique has often been applied to neurooncological patients in which the outlet catheter is placed in the resection cavity following surgical debulking of a brain tumor. Chemotherapeutic agents can be periodically injected into the subcutaneous reservoir and then delivered directly to the tumor bed. This technique, however, does not achieve truly continuous drug delivery.

More recently, several implantable pumps have been developed that possess several advantages over the Ommaya reservoir. Like the Ommaya reservoir, these pumps can be implanted subcutaneously and refilled by subcutaneous injection. Unlike the Ommaya reservoir, these pumps are capable of delivering drugs as a constant infusion over an extended period of time. Furthermore, the rate of drug delivery can be varied using external hand-held computer control units.

Currently each of the three different pumps available for interstitial CNS drug delivery operates by a distinct mechanism. The Infusaid pump (Infusaid Corp., Norwood, MA) uses the vapor pressure of compressed freon to deliver a drug solution at a constant rate; the MiniMed PIMS system (MiniMed, Sylmor, CA) uses a solenoid pumping mechanism; and the Medtronic SynchroMed system (Medtronic, Minneapolis, MN) delivers drugs via a peristaltic mechanism.

The Ommaya reservoir or infusion pumps have thus far been used in various clinical trials with brain tumor patients to interstitially deliver the chemotherapeutic agents BCNU or its analogs, methotrexate, adriamycin, bleomycin, fluorodeoxyuridine, cisplatin, L-2,4-diaminobutyric acid, interleukin 2 (IL-2), and interferon $\gamma$ (48). In most of these studies the intratumoral drug concentrations were often high, and the side effects of the therapy were mild. In all of these clinical studies, however, the patient population was small (<30), and the survival benefit was rarely significant. The success of these techniques is limited by catheter clogging or blocking by tissue debris, inadequate drug distribution throughout the tumor, and a high degree of burden to the patient.

Oldfield et al. have suggested that the distribution of large and small drug molecules in the brain could be enhanced using convection to supplement simple diffusion (49). The group proposes maintaining a pressure gradient during interstitial drug infusion to generate bulk fluid convection through the brain interstitium. This technique was employed in mongrel cats to increase the distribution of $^{111}$In-labeled transferrin and [$^{14}$C]sucrose over centimeter distances from the site of infusion. Neuwelt et al. have, however, recently argued that increasing the diffusion gradient by maximizing the concentration of the infused agent results in a higher volume of intraparenchymal drug distribution than is achieved by enhancing fluid convection (50). Further study is warranted to optimize the infusion techniques used in future clinical testing.

*Drug Delivery from Biological Tissues.* Another strategy to achieve interstitial drug delivery involves releasing drugs from biological tissue. With the dawn of molecular medicine, there has been great interest in using biological tissue as both the source and delivery vehicle for therapeutic proteins. This concept has been the subject of intense research for the past two decades.

The simplest approach to this technique is to implant into the brain a tissue that naturally secretes a desired therapeutic agent. This approach has been most extensively applied to the treatment of Parkinson's disease (22). The debilitating symptoms of Parkinson's disease are largely caused by diminished interstitial dopamine concentrations due to the progressive degeneration of the dopamine-releasing neurons of the substantia nigra. Transplantation of embryonic dopamine-releasing neurons is an attractive therapeutic strategy because these cells demonstrate good post-transplantation survival and growth characteristics in animal models (51). For example, Sladeck et al. have demonstrated that grafting fetal embryonic nerve cell or paraneural tissue to damaged areas of the brain leads to substantial functional recovery in both rodent and primate models of parkinsonism (52). The observed improvements correlated with the duration of graft survival and the extent to which the grafted neurons innervated the host striatum.

When this strategy was first applied clinically, the results were disappointing (53). Transplanted tissue often did not survive owing to a lack of neovascular innervation. Without rapid neovascularization, implanted solid grafts undergo irreparable ischemic injury leading to cell death. Recently, Rogers et al. have demonstrated enhanced vascularization and microvascular permeability in cell-suspension embryonic neural grafts relative to solid grafts (53), but improvements in clinical results using this technique have not yet been dramatic.

An attractive extension of this method is to use gene therapy to develop optimized biological tissue for interstitial drug delivery. Prior to implantation, cells can be genetically modified to synthesize and release specific therapeutic agents. Laterra et al. have demonstrated the therapeutic potential of this technique in the treatment of brain tumors. Owing to the presence of endothelial mitogens and a high rate of vessel proliferation in CNS tumors, genetically modified endothelial cells might efficiently transplant to brain tumor microvasculature (54). This hypothesis was supported by demonstrating high grafting efficiency of immortalized rat brain endothelial cells genetically engineered to express the $\beta$-galactosidase reporter following implantation into experimental gliomas.

The use of nonneuronal cells for therapeutic protein delivery to the CNS has recently been reviewed (55). Thompson et al., for instance, have recently used this technique to develop a novel immunotherapeutic approach for the treatment of CNS tumors (56). Nonreplicating murine melanoma cells genetically engineered to produce the cytokine granulocyte–macrophage colony-stimulating factor (GM-CSF) or interleukin 2 (IL-2) in a paracrine fashion were implanted in the flank and/or the brain of mice challenged with a lethal intracranial injection of wild-type melanoma cells. Systemic GM-CSF administration from cells implanted in the flank in conjunction with intracranial IL-2 delivery demonstrated a synergistic survival benefit.

The survival of foreign tissue grafts may be improved by advancements in techniques for culturing distinct cell types. For example, cultured human tumorigenic neuroblastoma cells synthesize several potentially therapeutic neurotransmitters (57). These cells can be chemically rendered amitotic without depleted neurotransmitter synthesis. Transplantation into the primate brain yielded graft survival for up to 270 days and no evidence of neoplastic growth. Also cografting cells engineered to release neurotrophic factors with cells engineered to release therapeutic proteins may enhance the survival and development of foreign tissue (52).

Ideally it would be possible to perform in vivo genetic engineering to cause specific endogenous brain tissue to express a desired protein, circumventing the ischemic and immunogenic complications encountered with the implantation of foreign tissue grafts. In the last few years there have been numerous attempts to apply this strategy to the treatment of various CNS disorders including Parkinson's disease (58,59), epilepsy (60), Alzheimer's disease (61), and malignant brain tumors (62–64). One technique that has been developed for the treatment of CNS malignancies involves in vivo tumor transduction with the herpes simplex thymidine kinase (HS-tk) gene followed by treatment with the antiherpes drug ganciclovir (65). HS-tk transduction specifically into replicating glioma cells is achieved in various animal models following intratumoral injections of retroviral vector-producing cells containing the HS-tk gene, rendering the transfected tumor cells susceptible to treatment with ganciclovir. Other vector systems used in CNS gene transfer studies include retroviruses, adenoviruses,

adeno-associated viruses, encapsulation of plasmid DNA into cationic liposomes, and neural and oligodendrial stem cells. Although this approach holds remarkable therapeutic potential in the treatment of CNS diseases, its efficacy has thus far been hindered by a number of obstacles: restricted delivery of vector systems across the BBB, inefficient transfection of host cells, nonselective expression of the transgene, and deleterious regulation of the transgene by the host CNS (64).

In conclusion, despite significant advancements in interstitial drug delivery to the CNS using injections, catheters, pumps, and biological tissue implants, these techniques have thus far had only modest clinical impact. However, the therapeutic potential of interstitial drug delivery may soon be realized using new advances in polymer technologies to modify the aforementioned techniques. The second half of this chapter will focus on our work and the work of other groups on delivering drugs directly to the brain interstitium using controlled-release polymer systems.

## INTERSTITIAL CNS DRUG DELIVERY USING CONTROLLED-RELEASE POLYMER SYSTEMS

### Development of Polyanhydrides for Controlled Drug Delivery

Polymeric or lipid-based devices that can deliver drug molecules at defined rates for specific periods of time are now making a tremendous impact in clinical medicine (66,67). For a long time, however, polymers used in medicine were simply adapted from materials originally developed for other applications. For example, the highly flexible material used to construct an artificial heart was borrowed from a woman's girdle. Dialysis tubing was built with sausage casing, and the material used in breast implants was adapted from mattress stuffing. As a result, the scope of polymeric medical devices was limited by the narrow range of physical characteristics found in conventional polymeric materials.

Recognizing the shortcomings in this approach, a program was launched at MIT in 1979 to rationally design polymeric materials specifically for drug delivery applications (68,69). The idea was to first decide what mechanical, chemical, and biological characteristics were required for a particular application and then to design, from first principles, a polymeric material that possessed these properties.

**Rational Design of Polyanhydride Polymers.** To develop polymeric materials specifically for drug delivery applications, the following characteristics were considered. First, the delivery device should be biodegradable to obviate the need for surgical removal following treatment. Polymeric degradation can occur either by bulk erosion throughout the polymeric matrix (like a sugar cube) or by surface erosion, whereby the matrix dissolves layer by layer (like a bar of soap). Bulk erosion, the mechanism by which polyester-based suture materials erode, is not desirable for drug delivery applications because it can lead to drug dumping and potential toxicity. Surface erosion of a homogeneous polymer/drug formulation, however, would lead to constant, zero-order release of the entrapped drug, as long as the surface area of the device remained fairly constant over time. One geometry that would achieve this constancy is a long flat slab as shown in Figure 3. Polymeric degradation in vivo can be achieved by either enzymatic or hydrolytic mechanisms. Although enzyme levels can vary from patient to patient as well as both regionally and temporally within the same patient, excess water is ubiquitous. Therefore, we decided that drug release profiles would be more consistent if the polymer degraded via hydrolytic degradation.

The challenge was then to find a combination of monomers and bonds that would yield a polymeric material that would erode from its surface via hydrolytic degradation. We predicted that hydrophobic monomers would minimize the penetration of water into the delivery device and allow only the bonds at the surface to be cleaved. The exclusion of water inside the device would also protect unstable drugs from hydrolytic degradation. Toxicologists at MIT believed that the hydrophobic monomers carboxyalkanoic acid and sebacic acid would be nontoxic and nonmutagenic. This strategy would only work, however, if the bonds linking the hydrophobic monomers were highly labile in the presence of water. After considering the stability of various polymeric linkages, we predicted that anhydride bonds would demonstrate the appropriate instability to promote surface erosion. A variety of polyanhydride polymers and copolymers were therefore synthesized by then graduate students or postdoctoral fellows Rosen and Linhardt; initial studies demonstrated that, indeed, these new materials displayed surface eroding properties (70). MIT postdoctoral fellow Leong subsequently showed that altering the ratio of monomers in several copolymer systems achieved variable degradation rates (71). For example, altering the monomer ratio in the copolymer system composed of sebacic acid (SA) and the more hydrophobic 1,3-bis-*para*-carboxyphenoxypropane (pCPP) results in degradation rates ranging from a few days to up to 5 years. (Fig. 4). When this new class of polymers was then applied to drug delivery, zero-order or near-zero-order release was achieved with many drug/polymer formulations (Fig. 5).

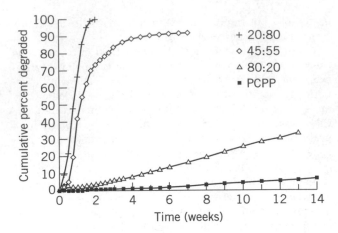

**Figure 4.** Rate of polyanhydride degradation versus time. pCPP and SA copolymers were formulated into 1.4-cm-diameter disks 1 mm thick by compression molding and placed into a 0.1 M pH 7.4 phosphate buffer solution at 37°C. The cumulative percentage of the polymer that was degraded was measured by absorption at 250 nm. *Source:* Reproduced with permission from Ref. 69.

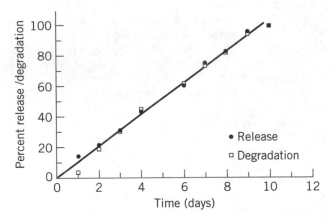

**Figure 5.** Release of *p*-nitroaniline and the degradation of poly-(carboxyphenoxyacetic acid) (P-CPA) versus time. Disks (1.4 cm in diameter and 1 mm thick) of PCPA were prepared containing 5% (w/w) *p*-nitroaniline by compression molding and degraded in 0.1 M pH 7.4 phosphate buffer at 37°C. The cumulative release of *p*-nitroaniline and degradation of PCPA were measured by absorbance at 380 and 235 nm, respectively. *Source:* Reproduced with permission from Ref. 69.

**Synthesis of Polyanhydrides.** Because the first polyanhydrides had low molecular weights (under 20,000 Da), they were brittle and not strong enough to be formed into practical drug delivery devices. However, Domb has developed several methods for synthesizing polyanhydrides with molecular weights up to 250,000 Da (72), including several solution methods such as catalyzed dehydrative coupling, dehydrochlorination, and one-step polymerization using phosgene- or diphosgene-coupling agents. The most successful synthesis, however, was achieved via polycondensation of melted monomers.

The optimized polycondensation synthesis scheme for the copolymer system composed of SA and pCPP in a 20:80 molar ratio is shown in Figure 6 (72). Highly purified SA

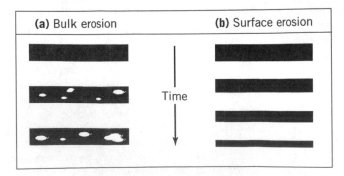

**Figure 3.** Polymers designed for drug delivery applications may degrade by (**a**) bulk erosion or (**b**) surface erosion. Surface erosion is generally preferred when constant drug release is desired. *Source:* Reproduced with permission from Ref. 67.

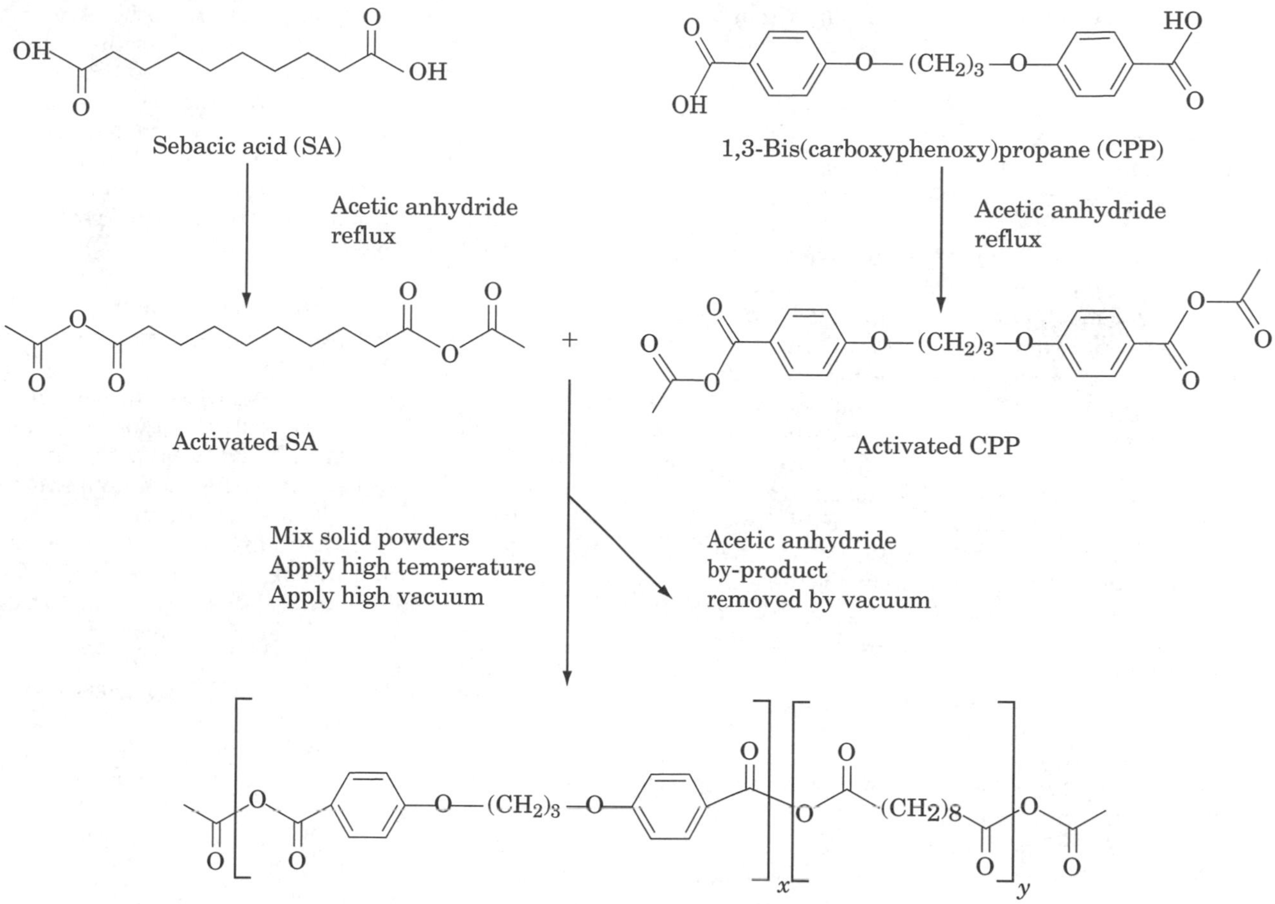

**Figure 6.** Synthesis scheme of pCPP:SA polyanhydrides by melt polycondensation. *Source:* Reproduced with permission from Ref. 67.

and pCPP monomers are first activated via reflux in acetic anhydride, leading to acylation of the terminal carboxylic acid functionalities. These activated monomers are then mixed together as solids in desired proportions and heated. As the monomers begin to melt, they start to react, forming a pCPP:SA copolymer with acetic anhydride as the by-product. As the reaction begins, the system is purged with nitrogen or argon, followed by the application of high vacuum, which leads to the removal of the volatile acetic anhydride. Without acylation in the first step, the byproduct of polymerization would be water, which is less volatile when acetic anhydride and could cause polymeric degradation. Because this synthetic technique does not require potentially toxic catalysts, the product is pure, high-molecular-weight polyanhydride polymer suitable for formulation into practical drug delivery devices.

**Techniques for Drug Incorporation.** Various methods have been developed for incorporating drugs into biodegradable polymers for controlled drug delivery applications. The trituration method involves physically mixing the polymer and drug in fine powder form (73). This method is universally applicable because it requires only that drugs may be formed into small particles. However, this method can result in inhomogeneous drug distribution leading to inconsistent drug release profiles. Also, release of loosely bound drug particles on the polymer surface can lead to a "burst effect," a disproportionately large burst of drug being delivered at the beginning of the release period. Another approach, the cosolution method, involves codissolving solvent and drug in an organic solvent and then evaporating the solvent in vacuo. This method tends to yield homogeneous drug/polymer formulations leading to highly reproducible release profiles. However, this method requires that the drug be soluble and stable in organic solvents such as methylene chloride or chloroform. Furthermore, solvent evaporation leaves pores and channels within the polymer matrix through which the drug can diffuse. A high rate of drug diffusion leads to release profiles that are non–zero order. A third method, the mix/melt method, involves heating the polymer above its melting point and then physically mixing the drug into the viscous polymer liquid (68). Cooling this mixture yields a dense and homogeneous polymer/drug matrix with minimal pores and channels for drug diffusion and highly reproducible release kinetics. Heating the polymer, however [about 80°C is required for pCPP:SA (20:80)], can potentially lead to interaction of the drug with the polymer and/or heat deactivation of temperature-sensitive drugs. The advantages and disadvantages of these drug-incorporating techniques must be considered for each individual polymer/drug formulation.

**Techniques to Fabricate Drug/Polymer Formulations into Drug Delivery Devices.** After a drug has been incorporated into a polyanhydride polymer matrix, a variety of techniques fabricate the drug/polymer formulation into a size and shape suitable for drug delivery (71). One such technique is compression molding: The drug/polymer formulation is ground or spray-dried into a fine powder, placed into a piston-type mold, and then compressed with a vice or hydraulic press into a flat slab or wafer. This method is applicable to a wide variety of polymer formulations. The melt-molding technique involves melting the drug/polymer formulation and molding the gel into a desired geometry. Another method, known as the solvent-casting method, involves pouring an organic solution containing polymer and drug into a flat, open mold chilled over dry ice. Slowly the solvent evaporates to yield a thin, flat film containing polymer and drug intimately mixed.

Various methods also exist for preparing drug-loaded polyanhydride microspheres that can be incorporated into a suspension and injected into tissue (74). In the hot-melt microencapsulation method, polymers and drug are cosuspended in a silicone or olive oil and the mixture is heated to 5°C above the polymer melting point with continuous stirring. When the solution is cooled, the polymer is solidified into microspheres while entrapping the drug within. Another method for preparing microspheres is the spray-drying microencapsulation technique. In this method, polymer and drug are codissolved in an organic solvent, and the solution is sprayed through an atomizer into a drying chamber where the drug-loaded microspheres are immediately dried by an upward flow of nitrogen. Drugs that are heat sensitive can be incorporated into polyanhydride microspheres via the solvent-removal microencapsulation method (75). In this method the polymer and drug are codissolved in a volatile organic solvent. This solution is then suspended in an organic oil that is immiscible with the organic solvent. A surfactant such as Span 85 is added, and the mixture is continuously stirred. Petroleum ether is then introduced, causing the condensation of polymer into the shape of microspheres with drug trapped inside.

With these various techniques, almost any drug, from small anticancer agents to large therapeutic macromolecules such as proteins, can be released in a controlled fashion from a biodegradable polymer matrix.

### Development of Drug-Impregnated Polyanhydride Wafers for the Treatment of Brain Tumors

In 1984, the Department of Neurosurgery at Johns Hopkins University and the Department of Chemical Engineering at MIT joined forces to apply this new drug delivery technology to the treatment of brain tumors. Despite substantial advances in imaging and surgical techniques, the prognosis for patients diagnosed with glioblastoma multiforme, the most malignant and deadly form of brain cancer, remains dismal. Survival is still measured in months, and rarely do these patient survive for more than 2 years. Because most malignant gliomas demonstrate a marked propensity for local recurrence rather than metastasizing to other areas of the brain or to the peripheral organs, we predicted that directing intense therapy at the site of the initial tumor would result in powerful control of postsurgical tumor growth. We believed that drug delivery directly to the brain interstitium using polyanhydride wafers would circumvent the BBB and release unprecedented levels of drug directly to an intracranial target in a sustained fashion for extended periods of time.

**Preclinical Studies.** With the lack of mutagenicity and carcinogenicity of the pCPP:SA polymers already established at MIT, we carried out a series of preclinical experiments in collaboration with Colvin et al. to determine the safety and efficacy of drug-loaded intracranial polymer implants. The alkylating agent BCNU was chosen for our initial studies because of its established antineoplastic activity against malignant gliomas. These preclinical studies investigated the biocompatibility of the pCPP:SA (20:80) polymer in the brain, the in vivo drug release profile of BCNU-loaded wafers, and the safety and efficacy of BCNU-loaded wafers in a murine model of malignant glioma.

The biocompatibility of the pCPP:SA (20:80) polymer was first tested by implanting placebo wafers in the corneas of rabbits. This assay has proven to be a highly sensitive system for determining the inflammatory effect of implanted agents (76). Over the course of a 6-week implantation period, the pCPP:SA (20:80) polymer implant did not cause neovascularization, corneal edema, or the infiltration of white blood cells (77). The biocompatibility of pCPP:SA polymers was then tested in the CNS of Fischer 344 rats and compared with that of other well-tolerated neurosurgical biomaterials: oxidized regenerated cellulose (Surgicel®) and absorbable gelatin sponge (Gelfoam®) (78). Rats underwent bilateral frontal-lobe implantation of pCPP:SA (20:80), Surgicel®, and Gelfoam® and were closely monitored for inflammatory responses upon serial sacrifice over a 5-week period. The pCPP:SA polymers elicited a minimal, transient, localized inflammatory response comparable to the well-tolerated clinical implants. The local response to pCPP:SA (20:80) wafers subsided as the polymer device degraded. Similar studies were performed in the rabbit brain which again demonstrated no differences from the inflammatory response elicited by pCPP:SA polymer and Gelfoam® implants (79).

To initially assess the therapeutic potential of intracranial drug delivery from a locally implanted polymer, the drug release profile was determined for intracranial implants of ethylene-vinyl acetate copolymer (EVAc) wafers impregnated with BCNU in Fischer 344 rats (80). EVAc was chosen as a delivery device because the in vitro and in vivo release kinetics of this nonbiodegradable polymer system are well established. The EVAc polymer used in these studies was obtained from the Du Pont Company (Wilmington, DE) and contained 40% vinyl acetate by weight. The polymer was first washed copiously with ethyl alcohol to remove inflammatory impurities. To prepare the BCNU-loaded EVAc polymers, drug and polymer (3:7 BCNU-to-EVAc w/w ratio) were codissolved in methylene chloride, and the solution was poured into cylindrical glass molds at $-70°C$. Solvent evaporation yielded cylindrical polymers homogeneously loaded with 30% BCNU that were transferred to glass plates at $-30°C$ and allowed to dry for 7 days. The drug-loaded EVAc cylinders were then cut into 10-mg disks, exposed to ultraviolet radiation for 1–2 hours, and implanted into the rat brain (81). Using a Bratton-Marshall assay to quantify BCNU concentrations in brain homogenates, the controlled release of BCNU in the implant hemisphere was observed for the entire 9 days of the study. Drug levels peaked at 49.6 $\mu g/g$ brain tissue. Much lower levels of BCNU were measured in the contralateral hemisphere and systemic circulation and only for 1 day.

Similar results were obtained with various [3]H-BCNU-loaded pCPP:SA (20:80) wafers implanted intracranially in New Zealand White rabbits (82). Drug-loaded polymers prepared by either the trituration or solution method were studied. Autoradiographic analysis of coronal sections bisecting the polymer implant was used to assess both the local drug concentrations as a function of distance from the polymer and the percent of the brain exposed to drug. In this study the results were compared to local injection of [3]H-BCNU. At days 3 and 7, BCNU concentrations of 6 mM were observed up to 10 mm from the polymer implant. In contrast, direct injection of [3]H-BCNU resulted in almost no detectable levels of drug in the brain on day 3. The percent of brain exposed as a result of polymeric drug delivery was determined at days 3, 7, and 10 to be 50%, 15%, and <10%, respectively. Both the average drug concentration and percent of brain exposed increased with greater [3]H-BCNU loading doses. The results obtained with the polymers prepared by solution and by trituration were not significantly different.

The ability of BCNU-loaded polymers to control the growth of established solid tumors was first evaluated in a flank model in Fischer 344 rats (83). BCNU-loaded EVAc wafers implanted adjacent to an established 9L gliosarcoma tumor resulted in a delay in tumor growth (defined as the time required for tumor volume to increase by 1 log) of + 16.3 days compared to controls ($p < 0.05$). In contrast, BCNU delivered systemically via intraperitoneal (IP) injections and BCNU delivered systemically from a contralateral EVAc implant resulted in significantly smaller growth delays of + 11.2 days and + 9.3 days, respectively. The efficacy of BCNU-loaded EVAc wafers in the treatment of an established intracranial tumor was then evaluated. Sixty Fischer 344 rats underwent a craniectomy followed by placement of a 1-mm³ 9L gliosarcoma implant in the cerebral cortex of the left parietal lobe. On the fourth day after tumor implantation, animals were randomized into one of four treatment groups receiving: (1) empty EVAc polymers implanted at the tumor site, (2) systemic BCNU (14 mg/kg) via a single IP injection, (3) interstitial delivery of BCNU (14 mg/kg) via EVAc, and (4) interstitial delivery of BCNU (14 mg/kg) via pCPP:SA (20:80) wafers. As shown in Figure 7, the systemic delivery of BCNU resulted in only a 2.4-fold increase in mean survival compared to controls, while the same drug dose delivered by EVAc and pCPP:SA (20:80) polymers resulted in significantly greater 7.3- and 5.4-fold increases, respectively ($p < 0.05$). Even more encouraging was that although there was a 100% mortality rate for the control animals and the animals receiving IP

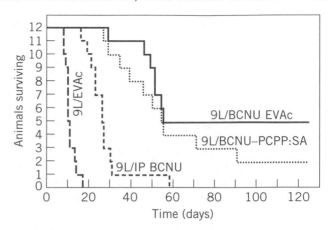

**Figure 7.** Survival curves of 9L glioma rats bearing intracranial tumors treated with BCNU either systemically (9L/IP BCNU) or polymerically delivered intracranially (9L/BCNU EVAc and 9L/BCNU–pCPP:SA). *Source:* Reproduced with permission from Ref. 83.

BCNU, the animals treated interstitially with drug-loaded EVAc and pCPP:SA (20:80) wafers had 42% and 17% long-term survival, respectively, to the scheduled date of sacrifice (125 days posttumor implantation). Pathological evaluation revealed no viable tumor in any of these long-term survivors, apparently indicating that the animals were cured.

The therapeutic benefit of interstitial delivery of BCNU from biodegradable pCPP:SA (20:80) polymer wafers was also compared with the intratumoral injection of the same dose of BCNU (84). In this study, Fischer 344 rats received intracranial implants of 9L gliosarcoma and were subsequently treated on day 5. The median survival for the animals receiving BCNU via polymer implants was extended 271% over controls, compared with only 36% for the same amount of BCNU injected intratumorally. These results demonstrate the advantages of sustained polymeric drug delivery over direct interstitial injection.

In preparation for clinical trials, the biocompatibility of pCPP:SA (20:80) wafers impregnated with BCNU was then evaluated in the primate brain (85). Eighteen *Macaca fascicularis* monkeys underwent a frontal lobectomy and received the following implants: (1) no treatment, (2) empty pCPP:SA (20:80) polymer wafers, (3) pCPP:SA (20:80) polymers impregnated with BCNU, and (4) empty pCPP:SA (20:80) polymer in one hemisphere and a BCNU-loaded polymer in the opposite hemisphere. The last group received concurrent external beam radiation therapy (XRT). Animals were closely monitored over a period of 80 days and showed no signs of neurological deficits. Extensive analysis of serially collected blood samples showed no signs of toxicity normally associated with systemic delivery of BCNU. Serial computerized tomography (CT) and magnetic resonance imaging (MRI) scans revealed a radio-opaque mass and minimal edema surrounding the site of polymer implantation. Detailed histological examination following serial sacrifice indicated a mild, localized, transient inflammatory reaction consistent with that observed in the rat and rabbit studies. Monkeys receiving concurrent radiation therapy demonstrated no further deleterious effects attributable to the polymers with or without BCNU.

From these experiments, we concluded that the intracranial implantation of BCNU-impregnated pCPP:SA (20:80) polymer wafers is safe in the brains of rats, rabbits, and primates. Furthermore, this intracranial drug delivery strategy is capable of delivering active drug in a sustained fashion directly to the site of an intracranial tumor, and the potential benefits of concurrent XRT are not lost. In animal glioma models, significant prolongation of survival and encouraging cure rates were consistently observed. Owing to the success of these preclinical studies, clinical trials to evaluate the efficacy of this therapy in human patients with malignant glioma were initiated.

**Clinical Studies.** The safety of implanting BCNU-loaded pCPP:SA (20:80) polymer wafers into the brains of neurooncology patients was first assessed in a Phase I–II clinical trial that began in 1987 (86). Twenty-one patients were enrolled at five institutions in the United States. The admission criteria included indication for reoperation for recurrent malignant glioma, a unilateral single focus of tumor in the cerebral cortex with an enhancing volume of at least 1.5 cm$^3$ on CT scans, and a Karnofsky Performance Scale (KPS) score of at least 60 (indicating that the patient is able to function independently). All patients had previously undergone XRT, had no significant hepatic or renal disease, and were not suffering from any other life-threatening illness.

Polymer wafers loaded with various doses of BCNU (1.93%, 3.85%, and 6.35% w/w) were prepared as follows. Clinical-grade pCPP:SA polymer was prepared by melt polycondensation of pCPP and SA monomers in a 20:80 molar ratio. Polymer and BCNU were codissolved in methylene chloride and spray-dried into microspheres. The drug-loaded microspheres were then compression-molded into cylindrical wafers measuring 1.4 cm in diameter and 1.0 mm thick (about the size of a dime).

In each patient, after maximum debulking of the recurrent tumor, the resection cavity was lined with up to eight BCNU-loaded polymer wafers (Figs. 8 and 9). The placement of the polymer wafers added approximately 3 minutes to the total time of surgery. For the first 7 weeks following polymer implantation, patients were closely monitored for signs of local and systemic toxicity by regular neurological examination, serial CT scans or magnetic resonance (MR) images, KPS score determination, hematologic and blood chemistry analysis, and urinalysis.

The treatment was well tolerated at all three dosing levels. CT scans and MR images revealed no direct toxicity in the brain attributable to the BCNU polymer implants. (Fig. 10). Furthermore, no patient experienced the significant reduction in blood cell counts normally associated with the systemic administration of BCNU. The overall median survival time following polymer implantation was 46 weeks, and 8 of the 21 patients (38%) survived more than 1 year. It was therefore concluded that the local delivery of BCNU from pCPP:SA (20:80) polyanhydride polymer wafers was safe in human patients and that further

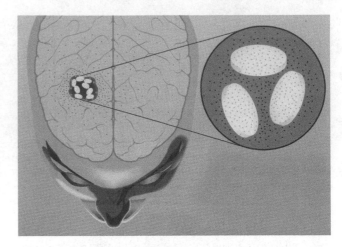

**Figure 8.** Up to eight polymer implants line a tumor resection cavity, where the loaded drug is gradually released as the implants dissolve. The inset shows conceptually how drug molecules diffuse away from these implants. *Source:* Reproduced with permission from Ref. 66.

clinical investigation was indicated to assess the efficacy of this new treatment modality.

A prospective, randomized, placebo-controlled study was carried out to evaluate whether pCPP:SA (20:80) polymer wafers impregnated with BCNU (3.85% w/w) could reduce the 6-month mortality in patients with recurrent malignant glioma (87). Two hundred and twenty-two patients were enrolled at 27 medical centers in the United States and Canada. The admission criteria were similar to those used for the Phase I–II study. All patients entered had failed conventional therapies and were at the end stages of their fatal disease.

At the time of reoperation, the tumor resection cavity was lined with up to eight polymer wafers with or without 3.85% BCNU. When the code was broken 6 months after the last patient was entered, 110 patients had received the BCNU-loaded implants, and 112 had received the placebo. The 6-month survival rate for patients receiving BCNU-loaded polymer implants was 60% compared with 47% for patients receiving placebo wafers ($p = 0.061$). For the subset of patients with glioblastoma multiforme, the 6-month survival for the patients receiving BCNU loaded wafers was 50% greater than in those treated with placebo (64% vs. 44%, $p = 0.020$). The overall survival curve after adjustment by the proportional hazards regression model is shown in Figure 11. The median survival was 31 weeks in patients receiving drug-loaded wafers and 23 weeks for the placebo group ($p = 0.006$, after accounting for the effects of prognostic factors). Again, no clinically important deleterious effects were attributable to the polymer implants. Based on these results, it was concluded that local delivery of BCNU from pCPP:SA polymer wafers is a safe and effective treatment for recurrent malignant glioma. On September 24, 1996, the U.S. Food and Drug Administration (FDA) approved the 3.85% BCNU wafers for the treatment of recurrent malignant glioma, making this the first new treatment for brain tumors to be approved by the FDA in 23 years. These polymer wafers are now manufactured by

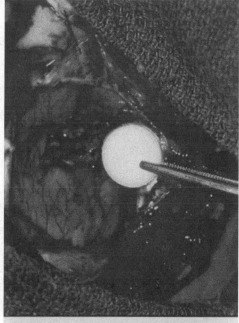
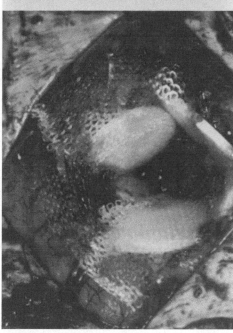

**Figure 9.** The surgical implantation of BCNU-loaded polyanhydride wafers within a tumor resection cavity. Placement of polymer wafers requires about 3 additional minutes of operation time. *Source:* Reproduced with permission from Ref. 66.

Guilford Pharmaceuticals in Baltimore, Maryland, under the name Gliadel® and are distributed by Rhone Poleunc Rorer worldwide except for Sweden, Norway, Denmark, and Finland, where it is distributed by Orion, Inc.

Based on the success of the 3.85% BCNU-loaded pCPP:SA wafers in the treatment of recurrent malignant glioma, we hypothesized that an even greater therapeutic benefit could be achieved using the BCNU polymer as a part of the initial treatment. To this end, a Phase I–II

study was carried out to evaluate the safety of BCNU polymer implantation in conjunction with XRT for the treatment of newly diagnosed malignant glioma (88). Twenty-two patients at three hospitals in the United States received up to eight 3.85% BCNU-loaded wafers and a standard course of postoperative XRT. Again, no systemic or local toxicity was attributable to the BCNU-loaded polymer. Neurotoxicity was equivalent to that typically associated with craniotomy and XRT. The patient population was older than in the two previous clinical studies (mean age = 60), which is typically a negative predictor of outcome. Nonetheless, the median postoperative survival was 42 weeks, and 4 patients survived more than 18 months. It was therefore concluded that implantation of 3.85% BCNU-loaded polymer wafers in conjunction with XRT is safe as the initial treatment for malignant glioma.

The first Phase III clinical trial evaluating the efficacy of 3.85% BCNU-loaded polymers in patients with newly diagnosed malignant glioma was carried out in Finland and Norway by Valtonen et al. (89). Thirty-two patients were enrolled in a prospective, randomized, placebo-controlled study in which 16 patients received 3.85% BCNU-loaded wafers and 16 patients received placebo wafers following the initial surgical resection of a malignant glioma. As shown in Figure 12, median survival for the treatment group was 58 weeks compared with 40 weeks for the placebo group. ($p = 0.001$). Ten of the 16 patients (63%) in the treatment group survived for 1 year after polymer implantation, while only 3 of 16 patients (19%) in the placebo group were still alive ($p = 0.029$). When the patients with glioblastoma were considered separately, median survival was 53 weeks for patients receiving BCNU and only 40 weeks for those receiving placebo ($p = 0.0083$). Two years after treatment 31% of the patients that received Gliadel were alive versus just 6% in the placebo group, and at 3 years 25% of the treatment group were still alive compared to 6% of those receiving placebo. Another clinical trial involving a greater number of patients is being carried out by the European Association of Neurosurgeons to further evaluate the efficacy of Gliadel® wafers implanted at the time of initial surgical resection in patients with glioblastoma multiforme.

Together, these clinical trials, summarized in Table 1, demonstrate that 3.85% BCNU-loaded pCPP:SA polymer wafers are safe and effective in the treatment of newly diagnosed as well as recurrent malignant glioma. This treatment provides an effective means for circumventing the BBB, resulting in sustained, high concentrations of drug in the region of an intracranial tumor. Furthermore, the systemic toxicities usually associated with chemotherapy are avoided, and the potential benefits of adjuvant XRT are still present.

**Future Clinical Studies with BCNU-Loaded Wafers.** We believe that the clinical success of the Gliadel® wafer serves as proof of principle and that the polymer wafer–mediated intracranial drug delivery strategy opens the door for a dramatic expansion in the armamentarium available for the treatment of brain tumors. Eight clinical trials have already begun, and many more are planned

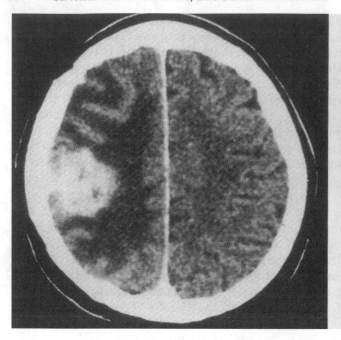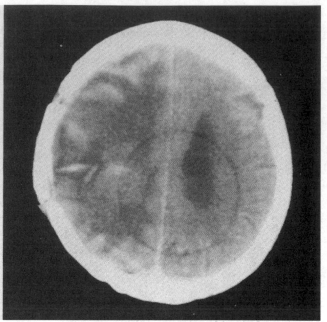

**Figure 10.** Computerized tomography (CT) scans show a recurrent malignant brain tumor preoperatively (left panel) and BCNU-loaded polymer wafers postoperatively (right panel). As the wafers dissolve and release the drug, they gradually disappear from the scans. *Source:* Reproduced with permission from Ref. 66.

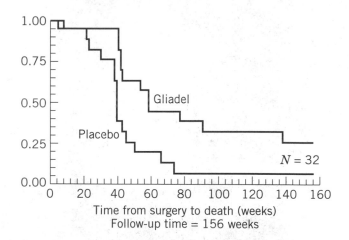

**Figure 12.** Kaplan-Meirer survival curves for patients receiving implantation of BCNU-loaded polymers (Gliadel®) or placebo controls at the time of initial operation for Grade III or Grade IV tumors. All patients with Grade III tumors were included in the group receiving Gliadel®. *Source:* Reproduced with permission from Ref. 89.

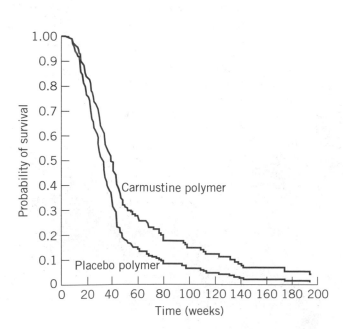

**Figure 11.** Overall survival for patients receiving implantation of carmustine (BCNU)-loaded polymers or placebo controls at the time of operation for a recurrent brain tumor after adjustment for prognostic factors. The curves illustrate the treatment effect expected if all patients were about age 48, white, had performance status >70, underwent >75% resection, had local irradiation, had not previously been exposed to nitrosoureas, and had glioblastomas pathologically classified as active. *Source:* Reproduced with permission from Ref. 87.

that aim to improve upon the demonstrated survival benefit of Gliadel®.

Recent laboratory investigations have shown that it may be possible to improve upon the efficacy of the 3.85% BCNU-loaded Gliadel® wafers simply by increasing the loading dose of BCNU (90). In the murine 9L glioma model, BCNU dose escalation from 3.8% to 32% polymer by weight revealed a dose-response relationship (hazard ratio 0.8354 for each mg/kg increase, $p < 0.001$) (Fig. 13). The best bal-

**Table 1. Clinical Studies with BCNU-Loaded pCPP:SA (20:80) Polymer Wafers at the Time of Publication**

| Title of clinical study | Principal investigator/ Funding source | Pts. | Participating institutions | Dates of study | Refs. |
|---|---|---|---|---|---|
| Interstitial chemotherapy with drug polymer implants for the treatment of recurrent gliomas. (Phase I–II—safety) | Henry Brem, M.D. Department of Neurological Surgery, Johns Hopkins Hospital, Baltimore, MD, USA/Nova Pharmaceutical Corporation, Baltimore, MD, USA | 21 | Johns Hopkins Hospital, Baltimore, MD; University of AL, Birmingham, AI; Northwestern University, Evanston, IL; University of CA, Los Angeles, CA; Duke University, Durham, NC | 9/87–3/90 | 86 |
| Placebo-controlled trial of safety and efficacy of intraoperative controlled delivery by biodegradable polymers of chemotherapy for recurrent gliomas. (Phase III—efficacy) | Henry Brem, M.D./ Guilford Pharmaceuticals Inc., Baltimore, MD, USA; Scios-Nova Corporation, Mountain View, CA, USA; NCDDG of the NCI of the NIH, USA | 222 | 27 medical centers in the USA and Canada | 3/89–9/93 | 87 |
| The safety of interstitial chemotherapy with BCNU-loaded polymer followed by radiation therapy in the treatment of newly diagnosed malignant gliomas. (Phase 1—safety) | Henry Brem, M.D./ Guilford Pharmaceuticals Inc., Baltimore, MD, USA; Scios-Nova Corporation, Mountain View, CA, USA; NCDDG of the NCI of the NIH, USA | 22 | Johns Hopkins Hospital, Baltimore, MD; Charlotte Memorial Hospital, Charlotte, NC; Columbia Presbyterian Medical Center New York, NY | 7/90–9/94 | 88 |
| Compassionate Use Study | N/A | 40 | Johns Hopkins Hospital, Baltimore, MD | | N/A |
| Treatment IND | Guilford Pharmaceuticals | 380 | 170 medical centers in the USA | | N/A |
| Interstitial chemotherapy with BCNU-loaded polymers for initial treatment of high-grade gliomas: a randomized double-blind study (Phase III—efficacy) | Simo Valtonen, M.D. Turku University Central Hospital, Turku, Finland/ Orion Pharma (UT, PT), Espoo, Finland | 32 | Turku University Central Hospital, Turku, Finland; Helsinki University Central Hospital, Helsinki, Finland; Tampere University Hospital, Tampere, Finland; University Hospital of Trondheim, Trondheim, Norway | 3/92–3/95 | 89 |

*Note:* Abbreviations: NCDDG, National Cooperative Drug Discovery Group; NCI of the NIH, National Cancer Institute of the National Institutes of Health, Bethesda, MD, USA.

ance between toxicity and antitumor efficacy, achieved with the 20% BCNU-loaded polymer wafers, resulted in a 75% long-term survival rate. Furthermore, 20% BCNU-loaded wafers are well tolerated in the non-human primate brain (90,91). Based on these preclinical findings, Olivi is leading an NIH-funded, open-label dose escalation clinical trial at 11 medical centers in the United States to evaluate the safety of Gliadel® wafers containing between 6.5% and 20% BCNU in patients with recurrent malignant glioma. A Phase III trial will subsequently evaluate the efficacy of the highest-tolerated BCNU loading dose.

Gliadel® may possess therapeutic potential in secondary CNS tumors as well. In addition to primary CNS tumors, many neurooncology patients present with systemic tumors that have metastasized to the brain (92). In fact, the majority of newly diagnosed intracranial tumors are the result of metastases from primary tumors outside the CNS. As the therapeutic options for patients with systemic malignancies are rapidly improving, fatalities due to CNS metastases are increasing. This is because many drugs with proven efficacy against the systemic aspects of these diseases do not sufficiently cross the BBB. At present the mainstays for the treatment of intracranial metastases are surgery in combination with XRT and/or systemic chemotherapy. Ewend proposed that in the treatment of intracranial metastases, local delivery of chemotherapy from polyanhydride wafers in combination with XRT might demonstrate enhanced efficacy over standard therapies. This hypothesis has been tested in a new series of murine models of the most common forms of intracranial metastases including melanoma, renal cell carcinoma, colon adenocarcinoma, lung carcinoma (15), and breast carcinoma (93). Treatment with BCNU-loaded wafers in combination with XRT was effective in prolonging median survival in all five metastatic tumor models. In most cases, the combination of XRT with local chemotherapy was more efficacious than either treatment alone. For example, the results achieved in the intracranial model of metastatic melanoma are presented in Figure 14. Based on these preclinical studies, a Phase II clinical trial has been initiated to evaluate the therapeutic potential of surgical resection

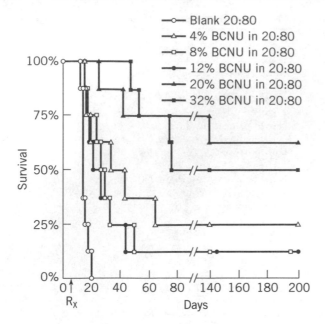

**Figure 13.** Survival curves for rats bearing 9L gliosarcomas treated 5 days after tumor implantation with pCPP:SA (20:80) polymers containing escalating doses of BCNU compared with control animals treated with empty polymers ($n = 8$ for each group). *Source:* Reproduced with permission from Ref. 90.

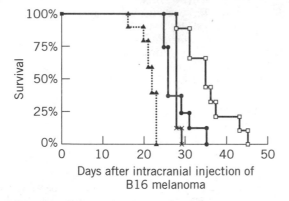

**Figure 14.** The efficacy of BCNU, delivered locally by biodegradable polymer, in treating a murine model of solitary intracranial metastases of B16 malignant melanoma. The treatment groups include blank polymer control (closed triangle), 20% loaded BCNU polymer implant (closed circle), radiation therapy with 300 cGy per day for 3 days ($\times$), and 10% loaded BCNU polymer implant followed by radiation therapy with 300 cGy per day for 3 days (open square). The combination therapy was superior to either BCNU alone or XRT alone ($p < 0.006$). *Source:* Reproduced with permission from Ref. 15.

in combination with Gliadel® and XRT in patients with operable CNS metastases from systemic malignancies.

Two additional clinical studies are underway to evaluate the therapeutic potential of Gliadel® in the treatment of recurrent supratentorial glial tumors. The first is a Phase I study of Gliadel® in children and adolescents, and the other is a Phase II study in adults. Another set of studies has been initiated that combines the local delivery of

BCNU with an additional anticancer agent delivered systemically. In one study, a Phase I trial, patients with glioblastoma multiforme or anaplastic astrocytoma are receiving intracranial implantation of Gliadel® wafers in conjunction with serially escalated systemic doses of carboplatin. In another study, patients with glioblastoma multiforme are receiving Gliadel® and XRT in conjunction with oral administration of the cytodifferentiating agent 13-*cis*-retinoic acid.

Another approach to improve the efficacy of Gliadel® in the treatment of brain tumors is to inhibit the neoplastic cellular resistance mechanisms to BCNU therapy. The cytotoxicity of BCNU against tumor cells depends on initial DNA alkylation at the $O^6$-position of guanine bases (94). The primary cellular chemotherapeutic resistance mechanism against BCNU therapy is the dealkylation reaction carried out by the enzyme alkylguanine-DNA alkyl transferase (AGAT). AGAT removes $O^6$-guanine adducts stoichiometrically, a reaction that renders this cytoprotective enzyme irreversibly inactive, making de novo protein synthesis the only means by which AGAT activity can be restored. Pegg et al. have shown that AGAT depletion with exogenous substrates such as $O^6$-benzyl guanine ($O^6$BG) enhances the sensitivity of various animal and human tumor cell lines to treatment with alkylating agents such as BCNU (94–97). Because many human gliomas express high levels of AGAT, $O^6$BG could potentiate the efficacy of Gliadel® in the treatment of human brain tumors. To test this hypothesis, a Phase I clinical trial is being initiated in which patients with recurrent malignant gliomas will receive Gliadel® wafers in combination with escalating doses of intravenous $O^6$BG.

Thirteen other clinical trials are also scheduled to begin soon. These studies include combining local BCNU with systemic chemotherapy, permanent I-125 seeds, other drug-resistance modulatory agents, various XRT protocols, and the use of Gliadel® in the treatment of nonglial tumors of the CNS. The current and planned clinical studies are summarized in Table 2.

**Laboratory Investigations with a Variety of Biodegradable Intracranial Controlled Drug Release Systems.** BCNU-loaded wafers possess great therapeutic potential in the treatment of brain tumors. However, many anticancer agents are substantially more potent than BCNU against neoplastic cells in vitro. Because many of these agents do not significantly cross the BBB, their therapeutic potential in the treatment of brain tumors remains unrealized. Finding ways to deliver a wide variety of anticancer agents to the brain is crucial: Malignant gliomas are noted for their cellular pleomorphism and associated potential for acquired resistance to a single chemotherapeutic agent, and multiple drug regimens are often required to eradicate all the malignant cells in a CNS tumor. To this end, we have investigated the potential efficacy of a wide variety of drug/polymer formulations in the treatment of brain tumors.

*pCPP:SA Polymers.* As described above, the benefits of the pCPP:SA wafers for the intracranial delivery of drugs include circumvention of the BBB, controlled drug release by surface erosion, and the protection of hydrolytically and/or metabolically unstable drugs. Because of these fa-

**Table 2. Future Clinical Applications of BCNU-Loaded Polyanhydride Wafers in the Treatment of Brain Tumors**

| Clinical trials currently underway | Clinical trials with letter of intent approved |
|---|---|
| Dose escalation study to evaluate the safety of Gliadel wafers containing between 6.5 and 20% BCNU in patients with recurrent malignant glioma | Phase I Gliadel and escalating doses of intravenous $O^6$-benzyl guanine trial in patients with recurrent malignant glioma |
| Phase II trial of surgery with Gliadel wafers and radiation in patients with operable CNS metastases from systemic cancer | Use of Gliadel with concomitant radiotherapy and IV BCNU in the treatment of patients with newly diagnosed glioblastoma multiforme |
| Phase III clinical trial to further evaluate the efficacy of 3.85% BCNU wafers implanted at the time of initial surgical resection in patients with glioblastoma multiforme | Phase I treatment of adults with recurrent supratentorial low-grade gliomas with Gliadel wafers |
| Phase I study of Gliadel in children and adolescents with recurrent supratentorial glial tumors | Phase II Gliadel, carboplatin, and XRT in newly diagnosed glioblastoma multiforme either as de novo or recurrent from low-grade gliomas |
| Phase II treatment of adults with recurrent supratentorial low-grade gliomas with Gliadel wafers | Phase I treatment of adults with recurrent supratentorial high-grade glioma with Gliadel wafers plus Irinotecan (CPT-11) |
| Phase I trial of Gliadel and carboplatin as initial treatment in patients with glioblastoma multiforme or anaplastic astrocytoma | Phase I/II pilot study using concurrent multimodality treatment for patients with relapsed glioblastoma multiforme using permanent I-125 seeds and Gliadel wafers |
| Use of Gliadel and 13-cis-retinoic acid in patients with primary glioblastoma multiforme | Use of Gliadel and permanent low-activity I-125 radioactive seeds for the treatment of recurrent malignant astrocytomas |
| | Drug resistance modulation with 6MP combined with Gliadel for the treatment of recurrent malignant gliomas; a Phase I study |
| | Use of Gliadel, fractionated radiation therapy, and stereotactic radiosurgery in patients with newly diagnosed malignant glioma |
| | Clinical and radiographic evaluation of patients receiving Gliadel wafers and stereotactic radiosurgery boost with the Gamma Knife in the initial treatment of malignant gliomas |
| | Phase I/II study of Gliadel wafer placement and intensity of modulated radiation therapy using the Peacock system for patients undergoing gross total resection for glioblastoma multiforme |
| | Phase I/II study of Gliadel, Gamma Knife radiosurgery, and postoperative fractionated radiation therapy in the treatment of patients with newly diagnosed high-grade malignant gliomas |
| | Phase II study of Gliadel wafers followed by RSR13 plus radiotherapy for the treatment of newly diagnosed malignant glioma |
| | Use of Gliadel in the management of aggressive pituitary adenomas and craniopharyngiomas |

vorable properties and the ease with which polymer/drug formulations can be prepared under mild conditions, the possibilities for intracranial drug release from pCPP:SA wafers are vast. We have studied several potentially therapeutic pCPP:SA/drug formulations in an attempt to improve upon the success of the BCNU-loaded wafers in the treatment of CNS tumors.

Taxol, an antitumor alkaloid, elicits antineoplastic activity by promoting tubulin polymerization and has demonstrated efficacy in clinical trials for ovarian and breast carcinoma, as well as for melanoma. Recently, Walter and Cahan have reported a quantitative analysis of brain tumor cell sensitivity to taxol in vitro (98). In this study, taxol demonstrated potent, exposure-time-dependent antitumor activity against human and murine malignant glioma. However, because taxol does not cross the BBB to an appreciable extent, its efficacy against intracranial tumors following systemic administration is severely limited. It was hypothesized that the therapeutic potential of taxol could be realized if the drug were delivered locally from biodegradable polymers.

This hypothesis was tested in the murine 9L gliosarcoma model (99). On the fifth day after intracranial tumor implantation in Fischer 344 rats, three separate groups were implanted with pCPP:SA (20:80) wafers loaded with 20%, 30%, and 40% taxol (w/w). A fourth group received placebo polymers. As shown in Figure 15, median survival was extended for all three taxol loadings (p < 0.02). In vivo studies were carried out to quantitatively assess the release of $^3$H-labeled taxol in the rat brain. It was found that polymeric delivery of taxol resulted in widely distributed intracranial drug concentrations. Similar in vitro and in vivo studies have recently been performed by Sampath with taxotere, a synthetic derivative of taxol. Intracranial controlled release of taxotere is also able to extend survival in the 9L glioma model (100). These studies demonstrate that pCPP:SA wafers loaded with taxol or taxotere may be efficacious in the treatment of gliomas in humans.

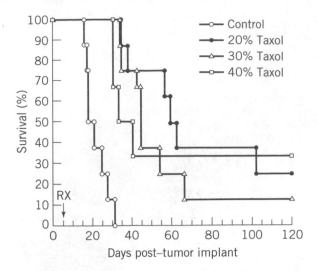

**Figure 15.** Survival curves for rats receiving an intracranial 9L gliosarcoma tumor implant on Day 0 followed by treatment on Day 5 with an intratumoral 10 mg pCPP:SA (20:80) polymer implant loaded with 20, 30, or 40% taxol by weight. Control animals received a 10-mg PCPP:SA (20:80) implant with no loaded drug. *Source:* Reproduced with permission from Ref. 99.

Camptothecin is an inhibitor of the DNA-replicating enzyme topoisomerase 1. This naturally occurring alkaloid was first isolated from the tree *Camptotheca acuminata* in China by Monroe Wall in 1966. Although this compound demonstrated extraordinary promise as a potentially new antineoplastic agent in preclinical studies, the clinical trials were abruptly derailed due to unexpected toxicity and low antineoplastic activity. Mary Wolpert of the National Cancer Institute proposed that the local delivery of camptothecin into the brain may achieve efficacious concentrations in brain tumors without systemic side effects. This hypothesis has been tested by Weingart and others in our laboratories (101,102).

In a series of in vitro studies, camptothecin demonstrated potent antineoploastic activity against several rat and human glioma cell lines. The therapeutic potential of camptothecin-loaded pCPP:SA (20:80) wafers was then tested in Fischer 344 rats. Sustained delivery of $^3$H-camptothecin from pCPP:SA wafers at tumoricidal concentrations was demonstrated for over 21 days. In rats challenged with an intracranial implant of 9L gliosarcoma, systemic administration of camptothecin did not extend survival compared to controls. However, local delivery via pCPP:SA wafers (50% camptothecin w/w) resulted in significant prolongation of survival (p < 0.05), and 40–79% of the animals were apparently cured.

More recently, Sampath and others have examined the in vitro cytotoxicity of twenty new analogs of camptothecin, synthesized by Monroe Wall's laboratory, against murine and human glioma cell lines (103). They concluded that several 10,11-methylene dioxy analogs of camptothecin are significantly more potent than the parent compound. The local delivery of these camptothecin analogs from pCPP:SA wafers holds great promise for future development in intracranial tumor models and clinical trials.

Adriamycin is an anthracycline antitumor antibiotic frequently used in the treatment of systemic malignancies. This compound intercalates with the DNA of dividing cells resulting in strand scission and double-stranded cross breaks (48). Although adriamycin demonstrates cytotoxicity against malignant gliomas in vitro, the drug is hydrophilic and therefore only minimally crosses the BBB, and when delivered systematically at high dosing regimens, it causes unacceptable cardiac toxicity. Therefore, adriamycin is a promising candidate for intracranial polymeric drug delivery.

Polymer-based drug delivery of adriamycin was first developed by Lin et al. using EVAc needles (104). In our laboratories Watts, Carter, et al. have investigated the intracranial delivery of adriamycin from pCPP:SA wafers for the treatment of malignant glioma (105,106). Initial in vitro studies of the release kinetics of adriamycin from pCPP:SA (20:80) wafers prepared by the solution method showed a large burst effect. This was attributed to the high solubility of this highly hydrophilic compound in aqueous media, which enhances the rate of drug diffusion from the polymer matrix. When adriamycin-loaded pCPP:SA wafers were prepared alternatively by the mix/melt method, it was found that the burst effect was minimized and more zero-order release was achieved in vitro. Polymers prepared by the mix/melt method normally have a denser polymer matrix than polymers prepared by the solution or trituration methods (68): A denser matrix slows the rate of drug diffusion through the polymer wafer. In Fischer 344 rats, adriamycin-loaded wafers prepared by the mix/melt method were also better tolerated and therefore more efficacious than those prepared by the solution method in vivo. In the 9L glioma model, rats receiving 3% and 5% adriamycin-loaded pCPP:SA wafers prepared by mix/melt method demonstrated a median survival of 29 and 45 days, respectively, compared to 19 days for untreated controls. Therefore, sustained delivery of adriamycin from pCPP:SA wafers prepared by the mix/melt method may provide another alternative for the treatment of malignant glioma.

Recently, the endogenous seco-steroid hormone, 1,25-dihydroxyvitamin $D_3$ (1,25 $D_3$) has demonstrated potent antiproliferative activity against a variety of systemic malignancies in vitro and in vivo. Although the mechanism by which 1,25 $D_3$ exerts antiproliferative activity is controversial, it is believed to involve the upregulation via the classic steroid receptor–mediated pathway of cell cycle control proteins that inhibit tumor cells from leaving $G_1$ and entering the phase of DNA synthesis (107). 1,25 $D_3$ also possesses potent antiangiogenic activity (108). The success of 1,25 $D_3$ in the treatment of cancer is limited, however, because the drug causes toxic hypercalcemia at supraphysiological systemic concentrations. The use of 1,25 $D_3$ against intracranial neoplasms is even more hindered by the drug's limited penetration of the BBB following systemic administration. Posner's group at Johns Hopkins University has designed and synthesized new analogs of 1,25 $D_3$ that demonstrate maintained or enhanced antiproliferative activity in vitro and dramatically minimized calcemic activity in vivo (109). Burke et al. have proposed that combining the polymer-wafer intracranial delivery method with the anticalcemic analogs may make it possible to achieve efficacious intracranial concentrations of 1,25 $D_3$ analogs for the treatment of intracranial tumors.

In a series of in vitro proliferation assays, several 1,25 $D_3$ analogs demonstrated potent antiproliferative activity against a variety of murine tumor cell lines frequently metastatic to the brain, including B16 malignant melanoma, EMT6 breast cell carcinoma, and RENCA renal cell carcinoma (110). Using quantitative high-performance liquid chromatography, in vitro release kinetics studies were carried out that demonstrated that these pH-sensitive 1,25 $D_3$ analogs can be successfully loaded and released from pCPP:SA (20:80) wafers with maintained structural integrity (111). These studies suggest that 1,25 $D_3$ analog–loaded pCPP:SA wafers may possess therapeutic potential in the treatment of intracranial metastases. In vivo studies to test this hypothesis are underway.

In the absence of an intratumoral blood supply, the three-dimensional growth of solid tumors is controlled by the limits of diffusion that restrict the transport of nutrients and waste products (112). Typically, avascular tumors cannot grow beyond about 4 mm$^3$, a size at which tumors are often asymptomatic and have little metastatic potential (104). To support exponential growth, solid tumors stimulate the process of intratumoral angiogenesis, the proliferation and migration of endothelial cells that bring about the formation of new blood vessels. Folkman first suggested that inhibiting the process of tumor angiogenesis may be a powerful strategy for controlling the growth of solid tumors (113).

Because malignant gliomas are among the most vascular of all solid tumors, we have investigated the application of this strategy to the treatment of primary brain tumors. Many inhibitors of angiogenesis have been discovered, most of which are not directly cytotoxic to tumor endothelial cells. Therefore sustained delivery, such as from pCPP:SA polymer wafers, appears to be crucial for the efficacy of these biological response modifiers. In our laboratories the well-known antiangiogenic agent cortisone acetate, as well as two novel angiogenic inhibitors, squalamine and minocycline, have demonstrated the ability to restrict the growth of 9L glioma in the rat flank (114–116). For example, the intratumoral implantation of pCPP:SA (50:50) wafers loaded with cortisone acetate resulted in a 2.3-fold reduction in the tumor growth rate (107).

The controlled release of the antibiotic minocycline for the inhibition of brain tumor angiogenesis has been studied extensively in our laboratory. The antiangiogenic potential of minocycline was first predicted and demonstrated in 1991 by Tamargo, Bok, and Brem using the rabbit cornea angiogenesis assay (114). We then showed that EVAc polymers loaded with 50% minocycline (w/w) could significantly inhibit the growth of 9L glioma in the rat flank when compared to blank polymer alone; subsequently, we investigated the ability of locally delivered minocycline to extend survival in the 9L glioma model with or without the concurrent intraperitoneal administration of BCNU to determine whether the retardation of tumor cell growth by angiogenesis inhibition could potentiate the

efficacy of chemotherapy (117). Although the local delivery of minocycline begun at the time of tumor implantation resulted in a 530% increase in median survival, no prolongation was seen when minocycline-loaded polymers were implanted on postchallenge day 5. However, treatment on postchallenge day 5 with intracranial minocycline-loaded polymer implants and concurrent intraperitoneal administration of BCNU resulted in a 93% increase in median survival over intraperitoneal BCNU alone. Systemic delivery of minocycline with intraperitoneal BCNU was not more effective than BCNU alone. These studies suggest that the local delivery of antiangiogenic agents such as minocycline can work synergistically with standard chemotherapy to control intracranial tumor growth.

To enhance the efficacy of XRT in the treatment of cancer, patients are often administered radiosensitizing agents prior to tumor irradiation. A radiosensitizing agent does not possess antineoplastic activity when administered alone but can greatly potentiate the efficacy of XRT by making tumor cells more sensitive to DNA damage. For example, the halogenated pyrimidines, such as 5-iodo-2′-deoxyuridine (IUdR) are readily incorporated into the replicating DNA of many human tumor cell lines, including malignant glioma. When malignant glioma cells are exposed to IUdR prior to irradiation, higher killing ratios are observed compared with treatment with radiation alone (118). However, because of rapid blood clearance and marrow suppression toxicity after systemic administration, the therapeutic potential of halogenated pyrimidines for the radiosensitization of human malignant gliomas has remained unrealized. Williams has proposed that the local delivery of IUdR and other halo-pyrimidines from biodegradable polymers implanted at the time of surgery could overcome the obstacles associated with systemic delivery and effect the radiosensitization of malignant glioma.

In a recent study, the controlled release of IUdR from pCPP:SA (20:80) wafers was characterized in vitro and in U251 human glioma xenografts grown in the flank of nude mice (118). Radiosensitization was then demonstrated in U251 tumors receiving IUdR from pCPP:SA (20:80) polymer implants. While minimal yet significant growth delay was observed for xenografts treated with placebo wafers and XRT ($p = 0.046$), extended growth delay and even tumor regression were observed when tumors were irradiated 5 days after the intratumoral implantation of pCPP:SA wafers loaded with 50% IUdR (w/w). This strategy holds promise for the radiosensitization of human gliomas and other tumors.

Saltzman has recently carried out more elaborate intracranial drug release studies in the rat, rabbit, and primate brain and used the data to develop mathematical models of intracranial drug delivery from biodegradable polymer wafers (91,119). A recent study determined the in vivo release profiles of BCNU, taxol, and the alkylating agent 4-hydroperoxycyclophosphamide (4-HC) from pCPP:SA (20:80) wafers implanted intracranially in adult *M. fasicularis* monkeys (91). Significant drug concentrations (0.4 $\mu$M for BCNU, 0.6 $\mu$M for taxol, and 3 $\mu$M for 4-HC) were measured up to ~5 cm from the polymer implant even 30 days after implantation.

With a mathematical model, this collection of data was used to evaluate the pharmacokinetics of polymerically delivered drugs in the brain. This model predicts that the fate of a drug delivered to the brain interstitium from a biodegradable polymer wafer depends on (1) rates of drug transport via diffusion and fluid convection; (2) rates of elimination from the brain via degradation, metabolism, and permeation through capillary networks; and (3) rates of local binding and internalization. A more detailed visual representation of this model is shown in Figure 16. These models are able to predict the intracranial drug concentrations that result from BCNU-loaded pCPP:SA (20:80) wafers as well as other drug–polymer combinations, paving the way for the rational design of drugs specifically for intracranial polymeric delivery.

*FAD:SA Polymers.* Since their introduction in the late 1970s, the polyanhydrides have found increasing use as vehicles for controlled delivery of many drugs, peptides, and proteins (69). Because various general synthetic techniques have been developed, the possibilities for altering both the monomers used as well as their respective ratios are vast. It is therefore possible to rationally design anhydride copolymers with desirable properties for the delivery of specific classes of drug molecules. pCPP:SA polymer wafers have demonstrated near zero-order in vivo controlled-release profiles with a wide range of drugs. It has been observed, however, that some drugs, especially highly hydrophilic compounds, undergo an initial burst of drug release. Furthermore, some water penetration does occur into pCPP:SA wafers (120), making very hydrolytically unstable drugs vulnerable to degradation. To improve upon these shortcomings, a new biodegradable anhydride copolymer of fatty acid dimer (FAD) and sebacic

acid (SA) was designed by Domb (121). FAD, a dimer derived from a naturally occurring oleic acid, is more hydrophobic than CPP. The FAD:SA copolymer was therefore predicted to provide more zero-order release of highly hydrophobic drugs and even greater protection of hydrolytically unstable compounds.

The FAD:SA copolymer is synthesized by mixing acylated FAD and SA monomers in various ratios and polymerizing the solid mixture by melt polycondensation at 180°C under vacuum. Polymerizing the FAD and SA monomer in a 1:1 ratio yields a copolymer with a melting point of 60–70°C and molecular weights up to 110,000 Da. Drug-loaded FAD:SA (50:50) polymer wafers are typically prepared by the mix/melt technique.

The tests of biocompatibility of the FAD:SA (50:50) polyanhydride polymer in the brain of Fischer 344 rats mirrored the tests used to establish the intracranial biocompatibility of pCPP:SA polymers (122). As before, the systemic and local reaction to FAD:SA polymer implants was compared to that of the clinically used Surgicel® and Gelfoam® implants. Rats receiving FAD:SA wafers demonstrated no neurological deficits or behavioral modifications, and the transient, minimal, local reaction in the area of the brain surrounding the implant was similar to that observed for the pCPP:SA wafers. It was therefore concluded that this novel polyanhydride polymer was as biocompatible as the pCPP:SA polymer in the rat brain.

The chemotherapeutic agent cyclophosphamide has been widely used for the treatment of systemic malignancies. Cyclophosphamide itself has no antineoplastic activity, but after systemic administration this compound is rapidly metabolized by the hepatic p450 microsomal enzyme system into active metabolites including 4-HC and

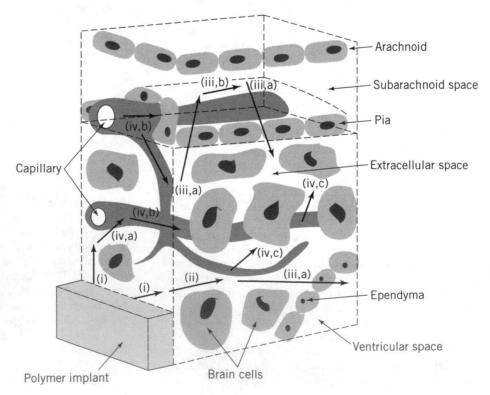

**Figure 16.** Fate of drug released from the polymer pellet to the brain. After drug molecules are released from the polymer matrix, they enter the extracellular space of the brain; they can be transported by diffusion due to drug concentration gradients (**i**); convection due to fluid pressure gradients (**ii**); drug migration into ventricular space via pial or ependymal surfaces (**iii,a**), circulation in the subarachnoid or ventricular spaces (**iii,b**), and subsequent diffusion back into the brain interstitium (**iii,c**); and permeation through the endothelium (**iv,a**), circulation in the cerebral blood vessels (**iv,b**), and reentry by permeation back into the brain interstitium (**iv,c**). *Source:* Reproduced with permission from Ref. 91.

phosphoramide mustard, both of which possess potent DNA-alkylating activity. These metabolites are, however, very hydrophilic and toxic at high systemic concentrations, so their efficacy in the systemic treatment of brain tumors is limited. It was proposed that the local delivery of the active metabolite 4-HC to the brain via biodegradable polymers may be a powerful treatment for malignant glioma.

Local delivery of 4-HC from biodegradable polymers was particularly challenging because the metabolite is not only very hydrophilic but extremely hydrolytically unstable as well. It was predicted that the highly hydrophobic FAD:SA polymer wafer would be able to deliver 4-HC for an extended period of time and protect the entrapped drug from hydrolytic degradation. The in vivo release kinetics of 20% 4-HC–impregnated FAD:SA (50:50) wafers prepared by the mix/melt method were studied in the brains of Fischer 344 rats by Buahin et al. (123). The 7-hydroxyquinoline assay to quantitate 4-HC levels in blood and brain tissue showed a biphasic release profile. An early burst of drug was followed by a period of slower release and a second larger peak between days 5–20. These results may be due to the coating of the polymer with oily fatty acid dimer degradation products. The levels of 4-HC in the blood were consistently lower than those measured in the brain.

The efficacy of 4-HC–loaded FAD:SA (50:50) wafers was then evaluated in the rat 9L gliosarcoma model and compared to the efficacy of an intratumoral injection of the highest tolerated dose of 4-HC (85). While the median survival of rats receiving a 2-mg intratumoral injection of 4-HC on postchallenge day 5 was extended by 36%, a 121% extension was observed in rats receiving the same amount of 4-HC released in a sustained fashion from FAD:SA (50:50) wafers. The FAD:SA polymer matrix seems to be able to protect the hydrolytically unstable 4-HC–alkylating agent and release the compound in a sustained fashion to effectively treat experimental brain tumors.

Another class of compounds that demonstrates promising antitumor activity in vitro against brain tumors is the platinum-based family of cisplatin derivatives. The platinum drugs elicit antineoplastic activity by binding to DNA, leading to interstrand cross-links. Like 4-HC, however, cisplatin and its analogs are highly water soluble and therefore only minimally cross the BBB. The efficacy of these agents delivered systemically for the treatment of intracranial tumors is limited by severe dose-dependent hematopoetic and gastrointestinal toxicity. The parent compound, cisplatin, is also highly neurotoxic and can lead to nephrotoxicity, emesis, ototoxicity, and peripheral neuropathy (124). The synthetic analog carboplatin is significantly less neurotoxic than cisplatin and retains antineoplastic activity in vitro (124). Olivi et al. have therefore investigated the toxicity and efficacy of carboplatin delivered intracranially from both FAD:SA and pCPP:SA polymer wafers (125). The in vitro drug release profiles of 5% carboplatin-loaded (w/w) FAD:SA (20:80) and pCPP:SA (20:80) wafers prepared by the mix/melt method were evaluated in buffered saline using flameless atomic absorption spectrophotometry. Carboplatin was released faster from the less hydrophobic pCPP:SA (20:80) wafers than from the polymers composed of the same ratio of FAD:SA (Fig.

17). By day 7 all of the drug had been released from the pCPP:SA wafers, and only 65% of the drug was released from the FAD:SA wafers during this period.

The efficacy of the two carboplatin–polymer formulations was then evaluated in the F98 rat gliosarcoma model. Both polymer formulations with 5% carboplatin loading extended median survival more than 3-fold over controls ($p < 0.004$). This was significantly better than the best systemic dose of carboplatin. These results suggest that carboplatin delivered from FAD:SA or pCPP:SA wafers may be efficacious against intracranial tumors in humans.

The extent of penetration into the brain parenchyma and therefore the therapeutic potential of antitumor agents delivered interstitially from biodegradable polymers depends on the rate of drug elimination from brain tissue (83,112). Hence, Saltzman has proposed that minimizing the tissue elimination rates of antineoplastic agents delivered from polymers will enhance their efficacy in the treatment of brain tumors (126). To test this hypothesis, the antineoplastic agent methotrexate (MTX), an antimetabolite, was covalently coupled to $M_r$ 70,000 dextran via a short-lived ester bond ($t_{1/2} \approx 3$ days in buffered saline) and a longer-lived amide bond ($t_{1/2} > 20$ days in buffered saline). A novel three-dimensional tissuelike matrix resembling a human brain tumor was developed by suspending H80 human glioma cells in agarose entrapment medium within a hollow fiber. The relative penetration distances for MTX and the two MTX–dextran conjugates was then evaluated by diffusing the drugs into the tumor cell matrix for 10 days and determining the extent of penetrating cytotoxic activity. Whereas the cytotoxicity of MTX and the MTX-ester-dextran conjugate penetrated about 10 mm, the cytotoxicity penetration of the MTX-amide-dextran conjugate was >20 mm. This enhanced penetration correlated with the enhanced stability of the MTX-amide-dextran linkage.

MTX and the MTX-amide-dextran conjugate were then loaded into FAD:SA (50:50) wafers, and the efficacy of these polymer formulations was evaluated in the 9L glioma model. Treatment with the two formulations

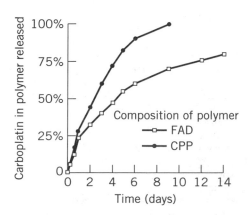

**Figure 17.** Drug release kinetics in vitro from two separate anhydride polymer formulations: pCPP:SA (CPP) and a more hydrophobic fatty acid dimer–sebacic acid copolymer (FAD). Each polymer weighed 10 mg and was loaded with 5% carboplatin by weight. *Source:* Reproduced with permission from Ref. 125.

yielded modest but significant increases in survival, and the MTX-amide-dextran conjugate seemed to shift the dose-response curve to a lower dosage. These studies support the proposal that conjugation of a polymerically delivered chemotherapeutic agent to a water-soluble macromolecule increases drug penetration into the brain by increasing the period of drug retention in brain tissue. This strategy may enhance the intracranial penetration distances of a wide variety of chemotherapeutic agents, potentiating their efficacy in the treatment of brain tumors.

**Polymer Microspheres for the Intracranial Delivery of Proteins.** The delivery of proteins from a polymeric matrix was first demonstrated in 1976 (127). Since that time, advancements in rational polymer design have made it possible to deliver a wide range of therapeutic proteins with retained tertiary structural integrity and catalytic activity (128). Hanes et al. have recently developed IL-2–loaded biodegradable polymer microspheres for local cytokine delivery to improve the immunotherapeutic approach to brain tumor treatment (129). In theory, polymeric cytokine delivery has several advantages over delivery from transduced cells, including obviating the need for transfecting cytokine genes, producing longer periods of cytokine release in vivo, and yielding more reproducible cytokine release profiles and total cytokine dose. Additionally, Sampath has recently shown that local cytokine therapy acts synergistically with local chemotherapy in a murine brain tumor model (130). As a strategy to treat CNS tumors, polymeric delivery may one day greatly enhance the efficacy of immunotherapy in the treatment of human brain tumors.

In conclusion, the capacity of the biodegradable polymer delivery methodology to deliver drugs directly to the brain interstitium is vast. Many of the polymer/drug formulations tested in our laboratories (summarized in Table 3) demonstrate greater therapeutic potential than the clinically approved 3.85% BCNU-loaded wafers in the murine 9L glioma model. It is our hope that these and other polymer drug combinations will soon be studied in clinical trials and lead to substantial improvements in the prognosis for patients diagnosed with malignant brain tumors. A proposed algorithm for incorporating polymer-mediated local delivery of bioactive agents into an overall regimen for the care of neurooncological patients is presented in Figure 18.

**Nonneoplastic Diseases of the CNS.** The strategy of controlled drug delivery to the brain interstitium with polymeric matrices has been applied experimentally to a variety of nonneoplastic diseases of the CNS as well. For example, Aebischer et al. have demonstrated that local delivery of NGF from EVAc rods can prevent the loss of choline acetyl transferase (ChAT) expression in basal forebrain neurons following a fimbria-fornix lesion, indicating the therapeutic potential of this strategy in the treatment of Alzheimer's disease (131). Ebner et al. then showed that NGF could be released into the brain for more than 4 weeks following injection of drug-loaded biodegradable polymer microspheres (132). Theoretically, NGF-releasing microspheres could be useful not only in Alzheimer's therapy but in prolonging graft survival and augmenting nerve regeneration as well.

Freese et al. at MIT have reported the controlled release of dopamine from a polymer brain implant as a potential novel therapy for patients with Parkinson's disease (133). Near-zero-order release for more than 60 days was

**Table 3. Preclinical Studies with Various Polymer/Drug Formulations for the Treatment of Brain Tumors**

| Polymer/Drug formulation[a] | Mechanism of drug's antineoplastic activity | Extension in Median Survival in 9L Glioma Model[b] | Refs. |
|---|---|---|---|
| pCPP:SA (20:80)/3.8% BCNU | DNA alkylation | 127% | 90 |
| pCPP:SA (20:80)/20% BCNU | DNA alkylation | >1,200% | 90 |
| pCPP:SA (20:80)/20% taxol | Inhibition of microtubule depolymerization | 215% | 99 |
| pCPP:SA (20:80)/50% camptothecin | Inhibition of the DNA-replicating enzyme topoisomerase 1 | 306% | 101,102 |
| pCPP:SA (20:80)/5% adriamycin | Intercalation with DNA leading to strand scission and double-stranded cross-breaks | 137% | 105,106 |
| pCPP:SA (20:80)/5% carboplatin | Binding to DNA leading to interstrand cross-links | Not yet tested in 9L model | 125 |
| pCPP:SA (20:80)/1,25 D$_3$ analogs | Upregulation of cell-cycle control proteins and inhibition of angiogenesies | Not yet tested in 9L model | 109,110 |
| EVAc/50% minocycline | Inhibition of tumor angiogenesis | 331% (in combination with IP BCNU) | 114,117 |
| pCPP:SA (20:80)/IUdR | Radiosensitization via incorporation into replicating DNA | Not yet tested in 9L model | 118 |
| FAD:SA (50:50)/20% 4-hydroperoxycyclophosphamide | DNA alkylation | 121% | 123 |
| FAD:SA (20:80)/5% carboplatin | Binding to DNA leading to interstrand cross-links | Not yet tested in 9L model | 125 |
| FAD:SA (50:50)/1% methotrexate-dextran conjugates | Antimetabolite | 140% | 126 |
| Polymer microspheres/cytokine IL-2 | Promotion of antitumoral immune response | Not yet tested in 9L model | 129 |

[a]The indicated drug loading dose, percent polymer by weight, represents the dose that yields the best efficacy/toxicity ratio in the murine brain; this does not apply to 4% BCNU.
[b]In the 9L glioma model, Fischer 344 rats are challenged on day 0 with an intracranial injection or implantation of 9L gliosarcoma cells. On day 5 animals are treated with polymer wafers loaded with the indicated amounts of drug (weight/weight). Extension in median survival expresses the difference in median survival between the treatment group and control animals receiving tumor challenge but no drug; ($p < 0.05$).

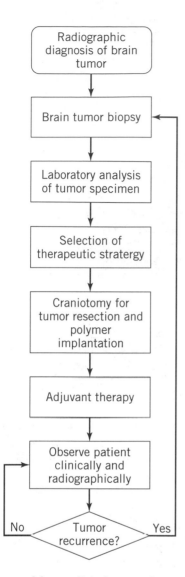

**Figure 18.** Proposed future clinical approach to management of patients with malignant brain tumors. *Source:* Reproduced with permission from Ref. 66.

achieved by coating a dopamine-loaded EVAc polymer matrix with an impermeable layer of EVAc and creating a single cavity for diffusional drug release in the impermeable layer.

Tamargo and others at Johns Hopkins University have also shown that intratumoral implantation of EVAc wafers loaded with the corticosteroid dexamethasone can effectively treat peritumoral brain edema with minimal systemic drug exposure (134,135). This finding is important because systemic administration of corticosteroids, the most common clinical approach to treating edema, is sometimes associated with serious systemic side effects.

### Use of Polymer Systems to Facilitate Drug Delivery to the CNS Interstitium from Biological Tissues

Several distinct approaches use polymer technology to enhance the survival and therapeutic potential of biological tissue implants as vehicles for drug delivery. One approach involves incorporating biologically active moieties into the backbone of biodegradable polymers and using these polymers to form highly porous scaffolds for cell transplantation (136,137). Synthetic polymers can be rationally designed so that the polymer scaffold guides cell organization and growth while allowing the diffusion of nutrients to the transplanted cells. For example, vascularization of the implant could theoretically be artificially induced by the slow release of angiogenic factors from the polymer matrix or the incorporation of capillary growth, promoting amino acid sequences such as the arginine-lysine-aspartic acid (RGD) sequence into the polymeric backbone (138). Furthermore, with recent advancements in cell culture techniques, large populations of cells genetically engineered to express desired proteins could be formed in vitro using a small sample of a patient's own cells, thus circumventing immunogenic complications (139). At MIT, we have demonstrated the feasibility of this technique in the engineering of liver, bone, tendon, skin, ureter, intestine, and cartilage (136). This technique could theoretically also be applied to the CNS to enhance the survival of tissue grafts designed to release therapeutic proteins.

An alternative strategy has been pursued by Aebischer et al. that involves the immunoisolation of transplanted tissue via encapsulation in a permselective polymeric membrane. In this approach, transplanted cells are kept selectively separated from the host tissue by a polymeric membrane with defined pore size that permits the passage of nutrients and neurotransmitters but is impermeable to cells, complement, antibodies, and viruses (140). This technique has been applied to deliver PC12 cells, a dopamine-secreting cell line derived from a rat pheochromocytoma, to the striatum of guinea pigs (141). The use of neoplastic cells is possible because the polymer capsule prevents tumor formation by physically sequestering the transplanted tissue. The therapeutic potential of this strategy was evaluated in the 6-hydroxydopamine unilaterally lesioned rat model of Parkinson's disease. Significant behavioral recovery was observed from 1 to 4 weeks following implantation of PC12 cell-loaded microcapsules as compared with empty microcapsules. Cell encapsulation technology holds promise as a strategy for long-term delivery of neurotransmitters to the brain (142).

### CONCLUSION

The treatment of CNS diseases is particularly challenging because the delivery of drug molecules to the brain is often precluded by a variety of physiological, metabolic, and biochemical obstacles that collectively comprise the BBB, BTB, and BCB. Two general strategies, drug modification and BBBD, have been employed to enhance the BBB penetration of drug molecules delivered via the cardiovascular route. Aided by an increased understanding of BBB and tumor physiology, significant progress has been made in these techniques. A third general strategy encompasses alternative routes to CNS drug delivery that attempt to circumvent the BBB altogether. The effectiveness of one of these routes, drug delivery directly to the brain interstitium, has recently been markedly enhanced through the rational design of polymer-based drug delivery systems.

The present outlook for patients suffering from many types of CNS diseases remains poor, but recent developments in drug delivery techniques warrant reasonable hope that the formidable barriers shielding the CNS may ultimately be overcome. Substantial progress will only come about, however, if continued vigorous research efforts to develop more therapeutic and less toxic drug molecules are paralleled by the aggressive pursuit of more effective mechanisms for delivering those drugs to their CNS targets.

## ACKNOWLEDGMENTS

The authors thank Eileen Bohan, Betty Collevecchio, and Betty Tyler for their generous research assistance as well as Ann Finkbeiner and Sarvenaz Zand for their thoughtful review of this manuscript. This work was supported by NIH Grant U01-CA52857.

## DISCLOSURE STATEMENT

Dr. Brem and Dr. Langer are consultants to Guilford Pharmaceuticals, Inc., and Dr. Brem is a consultant to Rhone-Poulenc Rorer. Guilford has provided a gift for research in Dr. Brem's laboratory. The Johns Hopkins University, Dr. Brem, and Dr. Langer own Guilford stock, the sale of which is subject to certain restrictions under University policy. The terms of this arrangement are being managed by the University in accordance with its conflict-of-interest policies.

## BIBLIOGRAPHY

1. N.J. Abbot and I.A. Romero, *Mol. Med. Today* March, pp. 106–113 (1996).
2. W.M. Pardridge, *Annu. Rev. Pharmacol. Toxicol.* **28**, 25–39 (1988).
3. R.J. Tamargo and H. Brem, *Neurosurg. Q.* **2**(4), 259–279 (1992).
4. W.M. Scheld, *Rev. Infect. Dis.* **2**(7), 1669–1690 (1989).
5. H. Brem et al., *Polymeric Site-specific Pharmacotherapy*, John Wiley and Sons, New York, 1994, pp. 117–139.
6. N.H. Grieg, *Cancer Treat. Rev.* **14**, 1–28 (1987).
7. G.W. Goldstein and A.L. Betz, *Sci. Am.* **255**, 74–83 (1986).
8. P.H. Hutson, G.S. Sarna, B.D. Kantamaneni, and G. Curzon, *J. Neurochem.* **44**, 1266–1273 (1985).
9. E.P. Sipos, R.J. Tamargo, J.D. Weingart, and H. Brem, *Ann. N.Y. Acad. Sci.* **732**, 263–272 (1994).
10. A.K. Sills, E.P. Sipos, J. Laterra, and H. Brem, *Adv. Neuro-Oncol.* **2**, 81–96 (1997).
11. R.K. Jain, *J. Natl. Cancer Inst.* **81**(8), 570–576 (1989).
12. R.K. Jain, *Sci. Am.* **271**(1), 58–65 (1994).
13. R.K. Jain, *Sci.* **271**, 1079–1080 (1996).
14. F. Yuan et al., *Cancer Res.* **54**, 4564–4568 (1991).
15. M.G. Ewend et al., *Cancer Res.* **56**, 5217–5223 (1996).
16. O. Merimsky, M. Inbar, B. Gerarg, and S. Chaitchik, *Melanoma Res.* **2**, 401–406 (1992).
17. M. Mastrangelo, R. Bellet, and D. Berd, *PPO Updates* **5**, 1–11 (1991).
18. A. Abner, *Chest* **103**(Suppl.), 445S–448S (1993).
19. A. Gerl et al., *Clin. Exp. Metastasis* **12**, 226–230 (1994).
20. V.A. Levin, C.S. Patlak, and H.D. Landahl, *J. Pharmacokinet. Biopharm.* **8**(3), 257–296 (1980).
21. V.A. Levin and P.M. Kabra, *Cancer Chemother. Rep.* **58**, 787–792 (1974).
22. Y. Madrid, L.F. Langer, H. Brem, and R. Langer, *Adv. Pharmacol.* **22**, 299–324 (1991).
23. N. Bodor and P. Buchwald, *Pharmacol. Ther.* **76**(1–3), 1–27 (1997).
24. N. Bodor, H.H. Farag, and M.E. Brewster, *Science* **214**(18), 1370–1372 (1981).
25. F.A. Omar, H.H. Farag, and N. Bodor, *J. Drug Target.* **2**(4), 309–316 (1994).
26. N. Bodor et al., *Science* **257**, 1698–1700 (1992).
27. Y. Takada et al., *Cancer Res.* **52**, 2191–2196 (1992).
28. P.M. Friden, *Neurosurgery* **35**(2), 294–298 (1994).
29. P.M. Friden et al., *Proc. Natl. Acad. Sci. U.S.A.* **88**, 4771–4775 (1991).
30. P.M. Friden et al., *Science* **259**, 373–377 (1993).
31. W.A. Jeffries et al., *Nature (London)* **312**(8); 162–163 (1984).
32. W.M. Pardridge, R.J. Boado, and Y.S. Kang, *Proc. Natl. Acad. Sci. U.S.A.* **92**, 5592–5596 (1995).
33. J. Huwyler, D. Wu, and W.M. Pardridge, *Proc. Natl. Acad. Sci. U.S.A.* **93**, 14164–14169 (1996).
34. R.A. Kroll and E.A. Neuwelt, *Neurosurgery* **42**(5), 1083–1099 (1998).
35. R.A. Kroll et al., *Neurosurgery* **43**(4), 879–886 (1998).
36. D.E. Bullard, S.H. Bigner, and D.D. Bigner, *Cancer Res.* **45**, 5240–5245 (1985).
37. E.A. Neuwelt et al., *Ann. Intern. Med.* **94**(1), 449–454 (1981).
38. E.A. Neuwelt et al., *Cancer Res.* **45**, 2827–2833 (1995).
39. M. Ohata et al., *Cancer Res.* **45**, 1092–1096 (1995).
40. E.A. Neuwelt et al., *Neurosurgery* **37**(1), 17–28 (1995).
41. E.A. Neuwelt et al., *Cancer Res.* **56**, 706–709 (1996).
42. T. Inamura, T. Nomura, R.T. Bartus, and K.L. Black, *J. Neurosurg.* **81**, 752–758 (1994).
43. S. Nakano, K. Matsukado, and K.L. Black, *Cancer Res.* **56**, 4027–4031 (1996).
44. K. Matsukado et al., *Neurosurgery* **39**(1), 125–134 (1996).
45. K.L. Black et al., *J. Neurosurg.* **86**, 603–609 (1997).
46. R.E. Harbaugh, R.L. Saunders, and R.F. Reedert, *Neurosurgery* **23**(6), 693–698 (1988).
47. R.G. Thorne, C.R. Emory, T.A. Ala, and W.H. Frey, *Brain Res.* **692**, 278–282 (1995).
48. K.A. Walter et al., *Neurosurgery* **37**(6), 1129–1145 (1995).
49. E.H. Oldfield et al., *Proc. Natl. Acad. Sci. U.S.A.* **91**, 2076–2080 (1994).
50. E.A. Neuwelt et al., *Neurosurgery* **38**(4), 746–754 (1996).
51. J. Sladek and D. Gash, *J. Neurosurg.* **68**, 337–351 (1988).
52. D.M. Yurek and J.R. Sladek, *Annu. Rev. Neurosci.* **13**, 415–440 (1990).
53. K. Leigh, K. Elisevich, and K.A. Rogers, *J. Neurosurg.* **81**, 272–283 (1994).
54. B. Lai et al., *Proc. Natl. Acad. Sci. U.S.A.* **91**, 9695–9699 (1994).
55. E.Y. Snyder et al., *Neurobiol. Dis.* **4**, 69–102 (1997).
56. R.C. Thompson et al., *J. Immunother.* **19**(6), 405–413 (1997).
57. D.M. Gash et al., *Science* **233**, 1420–1421 (1986).
58. P. Horellou and J. Mallet, *Mol. Neurobiol.* **15**(2), 241–256 (1997).
59. P. Horellou, O. Sabate, M.H. Buc-Caron, and J. Mallet, *Exp. Neurol.* **144**(1), 131–138 (1997).

60. W.M. O'Conner et al., *Exp. Neurol.* **148**(1), 167–178 (1997).

61. M.H. Tuszynski et al., *Gene Ther.* **3**(4), 305–314 (1996).

62. B.L. Maria and T. Friedman, *Semin. Pediatr. Neurol.* **4**(4), 333–339 (1997).

63. S. Benedetti et al., *Hum. Gene Ther.* **8**(11), 1345–1253 (1997).

64. B.V. Zlokovic and M.L. Apuzzo, *Neurosurgery* **40**(4), 805–812 (1997).

65. C. Raffel et al., *Hum. Gene Ther.* **5**, 863–890 (1994).

66. H. Brem and R. Langer, *Sci. Med.* **3**(4), 1–11 (1996).

67. J. Hanes and M. Burke, *Hosp. Pharm. Rep.*, December, pp. 1–11 (1997).

68. J. Tamada and R. Langer, *J. Biomater. Sci., Polym. Ed.* **3**(4), 315–353 (1992).

69. M. Chasin et al., in M. Chasin and R. Langer, eds., *Biodegradable Polymers as Drug Delivery Systems*, Dekker, New York, 1990, pp. 43–69.

70. K.W. Leong, V. Simonte, and R. Langer, *Macromolecules* **20**, 705–712 (1987).

71. K.W. Leong, B.C. Brott, and R. Langer, *J. Biomed. Mater. Res.* **19**, 941–955 (1985).

72. A.J. Domb and R. Langer, *J. Polym. Sci.* **25**, 3373–3386 (1987).

73. M. Chasin, D. Lewis, and R. Langer, *BioPharm. Manuf.* **1**, 33–46 (1988).

74. E. Mathiowitz and R. Langer, *J. Controlled Release* **5**, 13–22 (1987).

75. E. Mathiowitz et al., *J. Appl. Polym. Sci.* **35**, 755–774 (1988).

76. R. Langer, H. Brem, and D. Tapper, *J. Biomed. Mater. Res.* **15**, 267–277 (1981).

77. K.W. Leong, P. D'Amore, M. Marletta, and R. Langer, *J. Biomed. Mater. Res.* **20**, 51–64 (1986).

78. R.J. Tamargo et al., *J. Biomed. Mater. Res.* **23**, 253–266 (1989).

79. H. Brem et al., *Sel. Cancer Ther.* **5**, 55–65 (1989).

80. M.B. Yang, R.J. Tamargo, and H. Brem, *Cancer Res.* **49**, 5103–5107 (1989).

81. R.J. Tamargo, R. Langer, and H. Brem, *Methods Neurosci.* **21**, 135–149 (1994).

82. S.A. Grossman et al., *J. Neurosurg.* **76**, 640–647 (1992).

83. R.J. Tamargo et al., *Cancer Res.* **53**, 329–333 (1993).

84. K.G. Buahin and H. Brem, *J. Neuro-Oncol.* **26**, 103–110 (1995).

85. H. Brem et al., *J. Neurosurg.* **68**, 334 (1988).

86. H. Brem et al., *J. Neurosurg.* **74**, 441–446 (1991).

87. H. Brem et al., *Lancet* **345**, 1008–1012 (1995).

88. H. Brem et al., *J. Neuro-Oncol.* **26**, 111–123 (1995).

89. S. Valtonen et al., *Neurosurgery* **41**(1), 44–49 (1997).

90. E.P. Sipos et al., *Cancer Chemother. Pharmacol.* **39**, 383–389 (1997).

91. L.K. Fung et al., *Cancer Res.* **58**, 672–684 (1998).

92. J.H. Galicich, N. Sunderesan, and H.T. Thaler, *J. Neurosurg.* **53**, 63–67 (1980).

93. M.G. Ewend et al., *Am. Assoc. Neurol. Surg.*, Denver, Colo. 1997, Abstract.

94. S.R. Wedge and E.S. Newlands, *Br. J. Cancer* **73**, 1049–1052 (1996).

95. A.E. Pegg, *Cancer Res.* **50**, 6119–6129 (1990).

96. M.E. Dolan et al., *Cancer Res.* **51**, 3367–3372 (1991).

97. S.C. Schold et al., *Cancer Res.* **56**, 2076–2081 (1996).

98. M.A. Cahan, K.A. Walter, M.O. Colvin, and H. Brem, *Cancer Chemother. Pharmacol.* **33**, 441–444 (1994).

99. K.A. Walter et al., *Cancer Res.* **54**, 2207–2212 (1994).

100. P. Sampath et al., *Am. Assoc. Neurol. Surg.*, Denver, Colo. 1997, Abstract.

101. J.D. Weingart et al., *Int. J. Cancer* **62**, 605–609 (1995).

102. J.D. Weingart, B. Tyler, O.M. Colvin, and H. Brem, *3rd Drug Discovery Dev. Symp.* 1993, Abstract.

103. P. Sampath et al., *Soc. Neuro-Oncol.*, Charlottesville, Va., 1997, Abstract.

104. S.Y. Lin et al., *Biomater., Artif. Cells, Artif. Organs* **17**, 189–203 (1989).

105. V.L. Perry et al., *Cong. Neurol. Surg.*, 1998, Abstract.

106. M. Watts et al., *Am. Assoc. Neurol. Surg.*, Denver, Colo., 1997, Abstract.

107. D. Feldman, F.H. Glorieux, and J.W. Pike, *Vitamin D*, Academic Press, San Diego, Calif., 1997.

108. T. Oikawa et al., *Eur. J. Pharmacol.* **178**, 247–250 (1990).

109. G.H. Posner et al., *J. Med. Chem.* **41**(16), 3008–3014 (1998).

110. M. Burke et al., *Proc. 10th Workshop Vitam. D*, Strasbourg, France, May 24–29, 1997.

111. M. Burke et al., *5th Ann. Brown Univ. Symp. Vitam. D*, Providence, R.I. September 7–9, 1997, Abstract.

112. J. Folkman, *N. Engl. J. Med.* **333**, 1757–1763 (1995).

113. J. Folkman, *N. Engl. J. Med.* **285**, 1182–1186 (1971).

114. R.J. Tamargo, R.A. Bok, and H. Brem, *Cancer Res.* **51**, 672–675 (1991).

115. R.J. Tamargo, K.W. Leong, and H. Brem, *Neuro-Oncol.* **9**, 131–138 (1990).

116. A.K. Sills et al., *Cancer Res.* **58**(13), 2784–2792 (1998).

117. J.D. Weingart, E.P. Sipos, and H. Brem, *J. Neurosurg.* **82**, 635–640 (1995).

118. J.A. Williams et al., *J. Neuro-Oncol.* **32**, 181–192 (1997).

119. W.M. Saltzman et al., *Pharm. Res.* **13**(5), 671–682 (1996).

120. W. Dang et al., *J. Controlled Release* **42**, 83–92 (1996).

121. A. Domb et al., *Polym. Prep.* **32**(2), 219–220 (1991).

122. H. Brem et al., *J. Controlled Release* **19**, 325–330 (1992).

123. K.G. Buahin et al., *Polym. Adv. Technol.* **3**, 311–316 (1992).

124. A. Olivi et al., *Cancer Chemother. Pharmacol.* **31**, 449–454 (1993).

125. A. Olivi et al., *Cancer Chemother. Pharmacol.* **39**, 90–96 (1996).

126. W. Dang, O.M. Colvin, H. Brem, and W.M. Saltzman, *Cancer Res.* **54**, 1729–1735 (1994).

127. R. Langer and J. Folkman, *Nature (London)* **263**, 797–800 (1976).

128. R. Langer, *Nature (London)* **392**, 5–10 (1998).

129. J. Hanes et al., *Conf. Formul. Drug Delivery II*, La Jolla, Calif., October 5–8, 1997.

130. P. Sampath et al., *Congr. Neurol. Surg.*, New Orleans, La, September 27–October 2, 1997.

131. D. Hoffman, L. Wahlberg, and P. Aebischer, *Exp. Neurol.* **110**, 39–44 (1990).

132. P.J. Camarata et al., *Neurosurgery* **30**, 313–319 (1992).

133. A. Freese et al., *Exp. Neurol.* **103**, 234–238 (1989).

134. R.J. Tamargo et al., *J. Neurosurg.* **74**, 956–961 (1991).

135. C.S. Reinhard et al., *J. Controlled Release* **16**, 331–340 (1991).

136. R. Langer and J.P. Vacanti, *Science* **260**, 920–926 (1993).

137. L.G. Cima et al., *J. Biomech. Eng.* **113**, 143–151 (1991).

138. D.A. Barrera, E. Zylstra, P.T. Lansbury, and R. Langer, *J. Am. Chem. Soc.* **115**, 11010–11011 (1993).

139. R. Langer, *Ann. Biomed. Eng.* **23**, 101–111 (1995).

140. P. Aebischer, S.R. Winn, and P.M. Galletti, *Brain Res.* **448**, 364–368 (1988).

141. S.R. Winn et al., *Exp. Neurol.* **113**, 322–329 (1991).

142. P. Aebischer et al., *J. Biomech. Eng.* **113**, 178–183 (1991).

# CHARACTERIZATION OF DELIVERY SYSTEMS, DIFFERENTIAL SCANNING CALORIMETRY

PIERRE SOUILLAC
J. HOWARD RYTTING
University of Kansas
Lawrence, Kansas

## KEY WORDS

Differential scanning calorimetry (DSC)

Differential thermal analysis (DTA)

Drug purity

Heat conduction calorimetry

Isoperibol calorimetry

Polymorphism

Pseudopolymorphism

Thermogravimetric analysis (TGA)

## OUTLINE

## INTRODUCTION

Among the most useful and underused tools available to pharmaceutical scientists, in both research and development, are the various types of calorimeters. A large variety of calorimeters are currently available for the collection of information necessary for the preformulation and formulation of drugs: differential thermal analysis (DTA), differential scanning calorimetry (DSC), modulated differential scanning calorimetry (MDSC), thermogravimetric analysis (TGA), isoperibol calorimetry, and heat conduction calorimetry.

DSC has been one of the most widely used calorimetric techniques for the determination of various thermal parameters, which allow a better understanding of drug characteristics and provide a basis for drug formulation. DTA, often regarded as a more qualitative technique, is still widely used. Recently, the development of a novel application of DSC, MDSC, which uses a modulated (sinusoidal) heating rate, has permitted the resolution of overlapping thermal phenomena as well as the detection of higher order phase transitions such as glass transitions ($T_g$).

Often, other thermal analysis techniques are used concomitantly with DSC for the determination of parameters necessary for the formulation of drugs. TGA is widely used for the determination and characterization of pseudopolymorphs (hydrates and solvates). Heat conduction calorimeters (or microcalorimeters) are used to evaluate changes in heat due to chemical or physical reactions as well as to determine the rates of chemical reactions including degradation. Isoperibol calorimetry is used to measure various types of enthalpies (i.e., enthalpies of solution, mixing, hydration, or binding) as well as to determine the thermodynamics of binding.

## DTA AND DSC

Upon undergoing a physical change (i.e., phase transition or crystallization) or during a chemical reaction, a material usually absorbs or evolves heat. The principle of these two techniques is based on the differential detection of energy changes between the sample undergoing a thermal event (at a given temperature) and a reference (inert) when both are subjected to the same heating program. DTA is a technique in which the difference in temperature between the sample and a reference is recorded versus temperature, whereas DSC is a technique in which the difference in heat flow between the sample and a reference is recorded versus temperature.

## DTA

DTA has been widely used and is considered the simplest of the differential techniques. Typically, a DTA apparatus is composed of a sample holder which itself is comprised of two cells. One cell contains the sample and the other one an inert reference material (Fig. 1). Reference materials should not undergo thermal events in the temperature range of interest and should not react with either the thermocouples or the cells. Alumina and silica are generally used as reference materials (1). The sample holder is centered inside a furnace equipped with a temperature controller. Thermocouples can be either placed inside the sample and the reference (block type measuring system) or inserted in each cell (heat flux).

With the first type of apparatus design, the thermocouples are very sensitive to the nature of the material used. This particular design makes the method quite sensitive and allows for rapid detection of the thermal event associated with the sample (2). The signal is dependent on the

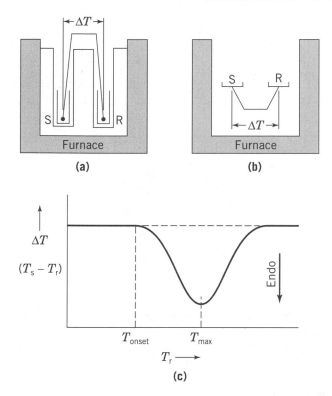

**Figure 1.** DTA apparatus. (**a**) Classical apparatus; (b) heat flux apparatus, (c) typical DTA curve. (S: sample, R: reference). *Source:* From Ref. 1.

difference in heat capacities between the sample and the reference, the heating rate, and the thermal resistance of the system.

$$\Delta T = R(dT/dt)(C_s - C_r) \qquad (1)$$

where $\Delta T$ is the differential temperature between the sample and the reference, $C_s$ and $C_r$ are the heat capacities of the samples and the reference, respectively, and $R$ is the thermal resistance. The thermal resistance is not only affected by the characteristics of the instrument but also by the properties of the sample and the reference (1).

With the second type of DTA apparatus the thermocouples are not inside the sample but attached beneath the cells. A slower response between the thermal event associated with the sample and its detection is observed. On the other hand, the response is less affected by the properties of the sample (1). In this case, the thermal resistance is only dependent on the characteristics of the instrument.

Independent of the apparatus design, both sample and reference are subjected to the same heating program. When the sample does not undergo a thermal event, the differential temperature is equal to zero. During a phase transition of the sample, an energy change will occur and the temperature of the sample will be different from the temperature of the reference (which does not undergo phase transition). The temperature differential is amplified, recorded, and plotted versus temperature. The minimum temperature differential that can be detected is about 0.01°C (3).

However, despite its simplicity and design, DTA has been progressively replaced by the more quantitative technique, DSC.

## DSC

DSC has provided qualitative information as well as quantitative determinations of physicochemical parameters of drugs. Differential scanning calorimeters measure the heat capacity ($C_p$) of the system as a function of temperature. Following the change in heat capacity of the sample as a function of temperature allows for the detection of phase transitions of various orders. DSC has been used in a variety of pharmaceutical applications: the characterization of solid crystals (polymorphism) (4–8); the study of phase changes (9); the evaluation of drug-excipient interactions (10,11); the study of thermal denaturation of biopolymers (12–16); and the influence of penetration enhancers on the lipid bilayer structure (17,18). The measurement of heat capacity is based on the differential monitoring of temperature or heat supplied (depending on the type of instrument) between a sample and a reference. The sample and the reference are submitted to the same preset temperature program in which temperature increases linearly at a predetermined rate. The thermal events associated with the samples are monitored versus temperature.

Two types of differential scanning calorimeters have been used to perform thermal analyses relevant to pharmaceutical science: heat flux DSC and power compensation DSC (19). A differential measurement of the heat flow is performed in both cases. This differential heat flow is deduced from either monitoring the difference in temperature between the sample and a reference (heat flux DSC) or monitoring the difference in the input of heating power necessary to maintain the sample and a reference at the same temperature (power compensation DSC). Most modern instruments allow for a wide range of scanning temperatures (from −170°C to 700°C) as well as for a wide range of heating rates (from 0.1°C/min to 200°C/min). However, the heating rates used for thermal analysis of most pharmaceutical compounds are typically between 5°C/min to 20°C/min. The choice of heating rate should be based on the characteristics of the particular compound (or system) of interest and the thermal parameters that are to be determined. Indeed, high scanning rates allow for relatively high heat flow sensitivity but low resolution, whereas low scanning rates permit high resolution but lower heat flow sensitivity. High scanning rates are often necessary for the determination of glass-transition temperatures. The increase in heat capacity of the system due to an increase in molecular mobility at the glass-transition temperature is relatively small and large heating rates are needed for its detection. On the other hand, for a system in which two different thermal phenomena are occurring in the same temperature range, a low heating rate is recommended to obtain a better resolution of the two thermal events. Unfortunately, a balance between resolution and sensitivity is often necessary.

**Types of Differential Scanning Calorimeters.** *Heat Flux DSC.* Two types of heat flux DSCs are encountered: the

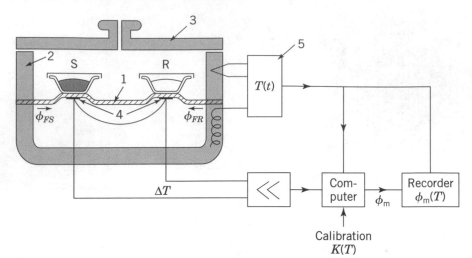

**Figure 2.** Heat flux DSC with disk-type measuring system. (1) Disk; (2) furnace; (3) lid; (4) differential thermocouples; (5) programmer and controller. (R: reference; S: sample; $\phi_{FS}$: heat flow rate from the furnace to the sample; $\phi_{FR}$: heat flow rate from the furnace to the reference; $\phi_m$: measured heat flow rate; $K$: calibration factor) *Source:* From Ref. 2.

disc type and the cylinder type. The disc type calorimeter is the most commonly used (Fig. 2). A sample and a reference are symmetrically placed on the same furnace (thermoelectric disc). Energy input is transferred to the sample and reference through the sample holder (or cell) and the sample pan. The temperature probes are placed within the disc and the temperature differential is detected as a voltage (2). Differential heat flow to the sample and reference is monitored as a function of temperature.

*Power Compensation DSC.* With power compensation DSCs, the sample and reference are placed on two different furnaces (Fig. 3). Both furnaces contain a temperature probe and a heating resistor. The same temperature is maintained at all times between the sample and the reference. The difference in the amount of heat supplied to both furnaces in order to maintain the temperature differential between the sample and the reference at zero is monitored and can be directly correlated to thermal events occurring to the sample (2).

This difference is followed as a function of temperature. When the sample does not undergo any thermal event, the same amount of heat is supplied to both furnaces; a horizontal baseline should be observed assuming that the change in heat capacity of the sample as a function of temperature is negligible. When the sample undergoes a thermal event, the differential temperature starts to deviate

from zero. A decrease or increase in the amount of electrical energy supplied to the reference will reestablish the temperature differential to zero. This heating power differential is monitored and after amplification a significant deviation from the horizontal baseline is observed.

### Practical Aspects

**Sample Preparation.** The amount of material necessary for these types of measurement are generally in the milligram range. The use of small amounts of sample presents the advantage of a better thermal gradient within the sample as well as the consumption of less material. The sample is generally placed in disposable a pan. Different types of sample pans (e.g., composition, geometry) are available. The reference usually consists of an empty pan of the same type as the pan containing the sample. Aluminum pans are the most common type. A lid can be crimped on the pans to avoid contamination of the DSC cell holder and to provide a better thermal gradient throughout the sample. When a sample is heated, depending on the size of the sample, a thermal gradient will exist within the sample. The thermal gradient throughout the sample can be reduced by reducing its size to a minimum or by compacting the sample with a lid. Hermetically sealed sample pans are necessary for volatile compounds or when liquid to gas transitions are studied. However, the pressure resistance of the pans should be carefully taken into account before designing any experiment. Gold or platinum sample pans can be used (*1*) for compounds reacting with aluminum or (*2*) when high temperatures are needed (above 600°C) (20). Gold also has the advantage over aluminum that it has a better thermal conductivity. However, due to the difference in cost between gold and aluminum, aluminum pans are most commonly used.

Determination of thermal parameters at temperatures below room temperature are sometimes necessary. With the increased use of lyophilized protein formulations, the determination of the glass-transition temperatures of frozen solutions ($T_g'$) for diverse additives is often useful for the development of the lyophilization cycle (21,22). The cooling system depends on the temperature range of inter-

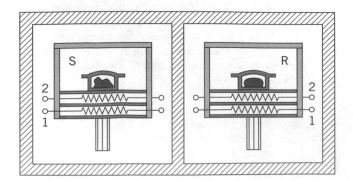

**Figure 3.** Power compensation DSC. (1) Heating wire; (2) resistance thermometer. (S: sample; R: reference) *Source:* From Ref. 2.

est. For the lowest temperatures ($-170°C$), liquid nitrogen is used. To avoid any moisture condensation that could interfere with the measurement, a purge gas system is required for measurements below room temperature. Dry nitrogen, argon, or helium are the recommended gases. The instrumental baseline can be affected by the flow rate of the purge gas, and therefore the flow rate should remain as stable as possible in the range of 20 to 50 mL/min (3). Nitrogen purge is also highly recommended at temperatures above room temperature when the sample of interest is prone to oxidative degradation pathways.

**Calibration.** Because the response of a calorimeter differs depending on the choice of several instrumental parameters, calibration of the instrument is necessary prior to analysis of a sample. Moreover, the transfer of heat between the furnace and the sample pans is not instantaneous. Even under ideal sample preparation conditions, there will be a difference between the temperature of the furnace and the temperature of the sample at any time during the temperature scan. The temperature lag between the furnace temperature and the sample (and reference) temperatures should be reproducible and should not depend on the type of sample analyzed. Usually, the thermal lag increases as the heating rate increases. To minimize the thermal lag problem, the calibration should be performed under the same conditions as those of the experiments; heating rate, type of sample pans, purge gas flow, and sample size should be identical during the calibration and the subsequent experimental measurements. The calibration of differential scanning calorimeters relies on chemical standards of high purity (metals or organic compounds). Calibration is performed for both the temperature scale and the amplitude of the signal obtained (baseline and thermal event). The material used for calibration should ideally (1) be pure, (2) have a thermal behavior that is perfectly described and reproducible, (3) be characterized by the presence of a thermal event in the same temperature range as the temperature range of interest, and (4) not react with the sample pans (2). Calibration can be performed using only one calibrant when the temperature range of interest is narrow, or with several calibrants for a large temperature range. In the case of a one-point calibration, indium is the most commonly used calibrant with a melting temperature at $156.6°C$ and an enthalpy of fusion equal to 28.7 J/g. For the newest instruments, the calorimetric response over the entire temperature range can be assumed constant and a multipoint calibration is usually not necessary. However, when a multipoint calibration is desired or when the temperature range of interest is far from the indium melting temperature, other substances may be used. Water can be used to calibrate calorimeters around $0°C$. For high temperature ranges, fusion of tin, lead, or zinc can be used (2). The melting temperature (and the enthalpies of fusion) for tin, lead, and zinc are, respectively, $231.9°C$ ($14.24$ cal.g$^{-1}$), $327.4°C$ ($5.55$ cal.g$^{-1}$), and $419.5°C$ ($26.6$ cal.g$^{-1}$) (23). The melting temperature of organic compounds such as cyclopentane ($-93.43°C$), 1,2-dichloroethane ($-32°C$), decane ($-30°C$), phenyl ether ($30°C$), o-terphenyl ($58°C$), or biphenyl ($69.3°C$) can also be used for calibration purposes.

**Detection of Thermal Events.** Different types of thermal events can be detected by DSC: (1) endothermic events (i.e., melting, dehydration), (2) exothermic events (i.e., crystallization, denaturation of proteins), or (3) changes in heat capacity (i.e., glass-transition temperature). Heat capacity is plotted versus temperature when dynamic measurements are performed. These types of plots are called thermoanalytical curves or thermograms (Fig. 4). Different software (depending on the calorimeter manufacturer) is used for the extraction of numerical data from the curves obtained.

Phase transitions of a substance (i.e., melting or crystallization) are detected over a range of temperatures and not at a single temperature, thus making the determination of a single transition temperature difficult. Different temperatures have been used to describe a phase transition (Fig. 5). In the case of a melting endotherm, the temperature at the apex of the peak is denoted $T_m$. The onset temperature ($T_0$) corresponds to the temperature at which the signal first deviates from the baseline. The extrapolated onset temperature ($T_e$) corresponds to the tempera-

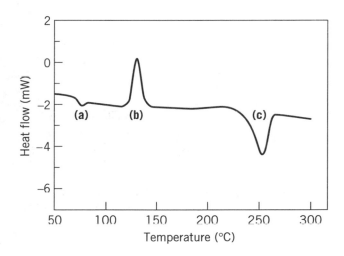

**Figure 4.** Typical DSC thermoanalytical curve. (**a**) Glass transition with associated enthalpic relaxation; (**b**) cold crystallization exotherm; (**c**) melting endotherm. *Source:* From Ref. 20.

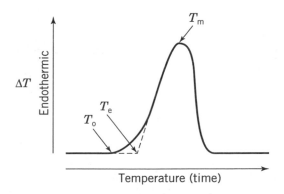

**Figure 5.** Typical endotherm. $T_0$: onset temperature; $T_e$: extrapolated onset temperature; $T_m$: peak temperature. *Source:* From Ref. 19.

ture at the intersection of the tangent to the maximum slope of the peak with the extrapolated linear base line. The extrapolated onset temperature is usually used as the melting temperature unless otherwise stated.

The same characteristic temperatures are used in the case of a crystallization exotherm. However, the temperature of the apex of the peak is usually denoted $T_c$.

Small sample size, homogeneous particle size, and good thermal contact between the sample pan and the calorimeter improve the quality of the data obtained. The quality of the data obtained is affected by the heating rate; the maximum temperature of the peak increases and the peak broadens as the scanning rate increases when the data are plotted versus temperature (Fig. 6).

The quality of the data obtained is also affected by the purity of the substance analyzed. The maximum temperature of the melting endotherm decreases and the melting endotherm broadens as the purity of the sample decreases. The change in shape of the melting endotherm due to a decrease in the purity of the sample provides the basis for sample purity determination by DSC and is discussed in a separate section.

The melting (or freezing) temperature of a substance should be determined by studying the phase transition from the solid state to the liquid state (increasing temperature program) rather than by studying the phase transition from the liquid state to the solid state (decreasing temperature program). Indeed, upon cooling a liquid can stay liquid below its freezing temperature to form a supercooled liquid. This phenomenon is frequent and causes a significant underestimation of the melting temperature (freezing temperature). Studying the transition temperature upon cooling and heating the sample should give relatively accurate results in the absence of supercooling phenomenon.

Determination of glass-transition temperatures ($T_g$) is slightly different due to the fact that glass transitions are not first-order phase transitions (such as melting or crystallization). The enthalpy change during higher order phase transition is equal to zero. A peak is not generally observed; instead, a change in heat capacity values causes a drift in the baseline. The extrapolated onset temperature

($T_f$) and the temperature obtained at half the height of the baseline shift ($T_{mid}$) have been used to define the glass-transition temperature (19) (Fig. 7). Due to the dramatic effect of the thermal history of the sample on the glass-transition determination for polymers, a cycle at high scanning rate (20°C/min) should be performed on the samples prior to analysis. This preliminary cycle would erase the thermal history of the sample (3).

Not only the temperature but also the enthalpies of transition can be estimated for the first-order transitions. If the sample weight is known and if the calorimeter has been correctly calibrated with a calibrant having a known enthalpy of transition (generally heat of fusion), the enthalpies of transition can be determined from the thermoanalytical curves. The heat absorbed by (endothermic events) or released from (exothermic events) the sample during a phase transition can be obtained from the area under the curve of the transition.

Very often a change in the baseline values before and after the phase transition peak are observed. This phenomenon occurs because the heat capacity of a substance in a given phase is generally slightly different from the heat capacity of the same substance in a different phase. Consequently, the true estimation of the baseline during the phase transition, necessary for the determination of the area under the curve, can be difficult (19). Different mathematical techniques of integration are used for the estimation of the enthalpy values (20,24) (Fig. 8).

When hermetically sealed pans are used (constant volume instead of constant pressure), internal energy of transition should be used rather than enthalpy of transition.

## Pharmaceutical Applications

**Drug Purity.** DSC can be used to determine the purity of pharmaceutical drug substances. The method is based on the melting point depression measurement of an impure material (25). The presence of an impurity in a sample depresses the melting temperature of the main component and broadens the melting endotherm (Fig. 9).

The Van't Hoff equation gives the melting point depression as a function of the mole fraction of the impurity:

$$T_0 - T_m = \frac{RT_0^2 \chi_2}{\Delta H_0} \qquad (2)$$

where $T_0 - T_m$ is the melting point depression, $R$ is the gas constant in cal/mol.K, $T_0$ is the melting temperature

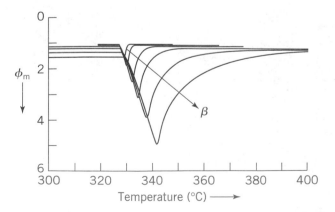

**Figure 6.** Measured curves showing the peak temperature maximum changing with the heating rate $\beta$. *Source:* From Ref. 2.

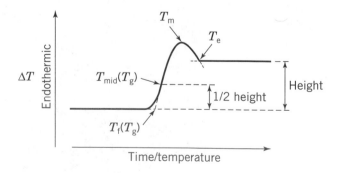

**Figure 7.** Typical transitions of a glass-forming material. $T_f$: extrapolated onset temperature; $T_{mid}$: glass-transition temperature. *Source:* From Ref. 19.

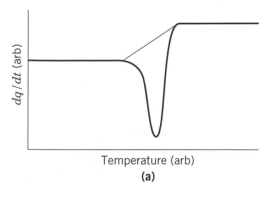

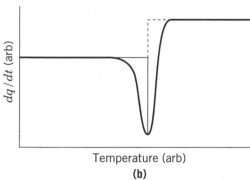

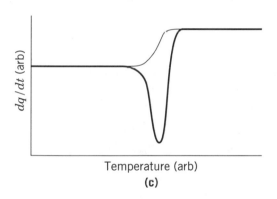

**Figure 8.** Methods of determining peak area by the construction of different baselines. (**a**) Tangential; (**b**) stepped; (**c**) sigmoidal. *Source:* From Ref. 20.

of the pure component in Kelvin, $T_m$ is the melting temperature of the impure material, $\chi_2$ is the mole fraction of the impurity, and $\Delta H_0$ is the enthalpy of fusion of the pure component in cal/mol.

Several assumptions and limitations apply to the calculation of the fraction of impurity present in the sample (27,28). It is assumed that the impurity forms an ideal solution with the main component and that no chemical reaction takes place in the molten state between the impurity and the main component. The melting temperature and the enthalpy of fusion of the pure compound of interest have to be known. The fraction of molten sample at any temperature, $F$, and the sample temperature at equilibrium, $T_S$, are given by the following equations:

$$F = (T_0 - T_m)/(T_0 - T_S) \qquad (3)$$

$$T_S = T_0 - RT_0^2\chi_2(1/F)/\Delta H_0 \qquad (4)$$

When $T_S$ is plotted versus $1/F$, a straight line with a slope of $RT_0^2\chi_2/\Delta H_0$ is obtained and the mole fraction of the impurity can be calculated. Figure 10 gives an example of such a plot for an impure sample of testosterone.

This method should be used only if the amount of impurity is less than about 2% (i.e., dilute solution) (19). Indeed, a significant deviation from linearity of the plots $T_S$ versus $1/F$ is observed for samples containing high levels of impurity.

**Polymorphism and Pseudopolymorphism.** Polymorphism is defined as the tendency for a compound to exist in the solid state with different crystal structures (or polymorphs). Despite their chemical identity, polymorphs generally have different physical characteristics (i.e., melting point, heat of fusion, solubility, X-ray pattern, density, hygroscopicity, dissolution rate, infrared spectrum). Pseudopolymorphs are defined as crystals containing solvent molecules. They can be further differentiated as hydrates when water is the solvent incorporated in the crystal pattern or solvates when other types of solvents (i.e., acetonitrile, methanol, or chloroform) are incorporated. By using various solvents, temperatures, and rates of crystal-

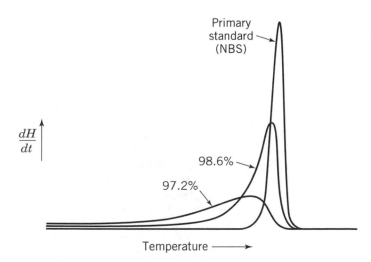

**Figure 9.** Effect of purity on melting endotherm of benzoic acid. *Source:* From Ref. 26.

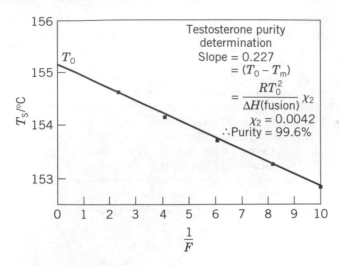

**Figure 10.** Graphical evaluation of testosterone purity. *Source:* From Ref. 26.

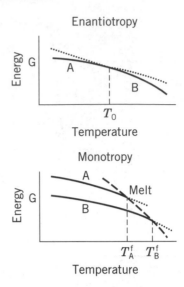

**Figure 11.** Relationship between Gibbs energy G and temperature for two modifications in the case of enantiotropy and monotropy. *Source:* From Ref. 5.

lization, different polymorphs may be obtained. Because different polymorphic forms of the same drug exhibit significant differences in their physical characteristics, therapeutic activity from one form to another may be different. Studying the polymorphism of a drug and the relative stability of the different polymorphs is a critical part of preformulation development.

Polymorphism of drugs has been extensively studied using DSC and other techniques. Numerous reviews have been published on the subject (5,29,30). Studies of the physical stability of polymorphic forms have led to the definition of two different types of polymorphic behavior, enantiotropy and monotropy; the difference is based on the reversibility of the transformation from one form to the other one (4,31). Enantiotropy is defined as the reversible change from one polymorphic form to another (31). Polymorphic changes can occur below the melting point of either polymorph and thus the stability of the different forms is dependent on temperature (4). At temperatures below the transition temperature, the form with the lower melting point is the most stable form, whereas the form with the highest melting temperature is the most stable at temperatures above the transition temperature (32). Monotropy, on the other hand, is defined as an irreversible transition from a metastable form to a stable form. The higher melting form is the most stable throughout the temperature range (31) (Fig. 11).

Differentiation between monotropy and enantiotropy is important for understanding the relative stability of different polymorphic forms of a given compound and can be achieved by DSC experiments (Fig. 12).

Other techniques (calorimetric or not) are useful in the differentiation process between monotropy and enantiotropy. Thermogravimetric analysis and thermomicroscopy are very often used and give complementary information necessary for a complete understanding of the polymorphic system. Determination of the transition temperature is also crucial and can be done by solubility studies based on the fact that at the transition temperature both forms have

the same solubility (33). Heat of solution data give valuable insight to polymorphic transitions.

**Amorphous Compounds: Glass-Transition Temperature.** Amorphous solids are often defined as supercooled liquids. Their molecules are not arranged in an orderly fashion but in a random manner, giving pharmaceutically interesting physicochemical characteristics to these types of solids. Dissolution rates and consequently bioavailability can be tremendously improved upon formation of an amorphous solid rather than the crystalline form of the same drug. This improvement is critical for many poorly soluble drugs. Amorphous pharmaceutical solids can be produced by several means: supercooling of the melt, vapor condensation, precipitation from solution, lyophilization, or milling of crystals (34). Unfortunately, due to their high internal energy, most of the metastable amorphous solids have a tendency to crystallize during manufacturing and storage. Knowledge of the (1) rate of crystallization, (2) glass-transition temperature ($T_g$), and (3) influence of temperature and humidity on the physical stability of amorphous solids is of importance in preformulation studies. The glass-transition temperature corresponds to the temperature where the characteristics of the solid change from the rubbery state (above $T_g$ and below $T_m$) to the glassy state (below $T_g$) (Fig. 13).

Molecular motion of amorphous solids is highly temperature dependent. At temperatures well below the glass-transition temperature, the molecules of an amorphous solid are kinetically frozen, whereas their motion significantly increases as the temperature gets closer to the $T_g$ or even above the $T_g$. Due to the thermodynamic instability of amorphous solids as compared to the crystalline state, spontaneous crystallization is always possible as soon as molecular mobility is above the threshold of nucleation. Molecular mobility, even at temperatures below $T_g$, is usually high enough to allow for nucleation and thus crystal-

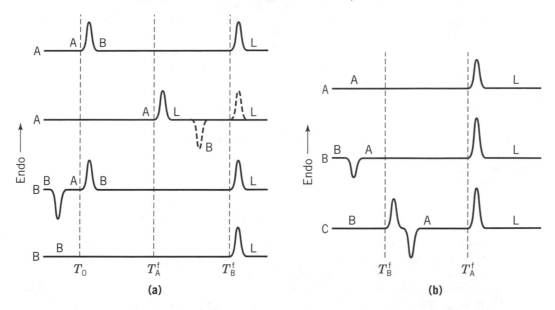

**Figure 12.** DSC curves for (a) enantiotropic and (b) monotropic polymorphism. *Source:* From Ref. 5.

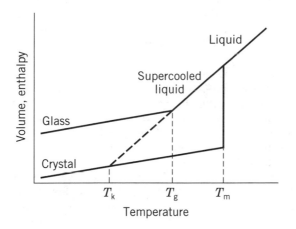

**Figure 13.** Schematic depiction of the variation of enthalpy (or volume) with temperature. *Source:* From Ref. 34.

lization (35). The determination of the glass-transition temperature of an amorphous solid is of primary importance for the formulation and prediction of stability of pharmaceutical solids. The determination of $T_g$ is possible by DSC because a solid above $T_g$ has a higher $C_p$ than a solid below $T_g$ (36). The strength or fragility of amorphous systems is indicated by the amplitude of the change in $C_p$ (36). The molecular mobility of strong glasses (i.e., proteins) is not temperature dependent near $T_g$. Strong glasses are characterized by small changes in heat capacity at $T_g$. On the other hand, fragile glasses are more sensitive to thermal excitation and are characterized by a larger change in heat capacity at $T_g$.

However, the change in heat capacity is often very small and thus difficult to detect with classical DSC. The presence of moisture in the sample is critical, because water can act as a plasticizer and thus reduce the $T_g$ of the sys-

tem (37,38). The glass-transition temperature is also very sensitive to the history of the sample and the heating and cooling rate used for its determination (39). The study of amorphous systems is also possible by the detection of crystallization, upon heating of the sample, as an exothermic peak. The extent of the amorphous character of a sample can be estimated by the value of the enthalpy of crystallization. Comparing the crystallization enthalpy value of a given sample to the enthalpy value of a purely amorphous sample of the same drug, as defined by an independent method (i.e., X-ray diffraction), gives the amorphous content of the sample (40,41).

Due to the recent interest in development of protein formulations in the pharmaceutical field and the search for "ideal" protectants during the lyophilization process, the glass-transition temperature of frozen amorphous solutions ($T'_g$) is often determined using DSC. Determining the $T'_g$ of frozen solutions is critical to optimize the freeze-drying cycle. Indeed, $T'_g$ represents the maximum allowable operating temperature during primary drying. If the product temperature exceeds the glass-transition temperature, the product could collapse due to the significant increase of the viscosity of the sample (42–44).

**Micro-DSC: Protein Applications.** Due to recent improvement of the sensitivity of differential scanning calorimeters, the study of thermal transitions in biological systems has been tremendously expanded. DSC is now routinely used to help predict the stability of proteins in liquid formulations. By determining the effect of formulation parameters (i.e., pH, salt concentration), as well as the effect of additives on the melting temperature of proteins, DSC has become a necessary analytical tool. DSC has also been used to study protein–ligand interactions (45–47). The complexity in interpreting the data obtained with macromolecules resides in the fact that it is difficult to separate the interactions specific to the protein from the solvent interactions (46).

To manifest their biological activity, proteins must be folded in a specific configuration. Covalent and noncovalent intramolecular bonds keep the protein in the folded state. The stability of the native state in aqueous solution is determined by amino acid sequence, temperature, pH, salt concentration, and presence of ligands (48). Due to their polymeric composition, proteins can be subject to physical denaturation without chemical degradation. Upon heating, covalent and noncovalent forces are disrupted and unfolding of the protein follows. Thermal denaturation of the protein can be classified either as reversible or irreversible depending on whether the protein returns to the native conformation upon cooling. In differential scanning experiments, complete reversibility of the unfolding process is often tested by reheating the same sample after cooling; both curves should be superimposable. In the case of a reversible denaturation, only two populations of conformations of the protein are present in solution: the native ($N$) and the unfolded ($U$) populations.

$$N \overset{K}{\leftrightarrow} U \qquad (5)$$

In the case of an irreversible denaturation, the reversibly unfolded population present in solution undergoes further reactions (i.e., aggregation, precipitation, or disulfide cross-linking) to form an irreversibly denatured population ($I$).

$$N \overset{K}{\leftrightarrow} U \overset{k}{\to} I \qquad (6)$$

For reversible denaturation, a two-state model can be used to represent the experimental phenomenon observed by DSC for most small monomeric proteins. The melting temperature ($T_m$) is the temperature at which 50% of the protein is in one of the native conformations and 50% is in one the unfolded conformations (49). Reversible denaturation of proteins can be represented by the following equilibrium expression:

$$K = \frac{[U]}{[N]} \qquad (7)$$

where [$U$] is the concentration of unfolded form, [$N$] is the concentration of native form, and $K$ is the equilibrium constant for the unfolding process (15). Equations 8 and 9 are used to interpret DSC data when the unfolding process is under thermodynamic control (no kinetic limitations):

$$\left(\frac{d \ln K}{dT}\right)_P = \frac{\Delta H_{VH}}{RT^2} \qquad (8)$$

$$\Delta H_{VH} = ART^2 \frac{C_{ex,1/2}}{\Delta h_{cal}} \qquad (9)$$

where $T$ is the absolute temperature at which the process is half completed, $\Delta H_{VH}$ is the Van't Hoff enthalpy, $R$ is the gas constant, $\Delta h_{cal}$ is the calorimetric specific enthalpy, $A$

is a constant with a value of 4.0 (two-state model), and $C_{ex,1/2}$ is the excess specific heat at $T$ (50) (Fig. 14).

Moreover, the calorimetric specific enthalpy ($\Delta h_{cal}$) is related to the calorimetric enthalpy ($\Delta H_{cal}$) by the following equation:

$$\Delta H_{cal} = \Delta h_{cal} M \qquad (10)$$

where M is the molecular weight of the protein. For a strict two-state process, $\Delta H_{HV} = \Delta H_{cal}$. The presence of intermediates during the unfolding process can be suspected when $\Delta H_{HV} < \Delta H_{cal}$. On the other hand, when $\Delta H_{HV} > \Delta H_{cal}$, the presence of intermolecular cooperation (subunits) can be suspected.

For irreversible denaturation, the analysis of the DSC data is still possible but should take into account the kinetics of formation of the irreversibly formed protein states (51–53).

Due to weakness of the forces involved in the folding process, highly sensitive differential scanning calorimeters are used in order to permit the accurate detection of the thermal phenomena of interest (i.e., denaturation or unfolding). These types of instruments are called high sensitive differential scanning calorimeters (HSDSC) or micro differential scanning calorimeters (micro-DSC). Most of the instruments commercially available are designed for these specific types of applications and thus have a limited temperature range ($-20°C$ to $130°C$), as well as a limited scanning rate range ($0.01°C/min$ to $2.5°C/min$). In most of these calorimeters, the reference and cell holder are placed in an adiabatic shield. The volume of the sample holders are usually larger (ca. 1 mL) than in conventional DSC. Due to the fact that the cells are fixed in place, cleaning of the cells between samples can be problematic. Indeed, upon unfolding and especially if the sample precipitates, proteins have a tendency to adsorb to the wall of the cells. Thorough cleaning, using various acidic and/or organic solutions, followed by extensive rinsing, should be performed. To assess the efficiency of the cleaning procedure, a baseline check using the same buffer solution in both the reference and sample cells should be performed prior to

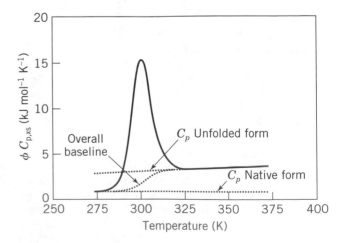

**Figure 14.** Typical melting thermogram for protein. *Source:* From Ref. 15.

any other experiments. In a typical experiment, the reference cell usually contains the placebo formulation (every component of the formulation but the macromolecule). The scope of applications for HSDSC is not limited to proteins (54–58). Numerous studies involving lipids and biological membranes (59,60) or polynucleotides (61) have been published.

## MODULATED DIFFERENTIAL SCANNING CALORIMETRY

As mentioned in the introduction, a novel extension of classical DSC was developed a few years ago. Its application to pharmaceutical science led to new insights on the determination of complex thermal parameters in multicomponent formulations. With classical DSC, depending on the scanning rate, either resolution or heat flow sensitivity are emphasized, but at the expense of one another. MDSC is a technique that allows for simultaneous resolution of overlapping thermal phenomena and provides good sensitivity for the detection of small thermal events. Indeed, by applying a modulated temperature program, low underlying heating rates are associated with a large instantaneous heating rate due to a large modulation amplitude (20). This novel approach permits the deconvolution of thermal phenomena into two different signals, reversible and irreversible. The reversibility of any thermal event is defined within the time scale of the temperature modulation. MDSC can be easily achieved using both types of instruments: power compensation and heat flux DSC (62). Two different instrument designs are commercially available: a Modulated DSC™ (TA instrument) pioneered by Reading (62), and a Dynamic DSC™ (Perkin-Elmer instrument) developed by Schawe (63,64). With the MDSC™ model, a sinusoidal ripple is overlaid on the standard linear temperature ramp (62), whereas with the DDSC™ model, the temperature program is composed of repeating units consisting of an isotherm followed by an isotherm or a heat segment followed by a cool segment (65).

With the TA instrument, a small sinusoidal term is added to the linear temperature program so the temperature can be separated into (1) a modulated temperature component and (2) an underlying linear component. The sinusoidal temperature profile can be described by the following equation (62):

$$Temp = T_0 + bt + B \sin(\omega t) \qquad (11)$$

where $T_0$ is the starting temperature, $b$ is the underlying heating rate, and $B$ and $\omega$ are the amplitude and the frequency of the sinusoidal function, respectively. The total heat flow out of the sample can be expressed as:

$$dQ/dt = C_p(b + B\omega \cos(\omega t)) + f'(t, T) + C \sin(\omega t) \qquad (12)$$

where $Q$ is the amount of heat absorbed by the sample, $f'(t, T)$ is the average underlying kinetic component of the calorimetric response, $C$ is the amplitude of the kinetic response to the sine wave modulation, and $(b + B\omega \cos(\omega t))$ corresponds to $dT/dt$ (linear temperature program). The

heat flow can be separated into (1) a heating rate–dependent heat flow that is influenced by heat capacity, and (2) an absolute temperature–dependent heat flow (or kinetic) which is influenced by the temperature at a given time. These two types of heat flow can be separated due to their different responses to the sinusoidal temperature program and the underlying temperature ramp. Indeed, the heat capacity contribution is present at all times (reversible and irreversible processes) whereas the kinetic contribution is predominant during irreversible processes. Discrete Fourier transform is used to separate both types of heat flow. The reversing heat flow component is calculated, and the nonreversing heat flow component is obtained by subtracting the reversing heat flow component from the total heat flow (20) (Fig. 15). The determination of heat capacity values is possible through the reversing component of the response, whereas the determination of enthalpies of crystallization, relaxation, and evaporation is possible through the nonreversing component of the response. Depending on the frequency and amplitude chosen for the temperature program, glass-transition temperature and melting are phenomena that can appear in either components (reversing and nonreversing) (66).

To avoid supercooling of the sample heating rate, frequency of modulation, and amplitude of modulation should be chosen in a such way that the overall heating rate is at all times positive. Moreover, the frequency of the sine function should be chosen such that at least six modulations are completed over any single thermal event, so that the frequency is not too slow in comparison to the rate at which the sample undergoes a thermal transition (62).

With the Perkin-Elmer instrument, the temperature program is composed of repeating units. The repeating units can either be composed of (1) an isotherm segment followed by a scan segment (Iso-Scan mode) or (2) a heat segment followed by a cool segment (Heat-Cool mode). The Iso-Scan mode should be used for the determination of thermal events in which supercooling of the sample could be detrimental to the quality of data obtained (enthalpy of crystallization). On the other hand, the Heat-Cool mode should be used for the determination of reversible thermal

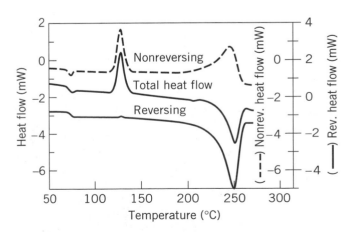

**Figure 15.** Thermograms of PET showing the total, reversing, and nonreversing heat flow signals. *Source:* From Ref. 20.

events (i.e., glass-transition temperature) that have to be separated from overlapping nonreversible events (i.e., enthalpy of relaxation). After correcting for the phase lag of time-dependent processes, the data analysis software separates the storage-specific heat (storage $C_p$) from the loss-specific heat (loss $C_p$). The storage-specific heat characterizes the molecular motion of the sample, whereas the loss-specific heat characterizes the dissipative properties of the sample (65) (Fig. 16).

MDSC is used in polymer sciences to determine glass-transition temperatures. Accurate determination of $T_g$ is achieved because the specific heat change (reversible thermal event) is separated from the enthalpy of relaxation (nonreversible thermal event). The change in heat capacity appears sigmoidal in shape (due to the absence of interference from relaxation phenomena) and is independent of the thermal history of the sample (66–70).

## HEAT CONDUCTION CALORIMETRY

Heat conduction calorimetry has been used in a wide range of applications including the stability of drugs in the solid state, the study of water adsorption by solids, the study of protein–ligand interactions, and various biological reactions (71–88). Heat conduction calorimeters are isothermal microcalorimeters and thus allow for the detection of reactions at ambient temperature rather than requiring that a reaction be studied at high temperatures (as is the case with DSC). Most of the microcalorimeters are of the twin design and are comprised of a reaction cell and a reference cell. The reference cell is usually filled with an inert material with preferably the same characteristics as the sample with regards to heat capacity and heat conductance. The sample cells can be of different designs depending on the type of experiment. For batch-type experiments, static cells are usually used. For very slow processes, such as biological processes, sealed ampoules are usually used. For titration experiments, a specific type of cell has to be used to allow for stirring and injection of one of the reactants (89). The heat generated or absorbed by the reaction taking place in the sample cell is usually detected by a thermopile

placed between the cell and a heat sink. The heat sink (thermostated water bath or a metal block) allows for thermal equilibrium. After an equilibration period, the temperatures of the heat sink and the sample cells are the same. As the reaction inside the cell takes place, heat will be conducted from the cell to the heat sink (for reactions evolving heat) or from the heat sink to the sample cell (for reactions absorbing heat). Because no reaction is taking place in the reference cell, a differential heat flow measurement is performed (90–92). However, because the heat generated by many of the processes studied by heat flux calorimeters is small, very sensitive detectors have to be used. The detection limit of the most sensitive calorimeters is in the nanowatt range. The working temperature of the thermostat ranges from $-40°C$ to $200°C$ with a precision of $\pm0.0001°C$. One of the disadvantages of heat conduction calorimeters is their large time constant. For multistep titration experiments, several hours might be needed to complete the experiment. However, a dynamic correction method can be used to shorten the experiment time (93,94). Calibration is usually performed by the use of electrical heaters. Electrical calibration is accurate but may not be representative of the reaction taking place during a biological or chemical reaction. Chemical calibration can be performed to better mimic the reaction taking place inside the cell. Different types of chemical calibration can be used depending on the type of reaction vessel (95). When the calorimeter is used below room temperature, dry nitrogen is flushed through the system to avoid condensation.

Among the various applications of heat conduction calorimetry, the study of drug degradation has been one the most interesting to pharmaceutical science. Kinetics of drug decomposition are usually estimated by following the rate of disappearance of the parent compound or the rate of appearance of decomposition products. However, for compounds relatively stable at room temperature, accurate determination of the rate of decomposition can be difficult due to insufficient amounts of degradants in the sample. The study of the rates of decomposition of drugs at elevated temperatures and the extrapolation of the rates at lower temperatures are possible using the Arrhenius relationship based on the following assumptions: (1) the mechanism of drug degradation is independent of temperature and (2) the energy of activation used in the Arrhenius equation is also independent of temperature (25). To assess the validity of these two assumptions, stability data at room temperature are usually necessary. However, the formation of sufficient amounts of degradation products (above their detection limits) might be a very slow process at room temperature. Moreover, no data are obtained characterizing the early stages of drug degradation due to insufficient amounts of degradants present in the sample (72,73). On the other hand, due to the production or consumption of heat associated with most chemical reactions, heat conduction calorimetry can be used to follow the rate of heat change in the sample as a function of time at a given temperature. Kinetics of chemical reactions can be subsequently deduced. By performing measurements at different temperatures, the activation energy can be determined.

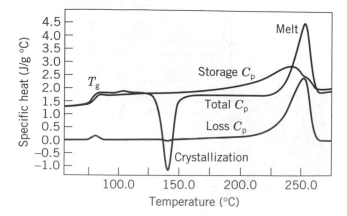

**Figure 16.** Thermograms of PET showing the storage, loss, and total specific heat curves. *Source:* From Ref. 65.

## THERMOGRAVIMETRY

TGA has been used in pharmaceutical research and development primarily to characterize pseudopolymorphism of drugs and to detect the presence of residual solvent in drug substances (i.e., from crystallization) or excipients. Decomposition of drugs can also be followed if a mass change accompanies drug degradation. Thermogravimetric analysis is a thermal technique in which the weight of a sample is monitored as a function of temperature. An electronic microbalance, which is thermally isolated from the furnace (to avoid interference), is used. Only the sample holder is surrounded by the furnace (Fig. 17). Most commercially available balances are null point balances; the sample is maintained in the same zone of the furnace as the mass of the sample changes. The balance is zeroed at its null position and deviations from this position are detected and monitored as a function of temperature (96). Electronic or optical methods are used to detect the deviation of the balance from its null position as the mass of the sample changes. Electronic detection uses the change in the capacity of a condenser, whereas optical detection uses the interception of a light beam to sense the change of the balance from its null position (97). Usually the restoring mechanism to the null position is electromagnetic. The furnace, which is usually a resistive heater, can be one of several different designs. The furnace can be inside, part of, or external to the housing (1). Because heating and weighing of the sample have to occur simultaneously, the sample cannot be in direct contact with the furnace. Thus, problems of heat transfer may arise (96). Temperature gradients can occur not only between the furnace and the sample but also within the sample. The latter can be improved by using a small sample size (generally in the order of a few milligrams). To avoid interference from the moisture content of air, the sample chamber should be purged using a dry inert gas such as nitrogen or helium at a flow rate of 15 to 25 mL/min (3).

Most current instruments permit a wide range of heating rates (from 0.1°C/min to 1,000°C/min). The most sensitive thermogravimetric analyzers are capable of detecting mass changes on the order of 0.1 $\mu$g.

The thermogravimetric data can be presented by plotting the (1) mass change of the sample (%) versus temperature or (2) rate of mass change of the sample versus temperature (differential thermogravimetry [DTG]) (Fig. 18).

A rapid initial mass loss is usually characteristic of a desorption phenomenon, whereas a mass loss after a flat portion is more characteristic of decomposition of the sample (i.e., dehydration). Depending on the stoichiometry of the reaction and on the presence of intermediates, decomposition can occur in one or several stages (1). On the other hand, a gain in mass is characteristic of a reaction of the sample with its surroundings (i.e., oxidation). Interpreting the data obtained by thermogravimetric analysis can be difficult; comparison of the data obtained with DSC and TGA for the same sample facilitates the data analysis (Fig. 19).

## ISOPERIBOL CALORIMETRY

Isoperibol calorimetry is used in a wide range of pharmaceutical applications in which solid–liquid or liquid–liquid interactions can be studied. Typical applications include measurements of solubility, heat and rate of dissolution, polymorphism, ligand interactions and binding, metal ion chelation and analysis as well as equilibrium and rate constants. The temperature of a reaction vessel is monitored as a function of time during mixing of two reactants. The instrument is comprised of (1) a reaction vessel (usually silvered inside and outside) containing the solvent, (2) a thermistor used as the temperature measuring device, (3) an electrical heater for calibrating and warming the solvent, (4) a stirrer allowing for continuous mixing, (5) a batch assembly or titration device containing the second

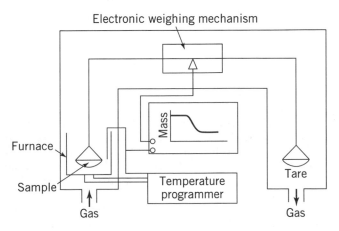

**Figure 17.** Thermobalance. *Source:* From Ref. 1.

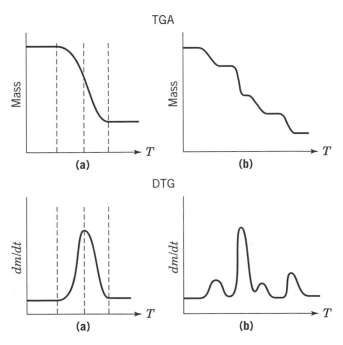

**Figure 18.** Comparison of TGA and DTG curves. *Source:* From Ref. 1.

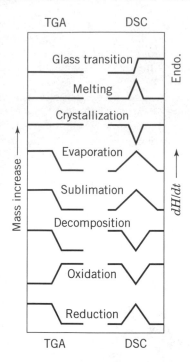

**Figure 19.** Comparison of the schematic TGA and DSC curves recorded for a variety of physicochemical processes. *Source:* From Ref. 3.

component (liquid or solid), and (6) a thermostated water bath acting as the constant surroundings (98) (Fig. 20).

The word *isoperibol* is used to describe that during the reaction phase the temperature of the surroundings stays constant (98,99). As the temperature inside the reaction vessel changes due to the reaction, heat exchange between the reaction vessel and the surroundings occurs. A correction factor must be applied to account for this heat exchange (98,100).

Because the value of the energy equivalent is necessary for the calculation of the heat of reaction taking place in the reaction vessel, determination of the energy equivalent of the system is performed before and/or after experiments. The energy equivalent value is obtained by electrical calibration. After submersion of the reaction vessel into the water bath and an initial equilibration period, a known current is passed through the heater for a known period of time. The increase in temperature of the solvent contained in the reaction vessel is recorded and corrected for heat leak with the surroundings. Subsequently, the energy equivalent of the system can be calculated using the following equation:

$$E_q = \frac{IVt}{\Delta T_1} \qquad (13)$$

where $E_q$ is the energy equivalent of the system, $I$ is the current, $V$ is the voltage, $t$ is the period of time during which the heater was on, and $\Delta T_1$ is the corrected increase in temperature.

After calculating the energy equivalent of the system, the heater is turned on to bring the temperature of the solvent to the same temperature as the water bath (surroundings). When the temperature differential between the reaction vessel and the surroundings is equal to zero, the heater is turned off and the experimental reaction procedure can be started.

The temperature inside the reaction vessel is recorded before the reaction occurs to give the slope of the initial (prereaction or lead) phase. Usually a slight positive slope is obtained due to heating from the stirring system. The reaction (mixing between the two reactants) is initiated, and the change in temperature due to the interaction between the two reactants is recorded. The temperature of the reaction vessel is recorded after the reaction period to give the slope of the final (postreaction or trail) phase. By extrapolating the slopes from the initial and final phases and after correcting for heat exchange with the surroundings, the corrected change in temperature due to the reaction is calculated. Knowing the energy equivalent of the system and the corrected change in temperature due to the reaction, the enthalpy of reaction can be determined using the following equation:

$$\Delta H_{reaction} = E_q \times \Delta T_2 \qquad (14)$$

where $\Delta H_{reaction}$ is the enthalpy of reaction, $E_q$ is the energy equivalent of the system, and $\Delta T_2$ is the corrected temperature change.

There are two different ways of initiating the reaction: the batch mode (or breaking ampoule mode) and the continuous titration mode. With the batch mode, the temperature of the reaction vessel is followed as a function of time. One reactant is sealed in a glass ampoule or contained in a stainless steel device, with glass seals; the device is secured in the reaction vessel containing the solvent (second reactant). At the beginning of the reaction phase, the ampoule or the two end pieces of glass are broken, releasing the sample into the reaction vessel and allowing for mixing of the sample with the solvent. The sample contained inside the glass ampoule or the stainless steel device can be a solid or a liquid. Batch solution calorimetry has been used to determine the enthalpy of solution, mixing, or hydration of numerous pharmaceutical compounds (76,101–104). Batch solution calorimetry has also been used to characterize polymorphism of drugs. Different values of enthalpy of solution are usually obtained for different polymorphic forms of a compound in a particular solvent. Enthalpy values for one given polymorphic form are usually different from one solvent to another. However, the difference in enthalpy values from one polymorphic form to another should be independent of the solvent (92). Batch solution calorimetry has also been used to determine the degree of crystallinity of a sample by comparison of the values obtained for a pure amorphous sample and a pure crystalline sample of the same compound (105). The enthalpy of solution for the pure crystalline sample corresponds to the most endothermic (or least exothermic) value, whereas the enthalpy of solution for the pure amorphous sample will be the most exothermic (or least endothermic) value. Partially crystalline samples are characterized by an enthalpy of solution between these two extreme values. A linear relationship between the en-

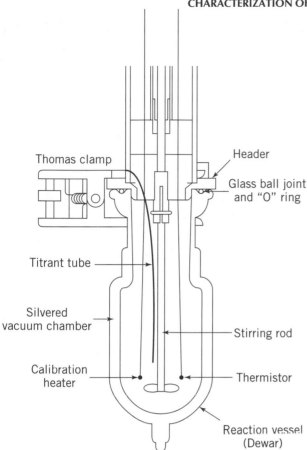

**Figure 20.** Isoperibol calorimeter reaction vessel. *Source:* Courtesy of Hart Scientific.

essary battery of tests for the development of new drug entities. Classical DSC is routinely used. However, (1) the accuracy and the ease with which thermal events such as $T_g$ are detectable and (2) the possibility of resolving overlapped complex thermal events using MDSC increases the interest for this new technique. Microcalorimetry techniques (isothermal microcalorimeters, micro-DSC, and isoperibol calorimeters) are not yet routinely used but are increasingly of interest in pharmaceutical science because of the valuable information they provide.

thalpy of solution and the degree of crystallinity is expected if (1) no reactions occur between the solvent and the sample and (2) the dissolution processes of the amorphous and crystalline samples are fast enough in the particular solvent to allow precise measurements.

With the titration mode, the temperature of the reaction vessel is followed as a function of the amount of titrant added. One of the reactants (titrant), placed in a thermostated motor-driven syringe, is added to the reaction vessel containing the other component through a continuous injection at a constant rate. The titration mode has been used to investigate the interactions between two components in solution (i.e., titration, complexation) (106–108). Equilibrium constants and enthalpy and entropy changes can be calculated using this method (109–111). Equilibrium constants can be obtained with precision only if they are relatively small (log K < 4) and if the enthalpy of reaction is large so that changes in temperature can be accurately observed. The free energy and entropy of binding can be calculated from the binding constant and the measured enthalpy change.

## CONCLUSIONS

Thermal analysis is successfully used in preformulation and formulation of pharmaceuticals and is part of the nec-

## BIBLIOGRAPHY

1. M.E. Brown, *Introduction to Thermal Analysis: Techniques and Applications*, Chapman & Hall, London, 1988.
2. G.W.H. Höhne, W. Hemminger, and H.J. Flammersheim, *Differential Scanning Calorimetry: An Introduction for Practitioners*, Springer-Verlag, Berlin, 1996.
3. T. Hatakeyama and F.X. Quinn, *Thermal Analysis: Fundamentals and Applications to Polymer Science*, Wiley, New York, 1994.
4. R.J. Behme, D. Brooke, R.F. Farney, and T.T. Kensler, *J. Pharm. Sci.* **74**(10), 1041–1046 (1985).
5. D. Giron, *Thermochim. Acta* **248**, 1–59 (1995).
6. R.K. Khankari, D. Law, and D.J.W. Grant, *Int. J. Pharm.* **82**, 117–127 (1992).
7. M.M.J. Lowes, M.R. Caira, A.P. Lotter, and J.G. Van Der Watt, *J. Pharm. Sci.* **76**(9), 744–752 (1987).
8. N.T. Nguyen, S. Ghosh, L.A. Gatlin, and D.J.W. Grant, *J. Pharm. Sci.* **83**(8), 1116–1123 (1994).
9. I. Hatta, H. Ichikawa, and M. Todoki, *Thermochim. Acta* **267**, 83–94 (1995).
10. F. Balestrieri et al., *Thermochim. Acta* **285**, 337–345 (1996).
11. S. Venkataram, M. Khohlokwane, and S.H. Wallis, *Drug Dev. Ind. Pharm.* **21**(7), 847–855 (1995).
12. J.F. Brandts, C.Q. Hu, and L.N. Lin, *Biochemistry* **28**, 8588–8596 (1989).
13. J.F. Brandts and L.N. Lin, *Biochemistry* **29**, 6927–6940 (1990).
14. G. Barone et al., *Pure Appl. Chem.* **67**, 1867–1872 (1995).
15. B. Chowdhry and S. Leharne, *J. Chem. Educ.* **74**(2), 236–241 (1997).
16. J. Lu and K.B. Hall, *Biophys. Chem.* **64**, 111–119 (1997).
17. H. Tanojo, J.A. Bowstra, H.E. Junginger, and H.E. Boddé, *Pharm. Res.* **11**(11), 1610–1616 (1994).
18. A. Martini, M. Crivellente, and R. De Ponti, *Eur. J. Pharma. Biopharm.* **42**(1), 67–73 (1996).
19. J.L. Ford and P. Timmins, *Pharmaceutical Thermal Analysis: Techniques and Applications*, Ellis Horwood Ser. Pharm. Technol. Wiley, New York, 1989.
20. N.J. Coleman and D.Q.M. Craig, *Int. J. Pharm.* **135**, 13–29 (1996).
21. F. Franks, *Cryo-Letters* **11**, 93–110 (1990).
22. L.M. Her and S.L. Nail, *Pharm. Res.* **11**(1), 54–59 (1994).
23. O. Kubaschewski and E.L. Evans, *Metallurgical Thermochemistry*, 4th ed., Pergamon, Oxford, 1958.
24. R.N. Goldberg and E.J. Prosen, *Thermochim. Acta* **6**, 1–11 (1973).

25. S. Lindenbaum, in J. Swarbrick and J. Boylan, eds., *Encyclopedia of Pharmaceutical Technology*, vol. 2, Dekker, New York, 1989, pp. 233–250.

26. J.L. McNaughton and C.T. Mortimer, *International Review of Science: Physical Chemistry Series 2*, Vol. 10, Butterworths, London, 1975, pp. 1–44.

27. E.E. Marti, *Thermochim. Acta* 5(2), 173–220 (1972).

28. A.A. Van Dooren and B.W. Muller, *Int. J. Pharm.* 20, 217–233 (1984).

29. L. Borka and J.K. Haleblian, *Acta Pharm. Jugosl.* 40(1–2), 71–94 (1990).

30. C. Rodriguez and D.E. Bugay, *J. Pharm. Sci.* 86, 263–266 (1997).

31. M. Kuhnert-Brandstatter, *Thermomicroscopy in the Analysis of Pharmaceuticals*, Pergamon, New York, 1971.

32. A.G. Mitchell, *J. Pharm. Pharmacol.* 37, 601–604 (1985).

33. W.I. Higuchi, P.K. Lau, T. Higuchi, and J.W. Shell, *J. Pharm. Sci.* 52, 150–153 (1963).

34. B.C. Hancock and G. Zografi, *J. Pharm. Sci.* 86(1), 1–12 (1997).

35. M. Yoshioka, B.C. Hancock, and G. Zografi, *J. Pharm. Sci.* 83(12), 1700–1705 (1994).

36. C.A. Angell, *Science* 267, 1924–1933 (1995).

37. J.T. Carstensen and K.V. Scoik, *Pharm. Res.* 7(12), 1278–1281 (1990).

38. B.C. Hancock and G. Zografi, *Pharm. Res.* 11(4), 471–477 (1994).

39. C.T. Moynihan, A.J. Easteal, J. Wilder, and J. Tucker, *J. Phys. Chem.* 78(26), 2673–2677 (1974).

40. M.L. Shively and S. Myers, *Pharm. Res.* 10, 1071–1075 (1993).

41. Y. Roos and M. Karel, *J. Food Sci.* 57(3), 775–777 (1992).

42. S.L. Nail and L.A. Gatlin, in K.E. Avis, H.A. Lieberman, and L. Lachman, eds., *Pharmaceutical Dosage Forms: Parenteral Medications*, vol. 2, Dekker, New York, 1993, pp. 163–233.

43. M.J. Pikal and S. Shah, *Int. J. Pharm.* 62, 165–186 (1990).

44. L.M. Her and S.L. Nail, *Pharm. Res.* 11(1), 54–59 (1994).

45. J.F. Brandts and L.N. Lin, *Biochemistry* 29, 6927–6940 (1990).

46. P.R. Connelly, *Curr. Opin. Biotechnol.* 5, 381–388 (1994).

47. G. Barone et al., *Pure Appl. Chem.* 67(11), 1867–1872 (1995).

48. B.A. Shirley, in T.J. Ahern and M.C. Manning, eds., *Stability of Protein Pharmaceuticals. Part A: Chemical and Physical Pathways of Protein Degradation*, Plenum, New York, 1992. Chapter 6.

49. D.B. Volkin and C.R. Middaugh, in T.J. Ahern and M.C. Manning, eds., *Stability of Protein Pharmaceuticals. Part A: Chemical and Physical Pathways of Protein Degradation*, Plenum, New York, 1992, Chapter 8.

50. B.Z. Chowdhry and S. Cole, *Trends Biotechnol.* 167, 11–18 (1989).

51. J.M. Sánchez-Ruiz, J.L. López-Lacomba, M. Cortijo, and P.L. Mateo, *Biochemistry* 27, 1648–1652 (1988).

52. E. Freire, W.W. van Osdol, O.L. Mayorga, and J.M. Ruiz, *Annu. Rev. Biophys. Biophys. Chem.* 19, 159–188 (1990).

53. M.L. Galisteo et al., *Thermochim. Acta* 199, 147–157 (1992).

54. P.L. Privalov, *Adv. Protein Chem.* 33, 167–241 (1979).

55. P.L. Privalov, *Adv. Protein Chem.* 35, 1–104 (1982).

56. G. Castronuovo, *Thermochim. Acta* 193, 363–390 (1991).

57. A. Cooper and C.M. Johnson, *Methods Mol. Biol.* 22, 109–124 (1994).

58. A. Cooper and C.M. Johnson, *Methods Mol. Biol.* 22, 125–136 (1994).

59. A. Blume, *Thermochim. Acta* 193(12), 299–347 (1991).

60. R.M. Epand and R.F. Epand, *Biophys. J.* 66(5), 1450–1456 (1994).

61. K.J. Breslauer, E. Freire, and M. Straume, *Methods Enzymol.* 211, 533–567 (1992).

62. M. Reading, A. Luget, and R. Wilson, *Thermochim. Acta* 238, 295–307 (1994).

63. J.E.K. Schawe, *Thermochim. Acta* 261(9), 183–194 (1995).

64. J.E.K. Schawe, *Thermochim. Acta* 260(8), 1–16 (1995).

65. M.P. DiVito, R.B. Cassel, M. Margulies, and S. Goodkowsky, *Am. Lab.* 27(12), 28–34 (1995).

66. P.S. Gill, R. Sauerbrunn, and M. Reading, *J. Therm. Anal.* 40, 931–939 (1993).

67. D.J. Hourston, M. Song, H.M. Pollock, and A. Hammiche, *J. Therm. Anal.* 49, 209–218 (1997).

68. A. Boller, C. Schick, and B. Wunderlich, *Thermochim. Acta* 266, 97–11 (1995).

69. C. Tomasi, P. Mustarelli, N.A. Hawkins, and V. Hill, *Thermochim. Acta* 278, 9–18 (1996).

70. J.M.E. Sarciaux and M.J. Hageman, *J. Pharm. Sci.* 86(3), 365–371 (1997).

71. L.D. Hansen et al., *Pharm. Res.* 6(1), 20–27 (1989).

72. L.D. Hansen, D.L. Eatough, and E.A. Lewis, *Can. J. Chem.* 68, 2111–2114 (1990).

73. L.D. Hansen, *Pharm. Technol.* 20, 64–74 (1996).

74. M.J. Pikal and K.H. Dellerman, *Int. J. Pharm.* 50, 233–252 (1989).

75. R. Oliyai and S. Lindenbaum, *Int. J. Pharm.* 73, 33–36 (1991).

76. K.C. Thompson, J.P. Draper, M.J. Kaufman, and G.S. Brenner, *Pharm. Res.* 11(9), 1362–1365 (1994).

77. G. Buckton, P. Darcy, and A.J. Mackellar, *Int. J. Pharm.* 117, 253–256 (1995).

78. Y. Aso, S. Yoshioka, T. Otsuka, and S. Kojima, *Chem. Pharm. Bull.* 43(2), 300–303 (1995).

79. T. Sebhatu, M. Angberg, and C. Ahlneck, *Int. J. Pharm.* 104, 135–144 (1994).

80. L.E. Briggner, G. Buckton, K. Bystrom, and P. Darcy, *Int. J. Pharm.* 105, 125–135 (1994).

81. G. Buckton, P. Darcy, D. Greenleaf, and P. Holbrook, *Int. J. Pharm.* 116, 113–118 (1995).

82. M. Angberg, C. Nystrom, and S. Castensson, *Acta Pharm. Suec.* 25, 307–320 (1988).

83. M. Angberg, C. Nystrom, and S. Castensson, *Int. J. Pharm.* 61, 67–77 (1990).

84. L.D. Hansen, J.W. Crawford, D.R. Keiser, and R.W. Wood, *Int. J. Pharm.* 135, 31–42 (1996).

85. M. Pudipeddi, T.D. Sokoloski, S.P. Duddu, and J.T. Carstensen, *J. Pharm. Sci.* 85(4), 381–386 (1995).

86. A. Chen and I. Wadsö, *J. Biochem. Biophys. Methods* 6, 307–316 (1982).

87. I. Wadsö, *Thermochim. Acta* 267, 45–59 (1995).

88. M.G. Nordmark et al., *J. Biochem. Biophys. Methods* 10, 187–202 (1984).

89. I. Wadsö, *Chem. Soc. Rev.* 26(2), 79–86 (1997).

90. I. Wadsö, *Thermochim. Acta* 294, 1–11 (1997).

91. G. Buckton and A.E. Beezer, *Int. J. Pharm.* 72, 181–191 (1991).

92. M.J. Koenigbauer, *Pharm. Res.* **11**(6), 777–783 (1994).

93. M. Bastòs, S. Hägg, P. Lönnbro, and I. Wadsö, *J. Biochem. Biophys. Methods* **23**, 255–258 (1991).

94. P. Bäckman et al., *J. Biochem. Biophys. Methods* **28**, 85–100 (1994).

95. L.E. Briggner and I. Wadsö, *J. Biochem. Biophys. Methods* **22**, 101–118 (1991).

96. W. Hemminger and G. Höhne, *Calorimetry: Fundamentals and Practice*, Verlag Chemie, Weinheim, 1984.

97. E.L. Charsley and S.B. Warrington, *Thermal Analysis: Techniques and Applications*, Royal Chemistry Society, London, 1992.

98. K.N. Marsh and P.A.G. O'Hare, *Solution Calorimetry, Experimental Thermodynamics*, vol. 4, Blackwell, Oxford, 1994.

99. L.D. Hansen and D.J. Eatough, *Thermochim. Acta* **70**, 257–268 (1983).

100. K. Grime, *Analytical Solution Calorimetry*, Wiley, New York, 1985.

101. M.J. Pikal, A.L. Lukes, J.E. Lang, and K. Gaines, *J. Pharm. Sci.* **67**, 767–773 (1978).

102. D.J.W. Grant and P. York, *Int. J. Pharm.* **28**, 103–112 (1986).

103. D. Gao and J.H. Rytting, *Int. J. Pharm.* **151**, 183–192 (1997).

104. S. Lindenbaum and S.E. McGraw, *Pharm. Manuf.* **2**, 26–30 (1985).

105. J.K. Guillory and D.M. Erb, *Pharm. Manuf.* **9**, 29–33 (1985).

106. R.M. Izatt, E.H. Redd, and J.J. Christensen, *Thermochim. Acta* **64**, 355–372 (1983).

107. P.R. Connelly, R. Varadarajan, J.M. Sturtevant, and F.M. Richards, *Biochemistry* **29**, 6108–6114 (1990).

108. W. Tong, J.L. Lach, T. Chin, and J.K. Guillory, *Pharm. Res.* **8**(7), 951–957 (1991).

109. J.J. Christensen, J. Ruckman, D.J. Eatough, and R.M. Izatt, *Thermochim. Acta* **3**, 203–218 (1972).

110. D.J. Eatough, J.J. Christensen, and R.M. Izatt, *Thermochim. Acta* **3**, 219–232 (1972).

111. D.J. Eatough, R.M. Izatt, and J.J. Christensen, *Thermochim. Acta* **3**, 233–246 (1972).

# CHARACTERIZATION OF DELIVERY SYSTEMS, GEL PERMEATION CHROMATOGRAPHY

CAMILLA A. SANTOS
Brown University
Providence, Rhode Island

## KEY WORDS

Cross-linked gels

Molecular weight

Polymer characterization

Polystyrenedivinylbenzene

Size exclusion chromatography

Gel permeation chromatography (GPC) is a form of size-exclusion chromatography used to separate molecules according to their size in solution and is a key element of polymer characterization in today's environment. It is used to quickly determine the molecular weight distribution of polymers, yielding information on the macromolecular nature of the polymer being studied. GPC has been in use since the late 1950s, when water-soluble polymers were separated according to size using cross-linked dextran gels, and commercial use of GPC began in the 1960s when rigid gels were developed for use with organic solvents (1). In GPC, molecular weight averages are determined by effecting size separation of a low concentration of polymer in organic solvent as the dilute polymer solution passes through a column packed with microporous gel. Size exclusion can also be effected in an aqueous mobile phase, and this is typically termed gel filtration chromatography. This article will concern itself with GPC, the size-exclusion chromatography utilizing an organic mobile phase.

The GPC system is composed of a solvent delivery system capable of delivering a constant volume of solvent, an injection port for injection of a small amount of a dilute polymer solution without disturbance of solvent flow, a column or series of columns packed with microporous gel, a detector, and a data system to automatically process data. The columns of a GPC system are typically packed with cross-linked polystyrenedivinylbenzene gel of a particle size range of 5–10 $\mu$m and a pore size range of 0.5–$10^5$ nm (2). Originally, column length was on the order of 4 feet, and the polystyrenedivinylbenzene gel particle size was approximately 40 $\mu$m or larger (1). Owing to improvements in the gels being used, column length is now typically 30–60 cm with a diameter of 7.5 cm (2). Columns can also be packed with other cross-linked polymers or porous glasses and silica, which are typically used in aqueous phase systems (1,3). Within each particle are numerous pores. As the dilute polymer solution passes through the column, it is separated by size because the low-molecular-weight chains of the polymer are able to migrate into and out of the gel particle pores, and the larger chains are forced to flow relatively unhindered through the interstices in the column, their larger size prohibiting diffusion into the pores of the gel particles. Thus, large molecular weights of a dilute polymer solution elute first, and smaller-molecular-weight chains elute much later. A mixed range of pore sizes in the packed column gel allows effective separation of a molecular weight distribution of $<100 - 4 \times 10^7$ (2).

The system should have the capability of operating at both ambient and high temperatures and must contain a device that can maintain a constant temperature along the length of the columns. The capability of operating at elevated temperatures is a function primarily of the packed gel. Effective size-exclusion separation can be achieved in shorter times at elevated temperatures as diffusion rates increase with increasing temperatures. Current GPC systems achieve separations on the order of minutes, making this characterization technique very popular. Typically, the organic solvents used in GPC system were nonpolar, but advances in the stability of the cross-linked gels allow some systems to now operate with more polar solvents.

There are many types of detectors used in GPC systems. The detector must be capable of continuous monitoring of solute elution and can be either a bulk property or solute property detector. Bulk property detectors measure a change in the bulk property of the eluting solvent, such as the refractive index of the solvent, which will change with the presence of a solute. Solute property detectors detect changes in the physical property of the solvent (occurring when solute is present), such as the ultraviolet absorption of the solvent. The limiting factor in any detector is the signal-to-noise ratio, and this must be considered when detecting minor concentrations of sample (this is not usually a problem in GPC as adequate sample mass is normally available) (3). Examples of common detectors are the refractive index detector, the ultraviolet detector, the viscometric detector, and the low-angle light-scattering detector.

The detector monitors the concentration of the eluting polymer. The molecular weight distribution is calculated by measuring the elution volume or peak retention volume, $V_R$. This is a function of the ability of the polymer in solution to diffuse into the pores of the gel. The retention volume is the volume of solution eluted as measured from the beginning of the injection into the column to the peak of the fraction of the particular molecular weight eluted as determined by the chromatogram. High-molecular-weight solutes will elute at a low $V_R$, and low-molecular-weight solutes will elute at a high $V_R$. $V_R$ is related to the interstitial volume, $V_O$, the internal pore volume, $V_i$, and the equivalent liquid volume of the gel, $V_S$, by the equation

$$V_R = V_O + kV_i + k_{LC}V_S$$

where $k$ is the coefficient of distribution, indicating the ease of the polymer solution to move into the pores. If $k = 0$, then the polymer cannot enter the pores of the gel at all, and if $k = 1$, then the polymer can freely enter the pores. Typically, $k_{LC}$, which is the ratio of the volume of solute in the stationary phase (within the gel) to the volume in the mobile phase, is negligible and the equation reduces to

$$V_R = V_O + kV_i$$

$V_O$ and $V_i$ are constant for a given column. Thus, $V_R$ is dependent on solute molecular weight and as such, a calibration curve can be generated using solutes of known monodisperse molecular weights. The molecular weight is related to the elution time and is plotted as a log (base 10) function against elution time (otherwise known as retention volume). This plot is typically linear over the range of molecular weights of the standards (1–3). From this calibration plot, the molecular weight of an unknown sample can be calculated in terms of weight average molecular weight ($M_w$), number average molecular weight ($M_n$), and the z-average molecular weight ($M_z$).

The limitation of the GPC system is the need for calibration with known molecular weight standards. Ideally, the standards should contain numerous monodisperse polymer chains of known molecular weight of the same polymer being tested. However, this is not usually the case. Polystyrene is the typical monodisperse polymer used to calibrate the GPC system, and there is some element of error introduced when other polymers are compared to the polystyrene standards. The system separates polymer chains by their effective size and does not account for branching of the polymer or for polymer–solvent interactions, both of which may affect the polymer size in solution. A detailed explanation of the calibration involved in GPC systems is beyond the scope of this article, and the reader is kindly referred to the texts listed in the reference section.

A survey of the current literature finds GPC used in many situations. For instance, the manner in which a polymer degrades can be described by GPC by evaluating samples throughout the degradation timeframe (4). Another use is to determine if a loaded agent (either a pharmaceutical drug or an excipient intended to enhance controlled release of a substance) has any effect of the degradation of a polymer, either increasing the speed of degradation or retarding it (5). Stability of a polymer in various storage conditions can be determined using GPC, yielding crucial information concerning shelf life. These are just a few uses of gel permeation chromatography, a method which has grown over the past 40 years to become one of the main tools utilized in polymer characterization.

## BIBLIOGRAPHY

1. A.R. Cooper, in L.S. Bark and N.S. Allen eds., *Analysis of Polymer Systems*, Applied Science Publishers, London, 1982, pp. 243–301.
2. D. Campbell and J.R. White, *Polymer Characterization: Physical Techniques*, Chapman & Hall, London, 1989.
3. J. Sherma et al., *CRC Handbook of Chromatography*, vol. II, CRC Press, Boca Raton, Fla., 1994.
4. C.A. Santos et al., *J. Controlled Release*, in press.
5. W.L. Webber, F. Lago, C. Thanos, and E. Mathiowitz, *J. Biomed. Mater. Res.* **41**, 18–29 (1998).

# CHARACTERIZATION OF DELIVERY SYSTEMS, MAGNETIC RESONANCE TECHNIQUES

PAUL A. BURKE
Amgen, Inc.
Thousand Oaks, California

## KEY WORDS

Biodegradable polymers

Drug delivery

EPR

Hydrogels

Magnetic resonance

Microspheres
MRI
NMR

## OUTLINE

## INTRODUCTION

Controlled release drug delivery systems are often characterized by heterogeneity of phase, consisting of a matrix and a drug that is in either solid or solution form or in transition between these states. This attribute complicates characterization of interior structure and microenvironment. Hydrogels, various implants, microspheres, and controlled release tablets represent quintessential controlled release technologies, and yet the internal features of many such systems are poorly understood. Invasive techniques, or indirect measures (such as the observed rate of drug release or the condition of drug following release), have substituted for direct observation. The unevenness of commercial success in the field has heightened the need to differentiate what works from what does not; as a result the use of sophisticated, noninvasive methods such as those based on magnetic resonance has grown. In some cases depot systems can be characterized in vivo, in real time. Detailed information regarding the internal microenvironments of a variety of matrix systems has resulted.

## PRINCIPLES AND SPECTROSCOPIC METHODS

Magnetic resonance refers to the absorption of electromagnetic radiation of a frequency corresponding to the energy difference between a multiplicity of states created by the application of a strong magnetic field. Experimental methods exploiting this phenomenon include solution and solid-state NMR spectroscopy, electron paramagnetic resonance spectroscopy (EPR), and magnetic resonance imaging (MRI). The semiclassical vector model of nuclear magnetization is a good starting point for understanding time domain (as opposed to frequency domain) methods, which predominate. Fortunately some excellent introductory texts make this subject accessible to anyone with a background in classical mechanics and elementary electromagnetic theory (1–3). Although the present discussion assumes familiarity with these concepts, the uninitiated will likely appreciate the potential of magnetic resonance methods. Specific techniques relevant to drug delivery system characterization are highlighted; most of these are practicable with commercial instrumentation. For a comprehensive overview of NMR methods (including imaging), the reader is referred to the Encyclopedia of NMR (4).

### Solution NMR

The most rudimentary solution NMR methods result in a one-dimensional spectrum of spin 1/2 nuclei such as $^1H$, $^{13}C$, $^{31}P$, $^{15}N$, or $^{19}F$ (1). Solution NMR can be used to determine pH by means of the Henderson-Hasselbalch equation. Given an acid with known $pK_a$, a chemical shift that is sensitive to protonation state can be used to determine directly the ratio of protonated and unprotonated species in a sample. This concept is illustrated in Figure 1 (5). The same principle applies in EPR, where hyperfine splittings are sensitive to protonation state. This experiment relies on the sensitivity of magnetic resonance methods to processes that are fast or slow relative to the angular frequency difference ($\Delta\omega$) between the species involved (1). Loosely speaking, $\Delta\omega$ is referred to as the time scale of the experiment and is on the order of milliseconds for most solution NMR experiments. Proton transfer, being comparatively rapid, results in a time-averaged chemical shift, weighted according to the relative proportion of protonated and unprotonated species. Under conditions of slow ex-

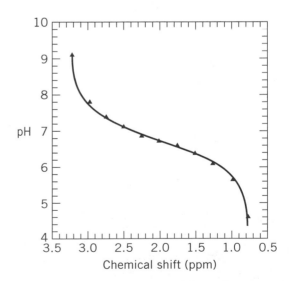

**Figure 1.** Solution NMR titration curve for inorganic phosphate solutions prepared with various ratios of monobasic and dibasic forms (155 mM ionic strength). Symbols depict the correspondence between experimentally determined chemical shifts and the measured pH. Also shown is the theoretical standard curve calculated using the following parameters: $pK_a = 6.75$; $\delta(HPO_4^{2-}) = 3.23$ ppm; $\delta(H_2PO_4^-) = 0.75$ ppm.

change, individual peaks are observed for each species; peak intensities are proportional to the population of each state. Examples of slow exchange include that between the cytosol and surrounding medium of living cells, and exchange within solids (6,7).

Diffusion in solutions can be measured using pulse field gradient (PFG) NMR, a variant of a spin-echo pulse sequence used to measure spin-spin relaxation times ($T_2$) (8). A transient magnetic field gradient pulse is applied to mark the positions of the nuclei of interest. Subsequently, a time interval is introduced during which diffusion is allowed to occur. The spins are refocused by reapplication of the original gradient pulse, and the attenuation in the echo amplitude related mathematically to the diffusion constant. PFG NMR is appropriate to the measurement of diffusion processes that are fast relative to $T_2$, such as water diffusion within hydrogels (9,10).

Solution NMR methods are of limited applicability to systems of heterogeneous phase. If the analyte lacks isotropic molecular motion, broad resonances will result. In the worst case the signal is not detectable. This is due to incomplete averaging of chemical shift anisotropy, dipole-dipole interactions, and (for nuclei of spin $>1/2$) quadrupolar interactions (7,11). Phase heterogeneity in a sample can be problematic even if the analyte is itself fully dissolved. In w/o emulsions where the magnetic susceptibilities of the dispersed and continuous phases differ, local magnetic field gradients result, giving broad lines and altering relaxation rates (12). Magnetic interactions between water and polymer protons in a hydrogel result in cross-relaxation, invalidating the conventional mathematical formalism used to interpret echo amplitude attenuation in PFG NMR (10).

## Solid-State NMR

In high-resolution, solid-state NMR, experimental techniques, as opposed to isotropic molecular motion, average spin interactions. Isotropic chemical shifts of dilute spin 1/2 nuclei are recovered by magic angle spinning (MAS), where the sample rotates about an axis which is at a fixed angle (the magic angle) with respect to the external magnetic field. Dipolar coupling to protons is removed by spin locking. When applied to $^{13}C$, which has a natural abundance of 1%, labeling is often unnecessary. For other nuclei, incorporation of spin 1/2 isotopes by synthetic means is an option; this can become a costly and time-consuming affair. An alternative is to use commercially available compounds as reporter groups (for example, $^{13}C$ or $^{15}N$ labeled amino acids (13)). Another limitation of solid-state NMR methods is poor sensitivity. Even with isotopic enrichment acquisitions can require hours or even days.

Rotational-echo double resonance (REDOR) is one of several solid-state NMR techniques used to determine internuclear distances (up to 5 Å) of heteroatoms, for example, $^{13}C$ and $^{15}N$ (14). The measurement is based on the distance-dependent heteronuclear dipolar coupling constant. Precise mathematical analysis requires that a given labeled site interacts with no more than one heteroatom; otherwise, an average dipolar coupling is obtained (13). Related techniques enable determination of homonuclear interatomic distances.

Solid-state deuterium NMR can elucidate molecular motions in solids using a quadrupole echo pulse sequence (15). The spectrum is dominated by the quadrupole interaction of the spin 1 nucleus and is sensitive to motions over a wide time scale range, from milliseconds to microseconds. Pharmaceutical applications of solid-state NMR techniques have been reviewed recently (16).

## Electron Paramagnetic Resonance

EPR differs from NMR in that the spins involved are those of unpaired electrons (17). EPR can detect free radicals in, for example, gamma-irradiated samples (18,19). Most analytes, however, lack unpaired electrons, and exogenous reporter groups such as paramagnetic metal ions or free radicals are required. Stable nitroxide spin labels can be introduced through chemical modification (20). EPR spectra, typically presented as the first derivative of the absorption profile, are sensitive to molecular motion on the microsecond time scale; the rate of label motion is determined by analysis of the line shape. This time frame is extended to the millisecond range using saturation transfer EPR (ST-EPR) (21). Other details, such as the reporter group protonation state or the polarity of its microenvironment, can be elucidated from the hyperfine splittings. The application of EPR to pharmaceutics has been reviewed by Mader et al. (22).

## Magnetic Resonance Imaging

In MRI, an image is constructed from the spatial distribution, and different relaxation times, of spins within a sample (23). As a medical diagnostic, MRI detects water proton spins, exploiting the variable water contents of body tissues and fluids. Paramagnet spins likewise can be used to construct images. One limitation of MRI as it relates to drug delivery systems is resolution. Even with advanced instrumentation the resolution limit is 20–50 $\mu$m, adequate for macroscopic observations but insufficient to illuminate microscopic architecture.

## APPLICATIONS

### Distribution of Water and of Active Agent

Water environments in hydrogel systems have been characterized with magnetic resonance techniques. Roorda et al. measured the relaxation times of $^{17}O$ labeled water in poly(hydroxyethyl methacrylate) (PHEMA) hydrogels (24). Advantages of $^{17}O$ include negligible natural abundance signals and absence of exchange with matrix oxygens. The rotational mobility of hydrogel water was reduced compared to pure water, and the degree of mobility restriction correlated with the hydrogel polymer content. Relaxation curves were fit with a single exponential decay, showing that the hydrogel contained a single water population on the millisecond time scale. This result contrasted with that obtained using invasive techniques.

Pitt et al. used nitroxide spin labels and spin probes to characterize the molecular environments of hydrogels and hydrophobic polyester matrices (25–27). The existence of distinct bulk water and lipophilic polymer phases within hydrogels was tested. A spin-labeled amino ligand—the lipophilicity of which depends on protonation state—was bound to the gel surface and the ligand molecular motions examined. Single-component EPR spectra were observed over a wide pH range, indicating a single, homogeneous water environment within the gel. Similar results were obtained in an investigation of PHEMA using spin-labeled solutes. In contrast, composite spectra were observed in several polyester-containing polymer-polymer blends, indicating phase separation.

Schmitt et al. used solid-state $^2$H NMR to characterize the water environments in pellets of poly(lactide-co-glycolide) (PLGA) copolymers as a function of lactide content, in order to provide a mechanistic explanation for the known decrease of hydrolysis rate with increasing lactide content (28). Using a quadrupole echo pulse sequence, the authors quantitated bound and free water populations in the hydrated pellets. Although the *total* water content was comparable for all lactide:glycolide ratios studied, the water distribution was not. With increasing lactide content, the fraction of bound water decreased. It was thus concluded that bound, but not free, water is reactive in the hydrolysis reaction.

Greenley et al. used REDOR to characterize the distribution of $^{13}$C- and $^{15}$N-labeled amino acids imbibed in cross-linked polyacrylic acid matrices (13). Since the resulting sample contains a distribution of carbon-nitrogen internuclear distances, an ensemble average dipolar interaction was observed. A significant dipolar interaction was present even at very low concentrations of imbibed compounds, indicating cluster formation. The amino acid clusters appeared to increase in size with increasing drug concentration in the matrix.

Goldenberg et al. used chemical shift analysis to characterize active agent distribution within micellar delivery systems (29). The chemical shift of a given nucleus is largely determined by the shielding influences of its substituent groups and atoms, but also can vary with the physical properties of the medium. Based on chemical shift analysis of both surfactant and active agents, some of the extremely water-insoluble agents studied were found to form clusters within the nonionic micelles, as opposed to being dispersed within either the hydrocarbon core or the surrounding poly(oxyethylene) mantle.

Tobyn et al. used EPR to study interactions between a polysaccharide matrix and incorporated nitroxide spin labels in an oral dosage form (30). Wet granulation, in contrast to dry mixing, was found to be effective for dispersing the crystalline spin labels.

### Internal Environment During Polymer Erosion

The internal environment of PLGA microspheres was characterized using solution $^{31}$P NMR (5). Microspheres were suspended in sheep serum (pH ~ 8), which contains endogenous inorganic phosphate at a concentration of a few millimolar. In the settled microsphere suspension two phosphate peaks were obtained, demonstrating the presence of a solution environment within the microspheres, with slow chemical exchange between internal phosphate and that in the surrounding medium, as well as a differ-

ence in pH between the two phosphate pools (Fig. 2). The $^{31}$P spectrum was monitored through day 45. The calculated internal pH changed modestly during this time frame, stabilizing at pH 6.4 (Fig. 3). The slight acidity of the interior is presumably due to the carboxylic acid degradation products of the polyester. That a more acidic value was not obtained suggests that the exterior medium effectively neutralized at least some of the acid generated on polymer hydrolysis; this scenario is consistent with the high porosity (by scanning electron microscopy) of the microspheres used in this study.

EPR has been used to measure pH in both w/o emulsions and polymer matrices, using spin probes sensitive to a wide pH range. Kroll et al. (31) monitored aqueous phase pH as a function of time and of temperature in w/o oint-

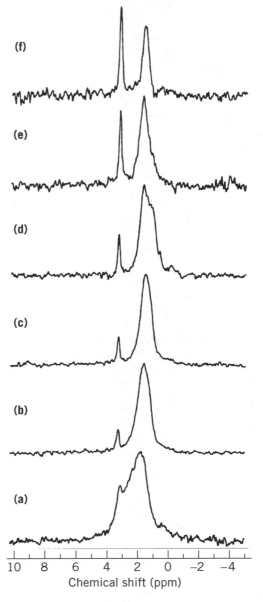

**Figure 2.** $^{31}$P NMR spectra of PLGA microspheres in sheep serum at (**a**) 3 hours; (**b**) 2 days; (**c**) 6 days; (**d**) 14 days; (**e**) 30 days; (**f**) 45 days.

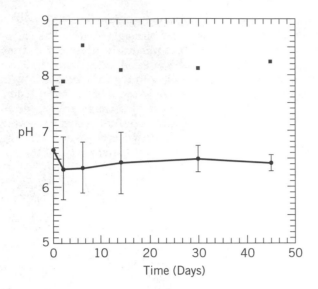

**Figure 3.** PLGA microsphere interior pH as a function of time, determined from the spectra in Figure 2. ●, pH of microsphere interior; ■, pH of suspension medium.

ments containing acetylsalicylic acid and benzocaine. Low-frequency ($\leq 1$ GHz) EPR has been used for in vivo characterization of the pH inside PLGA implants (32). Low frequency enables detection to depths of 1 cm or more using surface coils. Six nitroxide spin probes, with $pK_a$ values ranging from 1.0 to 7.2, were incorporated into monolithic PLGA matrices that were implanted subcutaneously in mice. While a dearth of endogenous paramagnets simplifies interpretation, precision in the pH determination was limited by an inability to resolve spectral components arising from spin probes of varying mobility (and possibly in regions of differing pH). It was concluded that the pH dropped from 4 to 2 over a three-week period.

The pH inside polyanhydride discs was monitored in vitro, with spatial resolution of roughly 20 $\mu$m, using two-dimensional (spectral-spatial) EPR imaging (33). The pH was monitored in a cross section of the 2-mm-thick wafer over a 7-day period. The outer disc layers exhibited an acidic environment (pH 4.7) after 2 days. The solvent penetration front proceeded with zero-order kinetics at a rate of 0.3 mm day$^{-1}$, with outer layers gradually reaching the pH of the outer medium (buffered with phosphate to pH 7.4), and the interior showing pH > 5.0. After 7 days, the pH gradient disappeared, and the pH throughout the disc matched that of the surrounding medium. This constituted the first direct observation of pH gradients within biodegradable polymers.

### Rates of Drug Release and Polymer Erosion in Vivo

Low-frequency EPR was coupled with MRI in studies of drug release and matrix erosion in vitro and in vivo (34,35). PLGA and polyanhydride implants were studied. In order to obtain spatial information, sandwich implants were fabricated with outer layers containing $^{14}N$-nitroxide and an inner layer containing $^{15}N$-nitroxide (34). (The splitting observed is determined by the nitrogen nuclear spin, and

therefore different for the layers.) The rate of nitroxide release was assessed by following the ratio of the two nitroxides remaining at the site. A zero-order decrease in the relative $^{14}N$ component (corresponding to the outer tablet layers) was observed. In vivo MRI studies indicated a surface erosion front mechanism, both in vitro and in vivo. Differences between in vitro and in vivo rates of both spin probe release and polymer erosion were noted, however (34,35). Surprisingly, a decrease in the spin probe mobility during the course of polyanhydride erosion was observed. This was assumed to derive from coprecipitation of soluble monomer (liberated during polymer erosion) with the encapsulated probe. Rates of bioerosion of polylactide bone screws in humans have been followed by MRI (36,37).

### Diffusion, Gelation, and Pore Size Distribution

Magnetic resonance imaging has been used to characterize diffusion, determine pore size distribution, and monitor gelation in hydrogels. Imaging imparts spatial resolution (in one or more dimensions) to diffusion analysis, and can be useful in the study of diffusion times longer than those accessible by PFG NMR. Null point MRI has been used to map Fickian diffusion of paramagnetic ions and spin probes through hydrogels and other heterogeneous media, via the effect of the paramagnet on water proton $T_1$s (38,39). The technique has been extended to two dimensions using fast inversion recovery (40), and has been combined with magnetization transfer in the characterization of alginate gelation in the absence of a paramagnetic tracer (41). Pore size distribution within calcium alginate gels has been determined from water proton $T_2$ maps, exploiting an inverse relationship between $T_2$ and alginate concentration (42,43). The temperature dependence of the diffusion constant of the tracer compound trifluoroacetamide was determined in a thermoresponsive poly($N$-isopropylacrylamide) hydrogel using one-dimensional $^{19}F$ MRI (44). Fyfe et al. investigated the hydration of hydroxymethylcellulose tablets, following concentrations and mobilities of both water and the polymer as functions of time and distance through a combination of techniques (45). Susceptibility to error introduced by entrapped gas bubbles within the tablet matrix was reported (46). NMR imaging was recently applied to compression-coated tablets in order to characterize the drug release mechanism (47).

Magnetic resonance methods have been used to characterize intracranial and transdermal drug transport. The delivery of paramagnetic contrast markers to the rabbit parenchyma by an osmotic pump was followed by MRI, and the results used to refine a theoretical model of local drug delivery to the brain. The model was used to predict distribution profiles of drugs delivered by implanted polymeric systems (48). Kinsey et al. used MRI to characterize water mobility in skin samples in vitro, with the aim of applying the results to transdermal drug delivery via passive diffusion or by electroporation (49). Under the highest resolution conditions studied, major anatomical structures such as sebaceous glands and hair follicles were well resolved.

### Interfaces and Colloidal Suspensions

Emulsions can exhibit sustained release effects and play a role in some microencapsulation processes (50). Emulsification of surface active drugs such as proteins can cause denaturation. Nevertheless, little is known about the physical environment under these conditions. For example, the ultimate fate of the surface denatured layer is not known. Stabilization of proteins toward double emulsion methods has been empirical (51). Recently, emulsions have been characterized by solution NMR methods. Lonnqvist et al. characterized double (w/o/w) emulsions by PFG NMR (12), which was used to determine droplet size distributions, and to distinguish oil-only from drug-containing droplets. Inner and outer water populations were distinguished and the average lifetime of an inner aqueous droplet determined. Ter Beek et al. used $^{13}$C and $^{31}$P solution NMR to characterize $\beta$-casein at an oil-water interface (52). They relied on the dependence of $^{13}$C chemical shift on protein secondary structure to conclude that the mobile portions of the surface-adsorbed protein derived from random coil, as opposed to $\alpha$-helical, portions of the protein. $^{31}$P signals from $\beta$-casein's five phosphoserine residues served as probes of motion in the molecule's highly charged N-terminal region, comparing surface-adsorbed versus solution (primarily micellar) $\beta$-casein. Finally, solution $^{31}$P NMR has been used to illuminate the structure of amphotericin B containing liposomes (53), as well as the fusogenic properties of cationic liposome-DNA complexes (54). The restricted molecular motions at interfaces make these systems amenable to solid-state NMR methods, which have been applied to lipid bilayers (15,55) and to protein-lipid interactions in emulsions (56).

## SUMMARY

Controlled release delivery systems are highly complex dosage forms. The past few years have witnessed the increased use of magnetic resonance techniques in the characterization of these systems. Although the body of literature to date is limited, early applications have demonstrated the power of magnetic resonance to address critical questions in the field, and to challenge dogmas based on invasive or indirect measurements. Despite the requirements of specialized instrumentation and training, such noninvasive analytical techniques will likely prove indispensable in the development of well-characterized, high quality drug delivery systems suitable for commercialization.

## BIBLIOGRAPHY

1. F.A. Bovey, *Nuclear Magnetic Resonance Spectroscopy*, 2nd ed., Academic Press, San Diego, Calif., 1988.
2. A.E. Derome, *Modern NMR Techniques for Chemistry Research*, Pergamon, Oxford, 1987.
3. T.C. Farrar and E.D. Becker, *Pulse and Fourier Transform NMR. Introduction to Theory and Methods*, Academic Press, New York, 1971.
4. D.M. Grant and R.K. Harris, *Encyclopedia of NMR*, Wiley, New York, 1996.
5. P.A. Burke, *Proc. Int. Symp. Control Relat. Bioact. Mater.* **23**, 133–134 (1996).
6. G. Navon, S. Ogawa, R.G. Shulman, and T. Yamane, *Proc. Natl. Acad. Sci. U.S.A.* **74**, 888–891 (1977).
7. C.A. Fyfe, *Solid State NMR for Chemists*, CFC Press, Guelph, Ontario, 1983.
8. P. Stilbs, *Prog. NMR Spectrosc.* **19**, 1–45 (1987).
9. W.R. Korsmeyer, E. von Meerwall, and N.A. Peppas, *J. Polym. Sci. Polym. Phys. Ed.* **24**, 409 (1986).
10. L.J. Peschier et al., *Biomaterials* **14**, 945–952 (1993).
11. C.S. Yannoni, *Acc. Chem. Res.* **15**, 201–208 (1982).
12. I. Lonnqvist, B. Hakansson, B. Balinov, and O. Soderman, *J. Colloid Interface Sci.* **192**, 66–73 (1997).
13. R. Greenley, H.S. Zia, J. Garbow, and R.L. Rodgers, *ACS Symp. Ser.* **469**, 213–236 (1991).
14. J.M. Griffiths and R.G. Griffin, *Anal. Chim. Acta* **283**, 1081–1101 (1993).
15. R.G. Griffin et al., in G.J. Long and F. Grandjean, eds., *The Time Domain in Surface and Structural Dynamics*, Kluwer Academic Publishers, Amsterdam, 1988, pp. 81–105.
16. D.E. Bugay, *Pharm. Res.* **10**, 317–327 (1993).
17. J.A. Weil, J.R. Bolton, and J.E. Wertz, *Electron Paramagnetic Resonance: Elementary Theory and Practical Applications*, Wiley, New York, 1994.
18. K. Mader, A. Domb, and A. Swartz, *Appl. Radiat. Isot.* **47**, 1169–1674 (1996).
19. L. Martini, J.H. Collett, and D. Attwood, *J. Pharm. Pharmacol.* **49**, 601–605 (1997).
20. L.J. Berliner, *Spin Labeling: Theory and Applications*, Plenum, New York, 1989.
21. J.S. Hyde, *Methods Enzymol.* **49**, 480–511 (1978).
22. K. Mader, H.M. Swartz, R. Stosser, and H.H. Borchert, *Pharmazie* **49**, 97–101 (1994).
23. D.D. Stark, *Magnetic Resonance Imaging*, Mosby, St. Louis, Mo., 1992.
24. W.E. Roorda, J. de Bleyser, H.E. Junginger, and J.C. Leyte, *Biomaterials* **11**, 17–23 (1990).
25. C.G. Pitt, X.C. Song, R. Sik, and C.F. Chignell, *Biomaterials* **12**, 715–721 (1991).
26. C.G. Pitt, J. Wang, C.F. Chignell, and R. Sik, *J. Polym. Sci. Part B: Polym. Phys.* **30**, 1029–1071 (1992).
27. C.G. Pitt et al., *Macromolecules* **26**, 2159–2164 (1993).
28. E.A. Schmitt, D.R. Flanagan, and R.J. Linhardt, *Macromolecules* **27**, 743–748 (1994).
29. M.S. Goldenberg, L.A. Bruno, and E.L. Rennwantz, *J. Colloid Interface Sci.* **158**, 351–363 (1993).
30. M.J. Tobyn, J. Maher, C.L. Challinor, and J.N. Staniforth, (1996) *J. Controlled Release* **40**, 147–155 (1996).
31. C. Kroll, K. Mader, R. Strober, and H.H. Borchert, *Eur. J. Pharm. Sci.* **3**, 21–26 (1995).
32. K. Mader, B. Gallez, K.J. Liu, and H.M. Swartz, *Biomaterials* **17**, 457–461 (1996).
33. K. Mader et al., *Polymer* **38**, 4785–4794 (1997).
34. K. Mader et al., *J. Pharm. Sci.* **86**, 126–134 (1997).
35. K. Mader et al., *Pharm. Res.* **14**, 820–826 (1997).
36. H. Pihlajamaki, J. Kinnunen, and O. Bostman, *Biomaterials* **18**, 1311–1315 (1997).
37. H.K. Pihlajamaki, P.T. Karjalainen, H.J. Aronen, and O.M. Bostman, *J. Orthop. Trauma* **11**, 559–564 (1997).
38. A.E. Fischer and L.D. Hall, *Magn. Reson. Imaging* **14**, 779–783 (1996).
39. B.J. Balcom, A.E. Fischer, T.A. Carpenter, and L.D. Hall, *J. Am. Chem. Soc.* **115**, 3300–3305 (1993).

40. A.E. Fischer et al., *J. Phys. D* **28**, 384–397 (1995).
41. J.J. Tessier, T.A. Carpenter, and L.D. Hall, *J. Magn. Reson. Ser. A* **113**, 232–234 (1995).
42. K. Potter, T.A. Carpenter, and L.D. Hall, *Carbohyd. Res.* **246**, 43–49 (1993).
43. K. Potter, T.A. Carpenter, and L.D. Hall, *Magn. Reson. Imaging* **12**, 309–311 (1994).
44. R. Dinarvand, B. Wood, and A. D'Emanuele, *Pharm. Res.* **12**, 1376–1379 (1995).
45. C.A. Fyfe and A.I. Blazek, *Macromolecules* **30**, 6230–6237 (1997).
46. C.A. Fyfe and A.I. Blazek, *J. Controlled Release* **52**, 221–225 (1998).
47. B.J. Fahie et al., *J. Controlled Release* **51**, 179–184 (1998).
48. S. Kalyanasundaram, V.D. Calhoun, and K.W. Leong, *Am. J. Physiol. (Regul. Integr. Comp. Physiol.)* **42**, R1810–R1821 (1997).
49. S.T. Kinsey, T.S. Moerland, L. McFadden, and B.R. Locke, *Magn. Reson. Imaging* **15**, 939–947 (1997).
50. S.D. Putney and P.A. Burke, *Nat. Biotechnol.* **16**, 153–157 (1998).
51. J.L. Cleland and A.J.S. Jones, *Pharm. Res.* **13**, 1464–1475 (1996).
52. L.C. Ter Beek et al., *Biophys. J.* **70**, 2396–2402 (1996).
53. A.S. Janoff et al., *Proc. Nat. Acad. Sci. U.S.A.* **85**, 6122–6126 (1988).
54. K.W. Mok and P.R. Cullis, *Biophys. J.* **73**, 2534–2545 (1997).
55. R.G. Griffin, *Methods Enzymol.* **72**, 108–174 (1981).
56. J.A.G. Areas et al., *Magn. Reson. Chem.* **35**, S119–S124 (1997).

# CHARACTERIZATION OF DELIVERY SYSTEMS, MICROSCOPY

JULES S. JACOB
Brown University
Providence, Rhode Island

## KEY WORDS

Drug delivery
Imaging
LM
Microstructure
SEM
TEM

## OUTLINE

## INTRODUCTION

The characterization of drug delivery systems using optical and electron optical methods provides important information about size, microstructure, surface topography and texture, porosity and physical state, and distribution of drug in the final product. All of these factors contribute to the performance and drug release characteristics of the delivery system and can offer insight into troubleshooting manufacturing problems as well as suggest ways for improving delivery.

The topics to be covered include optical or light microscopy (LM), transmission electron microscopy (TEM), and scanning electron microscopy (SEM). The advantages and disadvantages of each type of microscopy as well as preparation methods for drug delivery systems are discussed.

In practice, as characterization methods for drug delivery systems, LM, TEM, and SEM are used in an integrated approach. Preliminary information from LM is used to screen the sample and set initial processing parameters for electron microscopic examination. TEM is best suited for high-resolution analysis of the matrix of delivery systems, whereas SEM is applicable to surface topography and limited views of the interior (Table 1). However, the information derived from each method is continually checked against observations derived from the other methods to ensure validity of interpretation and to guard against introduction of artifacts inherent to sample preparation.

## OPTICAL MICROSCOPY

The light microscope is often used as a screening tool in the characterization of drug delivery systems. The stan-

**Table 1. Suitability of Imaging Techniques for Characterization of Drug Delivery Systems**

| Characterization | LM | TEM | SEM |
|---|---|---|---|
| Size | Yes | Yes | Yes |
| Size distribution | Yes | Yes | Yes |
| Macrostructure (0.1–10 mm) | Yes | Yes | Yes |
| Mesostructure (1–100 microns) | Yes | Yes | Yes |
| Microstructure (1–1,000 nm) | No | Yes | Yes |
| Internal morphology | Yes | Yes | Yes |
| External morphology | No | Replica only | Yes |
| Surface topography | No | Replica only | No |
| Porosity | No | Yes | Yes |
| Internal drug distribution | Yes | Yes | Yes |
| Surface drug distribution | No | No | Yes |

dard upright light microscope is capable of resolving structures that are 0.2 $\mu$m or larger as restricted by the wavelength of light. The components of a light microscope include a light source, condenser lenses to focus light on the specimen, a stage to move the specimen, objective lenses to focus and enlarge the transmitted light image, and eyepiece lenses to provide additional enlargement for viewing (Fig. 1). Most modern research microscopes are "compound" or equipped with multiple objective lenses, ranging from 0.5 to 100 enlargements in power and used mounted on a turret to provide a range of magnifications. At higher magnification, 40X and 63X and 100X, most objective optics are designed to be used with optically corrected immersion oil to reduce image degradation resulting from diffraction. With intermediate, objective, and eyepiece lenses summed together, the range of magnification spans from 3X to 1,500X. The images can be viewed directly, recorded photographically with a camera attachment, or digitally with a video camera attachment. Digital processing and image analysis of captured video microscope images has been widely used to characterize drug delivery systems in terms of size, porosity, and distribution.

## Imaging Modes

Bright field imaging, using differences in optical density and color of light transmitted through a sample, is the routine mode for viewing samples with a conventional light microscope. However, small particulates in solution can best be visualized using dark-field optics to increase specimen contrast by reversing background and sample optical densities. Another means of increasing contrast is to use a phase contrast condenser and matched lenses to increase sample edge contrast by introducing a phase shift between transmitted and scattered light passing through the specimen. Hoffman modulation optics can be used to produce

nearly three-dimensional images of light microscopic samples, using a special polarizer condenser to change amplitude rather than phase of scattered light transmitted through the sample, and a slit to regulate the level of illumination.

Polarized light microscopy is of great utility for analyzing the crystalline state of polymers comprising most drug delivery devices. When two polarizers are 'crossed' so that planes of transmitted light are perpendicular in orientation to each other, crystalline specimens appears birefringent against a totally dark background. Changes in crystallinity resulting from fabrication can be easily determined by direct observation.

Confocal light microscopes are specialized instruments, used to "optically section" specimens as thick as 1 mm by controlling the plane of focus of objective lenses with computer-regulated, stepper motors. Laser light sources are used to provide more powerful illumination to penetrate thicker samples. The laser can be used to excite fluorescent dyes, such as fluorescein isothiocyanate and rhodamine to name a few, and the emission from the fluorophores can be used to localize "labeled" molecules encapsulated in devices. Using image analysis software, it is possible to 'reconstruct' the optical slices and rebuild the sample in three dimensions, observing contour images of the slices for more information.

## Sample Preparation

Samples for light microscopy that can be observed directly include particulates dispersed in solution, small, dry particulates including microspheres, emulsions, fluids, thin films (either pressed, melted, or solvent-cast). Specimens thicker than 40–50 $\mu$m are normally sectioned to produce thinner specimens. Cryostat sectioning of samples, quickly frozen in liquid freon or liquid nitrogen can be used to rapidly produce 5–20-$\mu$m-thick sections suitable for observation. Samples may also be infiltrated with epoxy or acrylate-based embedding media (such as LR white and glycolmethacrylate [GMA]), cured and sectioned using glass knives to produce higher quality sections in the 1–10-$\mu$m thickness range. This procedure will be discussed in greater detail in the "Sample Preparation for TEM."

**Staining.** The contrast of specimens can often be improved using a variety of stains that have different chemical specificities. Many biological stains can be used to stain polymers and achieve contrast effects. Stains used for the characterization of drug delivery systems are divided into three basis categories. Acid stains are negatively charged stains containing hydroxyl, carboxyl, or sulphonic groups, such as eosin y, and will bind to positively charged polymers. Basic stains are positively charged molecules, such as toluidine blue, methylene blue, and Alcian blue, and bind to negatively charged polymers. Solvent stains or lysochromes are generally un-ionized, lipophilic substances, such as Sudan black, and oil red O, which partition by solubility into hydrophobic substances and stain them (Table 2).

Additionally, immunochemical and enzymatic techniques may be used on cryostat sections to localize bio-

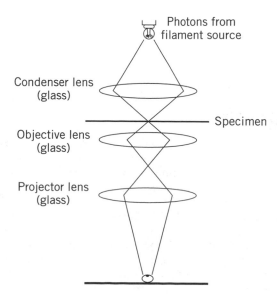

**Figure 1.** Simplified arrangement of optics in the upright light microscope.

Photons from filament source

Condenser lens (glass)

Specimen

Objective lens (glass)

Projector lens (glass)

**Table 2. Application of LM and TEM Stains for Characterization of Drug Delivery Systems**

| Stain | Imaging | Affinity | Application |
|---|---|---|---|
| Eosin Y | LM | Anionic dye with affinity to cationic moieties | Staining of positively-charged polymers |
| Alcian blue | LM | Cationic dye with affinity to anionic moieties | Staining of negatively-based polymers |
| Methylene blue | LM | Cationic dye with affinity to anionic moieties | Staining of negatively-charged polymers |
| Toluidine blue | LM | Cationic dye with affinity to anionic moieties | cationic dye with affinity to anionic moieties |
| Methyl green-pyronin | LM | Nucleic acids | Localization of encapsulated nucleic acids |
| Sudan black | LM | Lipid | Hydrophobic polymer stain |
| Oil red O | LM | Lipid | Hydrophobic polymer stain |
| Uranyl acetate | TEM | Phosphate/amino moieties | Negative/positive stain |
| Lead citrate | TEM | Negatively charged moieties | Positive stain |
| Ammonium molybdate | TEM | Cationic molecule with affinity to anionic moieties | Negative stain |
| Phosphotungstic acid | TEM | Anionic molecule with affinity to cationic moities | Negative stain |
| | | | Positive stain—polyamides and olypropylene |
| Osmium tetroxide | TEM | Unsaturated hydrocarbons, double bonds | Positive stain |
| Iodine | LM/TEM | Polyamides/polyvinylalcohol | Positive stain |
| Silver sulfide | LM/TEM | Cationic attraction to surface anions | Positive stain |

logical encapsulants. Sections are incubated with specific antibodies to the target molecule and reacted with secondary antibodies labeled with markers such as horseradish peroxidase (HRP; brown, insoluble, enzyme reaction product), alkaline phosphatase (AP; red or blue, insoluble enzyme reaction product), phycoerythrin (pink-colored, fluorescent), fluorescein isothiocyanate (green-fluorescent), and rhodamine (red-fluorescent). Other labeling schemes used at the light microscopic level include protein A–gold with silver enhancement and biotin-labeled antibodies, primary or secondary antibodies reacted with avidin-marker enzyme conjugate (either HRP or AP labels). The localization of biomolecules and enzymes will be discussed in greater depth in "Sample Preparation for TEM."

## TRANSMISSION ELECTRON MICROSCOPY

TEM extends the level of observation provided by the light microscope by orders of magnitude. Generally speaking, modern TEM is capable of resolving structures with dimensions smaller than 0.2 nm and provides magnification in the ranging from less than 500-fold to millionfold. Instead of focusing light rays with glass lenses, as in LM, TEM focuses electrons emitted by a heated filament to provides images restricted only by the wavelength of electrons or 0.003 nm.

The design of the TEM has many striking similarities to the optics employed in light microscopes (Fig. 2). Both systems use condenser lenses to focus incident electrons or light rays onto the sample and objective lenses to focus the electrons or light rays that interact and pass through the sample. However in the case of TEM the lenses are powerful electromagnets capable of bending the path of electrons. Electromagnets are also used to accelerate electrons as they are emitted from the filament source so that they have sufficient energy to penetrate the specimen. Typically energies in the range of 60 to 100 kV are used for this purpose. Even using powerful lenses and high accelerating voltages, the TEM must operate under high vacuum, generally less than $10^{-4}$ mmHg, to prevent collisions of electrons with air molecules and resultant loss in resolution.

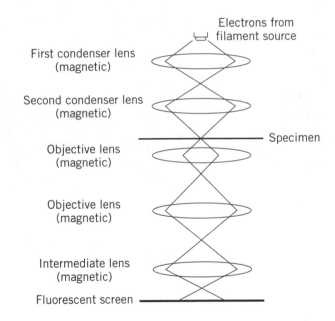

**Figure 2.** Simplified arrangement of optics in the transmission electron microscope.

The TEM has two sets of apertures or micron-sized holes laser-drilled into metal foil: condenser and objective. Condenser apertures serve to limit the diameter of the electron beam that reaches the specimen, thereby controlling to some extent beam damage. The incident electron beam strikes and interacts with atoms of the sample, particularly heavy metal stain atoms. Those electrons that emerge from the sample consequently have different energies resulting from their atomic interactions. The differences in energy result in differences in sample contrast. The contrast of the sample electrons may further be controlled by inserting different objective apertures. Smaller objective apertures tend to improve contrast by removing peripherally deflected electrons from the signal beam. Objective lenses focus the electron beam and projector lenses, like the eyepiece lenses of light microscopes, enlarge the focused image so that it can be observed. The final image is viewed through binoculars focused upon a phosphor-loaded screen inside the microscope chamber and may be photographically recorded by exposing sheet or roll film mounted in an automatic camera system situated in vacuo beneath the viewing screen.

## Sample Preparation for TEM

The requirement for high vacuum in all electron microscopy systems imposes important restrictions on the preparation of samples, namely that the samples be dehydrated to remove water, which would hinder the effectiveness of the vacuum system. In the case of biological electron microscopy, an important requirement is that the material be prepared so that it resembles the living state, with the introduction of as little artifact as possible. However, for the purpose of analyzing controlled release systems, it is vital that the sample be prepared with minimal losses of therapeutic agents or disruption of device morphology and material matrix. Borrowing from biological TEM, the following steps are necessary for most aspects of thin section TEM: (1) fixation or preservation of the sample to prevent disruption and extraction during subsequent manipulations; (2) dehydration; (3) infiltration with embedding material; (4) curing of the infiltrated sample; (5) sectioning and collection of the cured sample; and finally (6) staining of the sectioned material for observation.

**Fixation.** As with biological TEM, two types of fixing agents are popularly used: aldehydes and heavy metal tetroxides. In many instances both agents are used sequentially (1). Glutaraldehyde and formaldehyde are capable of cross-linking proteins and other agents containing free terminal amino groups. Additionally, glutaraldehyde has some limited reactivity with nucleic acids, carbohydrates, and lipids. For protein and peptide drugs encapsulated in polymer matrices, aldehydes quickly penetrate most systems (especially hydrogel-based devices) immobilizing the proteins by cross-linking within the device matrix, thus preventing extraction during subsequent steps. Fixation is generally complete within one to three hours and can be greatly shortened for systems measuring less than 1 mm, yet in many instances samples may be stored in fixative for years with minimal artifactual change. Aldehyde fixation is sufficiently gentle that many enzymes retain at least partial activity following fixation and suffer only minimal changes in antigenicity when reduced levels of fixatives are employed. This characteristic provides researchers with a powerful tool, since lightly fixed, controlled delivery systems can be processed, sectioned, and used to localize encapsulated enzymes after substrate addition, providing that the resulting reaction products are water-insoluble. Additionally, antibody histochemistry can be used to localize nearly any target material, assuming that a suitable primary antibody is available.

Osmium or, less commonly, ruthenium tetroxides are generally used as secondary fixatives, following aldehyde fixation, yet can be used as primary fixatives in protein-free systems. Tetroxides react with unsaturated double bonds commonly found in lipids and many polymers, cross-linking adjacent reactive moieties. The fixing agents also have some reactivity with proteins. Additionally, the heavy metal constituents of the fixative, either osmium, rubidium, or ruthenium are electron dense and serve as powerful contrast enhancing agents or TEM stains, which can be used to enhance polymer morphology. The fixatives have mordant properties that increase the binding of other stains, notably lead-based stains. Ruthenium tetroxide, particularly, has found favor among polymer chemists, owing to its heightened reactivity with ether, alcohol, amine, and aromatic functional moieties compared to osmium tetroxide. The ruthenium form is believed to enhance contrast in surface, crystalline regions of polymers relative to bulk, amorphous domains.

Aldehyde and tetroxide fixation are often used in a complementary fashion to offset the deficiencies of a single fixative (2,3). Unfortunately, artifacts are associated with the simultaneous use of the two types of fixatives. Since aldehyde fixatives offer greater penetrating power than the tetroxides, they are normally employed first, followed by a secondary fixation in tetroxide. A commonly used two-step protocol uses 3–4% (w/v) paraformaldehyde and 1–2% (v/v) glutaraldehyde in physiological buffer (PBS) for 2–5 h at 4°C with continuous agitation, followed by washing in PBS to remove unreacted aldehyde and a secondary fixation in 0.5–2% (w/v) osmium tetroxide in PBS for 2–4 h. For purposes of enzyme or antibody localization the secondary fixation is generally eliminated because of the harsher denaturing activity of tetroxides on proteins and most agents, and the concentrations of formaldehyde and glutaraldehyde are generally reduced to 1% and 0.5%, respectively.

The fact that these fixatives are aqueous-based imposes important limitations on their use for controlled delivery systems. Because many of the agents to be localized are water soluble, it is often necessary to quantitate losses of the agents during processing, especially during fixation, and determine if the final product is an accurate estimate of the original. As an alternative to aqueous fixation, in special cases nonaqueous and vapor-phase fixation may be employed. Osmium tetroxide has excellent solubility in many organic solvents as well as water and may be reacted with the sample, providing the organic solution is a nonsolvent for the delivery system. Lipid-free hexane, as an example, can be used as a solvent for osmium tetroxide and used to fix poly(lactide) polymer (PLA) and poly(lactide-co-glycolide) (PLGA) systems, since hexane is a nonsolvent for PLA and PLGA. Alternately, because of the high vapor pressures of both aldehydes and tetroxides, it is possible to fix very thin samples in a closed system containing vapors of the fixatives for prolonged periods of time at 4°C or for progressively shorter periods of time at higher temperatures (4). Vapor-phase fixation is inefficient, often unsatisfactory, and is most often used as a last alternative in difficult circumstances where the encapsulated agent is quickly extracted after brief immersion in fixative solution.

**Dehydration.** After fixation, the sample is sufficiently stable that water may be removed without disrupting morphology. Again, borrowing from biological TEM, nondestructive dehydration may be accomplished by three means: (1) treatment with dehydrating solvents; (2) critical point drying; or (3) lyophilization. Air-drying and heating have generally been avoided for TEM processing because of sample collapse associated with these techniques. Sample collapse is an artifact associated with the collapse of hollow structures to flat morphologies attributed to interfacial tensions during drying. In biological TEM, only the first treatment, employing dehydrating solvents, is used. Critical point drying and lyophilization are used for sample preparation during SEM. However, in the absence of biological tissue, as is the case with controlled delivery systems, all three methods may be employed (Table 3).

Dehydration with fluids is normally accomplished using graded series of either alcohols or acetones ranging from 30 to 100% (v/v) solvent in steps of 10%, starting from the lowest dehydrant concentration to pure dehydrant. It is important that the dehydrating fluid not be a solvent for the delivery system and that losses of encapsulated material during dehydration be minimal, as determined by assay of the dehydrant fluids (Table 4).

Critical point drying is normally performed after exchange with volatile solvents such as alcohol. The samples are sealed in a cooled, dehydrant-filled vessel, and the dehydrant is replaced with a transitional agent, such as carbon dioxide or Freon. The sample is then heated under pressure to the critical point for the transitional agent where the density of the vapor phase and liquid phase are the same. Critical point drying achieves specimen drying without sample collapse because the specimen is always immersed in a dense vapor phase and interfacial tensions resulting from the liquid phase are eliminated.

Lyophilization removes water without using dehydrant solutions and therefore eliminates an important source of encapsulant losses. Samples are quickly frozen in liquid nitrogen, dry ice, or Freon to eliminate damaging ice crystal formation. Water is removed via sublimation under reduced pressure as the sample slowly thaws. Lyophilized samples often suffer from a small degree of sample collapse, as compared to critical point-dried samples, but are clearly superior to air-dried or heated samples. It is often possible to prepare some samples without fixation, freeze, and lyophilize them directly and examine them by SEM. Portions of the sample can be infiltrated, sectioned, and observed by TEM, providing that the encapsulant and sample matrix are sufficiently stable during processing.

**Embedment and Infiltration.** Embedment is necessary for TEM preparation to provide solid support throughout the sample for purposes of sectioning. The media commonly used are either epoxy- or acrylate-based liquids having sufficiently low viscosity to be able to permeate the sample completely and then harden by polymerization after either chemical, thermal, or photochemical initiation (Table 5). Historically, the epoxy embedding media such as Epon 812 (now replaced by Embed 812), araldite, and mixtures of the two have been most widely used. Samples are typically infiltrated or soaked in the dehydrant-diluted epoxy resin solution for up to 48 h because of the relatively high viscosity of the media, followed by a supplemental treatment with full-strength epoxy. Unwieldy infiltration schedules prompted the search for lower viscosity epoxies, such as Spurr's media, which allowed times to be reduced by half. Epoxy-embedded samples are hardened by heat curing at 60–70°C and produce extremely hard blocks that may be cut, sawed, or machined, section easily, and have excellent stability under an electron beam (Table 6).

Acrylate-based embedding media are more hydrophilic than epoxies and offer important advantages for both general and specialized usages (Table 6). Commonly used ac-

**Table 3. Suitability of Dehydration Methods for Imaging Techniques**

| Dehydration | LM | TEM | SEM |
| --- | --- | --- | --- |
| Graded ethanol series (30, 50, 70, 80, 90, 95, 100% v/v) | Yes | Yes | Yes |
| Graded acetone series (30, 50, 70, 80, 90, 95, 100% v/v) | Yes | Yes | Yes |
| Air-drying | Yes | No | Yes |
| Lyophylization | Rarely | Rarely | Yes |
| Critical point drying | Rarely | Rarely | Yes |
| Fluorocarbon sublimation (Peldri) | No | No | Yes |
| Hexamethyldisilizane | No | No | Yes |

**Table 4. Suitability of Dehydration Media and Methods for Polymer Delivery Systems**

| Dehydration | Hydrogel | Celluloses | PLGA | PLA | PCL | Polyanhydride | Protein | EV Ac |
| --- | --- | --- | --- | --- | --- | --- | --- | --- |
| Ethanol/water | Excellent | Excellent | Good | Good | Good | Good | Good | Good |
| Acetone/water | Excellent | Excellent | Poor | Poor | Poor | Poor | Excellent | Poor |
| 2,2-Dimethoxypropane | Excellent | Excellent | Poor | Poor | Poor | Poor | Good | Poor |
| Peldri II | Excellent | Excellent | Poor | Poor | Poor | Poor | Excellent | Fair |
| Critical point drying | Excellent | Excellent | Good | Good | Good | Good | Excellent | Good |
| Lyophylization | Excellent | Excellent | Excellent | Excellent | Excellent | Excellent | Excellent | Excellent |
| Hexamethyldisilizane | Excellent | Excellent | Poor | Fair | Poor | Poor | Excellent | Poor |
| Air-drying | Poor | Poor | Good | Good | Good | Good | Poor | Good |
| Air-drying/60°C heat | Poor | Poor | Fair | Fair | Poor | Poor | Poor | Poor |

*Note:* PLGA, poly(lactide-*co*-glycoglide) 50:50; PLA, poly(lactide); PCL, poly(caprylactone); EVAc, ethylene-*co*-vinyl acetate.

**Table 5. Suitability of Embedment Media for Imaging Techniques**

| Embedment media | LM | Immunochemistry | Enzyme histochemistry | TEM | SEM |
|---|---|---|---|---|---|
| PEG/PVA | Good | Excellent | Excellent | No | Yes |
| GMA | Excellent | Good (with etching) | Good (with etching) | No | Yes (without infiltration) |
| Epon/araldite | Excellent | Poor (with etching) | Poor (with etching) | Yes | No |
| Spurr's low viscosity | Excellent | Fair (with etching) | Fair (with etching) | Yes | No |
| LR white/LR gold | Excellent | Good (LM/TEM) | Good (LM/TEM) | Yes | Yes (without infiltration) |
| Lowicryl | Excellent | Good (LM/TEM) | Good (LM/TEM) | Yes | Yes (without infiltration) |

**Table 6. Chemical Properties of Embedment Media**

| Embedment media | Low temperature | Hydrophilicity | Viscosity | UV cure | Heat cure | Chemical cure |
|---|---|---|---|---|---|---|
| PEG/PVA | $-40$ to $-20°C$ | Yes | High | No | No | No |
| GMA | No | Yes | Low | No | No | Yes |
| Epon/araldite | No | No | High | No | $70°C$ | No |
| Spurr's low viscosity | No | No | Medium | No | $70°C$ | No |
| LR white/LR gold | $-20$ to $4°C$ | Yes | Low | Yes | $60°C$ | Yes |
| Lowicryl | $-40$ to $4°C$ | Yes | Low | Yes | $60°C$ | Yes |

rylates, such as Lowicryl, LR white, and LR gold have extremely low viscosities, indeed much lower than epoxies, and can be used to infiltrate samples without dehydrant dilution, reducing the likelihood of solvent extraction. Acrylate monomers have superior infiltrative properties and often plasticize and fatigue brittle thermoplastics. They have been used with success in infiltrating difficult samples (Table 7). Samples can be infiltrated, for periods of 24 hours to years, and even cured at $-20$ to $4°C$ using blue-light photoinitiation. Since low-temperature treatment preserves the antigenicity and enzyme activity of protein encapsulants, it is not surprising that the acrylates have been widely used for enzyme and antibody localization studies at the TEM level. Histochemical experiments are facilitated by the hydrophilic properties of the hardened blocks, which are much more permeable to aqueous stains and solutions than epoxies (5). For general use, acrylate monomers may be cured by heating at $60°C$ or else exothermic catalysis with benzoyl peroxide. Acrylate-embedded samples produce hard blocks, softer than epoxies, that section easily and offer moderate electron-beam stability. Two shortcomings of acrylates and LR white in particular are the requirement to exclude oxygen from the sample during curing, necessitating the use of closed embedding capsules, and the inability of acrylates to infiltrate hard materials, causing samples to drop-out during sectioning.

**Sectioning.** Embedded sample blocks are first trimmed with razor blades to form a trapezoid with dimensions not exceeding 2–3 mm on a side. Semithin sections, ranging from 0.5 to 1 $\mu$m, are produced using glass knives mounted on an ultramicrotome. Once a suitable area has been identified by light microscopy, the block face may be reduced in size to 1–2 mm on a side, and from that point on sectioning is typically continued with a diamond knife. Sections, ranging in thickness from 70 to 90 nm, with interference colors of silver and gold, respectively, are produced as a floating ribbon on the surface of a liquid-filled trough or "boat" attached to the knife. Most commonly, the boat fluid is water; however, to reduce extraction of water-soluble encapsulants during sectioning, other fluids may be substituted. The sections are collected on copper grids, dried with filter paper, and may be examined directly or else stained with electron dense materials to improve contrast. In the case of fragile sections embedded in acrylate media, support-film-coated grids, composed of thin films of formvar or parlodian polymers are often used to improve the stability of the sections under the electron beam. Additionally, support films or else thin section-support film sandwiches on grids may be coated with a thin deposition of evaporated, electron-transparent carbon metal to improve thermal conductivity.

A specialized form of thin sectioning, performed with frozen samples is cryoultramicrotomy. Tissue is quickly

**Table 7. Suitability of Embedment Media for Infiltration of Polymer Delivery Systems**

| Embedment media | Hydrogel | Celluloses | PLGA | PLA | PCL | Polyanhydride | Protein | EV Ac |
|---|---|---|---|---|---|---|---|---|
| PEG/PVA | Excellent | Excellent | Fair | Fair | Fair | Fair | Good | Fair |
| GMA | Excellent | Excellent | Good | Good | Good | Good | Excellent | Poor |
| Epon/araldite | Excellent | Excellent | Poor | Poor | Poor | Poor | Good | Poor |
| Spurr's low viscosity | Excellent | Excellent | Good | Good | Fair | Fair | Good | Poor |
| LR white/LR gold | Excellent | Excellent | Excellent | Excellent | Good | Good | Excellent | Poor |
| Lowicryl | Excellent | Excellent | Excellent | Excellent | Excellent | Excellent | Excellent | Fair |

*Note:* PLGA, poly(lactide-*co*-glycoglide) 50:50; PLA, poly(lactide); PCL, poly(caprylactone); EV Ac, ethylene-*co*-vinyl acetate.

frozen by rapid contact with a liquid nitrogen-chilled copper block and sectioned using an ultramicrotome capable of maintaining temperatures at less than $-70°C$ during sectioning. Sectioning is normally performed on dry glass knives and sections are transferred to formvar-coated copper grids for TEM observation. This technique is not commonly used because of the technical proficiency required to obtain acceptable results; however, the method is unsurpassed for localizing bioactive agents at the TEM level, using histochemical methods, without danger of chemical denaturation from fixation and dehydration.

**Staining.** Most commonly used TEM stains are uranyl acetate and lead citrate used singly or in a two-step procedure (see Table 2). Lead and uranium both have high atomic numbers and are extremely electron dense. The heavy metal atoms interact with the incident electron beam to produce highly contrasted images. The citrate and acetate salts are most commonly used because of their relative resistance to precipitation compared with other chemical forms of the stains. Staining is normally performed in carbon dioxide–deprived environments to prevent formation of insoluble, heavy metal-carbonate precipitates, called cookies and needles.

Staining may be performed either on thin sections collected on grids or else en bloc on the sample before dehydration. In the first case, sample grids are floated on the surface of droplets of particle-free, filtered stain for periods of 10–60 min, followed by washing in distilled water. Typically, staining with uranyl acetate is performed first followed by staining in lead citrate. Alternately, immediately following fixation, samples may be immersed in solutions of 0.5–2% (w/v) uranyl acetate for periods of 2–5 h at 4°C. En bloc samples are then dehydrated, infiltrated, and embedded using normal protocols.

Vapor phase staining with osmium tetroxide vapors can be used to increase the contrast of samples embedded in acrylate mediums. Sections on sample grids are sealed in a closed vessel containing osmium tetroxide solution in a smaller open vessel or else as crystals and incubated at room temperature for periods of up to 6 h. This procedure will also preserve unfixed samples in addition to improving specimen contrast.

Additionally, material scientists have employed a number of stains to contrast polymeric materials (see Table 2). Mercuric trifluoroacetate has been used to contrast polystyrene and polybutadiene, either singly or as block copolymers (6). Iodine has been used to stain polyamides and poly(vinyl alcohols) (PVA) (7,8). Silver sulfide was used to increase contrast of microporous structures in polymers including cellulose, polyesters, acrylics, and PVA (9,10). Phosphotungstic acid has been used to contrast polyamides and polypropylene (11,12).

## Localization of Enzymes, Proteins, and DNA

Histochemistry may be used to localize biomolecules encapsulated in drug delivery devices using samples lightly fixed in 1% (w/v) formaldehyde and 0.5% (v/v) glutaraldehyde for periods of 1–2 h at 4°C. A suitable primary antibody, either monoclonal (reactive with a single specific epitope) or polyclonal (reactive with multiple specific epitopes) must first be obtained or developed within an animal such as a sheep, goat, horse, rabbit, or pig and established to be specific to the target molecule and not cross-reactive with other agents in the sample. The procedure may be performed either on thin sections on grids of samples embedded in LR white or Lowicryl or else using thin slices of the unembedded, fixed sample, which may then be processed for TEM (13). Chemical etching of epoxy-based embedding media, such as Spurr's and Epon/araldite, as well as acrylate-based media such as GMA, has been used to enable limited histochemistry (14). Samples are first incubated with blocking agents, such as serum albumin, gelatin, or glycine to reduce nonspecific binding of primary antibodies to free, reactive aldehyde groups.

Antibody staining may be done in two ways: either directly with the primary antibody or indirectly with a second agent directed against the primary antibody. In the first case, the primary antibody is chemically conjugated to an electron dense marker such as ferritin or colloidal gold to facilitate visualization at the TEM level. Typically, the primary antibody is used in a diluted form to increase specificity (1:100–1:500 for polyclonal antibodies versus 1:250–1:2,000 for monoclonal antibodies, as determined experimentally) and reacted with the blocked sample for periods of 1–24 h at room temperature. The sample is then washed and examined using TEM.

The second scenario, indirect labeling, has greater specificity and more general applicability. An untagged primary antibody is used, followed by secondary incubation with either a tagged secondary antibody directed against both the host animal and antibody class (IgG, IgM, etc.) for the primary antibody. The secondary antibody is normally available commercially and may be tagged with colloidal gold or ferritin markers. The second reaction is normally carried out at dilutions of 1:100–1:250 for periods of 1–3 h at room temperature. Alternately, protein A- or protein G-colloidal gold complexes may be used in place of the secondary antibody. Protein A and protein G are cell wall proteins isolated from staphylococcal and streptococcal bacteria, respectively, with specificity against the Fc portion of IgG molecules from most common laboratory animals.

Enzyme localization at the TEM level may be carried out for a number of enzymes producing an insoluble end product, including acetylcholinesterase (15), galactosidase, alkaline phosphatase (16), acid phosphatase (17), adenyl cyclase (18), arylsulfastase (19), ATPases (20), catalase (21), cytochrome oxidase (22), cytosine monophosphastase, esterases (23), glucose-6-phosphatase (24), nictotine adenine dinucleotide phosphatase (25), 5′-nucleotidase (26), peroxidase (27), and thiamine pyrophosphatase (28), to name a few. Typically, thin sections of lightly fixed samples, either embedded in acrylate medium or reacted as thin slices and then postembedded are reacted with excess substrate at optimal pH for the enzyme and with all necessary cofactors supplemented in the reaction buffer. Trapping agents, often insoluble metal salts or other enzymes such as peroxidase, are often employed to produce an electron dense, insoluble end product that may be visualized by TEM.

It is important to note that appropriate controls for all localization schemes should be employed at all times. In the case of immunochemical methods, nonspecific or nonimmune serum from the same animal species should be used as control for the primary antibody and in the case of enzyme methods, controls should include samples without added substrate or trapping agents. In all instances, LM should be used to confirm and target the extent and location of the reactions before carrying the analysis to the TEM level.

## Negative Staining

This method of visualization does not involve sectioning of embedded material and is best carried out on either fixed or unfixed material that is sufficiently thin to be be penetrated by the incident electron beam (~10 and 200 nm), although it can be used to crudely visualize and size larger particles such as bacteria, cells, and polystyrene latex particles (1–5 $\mu$m). Unlike positive staining, where the specimen reacts or binds heavy metal stains, the sample particles in negative stain are very often unstained and are instead embedded in a droplet of dried stain, resulting in negative contrast images. An important artifact is that images of hollow spherical objects are often collapsed because of interfacial tensions during drying. Commonly used negative stains are 1% (w/v) solutions of uranyl acetate, phosphotungstic acid, or ammonium molybdate. Plastic films, made from formvar or parlodian, on copper grids are used as supports for the particles. Specimens are applied to the grids in one of two ways: either mixed directly with the stain or else applied in a stepwise manner with the particles in suspension introduced onto the grid, followed by a droplet of stain. The method is simple and may be used to visualize a variety of biological entities such as large proteins, nucleic acids, viruses, ribosomes, and bacteria. Negative staining has also been useful in the structural analysis of liposomes for drug delivery (Fig. 3) and oily emulsions.

## Replica Techniques and Freeze-Fracture/Freeze-Etch

The use of metal shadowing techniques to prepare faithful replicas of surfaces is used extensively in materials sciences and can be used to collect high-resolution topographic details of drug delivery devices with resolutions of 3–4 nm. Replicas are prepared using a vacuum evaporator capable of routinely achieving vacuums of from $10^{-4}$ to $10^{-7}$ mmHg. Typically, the sample is fixed, applied to a clean surface such as freshly cleaved mica or a glass substrate, and placed under high vacuum. The most commonly used replica technique is the one-stage, fixed-angle replica: an electron dense platinum coating is evaporated onto the sample surface at an angle of 45°, followed by a second protective coating of electron-transparent carbon metal, applied at an angle of 80–90°, to conduct surface charge during TEM examination. The sample is removed from the instrument and the replica is displaced from the sample by floating at the interface of a chromic acid or bleach solution. Treatment with acid and bleach is used to destroy the sample and subsequent washing with water removes residue from the replica. The replica can then be mounted

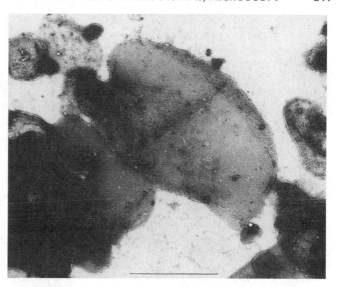

**Figure 3.** Negative stain TEM of liposomes. Phosphatidyl choline, unilamellar liposomes were prepared by vortex treatment and bath sonication of 10 mg/mL dispersion of the lipid in 50 mM phosphate buffer. A 10-microliter droplet of the suspension was placed on a carbon-formvar-coated grid. After 1 min the droplet was removed by wicking with filter paper and replaced with a microdroplet of 1% uranyl acetate (w/v) in distilled water for 30 sec. The stain was removed, and the grid was air-dried for 1 min and examined directly in the TEM at 80 kV. The vesicles appear collapsed after drying down with stain. This phenomenon is an accepted artifact of the negative staining. Bar = 1 $\mu$m.

on a copper grid and examined directly by TEM. By measuring the length of the shadow (area where metal deposition was occluded by sample interference) cast by samples, it is possible to determine the height of the sample using the angle of platinum deposition and a simple geometric relationship.

A variation of this method is rotary shadowing, in which case the sample is mounted onto a motor-driven stage in the evaporator and rotated at 10–50 rpm while platinum is evaporated at an angle of ~25°. The replica produced using this technique yields exquisite detail of all areas of the specimen because no shadow is created, as is the case with static evaporation.

A less commonly used form of replication is the two-stage negative replica. This method provides a lower resolution image than the one-stage replica, but has the advantage of not destroying the sample. Bulk samples are coated with several layers of 1% (w/v) collodion plastic in solvent. After drying, the plastic replica is stripped from the sample and a negative, platinum–carbon replica is prepared of the plastic replica, as already described. The plastic replica can be separated from the metal replica by dissolution in solvent.

Smaller samples (<10 $\mu$m in diameter) may be spray- or air-dried from a liquid suspension onto plastic-coated grids and simply shadowed with metal to reveal hidden surface detail without destroying the underlying sample with strong agents. Shadowing in this manner has many similarities to negative staining and can be used to render surface topography of nanoparticles and microspheres.

Finally, a variation of the replica technique widely used for evaluating liposomes and emulsions is freeze-fracture (Fig. 4a). Samples are frozen by immersion in liquid Freon on copper or gold supports and introduced into a refrigerated vacuum evaporator capable of cooling to $-170°C$. Vacuums of $10^{-4}$ mmHg are achieved, the sample is warmed to $-110°C$ and the sample is fractured by cutting with a liquid nitrogen–cooled razor blade mounted to a microtome inside the vacuum unit. The fractured surface is replicated using either single-sided or rotary platinum–carbon evaporation and the sample is separated from the replica by thawing in bleach or chromic acid. The replica is then washed with water and examined by TEM. The freeze-fracture replica can yield morphological information about the interior compartments of closed vesicles.

A variation of the freeze-fracture technique uses etching or sublimation of frozen water to reveal not only fracture planes, but also interior and exterior surfaces as well (Fig. 4b). Freeze-etching is performed immediately following fracture, for periods of 1–5 min at $-100°C$ instead of $-110°C$.

## SCANNING ELECTRON MICROSCOPY

SEM is an important supplement to LM and TEM. SEM is probably the most commonly used method for charac-

terizing drug delivery systems, owing in large part to simplicity of sample preparation and ease of operation. The three-dimensional information derived from the micrographs offers important information about macro- (0.1–10 mm), meso- (1–100 $\mu$m) and microstructure (10–1,000 nm), often within the same micrograph! SEM has been used to determine particle size distribution, surface topography, and texture and to examine the morphology of fractured or sectioned surfaces.

The modern SEM has a resolution of ~3 nm (Fig. 5) and provides magnifications ranging from less than 30-fold to 300,000-fold. Electrons emitted by a excitation of a lanthanum hexaboride crystal or heated filament are accelerated at voltages of 1–30 kV and focused onto the sample surface as a tight incident beam (Fig. 6). The 3–5 nm beam is translated about the probe area in a raster pattern. A variety of electrons and electromagnetic radiation with different energies are ejected from the sample surface in response to the incident electron beam. Secondary electrons having relatively low energy ranges of 0–50 eV are the result of inelastic scattering from atoms located near the surface of the sample. These electrons can be collected with a detector and used to produce three-dimensional images of the sample surface. Nearly all scanning electron micrographs are produced using secondary electrons. Backscat-

 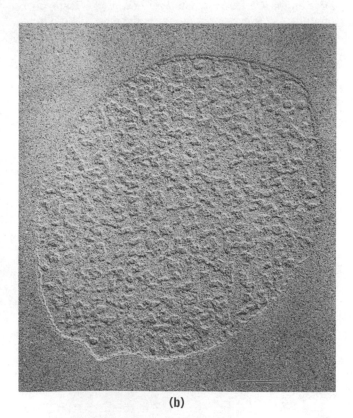

(a)                                  (b)

**Figure 4.** Freeze-fracture and freeze-fracture/etch of PLA microspheres prepared by solvent removal. (**a**) A single microsphere that has been frozen in chilled Freon 22, fractured at $-110°C$, and replicated with platinum and carbon evaporation while at high vacuum ($10^{-4}$ mmHg) in a Balzer BAF-400 instrument. This image shows a fracture plane in the polymer matrix and shows uniform crystallinity. (**b**) A similar microsphere that has been etched by sublimation of ice to water vapor at a slightly higher temperature ($-100°C$). The extreme crystallinity of the surface of the fracture plane is evident. Bar = 0.1 $\mu$m.

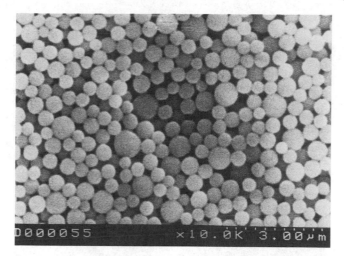

**Figure 5.** SEM of microspheres prepared by phase inversion nanoencapsulation (PIN). PLGA 50:50 microspheres range in size from 200 to 500 nm and have a high degree of monodispersity and smooth surface texture.

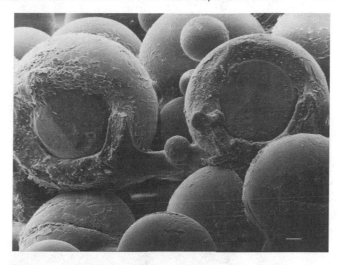

**Figure 7.** Three-dimensionality of SEM images. Double-walled microspheres composed of ethylene-*co*-vinyl acetate (EVA) in the inner layer and PLA in the outer layer. Two of the spheres were sectioned with a razor blade to reveal the inner cores. Bar = 50 $\mu$m.

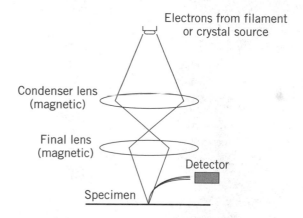

**Figure 6.** Simplified arrangement of optics in the scanning electron microscope.

tered electrons having higher energy ranges than secondary electrons are the result of elastic scattering with atoms located deeper beneath the surface. Backscattered electron images can be useful in distinguishing materials having large differences in atomic number and have been useful in gold-labeling experiments. Additionally, X-rays generated by sample interaction with the probe beam may be collected and used to evaluate the elemental composition of the sample.

The various signals collected by secondary, backscattered, and X-ray detectors at each point on the probed surface are processed, amplified, and displayed as a real-time, video image on a monitor. The images may be printed on photographic film or stored and printed digitally.

The SEM is most commonly used for generating three-dimensional surface relief images derived from secondary electrons (Fig. 7). The mechanism evoked to explain this phenomenon has been dubbed the edge effect. Simply put, secondary electrons from depressions or cavities on the sample surface are trapped within the surface topography

resulting in diminished emission from these features, whereas the converse situation is true for elevations or peaks on the sample surface where electron emission is enhanced.

A great advantage of SEM over other techniques is the tremendous versatility of the method. Samples ranging in size from nanometers, such as viruses and small microspheres, to centimeter-scale bulk samples, such as vascular grafts or osmotic pumps, may be analyzed, even on the same sample holder. Although the working resolution of the instrument is less than TEM, the ease of sample preparation and simplicity of use more than compensate for resolution limitations. The large sample chamber and goniometer accommodate extremely large samples compared to TEM and provide for nearly unlimited points of viewing, owing to translational, tilting, and rotary movements.

The SEM has a few disadvantages compared to LM and TEM. SEM is used primarily to observe surface topography and texture. SEM does not see through a sample, although it is possible to view inside a sample by examining fractured or sectioned specimens. SEM does not distinguish colors as does LM. The analysis of stained samples or polarized light can often be used to identify polymeric materials. Also, it is difficult to distinguish differences in materials. Identification of materials must be confirmed by comparison to pure reference samples fabricated using identical conditions and processed for SEM using the same protocols as the sample. Another disadvantage is that labeling techniques using immunochemicals or enzymes are limited to the use of electron-dense markers, such as gold or lead, and samples must be visualized using backscattered electron imaging. Additionally, the use of thermoplastic polymers or polymeric materials with low glass-transition temperatures for fabrication of drug delivery systems imposes limitations on the use of SEM. These materials require special precautions to avoid sample distortion, resulting from thermal changes during observation.

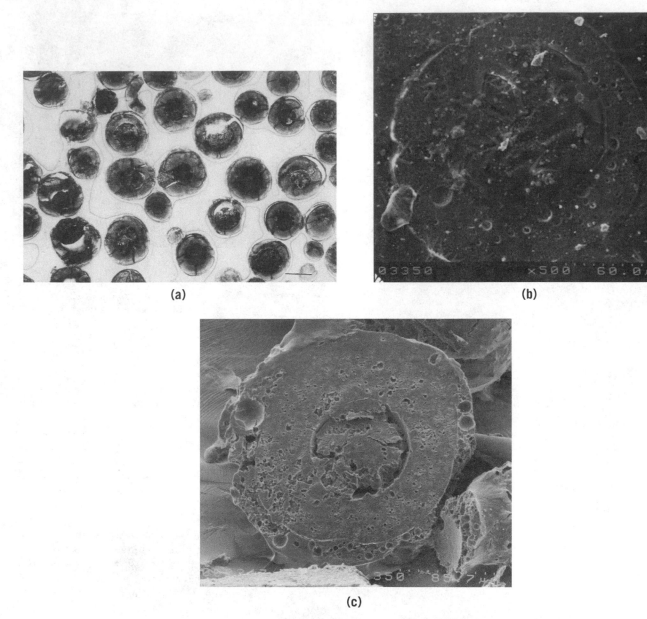

(a)

(b)

(c)

**Figure 8.** Analysis of double-walled microspheres using LM and SEM. (**a**) Double-walled microspheres made of polystyrene (inner layer) and PLA (outer layer) loaded with particles of ferric oxide (central core). Samples were embedded in GMA and sectioned with glass knives. LM indicates that nearly all the spheres are double-walled and shows the uniformity of loading. Bar = 100 $\mu$m. Courtesy of Benjamin Hertzog. (**b**) and (**c**) double-walled microspheres composed of polyanhydride and PLA. (**b**) A microsphere that was poorly embedded in GMA and sectioned with glass knives. The sections were mounted on stubs using carbon-impregnated, adhesive tape, coated, and examined by SEM. (**c**) A microsphere that was sectioned with a razor blade and prepared for SEM. SEM analysis shows the porosity of the inner layers and localization of the drug. For more information about double-walled microspheres refer to Ref. 29.

It is often necessary to use low accelerating voltages (less than 10 kV) to avoid excess heating, and heavy metal coating protocols must be optimized to reduce thermal artifacts. Finally, as in TEM, samples must in most cases be dehydrated because of the high vacuum requirement of most electron microscopes.

It is worth noting that a specialized version of the SEM, the environmental SEM, is commercially available and is applicable for the observation of hydrated, unfixed samples, although at lower resolutions than conventional high-vacuum SEM systems. This instrument functions by maintaining a differential vacuum; that is, a high vacuum environment is maintained near the electron gun while a lower vacuum environment is used for the sample chamber. Future developments will probably bring this specialized instrument

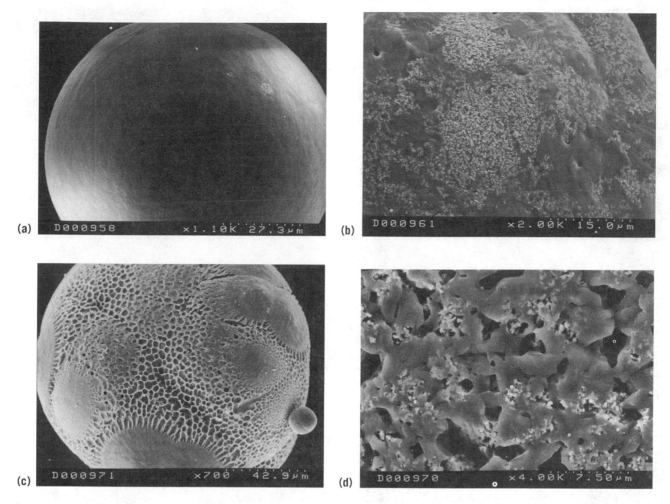

**Figure 9.** SEM analysis of topography and encapsulant distribution in solvent evaporation spheres. Poly(caprylactone) (PCL) (**a–b**) and PLA spheres (**c  d**) were prepared by dispersing a 20% (w/v) polymer solution in methylene chloride into a bath of 1% poly(vinyl alcohol) (w/v). The microspheres were stirred until they had hardened and were collected by filtration. Microspheres were prepared with and without 300 nm ferric oxide particle by including the particles in the polymer solution. (**a**) The smooth surface morphology of the unloaded microspheres. (**b**) The clustered distribution of iron oxide particle on the surface of the PCL spheres. (**c**) The extreme porosity and uneven distribution of pores on unloaded PLA spheres. (**d**) Higher magnification of the surface shows that the metal particles are associated with crystalline regions (spherulites).

into more widespread use for analysis of controlled release systems.

## Sample Preparation

**Fixation.** Fixation is generally not necessary, except where biological encapsulants are involved, such as biomolecules and cells, in which cases the general fixation procedures described for TEM are directly applicable. For biomolecules, these steps are only required when solvents are used for dehydrating the sample, not when lyophilization is used for dehydration.

Oftentimes, a postosmication step is employed to heighten contrast and increase the secondary electron signal derived from the sample. This treatment involves soaking the sample in thiocarbohydrazide, which binds to osmium tetroxide already on the sample and binds addi-

tional osmium tetroxide when the sample receives a second, supplemental osmium fixation. This additional coating of heavy metal greatly amplifies the number of secondary electrons emitted from the sample by providing additional atomic targets for the primary probe beam.

**Sectioning.** Although SEM is primarily used for analysis of surface topography, it is possible to look at interior surfaces by simply sectioning the bulk sample with a razor blade (Figs. 7, 8). This procedure suffers from artifacts introduced by compressing the sample downwards during cutting and is especially troublesome for soft samples. Cryosectioning obviates many of these concerns. Samples are quickly frozen by immersion in liquid Freon or liquid nitrogen and fractured by sectioning on a liquid nitrogen–chilled metal surface with a chilled razor blade. Samples

**Figure 10.** SEM of spray-dried microspheres. A 3% (w/v) solution of PLGA 50:50 in methylene chloride was spray-dried with a 1.0 mm atomizer nozzle at 50 psi atomizer pressure using a flow rate of 30 mL/min at 40°C. (**a**) Unloaded microspheres ranging in size from 0.1 to 15 μm with smooth surface morphology. (**b**) Microspheres loaded with 10% antibiotic (w/w) as encapsulant. The drug-loaded spheres have the same size range but extremely rough, porous surface texture.

**Figure 11.** (**a**) SEM of PLA microspheres prepared using PIN. The spheres are discrete, range in size from 0.1 to 6 μm and have smooth surface texture. (**b**) PLA particles prepared under similar solvent/nonsolvent conditions using the organic phase separation process. These particles range in size from 2 to 5 μm and have extremely coarse surface texture. The particles are individual spherulites of PLA. For detail of PIN, see Ref. 30, and for details of organic phase separation see Ref. 31. Bar = 1 μm.

as small as 300 μm in diameter may be sectioned using the razor blade technique.

Smaller samples may be prepared by cryostat sectioning of PVA-PEG embedded, frozen samples. Sections, between 10 and 50 μm in thickness, can be collected on a glass slide, dehydrated by either air-drying or freezing and lyophilization, and the entire slide mounted for observation by SEM. Alternately, smaller samples may be quickly embedded, without infiltration in acrylate-based embedding media such as LR white or GMA, sectioned with a glass knife and microtomed to produce 1–10-μm-thick sections mounted on glass slides for SEM analysis (see Fig. 8).

**Dehydration.** As with all electron microscopes (excluding the environmental SEM), high vacuum is a requirement for image formation, and samples must be thoroughly desiccated before entering the vacuum chamber. As mentioned previously (in the discussion of dehydration of TEM samples, care must be taken to avoid collapse. Samples are best prepared by critical point drying, lyophilization, or else by using solvent substitution with hexamethyldisilizane. Fluorocarbon solvents, such as Peldri II (a proprietary product of Ted Pella Co., California, now discontinued), that slowly sublimate from the solid state at room temperature have been widely used in the past. This latter procedure requires graded ethanol dehydration before solvent exchange, as does critical point drying.

**Mounting.** The dried sample is normally attached to an aluminum sample holder or stub using silver- or carbon-

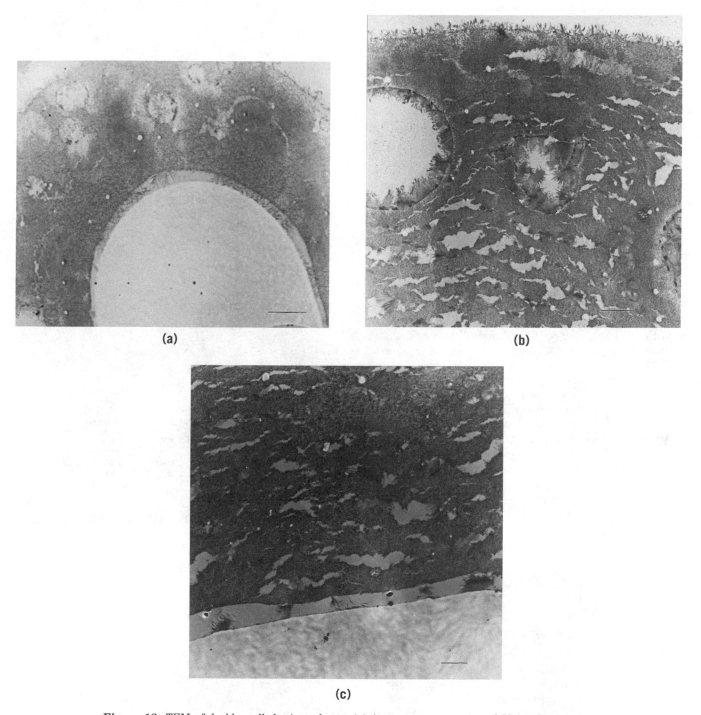

**Figure 12.** TEM of double-walled microspheres. (**a**) A survey cross section of PLA-polystyrene microspheres. Three distinct zones are identified: the outer PLA zone, an intermediate zone, and an inner core of polystyrene. Bar = 10 $\mu$m. (**b**) Details of the outer PLA zone and surface. The crystallinity of the polymer is obvious from observation of the inclusions and surface. Bar = 3 $\mu$m. (**c**) Details of the PLA-polystyrene zones and interfacial region. Bar = 3 $\mu$m. The polystyrene core at the bottom appears nearly featureless, which is consistent with the amorphous nature of the polymer. The microspheres were infiltrated with LR white and cured at 40°C for 4 days to minimize thermal changes to the polymer. Thin sections were stained with uranyl acetate to contrast the PLA layer.

based conductive cements that are commercially available for this purpose. A variety of double-sided adhesive tapes and tabs, impregnated with carbon, are often more convenient for smaller samples such as microspheres, which might become totally immersed in a layer of liquid cement.

A vexing artifact, commonly called charging, arises when SEM samples are poorly attached to the underlying substrate or contain residual moisture or a combination of both. Charging results from accumulation of electrons within the sample, imparted by irradiation from the incident electron beam. The excess electron load would normally be dissipated by conduction to the sample holder. Objects that are charging often appear to glow brightly on the viewing screen compared to objects that are optimally prepared. The artifact can often be eliminated by repeating the dehydration step with new sample or else by remounting and recoating the same sample.

**Coating.** Thin coatings (20–30 nm) of electron-dense metals, typically gold, palladium, platinum, or combinations, are applied to the mounted sample using either sputter-coating or thermal vacuum evaporation. The coatings serve two purposes: (*1*) to heighten sample signal by providing additional heavy metal atoms for interaction with the incident electron beam and (*2*) to conduct accumulated sample charge and heat to the sample holder.

Sputter-coating is most commonly used for SEM coating for reasons of simplicity and nondestructiveness. The mounted sample is placed in a vacuum chamber, the chamber is evacuated using a rotary pump, and an inert carrier gas, typically argon, is introduced to produce a partial vacuum of approximately $10^{-2}$ mmHg. The argon atmosphere is ionized by electrodes located near a metal foil, or target, composed of gold or gold–palladium metal. The interaction of ionized argon atoms with the metal foil causes heavy metal atoms to be ejected from the foil, covering the underlying sample with a finely dispersed, particulate coating. The process is complete within minutes and produces relatively little heating damage compared to thermal vacuum evaporation. Thermal vacuum evaporation is more commonly used for deposition of platinum and carbon coatings, as previously described in "Replica Techniques and Freeze Fracture."

### Imaging of Drug Delivery Systems

SEM is most often used to characterize drug delivery systems based on surface morphology, texture, and size. The examination of the surface of polymeric drug delivery systems can provide important information about the porosity and microstructure of the device (Figs. 5, 9–11). Increased porosity translates into rapid release of drug. The size and distribution of surface pores can be evaluated by direct measurement using the image scale marker or interpolation from micrographs using the magnification factor to calculate dimensions of objects (Fig. 9).

The distribution and morphology of drug on the surface and encapsulated in the matrix (fractured and sectioned samples) can also be directly observed (see Fig. 9). It is important to first examine pure samples of drug and excipients, prepared in exactly the same form used for en-

capsulation but with minimal SEM preparation (no exposure to solvents). These reference samples should be compared to drug samples that have been carried though the same SEM processing protocol as the delivery devices. Artifactual changes in morphology of encapsulant resulting from processing, notably, aggregation, coalescence, or dissolution should be investigated and distinguished from real changes resulting from manufacture. This information is best used in conjunction with data from size exclusion, high performance chromatography (SEC-HPLC) analysis of drugs extracted or released from delivery systems. SEC-HPLC describes the aggregation state of drug following encapsulation and SEM depicts the morphology of the drug particles in the system.

Examination of cross or tangential sections can similarly provide information about the uniformity of drug distribution throughout the delivery system (Figs. 7, 8b, 8c).

(a)

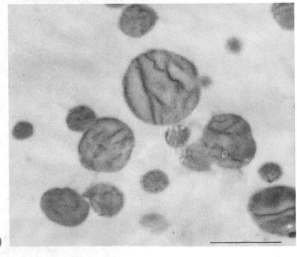

(b)

**Figure 13.** SEM and TEM of PLA microspheres prepared by solvent removal. (**a**) Microspheres dried on a glass coverslip and observed from a 75° tilt angle. (**b**) The same microspheres were prepared for thin-section TEM. The crystallinity of the inner core is revealed in the micrographs. Bar = 1 $\mu$m.

**Table 8. Suitability of TEM Techniques for Characterization of Drug Delivery Systems**

| TEM technique | Imaging | Application to drug delivery systems |
| --- | --- | --- |
| Thin-section | Interior morphology | Drug distribution, polymer morphology |
| Immunochemistry | Interior morphology | Drug distribution |
| Enzyme histochemistry | Interior morphology | Drug distribution |
| Negative stain | Exterior morphology | Small particles (<1 $\mu$m) liposomes, nanoparticles, lataxes |
| Freeze-fracture | Internal morphology | Drug distribution, polymer morphology (especially liposomes) |
| Freeze-etch | Internal morphology/internal and external surfaces | Drug distribution, polymer morphology (especially liposomes) |
| Replica | Exterior morphology | Small particles (<1 $\mu$m) liposomes, nanoparticles, lataxes |

Pores on the surface of the delivery vehicles can often be traced to the interior matrix and the interconnectivity of surface and interior voids can be estimated. As an example, the distribution of drug particles near porous microstructures can offer important information about drug release (Fig. 9d).

The morphology of the polymer matrix can be evaluated and differences between crystalline and amorphous regions and their locations in the delivery system can be discriminated and mapped. Crystalline regions can be identified by the presence of spherulites (Figs. 9d, 11b). This is important because release of drug from crystalline regions is slower than from amorphous domains. Therefore domains of crystalline and amorphous polymer contribute to a mixed release profile: the sum of slow- and fast-releasing polymer matrices in a single system. Changes in polymer morphology during manufacturing and storage can be monitored by SEM and used in concert with information about polymer molecular weight, from gel permeation chromatography and crystallinity, from differential scanning calorimetry, to provide a full profile of the polymer component of the delivery system.

## SUMMARY

In conclusion, LM, TEM, and SEM provide complementary morphological information about drug delivery systems and should be used in an integrated approach for purposes of characterization (Table 1, Figs. 4, 7, 8, 12, and 13). LM can be used to provide macro- and microstructural information about drug delivery systems, including size and morphology. Stains can be used to localize encapsulants and preferentially stain polymers based on ionic and lipophilic properties (see Table 2). TEM provides micro- and nanostructural data and extends the optical range of LM to the atomic level (Table 8). TEM analysis yields information about polymer morphology in the matrix of the delivery systems and the localization of encapsulant in the matrix (Figs. 12 and 13b). Using more advanced replica techniques, it is possible to examine the surface of delivery systems and observe the morphology of vesicles and nanoparticles (see Table 8).

Finally, SEM completes the characterization by providing a three-dimensional depiction of macro-, meso-, and nanostructure on exterior and interior (cut) surfaces of drug delivery systems. SEM provides a convenient means of studying porosity and interconnectivity of pores. Most importantly, the distribution of encapsulant in the polymer matrix and the interaction of drug with surrounding polymer can be analyzed, as well as crystallinity differences in the polymer matrix itself. The abundance of information derived from SEM analysis can be further supplemented by LM and TEM data, as well as information from other physical characterization methods such as differential scanning calorimetry, Fourier transform infrared, gel permeation chromatography, and high-performance liquid chromatography to gain a fuller understanding of the mechanisms underlying controlled release in drug delivery systems.

## BIBLIOGRAPHY

1. A.B. Maunsbach, *J. Ultastruct. Res.* **15**, 242–282 (1966).
2. M.A. Hayat, *Fixation for Electron Microscopy*, Academic Press, New York, 1981, p. 207.
3. A.M. Glauert, *Fixation, Dehydration and Embedding of Biological Specimens*, Elsevier, Amsterdam, 1975.
4. L.C. Sawyer and D.T. Grubb, *Polymer Microscopy*, Chapman & Hall, New York, 1987, p. 97.
5. P. Gerrits, R. Horobin, and D. Wright, *J. Microsc. (Oxford)* **160**, 279–290 (1990).
6. S.Y. Hobbs, V.H. Watkins, and R.R. Russell, *J. Polym. Phys. Ed.* **18**, 393–395 (1980).
7. K. Hess and H. Kiessig, *Naturwissenschaften* **31**, 171 (1943).
8. K. Hess and H. Mahl, *Naturwissenschaften* **41**, 86 (1954).
9. M. Sotton, *C.R. Hebd. Seances Acad. Sci.* **270B**, 1261 (1971).
10. W. Fredericks and K. Bosch, *Histochemistry* **100**, 297–302 (1993).
11. B.J. Spit, *Proc. 5th Int. Congr. Electron Microsc.*, Philadelphia, August 29–September 5, 1962, vol. 1, p. BB7.
12. C.W. Hock, *J. Polym. Sci. Part A2* **5**, 471–478 (1967).
13. W. Kuhlmann and P. Peschke, *Histochemistry* **75**, 151–161 (1982).
14. J. Litwin and K. Beier, *Histochemistry* **88**, 193–196 (1988).
15. R.M. Friedenberg and A.M. Seligman, *J. Histochem. Cytochem.* **20**, 771–792 (1972).
16. J. Hugson and M. Borgers, *J. Histochem. Cytochem.* **14**, 429–431 (1966).
17. A.L. Frank and A.K. Christensen, *J. Cell Biol.* **36**, 1 (1968).
18. L. Cutler, G. Rodan, and M.B. Feinstein, *Biochim. Biophys. Acta* **542**, 357–371 (1978).
19. J. Kawano and E. Akiwa, *J. Histochem. Chytochem.* **35**, 523–530 (1987).
20. S.A. Ernst, *J. Histochem. Cytochem.* **20**, 23–38 (1981).
21. A.B. Novikoff and R.J. Goldfischer, *J. Histochem. Cytochem.* **17**, 675–680 (1969).

22. A.M. Seligman et al., *J. Cell Biol.* **38**, 1–14 (1968).
23. R.A. Monahan, H.F. Dvorak, and A.M. Dvorak, *Blood* **58**, 1089–1099 (1981).
24. L.W. Tice and R.J. Barnett, *J. Histochem. Cytochem.* **9**, 635–636 (1961).
25. C.E. Smith, *J. Histochem. Cytochem.* **28**, 689–699 (1980).
26. G.W. Kreutzberg and S.T. Hussain, *Neuroscience* **11**, 857–866 (1984).
27. R.C. Graham and M.J. Karnovsky, *J. Histochem. Cytochem.* **14**, 291–302 (1966).
28. A.B. Novikoff and S. Goldfischer, *Proc. Natl. Acad. Sci. U.S.A.* **47**, 802 (1961).
29. K. Pekarek, J. Jacob, and E. Mathiowitz, *Nature (London)* **367**, 258–260 (1994).
30. E. Mathiowitz et al., *Nature (London)* **386**, 410–414 (1997).
31. U.S. Pat. 4,919,929 (April 24, 1990), L.R. Beck (to Stolle Research and Development Corp.).

# CHARACTERIZATION OF DELIVERY SYSTEMS, PARTICLE SIZING TECHNIQUES

K.L. WARD
M.A. TRACY
Alkermes, Inc.
Cambridge, Massachusetts

## KEY WORDS

Coulter principle
Electrical sensing zone
Electron microscopy
Image analysis
Laser diffraction
Light scattering
Optical microscopy
Particle sizing
Photon correlation spectroscopy
Quasielastic light scattering
Sieving
Static light scattering
Total intensity light scattering

## OUTLINE

## INTRODUCTION

Particle size is an important variable in drug delivery systems, affecting such parameters as rates of dissolution, rates of release, injectability, and dose delivery. Table 1 gives examples of the broad size ranges of particulate systems encountered. Unfortunately, no single method is capable of covering this large range of sizes. As a result, a wide variety of methods have been developed to measure particle size, using quite different physical principles. Because it is not possible to fully describe all of the methods in the space allotted here, we have focused this discussion on four of the techniques most widely used by drug delivery scientists, covering the millimeter to nanometer size range (sieving) microscopy (electron and optical), electrical sensing zone (Coulter principle), and laser light scattering methods. For a more comprehensive review of particle size analysis, including methods such as sedimentation that are not discussed here, see the work by Allen (1). Figure 1 gives the size ranges that can be measured by each of the techniques. Because each method has its strengths and limitations, the objective of this chapter is to provide the reader with an understanding of the physical principles and practical differences between these techniques which, it is hoped, will help in selecting the most appropriate method for the job. Within this work, examples are given which demonstrate how these methods can be applied to drug delivery systems.

## SIEVING METHODS

Sieving is one of the oldest methods of particle size determination. It involves passing sample through different sized meshes and determining the fraction captured by each mesh. This can be achieved with the powder in the dry or wet state. Sieving is most suitable for large particles (>75 $\mu$m), and it is often the pharmaceutical method of choice for sizing coarse powders, e.g., for nasal drug delivery (2). There is a U.S. Pharmacopeia (USP) method for accurately determining a particle size distribution by analytical sieving (3). There are limitations to this method, though, including a large sample requirement (>25 g) and difficulty in sieving cohesive and oily powders (3).

## MICROSCOPY METHODS

Examination of particles by microscopy is an absolute method for measuring particles, as it allows direct obser-

**Table 1. Particle Dimension Estimates in Drug Delivery and Pharmaceutical Systems**

| Particle Size ($\mu$m) | Examples |
|---|---|
| 0.001–0.5 | Proteins, polymers, oligonucleotides, DNA, micelles, microemulsions, colloidal suspensions |
| 0.5–10 | Suspensions, fine emulsions, micronized drugs |
| 10–50 | Coarse emulsions, flocculated suspensions |
| 50–150 | Fine powders |
| 150–1,000 | Coarse powders |
| 1,000–3,400 | Granules |

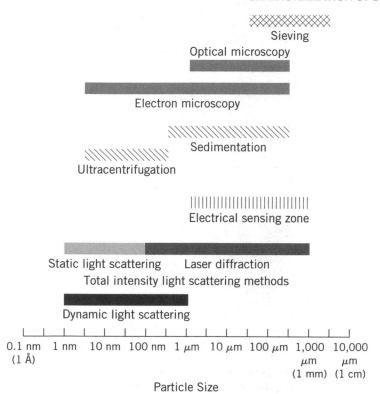

**Figure 1.** Approximate size ranges for methods used for particle size analysis.

vation of both the size and morphology of individual particles. Although it can be time consuming, microscopy is often used to validate the results of other sizing methods. Optical microscopy is most often used for particles between 3 and 150 $\mu$m. Larger than 150 $\mu$m, a magnifying glass is sufficient and sieving is the preferred method of particle size analysis. A lower limit of 3 $\mu$m is recommended due to gross overestimation of particle diameters close to the wavelength of light. Diffraction halos can result in oversizing a 1 $\mu$m particle by 15%, and even greater errors are introduced at smaller sizes (1). Another limitation of optical microscopy is the small depth of focus, which means that for wide-size distributions, only some particles in a particular field of view will be visible at any one time. For these reasons, it is recommended that particles from 20 nm to 3 $\mu$m be examined by scanning electron microscopy (SEM) (4), which can also be used to study larger (up to 1 mm) particles (5).

When viewed in an optical microscope, most particles appear as two-dimensional silhouettes, for which there are three major statistically acceptable measures of size:

- *Martin's diameter* is the length of a line which bisects the image of a particle into two equal areas. It can be measured in any direction relative to the particle's orientation, but a fixed direction must be used throughout the sample.
- *Feret's diameter* is the distance between parallel tangents on opposite sides of the particle. The orientation of the tangents is the same for the entire sample.
- The *projected area diameter*, i.e., the diameter of a circle having the same projected area as the particle,

is preferred over other measures such as Martin's or Feret's diameter since two dimensions are included as opposed to one.

Particle sizes as measured by microscopy tend to be greater than those measured by other methods because the third dimension, perpendicular to the viewing plane, is the smallest for nonspherical particles in stable orientations. A globe and circle eyepiece graticule, calibrated against a stage micrometer, is typically used to estimate projected areas. In manual techniques, untrained operators tend to overestimate the areas of irregular silhouettes, whereas experienced operators are able to compensate for this tendency (6). This somewhat subjective determination, however, leads to some degree of operator bias. Use of commercially available automatic image analyzers can eliminate these biases, as well as speed up the process and remove errors due to the slowness and tedium of manual methods. Image analyzers also make shape analysis more practical. Automatic systems, on the other hand, are less able to distinguish artifacts, such as abutting particles, without some intervention by a human operator. They are also unable to adjust focus within a field of view, making them less useful for samples with a wide distribution of sizes. Several excellent reviews have been written on the application of automatic image analysis (1,7).

Particle size analysis by microscopy is best suited to determining a size distribution by number. Statistical accuracy is determined by the number of particles counted, although time limitations make it impractical to examine every particle on a slide. A general rule is that counting 625 or more particles yields a standard error of 2% or less

(1). Extreme care should be taken if such a number distribution is to be converted to a weight or volume distribution. A single 50-$\mu$m particle in a weight distribution, for example, accounts for as much mass as 1,000 5-$\mu$m particles, and so miscounting the largest particles can drastically affect the overall results. Counting at least 25 particles in the largest size category helps to limit such errors but usually increases the time of analysis, as more of the sample must be examined to achieve this standard. Detailed procedures for determining particle size distributions by microscopy and reporting particle size data are described in the U.S. Pharmacopeia (8) and in the Pharmacopeial Forum (9), respectively.

## ELECTRICAL SENSING ZONE (COULTER PRINCIPLE) METHOD

The electrical sensing zone method, also known as the Coulter technique, was presented in 1956 as an automatic method for counting and sizing blood cells (10) and was soon applied to many kinds of particulate materials (11,12). The apparatus consists of two electrodes immersed in an electrolyte solution, separated by the wall of a glass tube with a single small orifice connecting the fluid inside the tube to the solution surrounding it. Particles suspended in the outer fluid are pulled through the orifice by applying vacuum to the fluid inside the tube. Electrical current flowing between the electrodes can only pass through the narrow orifice, and a particle that passes through the orifice, displacing a volume of electrolyte, momentarily increases the impedance of the circuit. The resulting change in resistance ($\Delta$R) is related to the volume (V) of a spherical particle by the following equation:

$$\Delta R = (16\rho_f V/\pi^2 D^4)F \qquad (1)$$

where $\rho_f$ is the resistivity of the fluid, $D$ is the diameter of the orifice, and $F$ is a function that incorporates certain correction factors, including a shape factor. In other words, $\Delta R$ is proportional to the particle volume, modified by the function $F$. Empirically, this function has little effect for particles much smaller than the orifice, and causes divergence from linearity in a predictable fashion as the particle size approaches the aperture diameter. Modern instruments are capable of reliably sizing particles between 2% and 60% of the orifice diameter, e.g., 2–60 $\mu$m for a 100-$\mu$m-diameter aperture. Aperture diameters from 30 to 2000 $\mu$m give the apparatus an overall range of 0.6 $\mu$m to 1,200 $\mu$m, but only a portion of this range can be measured with a given aperture. Consequently, a difficult two-tube technique is required for complete analysis of any sample with a broad-size distribution.

An electrical sensing zone instrument can count and size thousands of particles in a few seconds, and semiskilled operators can use it with reproducible results. Up to 256 size channels in a relatively narrow range leads to very high resolution, although with some increase in noise levels. The large number of particles analyzed, often many tens of thousands, greatly improves the statistical accuracy of a number-based distribution, and also improves the accuracy when the data are converted to a volume distribution. Care must be taken to control the concentration of suspended particles, though, because of the problem of coincidence. This is the passage of two particles through the orifice at roughly the same time, possibly detected as a single larger peak, which leads to oversizing when it becomes significant at higher concentrations. At low concentrations, the Coulter method is used for counting particulate contaminants in parenteral fluids. This application is described extensively in the British Pharmacopeia (13), and is also described in detail elsewhere (14). Other applications include the study of drug delivery systems such as liposomes (15), albumin microparticles (16), and microspheres for sustained release of proteins (17). A general method for all applications is described in British Standard 3406 (18).

### Dispersion of Samples

Whenever a powder is suspended in a liquid for purposes of particle size analysis, whatever the method, accurate results will only be obtained if the sample is properly dispersed. Usually, this is accomplished by the addition of a surfactant and possibly low power sonication. With the Coulter method, an additional factor is the requirement of an electrolyte solution. Some electrolytes, especially in nonaqueous media, can cause a sample to aggregate, thereby increasing the measured particle size. An example of a pharmaceutical with an electrolyte sensitivity is lyophilized particles of recombinant human growth hormone (rhGH), a water-soluble protein that has to be suspended in an organic solvent for particle size analysis. Lithium chloride or zinc chloride dissolved in acetone cause some agglomeration of the protein powder, although in the latter case the size distribution appears stable over time. When magnesium perchlorate is dissolved in 80/10/10 (% v/v) acetone/methanol/dimethyl sulfoxide, however, the measured particle size is smaller, as shown in Figure 2. Analysis on a laser diffraction instrument confirms that particles are larger in the presence of zinc chloride, relative to neat acetone or the magnesium perchlorate solution.

### LASER LIGHT SCATTERING METHODS

Instruments that use light scattering principles for particle size analysis have become popular because they allow

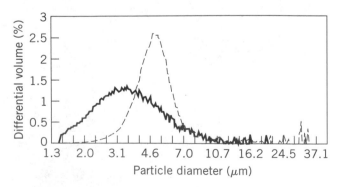

**Figure 2.** Size distribution of rHGH particles suspended in organic solvents with two different electrolytes. The solid line represents the data with magnesium perchlorate and the dotted line the data with zinc chloride.

quick and absolute determination of particle size, i.e., without the need for calibration. Although all of these approaches use a laser to measure particle size, there are really three major light scattering methods, each based on different principles. The first two techniques, static light scattering (SLS) and laser diffraction, are sometimes referred to as total intensity light scattering (TILS) methods and use time-averaged measurements of scattered light flux, whereas dynamic light scattering (DLS) instruments measure fluctuations in the intensity of scattered light and relate them to size-dependent parameters such as diffusion coefficients. Figure 1 shows the optimum particle size ranges for each method. SLS instruments are most often used for determining the molecular weights of macromolecules. Broad particle size distributions of 0.1–1000 $\mu$m can easily be measured with instruments based on laser diffraction principles. And for fine particulates from 2 nm up to about 2 $\mu$m, DLS is the method of choice. All these methods require dilute concentrations where interactions between particles and secondary scattering are minimized, although this must be balanced against the requirement for enough sample to obtain a sufficient signal-to-noise ratio. The discussion that follows briefly describes the theory behind each of these techniques and few practical issues. This does not represent the full extent of light scattering methods, though, that also can be used to obtain information on particle interactions, macromolecular solution dynamics, molecular relaxation, and phase changes.

### Static Light Scattering

The basic principles of light scattering are presented here to provide a basis for understanding how particle size information is obtained from light scattering methods. For more detailed theoretical discussion, see the work by van de Hulst (19). When incident light hits a particle, the particle acts as an oscillating dipole emitting scattered light in all directions. For homogeneous small particles, such as single macromolecules, the scattering particles act as point dipole oscillators. The scattering intensity for $N$ ($= \rho N_A / M$) particles per unit volume is then described by the Rayleigh equation:

$$I = N(8\pi^4(1 + \cos^2\theta)\alpha^2)\, I_0/r^2\lambda^4 \qquad (2)$$

where $r$ is distance from the particle, $\lambda$ is wavelength of incident light, $\alpha$ is particle polarizability, $I_0$ is intensity of incident light (unpolarized), $\theta$ is scattering angle, $\rho$ is density of the particles, $M$ is molecular weight of the particles, $N_A$ is Avogadro's number. Some of the underlying assumptions of this model are that the particles are dilute and therefore not interacting with each other and do not absorb light (19).

Another assumption of Rayleigh scattering is that interference effects are absent. It has been shown that this is a valid assumption only for molecules or particles with overall dimensions $<\lambda/20$ (19). This limit decreases as the particle refractive index approaches the refractive index of the medium (20). A larger molecule no longer acts as a point dipole oscillator but as a collection of many of them. Light scattered from different points within the molecule results in destructive interference. The Rayleigh-Gans-

Debye theory corrects the Rayleigh equation 2 for intramolecular interference effects, and the resulting equation is most commonly written as in equation 3 (20).

$$(Kc/R_\Theta) = [1/M_w + 2Bc][1 + (16\pi^2/3\lambda^2)(r_g^2 \sin^2(\theta/2))] \qquad (3)$$

where $K$ is $2\pi^2[n(dn/dc)]^2/\lambda^4 N_A$ (for unpolarized light), $R_\Theta = Ir^2/I_0(1 + \cos^2\theta)$, $c$ is the particle concentration, $n$ is the refractive index of the solvent, $M_w$ is the weight average molecular weight of the particle, $B$ is the thermodynamic second virial coefficient, and $r_g$ is the radius of gyration of the particle. This equation provides the basis for measuring size-related parameters, namely molecular weight and radius of gyration, which are commonly used to characterize polymers in solution, including those used in many drug delivery devices. Measuring molecular weights by light scattering is described in the USP (21). A method of extrapolating data from equation 3 to the limits $c = \theta = 0$, $\theta = 0$, and $c = 0$ to determine $M_w$, $r_g$, and $B$, respectively, in a single plot is called a Zimm plot and is described in detail by several authors (20,22). Several manufacturers make SLS instruments for the measurement of molecular weights, where the scattered light intensity is measured at multiple angles and equation 3 is applied in the limit as $\theta = 0$. This technique is sometimes called multiple angle laser light scattering (MALLS). SLS measured at single or multiple angles is even more powerful when used in conjunction with size exclusion chromatography (SEC) as a postcolumn detection method, as it allows absolute determination of polymer molecular weights without need for calibration (23–25).

### Laser Diffraction

If particles are sufficiently large compared to the wavelength of light, the particles diffract light in predictable patterns, according to the laws of geometrical optics (Fraunhofer diffraction). Large particles scatter more radiation but mostly at low angles, whereas smaller particles produce fainter, more complicated scattering patterns that extend to higher angles. In analogy to Young's single slit experiment, light diffracted from a large particle of diameter $d$ ($>1$ $\mu$m) produces an intensity pattern at a large distance from the particle with mimima at angles given by the equation below, where $m$ is any integer.

$$d \sin\theta = (m + 1/2)\lambda \qquad (4)$$

Thus, in principle, the particle size can be determined from the diffraction pattern. For some particles, though, such as transparent and submicron particles, the Fraunhofer approximation begins to break down. In such cases, an alternative is the Mie theory, which rigorously describes the scattering pattern for spheres of any size. This theory is quite complex, but with the help of modern computers, most laser diffraction instruments offer sizing results based on both Fraunhofer diffraction and Mie theory. In the case of Mie theory, a close estimate of the particle's refractive index is required to accurately compute particle size.

Instruments that use laser diffraction for particle size analysis are commercially available from a number of manufacturers. The chief attraction of these instruments is their ability to quickly and easily measure size distributions over a broad range, on average between 0.1 and 1000 $\mu$m, with the range subdivided into about 100 size intervals. Many instruments use information from additional detectors or polarization ratios to extend this range up to 2,000 $\mu$m or down to 0.04 $\mu$m (1). Some models are available with modules for sizing dry powders as well as liquid suspensions. The equipment typically consists of a transparent sample cell through which a fluid suspension is recirculated, or through which a dry powder is blown. A laser beam is directed through the cell, and one or more Fourier lenses on the other side focus the scattered radiation on an array of solid-state detectors, from low angles up to at least 40°. Using optical models based on Fraunhofer or Mie theory, the overlapping patterns of scattered light are deconvoluted in an iterative fashion to generate a volume-weighted size distribution (1). This method has been applied to a number of drug delivery systems, including powder aerosols for pulmonary delivery (26), and microspheres for sustained drug delivery (27,28).

Laser diffraction instruments in general provide very reproducible results, but the distributions obtained from different manufacturers' instruments can vary significantly, especially in the lower size range (1). One reason for these discrepancies is that the complete solution for the diffraction pattern using Mie theory is available only for spheres. For partially transmitting or nonspherical particles, converting the scattered light flux into a size distribution is less straightforward. Each manufacturer uses its own proprietary algorithms, leading to different results from different instruments (1). Nonspherical shapes can affect the distribution in other ways as well. One advantage of laser diffraction as compared with other techniques is that all orientations of a suspended particle can be presented to the laser beam, which prevents oversizing due to neglect of one dimension. Unfortunately, it may also produce a broader distribution, or a multimodal size distribution in the case of extremely anisotropic shapes such as rods or flakes (29). In such cases, combining this technique with microscopy may help to determine whether particle shape is affecting the appearance of the size distribution.

## Dynamic Light Scattering

For the techniques already discussed, particle size-related information is derived from time-averaged measurements of the intensity of scattered light at various angles. In contrast, DLS, also known as photon correlation spectroscopy (PCS) or quasielastic light scattering (QELS) involves measuring the variation in the scattered light intensity over time. Physically, the instantaneous scattered intensity can be thought of as the superposition of waves scattered from individual scattering centers subject to Brownian motion. As the scattering molecules or particles move toward or away from the detector, minute shifts in the frequency of the scattered light are introduced that generate beat frequencies proportional to the magnitude of the shift (30). As particle size decreases, particles exhibit increas-

ingly rapid motions, resulting in greater Doppler shifts and higher frequency fluctuations. These variations in the scattered light intensity can be related mathematically to size-dependent molecular transport properties such as diffusion coefficients. The theories describing that relationship are quite complex and beyond the scope of this discussion, but a thorough explanation may be found in the work of Berne and Pecora (31). Particle size for noninteracting spheres is related to the diffusion constant ($D$) through the Stokes-Einstein equation:

$$D = kT/6\pi\eta R \tag{5}$$

where $\eta$ is the solution viscosity, $R$ is the hydrodynamic radius of the particle, $k$ is Boltzmann's constant, and $T$ is the absolute temperature.

Manufacturers of DLS instruments claim an applicable size range from about 0.002 to 2 $\mu$m, although the technique is most reliable below about 0.5 $\mu$m (32). Measuring particles larger than 1 $\mu$m is complicated by their tendency to sediment, as well as inter- and intramolecular scattering effects. The apparatus for determination of particle size by DLS typically involves shining a polarized laser beam through a dilute suspension of the sample, and measuring scattered radiation at one or more angles from the incident light. An autocorrelator stores scattered light intensities at different times and calculates a correlation function that decays exponentially with time. For small, monodisperse particles undergoing Brownian motion

$$C(t) \propto (\exp(-qDt)) \tag{6}$$

where $C(t)$ is the value of the correlation function at time $t$, $q$ is the wave vector $= 4\pi n/\lambda \sin(\theta/2)$, $n$ is the solvent's refractive index, and $D$ is the diffusion constant. For multicomponent or polydisperse systems, the correlation function can be fit to a weighted sum of exponentials which can be deconvolved using computer programs such as CONTIN. From the components generated in this way, a size distribution can be determined, although multimodal distributions cannot always be resolved (1). A rule of thumb for spherical particles is that two modes differing by at least a factor of two in their mean diameter can be distinguished by CONTIN analysis. Results are usually given in terms of an average particle size and a polydispersity index. This method is particularly suitable for sizing drug delivery systems such as liposomes (33), microemulsions (34,35), and nanoparticles (36–39). DLS has also been used to examine the kinetics of aggregation for macromolecules such as insulin (40).

## COMPARING RESULTS FROM DIFFERENT PARTICLE SIZING METHODS

It is important to realize that each method of particle size analysis may produce a somewhat different result. These differences do not necessarily imply that one technique is more accurate than another, only that distinct principles are involved that measure particle size in different ways. For example, in the electrical sensing zone method the signal is proportional to the displacement volume of a particle. Porous particles, into which electrolyte solution can penetrate, appear smaller by this method than by a light scattering method, in which the diffraction pattern is dependent on the envelope volume of the particle. A number of studies have found that extremely nonspherical shapes also appear larger when measured by laser diffraction as compared with the Coulter method (41,42). The greatest differences between the techniques, about 40%, were observed for rods and plates, whereas the results for less extreme shapes differed by less than 10%. It has been suggested that the ratio of mean values obtained by these two methods could be used to predict shape, provided the particles are nonporous (42). In selecting a method for particle size analysis, one should consider not just whether a method may oversize some particles, but also the characteristics requirements of the material to be analyzed. For some applications, reproducibility and ease of use may be more important than absolute size determination.

## BIBLIOGRAPHY

1. T. Allen, *Particle Size Measurement*, 5th ed., Chapman & Hall, London, 1997.
2. A.D. Ascentiis et al., *Pharm. Res.* **13**, 734–738 (1996).
3. *Pharmacopeial Forum* **22**, 3240–3244 (1996).
4. *Scanning Electron Microscopy*, U.S. Pharmacopeia 25, U.S. Pharmacopeial Convention, Inc., Rockville, Md., 1995, pp. 1954–1957.
5. R. Falk et al., *J. Controlled Release* **44**, 77–85 (1997).
6. I.F. Nathan, M.I. Barnet, and T.D. Turner, *Powder Technol.* **5**, 105–110 (1972).
7. T.A. Barber, in J.Z. Knapp, T.A. Barber, and A. Lieberman, eds., *Liquid- and Surface-Borne Particle Measurement Handbook*, Dekker, New York, 1996, pp. 61–112.
8. *Optical Microscopy*, U.S. Pharmacopeia 25, U.S. Pharmacopeial Convention, Inc., Rockville, Md., 1995, pp. 2715–2717.
9. *Pharmacopeial Forum* **19**, 4640–4642 (1993).
10. W.H. Coulter, *Proc. Natl. Electron. Conf.* **12**, 1034 (1956).
11. H.E. Kubitschek, *Nature (London)* **182**, 234–235 (1958).
12. R.W. Lines and W.M. Wood, *Ceramics* **16**, 27–30 (1965).
13. *British Pharmacopoeia*, Vol. 11, HMSO, London, 1993, pp. A163.
14. R.W. Lines, in J.Z. Knapp, T.A. Barber, and A. Lieberman, eds., *Liquid- and Surface-Borne Particle Measurement Handbook*, Dekker, New York, 1996, pp. 113–154.
15. H. Talsma and D.J.A. Crommelin, *Pharm. Technol. Int.* **5**, 37 (1993).
16. I. Vural, H.S. Kas, A.A. Hincal, and G. Cave, *J. Encapsulation* **7**, 511 (1990).
17. O.L. Johnson et al., *Pharm. Res.* **14**(6), 730–735 (1997).
18. *British Standard 3406: Part 5*, British Standards Institution, London, 1983.
19. H.C. van de Hulst, *Light Scattering by Small Particles*, Dover, New York, 1981.
20. P.C. Hiemenz, *Polymer Chemistry: The Basic Concepts*, Dekker, New York, 1984.
21. *Spectrophotometry and Light Scattering*, U.S. Pharmacopeia 25, U.S. Pharmacopeial Convention, Rockville, Md., 1995, pp. 1830–1835.
22. M.B. Huglin, ed., *Light Scattering from Polymer Solutions*, Academic Press, New York, 1972.
23. P.J. Wyatt, C. Jackson, and G.K. Wyatt, *Am. Lab.* May-June (1988).
24. L. Jeng et al., *J. Appl. Polym. Sci.* **48**, 1359–1374 (1993).
25. U. Dayal and S.K. Mehta, *J. Liq. Chromatogr.* **17**(2), 303–316 (1994).
26. H.-K. Chan et al., *Pharm. Res.* **14**(4), 431–437 (1997).
27. A. Matsumoto et al., *J. Controlled Release* **48**(1), 19–27 (1997).
28. M.-K. Yeh, S.S. Davis, and A.G.A. Coombes, *Pharm. Res.* **13**(11), 1693–1698 (1996).
29. N. Gabas, N. Hiquily, and C. Lagueric, *Part. Part. Syst. Charact.* **11**(2), 121–126 (1994).
30. B.B. Weiner, in J.Z. Knapp, T.A. Barber, and A. Lieberman, eds., *Liquid- and Surface-Borne Particle Measurement Handbook*, Dekker, New York, 1996, pp. 155–171.
31. B.J. Berne and R. Pecora, *Dynamic Light Scattering*, Wiley, New York, 1976.
32. N. de Jaeger et al., *Part. Part. Syst. Charact.* **8**(9), 179–186 (1991).
33. E.C.A. van Winden, W. Zhang, and D.J.A. Crommelin, *Pharm. Res.* **14**(9), 1151–1160 (1997).
34. M.R. Lance, C. Washington, and S.S. Davis, *Pharm. Res.* **13**(7), 1008–1014 (1996).
35. F. Liu, J. Yang, L. Huang, and D. Liu, *Pharm. Res.* **13**(11), 1642–1646 (1996).
36. J.-Y. Chcrng et al., *Pharm. Res.* **13**(7), 1038–1042 (1996).
37. H. Heiati, N.C. Phillips, and R. Tawashi, *Pharm. Res.* **13**(9), 1406–1410 (1996).
38. R.H. Müller, S. Maassen, C. Schwarz, and W. Mehnert, *J. Controlled Release* **47**(3), 261–269 (1997).
39. M.T. Peracchia et al., *J. Controlled Release* **46**(3), 223–231 (1997).
40. M. Dathe et al., *Int. J. Pep. Protein Res.* **36**, 344–349 (1990).
41. J.A. Davies and D.L. Collins, *Part. Part. Syst. Charact.* **5**, 116–121 (1988).
42. A.T. Palmer, P.J. Logiudice, and J. Cowley, *Am. Lab.* November (1994).

# CHARACTERIZATION OF DELIVERY SYSTEMS, SPECTROSCOPY

GREGORY T. FIELDSON
ALZA Corporation
Palo Alto, California

## KEY WORDS

ATR spectroscopy

Characterization

Chemometrics

Infrared spectroscopy

Quantitative analysis

## Quantitative Analysis

The root of applied, quantitative spectroscopy is the basic relationship between the absorption of electromagnetic waves and the amount of the absorbing chemical in the infrared path. In transmission spectroscopy, at low absorbance, this relationship is expressed by the Beer-Lambert law

$$dl = \alpha I dz = -\epsilon C I dz \qquad (1)$$

where $I$ is the light intensity at position $z$, $\alpha$ is the absorption coefficient, $\epsilon$ is the molar extinction coefficient, and $C$ is the molar concentration of the absorbing group. This expression is strictly valid in the limit as $\alpha \to 0$ but is generally applicable unless the absorption coefficient, $\alpha$, is quite large. Integration of equation 1 through a sample of thickness $L$ gives the usual expression for percent transmittance

$$T = \frac{I}{I_0} 100 = \exp\left(\int_0^L \epsilon C dz\right) \times 100 \qquad (2)$$

where $T$ is the percent transmittance, $I$ is the intensity of the transmitted light, and $I_0$ is the intensity of the incident light. The transmittance is often converted into absorbance, $A$, by

$$A \equiv -ln\left(\frac{I}{I_0}\right) = \int_0^L \epsilon C dz \qquad (3)$$

When $C$ is constant throughout the sample, equation 3 reduces to

$$A = \epsilon C L \qquad (4)$$

As the link between light absorption and the molar concentration of an absorbing group, equation 4 is the basis for quantitative infrared spectroscopy. The value of the molar extinction coefficient is usually determined experimentally (27).

## APPLICATIONS

Infrared spectroscopy is applied to drug delivery research, development, and commercial product manufacture in a tremendous variety of ways. The applications range from fundamental research into protein–membrane interactions to on-line process monitoring in commercial production. The common thread that unites all of the applications is the need to investigate changes in covalent or intramolecular interactions in a nondisrupting manner.

## Research

**Basic Lipid Research.** One area of infrared-based research that has broad implications in many aspects of drug delivery is the study of lipid bilayers. Infrared spectroscopy is capable of revealing small changes in the orientation of a lipid bilayer without perturbing the sample (28–30). As a result, infrared spectroscopy has been used in numerous

studies. The effect of composition, temperature, and pressure on phase transitions has been closely studied, typically using a combination of infrared spectroscopy and differential scanning calorimetry (DSC) or nuclear magnetic resonance (NMR) spectroscopy (31–36). The resulting information may be used directly as a guide in the development of liposomal formulations (37–40). Understanding of lipid behavior also has a general impact on other drug delivery technologies, such as transdermal delivery and drug targeting (41–43).

**Materials and Formulations for Drug Delivery.** Materials and formulations for drug delivery systems are frequently developed using infrared spectroscopy. In controlled release formulations the relationship of raw materials and processing to dissolution rates and release rates may successfully be measured with infrared spectroscopy (44–47). For implant drug delivery devices, the interaction between biological proteins and materials is measurable, as well as the impact of altering these materials (for example, heparinization of an implant surface) (48–50). Strategies for preparing conformationally stable protein formulations for intravenous injections and implant devices are frequently based upon infrared spectroscopic results (51–54).

**Biological Structure and Composition.** Infrared tools developed for studying lipids can be extended to biological lipid membranes as well (55). Fundamental research into the perturbation of cell membranes by various proteins is conducted using infrared spectroscopy (56–59); cell membranes have been reconstructed on ATR cells for accurate measurements of protein–membrane interactions and liposome–membrane interactions (28,60).

Because the stratum corneum is commonly accepted as the primary barrier of the skin and these bilayers are quite amenable to infrared sampling, many researchers have used infrared spectroscopy to study the stratum corneum lipids (61). These studies have examined the normal structure of the stratum corneum (62–67) and the impact of hydration (68–72), chemical permeation enhancers, mechanical disruption, and electrical disruption upon the barrier (73–82).

It especially worth noting that infrared spectroscopy has been used repeatedly as a noninvasive, clinical tool for studying topical and transdermal formulations. These studies have investigated the accumulation of penetrant in the skin, the extraction of lipids from the skin, and the perturbation of the skin lipids when subjected to various treatments (65,68,70,71,80,83–85).

**Correlative Methods.** The infrared absorption spectrum of a chemical inherently contains a detailed description of the covalent structure and both intermolecular and intramolecular secondary interactions. As a result, infrared spectroscopy should be a successful tool for developing quantitative structure analysis relationships (QSARs), linking a simple, inexpensively measured property to a higher-level effect.

Outside of drug delivery, infrared spectroscopy was successfully applied to the prediction of polymer membranes' permeability (86,87). Drawing upon this methodology, a proprietary model, using infrared spectrum, is used to predict the effectiveness of chemical permeation enhancers for transdermal drug delivery. The resulting neural network models, based upon infrared spectral inputs, are the most accurate models developed to date (W. van Osdol, personal communication, 1998).

## Development

**Composition.** Infrared spectroscopy is a primary tool for compositional analysis of mixture components that are present in concentrations greater than 0.1 vol %. Although sample preparation or calibrations may sometimes represent a challenge, this application is routine.

**Stability.** In conjunction with composition analysis, infrared spectroscopy is also a rapid test of the gross chemical stability of any formulation, component, or material that can be sampled. The creation or disruption of covalent chemical bonds results in changes in the infrared absorption. Detection of significant changes (>0.1 vol %) by infrared spectroscopy may be used as an inexpensive screen that triggers the deployment of more elaborate techniques (such as liquid or gas chromatography and mass spectroscopy) to identify the precise nature of a stability problem (88,89).

## Manufacturing

**Quality Control.** Infrared spectra are routinely used as an acceptance criterion for receiving materials in the manufacturing environment. Gross errors, such as mislabeled materials, are almost instantly detected. Other changes in the infrared spectrum, when compared to a reference standard, can be used to detect minor errors, such as an incorrect mixture composition, excessive degradation, or significant impurities (90–93).

**Quality Assurance.** Infrared spectroscopy has been developed as an on-line tool for process control. Sampling methods exist for almost every type of process operation, and when accurate, rapid composition monitoring and control is desired, infrared spectroscopy can be applied (94–96).

Infrared spectroscopy may also play a role in product testing. Finished products may be compared to reference standards to assess compliance with composition specifications. Additionally, infrared spectroscopy may be used to monitor the long-term stability of product, detecting changes that result from both chemical and morphologic changes to the product (97).

**Defect Analysis.** A common application of infrared spectroscopy is the analysis of failed product or materials. An example is analysis of particles that appear in a finished drug delivery system. The use of infrared microscopy, combined with searches of infrared spectral libraries, will rapidly identify the nature of the contaminant (98). Other forensic applications can result from complex problems that are not as readily solved. Unexpected crystalline material in a product may be rapidly identified by infrared spectroscopy, but discovering the root cause of the crystals may

require extensive investigation and elimination of the problem may entail significant process changes (99,100).

## BIBLIOGRAPHY

1. J.I. Steinfeld, *Molecules and Radiation*, 2nd ed., MIT Press, Cambridge, Mass., 1985.
2. N.B. Colthup, *Introduction to Infrared and Raman Spectroscopy*, 3rd ed., Academic Press, Boston, 1990.
3. S. Johnston, *Fourier Transform Infrared*, Ellis Horwood, New York, 1991.
4. D.L. Pavia, G.M. Lampman, and G.S. Kriz, Jr., *Introduction to Spectroscopy: A Guide for Students of Organic Chemistry*, Saunders College Publishing, Orlando, Fla., 1979.
5. B.C. Smith, *Fundamentals of Fourier Transform Infrared Spectroscopy*, CRC Press, Boca Raton, Fla., 1996.
6. B. Stuart, *Modern Infrared Spectroscopy*, Wiley, Chichester, England, 1996.
7. B. Schrader, ed., *Infrared and Raman Spectroscopy*, Wiley-VCH, Weinheim, 1995.
8. P.B. Coleman, ed., *Practical Sampling Techniques for Infrared Analysis*, CRC Press, Boca Raton, Fla., 1993.
9. G. Socrates, *Infrared Characteristic Group Frequencies*, 2nd ed., Wiley, Chichester, England, 1994.
10. *Bio-Rad Sadtler IR Database*, Bio-Rad Laboratories, Philadelphia (software).
11. *Nicolet Aldrich FT-IR Libraries*, Nicolet Corporation, Madison, Wis. (software).
12. *GRAMS /32 Search*, Galactic Industries Corporation, Salem, N.H. (software).
13. D. Lin-Vien et al., eds., *Handbook of Infrared and Raman Characteristic Frequencies of Organic Molecules*, Academic Press, Boston, 1991.
14. C.J. Pouchert, ed., *The Aldrich Library of FT-Ir Spectra*, 2nd ed., Aldrich Chemical Company, Milwaukee, Wis., 1985.
15. N.J. Harrick, *J. Opt. Soc. Am.* **55**(7), 851–857 (1965).
16. F.M. Mirabella, Jr., ed., *Internal Reflection Spectroscopy: Theory and Applications*, Dekker, New York, 1993.
17. K. Ohta and R. Iwamoto, *Anal. Chem.* **57**, 2491–2499 (1985).
18. K. Ohta and R. Iwamoto, *Appl. Spectrosc.* **39**(3), 418–425 (1985).
19. J.B. Huang and M.W. Urban, *Appl. Spectrosc.* **46**(6), 1014–1019 (1992).
20. M. Yanagimachi, M. Toriumi, and H. Masuhara, *Appl. Spectrosc.* **46**(5), 832–840 (1992).
21. K.J. Kuhn, B. Hahn, V. Percec, and M.W. Urban, *Appl. Spectrosc.* **41**(5), 843–847 (1987).
22. K.S. Kalasinsky, G.R. Lightsey, P.H. Short, and J.R. Durig, *App. Spectrosc.* **44**(3) 404–407 (1990).
23. G.T. Fieldson and T.A. Barbari, *AIChE J.* **41**(4), 795–804 (1995).
24. G.T. Fieldson and T.A. Barbari, *Polymer* **34**(6), 1146–1153 (1993).
25. H.M. Reinl, A. Hartinger, P. Dettmar, and T.M. Bayerl, *J. Invest. Dermatol.* **105**, 291–295 (1995).
26. J.E. Harrison et al., *Pharm. Res.* **13**(4), 542–546 (1996).
27. American Society for Tasting Materials, *Standard Practices for General Techniques of Infrared Quantitative Analysis*, ASTM, Philadelphia, 1992.
28. U.P. Fringeli, *Chimia* **46**, 200–214 (1992).

29. D.G. Cameron, H.L. Casal, and H.H. Mantsch, *J. Biochem. Biophys. Methods* **1**, 21–36 (1979).

30. D.C. Lee and D. Chapman, *Biosci. Rep.* **6**(3), 235–256 (1986).

31. R.N.A.H. Lewis, R.N. McElhaney, M.A. Monck, and P.R. Cullis, *Biophys. J.* **67**, 197–207 (1994).

32. I. Cornut, B. Desbat, J.M. Turlet, and J. Dufourcq, *Biophys. J.* **70**, 305–312 (1996).

33. H.H. Mantsch, A. Martin, and D.G. Cameron, *Biochemistry* **20**, 3138–3145 (1981).

34. D.G. Cameron, H.L. Casal, and H.H. Mantsch, *Biochemistry* **19**, 3665–3672 (1980).

35. R.G. Snyder, G.L. Liang, H.L. Strauss, and R. Mendelsohn, *Biophys. J.* **71** 3186–3198 (1996).

36. M.P. Sanchez-Migallon, F.J. Aranda, and J.C. Gomez-Fernandez, *Biochim. Biophys. Acta* **1281**(1), 23–30 (1996).

37. J.H. Crowe, S.B. Leslie, and L.M. Crowe, *Cryobiology* **31**(4), 355–366 (1994).

38. K. Ozaki and M. Hayashi, *Int. J. Pharm.* **160**(2), 219–227 (1998).

39. Y. Setiawan, T. Rise, and D.E. Moore, *Pharm. Res.* **11**(5), 723–728 (1994).

40. P.L. Chong and P.T. Wong, *Biochim. Biophys. Acta* **1149**(2), 260–266 (1993).

41. Y. Takeuchi et al., *Chem. Pharm. Bull.* **40**(2), 484–487 (1992).

42. S. Zellmer et al., *Chem. Phys. Lipids* **94**(1), 97–108 (1998).

43. S. Zellmer, W. Pfeil, and J. Lasch, *Biochim. Biophys. Acta* **1237**, 176–182 (1995).

44. R. Ek et al., *Int. J. Pharm.* **125**, 257–264 (1995).

45. K.J. Pekarek et al., *J. Controlled Release* **40**(3), 169–178 (1996).

46. M.R. Kreitz, J.A. Domm, and E. Mathiowitz, *Biomaterials* **18**(24), 1645–1651 (1997).

47. S. Yamamura and J.A. Rogers, *Int. J. Pharm.* **130**(1), 65–73 (1996).

48. R. Barbucci and A. Magnani, *Biomaterials* **15**(12), 955–962 (1994).

49. R.W. Sarver, Jr. and W.C. Krueger, *Anal. Biochem.* **212**, 519–525 (1993).

50. G. Khang, H.B. Lee, and J.B. Park, *Bio-Med. Mater. Eng.* **5**(4), 245–258 (1995).

51. M.M. Tan, S.A. Corley, and C.L. Stevenson, *Pharm. Res.* **15**(9), 1442–1448 (1998).

52. J.M. Hadden, D. Chapman and D.C. Lee, *Biochim. Biophys. Acta* **1248**(2), 115–122 (1995).

53. M.J. Pikal and D.R. Rigsbee, *Pharm. Res.* **15**(2), 362–363 (1998).

54. M.J. Pikal and D.R. Rigsbee, *Pharm. Res.* **14**(10), 1379–1397 (1997).

55. H.H. Mantsch and D. Chapman, eds., *Infrared Spectroscopy of Biomolecules*, Wiley-Liss, New York, 1996.

56. J.H. Crowe, L.M. Crowe, and D. Chapman, *Arch. Biochem. Biophys.* **232**(1), 400–407 (1984).

57. E. Goormaghtigh et al., *Biochemistry* **32**, 6104–6110 (1993).

58. L.K. Tamm and S.A. Tatulian, *Biochemistry* **32**, 7720–7726 (1993).

59. H.L. Casal, D.G. Cameron, I.C.P. Smith, and H.H. Mantsch, *Biochemistry* **19**(3), 444–451 (1980).

60. H.M. Reinl and T.M. Bayerl, *Biochim. Biophys. Acta* **1151**, 127–136 (1993).

61. S. Nanda, S. Anand, and A. Nanda, *Indian J. Pharm. Sci.* **60**(4), 185–188 (1998).

62. N.A. Puttnam, *J. Soc. Cosmet. Chem.* **23**, 209–226 (1972).

63. K. Knutson et al., *J. Controlled Release* **2**, 67–87 (1985).

64. C.L. Gay et al., *J. Invest. Dermatol.* **103**(2), 233–239 (1994).

65. D. Bommannan, R.O. Potts, and R.H. Guy, *J. Invest. Dermatol.* **95**(4), 403–408 (1990).

66. B. Ongpipattanakul, M.L. Francoeur, and R.O. Potts, *Biochim. Biophys. Acta* **1190**, 115–122 (1994).

67. R.O. Potts and M.L. Francoeur, *Semin. Dermatol.* **11**(2), 129–138 (1992).

68. P.A.D. Edwardson, M. Walker, and C. Breheny, *Int. J. Pharma.* **91**, 51–57 (1993).

69. V.H.W. Mak, R.O. Potts, and R.H. Guy, *Pharm. Res.* **8**(8), 1064–1065 (1991).

70. M. Gloor, G. Hirsch, and U. Willebrandt, *Arch. Dermatol. Res.* **271**, 305–313 (1981).

71. R.O. Potts, D.B. Guzek, R.R. Harris, and J.E. McKie, *Arch. Dermatol. Res.* **277**, 489–495 (1985).

72. A. Alonso, N.C. Meirelles, V.E. Yushmanov, and M. Tabak, *J. Invest. Dermatol.* **106**, 1058–1063 (1886).

73. B. Ongpipattanakul, R.R. Burnette, R.O. Potts, and M.L. Francoeur, *Pharm. Res.* **8**(3), 350–354 (1991).

74. S.L. Krill, K. Knutson, and W.I. Higuchi, *J. Controlled Release* **25**, 31–42 (1993).

75. Y. Takeuchi et al., *Chem. Pharm. Bull.* **40**(7), 1887–1892 (1992).

76. R.P. Oertel, *Biopolymers* **16**, 2329–2345 (1977).

77. T. Kurihara-Bergstrom, K. Knutson, L.J. DeNoble, and C.Y. Goates, *Pharm. Res.* **7**(7), 762–766 (1990).

78. A.N.C. Anigbogu, A.C. Williams, B.W. Barry, and H.G.M. Edwards, *Int. J. Pharm.* **125**, 265–282 (1995).

79. K. Yoneto et al., *J. Pharm. Sci.* **85**(5), 511–517 (1996).

80. A. Naik, L.A.R.M. Pechtold, R.O. Potts, and R.H. Guy, *J. Controlled Release* **37**, 299–306 (1995).

81. S.-Y. Lin, K.-J. Duan, and T.-C. Lin, *Spectrochim. Acta, Part A* **52A**, 1671–1678 (1996).

82. M.J. Clancy, J. Corish, and O.I. Corrigan, *Int. J. Pharm.* **105**, 47–56 (1994).

83. D. Bommannan, R.O. Potts, and R.H. Guy, *J. Controlled Release* **16**, 299–304 (1991).

84. S. Thysman, D. Van Neste, and M. Préat, *Skin Pharmacol.* **8**, 229–236 (1995).

85. F. Pirot et al., *Proc. Natl. Acad. Sci. U.S.A.* **94**, 1562–1567 (1997).

86. M. Wessling et al., *J. Membr. Sci.* **86**(1–2), 193–198 (1994).

87. A. Bos, Doctoral Dissertation, University of Twente, Enschede, The Netherlands, 1993.

88. B.J. Tyler and B.D. Ratner, *J. Biomater. Sci. Polym. Ed.* **6**(4), 359–373 (1994).

89. M.C. Tanzi et al., *J. Biomed. Mater. Res.* **36**(4), 550–559 (1997).

90. J.P. Coates. *The Industrial Applications of Infrared Internal Reflectance Spectroscopy*, Dekker, New York, 1993, p. 374.

91. C. Wojciechowski et al., *Food Chem.* **63**(1), 133–140 (1998).

92. H. Peters, E. Unger, and F. Moll, *Pharmazie* **50**(1), 43–48 (1995).

93. Z. Bouhsain, S. Garrigues, and M. de la Guardia, *Analyst* **121**(12), 1935–1938 (1996).

94. Z. Bouhsain, S. Garrigues, and M. de la Guardia, *Analyst* **122**(5), 441–445 (1997).

95. L. Rintoul et al., *Analyst* **123**(4), 571–577 (1998).

96. E.D.S. Kerslake and C.G. Wilson, *Adv. Drug Delivery Rev.* **21**(3), 205–213 (1996).

97. Y. Wu et al., *J. Appl. Polym. Sci.* **46**, 201–211 (1992).

98. T. Gál and P. Tuth, *Can. J. Appl. Spectrosc.* **37**(2), 55–57 (1992).

99. H. Zhu et al., *J. Pharm. Sci.* **86**(12), 1439–1447 (1997).

100. J.E. Charbonneau, *Scanning* **20**(4), 311–317 (1998).

# CHARACTERIZATION OF DELIVERY SYSTEMS, SURFACE ANALYSIS AND CONTROLLED RELEASE SYSTEMS

BUDDY D. RATNER
CONNIE KWOK
University of Washington Engineered Biomaterials (UWEB)
Seattle, Washington

## KEY WORDS

Bioadhesion

Biocompatibility

Biodegradable polymers

Contact angle

Contamination

Deposition

Diffusion barriers

ESCA method

Infrared spectroscopy

Microsphere

Polymer colloids

Polymer matrix

Scanning probe microscopes

Secondary ion mass spectrometry (SIMS)

Surface analysis

## OUTLINE

## INTRODUCTION

The study of controlled-release and drug-release devices has been dominated by considerations of the bulk or average properties of materials or devices. Concentrations of drugs within a matrix, porosity, diffusion barriers, and erosion are important bulk matter concepts that are central to controlled release. Yet the outermost surface atoms of these systems play a central role in their performance. This article justifies surface studies for controlled-release systems, examines surface methods in a historical context, and then focuses on a few relevant examples.

Surface analysis is a term loosely used to describe the measurement of the outermost chemistry and structure of a material, with "outermost" ranging from 2 Å (an atomic surface monolayer) to a 10-$\mu$m region at the surface. There is no precise definition of surface in terms of depth into a material. Rather, *surface* can be operationally defined as the relevant, outermost zone that explains the performance of a material or device. Each of the major surface analysis tools probes to a different depth into a material. A few common methods and their penetration depths are summarized in Table 1. The specific reason why the surface region is important for controlled release is illustrated in Figure 1. Let us briefly examine each of the points highlighted in the figure.

### Diffusion Barriers

Diffusion barriers and protective layers are widely used in drug-release systems. These surface coatings can provide zero-order release, extended-release, or protection from environmental agonists such as stomach acids. Typically, these coatings are thin. Their nature can be explored using surface analysis methods, and new barrier layers can be developed and optimized. An example of the use of surface methods to develop controlled-release barriers is presented in the last section of this article.

### Biocompatibility

The word *biocompatibility* has been, to date, only loosely defined. However, it is clear that once toxicology issues (i.e., those associated with leachable substances toxic to cells) are addressed, biocompatibility is strongly influenced by surface properties. Because long-term controlled-release devices are rapidly encapsulated in a collagenous sheath after implantation and because the thickness of this sheath can be variable, achieving precise controlled release can be problematic. Contemporary implanted controlled-release devices address this by conservative en-

**Table 1. Methods to Characterize the Surfaces of Controlled-Release Systems**

| Method | Principle | Depth analyzed | Spatial resolution | Analytical sensitivity | Cost |
|---|---|---|---|---|---|
| Contact angles | The wetting of liquid drops on surface is used to estimate the surface energy. | 3–20 Å | 1 mm | Low or high depending on the chemistry | $ |
| ESCA | X rays stimulate the emission of electrons of characteristic energy. | 10–250 Å | 10–150 $\mu$m | 0.1 atom % | $$$ |
| Auger electron Spectroscopy[a] | A focused electron beam causes the emission of Auger electrons. | 50–100 Å | 100 Å | 0.1 atom % | $$$ |
| SIMS | Ion bombardment leads to the emission of surface secondary ions. | 10 Å–1 $\mu$[b] | 100 Å | Very high | $$$ |
| FTIR-ATR | IR radiation is adsorbed in exciting molecular vibrations. | 1–5 $\mu$m | 10 $\mu$m | 1 mol % | $$ |
| AFM | Measurement of the force of interaction between a sharp tip and a surface | 10 Å | 50 Å | Single molecules | $$ |
| SEM | Secondary electron emission caused by a focused electron beam is measured and spatially imaged. | 5 Å | 40 Å (typically) | High but hot quantitative | $$ |

*Note:* $—up to $5000
$$—$5000–$100,000
$$$—>$100,000
[a]Auger electron spectroscopy is damaging to organic materials and best used for inorganics.
[b]Static SIMS $\approx$ 10 Å; dynamic SIMS to 1 $\mu$m.

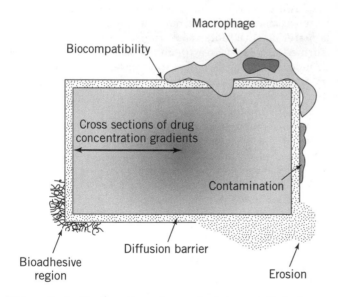

**Figure 1.** Applications for surface analysis in controlled release.

gineering so that even for the thickest expected capsule, drug concentrations in the efficacious range will diffuse to the body. Still, it is clear that control (or elimination) of the foreign body capsule would be desirable, and modern approaches to designing surfaces for implantation address this point (1,2). To design precision surfaces to control the healing, precision surface analysis is required.

### Surface Erosion

Some drug-release systems function through surface erosion. The kinetics of this process, the chemistry of the surface reactions occurring, and the morphological changes taking place at the surface can be probed with surface analytical methods.

### Drug Concentration Profiles in the Surface Region

Surface analysis methods can be used to measure drug concentrations in a surface zone, concentration gradients, and surface blooming of drugs. Also, using cross-sectional slices, drug profiles within a device can be directly measured. Thus, surface analysis can be used for studying release mechanisms and for quality control purposes.

### Adhesion

Many controlled-release systems are multilayer or multicomponent devices. Because adhesion is largely a surface phenomenon, the adhesion of one layer to another and the integrity of joints holding together components of a device can be assessed with surface analysis tools.

### Bioadhesion

The development of successful controlled-release strategies for mucosal sites requires that devices adhere to these slippery, hydrated surfaces. Such bioadhesion is often attributed to noncovalent chemical interactions (H-bonding), specific covalent bond formation, or interdiffusion of one polymeric species into another. The molecular size scale of these interactions is appropriate for study using surface analytical methods, and the analytical complexity associated with the interactions makes surface analysis methods particularly valuable.

### Contamination

Surfaces are readily contaminated. Two explanations for this are the intrinsic driving forces for all surfaces to reduce their free energy and the obvious accessibility (exposure) of a surface to the external world. An unintentional contamination layer (hydrocarbon, silicones, plasticizer, etc.) can influence the performance of a carefully engineered controlled-release device. The ability of surface analysis methods to detect and identify surface contaminant layers can help explain changes in performance and pinpoint the origin of surface contamination. The quantitative nature of surface analytical methods can be used to develop strategies for removal of surface contaminants and to establish quality control standards (e.g., minimum acceptable levels) for unavoidable contaminants.

### General Analysis

Although surface analysis methods exhibit their greatest strengths in their ability to measure the surface zone, they are also valuable general analytical tools. They can be used to identify unknowns and to perform quantitative analysis, often in situ and with little sample preparation (3).

## A HISTORICAL CONSIDERATION

The justification for the use of surface analysis methods to study controlled-release systems has been presented in the preceding section. It is useful to briefly review the history of these methods and then consider the principles behind them.

The appreciation of the importance and uniqueness of surfaces has been slow to arrive in the scientific community. This is, in part, because much of the instrumentation needed to study surfaces was perfected only with the last 30 years. Although contact angles and wettability considerations have been used to directly or indirectly investigate surfaces since antiquity, techniques such as electron spectroscopy for chemical analysis (ESCA) have much more contemporary origins.

The quantitative use of contact angles of liquid drops on solid surfaces to explain surface properties can trace its origins to the work of Thomas Young and others in the early nineteenth century. However, contact-angle measurement as a routine surface analysis tool is attributed to the work of Zisman at the Naval Research Laboratories starting in the 1940s (4). Zisman explored the relationships between contact angles, composition, and the thermodynamic parameter, surface free energy. This led to his development of the concept of critical surface tension, an approximation to the surface free energy.

The ESCA method (also called X-ray photoelectron spectroscopy [XPS]) is based on the photoelectric effect, first explained by Einstein in 1905. In 1925, an experimental demonstration was made of this method, but this was closer to a heroic physics demonstration than a practical analytical method. In the 1960s and 1970s, Professor Kai Seigbahn developed the ESCA technique and the instrumentation as we know it today (5,6). He defined its use in surface analysis and coined the ESCA acronym.

The principle of another surface analysis method widely used for inorganic surface, Auger electron spectroscopy (AES), is based on the cloud chamber observations of Pierre Auger in the 1920s. Auger electrons, given off from materials under bombardment with energetic surfaces, are different from photoemitted electrons measured in ESCA in that their energies are independent of the energy of the impinging energy source. Because an electron source, rather than an X-ray source, can be used to excite the emis-

sion and because electron sources are easier to design than X-ray sources, Auger instrumentation made more rapid strides than ESCA instrumentation. Still, at the end of the twentieth century, ESCA is used much more widely than AES (except in the semiconductor industry) because of its enhanced information content and lower sample damage.

Secondary-ion mass spectrometry (SIMS) developed in the 1950s and 1960s as a method to depth-profile materials. However, the pioneering work of Benninghoven revealed the potential of SIMS used in the static mode. Static SIMS permitted analysis of the outermost surface zone and offered rich molecular information about organic surfaces (7).

Surface analysis applied to biomedical problems evolved from the work of early visionaries and pioneers who appreciated both the unique nature of the surface zone and the relationship between the surface structure and biological response (8–11). The advent of modern ultrahigh-vacuum-surface analysis methods, spurred by the growth of the semiconductor industry and petrochemical catalysis, greatly advanced the ability to accurately study surfaces so that surface structures might be considered in the context of biological interactions (12,13). More recently, synchrotron methods, the scanning probe microscopics, and modern optical techniques have made impressive strides in permitting the study of the surface structure of biomedically important materials (14–16). Of special note is the ability of many of these methods to study the solid–water interface.

## A SHORT OVERVIEW OF METHODS

The principles behind a number of the commonly used surface analysis methods are illustrated in Figure 2. Characteristics and specifications for these methods are summarized in Table 1. A few comments on those surface methods especially applicable to drug-release systems follows. It is important to appreciate that although each of the methods can provide reliable information on surfaces, combinations of these methods allow the construction of a more complete picture of surface structure.

### ESCA

ESCA is probably the single most useful technique for studying the surfaces of devices and materials of technological importance. Many good review articles and books on this method are available (17–21). The instrumentation is well developed and is almost user-friendly. Sample preparation is generally simple—just insert the sample into the analysis vacuum chamber and commence the acquisition of data. Metals, polymers, ceramics, particulates, thin films, fabrics, and intact devices can all be readily examined. The information content, including all elements present, atomic stoichiometry, molecular detail, and depth-profile data in the outermost ~80 Å, is exceptionally high. Modern instruments offer good analytical sensitivity (approximately 0.1 atom %), good spatial resolution (approximately 25 m$\mu^2$), low sample damage, and good spectroscopic resolution. Finally, novel methods have been evolved to enhance the information content of ESCA. For example,

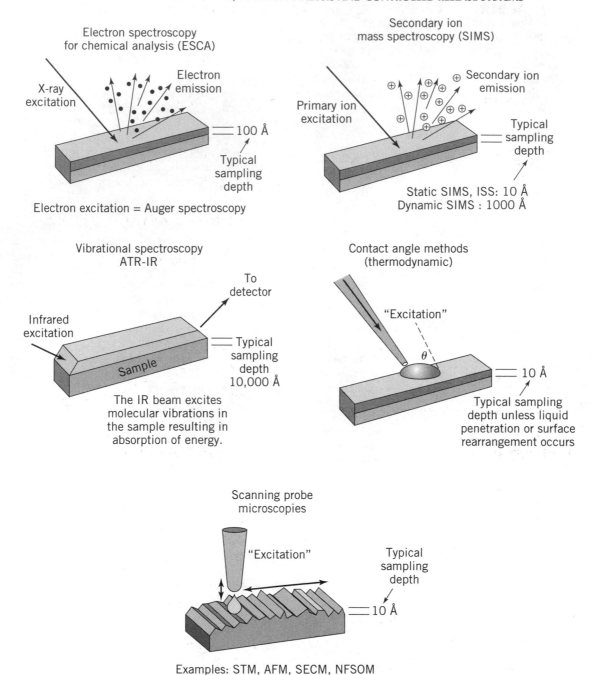

**Figure 2.** The principles of five surface analysis methods.

hydrated frozen samples can be studied to look at the polymer–aqueous interface (22). Also, chemical derivatization methods can enhance the accuracy of surface group identification and quantification (23).

**Contact-Angle Methods**

Contact-angle methods to probe surface properties are highly surface sensitive, low cost, readily performed in most laboratories, and reasonably easy to understand. In the simplest configuration, a small drop of a pure liquid of known surface tension is placed on the solid surface to be

measured. The angle of a tangent to the drop profile originating from the point of contact of drop and surface is measured, typically with a magnifying scope and protractor scale. The angle through the fluid phase is, by convention, the value reported. Analyses offering higher precision can be performed with a Wilhelmy plate apparatus. The measured force of insertion or removal of a flat plate of the material of interest through an air–liquid interface is measured and can be used to compute the contact angle between plate and liquid (24). Other techniques include underwater contact-angle measures with drops of an insoluble hydrocarbon or air and measurements of the angle of drop flow on a tilting stage. The relative simplicity of these methods must be tempered by a consideration of their limitations. The angles measured are influenced by operator experience, purity of the measurement liquid, penetration of the liquid into the substrate, sample roughness, drop size, surface swelling, and a number of other variables. Also, information about surface chemistry is inferred rather than directly measured. Still, with care, useful information can be obtained. Application of acid–base concepts to contact-angle measurements by Fowkes has significantly advanced specific information content obtainable from these methods (25). Useful review articles are available on contact-angle methodology (24–26).

## SIMS

In recent years, SIMS analysis of surfaces has approached ESCA in general utility and interpretability. In SIMS, the surface is bombarded with an energetic primary-ion beam, and the masses of charged atomic and molecular fragments sputtered into the vacuum region are measured. SIMS is performed in one of two modes. Dynamic SIMS, with a high primary-ion flux, gives elemental compositional information and provides composition as a function of depth over a range of roughly 100 to 10,000 Å. Because of high sample damage, dynamic SIMS is rarely useful for organic materials. In static SIMS (SSIMS), an extremely low primary-ion dose is played over the surface. This results in the erosion of less than 10% of the surface monolayer. The consequence is that only the outermost zone of a sample (approximately 10 Å deep) is probed, and complex molecular information can be read from the mass spectral output (7). In recent years, time-of-flight (TOF) mass analyzers for SIMS have revolutionized analytical sensitivity (detection down to $10^{10}$ molecules or less), mass accuracy (often masses can be expressed to four significant figures after the decimal point), and spatial resolution (elemental analysis can be performed in regions as small as $0.1~\mu m$ in diameter along with compositional mapping) (27). The downsides of the static SIMS method are expensive instrumentation (>500,000), the need for highly trained operators for quality data acquisition, the ability to probe only the outermost layer, and complex spectral interpretation. Still, if both ESCA and static SIMS can be performed, a detailed understanding of the surface chemistry and structure will be accessible.

### The Scanning Probe Microscopies

The scanning probe microscopies (SPMs) are the latest tools in the armamentarium of methods available to the surface analyst (16). An atomic or molecularly sharp tip is passed over a surface using extremely precise piezoelectric controllers, and the electrical or mechanical interaction of the tip with the surface is measured. The output is plotted as tip response per spatial location. SPM methods include the scanning tunneling microscope (STM), which is primarily applicable to electrically conducting samples, and the atomic force microscope (AFM), useful for conductors and insulators. AFM is most applicable to the types of samples likely to be encountered in controlled-release systems and is the focus here. AFM can produce maps of surface topography with a height ($z$) resolution of 1 Å and an $x,y$ spatial resolution of typically 50 Å. The sensitivity and accuracy in the $z$-plane permits surface textures to be observed with unprecedented accuracy and sensitivity compared with, for example, a scanning electron microscope (SEM). The microscope can be operated in air, underwater, or in vacuum, offering much flexibility for exploring real-world systems. Finally, it can be used in a number of specialized modes to enhance information content. The lateral force microscopy (LFM) mode permits surface friction to be measured at the nm scale. The tapping mode is useful for softer samples that could be readily damaged or moved by tip forces. The AFM can be used as a nanointendometer to measure hardness or modulus in nano-scale regions (16). Derivatization of the tip permits chemical imaging to be performed (28). Other variations of the SPM method include the scanning electrochemical microscope (SECM) for electrical information about surfaces in ionic media and the near-field optical scanning microscope (NFOSM) for optical information with spatial resolutions well below the defraction limit of light.

### Surface Infrared Methods

The infrared (IR) spectroscopy methods, part of a class of techniques referred to as vibrational spectroscopies, have a long history in surface analysis (29). IR spectroscopy is well understood and information rich. However, special methods must be applied to make it surface sensitive. Although vibration spectroscopy methods can measure surface zones as thin as 10 Å, the most generally useful of the techniques for drug-release systems, attenuated total reflectance (ATR) IR, probes $1–5~\mu m$ into a material (30). In ATR, the sample or device of interest is pressed against a trapezoidal crystal of high refractive index (often germanium), and an IR beam is passed through the crystal, allowing a portion of the IR beam extending beyond the face of the crystal to be absorbed by the sample. Hence, an IR absorption spectrum can be measured. ATR accessories are readily obtained for most laboratory IR spectrometers. Thus, if the high probe depth of this method is not a limitation for the specific question being asked, ATR remains a good method for surface analysis of materials and devices.

## SOME EXAMPLES FROM THE LITERATURE

Although the application of surface analysis methods to controlled release is relatively new, there are a surprisingly large number of published examples to cite on this subject. These examples will be organized into the following categories:

- Use of surface tools in site-specific delivery systems (ESCA, SIMS)
- Use of surface tools in biodegradable polymers and the study of surface erosion (ESCA, SIMS, AFM)
- Use of surface tools in establishing drug concentration profiles (ESCA, SSIMS, SIM imaging, TOF-SIMS, AFM)

- Use of surface tools in bioadhesion studies involving controlled-release systems (ESCA, AFM, FTIR-ATR)
- Use of surface tools to study a diffusion barrier (ESCA)

## Surface Tools for Studying Site-Specific, Microsphere Delivery Systems (ESCA, SIMS)

Polymer colloids have been shown to be useful in site-specific drug delivery systems, especially for pulmonary administration or intravenous drug delivery. However, one major drawback is their significant uptake by macrophages (phagocytosis) in the lung or liver/spleen (the reticuloendothelial system, RES), thereby reducing the delivery efficacy of the device. Davies et al. (31,32) have proposed that by copolymerizing poly(ethylene oxide) (PEO) of $M_r$ 2,000 with polystyrene, they were able to reduce the opsonization and engulfment of the particles by macrophages. Using both ESCA and SIMS, they reported that this observation might be due to the enrichment of ether carbon (C-O) on the particle surface. C-O surface density increased dramatically as the amount of the PEO in the polymerization mixture increased, and this increased C-O surface coverage correlated with decreased macrophage uptake and increased particle circulation time in the blood (33). Evora et al. (34) reported similar studies with 1,2-dipalmitoylphosphatidylcholine (DPPC), a natural surfactant in the lung, on the surface of microspheres during pulmonary drug delivery. ESCA indicates that DPPC dominated the surface chemistry, and its presence on the aerosol surface lowered the phagocytosis of the particles by alveolar macrophages. Both examples demonstrated the use of surface analytical tools, specifically ESCA and SIMS, to chemically characterize the surfaces and use the information obtained to understand biological responses.

## Surface Tools to Study the Surface Erosion of Biodegradable Polymers (ESCA, SIMS, AFM)

Biodegradable polymers have always been of interest for the controlled release of drugs, because the release device need not be removed after all of the drug has been delivered. However, the polymers must possess predictable erosion kinetics with mechanical and surface characteristics suitable for drug delivery systems. Mechanistically, biodegradable polymers resorb either by bulk or surface erosion. In this section, we focus on applying surface analysis techniques to study the kinetics of surface erosion. Specifically, the relationship between polymer erosion and drug release is examined. Shakesheff et al. (35,36) demonstrated the use of in situ AFM to visualize the dynamic surface morphology changes that result from the surface biodegradation of a poly(ortho ester) sample containing protein. This technique allowed the degradation process occurring at the polymer/water interface to be visualized directly at the nanometer lateral ($x$ and $y$) and vertical ($z$) scale on surfaces within a hydrolyzing (aqueous) medium. In the study, they noticed that the presence of bovine serum albumin (BSA) particles (a model drug) in the polymer film yielded a rougher surface morphology in comparison to a BSA-free film. A time-course study over 90 min on the

degradable polymer was conducted, and the vivid AFM images corresponding to each time point were presented. Using image analysis methods, the relative changes of volume of protein particles and polymer matrix over the time course of the experiment were calculated. Later, the researchers, using the same in situ AFM technique along with ESCA and SSIMS, examined more intensively the surface degradation mechanisms of various polymer blends of poly(sebacic anhydride) (PSA) and poly(D,L-lactic acid) (PLA) (37). The blends are normally immiscible. However, when the molecular weight of the PLA is less than 3,000 Da, the polymer blends showed limited miscibility. Using this intrinsic characteristic of the two polymers, both miscible (PLA, 2,000 Da) and immiscible (50,000 Da) blends of polymer with various PLA loading were prepared. In both cases, ESCA and SIMS suggested preferential concentration of PLA on the surface. Because PSA tends to form semicrystalline materials while PLA leads to an amorphous polymer, various amounts of each of the components on the surface, in additional to their different degradation rates (2–4 weeks for PSA versus 12–16 weeks for PLA), could dramatically influence the release characteristics. In situ AFM showed that as the PLA in the immiscible blend is increased from 30 to 50%, an inversion in the phase morphology occurs with PLA morphology changing from isolated granules in a network of PSA to a network of PLA separating PSA areas. When PLA reaches 70% in the blend, surface enrichment of PLA even produced a permeation barrier that retards penetration of the hydrolyzing solution into the polymer matrix. Thus, the surface becomes resistant to degradation even after 3 h of exposure to the solution, in comparison with 30% PLA where degradation was observed shortly after exposure of 5 min. At this 70% PLA blend composition, the *surface* dominance of PLA is the limiting factor in the degradation process of the polymer. In these examples, combining surface chemical information from ESCA and SIMS with surface morphology information from AFM, it is possible to characterize the complex surfaces of these polymer blends and record the effect of the surface organization on degradation. This offers important insights in designing biodegradable drug delivery devices.

## Measuring Drug Concentration Profiles (ESCA, SSIMS, SIM Imaging, TOF-SIMS, AFM)

As early as 1985, ESCA was used to determine drug distributions within polymer matrices in controlled-release systems (38–40). The drug concentration profile is important in understanding and designing systems with desired and predictable release characteristics. Carli and Garbassi (39,40) reported that with the use of ESCA, they could understand the distribution of drug molecules loaded in a polymer matrix, that is, whether the drug molecules are in excess on the surface or homogeneously distributed throughout the matrix. By using unique label atoms in the drug (Cl for griseofulvin) and the polymer (N for crospovidone), they proposed that the use of Cl/N ratio could identify the drug location. Cl/N values higher than the theoretical ones suggest a drug excess in the surface, whereas lower Cl/N values indicate drug entrapment in the inner core of the polymer matrix. With these criteria, they related the drug profiles to various loading techniques and the release kinetics observed. In most cases, they observed preferential location of the drug on the device surface.

Starting in the late 1980s, Davies et al. (41–43) began to apply SSIMS and SIMS imaging to controlled-release systems. SSIMS analysis offers high surface molecular specificity and is complementary to the chemical quantification provided by ESCA. Using SSIMS, this group observed molecular fingerprints of a drug molecule, indomethacin, in polymer beads (42), focusing on the intense signals of the 139 D and 359 D mass spectral peaks on the outermost surface of the beads. However, for the older SIMS instrument used, quantification and the detection limit for high-molecular-weight fragments posed major obstacles in applying SSIMS to peptide-containing reservoirs. To overcome these limitations, John et al. (44) applied TOF-SIMS to analyze a peptide/polymer drug delivery system. TOF-SIMS permits molecular weights to at least 10,000 Da to be measured. Furthermore, using SIMS and TOF-SIMS imaging analysis, the surface molecular ions on the polymer matrices can be chemically mapped. Specifically, John et al. used TOF-SIMS and its imaging techniques to look at the distribution of the drug molecules along the surfaces of cross sections (of thickness between 100 and 200 $\mu$m) prepared from different polymer formulations. Previous studies could determine only if the drug was concentrated on the surface or entrapped in the bulk. With TOF-SIMS imaging, the drug molecular distribution in each layer through a polymer matrix can be examined. Belu and Bryan (45) also demonstrated that they were able to chemically image cross-sectional layers of a drug capsule simultaneously and identified precisely the distribution of drug and coating materials. They even argue that one can easily gain insightful information on competitor's samples using the in situ TOF-SIMS analysis. zur Mühlen et al. (46) also applied AFM to image the drug distribution on the surface of lipid nanoparticles and attempted to relate the images to release characteristics. Similar to the study of Davies et al. (37) described earlier, zur Mühlen et al. examined the surface after elution studies. The authors concluded that the fast initial release was by the outer noncrystalline layers of the particles, while the subsequent sustained release was associated with the inner crystalline particle layers.

## Use of Surface Analytical Tools in Controlled-Release-System Bioadhesion Studies

Controlled-release systems made from bioadhesive polymers prolong the residence time of devices on a biological surface before being eliminated by the body and allow more time for drug molecules to penetrate into the tissue. Bioadhesive systems are especially important when dealing with mucous and highly hydrated tissues in dentistry, orthopedics, and ophthalmology. Peppas and Buri (47) offer an extensive review of bioadhesive polymers on soft tissues in the context of drug delivery, and they describe various surface analytical techniques. Recently, Westwood et al. (48) applied both ESCA and AFM to determine the chemical as well as topological nature of an ophthalmic drug delivery

system (Occumer). They concluded that both techniques provide direct evidence for the complexion behavior of drug and polymer (anionic/cationic) as well as information on the surface topology and chemical coverage of Occumer.

In addition to the commonly used ESCA and SIMS surface techniques, Fourier transform infrared attenuated total reflection spectroscopy (FTIR-ATR) is emerging as a complementary tool for drug delivery design (49). It is used mostly in transdermal delivery applications. Jabbari et al. (50) reported the use of ATR in studying the penetration of poly(acrylic acid) (PAA), an important mucoadhesive for controlled-release applications, into a mucin layer. ATR spectra showed that swelling of the PAA allowed chain penetration at the polymer/mucin interface over a 12-min interval, and the extent of the chain interpenetration determined the bioadhesiveness. Mak et al. (51) have examined the potential use of ATR in vivo to study percutaneous penetration enhancement using 4-cyanophenol (CP) as the model penetrant across a human subjects' stratum corneum, the skin's outermost barrier, in the presence and absence of a penetration enhancer (oleic acid). Using the distinct IR peaks of CP, the authors were able to conclude that CP permeated faster when codelivered with oleic acid and examined penetration time. This study demonstrates the potential use of ATR as a noninvasive, in situ technique in studying transdermal delivery in vivo.

### Surface Tools to Study Diffusion Barriers (ESCA)

Diffusion barriers on controlled-release systems have been proposed to achieve zero-order release kinetics as well as to extend the efficacy of drug release devices. Kwok et al. (52,53) focused on applying the radio frequency–glow discharge plasma deposition technique (RF-GDPD) to coat an ultrathin, rate-limiting barrier ($\sim$100 Å) on an antibiotic-containing polyurethane polymer. In the study, ESCA was used as a surface analytical tool to relate the RF–plasma operating conditions to the release characteristics observed.

The plasma operating conditions determine the degree of structural alteration and cross-linking for the thin plasma-deposited film (PDF), which, in turn, affects the release rate of antibiotics through this layer. In principle, higher cross-linking density (measured indirectly by the C/O ratio) and thicker films will give slower release rates. Usually, higher deposition power yields higher cross-linking PDF, while longer deposition time gives thicker PDFs. Kwok et al. (unpublished data) reported that the optimal plasma conditions to attain such a PDF was first to argon-etch and pretreat the matrices with an 80 W $n$-butyl methacryate (BMA) plasma for 1 min, followed by immediate BMA plasma deposition at 40 W and 150 mT for 20 min. Pretreatment was used to increase the adhesion between the PDF and the matrix, thereby preventing coating delamination. An ESCA study to develop this pretreatment protocol is described elsewhere (Kwok et al., unpublished data). Briefly, by using the specified plasma deposition protocols, the authors eliminated the initial burst effect, significantly reduced the release rates, and closely approached the zero-order release. The release rates were at the $N_{kill}$ (minimum required kill rate for *Pseudomonas aeruginosa*) for at least 5 days, compared to 1 day for uncoated matrices.

Figure 3 depicts a cross-plot of average release rates and C/O ratios versus deposition powers. Deposition powers higher than 40 W resulted in CASING (cross-linking by activated species of inert gases), which eroded the surface antibiotic, cross-linked the polyurethane surface and made it a barrier for the release of antibiotics yet not as effective a barrier as a deposited overlayer (54). This was also observed in ESCA by the presence of nitrogen (an element found only in the polyurethane matrix but not in BMA deposition) and unusual high C/O ratios at higher powers (50 and 60 W). Thus, the highest deposition power did not necessarily result in the lowest release rate—there was an optimum at 40 W in this case. Figure 4 is a cross-plot of average release rates and C/O ratios versus deposition times. An increased C/O ratio of 6.4 to 8.2 was observed as the deposition time increased from 5 to 15 min, in comparison with an uncoated polyurethane that had a C/O ratio of 4.3. Beyond 15-min deposition times, the C/O ratio is unchanged; therefore, the thickness of the film had exceeded the sampling depth of ESCA ($\sim$100 Å), and ESCA no longer detected the oxygen in the underlying polyure-

thane. From the 15-min deposition time onward, the spectra obtained described the chemical composition of only the uppermost PDFs, and the C/O ratio remained constant. This is also reflected in the measurements of the release rates, which were consistent when the PDFs were made at deposition times longer than 15 min. This study demonstrated the use of ESCA as a routine check tool to relate the surface chemistry of PDF and the plasma operating parameters. The information was used to optimize the plasma conditions so that the desired diffusional barrier could be obtained. The bactericidal efficacy of the released antibiotics was evaluated and showed that the released antibiotics were biologically active in preventing the bacteria from adhering to the polyurethane, thereby decreasing the chances of biofilm-related infection (55,56).

## CONCLUDING REMARKS

Possibilities for surface analysis to contribute to the development and understanding of bulk controlled-release devices have been highlighted in this article. Advances will be made as researchers begin to focus on the importance of surfaces and the contributions that surface analysis can make to these systems. Challenges include the problems in working with highly hydrated (swollen) systems, with systems immersed in aqueous environments and with systems that are delivering agents as the analysis is being performed ("moving targets").

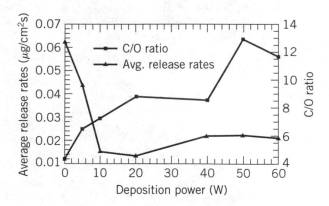

**Figure 3.** Correlation of release rates with C/O ratios and deposition powers. The system was at a pressure of 150 mT, and deposition time were 15 min throughout. *Note:* 0 W means uncoated polyurethane films.

## ACKNOWLEDGMENTS
The support of National Science Foundation (NSF) grant BES-9410429, National Institute of Health (NIH) grant RR01296 (NESAC/BIO), and NSF Engineering Research Center EEC-9529161 (University of Washington Engineered Biomaterials (UWEB) for some of the studies reported herein and during the writing of this article is appreciated.

## BIBLIOGRAPHY

1. B.D. Ratner, *J. Mol. Rec.* **9**, 617–625 (1997).
2. B.D. Ratner, *J. Biomed. Mater. Res.* **27**, 837–850 (1993).
3. B.J. Tyler, B.D. Ratner, D.G. Castner, and D. Briggs, *J. Biomed. Mater. Res.* **26**(3), 273–289 (1992).
4. W.A. Zisman, *Adv. Chem. Ser.* **43** (1964).
5. K. Seigbahn, *J. Electron. Spectrosc. Relat. Phenom.* **36**(2), 113–129 (1985).
6. K. Seigbahn, *Science* **217**, 111–121 (1982).
7. A. Benninghoven, *J. Vac. Sci. Technol., A* **3**, 451–460 (1985).
8. J.D. Andrade, *Med. Instrum.* **7**(2), 110–120 (1973).
9. R.E. Baier, *Bull. N.Y. Acad. Med.* **48**(2), 257–272 (1972).
10. E. Nyilas, E.L. Kupski, P. Burnett, and R.M. Haag, *J. Biomed. Mater. Res.* **4**, 369–432 (1970).
11. D.J. Lyman, J.L. Brash, and K.G. Klein, in R.J. Hegyeli, ed., *Proceedings of the Artificial Heart Program Conference*, 1969, pp. 113–121.
12. C.B. Duke, *J. Vac. Sci. Technol. A* **2**(2), 39–143 (1984).
13. G.A. Somorjai, *MRS Bull.* **23**(5), 11–29 (1998).
14. Y.R. Shen, *Nature (London)* **337**, 519–525 (1989).
15. D.G. Castner et al., *Langmuir* **9**, 537–542 (1993).

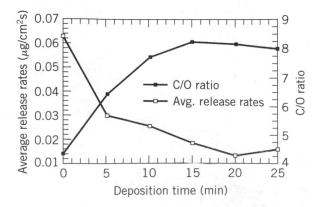

**Figure 4.** Correlation of release rates with C/O ratios and deposition times. The system was at a pressure of 150 mT, and deposition powers were at 40 W throughout. *Note:* 0 min means uncoated polyurethane films.

16. B.D. Ratner and V. Tsukruk, eds., *Scanning Probe Microscopy of Polymers*, ACS Symp. Ser. vol. 694, American Chemical Society, Washington, D.C., 1998.

17. D. Briggs and M.P. Seah, eds., *Practical Surface Analysis*, Wiley, Chichester, U.K., 1983.

18. B.D. Ratner and B.J. McElroy, in R.M. Gendreau, ed., *Spectroscopy in the Biomedical Sciences*, CRC Press, Boca Raton, Fla., 1986, pp. 107–140.

19. B.D. Ratner and D.G. Castner, in J.C. Vickerman, ed., *Surface Analysis: The Principal Techniques*, 1st ed., Wiley, Chichester, U.K., 1997, pp. 43–98.

20. J.J. Pireaux et al., *J. Electron. Spectrosc. Relat. Phenom.* **52**, 423–445 (1990).

21. G. Beamson and D. Briggs, *High Resolution XPS of Organic Polymers: The Scienta ESCA 300 Database*, 1st ed., Wiley, Chichester, U.K., 1992.

22. K.B. Lewis and B.D. Ratner, *J. Colloid Interface Sci.* **159**, 77–85 (1993).

23. A. Chilkoti and B.D. Ratner, in L. Sabbatini and P.G. Zambonin, eds., *Surface Characterization of Advanced Polymers*, VCH Publishers, Weinheim, 1993, pp. 221–256.

24. D.E. Gregonis et al., in R.B. Seymour and G.A. Stahl, eds., *Macromolecular Solutions*, Pergamon, New York, 1982, pp. 120–133.

25. F.M. Fowkes, in J.D. Andrade, ed., *Surface and Interfacial Aspects of Biomedical Polymers*, vol. 1, Plenum, New York, 1985, pp. 337–372.

26. E. Sacher, in B.D. Ratner, ed., *Surface Characterization of Biomaterials*, Elsevier, Amsterdam, 1988, pp. 53–64.

27. A. Benninghoven, *Angew. Chem., Int. Ed. Engl.* **33**, 1023–1043 (1994).

28. T. Boland and B.D. Ratner, *Proc. Natl. Acad. Sci. U.S.A.* **92**(12), 5297–5301 (1995).

29. J.T. Yates, Jr. and T.E. Madey, eds., *Vibrational Spectroscopy of Molecules on Surfaces*, Plenum, New York, 1987.

30. N.J. Harrick, *Internal Reflection Spectroscopy*, Interscience, New York, 1967.

31. M.C. Davies et al., *Polym. Prep. Am. Chem. Soc., Div. Polym. Chem.* **34**(2), 72–73 (1993).

32. A. Brindley, S.S. Davis, M.C. Davies, and J.F. Watts, *J. Colloid Interface Sci.* **171**(1), 150–161 (1995).

33. S.E. Dunn et al., *Pharm. Res.* **11**(7), 1016–1022 (1994).

34. C. Evora et al., *J. Controlled Release* **51**, 143–152 (1998).

35. K.M. Shakesheff et al., *Polym. Prepr., Am. Chem. Soc., Div. Polym. Chem.* **36**(1), 71–72 (1995).

36. K.M. Shakesheff et al., *Langmuir* **11**(7), 2547–2553 (1995).

37. M.C. Davies et al., *Macromolecules* **29**, 2205–2212 (1996).

38. F. Carli et al., *Polym. Sci. Technol.* **34**, 397–407 (1986).

39. F. Carli and F. Garbassi, *J. Pharm. Sci.* **74**(9), 963–967 (1985).

40. F. Garbassi and F. Carli, *SIA, Surf. Interface Anal.* **8**, 229–233 (1986).

41. M.C. Davies, A. Brown, J.M. Newton, and S.R. Chapman, *SIA, Surf. Interface Anal.* **11**(12), 591–595 (1988).

42. M.C. Davies and A. Brown, *ACS Symp. Ser.* **348**, 101–112 (1987).

43. M.C. Davies, M.A. Khan, A. Brown, and P. Humphrey. in A. Benninghoven, A.M. Huber, and H.W. Werner, eds., *Secondary Ion Mass Spectrometry VI*, Wiley, Chicheser, U.K., 1988, p. 667.

44. C.M. John, R.W. Odom, A. Annapragada, and M.Y. Fu Lu, *Anal. Chem.* **67**(21), 3871–3878 (1995).

45. A.M. Belu and S.R. Bryan, *Proc. Surf. Biomater. Meet.*, 1998.

46. A. zur Mühlen, E. zur Mühlen, and W. Mehnert, *Pharm. Res.* **13**(9), 1411–1416 (1996).

47. N.A. Peppas and P.A. Buri, *J. Controlled Release* **2**, 257–275 (1985).

48. A.D. Westwood, D.J. Leder, and D.H. Donabedian, *ACS Polym. Mater. Sci. Eng.* **76**, 130–131 (1997).

49. U.P. Fringeli, *Chimia* **46**(5), 200–214 (1992).

50. E. Jabbari, N. Wisniewski, and N.A. Peppas, *J. Controlled Release* **26**(2), 99–108 (1993).

51. V.H.W. Mak, R.O. Potts, and R.H. Guy, *Pharm. Res.* **7**(8), 835–841 (1990).

52. C. Kwok et al., in *Proceedings of 23rd International Symposium on Controlled Release of Bioactive Materials*, Controlled Release Society, Deerfield, Ill., 1996, pp. 230–231.

53. C.S.K. Kwok, B.D. Ratner, and T.A. Horbett. *Polym. Prepr. Am. Chem. Soc., Div. Polym. Chem.* **38**(1), 1077–1078 (1997).

54. F.Y. Chang, M. Shen, and A.T. Bell, *J. Appl. Polym. Sci.* **17**, 2915–2918 (1973).

55. C. Kwok et al., in *Fifth World Biomaterials Congress Program and Transaction*, Society of Biomaterials, University of Toronto Press, Toronto, Canada, 1996, p. 605.

56. J.D. Bryers et al., in *24th Annual Meeting for Society of Biomaterials*, Society for Biomaterials, Minneapolis, Minn., 1998, p. 347.

# CHARACTERIZATION OF DELIVERY SYSTEMS, XPS, SIMS AND AFM ANALYSIS

M.C. Davies
K.M. Shakesheff
C.J. Roberts
S.J.B. Tendler
University of Nottingham
Nottingham, United Kingdom

N. Patel
Molecular Profiles Ltd.
Nottingham, United Kingdom

S. Bryan
Physical Electronics Ltd.
Eden Prairie, Minnesota

## KEY WORDS

Chemical imaging
Controlled drug delivery
Secondary ion mass spectrometry
Surfaces
TOF-SIMS

## OUTLINE

Introduction
SIMS Process and Instrumentation

## INTRODUCTION

Surface chemistry plays a major role in the formation, stability, performance, and biointeractions of many pharmaceutical delivery systems. Surface coatings are widely used to protect the bulk drug carrier whether it be at the macroscopic level with micron-thick polymer film coatings or at the monolayer scale where polymer brush monolayers allow stealth liposomes and nanoparticles to avoid recognition by the reticuloendothelial system (1). Surface chemistry can promote selective interactions, as in the case of bioadhesive delivery systems that exploit the interaction between the hydrated polymer interface and the mucosal linings of the gastrointestinal tract.

Knowledge of the surface properties of delivery systems gives an important insight into both the development of conventional and novel delivery systems and also into their function and operation. Exactly which surface feature is important in the performance of delivery systems may be initially difficult to define (2). Surface topography, chemistry, hydration, charge, and bioactivity may all play an important role. In the chapter CHARACTERIZATION OF DELIVERY SYSTEMS, SURFACE ANALYSIS AND CONTROLLED RELEASE SYSTEMS the application of X-ray photoelectron spectroscopy (XPS or ESCA) was reviewed and the ability to obtain quantitative elemental and chemical state information was highlighted. Here, we describe the powerful complementary role of secondary ion mass spectrometry (SIMS) in defining the *molecular* chemical structure of interfaces.

## SIMS PROCESS AND INSTRUMENTATION

In the SIMS experiment, a surface is bombarded with stream of ions or atoms within an ultrahigh vacuum environment. This bombardment causes the emission or sputtering of electrons, neutral and charged species from the interface. The positive and negative ions of varying mass are collected and analysed within a mass spectrometer. Figure 1 illustrates a simple schematic of the process. The data are presented as a mass spectrum that may be interpreted using conventional mass spectrometry rules. A number of reviews (3–7) and reference databases (8,9) are available to aid the novice and the experienced operator alike in spectral interpretation. In certain acquisition modes, it is possible to produce chemical maps of surfaces

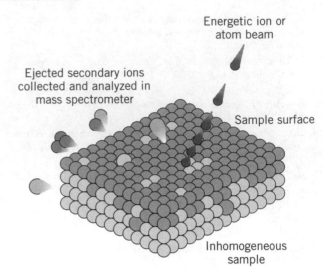

**Figure 1.** Schematic of the SIMS process.

by selecting a specific ion within the spectrum. An excellent detailed overview on the current state of the art of SIMS instrumentation and the physics of the SIMS process has been presented by Briggs recently (10) and, therefore, only relevant summary details are described here.

Current SIMS instruments have time-of-flight (TOF) mass spectrometers that are capable of routine high mass resolution in excess of m/M >3000, mass ranges from m/z 0 to 10,000 and spatial imaging resolutions at the submicron level. For most applications in this field, a mass spectrum may be acquired in one of two ways:

1. Where there is not a great need to define differences in chemical heterogeneity across the surface, a defocused beam can be used to obtain the spectrum.
2. Where the identification of different components, regions, or features across the surface is a key aim of the analysis, the primary ion or atom beam can be rastered over a predefined area and an entire mass spectrum acquired at each pixel point. The overall mass spectrum from that area is the summation of all the spectra acquired at all pixel points. This mode of data acquisition may be used to undertake retrospective image analysis (11) (Fig. 2). Ions in the summed spectrum that are known to be diagnostic for different chemical species or regions can be selected and an image formed from their intensity at each pixel. In addition, a region of interest from the resultant image, e.g., a coating film, may be selected and a mass spectrum formed from the summation of spectra from all the pixels in the defined region. This is a very powerful form of retrospective microanalysis of surfaces.

## GENERAL COMMENTS ON SIMS SPECTRA OF PHARMACEUTICAL MATERIALS

There is a growing SIMS literature on inorganic and polymeric compounds employed in many fields of materials sci-

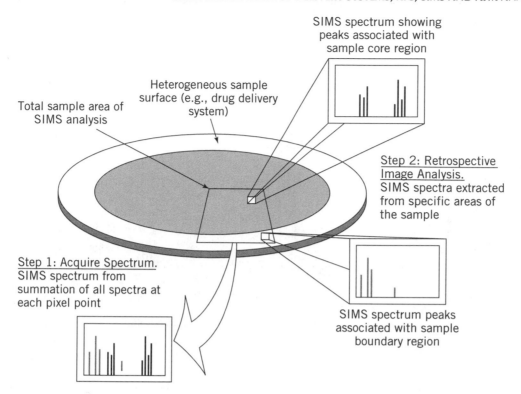

**Figure 2.** Schematic of the TOF-SIMS analysis of heterogenous surfaces. Retrospective data analysis is used to derive data from different regions of the sample.

ence. Although some activity has been within the drug delivery area, the majority of work within the medical sciences to date has focused on the SIMS analysis of biomaterials, which is reviewed elsewhere (2,5,6,12). In this chapter, we provide an overview of the general features of SIMS spectra of typical materials employed in drug delivery systems. A number of application areas are then discussed in more detail in the latter half of this article.

**Small Organic Molecules**

Conventional mass spectrometry is widely used in the elucidation of the molecular structure of organic molecules (13). The same principles may be employed to detect the presence of low molecular weight molecules (typically m/z <1000) within surfaces of delivery systems. Many drug molecules form quasimolecular ions $(M + H)^+$ or $(M - H)^-$ with a distinctive fragmentation pattern, which may be readily interpreted using mass spectrometry rules. For example, the presence of theophylline within the surface of cellulose spheroid beads was detected using SIMS where the molecular ions $(M + H)^+$ and $(M - H)^-$ ions and associated fragments were readily observed within both the negative and positive ion mass spectrum, respectively (14). Such an approach can be employed to detect the presence of many other types of pharmaceutical materials within dosage forms such as excipients and additives (5,15), e.g., plasticizers, inorganic fillers, pigments, surfactants and, to be discussed later, contaminants.

**Polymeric Materials**

A diverse range of polymers are currently employed in conventional and advanced drug delivery. Over the last 15

years, a rich literature has been compiled on the SIMS analysis of polymers (see Ref. 7 for most recent overview of area). The systematic examination of different classes of materials, including homologous series of some of the more technologically important ones, has made a significant contribution to our understanding of SIMS polymer data. From this work it has emerged that the SIMS spectra of polymers are dominated by ions that arise from cleavage of polymer chain that reflects the monomer structure (i.e., $(M + / - H)^{+/-}$, where M is monomer molecular weight) and any pendent or substituent side groups. This is illustrated in the negative ion SIMS spectrum in Figure 3 for the biodegradable polymer polylactic acid (PLA). Here, the major anions arise from $(nM - H)^-$ and $(nM + O - H)^-$ for n = 1–2 at m/z 71 and 143, and 87 and 159, respectively. Similar fragmentation patterns have been observed for many other classes of biomedical polymers that are used in drug delivery and biomaterials (2,5).

For copolymer systems, it is possible to detect ions that arise from all the monomers present. For example, ions diagnostic of both glycolide (GA) and lactide (LA) acids are present in the SIMS spectrum of the poly(lactide-*co*-glycolide) copolymer (PLGA) (16,17). From an analysis of the relative intensity of the diagnostic peaks, one can estimate the surface contribution of each component of the copolymer. Such data is not quantitative, because the surface environment, i.e., the so-called matrix effects (18), influence the intensity of the ions in the SIMS data. However, numerous examples exist in the literature that have shown that such an approach can be employed to derive relationships that reflect the surface chemistry in a quantitative manner (10).

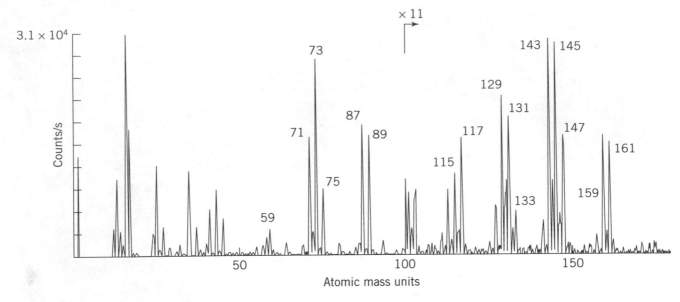

**Figure 3.** Negative ion SIMS spectrum of a PLGA film.

It is possible to deduce some important structural information from such semiquantitative analyses. One can detect ions in the SIMS spectra that arise from the sequence (i.e., dimers, trimers, etc.) of the copolymer chain. A number of groups have employed statistical analyses of the intensities of these ions to determine the level of random versus block nature of a copolymer, properties that may affect material performance. Such an approach has been employed on copolymers of polyalkyl methacrylates (19) and also for the biodegradable polymer PLGA (17). An examination of ion intensities has allowed the detection of the presence and level of polymer endgroups necessary for the stability of colloidal nanoparticle systems (20). Other surface properties such as polymer tacticity (21), cross-linking, and molecular weight as either oligomers (22) or as polymer brush (23) have been addressed using SIMS.

### Biomolecules and Cells

The molecular characterization of biomolecules is better suited to other powerful forms of mass spectrometry such as MALDI (24,25) due to the limited mass range of TOF-SIMS (m/z <10,000). However, the technique does have some value in the detection of biomolecules adsorbed or grafted to interfaces. Chemometrics has been used to study the low molecular weight regions (m/z 0–200) of TOF-SIMS spectra of protein films to identify diagnostic peaks and clusters for different protein molecules (26). Further studies suggest that the technique is sensitive to protein orientation and/or conformation at surfaces (27). SIMS imaging has revealed coexisting phases in lipid monolayers (28). The ability to detect surface-immobilized peptides (29) and oligonucleotides (30) suggests a considerable future potential in the use of TOF-SIMS analysis in identifying and estimating the surface density of targeting biomolecules grafted at the interface of advanced delivery systems.

Given a rich literature in the high-resolution elemental imaging of cells using SIMS (31), there has been relatively little activity in the pharmaceutical sector. Very recently, Winograd and his group (32) have demonstrated that it possible to get *molecular* TOF-SIMS images across a cell treated with a number of dopants, including the drug cocaine.

### APPLICATION OF TOF-SIMS TO THE ANALYSIS OF PHARMACEUTICAL DELIVERY SYSTEMS

#### Controlled Release Dosage Forms

Over the last 10 years a number of reports have demonstrated the detection of drugs (5–14) including peptides (33) in drug delivery systems surfaces using SIMS. Current instrumentation allows not only the chemical identification of surface species but also their spatial location within an interface. This ability may be exploited in the characterization of multicomponent formulations, a good example of which is shown in Figure 4. A polymer-coated bead (circa 1 mm) has been sectioned in half and the exposed cross-sectional surface has been analyzed by TOF-SIMS. The positive ion spectrum in Figure 4a is dominated by the signals of the drug, metoprolol at m/z 72 and 116 and the protonated molecular ions, $MH^+$ at m/z 268. It is possible to generate chemical images from such surfaces by identification of ions that are diagnostic of the drug, any excipients present, and the polymer coating. The $MH^+$ image in Figure 4b shows that the drug is well distributed throughout the bead with the exception of the central region. This area is dominated by $Si^+$ cations arising from a silica core. The polymer film is clearly distinguished by the ethyl substituent on the polymer chain from the ethylcellulose coating. Using the retrospective image analysis (11) already described, it is possible to generate SIMS

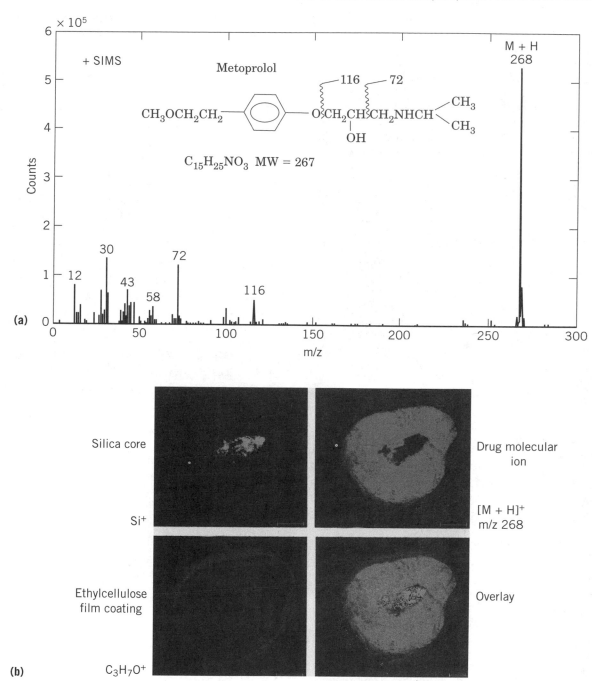

**Figure 4.** TOF-SIMS analysis of sectioned bead: (**a**) SIMS spectrum of drug; (**b**) SIMS images of drug, core, and coating. *Source:* The authors are grateful to Scott Bryan of Physical Electronics for kindly allowing the use of this figure.

spectra solely from each of the silica core, main drug-laden body of the bead, and the outer film coating regions. These spectra may be compared to standard control spectra of the materials. Such capability plays a powerful role not only in the characterization of the integrity of delivery systems but also in the use of SIMS in investigations within the patent protection field.

## Contaminants

One of the most significant areas for the routine exploitation of SIMS surface analysis in mainstream material science is in the detection and identification of surface contaminants. The presence of contaminants such as residual catalysts, mold-releasing agents and surfactants may also indeed feature within the interface of biomaterials and drug delivery systems. Ratner and co-workers have used XPS and SIMS to show that the surfaces of two different batches of a commercially available biomedical polymer Biomer were not composed of pure polyurethane but also contained a phenol antioxidant, Irganox 245, in one case and a di-isopropyl-amino-ethyl-methacrylate in the other (34). A common surface contaminant of SIMS polymer spectra is poly(dimethylsiloxane), a mold lubricant (3,4). SIMS has also been used to detect and monitor the efficiency of various techniques for the removal of residual surfactants such as sodium lauryl sulfate (35) and poly(vinyl alcohol) (PVA) (36) from the surface of biodegradable polyhydroxybutyrate and PLGA nanoparticles, respectively.

### Nano- and Micro-Particle Delivery Systems

There has been considerable interest in the application of colloidal formulations in drug delivery. SIMS (and XPS) are powerful tools for the characterization of the surface chemistry of nanoparticles and microcapsules. The surface enrichment of both polymer endgroups responsible for colloid stability (20), and one or more of the components of a copolymer colloid formulation have been detected by SIMS (37,38). The technique has also provided a detailed analysis of the interfacial chemistry of colloids with surface-grafted poly(ethylene oxide) (PEO) chains intended as model systems for site-specific drug delivery (39). The detection of ions diagnostic of the PEO within the SIMS spectra and their increase in relative intensity with rising PEO bulk levels provided clear evidence of the surface enrichment of PEO polymer brush. The analysis of sugars coupled to model colloid surfaces (40) suggests SIMS may also have considerable potential for the analysis of targeting molecules on colloid surfaces.

### Biodegradable Polymeric Systems

SIMS and XPS have been used to characterize the surface chemistry of homopolymers and copolymers of most biodegradable polymeric systems including the polyesters, poly(ortho esters), polyanhydrides, polyorganophosphazines, poly($\beta$-malic acids) and some of natural origin such as gelatin and modified hyaluronic acids, which are reviewed elsewhere (41). In most cases there was no evidence of phase separation and/or preferential orientation of one or more component within the copolymer films and the surface chemistry mirrored the bulk. Such a situation may not arise where polymer blends are used to develop materials with novel and controllable biodegradation kinetics as was shown for a multitechnique study of films formed from blends of PLA and poly(sebacic anhydride) (PSA) (42). Diagnostic ions of both polymers were observed in the SIMS spectra and an analysis of the relative intensities of each set of ions revealed that there was a surface excess of PLA in all of the blend compositions. To date, limited work has been done on the SIMS analysis of the surface degradation of biodegradable polymers. Leadley et al. successfully exploited the molecular specificity of SIMS to confirm the mechanism of hydrolysis of some poly(ortho ester) polymers (43) and the high mass range of ToF-SIMS has been employed to detect the presence of low molecular weight oligomers formed by the hydrolytic cleavage of surface chains (J.R. Gardella, personal communication).

### CONCLUDING REMARKS

There is rich potential for the exploitation of SIMS in the surface characterization of pharmaceutical delivery systems. The most fruitful information will arise when the technique is employed in tandem with other surface and biophysical techniques as part of a comprehensive strategy for the analysis of the surfaces of drug delivery devices. Such an approach cannot only aid our understanding of the physicochemical properties of such systems but also makes a valuable contribution to our understanding of their function and biological interactions.

### BIBLIOGRAPHY

1. L. Illum and S.S. Davis, *Life Sci.* **40**(16), 1553–1560 (1987).
2. M.C. Davies, C.J. Roberts, S.J.B. Tendler, and P.M. Williams, in J Braybrook, ed., *Biocompatibility: Assessment of Materials and Devices for Medical Applications*, 1997, pp. 64–100.
3. D. Briggs, SIA, *Surf. Interface Anal.* **9**, 391–404 (1986).
4. D. Briggs, *Br. Polym. J.* **21**, 3–15 (1989).
5. M.C. Davies and R.A.P. Lynn, *Crit. Rev. Biocompat.* **5**(4), 297–341 (1990).
6. M.C. Davies, in W.J. Feast, H.S. Munro, and R.W. Richards, eds.; *Polymer Surfaces and Interfaces II*, Wiley, New York, 1993, pp. 203–225.
7. D. Briggs, *Surface Analysis of Polymers by XPS and Static SIMS*, Cambridge University Press, Cambridge, U.K., 1998, pp. 119–184.
8. D. Briggs, A. Brown, and J.C. Vickerman, *Handbook of Static Secondary Ion Mass Spectrometry*, Wiley, London, 1989.
9. Static SIMS Library, *Surface Spectra*, Static SIMS Library, Manchester, U.K., 1998.
10. D. Briggs, *Surface Analysis of Polymers by XPS and Static SIMS*, Cambridge University Press, Cambridge, U.K., 1998, pp. 88–118.
11. S. Reichlmaier, S.R. Bryan, and D. Briggs, *J. Vac. Sci. Technol., A* **13**, 1217–1223 (1995).
12. B.D. Ratner, SIA, *Surf. Interface Anal.* **23**, 521–528 (1995).
13. F.W. McLafferty and F. Turecek, *Interpretation of Mass Spectrometry*, 4th ed., University Science Books, Mill Valley, Calif.
14. M.C. Davies et al., SIA, *Surf. Interfac. Anal.* **14**, 115–120 (1989).
15. M.C. Davies and A. Brown, *ACS Symp. Ser.* **348** (1987).
16. M.C. Davies, M.A. Khan, A. Brown, and P. Humphrey, in A. Benninghoven, A.M. Huber, and H.W. Werner, eds., *Secondary Ion Mass Spectrometry VI*, Wiley, Chichester, U.K., 1988.
17. A.G. Shard, C. Volland, M.C. Davies, and T. Kissel, *Macromolecules* **29**(2), 748–754 (1996).
18. P. Williams, *Annu. Rev. Mater. Sci.* **15**, 517–548 (1985).
19. D. Briggs and B.D. Ratner, *Polym. Commun.* **29**, 6–8 (1988).
20. M.C. Davies et al., *J. Colloid Interfac. Sci.* **156**, 229–239 (1993).
21. X.V. Eynde, L.T. Weng, and P. Betrand, SIA, *Surf. Interface Anal.* **25**(1), 41–45 (1997).
22. I.V. Bletsos, D.M. Hercules, D. Griefendorf, and A. Benninghoven, *Anal. Chem.* **57**, 2384–2388 (1985).

23. D. Briggs and M.C. Davies, *SIA, Surf. Interface Anal.* **25**(9), 725–733 (1997).

24. K.J. Wu and R.W. Odom, *Anal. Chem.* **68**, 456–461 (1998).

25. C. Fenselau, *Anal. Chem.* **69**, A661–A665 (1997).

26. D.S. Mantus, B.D. Ratner, B.A. Carlson, and J.F. Moulder, *Anal. Chem.* **65**(10), 1431–1438 (1993).

27. J.B. Lhoest, E. Detrait, P. VandenBosch de Aguilar, and P. Betrand, *J. Biomed. Mater. Res.* **41**(1), 95–103 (1998).

28. K. Leufgen et al., *Prog. Biophys. Mol. Biol.* **65** (S1), PC435.

29. C. Drouot et al., *Tetrahedron Lett.* **38**(14), 2455–2458 (1997).

30. J.S. Patrick, R.G. Cooks, and S.J. Pachuta, *Biol. Mass. Spectrom.* **23**(11), 653–659 (1994).

31. C. Chassard-Bouchaud, *Conf. Ser. Inst. Phys.*, 793–798 (1990).

32. T.L. Colliver et al., *Anal. Chem.* **69**(13), 2225–2231 (1997).

33. C.M. John et al., *Anal. Chem.* **67**(21), 3871–3878 (1995).

34. B.J. Tyler, B.D. Ratner, D.G. Castner, and D. Briggs, *J. Biomed. Mater. Res.* **26**, 273–289 (1992).

35. F. Koosha, R.H. Muller, S.S. Davis, and M.C. Davies, *J. Controlled Release* **94** 149–157 (1989).

36. P.D. Scholes et al., *J. Controlled Release*, in press.

37. M.C. Davies et al., *Langmuir* **12**, 3866–3875 (1996).

38. M.C. Davies et al., *Langmuir* **10**, 1399–1409 (1994).

39. A. Brindley, S.S. Davis, M.C. Davies, and J.F. Watts, *J. Colloid Interfac. Sci.* **171**, 150–161 (1995).

40. M.C. Davies et al., *Langmuir* **9**(7), 1637–1645 (1993).

41. A.G. Shard et al., in A.J. Domb, J. Kost, and D.M. Wiseman, eds., *Handbook of Bioerodable Polymers*, Gordon & Breach, Amsterdam, 1997, pp. 417–450.

42. K.M. Shakesheff et al., *Macromolecules* **29**(6), 2205–2212 (1996).

43. S.R. Leadley et al., *Biomaterials* **19**, 1353–1360 (1998).

# CHARACTERIZATION OF DELIVERY SYSTEMS, X-RAY POWDER DIFFRACTOMETRY

Suneel K. Rastogi
Raj Suryanarayanan
University of Minnesota
Minneapolis, Minnesota

## KEY WORDS

Amorphous
Crystalline
Crystallinity
Drug delivery system
Hydrate
Interaction
Internal standard
Phase identification
Phase transformation
Polymorph
Processing
Quantitative analysis
Solid dispersion
Storage
X-ray powder diffractometry

## OUTLINE

## INTRODUCTION

Discovered in 1895 by the German physicist Roentgen, X-rays are electromagnetic radiation, and their wavelength can range between $10^{-2}$ and $10^2$ Å. When X-rays pass through matter, they interact with the electrons and are scattered in all directions. Crystalline substances are characterized by long-range periodicity of the constituent molecules (or atoms). Because the interplaner distances are similar in magnitude to the wavelength of X-rays, their interaction results in diffraction. This was hypothesized and proven by Max von Laue in 1912. The condition under which diffraction occurs is described by the Bragg equation (1,2):

$$n\lambda = 2d \sin\theta \tag{1}$$

where $d$ = the distance between the successive planes of a crystal, $n$ = order of reflection (an integer), $\lambda$ = the wavelength of the X-rays, and $\theta$ = the angle that the incident beam of X-rays make with the sample.

The technique of X-ray powder diffractometry (XRD) was devised independently by Debye and Scherrer (1916) in Germany (3) and Hull (1917) in the United States (4). X-ray diffraction data show peak intensity as a function of diffraction angle ($\theta$) and can be presented in several ways (Fig. 1). Crystalline materials exhibit characteristic XRD patterns in which the positions and relative intensities of peaks are well defined and reproducible. The technique is usually nondestructive and is suited for analysis of pharmaceuticals both in pure state as well as in finished dosage forms. In recent years, the U.S. Food and Drug Administration (FDA) has realized the potential utility of XRD as a fingerprinting tool for characterization of pharmaceuticals (6). A brief description of the principles and applications of XRD is provided in the U.S. Pharmacopeia (7). In 1995, Byrn et al. developed some strategic approaches to regulatory issues involving solid pharmaceuticals in the form of flow charts/decision tree (8). From one such flow chart (Fig. 2), it is evident that XRD can be utilized for identifying and characterizing polymorphs. If multiple polymorphic forms exist, their quantitation can also be performed by XRD.

## PHASE IDENTIFICATION

XRD finds widespread application for the identification of crystalline solid phases. The diffraction pattern of every crystalline form of a compound is unique, which is the basis of its identification. The FDA requires proper control of the solid form of the drug in the formulation (6,8–10).

Phase identification is based on a comparison of the experimental data with standard powder patterns published in a database known as the *Powder Diffraction File* (PDF). This database was first introduced by Hanawalt et al. (11) in 1938 and has been issued annually since 1950. It is currently administered and distributed by the International Centre for Diffraction Data (ICDD). The database contains a collection of single-phase XRD patterns where the $d$-spacings (in Å) and relative intensities are tabulated. Each powder pattern is assigned a quality mark, the details of which can be found in the ICDD literature. This process of phase identification has been revolutionized by the introduction of compact disks for storage of standard data and faster search/match software algorithms for accessing and searching databases (12).

### Raw Materials

XRD is widely used for the identification and characterization (polymorphic forms, state of solvation, degree of crystallinity) of solid phases (13–15). The solid state of several forms of acemetacin were characterized by thermomicroscopy, differential scanning calorimetry (DSC), thermogravimetry, infrared spectroscopy (IRS), XRD, and pycnometry (13). Six crystalline forms of the drug were identified, of which five were anhydrates and one was a monohydrate. An amorphous phase was also identified. Each crystalline phase was characterized by a unique XRD pattern (Fig. 3). Similarly, the XRD patterns of seven crystalline phases (three anhydrate forms, two monohydrate forms, a quarter-hydrate and a methanol hemisolvate) of dehydroepiandrosterone revealed clear differences (14). XRD can also be used to identify and differentiate the different solid phases of pharmaceutical excipients. For example, the different polymorphs of mannitol exhibited differences in their diffraction patterns (15). These studies demonstrate the utility of XRD for the rapid identification of the various solid forms of pharmaceutical compounds.

(a)

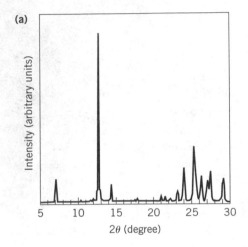

(b)

| $2\theta$ (degree) | $d$-spacings (A) | Relative intensity (%) |
|---|---|---|
| 7.11 | 12.4 | 13 |
| 12.6 | 7.02 | 100 |
| 14.25 | 6.21 | 10 |
| 17.67 | 5.02 | 1 |
| 20.88 | 4.25 | 2 |
| 21.45 | 4.14 | 2 |
| 22.11 | 4.02 | 2 |
| 23.19 | 3.83 | 6 |
| 23.85 | 3.73 | 16 |
| 23.94 | 3.71 | 20 |
| 25.29 | 3.52 | 33 |
| 25.38 | 3.51 | 32 |
| 26.19 | 3.40 | 15 |
| 27.03 | 3.30 | 12 |
| 27.39 | 3.25 | 18 |
| 29.04 | 3.04 | 15 |

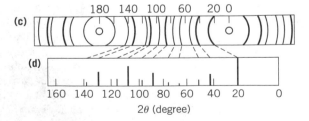

**Figure 1.** Different ways of presentation of X-ray powder diffraction data. (**a**) Diffraction pattern recorded experimentally (raw data), (**b**) tabular form of the experimentally obtained data, (**c**) Debye–Scherrer diffraction pattern, and (**d**) reduced diffraction pattern. (**a**), (**b**), and (**d**) are obtained by using a powder diffractometer with Bragg-Brentano geometry, whereas (**c**) is obtained using a Debye–Scherrer camera. *Source:* (**c**) and (**d**) are reproduced with permission of the copyright owner, John Wiley & Sons, Inc., from Ref. 5.

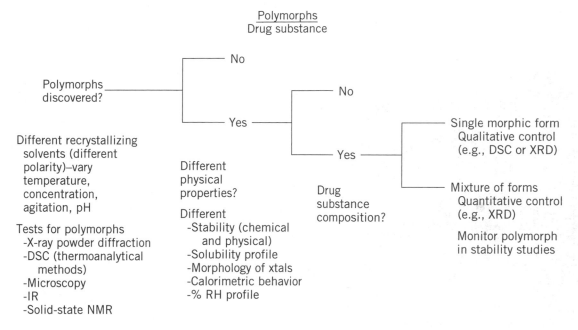

**Figure 2.** Flow chart/decision tree describing a logical sequence for collecting data for the identification and characterization of polymorphs. *Source:* Reproduced with permission of the copyright owner, Plenum Publishing Corporation, from Ref. 8.

### Dosage Forms

Solid dosage forms are usually multicomponent systems and contain one or more excipients in addition to the active ingredient. In such cases, each phase produces its XRD pattern independently of the others. Thus, simultaneous identification of multiple ingredients in a dosage form is possible (16). The technique may permit the detection of crystalline impurities. The method is nondestructive, and active ingredients in dosage forms can be analyzed with minimal or no sample preparation. However, in some formulations, the diffraction peaks of the ingredients overlap to such an extent that unambiguous identification of one or more ingredients may become difficult, if not impossible. Phadnis et al. (16) have described a pattern-subtraction technique that permitted selective subtraction of the XRD pattern of constituents of a formulation from the overall XRD pattern. The same approach was also used to improve the sensitivity of the technique. Chlordiazepoxide hydrochloride is available as an injectable and as oral (tablet and capsule) formulations. Because this is a low-dose drug, the content per tablet or capsule ranges between 5 and 25 mg. The drug is crystalline and has a characteristic XRD pattern (Fig. 4a). In physical mixtures of drug and microcrystalline cellulose, the presence of chlordiazepoxide hydrochloride was not readily discernable (Fig. 4b) because the drug constituted only 5% w/w of the mixture. However, using the pattern-subtraction technique, it was possible to subtract the XRD pattern of microcrystalline cellulose from that of the mixture. Identification of the drug was possible because the subtracted pattern (Fig. 4d) clearly revealed the peaks due to chlordiazepoxide hydrochloride. It should be pointed out here that conventional pharmacopeial identification techniques such as high-performance liquid chromatography (HPLC) require extraction of the active ingredient from the formulation, resulting in loss of valuable information about the solid state of the drug.

### Limitations of XRD

This technique can only be used if the analyte is crystalline. Noncrystalline (amorphous) materials produce diffraction patterns with one or more broad diffuse maxima. Because the technique is not very sensitive, crystalline impurities in low concentrations may be undetected. Other potential problems in the analyses of organic solids have been discussed comprehensively and elegantly by Jenkins (17).

### DETERMINATION OF CRYSTALLINITY

Solids are classified as either crystalline or noncrystalline (amorphous). Although the crystalline state is characterized by an ordered lattice, the amorphous state is characterized by the lack of long-range periodicity of the constituent molecules or atoms. The crystalline and amorphous states represent the two extremes of lattice order and intermediate states are possible. The term *degree of crystallinity* is useful in attempts to quantify these intermediate states of lattice order. During pharmaceutical processing, a drug is subject to a number of stresses such as milling and compression, which can cause a decrease in lattice order (18,19). Many properties, including chemical stability, water uptake and loss, and dissolution rate, depend on the degree of crystallinity of the drug.

Several XRD methods are available for the determination of degree of crystallinity. Ruland described a rigorous

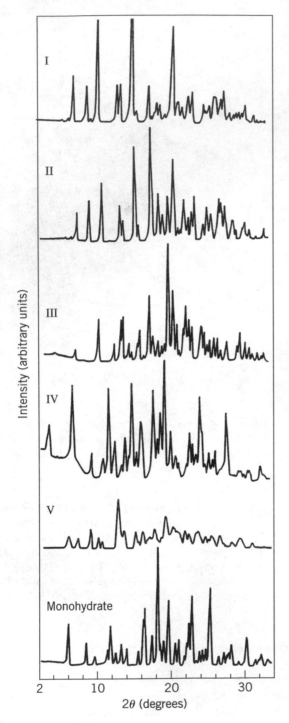

**Figure 3.** XRD patterns of different solid forms of acemetacin. I to V are different polymorphs of the anhydrate. *Source:* Reproduced with permission of the copyright owner, Govi-Verlag Pharmazeutischer Verlag GmbH, from Ref. 13.

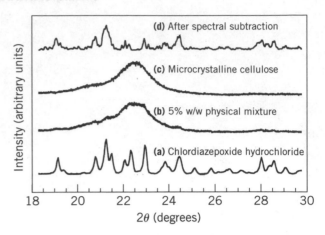

**Figure 4.** (**a**) The XRD pattern of chlordiazepoxide hydrochloride powder, (**b**) the XRD pattern of a powder mixture of chlordiazepoxide hydrochloride (5% w/w) and microcrystalline cellulose (95% w/w), (**c**) the XRD pattern of microcrystalline cellulose powder, **d**) the residual pattern after proportional subtraction of the microcrystalline cellulose XRD pattern from that of the physical mixture (**b**). The full scale in (**d**) is different from that of the other three XRD patterns. *Source:* Reproduced with permission of the copyright owner, Elsevier Science B. V., from Ref. 16.

tensive and time-consuming calculations, it has rarely been applied to evaluate the crystallinity of pharmaceuticals.

A simpler yet reliable method was developed by Hermans and Weidinger (23). This method is based on three assumptions: (1) the XRD pattern can be divided to demarcate between crystalline intensity ($I_c$) and amorphous intensity ($I_a$), (2) the experimentally measured crystalline intensity is directly proportional to the crystalline weight fraction ($x_c$) in the sample, and (3) the experimentally measured amorphous intensity (background intensity) is proportional to the amorphous weight fraction ($x_a$) in the sample. Thus,

$$x_c = pI_c \qquad (2)$$

$$x_a = qI_a \qquad (3)$$

where $p$ and $q$ are proportionality constants. If $x$ is the sum of the crystalline and amorphous weight fractions, then it can be shown that

$$I_a = \frac{x}{q} - \frac{p}{q}I_c \qquad (4)$$

Samples of varying degrees of crystallinity can be prepared and analyzed to obtain the values of $I_c$ and $I_a$. Based on equation 4, a plot of $I_a$ versus $I_c$ must be a straight line with a slope of $p/q$. The degree of crystallinity ($x_{cr}$) is related to XRD intensities $I_c$ and $I_a$ through the relation

$$x_{cr} = \frac{x_c}{x_c + x_a} \times 100 = \frac{I_c}{I_c + I_a \dfrac{q}{p}} \times 100 \qquad (5)$$

For a given unknown sample, the value of $x_{cr}$ can be obtained by substituting the experimentally obtained val-

method of determination of crystallinity (20). The method is based on Hosemann's paracrystal theory and takes into account all forms of lattice imperfections, including those produced by paracrystalline disorder. The applicability of this method to pharmaceutical systems has been demonstrated. (21,22). However, because the method requires ex-

ues of $I_c$ and $I_a$ into equation 5. Black and Lovering (24), simplified equation 5 to

$$x_{cr} = \frac{I_c}{I_c + I_a} \times 100 \qquad (6)$$

and determined the degree of crystallinity of several compounds including digoxin. In order to determine the degree of crystallinity using equations 5 and 6, it is necessary to separate the crystalline ($I_c$) and amorphous ($I_a$) intensities. This is at best arbitrary and based on subjective judgment, making it a potentially serious source of error.

## Use of Internal Standard

The use of an internal standard will assist in minimizing some of the errors in the measurement of intensities. Saleki-Gerhardt et al. used this approach to assess the percent disorder (percent disorder = $100 - x_{cr}$) of sucrose (25). Sucrose samples were prepared with percent disorder ranging from 0 to 100 to which a constant proportion of lithium fluoride (internal standard) was added. The integrated intensities of a peak unique to sucrose and lithium fluoride were determined. A linear standard curve was obtained when the intensity ratio (intensity of sucrose peak/intensity of lithium fluoride peak) was plotted as a function of percent disorder.

## Use of Peak Intensity

It is possible to use the peak height (instead of integrated intensity) as a measure of diffraction line intensity, provided the variations in disorder and particle size do not significantly influence the shape of the XRD peaks. Imaizumi et al. (26) used this method for the evaluation of degree of crystallinity of several samples of indomethacin, again using lithium fluoride as an internal standard.

## Other Methods

Ryan described a method to calculate the relative degree of crystallinity (RDC) of imipenem in a lyophilized product containing imipenem and amorphous cilastatin sodium (27). The full width at half the maximum intensity (FWHM) of the diffraction lines can be used to rank-order samples of varying degrees of crystallinity (28). Assuming that there are no particle size effects, an increase in crystallinity will result in an increase in peak height and a decrease in FWHM.

In spite of the popularity of XRD for crystallinity evaluation, the technique lacks precision and is subject to artifacts. It is also unsuitable for the detection and quantitation of small changes in lattice order. Pikal et al. performed XRD and solution calorimetry of spray-dried and freeze-dried cephalothin sodium (29). The differences in crystallinity between these could be evaluated by solution calorimetry but not by XRD. For the quantitation of small levels of disorder in highly crystalline samples, water sorption (25) and microcalorimetry (30) appear to be the techniques of choice. Recently, using FT-Raman spectroscopy, it was possible to detect down to either 1% amorphous or 1% crystalline content (31). This technique can therefore be used to quantify small levels of disorder in highly crystalline samples and also small levels of order in highly disordered samples.

## QUANTITATIVE ANALYSIS

### Theory

Alexander and Klug derived the theoretical basis for quantitative analysis by XRD (32). When a beam of X-ray passes through any homogenous substance, the fractional decrease in the intensity, $I$, is proportional to the distance traversed, $x$. Thus,

$$-\frac{dI}{I} = \mu dx \qquad (7)$$

where $\mu$ is a proportionality constant known as the *linear attenuation coefficient*. The value of $\mu$ depends on the density of the substance and the wavelength of the X-rays. The *mass attenuation coefficient*, $\mu^*$, obtained by dividing $\mu$ by the density of the substance, is a constant independent of the physical state. The mass attenuation coefficient of a substance is the weighted average of the mass attenuation coefficients of that substance's constituent elements. Thus,

$$\mu^* = \sum_{k=1}^{n} w_k \mu_k^* \qquad (8)$$

where, $w_1, w_2, \ldots, w_n$ are the weight fractions of elements $1, 2, \ldots, n$ in the substance and $\mu_1^*, \mu_2^*, \ldots, \mu_n^*$ are their respective mass attenuation coefficients.

A solid mixture containing several components can be regarded as being composed of only two components—the analyte (component 1) and the sum of the other components (the matrix, subscripted as M). The relationship between the intensity of diffraction peak $i$ of the unknown component ($I_{i1}$) and its weight fraction ($x_1$) in the mixture is described by the equation:

$$I_{i1} = \frac{Kx_1}{\rho_1(x_1(\mu_1^* - \mu_M^*) + \mu_M^*)} \qquad (9)$$

where K is a constant, $\rho_1$ is the density of component 1, and $\mu_1^*$ and $\mu_M^*$ are the mass attenuation coefficients of components 1 and the matrix, respectively.

### The Direct Method

The direct method is applicable to two component mixtures where $\mu_1^* \neq \mu_M^*$. Such cases are frequently encountered in pharmaceutical systems where a solid drug might be contaminated by pseudopolymorphs or decomposition products. It is necessary to determine the intensity of peak $i$ of a sample consisting of only the analyte [$(I_{i1})_0$]. Equation 9 is modified so that the relative intensity of the XRD peak of component 1 [expressed as $I_{i1}/(I_{i1})_0$] is given by

$$\frac{I_{i1}}{(I_{i1})_0} = \frac{x_1\mu_1^*}{x_1(\mu_1^* - \mu_M^*) + \mu_M^*} \qquad (10)$$

As mentioned previously, it is possible to calculate the mass attenuation coefficients of the analyte and the matrix based on their chemical compositions. It is then possible to calculate the relative intensity $[I_{il}/(I_{il})_0]$ as a function of the weight fraction of the analyte in the mixture and thus generate a theoretical standard curve. This eliminates the need for the preparation of experimental standard curves.

A simpler case of quantitative analysis arises when $\mu_1^* = \mu_M^*$. This occurs when the analyte and the matrix have the same molecular formula. Two examples are polymorphic mixtures and enantiomer–racemic compound mixture. In such cases, equation 10 reduces to

$$\frac{I_{il}}{(I_{il})_0} = x_1 \quad (11)$$

A plot of the relative intensity as a function of the analyte weight fraction will be linear.

### The Internal Standard Method

In this method, an internal standard is added, and its weight fraction in the mixtures is maintained constant in all the samples. The integrated intensities of line 1 of the analyte ($I_1$) and line $s$ of the internal standard ($I_s$) are determined. The relationship between the intensity ratio and the analyte weight fraction ($x_1$) in the mixture is expressed as

$$x_1 = k\,\frac{I_1}{I_s} \quad (12)$$

Thus, the intensity ratio is linearly related to the analyte weight fraction in the mixture.

### Choice of Internal Standard

Shell (33) suggested that an ideal internal standard should have high crystal symmetry so that its XRD pattern consists of strong but few peaks. It must be chemically stable and should not interact with the system ingredients. The high-intensity lines of the analyte and the internal standard should not overlap with one another but should be close to each other. To facilitate the preparation of a homogeneous mixture, the density of the internal standard should not be very different than those of the system ingredients. Corundum($\alpha$-Al$_2$O$_3$), silicon, lithium fluoride, and zinc oxide have been used as internal standards. Among organic compounds, adamantane has several of the properties desired in an internal standard. However, it has not found widespread use in quantitative XRD.

Besides the methods mentioned above, several others have been described in the literature. These include (1) the method of standard additions, (2) the reference intensity ratio (RIR) method, and (3) Reitveld analysis. These methods are described in standard texts (34,35) but have found little use in analyses of pharmaceuticals.

The application of XRD for the quantitative analyses of pharmaceutical systems was pioneered by Shell (33). Early quantitative XRD studies were based on the heights of XRD peaks (36,37). It is well known that variations in particle size and microstrain can affect the peak heights. How-

ever, the integrated peak intensities are much less affected by these factors. Thus, for quantitative XRD, it is desirable to measure the integrated intensities (2).

### Preformulation

The direct method described above has been successfully used to determine the weight fractions of anhydrous carbamazepine (C$_{15}$H$_{12}$N$_2$O) and carbamazepine dihydrate (C$_{15}$H$_{12}$N$_2$O · 2H$_2$O) in their mixtures (38). Based on the mass attenuation coefficients of carbamazepine anhydrate and dihydrate, the intensity ratios $(I_{il}/(I_{il})_0)$ were calculated as a function of the sample composition (Fig. 5). These were in good agreement with the experimentally obtained values of $I_{il}/(I_{il})_0$. The relative amounts of $\alpha$- and $\beta$-carbamazepine in a mixture have also been quantified using this method (39). The internal standard method has been used for the simultaneous quantification of the $S(+)$-enantiomer and the racemic compound of ibuprofen (40).

### Dosage Forms

XRD has been used for quantification of active ingredients in intact formulations. As early as 1964, Papariello et al. determined the glutethimide content in intact tablets by XRD (41). However, the authors were somewhat limited by the fact that computer-based data collection and analysis was not possible at that time. Using the direct method (equation 10), the carbamazepine content in intact, multicomponent tablet formulations was determined (42). Though the experimentally observed intensity ratios were in good agreement with the theoretical intensity ratios, the

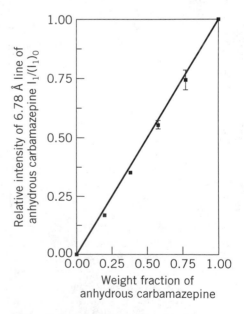

**Figure 5.** The relative intensity of the 6.78-Å line of anhydrous carbamazepine as a function of its weight fraction in a binary mixture of anhydrous carbamazepine and carbamazepine dihydrate. The line is based on theoretical values, while the data points are experimental measurements. Error bars represent standard deviations. *Source:* Reproduced with permission of the copyright owner, Plenum Publishing Corporation, from Ref. 38.

relative error in the determination of carbamazepine content was high ($\pm$10%). When the tablets were crushed and the powder sample was analyzed after the addition of an internal standard, the precision and accuracy improved substantially (43). The experimentally determined weight fraction of carbamazepine in the tablets ranged between 98.7 and 102% of the true weight fraction, and the highest coefficient of variation value was 2.2%. The relative error in the determination of carbamazepine content was reduced to $\pm$4%.

Kuroda and coworkers used the technique for the quantitative analysis of active components in a variety of formulations including ointments (44), vaginal tablets (45), oral suspension (46), plasters (47) and rectal suppositories (48). Both the direct as well as the internal standard method were used, and the active ingredient was quantified even at concentrations down to 0.05% w/w. There was a good agreement of the results obtained by XRD and by chemical analysis.

### Sources of Error in Quantitative XRD

There are numerous sources of error in quantitative XRD. These are discussed comprehensively in the literature (1,49).

## PHASE TRANSITIONS INDUCED DURING PROCESSING OR STORAGE

The active ingredient as well as the excipients in a dosage form can undergo a variety of physical transformations during processing or storage. These include polymorphic transformations, alterations in the degree of crystallinity, and changes in the state and degree of hydration. Such processing- and storage-induced phase transformations can profoundly influence the performance of the dosage form (50). XRD has been used to identify crystalline reactants, products, and intermediates (if any). The technique has been used to monitor phase transitions both qualitatively and quantitatively. An added advantage of the technique is that it provides quantitative information about the degree of crystallinity.

The transition of anhydrous carbamazepine to carbamazepine dihydrate in aqueous suspensions was observed to follow first-order kinetics (51). The dehydration of 6-mercaptopurine monohydrate was studied both in the presence and absence of 6-thioxanthine (52). Whereas pure 6-mercaptopurine monohydrate converted directly to polymorphic form II of the anhydrate, in the presence of 6-thioxanthine it converted to a different anhydrate (form III) through an intermediate. The effect of processing operations such as grinding and compression on the crystallinity of an experimental drug, TAT-59, was monitored by XRD (53). An increase in compression pressure or grinding time caused a decrease in drug crystallinity. This was accompanied by a decrease in the stability of TAT-59.

### Special Instrumentation

Recent advances in instrumentation and data analysis software have greatly facilitated the study of pharmaceutical systems. Variable-temperature powder diffractometry (VTXRD) is a technique whereby XRD patterns are obtained while the sample is subjected to controlled temperature program. During such studies, it is also possible to control the environment and maintain the sample under the desired relative humidity (54,55). Conventionally, thermoanalytical techniques such as DSC or themogravimetric analysis (TGA) have been used for the investigation of reactions in solid drugs. These techniques have several drawbacks. They do not unambiguously identify crystalline phases and provide little or no information about the degree of crystallinity. Because intermediate phases (if any) may not be readily identified, these techniques are not necessarily useful for discerning the reaction mechanism. These drawbacks can be overcome by using VTXRD, which provides an excellent compliment to thermoanalytical techniques. Because the crystalline intermediates can be easily identified, the technique provides critical information about the reaction mechanism. For example, VTXRD was used to investigate the dehydration of theophylline monohydrate (Fig. 6). Dehydration of the monohydrate resulted in the formation of anhydrous theophylline (50). However, the VTXRD study revealed that theophylline monohydrate dehydrated to a metastable anhydrous phase that then transformed to the stable anhydrate. Fawcett et al. (56) have built an instrument that allows simultaneous XRD and DSC studies on the same sample. Another unique feature of the instrument is that the gaseous reaction product can be subjected to mass spectrometric analysis (57).

By attaching a low-temperature stage to an XRD, characterization of solutes in frozen aqueous solutions has been accomplished (58). Sodium nafcillin does not crystallize when its aqueous solution is frozen. However, when the sample is annealed at $-4°C$, the solute gradually crystallizes. Using XRD, it was possible to monitor the amount of sodium nafcillin crystallizing as a function of the annealing time. Recently, the low-temperature stage of an XRD was modified so that the sample chamber could be evacuated using a vacuum pump. As a result, it was possible to carry out the entire freeze-drying process in situ in the sample chamber of the XRD (59).

### Interactions between Solids

In a powder mixture, the powder pattern of each crystalline phase is produced independently of the other constituents. This makes it feasible to study solid–solid interactions such as those between the drug and excipients in a formulation. If there is no interaction, then the diffraction pattern of a solid mixture will be the summation of diffraction patterns of individual constituents. If the interaction between the constituents results in the formation of crystalline product, then this will be characterized by the appearance of new peaks in the powder pattern. However, if the interaction results in amorphous products, this will be evident from the broad halos in the XRD pattern. Thus, irrespective of the nature of product phase, XRD is capable of revealing solid–solid interactions. Position-sensitive detectors (PSD) allow very rapid data collection and are particularly useful for investigating fast reactions. The com-

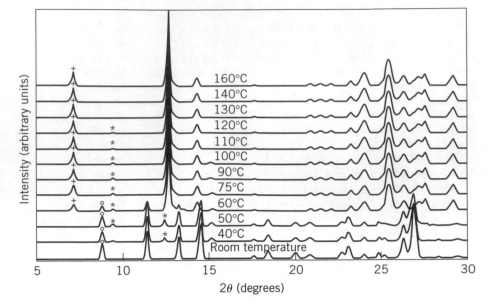

**Figure 6.** VTXRD of theophylline monohydrate. XRD patterns were taken at all the temperatures indicated in the figure. The *, +, and o marks indicate peaks unique to metastable anhydrous theophylline, stable anhydrous theophylline, and theophylline monohydrate, respectively. *Source:* Reproduced with permission of the copyright owner, American Chemical Society and American Pharmaceutical Association, from Ref. 50.

ponents of a formulation can undergo a variety of interactions resulting in formation of various dispersion systems such as glass dispersions, solid solutions, eutectic mixtures, and inclusion complexes. XRD is useful for characterizing such interactions.

## CHARACTERIZATION OF DRUG DELIVERY SYSTEMS

The principles and procedures discussed so far can be extended for the characterization of complex drug delivery systems. In sustained- and controlled-release formulations, the release of the active ingredient is controlled through the use of excipients that are often polymeric in nature. XRD has been used for the characterization of a wide variety of formulations, including microspheres (60–62), hydrogel matrices (63), solid liposphere (64), microcapsules (62), solid-dispersion polymer films (65,66), solid-dispersion granules (67), waxy matrix tablets (68), and nanoparticles (69,70).

### Physical State of the Active Ingredient

There are a variety of polymer-based delivery systems. Hydrogels are polymeric materials that can take up and retain water within their structure. This ability, combined with their biocompatibility, makes them useful for drug delivery applications. The physical state of the drug in such delivery systems depends on the solubility of the drug in the matrix. The drug could either be completely dissolved in the matrix, or a fraction of the incorporated drug could be dissolved, and the rest could be dispersed in the matrix. To model the kinetics of drug release from these dosage forms, it is necessary to know the physical state of the drug in the matrix. XRD has been used to determine the physical state of salicylic acid in a hydrogel topical formulation (63). Formulations were prepared with high salicylic acid concentrations so that a fraction of the incorporated salicylic acid was undissolved. A fixed weight of each formulation was subjected to XRD, and the integrated

intensities of two peaks of salicylic acid were determined. The intensity of each of these peaks was linearly related to the weight percent of salicylic acid in the formulations. The intercept on the x-axis was the solubility of salicylic acid in the matrix. The results obtained by XRD were confirmed by scanning electron microscopic studies. Many transdermal dosage forms are monolithic systems where the active ingredient is incorporated in a rate-controlling polymeric matrix. The XRD method described here can find application in the characterization of the physical state of the drug in such systems.

### Solid Dispersions

XRD has been used to differentiate between various types of solid-dispersion systems (71,72) such as solid solutions (73), glass dispersions (74), binary eutectics (74), solid-surface dispersions (75,76), physical mixtures (66), and amorphous precipitates of drug in the crystalline carrier. The identification of the polymorphic form of drug in solid-dispersion systems is readily possible (60,64). XRD can also been used to generate the phase diagram of binary solid systems. In the case of oxezapam (77) and ketoprofen (78), an increase in dissolution rate was brought about by rendering the drugs amorphous. XRD has been used for the quantification of crystalline frusemide in solid dispersions with polyvinylpyrrolidone (79).

XRD has been used to monitor solid-state phase transformations and decomposition reactions of crystalline ingredients in solid delivery systems. Marini et al. described the time-dependent phase changes in β-cyclodextrin when stored under dry nitrogen at room temperature (80). The technique can be used to detect phase transformations during processing (81) and storage (61,75,82) of solid-dispersion systems.

### Influence of Drug Load

Lee et al. prepared transdermal formulations of estradiol by dispersing the drug in a polymeric matrix and heating to temperatures ranging from 110 to 165°C (83). After complete dissolution of estradiol, the system was cooled, which resulted in crystallization of the drug. Interestingly, XRD patterns of the final product revealed that the solid state of the crystallized drug was dependent on the concentration of drug in the polymer. Whereas estradiol hemihydrate crystallized at low drug concentrations, high concentrations favored the formation of a new solid form (so far unreported and termed "form X"). Flux measurements revealed that the release rate of form X was significantly higher than that of the hemihydrate.

### Drug–Cyclodextrin Complexes

Many drugs form complexes with cyclodextrins, and the XRD pattern of these solid complexes is usually markedly different from that of pure drug or pure cyclodextrin (84,85). However, in some cases, complexation results in an amorphous product (86). XRD was used to determine the degree of crystallinity of indomethacin complexed with $\beta$-cyclodextrin (87). The colyophilization of naproxen and $\beta$-cyclodextrin resulted in the formation of an amorphous solid with increased solubility (88). In this case, it was not possible to ascertain whether increased solubility was a result of molecular encapsulation of drug inside the cyclodextrin cavity or due to formation of an amorphous solid dispersion. It was concluded that complementary techniques such as IR spectroscopy should be used to develop a more complete understanding of the system. Similarly, the solid complexes of tretinoin with $\beta$-cyclodextrin and dimethyl $\beta$-cyclodextrin were amorphous to X-rays (89). The formation of an inclusion complex was proven by NMR spectroscopy and DSC.

### Degree of Crystallinity

The degree of crystallinity of the active ingredient can have a significant influence on the performance of a dosage form. Interestingly, the crystallinity of the excipients in a formulation can also influence product performance. XRD has been useful in explaining the alterations in the release rate of drugs from different delivery systems due to changes in crystallinity of both the drug and the excipients. Izumikawa et al. have demonstrated that the drug release rate from progesterone-loaded poly(L-lactide) microspheres was closely correlated with the crystallinity of the polymer matrix (90). Chan et al. investigated the dispersion properties of dry powder aerosols for inhalation delivery of recombinant human deoxyribonuclease (rhDNase) (91). Powder blends of rhDNase with varying amounts of carrier material (NaCl) were prepared, and the crystallinity of these blends was determined using XRD. The study revealed that the crystallinity of the powder blends increased linearly with increase in NaCl content (Fig. 7). This increase in crystallinity caused a decrease in the cohesive force between particles, leading to a proportional increase in the respirable fraction (i.e., % wt. of particles <7 $\mu$m in the aerosol). Thus, XRD can be used to select the weight fraction of NaCl that provides the optimum formulation performance. These results were supported by scanning electron microscopic studies.

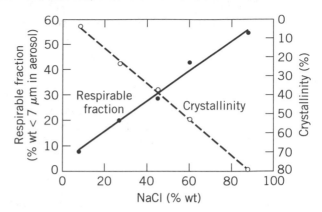

**Figure 7.** The relationship between sodium chloride content, the percent crystallinity, and the dispersing properties (expressed here as "respirable fraction") of rhDNase powders. Pure rhDNase powder is amorphous. *Source:* Reproduced with permission of the copyright owner, Plenum Publishing Corporation, from Ref. 91.

### Limitations of XRD in the Characterization of Drug Delivery Systems

The major limitation of XRD technique is its very limited utility in the characterization of amorphous systems. This is because amorphous materials exhibit one or more broad diffuse maxima. For example, XRD revealed the formation of an X-ray amorphous solid dispersion between crystalline ibuprofen and PVP (92). However, the exact nature of interaction between the drug and polyvinylpyrrolidone (PVP) could only be determined using infrared spectroscopy and thermal analysis. In drug–PVP solid dispersions, increasing the amount of PVP in the system resulted in a decrease in the intensity of the XRD peaks of the drug (93,94). This has often been interpreted as being due to a decrease in drug crystallinity. However the decrease in peak intensity may have resulted from partial dissolution of drug in PVP. Thus, XRD alone is often insufficient for the characterization of multicomponent delivery systems.

### OTHER APPLICATIONS

There are several other potential applications of XRD, though these find use only infrequently. These include (1) evaluation of preferred orientation in powders and compacts (95,96), (2) determination of crystal structure (97,98), (3) determination of crystallite size and local lattice distortion (microstrain) (99), and (4) determination of density of crystalline substances.

### SUMMARY

The importance of solid-state characterization of drugs and delivery systems is increasingly recognized by the pharmaceutical community. Because XRD provides direct and unequivocal information about the lattice structure of a crystalline solid, it is a valuable technique for solid-state characterization studies. XRD is an excellent complement to other techniques such as DSC, TGA, solid-state NMR spectroscopy, and microscopy. Recent advances in instrumentation and computerized data analysis have improved the capability of both qualitative and quantitative XRD. XRD can be used to develop a mechanistic understanding of the phase changes that occur during pharmaceutical processing and storage. The technique finds application in several stages of the design and development of delivery systems, including preformulation, formulation, during manufacture (in process control), and in the evaluation of the finished product (quality control).

## BIBLIOGRAPHY

1. H.P. Klug and L. Alexander, *X-ray Diffraction Procedures for Polycrystalline and Amorphous Materials*, 2nd ed., Wiley, New York, 1974, pp. 120–174.
2. B.D. Cullity, *Elements of X-ray Diffraction*, 2nd ed., Addison-Wesley, Reading, Mass., 1978, pp. 3–31, 81–106.
3. P. Debye and P. Scherrer, *Phys. Z.* 17, 277–283 (1916).
4. A.W. Hull, *J. Am. Chem. Soc.* 41, 1168–1175 (1917).
5. R. Jenkins, *X-Ray Fluorescence Spectroscopy*, Wiley, New York, 1988, p. 39.
6. T. Layloff, *Pharm. Tech.* 15, 146–148 (1991).
7. *The United States Pharmacopeia*, 23rd rev., U.S. Pharmacopeial Convention, Rockville, Md., 1994, pp. 265–267, 1843–1844, 2216.
8. S. Byrn et al., *Pharm. Res.* 12, 945–954 (1995).
9. *Preliminary Draft Guidance for Industry: CMC Content and Format of INDs for Phases 2 and 3 Studies of Drugs, Including Specified Therapeutic Biotechnology-Derived Produces*, Food and Drug Administration, Washington, D.C., 1997.
10. Food and Drug Administration, *Fed. Regist.* 62, 62889–62890 (1997).
11. J.D. Hanawalt, H.W. Rinn, and L.K. Frevel, *Ind. Eng. Chem., Anal Ed.* 10, 457–512 (1938).
12. J.I. Langford and D. Louer, *Rep. Prog. Phys.* 59, 131–234 (1996).
13. A. Burger and A. Lettenbichler, *Pharmazie* 48, 262–272 (1993).
14. L. Chang, M.R. Caira, and J.K. Guillory, *J. Pharm. Sci.* 84, 1169–1179 (1995).
15. B. Debord et al., *Drug Dev. Ind. Pharm.* 13, 1533–1546 (1987).
16. N.V. Phadnis, R.K. Cavatur, and R. Suryanarayanan, *J. Pharm. Biomed. Anal.* 15, 929–943 (1997).
17. R. Jenkins, *Adv. X-Ray Anal.* 35, 653–660 (1992).
18. Y. Takahashi, K. Nakashima, H. Nakagawa, and I. Sugimoto, *Chem. Pharm. Bull.* 32, 4963–4970 (1984).
19. N. Kaneniwa, K. Imagawa, and M. Otsuka, *Chem. Pharm. Bull.* 33, 802–809 (1985).
20. W. Ruland, *Acta Crystallogr.* 14, 1180–1185 (1961).
21. Y. Nakai, E. Fukuoka, S. Nakajima, and M. Morita, *Chem. Pharm. Bull.* 30, 1811–1818 (1982).
22. M. Morita and S. Hirota, *Chem. Pharm. Bull.* 30, 3288–3296 (1982).
23. P.H. Hermans and A. Weidinger, *J. Appl. Phys.* 19, 491–506 (1948).
24. D.B. Black and E.G. Lovering, *J. Pharm. Pharmacol.* 29, 684–687 (1977).
25. A. Saleki-Gerhardt, C. Ahlneck, and G. Zografi, *Int. J. Pharm.* 101, 237–247 (1994).
26. H. Imaizumi, N. Nambu, and T. Nagai, *Chem. Pharm. Bull.* 28, 2565–2569 (1977).
27. J.A. Ryan, *J. Pharm. Sci.* 75, 805–807 (1986).
28. R. Suryanaranan and S. Venkatesh, *Pharm. Res.* 7, S-104 (1990).
29. M.J. Pikal, A.L. Lukes, J.E. Lang, and K. Gaines, *J. Pharm. Sci.* 67, 767–773 (1978).
30. T. Sebhatu, M. Angberg, and C. Ahlneck, *Int. J. Pharm.* 104, 135–144 (1994).
31. L.S. Taylor and G. Zografi, *Pharm. Res.* 15, 755–761 (1998).
32. L. Alexander and H.P. Klug, *Anal. Chem.* 20, 886–889 (1948).
33. J.W. Shell, *J. Pharm. Sci.* 52, 24–39 (1963).
34. R. Jenkins and R.L. Snyder, *Introduction to X-ray Powder Diffractometry*, Wiley, New York, 1996, pp. 355–387.
35. R.L. Snyder and D.L. Bish, in D.L. Bish and J.E. Post, eds., *Modern Powder Diffraction*, Mineralogical Society of America, Washington, D.C., 1989, pp. 101–144.
36. C.L. Christ, R.B. Barnes, and E.F. Williams, *Anal. Chem.* 20, 789–795 (1948).
37. A.J. Aguiar, J. Krc, Jr., A.W. Kinkel, and J.C. Samyn, *J. Pharm. Sci.* 56, 847–853 (1967).
38. R. Suryanarayanan, *Pharm. Res.* 6, 1017–1024 (1989).
39. R. Suryanarayanan, *Powder Diffr.* 5, 155–159 (1990).
40. N.V. Phadnis and R. Suryanarayanan, *Pharm. Res.* 14, 1176–1180 (1997).
41. G.J. Papariello, H. Letterman, and R.E. Huettemann, *J. Pharm. Sci.* 53, 663–667 (1964).
42. R. Suryanarayanan and C.S. Herman, *Int. J. Pharm.* 77, 287–295 (1991).
43. R. Suryanarayanan, *Adv. X-Ray Anal.* 34, 417–427 (1991).
44. K. Kuroda and G. Hashizume, *Yakugaku Zasshi* 87, 148–158 (1967).
45. K. Kuroda, G. Hashizume, and K. Fukuda, *Yakugaku Zasshi* 87, 1175–1183 (1967).
46. K. Kuroda, G. Hashizume, and K. Fukuda, *J. Pharm. Sci.* 57, 250–254 (1968).
47. K. Kuroda, G. Hashizume, and F. Kume, *Chem. Pharm. Bull.* 17, 818–821 (1969).
48. K. Kuroda and G. Hashizume, *Chem. Pharm. Bull.* 20, 2059–2062 (1972).
49. R. Suryanarayanan, in H.G. Brittain, ed., *Physical Characterization of Pharmaceutical Solids*, Dekker, New York, 1995, pp. 187–221.
50. N.V. Phadnis and R. Suryanarayanan, *J. Pharm. Sci.* 86, 1256–1263 (1997).
51. W.W.L. Young and R. Suryanarayanan, *J. Pharm. Sci.* 80, 496–500 (1991).
52. H. Nakamachi et al., *Chem. Pharm. Bull.* 29, 2956–2965 (1981).
53. Y. Matsunaga, N. Bando, H. Yuasa, and Y. Kanaya, *Chem. Pharm. Bull.* 44, 1931–1934 (1996).
54. H. Hashizume et al., *Powder Diffr.* 11, 288–289 (1996).
55. S.K. Rastogi, M. Zakrzewski, and R. Suryanarayanan, *47th Annu. Denver X-Ray Conf.*, Colorado Springs, August 3–7, 1998.
56. T.G. Fawcett et al., *Adv. X-Ray Anal.* 29, 323–332 (1986).
57. T. Fawcett, *Chemtech (Heidelberg)* 17, 564–569 (1987).
58. R.K. Cavatur and R. Suryanarayanan, *Pharm. Res.* 15, 194–199 (1998).
59. R.K. Cavatur and R. Suryanarayanan, *47th Annu. Denver X-Ray Conf.*, Colorado Springs, August 3–7, 1998.
60. V. Rosilio et al., *J. Biomed. Mater. Res.* 25, 667–682 (1991).
61. Y. Aso, S. Yoshioka, and T. Terao, *Int. J. Pharm.* 93, 153–159 (1993).

62. A.K. Dash, *J. Microencapsul.* **14**, 101–112 (1997).

63. R. Suryanarayanan, S. Venkatesh, L. Hodgin, and P. Hanson, *Int. J. Pharm.* **78**, 77–83 (1992).

64. D. Aquilano, R. Cavalli, and M.R. Gasco, *Thermochim. Acta* **230**, 29–37 (1993).

65. Y. Kohda et al., *Int. J. Pharm.* **158**, 147–155 (1997).

66. K. Danzo, F. Higuchi, and A. Otsuka, *Chem. Pharm. Bull.* **43**, 1759–1763 (1995).

67. K. Goracinova et al., *Drug Dev. Ind. Pharm.* **22**, 255–262 (1996).

68. M. Otsuka and Y. Matsuda, *J. Pharm. Sci.* **83**, 956–961 (1994).

69. J. Kreuter, *Int. J. Pharm.* **14**, 43–58 (1983).

70. R. Cavalli, D. Aquilano, M.E. Carlotti, and M.R. Gasco, *Eur. J. Pharm. Biopharm.* **41**, 329–333 (1995).

71. G.P. Bettinetti, *Boll. Chim. Farm.* **128**, 149–162 (1989).

72. W.L. Chiou and S. Reigelman, *J. Pharm. Sci.* **60**, 1281–1302 (1971).

73. H.H. El-Shattawy, D.O. Kildsig, and G.E. Peck, *Drug Dev. Ind. Pharm.* **10**, 1–17 (1984).

74. A. Hoelgaard and N. Moller, *Arch. Pharm. Chemi, Sci. Ed.* **3**, 34–47 (1975).

75. T. Ishizaka, H. Honda, and M. Koishi, *J. Pharm. Pharmacol.* **45**, 770–774 (1993).

76. I. Sugimoto et al., *Chem. Pharm. Bull.* **30**, 4479–4488 (1982).

77. R. Jachowicz and E. Nurnberg, *Int. J. Pharm.* **159**, 149–158 (1997).

78. K. Takayama, N. Nambu, and T. Nagai, *Chem. Pharm. Bull.* **30**, 3013–3016 (1982).

79. C. Doherty, P. York, and R. Davidson, *J. Pharm. Pharmacol.* **37**, 57P (1985).

80. A. Marini et al., *Solid State Ionics* **63–65**, 358–362 (1993).

81. H. Lin, Y. Huang, L. Hsu, and Y. Tsai, *Int. J. Pharm.* **112**, 165–171 (1994).

82. J.D. Ntawukulilyayo, S. Bouckaert, and J.P. Remon, *Int. J. Pharm.* **93**, 209–214 (1993).

83. S.M. Lee et al., in N.A. Peppas, D.J. Mooney, A.G. Mikos, and L. Brannon-Peppas, eds., *Proceedings of the Topical Conference on Biomaterials, Carriers for Drug Delivery, and Scaffolds for Tissue Engineering*, American Institute of Chemical Engineers, New York, 1997, pp. 196–198.

84. F. Giordano et al., *Farmaco, Ed. Prat.* **43**, 345–355 (1988).

85. D. Watanabe et al., *Chem. Pharm. Bull.* **44**, 833–836 (1996).

86. P. Mura, G. Bettinetti, F. Melani, and A. Manderioli, *Eur. J. Pharm. Sci.* **3**, 347–355 (1995).

87. R. Casella, D.A. Williams, and S.S. Jambhekar, *Int. J. Pharm.* **165**, 15–22 (1998).

88. G.P. Bettinetti et al., *Farmaco* **44**, 195–213 (1989).

89. P. Montassier, D. Duchene, and M. Poelman, *Int. J. Pharm.* **153**, 199–209 (1997).

90. S. Izumikawa, S. Yoshioka, Y. Aso, and Y. Takeda, *J. Controlled Release* **15**, 133–140 (1991).

91. H. Chan et al., *Pharm. Res.* **14**, 431–437 (1997).

92. N.M. Najib, M.A. El-Hinnawi, and M.S. Suleiman, *Int. J. Pharm.* **45**, 139–144 (1988).

93. G.P. Bettinetti et al., *Farmaco, Ed. Prat.* **43**, 331–343 (1988).

94. A.S. Kearney, D.L. Gabriel, S.C. Mehta, and G.W. Radebaugh, *Int. J. Pharm.* **104**, 169–174 (1994).

95. E. Fukuoka, M. Makita, and S. Yamamura, *Chem. Pharm. Bull.* **39**, 3313–3317 (1991).

96. E. Fukuoka, M. Makita, and S. Yamamura, *Chem. Pharm. Bull.* **41**, 595–598 (1993).

97. R.J. Cernik et al., *J. Appl. Crystallogr.* **24**, 222–226 (1991).

98. M. Tremayne et al., *J. Solid State Chem.* **100**, 191–196 (1992).

99. E. Fukuoka, K. Terada, M. Makita, and S. Yamamura, *Chem. Pharm. Bull.* **41**, 1636–1639 (1993).

# CHEMICAL APPROACHES TO DRUG DELIVERY

Nicholas Bodor
University of Florida
Gainesville, Florida

James J. Kaminski
Schering-Plough Research Institute
Kenilworth, New Jersey

## KEY WORDS

Blood-brain barrier

Brain targeting

Chemical delivery systems

Eye targeting

Metabolism

Molecular packaging

Peptide delivery

Soft anticholinergics

Soft drugs

Soft steroids

## OUTLINE

## INTRODUCTION

The search for novel chemical entities which may become effective therapeutic agents for the treatment of various diseases traditionally involves identifying a lead compound of well-defined structure that exhibits the desired biological activity in a relevant assay. In most cases, the primary test considered relevant is a mechanism-based in vitro assay rather than an in vivo animal model. The in vitro assay may involve inhibition of a specific enzyme or may involve antagonism of a particular receptor thought to be involved in the pathogenesis of the disease.

The source of the lead compound could be from a corporate collection of compounds, from third-party databases of chemical entities and natural products offered for sale, or from libraries of compounds prepared using the techniques of combinatorial chemistry. In any case, the screening of these compound collections using random, focused or high-throughput screening technologies could identify a lead compound of interest.

Once identified, the structure of the lead compound is systematically, and in some cases randomly, modified to learn as much as possible about the relationship that may exist between its structure and the observed biological activity. This empirically derived knowledge, if it exists, is then used to design other novel compounds that might exhibit even greater biological activity.

In some instances, the structure of the enzyme (protein) or receptor of interest, in the presence of the lead structure or one of its analogues (ligand), might be known from an X-ray crystal structure of the ligand–protein complex. In these cases, structure-based design methods can be invoked to propose compounds that could exhibit maximum biological activity by optimizing the intermolecular interactions occurring between the ligand and specific amino acids in the active site of the enzyme, or at the receptor.

Despite all these attempts to introduce rationale and logic into the drug discovery paradigm, the structural (chemical) modifications made to the lead compound in order to optimize biological activity rarely consider their effect on determining the physicochemical and metabolic characteristics of the candidate selected. As a result, very few compounds with maximal or optimal biological activity have become clinically useful drugs. The main, and most frequent, reason for this being the toxicity observed in the development of these compounds. As this situation occurs so frequently, we have to recognize that the main objective of the drug discovery (design) process should not be to optimize the activity (potency) of the drug, but rather to optimize its therapeutic index (i.e., to optimize the ratio between the toxic and therapeutic doses of the drug). Ideally, drugs should have rather large therapeutic indices, indicating a significant and safe separation between these two important dose levels. However, most of the time, undesired and toxic side effects of novel biologically active compounds run parallel with the desired activity. Thus, even if more potent drugs are obtained, their therapeutic index remains essentially unchanged.

Side effects are usually related to the intrinsic receptor affinity responsible for the desired biological activity. Therefore, it is possible that changes in the pharmacokinetic and pharmacodynamic properties of a drug can parallel biological activity changes. Furthermore, when introduced into the body, drugs can be extensively metabolized. Most drugs generate multiple metabolites whose chemical structures are closely related to each other. These metabolic products can exhibit an enhanced or different biological activity, or they can exhibit their own toxicity.

Data accumulated on metabolic activation, toxification, and/or detoxification of biologically active compounds demonstrate the necessity to include metabolic considerations as an integral part of the drug discovery (design) process. Rather than determining the metabolic profile of only the candidate selected for preclinical evaluation, structure–activity relationship studies should be conducted concomitantly with structure–metabolism (toxicity) relationship studies throughout the drug discovery (design) process. This combination of structure–activity, structure–metabolism and structure–toxicity relationship studies is the principal tenet of the retrometabolic drug design concept, a cycle which includes soft drug design (SD) and chemical delivery system (CDS) design (Fig. 1) as distinct from prodrug design. Chemical approaches for drug delivery using prodrug strategies have been extensively reviewed recently (1–3) and the reader is directed to these comprehensive references.

Retrometabolic drug design approaches include two distinct methods to improve the therapeutic index of a drug and form the subject matter of this article. One is the general concept of CDSs. A CDS is defined as a biologically inert molecule that requires several steps in its conversion to the active drug, and that enhances drug delivery to a particular organ or site. SD design strategies represent the alternate method in the retrometabolic drug design cycle. Although both approaches involve chemical modifications to obtain an improved therapeutic index and both require enzymatic reactions to fulfill drug targeting, there is not much in common between the principles of SD design and those of CDS-based retrometabolic drug design. Whereas the CDS is inactive by definition and sequential enzymatic reactions provide the differential distribution and drug activation, SDs are active isosteric/isoelectronic analogues of the lead drug that are deactivated in a predictable and controllable manner after achieving their therapeutic role. In the extreme case for a CDS, enzymatic processes produce the drug at the site and nowhere else in the body, whereas in the SD case, the drug is present at the site and absent in the rest of the body due to enzymatic activity at those sites.

## SD DESIGN

A SD by definition is a biologically active species. However, in general, the SD behaves very differently from the drug itself, and this behavior is by design. The SD undergoes a predicted and controlled singular transformation to a nontoxic and inactive metabolite, or in some cases, is degraded to more than one nontoxic and inactive metabolite. The

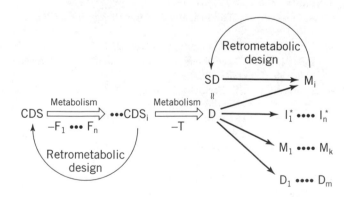

**Figure 1.** The retrometabolic drug design cycle. CDS: chemical delivery system; SD: soft drug; CDS$_i$: CDS metabolites; D: lead drug; M$_i$: inactive metabolites; I$_n^*$: reactive intermediates; M$_n$: other metabolites; D$_n$: analogue metabolites; F$_n$: modifier functions; T: targetor moiety.

main point is that by design, the SD simplifies the transformation, distribution, and activity profile that most drugs exhibit. SD design aims to produce biological activity of a specific, desired kind and then predicted processes deactivate or metabolize the SD in one step to an inactive species. When applied at the desired site of action, be it topical or internal, the SD elicits the desired biological effect locally, but as soon as it is distributed from the site, it is susceptible to deactivation, thereby preventing any undesirable (toxic) side effect, which would be otherwise characteristic for this class of drug.

The SD concept was introduced in 1980, and since then five distinct SD classes have been identified:

1. *Soft analogues*. These SDs are close structural analogues of known active drugs but have a specific metabolically sensitive site built into their structure that provides a one-step controllable detoxification.

2. *SDs based on the inactive metabolite approach*. An inactive metabolite of a lead drug is chemically modified to provide an isosteric–isoelectronic analogue of the lead drug that exhibits the desired biological activity. The biologically active analogue is metabolically converted to the inactive metabolite in one step.

3. *Controlled release SDs*. Controlled release endogenous agents are derivatives of natural hormones or neurotransmitters (natural SDs) for which a local or site-specific delivery is developed.

4. *Activated SDs*. These compounds are not analogues of known drugs but are derived from nontoxic chemical compounds activated by introduction of a specific pharmacophore whose presence in the molecule is responsible for eliciting the desired biological activity.

5. *SDs based on the active metabolite*. A metabolite that significantly retains the activity of the parent drug can be chemically transformed to a potent drug, but undergoes a one-step deactivation process, as it is already at its highest oxidation state.

Examples of each of these classes of SDs are described in the literature. However, the most useful and successful strategies for designing safe and selective drugs have been the *inactive metabolite* and *soft analogue* approaches.

## Soft Analogue Approach

Research on soft analogues began after the discovery of "soft quaternary salts", **1** (Fig. 2). These compounds undergo a facile hydrolysis via a short-lived intermediate to form an aldehyde, an amine, and a carboxylic acid upon deactivation. Structures exemplified by **1** were first used for the improved delivery of drugs containing tertiary amines such as pilocarpine (4) and erythromycin (5). The term "soft" in the context of these quaternary salts was adopted to reflect their relative ease to biodegrade as compared with conventional "hard" quaternary salts that generally require multiple oxidative steps to be metabolized.

The simplest example of a useful SD is provided by the isosteric analogue **2** of cetylpyridinium chloride **3** (Fig. 3), a well-known hard quaternary antimicrobial agent that generally requires several oxidative steps to lose its surface active antimicrobial properties. The structure of **2** and **3** are very similar in that both side chains are essentially 16 atoms in length. Accordingly, their physical properties, as measured by their respective critical micelle concentration values, are also very similar (6). Whereas both compounds exhibit comparable antimicrobial activity against a wide variety of microorganisms, **2** undergoes facile hydrolytic cleavage leading to its deactivation. Because of this facile degradation, **2** is approximately 40 times less toxic than **3** (i.e., the median oral lethal dose for **2** and **3**, determined using white Swiss mice, was 4,110 mg/kg and 108 mg/kg, respectively).

Other examples for the design of soft analogues based on soft quaternary salts were provided by the soft anticholinergic agents (7) and the soft cholinesterase inhibitors (8).

The soft anticholinergic agents were designed to exhibit high local, but practically no systemic, anticholinergic activity. It is well-established that anticholinergic agents exhibit many useful clinical effects. For example, the local antisecretory activity of these compounds was long thought to be beneficial for inhibiting eccrine sweating (perspiration). Indeed, a number of anticholinergic agents such as atropine, scopolamine, or their quaternary ammonium salts inhibit perspiration. However, these compounds are also responsible for producing the many side effects observed (e.g., dry mouth, urinary retention, drowsiness, and mydriasis). Therefore, these compounds are not useful as antiperspirants, but their soft analogues may provide a viable alternative.

Although the structural differences between hard and soft anticholinergics agents of this type are relatively small, they are nonetheless profound (Fig. 4). Classical quaternary ammonium anticholinergics, **4** and **5**, have two or three carbon atoms separating the quaternary head from the ester function. For many years, the dogma was that this separation was critical for effective receptor binding. The corresponding soft analogue **6** would contain just one carbon atom separating the quaternary head from the ester function allowing facile hydrolytic deactivation. Several compounds of type **6** were synthesized and evaluated; many were found to be at least as potent as atropine (7). The example compound in the following diagram, **7**, is equipotent with atropine in a number of in vitro and in vivo anticholinergic tests, but it is extremely short acting following intravenous administration due to its facile hydrolytic deactivation. When applied topically to humans, **7** exhibited high local antisecretory activity with no observation of systemic toxicity.

**7**                    Atropine

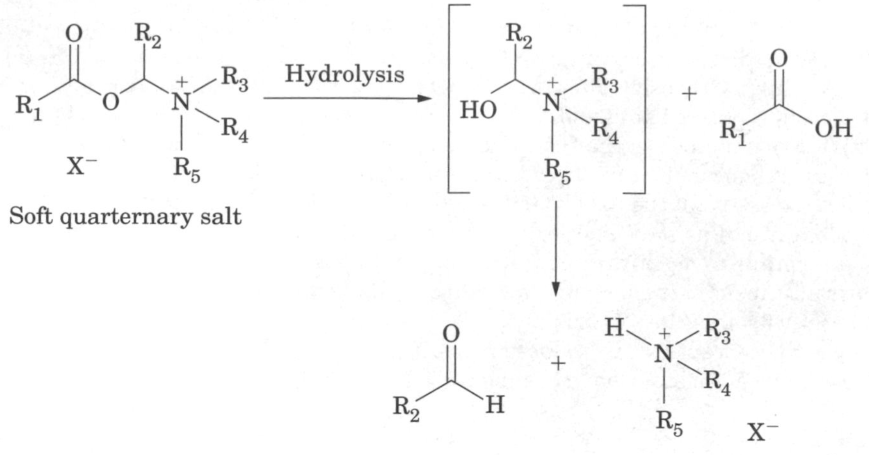

**Figure 2.** Hydrolytic deactivation of soft quaternary salts, **1**.

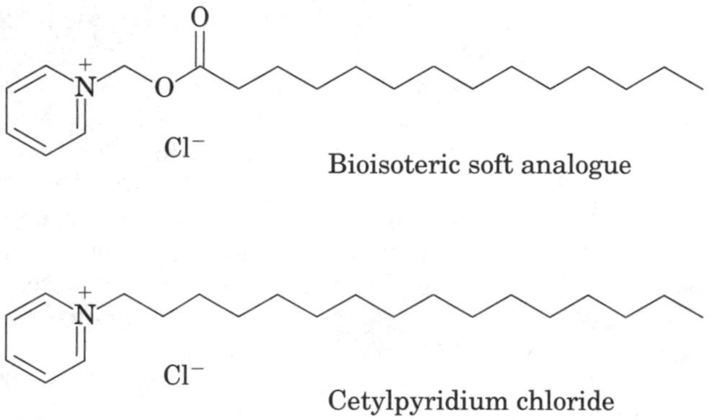

Soft quarternary salt

Bioisoteric soft analogue

Cetylpyridium chloride

**Figure 3.** Bioisosteric soft analogue, **2**, of cetylpyridinium chloride, **3**.

These examples clearly illustrate the application of replacing neighboring methylene groups with an ester or "reverse ester" function resulting in a bioisosteric soft analogue. The method as described is in fact general and not restricted to quaternary salts. Some of the basic isosteric and isoelectronic design concepts of SD design are integrated as part of the inactive metabolite approach.

### Inactive Metabolite Approach

This is one of the most promising and versatile methods for developing new and safer drugs. The main objective is to include the metabolism of the active species into the design process. The following strategies can be used in developing new drugs based on the inactive metabolite approach:

- The design process starts with a known inactive metabolite of a drug as the lead compound.
- Novel structures are designed from the inactive metabolite; these structures are isosteric and/or isoelectronic with the drug from which the lead inactive metabolite was derived.
- New structures are designed so that their metabolism produces the starting inactive metabolite in one

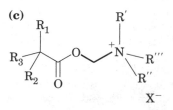

**(a)**

**(b)**

**(c)**

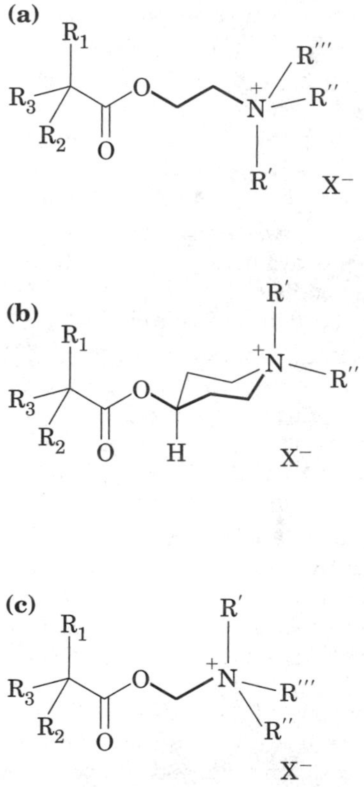

**Figure 4.** Traditional quaternary ammonium anticholinergic agents, **4** and **5**, and a soft anticholinergic analogue, **6**.

step without any other metabolic processes occurring (predicted metabolism).

- The specific binding and transport properties and metabolic rates of the SD can be controlled by molecular manipulation in the activation stage of the process (controlled metabolism).

With respect to these considerations, two points are worthy of elaboration. The inactive metabolite selected does not have to be isolated and identified. In addition, there are two classes of SDs based on inactive metabolites. In one class, the inactive metabolite formed involves the pharmacophore; in the other class, it does not. Enzymatic modifications affect only the portions of the molecule that are not directly responsible for eliciting activity. Understandably for the second class, there is more freedom to modify inactive metabolites in the activation stage.

Many examples of potential drugs designed using this approach are available, which include several dermatological and ophthalmic products under development, but the principles can be illustrated with some general examples. One of the few instances in which these principles have been accidentally used for the design of nonpharmaceutical products is the pesticide dichlorodiphenyltrichloromethane (DDT), 8, along with its analogues and metabolites (Fig. 5) (9). In 1939, DDT was discovered as a powerful insecticide. During World War II, it was also demonstrated that DDT could also be used to treat typhus and malaria. DDT undergoes several in vivo oxidative dehydrohalogenations and oxidations to produce the inactive metabolite 9. This inactive acidic metabolite is of low toxicity and can be excreted as a water-soluble species. Therefore, it is an ideal lead compound for the inactive metabolite approach.

It was found that ethyl 4,4'-dichlorobenzilate, 10, is highly active as a pesticide, but is less carcinogenic than DDT or dicofol (kelthane, 11). The ethyl ester moiety apparently functions similarly to the trichloromethyl group in DDT imparting pesticidal activity to 10 however, 10 is considerably less toxic than 8 because metabolism of the labile ethyl ester generates the corresponding nontoxic carboxylic acid, 9, the inactive metabolic of Figure 5.

Another good example of the inactive metabolite approach is provided by the antiinflammatory corticosteroids known to have multiple adverse effects that often limit their usefulness. Corticosteroids undergo a variety of oxidative and reductive metabolic conversions that lead to the formation of numerous steroidal metabolites. A major metabolic route of hydrocortisone, 12, is oxidation of the dihydroxyacetone side chain, which ultimately leads to the formation of cortienic acid (Fig. 6). Cortienic acid 13 is a major metabolite excreted in human urine and lacks corticosteroid activity, and therefore, is an ideal lead candidate for the inactive metabolite approach (10–12). The design process can directly involve the important pharmacophoric features found in the 17-$\alpha$ and 17-$\beta$ side chains. Suitable isosteric/isoelectronic substitution of the $\alpha$-hydroxy and $\beta$-carboxy substituents with esters or other functional groups should restore the original corticosteroid activity without imparting the potential to produce adverse effects. Modification of the 17-$\beta$ carboxyl function, in addition to changes in the 17-$\alpha$ position and retention of the customary activity enhancing structural modifications (e.g., introduction of $\Delta^1$, fluorination at 6-$\alpha$ and/or 9-$\alpha$, methyl introduction at 16-$\alpha$ or 16-$\beta$) lead to a host of active analogues exemplified by the general structure of soft steroids, 14. Critical functions for activity are clearly the haloester in the 17-$\beta$ position and the novel carbonate (13) and ether (14) substitutions in the 17-$\alpha$ positions, which provided the best activity.

14

$R_1$ = alkyl, haloalkyl
$R_2$ = alkyl, alkoxyalkyl, COOalkyl
$R_3$ = H, $\alpha$- or $\beta$-CH$_3$
$X_1$, $X_2$ = H, F
$\Delta^1$ = double bond (absent or present)

One of these compounds derived from prednisolone, loteprednol etabonate, 15, was selected for clinical development. Based on receptor binding studies, 15 is approximately fivefold more potent than dexamethasone and was successfully developed as a unique ophthalmic antiinflammatory and antiallergic agent. The compound is highly active pharmacologically, but lacks the intraocular pressure (IOP)-elevating side effect characteristic of the other steroids in use. The FDA has recently approved 15 for marketing in the United States.

15, loteprednol etabonate

An example in which SDs based on the inactive metabolite approach can be obtained by introducing the hydrolytically sensitive function at a position remote to the pharmacophore is the case of soft $\beta$-blockers. In this case, there is more freedom to choose the structural modification; consequently, transport and the rate of metabolism can be controlled more easily. The $\beta$-amino alcohol is the main pharmacophore in $\beta$-adrenergic antagonists. The $\beta$-blockers metoprolol, 16, and atenolol, 17, undergo metabolism involving the $\beta$-amino alcohol pharmacophore, but they are also metabolized to a significant extent at the *para* position of the phenol ring producing the corresponding inactive phenylacetic acid derivative, 18. Esterification of this phenylacetic acid lead to the soft $\beta$-blockers, which exhibited different transport, rate of cleavage, and metabolic properties (15,16).

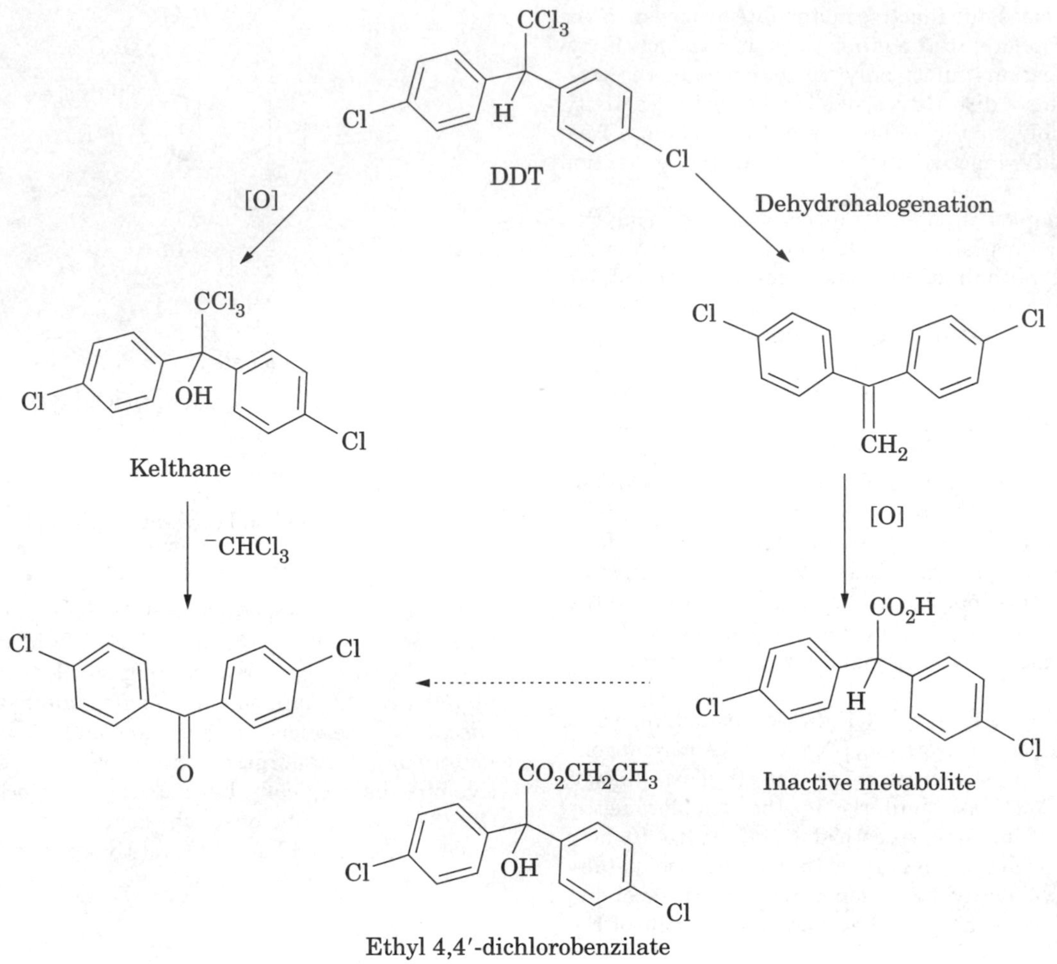

**Figure 5.** Metabolism of DDT.

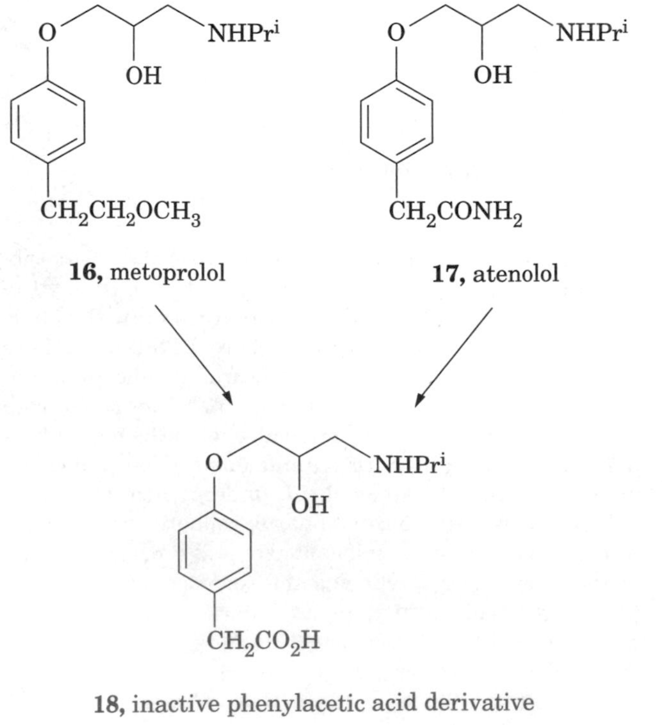

**16,** metoprolol

**17,** atenolol

**18,** inactive phenylacetic acid derivative

If for the desired pharmacological activity, membrane transport (lipophilicity) and relative stability are important, the R group of the ester moiety should be relatively lipophilic and stable. Among the various soft $\beta$-blockers the adamantane ethyl derivative, adaprolol, **19,** is being developed as a topical antiglaucoma agent with prolonged and significant reduction of IOP and rapid hydrolysis in human blood. Thus, it produces a significant separation between local activity and undesired systemic cardiovascular/pulmonary activity. On the other hand, if short systemic action is the objective, then R groups that make the ester moiety susceptible to rapid hydrolysis should be used. Methyl thiomethyl and related esters, **20,** were recently discovered as ultra short acting $\beta$-blockers (17).

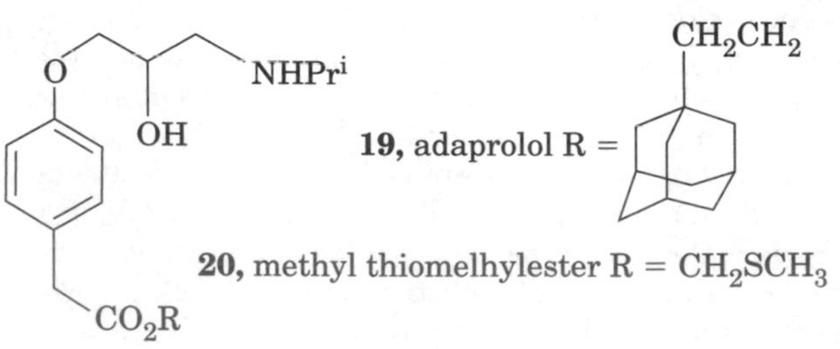

**19,** adaprolol R =

**20,** methyl thiomelhylester R = $CH_2SCH_3$

Finally, the flexibility and potential of SD design is clearly illustrated by the possibility of developing two distinct classes of soft anticholinergic agents. Whereas soft anticholinergics, exemplified by **6**, have been described previously (see Fig. 4c), soft anticholinergic agents based on the inactive metabolite approach have more recently been disclosed (18–22). Here the lead is atropine or the closely related scopolamine. The benzylic hydroxymethyl function in atropine can be oxidized to the corresponding carboxylic acid, a hypothetical inactive metabolite, **21**. Esterification of the carboxyl group in **21** yields the inactive metabolite–based soft anticholinergics, **22**.

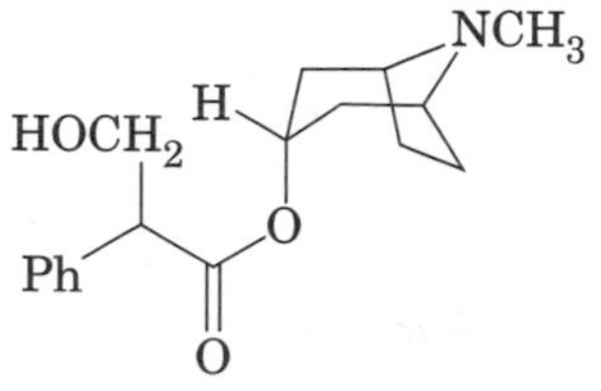

Atropine

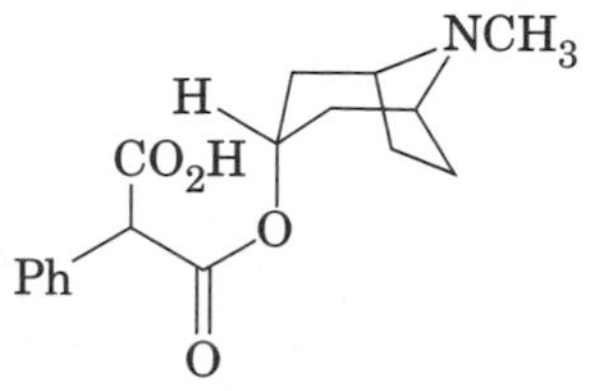

**21,** inactive metabolite

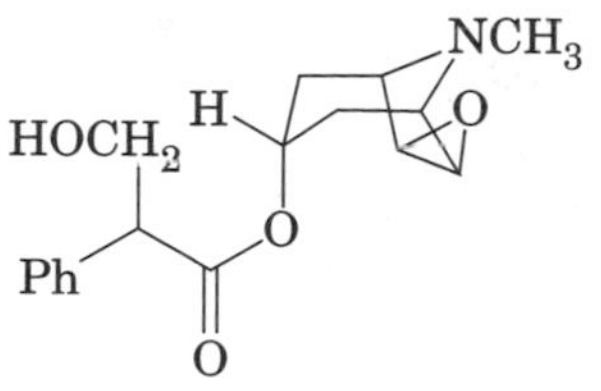

Scopolamine

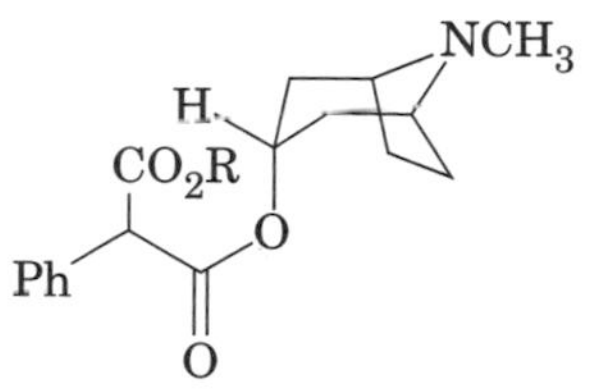

**22,** R = alkyl
**23,** R = $CH_2CH_3$

Soft anticholinergics of this type exhibited good intrinsic activity, with antispasmodic $pA_2$ values of up to 7.85 relative to 8.29 for atropine. However, systemic in vivo activities were much shorter compared to the "hard" atropine. Accordingly, when equipotent mydriatic concentrations of atropine and tematropium, **25**, the corresponding methyl quaternary salt of the ethyl ester **23** were compared, the same maximal mydriasis was obtained, but the area under the mydriasis versus time curve for **25** was only 11–19% of that for atropine (23,24). This observation is consistent with the facile hydrolytic deactivation of the SD. Similarly, the cardiovascular activity of **25** exhibited ultrashort duration. The effect of **25** on the heart rate and its ability to antagonize the cholinergic cardiac depressant action induced by acetylcholine injection, or by electrical vagus stimulation, was determined in comparison to atropine. It was found that a dose of 1 mg/kg of atropine could completely abolish the bradycardia induced by acetylcholine injection or electrical vagus stimulation for more than two hours following intravenous injection. On the other hand, similar doses of **25** exerted muscarinic activity for only 1–3 minutes following intravenous injection. Even a ten-fold increase of **25** to 10 mg/kg did not lead to any significant prolongation or duration of anticholinergic activity.

Further improvement was obtained by using the corresponding ethyl ester soft analogue of methscopolamine,

**24**, which proved significantly more potent than the similar soft analogue, **25** (19).

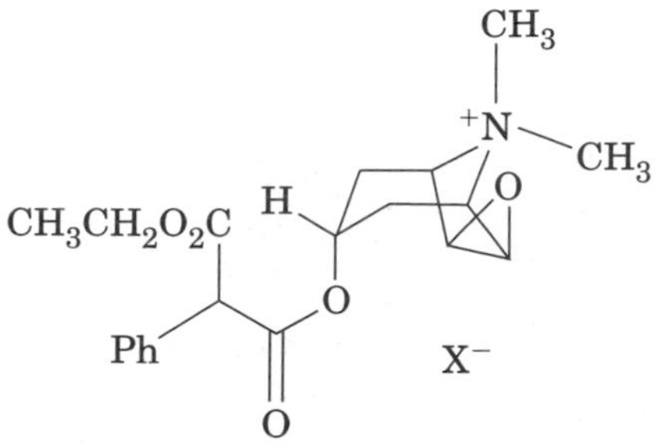

**24,** PMSC. Et

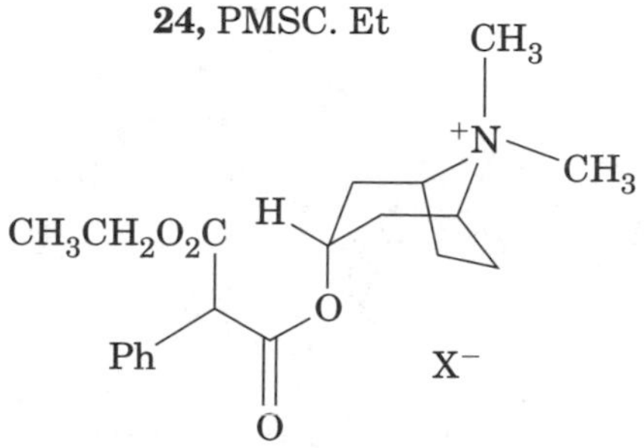

**25,** Tematropium (PMTR. Et)

Meanwhile, the SD **24** was shorter acting than even tropicamide, and consistent with the SD approach, the untreated eye did not exhibit mydriasis as compared to administration of methscopolamine, which did. All of these results provide further support for the ultrashort duration of action, and for the potential of use of these agents as safe topical muscarinic drugs. Their ultrashort systemic activity provides the basis for their successful SD-based targeting.

## CHEMICAL DELIVERY SYSTEMS (CDSS)

CDSs refer to inactive chemical derivatives of a drug where one or more chemical modifications were carried out on the drug by design. The moieties attached to the drug in these modifications are monomolecular units generally comparable in size to the drug. Importantly, these modifications provide a site-specific, or site-enhanced delivery of the drug. Generation of the drug requires multistep enzymatic and/or chemical transformations.

During the chemical manipulations, two types of bioremovable moieties are introduced to convert the drug into an inactive precursor form: a targetor (T) responsible for targeting, site-specificity and imparting "lock-in" character, and modifier functions ($F_1$, $F_2$, . . . , $F_n$) that serve to protect certain functionality, alter lipophilicity, or modulate other molecular properties to prevent premature, undesirable metabolic conversions. Targeting is achieved by design. The CDS undergoes sequential metabolic conversions that remove the modifier functions and finally after fulfilling its site- or organ-targeting role, the targetor moiety is removed. The CDS concept evolved from strategies used to design prodrugs but is very different and significantly more complex. Within the present formalism, prodrugs contain one or more modifier functions (F) for pro-

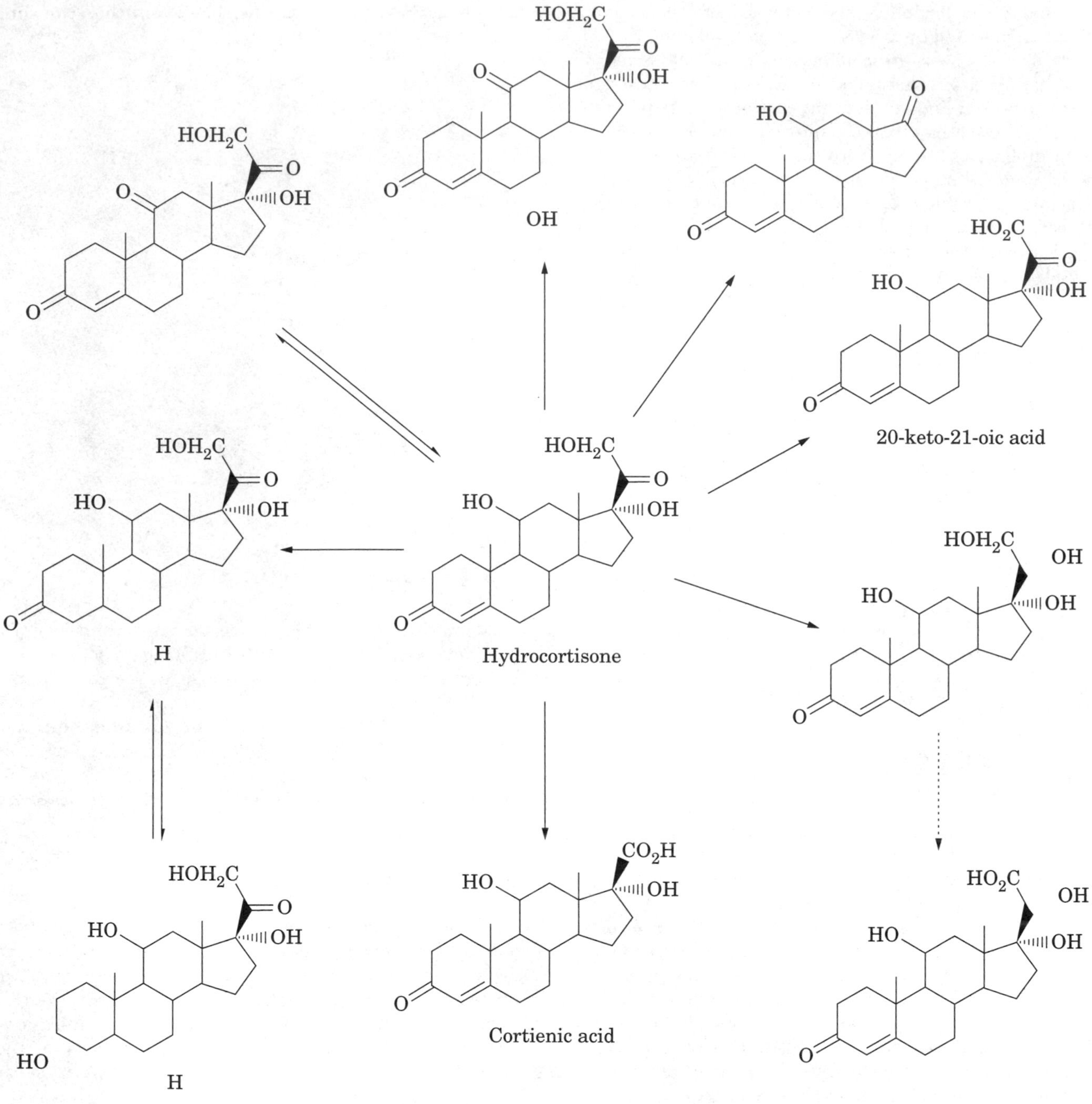

**Figure 6.** Metabolism of hydrocortisone, **12**.

tected or enhanced overall delivery, but they do not contain any targetor function (T). In general, prodrugs do not achieve significant drug targeting, which is the major way to improve the therapeutic index of a drug. It is important to reiterate that the CDS, and all its intermediary metabolic products leading to the active drug, are biologically inactive and do not produce any undesirable side effects. Chemical delivery systems will only be activated to the desired drug by a designed metabolic conversion at the site of action, and no other metabolism should occur.

With a chemical delivery system, targeting is achieved by design, capitalizing on specific enzymes found primarily,

exclusively, or exhibiting higher levels of activity at the site of action, or by exploiting site-specific trafficking properties at the site of action (e.g., the blood-brain barrier). The strategically [BBB] predicted, multienzymatic transformations result in the differential distribution of the drug.

Chemical delivery systems have been divided into the following classes:

1. *Enzymatic physical-chemical–based CDS*. These chemical delivery systems exploit site-specific trafficking properties by sequential metabolic conver-

sions at the site of action, which result in modified transport properties.

2. *Site-specific enzyme activated CDS.* These chemical delivery systems exploit specific enzymes found primarily, exclusively, or exhibiting higher levels of activity at the site of action.

3. *Receptor-based CDS.* These chemical delivery systems capitalize on the transient and reversible nature of the drug binding to its receptor to enhance selectivity and activity.

## Enzymatic Physical-Chemical–Based CDSs

In this approach, the drug is chemically modified to introduce the protective functions ($F_n$) and the targetor moiety (T). Upon administration, the resulting CDS is distributed throughout the body. Predictable enzymatic reactions convert the original CDS by removing some of the protective functions and by modifying the targetor moiety, leading to a precursor form that is still active but has significantly different physicochemical properties. Although these intermediates are continuously eliminated from the rest of the body, efflux–influx processes at the targeted site are not the same due to the presence of a specific membrane or other distributional barriers. These processes ultimately allow release of the active drug in a specific concentration at the site of action.

For example, the BBB can be regarded as a biological membrane that is permeable to most lipophilic compounds but not to hydrophilic molecules. In most cases, the transport criteria apply to both sides of the barrier. Thus, if a lipophilic CDS that can enter the brain is converted there to a hydrophilic molecule, one can assume that it is "locked-in" and is no longer be able to come out. The same conversion taking place in the rest of the body actually accelerates peripheral elimination and further contributes to brain targeting. In principle, many targetor moieties are possible for a general system of this type, but the one developed some years ago that proved most useful was based on the 1,4-dihydrotrigonelline ↔ trigonelline system (Fig. 7) (25–27). Here the targetor of this redox system is the lipophilic 1,4-dihydro form, which is converted in vivo to the highly hydrophilic quaternary form. Because this system is closely related to the ubiquitous NAD⁺ ↔ NADH coenzyme system, this conversion occurs very easily everywhere in the body. The resulting charged species is locked into the brain but is easily eliminated from the body due to the acquired positive charge. After a relatively short time, a sustained, brain-specific release of the active drug is achieved.

Examples of the use of this system for a wide variety of drug classes have been described: various steroid hormones (28,29), antiinfective agents (30,31), antivirals (32,33), anticancer agents (34), antiretroviral agents (35–38) neurotrophomodulator catechol derivatives (39,40), and many others.

Most recently, successful brain deliveries of enkephalin (41,42) and thyrotropin-releasing hormone (TRH) analogues using a combination of approaches including the redox targetor system were reported (43). Delivery of such peptides through the BBB is an even more complex prob-

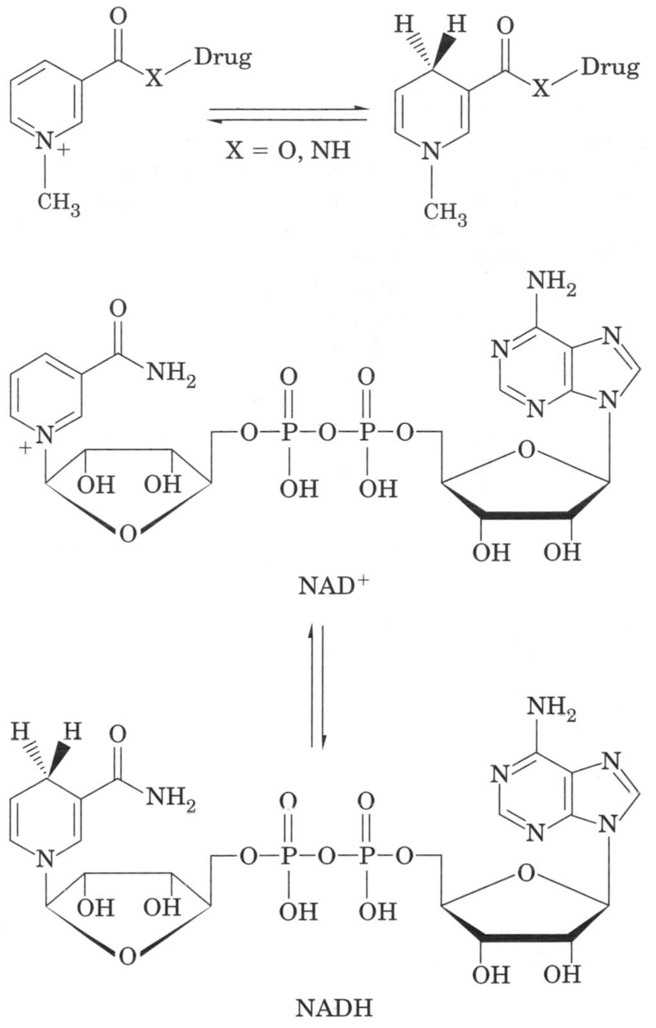

**Figure 7.** The pyridinium ↔ 1,4-dihydropyridine redox and NAD⁺ ↔ NADH coenzyme systems.

lem because endogenous enzymes recognize and rapidly degrade most naturally occurring neuropeptides (44). In the applied "molecular packaging" strategy (45), the peptide unit in the delivery system appears as a perturbation on a bulky molecule which is dominated by lipophilic modifying groups that direct BBB penetration and prevent recognition by peptidases. A brain-targeted packaged peptide delivery system contains the following major components: the redox targetor (T); a spacer function (S) consisting of specific amino acids selected to ensure timely removal of the charged targetor (T) and spacer (S) from the peptide; the peptide itself; and a bulky lipophilic moiety (L) attached through an ester bond, or occasionally through a C-terminal adjuster (A), at the carboxy terminus to impart lipid solubility and to disguise the peptide nature of the molecule. To achieve delivery using such a complex system, it is imperative that the designated enzymatic reactions occur in a specific sequence. Upon delivery, the first step must be oxidation of the targetor (T) to assure the "lock-in" character. This step is followed by removal of the lipophilic function (L) to form a direct precursor of the peptide that is still attached to the targetor. Subsequent cleavage of the targetor–spacer moiety (T–S) finally produces the active peptide.

Delivery of a TRH analogue, **26**, to the central nervous system using this approach is summarized in Figure 8. In this case, a precursor sequence (Gln-Leu-Pro-Gly) is in fact delivered, necessitating two additional steps. One where the C-terminal adjuster glycine (Gly) is cleaved by peptidyl glycine α-amidating monooxygenase to form the terminal prolinamide, and one where the N-terminal pyroglutamyl is formed from glutamine by glutaminyl cyclase. To improve and to optimize this delivery strategy, an efficient synthetic route for the peptide chemical delivery system was established, and the role of the spacer (S) and lipophilic (L) functions were investigated recently for Tyr-D-

Ala-Gly-Phe-D-Leu (DADLE), an analogue of leucine enkephalin that binds to opioid receptors (42). Intravenous injection of several chemical delivery systems synthesized by a segment-coupling method produced a significant and long-lasting response in rats monitored by tail-flick latency measurements. Compared to the delivered peptide (DADLE), the packaged peptide and the intermediates formed during the sequential metabolism had weak affinity to opioid receptors. The antinociceptive effect was naloxone reversible and methylnaloxonium irreversible, suggesting that central opiate receptors must be solely responsible for mediating the induced analgesia. These observations support the hypothesis that the peptide chemical delivery system successfully delivered, retained and released the peptide in the brain. Changing the spacer (S)

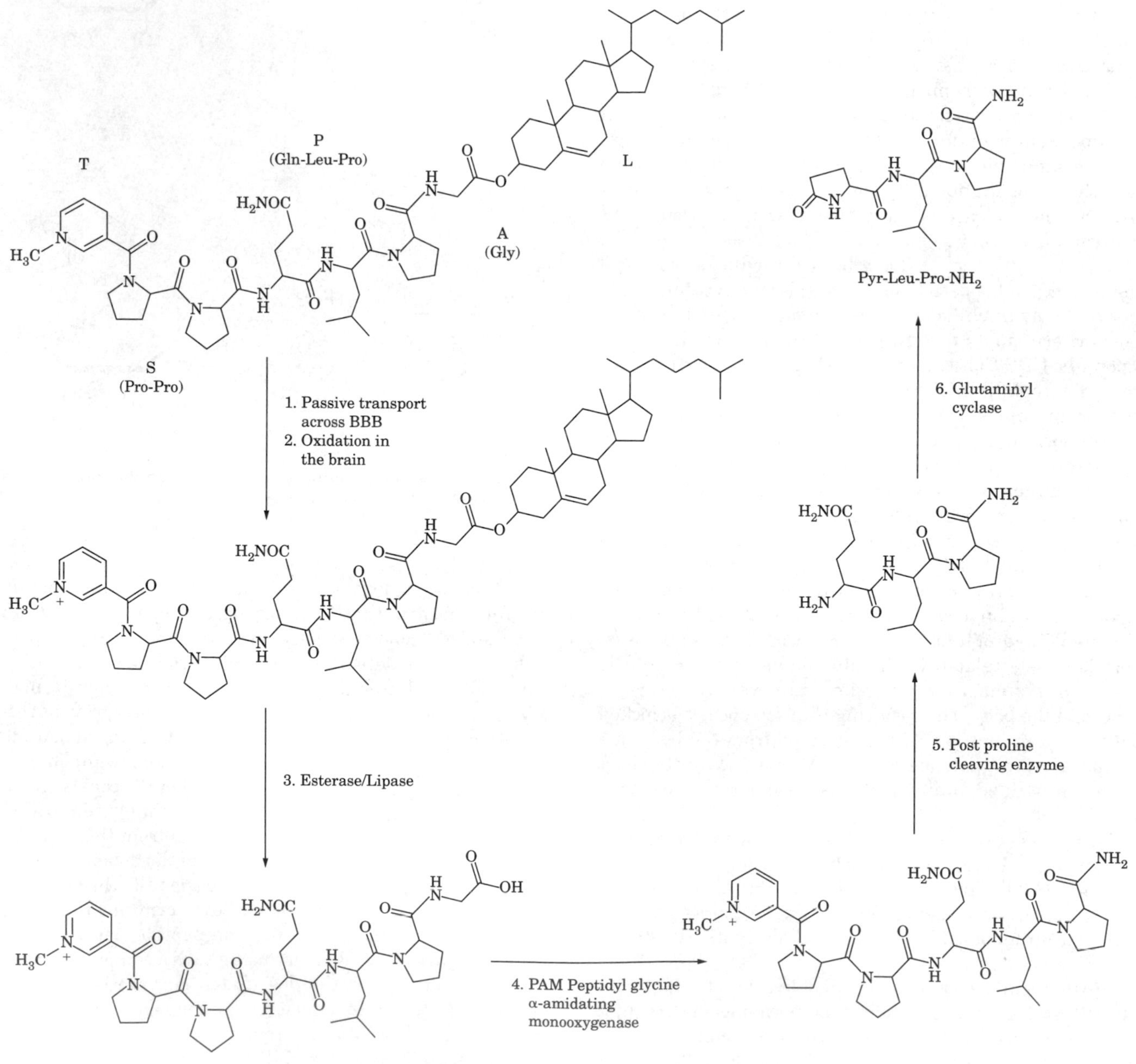

**Figure 8.** CDS for brain targeting of a TRH analogue.

and lipophilic (L) components can greatly influence the efficacy. The bulkier cholesterol group exhibited a greater efficacy than the smaller 1-adamantane ethyl group, but the most important factor which modulated DADLE release and pharmacologic activity was the identity of the spacer (S) function. Proline as the spacer produced more potent analgesia compared to alanine. Selection of the optimum spacer proved also important for the efficacy of the TRH chemical delivery system as measured by the decrease in barbiturate-induced sleeping time in mice following intravenous injection. At equimolar doses the TRH analogue exhibited only a marginal decrease, whereas the chemical delivery system with L-Ala as the spacer produced a 30% reduction, and the one with L-Ala–L-Ala as the spacer produced a decrease greater than 50% (45).

## Site-Specific, Enzyme-Activated CDSs

This second type of CDS is based on enzymatic conversions of a CDS that take place only at the site of action as a result of the differential distribution of certain enzymes within the body. If some specific enzymes are present only around the designated site of action, their use in a strategically designed system can lead to high site specificity. The drug is chemically modified to introduce the targetor (T) moiety and the necessary protective function(s). Following administration, the resulting chemical delivery system is distributed throughout the body. Although it is continuously eliminated from the rest of the body without producing any pharmacologic effects, predictable enzymatic reactions convert the original CDS to the active drug only at the targeted site where the necessary enzymes are found primarily, exclusively, or at a higher level of activity.

Certainly, the system depends upon the existence of such enzymes. The basis of the retrometabolic design that resulted in a successful site- and stereospecific delivery of IOP-reducing β-adrenergic blocking agents to the eye is a general metabolic process that applies to all lipophilic ketone precursors of β-amino alcohols (46–51). Recognizing that lipophilic esters of adrenolone are effectively reduced within the eye to epinephrine, whereas oral administration does not result in epinephrine formation (52), the ketone function was identified as the targetor that can be reduced to produce the β-amino alcohol pharmacophore necessary for action. Since the approach could be extended to other β-agonists (53), the possibility arose to produce ophthalmically useful β-adrenergic antagonists within the eye from the corresponding ketones. However, ketones of the phenol ether type β-blockers are not good precursors to produce the β-amino alcohols because they decompose to their corresponding phenols in solution. To stabilize them, they were converted to oximes, which require a facile enzymatic hydrolysis to the bioreducible ketone. Although the oximes are much more stable relative to their corresponding ketones, their aqueous stability does not provide an acceptable shelf-life. However, significant stabilization can be obtained using the corresponding methoximes. A variety of oxime and methoxime analogues have been synthesized and studied (46–51). They were all shown to undergo the predicted specific activation within, and only

within, the eye. Highest concentrations of the active β-agonists were observed in the iris-ciliary body. When applied topically in rabbits, both the oximes and the methoximes exhibited significantly higher IOP-reducing activities than there corresponding alcohols. On the other hand, topical or intravenous administration did not produce any cardiovascular effects in normal rabbits, rats, and dogs, nor after inducing tachycardia. Following intravenous administration, the oxime disappeared rapidly from the blood, and at no time could the corresponding alcohol be detected (52). This observation suggests that the required enzymatic hydrolysis–reductive activation sequence that occurs in the eye does not take place in the systemic circulation.

To summarize, in these chemical delivery systems, a β-aminooxime or β-methoxime function replaces the corresponding β-amino alcohol pharmacophore in the original molecules (Fig. 9). This oxime derivative is enzymatically hydrolyzed within the eye by enzymes located in the iris-ciliary body, and subsequently, reductive enzymes stereospecifically produce only the S-(-) stereoisomer of the β-blocker (18).

## Receptor-Based CDSs

This last class of CDS is based on formation of a reversible covalent bond between the delivered entity and some part of the targeted receptor site. The targetor moiety in these CDSs is expected to undergo a modification that allows transient anchoring of the active agent. This class of CDS has been developed to a lesser extent than the previous ones.

The possibility of a receptor-based CDS for naloxone, **31**, a pure opioid antagonist, was explored. The presence of an essential sulfhydryl group near the binding site in opioid receptors was long suspected because agonist binding capacity is destroyed by sulfhydryl scavenging agents such as N-ethylmaleimide (54). Recent mutagenesis and other studies began to elucidate the mechanism for this inhibition of binding at opioid receptors (55,56). There is evidence that Leu-enkephalin, or even morphine analogues, containing sulfhydryl groups can form mixed disulfide linkages with opioid receptors (57). Chlornaltrexamine, an irreversible alkylating affinity label on the opiate receptor, can maintain its antagonist effect for 2–3 days, suggesting that covalent binding can increase the duration of action at these receptors. A more reversible spirothiazolidine system was investigated (Fig. 10). Spirothiazolidine derivatives were selected because they should be subject to biological cleavage of the imine formed after spontaneous cleavage of the carbon-to-sulfur bond (58), and they have been investigated previously in CDSs for the controlled release of endogenous agents such as hydrocortisone or progesterone (59,60). As modifications at the 6-position are possible without the loss of activity, the spirothiazolidine ring was attached at this site. Opening of the thiazolidine ring allows oxidative anchoring of the delivered agent to sulfhydryl groups nearby via disulfide bridging. Blocking of the receptor is reversible, because endogenous sulfhydryl compounds can presumably regenerate the sulfhydryl group at the receptor. If a favorable conformation can be attained, greater affinity at the opioid

Oxime
hydrolase
Ketone
reductase
S - (-)
Alprenolol
Propanolol
Betaxolol
Timolol


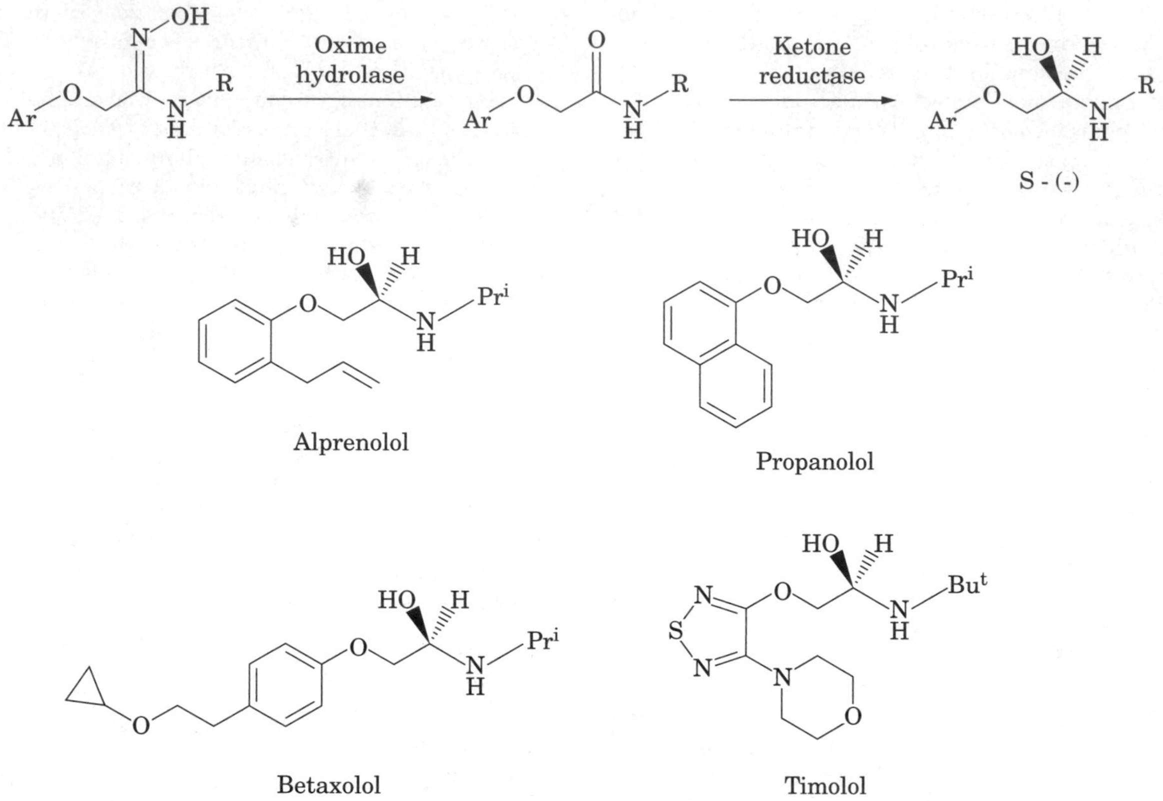

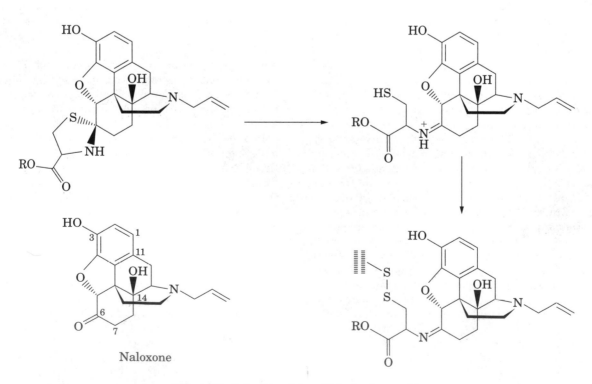

**Figure 9.** Site- and stereospecific delivery of $\beta$-adrenergic antagonists to the eye.

Naloxone


**Figure 10.** Spirothiazolidine CDS of naloxone, **31**.

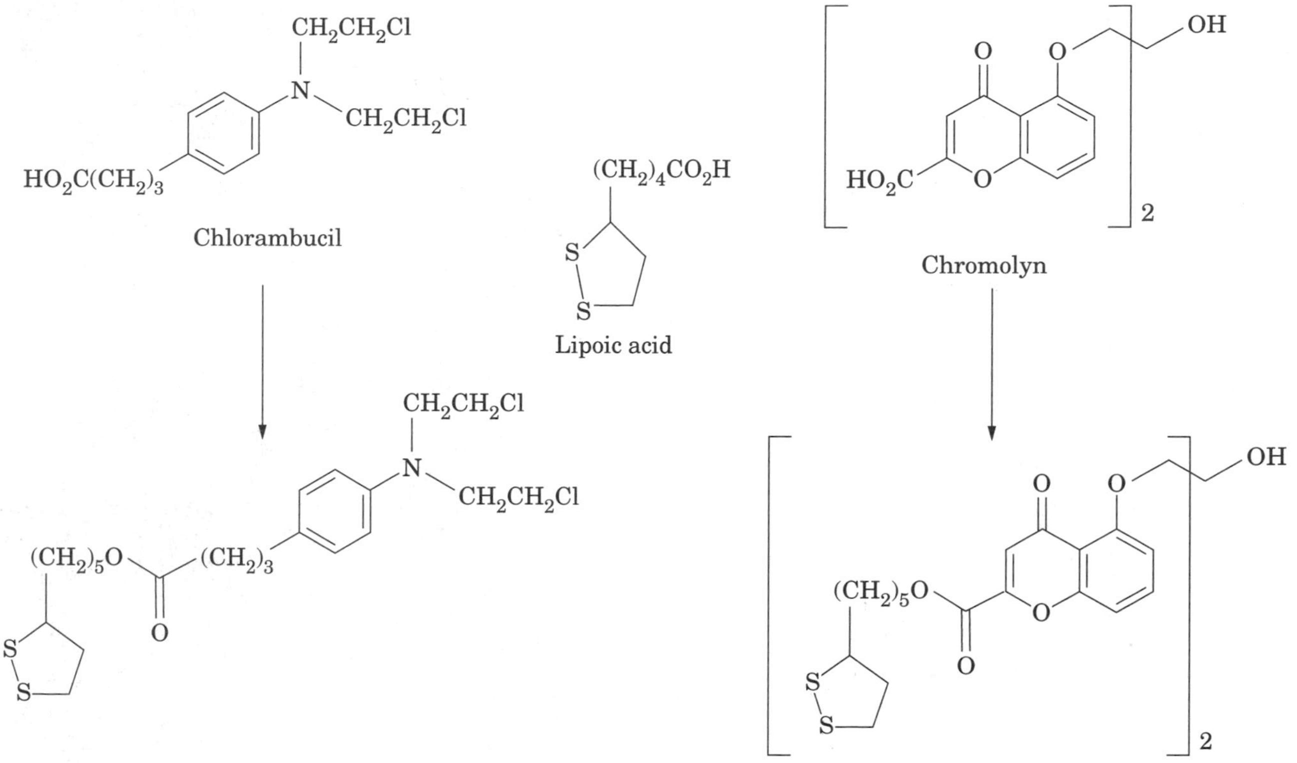

**Figure 11.** Lipoic acid–based CDS for lung targeting.

receptor can be expected. Indeed, naloxone-6-spirothiazo-lidine exhibits an increased affinity at opioid receptors. The increase is especially significant at the $\kappa$-receptors known to be involved in sedative analgesia. The intrinsic activity for guinea pig ileum of the naloxone chemical delivery system also demonstrated a 25-fold increase compared to naloxone itself.

Because several other G-protein–coupled receptors (e.g., TRH, $D_1$ and $D_2$ dopamine, vasopressin, follicle-stimulating hormone, and cannabinoid receptors) are also sensitive to sulfhydryl reagents (56), there is considerable potential for this class of CDS if modification of the corresponding ligands at appropriate sites can be performed. In addition, as earlier evaluations of spirothiazolidine-based systems indicated an enhanced deposition to the lung, the possibility to use similar mechanisms in developing other lung tissue targeting chemical delivery systems presented itself. Because of its potential to form disulfide bonds, lipoic acid, **32**, a nontoxic coenzyme for acyl transfer and redox reactions, was investigated as a targetor for selective delivery to lung tissue (61). The corresponding chemical delivery systems for chlorambucil, **33**, an antineoplastic agent, and chromolyn, **34**, a bischromone used in antiasthmatic prophylaxis, were synthesized via an ester linkage (Fig. 1). In vitro kinetic and in vivo pharmacokinetic studies demonstrated that the respective CDSs were sufficiently stable in buffer and biological media, hydrolyzed rapidly into the respective active parent drugs, and, relative to the underivatized parent compounds, significantly enhanced delivery and retention of the active drug in lung tissue.

## CONCLUSIONS

Several novel metabolic-based drug design approaches have been described. The particular advantage of these approaches is to enhance drug targeting to the site of action. Retrometabolic approaches include two major classes: CDSs and SDs. They are opposite in terms of how each approach achieves their drug targeting role, but they have in common the basic concept of designed metabolism to control drug action and targeting. For CDSs, the molecule is designed to be inactive and to undergo strategic enzymatic activation in order to release the active agent only at the site of action. Delivery of this kind was successfully made to the brain, to the eye, and to other organs such as the lung. On the other hand, SDs are intrinsically potent new chemical entities that are strategically deactivated after they achieve their therapeutic role. These two approaches are general in nature and can be applied essentially to all drug classes.

## BIBLIOGRAPHY

1. J.J. Kaminski and N. Bodor, in C.R. Clark and W.H. Moos, eds., *Drug Discovery Technologies*, Ellis Horwood, Chichester, U.K., 1990, pp. 65–91.

2. C.G. Wermuth, J.C. Gaignault, and C. Marchandeau, in C.G. Wermuth, ed., *The Practice of Medicinal Chemistry*, Academic Press, London, 1996, pp. 672–696.

3. C.G. Wermuth, in C.G. Wermuth, ed., *The Practice of Medicinal Chemistry*, Academic Press, London, 1996, pp. 697–715.

4. U.S. Pat. 4,061,722 (December 6, 1977), N. Bodor (to INTER$_x$ Research Corp.).

5. U.S. Pat. 4,264,765 (April 28, 1981), N. Bodor and L. Freiberg (to Abbott Laboratories).

6. N. Bodor, J.J. Kaminski, and S. Selk, *J. Med. Chem.* **23**, 469–474 (1980).

7. N. Bodor et al., *J. Med. Chem.* **23**, 474–480 (1980).

8. Jpn. Pat. 21201 (February 10, 1983), N. Bodor and Y. Oshiro.

9. M.B. Abou-Donia and D.B. Menzel, *Biochem. Pharmacol.* **17**, 2143–2161 (1968).

10. N. Bodor, in H.P. Tipnis, ed., *Advances in Drug Delivery Systems*, MSR Foundation, Bombay, 1985, pp. 111–127.

11. N. Bodor, in E. Christophers, A.M. Kligman, E. Schopf, and R.B. Stoughton, eds., *Topical Corticosteroid Therapy: A Novel Approach to Safer Drugs*, Raven Press, New York, 1988, pp. 13–25.

12. N. Bodor, *Curr. Probl. Dermatol.* **21**, 11–19 (1993).

13. P. Druzgala, G. Hochaus, and N. Bodor, *J. Steroid Biochem.* **38**, 149–154 (1991).

14. P. Druzgala and N. Bodor, *Steroids* **56**, 490–494 (1991).

15. N. Bodor et al., *Pharm. Res.* **3**, 120–125 (1984).

16. N. Bodor, A. El-Koussi, M. Kano, and M. Khalifa, *J. Med. Chem.* **31**, 1651–1656 (1988).

17. H. Yang, W.M. Wu, and N. Bodor, *Pharm. Res.* **12**, 329–336 (1995).

18. N. Bodor and L. Prokai, *Pharm. Res.* **7**, 723–725 (1990).

19. G.N. Kumar, R.H. Hammer, and N. Bodor, *Drug Des. Discovery* **10**, 11–21 (1993).

20. G.N. Kumar, R.H. Hammer, and N. Bodor, *Drug Des. Discovery* **10**, 1–9 (1993).

21. G.N. Kumar, R.H. Hammer, and N. Bodor, *Bioorg. Med. Chem.* **1**, 327–332 (1993).

22. G.N. Kumar, R.H. Hammer, W.M. Wu, and N. Bodor, (1993) *Curr. Eye Res.* **12**, 501–506 (1993).

23. R. Hammer et al., *Novel Soft Anticholinergic Agents. Drug Des. Delivery* **2**, 207–219 (1988).

24. R. Hammer, W.M. Wu, J.S. Sastry, and N. Bodor, *Curr. Eye Res.* **10**, 565–570 (1991).

25. N. Bodor, H. Farag, and M.E. Brewster, *Science* **214**, 1370–1372 (1981).

26. N. Bodor and M.E. Brewster, *Pharmacol. Ther.* **19**, 337–386 (1983).

27. N. Bodor and H. Farag, *J. Med. Chem.* **26**, 313–318 (1983).

28. N. Bodor and J.J. Kaminski, *Annu. Rep. Med. Chem.* **22**, 303–313 (1987).

29. M.E. Brewster, K.S. Estes, and N. Bodor, *J. Med. Chem.* **31**, 244–249 (1988).

30. E. Pop, W.M. Wu, and N. Bodor, *J. Med. Chem.* **32**, 1789–1795 (1989).

31. E. Pop, W.M. Wu, E. Shek, and N. Bodor, *J. Med. Chem.* **32**, 1774–1781 (1989).

32. K. Rand et al., *J. Med. Virol.* **20**, 1–8 (1986).

33. M.E. Brewster and N. Bodor, *Adv. Drug Delivery Rev.* **14**, 177–197 (1994).

34. K. Raghavan, E. Shek, and N. Bodor, *Anticancer Drug Des.* **2**, 25–36 (1987).

35. P.T. Torrence et al., *FEBS Lett.* **234**, 135–140 (1988).

36. S. Gogu, S.K. Aggarwal, S.R.S. Rangan, and K.C. Agrawal, *Biochem. Biophys. Res. Commun.* **160**, 656–661 (1989).

37. E. Palomino, D. Kessel, and J.P. Horowitz, *J. Med. Chem.* **32**, 622–625 (1989).

38. R. Little et al., *Biopharm. Sci.* **1**, 1–18 (1990).

39. A. Kourounakis, N. Bodor, and J. Simpkins, *Int. J. Pharm.* **141**, 239–250 (1996).

40. A. Kourounakis, N. Bodor, and J. Simpkins, *J. Pharm. Pharmacol.* **49**, 1–9 (1997).

41. N. Bodor and M.J. Huang, *J. Pharm. Sci.* **81**, 954–960 (1992).

42. K. Prokai-Tatrai, L. Prokai, and N. Bodor, *J. Med. Chem.* **39**, 4775–4782 (1996).

43. L. Prokai, X.D. Ouyang, W.M. Wu, and N. Bodor, *J. Am. Chem. Soc.* **116**, 2643–2644 (1994).

44. J. Brownlees and C.H. Williams, (1993) *J. Neurochem.* **60**, 793–803 (1993).

45. N. Bodor and L. Prokai, in M. Taylor and G. Amidon, eds., *Peptide-Based Drug Design: Controlling Transport and Metabolism*, American Chemical Society, Washington, D.C., 1995, pp. 317–337.

46. A. El-Koussi and N. Bodor, *Int. J. Pharm.* **53**, 189–194 (1989).

47. N. Bodor, H. van der Goot, G. Domany, L. Pallos, and H. Timmerman, eds., *Trends in Medicinal Chemistry*, Elsevier, Amsterdam, 1989, pp. 145–164.

48. N. Bodor, *Adv. Drug Delivery Rev.* **16**, 21–38 (1995).

49. L. Prokai, W.M. Wu, G. Somogyi, and N. Bodor, *J. Med. Chem.* **38**, 2018–2020 (1994).

50. N. Bodor et al., *J. Ocul. Pharmacol. Ther.* **13**, 389–403 (1997).

51. H. Farag et al., *Drug Des. Discovery* **15**, 117–130 (1997).

52. N. Bodor, A. El-Koussi, M. Kano, and T. Nakamuro, *J. Med. Chem.* **31**, 100–106 (1988).

53. I.K. Reddy and N. Bodor, *J. Pharm. Sci.* **83**, 450–453 (1994).

54. J.R. Smith and E.J. Simon, *Proc. Natl. Acad. Sci. U.S.A.* **77**, 281–284 (1980).

55. B.D. Joseph, and J.M. Bidlak, *Eur. J. Pharmacol.* **267**, 1–6 (1996).

56. M. Shahrestanifar, W.W. Wang, and R.D. Howells, *J. Biol. Chem.* **271**, 5505–5512 (1996).

57. K. Kanematsu et al., *Chem. Pharm. Bull.* **38**, 1438–1440 (1990).

58. W.M. Schubert and Y. Motoyama, *J. Am. Chem. Soc.* **87**, 5507–5508 (1965).

59. N. Bodor et al., *Int. J. Pharm.* **10**, 307–321 (1982).

60. N. Bodor and K.B. Sloan, *J. Pharm. Sci.* **71**, 514–520 (1982).

61. M. Saah et al., *J. Pharm. Sci.* **85**, 496–504 (1996).

See also CANCER, DRUG DELIVERY TO TREAT—PRODRUGS; PENDENT DRUGS, RELEASE FROM POLYMERS.

# COATINGS

C.T. RHODES
PharmaCon, Inc.
West Kingston, Rhode Island

S.C. PORTER
Colorcon
West Point, Pennsylvania

## KEY WORDS

Cellulosic
Compaction coating
Controlled release
Drug master file
Enteric coatings
Excipient
Fluidized bed coating
Food and Drug Administration
Melt coating
Microsphere
Noncellulosic
Pan coating
Plasticizer
Polymer
Product validation

## OUTLINE

The coating of pharmaceutical products goes back many years. For example, Shakespeare refers to gilding (gold coating) pills. Certainly, even as recently as the 1950, pharmacists in England and the United States were, for example, trained to coat pills with silver or gold leaf or shellac. To what extent such drug delivery systems provided controlled release is a matter of conjecture. Probably the first reliable coating products were enteric-coated tablets designed to protect acid labile drugs from degradation in the relatively low pH conditions of the stomach. However, even here there is reason to doubt the efficacy of many of the enteric-coated products that were marketed as recently as the 1960s.

It is probably true that coated pharmaceuticals that can be relied on to provide controlled release are less than 50 years of age. The development of such products has depended, to a very considerable extent, on the availability of chemically modified coatings (especially cellulose derivatives), which are supplied to the pharmaceutical industry as materials of reliable quality. Also, the improvements in coating equipment technology, especially the invention of the Wurster film coating device, has been essential to the improvement in coating technology.

This article is focused on coatings and the method by which coatings can be applied to manufacture controlled release products. It will be appreciated that coatings are sometimes applied to pharmaceutical products for reasons separate from a desire to modify drug release. For example, coatings are applied to some tablets, e.g., ibuprofen, which have an unpleasant taste or to improve shelf life stability, e.g., conjugated estrogens. The thrust of this chapter is controlled release, although in some instances coatings designed for control of drug release may also confer other benefits, such as improving ease of swallowing tablets by geriatric patients. Thus in this chapter, sugar coating, which is not normally used primarily as a method of control of drug release, receives only minimal attention.

The essential elements of a coated pharmaceutical product is that a core (consisting of drug, or drug plus excipients) is encased by a layer, or layers, of materials that regulate the rate at which drug is released into the surrounding medium. The primary coating materials, usually polymeric, often require the addition of other excipients such as plasticizers, pore formers, or antiaggregation agents for the product to be conveniently manufactured or for the coating to perform in the desired fashion. It is therefore appropriate to give attention to some of these types of excipients. The demarcation between a coated and a matrix-type pharmaceutical, controlled release product is not always clear. Thus, as Tice and Tabibi have pointed out (1), a microcapsule has its drug:

... centrally located within the particle where it is encased within a unique polymer membrane. ... Whereas a microsphere has its drug dispersed throughout the particle; that is the internal structure of a matrix of drug and polymer excipient.

Obviously, therefore, some of the materials used as coatings to control drug release may also be used for a similar function in matrix-type products, and are discussed elsewhere in this encyclopedia. It is also true that some excipients may be used in coated pharmaceuticals for one purpose and be used in other pharmaceuticals to fulfill a different function. Thus, magnesium stearate, most commonly used in pharmaceutical technology as a tablet lubricant, has been used to prevent agglomeration of coated microparticles (2).

## FUNDAMENTAL REQUIREMENTS

There are three preeminent requirements for any coating intended for use in controlling the rate of release of drug from a commercially available product:

1. The coating must control the release pattern of the drug such that the release profile, zero-order, first-order pulsatile, or other is within acceptable limits so as to comply with the pharmacokinetic and/or pharmacodynamic requirements of the particular drug product.
2. All components of the coating must be acceptable to the relevant regulatory agencies for the route of administration and species for which the product is intended.
3. The coating materials must be commercially available at a reasonable price, reproducibly meeting functionally relevant standards so that process validation can clearly demonstrate that the manufacturing process is fully under control and that the product is likely to show an acceptably low level of batch-to-batch variation.

The first of these three desiderata is considered in other parts of this encyclopedia, and thus will not be discussed in this chapter. It is, however, essential before considering specific coating components to give some attention to regulatory concerns.

### Regulatory Concerns

There are many polymeric materials that research workers have identified as having potential for control of drug release. Unfortunately, relatively few have been approved for use in either human or veterinary pharmaceutical products. It would be most imprudent for any controlled release project involving the use of a coating to be initiated without some attention being given to the regulatory status of the components being considered, if it is hoped that ultimately a product can be marketed. It would be attractive if a simple list of "acceptable" components could be provided to readers of this encyclopedia. However, there are several reasons that prevent the preparation of such a

list. The processes of globalization and harmonization are far from being complete. Although significant progress is being made, there are still important differences in the regulatory requirements of the United States, the European Union, and Japan (3–5). Thus, what may be acceptable in one jurisdiction is not necessarily acceptable in another.

Even within one jurisdiction, different levels of acceptance may apply to an excipient intended for the oral or intramuscular route or for use in humans or feed animals. Also, of course, the approval of a coating component depends on the quality and quantity of minor constituents. Thus, approval may well be source specific. In the United States, the mechanism of Drug Master Files provides a most useful mode, whereby the manufacture of an excipient can establish, with the U.S. FDA, a detailed record of test data pertaining to its own marketed product. Upon authorization by the sponsor of the Drug Master File, FDA personnel can access the file when considering a marketing submission document (NDA, PLA, etc.) that involves use of the excipient. Information from excipient suppliers as to the current regulatory acceptability of one of their products for administration to humans or animal species by a specified routes of administration can be most useful.

In the recent past, there was considerable gloom in the pharmaceutical industry about the possibilities of obtaining regulatory approval for new excipients. The formation of the International Pharmaceutical Excipient Council (IPEC) has raised hopes that in the future this process will be less difficult than previously (6). (Call 201-628-3231 for information on IPEC.)

At the time marketing permission is requested from a regulatory agency such as the FDA, the sponsor is required to demonstrate that the manufacturing process is fully validated (7). Validation of process can only be successful if there are reliable, functionally relevant standards for the raw materials so that batch-to-batch variability of such raw materials is low. This requirement can often create problems for excipients of natural origin that commonly show considerable batch, seasonal, and source variability. Thus, unless a new coated, controlled release product is being developed for a nutraceutical for sale through health food stores that insist on all components being "natural," any group considering a new product is probably well advised to avoid the use of natural products, such as shellac. Although shellac is still used quite substantially in coated products, it is very difficult to obtain reliable, quality material. Substantial batch-to-batch variability of shellac has been the source of difficulties with the FDA because of a lack of sufficiently precise control over drug release.

### Sources of Information on Coating Materials and Processes

It is hoped that this chapter provides a useful coverage of the fundamentals of both coating materials and processes. However, since this field is so vast, and also because it is in a state of flux, those readers interested in details on a particular coating material or piece of coating equipment may find it useful to supplement this chapter with information derived from other sources.

For general coverage of the selection of formulation and processing factors in the development of a coated, controlled release products, the discussions presented by Rudnic and Kotke (8) and Schwartz and O'Connor (9) are recommended.

It is generally recognized that national pharmacopoeias (e.g., USP, BP, PF, JP) have done an excellent job in setting standards for drug substances. Unfortunately, until quite recently, comparatively little attention had been given by national or professional bodies to developing chemical, physical, or biological standards for excipients, including materials used in coatings; thus, USP/NF does not have monographs on all the commonly used coating excipients. Although this failing is now recognized, there is still no official compendium that might be used by those seeking standards for pharmaceutical excipients. However, the *Handbook of Pharmaceutical Excipients* (10) has attained quasiofficial status in this regard. Although this book is not perfect (some feel that there is an uneven level of treatment in the different monographs, and others feel that additional substances should be included), this text is often used by those seeking standards for excipients.

Another book likely to find increasing use by those looking for specifications for excipients is that written by Ash and Ash (11), which provides a global survey of more than 6,000 products, which can be accessed by trade name, chemistry, function, and manufacturer. However, for those who become directly concerned with developing a controlled release coated product, there is much to be said for consulting the manufacturers' literature. Although there is considerable variation in the content and detail of such material, much is comprehensive, detailed, and objective. Reputable manufacturers of excipients, which derive a substantial portion of their income from the pharmaceutical industry, have an obvious interest in disseminating reliable information about their products.

Additional sources of specialized information are consultants or the services of contract houses that provide formulation and processing development assistance. The Coating Place in Verona, Wisconsin, is one such organization. Those interested in locating such contract houses may find the Drug Information Association's *Pharmaceutical Contract Support Organizations Register* of value (12).

For those who are interested in becoming current with the most recent research developments in controlled release coating technology, perusal (manual or computer-assisted) of abstracting journals such as *Current Contents* and the primary literature in research journals is, of course, essential. Among those journals that are particularly likely to publish research or review papers in this area are the *Journal of Controlled Release, International Journal of Pharmaceutics, Drug Development and Industrial Pharmacy, Pharmaceutical Research, Journal of Microencapsulation, Journal of Pharmacy and Pharmacology,* and *Journal of Pharmaceutical Sciences.*

## Mechanisms by Which Coatings Can Effect Controlled Release

Elsewhere in this book, detailed consideration is given to the mathematical definition of various release profiles and how such profiles can be obtained. This brief summary of mechanisms by which coatings and the components

therein can function to control release is only intended so that the properties of the components described in later sections of this chapter can be related to the overall objective for which the product is designed.

Although there is considerable semantic confusion concerning the various terms (sustained, extended, modified) applied to controlled release products, the present authors follow Berner and Dinh (13) who state that a controlled-release product "involves control of either the time course or location of drug delivery." The simplest mechanism by which a coating can control temporal aspects of drug release is to apply layers of a single material that can erode, melt, or become permeable at body temperature and thus allow release of drug. By a judicious selection of blends of pellets with varying thicknesses of coating, it may be possible to construct a dissolution profile of the form required for a particular drug. A number of waxes and similar materials are used for controlled release products that function in this manner, and some of these products have been successfully marketed for many years. Recently, considerable interest has been directed to the use of hot-melt coating methods, some of which incorporate modified, fluidized-bed coaters, to the manufacture of such formulations (14).

Enzymes present in the gastrointestinal tract, or elsewhere, have considerable potential for exploitation as agents to stimulate the breakdown of the coating of a controlled release product, and thus release the drug at an appropriate time after ingestion or at a particular location in the tract. For example, there is evidence that short, linear chains of amylose are resistant to enzymatic attack in the small intestine (15), but can be degraded by colonic microflora. There is thus considerable interest in coating drug delivery systems with a mixture of an appropriate form of amylose and other coating materials, such as ethylcellulose so that the resultant drug delivery system is capable of releasing drug in the colon. The coatings described by Milojevic and associates (16) may prove to be precursors of coatings that will eventually be commercialized. Also, the possible use of coatings containing azopolymers that are impermeable in the stomach but are converted to amides in the colon by bacteria, such conversion resulting in coating collapse, also may have potential (17).

The variation in pH that occurs as an orally administered drug delivery system moves down the gastrointestinal tract and has been very widely used as the trigger to cause the release of drug in enteric-coated drug delivery systems. Polymers, often esters of phthalic acid, used as enteric coatings commonly possess pendent carboxylic acid groups that are un-ionized in the relatively low pH of the stomach (normally about 1.5 to 4.5), but which ionize and thus repel one another, causing coating disruption as the pH rises when the delivery system enters the small intestine.

Diffusion of water from the exterior of the product across the coating, followed subsequently by diffusion in the reverse direction of dissolved drug, is very commonly the major mechanism by which coated products achieve control of drug release. In the least complicated limit case, the rate of release of drug might be regarded as a simple function of the diffusion coefficient of water or the solvated drug in the polymer, which is regarded as invariant in properties. In reality, such simple conditions rarely exist in commercially available products. Coatings often contain several components in addition to the primary component polymeric species, which provides the backbone of the coating. Some of the secondary coating components may be deliberately added in order to modify the permeability of the primary polymer by providing channels or pores within the coating (18,19). Other materials, such as plasticizers, although they are added for entirely different reasons, can, in some instances, significantly modify drug release rates (20). Also, it must be kept in mind that changes in processing variables, such as inlet temperature in a fluidized-bed coater, which may be required during the scale-up from laboratory to pilot scale to manufacturing scale, may result in significant changes to dissolution profiles. Thus, it is wise not to rigidly define release rate profiles for coated, controlled release products at an early stage of the research and development process.

It should also be kept in mind that the various theoretical models that have been advanced to rationalize the release of drug from coated, controlled release products are often quite insufficient to reliably and precisely predict in vivo dissolution rates. Although we may find it intellectually convenient to classify coatings in terms of their purported mechanism of control, the reality probably is that in many formulations the actual behavior of the product in the patient is a hybrid of several models with surface erosion, enzymatic degradation, pH-induced changes, swelling of coating components, etc., all playing roles in the control of drug release. Finally, one must be careful not to fall into the error of assuming that in vitro dissolution tests (even those performed using the flow-thru cell) are necessarily always predictive of pharmacokinetic or pharmacodynamic parameters (21).

The osmotic pump approach to the design of coated, controlled release drug products is one of the most exciting developments that has occurred in controlled release coating technology, and the background of this concept is described elsewhere in this encyclopedia.

Multilamellar coating products in which there are clear differences in composition of two or more sequentially applied coats are available and can offer special advantages. For example, Shah and his associates teach in their patent (22) the use of a core-containing drug that is coated with an erodible polymer coating which, in turn, is surrounded by an enteric-release type coating. In example 1 of their patent, Shah and his co-workers describe compaction coating of granules by an erodible coating comprising hydroxypropylmethylcellulose, microcrystalline cellulose, and polyvinyl pyrrolidone. These coated granules were then enteric coated by a spray method with a mixture of hydroxypropylmethylcellulose phthate and acetylated monoglyceride. The outer coating protects the drug until the small intestine is reached. The inner coating (proximal to the core) is of such a thickness and composition that the drug is released in the colon. Multilamellate coatings in which each coating has a different function to perform offer the potential of developing very sophisticated, controlled release coating system.

If a coating is prepared around a drug delivery system in which the outermost layer contains a bioadhesive material, then the control of the locus at which drug release occurs becomes possible. A number of materials, including hydroxypropylcellulose, polyacrylic acid, and sodium carboxymethycelluse have been examined for this application (23,24). Bioadhesive coatings have potential for topical buccal, periodontal, gastric, intestinal, ophthalmic, and vaginal delivery (25–28).

## MANUFACTURING METHODS FOR THE APPLICATION OF COATING MATERIALS FOR CONTROLLED RELEASE

Although considerable ingenuity has been applied to developing manufacturing methods which might be used for the application of coating materials for control release products, the four basic approaches are (1) pan coating using solvent evaporation, (2) fluidized-bed coating using solvent evaporation, (3) compaction coating, and (4) melt coating.

For those coating methods that involve the use of a solvent system (i.e., methods 1 and 2), it is convenient to define whether the solvent is aqueous or organic since the selection of solvent not only has regulatory implications (e.g., EPA), but also affects the nature or type of materials used in the coating formulation.

It is possible that other coating methods, such as various types of coacervation, may come to be of greater commercial importance for controlled release, coated products in the future.

### Pan Coating

The oldest form of pharmaceutical coating—sugar coating—was an art rather than a science. The basis of the process was, in effect, to use giant saucepans, which were continually agitated, as the cook supervised the addition of coating fluid and the subsequent solvent evaporation. Film coating (using cellulosic-type materials for controlled release) naturally evolved from practices and equipment that had been used in sugar coating of tablets.

Conventional coating pans were top-open-ended truncated spheroids mounted on a power unit, which allowed the pan to be rotated at an angle of about 45° to the horizontal at a speed of between 20 and 50 rpm. Both ingress of coating fluid and the egress of solvent vapors was via the top of the pan.

The basic conventional pan design has been substantially improved by a number of modifications which have inter alia improved air flow, throughput, and coating uniformity. Different shaped pans, cylinders, hexagons, and pear shapes can offer advantages. Improvement of air flow patterns have been made by using side-vented cylindrical pans and perforated pans. In the Hi-Coater (Vector) and Acela-Cata (Thomas), airflow is through the tablet bed and then out of the perforated walls of the pan. The Glatt coater allows airflow with or against the direction in which coating spray is being applied.

In the old days of manual sugar coating, the coating fluid was ladled into the pan. With modern film coating, the coating is sprayed into the tablet bed, often with the assistance of an air jet. Specialized sword and plough spray devices are available. Modern pan coating certainly represents a great improvement over traditional methods, but there are many in the industry who believe that for the most reproducible, controlled release coatings showing minimal defects and tablet-to-tablet variability fluidized-bed methods are, for many formulations, probably to be preferred.

### Fluidized-Bed Coating

The fundamental principle behind fluidized-bed coating in general and the Wurster technique in particular is to suspend tablets in an upward moving column of warm air during the coating process. This minimizes tablet abrasion and unevenness of film distribution caused by tablet-to-tablet contact. The basic Wurster fluidized-bed coating chamber is essentially cylindrical in shape with the axis of the cylinder in the vertical plane and the top of the chamber having a convex exterior. Drying air is forced into the base of the cylinder with the coating materials being sprayed by an atomizer into the center of the base of the cylinder. An open-ended, inner cylinder acts as a partition guiding the air flow and the drug delivery units to be coated up the center of the unit across the inside of the roof and then down the outer space of the cylinder to the base of the unit where the spraying and drying cycle recommences. Each time a unit of product reaches the bottom of the coating chamber, it receives an additional dose of coating material. The solvent is removed when the unit rises up the center of the chamber and descends down the periphery. The coating is therefore built up in a series of incremental steps and thus, from a processing point of view, although not in terms of composition or function, the coating is multilayered. This cyclic process of spraying, drying, spraying, and drying can allow optimum conditions for gradual deposition of a coating of uniform thickness and structure. Batches of product from about 0.5 to 500 kg can be coated, and particles as small as 50 μm up to conventional tablets can be coated on this type of equipment. The original Wurster design has, in recent years, been significantly improved by, for example, introducing improved spray-nozzle design (Wurster HS) and improved airflow (Swirl Accelerator). There are various types of fluidized-bed coating equipment now commercially available (tangential spray, top spray, etc.). Different operators have their own preferences and, of course, the individual requirements of each controlled release coating formulation must be considered as well as equipment availability. Probably for controlled release coating products with aqueous coatings, bottom or tangential spray equipment is likely to be preferred. A detailed, comprehensive and authoritative review of Wurster and other fluidized-bed coating technologies has been published by Christensen and Bertelsen (29).

### Compression Coating

It is possible to compress a coating around a preformed (relatively soft) core by using a special tablet press (e.g., Drycota by Manesty or Precoter by Killian). The process basically consists of compression of the core to give a relatively soft compact, which is then fed into a die of a tablet press that already has received half of the coating material. The core is centered within the die, the remainder of the coating material added, and the product then compressed (9). Compaction coating of tablets is not as common as perhaps some of us expected when the equipment was introduced. Perhaps there is still a need for coating mixtures of improved compressibility and flow. Certainly the very high production rates that can be obtained on a tablet press make this coating approach very attractive.

### Melt Coating

For those controlled release coating materials that have a relatively low melting point and have acceptable thermostability, melt coating is possible. Materials applied in this manner include polyglycolyzed glycerides, hydrogenated vegetable oils, glyceride, ethylene glycol polymers, Carnauba wax, and synthetic wax. Top spraying, fluidized-bed coaters are probably the most commonly used type of equipment for this purpose, and operating temperatures as high as 200–250°C have been employed. Jozwiakowski, Jones, and Franz (30), and Jones and Percel (31) reviewed the processes.

### Development of New Controlled Release Coating Products

If a new, controlled release coating product is developed, it is quite likely that it uses a synthetic polymer applied to the core as an aqueous dispersion, rather than a material of natural origin dissolved in an organic solvent. There are certainly advantages in terms of process validation, and likely response from regulatory agencies from the use of aqueous-based coatings of synthetic polymers. However, even with these systems, scale-up from laboratory to pilot scale to manufacturing scale may well require modification to either formulation or processing variables. Even though such changes may seem to be inconsequentially minor, it is still possible that they may cause significant changes to the release profile. Similarly, changes in formulation or processing variables that may occur during clinical trials can modify the behavior of the product (32). In a marketing approval document submitted to a regulatory agency, it may be necessary to justify reliance on clinical trials results that derive from formulations which are not entirely identical. Also, regulatory agencies require data demonstrating that the manufacturing process is validated (8). The FDA, for example, normally require data on at least three production-scale batches before marketing approval is given. It must be kept in mind that a formulation and process that may yield coated products commendably free of coating defects when prepared at the laboratory scale may at a later point in the development process exhibit coating defects such as those described by Rowe (33). Rudnic and Kottke (9, p. 379) have presented a useful table to use in diagnosing coating problems.

## TRENDS IN THE FORMULATION AND MANUFACTURE OF COATED, CONTROLLED-RELEASE PRODUCTS

The trends of major importance in the current development of the formulation and manufacture of coated, controlled release dosage forms may be divided into two categories. Firstly, there are trends common to the manufacture of all types of drug delivery systems and, secondly, there are trends specific to this type of product. In the first class we include:

1. More rigorous evaluation of data by the FDA before marketing approval is granted
2. Increasing interest and regulatory concern about validation and optimization

In the second group we include:

1. Greater knowledge of, and control over, enteric-coated drug delivery systems
2. Lively interest in coated products for colonic drug delivery
3. Improvements in both materials and equipment used in controlled release, coated products

### Rigorous Evaluation of Data by FDA Before Marketing Approval Is Granted

The so-called Generic Drugs scandal (34) was a major factor in causing the FDA to become increasingly rigorous in its perusal of requests to market new pharmaceutical products. The introduction of the Application Integrity Policy, designed to discover or discourage fraud, has undoubtedly had significant effects on the research and development process in many companies. The requirement that the FDA District Office conducts a Preapproval Inspection before a new product can be marketed has caused some dramatic changes in both the pharmaceutical industry per se and also in contract houses that perform services to our industry in research, development, manufacturing, or product evaluation.

### Optimization, Validation, and Product Evaluation for Coating Formulation and Processing

During the past decade or so, a number of most useful reports have been published describing various approaches to optimizing the selection of formulation and processing variables for coatings. Johansson, Ringberg, and Nicklasson (35) described the use of sequential, reduced factorial studies on the fluidized bed coating process for organic solvent–dissolved ethylcellulose. Voight and Wunsch (36) reported an optimization study for the coating of granules by spray-coating. Optimization of a Wurster-type spray-drying process, including different thicknesses of film, was reported by Bianchini and Vecchio (37). Losa and co-workers (38), Bodmeier and Paratakul (39), and Parikh and associates (40) have also described studies relevant to optimization of coated pharmaceuticals. The paper by Parikh and associates is of special interest to those wishing to optimize and validate the aqueous spray-coating process. Farivar and co-workers (41) conducted a factorial design-type optimization study of ethycellulose-coated microcapsules, which included the evaluation of dissolution as a response parameter. Hsieh and associates (42) have reported a novel approach to coating controlled release products in

which the coating is incomplete, i.e., the drug delivery system has an open, uncoated face. Theoretical and experimental work indicated the potential of this approach. Brook and co-workers (43) have reported a separate study of the utility of partial coating in attaining controlled release. Optimization of this approach is possible. The Li team (44) have reported interesting studies of the rational development of a diltiazem, controlled release product in which a two-step layering process using a Wurster column is used.

Although there is often overlap between optimization and validation, and some workers tend to use the terms almost interchangeably, it is probably useful to maintain a distinction between their meanings. Optimization relates to studies of the influence of formulation and processing variables on the properties of a drug delivery system. Such studies often involve the application of empirical mathematical models to defining the interrelationship of such variables. Validation has taken on a special regulatory meaning and is now generally used to refer to the process whereby the sponsor of an NDA, ADA, or some other marketing approval document tries to convince a regulatory authority, such as the FDA, that the formulation and process are under control, that batch-to-batch variability is at an acceptably low level, and that the process is sufficiently robust so that minor perturbation of any one formulation or processing variable do not have a catastrophic effect on the product. Obviously, therefore, the validation of the coating segment of the manufacturing process normally involves evaluation of such basic parameters as coating temperature, coating times, equipment loading, and coating composition, together with such other values as may be important for an individual product. Before commencing a process validation study for a coating process, it can be helpful to examine published reports of other workers. However, some caution is advisable. Studies that might have been deemed acceptable by regulatory agency A in 1990 will not necessarily be regarded as acceptable by regulatory agency B in 2000.

Dietrich and Brausse (45) reported a validation study in which the effects of spray-drying temperature, air velocity, and relative humidity on release rates of a controlled release, coated product was explored. The starting point for any current validation study should be the definition of all functionally relevant attributes of the raw materials. Depending on the individual excipient, USP/NF, *Handbook of Pharmaceutical Excipients*, or the manufacturer's data may well provide most, if not all, of what is needed in this respect. However, depending on the nature and function of the excipient, it also may be prudent for the manufacturer of the drug delivery system to consider the use of additional test methods for excipients, intermediates, or final product. Such test methods may not only be useful during process validation, but may also be of value in routine testing of marketed product or during troubleshooting. For test data to be submitted to the FDA it is, of course, essential that the data be derived by objective test methods applied in accordance with an acceptable standard operating procedure. Thus, manual flexing of a cast polymer film between the two hands of an experienced polymer technologist, although perhaps of use in troubleshooting a plasticization problem, is not likely to be regarded as a test of pivotal value for documentation submitted to the FDA.

Microscopic examination of films, either cast or in situ on the drug delivery system, can provide useful data on the fine structure of the coating, which may well be related to function. For example, Hasirci and colleagues (46) used optical and scanning electron microscopy to demonstrate the pore structure of a hydrophobic tablet coat. Evaluation of pharmaceutical polymer films using a confocal laser scanning microscope (47) may well become of increasing importance. Gopferich, Alsonso, and Langer (48) have published a most useful paper in which a variety of techniques—polarized light microscopy, electron microscopy, specialized chemical analysis, and elemental analysis—were employed to characterize coating structure. Pourkavoos and Peck (49) have characterized coated tablets using thermogravimetric analysis, differential scanning calorimetry, and mercury intrusion porisimetry.

Zhou and co-workers (50) have described an approach for determining tablet coating distribution from the weight distribution of uncoated and coated tablets. The authors indicate the potential value of the method in scale-up and validation studies.

Fukumori and associates (51) have published a most interesting paper in which a computer simulation of agglomeration in a Wurster coating process is described. Results were compared with experimentally generated data. Voigt and Wunsch (52) have reported the use of equipment for evaluating drug release from film-coated pellets. Wu, Jean, and Chen (53) have used scanning electron microscopy in their investigation of the relationship between coating structure and dissolution rate. Murthy and Ghebre-Sellasie (54) have investigated dissolution stability of coated pharmaceuticals. Pathak and Dorle (55) have examined dissolution of coated microcapsules; their results indicate that the release mechanism does not follow a simple process. Lee and Liao (56) have presented an interesting theoretical study on the effect of coating deformation on dissolution. Dissolution studies of a coated theophylline product have been reported by Lavasanifar and co-workers (57).

Malamataris and Pilpel (58) have studied compaction properties of coated products using the Heckal and the Cooper and Eaton equations to assist in the interpretation of the data. Since there is an increasing interest in the use of drug delivery systems that consist of controlled release, coated microcapsules incorporated into compressed tablets (that can, if desired, also be coated), there is substantial value in examining compaction behavior of coated products. The combination of a multiparticulate drug delivery and a compressed tablet yields a hybrid product that can have the advantages of the reliable controlled delivery of coated microcapsules and the convenience, stability, and patient acceptability of compressed tablets (59). Although not a large amount of experimental data pertains directly to the resistance of polymer-coated microcapsules to coating fracture during the process of tablet compaction, it seems likely that the following factors are of importance in delineating the extent of such fracture:

1. Shape, size, and other characteristics of the particles to be coated
2. Composition and thickness of the polymer coating applied to the particles
3. Nature of the tablet matrix and the ratio of coated particles to total tablet weight
4. Compression force applied when the tablet is produced

Chang and Rudnic (60) have shown that not all controlled release coatings applied to cubic KCl crystals are equally resistant to compaction fracture.

### Enteric-Coated, Controlled Release Products

During the past decade or so there have been significant advances in the reliability of enteric coatings and their sensitivity to the pH at which drug is released. Agyilrath and Banker (61) indicated the various factors that influence the dissolution of enteric coatings:

1. The $pK_a$ of the polymer
2. Total free carboxylic and content of the polymer
3. Nature of the core material
4. Ionic strength of the dissolution fluid
5. Coating thickness
6. Presence or absence of plasticizers and other nonenteric components in the coating layer

Porter and Ridgeway (62) used X-ray crystallography in an investigation of two enteric coating polymers, cellulose acetate phthalate and polyvinyl acetate phthalate, which were found to be essentially amorphous in nature. Lin and Kawashima (63) reported on the development of an enteric-coated formulation using cellulose acetate phthalate. Malfertheiner and co-workers (64) have described clinical evaluation of some enteric-coated aspirin products. The topic of how effective enteric-coated aspirin is in preventing blood loss caused by ulceration of the gastrointestinal tract is still a topic of lively interest. Ebel and associates (65) have applied an empirical mass transfer model for enteric coat dissolution that uses in vivo dissolution data to characterize the pH-dependent solubility properties of the polymer film and estimation of a mass transfer coefficient. The authors presented data comparing predicted and experimentally determined mass transfer coefficients. Taniguchi and associates (66) have described enteric coating of an enzyme. Lin and co-workers (67) have reported the enteric coating of a vaccine using cellulose acetate phthalate as coating material. Kiriyama and colleagues (68) have reported on the bioavailability of a tripeptide protease inhibitor, enteric-coated by hydroxypropyl methylcellulose phthalate. Duodenal absorption appears satisfactory. Arica and co-workers (69) have described successful enteric coating and evaluation of diclofenac sodium.

### Coated Pharmaceuticals for Colonic Delivery

The controlled release of drugs in the colon has, in recent years, become a topic of great interest to many pharmaceutical scientists. This interest derives primarily from two causes. Firstly, there is a desire to develop therapies allowing specific delivery of drugs to the colon for the treatment of such conditions as ulcerative colitis. Secondly, the exciting possibility of using the colon as a portal entry for polypeptides and small proteins into the vascular system has stimulated many research groups to work in this field. (There are, of course, many matrix-type approaches also being examined for colonic drug delivery.) In earlier sections of this chapter, some references to coated drug delivery systems designed for drug release in the colon have been given (15–17,22). The references given hereafter are samples of some of the many varied strategies being employed for coated products for colonic drug delivery.

Van-den-Mooter and co-workers (70,71) have published two papers dealing with the use of copolymers of 2-hydroxyethylmethacrylate (HEMA) and methylmethacrylate (MMA), and terpolymers of HEMA, MMA, and methacrylic acid were synthesized in the presence of an azobenzene derivative for use as coatings for pharmaceutical products designed for colonic drug delivery. In vivo testing of the coating was reported. Ashford and Fell (72) have published a useful review of oral delivery of drugs to the colon. Niwa and co-workers (73) have described an ethylcellulose capsule that only ruptures and allows the drug within to be released when a swellable polymer, such as low-substituted hydroxypropyl cellulose, has expanded so much as to destroy capsule integrity. The lag time during which the pressure within the capsule is developing is such that the drug is released in the colon. The release time of the drug was mainly dependent on the thickness of the capsule. Wakerly and associates (74) have investigated the use of film coatings consisting of ethylcellulose and pectin for colonic drug delivery. In vitro dissolution studies in a flow-thru cell indicate the potential of this approach. Gardener and co-workers (75) have reported on the use of a pig model for studying colonic drug delivery systems.

### Developments in Materials and Processing

Published studies, from both academia and industry, clearly indicate that much energy and skill is being devoted to improving existing coating materials and processing methods as well as exploring the potential of new materials and coating techniques. Fukimore and co-workers (76) described the potential of aqueous coats of ethylacrylate methymethacrylate 2-hydroxyethyl methacrylate copolymer. Park, Cohen, and Langer (77) have presented a most interesting report describing the use of aqueous coatings of an adsorptive water-soluble polymer polyethyleneimine for control of protein release. Bhagat and associates (78) have described a novel coating process leading to the formation of a uniform defect-free coating. This coating process, termed diffusion controlled interfacial complexation, involves a chemical reaction as part of the coating process. The technique can be applied to tablets that may be characterized by zero-order release. Ramade (79) has presented a review on the role of polymers in drug delivery that contains some useful data for any worker considering testing polymers for inclusion in drug delivery systems. Bianchini and Vecchio (80) have described the coating of

dose units prepared by extrusion-spheronization. Pourkavoos and Peck (81) reported on the mechanism and extent of sorption of water following aqueous film coating. Ashton and co-workers (27) have described the design and evaluation of a coated drug delivery device intended for implantation in the eye. Deasy (82) has described, in a helpful review article, some recent advances in the design and evaluation of coated products.

Coating of pharmaceuticals as a method of assuring reproducible controlled release has come of age. The uncertainties about reliability of the natural-origin raw materials used for coating has been replaced by quiet confidence in the standardization of the synthetic materials now commonly used. The difficulties encountered during the conversion from organic solvents to water have been resolved, and prospects for future developments are good.

What factors should be taken into account when a pharmaceutical research and development group selects a formulation and processing strategy for a controlled release product? More specifically, what are the advantages of coated, vis-à-vis noncoated, products? It is not possible to give a precise answer to this question of general applicability. Obviously, the physicochemical, pharmacokinetic, and pharmacoeconomic attributes of the drug substance, together with such factors as the formulation group's experience in previous controlled release projects, the availability of equipment, and patents may all play legitimate roles in this type of decision. It would certainly be unreasonable to claim that coated, controlled release approaches should always be selected in preference to noncoated methods. However, there are certain strong advantages of coating procedures that do merit careful consideration:

1. The supply of a variety of reliable-quality excipients, which are acceptable to regulatory agencies
2. A successful track record of products being taken from research and development through pilot-scale to large-scale manufacturing with, in many cases, surprisingly few problems
3. Manufacturing methods that normally do not require the fabrication of any new manufacturing equipment but rely on well-tested manufacturing methods, which in many instances, are readily validated
4. The availability of a remarkably flexible range of formulation and processing variables, which can be used to custom build controlled release profiles for a plethora of purposes

## MATERIALS USED IN CONTROLLED-RELEASE PRODUCTS

The following section presents data on some of the more common materials used in pharmaceutical, controlled release coatings. This list is not represented as a comprehensive statement of all materials which have ever been proposed for use in such products. Generally, the materials outlined in Table 1 and described in more detail in this section are used for one or more of three main functions: (1) provisions of the backbone structure of the coating, (2) facilitation or control over transport of drug across the

**Table 1. Materials Listed as Components in Controlled Release Coating**

*Enteric coatings*

A. *Cellulosic*
 1. Cellulose acetate phthalate (CAP)[a]
 2. Hydroxypropylmethylcellulose phthalate (HPMCP)[a]
B. *Noncellulosic*
 1. Methacrylic acid polymers[a,b]
 2. Polyvinylacetate phthalate (PVAP)
 3. Shellac[a]

*Nonenteric coatings*

A. *Cellulosic*
 1. Ethylcellulose (EC)[a]
 2. Hydroxyethylcellulose (HEC)[a]
 3. Hydroxypropylmethylcellulose (HPMC)[a]
 4. Methylcellulose (MC)[a]
 5. Sodium caroboxymethylcellulose (NaCMC)[a]
B. *Noncellulosic*
 1. Carnauba wax[a]
 2. Castor oil[a]
 3. Cetyl alcohol[a]
 4. Ethylene vinyl acetate copolymer
 5. Hydrogenated vegetable oils[a]
 6. Polyvinyl alcohol[a]
 7. Silicon-based polymers

*Plasticizers*

 1. Benzyl benzoate[a]
 2. Chlorobutanol[a]
 3. Dibutyl sebacate[a]
 4. Diethyl phthalate[a]
 5. Glycerol[a]
 6. Polyethylene glycol[a]
 7. Sorbitol[a]
 8. Triacetin[a]
 9. Triethyl citrate[a]

[a]Monograph in USP23/NF18.
[b]Also used as nonenteric coatings.

membrane, or (*3*) plasticization of the coating. Coatings may well contain other components, such as colors or antioxidants, but specific attention is not given to components the primary function of which is not covered by one of these three categories. The functions of the first two classes is self-evident; however, it is appropriate to define what is meant by a plasticizer, since this term is sometimes used as a catchall to cover any excipient other than the primary coating material regardless of what function the excipient is expected to fulfill in the coating. A plasticizer, in terms of its functional definition, is an excipient, usually a low molecular weight organic solvent, added to coating material to assist in the processing by rendering the film more flexible. It is believed that such materials act by lowering the polymer glass-transition temperature. Plasticizers must be compatible, in terms of solubility parameter, with the polymer to which they are added to make it more flexible. Thus, plasticizers are not general purpose but must be chosen from materials that have been shown to be useful for a particular polymer.

Examination of the substances listed in this section indicates that a substantial number of the materials are cellulose derivatives. The review of chemically modified cellulose by Kumar and Banker (83) provides most useful background data in this area. Also, for more detailed information on polymers used in controlled drug delivery, the book edited by Tarcha (84) is likely to be of value. Also, for those excipients listed in the USP/NF, it is appropriate to check standards given in this official compendium. The materials described here are presented in a sequence which relates to function and alphabetical order. The list of suppliers for the various materials is not necessarily exhaustive.

**Enteric Coatings: Cellulosic**

**Cellulose Acetate Phthalate (CAP).** CAP is a cellulose ester referred to as cellacephate in BP and as Cellulosi acetas phthalus in Ph. Eur. It is soluble at pH values above about 6.0, and thus will tend to release drug toward the distal end of the small intestine. CAP is soluble in acetone and ethyl acetate. Plasticizers, such as triacetin or castor oil, are usually incorporated into this polymer. CAP is approved for oral use in the U.S. and European Union. It has been reported that CAP may be incompatible with ferrous sulfate, ferrous chloride, silver nitrate, and other inorganic salts. Also, since CAP has free carboxylic acid incompatibilities with acid-sensitive drugs (e.g., omeprazole) are possible. Application from aqueous dispersion may be possible (85). Suppliers include Eastman Kodak and FMC.

**Hydroxypropylmethylcellulose Phalate (HPMCP).** HPMCP is a monophthalic ester of hydroxypropylcellulose. The NF specifies two grades viz 200731 and 220824. BP refers to HPMCP as hypromelluse phthalate and Ph. Eur. as methylhydroxypropylcellusoi phthalus. It normally releases drug in the proximal part of the small intestine. Shin-Etsu (86) markets three grades that vary in molecular weight (higher values generally yield tougher films) and in the pH at which drug release normally occurs (from about 5.0 to 5.5). Plasticizers, such as castor oil, diacetin, diethyl and dibutylphthalate and polyethylene glycols, are commonly used with this polymer. Aqueous coating is possible (87). HPMCP is approved for oral use in the European Union, U.S., and Japan. Possible incompatibilities include strong oxidizing agents. Also, inclusion of more than about 10% titanium dioxide as a color may adversely affects physical stability or gastric resistance of the film coat. This material is also made by Eastman.

**Enteric Coatings: Noncellulosic**

**Methacrylic Acid Polymer and Other Polymeric Methacrylates.** There are three grades specified in NF18 of these materials, often referred to as polymeric methacrylates or Eudragit®. (Polyacrylates in NF18 are listed under ammonio methacrylate copolymer and methacrylic acid copolymer.) Eudragit E, which is soluble below pH 5.0, is used as a nonenteric coating. Eudragit L100-55 is used as an enteric coating for drug release at a pH of above 5.5, NE 30 D is used as a permeable, controlled release coating, RL 30 D is used in controlled release coatings where a high permeability film is required. The manufacturer of Eudragit has published a guide to the use and properties of the whole range of Eudragits (88). Aqueous enteric coating is possible with Eudragit. By blending different grades of Eudragit, the formulator may be able to control release pH from about 5.5 to 7.0, and also the permeability of the film at constant pH (89). Lehman has presented authoritative reviews of polymethacrylate coating systems (90). Ghebre-Sellasie and co-workers (91) have reviewed the use of aqueous dispersions of Eudragits for controlled release. Eudragits have been used in topical, oral, parenteral, ophthalmic, and other types of pharmaceutical products.

**Polyvinylacetate Phthalate (PVAP).** PVAP is a reaction product of phthalic anhydrode and partially hydrolyzed polyvinyl acetate. As an enteric film coating it releases drug at pH values above about 5.0 so that absorption may occur throughout the small intestine. It is approved for oral use in the European Union and the U.S. Trade names include Opaseal®, pHthalvin®, and Sureteric®. The Colorcon literature (92) provides quite specific advice on processing conditions using, for example, an Acela-Cota. Porter (93) has described aqueous film coating using PVAP.

**Shellac.** Shellac is obtained from a gummy exudation produced by female insects. *Laccifer lacca Kerr.* Commercially, it is available in several grades (bleached, orange, white, etc.). Although its precise composition is both unknown and variable, its main component is a resin. The quality of this material shows significant variation, both with respect to source and storage time. The pH at which drug is released is about 7.0, which may well be somewhat too high for most enteric-coated products. Unless there is some specific requirement that an enteric coating shall be "natural," this material is *not* recommended for anyone developing a new product. Depending on the grade, it is acceptable for oral use in the European Union, U.S., and Japan. Suppliers include Classic Flavors, H.E. Daniel, Ikeda, and Ruger.

### Nonenteric Coatings: Cellulosic

**Ethylcellulose (EC).** EC is a cellulose ether derivative with three hydroxy groups available for substitution. It is sometimes referred to as cellulose ethyl ether. Grades with differing degrees of substitution and/or molecular weight are available commercially. This polymer is virtually insoluble in water, but is freely soluble in organic solvents. Higher viscosity grades (e.g., Ethocel® 100) tend to produce tougher films. The permeability of ethylcellulose films can be increased by adding materials such as hydroxypropyl cellulose, polyethylene glycol, etc. (94,95). Plasticizers used with ethylcellulose include dibutyl phthalate, diethyl phthalate, dimethylphthalate, benzyl benzoate, cetyl alcohol, castor oil, and corn oil. Oxidative degradation can be reduced by adding stabilizers such as octyl phenol or butylated hydroxyphenal. An ultraviolet light absorber, such as 2,4,dihydroxybenzophenone (at about 0.7%) has also been used as a stabilizer. Aqueous dispersion such as Aquacoat® (FMC) and Surelease® (Colorcon) are available. The Surelease literature (96) contains a number of cookbook-type recipes. Ethylcellulose is accepted for oral use in many parts of the world. Suppliers include Colorcon, FMC, Dow, and Hercules.

**Hydroxyethylcellulose (HEC).** HEC, Ph. Eur. hydroxyethylcellulosum, is cellulose 2-hydroxyethyl ether. This water-soluble polymer, available under such trade names as Cellosize® and Natrosol®, is available in several grades (97) and is approved for oral use in the European Union, U.S., and elsewhere. Suppliers include Allchem, Amerchol, Aqualon, Sumisho, and Union Carbide.

**Hydroxypropyl Methylcellulose (HPMC).** HPMC (hypromelose, BP, hydroxypropylcellulosum, Ph. Eur.) is cellulose 2-hydroxypropyl methyl ether. HPMC is an excellent water-soluble film former, low viscosity grades, e.g., Methocel® HG (Dow), are commonly used in controlled release, coated pharmaceuticals. It is approved for oral and topical use in the European Union, U.S., and Japan. *The Handbook of Pharmaceutical Excipients* (10) reports that it may be incompatible with oxidizing agents, metallic salts, and ionic organics (98). Suppliers include Aldrich, Dow, Aqualon, Cortaulds, and Hercules.

**Methylcellulose (MC).** MC (methylcellulosum, Ph. Eur.) is cellulose methyl ether. It is soluble in water and organic solvents and used in a number of controlled release, coated products (99). It is approved (at varying levels) in the European Union, U.S., and Japan for use in oral, topical, buccal, vaginal, and parenteral pharmaceutical use. This material can be obtained from Dow.

**Sodium Carboxymethylcellulose (NaCMC).** NaCMC (carboxymethylcellulose sodium, USP/NF; carmelose sodium, BP; carboxylmethylcellulosum natricum, Ph. Eur.) is the sodium salt of cellulose carboxycellose ether. NaCMC is soluble in water and polar organic solvents (100). A number of different grades are commercially available with different degrees of substitution and viscosity. NaCMC is approved in the European Union, U.S., and Japan for dental, oral, topical, and parenteral use. Suppliers include Hercules and Cortaulds.

### Nonenteric Coatings: Noncellulosic

**Carnauba Wax.** This material, sometimes referred to as Brazil Wax or Sera Carnauba, is a gummy exudate obtained from the Brazilian wax palm. It is photlabile. It has been accepted (for different purposes) for inclusion in foods and/or oral pharmaceuticals in the European Union, U.S., and Japan. Suppliers include Spectrum, Ruger, Aldrich, and Penta. Sterotex® is a trade name.

**Castor Oil.** There is an NF monograph (hydrogenated castor oil) for this material in which it is defined as mainly consisting of the triglyceride of hydrosteric acid. It has been used in topical and oral pharmaceutical products in the European Union, U.S., and Japan.

**Cetyl Alcohol.** The NF monograph defines cetyl alcohol as containing not less than 90.0% of cetyl alcohol with the remainder consisting of related alcohols. It is referred to as alcohol cetylicus in Ph. Eur. *The Handbook of Pharmaceutical Excipients* (10) states that cetyl alcohol is incompatible with oxidizing agents and also suggests that it may cause allergic reactions. Cetyl alcohol has been used in topical and oral pharmaceutical products in the European Union, U.S., and Japan. Suppliers include Aldrich, Proctor and Gamble, Ruger, and Spectrum. This material is marketed under a number of trade names, including Dyhdag®, Lanette®, and Nacol®.

**Ethylene Vinylacetate Copolymer.** This material, sometimes referred to as EVA copolymer, is used as a coating material in transdermal products in the U.S. Suppliers include Allied Signal, Bayer, and Focus.

**Hydrogenated Vegetable Oils.** Oils derived from cottonseed, palm, and soybean are covered by this title. Trade names include Lubritab®, Sofisan®, and Sterotex®.

**Polyvinyl Alcohol.** This material is commercially available in at least three different viscosity grades. It is used in the U.S. in ophthalmologic, parenteral, topical, and oral products.

**Silicon-Based Polymers.** These relatively new materials for use in controlled release coatings are presently attracting considerable attention. They can be applied as a coating from aqueous dispersions (101).

## Plasticizers

**Benzyl Benzoate.** Benzyl benzoate, which has been accepted for oral use in both the U.S. and U.K., has been used as a plasticizer for several cellulosic-type films. Suppliers include Ruger, Aldrich, and Spectrum.

**Chlorobutanol.** Chlorobutanol has many uses in pharmaceutical technology, including that of plasticizing some cellulose esters and ethers. It has been widely used in topical, oral, and parenteral pharmaceutical products throughout the world. Suppliers include R. W. Greef, Aldrich, and Ruger.

**Dibutyl Sebacate.** This plasticizer is used in a number of oral products. Suppliers include Union Carbide and Unitex.

**Diethy Phthalate.** This material has been used as a plasticizer (in the range of 10 to 30%). It is included in oral products in the U.K. and U.S. Suppliers include Allchem, Daihachi Chem., Eastman, Aldrich, and Spectrum.

**Glycerol (Glycerin).** This multipurpose excipient has been used as a plasticizer. It is used in many types of pharmaceutical products throughout the world. Suppliers include Ruger, Asland, Croda, Compton and Knowles, Proctor and Gamble, and Ellis and Everard.

**Polyethylene Glycol.** Polyethylene glycol (also known as Macrogol) is commercially available in a variety of molecular weight ranges. It is widely used in many parts of the world in parenteral, topical, and oral pharmaceutical products. Suppliers include BASF, Dow, duPont, Texaco, and Union Carbide.

**Sorbitol.** Sorbitol has many pharmaceutical uses including that of plasticization. It is available in many different grades and is included in many oral pharmaceutical products used in many parts of the world. Suppliers include Aldrich, ICI, Penta, Ruger, and Spectrum.

**Triacetin.** This plasticizer has been included in oral pharmaceuticals in both the U.K. and U.S. Suppliers include Aldrich, Eastman, and Ruger.

**Triethyl Citrate.** This plasticizer has been used in oral pharmaceuticals, including coated products prepared using aqueous coating methods. Suppliers include Aldrich, Penta, H.E. Daniels, and Sharon. Hydagen® is a trade name for this material.

## BIBLIOGRAPHY

This chapter was previously published in *Drug Development and Industrial Pharmacy*, Volume 24, Issue 12, by Marcel Dekker, Inc. (November 11, 1998). Reprinted with permission.

1. T.R. Tice and S.E. Tabibi, in A. Kydonieus, ed., *Treatise on Controlled Drug Delivery*, Dekker, New York, 1991, p. 320.
2. U.S. Pat. 4,863,743 (September, 1989), D. Hsiao and T. Chou (to Key Pharmaceuticals, Inc., Kenilworth, N. J.).
3. A.C. Cartright and B.R. Matthews, eds., *International Pharmaceutical Product Registration*, Ellis Horwood, New York, 1994.
4. G.E. Peck, in G.S. Banker and C.T. Rhodes, eds., *Modern Pharmaceutics*, 3rd ed., Dekker, New York, 1995, Chapter 19.
5. B.A. Matthews, in G.S. Banker and C.T. Rhodes, eds., *Modern Pharmaceutics*, 3rd ed., Dekker, New York, 1995, Chapter 20.
6. A. Merrill, *Pharm. Technol.*, December, pp. 34–40 (1995).
7. I.R. Berry and R.A. Nash, *Pharmaceutical Process Validation*, 2nd ed., Dekker, New York, 1994.
8. E.M. Rudnic and M.K. Kottke, in G.S. Banker and C.T. Rhodes, eds., *Modern Pharmaceutics*, 3rd ed., Dekker, New York, 1995.
9. J.R. Schwartz and R.E. O'Connor, in G.S. Banker and C.T. Rhodes, eds., *Modern Pharmaceutics*, 3rd ed., Dekker, New York, 1995.
10. *Handbook of Pharmaceutical Excipients*, 2nd ed., Royal Pharmaceutical Society of Great Britain, London, and the American Pharmaceutical Association, Washington, D.C., 1994.
11. M. Ash and I. Ash, *Handbook of Pharmaceutical Additives*, Gower, Aldershot, England, 1995.
12. *Pharmaceutical Contract Support Organizations Register*, Drug Information Association, Washington, D.C., 1996.
13. B. Berner and S. Dinh, in A. Kydonieus, ed., *Treatise on Controlled Drug Delivery*, Dekker, New York, 1991, p. 2.
14. A.S. Achanta, P.S. Adusmili, K.W. James, and C.T. Rhodes, *Drug Dev. Ind. Pharm.* **23**, 441–450 (1997).

15. R.L. Botham et al., in G.O. Phillips, P.A. Williams, and D.J. Wedlock, eds., *Gums and Stabilizers for the Food Industry*, Oxford University Press, London, 1994, pp. 187–195.

16. S. Milojevic et al., *J. Controlled Release* **38**, 75–84, 85–94 (1996).

17. M. Saffran et al., *Science* **233**, 1081 (1986).

18. M. Donbrow and D.Y. Samuelov, *J. Pharm. Pharmacol.* **32**, 461–470 (1979).

19. U.S. Pat. 4,587,118 (May 6, 1986), C.M. Hsiao (to Key Pharmaceutical, Inc., Miami, Fla.).

20. T.A. Wheatley and C.R. Steuernagel, in J.W. McGinity, ed., *Aqueous Polymeric coatings for Pharmaceutical Dosage Forms*, 2nd ed., Dekker, New York, 1997, p. 50.

21. *USP 23/NF18*, U.S. Pharmacopoeial Convention, Inc., Rockville, Md., 1994, pp. 1793–1799.

22. U.S. Pat. 5,482,718 9 (January 9, 1996), N.H. Shah, W.P. Kearney, and A. Railkar (to Hoffman-LaRoche, Inc., Nutley, N.J.).

23. V. Lenaerts and R. Gurny, *Bioadhesive Drug Delivery Systems*, CRC Press, Boca Raton, Fla., 1990.

24. H.R. Chueh, H. Zia, and C.T. Rhodes, *Drug Dev. Ind. Pharm.* **21**, 1725–1748 (1995).

25. A. Joglekar, C.T. Rhodes, and M. Danish, *Drug. Dev. Ind. Pharm.* **17**, 2103–2153 (1991).

26. N.A. Pappas and J.J. Sahlin, *Biomaterials* **17**, 1553–1561 (1996).

27. P. Ashton et al., *J. Ocul. Pharmacol.* **10**, 691–701 (1994).

28. J.G. Needleman, *Br. Dent. J.* **170**, 405–408 (1991).

29. F.N. Christensen and P. Bertelsen, *Drug. Dev. Ind. Pharm.* **23**, 451–463 (1997).

30. M.J. Jozwiakowski, D.M. Jones, and R.M. Franz, *Pharm. Res.* **7**, 1119–1135 (1990).

31. D.M. Jones and P.J. Percel, in I. Ghebre-Sellasic, ed., *Multiparticulate Oral Drug Delivery*, Dekker, New York, 1994, pp. 113–142.

32. D.C. Monkhouse and C.T. Rhodes, eds., *Drug Products for Clinical Trials*, Dekker, New York, 1997.

33. R.C. Rowe, in J.W. McGinity, ed., *Aqueous Polymeric Coatings for Pharmaceutical Dosage Forms*, 2nd ed., Dekker, New York, 1997, Chapter 12.

34. M.R. Hamrel, *Clin. Res. Drug Regul. Affairs* **14**, 139–154 (1997).

35. M.E. Johansson, A. Ringberg, and M. Nicklasson, *J. Microencapsul.* **4**, 217–222 (1987).

36. R. Voigt and G. Wunsch, *Pharmazie* **40**, 772–776 (1985).

37. R. Bianchini and C. Vecchio, *Boll. Chim. Farm.* **128**, 373–379 (1989).

38. C. Losa et al., *Pharm. Res.* **10**, 80–87 (1993).

39. R. Bodmeier and O. Paratakul, *Pharm. Res.* **11**, 882–888 (1994).

40. N.H. Parikh, S.C. Porter, and B.D. Rohera, *Pharm. Res.* **10**, 535–534 (1993).

41. M. Farivar, H.S. Kas, L. Omer, and A.A. Hineal, *J. Microencapsul.* **10**, 309–317 (1993).

42. D.S. Hsieh, W.D. Rhine, and R. Langer, *J. Pharm. Sci.* **72**, 17–22 (1983).

43. I.M. Brook, C.W. Douglas, and R. vanNoort, *Biomaterials* **7**, 292–296 (1986).

44. S.P. Li et al., *Pharm. Res.* **12**, 1338–1342 (1995).

45. R. Dietrich and R. Brausse, *Arzneimi. Forsch.* **38A**, 1210–1219 (1988).

46. V.M. Hasirci and I.L. Kamel, *Biomaterials* **9**, 42–48 (1988).

47. K. Carlsson et al., *Opt. Lett.* **10**, 53–55 (1985).

48. A. Gopferich, M.J. Alsonso, and R. Langer, *Pharm. Res.* **11**, 1568–1574 (1994).

49. N. Pourkavoos and G.E. Peck, *Pharm. Res.* **10**, 1212–1218 (1993).

50. J. Zhou, T. Williams, H. Swopes, and T. Hale, *Pharm. Res.* **13**, 381–386 (1996).

51. Y. Fukimora et al., *Chem. Pharm. Bull.* **40**, 2159–2163 (1992).

52. R. Voigt and G. Wunsch, *Pharmazie* **41**, 114–117 (1986).

53. J.C. Wu, W.J. Jean, and H. Chen, *J. Microencapsul.* **11**, 507–518 (1994).

54. K.S. Murthy and I. Ghebre-Sellassie, *J. Pharm. Sci.* **82**, 113–126 (1993).

55. Y.V. Pathak and A.K. Dorle, *J. Microencapsul.* **7**, 185–190 (1990).

56. D.J. Lee and Y.C. Liao, *J. Pharm. Sci.* **84**, 1366–1373 (1995).

57. A. Lavanisifar et al., *J. Microencapsul.* **14**, 91–100 (1997).

58. S. Malamataris and N. Pilpel, *J. Pharm. Pharmacol.* **35**, 1–6 (1983).

59. M.K. Kottke, G. Stetsko, S.E. Rosenbaum, and C.T. Rhodes, *J. Geriatr. Drug Ther.* **5**, 77–92 (1990).

60. R.K. Chang and E.M. Rudnic, *Int. J. Pharm.* **70**, 261–270 (1991).

61. G.A. Agyilrath and G.S. Banker, in P.J. Tarcha, ed., *Polymers for Controlled Drug Delivery*, CRC Press, Boca Raton, Fla., 1991, p. 37.

62. S.C. Porter and K. Ridgeway, *J. Pharm. Pharmacol.* **35**, 341–344 (1983).

63. S.Y. Lin and Y. Kawashima, *Pharm. Res.* **4**, 70–74 (1987).

64. P. Malfertheiner, A. Stanescu, W. Rogatti, and H. Ditschuneit, *J. Clin. Gastroenterol.* **10**, 269–272 (1988).

65. J.P. Ebel, M. Jay, and R.M. Beihn, *Pharm. Res.* **10**, 233–238 (1993).

66. M. Taniguchi, S. Tanahashi, and M. Fujii, *Appl. Microbiol. Technol.* **33**, 629–632 (1990).

67. S.Y. Lin, Y.L. Tzan, C.J. Lee, and C.N. Weng, *J. Microencapsul.* **8**, 317–325 (1991).

68. A. Kiriyama et al., *Biopharm. Drug Dispos.* **17**, 125–134 (1996).

69. B. Arica et al., *J. Microencapsul.* **13**, 699 (1996).

70. G. Van-den-Mooter, C. Samyn, and R. Kinget, *Pharm. Res.* **11**, 1737–1741 (1994).

71. G. Van-den-Mooter, C. Samyn, and R. Kinget, *Pharm. Res.* **12**, 244–247 (1995).

72. M. Ashford and J.T. Fell, *J. Drug Target.* **2**, 241–257 (1994).

73. K. Niwa, T. Takaya, T. Morimoto, and K. Takada, *J. Drug Target.* **3**, 83–89 (1995).

74. Z. Wakerly, J.T. Fell, D. Atwood, and D. Parkins, *Pharm. Res.* **13**, 1210–1212 (1996).

75. N. Gardener et al., *J. Pharm. Pharmacol.* **48**, 688–693 (1996).

76. Y. Fukimore et al., *Chem. Pharm. Bull.* **36**, 3070–3078 (1988).

77. T.G. Park, S. Cohen, and R. Langer, *Pharm. Res.* **9**, 37–39 (1992).

78. H.R. Bhagat, R.W. Mendes, E. Mathiowitz, and H.N. Bhargava, *Pharm. Res.* **8**, 576–583 (1991).

79. V.V. Ramade, *J. Clin. Pharmacol.* **30**, 10–23 (1990).

80. R. Bianchini and C. Vecchio, *Farmaco* **44**, 645–654 (1989).

81. N. Pourkavoos and G.E. Peck, *Pharm. Res.* **10**, 1363–1371 (1993).

82. P.B. Deasy, *J. Microencapsul.* **11**, 487–505 (1994).

83. V. Kumar and G.S. Banker, *Drug Dev. Ind. Pharm.* **19**, 1–32 (1993).

84. P.J. Tarcha, ed., *Polymers for Controlled Drug Delivery*, CRC Press, Boca Raton, Fla., 1991.

85. A.M. Ortega, Ph.D. Thesis, Purdue University, West Lafayette, Ind., 1977.

86. Shin-Etsu Chemical Company, Ltd., technical literature on HPMCP, 1993.

87. N.A. Muhammad, W. Boisvert, M.R. Harris, and J. Weiss, *Drug Dev. Ind. Pharm.* **18**, 1787–1797 (1992).

88. Eudragit literature, Rohm. Pharm. GmBh, 1990.

89. K.O.R. Lehman, *Drug Dev. Ind. Pharm.* **12**, 265–287 (1986).

90. K.O.R. Lehman, in J.W. McGinity, ed., *Aqueous Polymeric Coating for Pharmaceutical Dosage Forms*, 2nd ed., Dekker, New York, 1997, Chapter 4.

91. I. Ghebre-Sellasi, in J.W. McGinity, ed., *Aqueous Polymeric Coatings for Pharmaceutical Dosage Forms*, 2nd ed., Dekker, New York, 1997, Chapter 7.

92. Sureteric®, Colorcon, West Point, Pa., 1995.

93. S.C. Porter, in J.W. McGinity, ed., *Aqueous Polymeric Coatings for Pharmaceutical Dosage Forms*, 2nd ed., Dekker, New York, 1997, Chapter 9.

94. *Ethocel*, Dow Chemical Company, 1993.

95. *Ethocel Bibliography*, Dow Chemical Company, 1993.

96. Surelease technical literature, Colorcon, West Point, Pa., 1990.

97. B. Kovacs and G. Merenyi, *Drug Dev. Ind. Pharm.* **16**, 2302–2333 (1990).

98. G. Banker and G. Peck, *Drug Dev. Ind. Pharm.* **7**, 693–716 (1981).

99. N.M. Sanghavi, P.R. Kamath, and D.S. Amin, *Drug Dev. Ind. Pharm.* **16**, 1843–1848 (1990).

100. G. Banker et al., *Drug Dev. Ind. Pharm.* **8**, 41–51 (1982).

101. L.C. Li and G.E. Peck in J.W. McGinity, ed., *Aqueous Polymeric Coatings for Pharmaceutical Dosage Forms*, 2nd ed., Dekker, New York, 1997, Chapter 14.

See also MICROENCAPSULATION; MICROENCAPSULATION FOR GENE DELIVERY.

# D

## DIAGNOSTIC USE OF MICROSPHERES

MARGARET A. WHEATLEY
PADMA NARAYAN
Drexel University
Philadelphia, Pennsylvania

**KEY WORDS**

Biodegradable

Contrast agent

Diagnostic imaging

Imaging

Liposome

Magnetic resonance imaging

Microbubble

Microcapsule

Microparticle

Microsphere

Radiopaque

Therapeutic

Ultrasound

**OUTLINE**

In 1896 a physicist, Hascheck, and a physician, Lindenthanl, working together, injected a "heavy mixture" into the hand veins of a cadaver for X-ray analysis during autopsy (1). Although not involving microcapsules at this early stage, it was the beginning of what has become a large and rapidly developing enterprise to produce contrast agents (CA) that can be safely injected into a patient to assist the physician in diagnosis, by improving the quality of information obtained during noninvasive diagnostic procedures. The search for diagnostic CA has spread from X-ray to other imaging modalities, principally to magnetic resonance imaging (MRI) and ultrasound (2). Microencapsulated agents have been developed using a wide variety of methods, from liposomal entrapment (3) to use of biodegradable polymers (4). In addition, microencapsulated agents have been used in such areas as targeted imaging (5), biodistribution studies (6), therapy involving embolization (7), and combinations of these, such as concomitant drug delivery and imaging (8).

Microencapsulation of imaging agents offers many advantages that parallel those found with encapsulated drugs. Advantages of encapsulated imaging agents include increased stability, prolonged in vivo half-life, reduction of possible adverse side effects, concentration of the agent resulting in lower required doses, ease of administration, and the possibility of agent targeting. The next generation of contrast agents is expected to facilitate areas such as highly specific targeting and functional imaging.

Each imaging modality has certain requirements to obtain good contrast, for example X-ray CA need to be radiopaque, whereas ultrasound CA need to produce a large impedance mismatch between themselves and the suspending fluid (usually blood). General requirements that encompass all agents include lack of toxicity, stability during imaging, ease of elimination from the body and most importantly, being amenable to intravenous injection, requiring the agent to be small enough to pass the pulmonary bed (preferably less than 6 $\mu$m diameter).

## X-RAY CONTRAST AGENTS

### Physical Principles

Contrast in X-ray imaging is produced by use of a radiopaque material that strongly absorbs incident radiation. At imaging energies, the absorption is due to electron density. Electron density relates to the number of electrons per atom multiplied by the number of atoms per volume, or approximately the density of the material. The most effective agents contain a heavy element such as the barium in barium sulfate ($BaSO_4$). $BaSO_4$ was one of the first agents to be used, and is still in use today, predominantly for gastrointestinal (GI) imaging. The most commonly used heavy element is iodine, generally covalently bound to an organic moiety such as in diatrizoate, which has three iodines on a benzene ring (Table 1.).

**Table 1. Main Structures of Typical Monomeric Iodinated X-Ray CA**

| Basic monomer structure | Group | Structure/significance | | |
|---|---|---|---|---|
| | I | Radiopaque marker | | |
| | | | Ionic | Nonionic |
| | –COR | | R = Kat$^+$; Na$^+$; Meg$^+$ | R = –N(R)$_2$ amine carrying hydroxyl groups |
| | | Affects osmolarity, solubility | | |
| | –R$_{1/2}$ | | R-OC-HN- | R-OC-N-<br>R |
| | | Influences elimination and toxicity | | |

Basic monomer structure diagram: benzene ring with COR at position 1, I at positions 2 and 6, R$_2$ at position 3, I at position 4, R$_1$ at position 5.

## Early CA for X-Ray Imaging

Intravascular agents became available with the development of liquid iodinated media. There was a steady progression from insoluble oils such as Lipiodol, a halogenated poppy seed oil (9) to water-soluble ionic media such as meglumine (Meg$^+$) diatrizoate, also known as Renografin®. The early 1980s saw rapid and significant changes. The agents at the time were ionic monomers, which had high osmolarity, with an in-solution iodine to particle ratio of 3:2. The first change involved the development of ionic dimers such as ioxalgic acid (Hexabrix), which in-solution has an iodine to particle ratio of 6:2 and therefore a lower osmolarity for an equivalent iodine dose. The next improvement came with the development of nonionic agents (monomers such as Iohexol known as Omnipaque and dimers such as Iodixanol or Acupaque). The monomer and dimers showed a large reduction in both osmolarity and in the associated side effects together with even higher iodine to particle ratios. Improved toxicological characteristics are associated with key alterations in structure and properties, namely: the absence of acid groups ($-COO^-$), low osmolarity, symmetry of the molecule and a high number of $-OH$ groups. However, as the molecular size increases there is a concomitant increase in solution viscosity. Because of the high viscosity of many nonionic contrast agents, they are usually warmed to body temperature before intravascular injection. These more recent aqueous iodinated CA have been reviewed (10,11). Table 1 summarizes the main structures of monomeric iodinated agents, Table 2 summarizes the main structures of dimeric iodinated agents, and Table 3 reviews the salient properties of each.

## Main Areas of Use of Microencapsulated X-Ray Agents

Microencapsulation of X-ray CA has advantages over use of unencapsulated agents, primarily in the area of targeting to the liver and embolization:

1. Computed tomography (CT) scans of the liver are currently one of the most important techniques for examination of metastatic lesions. Without the use of contrast, greater than 10% of focal hepatic lesions can go undetected (12). The use of iodinated CA such as meglumine diatrizoate, iothalamate, and metrizoate also does not, however, significantly improve attenuation differences between parenchyma and lesions, because the agent distributes uniformly throughout the vasculature and the interstitial spaces of the tissue (12). In addition, the agents are rapidly cleared from the liver, with optimal conditions for image acquisition occurring only within 2–3 minutes after injection, giving insufficient time for an entire liver scan. Particulates such as microencapsulated CA, however, are usually taken up by the reticuloendothelial system (RES) and with time become concentrated in the liver and spleen. Since few tumor cells have an RES, they are not enhanced by CA and appear as areas of decreased density (13–15). Microencapsulation concentrates the agent in the liver, and dramatically increases the residence time.

2. An additional area in which microencapsulated X-ray CA have been widely investigated is for the therapeutic use of transient or permanent embolization. Embolization is used for treatment of hemorrhages and hemoptysis, arteriovenous malformations, and malignant neoplasmas, and has been reviewed by Benoit and Puisieux (16). Puisieux holds a patent describing the use of cellulose embolization gel (17). Usually postoperative follow-up is conducted by angiography, but if the embolus possesses X-ray contrast properties, angiography can be avoided and the embolus itself can be followed (18–21).

## Microcapsule Types Developed for X-Ray CA

**Hydrogels.** The use of poly(2-hydroxyethyl methacrylate) (PHEMA) microspheres to which a radiopaque marker was chemically attached was suggested by Horak

**Table 2. Main Structures of Typical Dimeric Iodinated X-Ray CA**

| Basic dimer structure |
|---|

| Group | Structure/significance |
|---|---|
| X | $-CH_2-$; $-CH_2-O-CH_2-$ |
| $R^a$ | D,L, Threitol: $HOCH_2-CHNH-CHOH-CH_2OH$ |

[a]Ionic or nonionic depends on acid group substitution of both benzene rings

**Table 3. Properties of Iodinated CA of 300 mgI₂/mL at 37°C**

| Class of CA | $I_2$:particle ratio | Typical osmolarity mOsmols/kg water | Viscosity (cps) |
|---|---|---|---|
| Ionic monomer | 3:2 | 1,500–2,000 | 2.4–5.0 |
| Ionic dimer | 6:2 | 690 | 6.0 |
| Nonionic monomer | 3:1 | 580 | 4.6–5.7 |
| Plasma | — | 300 | 1.2 |
| Nonionic dimer | 6:1 | 272 | 9.0 |

et al. for preparation of contrast-containing embolic material (19). Using 3-acetylamino-2,4,6-triiodobenzoic acid attached to the reactive hydroxyl functions, they prepared microcapsules with greater than 40 wt % iodine content. However, the resulting capsules were hard, brittle, and nonswelling. Jayakrishman et al. attempted to circumvent these disadvantages by preparing highly porous microcapsules using suspension polymerization of 2-hydroxy-ethyl methacrylate in the presence of polymeric diluents such as poly(methyl methacrylate) (PMMA) and poly(tetraethyl glycol) (18). The radiopaque markers iothalamic and iopanoic acid were esterified to the free hydroxyl groups, yielding greater than 30 wt % bound iodine. The same group has prepared similar PMMA capsules by alkaline hydrolysis of ethylene glycol dimethacrylate cross-linked PMMA, which were then impregnated with BaSO₄ by precipitation using barium chloride and sodium sulfate solution. More than 70 wt % BaSO₄ was incorporated into the capsules, which demonstrated good contrast properties. A loading of 40–50 wt % was reported in PHEMA micro-

spheres with a size range of 90 to 1500 μm, produced by a solvent evaporation method. Porosity was imparted to the microspheres by incorporation of the porogen sodium chloride (NaCl), and the BaSO₄ was trapped inside by cross-linking the capsule surface using γ-irradiation or homomethylene diisocyanate in n-heptane.

**Polysaccharides.** Starch microcapsules have been studied for many years as potential drug delivery and embolytic devices, and their synthesis has been extensively characterized (22). Rongved and co-workers describe the synthesis of the succinic acid derivatives of iohexol and iodixanol (monomeric and dimeric nonionic CA) bound to starch particles through an amino acid–ester bond (23). They used the method of Leegwater to prepare the starch particles from potato starch (24). Glycine derivatives had a shorter (4.5 h) half-life than alanine derivatives (8.6 h) in human blood serum, indicating that they would be superior CA for imaging of the liver. Mir et al. caused hepatic embolism by injecting starch microcapsules into the renal artery followed by injection of Renografin-60 in dog (25). They also noted a negative effect on the degradation rate of the starch capsules. Epenetos and co-workers used starch in a similar way to cause hepatic embolism, but this was to promote enhanced uptake of [131]I, which had been labeled with monoclonal antibodies to carcinoembryonic antigen (26). The iodine was used not only as the curative agent but also for imaging to monitor tumor uptake. Clinical improvement and a drop in circulating carcinoembryonic antigen was noted.

Another example of use of polysaccharide microspheres for embolization in conjunction with contrast material is presented by Dion et al. (27) with the use of dextran microspheres (Sephadex G-25, 40–150 $\mu$m, Sephadex G-50, 100–300 $\mu$m, Pharmacia Laboratories) suspended in 10 mL of equal parts of saline and contrast material. The resulting mixture was isogravitational, which facilitated injection, and the microspheres reached more distally, producing more complete occlusion. Ethyl cellulose microspheres containing 20% $BaSO_4$ (5–301 $\mu$m) were used by Yang and others (28) as long-term peripheral emboli for percutaneous maxillofacial arterial embolization in dog. At 2 and 6 months the microspheres were neither degraded nor resorbed and the occluded peripheral arterioles were not recannulized. Absence of focal necrosis of the bones and soft tissue indicates that these microspheres have the potential for maxillofacial artery embolization without significant cosmetic problems.

**Polyesters, Polyamines, Ethyl Ester, Polyanhydrides, and Polyurethane.** Biodegradable poly(DL-lactide) microcapsules (0.5–3 $\mu$m) containing either ethyliopanoate or ethyldiatrizolate have been evaluated for detection of liver metastases using CT (4). High liver uptake was reported, and capsules were actively taken up by the Kupffer cells. The same group has also investigated the use of poly(benzyl L-glutamate) microcapsules (PBLG), which possess available side chains for modification with functional groups (5). Maximum uptake of the capsules in the liver was 15–60 minutes postinjection, and contrast effect lasted up to 2 hours. PBLG-estrone conjugates of the same capsules displayed potential in estrogen receptor targeted therapy. The liver was also the target of a study involving microcapsules prepared from iodipamide ethyl ester (6,29). Using rats, they showed that the agent was actively accumulated in the liver and contrast remained high for over 2 hours, then gradually cleared in approximately 2 days. Leander et al. used the VX2 carcinoma model in rabbit to show the efficacy of a biodegradable particulate of (1'-Ethyloxycarbonyloxy)-ethyl-5-acetylamino-3-(N-methyl-acetylamino)-2,4,6-triiodo-benzenecarboxylate, based on a prodrug ester design of metrizoic acid, which accumulates rapidly in the liver (30). Some 40% of small tumors (2–4 mm) were detected, compared with 27% for iohexol and 15% for controls. Mathiowitz et al. have investigated radiopaque ($BaSO_4$) microcapsules prepared from alginate and polyanhydrides for nonhepatic use in a study of GI transit of capsules (31). The method allowed the investigators to pinpoint regions of the GI tract where high concentrations of capsules adhered. Nonbiodegradable capsules such as those made from polyurethane have been used to prepare microcapsules of 150–1500 $\mu$m for embolization, which contained tantalum as an imaging agent (32). The high atomic number of tantalum renders it considerably more radiopaque than either barium or iodine, and it is well tolerated by the body. The capsule surfaces were modified with methacrylic acid in order to increase hydrophilicity and improve ease of passage through micro catheters. The lack of biodegradability makes the embolism more permanent.

**Liposomes.** Liposomes, bilayered vesicles composed of phospholipids entrapping an aqueous space, have been ex-

tensively investigated for drug delivery. Being nontoxic, easily injected, and readily taken up by the RES, they are also a natural choice for development as CA, especially to target organs such as the liver (33,34). The group of Seltzer has studied the use of multilamella vesicles containing diatrizoate (13,35,36). The vesicles were 1.4–2 $\mu$m in size and contained 55.5 mg of iodine per mL. Maximum spleen enhancement in rat was 6 times that for unencapsulated agent. Marked liver and spleen opacification was measured 60–90 minutes after injection. In another study from the same laboratory, the liver, bile ducts, gallbladder, spleen, and GI tract were enhanced with use of entrapped Iosefamate, which is excreted through the hepatobiliary system when it is released from the liposomes (37). Later, Seltzer's group showed that the phospholipid composition (phosphatidylcholine/cholesterol/stearylamine [PC/C/S] compared with distearoyl phosphatidylcholine/sphingomyelin [DSPC/SM]) but not the nature of the entrapped iodinated species (ionic[$^{125}$I] diazotriate or nonionic labeled iotrol) could influence the in vivo biodistribution and the in vitro permeabilities of the liposomes in the presence of serum (38). The leakage of $^{125}$I activity was 2 to 3 times greater from PC/C/S liposomes and after injection into rat, the clearance half times for $^{125}$I activity from the liver, spleen, and whole body were 4.4, 4.5, and 2.8 hours, respectively. For DSPC/SM the clearance was significantly slower, at 24.0, 18.4, and 17.2 hours, respectively. Other groups have used multilamella vesicles (MLV) (39) with similar results as well as large unilamella vesicles which have a higher entrapment capacity (40,41). An alternative approach employed liposomes made with labeled lipids. The use of brominated lipids (42) and iodocholesterol (43) has demonstrated poor opacification, but ethiodol was more successful (43).

**Chylomicron Remnant-Like Polyiodinated Triglycerides.** Weichert and his group at Michigan have reported on a particulate contrast agent which exploits the natural lipid transport system to specifically target the liver (44–48). The approach involves synthesis of a polyiodinated triglyceride (ITG) which is then assembled into the core of a synthetic chylomicron remnant by processing in a microfluidizer. The resulting microemulsion (ITG-LE) has a mean diameter of between 200 and 300 nm. Cholesterol was found to produce smaller, more stable emulsions, and inclusion of $\alpha$-tocopherol helped protect the phospholipids that were present. The ITG-LE is metabolized in the hepatic parenchyma and eliminated through the bile. This allows thorough examination of the hepatobiliary system using standard or spiral CT. In a rabbit/VX2 carcinoma model, at CT enhanced with triglyceride (especially when combined with iohexol), sensitivity values and liver to lesion attenuation differences were greater with lower iodine doses than with iohexol or at CT during arterial portography.

**Ethyl Ester Derivatives of Diatrizoic Acid.** Another nanometer scale particle of 200–400 nm has been developed for blood-pool and liver imaging. It consists of ethyl ester derivatives of diatrizoic acid which are milled in the presence of different surfactants (49). The balance between blood-pool and liver-spleen enhancement could be controlled by modifying the surfactant used. Complete elimination of the agent was within a week of administration. All formulations produced superior enhancement of the blood-pool and liver-spleen compared with iohexol, and persisted for at least 30 minutes after injection. These nanoparticles have also been used to advantage for passive targeting of the lymphatic system, with demonstration of intranodal architecture (50–52). In a 1995 study, Wisner et al. examined the anatomic and temporal distribution of lymph node opacification after subcutaneous, submucosal, or intestinal injection of this agent for indirect lymphography of subdiaphragmatic lymph nodes in normal dogs (53). CT examinations revealed excellent opacification and visibility of lymph nodes after contrast administration. Advantages include visualization of some internal architectural detail because of the nonuniform distribution in larger nodes. The mechanism is not known, but is presumed to involve particle uptake by lymph node macrophages.

## Contrast Imaging with Therapeutic Modalities

Many innovations in particulate CA development have their routes in techniques developed for drug delivery. It is an attractive concept that a therapeutic agent could be injected at the same time as administration of the CA, and even that the mechanisms involved in imaging could facilitate and/or enhance drug or even gene delivery at a particular site. One example of concurrent CT imaging and treatment, the practice of chemoembolization (CE) using radiopaque microspheres, has already been mentioned for treatment of hemorrhage (16,17). Radiopaque emulsion of iodized oil (lipiodol) have also been described in the literature for this purpose. Rougier and co-workers report Phase I–II French trials for treatment of primary metastatic liver carcinomas with CE performed during hyperselective arteriography using gelfoam and hyperfluid lipiodol with interarterial adriamycin (54). Two Japanese groups reported favorable results, one using transcatheter hepatic subsegmental arterial chemoembolization (TAE) with iodized oil (Lipiodol-TAE) for small hepatocellular carcinomas (55), and the second using selective arterial infusion of styrene maleic acid neocarzinostatin/lipiodol for renal cell carcinoma (56). In both cases tumor growth was inhibited, and the treatment was followed closely with the contrast-enhanced imaging. However, other studies show that not all drug–CA combinations in chemoembolization are beneficial. Dehmer et al. showed that ipohexol impaired the action of fibrinolysis by streptokinase, urokinase, and by recombinant tissue-type plasminogen activator (57).

Investigators at Massachusetts General Hospital and Harvard Medical School have been exploiting the unique characteristics of micellar carriers for therapeutic and diagnostic agents (58,59). They synthesized amphiphilic AB-type block copolymers of methoxy poly(ethylene glycol) (MPES) (12 kDa) and poly[$\epsilon,N$-(triiodobenzoyl)]-L-lysine which formed ~80 nm micelles in water, with a 44% iodine content. Rabbit studies showed that IV injection (250 mg iodine/kg) gave significant enhancement of blood-pool,

liver and spleen, and a prolonged half-life of about 24 h. Biodistribution and CT-imaging studies in rats have been described (60). Significant and sustained enhancement of blood-pool (aorta and heart), liver and spleen were observed for a period of at least 3 hours. Figure 1 shows CT images demonstrating the effects of iodine-containing polymeric micelles in rat.

## MRI CONTRAST AGENTS

### Physical Principles

The nuclei of certain atoms can act like minute magnets and become lined up when placed in a magnetic field. MRI measures the energy required to change the alignment of the nuclei in a magnetic field. An external field is applied across a part of the body to align the magnetic nuclei in the longitudinal direction, and a radio wave is broadcast into the tissue in the transverse direction, which induces some of the magnets to tilt over, usually by either 90° or 180° depending on the duration of the pulse (61). The radio wave is turned off and after some time the nuclei retransmit the signal at the same frequency as it was received, as they return to their equilibrium orientation. The time required for the net tissue magnetization vectors to return to equilibrium conditions in the external magnetic field after a 180° resonant frequency pulse is known as the $T_1$ or longitudinal relaxation time. $T_1$ is also known as the spin lattice relaxation time since during the relaxation process relaxation energy in the form of heat is lost through molecular collisions to the surrounding tissue (lattice). A second parameter $T_2$, the transverse or spin-spin relaxation time, is a measure of a dephasing phenomenon taking place between the protons as they precess about the transverse plane. It is somewhat analogous to the tone that one would hear if a number of tuning forks, all with a note close to but slightly different from each other, were struck simultaneously. There would be an initial loud tone, followed by a waxing and waning muted tone that was rapidly damped due to the destructive interference of the different wave forms. Following a 90° resonant frequency pulse the protons precess together in the x–y plane. If all the protons had exactly the same resonant frequency they would produce a strong signal proportional to the number of nuclei. Small differences in resonant frequency, however, result in a rapid dephasing of the signal with a time constant $T_2$. $T_2$ is a measure of how long the substance maintains the transverse magnetization, which is perpendicular to the external field (61). $T_2^*$ is an effective time constant governing decay in an inhomogeneous external field, and it is smaller than $T_2$.

In MRI the majority of the signals arise from protons of cellular water and lipids. Natural contrast is a direct result of inherent differences in the various microenvironments of the constituent protons. CA can alter this contrast by affecting any of three parameters, $T_1$, $T_2$ and $\rho$, the nuclear spin density. The majority of CA for MRI change $T_1$ and $T_2$, although the spin density has been shown to be changed by injection of furosemide, olive oil, or clomiphene (62).

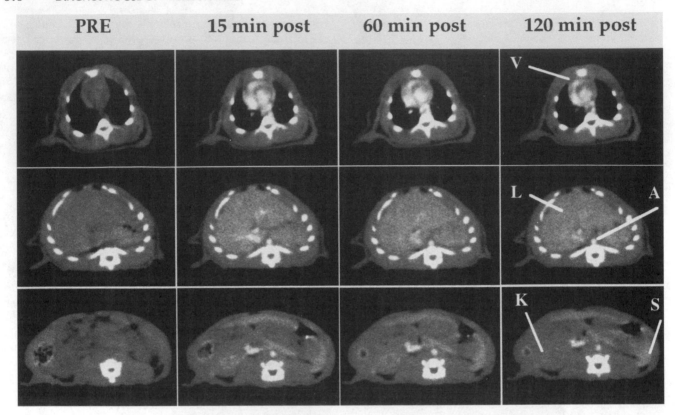

**Figure 1.** CT images demonstrating effects of iodine-containing polymeric micelles in various rat organs (V, ventricle; L, liver; A, aorta; K, kidney; S, spleen) at different postinjection times. MPEG-iodomicelles were injected into the rat tail vein at a dose of 170 mg I/kg. *Source:* From Ref. 60 with permission of the publishers.

There are three types of CA for MRI: paramagnetic, superparamagnetic, and ferromagnetic. Addition of small amounts of paramagnetic material to the microenvironment of the proton brings about a contrast effect. Paramagnetics affect both $T_1$ and $T_2$. At the low concentrations usually used in imaging the $T_1$ shortening effect predominates and the change in tissue is best seen on $T_1$-weighted images. The magnetic fields induced in paramagnetics by the external field are additive to that of the applied field and the increased signal intensity is known as positive enhancement. Paramagnetics also produce local magnetic inhomogeneities and thus shorten $T_2$ relaxation times as well. At high concentrations such as would be found in the urine, the $T_2$ shortening predominates and $T_2$-weighted images are preferred. Certain metals, including manganese(II), iron(III), and gadolinium(III), and stable free radicals such as nitroxides and melanin contain unpaired electrons which impart strong paramagnetic properties. Paramagnetic ions have magnetic dipoles 1,000 times that of protons, and the predominant mechanism of proton relaxation is through dipole–dipole interaction with the surrounding spin lattice (62). Appropriately fluctuating local magnetic fields generated in the magnetic spin lattice by random rotational and translational motions promote relaxation. Unpaired electrons can greatly influence relaxation times because the magnetic moment of an electron is 657 times larger than the magnetic moment of a proton. The increase in relaxation rate is proportional to the concentration of the paramagnetic agent and to the square of the magnetic moment and inversely proportional to the sixth power of the distance between the paramagnetic center and the proton to which it is bound (11). The formula for relaxation rate is given by Gadian and Bydder (63):

$$1/T_{1,\text{obs}} = 1/T_{1,\text{o}} + 1/T_{1,\text{p}} \qquad (1)$$

where $T_{1,\text{obs}}$ is observed relaxation time, $T_{1,\text{o}}$ is intrinsic relaxation time, and $T_{1,\text{p}}$ is the contribution the CA makes to the relaxation rate.

### Early CA for MR Imaging

First-row transition metals with partially filled 3d electron shells and lanthanide metal ions with partially filled 4f electron shells are possible paramagnetic ions. Mn(II), Cr(III), and Fe(III) have 5, 3, and 5 unpaired 3d electrons, respectively; however, Gd(III) of the lanthanides with 7 unpaired electrons has been found to be most effective. Because of its toxicity, gadolinium has to be chelated to, for example, diethyltriamine pentaacetic acid (DTPA), making Gd DTPA. The first clinically approved agent in the U.S. was gadopentetate dimeglumine (Magnevist®; Berlex Laboratories, Cedar Knolls, NJ), approved in 1988. The agent moves from the blood vessels to the tissue by diffusion, except in the brain and spinal cord. The plasma half-life is 90 minutes, and it is excreted intact by glomerular filtration (64). The structure is shown in Figure 2. The chelating agent on its own is also toxic, because of its high affinity for cations.

Development of paramagnetic MRI contrast agents has followed a similar pattern to that for X-ray agents, with a search for less ionic chelating moieties. Some examples include Gd DTPA-BMA (Omniscan; Salutar, Sunnyvale, CA) and Mn-DPDP (Salutar, Sunnyvale, CA), Gd-DOTA meglumine (Gadoterate; Guerbet Laboratories, Aulnay-sous Bois, France), Gd-BOPTA (Bracco, Milan, Italy), EOB-DTPA (Schering, Berlin, Germany), and HP-DO3A (Gadoteridol; Squibb Diagnostics, Princeton, NJ). The more lipophilic agents are excreted by both the kidney and liver.

Ferromagnetic materials consist of a crystalline array of many paramagnetic elements that behave in synchrony when placed in an external magnetic field. A subset of these are superparamagnetic particles that lack magnetic memory. Ferrites and magnetites ($Fe_3O_4$) belong to this class. The array of paramagnetic elements produces a localized disturbance in magnetic field homogeneity, which affects the proton relaxation, causing a rapid dephasing of the spins. Superparamagnetic and ferromagnetic agents reduce $T_2$ relaxation much more than $T_1$ relaxation, which can be a disadvantage because it causes a decrease in signal intensity on spin-echo images (65). Although $T_2$ is reduced, their main reduction is in $T_2^*$, hence the effects are best observed on gradient echo sequences where $T_2^*$ effects are retained. Dysprosium and its chelates have a similar effect. Ferromagnetics are referred to as negative enhancers because they cause a local blackening of the image in $T_2$-weighted and gradient-echo pulse sequences. Both superparamagnetic and ferromagnetic agents are in the form of particulates and are therefore not suitable for neuroradiological uses. Superparamagnetic agents have a smaller particle size; however, if ferromagnetics are reduced in size to below 10 nm they lose their permanent magnetic characteristics and become superparamagnetic. It is ferro- and superparamagnetic particles that have been the subject of the majority of microencapsulation research in MRI.

The state of the art in MRI contrast has been reviewed (66–68). Specific areas of contrast use have been reviewed for blood-pool imaging (69), metastases to the liver (70), myocardial perfusion (71), cerebral perfusion (72), pediatric neuroimaging (11), and the brain (73).

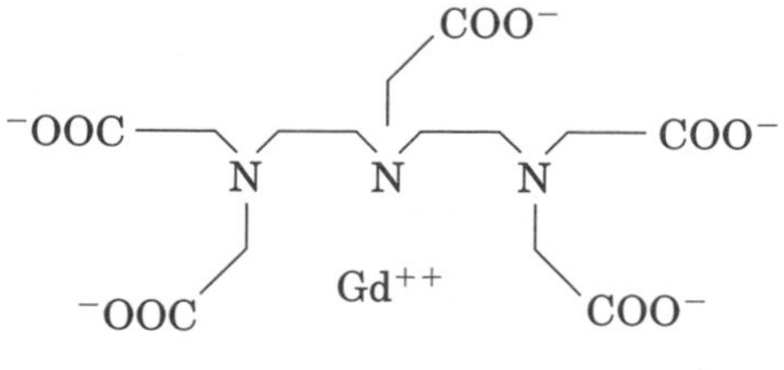

**Figure 2.** The structure of Gd DTPA.

## Main Areas of Use for MRI CA

Brasch (66) defines seven main areas of use for MRI CA. Many parallel areas of use for CT agents and current research are ever expanding the applicability of CA in MRI. The main areas of all contrast use in MRI are briefly reviewed here.

1. Extracellular agents (small molecular weight agents such as Magnevist described previously) comprise the gadolinium-based agents and are used primarily to define extracellular fluid spaces and defects in the blood-brain barrier on $T_1$-weighted images (66,72). After intravenous injection, the agents diffuse from the blood vessels into the surrounding extravascular spaces. Encapsulation of these agents can increase the blood-pool and total residence time and alter the biodistribution.

2. Blood-pool and capillary-integrity markers rely on high (>20,000 daltons) molecular weight agents (69) which escape only slowly from the vascular compartments. Approaches to macromolecular contrast media (MMCM) include ex vivo labeling of blood cells or plasma components or use of modified agents in the form of macromolecules (albumin-(Gd-DTPA)$_{30}$, dextran-(Gd-DTPA)$_{15}$, polylysine-(Gd-DTPA)$_{30}$), colloids, and liposomes. They have been used to define tissue ischemia and/or the adequacy of reperfusion procedures in the heart, lungs, kidney, and brain. One advantage of these agents is that they can be synthesized with a high number of reporter groups (i.e., the paramagnetic agent such as Gd).

3. MRI is used in the heart to assess myocardial perfusion (71). This is not possible without the use of CA. Newly developed ultrafast imaging sequences are expected to greatly assist in this measurement.

4. In the same way as in CT, there is extensive research into developing CA for the liver (70). Agents can be directed either to the hepatocytes or the RES. Certain paramagnetic complexes of Fe(III) and Gd(III) have been shown to be taken up by hepatocytes through a nonspecific anion transporter and subsequently secreted in the bile (74). Other agents with hepatobiliary excretion include Gd-BOPTA, Gd-EOB-DTPA and Mn-DPDP. Superparamagnetic iron oxide particles coated with asialoglycoprotein (ASG) have also been specifically targeted to normal hepatocytes (75). Particles that are targeted to the RES include paramagnetic liposomes (76) and dextran-coated superparamagnetic iron oxide particles (77). Enhancement is achieved in less than an hour with iron oxide particles and lasts for more than a day. This advance has reduced the size threshold for detection from a 10 mm lesion down to 3 mm. Some concerns do, however, exist about reports of hypotension at high CA doses.

5. Again, by the same mechanisms as with CT agents, very small particles can travel to the lymph and act as CA. One route is via subcutaneous injection (78); the more usual route is intravenous injection (79). Weissleder (79) used ultrasmall superparamagnetic

iron oxide particles (USIPO) with a mean diameter of 11 nm. Phagocytic uptake in the normal nodes results in low signal intensity in $T_2$-weighted MRI, compared to high intensity in nodal metastases.

6. Tumor-specific and antigen-specific enhancement is also under investigation. Studies report on the attachment to monocrystalline iron oxide nanoparticles (2.9 nm) of antimyosin Fab fragment directed against infarcted myocardium and human polyclonal immunoglobulin G directed against abscesses, and of monoclonal anti-Burkitt lymphoma antibody attached to Gd-DTPA (66). An interesting targeted molecule has been investigated which uses the observation that phosphate-containing compounds accumulate in bone and soft-tissue calcifications (80). In this study, gadopentetate dimeglumine modified with 1-hydroxy-3 aminopropane-1,1-diphosphonate was used to image infarcted myocardium with a high degree of success.

7. Finally, agents are being developed for use in the GI tract. Positive enhancers have been administered both orally and rectally in the presence of mannitol to increase the osmolarity and minimize dilution by intestinal contents, for example Gd-DTPA (81). An inexpensive agent consisting of ferric ammonium citrate (Geritol) has also been suggest for oral and rectal administration (66,82). In other studies negative enhancers have been used in the bowel and including $BaSO_4$ (83), kaolin (84) and perfluorooctylbromide (Imagent®; Alliance Pharmaceuticals, San Diego, CA) (85). Iron oxide particles and crystalline iron oxide on an organic polymer matrix have also been studied (86). An oral agent consisting of 3.5-$\mu$m-size sulphonated-divinylbenzene copolymer coated with superparamagnetic iron oxide has been developed by Nycomed and reported on by Van Beers and his group (87). They reported significantly improved delineation of lesions, small bowel and paraaortic regions post contrast. An agent consisting of 2–5 $\mu$m magnetic albumin microspheres (MAM) has been shown to be orally stable over 24 hours in rabbits and dogs (88). Possessing a larger magnetic moment than paramagnetic agents, the transverse relaxation rate (R2) can be as much as 40 times the longitudinal rate (R1). MAM cause marked signal loss in the stomach and small bowel on both $T_1$- and $T_2$-weighted pulse sequences. They are promoted as reducing motion artifact throughout the GI tract. Undoubtedly the best tasting oral CA that has been proposed recently is blueberry juice, which is high in Mn(II) and acts as a positive CA on $T_1$-weighted images and negative CA on $T_2$-weighted images (89).

## Microcapsule Types Developed as CA for MRI

**Macromolecular Contrast Media.** MMCM inhibit diffusion of the imaging species and slow renal filtration, greatly reducing the stringent requirement for timing of image acquisition and permitting imaging of multiple body regions without need for further dosing (90). The majority of the agents being investigated involve chemical linking of chelated imagers such as Gd-DTPA to either exogenous macromolecules such as poly-L-lysine, and dextran, or endogenous blood components such as red blood cells or human serum albumin. The resulting complexes have very high dosing efficiency due to the multiplicity of imaging moieties per molecule (>60). In addition, the combination with a macromolecule slows the molecular rotation so that it more closely resembles that of water.

Poly-L-lysine is an attractive candidate for MMCM because of the large number of $\epsilon$-lysine residues available for substitution with either agent or a targeting moiety. Low molecular weight polymers (e.g., PL-Gd-DTPA$_{40}$) are not satisfactory because of rapid clearance (91). Higher molecular weight agents have longer circulation times, but also demonstrate renal accumulation, which could limit their use (92). PL-DTPA has been investigated for cardiac (93–95), pulmonary (96–99), and renal (100,101) use. Grandin et al. (101) compared the imaging characteristics of Magnevist with the Schering poly-L-lysine conjugate using $T_1$-weighted spin-echo images in the rat hepatic tumor model. Peak intensity of the MMCM was reached at 3 minutes and remained high at 60 hours, whereas at 3 minutes tumor-to-liver contrast of the Magnevist was insignificant and baseline was already achieved by 60 hours. The same group conducted a similar study to investigate reperfused ischemia of the rat intestine (102). They reported that contrast between reperfused and normal intestine was higher on PL-DTPA-enhanced images than on unenhanced or Gd-DTPA-enhanced $T_1$ and $T_2$-weighted images. Because previous studies showed that no enhancement was evident during total vascular occlusion, they suggest this as a potential method of detecting reperfused ischemia of the intestine.

Weissleder et al. (103) studied the uptake of partially derivatized Gd-DTPA by the adrenal glands, based on the theory that positive charge is essential for adrenal uptake. The highest uptake was in the liver, followed by the adrenal glands and the spleen. At two hours the adrenal uptake of the partially derivatized agent was 3.4 times higher than the fully derivatized one. The same group has synthesized and evaluated the long-circulating copolymer, methoxy poly(ethylene) glycol-succinyl-($N$-$\epsilon$-poly(L-lysyl) DTPA) (MPEG-PL-DTPA) (52,104–106). The agent was found to be nontoxic, and although it binds fewer Gd per molecule compared with PL-Gd-DTPA with the same degree of polymerization, this is partially compensated for by an increase in atomic relaxivity, ($R_1 = 12$ (mmol s 1)$^{-1}$ for PL-Gd-DTPA and $R_1 = 18$ (mmol s 1)$^{-1}$ for MPEG5-PL-Gd-DTPA at 20 MHz, 37°C. High-quality, high-resolution images were obtained in rabbit and rat using a combination of three-dimensional time of flight angiographic technique and MPEG-PL-Gd-DTPA injection. Postinjection images clearly show blood vessels that were absent in precontrast images (69). Equally significant results were obtained in pulmonary magnetic resonance angiography and imaging of tumor neovascularity. The copolymer has also been used for passive targeting to solid tumors (106). A group at Cal Tech has even used poly-L-lysine with bound DNA and Gd-DTPA to develop a particle that delivers both the transfecting gene and an MRI CA to enable the monitoring of DNA for gene therapy in real time (107). PL-Gd-

DTPA has also been conjugated with antihuman antibodies to demonstrate increased signal intensity from experimental human tumors implanted in nude mice (108). Finally poly(glutamic acid, lysine, tyrosine) (PEKY) has also been used to coat superparamagnetic particles followed by attachment of carcinoembryonic antigen (final radius ~50 nm) for tumor-specific targeting (109).

Some groups have chosen dextran linked to Gd-DTPA to increase the CA molecular weight and prolong its residence time in the blood (110,111). Blustajn and co-workers investigated carboxymethyldextran-Gd-DTPA (Guerbet, Aulnay, France) to measure liver blood volume with $T_1$-weighted spin-echo sequences (112). They reported remarkable stability of signal enhancement over time, allowing the summation of a large number of measurements. However, not all reports involve chemical linkage of the high molecular weight polymer to the chelate. Simple mixing of dextrose or poly(ethylene glycol) 3350 or 8000 has been shown to improve the relaxivity of Gd-DTPA and Gd-HP-DO3A, and this improvement followed a geometric function of concentration (113).

Albumin-labeled Gd-DTPA CA have been reported with long half-lives (3.6 h) in rats, and have been found to be more potent than the corresponding individual chelates (114–117). However their usefulness is limited due to concerns over cardiovascular toxicity, immunogenicity, and retention of Gd for long periods of time in the liver and bone (118). More recently a related agent, MS-325 (Epix Medical, Inc, Cambridge, MA) has been described in preclinical evaluations in rabbits and nonhuman primates (119). MS-325 consists of a diphenylcyclohexyl group attached to Gd chelated by a phosphodiester linkage which binds reversibly to serum albumin. A 5 to 10 times higher reduction in monomolecular tumbling of the chelate upon binding compare to Gd-DTPA is reported. The reversible binding results in a small volume of distribution, increased plasma half-life, and reduced extravasation while preserving efficient excretion due to the reversible nature of the binding.

**Starch Microspheres.** Starch has been used to microencapsulate superparamagnetic oxide particles (SPIO) and USPIO, usually for passive targeting to macrophages (120–123). Nycomed (Oslo, Norway) has been active in this area. Magnetic starch microspheres (MSM), with a mean size of 0.3–0.4 $\mu$m have been shown to be well suited as $T_2$ CA at 0.5T and 1.5T using conventional spin-echo and fast turbo spin-echo (120). The same MSM were used by Kreft to differentiate diffuse and focal splenic disease (121). One hour after injection of 20 $\mu$m/kg, lymphomatous spleen showed significantly ($P < 0.001$) reduced enhancement relative to normal splenic tissue. Diffuse lymphoma (signal-to-noise ratio, SNR: 10.3 $\pm$ 1.7) could easily be differentiated from control animals (SNR: 5.5 $\pm$ 0.6) on $T_2$-weighted images. Rongved and co-workers used Gd-DTPA-loaded starch particles cross-linked with epichlorohydrin and found the in vitro spin-lattice relaxivities were 40–260% higher than that for Gd-DTPA (122). The relaxivity increased with decreasing Gd content. The in vitro degradation rate of the particles decreased with decreasing Gd content, with half-lives of from 24 h with 12.2 wt % Gd(III)

to 0.67 h with 1.1 wt %. The exact mechanism was not clear. In in vitro studies, MSM have been shown to accumulate in macrophages were they were localized to the lysosomes (123). The accumulation was highly competed by regular microspheres, suggesting a receptor-mediated mechanism. In contrast to unencapsulated SPIO, the MSM were found to be nontoxic to the cells. Colet et al. (124) expanded on this work using perfused rat liver to show that binding to Kupffer cells is caused by membrane lectins.

**Dextran-Coated SPIO.** Dextran-coated SPIO have been widely studied, especially for targeted CA. Many patents have appeared such as that assigned to Advanced Magnetics, Inc. (125), and monocrystalline iron oxide from Weissleder (126). A French group at Pouliquen's laboratory has used dextran-coated SPIO encapsulated in isobutyl cyanoacrylate to target the liver and spleen, where they report a sevenfold decrease of $T_2$ relaxivity rate and a halving of the $R_2/R_1$ ratios (127). Weissleder's group at Massachusetts General have studied the effect of particle surface on the biodistribution and have synthesized targeted CA including agents to pancreatic receptors (128–130). Dextran-coated SPIO with diameters of 4.4–6 nm were modified with poly-L-lysine (PL), and in some cases the PL was then succinilated. They found that negatively charged and neutral particles had the longest circulation times. Linking cholecystokinin to 430 nm spheres by a non-covalent method resulted in a 50% drop in blood half-life to 20 minutes, and shortened in vitro $T_2$ relaxation times (129). Pancreatic $T_2$ relaxation times decreased in a dose-dependent fashion as demonstrated by ex vivo imaging. Similar dextran-coated SPIO were demonstrated to be taken up by receptor-mediated endocytosis when opsonized with albumin (130). In angiography, their dextran-coated monocrystalline iron oxide (MION-46) nanoparticles resulted in a three to fourfold increase of aortic SNR and markedly improved the quality of images of the vasculature of the lungs, abdomen, and extremities (131). This was attributed to the long plasma half-life (180 minutes in rats). Long-circulating iron oxide particles for MRI have been reviewed (132). The particles are particularly useful for lymph node imaging, as seen in Figure 3 which shows

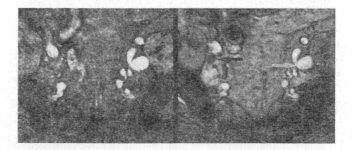

**Figure 3.** SPGR images of the pelvis before (left) and 24 h after (right) intravenous administration of long-circulating iron oxide particles in a patient with primary prostate cancer. The lymph node metastasis (white arrow) is of much higher signal than the contralateral normal lymph node (black arrow). *Source:* Courtesy of Dr. R. Weissleder.

the spoiled grass (SPGR) images of the pelvis before and after MION injection. The same MION agent has recently also been described for use in in vivo cell trafficking studies (133). One advantage of the use of MION lies in the fact that the cells internalize the label, leaving the cell surface receptors free for adhesion. Roughly $5 \times 10^6$ particles were taken up per lymphocyte, and they were retained for at last 3 days. Labeling did not affect viability or biodistribution. Detection threshold was determined to be $2.5 \times 10^6$ cells/30 $\mu$L sampling volume. Lymphocyte in the spleen resulted in MRI signal intensity changes that were readily detectable by MRI.

Other groups have developed dextran-coated SPIO linked to antibodies directed against leukocyte common antigen (CD45) by biotin-streptavidin binding to successfully target brain tumors (134). A recent report of the use of dextran-coated USPIO (AMI-227 Combidex; Advanced Magnetics Inc., Cambridge, MA; also known as ferumoxtran) in humans describes improved myocardial/blood-pool contrast with a trend toward contract-to-noise ratio improvement in the short axis and significant improvement in the long axis cine by an average of 128% (135).

**Actively Targeted Microspheres.** In a 1992 review of drug targeting in MRI, Weissleder et al. conclude that "target selectivity and receptor and/or antigen specificity is the most important issue if functional MRI is to establish itself in the next decade" (136). Although the review includes passive targeting, they list carriers including antibodies, proteins and polysaccharides, cells and liposomes, and targets ranging from colorectal tumor, heart (myosin), liver (ASG receptor), autonomic nervous system, peripheral nervous system (axonal transport), inflammatory cells, receptors (fucose, manose, ASG), inflammation, lymph nodes, cell traffic, and liver and spleen. Many actively targeted imaging microspheres have been described elsewhere in the literature (101,104–108,130,134). In addition, magnetite nanoparticles with a PEKY coating (see also Ref. 109) have been developed which have then been cross-linked with the homobifunctional regent ethleneglycol bis(succinimidylsuccinate) followed by coupling to carcinoembryonic antigen (137). Poly(ethylene glycol) was used to prolong the particle circulation time as were mucin, albumin, $N$-acetyl neuraminic acid, glucuronic acid, or glycophorin. The ability of these protective agents to prevent rapid elimination rates was disappointing, although the relaxivity was very good, between 300 and 500 mM$^{-1}$ sec$^{-1}$. Work on an ASG receptor–targeted agent consisting of arabinogalactan-coated USPIO made by coating BMS 180550, (developed by Advanced Magnetics, Inc., Cambridge, MA) and obtained here from Bristol-Myers-Squibb, has been described by Leveille-Webster and co-workers (138). The particles were targeted specifically to hepatocytes, as opposed to Kupffer cells, which are the usual target in the liver. The goal was to develop an agent that could be used to monitor the regenerative responses after acute liver injury. A similar ASG-USPIO was reported by Reimer and co-workers in a comparative study of three agents, the hepatocyte targeting agent, a USPIO targeted to the lymph and bone marrow, and an RES-directed iron oxide agent (139). They showed that signal intensity changes,

which reflect local change on relaxivity, were not affected by intracellular redistribution such as clustering.

The nanomolar, or high affinity, folate receptor is expressed or up-regulated on tumors of epithelial origin including ovarian, endometrial, and breast cancers. This fact has been exploited by Wiener and co-workers who developed Gd complexes of folate-conjugates dendrimer-conjugates (140). They showed that cells accumulate the conjugate in a receptor-specific manner. Although no imaging studies were reported, the approach looks promising for tumor targeting of MRI CA.

An abscess-specific CA has been investigated which involves the labeling of neutrophils with SPIO particles (141). The particles had to be coated with polystyrene (Estapoe® M1-0.07/60, 0.8 $\mu$m; Rhone-Poulenc, Combevoie, France) to prevent strong, irreversible cell aggregation. The coated particles became engulfed by the neutrophils and permanently sealed inside of them. After 48 hours postinjection, $T_2$-weighted images of an artificially induced abscess in rabbit muscle produced dark contrast around a characteristically bright center of the inflammation. However, care had to be taken to balance the number of ingested particles needed for adequate imaging with a noted decrease in chemotactic potential upon uptake of the particles.

Metaloporphyrins have been evaluated as positive MRI CA for a variety of tumors including human carcinoma, lymphoma, and fibrosarcoma (142–144). Tumor enhancement was evident in all cases. The long-circulating copolymer MPEG-PL-DPTA has also been investigated as a carrier for diagnostic and therapeutic drug delivery to tumors in vivo (106). The copolymer was labeled with both Gd and $^{111}$In. The adducts were succinilated to test the theory that a hydrophilic polymer may be able to overcome diffusion barriers in a solid tumor which are thought to be caused by interstitial polysaccharides which impede motion across tumor interstitium (145). The copolymer accumulates in solid tumors at the level of 1.5–2% injected dose/g of tumor in 24 h. The polycarboxylated copolymer was also associate with cis-diamminedichloroplatinum(II) by coincubation, and the resulting drug–copolymer complex showed a cytotoxic effect in mouse F9 teratocarcinoma, and induced reversal in tumor growth after intravenous administration. In vitro the release profile in the presence of blood components showed a burst effect over the first 23 minutes (20% release) followed by a slow, essentially linear release with a half-life of 79 h. Despite a fourfold reduction in activity as an inhibitor of F9 proliferation in cell culture, the adduct was remarkably successful in vivo, with all animals surviving 6 consecutive treatments with a resulting reduction of tumor volume to an average of 90% of pretreatment values. Two of the eight animals showed complete tumor regression. The pronounced toxic side effects observed in control animals treated with uncomplexed drug were notably absent in the treated population.

Pislaru reported on the use of bis-Gd-mesoporphyrin to evaluate the possibility of noninvasive measurement of infarct size after thrombolysis in a dog using the Cu-coil technique (146). The agent demonstrates a high affinity for necrotic tissue. In the animal model, strong enhancement (150–280%) was observed on $T_1$ weighted images, which closely correlated with postmortem triphenyltetrazolium chloride staining results.

Finally, reports have appeared in the literature that propose both CT and MRI imaging of antibody-labeled iron oxide particles and physical movement of these microcapsules in response to a magnetic field gradient (147). The antibody is cross-linked to silicone-coated iron oxide to give particles similar to BioMag™ (148). In rats, the injected particles could be transported through the cerebrospinal fluid and localized to the medial aspect of one or other cerebral hemisphere using an external magnet.

**Liposomes.** A paramagnetic species in free solution can directly relax only adjacent water protons. It follows that use of liposomes can only be effective with entrapped CA if protons can pass across the membrane or if the membrane is sufficiently leaky. This is usually achieved by use of small unilamella liposomes (SUV) with a small internal volume compared with the surface area, which facilitates water exchange (149). Theoretical expressions have been derived that allow calculation of water permeabilities of liposomes for use in CA preparation (150). Variations of liposome size and permeability have been shown to influence changes in relaxation rate in bulk water (151,152). In rats, liposomal Gd-DTPA caused significant improvement in contrast between liver and tumor, on $T_1$-weighted MR images, and smaller 70 nm liposomal Gd-DTPA caused the greatest contrast enhancement (153). This reflects the larger surface-area-to-volume ratio of smaller vesicles. Alternatively, the imaging moiety can be incorporated into the bilayer, so that that it is exposed on the outer surface, which is in contact with the aqueous environment (149). In either case, whether the agent is entrapped or exposed on the surface, liposomes act as blood-pool agents to increase the circulation time of the agent and to target it to the liver and spleen. Specific target molecules can also be attached to the liposome bilayer to create an immunoliposome, adding a further degree of sophistication to the agent.

Karlik and coworkers prepared SUV from phosphatidylcholine and dipalmitoyl phosphatidylethanolamine-DTPA, chelating either Gd(III) or the spin label, nitroxide free radical (154). In vivo in rats, the spin label significantly highlighted liver and spleen compared with noncontrast liposomes, and the enhancement increased with time. However, the Gd-containing liposomes were far more effective. The spin label had a short tissue half-life, compared with the Gd-liposomes which had a 24 h half-life in liver. The results are similar to those reported by Kabalka and by Schwendener, both using a sterylamine DTPA-Gd chelate (155,156). Unger claims the highest relaxivity for SUV prepared with Mn(II) complexes of DTPA, which were derivatized with hydroxylated acyl chains (149). They reported $R_1$ and $R_2$ values of over 20 mM$^{-1}$ sec$^{-1}$, some five times higher than Gd-DTPA.

Polymerizable paramagnetic liposomes have been proposed by Storrs et al. which have derivatized gadopentetate dimeglumine as the hydrophilic head group and diacetylene groups in the hydrophobic acyl chains which cross-link when irradiated with ultraviolet light (157). About 2–5% of the head groups are left uncoordinated with

Gd(III) to ensure that there is no free ion present. The liposomes were well tolerated by rats, and images showed an increase in the $T_1$-weighted signal intensity which persisted through the 90-minute experiment (20% in the liver and 34% in the kidney at one-sixth the clinical dose of Gd(III)).

Finally, liposomes that are sensitive to pressure have been suggested for use in cardiovascular MRI (158). They consist of gas-filled liposomes between 4 and 9 $\mu$m in size. Various gases have been investigated including nitrogen, argon, air, xenon, neon, pefluoropentane, and sulfur hexafluoride. Assuming ideal gas behavior, it can be shown that magnetic field perturbation is inversely proportional to the pressure. Pressure changes could cause the transverse relaxation of the surrounding medium to change. This does not take into account effects of any membrane around the bubble. The $1/T_2$ for gas-filled liposomes decreased linearly with pressure, but the effect was different for different gases. Oxygen (or air) had the greatest effect on $T_2$ relaxation, which was attributed to either bubble stability or the inherent paramagnetic properties of oxygen. Although the effects are far smaller than those noted in ultrasound, these gas-filled bubbles are considered to be a first step in developing pressure-sensitive MRI agents.

**Microcapsules for Fluorine MRI.** The use of $^{19}$F in MRI allows for contrast-only imaging since there is no background signal from the tissue. Fluorine is highly sensitive and the CA, usually perfluorocarbons (PFC), are low in toxicity. Different mixtures of PFC with melting points around 37°C can be used for temperature mapping due to the absence of signal in the solid phase. Further, because oxygen is highly soluble in PFC, and because dissolution is accompanied by a decrease in $T_1$, fluorine MRI has been used to measure the partial pressure of oxygen, an area that has been reviewed by Clark (159). Most imaging agents are fluorocarbon emulsions such as perfluorotributylamine. Sotak suggested the use of perfluoro-2,2,2',2'-tetramethyl-4,4'-bis(1,3-dioxalane) (PTBD) to overcome some of the problems of earlier agents such as multiple chemical shifts and short $T_2$ (160). Despite the superior properties of PTBD, PFC emulsions can not be targeted away from the RES, and only contain 10–40% (w/v) of active fluorocarbon. This means that multiple doses are often required, and the SNR is frequently insufficient to measure oxygen pressure.

Webb and co-workers have reported the development of sonochemically produced protein microspheres to overcome these problems and to allow for the flexibility of surface modification to redirect the biodistribution of the agent (161). The bovine serum albumin spheres have a mean diameter of 2.5 $\mu$m, are filled with liquid perfluorooctane, and are similar to the air-filled microspheres used in ultrasound known as Albunex®. They are produced by high power ultrasound irradiation at the interface of an aqueous protein solution and a nonpolar liquid. They target the RES and give high SNR (up to 300% compared with commercially available emulsions), and do not induce an immune response. High encapsulation efficiencies result in a sixfold increase in the volume of PFC per dose.

An area where microencapsulation has produced superior agents to unencapsulated examples is in temperature mapping. Well-defined mixtures of PFC with appropriate melting points can be encapsulated for temperature mapping and the surface properties of the capsules can be modified. When modified with PEG 2000, the microspheres were retained in the circulation 30 times longer than the unmodified capsules (circulation half-lives of 70 and 2.5 minutes, respectively). These capsules represent a class of fluorine imaging agents that can give good quality fluorine images in a reasonable data acquisition time, with high SNR.

### Contrast Imaging with Therapeutic Modalities

As was the case with CT CA, the possibility of concomitant imaging and therapy is an attractive possibility. A recent review dealing with macromolecular systems for chemotherapy specifically concentrated on this concept for MRI (162). Although acknowledging the current drawbacks of limited tissue penetration and relatively slow rates of internalization by endocytosis, the authors pointed out that the field is yet in its infancy, and much is yet to be learned about how macromolecular prodrug chemistry affects their biological properties. Torchilin and co-workers describe the use of micelles for both drug delivery and imaging for CT, MRI, and $\gamma$-scintigraphy (58,59). Agent was prepared from amphiphilic AB-type copolymers to produce micelles in which poly(ethylene glycol) groups surround and protect a hydrophobic core that contains the drug and/or imaging agent. The entrapped species dictates the imaging modality for example, iodine for CT, Gd-DTPA-PE for MRI, and [111]In-DTPA-SA for scintigraphy. The important feature of Gd-DTPA-PE is its amphiphilic nature, which facilitates incorporation into a micelle. Excellent MRIs can be obtained with these agents (58). An interesting feature of these agents is the ability to gently massage the 20 nm particles from the subcutaneous injection site in the rabbit hind paw, down the lymphatic pathway, to the thoracic duct. In the case of $\gamma$-scintigraphy, the investigators used direct massage of the popliteal node to "squeeze" the [111]In-labeled micelles to the systemic circulation. In MRI, axilliary lymph nodes became visible only 4 minutes after administration of Gd-containing micelles.

Reports are starting to appear in the literature addressing issues in imaging and gene/DNA transfer. As mentioned earlier, Meade and co-workers at the California Institute of Technology report on the elegant use of MRI as a method of noninvasive monitoring of gene transfer (107). They synthesized a new class of agents and demonstrated in vitro the ability of these agents to transfect genes into cells while enhancing the MRI contrast of the targeted cells. The strategy involved preparation of multimodal particles composed of DNA (luciferase plasmid) condensed with poly-D-lysine (PDL) conjugates with Gd-DTPA-PDL, and poly-L-lysine (PL) conjugates with transferin-PL. Approximately 1,500 Gd ions were taken up per cell. Significantly, higher gene expression (over twofold) was observed compared with particles without MRI contrast. A possible explanation could lie in the higher degree of neutralization of the PL backbone when Gd-DTPA-PDL was present.

Nantz et al. have reported imaging a DNA transfection event using a Gd-chelated liposomal CA formulated with the cationic transfection lipid DOTAP, complexed with the reported gene encoding for luciferase (163).

Again in an attempt to use MRI as a noninvasive tool for monitoring in vivo events, drug release from an implanted hydrogel has been followed by depicting the three-dimensional structure of the implanted biodegradable drug delivery matrix with time (164). Release was from bis-hydroxysuccinimide ester of poly(ethylene glycol)disuccinate cross-liked with albumin, to which gentamicin and Gd-DTPA were covalently attached in stoichiometric quantities. Impressive three-dimensional reconstructions and MRI images of the implant (Fig. 4) are shown, and a correlation coefficient of 0.965 for concentration of released gentamicin and hydrogel volume are reported. The degradation of the hydrogels was followed by serial MRI. Complete resorbtion was judged to take 4 weeks, and the half-life of a 1-mL "button" containing 18 mg of gentamicin was approximately 7 days (Fig. 5).

Bacterial glycopeptides have been used for penetration of the blood-brain barrier with enhancement of drug delivery and imaging (165). In rabbit, Gd-DTPA, which is normally excluded from the brain parenchyma, showed enhancement 4 hours after administration of 1 $\mu$g of pneumococal cell wall and became prominent, particularly in the region of the occipital cortex, 5 hours after challenge.

Finally, in an interesting combination of MRI and ultrasound, the use of target-specific delivery of macromolecular agents with MR-guided focused ultrasound has been reported (166). In this study MRIs of a Gd-labeled liposome were used to select a target region in rabbit muscle for application of pulsed-focused ultrasound to deliver biotin-labeled liposomes. Tissue changes observed in $T_2$-

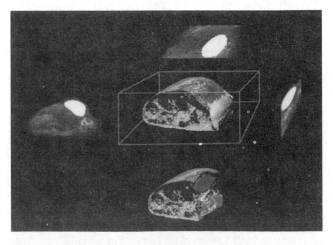

**Figure 4.** MRI of gentamicin-hydrogel. A gentamicin-hydrogel was implanted into a paraspinal abscess in a rat. Sixty individual MRI slices (individual slice thickness of 700 $\mu$m to render a slab of 4.2 cm) through the midabdominal of a rat were acquired and are displayed as axial (left), coronal (top), and sagittal (right) images or as colorized 3D reconstructions in which hydrogel is segmented in green (see color plate). The calculated gel volume by MRI (0.94 mL) corresponds to 16.292 mg of gentamicin. *Source:* From Ref. 164, courtesy of Dr. R. Weissleder and the publishers.

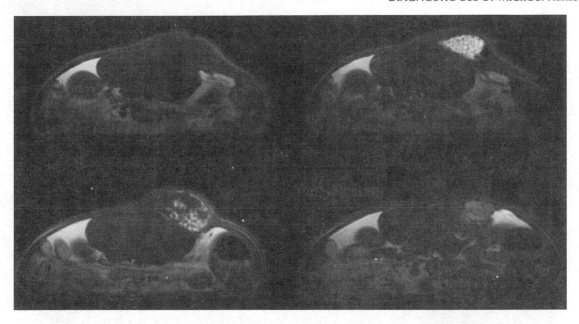

**Figure 5.** Degradation of gentamicin-hydrogel. MRI of an implanted hydrogel is shown. The brightness of the hydrogel is due to the paramagnetic Gd label that makes it better distinguishable from muscle signal intensity. Time point: precontrast (top left), postcontrast (top right), 5 days (bottom left), and 18 days (bottom right). *Source:* From Ref. 164, courtesy of Dr. R. Weissleder.

weighted images and the accumulation of Gd-liposomes in the regions treated with pulsed-focused ultrasound were a function of the total energy deposition, the duty cycle, and the power of the focused ultrasound pulse. The changes seemed reversible in the case of low-energy deposition.

## ULTRASOUND CA

### Physical Properties

An ultrasound CA must modify one or more of the acoustic properties of tissue which determine the ultrasound imaging process (167). Essentially there must be an acoustic impedance mismatch between the CA and the suspending fluid, usually blood. Acoustic impedance is the product of the density of the medium and the speed of sound in that medium. The most important properties that are affected are backscatter cross section ($\sigma$), attenuation, and acoustic propagation velocity, although the latter is less frequently used. The backscatter cross section is related to the received ultrasound intensity $I_s$, the incident intensity $I_i$, and the distance $R$ between the transducer that emits the radiation and a single small scatterer (168):

$$I_S = \frac{I_i \sigma}{4\pi R^2} \qquad (2)$$

For multiple scatterers this value is multiplied by the number of scatterers. The relationship assumes that the scatter is much smaller than the wavelength. This is usually the case because the size of the agent is in the order of 1–6 $\mu$m, much smaller than the wavelength of the acoustic field (at 3 MHz the wavelength in water is 0.5 mm). The

scattering cross section $\sigma$ (m$^2$) can be described by the Born approximation (169):

$$\sigma = \frac{4\pi}{9} \, k^4 r^6 \left[ \left( \frac{\kappa_S - \kappa}{\kappa} \right)^2 + \frac{1}{3} \left( \frac{3(\rho_S - \rho)}{2\rho_S + \rho} \right)^2 \right] \qquad (3)$$

where k is wave number = $2\pi/\lambda (\text{m}^{-1})$; $\lambda$ is wavelength (m); $\kappa_s$ is compressibility of the scattering particle; $\kappa$ is compressibility of the medium; $\rho_s$ is density of the scattering particle (kg/m$^3$); $\rho$ is density of the medium (kg/m$^3$). This can be expressed as

$$\sigma = A \left[ B^2 + \frac{1}{3} \, C^2 \right] \qquad (4)$$

where A is amplitude term; B is compressibility term; and C is density term.

The objective in developing an ultrasound CA is to maximize the scattering cross section. It is nonlinearly related to the geometric cross section. It can be seen from the amplitude term in equations 3 and 4 that the cross section is related to the sixth power of the radius and the fourth power of the frequency. For a given radius, the higher the applied frequency, the greater the cross section, and for a given frequency, the larger the radius the higher is the cross section. We know, however, that particles that are to cross the pulmonary capillary bed must have diameters less that about 6 $\mu$m if they are to avoid being trapped, so there is an upper limit on size.

The compressibility term and the density term of equation 3 demonstrate that the physical properties of the scatterer compared with those of the medium are paramount. Of the two terms the compressibility difference is dominant. For example, the compressibility of air and water are $2.3 \times 10^{-4}$ cm$^2$/dyne, and $4.6 \times 10^{-11}$ cm$^2$/dyne respectively. In comparison, the compressibility of a rigid substance such as nickel, is $5.0 \times 10^{-13}$ cm$^2$/dyne. Using a radius of 5 $\mu$m, and a frequency of 5 MHz (the medical imaging range is 3–10 MHz), the scattering cross section of a liquid, solid, and gas can be calculated as approximately 0 m$^2$, $6.65 \times 10^{-15}$ m$^2$, and 0.38 m$^2$, respectively.

Gas bubbles have an added advantage when being considered as ultrasound CA because they can act as harmonic oscillators and resonate when insonated at their resonant frequency. Resonance increases the backscatter cross section by three orders of magnitude (168). An approximate relationship between resonant frequency f(s$^{-1}$) and particle radius r(m), which ignores surface tension, thermal effects, and effects due to viscous damping by the surrounding medium, is given by equation 5 (170):

$$f_r = \frac{1}{2\pi r} \sqrt{\frac{3\gamma P}{\rho}} \tag{5}$$

where $P$ is pressure (N/m$^2$); $\gamma$ is $C_p/C_v$ (heat capacity ratio equals 1.4 for an adiabatic process in air); and $\rho$ is density of medium (kg/m$^3$). Using values for air and ambient pressure we obtain a relationship for resonant frequency $f_o$ in kHz for a diameter $d$ in $\mu$m as:

$$f_o = \frac{6500}{d} \tag{6}$$

which indicates that the resonant frequency falls right in the medical imaging frequency range (171). Expressions have also been derived which account for the damping of the resonance frequency caused by the encapsulation of the bubble which is used to increase bubble stability (172,173).

An interesting property of an oscillating bubble has recently gained increasing attention as a possible new imaging technique (174,175). At slightly higher intensities of the sound field the oscillations of the bubbles are driven to a point where the alternate expansions and contractions of the bubble's size are not equal. This nonlinear response gives rise to the generation of harmonics of the insonating frequency which are emitted by the bubbles. Harmonic images are created by transmitting at the fundamental and receiving at the second harmonic, twice the fundamental frequency. Since the tissues and vessel walls are not oscillating in this fashion, this allows for contrast-specific imaging. Machines are currently available that can use this imaging technique in a clinical setting. Even though the harmonic signals are considerably weaker than the fundamental, the signal-to-clutter ratio enhancement is over 1,000 (176).

As already derived, the best medium for ultrasound CA is a gas in the form of very small bubbles with diameters less than 10 $\mu$m, preferably less than 6 $\mu$m. When considering the approach to be taken in developing these microbubbles, it is important to review some of the physics describing the stability of gas bubbles of this size in a liquid. The expression for the pressure difference $\Delta P$ between the outside and inside of a bubble of radius $r$ in a liquid with surface tension $\gamma$ is give by the Laplace equation:

$$\Delta P = \frac{2\gamma}{r} \tag{7}$$

It is derived by noting that at equilibrium the tendency for the surface area to decrease due to the sum of the external pressure and the surface tension, is balanced by the rise of internal pressure which would result from this area decrease. If we consider a 0.1 mm radius bubble in champagne, the pressure difference would be $2(7.4 \times 10^{-2}$ Nm$^{-1})$ $(1.0 \times 10^{-4}$ m) which is 1.5 kPa. Unfortunately, such bubbles are too large for CA use! The pressure inside a more suitable 10-$\mu$m bubble would be 15 kPa, enough to rapidly collapse any bubble that is not well stabilized. In fact free bubbles of this size have extremely short half-lives. Epstein and Plesset derived a relationship (equation 8) for the shrinkage (or enlargement) of bubbles in a still liquid, which took into account the diffusion coefficient of the gas and the gas solubility in the suspending medium (177):

$$\frac{\delta r}{\delta t} = \frac{-Dy}{r} \left[ \frac{1 - \dfrac{C_i}{C_S} + \dfrac{2\gamma}{rP_h}}{1 + \dfrac{4\gamma}{3rP_h}} \right] \left( 1 + \frac{r}{\sqrt{\pi D t}} \right) \tag{8}$$

where $r$ is radius of bubble (m); D = diffusion coefficient (m$^2$/s); $C_i/C_s$ = ratio of dissolved gas concentration $C_i$ to saturated concentration $C_s$; $\gamma$ = surface tension (N/m); $P_h$ = hydrostatic pressure (N/m$^2$); $t$ = time (s); y = (RTC$_0$)/P$_h$M; $M$ = gas molecular weight; R = gas constant; $T$ = absolute temperature.

Using this equation, de Jong (178) calculates the lifetime of a free air bubble in water saturated with air to be $5 \times 10^3$ seconds for a 100-$\mu$m bubble and only 6.0 seconds for a 10-$\mu$m bubble. Because it takes 2 s for an injected bubble to travel to the right side of the heart and 10–27 s to travel to an end organ, free air bubbles are clearly inadequate. Gases with lower solubility and higher molecular weights have much greater persistence. Recently encapsulated gases such as PFC and sulfur hexafluoride (SF$_6$) have gained preeminence for use as ultrasound CA.

Before discussing the evolution of CA from their beginning in 1968 (179), it is important to briefly review the different modes used in ultrasound imaging. The transducer emits the ultrasound in short pulses and also receives the reflected signal, or echo. The simplest instrument is an A-mode scanner, which depicts the amplitude of the received echo against time. M-mode, or sometimes TM (time motion) scanning depicts the movement of tissue with time, and is most often used in cardiac imaging. The time it takes for a transmitted disturbance to return to the transducer depends on the depth of the reflector, and, if the reflector is moving, this depth changes with time. The M-mode display depicts this change of depth with time and is very useful in analyzing dynamics of heart valves. It is composed of a rastor of A-mode lines. The most familiar ultrasound scan is the B-mode, which gives a two-dimensional image through the tissue, for example, an image of a fetus in the womb. The ultrasound beam is scanned through the tissue either mechanically or, more usually, by electronic beam steering using a multielement transducer. The time of arrival of the echoes indicates the depth from which the beam was reflected, and signals are displayed as spots with a brightness proportional to their intensity (known as gray scale). Tissue movement can be seen as the beam is periodically swept through the region if the image is updated at a sufficiently high rate. These scans are known as real-time images. In addition to B-mode, Doppler ultrasound is also an area where CA are having an impact. The usual scatterers in blood are the red blood corpuscles, and these are, of course, moving with the flow of blood. The frequency of echoes reflected from these, and any other moving species (for example CA), have an altered frequency, depending on their velocity and direction relative to the transducer. This is the familiar Doppler effect, noticed by trainspotters as they stand by the side of the rails and hear a train whistle shift in pitch as the train approaches and then passes them. The cyclical variation of blood velocities in arteries (and veins) can be monitored as the heart beats, and in fact can be heard because the frequency is shifted into the audible range. Duplex scanners are available in which both the B-mode image and Doppler image are displayed simultaneously one beside the other. A further refinement, known as color flow mapping (CDI), superimposes a color-coded Doppler signal onto a real-time B-mode image. The color coding can indicate both direction and speed of flow, and regions of abnormal flow can be detected. (Usually red indicates motion toward and blue indicates motion away from the transducer, with gradation of color depending on velocity.) A new modality that is gaining interest because of CA is color amplitude imaging (CAI, also known as power Doppler). In CAI the density as opposed to the velocity of moving scatters is displayed. This is achieved because the amplitude, and thus the power and intensity of the echo, depends on the number of particles. In fact the Doppler frequency shift is not used so the signal is not dependent on direction. Finally, as already mentioned, harmonic imaging is now being developed in which the nonlinear properties of oscillating bubbles are being used to gain contrast-specific images with high SNR, free from clutter of surrounding tissue (180,181). The tissue is interrogated at the fundamental frequency and the signal is received at the second harmonic (twice the fundamental).

### Early CA for Ultrasound Imaging

The first ultrasonic detection of bubbles in the blood was in 1963 during investigation of microbubbles formed during decompression sickness (182). However, it was not until Gramiak noticed a cloud of echoes during injection of indocyanine green into the left heart in 1968 that the possibility of their use for ultrasound contrast was considered (179). The realization that the signal originated from microbubbles created by cavitation at the catheter tip upon injection was followed by development of the first-generation agents consisting of hand-agitated solutions of compounds such as indocyanine green, saline, Renografin, 5% dextrose, and even the patient's own blood (183–185). The bubbles that were produced were extremely short lived and too large to avoid clearance by the lung. Later,

Lang reported production of microbubbles bubbles in Renografin by sonication (186). The early work on CA has been reviewed (187).

### Main Areas of Use of CA in Ultrasound Imaging

Ultrasound has many advantages over other imaging modalities, primarily in the areas of safety, low cost, convenience, and the possibility of real-time imaging. However, without contrast ultrasound is limited in the detection of small tumors and in Doppler evaluation of low-velocity blood flow either in small, deep vessels, or in areas of low flow due to pathology. The initial emphasis for CA development was in echocardiography, and it remains an active area of interest today. Since the initial investigations, the use of contrast has spread to all areas and modes of ultrasound imaging with the possible exception of fetal monitoring. The main areas are reviewed here.

**Cardiac and Blood Vessel Imaging.** de Jong defines the main areas as (178):

- Identification of cardiac structures (e.g., endothelial boarder)
- Detection of intracardiac shunts
- Visualization of blood flow in M-mode
- Detection of valvular regurgitation
- Analysis of complex congenital heart disease (e.g., arterial and ventricular septal defects, transpositions)
- Cardiac output determination by indicator dilution curves, left ventricular ejection fraction
- Perfusion studies
- Patency

Fan and co-workers have reported on in vitro studies comparing the commercially available Albunex and Echovist® with sonicated indocyanine green, Renografin-60, normal saline, and mannitol for potential CDI echocardiography (188). The commercial preparations were clearly superior. Reviews have appeared for cardiac use of Albunex (189), and the use of the gas-filled, lipid-coated preparation MRX-115 (now DMP-115 or Definity) for myocardial perfusion (190). Contrast has been investigated to alleviate problems of attenuation due to overlying tissue such as the skull, for example in transcranial Doppler (191). Using Levovist®, Bauer reports visualization of the circle of Willis in 100% of patients as opposed to only 80% in the absence of contrast. Levovist® has also been used by Sitzer to study renal arterial stenosis using contrast and CDI (192). They report a maximum enhancement of 21 dB after 15 s, and a clinically useful increase that lasted three minutes. Langholz reports that leg veins have been successfully imaged with Levovist after injection into the dorsal foot vein (193).

**Liver Imaging.** As with X-ray and MRI, liver imaging is an area in which strong interest in contrast enhancement of ultrasound images exists. Ultrasonography is the initial modality of choice for liver scans in clinical practice, for the reasons already given. Use of contrast adds greatly to these investigations (53,194). Kudo and co-workers report on sonographic angiography in which 5–20 mL of a vigorously agitated mixture of carbon dioxide, heparinized normal saline, and the patients own blood were injected through a catheter placed within the hepatic artery. Hepatocellular carcinomas could be distinguished from focal nodular hyperplastic nodules by distinguishing the hypervascular pattern. Adenomatous hyperplasia, hemangiomas, and metastases displayed yet another hypervascular pattern (195). The same technique was used to identify tumors of less than 1 cm in diameter, which had been missed by other methods (196). Finally, a new polymeric agent, Sonovist®, is reported, which is taken up by the Kupffer cells and upon ultrasonic radiation is stimulated to resonate and collapse, producing a strong Doppler signal (197). This signal is absent in tumors that lack an RES.

**Oncology.** In addition to hepatic imaging, CA are used in many other areas of oncology, including central nervous system tumors (198), prostate (199), and breast (200). Cosgrove has performed extensive studies into identifying key features of tumor vasculature to aid in predicting prognostic outcome (201). Doppler has been used to attempt to predict the metastatic propensity of tumors (202). A quite dramatic demonstration of the enhanced power of CA Doppler was demonstrated by the same team who identified connections between adjacent vessels in breast tumors that were not seen in benign tumors and which presumably represented arterio-arterial shunts (203). Identification or absence of these shunts has been used to change the diagnosis of breast lumps, and those with shunts were confirmed to be malignant upon biopsy, whereas those without were found to be benign (204).

**Gynecology.** Ultrasound CA have been used in gynecological examination for both sonohysterography (transvaginal examination of the endometrium) and sonohysterosalpingography (investigation of tubular patency) (205). Both hyper- and hypoechoic agents are becoming widely used.

**GI Tract.** Ultrasonic investigation of the GI tract is severely hampered by the natural presence of gas, which acts as a reflector. A cellulose-based agent for oral administration has been developed by ImaRx Pharmaceutical Corp. and licensed by Bracco Diagnostics (Princeton, NJ). The cellulose serves to disperse and displace gas bubbles, allowing good GI imaging (206,207).

### Microcapsule Types Developed as CA for Ultrasound

Whereas nearly all CA for ultrasound use microbubbles as the reflective agent, not all can be considered as true microcapsules. One of the earliest examples, Echovist (Schering, Germany), consists of a saccharide crystal base from which microbubbles are liberated upon dissolution in saline, just prior to injection. Other agents, for example, the phase-shift agent EchoGen® (Sonus, CA), consist of a liquid emulsion that essentially boils at body temperature. All of these agents can be described under the heading of "microcapsule" for consistency, however, micro*bubble* would perhaps be a more suitable title. The last 10 years have seen a flurry of activity in the ultrasound contrast area, with many new companies and agents appearing on the scene. Table 4 is presented to orient the reader as to the compositions and current names of some of the recent agents.

Several reviews have appeared describing recent ultrasound CA (167,176,178,194,208–211).

**Gelatin, Collagen, and Alginate.** In one of the earliest reports of capsules for ultrasound contrast, 80 $\mu$m gelatin-encapsulated nitrogen bubbles were reported to produce rim enhancement in VX2 carcinomas in rabbit, which lasted for several minutes (212). The large diameter required arterial injection of the agent close to the tumor area to avoid loss of the agent by entrapment in capillaries. The following year solid collagen microspheres of 3-$\mu$m diameter were prepared by an emulsion technique in which warm collagen was dispersed in cottonseed oil followed by cooling. However, in vitro testing in dog indicated that the strong enhancement in liver was due to microcapsule aggregation (213). We have reported on two methods for producing alginate-encapsulated gas bubbles using ionotropic gelation with calcium ions. In one case we designed a dual-barrel injection method in which gas was injected into the nascent alginate capsules as they emerged from the spray nozzle, and in the other a pregassed sodium alginate solution was fed to a single-barrel spray nozzle (214). Both methods produced echogenic gas-filled capsules, but in the first method the size was around 100–150 $\mu$m and in the second 30–40 $\mu$m.

**Galactose Particles.** Another early experimental agent, Echovist (Schering, Germany), consisted of crystalline galactose particles that contained numerous sites to entrap gas (215). The gas was released as bubbles with a mean diameter of 3 $\mu$m (but a size range up to 15 $\mu$m) when the particles were dissolved in a saturated galactose solution. Despite giving good right heart opacification, this agent was replaced by a similar agent, Levovist, with improved stability and narrower size range (90% < 8 $\mu$m), which passes the lungs and can be used to image the left heart (216). Levovist differs from Echovist not only in the way that the galactose particles are produced but also because it contains 0.1% palmitic acid, which stabilizes the bubbles. The bubbles are released by addition of sterile water. Levovist has undergone extensive animal and human testing, is approved in Germany, and is awaiting approval in the United States. Although no grayscale (B-mode) tissue enhancement is reported, woodchuck hepatic tumors and large and small vessel enhancement has been seen in color and spectral Doppler, and enhanced echogenicity was demonstrated for ophthalmic and retinal vessels (217,218). Multicenter European trials report the agent to be well tolerated, and Phase II clinical trials in the U.S. have shown significant improvement in vessel enhancement (219). Breast tumors have shown distinct enhancement with agent on color-coded Doppler and the signals have been reported to allow reliable demonstration of vascularization characteristics (200). The same study suggested

**Table 4. Examples of Agents Currently of Interest**

| Investigatory Name (Company) | Current Name | Composition/Gas |
|---|---|---|
| SHU 454 (Schering) | Echovist® | Galactose/air |
| SHU 508 (Schering) | Levovist® | Galactose/fatty acid–stabilized PFC (German approved) |
| Albunex® (MBI) | Albunex® | Sonicated HSA/Air (FDA approved) |
| FS069 (MBI) | Optison® | HSA/PFC |
| Quantison (Andaris) | Quantison® | HSA/spray-dried |
| PESDA (U. Nebraska) | PESDA | HSA/dextrose/PFC |
| AF0150 (Alliance) | Imagent® | Surfactant/PFC |
| QW3600 (Sonus) | Echogen® | PFC emulsion |
| MRX-115 (ImaRx) | Aerosomes® | Phospholipid/PFC |
| Now DMP 115 or Definity® (DuPont Merck) | | |
| BR1(Bracco) | Sonovue® | PEG4000/phospholipids/$SF_6$ |
| IDE particles (Univ. of Rochester) | Bubbicles | Iodapamide ethyl ester/air |
| SH U 563A (Schering) | Sonovist® | Polybutyl-2-cyanoacrylate/PFC |
| SonoRx (Bracco) | SonoRx® | Semithicone-coated cellulose (GI applications) |
| ST68,ST44 (Drexel Univ.) | ST68,ST44 | Surfactants/PFC |

that arrival, retention, and washout studies could give information about the growth dynamics of the tumor. Recently Levovist has been demonstrated to act as an active sound source during disintegration (stimulated acoustic emission or SAE) in myocardial contrast echocardiography (220). An advantage is that Doppler signals can be received even from stationary bubbles or ones with very low flow velocities. Myocardial contrast (MC) was studied in 36 patients using harmonic power Doppler investigations. Delineation of MC from contrast signals of the cavities was possible, whereas signal intensities in the cavities, even at the endocardial border, were at least 20 times higher than those of MC. In other studies in Europe, Levovist has been used as a continuous infusion for quantitative assessment using color power Doppler (221). Relative microbubble numbers could be accurately quantified in vivo, opening the door to the application of formal quantitative enhanced ultrasound in patients.

**Protein Microcapsules.** Another agent that was on the scene early is Albunex (MBI, California), which is the first agent to be approved by the FDA. Albunex® consists of air-filled albumin spheres with a mean diameter of 3–5 $\mu$m, produced by sonicating a 5% solution of human serum albumin. The usual concentration is 3–5 × $10^8$ microspheres/mL of sterile, isotonic buffer at pH 6.4–7.4. The wall thickness is estimated to be around 15 nm. Its refrigerated shelf life is around 1 year (190). The half-life after injection is around 1 minute, and after 3 minutes more than 80% of the contrasts is found in the liver. Goldberg has demonstrated that Albunex can produce dramatic Doppler signal enhancement in both normal and tumor vessels in woodchuck with a naturally occurring hepatocellular carcinoma (222). Doppler signal enhancement recorded via a cuff transducer around the celiac artery gave an optimal signal gain of approximately 10 dB. Albunex has been proven effective for enhancing left ventricular endocardial definition in Phase III multicenter clinical trials (223). It has also been used to measure cardiac output (190), in dobutamine (a vasodilator) stress echocardiography (224), fallopian tube patency (225), vesicoureteral reflux (a common pediatric urinary tract abnormality), and detection of thrombus (226). In the thrombus study, the Albunex appeared to adhere to the thrombi. One of the chief disadvantages of Albunex is its short half-life, and Vandenberg conducted a study that indicated that Albunex is acoustically labile to acoustic pulse pressures greater than 0.33 MPa (227).

A second generation agent, Optison® (originally FS069), has been developed, which has greater in vivo half-life than Albunex (190). As mentioned previously, one of the factors in the stability of gas bubbles is the solubility of the entrapped gas in the blood (see equation 8). In Optison, the relatively soluble, low molecular weight components of air are replaced by perfluoropropane, an insoluble, relatively high molecular weight polyfluorinated hydrocarbon. The capsules are taken up by the RES in a similar fashion to Albunex, and the PFC is expired from the lungs after it is released when the protein coat is metabolized. In addition to left heart opacification, Optison has shown promise in use in myocardial perfusion and harmonic imaging (228,229).

Quantison® (Andaris Ltd., Nottingham, UK) also has a shell composed of human serum albumin (recombinant), but these capsules are produced by spray-drying and are thought to have a 200-nm shell. They are considerably more resistant to pressure than Albunex (178).

Porter in Nebraska has used a sonicated mixture of 1 part dextrose and 5 parts human serum albumin to stabilize perfluoropropane in an agent called PESDA, which has a particle size of 4.7 $\mu$m and a concentration of 6.0 × $10^9$/ml (230). Porter has also encouraged the use of so-called transient response, or intermittent imaging, to improve the amount of MC observed from gaseous agents (231). The technique, which uses ultrasound in an interrupted mode such as 1 frame per cardiac cycle as opposed to the usual frame rate of 30 frames per second or greater, causes less overall acoustic destruction of the bubbles. Also, if this is combined with a high acoustic output, it may induce an emission from the microbubble when it is destroyed during the less frequent insonation, similar to the SAE noted previously with Levovist. In vitro this SAE was measured with a 20-MHz probe transducer positioned confocally with a 1-MHz cavitation transducer (232). The average internal cavitation activity thus measured was higher for a pulse repetition frequency (PRF) of 10 Hz compared with one of 100 Hz, and both decreased with time. At PRF of 1,000 Hz and 5,000 Hz, the overall cavitation activity was higher than at either 10 or 100 Hz and did not decrease with time, indicating sustained bubble reseeding. The same effect was demonstrated in vivo in dog with intravenous injection of agent (233). The heart was simultaneously imaged with a 1.7-MHz harmonic transducer using a peak negative pressure (PNP) of 0.3–0.9 MPa and frame rates of 43–10 Hz, and one frame rate every one (1:1) to three (1:3) cardiac cycles. Cavitational activity increased as frame rate decreased, and higher PNP increased intramyocardial contrast. PESDA has been used in a human trial (92 patients) to detect coronary artery disease(CAD) (234). A perfusion defect was observed with myocardial contrast echography at rest or after stress that strongly correlated with the presence of CAD.

**Stabilized PFC Emulsions.**  Perfluorooctyl bromide is radiopaque due to the terminal bromine. When a phospholipid-stabilized emulsion with particle size 0.1–0.4 $\mu$m, is administered intravenously, it can act as both an X-ray and ultrasound CA (235,236). The chief mechanism of ultrasound enhancement is from the compressibility difference between the emulsion droplets and blood. Use of these agents caused several side effects, and emphasis has now switched to a phospholipid-stabilized gaseous agent, known as Imagent US (237). Imagent is a powder consisting of buffer salts and phospholipid shells encapsulating perfluorohexane gas and nitrogen. Sterile water is added to suspend the capsules prior to injection. The agent has been shown to opacify the left ventricle, enhance the myocardium, and visualize regions of infarct in rabbit (238). Imagent has also been shown to enhance the placenta in rats (239). Recent second harmonic studies appear to show linear branching and discrete punctuated densities within the myocardium consistent with intramyocardial coronary vessels (240).

A different type of PFC emulsion is described by Correas and co-workers at Sonus Corporation, who describe a so-called 'phase shift' CA (241). The agent, Echogen, exists at room temperature as a liquid emulsion of the isomers dodecafluoro-$n$-pentane and dodecafluoroisopentane, with a mean diameter of about 0.3 $\mu$m. Upon injection, it vaporizes at body temperature. After rapid diffusion of dissolved gases from the blood into the PFC microbubbles the diameter is thought to increase to between 2 and 5 $\mu$m. The gas is exhaled through the lungs. The agent has been shown to exhibit parenchymal enhancement and tumor visualization (242). Vascular enhancement lasted 2 to 3 minutes, and liver and kidney enhancement was observed for up to 20 minutes. A maximum enhancement of 18.7 dB for a 0.6 mL/kg dose was reported. Detection of flow characteristics have been demonstrated to be markedly enhanced on spectral Doppler, color flow Doppler and CAI (243). Echogen is currently undergoing clinical trials. A clinical trial of perflenapent (2% dodecafluoropentane emulsion; Echogen Sonus Pharmaceuticals, Bothell, WA) has re-

cently been described (244). Contrast enhancement was observed in 94% of patients, compared with 8% of placebo group, and lasted between 5 and 15.4 minutes. It was concluded that perflenapent emulsion is a safe and effective CA for ultrasound of the liver, kidneys, and vasculature.

**Liposomes and Phospholipid-Stabilized Gas Bubbles.**  Unger and co-workers have described a dipalmitoylphosphatidyl choline liposome–containing gases such as nitrogen, oxygen, xenon, and PFC (originally Aerosomes®, now DMP-115 or Definity) (3,190). They have shown that the stability in the blood is very dependent on the solubility of the entrapped gas and the relative saturation of the blood. For example, air- or nitrogen-containing liposomes give far less contrast in the left heart in dogs ventilated on 100% oxygen, compared with those on atmospheric oxygen, whereas PFC-containing liposomes showed no such sensitivity (245). This result was attributed to the greater solubility of oxygen compared with nitrogen. Oxygen is thought to diffuse across the Aerosomes wall to exchange with nitrogen and hasten bubble loss. The agent can be prepared in an extruder, dried under vacuum, and filled with the gas of choice (246). Commercial preparation currently involves agitation using a mechanical shaking device to agitate sealed vials with the appropriate head space of gas (247). The resulting vesicles are around 2.5 $\mu$m in diameter. Acoustic measurements gave maximal attenuation at around 2 MHz, indicating that the bubbles were around 3 $\mu$m in diameter. The lipid-coated microbubbles have been used in a Phase II clinical study involving patients with acute myocardial infarction. The material (DMP-115 now Definity) was injected as an intravenous bolus. Images were obtained with second harmonic mode using commercially available ultrasound equipment, and the data were recorded on SVHS videotape and later analyzed off-line with software from Dr. Sanjiv Kaul's laboratory. The images were obtained at end systole of every heart cycle. Figure 6a is a composite of three images obtained at baseline in the two chamber view. Figure 6b is a composite of three images obtained after injection of the contrast and after attenuation was no longer present, in the same view. Figure 6c is the myocardial portion of Figure 6a subtracted from the myocardial portion of Figure 6b and the remainder subjected to a "heated object" lookup table. That is, white represents the most intense contrast and black would represent no contrast. Red therefore would be some limited amount of contrast. Figure 6d is a sestimibi scan of the same patient in the view analogous to the ultrasound image. The sestimibi and ultrasound contrast images (Figure 6c and 6d) demonstrate decreased myocardial perfusion in the anterior-apical regions of the left ventricle. The images were obtained from a 36-year-old male patient who had an acute anterior myocardial infarction 3 days prior to the ultrasound and sestimibi studies.

An agent identified as BY963 and described as being phospholipid-stabilized microbubbles has been reported for use in transcranial Doppler. The lipid in this case is a saturated stearic acid containing phospholipid of "plant origin," and the size of the resulting microbubbles is 3.9 $\mu$m (248,249).

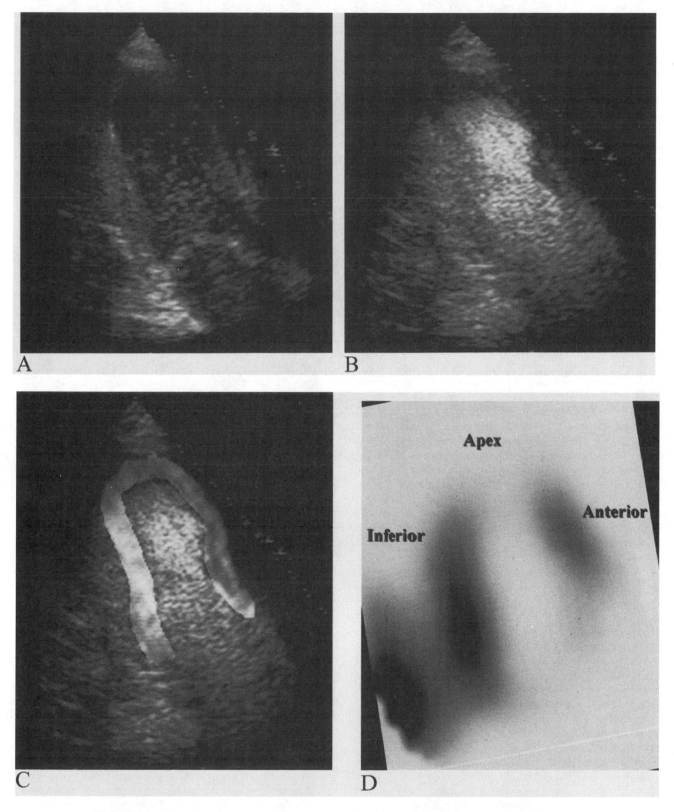

**Figure 6.** Images from a 36-year-old male patient with anterior wall myocardial infarction 3 days prior to the myocardial ultrasound study. (**A**) Three aligned unenhanced images of the left ventricle in the 2-chamber long axis view. (**B**) Three aligned DMP 115–enhanced images of the left ventricle in the 2-chamber long axis view. (**C**) Digital subtracted color enhancement of the left ventricle in the 2-chamber long axis view enhanced with DMP 115. (**D**) $^{99m}$Tc sestimibi image of the heart of the patient. See color plate. *Source:* Courtesy of Dr. Unger, Imarex Pharmaceutical.

Lipid-coated microbubbles have been described for detection of brain tumors, neurosonography, and internalization in tumors by D'Arrigo and co-workers (250–253). The proprietary lipid mixture contains mostly glycerides and cholesterol esters, and the capsules are 99% <4.5 $\mu$m in size.

Bracco Research in Switzerland has reported on an agent, Sonovue® (was BR1), consisting of $SF_6$ stabilized with a lyophilized coat of poly(ethylene glycol) (PEG) 4000 and the phospholipids distearoylphosphatidylcholine and dipalmitoylphosphatidylglycerol (254). Reconstitution of the freeze-dried agent in 5 mL of sterile water yields a milky suspension of $2 \times 10^8$ bubbles/mL with a mean diameter of 2.5 $\mu$m with 90% < 8 $\mu$m. The agent was found to be highly echogenic in vitro over the range 1–10 MHz and to opacify the left heart. In a study in dogs, the agent was tested for its ability to detect myocardial perfusion defects as a result of coronary occlusion and myocardial reperfusion after thrombosis and intravenous administration of CA (255). Contrast echo detected myocardial perfusion defects in all the injections performed during coronary occlusion and detected myocardial perfusion with a sensitivity of 50% versus microspheres, and the extent showed a good correlation with microspheres ($r = 0.73$).

**Iodipamide Ethyl Ester "Nano-Sponges."** A group at Rochester University has taken a novel approach to the concept of bubble stabilization by developing porous particles, 1–2 $\mu$m in diameter, primarily for hepatic imaging (256). Initially the microspheres were relatively smooth amorphous spheres of iodipamide ethyl ester (IDE) prepared by precipitation of IDE in dimethylsulfoxide/ethanol by addition of an aqueous solution of poly(vinyl pyrrolidone). After intravenous injection they were taken up by the Kupffer cells and accumulated in the liver within 10 to 20 minutes (257). The rather low enhancement (~3 dB) has been improved by modifying the particle surface to give many crevices for stabilization of minute gas bubbles (258). These modified microspheres have been named Bubbicles. The iodine in the IDE allows for dual ultrasound and X-ray imaging.

**Porous Inorganic Particles.** A similar concept to a nanosponge (and to Ecovist described in "Galactose Particles") is reported by Glajch and co-workers in a patent that describes the preparation of porous inorganic particles for use as ultrasound CA (259). The inorganic particles, for example alkali salts of carbonates or phosphates, are prepared in such a way as to make them highly porous, and the pores trap the gas of choice. Once inside the body, the particles dissolve within one to two hours, depending on fabrication conditions. Unlike Bubbicles, these particles would be less likely to be retained in the liver for extended periods of time, and the slow dissolution rate renders them more long lived than Echogen.

**n-Butyl-2-cyanoacrylate.** A different polymeric particle has been developed at Schering (SH U 563, or Sonovist), which consists of n-butyl-2-cyanoacrylate shells with a mean size of 1 $\mu$m and a thickness of about 100 nm (260). The thin shell allows the capsule to resonate when stimulated by ultrasound and to produce a strong harmonic signal (261). When the capsules break they emit a broadband frequency spectrum, which can be used in SAE. They are prepared by emulsion polymerization of the monomer and water in a surfactant-stabilized gas–water system. In vivo they strongly enhance spectral and color Doppler and B-mode, whereas in the vascular phase, and once taken up by the liver, allow dramatic demarcation of VX2 tumors (261).

**Poly(t-butyloxycarbonylmethyl glutamate).** Polymeric microballoons with a shell composed of the biodegradable polymer poly(t-butyloxycarbonylmethyl glutamate) have been described for use as CA (262–264). The mean size is around 3 $\mu$m, with 99% <8 $\mu$m. The microballoons are prepared by interfacial polymer deposition. An in vitro backscatter coefficient of 7.5 MHz was measured, and liver parenchyma was enhanced in rats after intravenous injection of $7.8 \times 10^9$ microballoons/kg. Enhancement reached a maximum 2 minutes after injection and lasted 1 to 2 hours (264).

**Cellulose Compounds as GI CA.** Initial investigations by Lund and co-workers, using simethicone and methylcellulose to diminish the gas artifact encountered during abdominal imaging, have led to the development of a GI CA, SonoRx® (Bracco Diagnostics, Princeton, NJ) (265,266). The agent consists of 22 $\mu$m cellulose particles with a 0.25% simethicone coating in a suspension containing an antifoaming agent (sodium lauryl sulfate), flavoring, and coloring. Clinical trials involved comparison of scans before and within 15 minutes of consuming agent or placebo. Improvement of imaging of cases with upper abdominal pathologies was dramatic (208). A combination of SonoRx and an intravenous agent has been suggested for detection of pancreatic tumors.

**Poly(DL-lactide-co-glycolide) Microcapsules.** In our laboratory we are developing a biodegradable polymeric agent with a shell composed of poly(DL-lactide-co-glycolide) (PLGA) (P. Narayan and M.A. Wheatley, unpublished data, 1998). The capsules, which can be made from any polymer amenable to microcapsule formation, are prepared by encapsulating a solid, volatilizable core, which is removed by lyophilization after the capsules are hardened. The evacuated capsules are filled with the gas of choice (usually PFC or $SF_6$) by venting the lyophilization chamber with the appropriate gas. Depending on the method of encapsulation we have achieved populations with a mean diameter of 6.7 $\mu$m (coacervation) or 4.9 $\mu$m (solvent evaporation). Figure 7 shows scanning electron micrographs of (1) whole capsules and (2) sectioned capsules made by coacervation. When 0.2 mL/kg of PLGA microcapsules containing decafluorobutane (concentration of approximately $5 \times 10^8$ microcapsule/mL) was injected through a shunt in the jugular vein of rabbit, they produced noticeable enhancement on CDI (Fig. 8). Studies are ongoing.

**Surfactant-Stabilized Microbubbles.** We have also developed a surfactant-stabilized microbubble, which has a far more flexible shell than the polymer-encapsulated agent.

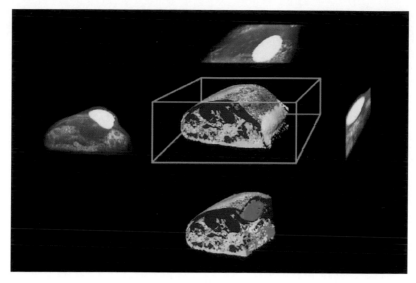

**Figure 4.** MRI of gentamicin-hydrogel. A gentamicin-hydrogel was implanted into a paraspinal abscess in a rat. Sixty individual MRI slices (individual slice thickness of 700 $\mu$m to render a slab of 4.2 cm) through the midabdominal of a rat were acquired and are displayed as axial (left), coronal (top), and sagittal (right) images or as colorized 3D reconstruction in which hydrogel is segmented in green. The calculated gel volume by MRI (0.94 mL) corresponds to 16.292 mg of gentamicin. *Source:* From Ref. 164, courtesy of Dr. R. Weissleder.

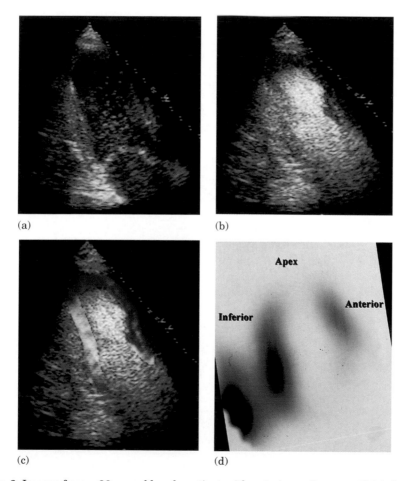

(a)

(b)

(c)

(d)

**Figure 6.** Images from a 36-year-old male patient with anterior wall myocardial infarction three days prior to the myocardial ultrasound study. **(a)** Three aligned unenhanced images of the left ventricle in the 2-chamber long axis view. **(b)** Three aligned DMP 115-enhanced images of the left ventricle in the 2-chamber long axis view. **(c)** Digital subtracted color enhancement of the left ventricle in the 2-chamber long axis view enhanced with DMP 115. **(d)** $^{99m}$Tc sestimibi image of the heart of the patient. *Source:* Courtesy of Dr. Unger, Imarex Pharmaceutical.

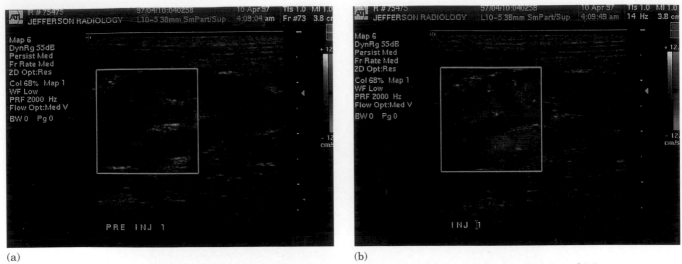

(a)  (b)

**Figure 8.** Color Doppler image of the right kidney. **(a)** Preinjection and **(b)** postinjection of 0.2 mL/kg of PLGA-decafluorobutane capsules. *Source:* Images obtained in collaboration with F. Forsberg and B.B. Goldberg at Thomas Jefferson University Hospital and funded in part by grants number NIH HL52901 and NIH CA52823.

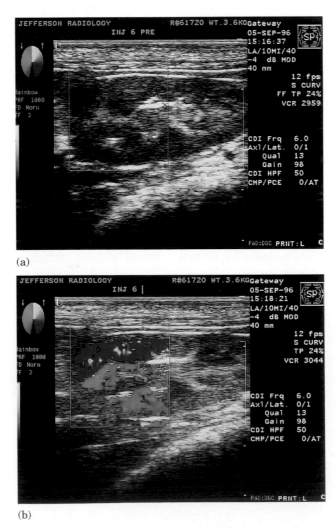

(a)

(b)

**Figure 9.** Color Doppler image of the right kidney. **(a)** Preinjection and **(b)** postinjection of 0.1 mL/kg of ST68-containing PFC. The color reaches the kidney capsule, and marked parenchymal enhancement is seen on the B-mode (outside the color box). *Source:* Images obtained in collaboration with F. Forsberg and B.B. Goldberg at Thomas Jefferson University Hospital and funded in part by grants number NIH HL52901 and NIH CA52823.

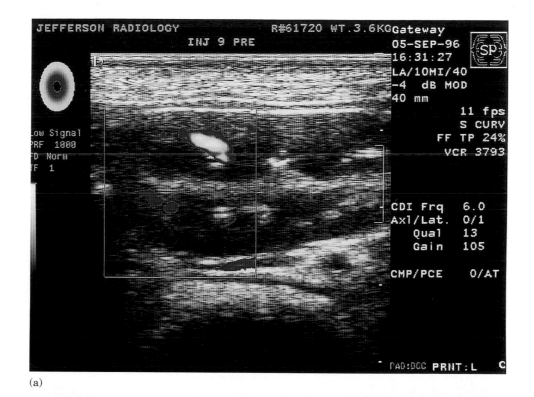

(a)

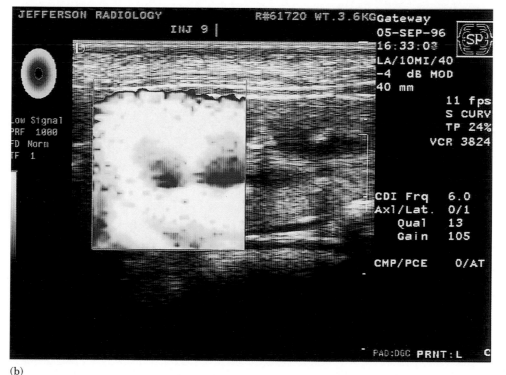

(b)

**Figure 10.** CAI of right kidney. **(a)** Preinjection and **(b)** postinjection of 0.1 mL/kg of ST68-containing PFC. Color blooming and flash artifact subsided after a few passes. Enhancement lasted 15 minutes. *Source:* Images obtained in collaboration with F. Forsberg and B.B. Goldberg at Thomas Jefferson University Hospital and funded in part by grants number NIH HL52901 and NIH CA52823.

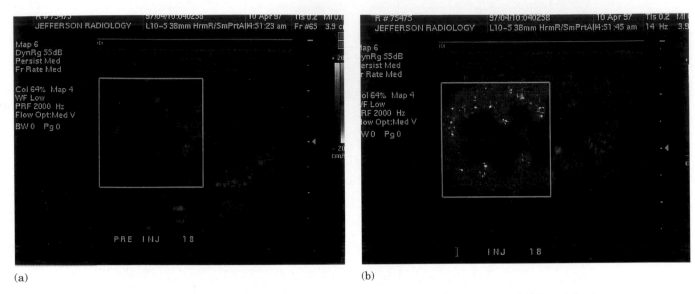

(a)

(b)

**Figure 11.** Color Doppler harmonic image of the right kidney. **(a)** Preinjection and **(b)** postinjection of 0.1 mL/kg of ST44-containing PFC. With a medical index of 0.6 the precontrast image is poorly defined. Strong color and B-mode enhancement are obtained with contrast. *Source:* Images obtained in collaboration with F. Forsberg and B.B. Goldberg at Thomas Jefferson University Hospital and funded in part by grants number NIH HL52901 and NIH CA52823.

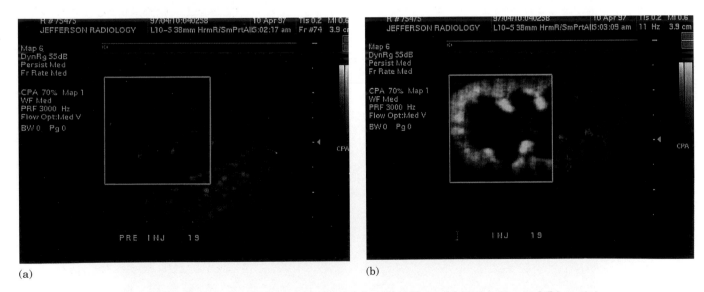

(a)

(b)

**Figure 12.** Color amplitude harmonic image of the right kidney. **(a)** Preinjection and **(b)** postinjection of 0.1 mL/kg of ST68-containing PFC. The dark vertical lines in the lower kidney indicate strong shadowing. *Source:* Images obtained in collaboration with F. Forsberg and B.B. Goldberg at Thomas Jefferson University Hospital and funded in part by grants number NIH HL52901 and NIH CA52823.

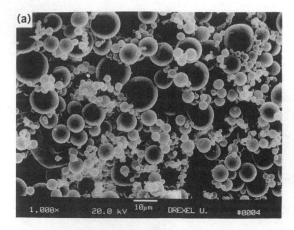
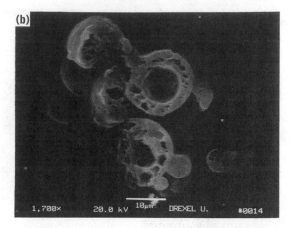

**Figure 7.** Scanning electron micrographs of hollow PLGA microcapsules prepared by a coacervation method. (**a**) Surface view of whole capsules; (**b**) Cross section of capsules. Bar = 10 $\mu$m.

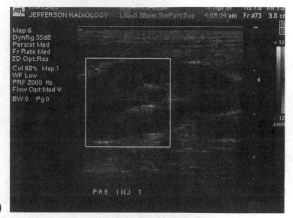
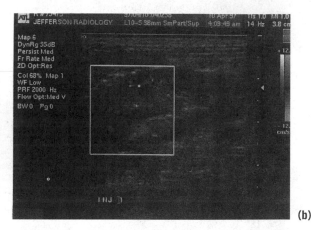

(a)

(b)

**Figure 8.** Color Doppler image of right kidney (see color plate). (**a**) preinjection and (**b**) postinjection of 0.2 mL/kg of PLGA-decafluorobutane capsules. *Source:* Images obtained in collaboration with F. Forsberg and B.B. Goldberg at Thomas Jefferson University Hospital and funded in part by grants number NIH HL52901 and NIH CA52823.

It is composed of two nonionic surfactants; for example ST68 contains Span 60 and Tween 80, whereas ST44 contains Span 40 and Tween 40 (267). The agent is produced by sonication of a phosphate-buffered saline solution of the two surfactants (268). The resulting microbubbles have a mean diameter of roughly 3.8 $\mu$m and a concentration of 7 $\times$ 10$^8$ microbubbles/mL. The agent can be made with a variety of entrapped gases, for example air, PFC, and SF$_6$, by purging the surfactant mixture with the gas of choice prior to and during sonication. The exact size distribution is dependent on the gas used and the preparation technique, but in all cases 95% of the bubbles are less than 10 $\mu$m in diameter. In vitro a pulsatile flow system was used to acquire digitized RF A-lines (269). For the air-filled agent backscatter changes nonlinearly with dose; the maximum was 13.1 dB $\pm$ 1.0 dB for a dose of 0.30 $\mu$m/mL of suspending fluid. Attenuation reached approximately 11 dB/cm for dosages above 0.27 $\mu$l/mL and for frequencies between 2.5 and 6.0 MHz. In vivo, intravenous injections of ST68 were given to rabbits at doses from 0.01 to 0.23

mL/kg. A clear increase in flow signal was observed for 1 to 2 minutes. An in vivo dose response curve was calculated from audio Doppler signals obtained with a 10 MHz cuff transducer placed around the distal aorta. Maximum enhancement was 18.3 dB $\pm$ 3.13 dB for a 0.13 mL/kg dose (269). In color Doppler and CAI all agents produced remarkable enhancement, but PFC-filled agents gave the most dramatic and long lasting effect (up to 15 minutes). Figure 9 compares (1) the preinjection Doppler image (2) with the image after injection of 0.1 mL/kg of ST68-containing PFC. Injections were through the catheterized jugular vein. In addition to the Doppler enhancement, parenchymal enhancement can be observed in the B-mode. Even more dramatic enhancement was obtained with a similar dose but using CAI as shown in Figure 10.

Recently we have investigated the use of the agent for harmonic imaging. Again the PFC-filled agent gave the best results, but the SF$_6$ was a close second. Figure 11 shows the poorly defined harmonic color Doppler image of a New Zealand white rabbit kidney (1) before contrast, and

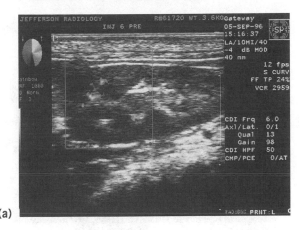
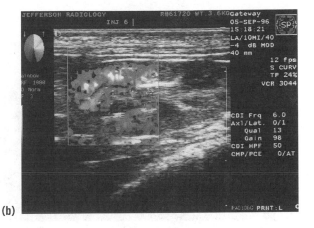

**(a)** **(b)**

**Figure 9.** Color Doppler image of right kidney (see color plate). (**a**) Preinjection and (**b**) postinjection of 0.1 mL/kg of ST68-containing PFC. The color reaches the kidney capsule, and marked parenchymal enhancement is seen on the B-mode (outside the color box). *Source:* Images obtained in collaboration with F. Forsberg, and B.B. Goldberg at Thomas Jefferson University Hospital and funded in part by grants number NIH HL52901 and NIH CA52823.

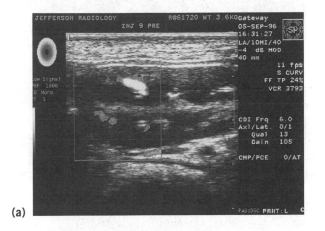
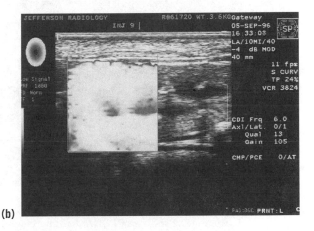

**(a)** **(b)**

**Figure 10.** CAI of right kidney (see color plate). (**a**) Preinjection and (**b**) postinjection of 0.1 mL/kg of ST68-containing PFC. Color blooming and flash artifact subsided after a few passes. Enhancement lasted 15 minutes. *Source:* Images obtained in collaboration with F. Forsberg and B.B. Goldberg at Thomas Jefferson University Hospital and funded in part by grants number NIH HL52901 and NIH CA52823.

(2) the strong color and greatly enhanced B-mode image after intravenous contrast, using ST44-PFC. After contrast the kidney is clearly outlined, and shadowing can be observed in the lower edge of the kidney due to the high reflectance of the ultrasound wave by the agent closer to the transducer. Enhancement lasted from 10 to 15 minutes depending on the surfactant mixture that was used. The results are again more dramatic in color amplitude mode, as shown in Figure 12. These results are extremely encouraging, and are the subject of ongoing research.

**Actively Targeted Ultrasound CA.** Developments in ultrasound CA have followed those in MRI, with development of agents which contain a targeting moiety specific for a site at a given pathology. Active targeting to a thrombus is being pursued by several groups. One such agent is being developed at Imarex (Tucson, Arizona). The agent,

MRX-408 consists of an Aerosome-bearing ligands (RGD analogues) specific to the GPIIBIIIA receptor of activated platelets in thrombus (270–273). Compared with injections of the nontargeted Aerosomes MRX-113 or MRX-115, significant increases in intensity of ultrasound postcontrast with MRX-408 were shown using videodensitometry. Thrombus was created in dog by advancing a guidewire-binding thrombin-coated thread into various vessels, including the aortobifemoral system, inferior vena cava, and left auricular appendage.

Others are exploiting the strong binding between avidin and biotin in a targeting method reminiscent of site-specific drug delivery (274,275). The thrombus was first pretargeted with a biotin–antibody conjugate (biotin–antifibrin monoclonal antibody), followed by injection of excess avidin. This resulted in a clot labeled with avidin via an antibody–biotin bridge. Biotinylated, lipid-coated, PFC

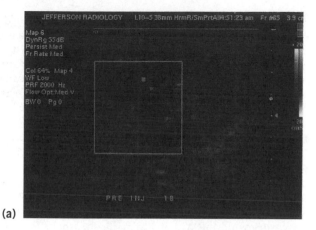

(a)

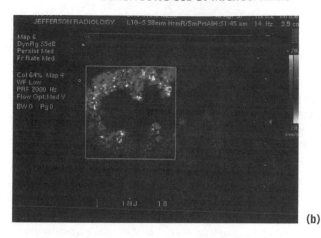

(b)

**Figure 11.** Color Doppler harmonic image of right kidney (see color plate). (**a**) Preinjection and (**b**) postinjection of 0.1 mL/kg of ST44-containing PFC. With a mechanical index of 0.6 the precontrast image is poorly defined. Strong color and B-mode enhancement are obtained with contrast. *Source:* Images obtained in collaboration with F. Forsberg and B.B. Goldberg at Thomas Jefferson University Hospital and funded in part by grants number NIH HL52901 and NIH CA52823.

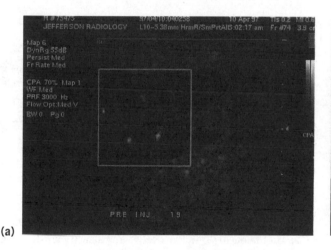

(a)

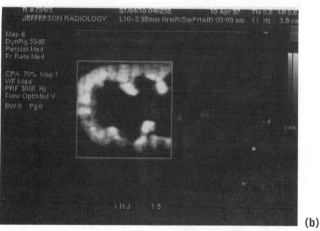

(b)

**Figure 12.** Color amplitude harmonic image of right kidney (see color plate). (**a**) Preinjection and (**b**) postinjection of 0.1 mL/kg of ST68-containing PFC. The dark vertical lines in the lower kidney indicate strong shadowing. *Source:* Images obtained in collaboration with F. Forsberg and B.B. Goldberg at Thomas Jefferson University Hospital and funded in part by grants number NIH HL52901 and NIH CA52823.

emulsion was then administered, which bound to the avidin exposed on the thrombus surface. They demonstrated that in dog, thrombi exposed to antifibrin-targeted contrast exhibited significantly increased echogenicity ($P <$ .05). The great advantage of this method is that a single biotinylated agent could be used in a variety of targeting situations. The initial treatment with a unique biotin–antibody conjugate defines the target site by the choice of antibody. The same biotinylated agent targets any site once the avidin in the follow-up injection has bound to the specific biotin–antibody conjugate at the site.

Antibody is also used in an agent targeted to atherosclerosis as well as thrombi (276). The CA was again a liposome-based agent composed of phosphatidylcholine, 4-(p-maleimidophenyl)butyrylphosphatidyl ethanolamine, phosphatidylglycerol, and cholesterol. Antibody was con-

jugated via the amine group on the phosphatidyl ethanolamine (PE) using 3-(2-pyridyldithiolpropionic acid) *N*-hydroxysuccinimide ester. Electron microscopy revealed attachment of the agent to in vitro thrombi, and attachment was demonstrated ex vivo by ultrasound imaging of postmortem samples from standard Yucatan miniswine model of induced atherosclerosis.

## Ultrasound Contrast Imaging with Therapeutic Modalities

The therapeutic applications of ultrasound have primarily used the potential of ultrasound to induce cell damage, both lethal cell lysis induced by ultrasonic cavitation (277) and transient permeabilization of cell membranes (278). Cavitation can be either unstable (also called transient) or stable. Transient cavitation is described as the generation, growth, and disruption of microbubbles. Bubble disruption produces a very high local pressure change and extremely high, transient temperatures. Stable cavitation gives rise to oscillation of the microbubbles, which itself is accompanied by large shock waves that produce microstreaming of liquids. Catheter delivery of ultrasound has been demonstrated to dissolve clots (279,280) and transcutaneous ultrasound has been reported to enhance the effect of thrombolytic agents (281). It is a logical extension to investigate the effect of the presence of CA on these and similar effects, because the microbubbles can act as cavitation nuclei. Tachibana and Tachibana demonstrated in vitro that Albunex could accelerate the fibrinolytic effect of urokinase in the presence of 170 kHz ultrasound (282). Porter has shown, again in vitro, that PESDA can enhance clot disruption (283). Finally, Siegel and co-workers have demonstrated both in vitro and in vivo in rabbit that the phase shift agent Echogen (dodecafluoropentane emulsion) significantly enhances the clot disrupting effect of low frequency ultrasound (284).

In the area of drug and gene delivery, ultrasound has not lagged behind progress already described for MRI. The Unger group has been in the forefront, with initial investigations into the successful use of microbubbles to increase the rate of heating during ultrasound hyperthermia as early as 1992 (285). They followed this with the use of gas-filled liposomes (286), and in both cases found that the microbubbles were destroyed. This gave the idea of incorporation of drug into the lipid bilayer of the agent, and the use of ultrasound to first release the drug and then facilitate its delivery (287). Using 1 MHz continuous wave ultrasound, they have demonstrated an increased rate of expression from liposomal agents in vitro in cell culture (288,289). Gene expression was assessed 48 hours after transfection with cationic liposomes complexed to the gene for chloramphenicol acetyltransferase. Interestingly, ultrasound given before during or after addition of the gene to the cells increased the rate of gene expression. However, perhaps the most significant observation came when they used fluorinated gas precursors in the liposomes and achieved even higher gene expression. Miller and co-workers have demonstrated the use of Albunex in cultured Chinese hamster ovary cells exposed to 2.25 MHz ultrasound to produce cavitation nuclei to facilitate transient permeabilization (sonoporation) at spattila peak pressure amplitudes as low as 0.1 MPa (290). Above this pressure considerable lysis was observed, increasing rapidly above the cavitation threshold of 0.4 MPa, as determined by $H_2O_2$ production test. In DNA transfection tests, Luciferase production reached a maximum of 0.33 ng per $10^6$ cells at 0.8 MPa exposure.

## CONCLUSION

The use of CA has greatly enhanced the diagnostic potential of the major noninvasive imaging techniques, whether they are an inexpensive agent such blueberry juice, used with the most expensive imaging technology of MRI (89), or highly sophisticated agents such as those enabling gene transfection to be followed by MRI (107). Microencapsu-

lation adds further dimensions to these agents. Indeed in some case, such as gaseous ultrasound CA, microencapsulation affords the imaging moiety enough stability to pass the lung to be used in the left heart and other areas of the body. Examples of the advantages of microencapsulation in CA can be chosen for all modalities. In many cases microencapsulation decreases inherent toxicity of the agent, as seen with microencapsulated SPIO (123), and frequently the natural biodistribution of the unencapsulated agent is altered. Using much of the experience acquired in drug delivery research, targeting by passive and active targeting can be directed to numerous sites: for example, tumors (201), lymph (79), blood clots (226), and even different cell types in the liver (75,76). In addition, microencapsulation has reduced the threshold for detection of hepatic lesions and continues to do as the field advances. The future will see many more advances as microencapsulated agents are developed that will demonstrate even more exquisitely specific targeting, and that will differentiate normal from abnormal physiological activity in the developing field of functional imaging.

## BIBLIOGRAPHY

1. E.R.N. Grigg, *The Trail of the Invisible Light: From X-Strahlen to Radio(bio)logy and Therapeutics*, Sec. 76, vol. 1, Pergamon, Oxford, 1993.

2. D.D. Shaw, *Invest. Radiol.* **28**(3), S138–S139 (1993).

3. E.C. Unger et al., *Radiology* **185**, 453–456 (1992).

4. D.J. Yang et al., *ACS Symp. Ser.: Polymeric* **520**, 371–381 (1993).

5. D.J. Yang et al., *J. Pharm. Sci.* **83**(3), 328–331 (1994).

6. M.R. Violante, K. Mare, and H.W. Fischer, *Invest. Radiol.* **16**, 40–45 (1981).

7. O. Laccourreye et al., *Invest. Radiol.* **28**, 150–154 (1993).

8. H. Taniguchi, T. Takahashi, T. Yamaguchi, and K. Sawai, *Cancer (Philadelphia)* **64**, 2001–2006 (1989).

9. S. Batnitzky, in R. Miller and J. Skucas, eds., *Radiographic Contrast Agents*, University Park Press, Baltimore, Md., 1977, pp. 429–469.

10. G. Marshall, *Radiogr. Today* **61**(694), 9–11 (1995).

11. S. Chuang, *Am. J. Neuroradiol.* **13**, 785–791 (1992).

12. M.E. Bernardino, in Z. Parver, ed., *Contrast Media: Biological Effects and Clinical Applications*, CRC Press, Cleveland, Ohio, 1987, pp. 25–39.

13. S.E. Seltzer et al., *Invest. Radiol.* **19**, 142–151 (1984).

14. M.R. Violante and H.W. Fischer, in Z. Parver, ed., *Contrast Media: Biological Effects and Clinical Applications*, CRC Press, Cleveland, Ohio, 1987, pp. 89–103.

15. M.S. Sands, M.R. Violante, and G. Gadenolt, *Invest. Radiol.* **22**, 408–416 (1987).

16. J.B. Benoit and F. Puisieux, in P. Guiot and P. Couvreur, eds., *Polymeric Nanoparticles and Microspheres*, CRC Press, Boca Raton, Fla., 1986, pp. 137–174.

17. Fr. Demande 2,548,902 (January 18, 1985), P. Trampont, P. Madoule, B. Bonnemain, and F. Puissieu (to Guerbet).

18. A. Jayakrishnan, B.-C. Thanoo, K. Rathinam, and M. Mohanty, *J. Biomed. Mater. Res.* **24**, 993–1004 (1990).

19. D. Horak et al., *Biomaterials* **8**, 142–145 (1987).

20. B.-C. Thanoo and A. Jayakrishnan, *J. Microencapsul.* **66**(2), 233–244 (1989).

21. B.-C. Thanoo and A. Jayakrishnan, *Biomaterials* **11**, 477–481 (1990).

22. B. Lindberg, K. Lote, and H. Teder, in S.S. Davis, L. Illum, J.G. McVie, and E. Tomlinson, eds., *Microspheres and Drug Therapy*, Elsevier, Amsterdam, 1984, pp. 153–188, 189–203.

23. P. Rongved, J. Klaveness, and P. Strande, *Carbohydr. Res.* **297**, 325–331 (1997).

24. D.C. Leegwater and J.B. Luten, *Staerke* **23**, 430–432 (1971).

25. S. Mir et al., *Invest. Radiol.* **120**, 119–122 (1984).

26. A.A. Epenetos et al., *Nucl. Med. Commun.* **8**, 1047–1058 (1987).

27. J.E. Dion et al., *Radiology* **160**, 717–721 (1986).

28. J.-Y. Yang et al., *Invest. Radiol.* **30**(6), 354–358 (1995).

29. M.R. Violante and P.B. Dean, *Radiology* **134**, 237–239 (1980).

30. P. Leander et al., *Invest. Radiol,* **28**, 513–519 (1993).

31. D.E. Chickering, J.S. Jacob, and E. Mathiowitz, *Proc. Int. Symp. Controlled Relat. Bioact. Mater.* **20**, 244–245 (1993).

32. B.-C. Thanoo, M.C. Sunny, and A. Jayakrishnan, *Biomaterials* **12**, 526–529 (1991).

33. O.A. Rozenberg, K.P. Hanson, and E.A. Zherbin, *Radiology* **149**(3), 877–878 (1983).

34. G.L. Kendrasiak, G.D. Frey, and R.C. Heim, *Invest. Radiol.* **20**, 995–1002 (1985).

35. A. Havron, S.E. Seltzer, M.A. Davis, and P.M. Shulkin, *Radiology* **149**, 507–511 (1981).

36. S.E. Seltzer, *Radiology* **171**, 19–21 (1989).

37. S.E. Seltzer et al., *AJR, Am. J. Roentgenol,* **43**(3), 575–579 (1984).

38. M. Zalutsky, A. Noska, and S.A. Seltzer, *Invest. Radiol.* **22**(2) 141–147 (1987).

39. F.A. Zherbin et al., *Vestn. Akad. Med. Nauk SSSR* **4**, 84–89 (1982).

40. P.J. Ryan, M.A. Davis, and D. Melchior, *Biochim. Biophys. Acta* **756**(1), 106–110 (1983).

41. P.J. Ryan et al., *Radiology* **152**, 759–762 (1984).

42. V.J. Caride et al., *Invest. Radiol.* **17**, 381–385 (1982).

43. G.L. Jendrasiak, G.D. Frey, and R.C. Heim, *Invest. Radiol.* **20**, 995–1002 (1985).

44. J.P. Weichert et al., *Invest. Radiol.* **28**, S284–S285 (1994).

45. M.A. Longino, D.A. Bakan, J.P. Weichert, and R.E. Cousell, *J. Pharm. Res.* **13**(6) 875–879 (1996).

46. D.A. Bakan et al., *Hepatology* **24**(4) Pt.2,481A (1996).

47. D.A. Bakan, M.A. Longino, J.P. Weichert, and R.E. Cousell, *J. Pharm. Sci.* **85**(9), 908–914 (1996).

48. F.T. Lee et al., *Radiology* **203**, 465–470 (1997).

49. D.L. Rubin et al., *Invest. Radiol.* **29**(Suppl. 2), S280–S283 (1994).

50. G.L. Wolf and G.S. Gazelle, *Invest. Radiol.* **28**(Suppl. 4), S2–S4 (1993).

51. G.L. Wolf, *Radiology* **185**, 115 (1992).

52. G.L. Wolf et al., *Acad. Radiol.* **1**(4), 352–357 (1994).

53. E. Wisner et al., *Acad. Radiol.* **2**, 405–412 (1995).

54. P. Rougier et al., *Proc. 12th Annu. Meet. Eur. Soc. Med. Oncol., Cancer Immunother.* Nice, France, November 28–30 1986.

55. R. Murakami et al., *Acta Radiol.* **35** (Fasc. 6), 576–580 (1994).

56. Y. Tsushinma et al., *Cardiovasc Intervent. Radiol.* **16**(3) 189–192 (1993).

57. G.J. Dehmer et al., *J. Am. Coll. Cardiol.* **25**(5), 1069–1075 (1995).

58. V.P. Torchilin and V.S. Trubetskoy, *Polym. Prepr.* **38**(1), 545–546 (1997).

59. V.P. Torchilin et al., *Acad. Radiol.* **3**, 232–238 (1996).

60. V.P. Torchilin, M.D. Frank-Kamentsky, and G.L. Wolf, *Acad. Radiol.* **6**, 61–65 (1999).

61. S.W. Young, *Magnetic Resonance Imaging: Basic Principles*, Raven Press, New York, 1988, p. 24.

62. R.C. Brasch, M.D. Ogan, and B.L. Engelstad, in Z. Parvez, R. Moncada, and M. Sovak, eds., *Contrast Media: Biological Effects and Clinical Applications*, vol. III, CRC Press, Boca Raton, Fla., 1987, pp. 131–143.

63. D.G. Gadian and G.M. Bydder, *Contrast Agents Magnet. Reson. Imaging, Proc. Int. Workshop*, San Diego, Calif., January 1986, pp. 14–18.

64. H.J. Weinmann, R.C. Brasch, W.R. Press, and G.E. Wesbey, *AJR, Am. J. Roentgenol.* **142**, 619–624 (1984).

65. G.L. Wolf, *Contrast Media Magn. Reson. Imaging, Proc. Workshop*, Berlin, February 1–3, 1990, pp. 15–17.

66. R.C. Brasch, *Radiology* **183**, 1–11 (1992).

67. G. Marshall, *Radiogr. Today* **61** (697) 9–12 (1995).

68. G. Marshall, *Radiogr. Today* **61** (698), 9–10 (1995).

69. A. Bogdanov, Jr., R. Weissleder, and T.J. Brady, *Adv. Drug Delivery Rev.* **16**, 335–348 (1995).

70. A.-E. Mahfouz, B. Hamm, and D. Mathieu, *Eur. Radiol.* **6**, 607–614 (1996).

71. J. Crnac, M.C. Schmidt, P. Theissen, and U. Sechtem, *Herz* **22**, 16–28 (1997).

72. E.C. Unger, K. Ugurbil, and R.E. Latchaw, *J. Magn. Reson. Imaging* **4**, 235–242 (1994).

73. W. Bradley, Jr., W.T.C. Yuh, and G.M. Bydder, *J. Magn. Reson. Imaging* **3**, 199–218 (1993).

74. R.B. Lauffer et al., *Magn. Reson. Med.* **4**, 582–590 (1987).

75. P. Reimer et al., *Radiology* **177**, 729–734 (1990).

76. E.C. Unger, P. MacDougall, P. Cullis, and C. Tilcock, *Magn. Reson. Imaging* **7**, 417–423 (1989).

77. D.D. Stark et al., *Radiology* **168**, 297–301 (1988).

78. R. Weissleder et al., *Radiology* **171**(3), 835–839 (1989).

79. R. Weissleder et al., *Radiology* **175**, 494–498 (1989).

80. I. Adzamli, D. Johnson, and M. Blau, *Invest. Radiol.* **29**, 243–248 (1991).

81. S. Kaminsky et al., *Radiol. Diagn.* **30**, 541–548 (1989).

82. P. Ang et al., *Radiology* **172**(P), 51 (1989).

83. R. Tart, et al., *Radiology* **172**(P), 517 (1989).

84. J.J. Linskey and R.G. Bryant, *Magn. Reson. Med.* **8**, 285–292 (1988).

85. R. Mattrey, *AJR, Am. J. Roentgenol.* **152**, 247–252 (1989).

86. P.F. Park et al., *Radiology* **164**, 37–41 (1987).

87. B. Van Beers et al., *Eur. J. Radiol.* **14**, 252–257 (1992).

88. D. Widder, R.R. Edelman, W.L. Grief, and L. Monda, *AJR, Am. J. Roentgenol.* **149**(4), 839–843 (1987).

89. K. Hiraishi et al., *Radiology* **194**, 119–123 (1994).

90. R.C. Brasch, *Magn. Reson. Med.* **22**, 282–287 (1991).

91. G. Schumann-Crampion et al., *Invest. Radiol.* **26**, 969–974 (1991).

92. V.S. Vexler, O. Clement, H. Schmitt-Willich, and R.C. Brasch, *J. Magn. Reson. Imaging* **4**, 381–388 (1994).

93. M. Saeed et al., *Radiology* **180**, 153–160 (1991).

94. V.S. Vexler et al., *Invest. Radiol.* **27**, 935–942 (1992).

95. Y. Berthezene et al., *Radiology* **181**, 773–777 (1991).

96. Y. Berthezene et al., *Invest. Radiol.* **27**, 346–351 (1992).

97. Y. Berthezene et al., *Radiology* **183**, 667–672 (1992).

98. Y. Berthezene et al., *Radiology* **185**, 97–103 (1992).

99. J.C. Bock, U. Pison, P. Kaufman, and R. Felix, *J. Magn. Reson. Imaging* **4**, 473–476 (1992).

100. V.S. Vexler et al., *J. Magn. Reson. Imaging* **2**, 311–319 (1992).

101. C. Grandin et al., *Invest. Radiol.* **30**(10), 572–582 (1995).

102. I. Mottet, et al., *Magn. Reson. Med.* **35**, 131–135 (1996).

103. R. Weissleder et al., *J. Magn. Reson. Imaging* **3**(1), 93–97 (1993).

104. A.A. Bogdanov, Jr. et al., *Radiology* **187**, 701–706 (1993).

105. H.W. Frank, R. Weissleder, A.A. Bogdanov, Jr., and T.J. Brady, *AJR, Am. J. Roentgenol.* **162**, 1041–1046 (1994).

106. A.A. Bogdanov, Jr. et al., *J. Drug Target.* **4**(5), 321–330 (1997).

107. J.F. Kayyem, R.M. Kumar, S.E. Fraser, and T.J. Meade, *Chem. Biol.* **2**, 615–620 (1995).

108. S. Gohr-Rosenthal, H. Schmitt-Willich, W. Ebert, and J. Conrad, *Invest. Radiol.* **28**, 789–795 (1993).

109. L.X. Tiefenauer, G. Kuhne, and R.Y. Andres, *Bioconjugate Chem.* **4**, 347–352 (1993).

110. W.A. Gibby, J. Billings, J. Hall, and T.W. Ovitt, *Invest. Radiol.* **25**, 164–172 (1990).

111. S.C. Wang et al., *Radiology* **175**(2), 483–488 (1990).

112. J. Blustajn et al., *Magn. Reson. Imaging* **15**(4), 415–421 (1997).

113. C. Tilcock et al., *J. Magn. Reson. Imaging* **1**, 463–467 (1991).

114. R.B. Lauffer and T.J. Brady, *Magn. Reson. Imaging* **3**, 11–16 (1985).

115. U. Schmiedl et al., *Radiology* **163**, 205–219 (1987).

116. U. Schmiedl, R.C. Brasch, M. Ogan, and M.E. Moseley, *Acta Radiol., Suppl.* **374**, 99–102 (1990).

117. H.C. Schwickert et al., *Acad. Radiol.* **2**, 851–958 (1995).

118. A.B. Baxter, S. Melnikoff, D.P. States, and R.C. Brasch, *Invest. Radiol.* **26**, 1035–1040 (1991).

119. D.J. Parmelee, R.C. Walovitch, H.S. Oullet, and R.B. Lauffer, *Invest. Radiol.* **32**(12), 741–747 (1997).

120. B. Kreft et al., *J. Magn. Reson. Imaging* **6**(2), 378–383 (1996).

121. B. Kreft et al., *J. Magn. Reson. Imaging* **4**(3), 373–379 (1994).

122. P. Rongved, B. Lindberg, and J. Klaveness, *Carbohydr. Res.* **214**, 325–330 (1991).

123. A. Fahlvik, P. Artursson, and P. Edman, *Int. J. Pharm.* **65**, 249–259 (1990).

124. J.-M. Colet, Y. Van Haverbeke, and R.N. Muller, *Invest. Radiol.* **29**(Suppl. 2), S223–S225 (1994).

125. U.S. Pat. 5,314,679 (May 24, 1994), J.M. Lewis et al. (to Advanced Magnetics Inc.).

126. U.S. Pat. 5,492,814 (February 29, 1996), R. Weissleder (to The General Hospital Corporation).

127. D. Pouliquen et al., *Magn. Reson. Imaging* **7**, 619–627 (1989).

128. M.I. Papisov et al., *J. Magn. Reson. Mater.* **122**, 383–386 (1993).

129. P. Reimer et al., *Radiology* **193**, 527–531 (1994).

130. E. Schulze et al., *Invest. Radiol.* **30**(10), 604–610 (1995).

131. H. Frank, R. Weissleder, and T. Brady, *AJR, Am. J. Roentgenol.* **162**, 209–213 (1993).

132. R. Weissleder, A. Bogdanov, E.A. Neuwelt, and M. Papisov, *Adv. Drug Delivery Rev.* **16**, 321–334 (1995).

133. U. Schoepf et al., *BioTechniques* **24**, 642–651 (1998).

134. K.G. Go et al., *Eur. J. Radiol.* **16**, 171–175 (1993).

135. A.E. Stillman, N. Wilke, and M. Jerosch-Herold, *J. Magn. Reson. Imaging* **7**, 765–767 (1997).

136. R. Weissleder, A. Bogdanov, and M.I. Papisov, *Magn. Reson. Q.* **8**(1), 55–63 (1992).

137. L.X. Tiefenauer, G. Tschirky, G. Kuhne, and R.Y. Andres, *Magn. Reson. Imaging* **14**(4), 391–402 (1996).

138. C. Leveille-Webster, J. Rogers, and I.M. Arais, *Hepatology* **23**, 1631–1641 (1996).

139. P. Reimer et al., *J. Magn. Reson. Imaging* **2**, 177–181 (1992).

140. E. Weiner, S. Konda, A. Shadron, M. Brechbiel, and O. Gansow, *Invest. Radiol.* **32**(12), 748–754 (1997).

141. F.M. Krieg, R.Y. Andres, and K.H. Winterhalter, *Magn. Reson. Imaging* **13**(3), 393–400 (1995).

142. J.A. Nelson, U. Schimedl, and E.G. Shankland, *Invest. Radiol.* **25**, S71–S73 (1990).

143. M.D. Ogan, D. Revel, and R.C. Brasch. *Invest. Radiol.* **22**, 822–828 (1987).

144. D.A. Place et al., *Invest. Radiol.* **25**, S69–S70 (1990).

145. R. Jain, *Cancer Metastasis Rev.* **9**, 253–266 (1990).

146. S. Pislaru et al., *Eur. Heart J.* **18**, 572 (1997).

147. H.H. Engelhard and D.A. Petruska, *Cancer Biochem. Biophys.* **13**, 1–12 (1992).

148. V.E. Groman et al., *BioTechniques* **3**, 156–160 (1985).

149. E.C. Unger, D.K. Shen, G. Wu, and T. Fritz, *Magn. Reson. Med.* **22**, 304–308 (1991).

150. D. Barsky, B. Putz, K. Schulten, and R.L. Magin, *Magn. Reson. Med.* **24**, 1–13 (1992).

151. G. Bacic, M.R. Niesman, R.L. Magin, and H.M. Swartz, *Magn. Reson. Med.* **13**, 44–61 (1990).

152. C. Tilcock, E. Unger, P. Cullis, and P. MacDougall, *Radiology* **171**, 77–80 (1989).

153. E. Unger et al., *Radiology* **171**, 81–85 (1989).

154. S. Karlik, E. Florio, and C.W.M. Grant, *Magn. Reson. Med.* **19**, 56–66 (1991).

155. G.W. Kabalka et al., *Magn. Reson. Med.* **8**, 89–95 (1988).

156. R. Schwendener et al., *Invest. Radiol.* **25**(8), 922–932 (1990).

157. R. Storrs et al., *J. Magn. Reson. Imaging* **5**, 719–724 (1995).

158. A.L. Alexander et al., *J. Magn. Reson. Imaging* **35**, 801–806 (1996).

159. L.C. Clark, J.L. Ackerman, and S.R. Thomas, *Adv. Exp. Med. Biol.* **180**, 835–845 (1984).

160. C.H. Sotak et al., *Magn. Reson. Med.* **29**, 188–195 (1993).

161. A.G. Webb et al., *J. Magn. Reson. Imaging* **6**(4), 675–683 (1996).

162. S.E. Mathews, C.W. Poulton, and M.D. Threadgill, *Adv. Drug Delivery Rev.* **18**, 219–267 (1996).

163. E.R. Wisner et al., *J. Med. Chem.* **40**, 3992–3996 (1997).

164. R. Weissleder et al., *Antimicrob. Agents Chemother.* **39**(4), 839–845 (1995).

165. B. Spellerberg et al., *J. Exp. Med.* **182**, 1037–1044 (1995).

166. M.D. Bednarski, J.W. Lee, M.R. Callstrom, and K.C.P. Li, *Radiology* **204**, 263–268 (1997).

167. P.N. Burns, *Radiol. Med.* **87**, 71–82 (1994).

168. J. Ophir and K.J. Parker, *Ultrasound Med. Biol.* **15**, 319–333 (1989).

169. P.M. Morse and K.V. Ingard, *Theoretical Acoustics*, McGraw-Hill, New York, 1968.

170. A.L. Anderson and L.D. Hampton, *J. Acoust. Soc. Am.* **67**, 1865–1889 (1980).

171. E.G. Tickner and N.S. Rasnor, *Adv. Bioeng.*, pp. 101–103 (1978).

172. N. de Jong, L. Hoff, T. Skotland, and N. Bom, *Ultrasonics* **30**, 95–103 (1992).

173. C.C. Church, *J. Acoust. Soc. Am.* **97**(30), 1510–1521 (1995).

174. B.A. Schrope and V.L. Newhouse, *Ultrasound Med. Biol.* **19**, 567–570 (1993).

175. P.N. Burns, J.E. Powers, and T. Fritzsch, *Radiology* **182**, 142 (Abstr.) (1992).

176. B.B. Goldberg, J.-B. Liu, and F. Forsberg, *Ultrasound Med. Biol.* **20**, 319–333 (1994).

177. P.S. Epstein and K.J. Plesset, *J. Chem. Phys.* **18**, 1505–1509 (1950).

178. N. de Jong, *IEEE Eng. Med. Biol.* 72–82 (1996).

179. R. Gramiak and P.M. Shah, *Invest. Radiol.* **3**, 356–366 (1968).

180. P. Burns, *Clin. Radiol.* **51**(Suppl. 1), 50–55 (1996).

181. F. Forsberg et al., *Int. J. Imaging Syst. Technol.* **8**, 69–81 (1997).

182. R.S. McKay, *Proc. 2nd Symp. Underwater Physiol.*, National Academy of Sciences, Research Council Number 11181, 41 (1963).

183. H. Feigenbaum et al., *Circulation* **41**(4), 615–621 (1970).

184. M.C. Ziskin, A. Bonakdapour, D.P. Weinstein, and P.R. Lynch, *Invest. Radiol.* **7**, 500–505 (1972).

185. S.B. Feinstein et al., *J. Am. Coll. Cardiol.* **3**, 14–20 (1984).

186. R.M. Lang et al., *J. Am. Coll. Cardiol.* **8**, 232–235 (1986).

187. R.S. Meltzer and J. Roelandt, eds., *Contrast Echocardiography*, Martinus Nijhoff, Dordrecht, The Hague, The Netherlands, 1982.

188. P. Fan et al., *Ultrasound Med. Biol.* **19**, 45–57 (1993).

189. A. Killam and H.C. Dittrich, in B.B. Goldberg, ed., *Ultrasound Contrast Agents*, Martin Dunitz Ltd., London, 1997, pp. 43–55.

190. E. Unger et al., in B.B. Goldberg, ed., *Ultrasound Contrast Agents*, Martin Dunitz Ltd., London, 1997, pp. 57–74.

191. F. Bauer, C.M. Allen, and C. Missouris, *Adv. Echo-Contrast* **4**, 56 (Abstr. 50) (1995).

192. M. Stizer, *Adv. Echo-Contrast* **4**, 55 (Abstr. 55) (1995).

193. J. Langholz, *Adv. Echo-Contrast* **4**, 56 (Abstr. 51) (1995).

194. E. Leen and C.S. McArdle, *Clin. Radiol.* **51**(Suppl. 1), 35–39 (1996).

195. M. Kudo et al., *AJR, Am. J. Roentgenol.* **158**(1), 65–74 (1992).

196. M. Kudo et al., *Radiology* **182**(1), 155–160 (1992).

197. T. Fritzsch, D. Heidmann, and M. Reinhardt, in B.B. Goldberg, ed., *Ultrasound Contrast Agents*, Martin Dunitz Ltd., London, 1997, pp. 169–192.

198. U. Bogdahn et al., *Radiology* **192**, 141–148 (1994).

199. D.O. Cosgrove, *Proc. 4th Int. Symp. Recent Adv. Urol. Cancer*, Paris, 1995, p. 13.

200. V.F. Duda, G. Rode, and R. Schlief, *Ultrasound Obstet. Gynecol.* **3**, 191–194 (1993).

201. D. Cosgrove, *Adv. Echo-Contrast* **3**(38), 45 (1994).

202. R.P. Kedar et al., *Radiology* **181**(P), 202 (1991).

203. R.P. Kedar et al., *Radiology* **198**, 679–686 (1996).

204. D.O. Cosgrove and M.J.K. Blomley, in B.B. Goldberg, ed., *Ultrasound Contrast Agents*, Martin Dunitz Ltd., London, 1997, pp. 159–168.

205. A. Fleischer, A.K. Parson, and J.A. Cullinan, in B.B. Goldberg, ed., *Ultrasound Contrast Agents*, Martin Dunitz Ltd., London, 1997, pp. 137–148.

206. A.S. Lev-Toaff, D.A. Merton, J. Dumsha, and B.B. Goldberg, *Radiology* **185**, 783–788 (1992).

207. A.S. Lev-Toaff and B.B. Goldberg, in B.B. Goldberg, ed., *Ultrasound Contrast Agents*, Martin Dunitz Ltd., London, 1997, pp. 121–135.

208. B.B. Goldberg, ed., *Ultrasound Contrast Agents*, Martin Dunitz Ltd., London, 1997.

209. D. Cosgrove, *Clin. Radiol.* **51**(Suppl. 1), 1–4 (1996).

210. L. Needleman and F. Forsberg, *Ultrasound Q.* **13**, 121–138 (1996).

211. G. Marshall, *Radiogr. Today* **61**(696), 9–10 (1995).

212. B.A. Carroll et al., *Invest. Radiol.* **3**, 260–266 (1980).

213. J. Ophir, A. Gobuty, R.E. McWhirt, and N.F. Maklad, *Ultrason. Imaging* **2**, 67–77 (1980).

214. M.A. Wheatley, B. Schrope, and P. Shen, *Biomaterials* **11**(9), 713–717 (1990).

215. H. Becher and R. Schlief, *Am. J. Cardiol.* **64**, 374–377 (1989).

216. T. Fritzsch et al., *Invest. Radiol.* **25**, S160–S161 (1990).

217. B.B. Goldberg et al., *J. Ultrasound Med.* **14**, S7 (1995).

218. B.B. Goldberg et al., *Radiology* **177**, 713–717 (1990).

219. R. Schlief, *Proc. 1st Symp. Contrast Agents Diagn. Ultrasound, Leading Edge Diagn. Ultrasound*, Atlantic City, N.J., May 9, 1995, pp. 25–26.

220. H. Becher et al., *J. Am. Coll. Cardiol.* **31**(2), Suppl. A, Abstr. 907-1, p. 400A (1998).

221. M.J. Blomley et al., *Radiology* **205**(P), 418 (1997).

222. B. Goldberg et al., *Radiology* **177**(3), 713–717 (1990).

223. L.J. Crouse et al., *J. Am. Coll. Cardiol.* **22**(5), 1494–1500 (1993).

224. G. Galanti et al., *J. Am. Soc. Echocardiogr.* **6**, 272–278 (1993).

225. J.H. Wible, Jr. et al., *Invest. Radiol.* **29**, S145–S148 (1994).

226. L. Needlemann, T.L. Nack, and R.I. Goldberg, *Radiology* **185**(P), 143 (1992).

227. B.F. Vandenberg and H.E. Melton, *J. Am. Soc. Echocardiogr.* **6**, 582–589 (1994).

228. H.C. Dittrich et al., *J. Am. Soc. Echocardiogr.* **7**, S16 (1994).

229. F. Forsberg et al., *J. Ultrasound Med.* **15**, 853–860 (1996).

230. T.R. Porter, A. Kricsfeld, and K. Kilzer, *J. Am. Coll. Cardiol.* **26**, 33–40 (1995).

231. T.R. Porter, *Proc. 3rd Ann. Symp. Contrast Agents Diagn. Ultrasound, Leading Edge Diagn. Ultrasound*, Atlantic City, NJ, May 13, 1997, pp. 45–46.

232. T. Porter et al., *Circulation* **94**(Suppl.), I-319, (Abstr. 1858) (1996).

233. T. Porter, S. Li, and C. Everbach, *J. Am. Coll. Cardiol.* **31**(2), Suppl. A, Abstr. 907-3, p. 400A (1998).

234. A. Morares et al., *J. Am. Coll. Cardiol.* **31**(2), (Suppl. A), Abstr. 907-4, p. 400A (1998).

235. R.F. Mattrey et al., *Radiology* **163**, 339–343 (1987).

236. S.F. Flaim, D.R. Hazard, J. Hogan, and R.M. Peters, *Invest. Radiol.* **26**, S122–S124 (1991).

237. R. Mattrey and T.J. Pelura, in B.B. Goldberg, ed., *Ultrasound Contrast Agents*, Martin Dunitz Ltd., London, 1997, pp. 83–99.

238. R. Satterfield et al., *Invest. Radiol.* **4**, 325–331 (1993).

239. R.F. Mattrey et al., *Acad. Radiol.* **3**, S320–S321 (1996).

240. S.L. Mulvagh et al., *J. Am. Coll. Cardiol.* **25**, 288A (1995).

241. J.-M. Correas, D. Kessler, D. Worah, and S. Quay, in B.B. Goldberg, ed., *Ultrasound Contrast Agents*, Martin Dunitz Ltd., London, 1997, pp. 100–120.

242. F. Forsberg et al., *Int. J. Ultrasound Med.* **14**, 949–957 (1995).

243. C.R. Pugh, P.H. Arger, and C.M. Sehgal, *J. Ultrasound Med.* **15**, 843–852 (1996).

244. M.L. Robbin and A.J. Eisenfeld, *Radiology* **207**, 717–722 (1998).

245. T. Barrette et al., *Abstr., Proc. Symp. Am. Inst. Ultrasound Med.*, San Francisco, March 26–29, 1995.

246. U.S. Pat. 5,580,575 (December 3, 1995), E. Unger et al. (to ImaRx Pharmaceutical Corp., Tucson, Ariz.).

247. U.S. Pat. 5,656,211 (August 12, 1997), E. Unger, T. McCreer, D. Yellowhair, and T. Barrette (to Imarx Pharmaceutical Corp., Tucson, Ariz.).

248. M. Kaps et al., *Stroke* **5**, 1006–1008 (1997).

249. P. Sollender, K.D. Beller, and R. Linder, *Acad. Radiol.* **3**(Suppl. 2), 194–197 (1996).

250. J.S. D'Arrigo, R.H. Simon, and S.-Y. Ho, *J. Neuroimaging* **1**, 134–139 (1991).

251. R.H. Simon et al., *Ultrasound Med. Biol.* **19**(2), 123–125 (1993).

252. R.H. Simon et al., *Invest. Radiol.* **25**, 1300–1304 (1990).

253. E. Barbarese, S.-Y. Ho, J.S. D'Arrigo, and R.H. Simon, *J. Neuro-Oncol.* **26**, 25–34 (1995).

254. M. Schneider et al., *Invest. Radiol.* **30**(8), 451–457 (1995).

255. D. Rovai et al., *J. Am. Soc. Echocardiogr.* **11**(2), 169–180 (1998).

256. D.P. Phillips et al., in B.B. Goldberg, ed., *Ultrasound Contrast Agents*, Martin Dunitz Ltd., London, 1997, pp. 149–158.

257. K.J. Parker, T.A. Tuthill, R.M. Verner, and M.R. Violante, *Ultrasound Med. Biol.* **13**(9), 555–566 (1987).

258. K.J. Parker et al., *Invest. Radiol.* **21**, 1135–1139 (1990).

259. U.S. Pat. 5,147,631 (September 15, 1992), J.L. Glajch, G.L. Loomis, and W. Mahler (to Du Pont Merck Pharmaceutical).

260. J.R. Harris, F. Depoix, and K. Urich, *Micron* **26**(2), 103–111 (1995).

261. T. Fritzsch, D. Heldmann, and M. Reinhardt, in B.B. Goldberg, ed., *Ultrasound Contrast Agents*, Martin Dunitz Ltd., London, 1997, pp. 169–176.

262. M. Schneider et al., *Invest. Radiol.* **26**, S190–S191 (1991).

263. M. Schneider et al., *Invest. Radiol.* **27**, 134–139 (1992).

264. M. Schneider, A. Broillet, P. Bussat, and R. Ventrone, *Invest. Radiol.* **29**, S149–S151 (1994).

265. P.J. Lund et al., *Radiology* **185**, 783–788 (1992).

266. M.G. Harisinghani et al., *Clin. Radiol.* **52**, 224–226 (1997).

267. U.S. Pat. 5,352,436 (October 4, 1994) M.A. Wheatley, S. Peng, S. Singhal, and B.B. Goldberg, (to Drexel University).

268. M.A. Wheatley and S. Singhal, *React. Polym.* **25**, 157–166 (1995).

269. F. Forsberg et al., *Ultrasound Med. Biol.* **23**(8), 1201–1208 (1997).

270. T.P. McCreary et al., *Circulation*, **96** (Suppl. 8), I-213 (1997).

271. M. Takeuchi et al., *J. Am. Coll. Cardiol.* **31**(2), Suppl. A, Abstr. 907–2, p. 400A (1998).

272. M. Takeuchi et al., *J. Am. Coll. Cardiol.* **31**(2), Suppl. A, Abstr. 803–6, p. 57A (1998).

273. W.H. Wright et al., *Radiology* **205**(P), 418 (1997).

274. G.M. Lanza et al., *Circulation* **94**(12), 3334–3340 (1996).

275. G.M. Lanza et al., *Circulation* **94** (Suppl.), I-319, (Abstr. 1860), (1996).

276. S.A. Demos et al., *J. Pharm. Sci.* **86**(2), 167–171 (1997).

277. M.W. Miller, D.L. Miller, and A.A. Brayman, *Ultrasound Med. Biol.* **22**, 1131–1154 (1996).

278. M. Frechheimer et al., *Proc. Natl. Acad. Sci. U.S.A.* **84**, 8463–8467 (1987).

279. W. Steffen et al., *J. Am. Coll. Cardiol.* **24**, 1571–1579 (1994).

280. C.W. Hamm et al., *Lancet* **343**, 605–606 (1994).

281. A. Blinc, C.W. Francis, J.L. Trudnowski, and E.L. Carstensen, *Blood* **81**, 2636–2643 (1993).

282. K. Tachibana and S. Tachibana, *Circulation* **92**(5), 1148–1150 (1995).

283. T.R. Porter et al., *Am. Heart J.* **132**, 964–968 (1996).

284. T. Noshihiko et al., *J. Am. Coll. Cardiol.* **30**(2), 561–568 (1997).

285. U.S. Pat. 5,149,319 (September 22, 1992), E. Unger.

286. U.S. Pat. 5,209,720 (May 11, 1993), E. Unger.

287. U.S. Pat. 5,580,575 (December 3 1996), E. Unger et al. (to Imarx Pharmaceutical Corp., Tucson, Ariz.).

288. E.C. Unger, *2nd Thoraxcent. Eur. Symp. Ultrasound Contrast Imaging*, Rotterdam, The Netherlands, January 23–24, 1997.

289. R. Sweitzer, T. McCreary, and E. Unger, *Ultrasound Contrast Medium Res. Radiol.*, San Diego, Calif., February 7–9, 1977.

290. S. Bao, B.D. Thrall, and D. Miller, *Ultrasound Med. Biol.* **23**(6), 953–959 (1997).

# E

## ECONOMIC ASPECTS OF CONTROLLED DRUG DELIVERY

Mark Speers
Clarissa Bonnano
Health Advances, Inc.
Wellesley, Massachusetts

### KEY WORDS

Commercialize
Competitors
Economics
Markets
Pricing
Prioritize
Strategy
Transactions
Unmet needs
Value

### OUTLINE

### OVERVIEW

Many drugs are launched with less than ideal delivery characteristics—either in the route of administration, the frequency of administration, and/or the incidence of side effects. Pharmaceutical companies have traditionally focused on developing novel compounds with documented safety and efficacy in treating a disease. To a large extent, these companies have relied upon the monopoly position their patent positions afford. These companies have known that most patients and clinicians tolerate inconvenience and side effects to benefit from the therapeutic benefits.

A separate industry has arisen that focuses on the improvement of drug delivery systems. Drug delivery systems are technologies that aid or enable the administration of therapeutic compounds. These systems include devices, such as inhalers or transdermal patches, as well as formulation technologies. Since R.P. Scherer and KV Pharmaceuticals were founded in the pre–World War II era, more than 100 companies have become actively involved in developing drug delivery systems, and the industry is growing at a considerable pace. According to the investment bank Dillon Read & Company, the drug delivery market will grow from $11.5 billion, or 12% of the total pharmaceutical market in 1996, to $35 billion, or 20% of the total pharmaceutical market in 2005.

Pharmaceutical companies have traditionally partnered with drug delivery companies to codevelop second-generation proprietary products. For example, Alza has developed several sustained-release cardiovascular products, arguably the most successful drug delivery products introduced to date, with the pharmaceutical firms Bayer AG and Pfizer. Drug companies need to use a range of technologies to reformulate their products, as each drug represents different technical challenges and different unmet clinical needs. Drug delivery companies, therefore, apply their technologies across a range of clinical segments. Although they are founded as focused niche players (e.g., depot injectables of small molecules, inhaled forms of peptides), a few of the older, established drug delivery companies have diversified their technology platforms so as to expand their commercial applicability.

Drug delivery deals with pharmaceutical companies are announced daily. Although the specific terms of these transactions are seldom disclosed, the headline deal values appear, initially, to be inconsistent and random. Table 1 lists some recent transactions. In our experience, however, the deals generally make sense when considered in a market context.

The terms of the relationship between pharmaceutical and drug delivery companies are ultimately determined by the value that the reformulation creates in the marketplace. The sources of value in reformulation include extending patent life, compliance improvement, improved therapeutic efficacy, reduced manufacturing costs and market share expansion. In today's managed care environment, with restricted pricing power for the pharmaceutical firms and managed care organizations' use of drug formularies, drug reformulations only command premium prices when they actually deliver value on these dimensions.

### SOURCES OF ECONOMIC VALUE

#### Value of Patent Extension

Generic drugs have assumed a prominent role in the pharmaceutical industry. The modern generics industry only

**Table 1. Drug Delivery Transactions**

| Date | Drug delivery company | Pharmaceutical company | Therapeutic(s) | Transaction value ($ millions) |
|------|----------------------|------------------------|----------------|--------------------------------|
| 3/96 | Inhale | Baxter | Heparin, Factor VIII analogs | 80, 20 in equity |
| 8/96 | Sano | Bristol-Myers Squibb | BuSpar | 40 |
| 1/97 | OSI | Hoechst Marion Russel | EPO | 30 |
| 2/97 | Emisphere | Eli Lilly | Insulin, growth hormone | 60 |
| 7/97 | Unigene | Warner Lambert | Calcitonin | 54.5, 3 in equity |
| 9/97 | Aradigm | SmithKline Beecham | Morphine sulfate | 40, 5 in equity |
| 9/97 | Alkermes | Alza | RMP-7 bradykinin | 60, 50% royalties |
| 12/97 | TheraTech | Proctor & Gamble | Estradiol/testosterone | 35 |

*Source:* Ref. 1.

began in 1984 with the passage of the Drug Price Competition and Patent Term Restoration Act. However, the pace and level of generic substitution has increased dramatically as the generics companies gained legitimacy and as managed care rewarded the practice.

Because patents guarantee market exclusivity and artificially high premiums, patent expiration translates into rapidly declining sales for brand-name pharmaceuticals. Generic versions of drugs are typically introduced at 20–25% of the branded drugs' prices. The branded drug's market erodes rapidly and loses 60–80% of the total days of therapy within 6 months—the influence of managed care and mandatory substitution laws. Figure 1 demonstrates a typical branded drug's precipitous loss of market share. For example, after Tagamet went off-patent in May of 1994, SmithKline Beecham retained only 34% of the sales of cimetidine in 1995.

Reformulating a branded drug may create an improved version that is preferred by clinicians and patients over the less expensive generic versions of the original branded product. Patents on the reformulated product allow the company to effectively extend the patent life of the drug. This strategy can dramatically affect the branded drug's market share (Fig. 2). In this scenario, one year before patent expiration, the company with the branded product launches a reformulation that quickly preserves market share. Generic versions capture only a small portion of the

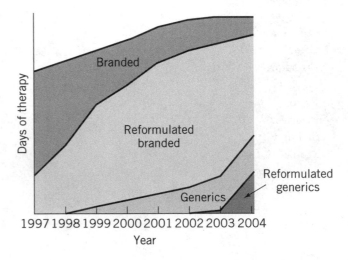

**Figure 2.** Generic penetration with reformulation.

market, until the patent expires on the reformulated product. When Hoechst Marion Roussel reformulated Cardizem to Cardizem CD, a once-daily version of the drug, the company retained 86% of sales of diltiazem after patent expiration, a marked contrast to the market evolution for Tagamet.

Generic drugs are expected to represent an increasing percentage of the pharmaceutical market. One of the trends driving growth of this industry is the large number of upcoming patent expirations on blockbuster products. Table 2 lists the drugs among the top 100 best-selling drugs in the U.S. that will be going off-patent between 1998 and 2005. Figure 3 summarizes the value of these drugs to several major pharmaceutical firms. An important strategic move for these companies is to reformulate these branded drugs with drug delivery technologies. Likewise, generic companies are targeting these drugs as their opportunity to develop premium-priced products. Finally, well-capitalized drug delivery companies are targeting some of these generics as their first candidates to develop products for their own account.

### Value of Compliance Improvement

One of the most significant impediments to keeping patients healthy and curing disease is noncompliance with prescribed medication regimens. Noncompliance explains

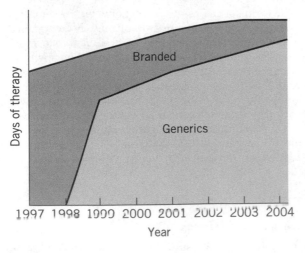

**Figure 1.** Generic penetration without reformulation.

**Table 2. Major Drugs Going Off-Patent 1998–2005**

| Patent expiration | Drug | Manufacturer | Indication | 1997 Worldwide sales ($ millions) |
|---|---|---|---|---|
| 1998 | Atrovent (ipratropium bromide) | Boehringer Ingelheim | Asthma | 691.7 |
|  | Omnipaque (iohexol) | Nycomed Amersham | Contrast enhancement | 393.0 |
| 1999 | Beclovent (beclomethasone dipropionate) | Glaxo Wellcome | Asthma | 542.8 |
|  | Versed (midazolam) | Hoffmann-LaRoche | Depression | 431.3 |
|  | Fortaz (ceftazidime) | Glaxo Wellcome | Infections | 426.4 |
|  | Nizoral (ketoconazole) | Johnson & Johnson | Fungal infections | 364.0 |
| 2000 | Vasotec (enalapril) | Merck | Hypertension | 2,510.0 |
|  | Pepcid (famotidine) | Merck | Ulcers | 1,180.0 |
|  | Rocephin (ceftriaxone sodium) | Hoffmann-LaRoche | Infections | 1,011.4 |
|  | Humulin (insulin) | Eli Lilly | Diabetes | 939.0 |
|  | Procardia XL (nifedipine) | Pfizer | Hypertension | 822.0 |
|  | Ceftin (cefuroxime axetil) | Glaxo Wellcome | Infections | 649.4 |
|  | Glucophage (metformin) | Bristol-Myers Squibb | Diabetes | 579.0 |
|  | Sporanox (itraconazole) | Johnson & Johnson | Fungal infections | 537.0 |
|  | BuSpar (buspirone) | Bristol-Myers Squibb | Anxiety | 443.0 |
| 2001 | Losec (omeprazole) | Astra AB | Ulcers | 2,815.8 |
|  | Prozac (fluoxetine) | Eli Lilly | Depression | 2,559.0 |
|  | Prilosec (omeprazole) | Astra Merck | Ulcers | 2,240.0 |
|  | Mevacor (lovastatin) | Merck | Cholesterol reduction | 1,100.0 |
|  | Zestril (lisinopril) | Zeneca | Hypertension | 1,035.0 |
|  | Prinivil (lisinopril) | Merck | Hypertension | 585.0 |
|  | Accutane (isotretinoin) | Hoffmann-LaRoche | Acne | 451.3 |
|  | Ambien (zolpidem tartrate) | Searle | Insomnia | 396.0 |
|  | Accupril (quinapril) | Warner-Lambert | Hypertension | 378.0 |
| 2002 | Augmentin (amoxicillin) | SmithKline Beecham | Infections | 1,517.0 |
|  | Mevalotin (pravastatin) | Sankyo | Cholesterol reduction | 1,406.7 |
|  | Pulmicort Turbuhaler (budesonide) | Astra AB | Asthma | 643.9 |
|  | Intron A ($\alpha$-2B interferon) | Schering-Plough | Cancer and viral infections | 598.0 |
|  | Axid (nizatidine) | Eli Lilly | Ulcers | 526.5 |
|  | Nolvadex (tamoxifen) | Zeneca | Breast cancer | 501.1 |
|  | Relafen (nabumetone) | SmithKline Beecham | Arthritis | 489.0 |
| 2003 | Biaxin (clarithromycin) | Abbott | Infections | 1,300.0 |
|  | Cardura (doxazosin) | Pfizer | Benign prostatic hypertrophy | 626.0 |
|  | Flovent (fluticasone propionate) | Glaxo Wellcome | Asthma | 516.6 |
|  | Lotensin (benazepril hydrochloride) | Novartis | Hypertension | 456.1 |
| 2004 | Cipro (ciprofloxacin) | Bayer | Infections | 1,441.1 |
|  | Procrit (epoetin alfa) | Johnson & Johnson | Red blood cell enhancement | 1,169.0 |
|  | Taxol (paclitaxel) | Bristol-Myers Squibb | Ovarian cancer | 941.0 |
|  | Diflucan (fluconazole) | Pfizer | Fungal infections | 881.0 |
|  | Ortho-Novum | Johnson & Johnson | Contraception | 658.0 |
|  | (norethindrone/mestranol) | SmithKline Beecham | Hepatitis B | 584.0 |
|  | Engerix-B (hepatitis B vaccine) | Rhone-Poulenc Rorer | Deep vein thrombosis | 462.2 |
|  | Lovenox (enoxaparin sodium) | Bristol-Myers Squibb | Ovarian cancer | 437.0 |
|  | Paraplatin (carboplatin) |  |  |  |
| 2005 | Zocor (simvastatin) | Merck | Cholesterol reduction | 3,575.0 |
|  | Zoloft (sertraline) | Pfizer | Depression | 1,507.0 |
|  | Paxil (paroxetine) | SmithKline Beecham | Depression | 1,474.0 |
|  | Pravachol (pravastatin) | Bristol-Myers Squibb | Cholesterol reduction | 1,437.0 |
|  | Zithromax (azithromycin) | Pfizer | Infections | 821.0 |
|  | Prevacid (lansoprazole) | TAP Pharmaceuticals | Ulcers | 730.0 |
|  | Lamisil (terbinafine) | Novartis | Fungal infections | 628.4 |
|  | Zofran (ondansetron) | Glaxo Wellcome | Nausea and vomiting | 619.9 |
|  | Zoladex (goserelin) | Zeneca | Breast cancer | 569.9 |
|  | Retrovir (zidovudine) | Glaxo Wellcome | HIV infection | 470.7 |

*Source:* Ref. 2.

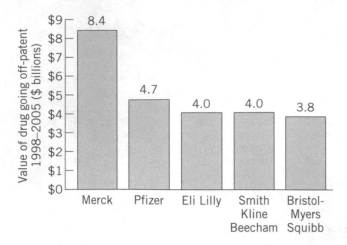

**Figure 3.** Market pressures to discover new drugs or reformulate old drugs. Value based on worldwide revenues.

why therapies often demonstrate greater efficacy in closely monitored clinical trials than in routine medical practice. Although compliance with prescribed regimens is not a significant issue in the closely supervised hospital setting, it is one of the most important issues for outpatients. With the average length of stay in hospitals decreasing and patients being released in a less stable state, compliance is becoming an even more significant issue. According to the Washington-based Center for the Advancement of Health, noncompliance is estimated to cost the U.S. health care system at least $100 billion annually in direct and indirect costs, significantly exceeding the $30 billion cost of prescriptions themselves.

Regardless of any therapy's potential benefits, adherence to the prescribed regimen—the correct timing, dosage, method of delivery, physical status—determines the drug's ultimate success. Clinicians often have to make predictions about expected patient compliance in deciding which drug to prescribe, if any. Clinicians may choose a suboptimal drug for fear that compliance on the more appropriate therapy will be low.

Many factors influence patient compliance, including the nature of the disease and disease symptoms, cognitive or functional ability, and financial resources. Some important factors influencing compliance, the frequency and mode of administration and the extent of drug-related side effects, can be modified through drug reformulation.

**Frequency and Mode of Administration.** Clinicians have learned that to achieve high patient compliance, in the absence of serious noncompliance penalties, drug regimens must be convenient and uncomplicated. Inconvenient (injectables) or complex (many dosages per day) regimens lead to poor compliance.

The oral formulation is the most preferred mode of administration as it is the easiest form for patients to tolerate. In terms of the frequency of administration, less is more. Drugs that must be taken only once per day are ideal, because they gain the highest compliance. Compliance has been shown to drop off sharply for drugs that have

to be taken more than three times per day, thus drugs with more frequent dosing schedules are generally considered unacceptable for therapies that must be taken chronically. Figure 4 illustrates how clinicians rank the main alternative modes and frequencies of administration.

A sustained release oral reformulation that allows for once per day dosing is the preferred form of drug delivery. Aerosol formulations have not been as convenient as oral because they have often required frequent (3 or more times per day) dosing. In addition, the effectiveness of aerosol formulations has been hampered by the inconsistency of inhalers, which results in inadequate or varying levels of drug absorption. Nasal delivery has also lacked consistency in dosage absorbed due to backflow of drug after administration and variations in nasal architecture and volume of mucus between patients. Transdermal systems, although usually providing dosing from three to seven days, are perceived by patients as less attractive because patches can result in skin irritation and may not adhere to the skin efficiently. Depot injections offer significant improvement over frequent injections or intravenous infusions.

There are exceptions to Figure 4. For example, some patients whose compliance on daily oral medications is poor may even prefer depot injections. Contraceptives are a good example. Yet in general, patient safety may be compromised because the drug cannot be discontinued quickly. In addition, rarely will patients self-administer injectable drugs, adding to the inconvenience and cost of the therapy because patients may have to use home care or visit the doctor's office for their medication.

The most successful drug delivery formulations have been, as would be expected, oral sustained release reformulations. Currently, Bayer AG's Adalat CC product for hypertension leads the oral sustained-release market. Reformulated in 1993 from a 3–4/day to 1/day, Adalat CC has climbed to sales of $1.1 billion worldwide in 1997. Another highly successful reformulation has been TAP Pharmaceuticals' Lupron Depot for prostate cancer and endometriosis. Reformulated from a daily injection in 1989 to a once per month injection, Lupron Depot has climbed to worldwide sales of $990 million in 1997.

**Extent and Nature of Side Effects.** Side effects are common with many drugs and eliminating them can significantly increase the value of the therapy. For example, amphotericin B is the most powerful antifungal available, but its use is limited by a severe nephrotoxic side effect profile. Up to 40% of the patients who require amphotericin do not receive the full therapeutic course as a result of the risk of kidney failure. The clinical value of reducing the severity of this side effect is huge, and three companies (NeXstar, the Liposome Company, and Sequus) have developed liposomal formulations that protect the kidney while maintaining equivalent efficacy to the generic formulation. The improved formulations command a price premium of 10 times the generic and after less than 2 years on the market they have increased dollar sales of amphotericin 14-fold.

Side effects may be caused by the action of the drug's active ingredient and therefore are unavoidable. However,

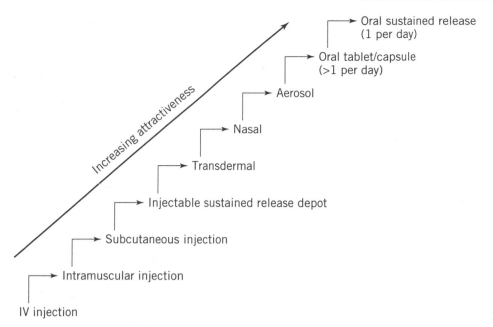

**Figure 4.** Hierarchy of reformulation.

drug delivery technologies can reduce or eliminate side effects such as nephrotoxicity and improve compliance.

### Value of Improved Therapeutic Efficacy

Drug delivery technologies can improve medical outcomes not only by affecting compliance but also by improving therapeutic efficacy. Improvements in bioavailability may help drugs work more effectively, as can technologies that allow the release of the drug at specific times. For example, Covera HS, an Alza reformulation of hypertensive drug initially launched by Searle in 1996, is designed to deliver peak concentrations when blood pressure and heart rate are at their highest. In the field of diabetes, a number of companies are working on technologies to help diabetics maintain stable, close to normal physiologic levels of insulin, because closely titrated patients have been shown to incur fewer long-term complications from the disease than those with widely fluctuating levels.

### Value in Reduced Manufacturing Costs

One method of increasing profitability on a drug is decreasing manufacturing costs. Often orally formulated drugs with poor bioavailability must be administered in high doses, because only a small percentage of the active ingredient is absorbed by the body. Drug reformulations that improve the bioavailability of the drug require less active ingredient to produce an equivalent therapeutic effect, thereby reducing manufacturing costs.

Although the manufacturing cost of a small molecule is low, active ingredients such as proteins and peptides are very expensive. Maximizing the bioavailability of these types of agents will be rewarded in the market.

### Value of Market Share Expansion

Due to the size and delicate structure of peptides and proteins, these molecules have traditionally been delivered via injection. Although this format is rarely desirable, injections may be acceptable for some indications. Some indications require periodic physician visits or admission to the hospital, during which injection therapy does not significantly increase cost or inconvenience to the patient.

However, patients will simply not tolerate injections for many chronic conditions. More patient-friendly formats would therefore allow pharmaceutical firms to apply their drugs to a much broader patient population. In addition, it would expand the range of chemical compounds that may be considered as drugs.

Although much of the value in drug delivery currently lies in the reformulations of small molecules, the potential market for reformulating biotech drugs is considerable. The vast majority of these drugs remain in development, and many are appropriate for chronic diseases. Most of these drugs, however, will never reach their potential as therapeutics until they are reformulated into formats that are more acceptable to patients.

The oral delivery of large molecules constitutes the biggest opportunity within the drug delivery industry. The challenges lie in protecting these molecules from inactivation by peptidases in the gastrointestinal tract while gaining absorption from the intestine to the bloodstream. Because the oral route has been considered a longer-term option, companies have also been investigating transdermal, parenteral depot, and transmucosal delivery systems.

## PRIORITIZING REFORMULATION CANDIDATES

One of the most critical components of a drug or drug delivery company's strategy is the choice of which compounds to reformulate. This process should include the following explicit steps: developing a starting list of drug candidates, examining the delivery technology, assessing the therapeutic or administrative unmet needs, performing a competitive screen, and sizing the market (Fig. 5).

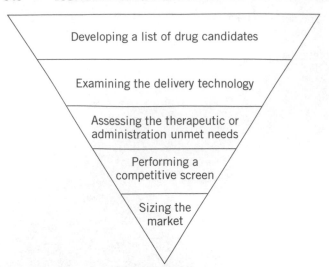

**Figure 5.** Prioritizing reformulation candidates.

### Developing a Starting List of Drug Candidates

Because there is considerable value in extending the patent life of a compound, drug companies often focus on their blockbuster drugs that are coming off-patent. The beginning to that list is Table 2. The pharmaceutical industry's "graveyard" is another source of drug candidates that have clinical potential but have failed in clinical trials due to side effects or administrative problems. There may be even more value in resurrecting these drugs that have consumed costly drug discovery dollars but would not gain regulatory approval and/or market acceptance without reformulation.

Drug delivery companies are often faced with the choice of reformulating an existing product or collaborating with a drug company to reformulate a product that is not yet FDA-approved. The latter choice obviously bears far more risk because the drug is unproven. In addition, reformulating an existing product is particularly appropriate for delivery technologies that have never been reviewed by the U.S. FDA. Thus, the main variable under consideration when the FDA examines the product is the delivery technology. The review process is considerably shorter than if the FDA has to scrutinize the efficacy of a new drug as well. After the FDA is comfortable with the safety of the delivery system, the company might consider reformulating novel compounds. Another important consideration for drug delivery companies is whether the compound is easily sourced. For instance, amoxicillin is certainly easier to come by than paclitaxel.

### Examining the Technology

Each delivery technology has different technical constraints, and companies must evaluate the drugs under consideration against these constraints. For example, many technologies have a maximum payload or molecular weight that can be delivered. Other systems will work more effectively with water-soluble or water-insoluble compounds or with compounds that can withstand a wide range in temperature.

Some drugs and indications present technical challenges as well. Insulin has a narrow therapeutic index; the dose absorbed by the body must be accurate, or hypoglycemic events may occur. This does not allow significant variability in absorption between doses or patients. For some depot formulations, an irretrievable format must be acceptable to clinicians. Drugs that are reformulated from injectable drugs to oral drugs inevitably demonstrate different pharmacodynamic and pharmacokinetic profiles. Some biologics cannot tolerate harsh manufacturing (e.g., encapsulation) techniques.

As more drug delivery technologies compete for a limited number of drug reformulations, it is critical that the drug delivery company honestly assesses its strengths and weaknesses. This screen should dramatically reduce the relevant list of drugs for further screening.

### Assessing the Therapeutic or Administrative Unmet Needs

A drug reformulation must satisfy an unmet need to succeed in the marketplace. For the set of reformulation candidates that have been selected, clinicians can help identify therapeutic or administrative unmet needs. Their input can assist companies in determining what type of reformulation technology would be appropriate.

The cardiovascular and respiratory categories are currently the two biggest therapeutic segments of the drug delivery industry. Reformulations in these areas have generally met administrative unmet needs. In the cardiovascular segment, many drugs have been reformulated into sustained-release products, which increase compliance for the asymptomatic conditions of hypertension and high cholesterol. In the asthma segment, many of the reformulations have used new inhaler systems, exploiting the ease of use of these technologies.

Therapeutic reformulations have generally offered improvements on the side effect profile of drugs. Many nonsteroidal antiinflammatory drugs have been reformulated with delayed release technologies that reduce gastric irritation. Kos' new Niaspan product is a reformulation of the cholesterol product niacin. Long hailed by clinicians as an effective therapy for cholesterol and triglyceride management, niacin is difficult for patients to tolerate because of the side effect of flushing and administration multiple times per day. Niaspan improves upon this product by creating a 1/day controlled-release product to be taken at night. Niaspan also reduces the liver toxicity and flushing side effect associated with niacin. Although patients may still experience some flushing, administration of the controlled release product at night minimizes the discomfort of this side effect.

### Performing a Competitive Screen

Another important step in selecting reformulation candidates is evaluating potential competitive advances. Drug reformulations of compounds going off-patent are prime candidates for significant competitive activity. In addition, many successful drugs are part of successful drug classes. Reformulations of drugs within the same class present a significant competitive threat as well.

In order to perform a comprehensive competitive screen, companies must also analyze drugs in the pipeline and other novel therapies. No matter how successful a reformulation is, it will not meet with commercial success if a therapy in a competitor's pipeline will soon cure or dramatically alter the treatment of a disease. Even experimental surgical remedies should be viewed as competitive with a drug reformulation, if the procedure is expected to change treatment and gain widespread acceptance.

### Sizing the Market

In order to negotiate a successful deal between pharmaceutical and drug delivery companies, both sides need to understand the market potential of the product. The components of market size are the relevant patient population, the price of the therapy, and penetration and market share estimates.

Estimates of most patient populations are readily available in clinical literature. However, the relevant "subset" of the patient population is a more difficult estimate. For example, depending upon the drug candidate, the subset may be defined as those patients with concomitant diabetes, or those patients seen as outpatients, or those patients insured by indemnity insurers. These subset estimates require clinician interviews, secondary research of the literature, and/or mining providers' databases and patient records.

The price of therapy, which includes the price per day and days of therapy required, can initially be estimated using analogs from the product class. With the scrutiny of managed care, few reformulations command a price premium unless they offer extraordinary improvements over existing products. The increasingly common pricing strategy for many companies is to lower the price to gain market share.

Market share estimates should reflect the degree of competition in the category and the marketing partner's previous success in this category. In addition, the pace of penetration is influenced by the degree of fragmentation of the market, indicating how easily physicians can be targeted through a marketing campaign.

Another important benchmark in preparing for negotiations is to identify analogous deals that have been conducted on similar products. Example deals can not only help in determining the value of a technology, but can also suggest creative deal structures.

## THE FUTURE

### Some Aspects of the Industry Will Not Change

We anticipate that the drug delivery industry will continue to constitute its own segment of the pharmaceutical industry. Vertical integration of pharmaceutical companies into drug delivery will not be economically viable as each pharmaceutical company's drug portfolio requires a wide variety of drug delivery technologies. In addition, pharmaceutical companies would be reluctant to partner with the drug delivery subsidiaries of their archrivals.

Although there are undoubtedly complementary drug delivery technologies that will be required to solve specific reformulation challenges, we believe that the boutique nature of the drug delivery industry will continue for the foreseeable future. The pharmaceutical companies will remain open to funding much of the necessary research, and drug delivery companies will remain open to collaborating as necessary.

### Other Aspects Will Undoubtedly Change

We expect a large shift in the mix of drugs that are being reformulated. Historically, the vast majority of reformulations have occurred just before a branded drug was to lose its patent. The branded pharmaceutical companies developed these products.

In the near term, we expect to see generic drug companies and drug delivery companies competing to reformulate generic drugs and, effectively, introduce improved drugs. These reformulation opportunities only exist because drug delivery technologies were not harnessed before the branded products' patent expirations.

The sophistication of drug delivery technologies is advancing rapidly and companies' success rates are growing. As a result, in the longer term, pharmaceutical companies will proactively elect to reformulate drugs well before they reach their patent maturities. As a recent example, Pfizer announced that it was teaming with R.P. Scherer to reformulate a faster-acting form of its blockbuster Viagra less than 4 weeks after it launched the drug. In fact, as drug delivery matures as a science, pharmaceutical companies will involve the technologies in their initial formulations.

## BIBLIOGRAPHY

1. R. Longman, *Windhover's In Vivo* **15**, 49–60 (1997).
2. *MedAdNews* **17**, 94 (1998).

# F

## FABRICATION OF CONTROLLED-DELIVERY DEVICES

JAMES P. ENGLISH
Absorbable Polymer Technologies
Pelham, Alabama

WENBIN DANG
ZHONG ZHAO
Guilford Pharmaceuticals
Baltimore, Maryland

### KEY WORDS

Capsule filling
Drug delivery
Drug–polymer conjugates
Encapsulation
Extrusion
Fabrication
Fluid-bed coating
Injection molding
Manufacturing processes
Polymer
Solvent evaporation
Tablet coating
Tableting
Transdermal patches

### OUTLINE

Manufacturing or fabrication of controlled delivery devices can be defined as "the processing of a bioactive agent into a finished product such that the finished product exhibits controlled delivery of the bioactive agent to a host site." Controlled delivery can be continuous delivery, extended delivery, delayed delivery, pulsatile delivery, or triggered delivery. Using this definition, the bioactive agents may be human or veterinary pharmaceuticals, cosmetic agents, or biocidal agents including pesticides, herbicides, fungicides, and parasiticides. The materials used to sequester and subsequently deliver the bioactive agent are typically polymeric; however, other materials such as waxes and fats are also used. Several physicochemical factors controlling the delivery of a bioactive agent to the host:

- Local pH of the host site
- Hydrophilic or hydrophobic nature of the active agent
- Solubility of the active in the local environment
- Solubility of the active in the delivery matrix
- Permeability of the delivery matrix to water
- Permeability of the delivery matrix to the active agent
- Biostability of the delivery matrix

### APPLICATIONS

Because of the significant advantages of controlled delivery, which include enhanced efficaciousness, reduced toxicity, and improved product stability, the list of products using controlled delivery continues to grow, particularly in cost-neutral or cost-advantageous situations. There are some cases where controlled delivery is actually an enabling technology for a product. For example, the active agent is unstable or is highly toxic without the controlled delivery device. The most active areas for development, marketing, production, and sales of controlled delivery products are human pharmaceuticals, veterinary pharmaceuticals, cosmetics, and agricultural products.

#### Human Pharmaceuticals

**Oral Drug Delivery.** Oral drug delivery is considered to be the holy grail of drug delivery because convenience results in high patient compliance. In the area of human pharmaceuticals, controlled drug delivery had its beginnings in simple wax coatings, which prolonged the delivery of drugs taken orally. Combining many very small encapsulated drug pellets having variations in the solubility and thickness of the coating in a gelatin capsule resulted in the first extended-release cold formulation. Several enteric polymers have been developed which are insoluble in a low pH environment, but swell or dissolve in a high pH environment. Coating a drug with an enteric polymer protects the drug through the low pH of the stomach while allowing it to be released later in the higher pH environment of the intestines (1). Many drugs that are otherwise suitable for oral delivery have a very bad taste, e.g., acetaminophen.

Therefore, coating the drug to mask the taste is often necessary. Taste-masking is sometimes an added benefit obtained in coating oral drugs for other controlled release reasons.

Although oral drug delivery is very desirable, there are many problems encountered in oral delivery that severely restrict the number of drugs that can be delivered orally. The chemical environment of the gastrointestinal tract is very severe. The pH varies from a low of 0.9–1.5 in the stomach to a high of 8.1–9.3 in the intestine. A variety of adverse chemical entities, ranging from salts to digestive enzymes, are present at varying concentrations throughout the gastrointestinal tract (2). And, these chemical entities may react with or otherwise denature many drugs. Therefore, most product development activities in oral drug delivery are focused on protecting a particular drug in the gastrointestinal tract until it reaches its site of most active absorption and prolonging its residence at that site while the drug is being delivered. Very elaborate products have been developed combining features such as enteric coating, particle size, and bioadhesion for delivery of actives to targeted areas of the gastrointestinal tract, particularly the small intestine. An aspirin product has been developed and recently approved for use in some countries that is specifically designed for prophylaxis of cardiovascular disease (3).

**Transdermal Drug Delivery.** Again, because convenience results in high patient compliance, transdermal drug delivery is another highly desirable means of controlled drug delivery. In transdermal drug delivery, the drug delivery device can be a reservoir-type or a matrix-type device. In a reservoir-type device, the device has an impermeable backing film on the outer side, followed by a reservoir containing the drug, then a semipermeable, rate-controlling membrane, followed by an adhesive layer for attachment to the skin, and a final protective, removable inner film (4). Alternatively, for a matrix-type device, the drug can be dispersed in a polymeric matrix, laminated to the backing film and coated with an adhesive layer, followed by a protective, removable inner film. Some devices actually have the drug combined directly with the adhesive (5). The matrix serves as the drug reservoir, and in combination with the skin itself, the rate-controlling portion of the device, thus eliminating the rate-controlling membrane and simplifying the manufacturing of the product.

The skin is designed as a protective membrane for the body and is composed of several complex layers. The most impermeable of these layers is the stratum corneum (the outermost layer of the epidermis). It is difficult to effectively penetrate this barrier; skin penetration is highly dependent on the particular drug. Therefore, as with oral drug delivery, there are limitations to the number of drugs that can be effectively delivered transdermally. Several penetration enhancers, such as oleic acid, oleic acid esters, and poly(ethylene glycol) (PEG) have been discovered and are often added to transdermal formulations as an aid in crossing this barrier (6). Another drawback to transdermal delivery is that it typically takes several hours for a drug to reach a systemic equilibrium. Products containing drugs such as nicotine, glyceryl trinitrate, scopolamine, fentanyl, flurbiprofen, and several steroidal hormones are being marketed (7–10). Several others are under development.

**Parenteral Delivery.** Perhaps the most complex of the controlled drug delivery systems are the human parenteral systems. Biodegradable microsphere and implantable-rod systems which deliver peptides for treatment of prostate cancer have been developed and approved in several countries. Implantable osmotic pumps are used in laboratory animals to conveniently evaluate the controlled delivery of active agents under a variety of conditions. Implantable silicone rods have also been developed and marketed for delivery of steroidal hormones (11–16).

**Dental.** A biodegradable, in situ–forming implant containing doxycycline has been approved in the U.S. for treatment of periodontal disease. The polymer and drug are both dispersed in a water-soluble solvent. When injected into the periodontal pocket, the mixture sets by extraction of the solvent. The implant then delivers its payload and subsequently biodegrades (17,18). Nondegradable fibers containing tetracycline are also used to treat periodontal disease. The fibers are placed in the periodontal pocket where they release tetracycline. The fibers are removed by the dentist once their payload is spent (19). Fluoride salts are routinely added to dental composites, pulp liners, pit and fissure sealants, and bases to aid in recalcification following dental restoration.

## Veterinary Pharmaceuticals

Veterinary pharmaceuticals are replete with controlled delivery products. Products marketed include parasiticides, pesticides, fungicides, vaccines, nutritional supplements, growth hormones, and fertility and estrus regulators. Types of delivery include rumen-bolus delivery, parenteral delivery, and topical delivery. Rumen-bolus delivery is used primarily for delivery of nutritional supplements and parasiticides to ruminant animals. Rumen boluses are designed to be heavy and typically have a complex geometry to prevent their regurgitation. Parenteral delivery is the favored route for delivery of growth hormones, some parasiticides, and antibiotics. Microspheres and implants (pellets and rods) are typical forms. Topical delivery is used primarily for pesticides and ectoparasiticides. Molded or extruded ear tags and collars are the primary forms used (20–25).

## Agricultural Products

The typical agricultural applications of controlled delivery technology are encapsulated fertilizers, pesticides, and herbicides. For agricultural products, cost effectiveness is a major consideration; therefore, process and coating-material selection are limited to the simple and inexpensive. Spray-coating is very common. Interfacial polymerization processes are sometimes used where the coating forms as the product is being sprayed on the host at the time of use thus eliminating isolation of the microcapsules. Some pheromone products are also available for trapping or confusing certain insects (26).

## Cosmetics

Numerous controlled delivery cosmetic products are marketed ranging from encapsulated fragrances to topical insect repellants. The most common controlled delivery cosmetics are skin-cream preparations made with liposomes containing various moisturizers and antioxidant vitamins such as vitamin C and E. Liposomes are small bilayer lipid vesicles formed by phospholipids and similar amphipathic lipids. They were originally thought to be an almost perfect drug delivery system for targeted delivery, but because of numerous problems with bioavailability and formulation stability, they have not yet found widespread use in human pharmaceuticals. However, they are being successfully used in a variety of cosmetic formulations. Methods have recently been developed to improve liposome stability by lyophilization, and liposomes continue to be tested for controlled delivery of a wide variety of pharmaceuticals.

Liposomes are formed when the lipid comes in contact with water. The lipids rearrange into concentric bilayers around aqueous inner compartment(s). Actives can be entrapped in the aqueous or the bilayer regions of the liposomes. Liposomes range in size from nanometers to micrometers and can have one (unilamellar) to many (multilamellar) lipid bilayers. Liposomes with fewer numbers of bilayers have more encapsulated volume to lipid volume and are more permeable. The liposome formation process can be simple, where thin, dry lipid films are simply hydrated, or more complex emulsification processes. Because laboratory processes are not always representative of large-scale processes, product development can often be difficult and costly. High-pressure microfluidization is one useful technology for manufacturing liposomes. Here, high pressure streams are caused to collide at high velocities in microchannels inside an interaction chamber resulting in high shear, impact, and cavitation forces and resulting in uniform, fine-particle emulsions (27–32).

## REGULATORY CONSIDERATIONS

Most controlled delivery devices will fall under some kind of regulatory agency control and/or monitoring depending on the active agent. Human and veterinary pharmaceuticals are regulated by the U.S. FDA and by similar agencies in foreign countries. Pesticides, herbicides, and fungicides are regulated by the U.S. Environmental Protection Agency (EPA) or its equivalent in foreign countries. In the following descriptions of manufacturing processes for controlled delivery of bioactive agents, the reader is likely to be subject to such governmental regulations in manufacturing products of this type.

## MANUFACTURING PROCESSES

There are numerous processes used in manufacturing pharmaceutical controlled drug delivery devices. In the United States, the manufacturing of controlled delivery devices for drug substances is subject to the regulations set forth in Parts 210–226 of the Code of Federal Regulations. These regulations contain the minimum current good manufacturing practice for methods to be used in, and the facilities and controls to be used for, the manufacturing, processing, packaging, or holding of a drug. These regulations assure that the drug meets the FDA requirements for safety, identity, and strength, and that the drug meets the quality and purity characteristics that it is represented to possess. Failure to comply with the regulations renders a drug to be adulterated; and such drugs, as well as the person who is responsible for the failure to comply, are subject to regulatory action (33). Similar regulations exist in other countries.

## Polymeric Materials

Because most controlled delivery devices use polymeric materials to sequester and subsequently deliver an active agent, most controlled delivery manufacturing processes involve the processing of polymeric materials.

A polymer is a material where many smaller molecules are joined together to form a much larger molecule, a macromolecule, often having a molecular weight in the millions. The smaller molecules joined together to form the polymer are called monomers, and the process of joining them together to form the polymer is called polymerization.

### Polymer Structure and Molecular Weight.

The molecular weight of a polymer is quite different from the molecular weight of a small-molecule organic compound. Polymers are polydisperse species that are heterogeneous in molecular weight. A polymer is actually made up of many molecules of differing size. For polymers, the term *molecular weight* means an average molecular weight. And, the average molecular weight and the degree of polydispersity, or molecular weight distribution (MWD), are both required to completely characterize its molecular weight. Analytical techniques are available to accurately determine the number-average molecular weight (Mn), the weight-average molecular weight (Mw), and the polydispersity (MWD) of a polymer. Mn, determined by measurement of colligative properties, is defined as

$$Mn = \Sigma N_x M_x \qquad (1)$$

Where $N_x$ is the mole fraction of molecules of size $M_x$. Mw, determined by light-scattering measurements, is defined as:

$$Mw = \Sigma w_x M_x \qquad (2)$$

Where $w_x$ is the weight fraction of molecules whose molecular weight is $M_x$. The polydispersity or MWD is defined as:

$$MWD = Mw/Mn \qquad (3)$$

Although this value may be determined by measuring the actual values of Mw and Mn, more often it is determined by a process known as size exclusion chromatography (SEC), or gel permeation chromatography (GPC). SEC and GPC are two names for the same analytical process. SEC or GPC is an analytical process whereby a solution of the polymer is passed through a column containing micrometer-sized particles having angstrom-sized pores on their surfaces. The smaller molecules of the polymer are retained on the column longer than the larger molecules and a fractionation of the polymer molecules occurs. The mobile phase is then continually analyzed by various methods, usually refractive index or ultraviolet absorption, to quantify the fraction of the polymer sample as it is eluted from the column. The molecular weight is determined by comparison of the elution time of the polymer fractions to the elution time of known molecular-weight standards. These standards are usually narrow-molecular-weight-range polystyrene. The retention times of several (5–10) polymer standards are determined to obtain a molecular-weight-calibration curve. This method provides a Mw and Mn relative to the standards, and the MWD of the polymer sample.

There are several important factors to consider in selecting or developing a polymer for controlled delivery:

- Biocompatibility and toxicology
- Regulatory acceptance or concerns
- Degradation rate and degradation products and their biocompatibility and toxicology, if biodegradable
- Cost
- Chemical, physical, and mechanical properties
- Suitable solvents
- Processing requirements
- Compatibility limits of the active agent with the polymer
- Required sterilization methods
- Thermal transition temperatures

The relative importance of these factors varies depending on the particular application. For example, if the product contains a thermally sensitive active agent, a polymer requiring thermal processing might not be acceptable.

In developing polymers for use in controlled delivery, manufacturers carefully consider each of the aforementioned factors. The polymers' suitability for particular applications is determined, and the polymer is marketed based on its advantages. The manufacturing requirements for the polymer vary depending on the intended applications and applicable regulations. Manufacturers seek market advantages based on their polymer's suitability for an intended application, its quality, consistency, and cost, most of which depend on their particular manufacturing process. Patented or proprietary processes are often used.

**Polymer Classification.** There are several different ways of classifying polymers. The most common are addition and condensation polymers, referring to composition or structure of the polymer, and step and chain polymers, referring to the polymerization reaction mechanization. Polymers may be organic (containing carbon in the main chain), semi-inorganic (containing no carbon in the main chain, but containing carbon in side groups or chains), or inorganic (containing no carbon at all). Organic and semi-inorganic polymers are used extensively in controlled drug delivery; however, there are no completely inorganic polymers that have found utility. Numerous types of organic polymers are used for controlled drug delivery, but only one class of semi-inorganic polymer, the polysiloxanes or silicone elastomers, has found extensive use.

Polymer molecules may be linear (with all the monomer units linked together continuously), branched (where part of the monomer units have linked together to form side branches extending from the main polymer chain with various degrees of subordinate branching), or cross-linked (where the main polymer chains are covalently interconnected at various points). Extensively cross-linked polymers essentially form a three-dimensional matrix where the polymer becomes one extremely large molecule. A high degree of cross-linking results in insoluble, intractable materials. Polymers may be thermosetting (set irreversibly when heated, usually because of cross-links forming between the polymer chains) or thermoplastic (soften when heated but return to their original condition on cooling). Thermoplastic polymers are by far the most often used for controlled drug delivery. Polymers may also be flexible (having moderate to high elongations when stressed with little reversibility in strain), rigid (having very low elongations when stressed with essentially no reversibility in strain), or elastomeric (having high elongations when stressed with a high reversibility in strain). Polymers may be amorphous or semicrystalline. In typical use polymers are never completely crystalline because they are polydisperse. Even highly crystalline polymers always have some amorphous zones, and thus have a glass-transition temperature ($T_g$) in addition to their crystalline-melting temperature ($T_m$). The $T_g$ is a temperature region below which a significant reduction in segmental and translational molecular mobility occurs resulting in solidification of the polymer.

Elastomeric polymers typically have highly flexible, kinked chains with a low degree of secondary interchain attractive forces and a low degree of cross-linking. When a stress is applied, the chains unkink to the point of restriction by the cross-links and rekink when the stress is removed. Fracture or permanent deformation occurs at strains beyond the point of significant resistance. "Thermoplastic elastomers" are materials where the polymer chains form regions having a high degree of secondary bonding or undergo partial crystallization, when stretched, forming pseudo cross-links (noncovalent). These pseudo cross-links provide resistance to straining similar to covalent cross-links, but because of the absence of covalent crosslinks, these materials remain thermoplastic and can be melt processed. Elastomeric polymers are typically above their $T_g$ at use temperatures. Flexible polymers are usually semicrystalline with low to moderate crystallinity at their use temperatures. The $T_g$ of flexible polymers vary, but are usually <100°C. Rigid polymers, on the other hand, are usually amorphous, have a high $T_g$, have very rigid polymer chains, are extensively cross-linked, or possess various combinations of these characteristics at their use temperatures (34).

Polymers useful in controlled drug delivery may be further characterized as biodegradable or nonbiodegradable, depending on whether or not the polymer is degraded in a biological environment. Both types are used extensively. Some water-soluble polymers such as PEG and poly(vinyl pyrollidone) are absorbed and excreted by a biological host without degradation of the polymer. Such polymers are said to be bioabsorbable to distinguish them from biodegradable polymers. Synthetic biodegradable polymers that have found use in drug delivery are polylactide (PLA), poly(lactide-co-glycolide) (PLGA), poly(caprolactone) and polyanhydrides. Some synthetic bioabsorbable polymers that have found use in drug delivery are poly(vinyl pyrollidone) and PEG (35). Silicone elastomers, polyurethanes, poly(vinyl chloride) (PVC), and numerous other nondegradable, nonabsorbable polymers are also used.

## Manufacturing of Polymeric Controlled-Delivery Products by Melt Processing

**Polymer Rheology.** Manufacturing of polymer-based, controlled delivery products by melt processing involves heating the polymer to a point where melt flow can occur. Depending on the molecular structure of the polymer, this point may be well above its melting point. However, at a sufficiently high temperature, the polymer is soft enough for flow to occur. Viscous flow can be modeled by a dashpot where a piston is displaced against the resistance of an oil (a Newtonian fluid) having an absolute or dynamic viscosity of $\eta$. When a shear force is applied to the fluid medium, the fluid continues to deform and the shear stress $\tau$ is proportional to the shear strain rate $d\gamma/dt$:

$$\tau = \eta(d\gamma/dt) \qquad (4)$$

On cooling, a polymer may solidify either by crystallization or by passing through its glass-transition temperature, $T_g$. The route of solidification is determined by the molecular structure of the polymer. Crystal formation is favored by linear, unbranched polymer structures. Crystal formation is also assisted by the presence of polar groups in the chain. Branching, and stereoirregularity in general, disrupt chain packing and crystallinity. Thus, highly stereoirregular polymers are typically amorphous.

At some temperature well below the $T_g$, molecular motion becomes sufficiently slow enough that the behavior of the polymer can be modeled by a spring. When a tensile stress $\sigma$ is applied, the tensile strain $e_t$ is proportional to a constant $E$, called Young's modulus:

$$e_t = \sigma/E \qquad (5)$$

From some point below $T_g$ to some point above $T_m$, thermoplastic polymers exhibit what is known as viscoelastic behavior where the polymer exhibits both viscous flow and elastic properties in response to an applied stress. Therefore, the total strain is the sum of the elastic or instantaneous strain and the viscous (time-dependent and strain-rate-sensitive) strain components. The strain in response to an applied stress in this region is given by:

$$\epsilon_{total} = \sigma/E + \sigma t/K \qquad (6)$$

where $E$ and $K$ are both constants. The viscous and viscoelastic behavior of the polymer are both very important in the

fabrication of polymeric devices by melt processing. Viscoelastic behavior changes with temperature leading to time-temperature equivalence particularly at very low strain rates. Thus, an increase in temperature is equivalent to a decrease in strain rate. Viscous flow is highly strain-rate-sensitive and becomes increasingly important as the temperature continually increases above $T_g$. When the material is stretched at a temperature above $T_g$, the molecules flow and align. The material in this aligned region becomes stronger, allowing more molecular flow to occur in the less aligned regions. Ultimately, this alignment gives the polymer directional properties with increased strength in the direction of alignment. This characteristic, known as orientation, is very important in the processing of fibers and films. Polymers with less symmetrical or irregular structures, polymers with compatible fluid additives or contaminants, or polymers with lower molecular weight species present, have a greater tendency for viscous flow. Some polymers are purposely formulated with plasticizers, which are essentially high boiling solvents for the polymer, that provide interchain lubricity, effectively lowering the $T_g$. Plasticized PVC is used in pesticidal collars and ear tags (36–38).

Scale-up of a manufacturing process to prepare a polymer for controlled delivery applications is somewhat different than for most commodity applications. For most commodity applications, the specifications of the final polymer are far less stringent than for polymers used in controlled delivery with far fewer regulatory requirements. And often there are numerous final products that can be made from a particular grade of such a polymer. Thus, the market size for a particular type and grade of a polymer used to make commodity products is typically very large. Here, cost is normally a key factor. On the other hand, final controlled delivery products are usually high value-add and their performance in the intended application usually requires very exacting specifications. Therefore, the specifications of all raw materials, including any polymers used to manufacture these products, are also very exacting. Also, the market for a controlled release product is typically a relatively small niche market requiring much smaller volumes of raw materials. Thus, for many controlled release applications the polymers used are custom made in relatively small size lots. A production lot of such polymers is therefore often very small by comparison and may range in size from <1 kg to 25 kg, especially for biodegradable controlled drug delivery products for human use. Production quantities of such polymers are therefore usually prepared in small-scale stirred batch reactors ranging in size from 1 to 50 gallons in a clean-room environment with close adherence to applicable regulatory guidelines. There are, however, some polymers used for human drug delivery that are prepared in relatively large quantities. Examples are the cellulose ethers and esters that are prepared in large-scale batch reactors. Typical batch sizes are 1,000–2000 kg. These polymers are used in enteric-coated oral formulations and currently represent one of the largest volume polymeric raw materials used in human controlled drug delivery.

**Extrusion and Injection Molding.** Extrusion is a process used for melting, blending, and forming a polymeric material into a desired finished product. Postforming operations such as orientation, pressing, or final molding may also be coupled with extrusion. Rod, tubing, film, channels, and filaments are examples of shapes that can be continuously extruded. Coatings and coextruded shapes (two different polymers extruded through a die and combined into the final product shape) of all of the above can also be produced. Extruders are also used for compounding and pelletizing materials to be later molded by various processes.

An extruder consists of a heated barrel having one or more rotating extrusion screws through it. The single screw variety is the most common. The screw turns and the material moves forward through the extruder in a fashion similar to the action of a progressive-cavity pump. The barrel is often vented to remove volatiles (residual monomers, solvents, moisture, and entrapped air), thus preventing defects in the finished product. A screen pack and breaker plate (support for the screen) for filtering the material are located at the barrel exit. The pressure is measured and controlled at the exit by a feedback control loop to the screw-rotation-speed controller. Heated dies are attached to the end of the extruder to form the polymeric material into the desired shape. A melt-metering pump usually precedes the forming die to provide precise flow control.

Extruders are sized by barrel diameter and can be as small as one-half inch to as large as about 8 inches; although, for most controlled delivery applications the smaller size machines will most likely be used. The length of the flighted portion of the screw to the inside diameter of the barrel determines the available surface area of the barrel and the average residence time of the material. The barrel and screw are designed of materials suitable for the temperature, pressure, and chemical aggressiveness of the material being extruded. Typical process pressures are <35 mPa; however, pressures up to 70 mPa are not uncommon. The materials used are usually surface-hardened, high-strength alloys. They are sometimes chromeplated for added corrosion resistance. Barrel liners of highly corrosion-resistant materials are also available.

Extruders are usually heated by external electrical heater bands controlled in various zones over the length of the barrel. Heating is controlled through a feedback controller loop which actuates an electrical contacter to activate the heating elements. Because of the high shear forces involved in extrusion, heat can begin to build once the materials begin to be extruded, and it often becomes necessary to cool the extruder. Extruders are typically cooled by passing a heat transfer fluid through internal cores or jackets in the barrel, or cooling coils surrounding the barrel. Cooling is controlled through a feedback controller loop which actuates a valve to the circulating heat-transfer-fluid system.

The geometry of the screw can vary considerably depending on the material and the final product desired. The screw can have a constant pitch or "lead" or variations in screw pitch or lead beginning larger at the feed throat (point of introduction of material) and getting progressively smaller as the material progresses through and exits the barrel. The later is usually used when an intense mixing action is desirable. However, each application typically

has its own requirements. The major or outer diameter of the screw is as close as possible to the barrel diameter to prohibit material from passing over the screw flights. The minor or "root" diameter of the screw will typically vary in the first screw type already described, having a smaller minor or root diameter at the feed throat and getting progressively larger and becoming constant as the later portion of the barrel (metering section) is approached. This provides a deep "channel" on material entry to accommodate incoming unmelted and uncompressed material and floods the entry zone with material to prevent starving of the extruder. Most screws are also bored, at least in the feed section, to provide entry of heat-transfer fluids for cooling.

The screen pack is located at the exit of the barrel and consists of several screens of different mesh size to filter rough contaminants from the melt. The finest mesh screen is located in the middle of progressively coarser mesh screens. The ones preceding it are for progressively finer filtering of the melt, and the ones after it are for support. The breaker plate is a thick metal plate with numerous large (~1/8-inch diameter) holes located past the screen pack and before the melt-metering pump and the forming die; it serves to support the screen pack and equilibrate the melt pressure to the pump. The metering pump is a positive displacement pump, which is controlled separately from the extruder screw to provide precise flow to the forming die (39–42).

Injection molding is probably the most widespread molding technique for quickly and easily forming a polymer melt into a finished product. In this process, the mold is split to allow part removal. It is kept closed during injection by an appropriate clamping force. The mold is filled by forcing a precompounded (containing all additives), molten polymer formulation into the mold. The injection-molding machine may be a simple piston (ram) injector design, or a more complex reciprocating-screw design. Either type consists of a feed hopper attached to a heated cylindrical barrel with an injection nozzle attached to the end of the barrel. The reciprocating screw or ram are typically capable of applying 70–140 mPa of pressure to the melt during the injection cycle. This is the operation sequence of the reciprocating-screw machine:

1. With the reciprocating screw forward, unmelted material is fed from the hopper.
2. The material is then plasticated and forced to the front of the barrel by the rotating screw, which simultaneously moves backward against a hydraulic cylinder (reciprocates) as the front of the barrel fills.
3. The mold clamp is released, the mold opens, and the part, formed on the previous cycle, is ejected.
4. The mold closes and the clamp pressure is reapplied.
5. The screw moves forward, as a ram, injecting the melt into the mold, and remains forward to begin the next cycle.

The mold temperature is kept warm but held at a suitable solidification temperature for the material being injected. Too cold a mold can lead to material freezing before the mold is filled. Materials used to construct injection molding equipment are similar to those used to build extruders. The molds must also be constructed of rugged materials to avoid both the erosive and corrosive forces of material flow. The major advantages to injection molding are speed and the ability to simultaneously form multiple complex geometric parts. These are the disadvantages:

- The high temperature and shear require the polymers and actives to be very stable.
- The process wastes materials in runners and sprues.
- The mold and equipment costs are high.
- Mold erosion occurs from material flow.
- Runners and sprues are difficult to clean from the final products.
- The directional flow patterns, inherent in the process, can leave residual internal stresses in the part.

Some examples of controlled delivery products produced by extrusion and injection molding are antibiotic periodontal fibers and insecticidal collars and ear tags (43,44). Extrusion and injection molding processes are very efficient and available equipment is capable of producing several pounds to several hundred pounds of final product per hour.

**Compression Molding.** Compression molding is usually used to process thermosetting polymers and is a simple and economical process. However, it can also be used to process thermoplastic materials where it is advantageous in producing the product and provided that appropriate molding conditions are used. In compression molding, the material to be molded is placed in the preheated mold in the form of a loose powder or prill. An excess of several percent is usually added to the mold to ensure complete filling. The mold is closed and sufficient pressure (several mPa) is applied to force the material into the mold cavity. The pressure is dependent on the flow characteristics of the molding material and the complexity of the part being molded. Excess material is forced out of the mold as flash or through a vent. For thermosetting materials the pressure is maintained long enough for the part to cure. The molds are generally heated and cooled by passing a heat transfer fluid through internal cores; however, internal or external electrical heating elements can also be used if precautions are taken to avoid hot spots. The advantages of compression molding are several:

- Waste is low because no runners or sprues are used.
- The final parts are easier to clean because no runners or sprues are used.
- Mold erosion from material flow is minimized.
- Residual internal stresses are low because of the short, multidirectional flow patterns of the material.
- The mold and equipment costs are low.
- Low process temperatures can be used.

There are some disadvantages to compression molding:

- It is not suitable for intricate parts because the flow is minimal.
- Because of polymer viscoelasticity, thermoplastic parts are difficult to mold without distortion.
- It is best suited for fairly thin (≤1/4-inch) products.

**Solvent Processing of Polymeric Controlled Delivery Products**

**Solvent Casting of Films.** Solvent casting has been an established process for the preparation of polymeric films for decades. The polymer and soluble or dispersable additives are first dissolved and dispersed in a suitable solvent. The solution is then cast onto a continuous, release-coated belt or web-supported film and passed through an oven to drive off the solvent. The solvent is usually reclaimed. The dried film is continuously removed from the belt and wound as it passes from the oven. Care must be taken in design of the process line to ensure that it is particularly suited to the product being manufactured. Consideration should be given to the flow characteristics of the casting solution, the evaporation rate of the solvent, and the changing flow characteristics of the polymer solution throughout the drying process to ensure a uniform film. The thermal stability of the polymer and any additives or active agents must also be considered. Solvent casting of films containing active agents for controlled delivery can be advantageous to melt processing if the active is thermodynamically unstable. However, because of the high initial process investment and the inherent difficulty in process control, solvent casting should only be considered for manufacturing controlled drug delivery devices when absolutely necessary. Solvent casting of films is often used as a part of the overall manufacturing process in manufacturing transdermal patch products.

**In Situ–Forming Implant Depots.** The first in situ–forming implant depot formulation to be approved for delivery of a drug in humans is ATRIDOX®. The product is designed for controlled delivery of an antibiotic for treatment of periodontal disease. It is supplied as an injectable liquid. When injected into the periodontal cavity, the formulation sets forming a drug-delivery depot that delivers the antibiotic doxycycline to the cavity. ATRIDOX is based on the ATRIGEL® drug-delivery technology developed by Atrix Laboratories, Inc. The technology consists of a biodegradable polymer, a bioactive agent, excipients, and additives dissolved in a water-soluble, bioabsorbable, and biocompatible solvent. The formulations may be liquids, pastes, or putties; however, the liquid injectable is the preferred form for most applications (17,18). Products based on the ATRIGEL technology are typically made using a high-intensity mixer to dissolve the polymer and dissolve or disperse the active agent and any excipients.

Another company, Matrix Pharmaceutical, Inc., has developed a proprietary biodegradable gel matrix consisting of purified bovine collagen for the targeted, intralesional delivery of chemotherapeutic agents (45,46). The system is prepared by dispersing chemotherapeutic agents in an aqueous solution of a protein such as purified bovine collagen. The chief advantage of this system is that the bovine collagen gel allows one to optimize the retention and release of the drug at a targeted injection site, thus minimizing systemic toxicity. The therapeutic effect of the delivery is further enhanced by the inclusion of a vasoconstrictor such as epinephrine. The tumor or other diseased tissues are exposed to high concentrations of the drugs for prolonged periods of time because of the site-specific and sustained release of the drug. In one study, cisplatin either in a single solution (CDDP suspension) or within the Matrix gel (cisplatin/epinephrine gel) are injected into a mouse tumor (100 mm$^3$). The free cisplatin was cleared from the site within 1 h, whereas the gel retained the drug between 24 to 72 h (47).

One product based on this concept, AccuSite (fluorouracil/epinephrine) Injectibel Gel for the treatment of recurrent genital warts has been approved in seven European countries and has completed Phase III studies in the United States (48). Here the gel matrix is a viscous, aqueous gel consisting of fluorouracil (30 mg/mL) and epinephrine (90.1 mg/mL) and other inactive buffering and osmotic excipients, within a purified bovine collagen matrix gel. This technology is covered by several U.S. and European patents (49–52).

Another Matrix product, IntraDose Injectable Gel, has advanced into a second level of Phase II clinical trials in the U.S. for treatment of inoperable liver cancer. The product consists of a dispersion of cisplatin and epinephrine in a purified bovine collagen matrix.

The protein gel matrix is a simple yet effective delivery system that permits direct delivery of chemotherapeutic agents to a tumor and provides a high local concentration of drug while minimizing systemic toxicity. The major disadvantage of the system is that it cannot provide long-term release of the drug (usually less than 1–3 days) because the drug can easily diffuse out of the gel matrix.

**Encapsulation.** Encapsulation, especially microencapsulation (particles ranging in size from a few to several hundred micrometers), is a process whereby particles of an active agent are surface coated to provide changes in the physicochemical properties of the active agent. There are many different processing techniques used depending on the desired properties of the final product, the properties of the agent being coated, and the properties of the coating material. The term microsphere is often used synonymously with microcapsule; however, a distinction should be made between the two terms as they are used in controlled delivery, because the final products produced and their release characteristics are quite different. Microcapsules are essentially discontinuous microspheres where the active core material is completely covered with a nonactive surface coating. The coating thickness may be varied depending on the characteristics desired in the final product. The surface coating of a microcapsule sequesters the active and serves as a protectant and/or a sustained release, rate-controlling membrane. The release mechanism of the active agent is usually mediated by diffusion of the active agent through the coating. On the other hand, microspheres are micrometer-sized homogeneous, monolithic spheres containing the active agent dispersed in a nonactive matrix material. The matrix material is often

biodegradable. And in this case, the release mechanism of the active agent is usually mediated by degradation of the matrix. Although the final products are quite different, some of the processes used to prepare microspheres and microcapsules are very similar.

Some of the more common processes used to form microspheres and microcapsules are as follows:

- Spray-drying
- Fluid-bed coating
- Phase separation
- Solvent evaporation
- Solvent extraction
- Cryogenic solvent extraction

Several proprietary processes also exist.

***Spray-Drying.*** Spray-drying is a process that transforms the feed material from a fluid state into a dried particulate form by spraying the feed into a hot drying medium. It is a one-step, continuous, particle-drying process. The feed material can be in the form of a solution, suspension, emulsion, or paste. The resulting product can be powdered, granular, or agglomerated particles, depending upon the physical and chemical properties of the feed material, the drier design, and its operation. Spray-drying is used in all major industries where particle drying is required, ranging from food and pharmaceutical manufacturing to chemical industries such as mineral ores and clays (53).

Spray-drying involves atomization of the feed into a drying medium, resulting in the evaporation of the solvent and the formation of dried particles. Atomization is a process that breaks up the bulk liquid into millions of individual spray droplets. The energy necessary for this process is supplied by centrifugal force (rotary atomizer), pressure (pressure nozzle), kinetic (two-fluid nozzle), or sonic vibration (ultrasonic nozzle). The selection of the atomizer type depends on the nature of the feed and the desired characteristics of the final product. For all atomizer types, increasing the amount of energy available for atomization results in smaller droplet sizes. If the atomization energy is held constant and the feed rate is increased, larger particles result. Atomization also depends upon the fluid properties of the feed material, where higher viscosity and surface tension result in larger droplet sizes at the same atomization energy.

In most cases, air is used as the spray-drying medium; however, dry nitrogen can be used for moisture-sensitive compounds. Contact with the spray-drying medium causes evaporation of the solvent (water or organic solvent) from the droplet surfaces. The evaporation is rapid due to the vast surface area of the droplets in the spray. The manner in which the spray contacts the drying medium is an important design factor. It influences droplet behavior; and therefore, has a great effect on the properties of the dried product. Contact with the spray drying medium is determined by the position of the atomizer in relation to the drying air inlet. Co-current flow (the product and air pass through the dryer in the same direction), counter-current flow (the spray and air enter the dryer at the opposite direction), and mixed flow driers are available. The selection of the appropriate design is based on the required particle size, the required dried particle form and the temperature to which the dried particle can be subjected. For example, if a fine-particle product (mean size 20–120 $\mu$m) is required, but a low product temperature must be maintained at all times during the drying operation, a co-current, rotatory-atomizer spray-drier is selected (53).

Product separation from the drying air follows completion of the drying stage. Primary separation of the dried product takes place at the base of the drying chamber. Small fractions can be recovered in separation equipment such as a cyclone.

Spray-drying is a useful method for the processing of pharmaceuticals since it offers a means for obtaining powders with predetermined properties, such as particle size and shape. In addition a number of formulation processes can be accomplished in one step in a spray-drier; these include encapsulation, complex formation, and even polymerization (54). Spray-drying (55) and spray-congealing (56) processes can be used for preparing microparticles for controlled release applications. In the spray-congealing process, no solvent is used. The feed, which consists of the coating and core materials, is fed to the atomizer in the molten state. Microparticles form when the droplets meet cool air in the drying chamber and congeal.

Oil-soluble vitamins, such as A and D, have been microencapsulated by spray-drying an emulsion of the oil in a gum arabic or gelatin solution. Spray-drying has also been used in the preparation of polymer-coated microcapsules for the purposes of taste masking (57).

Biodegradable microparticles have also been prepared by spray-drying. PLA and PLGA microspheres have been prepared from solutions or suspensions of a number of drugs dissolved or dispersed in methylene chloride (58,59). Microcapsules of progesterone and PLA were formed with diameters of less than 5 $\mu$m. Crystallization of the drug occurred in the aqueous phase when microspheres were prepared by a solvent evaporation method, but spray-drying avoided this problem. The main difficulty encountered in preparing spray-dried microcapsules is the formation of polymer fibers as a result of inadequate forces to disperse the feed liquid into droplets; the successful atomization into droplets is dependent on both the type of polymer used and the viscosity of the spray solution (60).

Spray-drying is also used as one of the steps in the multistep manufacturing process for GLIADEL®. GLIADEL, a product of Guilford Pharmaceuticals, Inc., is a biodegradable implant containing the anticancer agent BCNU. It is approved in the U.S. and several other countries for the treatment of malignant glioma. After the tumor has been removed from the patient's brain, up to 8 wafers are implanted into the cavity created by the removal of tumor mass. The drug BCNU is slowly released from the polymer matrix as the polymer degrades. BCNU is incorporated into the biodegradable polyanhydride matrix at 3.85% loading by spray-drying. Microspheres with a size ranging from 1 to 20 $\mu$m are obtained (61). The microspheres are subsequently compressed into small wafers to prepare the final dosage form. The spray-drying method is especially useful in this application since the drug BCNU is both heat- and moisture-sensitive.

The feasibility of encapsulating proteins into polymers has also been demonstrated. Bovine serum albumin (BSA) was incorporated into poly(DL-lactide)(DL-PLA) microspheres by a spray-drying method. Successful incorporation of the protein depends on the selection of an appropriate solvent system. Dichloromethane, ethyl acetate, and nitromethane proved to be the most suitable solvents for the DL-PLA/BSA system (62).

Spray-polycondensation is another method that can be used in the preparation of microcapsules. In this technique, polymer formation from reactive monomers, encapsulation, and product separation from the vehicle are all accomplished in a one-stage process. The feed consists of a dispersion of the core material and monomers, or precondensates of relatively low molecular weight, in addition to other film-forming agents and the catalyst. This technique was used to produce microcapsules that developed slow release properties after curing (55).

Spray-drying has proven to be a useful technique in producing controlled release products. The main advantage of this process is that it is convenient, fast, and allows the use of mild conditions. However, considerable loss of material can occur during spray-drying from adhesion of the particles to the inside wall of the spray-drier and agglomeration of the microparticles if improper process conditions are used. In order for the spray-drying process to be successful, appropriate selection of the spray-drying equipment and optimization of the spray-drying conditions are essential.

***Fluid-Bed Coating.*** Fluid-bed coating is a process whereby particulates are suspended in a column of heated air or inert gas while a solution or emulsion of a polymer or other film-forming coating material is applied to the particles through spray nozzles. High-quality microcapsule products are economically produced by this process. Typical products are taste-masked drugs, enteric-coated drugs, and sustained-release drugs (63,64).

Fluid-bed coating is a complex process consisting of three major operations: fluidization, atomization, and drying. The coating chamber has a high volume of flow to suspend, agitate, and dry the coated particles. The spraying nozzles can be located at various positions in the coating chamber providing top, bottom, and side or tangential spraying of the particles. Bottom spraying, or Würster coating as it is often called, is the most common technique used for encapsulation particles as small as 30–40 $\mu$m (65). A detailed review article on this process was provided by Christensen and Bertelsen (66). The Würster technique is a circulating fluid-bed specifically designed for small particle coating. It consists of a container with one or more Würster partition inserts, typically one partion for lab-scale equipment and three for production-scale equipment. The product load ranges from a few kilograms for laboratory scale equipment (e.g., Glatt, GPCG 3 and STREA-1, Niro) to a few hundred kilograms for production-scale equipment (e.g., Glatt, GPCG 200). The coating process consists of three phases: the start-up phase, the coating phase, and the drying phase. The start-up phase consists of an optional preheating of the equipment and the substrate to either shorten the heating period, prevent over-wetting of the substrate during the initial coating application, or both. In the coating phase, several processes take place simultaneously: atomization of the coating solution or suspension, transport of the film droplets to the substrate, adhesion of the droplets to the substrate, and film formation onto the substrate. The properties of the substrate such as particle density, size, and tackiness are very important in determining the coating process parameters during this phase. The coated product is finally dried to prevent adherence of the individual particles and remove residual solvents. High-quality microcapsule products are economically produced by this process. Typical products are taste-masked drugs, enteric-coated drugs, and sustained-release drugs (63–66).

***Phase Separation.*** The phase separation process, or coacervation process as it is sometimes called, involves:

- Preparing an organic solution of a water-insoluble, matrix material (usually polymeric)
- Addition of an aqueous solution of the active agent or dispersion of particulates of an active agent to the organic solution with vigorous agitation
- Introduction of a coacervating agent or event for the matrix material to the matrix solution/active agent emulsion or dispersion
- Depending on the means of coacervation, the coacervated matrix solution/active agent emulsion or dispersion is then added to an appropriate hardening agent to extract the excess matrix solvent
- collecting, washing, and drying of the final product

Coacervation may be brought about by various means. It can be induced by:

- A change in the temperature of the system
- A change in the pH of the system
- A change in electrolyte balance
- Addition of nonsolvents
- Addition of other materials which are incompatible with the polymer solution

The term coacervation used in this context is borrowed from the description of an equilibrium state of colloidal systems. In the phase separation process for encapsulation the term refers to a state of partial precipitation of the matrix phase by the coacervation agent. The matrix material at this point is still solvent rich, and not completely precipitated, and thus maintains a high degree of flow necessary for good film- and particle-forming properties. Several of the coacervation methods described here are being successfully used to prepare a variety of products from peptide microcapsules to encapsulated electrolytes (67–72).

***Solvent Evaporation.*** The emulsification-solvent-evaporation technique for preparation of microspheres was described by Beck et al. for the delivery of contraceptive steroids (73). It is the most widely used manufacturing technique for biodegradable microspheres. The microsphere formation process consists of three stages.

1. Droplet formation
2. Droplet stabilization
3. Microsphere hardening

First, a dispersed phase containing the polymer is emulsified in an immiscible continuous phase containing a stabilizing agent. The second phase involves the diffusion of the solvent from the emulsion droplet into the continuous phase and its subsequent evaporation. Simultaneous inward diffusion of the nonsolvent into the droplet causes polymer precipitation, microsphere formation, and hardening. Depending on the nature of the two phases, the process may be termed oil-in-water (o/w) or water-in-oil (w/o) method. The solvent evaporation process requires the use of a surfactant to stabilize the dispersed-phase droplets formed during emulsification and inhibit coalescence. Surfactants are amphipathic in nature and therefore align themselves at the droplet surface, thereby promoting stability by lowering the free energy at the interface between the two phases. Furthermore, the creation of a charge or steric barrier at the droplet surface confers resistance to coalescencing and microsphere flocculation. Surfactants employed in the o/w process tend to be hydrophilic in nature and by far poly(vinyl alcohol) is the most widely used. The emulsification systems used for microparticle production have included both low- and high-speed mechanical stirring, sonication, and microfluidization. The particle size and size distribution can be controlled by the emulsification speed and mixing vessel design. Following emulsification, the removal of remanent solvent and complete microsphere hardening is usually accomplished by gentle agitation of the suspension. After evaporation of the solvent, the final stage of the emulsification-solvent-evaporation process is the isolation of microspheres from the dispersed phase containing surfactant. This has generally been achieved by centrifugation and filtration, and it is usually followed by a further cleaning process in which the particles are washed several times with distilled water. The microspheres are finally dried using lyophilization or fluid-bed drying.

***Solvent Extraction.*** Solvent extraction is similar in nature to o/w solvent evaporation process except that following emulsification the preformed microspheres are poured into a large volume of nonsolvent to extract the remaining organic solvent, thus causing rapid hardening of the particles.

Microencapsulation by either the solvent evaporation or the solvent extraction process is normally achieved by first emulsifying or dispersing a drug-containing polymer solution in an immiscible solvent by stirring, agitation, or some other vigorous mixing technique followed by extraction or evaporation of the polymer solvent. This process is difficult to scale-up for several reasons. The key limiting factor in reducing overall process time is the time required to form a uniform emulsion. Increased batch sizes in large tanks require a longer time to form the emulsion, resulting in longer overall production times. Longer exposure times of the active agent to the process solvents can lead to the degradation or destruction of the active agent. Furthermore, as the emulsion tank size increases, stir speeds and viscosity within the larger tank have to be empirically optimized by the trial and error at each stage of the scale-up. Control of the particle size can become less reliable as batch size is increased because it is difficult to maintain the same shear while still providing uniform mixing in larger vessels. Herbert et al. (74) and Ramstack et al. (75) described a process that potentially overcomes this problem. This process uses a static mixer instead of a dynamic mixer to prepare the emulsion. In this process, a first phase comprising the active agent and the polymer and a second phase comprising the dispersing medium are pumped through a static mixer into a quench medium where the polymer solvent is either extracted or evaporated to form microparticles containing the active agent. The key advantage of this process is that accurate and reliable scaling from laboratory to commercial batch sizes can be achieved while obtaining a narrow and well-defined particle size distribution. One other significant advantage is that the process can be made continuous.

*Cryogenic Solvent Extraction.* A cryogenic solvent extraction process was developed (76) for preparing microspheres. This process proves to be especially useful in encapsulation of biologicals such as proteins and peptides. In this process, a solution of the polymer and the active agent to be encapsulated are atomized into a liquified gas (e.g., liquid nitrogen, liquid argon). The atomized particles freeze when they contact the liquified gas (liquid nitrogen), forming frozen spheres. These frozen spheres then sink to the surface of a frozen nonsolvent below the liquified gas. The liquid gas is evaporated and spheres begin to sink into the nonsolvent as the nonsolvent thaws. The solvent in the spheres is extracted into the nonsolvent to form microspheres containing the agent to be encapsulated. In this process, it is important that the polymer/active agent freeze immediately upon contacting the cold liquid, and then be slowly thawed while the polymer solvent is extracted from the microspheres. The thawing rate is dependent on the choice of solvents and nonsolvents. It is important to select a solvent for the polymer having a higher melting point than the nonsolvent for the polymer so that the nonsolvent melts first, allowing the frozen microspheres to sink without agglomerating into the nonsolvent where they later thaw. The solvent and nonsolvent for the polymer must also be miscible to allow extraction of the solvent from the microspheres. The microspheres are collected by filtration and lyophilized.

## Tableting and Tablet Coating

Tableting is perhaps the simplest, most common, and most economical method of processing an active agent into a drug-delivery product. Tableting is typically reserved for oral dosage forms of an active agent; however, there are some tableted or compressed wafer, controlled release implant depots produced, such as the GLIADEL wafer discussed earlier. The tableting process involves feeding a metered amount (usually from ~0.5 to 5 g) of a "granulation" (large particle blend) of an active agent to the dies of a tableting press where the granules are compressed into a tablet of the desired shape. Some manufacturers use intricate shapes for product identification. The premixed

granulation contains the active and various excipients to modify the flow, compaction, die release, and dissolution characteristics of the tablet. Granulation is accomplished wet (using solvent) or dry (using no solvent) by high-intensity mixers or fluidized-bed granulators. Although single-cavity presses can be used for specialized applications and testing, high-speed rotary presses capable of producing thousands of tablets per minute are typical. The pressures used in tablet compression are quite high, ranging from ~50 to 500 mPa; therefore, the dies used in tableting are typically made with high-strength, surface-hardened alloys.

Tableting has become much more sophisticated as the demand for controlled release, oral drug delivery has grown. Technology now exists that produce bilayer tablets providing two different release rates for a drug. The technology uses different polymeric excipients in the tablet providing both immediate and sustained release. Various over-the-counter analgesics use this technology. Tablets composed of compacted microencapsulated beads have also been developed to provide both sustained release and protection from the irritating effects of certain drugs such as non-steroidal anti-inflammatory drugs. The drug is microencapsulated in a separate process, such as the fluidized-bed process, and then incorporated into the tableting formulation. Effervescent tablets are also produced to provide chewable and fast-dissolving tablets for patients who have difficulty swallowing. The drug is again typically incorporated into the effervescent tablet formulation as microencapsulated beads.

Many tableted products are also subsequently coated to provide controlled release products. Coating usually takes place in a pan coater, which is a rotating drum, similar to a clothes drier, which tumbles the tablets in front of a spray nozzle for application of the coating. The coating is dried by a continuous flow of process air. The process air is often heated and solvent reclamation is used for organic-solvent-based coatings. The batch sizes for pan coating can vary between ~100 and 2,000 kg. Small laboratory models are also available. Various wax and enteric coatings are applied in this manner. For very small tablets, fluidized-bed coating may also be used.

## Capsule Filling

By far the most versatile of all dosage forms is the gelatin capsule. Gelatin capsules can be filled with powders, small pellets or small tablets, liquids, or semisolids, as well as some combinations of these forms. The capsule also provides efficient taste-masking. The drug product is released by dissolution of the gelatin in the stomach. Gelatin capsules can be prepared from hard or soft gelatin; however, hard-gelatin capsules are more versatile for controlled drug delivery.

Hard-gelatin capsules are available in several different sizes, and high-speed filling machinery, capable of filling ~1,500 capsules per minute, is available. The machines typically consist of a drug hopper, a capsule hopper, a dose-metering device, a dose chamber, a filling or tamping pin, a capsule tray, and a finished-capsule collection bin. The capsules are automatically opened, filled, and closed during the manufacturing process. Technologies have been developed whereby controlled release beads and minitablets are used to fill a gelatin capsule for convenient administration of an oral, controlled release dosage form. Examples of such products are the sustained release cold medications, where sustained release antihistamines, antitussives, and analgesics are first preformulated into extended release microcapsules or microspheres and then placed inside a gelatin capsule. Another example is enteric-coated lipase minitablets that are placed in a gelatin capsule for more effective protection and dosing of these enzymes (77).

### Transdermal and Iontophoretic Devices

Transdermal patches have proven to be an effective means of delivering drugs. Because of the different physicochemical characteristics of individual drugs, each drug typically has a particular delivery formulation and optimal type of transdermal delivery device. The drug is typically sequestered in a device that has an impermeable backing film on the outer side, followed by a reservoir or matrix containing the drug, then a semipermeable, rate-controlling membrane, and an adhesive inner layer for attachment to the skin (4). Preferably, the device can be a monolithic, matrix-type design with the drug dissolved or dispersed in a polymeric matrix, which is then applied directly to the impermeable membrane, and is finally coated with a pressure-sensitive adhesive (5). The active agent diffuses through the polymeric matrix, through the adhesive layer to the skin and, in combination with the skin itself, serves as the rate-controlling portion of the device. In a simpler variant of the matrix-type device, the drug is dispersed directly in the adhesive. This matrix-type device eliminates the rate-controlling membrane and simplifies the manufacturing of the product. In practice, the drug–polymer matrix is usually formed from a solution or melt blend of the polymer and drug, or a solution or melt of the polymer with the drug dispersed in it. The polymer–drug melt blend or solution is then coated onto the impermeable base film.

The impermeable base film is usually a multipolymer film composite or metalized film. Material combinations such as polyethylene/poly(ethylene terephthalate), poly-(vinylidene chloride)/poly(ethylene terephthalate) laminates, or polyethylene/metalized poly(ethylene terephthalate) are typically used for the backing film. For a monolithic device, the matrix polymer is selected from a variety of flexible or elastomeric thermoplastic polymers suitable for sequestration and diffusion of the active agent and any penetration enhancers. If the matrix backing film is to be coated by solution casting of the matrix film, the matrix polymer should be soluble in a low-boiling organic solvent. Thermoplastic poly(ether urethane)s and several other polymers are used as matrix materials. Melt blending and extrusion coating is preferable if the drug is thermally stable. After the monolithic drug–polymer matrix is applied to the backing film, a drug-permeable adhesive for adhering the patch to the skin, and a final protective film laminate that is impermeable to the active ingredients are applied. The final protective film usually has a release

coating of Teflon® or silicone to permit easy removal at the time of use. The adhesive is a low-irritating, pressure-sensitive type such as a pressure-sensitive acrylic or silicone adhesive. The final laminate is die cut to the appropriate size for the desired release profile of the drug, and the product is packaged in protective foil-lined pouches. Various proprietary versions of conventional continuous film-coating and laminating equipment are used. In a typical process, the drug–polymer matrix material is applied to the backing film by continuous direct extrusion or solvent casting onto the film which is supported on a continuous web. In the case of solvent casting, the composite is passed through an oven to remove the solvent. All the problems associated with solution casting of films, discussed earlier, apply. An adhesive film layer and protective release film are finally applied by transfer coating to the backing-film–drug-matrix laminate. Transfer coating is a process whereby two materials are laminated by passage through a pair of rollers where a low-pressure nip is applied.

Reservoir-type devices are much more complicated to manufacture. For a typical reservoir device, a drug-impermeable, protective-release film is first coated with a pressure-sensitive adhesive. The adhesive-coated protective film can then be laminated to semipermeable, rate-controlling membrane by a transfer-coating process. Reservoirs of the premixed drug formulation are then applied to the membrane side of the product-build laminate. A suitable heat- or pressure-sealable, drug-impermeable backing film is then applied and sealed at the periphery of the reservoir. The final laminate is then die cut to form the final patch. From a manufacturing standpoint, the monolithic- or matrix-type device is preferable whenever the appropriate sequestration and delivery profile for the drug can be met (78–82).

Iontophoresis is a process used to enhance the skin permeation of certain ionized drugs. Here the driving force to skin penetration of the drug is an applied electric potential. The current generator is typically a low-voltage source of direct current with a low current flow of less than one milliampere per square centimeter of skin surface area. The two electrodes are placed in contact with the skin through electrolyte-containing hydrogels or foams. The anode and cathode are typically placed within a few centimeters of each other. The anode, or positive electrode, serves as the drug reservoir; and when the voltage potential is applied, ionized species of drug are transported across the skin membrane (83). Pilocarpine is delivered in this manner in an approved test method for cystic fibrosis. Here the administered pilocarpine serves to cause local sweating which is then collected and tested for chloride-ion content. The presence of abnormally high chloride ion in body fluids is a strong indicator for the disease.

### Drug–Polymer Conjugates

Extensive research and development efforts have concentrated on the alteration of drug properties through drug–polymer conjugation. Efforts have primarily been focused on enhancing the solubility and bioavailability of the drug, or on targeted delivery of the drug by creating sites for natural recognition. The most investigated and successful class of polymers that have been used for drug–polymer conjugation to date are the PEGs.

PEG is a linear or branched neutral polyether with the following structure:

$$HO\text{-}(CH_2CH_2O)_n\text{-}CH_2CH_2OH \qquad (7)$$

Despite its apparent simple structure, PEG has been used in a variety of bioanalytical and biomedical applications (84–89). The versatility of PEG stems mainly from the fact that it is nontoxic (approved by FDA for internal consumption), soluble in both water and most organic solvents, and available in a wide range of molecular weights. In addition, PEG can be readily linked covalently to other molecules through its terminal primary hydroxyl groups. This particular area of PEG's applications is commonly known as PEGylation. In general, covalent attachment of PEG to other molecules:

- Enhances their solubility
- Alters their pharmicokinetics by prolonging plasma circulation half-life
- Reduces their clearance rate through the kidneys
- Renders proteins nonimmunogenic

The use of PEG to alter the properties of drugs began with the early observations by Davis and Abuchowski that the covalent attachment of PEG to proteins resulted in protein–polymer conjugates that were biologically active, nonimmunogenic, and had prolonged plasma circulation half-lives (90). Subsequent researchers in this area have established that molecules (mostly proteins such as enzymes) retain most of their bioactivity after the PEGylation process, indicating that covalent attachment of PEG does not interfere with the active sites of these molecules (91,92). A molecular-weight range for PEG between 2,000 and 20,000 daltons is commonly used in the modification of proteins and peptides. Monofunctional PEGs, such as the ones end-capped by a methyl group (mPEGs) are preferred for the modification of proteins to eliminate unwanted cross-linking that accompanies the use of difunctional PEGs (93–95).

**Methods of Forming a Covalent Linkage Between PEG and Proteins.** PEG can be derivatized and covalently linked to various functional groups of a protein such as the amino and carboxylic acid groups. However, the preferred route of PEGylation is through the amino groups. The derivitization process begins with an activation step wherein the terminal hydroxyl group of PEG is replaced by a nucleophile which subsequently reacts with the amino groups of a protein.

One commonly used reagent is trichloro-s-triazine (cyanuric chloride). This is the reagent originally used by Davis and coworkers (90) and has recently been improved to ensure reproducibility and complete conversion of mPEG to the intermediate (96). The hydroxyl group of PEG displaces one chloride attached to the triazine ring and the remaining two chlorides are used for reaction with the amino groups of a protein. This simple reaction provides

an effective one-step activation of PEG; however, it is not optimal in that cyanuric chloride released from the reaction is toxic. And, because of its nonselective reactivity toward other functional groups such as cysteinyl and tyrosyl groups other, less desirable, derivatives can be formed (97). It is not surprising that substantial loss of protein activity have been observed after such modification (98,99).

Another popular way to activate PEG is through a PEG-succinimidyl succinate intermediate (SS-PEG) (100,101). The intermediate is normally prepared first by succinylation of the hydroxyl end group of PEG followed by coupling with N-hydroxy succinimide (NHS) in the presence of dicyclohexylcarbodiimide. This derivative can react rapidly with proteins under mild conditions (near neutral pH, room temperature), and leads to modified proteins with high bioactivity (100). However, the PEG-succinate aliphatic ester linkage is susceptible to hydrolysis in an aqueous environment leading to a gradual dissociation of PEG from the protein (94).

The hydrolytic stability of the linkage between PEG and a protein can be enhanced by using a urethane linkage instead of an ester linkage. Such coupling can be achieved through a one-step process using carbonyldiimidazole (102), but such PEG derivatives have only mild reactivity with proteins and the modification process usually takes a long time. However, the resulting modified proteins, retain high bioactivity (103,104). Similar activations can also be accomplished in a single step using the chloroformate of 4-nitrophenol and 2,4,5-trichlorophenol (105). The major drawback of these processes is that 4-nitrophenol and 2,4,5-trichlorophenol in the reaction are toxic and hydrophobic and may be retained by the protein.

An activation process developed more recently uses succinimidyl carbonate as the intermediate (106). In this process PEG is first treated with phosgene to generate a chloroformate of PEG which subsequently reacts with NHS. The by-product, NHS, generated during the protein coupling step is not toxic and can be easily separated. However, caution must be exercised in the coupling process because of the use of phosgene.

Other less common derivatives of PEG that are reactive with amino groups have been reported in the literature with varying degrees of success. Some examples are PEG-carboxymethyl azide (107), PEG-tosylate (108,109), and PEG-imidoester (110). Currently the most frequently used PEG derivatives for amino coupling are the NHS active esters such as SS-PEG, succinimidyl carbonate, and succinimidyl propionate (111).

PEG derivatives that are reactive with other amino acid residues of a protein, such as arginine residues, can be achieved by phenylglyoxal (112), cysteine residues by maleimide containing reagents (91,97), and carboxylic acids by amino PEG (113–115).

**Examples of PEGylated Proteins and Peptides.** Covalent attachment of PEG has been applied to a large number of proteins and peptides with varying degrees of success. For instance, PEG-adenosine deaminase has been approved by the FDA as a replacement therapy for patients with severe combined immunodeficiency disease who are not suitable candidates for bone marrow transplantation. PEGylated recombinant interleukin-2 (rIL-2) has been tested in randomized human clinical trials (116–118). Covalent attachment of PEG enhances solubility, increases circulatory half-life, and eliminates the possibility of potential aggregates of rIL-2, which may elicit an immune response. Other PEG-modified proteins that have been investigated include brain-derived neurotrophic factor (119,120), interleukin-15 (121), interleukin-6 (122), tumor necrosis factor-alpha III (123), human megakaryocyte growth and development factor (124), human granulocyte colony stimulating factor (125), and recombinant human interferon-gamma (126). As more and more proteins become available one can expect to see an increasing number of proteins and peptides conjugated with PEG for various bioanalytical and biomedical applications. Covalent coupling of PEG to proteins and peptides represents an efficient way of improving the physicochemical properties of the proteins such as the reduced immunogenicity, improved plasma pharmacokinetics, increased resistance to proteolysis, and reduced renal excretion rate. The intensive research in the area during the past two decades has led to many exciting developments as evidenced by the increasing number of efficient and specific reagents for PEGylation, as well as more and more protein and peptide drugs being successfully PEGylated and advancing into human clinical trials. With the increasing number of recombinant proteins made available by genetic engineering and the increasing need to extend the patent protection of existing proteins and peptides, one can expect to see more and more applications of PEGylation.

Currently, most commercial PEGylation processes are batch processes of less than 1 kg, where the larger-scale synthetic procedures are merely the smaller-scale procedures proportionally enlarged to the larger batch size. This method of scale-up can result in problems in that suitable PEGylating agents need to be individually selected depending on the properties of the protein being used and on the particular functional group or groups on the protein to be PEGylated. The reaction conditions need to be optimized so that an even mixing between the PEGylating agent and the protein is achieved to ensure a uniform degree of PEGylation of all the protein molecules. Excessive PEGylation can adversely affect the binding sites of the particular protein, leading to a loss of bioactivity. Precise control of reaction temperature is also important in achieving successful PEGylation while retaining the bioactivity of the protein. Also, because the binding affinity varies among functionalized PEGylating agents, the final purification procedure must be carefully tailored to ensure complete removal of any unreacted agents. Dialysis is the purification method routinely used in small-scale PEGylation processes, whereas ultra-high-speed membrane centrifugation is the standard method used in larger scale processes. If required, further purification of the PEGylated protein can be achieved by passing the product through ion exchange columns. One can expect to see newer and better analytical methods for PEGylated proteins and more commercially available functionalized PEGs in the future.

**Combination Processes**

A controlled release dosage form may need to be prepared using a combination of processes. One example is the GLIADEL wafer described earlier. The production of GLIADEL involves both a spray-drying and a compression molding process. Spray-drying is used in the incorporation of BCNU into polymer matrix. The polymer—poly(carboxyphenoxypropane: sebacic acid)—is first dissolved in methylene chloride. After the polymer is completely dissolved, an appropriate amount of BCNU is added to the polymer solution. The drug–polymer solution is then fed into a spray-drier to form microspheres of the drug–polymer mixture. The feed rate, inlet air temperature, and vacuum are carefully controlled during the spray-drying process to ensure uniformity and lot-to-lot consistency. To produce the final GLIADEL wafer, approximately 200 mg of the spray-dried microsphere powder is weighed into a punch and die assembly and compressed into wafers using a press. The wafer is approximately 14 mm in diameter and 1 mm thick. The wafers are then inspected, packaged, and sterilized by irradiation.

Several tableted and gelatin-capsule oral dosage forms are also prepared using combined processes. In a typical combination process for a tableted oral dosage form, the active is first encapsulated in a fluidized-bed or spray-drying process. The microspheres or microcapsules thus formed are subsequently tableted. The tablets may also be coated for added protection in the stomach or to obtain the final release profile desired.

## BIBLIOGRAPHY

1. *Enteric Polymer Literature*, EFC-205C, Eastman Chemical Company, Kingsport, Tenn., 1998.
2. L.C. Junqueira, J. Carneiro, and R.O. Kelley, *Basic Histology*, 6th ed., Appleton & Lange, Norwalk, Conn., 1989.
3. U.S. Pat. 5,603,957 (February 18, 1997), O. Burquiere et al. (to Flamel Technologies, France).
4. U.S. Pat. 5,008,110 (April 16, 1991), A.G. Benecke et al. (to The Procter & Gamble Company.).
5. U.S. Pat. 5,676,969 (October 14, 1997), J. Wick et al. (to Bertek, Inc.).
6. U.S. Pat. 5,227,169 (July 13, 1993), S. Heiber et al. (to TheraTech, Inc.).
7. U.S. Pat. 5,023,084 (June 11, 1991), Y.W. Chien et al. (to Rutgers, The State University of New Jersey).
8. U.S. Pat. 5,126,144 (June 30, 1992), H. Jaeger et al. (to LTS Lohmann Therapie-Systems GmbH & Co. KG.).
9. U.S. Pat. 4,839,174 (June 13, 1989), R.W. Baker et al. (to Pharmetrix Corp.).
10. U.S. Pat. 5,364,630 (November 15, 1994), J.L. Osborne (to ALZA Corporation).
11. T.R. Tice and D.R. Cowsar, *Pharm. Tech.*, 26–34 (1994).
12. U.S. Pat. 4,320,758 (March 23, 1982), J.B. Eckenhoff et al. (to ALZA Corporation).
13. U.S. Pat. 5,023,088 (June 11, 1991), P.S.L. Wong et al. (to ALZA Corporation).
14. A.B. Berenson and C.M. Wiemann, *Am. J. Obstet. Gynecol.* **172** (4), Pt. 1, 1128–1135; discussion: pp. 1135–1137 (1995).
15. A.M. Kaunitz, *Int. J. Fertil. Menopausal Stud.* **41**(2), 69–76 (1996).
16. U.S. Pat. 5,660,848 (August 26, 1997), A.J. Moo-Young (to The Population Council, Center for Biomedical Research).

17. *Annual Report*, Atrix Laboratories, Inc., Fort Collins, Colo., 1996.

18. U.S. Pat. 4,938,763 (July 3, 1990), R.L. Dunn, D.R. Cowsar, J.P. English, and D.P. Vanderbilt (to Southern Research Institute).

19. J.M. Litch et al., *J. Periodont. Res.* **31**(8), 540–544 (1996).

20. U.S. Pat. 5,733,566 (March 31, 1998), D.H. Lewis (to Alkermes Controlled Therapeutics Inc.).

21. U.S. Pat. 4,166,107 (August 28, 1979), J.A. Miller et al. (to The United States of America as represented by the Secretary of Agriculture).

22. U.S. Pat. 5,018,481 (May 28, 1991), W.B. Rose (to Mobay Corp.).

23. U.S. Pat. 4,195,075 (March 25, 1980), W.V. Miller (to Shell Oil Co.).

24. U.S. Pat. 4,150,109 (April 17, 1979), P.R. Dick et al.

25. U.S. Pat. 4,879,117 (November 7, 1989), M.A. Rombi.

26. C. Thies, *Today's Chemist at Work*, November, 40–43 (1994).

27. U.S. Pat. 5,614,215 (March 25, 1997), A. Ribier et al. (to L'Oreal).

28. U.S. Pat. 4,217,344 (August 12, 1980), G. Vanlerberghe et al. (to L'Oreal).

29. U.S. Pat. 4,241,046 (December 23, 1980), D.P. Papahadjopoulos.

30. U.S. Pat. 5,077,056 (December 31, 1991), M.B. Bally et al. (to The Liposome Company, Inc.).

31. U.S. Pat. 4,235,871 (November 25, 1980), D.P. Papahajopoulos et al.

32. U.S. Pat. 5,616,334 (April 1, 1997), A.S. Janoff et al. (to The Liposome Company, Inc.).

33. *Code of Federal Regulations*, Title 21, Food and Drug Administration, Washington D.C., September 29, 1978, Pt. 210–211.

34. G. Odian, *Principles of Polymerization*, McGraw-Hill, New York, 1970.

35. A. Domb, J. Kost, and D. Wiseman, eds., *Handbook of Biodegradable Polymers*, Harwood Academic Publishers; Overseas Publishers Association, Amsterdam, 1997.

36. J.D. Ferry, *Viscoelastic Properties of Polymers*, 2nd ed. Wiley, New York, 1970.

37. J.E.A. John and W.L. Haberman, *Introduction to Fluid Mechanics*, 2nd ed., Prentice-Hall, Englewood Cliffs, N.J., 1980.

38. G.M. Barrow, *Physical Chemistry*, 2nd ed. McGraw-Hill, New York, 1966.

39. G. Schenkel, *Plastics Extrusion Technology and Theory*, American Elsevier, New York, 1966.

40. J.A. Schey, *Introduction to Manufacturing Processes*, McGraw-Hill, New York, 1977.

41. S. Gross, *Modern Plastics Encyclopedia*, McGraw-Hill, New York, 1970.

42. D.F. Mielcarek, *Chem. Eng. Process*, June, 59–67 (1987).

43. I.I. Rubin, *Injection Molding Theory and Practice*, Wiley, New York, 1972.

44. N.M. Bikales, ed., *Encyclopedia of Polymer Science and Technology*, Vol. 9, Wiley-Interscience, New York, 1968.

45. E. Luck and D. Brown, *Proc. Vet. Cancer Soc.*, Purdue University, Lafayette, Ind., May 6–8, 1985, pp. 22–23.

46. N. Yu, F. Conley, E. Luck, and D. Brown, *NCI Monogr.* **6** (1988).

47. Matrix Pharmaceutical, Inc., web page, available at: *www.matx.com*

48. J.M. Swinehart et al., *Arch Dermatol.* **133**, 67–73 (1997).

49. U.S. Pat. 4,978,332 (December 18, 1990), E. Luck (to Matrix Pharmaceutical).

50. U.S. Pat. 4,619,913 (October 28, 1986), E. Luck and D. Brown (to Matrix Pharmaceutical).

51. U.S. Pat. 5,290,552 (March 1, 1994), D. Sierra, E. Luck, and D. Brown (to Matrix Pharmaceutical).

52. U.S. Pat. 5,750,146 (May 12, 1998), R. Jones and M. Li (to Matrix Pharmaceutical).

53. K. Masters, *Spray-drying Handbook*, Wiley, New York, 1984.

54. J. Broadhead, S.K. Edmond Rouan, and C.T. Rhodes, *Drug Dev. Ind. Pharm.* **18**(11 and 12), 1169–1206 (1992).

55. C. Voellmy, P. Speiser, and M. Soliva, *J. Pharm. Sci.* **66**(5), 631–634 (1977).

56. P.B. Deasy, *Microencapsulation and Related Drug Processes*, Dekker, New York, 1984, pp. 181–193.

57. F. Nielson, *Manuf. Chem.*, 38–41 (1982).

58. R. Bodmeier and H.J. Chen, *Pharm. Pharmacol.* **40**, 754–757 (1988).

59. D.L. Wise, G.J. McCormick, and G.P. Willet, *Life Sci.* **19**, 867–874 (1976).

60. F. Pavanetto, I. Genta, P. Giunchedi, and B. Conti, *J. Microencapsul.* **10**(4), 487–497 (1993).

61. W. Dang, T. Daviau, and H. Brem, *Pharm. Res.* **19**, 683–691 (1996).

62. B. Gander, E. Wehrli, R. Alder, and P. Merkle, *J. Microencapsul.* **12**(1), 83–97 (1995).

63. U.S. Pat. 5,552,152 (September 3, 1996), R.W. Shen (to The Upjohn Company).

64. U.S. Pat. 5,498,447 (March 12, 1996), H. Nishii (to Sumatomo Pharmaceuticals Co., Ltd.).

65. Y. Fukomori et al., *Chem. Pharm. Bull.* **36**, 1491–1501 (1988).

66. F.N. Christensen and P. Bertelsen, *Drug Dev. Ind. Pharm.* **23**(5), 451–463 (1997).

67. C. Migliaresi, L. Nicolais, P. Ginisti, and E. Chiellin, *Polymers in Medicine III*, Elsevier, Amsterdam, 1987.

68. U.S. Pat. 5,000,886 (March 19, 1991), J.R. Lawter et al. (to American Cyanamid Company).

69. U.S. Pat. 4,673,595 (June 16, 1987), P. Orsolini et al. (to Debiopharm, S.A.).

70. U.S. Pat. 4,166,800 (September 4, 1979), J.W. Fong (to Sandoz, Inc.).

71. U.S. Pat. 5,503,851 (April 2, 1996), R. Mank et al. (to Ferring Arzneimittel GmbH.).

72. M.A. Tracy, *Biotechnol Prog.* **14**(1), 108–115 (1998).

73. L.R. Beck et al., *Fertil. Steril.* **31**, 545–551 (1979).

74. U.S. Pat. 5,654,008, P.F. Herbert and A.M. Hazrati (to Alkermes, Inc.).

75. U.S. Pat. 5,650,173, J.M. Ramstack, P.F. Herbert, J. Strobel, and T.J. Atkins (to Alkermes, Inc.).

76. U.S. Pat. 5,019,400 (May 28, 1991), W. Gombotz et al. (to Enzytech, Inc.).

77. *Product Literature*, Scandipharm, Inc., Hoover, Ala., 1998.

78. U.S. Pat. 5,679,373 (October 21, 1997), J. Wick et al. (to Bertek, Inc.).

79. U.S. Pat. 5,006,342 (April 9, 1991), G.W. Cleary et al. (to Cygnus Corp.).

80. U.S. Pat. 4,710,191 (December 1, 1987), A. Kwiatek et al. (to Jonergin, Inc.).

81. U.S. Pat. 5,662,925 (September 2, 1997), C.D. Ebert et al. (to TheraTech, Inc.).

82. G.A. Van Buskirk et al. *Pharm Res.* **14**(7), 848–852 (1997).

83. U.S. Pat. 5,445,607 (August 29, 1995), S. Venkateshwaran et al. (to TheraTech, Inc.).

84. J.M. Harris, ed., *Biotechnical and Biomedical Applications*, Plenum, New York, 1992.

85. C. Delgado, G.E. Francis, and D. Fisher, *Ther. Drug Carrier Syst.* **9**(3–4), 249–304 (1992).

86. H.F. Gaertner and R.E. Offord, *Bioconjugate Chem.* **7**(1), 38–44 (1996).

87. N.L. Burnham, *Am. J. Hosp. Pharm.* **51**(2), 210–218 (1994).

88. S. Tsunoda, Y. Tsutsumi, and T. Mayumi, *Nippon Rinsho* **56**(3), 573–578 (1998).

89. G.E. Francis et al., *J. Drug Target.* **3**(5), 321–340 (1996).

90. A. Abuchowski, T. van Es, N.C. Palczuk, and F.F. Davis, *J. Biol. Chem.* **252**(11), 3578–3581 (1997).

91. J. Glass et al., *Biopolymer* **18**, 383–393 (1979).

92. K. Ulbrich, J. Strohalm, and J. Kopeck, *Makromol. Chem.* **187**, 1131 (1986).

93. A. Abuchowski and F. Davis, in J. Holcenbery and J.P. Roberts, eds., *Enzymes as Drugs*, Wiley, New York, 1981, p. 367–383.

94. S. Dreborg and E. Akerblom, *Crit. Rev. Ther. Drug Carrier Syst.* **6**, 315–365 (1990).

95. J. Harris, *J. Macromol. Sci., Rev. Macromol. Chem. Phys.* **C25**, 325 (1985).

96. S. Schafer and J. Harris, *J. Polym. Sci., Part A: Polym. Chem.* **24**, 375–378 (1986).

97. M. Atassi, ed., *Immunochemistry of Proteins*, vol. 1, Plenum, New York, 1977, p. 1.

98. S. Davis, A. Abuchowski, Y.K. Park, and F.F. Davis, *Clin. Exp. Immunol.* **46**(3), 649–652 (1981).

99. K.J. Wieder, N.C. Palczuk, T. van Es, and F.F. Davis, *J. Biol. Chem.* **254**(24), 12579–12587 (1979).

100. A. Abuchowski et al., *Cancer Biochem. Biophys.* **7**(2), 175–186 (1984).

101. U.S. Pat. 4,101,380 (July 18, 1978), M. Rubinstein (to Research Products Rehovot, Ltd.).

102. L. Tondelli, M. Laus, A. Angeloni, and P. Ferruti, *J. Controlled Release* **1**, 251 (1985).

103. H. Berger, Jr. and S.V. Pizzo, *Blood* **71**(6), 1641–1647 (1998).

104. K. Yoshinaga and J. Harris, *J. Bioact. Compat. Polym.* **4**, 17–24 (1989).

105. F. Veronese, R. Largajolli, C. Boccu, and O. Schiavon, *Appl. Biochem. Biotechnol.* **11**, 141 (1985).

106. S. Zalipsky, R. Seltzer, and S. Menon-Rudolph, *Biotechnol. Appl. Biochem.* **15**(1), 100–114 (1992).

107. U.S. Pat. 4,179,337 (December 18, 1979), F. Davis, T. van Es, and N. Palczak.

108. F. Kawai, *CRC Crit. Rev. Biotechnol.* **6**, 273 (1987).

109. G. Royer and G. Ananthramaiah, *J. Am. Chem. Soc.* **101**, 3394 (1979).

110. Eur. Pat. Appl. 0,236,987 (September 16, 1987), H. Ueno and M. Fujino (to Takeda Chemicals, Ltd.).

111. Shearwater Polymers, URL: *www.swpolymers.com*.

112. Eur. Pat. Appl. 0,340,741 (November 8, 1989), A. Sano et al. (to Sumitano Pharmaceuticals Co., Ltd.).

113. S. Romani et al., eds., *Chemistry of Peptides and Proteins*, Vol. 2, de Gruyter, Berlin, 1984, p. 29.

114. A. Pollak and G. Whiteside, *J. Am. Chem. Soc.* **98**, 289–291 (1976).

115. M. Kimura, H. Matsumuro, Y. Miyauchi, and H. Maeda, *Proc. Soc. Exp. Biol. Med.* **188**, 364 (1988).

116. A. Carr et al., *Int Conf AIDS* **7–12**(11), 2–23 (Abstr. No. We.B.292) (1996).

117. N. Katre, *J. Immunol.* **144**(1), 209–213 (1990).

118. R.J. Goodson and N.V. Katre, *Bio Technology* **8**(4);343–346 (1990).

119. W.M. Pardridge, D. Wu, and T. Sakane, *Pharm. Res.* **15**(4);576–582 (1998).

120. T. Sakane and W.M. Pardridge, *Pharm. Res.* **14**(8);1085–1091 (1997).

121. D.K. Pettit et al., *J. Biol. Chem.* **272**(4);2312–2318 (1997).

122. Y. Tsutsumi et al., Thromb. Haemostasis **77**(1);168–173 (1997).

123. Y. Tsutsumi et al., *J. Pharmacol. Exp. Ther.* **278**(3);1006–1011 (1996).

124. L.A. Harker et al., *Blood* **88**(2);511–521 (1996).

125. K.E. Jensen-Pippo et al., *Pharm. Res.* **13**(1);102–107 (1996).

126. Y. Kita et al., *Drug Des Delivery* **6**(3), 157–167 (1990).

See also COATINGS; MICROENCAPSULATION.

# FERTILITY CONTROL

CAMILLA A. SANTOS
Brown University
Providence, Rhode Island

## KEY WORDS

Biodegradable

Contraception

Depo-Provera

EVA

Hormones

Hydrogels

Implants

Injectable

IUD

Microspheres

Norplant

PLA

Silicone

Transdermal

Vaginal rings

## OUTLINE

## BACKGROUND AND HISTORY

Many countries have experienced a population growth during this century. The societal and economic impacts of an increased population are enormous. There is concern that the world will not be able to meet the increasing needs of feeding and supporting an ever-growing population. Controlling the population growth through education and improved contraceptive devices available to all is perhaps the main driving force behind today's contraceptive research. Research in the field of fertility control promotes the use of a contraceptive agent or device that requires minimal medical intervention and provides safe, reliable contraception for an extended time. Providing contraceptive technology to third world countries, which have an immediate need for population control at a low cost with minimal medical intervention, is the goal behind research in this field.

The main objective of many fertility control devices is to achieve contraception by inhibiting ovulation. This can be accomplished with the administration of progesterone, which inhibits the preovulatory surge of luteinizing hormone (1). Additionally, many progesterones cause cervical mucus thickening and a thinning of the endometrial lining, reducing capacity of the endometrium to support implantation. Common side effects of excessive hormone dosage include irregular bleeding patterns, weight gain, breast tenderness, and acne. Contraceptive devices being developed today need to optimize hormone delivery and minimize side effects, while maintaining beneficial contraceptive effects. Another issue to consider is whether to deliver natural hormones or synthetic steroids. Natural hormones are synthesized from blood cholesterol, transported in the blood bound to albumin and specific binding globulins, and are rapidly degraded by the liver (1). Natural hormones are advantageous because their effect on the body is known, but their main disadvantages are the requirement for production from a pure source, their rapid degradation in the body, and the fact that they are difficult to deliver orally. The development of synthetic hormones has overcome these primary disadvantages of natural hormone therapies and has greatly improved contraceptive systems. Synthetic hormones avoid hepatic degradation and can thus be administered in much lower quantities, decreasing the numerous side effects normally encountered with natural hormones. Advancements in different types of systems for delivering hormones (both synthetic and natural) greatly aid in reducing unwanted side effects, as have improvements in the development of synthetic steroids themselves. The aim of this article is to acquaint the reader with the different types of fertility control devices commercially available today, as well as those in the preclinical and research stages. Developments in synthetic steroids have allowed the subsequent development of many of the contraceptive systems to be discussed. It should be noted that this article emphasizes the vehicle of delivery rather than the variety of steroid delivered.

A brief pharmacological analysis of the estrogens and progestins used in contraceptive devices is presented in Goodman and Gilman's *The Pharmacological Basis of Therapeutics* and is helpful toward understanding the action of many fertility control devices (2). One of the most potent naturally occurring estrogens is 17$\beta$-estradiol, as well as estrone and estriol. The most common synthetic estrogen is ethinyl estradiol. All of the estrogens are 18-carbon steroids, which contain a phenolic A ring necessary for binding to estrogen receptors. The estrogens are produced in the ovaries and placenta, and in the adipose tissue of postmenopausal women and men. Estrogens are readily absorbed through the skin, mucus membranes, and gastrointestinal tract and passively diffuse through cell membranes to reach nuclear estrogen receptors. The steroids are rapidly degraded by the liver and therefore have limited oral effectiveness. Synthetic estrogens have been modified to obtain slow hepatic degradation. Estrogens are not considered contraceptives except when administered in high doses; they are usually coadministered with progestins to maintain the endometrium and prevent irregular bleeding. The synthetic estrones are 17$\alpha$-ethinyltestosterone (ethisterone), 17$\alpha$-ethinyl-19-nortestosterone (norethindrone, norethisterone), and esters of 17$\alpha$-hydroxyprogesterone. The potent contraceptive steroids are the progestins. The progestins diffuse freely across cell membranes and depend on a $\Delta^4$-3-one A ring structure in an inverted 1$\beta$,2$\alpha$-conformation to bind to nuclear receptors, but this structure also binds to receptors for glucocorticoids, androgens, and mineralocorticoids. Progesterone is secreted by the ovaries in the luteal phase of the menstrual cycle and also in the testis, adrenal cortex, and placenta. Progesterone therapy extends the luteal phase of the cycle and mimics endometrial changes similar to early pregnancy. Secretion or administration of progesterone leads to the development of a secretory endometrium, and the abrupt decline in progesterone leads to menstruation. Continuous administration of progestins disrupts the cycle and leads to endometrial and ovarian atrophy. Low doses of the steroid can alter the endometrium and cervical mucus, inhibiting implantation but not affecting ovulation. Like the estrogens, the progestins are metabolized in the liver and excreted in the urine.

Synthetic steroids for oral contraception were not tested on a large scale until the early 1950s. After the development and testing of the numerous synthetic steroids, many contraceptive devices were designed. The main intent of many of these devices was to stem the fast-growing world population, and emphasis was placed on long-term contraceptive action with minimal medical intervention. The main forces behind research initiatives in this area were, and still remain, the World Health Organization (WHO), the Population Council, and pharmaceutical companies. By 1953, the injectable depot medoxyprogesterone acetate (Depo-Provera®) had been developed for use in cancer treatment and underwent clinical trials by the end of the decade (Table 1). Depo-Provera was tested as a contraceptive agent in several countries in the 1960s and marketed in the 1960s in Europe and Asia. During this same time frame another contraceptive device, vaginal rings made of silicone releasing progestins, began to be investigated. Norplant® clinical trials began in the early 1970s, as did clinical trials of the vaginal rings and the Copper T 380A intrauterine device (IUD). By 1973, research advanced to include biodegradable systems such as Capronor®, a

**Table 1. History of Fertility Control Development**

*1950s*

| | |
|---|---|
| early | Beginning development of oral contraceptives |
| 1953 | Depo-Provera developed for cancer treatment |
| 1957 | Noristerat two-month injectable developed |
| 1958 | Clinical trials with Depo-Provera as a cancer treatment begins |

*1960s*

| | |
|---|---|
| 1963 | First clinical trials with Depo-Provera as contraceptive, a three-month injectable suspension of microcrystalline steroid |
| 1968 | Cylofem (one-month injectable) research initiated |
| | Research begins on vaginal rings made of silicone-rubber-releasing progestins |

*1970s*

| | |
|---|---|
| 1972 | Norplant preliminary trials in humans, testing continued throughout the 1970s |
| | Clinical trials begin with Copper T 380A IUD |
| 1973 | In vitro research with Capronor |
| | Progestasert IUD developed |
| 1974 | Mesigyna developed (one month injectable) |
| 1976 | Implanon developed |
| | Progestasert IUD marketed in the U.S. by Alza Pharmaceuticals |
| | Copper T 200 approved by U.S. FDA |
| 1978 | Work begins with hydrogels releasing progesterone (hydroxyethyl methacrylate) |
| | IPCS trials begin |
| | Vaginal Contraceptive Film (VCF) marketed in Europe |
| 1979 | Preliminary trials with vaginal rings made of Silastic-releasing progestins for 90 days |
| | Research begins with vaginal rings releasing spermicide, continued in the 1990s |
| | Poly(l-actic acid) (PLA) microspheres containing NET (norethisterone), trials in baboons |

*1980s*

| | |
|---|---|
| 1981 | Development of vaginal rings releasing progestins and estrogen for one month |
| | Human trials with PLA microspheres releasing NET |
| 1982 | Capronor trials in humans |
| | Injectable norethisterone enanthate (NET-EN) |
| 1983 | Norplant-2, in vitro research |
| | Depo-Provera marketed in Finland as a contraceptive |
| 1984 | Clinical studies began with Cyclofem |
| | Clinical studies began with NET release from poly(dl-lactide-co-glycolide) (PLGA) (3–6 mos application) |
| | Phase III trials with vaginal rings releasing progestin (LNG) |
| | Biodegradable pellets fused with cholesterol and releasing NET developed |
| | FDA approves Copper T 380A IUD |
| 1985 | Norplant clinical trials in the United States |
| 1986 | Polyethylene oxide-based hydrogel strips releasing methyl ester developed for early abortion (24 h) |
| | Norplant-2 humantesting began |
| | VCF available in the U.S. |
| 1988 | Ethylene–vinyl acetate (EVA) strips releasing MSGB developed to prevent fertilization |
| | ParaGard380/Copper T 380A IUD marketed in the U.S. |
| 1989 | Transdermal delivery of LNG from an osmotic pump with an EVA membrane tested in rabbits |
| | Phase II clinical trials with PLGA microspheres releasing NET |

*1990s*

| | |
|---|---|
| 1990 | FDA approval of Norplant |
| 1991 | Research continues on vaginal rings releasing LNG and rings releasing spermicide (silicone-covered EVA, releasing for 30 days) |
| | Chitosan matrix for steroid release tested in vitro |
| 1992 | Depo-Provera approved by FDA |
| | LNG ROD device approved for use in Finland |
| 1993 | FDA extends effective duration of Copper T 380A to 10 years |
| 1996 | Implantable 2-rod system (LNG ROD) approved by FDA |

poly($\epsilon$-caprolactone) polymer encasing the synthetic progestin levonorgestrel (LNG). Biodegradable injectable systems were developed, specifically the forerunners of the poly(lactide-co-glycolide) and steroid microspheres that are still in the testing stages today. The 1970s also saw the development and production of IUDs designed to deliver steroids over the course of one year, such as the Progestasert® device. In the 1980s, injectables with an estrogen as well as a progestin were tested clinically, mainly Cyclofem®. The vaginal rings were still being researched, as were several injectable formulations. Optimization of the type of synthetic steroid and the exact amount of release were the main goals of many of the new systems. Transdermal delivery of steroids was also investigated. In the 1990s, the U.S. FDA approved two contraceptive methods, Norplant and Depo-Provera, and research continued on the many other forms of contraceptive devices, including new systems designed for male contraception.

Although Depo-Provera and other depot injections of pure steroid are not considered controlled release per se, they have been included in this section because they are among the few commercially available long-term contraceptives. There are many formulations of crystalline steroid injection currently in the testing phases, but these have not been mentioned in this article as the release profiles are first order and depend only on the solubility of the crystalline steroid involved, not on controlled release from a polymeric delivery vehicle.

In addition to the numerous contraceptive devices both on the market and in the design stages, investigators are creating devices used to induce abortion. Recent developments in controlled release have allowed the design of devices to release abortifacients over 24 hours, with the hope of reducing the side effects of potent drugs currently administered in a bolus form.

The aim of this article is to familiarize the reader with the different types of controlled release systems designed for fertility control purposes on the market and in the research stages. Basic mechanisms of release are discussed, and examples of each type of device are provided wherever possible.

## CONTRACEPTIVE DEVICES

### Female Contraceptives

**Marketed Products.** There are several marketed contraception products that rely on constant release of steroid to achieve their goal. Such products include Norplant, Depo-Provera, and the Progestasert IUD.

*Norplant.* Norplant is a constant, low dose, progestin-only contraceptive device that was developed by the Population Council and first marketed by Leiras Pharmaceuticals, Turku, Finland, in 1983. Today it is used worldwide by more than 3 million women (3). In the U.S. the device is marketed by Wyeth-Ayerst Laboratories, Philadelphia, PA. The original Norplant consists of six capsules of crystalline LNG encapsulated in Silastic®. The capsules are implanted in the inner region of the upper arm, with the use of a 10-gauge trocar, and are designed for 5 years of contraceptive efficacy. Each rod contains 36 mg of the synthetic progestin LNG, with a total daily release rate from the six capsules of 70 $\mu$g/day initially, dropping to 30 $\mu$g/day after a few months with a slow decline over five years of action. Release from the capsules has been reported to continue for ten years, but the high level of contraceptive efficacy is maintained for only the first five years (4). Efficacy of the device is dependent upon its steroid release to achieve 0.25 ng/mL serum concentration of LNG, which is sufficient to inhibit pregnancy (3). The capsules are 34-mm long and have an outer diameter of 2.41 mm. The inner diameter of the Silastic tubing is 1.57 mm. The core of the device is 30 mm in length and consists solely of crystalline levonorgestrel (5–8). The main advantages of the Norplant system are its ease of use, five-year duration of action, reversibility at any time by simple removal of the capsules, and its low dose of daily steroid when compared with oral contraceptives. Its main disadvantages are the need for surgical implantation and removal and irregular bleeding patterns (9,10).

In terms of a polymer delivery vehicle, Norplant is a membrane limited device in which the release is governed by diffusion. The membrane-limiting barrier of Norplant is made of the common medical elastomer, Silastic. Silastic is the commercial term for poly(dimethylsiloxane)(PDMS) manufactured and patented by Dow Corning. It is a silicone elastomer, which is a lightly cross-linked macromolecular network that will burn upon the application of high heating. It has a glass-transition temperature of $-123°C$, is an amorphous polymeric material, is flexible at room temperatures, and has low mechanical strength. In general, silicone has the physical capability to be stretched reversibly between temperatures of $-75°C$ and $240°C$. The ultimate tensile strength of silicone polymers can reach 110 kg/cm$^2$ (the ultimate tensile strength is the strength required to rupture in tension), and the ultimate elongation can reach 700%. Silicones are also resistant to compression and return to their original dimensions upon the removal of any compressive forces. Silicone elastomer is a blend of silicone gum (uncross-linked), filler, and a catalyst that induces cross-linking. Once reacted, these ingredients yield PDMS. It is a nonporous, hydrophobic material, impermeable to ions. Water diffuses easily into the matrix. A filler is normally added for strength and consists of silica particles of approximately 30-$\mu$m diameter, composing 15–20% of the volume. Strength is increased by decreasing the freedom of the polymer chains as they form around the silica filler particles. In addition to increasing the tensile strength, the filler increases the tear strength and abrasion resistance. The silica particles are porous, being composed of diatomaceous earth and have a negative effect on diffusion of active agents (i.e., steroids) due to adsorption between the filler and these solid materials (11).

PDMS has a backbone of silicone and oxygen. The Si-O bond is very strong and resistant to hydrolytic attack. Silicone polymers are also resistant to aqueous systems (i.e., acids, bases, salts) such as those found in the body. They have a high permeability to organic solvents and can be sterilized by gas ethylene oxide. For these reasons, PDMS is very biocompatible and is used in a variety of medical applications. Silastic is known to be one of the most easily diffused polymers and lipophilic compounds, such as progesterone, are soluble in it and diffuse at a high rate through a solid matrix of Silastic 382 (7,12).

There are two types of diffusion-controlled polymeric release systems, reservoir and matrix. It is also possible to have a combination of both types in one system. Reservoir systems, such as Norplant, ideally have zero-order release, where the drug is released independent of time, and the kinetics of the release are controlled by the design of the device. The system depends on surface area, concentration difference, diffusion coefficient, partition coefficient, and thickness of the device. Flux in a reservoir system is described by the equation

$$\text{flux} = -D\Delta C_\text{m}/\Delta x \qquad (1)$$

where $D$ is the diffusion coefficient of the drug, $\Delta C_\text{m}$ is the concentration difference, and $\Delta x$ is the diffusion distance (thickness). The crystalline steroid in Norplant is rather large and has a low solubility in water, causing it to aggregate as water diffuses into the capsule over time. The aggregation of steroid crystals acts to decrease the release rate, necessitating the implantation of six capsules to maintain adequate drug delivery over five years (6,7,13–15).

The Norplant system was approved by the FDA in 1990 and has a reported pregnancy rates of 0.7 per 100 over the course of 4 years (16) and 1.5 per 100 (over 5 years), significantly below the pregnancy rates for users of non-medicated IUDs (6,17). Continuation rates for Norplant users are high, at 76–90% at the end of the first year as compared with 50–75% for IUD and oral contraceptive users (3). Main reasons for discontinuation are common problems associated with progestin-only contraceptive methods, mainly abnormal bleeding. After removal of the implant, fertility is 90% at 2 years (16). A fibrotic capsule commonly forms around the Norplant implants but it does not hinder release of the steroid as diffusion of the steroid through the membrane is the limiting factor and not the solubility of the drug in the body (15).

*Other Implantable Capsules.* Contraceptive systems comprised of only two capsules constitute the second generation of the Norplant system. The Norplant-2 system was designed as an improvement over the original Norplant system, which consists of six capsules implanted under the skin of the upper arm (7). The second-generation device, Norplant-2, consists of two capsules implanted in a similar fashion in the same location. The two capsules of Norplant-2 are solid 1:1 weight ratio of micronized crystalline LNG and PDMS without filler, encapsulated within an ultrathin Silastic tube. Norplant-2 covered capsules are 44 mm in length and 24 mm in diameter, with each rod containing 70 mg of steroid. The release of steroid from Norplant-2 is much more efficient than the release from Norplant mainly because the steroid is micronized and in constant contact with the polymer matrix, increasing its diffusional capability. In the Norplant system, the steroid is crystalline and rather large, so as water enters the capsule over time the steroid aggregates due to its low solubility in water. This clumping tends to decrease its release from within the capsule. Norplant-2 release is membrane controlled, much like the six capsules of Norplant (6,7,13–15).

Norplant-2 was delayed in clinical trials due to procurement problems, as production was ceased for the Silastic elastomer used as the inner core for Norplant-2 (3,10). However, because initial testing of Norplant-2 was promising, the Population Council sought to continue development with a similar device. The system was redesigned with another elastomer as the inner matrix, a copolymer of dimethylsiloxane and methylvinylsiloxane (18). This copolymer is cured in a physical mixture with the steroid LNG. The total content of steroid in the two-rod system is 150 mg. This system is known as the LNG ROD implant and is approved for use in Finland and China. It is marketed in Finland by Leiras Pharmaceuticals, Turku. The two-rod system was also approved by the FDA in 1996 (19). Testing done to date has shown the LNG ROD system to be comparable to Norplant-2. Three-year studies of LNG ROD implants in 398 women showed no pregnancies. Serum concentrations of LNG are above 300 pg/mL for the first year, dropping well below this during the third year. The lowest serum level detected was 151 pg/mL, and the device still achieved its contraceptive efficacy at this low dose (18). The main advantages of LNG ROD over Norplant are the reduced number of capsules needed to achieve contraceptive efficacy. Additionally, the LNG RODs are more rigid than those in the original Norplant system, allowing easier implantation and retrieval of the system (3). Advantages of both systems are convenience, long-term action, high effectiveness, voluntary reversibility, and low dose of steroid. The disadvantages of the original Norplant system also apply to the LNG ROD system, mainly the need for surgical implantation and removal, and bleeding irregularities (5,9,10). In three years of use, there were no differences in menstrual irregularity or efficacy of the two-rod system as compared to the six-rod system, but the two-rod system was easier to insert and remove (16). In both systems, the average steroid serum level is affected by duration of use and by weight of the patient (18). The LNG ROD system is approved by the FDA for 3 years of use, even thought it remains effective for 5 years.

*Depo-Provera.* Depo-Provera is an intramuscular injection of 150 mg of microcrystalline depot medroxyprogesterone acetate (DMPA), which is administered in an aqueous solution. It has a much higher serum concentration of MPA than the LNG levels with Norplant, as shown in Figure 1 (3,20). The steroid is absorbed into the bloodstream over the course of several months. Another injection is given at 3 months. It is the most common injectable contraceptive system and was FDA approved in 1992 (21). Depo-Provera is currently marketed by the Pharmacia and Upjohn Company in the United States as well as in many other countries. An identical formulation is also marketed by Organon under the brand name Megestron (21,22). The mode of contraceptive action is the same as that of any progestin; its contraceptive effect lies primarily in its capacity to inhibit ovulation, as well as in its ability to cause cervical mucus thickening and reduce endometrial capacity to support implantation (21,23).

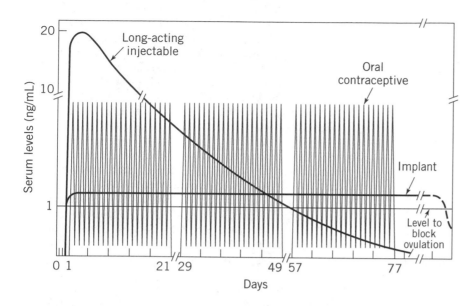

**Figure 1.** Release profiles of hormone levels in blood serum for three different types of contraceptive devices: daily oral contraceptives, Norplant implants and Depo-Provera 3-month injectable. *Source:* From Ref. 20.

The release profile of Depo-Provera is first-order, and serum levels of the drug are seen as one large bolus which drops off continuously over the time course of action. The high initial serum level peaks occur within the first three weeks postinjection and are most likely due to rapid absorption of loose particulates of drug (21). Advantages of the Depo-Provera system are its ease of administration (single injection every three months), high user compliance, high effectiveness, length of action, and reversibility. Side effects are the same as those common for progestin-only contraceptive series, mainly irregular bleeding patterns and weight gain. An additional disadvantage of the system is its irretrievability after injection (22,23).

Recent analysis has shown the pharmacokinetics of DMPA are affected by particle size, and recipient's diet, body mass, and fat distribution (9). The WHO is sponsoring a pharmacokinetic and pharmacodynamic study to determine the effects of three compositions of particle size, 3–10, 10–20, and 20–40 $\mu$m, to determine the effect of particle size of DMPA on pregnancy rate.

*Cyclofem.* An attempt to offer an alternative to Depo-Provera resulted in Cyclofem (formerly known as Cycloprovera), marketed by the Pharmacia and Upjohn Company, and HRP 102, manufactured by Schering AG. Cyclofem is a one-month injectable similar to Depo-Provera and is available for use in several countries but has not yet been FDA approved in the United States (9,22,24). In both Cyclofem and HRP 102, the steroid is in microcrystalline form and contains an estrogen compound to simulate more natural hormone levels and reduce the irregular bleeding side effects common to most progestogen-only systems (Depo-Provera is a crystalline progestin-only formulation) (22,25). Cyclofem contains 25 mg DMPA and 5 mg estradiol cypionate, and HRP 102 contains 50 mg norethindrone enanthate (NET-EN) and 5 mg estradiol valerate (25). There are several one- or two-month injectables currently marketed outside the United States under varying brand names, such as Mesigyna, Doryxus, and Cyclo Geston (22). Each of these systems employs the same method of drug delivery as Depo-Provera,

namely a slow uptake of the crystalline steroid injection due to the low solubility of the steroid in the aqueous environment of the body. The systems are composed of a variety of synthetic progestins each with an estrogen component. Release from the one- and two-month injectable systems is typically similar to Depo-Provera, an initial burst that gradually declines.

Another variation on Depo-Provera is the injectable Noristerat. Noristerat is a two-month injectable composed solely of the synthetic progestin NET-EN and is manufactured by Schering AG. It is a crystalline steroid, and its use is approved for use in over 40 countries (10).

*Intrauterine Devices.* There are several marketed IUDs that employ release of a substance to achieve their contraceptive efficacy. The initial IUDs relied on the slow release of copper to inhibit pregnancy, and later devices employed the use of steroids delivered directly to the uterus to prevent conception. IUDs are typically designed in the shape of a T, with drug release normally occurring from the stem portion of the device.

There are several designs of copper-releasing IUDs, and the most common is the Copper T IUD, which releases copper through corrosion over the course of 12 years. The copper particles act to cause a sterile inflammation of the endometrial lining, thus prohibiting contraception. In the many copper-releasing IUDs, the copper wire thickness varies from 0.2 to 0.4 mm with a surface area of exposed copper that varies correspondingly from 200 to 380 mm$^2$. The corrosion is time dependent and exhibits a linear release pattern, with release directly proportional to the surface area of exposed copper. An exposed area of 375 mm$^2$ releases 37.5 $\mu$g/day of copper and an exposed area of 250 mm$^2$ releases 25 $\mu$g/day (26). Examples of copper-releasing IUDs include the Copper T 380A developed by the Population Council, the polyethylene Nova-T (Cu T Ag 200) developed by Leiras and Outokumpo Oy, Finland, and the polyethylene Multiload developed by Multilan, Switzerland, which is available in several sizes to accommodate different body types (26).

The current version of the Copper T IUD is the ParaGard® T 380A intrauterine copper contraceptive, and has 380 mm² of exposed copper wire surface area. It is marketed in the U.S. by Ortho-McNeil Pharmaceutical (27). The IUD is comprised of a polyethylene T, with 176 mg of copper wire on the stem of the device and an additional 66.5 mg of copper wire on the arms. The IUD has a monofilament polyethylene thread attached at the tip of the T to aid in retrieval. In 1994, the FDA approved this device for 10 years of contraceptive use. Its original design was intended for a shorter time frame but recent clinical analysis has shown its contraceptive effect to extend to 10 years. The copper-releasing Copper T 380 Ag IUD manufactured by Leiras is similar to the Copper T 380A, but differs in that the copper wire has a silver core. The silver wire acts to slow the rate of corrosion and fragmentation of the copper. Ortho Pharmaceuticals of Canada markets Gyne T 380 Slimline, a device similar to the original Copper T device. It differs in its structural design; the side arms of the Slimline IUD are designed to fit into the inserter tube more easily for easier administration (19).

There are several marketed IUDs releasing steroid available today. All of the progestin-releasing IUDs produce a local sterile inflammation in the endometrial cavity much like other IUDs. However, the steroid-releasing IUDs do not rely on endometrial inflammation to achieve contraception but instead depend on the effect of the steroid delivered. The only one manufactured in the U.S. is the Progestasert IUD, illustrated in Figure 2 (28). Progestasert releases progesterone and has been marketed in the U.S. since 1976. It was the first contraceptive device to use the natural hormone progesterone, and is currently the only IUD marketed in the U.S. that delivers a steroid (28). The device has solid poly(ethylene vinyl acetate) (EVA) side arms and a hollow core. The microcrystalline progesterone steroid is suspended in the core in medical grade silicone oil along with barium sulfate (for radiopaque properties). The drug reservoir is encased by the EVA (9% w/w vinyl acetate) copolymer membrane. The EVA membrane is 0.25 mm thick and has a surface area of 2 cm². Release is governed by diffusion through this rate-limiting membrane. Progestasert is loaded with 38 mg of progesterone and releases at a rate of 65 μg/day over the course of 1 year (28). Release begins at 70 μg/day and decreases over the course of 1 year to 55 μg/day, at which point 60% of the loaded drug has been released (26). The device was originally designed for use as a 1-year contraceptive agent, but testing has shown the same level of efficacy for 18 months. The pregnancy rate at this time is reported as 1.8/100 for parous women and 2.5/100 for nulliparous women (28), whereas the pregnancy rate at 2 years of use is dramatically higher at 7.4/100 (29). Progestasert does not achieve its contraceptive effect by inhibiting ovulation, but instead prevents implantation in the endometrium by suppressing the normal cyclic changes of the endometrium lining as well as thickening cervical mucus (20,29). Progesterone released from the IUD is completely metabolized in the uterine wall and does not reach systemic circulation (29). Ovulation is not inhibited with the Progestasert device. The advantages of Progestasert over other nonsteroidal IUDs is its increased effectiveness, decrease in menstrual blood loss, and decreased dysmenorrhea. Disadvantages include its need to be replaced on a yearly basis, intermenstrual bleeding and spotting, and the possibility of increased ectopic pregnancies (24,29,30).

Other medicated IUDs use the more potent synthetic steroids to achieve contraception. The LNG Nova-T IUD has a polyethylene stem and side arms. The stem is covered by a matrix consisting of Silastic and LNG (2:1 weight ratio) covered with a nonmedicated Silastic membrane. It is designed to release LNG at a rate of 20 μg/day for 5 years. Early designs of this system had a stem outer diameter of 2.5 mm, which consisted of a thinned vertical stem of polyethylene (0.9-mm diameter) encased in the drug-loaded matrix and covered with Silastic tubing (0.41-mm thick) (31). In vivo release rates show an initial release of 24 μg/day, decreasing to 16 μg/day at the end of 2 years. By the end of the fifth year, 60% of the loaded drug is released (26). The synthetic steroid is released into the uterus much like the progesterone released by Progestasert, but avoids complete degradation and is systemically absorbed. The hormone acts to suppress the endometrium as well as inhibit ovulation (26).

***Vaginal Contraceptive Film®.*** The Vaginal Contraceptive Film (VCF®) is a small, thin, semitransparent square composed of 28% nonoxynol 9 in a base of poly(vinyl alcohol) and glycerin. It is a 2-inch by 2-inch square film that is 50 μm thick (32). VCF is the same device as C-Film, marketed in Great Britain since the mid-1970s. VCF contains a total of 70 mg of nonoxynol-9, delivering an average of 56–73 mg of drug (33). Nonoxynol-9 is a nonionic detergent that acts as a spermicide by creating a chemical barrier to spermatozoa in the vagina and disrupting the lipid membrane of the sperm (34). VCF is intended to be placed in the vagina against the cervix and dissolves into a gel upon contact with the vagina, in the time course of 15 minutes to 1 hour. The film and its contents are completely washed away by vaginal and cervical mucus discharge (35). It has a 96% efficacy rate and is an effective spermicide for up to 1 hour, although preliminary studies indicate that a large enough dose to inhibit spermatozoa (0.100 mg/mL) remains in the vagina for up to 4 hours (34). VCF has been available in

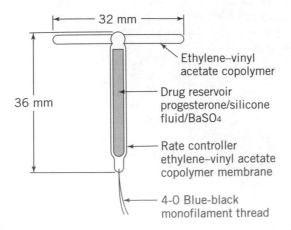

**Figure 2.** Progestasert IUD. *Source:* ALZA Pharmaceuticals, from Ref. 28.

the United States since 1986 and is marketed by Apothecus Pharmaceutical Corporation (36).

**Precommercial and Preclinical Products.** There are many types of contraceptive devices in the design and preclinical stages. Development of new synthetic steroids has led to development of optimal systems, and current regulations require extensive testing of these systems prior to clinical use. This section describes the current formulations still in the research phase.

*Vaginal Rings.* Development of contraceptive vaginal rings (CVRs) began in the late 1960s and early 1970s. Both the Population Council and the WHO have promoted research of this promising device. Although there are several designs of vaginal rings, the main features are essentially the same. The rings are donut-shaped with an outer diameter of 50–58 mm. They release either a progestin or a combination of progestin and estrogen. The progestin-only rings typically are designed to remain in place for the entire duration of action (normally 3 months), whereas the combined progestin–estrogen delivery system is removed every fourth week to simulate menstrual bleeding. When compared with oral tablets, vaginal rings deliver a lower dosage of hormones to achieve the same contraceptive effect. The rings are simple to self-administer and remove, produce little or no local irritation, provide a near zero-order sustained release of steroid, and the treatment can be ceased at any time to provide a reversible contraceptive effect. The rings are held in place by the vaginal muscles, and it is not necessary to fit the CVR to each user or to position the CVR in a certain place in the vagina. The rings can be removed for 3 hours without decreasing the efficacy of the device (9,37,38).

The mechanism of release is very similar among the many CVR designs. Steroids delivered vaginally are absorbed across the vaginal epithelium directly into the systemic bloodstream without any metabolic alteration in the digestive tract, in specific, within the liver. The release kinetics of the vaginal rings optimize the efficacy of the steroid while minimizing its required loading, aspects that benefit the user by decreasing common side effects of hormone therapy. Vaginal rings are an optimal delivery system for natural progesterone, which is a safe contraceptive for lactating women because it is not harmful to infants but cannot be administered orally because it is metabolized too quickly (39,40). The main disadvantage of the vaginal ring containing progestin only is irregular bleeding; vaginal rings with a combined estrogen and progestin result in better bleeding patterns.

The vaginal rings are made of Silastic and loaded with a steroid. Several different designs releasing both natural progesterone and several varieties of synthetic progestins with or without an estrogen component have been designed (Table 2). The intent of the many designs is to optimize the type of steroid and its daily dosage in a form that minimizes the side effects while still maintaining the contraceptive efficacy of the system. The main mechanism of drug release is diffusion. Diffusion depends on concentration gradients, where one element diffuses from an area of high concentration to an area of low concentration. Silastic is known to be one of the most easily diffused polymers, and lipophilic compounds, such as progesterone, are soluble in and diffuse at a high rate through a solid matrix of Silastic 382 (7,12). There are two types of diffusion controlled polymeric release systems, reservoir and matrix. It is also possible to have a combination of both types in one system, such as in some designs for the vaginal rings.

The best tolerated vaginal rings in clinical studies have an outer diameter of 50–58 mm and a thickness of 7–9.5 mm (41). There are four types of CVR that have been developed to date. The initial design consisted of a homogenous matrix of drug-loaded elastomer. The solid matrix of drug-loaded polymer was not encased within a rate-limiting membrane. Steroid diffused both through the polymer matrix and through pores left by released drug. There was a high burst effect with this design and a rapid decline in release due to depletion of the drug from the outer regions of the matrix, and subsequent testing was abandoned (38,42). Another design is the band design, in which a band of drug-loaded collagen is placed in a groove on the outer surface of a solid ring of polysiloxane (38). The two main types of ring design that have shown reliable and effective release rates over the course of 3 to 6 months are the core and shell designs. In the core design, a 3.5 mm homogenous core of drug-loaded PDMS is covered with a 5 mm layer of nonmedicated polysiloxane (38,42). The shell ring, depicted in Figure 3, is a design in which an inner layer of solid elastomer provides support, the middle layer is a matrix of polymer and drug, and the outer layer is a thin rate-limiting membrane of Silastic (38,40,42,43).

In the core ring, the matrix is in the form of an inner core of polymer loaded with steroid, and the reservoir boundary is an outer layer of pure polymer, generally Silastic. The outermost layer of progesterone in the matrix core diffuses away and leaves pores or channels for the remaining progestin to diffuse out. Because each molecule must still diffuse out of the unloaded outer layer of pure Silastic, the outer membrane maintains constant release rates. It was determined that the core ring device releases drug according to the diffusion model, which can be described by the equation

$$Q = KCDt/h \qquad (2)$$

where $Q$ is the amount of drug released, $t$ is the time, $K$ is the partition coefficient of drug between the polymeric device and surrounding fluid, $D$ is the diffusion coefficient of the drug in the polymer, and $h$ is the thickness of the hydrodynamic diffusion layer (12). After several weeks, the rings deviate from their expected release profile. This is due to a depletion zone that is created as the outermost layer of steroid in the matrix is diffused away, increasing the diffusion distance for the remaining steroid (12). The release of progesterone from core-loaded vaginal rings made of Silastic 382 show an inverse dependence on diffusion distance. This is characteristic of membrane-controlled drug delivery. Release rates can be varied with parameter changes of the inner core to yield high, medium, or low loading depending on the core diameter (12). The natural progesterone is more soluble in Silastic than its synthetic counterpart LNG, and tends to diffuse faster through the membrane than through the inner core, which

**Table 2. Designs of Contraceptive Vaginal Rings**

| Design type | Progestin (or spermicide) | Estrogen | Length of use | Outer layer | Middle layer | Inner layer | Developer/ Manufacturer | Refs. |
|---|---|---|---|---|---|---|---|---|
| Homogenous | Progesterone 20 mg/day | None | 90 days | PDMS and 22.5% w/w steroid | None | None | WHO/Dow Corning, Population Council | (12,40,42) |
| Core | LNG 20–25 µg/day | None | 90 days | PDMS | None | 5 mg LNG and PDMS | WHO | (26,38) |
| Core | Progesterone 15 mg/day | None | — | PDMS | None | PDMS with 20% w/w progesterone | — | (16,42) |
| Core | Progesterone 5 mg/day | None | 90 days | Thin Silastic | None | Silastic with 25% w/w steroid | WHO/Dow Corning | (12,40) |
| Core | 3-ketodesogestrel 75 or 150 µg/day | Ethinylestradiol 15 µ/day | 120 days | Silastic | None | Silastic and drug | Organon | (24,32) |
| Core | ST-1435 60 mg loaded | Ethinylestradiol 120 mg loaded | | Silastic | None | Silastic and drug | — | (26,38) |
| Shell | Medroxyprogesterone acetate 100 mg | None | 3–6 mos | PDMS | PDMS and drug | PDMS | — | (38) |
| Shell | Norgestrel 50 or 100 mg | None | 3–6 mos | PDMS | PDMS and drug | PDMS | — | (38) |
| Shell | Norethindrone 100 or 200 mg | None | 3–6 mos | PDMS | PDMS and drug | PDMS | — | (38) |
| Shell | Norethindrone 850 µ/day | Estradiol 200 µ/day | 3 mos | PDMS | PDMS and drug | PDMS | — | (38) |
| Shell (suspended) | LNG 230–290 µ/day | Estradiol 150–180 µ/day | | PDMS tubing | PDMS, LNG (30–77 mg), estradiol (20–66 mg) | PDMS | Population Council | (26,38,42) |
| Shell | LNG 25 µ/day | None | 90 days | Thin Silopren® membrane | Drug-loaded Silopren matrix | Silopren support structure | Schering AG | (40) |
| — | NET acetate | Ethinylestradiol 20–65 µg/day | | — | — | | — | (38) |
| Core | Nonoxynol-9 600 µg or 1 mg/day | None | 30 days | Silastic | None | EVAc and drug | — | (45) |
| Homogenous | RS-37367 spermicide 48 µg/cm/h | None | 8 days | PDMS and 1.37% w/w drug | None | None | — | (46) |

*Note:* Table cells containing a dash indicate that data for this feature was not specified.

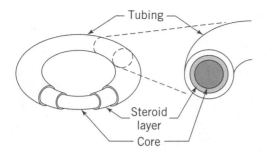

**Figure 3.** Cross section of the shell design contraceptive vaginal ring. *Source:* From Ref. 43.

accounts for the decreased release rate over time (42). After insertion, peak plasma levels of LNG from a core CVR design were detected within 1–8 hours, reaching 200% of their constant-release level. Release then becomes fairly constant after the initial burst but decreases by 23% to 26% over the course of 90 days (26). The factors governing release are the dissolution of the drug within the drug-loaded layer, diffusion of the drug through the rate-limiting Silastic membrane, and dissolution of the steroid within the body fluids.

Core ring designs are also being evaluated by Olsson and Odlind for the release of 3-keto desogestrel and ethinylestradiol in clinical trials (44). The core ring for delivery of two steroids is designed a bit differently; the ring is made of two Silastic tubes connected by glass stoppers. The outer diameter of the ring is 6 cm and the cross-sectional diameter is 4.9 mm. This type of CVR was developed by Organon International (Oss, The Netherlands). The rings have a daily release rate of 150 $\mu$g 3-keto desogestrel and 15 $\mu$g ethinylestradiol. The rings maintain their contraceptive efficacy for three consecutive cycles of 21 days each, with a removal period of 7 days in between cycles. Upon removal, the rings should be rinsed in water and stored in a dry location. Preliminary trials with these two-compartment rings indicate their usefulness as a contraceptive device, but also demonstrate that not all women find the CVR a comfortable means of birth control as there was some discomfort associated with use of the rings as well as some irregular bleeding patterns.

The shell rings are the most successful CVR design (38). The steroid is kept close to the rate-limiting outer Silastic membrane, acting to decrease the diffusion distance from within the matrix to the membrane. The release in these systems is governed by the outer layer of pure polymer, and release is more constant than the core ring design due to the near constant diffusion distance within the drug-loaded matrix. If the outer membrane is doubled in thickness, the release is decreased by one-half, indicating that the outer shell is the rate-limiting membrane (42). The shell CVR device provides a fairly constant diffusion distance due to the narrow inner drug-loaded polymer matrix, which is surrounded by a nonmedicated polymer membrane. The outer membrane is the rate-limiting device, and a thin enclosed ring of drug-loaded polymer provides a near-constant diffusion distance to the outer limiting membrane. These two aspects work together to provide the shell device with fairly constant release kinetics.

The manufacture of the early shell rings was a three-step injection molding process done by hand. With the increase in promising clinical trials, a new technique has emerged that produces uniform shell rings with decreased cost. In the new manufacturing process, a rod of Silastic is coated with a drug-polymer layer. After this first layer has cured, an additional layer of pure polymer is applied. The rod is then cut into sections and formed into a ring with a band of polymer to secure it (42).

Vaginal rings have also been explored for use in delivering spermicides. Vaginal rings of ethylene–vinyl acetate copolymer (EVAc) loaded with 33% of the spermicide nonoxynol-9 and covered with Silastic have been shown to release the agent over 30 days (45). This device can be used intermittently or continuously for a total of 30 days. Additionally, vaginal rings are being investigated as a delivery vehicle for the spermicide RS-37367, tested in macaque monkeys (46). This spermicide is more potent than nonoxynol-9, and has been shown in primates to be effective even when the ring is removed immediately prior to coitus. The RS-37367-containing vaginal rings are 2.3 cm in outer diameter, 1.5 cm in inner diameter, and approximately 0.4 cm in cross section. They are a homogenous matrix of 1.37% w/w drug with Silastic, formed by the injection molding technique described earlier. The devices were tested for up to 8 days in vivo.

There was some initial concern about infection with use of the CVR as it is a foreign device that is removed from and replaced into a body cavity. In 2-year studies, there was no increased incidence of vaginal infection with use of the CVRs, although there was a noted increase in vaginal secretions (38). No cervical changes were noted over this time course either.

The CVR appears to be a very promising delivery vehicle for contraceptive steroids. It is easy to use, requires minimal supervision, maintains constant serum levels of steroid, and there is minimal user requirement (once-a-month insertion). Further testing is needed, however, to optimize drug loading and provide additional safety data.

***Implantable Devices.*** A single implanted rod of nonbiodegradable EVAc similar to Norplant is being developed by Organon to deliver ST-1435, which is 3-keto desogestrel, the main metabolite of desogestrel. The device is called Implanon® (47). One 4-cm rod releases approximately 25 $\mu$g/day of steroid for 2 years (3,16,24,48). Due to the potency of the progestin used in this system, less drug than LNG is needed to achieve contraceptive efficacy (minimum serum concentrations of 0.09 mg/dL), and the system has reportedly fewer side effects than systems using LNG (3). Implanon consists of a drug-loaded EVAc matrix core surrounded by a rate-limiting EVAc membrane, with final dimensions of 40 mm in length and 2 mm in diameter. Testing with this contraceptive device has been performed in beagle dogs, and clinical studies have been performed in women. Implanon is manufactured in a coaxial-melt-extrusion process. The steroid is in micronized form, and 68.3 mg of steroid are contained in each rod.

The Nestorone® implant is a single-rod implant, much like the Norplant implants. It is a 4-cm-long device, 2.5 mm in diameter, that contains a core matrix of Nestorone progestin, which is a potent 19-norprogesterone derivative

(ST-1435) that is inactive when delivered orally and Silastic in a 2:1 weight ratio. Much like Norplant and Implanon, Nestorone is a constant, slow release device, and was designed to improve the insertion and removal problems encountered with Norplant (3,4,49,50). Nestorone contains 78 mg of synthetic steroid and is implanted in the ventral aspect of the upper left arm. Nestorone is advantageous because it is a potent steroid that inhibits ovulation but is not effective when delivered via the oral route. Thus, it is safe contraceptive agent for lactating women, as the steroid does not affect breastfeeding infants. The Population Council is developing this device, designed for 2 years of contraceptive efficacy. The device consists of a core of polymer-steroid, covered in a rate-limiting cellulose membrane, then encased within Silastic tubing. The ends of the rod are sealed with Silastic tape. Phase II clinical testing has demonstrated the effective release to be 45–50 $\mu$g/day (4).

Another single-rod implant under development is called Uniplant (3,4). This rod contains 38 mg of the progestin nomegestrol acetate, and preliminary clinical trials demonstrate a daily release rate of 75 $\mu$g/day. The single capsule is 3.5 cm long and is designed for 1 year of contraceptive efficacy.

Other types of devices still in the clinical development stages are the biodegradable systems. With biodegradable systems, a concern is the amount of time the degraded device remains in the body after release of drug is complete. The longer a device remains in situ after drug delivery, the greater the potential for tissue injury (51). The most common mechanisms of drug delivery through implantable devices is diffusion, biodegradation due to hydrolysis of chemical bonds within the polymer chain or backbone, or a combination of both. The implantable biodegradable systems consist of polyesters, poly(ortho esters) and poly(acyrlic acids). By-products of degradation from these polymers are nontoxic and induce little to no adverse reaction at their location within the body. An additional requirement of biodegradable contraceptive systems is the need to retain structural integrity throughout the length of time of drug delivery in case removal is necessary (3).

Capronor is a system using poly($\epsilon$-caprolactone) (PCL), which is of the polyester class, and was reported to be clinical testing (9) but no current data could be found regarding present development of this device. The biodegradable polymer is semicrystalline and rigid, and is able to withstand gamma irradiation for sterilization purposes. It is permeable to lipophilic steroids, such as LNG. The contraceptive device consists of a hollow capsule of PCL, which encases about 16 mg of LNG and 61 mg of ethyl oleate as a suspending vehicle. Capronor is 2.4 mm in diameter and 2.5–4 cm in length depending on the amount of drug to be encapsulated (12 or 21.6 mg LNG), with heat-sealed ends (8,26). Release rates of this device are about 45 $\mu$g/day, and release is governed by diffusion as the polymer degradation rate is relatively slow (18–24 months) (9). This release is approximately 10 times faster than release of LNG from the Silastic Norplant system (3,8,52). The device is intended to be placed subcutaneously in the ischial area for drug release up to one year (20,53,54). The ester bonds of PCL undergo hydrolytic degradation, which occurs throughout the bulk of the material and is independent of geometry. PCL degrades first to $\epsilon$-hydroxycaproic acid, and then to $H_2O$ and $CO_2$. The contraceptive is a reservoir-type device in which the drug is encased within a polymer wall. The wall must be thick enough to limit the release of the drug and prevent the smaller drug particles from diffusing out of the device rapidly, leaving the larger particles behind. With a wall of appropriate thickness, the device can approximate zero-order release. Release is modulated with excipients placed in the crystalline steroid region (3,8). The suspending medium for the drug acts to keep the drug in constant contact with the capsule wall, allowing for dissolution of the drug through the wall. The polymer wall degrades slowly via hydrolysis, and there is a long induction period between the time the ester linkages are cleaved and the molecular weight of the material drops off. This allows the polymer wall sufficient time to act as the limiting membrane for diffusion of the drug (53). Studies of the degradation of an implanted device show the half-life for molecular weight to be 10.6 months, as determined by intrinsic viscosity in toluene. The crystallinity over this time frame increased from 46.8 to 60.0%, which is indicative of the structural integrity of the system. Clinical studies of steroid release from Capronor were done on two systems, one comprised of a single 2.5-cm device and the other of a 4.0-cm device. In the trials with the 4.0-cm device, 80% of menstrual cycles showed suppressed ovulation, whereas all cycles with the 2.5-cm device were ovulatory. Thus, the 4.0-cm device is the optimal contraceptive system. It delivers an average of 33 $\mu$g/day over the course of one year, with release decreasing with time (3,8). One of the disadvantages of the Capronor system is the possible rupture of the device once it is implanted. Breakage of the PCL membrane would allow the remaining load of steroid to be released as a bolus (55).

Another type of bioerodible system investigated in the 1980s (no data could be found regarding current studies) is the poly(ortho ester) rod device in which the hydrolysis rates are pH dependent (56–58). The cross-linked poly(ortho ester) matrix is formed at room temperature in a reaction that is nearly instantaneous. The cylindrical devices tested in vitro and in vivo in rabbit models are in the form of molded capsules of 2.4 mm in diameter and 20 mm in length, where the drug is suspended within a polymer matrix. In vitro release rates averaged 20 $\mu$g/day, whereas the in vivo release was significantly higher at 33 $\mu$g/day. The release of LNG is governed in part by surface erosion of the device, but mainly by solubility of the steroid. The polymer hydrolyzes faster than the drug is solubilized, creating a foamed drug layer around the implant, a layer that also contains relatively insoluble degradation products. The in vitro release appears lower due to the aqueous environment used as the sink, whereas the in vivo release is higher due in part to lipophilic carriers within the body that act to solubilize the drug and the degradation products more quickly. As the osmotic pressure and polymer erosion increase, the matrix is weakened and eventually ruptures, leaving an open cell. The open cells gradually increase the surface area and drug release is accelerated. Slightly acidic salts with low water solubility are incorporated into the matrix to control the swelling. Another way in which the swelling is controlled is by incorporating an acidic monomer in the polymerization process. The fact that the polymer erodes faster than the drug is solubilized obviates the need for this device; a pellet of solid steroid would achieve similar release. Therefore, the poly(ortho ester) devices are still in the developmental stages.

Rice-sized norethindrone pellets fused with cholesterol are another biodegradable contraceptive system (24,26). The ellipsoidal pellets are composed of 85–90% steroid and 10–15% cholesterol, with each pellet containing about 35 mg of drug. Two to four pellets are implanted subdermally in the upper arm. Release is dependent upon surface area of the pellets, with reported release rates of 110 or 210 $\mu$ grams per day, to provide contraceptive efficacy for up to 12 months. Release is dependent solely upon solubility of the steroid in this type of device. It is possible to remove the pellets before they biodegrade, but there is a time limit within which removal can be accomplished. The pellets degrade completely within 24 months. Primary results show the four-pellet system to be most effective, but studies are still underway to determine the number of pellets needed, their size, and the cholesterol–hormone ratio (3).

Hollow fibers made of poly(L-lactic acid) (PLA) have been investigated in rabbits for release of LNG (59). The hollow fibers are manufactured with dry-wet phase inversion spinning technology and form the rate-limiting barrier for a solution of micronized steroid in castor oil. The polymer is extruded through a spinneret into a nonsolvent bath to form a porous membrane. Porosity of the fiber (and thus steroid permeability through the membrane) is greatly affected by the parameters of the solvent-nonsolvent system and process conditions (26). The steroid solution is loaded into the hollow fiber in a series of steps. First, the ends of the hollow fiber are heat sealed with Silastic. The steroid solution is then injected into the fiber through the Silastic end, and finally, the device is heat-sealed at the desired length and the silicone ends are discarded. The PLA hollow fiber is a reservoir device and provides a zero-order release of LNG. Release of steroid is governed by the dimensions of the hollow fiber (length, thickness, porosity, etc.), with the ideal release falling between 0.1 and 10 $\mu$g/cm/day for 210 days. PLA provides a useful vehicle for steroid delivery. It is biodegradable but has a low rate of degradation and also has a low permeability for steroids. These factors act to make the hollow fiber device ideal for long-term administration of low doses of steroid and obviate the need for surgical removal. The devices can be administered in a simple injection and have the advantage over other injectable contraceptives in that they can be removed if desired. The dimensions of the device are a maximum length of 3 cm and maximum diameter of 1.5 mm. Experimental data showed release in rabbits at an average of 2 $\mu$g/cm/day. The hollow fibers can be sterilized with $\gamma$ radiation.

PLA has also been investigated for use in a contraceptive delivery device composed of a block copolymer of PLA and PCL. Release from the block copolymer of PCL:PLA is controlled more easily than release from the random copolymer system (55). Ye and Chien (55) describe several devices made from the block copolymer of PCL:PLA. One system is a disc laminate system and the other is a cylinder

system for subdermal implantation. Both systems release a combination of LNG and estradiol ($E_2$). The copolymer in each system is prepared by sequence polymerization in dry toluene at 90°C in the presence of an AL/Zn bimetallic alkoxide catalyst. Release studies of the disc and cylinder systems were conducted in vitro in Valia-Chen permeation cells.

Initially there were two disc systems, one made by solvent casting and another made by compression molding. The solvent cast system exhibited faster release than the compression-molded system due to the channels formed when the solvent evaporated. Additionally, release was also different on each side of the solvent cast film, with release from the side in contact with air slower than release from the side of the film in contact with the casting surface. For these reasons, the solvent cast film system was abandoned. The disc system made by compression molding consisted of combining two matrix systems. The outer matrix system was in contact with the release medium and consisted of the block copolymer loaded with LNG, whereas the inner matrix was loaded with ethinyl estradiol ($E_2$). The intent of this design was to slow the release of $E_2$ from the combined system, as $E_2$ has faster release kinetics than LNG. The investigators found release of LNG from the outer disc to be matrix diffusion controlled, linear with $Q$ vs. $t^{1/2}$. The inner matrix release was found to be quite different, controlled instead by a membrane barrier and linear with $Q$ vs. $t$. The barrier membrane for release of $E_2$ from the inner matrix is the outer matrix of block copolymer loaded with LNG. Release from both components of the disc system was found to increase with PCL content of the block copolymer. As the weight fraction of PCL was increased from 0.6 to 0.9, release of LNG increased from 2.7 to 37.5 $\mu$g/cm$^2$/h$^{1/2}$ and release of $E_2$ increased from 9.3 to 32.9 $\mu$g/cm$^2$/h per unit thickness of the device. The increase in release with an increasing PCL content is attributed to an overall increase in free volume of the system. Matrix thickness was also investigated for its effect on release, and it was found to affect only release of $E_2$ from the inner matrix. LNG release was not affected. $E_2$ release decreased as the outer membrane thickness increased, and the permeation rate was found to be inversely proportional in a linear manner with the outer membrane thickness. An increase in loading of LNG in the outer matrix increased release of LNG but did not affect release of $E_2$ from the inner matrix (55).

The second type of PCL:PLA block copolymer system investigated by Ye and Chien was a cylinder of the block copolymer. PCL:PLA loaded with both LNG and $E_2$ was extruded through Teflon tubing (inner diameter 1.6 mm). The device was cut to 0.5 cm in length and coated by dipping into a polymer solution containing 20% LNG (w/w). Various copolymer ratios and drug loadings were evaluated, and all exhibited the typical burst effect with steady release thereafter. A system made with a single coat of LNG-polymer exhibited zero-order release, whereas systems made with additional coats had decreased release rates. LNG release was found to be dependent upon the core loading, the number of external coats, and the loading of each coat, as well as the thickness of the outer coats.

Release of $E_2$ from the core was dependent upon the core loading and the thickness of the outer coating layers (55).

In both systems described by Ye and Chien, release of LNG and $E_2$ can be controlled by modulating the parameters of the system, allowing an optimal drug delivery device to be developed for the purpose of contraception. Further research is needed, however, to identify the optimal design and determine in vivo release rates of both LNG and $E_2$.

*Injectable Devices.* The advantages of an injectable contraceptive microsphere delivery system are the ease of administration via single injection, biodegradation of the polymer device, low release of steroid with minimal loading, and long duration of release. The disadvantages of any microsphere delivery system are its relative irretrievability after administration (9,60). Microspheres made of poly(DL-lactide-*co*-glycolide) (PLGA) 85:15 exhibit zero-order release over a 3-month period (61–63). The norethindrone–PLGA microspheres are manufactured in a solvent evaporation process. Clinical trials with PLGA spheres loaded with 47% norethindrone (a synthetic progestin) have been conducted in the attempt to optimize the dosage. A release of 100 mg/day appears adequate. The injection is administered with a 21-gauge needle. Serum levels of the progestin are 2 ng/mL, about 15–30 times lower than serum levels for Depo-Provera. The system approximates zero-order, with steadier release of the drug at a concentration low enough to minimize side effects while still maintaining efficacy.

The PLGA-steroid microspheres are a matrix device from which release occurs in two phases. The first phase of release is through diffusion. This occurs for about 2 months, with an initial burst and gradual decline in the amount of drug released. At the end of 2 months, the PLGA spheres begin to degrade, and the drug release gradually increases and peaks as trapped drug particles are made available for diffusion. At the 3-month mark there is very little drug left to maintain adequacy as a contraceptive (26,62).

The principle of norethindrone release from PLGA spheres can also be applied to the release of other steroids, such as norgestimate, another synthetic progestin (64). This system, which was evaluated by Hahn et al. in the 1980s (no current data could be found), has a 20–70 weight % loading, depending on need. These solvent-evaporation particles appear as a "free flowing powder" and encapsulation efficiency is 81.8–91.2%. As the spheres release drug via diffusion, the rate of release can be controlled by geometry as well as crystallinity and molecular weight of the polymer. Generally, a more crystalline polymer releases drug more slowly (due to a decrease in diffusion of the drug through the polymer matrix and increased degradation time of the polymer). A higher molecular weight polymer also tends to decrease the release rate. However, smaller spheres increase the release rate due to the increased surface area available for diffusion of drug. There is also a membrane thickness effect, where higher loading increases the release rate due to smaller effective membrane thickness, reducing the area over which diffusion must take place. In vivo testing on the PLGA–norgestimate microspheres, as well as PLA–steroid microspheres, has been conducted in rats and baboons to determine the optimal loading and dosage necessary for contraceptive efficacy.

PLGA is also the delivery vehicle of choice for administration of natural progesterone in microsphere form. The PLGA-progesterone system was developed by Biotek, Inc., in conjunction with CONRAD. The microspheres released progesterone over a 90-day period in vitro and was further tested in vivo in 77 rabbits. However, the FDA is questioning the use of solvents in the manufacturing process for the PLGA microsphere formulation, and thus testing has been suspended (65).

Sterility is also important in any injectable system. PLGA microspheres can be sterilized via $\gamma$ irradiation, although this does cause a slight increase in the release rate. Storage is also a major concern for marketability, and the PLGA systems can be stored for up to 12 months in a cool, dry location.

PLGA has also been used as the delivery vehicle for luteinizing hormone-releasing hormone (LH-RH) analog (66,67). The LH-RH analog is a profertility agent when administered in pulsatile manner, but has a contraceptive effect when administered continuously. LH-RH has poor oral bioavailability, so the PLGA system greatly increases its potential as a contraceptive agent. The LH-RH analog is hydrophilic, as opposed to the lipophilic nature of the other steroids commonly used for contraception, and has a rather low diffusivity through PDMS, thus the choice of PLGA as the delivery vehicle (67). Spheres of PLGA and LH-RH analog have shown continuous release for over 40 days in rats. Disks made of the same materials have shown continuous release longer than the microspheres. PLGA was chosen over PLA due to its faster degradation rate of 6 months versus 12 months (66). It is not desired for the spheres to remain in the body much longer than the 3-month release time.

Another contraceptive system, designed by Nuwayser et al., employs PLA to coat steroids (68). The polymer-coated steroid systems are manufactured via air-suspension coating. The thickness of the polymer wall can be adjusted so as to control the diffusion of the drug and hence control the release characteristics of the system. The coated devices are similar to reservoir devices, and release is controlled by geometry, membrane permeability to the drug, and the thermodynamics of the encapsulated steroid. Increasing the thickness of the polymer wall decreases the release rate, and decreasing the capsule size while maintaining the same wall thickness also decreases the release rate. In order to administer these spheres via injection with a 15–22 gauge needle, the particle size must be kept below 500 $\mu$m. With the air suspension coating, the wall thickness can easily be controlled as the coating is applied incrementally. The devices under investigation have a high core-to-polymer ratio (on the order of 90:10). Using a high molecular weight PLA also ensures that the drug is released through diffusion before the polymer begins to degrade. The air-suspension-coated devices were tested in rabbits using an intramuscular injection as the route of administration. The release was mainly zero-order with a slight initial burst, and constant release of 2 $\mu$g/day/mg spheres continued for over 200 days.

One disadvantage with the PLGA microspheres is their rapid degradation when compared with other polymers, such as PCL. Research has been conducted into developing a zero-order release system that is composed of microspheres made of PLA-PLGA-PCL block copolymers (60,69). The copolymer microspheres are manufactured via solvent evaporation, and zero-order release is obtained with a balance of the ratio of PCL to PLA or PLGA. It was found that microspheres made of 60–65% PCL blocks exhibit zero-order release for approximately 5 months. The release of steroid (LNG) from the PCL blocks is controlled by diffusion because the PCL polymer degrades slowly. Steroid release from the PLA-PLGA block is controlled by erosion due to the faster degradation of these polymers. Thus, the combined system produces a long-acting device in which the release is balanced between the quick zero-order release from PLA-PLGA and the first-order release from PCL, yielding a constant release of drug for an extended length of time.

Another type of injectable contraceptive system uses a natural hydrogel. In vitro work using chitosan as the delivery vehicle began in 1991 (70). Chitosan is a polysaccharide that is hydrophilic, bioabsorbable, and nontoxic. Bioabsorption is due to lysosomal enzymes digesting the partially acetylated chitin derivatives of the chitosan polymer. The chitosan systems investigated are in the form of beads and films. The films are prepared by mixing the steroid and polymer in a solvent and evaporating the solvent, whereas the beads were prepared by a solvent removal technique using compressed air. Chitosan is a good vehicle for release of drugs because it is biocompatible, easy to control, and hydrophilic. The chitosan films showed first-order release due to the large surface area over which swelling occurs rapidly and uniformly. The beads showed zero-order release due to the swelling front, which begins at the outer core and gradually advances to the center of the bead as the pores open up. The drug is released as the matrix swells.

*Transdermal Devices.* Transdermal delivery is useful when a small amount of drug is needed to accomplish the desired medical purpose. This type of system is beneficial for steroid delivery because the drugs are very potent at low dosages. The main obstacle in steroid delivery is the low permeability of the skin to many compounds. Factors to be considered in transdermal delivery are drug and formulation stability, irritability of the skin, storage properties of the device, and cosmetic appearance (71). Transdermal delivery of steroids is currently being used in estrogen replacement therapy (10,72). In order to use this system for contraceptive purposes, the release rate of drug must be carefully optimized. Transdermal systems are reservoir-type devices that deliver drugs via diffusion. Release is controlled by the membrane through which the drug permeates, the diffusional resistance of the skin, and the thermodynamic activity of the drug (which affects its interaction with the delivery device) (71). Permeation enhancers are often incorporated within the reservoir to alter the drug's thermodynamic activity or the skin permeability.

One type of patch designed for contraceptive use is 4.91 $cm^2$, made with an impermeable backing of clear polyester film coated with heat-sealable polyethylene, and is being investigated by Friend et al. (72). The reservoir of the device is filled with 6.4 mg of LNG suspended in a hydroxypropyl cellulose and ethyl acetate/ethanol mixture. The membrane of the device is EVAc, adhering to the skin via a silicone-based adhesive. An optimal delivery of 30–35 $\mu g$/day of LNG is desired. For the steroid to penetrate the somewhat impermeable barrier of skin, a penetration enhancer must be incorporated into the design of the system. The penetration enhancer that functions most efficiently with a transdermal patch comprised of an EVA membrane is a mixture of ethyl acetate and ethanol. The devices are designed to be worn for 24 hours, and in vivo tests in rabbits indicate that this system may be suitable for contraceptive use in humans. Other transdermal patches are being developed for weekly use, where the user wears three medicated patches over the course of three consecutive weeks and a placebo patch during the fourth week to simulate menstruation through withdrawal bleeding (71).

Experiments delivering a combination of progestin and estrogen transdermally have also been conducted (73). Delivery of LNG and estradiol through a transdermal delivery system with an EVA membrane and ethanol as the permeation enhancer provide promising results. Experiments were performed in vitro on rat and human cadaver skin. It was found that release of steroids was controlled by a variety of factors, namely the composition of the reservoir formulation, the weight fraction of vinyl acetate in the EVA membrane, the membrane thickness, and the loading of steroid. It was also determined that the limiting factor in this system is the polymeric membrane and not the skin.

*Intrauterine Devices.* Intrauterine devices that release progestins are currently being used in the United States, such as Progestasert. Other IUDs were developed for 3-year use, such as the Intrauterine Progesterone Contraceptive System (IPCS) developed by Alza Corporation. The main difference between the Progestasert and the IPCS is the increased drug loading in the IPCS, which holds 52 mg of progesterone instead of 38 mg. The IPCS is 36 mm long, 32 mm wide, with an 8 mm insertion width. The device is a reservoir-type system, with EVAc encasing a solution of progesterone in silicone oil. The stem holds the drug, while the cross arms are solid copolymer (30). Trials with the IPCS were discontinued due to an increased pregnancy rate at 20–22 months, fully a year before its designed contraceptive efficacy was reached (20). Additionally, there is the Alza UTS system in which drug is released from the side arms as well as from the stem (74).

Another delivery vehicle being investigated is a fibrous system, in which fibers are loaded with steroid and wrapped around IUDs or vaginal rings (75). The fibers can be either biodegradable PLA, PLGA, or PCL, or nonbiodegradable polyethylene and polypropylene. It is possible to have a monolithic device in which the polymer and drug are mixed before spinning the fibers, or a coaxial fiber where the drug is the core of the device and is encased in a polymer sheath. Testing to date has been successful. The fibers are easily manufactured and stored and can be used in a variety of applications.

Another device similar to the IUD is an intracervical device (ICD) that contains a synthetic progesterone (either

LNG or Nestorone) (20,76). It was initially thought that an ICD would be more beneficial than an IUD because placement of the device in the cervix would decrease the basic problems of IUDs associated with intrauterine muscular contractility. The ICD is designed to release 10–25 $\mu$g of the progesterone daily over 5 years. The steroid is contained within a Silastic rod affixed to the vertical arm of a frame similar to the Nova-T IUD. In human testing, however, the ICD showed unsatisfactory release of the steroid through the Silastic carrier and an increase in expulsion rate over traditional IUDs.

Hydrogels are being investigated for their role in vaginal spermicide delivery. The spermicide nonoxynol-9 and ethylendiaminetetraacetic acid (EDTA) are loaded into a gel matrix composed of Carbopol 934P (77). Initial testing demonstrated a synergistic effect between EDTA and Carbopol in increasing the potency of nonoxynol-9, thus requiring lower dosage levels. Carbopol 934P is a high molecular weight poly(acrylic acid) polymer produced by B.F. Goodrich (78). It is made by cross-linking acrylic acid with allyl sucrose and polymerizing in benzene. The gel network is highly cross-linked and is typically in the form of a resin. Carbopol has a glass-transition temperature of 105°C when dry, but the glass-transition temperature drops dramatically when water is introduced to the system. Cross-linked networks swell to 1000 times their original volume in pH > 6, and Carbopol 934P is a flexible gel when hydrated. When the polymer is dry, drug is entrapped in the "glassy core." As the water penetrates the external surface of the polymer–drug matrix, the Carbopol polymer forms a gelatinous layer and drug becomes entrapped in the hydrogel domain. Carbopol is unique in that it forms "discrete microgels" of the polymer, and water forms tiny channels between the microgels as the system is hydrated. The water eventually forces the microgels to separate due to osmotic pressure, and the microgel–drug complex is sloughed off. Release of drug continues from within the stable gel layer of the microgel complex. Drug release from Carbopol is affected by the solubility of the drug, the rate of hydration of the polymer network, the degree of cross-linking and swelling of the matrix, as well as ionic interactions between the polymer and the drug. Polymer swelling is affected by the pH of the local environment, which affects drug release as well. Highly soluble drugs typically exhibit Fickian release and the release profile is dependent mainly upon the solubility and diffusion kinetics of the drug. The Carbopol matrix is the limiting membrane in the delivery of poorly soluble drugs, and release is typically near zero-order.

*Contraceptive Films.* Allendale Labs, Inc., is developing a vaginal contraceptive film much like VCF (marketed by Apothecus Pharmaceutical Corp) (33). The films differ in the composition of the polymeric carrier; VCF contains 70 mg of nonoxynol-9 in a base of poly(vinyl alcohol), whereas the Allendale-N9 film contains 100 or 130 mg nonoxynol-9 in a hydroxypropylmethylcellulose matrix. It is thought that the Allendale-N9 film will be less irritable to the mucosal tissues of the vagina, as well as enhance the stability of the device in warm, humid climates of third world countries. In Phase I clinical trials both Allendale-N9 films inhibited the action of sperm in the vagina, and there were no significant differences detected between VCF and the Allendale-N9 films. Further testing is needed to determine the efficacy of each film.

Another device that is undergoing testing is comprised of EVAc film (40% vinyl acetate by weight) loaded with 65% 2′-carbomethoxyphenyl 4-guanidinobenzoate (MSGB) (79). The EVAc-MSGB film is a nonbiodegradable and nonswellable matrix system that is highly flexible and can be molded into several different forms. MSGB is released from the system through diffusion through interconnecting pores. Its design is for in utero release, as MSGB is a sperm acrosin inhibitor. MSGB is advantageous to hormonal contraceptive systems as it does not interfere with the normal hormonal regulation of the endometrium. It is designed for an application of 1 to 2 years, but no test results on humans have been published.

### Male Contraceptives

Very few male contraceptives are available today. However, research is progressing to allow men a wider choice of contraceptive systems. One such system involves delivery of testosterone through an implanted device. In this system, crystalline testosterone is fused into a pellet formation and six 200-mg pellets are implanted in the abdominal wall (80). The pellets are biodegradable and the release of drug suppresses sperm output for about 6 months. The release kinetics have been shown to approach zero-order. Another system that could possibly act as a contraceptive agent in males is the injectable LH-RH-containing PLGA spheres mentioned previously (67). Developments are being made for an injectable system much like Depo-Provera, in which testosterone enanthate would be administered along with DMPA (81). Sundaram et al. report on the use of a synthetic testosterone, $7\alpha$-methyl-19-nortestosterone (MENT) for use in subdermal contraceptive implants in men (82). MENT is more potent than testosterone, maintains normal muscle mass, and does not hyperstimulate the prostate because it does not undergo $5\alpha$-reduction in the prostate. In conjunction with an LH-RH analogue, this potent steroid could be used in a system designed for 1 year of contraceptive use.

A further development in male contraceptives is the navel administration of testosterone via a transdermal device (83). The device is designed to administer testosterone for one month or longer, and initial testing shows a zero-order release both in vitro and in vivo testing on rhesus monkeys. The system is a disk-shaped bandage-type device that contains microspheres of hormone suspended in liquid and immobilized in a cross-linked polymer. (Further description of the liquid or polymer were not available.) In vitro release rates were 40.25 $\mu$g/cm$^2$/day and in vivo release rates were 27.66 $\mu$g/cm$^2$/day. The decrease in vivo release rate is most likely due to the additional barrier of skin.

Cylindrical capsules of PLGA have also been investigated for their use in delivering buserelin acetate (84). Buserelin is a gonadotropin-releasing hormone (GnRH) agonist (D-Ser(Bu$^t$)$^6$-GnRH-(1-9)nonapeptide-ethylamide) GnRH previously known as LH-RH, is one of the hypothalamic peptides and promotes pituitary secretion of follicle stimulating hormone (FSH) and luteinizing hormone (LH).

FSH induces spermatogenesis and LH causes the testicular Leydig cells to produce testosterone. A GnRH agonist will inhibit FSH and LH secretion from the pituitary, thus decreasing testosterone secretion and suppressing spermatogenesis. The PLGA capsules are 10 mm in length and contain 3.3 mg of buserelin acetate. The devices can be sterilized by $\gamma$-irradiation and are intended to be implanted subcutaneously in the abdominal wall lateral to the rectus abdominis muscle. Preliminary testing of the device as a contraceptive has been inconclusive as the delivery of GnRH agonists must be carefully combined with the administration of androgens so as to inhibit only spermatogenesis and avoid adversely affecting other androgen-supported functions in the body.

## ABORTIFACIENTS

In the past, prostaglandin analogs have commonly been used as abortion-inducing agents. However, there are many side effects of prostaglandin use, especially to the gastrointestinal tract; vomiting and diarrhea are very common. To avoid the side effects, 24-hour delivery systems for the prostaglandin were developed. One such device is a polyethylene-oxide based hydrogel strip into which 3 mg of drug have been loaded (85). The strip measures $1.3 \times 10 \times 30$ mm, and releases an average of 53% of the drug over 24 hours. The hydrogel is cross-linked with aliphatic difunctional isocyanate and an aliphatic triol. It swells to three times its dry volume and releases the drug through diffusion at zero-order. It has been demonstrated to be 85% effective, yet some of the side effects still persist. Another abortifacient device is a reservoir-type system (86). The device consists of three layers, a drug-loaded membrane surrounded by an outer rate-controlling membrane on one side and a drug-impermeable membrane on the other. The system is mounted on a holder and delivers from 50 to 100 $\mu$g/hour depending on loading. The device is 1.5 cm in diameter and 3.7 cm in length.

## SUMMARY

There are many types of contraceptive agents both in use and being developed today. This article was limited to discussing the contraceptive devices involving controlled release and does not touch upon systems involving vaccines or pure polymer systems without drug release (such as a plug of silicon placed in the vas deferens for male contraception). There is quite a variety of controlled release systems, and as research progresses more and more of these systems will become available to the general public. Additionally, advances in the types of synthetic steroids used in contraceptive devices act to minimize the numerous side effects of current systems and will greatly improve the efficacy and acceptability of future devices.

## BIBLIOGRAPHY

1. A.C. Guyton, *Textbook of Medical Physiology*, Saunders, Philadelphia, 1991.
2. F. Murad and J.A. Kuret, in A.G. Gilman, T.W. Rall, A.S. Nies, and P. Taylor, eds., *Goodman and Gilman's The Pharmacological Basis of Therapeutics*, 8th ed., Pergamon, New York, 1990, Chapter 58.
3. P.D. Darney, *Am. J. Obstet. Gynecol.* **170**, 1536–1543 (1994).
4. O. Peralta, S. Diaz, and H. Croxatto, *J. Steroid Biochem. Mol. Biol.* **53**, 223–226 (1995).
5. S. Koetsawang, S. Varakamin, S. Satayapan, and N. Dusitsin, in G.I. Zatuchni, A. Goldsmith, J.D. Shelton, and J.J. Sciarra, eds., *Long-Acting Contraceptive Delivery Systems*, Harper & Row, Philadelphia, 1984, pp. 459–470.
6. I. Sivin, in S.S. Ratnam, E.S. Teoh, and S.M. Lim, eds., *Contraception*, The Parthenon Publishing Group, Park Ridge, N.J., 1986, pp. 121–126.
7. S. Segal, *Am. J. Obstet. Gynecol.* **157**, 1090–1092 (1987).
8. P.D. Darney et al., *Fertil. Steril.* **58**, 137–143 (1992).
9. H.L. Gabelnick and P.E. Hall, *J. Controlled Release* **6**, 387–394 (1987).
10. L. Mastroianni, Jr., P.J. Donaldson, and T.T. Kane, eds., *The Current Status of Contraceptive Research*, National Academy Press, Washington, D.C., 1990, pp. 30–40.
11. M.R. Brunstedt, in D.F. Williams, ed., *Materials Science and Technology: A Comprehensive Treatment*, VCH Publishers, Mannheim, Germany, 1992, Chapter 11.
12. S.A. Matlin, A. Belenguer, and P.E. Hall, *Contraception* **45**, 329–341 (1992).
13. S. Kuldip, O.A.C. Viegas, and S.S. Ratnam, *Contraception* **45**, 453–461 (1992).
14. S. Koetsawang et al., in S.S. Ratnam, E.S. Teoh, and S.M. Lim, eds., *Contraception*, The Parthenon Publishing Group, Park Ridge, N.J., 1986, pp. 133–142.
15. D.N. Robertson, in G.I. Zatuchni, A. Goldsmith, J.D. Shelton, and J.J. Sciarra, eds., *Long-Acting Contraceptive Delivery Systems*, Harper & Row, Philadelphia, 1984.
16. J. Newton, *British Medical Bulletin* **49**, 40–61 (1993).
17. I. Sivin, in G.I. Zatuchni, A. Goldsmith, J.D. Shelton, and J.J. Sciarra, eds., *Long-Acting Contraceptive Delivery Systems*, Harper & Row, Philadelphia, 1984, pp. 488–500.
18. I. Sivin et al., *Contraception* **55**, 81–85 (1997).
19. *Reproductive Health Product Development*, available at: *http://www.popcouncil.org/*, 1997, updated October 7, 1998, accessed October 14, 1998.
20. L. Liskin and W.F. Quillin, *Popul. Rep. Ser. K* **2** (XI), K19–K47 (1983).
21. *Depo-Provera*, available at: *http:www.depo-provera.com*, 1997, accessed October 14, 1998.
22. R. Lande, *Popul. Rep. Ser. K* **5**, 1–5 (1995).
23. H. Nash, *Contraception* **12**, 377–393 (1975).
24. L. Liskin, R. Blackburn, and R. Ghani, *Popul. Rep. Ser. K* **3** (XV), K57–K84 (1987).
25. P.E. Hall, in G.I. Zatuchni, A. Goldsmith, J.D. Shelton, and J.J. Sciarra, eds., *Long-Acting Contraceptive Delivery Systems*, Harper & Row, Philadelphia, 1984, pp. 515–522.
26. A.P. Sam, *J. Controlled Release* **22**, 35–46 (1992).
27. *PARAGARD T 380A Intrauterine Copper Contraceptive*, available at: *http://ortho-mcneil.com/*, 1998, updated October 12, 1998, accessed October 14, 1998.
28. ALZA Pharmaceuticals, *Progestasert Intrauterine Contraceptive System: Clinical Evidence*, Palo Alto, Calif., 1994.
29. R. Aznar and J. Giner, in G.I. Zatuchni, A. Goldsmith, J.D. Shelton, and J.J. Sciarra, eds., *Long-Acting Contraceptive Delivery Systems*, Harper & Row, Philadelphia, 1984, pp. 613–620.

30. D.A. Edelman, L.P. Cole, R. Apelo, and P. Lavin, in G.I. Zatuchni, A. Goldsmith, J.D. Shelton, and J.J. Sciarra, eds., *Long-Acting Contraceptive Delivery Systems*, Harper & Row, Philadelphia, 1984, pp. 621–627.

31. S. El-Mahgoub, *Contraception* **22**, 271–286 (1980).

32. A.S. Lichtman, V. Davajan, and D. Tucker, *Contraception* **8**, 291–297 (1973).

33. C.K. Mauck et al., *Contraception* **56**, 97–102 (1997).

34. C.K. Mauck et al., *Contraception* **56**, 103–110 (1997).

35. Apothecus Pharmaceutical Corporation, *VCF Vaginal Contraceptive Film: A Compendium of Medical and Technical Data*, Oyster Bay, New York, 1995.

36. *Apothecus Pharmaceutical Corporation VCF Vaginal Contraceptive Film*, available at: *http://www.apothecus.com/*, accessed October 14, 1998.

37. E. Diczfalusy and B.M. Landgren, in G.I. Zatuchni, A. Goldsmith, J.D. Shelton, and J.J. Sciarra, eds., *Long-Acting Contraceptive Delivery Systems*, Harper & Row, Philadelphia, 1984, pp. 213–227.

38. D.R. Mishell, Jr., *Ann. Med.* **25**, 191–197 (1993).

39. S. Diaz et al., *Contraception* **32**, 603–621 (1985).

40. B.M. Landgren, P.E. Hall, and S.Z. Cekan, *Contraception* **45**, 343–349 (1992).

41. S. Roy and D.R. Mishell, Jr., in G.I. Zatuchni, A. Goldsmith, J.D. Shelton, and J.J. Sciarra, eds., *Long-Acting Contraceptive Delivery Systems*, Harper & Row, Philadelphia, 1984, pp. 581–594.

42. T.M. Jackanicz, in G.I. Zatuchni, A. Goldsmith, J.D. Shelton, and J.J. Sciarra, eds., *Long-Acting Contraceptive Delivery Systems*, Harper & Row, Philadelphia, 1984, pp. 201–212.

43. D.R. Mishell, Jr., et al., *Am. J. Obstet. Gynecol.* **130**, 55–62 (1978).

44. S.-E. Olsson and V. Odlind, *Contraception* **42**, 563–572 (1990).

45. W.M. Saltzman and L.B. Tena, *Contraception* **43**, 497–505 (1991).

46. B.H. Vickery, J.C. Goodpasture, and L.Y.W. Lin, in G.I. Zatuchni, A. Goldsmith, J.D. Shelton, and J.J. Sciarra, eds., *Long-Acting Contraceptive Delivery Systems*, Harper & Row, Philadelphia, 1984, pp. 228–240.

47. *A Tradition in Reproductive Medicine*, available at: *http://www.organon.com/*, 1998, updated accessed October 14, 1998.

48. J.A.A. Geelen et al., *Contraception* **47**, 215–226 (1993).

49. M. Laurikka-Routti and M. Haukkamaa, *Fertil. Steril.* **58**, 1142–1147 (1992).

50. S. Diaz et al., *Contraception* **51**, 33–38 (1995).

51. L.R. Beck et al., in G.I. Zatuchni, A. Goldsmith, J.D. Shelton, and J.J. Sciarra, eds., *Long-Acting Contraceptive Delivery Systems*, Harper & Row, Philadelphia, 1984, pp. 406–417.

52. G.I. Dhall, U. Krishna, and R. Sivaram, *Contraception* **44**, 409–417 (1991).

53. C.G. Pitt and A. Schindler, in G.I. Zatuchni, A. Goldsmith, J.D. Shelton, and J.J. Sciarra, eds., *Long-Acting Contraceptive Delivery Systems*, Harper & Row, Philadelphia, 1984, pp. 48–63.

54. S.J. Ory et al., *Am. J. Obstet. Gynecol.* **145**, 600–605 (1983).

55. W.-P. Ye and Y.W. Chien *J. Controlled Release* **41**, 259–269 (1996).

56. J. Heller, D.W.H. Penhale, B.K. Fritzinger, and S.Y. Ng, in G.I. Zatuchni, A. Goldsmith, J.D. Shelton, and J.J. Sciarra, eds., *Long-Acting Contraceptive Delivery Systems*, Harper & Row, Philadelphia, 1984, pp. 113–128.

57. J. Heller, B.K. Fritzinger, S.Y. Ng, and D.W.H. Penhale, *J. Controlled Release* **1**, 225–232 (1985).

58. J. Heller, B.K. Fritzinger, S.Y. Ng, and D.W.H. Penhale, *J. Controlled Release* **1**, 233–238 (1985).

59. M.J.D. Eenink and J. Feijen *J. Controlled Release* **6**, 225–247 (1987).

60. Z. Gu et al., *J. Controlled Release* **22**, 3–14 (1992).

61. G.S. Grubb et al., *Fertil. Steril.* **51**, 803–810 (1989).

62. L.R. Beck et al., *Am. J. Obstet. Gynecol.* **147**, 815–821 (1983).

63. L.R. Beck et al., *Am. J. Obstet. Gynecol.* **135**, 419–426 (1979).

64. D.W. Hahn et al., in G.I. Zatuchni, A. Goldsmith, J.D. Shelton, and J.J. Sciarra, eds., *Long-Acting Contraceptive Delivery Systems*, Harper & Row, Philadelphia, 1984, pp. 96–112.

65. *Systemic Hormonal Methods for Women*, available at: *http://www.reproline.jhu.edu/*, 1997, updated August 20, 1997, accessed October 16, 1998.

66. J.S. Kent, L.M. Sanders, T.R. Tice, and D.H. Lewis, in G.I. Zatuchni, A. Goldsmith, J.D. Shelton, and J.J. Sciarra, eds., *Long-Acting Contraceptive Delivery Systems*, Harper & Row, Philadelphia, 1984, pp. 169–179.

67. B.H. Vickery et al., in G.I. Zatuchni, A. Goldsmith, J.D. Shelton, and J.J. Sciarra, eds., *Long-Acting Contraceptive Delivery Systems*, Harper & Row, Philadelphia, 1984, pp. 180–189.

68. E.S. Nuwayser et al., in G.I. Zatuchni, A. Goldsmith, J.D. Shelton, and J.J. Sciarra, eds., *Long-Acting Contraceptive Delivery Systems*, Harper & Row, Philadelphia, 1984, pp. 64–76.

69. Y.-X. Li and X.-D. Feng, *Makromol. Chem., Macromol. Symp.* **33**, 253–264 (1990).

70. T. Chandy and C.P. Sharma, *Biomater. Artif. Cells, Immobil. Biotechnol.* **19**, 745–760 (1991).

71. V.V. Ranade and M.A. Hollinger, *Drug Delivery Systems*, CRC Press, Boca Raton, Fla. 1996.

72. D.R. Friend, P. Catz, and S. Phillips, *Contraception* **40**, 73–80 (1989).

73. G. Chen, D. Kim, and Y. Chien, *J. Controlled Release* **34**, 129–143 (1995).

74. T.G. McCarthy, E.L. Cheng, A. Loganath, and S.S. Ratnam, in S.S. Ratnam, E.S. Teoh, and S.M. Lim, eds., *Contraception*, The Parthenon Publishing Group, Park Ridge, N.J., 1986, pp. 95–98.

75. D.R. Cowsar and R.L. Dunn, in G.I. Zatuchni, A. Goldsmith, J.D. Shelton, and J.J. Sciarra, eds., *Long-Acting Contraceptive Delivery Systems*, Harper & Row, Philadelphia, 1984, pp. 145–163.

76. P. Lahteenmaki et al., in G.I. Zatuchni, A. Goldsmith, J.D. Shelton, and J.J. Sciarra, eds., *Long-Acting Contraceptive Delivery Systems*, Harper & Row, Philadelphia, 1984, pp. 595–600.

77. C.-H. Lee, R. Bagdon, and Y.W. Chien, *J. Pharm. Sci.* **85**, 91–95 (1996).

78. *CARBOPOL: The Proven Polymers in Pharmaceuticals. Controlled Release Tablets and Capsules*, B.F. Goodrich, Cleveland, Ohio, 1994, Bull. 17.

79. J.W. Burns, A.T. Fazleabas, I.F. Miller, and L.J.D. Zaneveld, *Contraception* **38**, 349–364 (1988).

80. D.J. Handelsman, A.J. Conway, and L.M. Boylan, *J. Clin. Endocrinol. Metab.* **75**, 1326–1332 (1992).

81. G.M.H. Waites, *Br. Med. Bull.* **49**, 210–221 (1993).

82. K. Sundaram, N. Kumar, and C.W. Bardin, *Ann. Med.* **25**, 199–205 (1993).

83. Y.W. Chien, *J. Pharm. Sci.* **73**, 1064–1067 (1984).

84. H.M. Behre, D. Nashan, W. Hubert, and E. Nieschlag, *J. Clin. Endocrinol. Metab.* **74**, 84–90 (1992).

85. I.T. Cameron and D.T. Baird, *Contraception* **33**, 121–125 (1986).

86. M. Bygdeman et al., *Contraception* **27**, 141–151 (1983).

# FOOD AND DRUG ADMINISTRATION REQUIREMENTS FOR CONTROLLED RELEASE PRODUCTS

HENRY J. MALINOWSKI
PATRICK J. MARROUM
U.S. Food and Drug Administration
Rockville, Maryland

## KEY WORDS

Area under the curve (AUC)

Bioavailability

Bioequivalence

Controlled release

Delayed release

Dissolution

Extended release

In vitro in vivo correlation (IVIVC)

Modified release

Peak plasma concentration ($C_{max}$)

Pharmacokinetics

## OUTLINE

## INTRODUCTION

Controlled-release pharmaceutical dosage forms may offer one or more advantages over conventional (immediate release) dosage forms of the same drug, including a reduced dosing frequency, a decreased incidence and/or intensity of adverse effects, a greater selectivity of pharmacologic activity, and a reduction in drug plasma fluctuation resulting in a more constant or prolonged therapeutic effect. In some cases, controlled-release products may be therapeutically advantageous primarily for certain subpopulations of patients. For example, a controlled-release drug product may allow a child to attend school without drug administration during the school day. In other instances, controlled-release products may have no significant advantages or may actually be less effective or more hazardous than conventional dosage forms of the same drug. Therefore, not all drugs are good candidates for formulation as controlled-release drug products (1). For example, some drugs are more effective if fluctuation in plasma concentrations occur. For such drugs, tolerance to drug effect may develop with the constant levels seen with controlled-release drug products. Ordinarily, oral controlled-release dosage forms result in a longer recommended dosing interval for the controlled-release dosage form, usually twice as long, compared with the dosing interval for the immediate release dosage form. Also, a controlled-release drug product may be warranted if significant clinical advantages for the controlled-release dosage form can be demonstrated, for example, decreased side effects resulting from a lower peak plasma concentration with the controlled-release dosage form relative to the immediate-release dosage form.

## TYPES OF CONTROLLED-RELEASE PRODUCTS

The most common type of controlled-release products are oral dosage forms. These products normally provide for a 12- or 24-h dosing interval. Dosing intervals for oral controlled-release products beyond once-a-day dosing are limited by physiologic characteristics of the human gastrointestinal tract. Gastrointestinal transit time, which normally averages 24 h but can vary from a few hours to several days, prevents oral controlled-release products with a dosing interval beyond 24 h. Other types of controlled-release products include transdermal patches that are applied to the skin for periods of 1 day to perhaps 1 week. In addition, controlled-release implants are dosage forms that are implanted below the skin surface and have been developed for continuous therapy for as long as 5 years.

## DEFINITIONS

Before beginning a discussion of the regulatory requirements of controlled-release products, it is useful to understand several commonly used definitions for these types of products:

*Controlled-release dosage forms.* A class of pharmaceuticals or other biologically active products from which a drug is released from the delivery system in a planned, predictable, and slower-than-normal manner (2).

*Modified-release dosage form.* This refers, in general, to a dosage form for which the drug-release characteristics of time course and/or location are chosen to accomplish therapeutic or convenience objectives not offered by conventional dosage forms (2).

*Extended-release dosage form.* This is a specific type of modified-release dosage form that allows at least a two-fold reduction in dosage frequency as compared to that drug presented as an immediate- (conventional-) release dosage form (2).

*Delayed-release dosage form.* This is a specific type of modified-release dosage form that releases a drug at a time other than promptly after administration. An example is enteric-coated tablets (2).

The requirements discussed in this article cover all types of controlled-release dosage forms. The primary focus will be on oral controlled-release drug products, which are most common. Requirements for other types of controlled-release drug products, such as transdermal patches or implants, are similar to those described in this article.

## LAWS, REGULATIONS AND GUIDANCES FOR CONTROLLED-RELEASE PRODUCTS

### Need for Clinical Studies

Premarketing evaluation of a controlled-release product should include consideration of the possible development of tolerance to the drug, the occurrence of sensitivity reactions or local tissue damage due to dosage-form-dependent persistence or localization of the drug, the clinical implications of dose dumping or of an unexpected decrease in bioavailability by physiological or physicochemical mechanisms, and a quantitative alteration in the metabolic fate of the drug related to nonlinear or site-specific disposition.

Specific claims for all therapeutic advantages of a controlled-release product over the conventional dosage forms should be based on adequate clinical studies, the results of which should be available to health professionals upon request. Where no therapeutic advantage is claimed, the need for clinical studies may be lessened.

An important consideration for the development of controlled-release products as original new drugs is the quantity of evidence needed in particular circumstances to establish substantial proof of effectiveness. The usual practice for all new molecular entities is to accept as proof two or more clinical studies that conclusively define the safety and efficacy of the drug. Within the U.S. Food and Drug Administration (FDA) Modernization Act of 1998 are described situations in which alternative approaches regarding the quantity of evidence to support effectiveness may be possible (3). These include (1) situations in which effectiveness of a new use may be extrapolated entirely from existing efficacy studies; (2) situations in which a single adequate and well-controlled study of a specific new use can be supported by information from other related adequate and well-controlled studies, such as studies in other phases of a disease; in closely related diseases; or other conditions of use (different dose, duration of use, regimen), of different dosage forms, or of different endpoints; and

(3) situations in which a single multicenter study, without supporting information from other adequate and well-controlled studies, may provide evidence that a use is effective. In each of these situations, it is assumed that any studies relied on to support effectiveness meet the requirements for adequate and well-controlled studies in 21 CFR 314.126 (4). It should also be appreciated that reliance on a single study of a given use, whether alone or with substantiation from related trial data, leaves little room for study imperfections or contradictory (nonsupportive) information. In all cases, it is presumed that the single study has been appropriately designed; that the possibility of bias due to baseline imbalance, unblinding, post hoc changes in analysis, or other factors is judged to be minimal; and that the results reflect a clear prior hypothesis documented in the protocol. Moreover, a single favorable study among several similar attempts that failed to support a finding of effectiveness would not constitute persuasive support for a product use unless there were a strong argument for discounting the outcomes in the studies that failed to show effectiveness (e.g., if the study was obviously inadequately powered or there was a lack of assay sensitivity as demonstrated in a three-arm study by failure of the study to show efficacy of a known active agent).

Whether to rely on a single study to support an effectiveness determination is not often an issue in contemporary drug development. In most drug development situations, the need to find an appropriate dose, to study patients of greater and lesser complexity or severity of disease, to compare the drug with other therapy, to study an adequate number of patients for safety purposes, and to otherwise know what needs to be known about a drug before it is marketed will result in more than one adequate and well-controlled study upon which to base an effectiveness determination.

In certain cases, effectiveness of a new controlled-release drug product may be demonstrated without additional adequate and well-controlled clinical efficacy trials. Ordinarily, this will be because other types of data provide a way to apply the known effectiveness to a new population or a different dose, regimen, or dosage form. Controlled-release dosage forms may be approved on the basis of pharmacokinetic data linking the new dosage form to a previously studied immediate-release dosage form. Because the pharmacokinetic patterns of modified- and immediate-release dosage forms are not identical, it is generally important to have some understanding of the relationship of blood concentration to response, including an understanding of the time course of that relationship, to extrapolate the immediate-release data to the modified-release dosage form (3).

### Bioavailability Study Requirements for Controlled Release Products

The bioavailability requirements for controlled-release products are covered in the U.S. Code of Federal Regulations under 21 CFR 320.25(f) (5). The aims of these requirements are to determine that the following conditions are met:

- The drug product meets the controlled-release claims made for it.
- The bioavailability profile established for the drug product rules out the occurrence of clinically significant dose dumping. This is usually achieved by the conduct of a food effect study whereby the drug is administered with and without a high-fat breakfast.
- The drug product's steady-state performance is equivalent to a currently marketed noncontrolled- or controlled-release drug product that contains the same active drug ingredient or therapeutic moiety and that is subject to an approved full new drug application.
- The drug product's formulation provides consistent pharmacokinetic performance between individual dosage units.

The reference material for such a bioavailability study shall be chosen to permit an appropriate scientific evaluation of the controlled-release claims made for the drug product. The reference material is normally one of the following:

- A solution or suspension of the active drug ingredient or therapeutic moiety
- A currently marketed immediate-release drug product containing the same active drug ingredient or therapeutic moiety and administered according to the dosage recommendations in the labeling of immediate-release drug product
- A currently marketed controlled-release drug product subject to an approved full new drug application containing the same active drug ingredient or therapeutic moiety and administered according to the dosage recommendations in the labeling proposed for the controlled-release drug product

Guidelines for the evaluation of controlled-release pharmaceutical dosage forms provide assistance to those designing, conducting, and evaluating studies. However, a drug may possess inherent properties that require considerations specific to that drug and its dosage form that may override the generalities of these guidelines. Guidances related to the evaluation of controlled-release drug products as well as many other types of guidances are available on the Internet at the Center for Drug Evaluation and Research Web site (*http://www.fda.gov/cder/*).

### Controlled-Release New Drug Applications

As mentioned earlier, a fundamental question in evaluating a controlled-release product is whether formal clinical studies of the dosage form's safety and efficacy are needed or whether only a pharmacokinetic evaluation will provide adequate evidence for approval. A rational answer to this question must be based on evaluation of the pharmacokinetic properties and plasma concentration/effect relationship of the drug. Where there is a well-defined predictive relationship between the plasma concentrations of the drug and the clinical response (regarding both safety and

efficacy), it may be possible to rely on plasma concentration data alone as a basis for the approval of the controlled-release product. In the following situations, it is expected that clinical data be submitted for the approval of the controlled-release New Drug Application (NDA):

- When the controlled-release product involves a drug that is an unapproved new molecular entity, because there is no approved reference product to which a bioequivalence claim could be made
- When the rate of input has an effect on the drug's efficacy and toxicity profile
- When a claim of therapeutic advantage is intended for the controlled-release product
- When there are safety concerns with regards to irreversible toxicity
- Where there are uncertainties concerning the relationship between plasma concentration and therapeutic and adverse effects or in the absence of a well-defined relationship between plasma concentrations and either therapeutic or adverse clinical response
- Where there is evidence of functional (i.e., pharmacodynamic) tolerance
- Where peak-to-trough differences of the immediate-release form are very large

In all of these instances except for the first, a 505(b)(2) NDA could be submitted for approval to the FDA. The regulations for such application are covered under 21 CFR 314.54. These regulations state that any person seeking approval of a drug product that represents a modification of a listed drug, for example, a new indication or a new dosage form, and for which investigations other than bioavailability or bioequivalence studies are essential to the approval of the changes may submit a 505(b)(2) application. However, such an application may not be submitted under this section of the regulations for a drug product whose only difference from the reference-listed drug is that the extent of absorption or rate of absorption is less than that of the reference-listed drug or if the rate of absorption is unintentionally less than that of the reference-listed drug (6).

## INFORMATION TO CHARACTERIZE THE DRUG ENTITY

### Physicochemical Characterization

Although the required physicochemical information to characterize the drug entity in a controlled-release dosage form should generally be no different from that for the drug entity in an immediate-release dosage form, additional physicochemical information related to solubility, dissolution, stability, and other release-controlling variables of the drug under conditions that may mimic the extremes of the physiologic environment experienced by the dosage form is necessary (7).

### Pharmacokinetic Characterization

**Absorption.** It is necessary to characterize the relative bioavailability and the fractional absorption profile versus time for the drug entity in the controlled-dosage form to evaluate the input profile of the controlled-release dosage form. This can be achieved by using deconvolution techniques that allow one to determine the fraction of drug absorbed versus time. Two commonly used techniques are the Wagner–Nelson method, which is used for drugs exhibiting one-compartment pharmacokinetics, and the Loo–Riegelman method, for drugs with two-compartment pharmacokinetic characteristics. The fractional absorption input profile can serve to assess the drug-release claims for that formulation. Further drug-release-rate and release-rate constants can be determined from the fractional absorption data. Moreover, this information, together with the disposition characteristics for the drug entity, can be used to characterize and predict changes in the bioavailability of the drug entity when input is modified for the controlled-release dosage forms. For example, the drug may exhibit saturable first-pass hepatic metabolism, which could result in decreased systemic availability when the input rate is decreased.

Although it may not be a regulatory requirement in designing a controlled-release dosage form, it may be useful to determine the absorption characteristics of the drug entity in various segments of the gastrointestinal tract (particularly the colon for dosage forms that may release drug in the color) (8).

**Disposition.**  The information required to characterize the disposition processes for the drug entity in a controlled-release dosage form should include those generally determined for the drug entity in an immediate-release dosage form. This may include the following:

- Disposition parameters—clearance, volume of distribution, half-life, mean residence time, or model dependent or noncompartmental parameters
- Linearity or characterization of nonlinearity over the dose and/or concentration range that could possibly be encountered
- Accumulation of drug in the blood following multiple doses
- Ratio of parent drug to metabolites if different from immediate-release products

The following information is usually not included in a controlled-release NDA but is known from the development of the molecular entity as an immediate-release formulation:

- Metabolic profile and excretory organ dependence with special attention to active metabolites and active enantiomers of racemic mixtures
- Enterohepatic circulation
- Protein-binding parameters and dialyzeability
- Effect of age, gender, race, and relevant disease states on drug disposition from the controlled-release drug product.
- Plasma/blood ratio

In addition, in cases where the drug can be given in the morning or at bed time or where there is evidence that the clinical response varies significantly as a function of the time of the day, it is recommended that circadian variability in the drug's disposition parameters and pharmacodynamics be characterized to determine whether changes in the rate of drug input with time have an impact on safety and efficacy.

## INFORMATION TO CHARACTERIZE THE DOSAGE FORM

### Physicochemical Characterization

Similar characterization of the physicochemical properties of the dosage form should be undertaken as for the drug entity. Dissolution profiles from pH 1 to pH 7.4 should be obtained, with particular attention to the effect of the formulation changes on dissolution characteristics. A determination of whether the release of the drug from the formulation is dependent on the dissolution conditions such as pH, rotation speed, and type of apparatus and medium. Characterization of formulations that are highly insoluble in purely aqueous systems may require the addition of small amounts of surfactant to mimic in vivo conditions, where bile salts act as surfactants, more closely.

### Bioavailability/Bioequivalence Studies

The type of bioavailability studies that need to be carried out depends upon how much is known about the drug, its clinical pharmacokinetics and biopharmaceutics, and whether bioequivalence studies are intended to be the sole basis for product approval. Several dosage strengths of the controlled-release dosage form should normally be developed to allow flexibility for clinicians to titrate the patient over the recommended therapeutic dose range of the marketed immediate-release dosage form. Scoring of the tablet is sometimes possible in the case where it does not affect the controlled-release mechanism of the formulation.

Each strength of the controlled-release product should be included in a single-dose crossover that involves a reference treatment that can consist of an approved formulation whether immediate or controlled-release, a solution or a suspension. In certain cases for substitution claims, the bioequivalence of the different strengths should be assessed. Also, a multiple-dose, steady-state study using the highest dose of the controlled-release dosage form versus a reference product is required for NDA approval (9,10). Comparison of such parameters such as AUC, $C_{max}$, $T_{max}$, and fluctuation index as well as $C_{min}$ at steady state should be made between the controlled-release product and the reference product. If the drug exhibits nonlinear characteristics in its absorption or disposition, then a steady-state study would also be required for the lowest and highest strength.

In the case of a controlled-release dosage form, where the dosage strengths differ from each other only in the amount of identical beaded material contained, a single-dose and a multiple-dose steady-state study at the highest dosage strength will be sufficient for NDA/Abbreviated New Drug Application (ANDA) approval as long as supported by dissolution data. The types of studies needed can be categorized in the following sections.

**Case I: Controlled-Release Oral Dosage Form of a Marketed Immediate-Release Drug for which Clinical Studies Have Been Conducted.** The following studies would be needed for most controlled-release dosage forms. If approval is to be sought without clinical trials, it is recommended that there be preconsultation with the regulatory authorities to ensure that an adequate database exists for such approval.

*Single-Dose Crossover Study.* A single-dose crossover study would include the following treatments: the controlled-release dosage form administered under fasting conditions, an immediate-release dosage form administered under fasting conditions, and the controlled-release dosage form administered immediately after completion of a high-fat meal typically consisting of the following: two eggs fried in butter, two strips of bacon, two slices of toast with butter, 4 oz of hash brown potatoes, and 8 oz of whole milk (11). Alternatively, other meals with 1,000-calorie content, with 50% of the calories derived from fat, could be used. The dosage form should be administered immediately following the completion of the breakfast or meal. Absence of food effect will be concluded when the 90% confidence interval for the ratio of means (population geometric means based on log transformed data) of fed and fasted treatments fall within 80–125% for AUC and 70–143% for $C_{max}$ (11). If there are no significant differences seen in the rate ($C_{max}$) or extent of bioavailability of the controlled-release product in this food effect study, then no further food effect studies are necessary. If significant differences in bioavailability are found, it would be useful to define the cause of the food effect on the controlled-release dosage form as well as the effect of time between meal and drug administration on the food–drug effect.

The purposes of this study are to assess bioavailability, rule out dose dumping, determine whether there is any need for labeling specifications regarding special conditions for administration with respect to meals, and provide information concerning the pattern of absorption of the controlled-release dosage form (12). The drug input function should be defined for controlled-release dosage forms by an appropriate method, for example, the Wagner–Nelson, Loo–Riegelman, or other deconvolution methods. This will aid in assessing the release-rate characteristics of the controlled-release product as well as in the development of an appropriate in vitro dissolution test. For dosage forms that exhibit high intrasubject variability, replicate design studies, in which subjects receive each treatment twice, are suggested.

An illustration of the importance of the study of the effect of food on the bioavailability of a drug from a controlled-release formulation is the case of nisoldipine (13). Nisoldipine is formulated as a once-a-day controlled-release formulation of a dihydropyridine calcium channel blocker that is approved in the United States for the treatment of hypertension.

During the course of development, a food effect study was conducted that revealed that a high-fat breakfast had a profound effect on the bioavailability of nisoldipine from this formulation. As can be seen from Figure 1, food increased the peak plasma concentration ($C_{max}$) from 2.75 to 7.5 ng/mL, an increase of 2.75-fold, while the area under

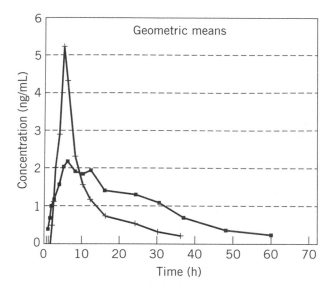

**Figure 1.** Nisoldipine plasma concentration time profile with and without the administration of a high-fat breakfast.

the curve (AUC) decreased from 70.4 to 53 ng × h/mL. These results raised a question related to the safety and efficacy of this controlled-release formulation when taken with food.

However, an $E_{max}$ pharmacokinetic pharmacodynamic (PK/PD) model established the relationship between nisoldipine plasma concentrations and the reduction in blood pressure. This model was constructed using data from a study in which 23 patients with essential hypertension were given, on an escalating basis, 30, 60, 90, and 120 mg of nisoldipine as could be tolerated by the patients. The $EC_{50}$ was estimated to be 3.54 ng/mL, and the $E_{max}$, expressed as the maximum drop in diastolic pressure, was 23.9 mm of Hg (14). The PK/PD model enabled the assessment of the clinical impact of the rapid increase in the rate of bioavailability of nisoldipine from this controlled-release formulation when taken with food. Because the relationship between dose and blood pressure lowering was log linear (the dose of drug would have to be increased greatly to observe an effect on the blood pressure) and the maximal drop in diastolic blood pressure would not be expected to pose any harmful hemodynamic effect to the patient and, with some additional experience in the fed state available from the clinical trials, the effect of food on the bioavailability of nisoldipine was not deemed to pose any safety concerns to the patients. However, in the labeling of this drug, it was recommended that Sular® be administered on an empty stomach for optimal efficacy (15).

*Multiple-Dose, Steady-State Studies.* When data exist for the immediate-release product establishing linear pharmacokinetics, a steady-state study with the controlled-release product at one dose level (at the high end of the dosage range) using an immediate-release formulation as a control should be conducted. At least three consecutive trough concentrations ($C_{min}$) taken at the same time of the day should be taken to ascertain that the subjects are at steady state. Concentrations over at least one dosage interval of the controlled-release product should be measured in each leg of the crossover study. In cases where the controlled-release dosage interval is not 24 h, it may be preferable to measure concentrations over an entire day in each leg for purposes of assessing diurnal variation.

Where it exists, consideration must be given to the "therapeutic window." of the drug. The occupancy time, which is defined as the percentage of time over a dosage interval where the concentration lies within the therapeutic window, should be determined.

Appropriate concentration measurements should include unchanged drug and major active metabolites. For racemic drugs, consideration should be given to specific measurement of the active enantiomers. However, approval of the controlled-release product is based on the results of the clinical trials. The pharmacokinetic data is used to establish the pharmacokinetic and bioavailability characteristics of the controlled-release formulation.

If the controlled-release product is aimed at a specific subpopulation, for example, children, it should be tested in that subpopulation. A controlled-release dosage form does not necessarily have to be administered at the same total daily dose as for the immediate-release reference. For example, if first-pass metabolism was greater for the controlled-release dosage form, the total daily dose may be more than the total daily dose of the immediate release product. Also, transdermal patches may contain a reservoir of drug that will not be absorbed during the time of use of the patch. The goal of differences in treatment dosage is to achieve equal exposure.

Steady-state studies in selected patient population groups and/or drug interaction studies may also be necessary, depending upon the therapeutic use of the drug and the type of individuals for which the controlled-release product will be recommended. In such studies, it may be advisable to carry out more than one interval measurement per patient to assess variability with both the controlled-release and the immediate-release dosage forms.

***Example of an NDA with Clinical Data.*** This NDA for a once-a-day diltiazem formulation is a typical example of an NDA application with clinical data. The clinical data consisted of two clinical trials. The first study was a multicenter placebo-controlled randomized-dose-ranging study in which 198 patients participated. This clinical trial combined a dose-escalation arm with 180 to 540 mg doses and a fixed-dose parallel arm involving doses of 90 to 360 mg. In the fixed-dose arm, those patients randomized to 90 or 180 mg remained on these doses throughout the study. Those randomized to 360 mg received 180 mg for the first 2 weeks before receiving the 360-mg dose. The second clinical study was also a randomized double-placebo-controlled-dose escalation study, with doses from 120 to 540 mg in 56 patients with mild to moderate hypertension.

Figure 2 shows the changes in blood pressure among the different doses compared with the placebo (16). The biopharmaceutics and clinical pharmacology portion of the NDA consisted of six studies, including (17)

- A single-dose relative bioavailability study comparing the controlled-release product to an approved immediate-release diltiazem

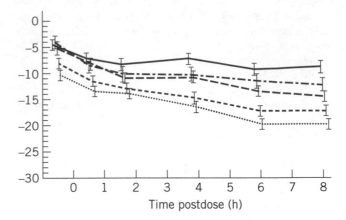

**Figure 2.** Change in supine diastolic blood pressure from baseline predose levels (mm Hg) compared with placebo across all dose levels.

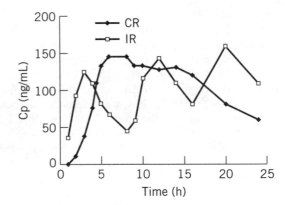

**Figure 3.** Comparative single-dose plasma concentration time profile of diltiazem from a once-a-day controlled-release (CR) formulation and three doses of the immediate-release (IR) formulation.

- A multiple-dose relative bioavailability study comparing the controlled-release product to an approved immediate-release diltiazem
- A food effect study

Figure 3 shows the comparative plasma concentration time profile for this controlled release formulation relative to the approved immediate release formulation. These three studies, along with the clinical data, fulfilled the regulatory requirements for this NDA.

It is to be noted that the trough level at the end of the dosing interval for the controlled-release product was lower than that of the immediate release formulation as can be seen in Figure 4. However, because the efficacy of the controlled-release product was established in the two clinical trials, the lower trough levels were of no clinical concern.

The sponsor also conducted the following:

- A dose proportionality study, because diltiazem is known to undergo saturable first-pass metabolism, and thus the degree of nonlinearity in pharmacokinetics needed to be characterized for the controlled-release product.

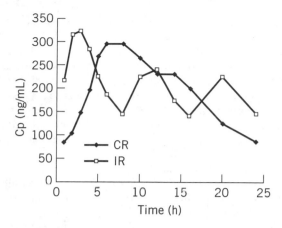

**Figure 4.** Comparative steady-state plasma concentration time profile of diltiazem from a once-a-day controlled-release formulation and three doses of the immediate-release formulation given during a 24-h dosing interval.

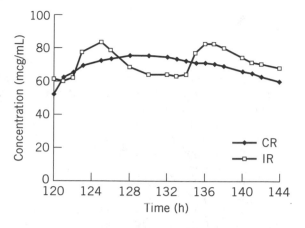

**Figure 5.** Example of a controlled-release product with steady-state $C_{min}$ values at the end of the dosing interval lower than those for the corresponding immediate-release product.

- A bioequivalence study comparing this once-a-day product with the two other once-a-day diltiazem products that are already approved for marketing. This study was undertaken for marketing purposes by the sponsor of the NDA and was not required from the regulatory point of view.

- An in vivo/in vitro correlation study that would enable the sponsor to obtain in vivo bioavailability waivers and possibly wider dissolution specifications. Such a study is highly desirable in an NDA but is not a requirement.

**Case II: Controlled-Release NDA with No Clinical Studies or PK/PD Information.** It may be possible to obtain approval of an NDA for a controlled-release drug product based on bioequivalence data alone. In this situation, there is an absolute need to establish bioequivalence conclusively. To allow marketing of a controlled-release drug product in the absence of clinical data or PK/PD information for the controlled-release product requires that all parameters related to bioequivalence be shown to meet the equivalence criteria under steady-state conditions. These parameters include $C_{max}$, AUC, and $C_{min}$ (absolute lowest concentration over an interval).

Similar $C_{max}$ comparisons for the test and reference products relate to establishing comparable safety and efficacy. Safety relates to the fact that the controlled-release product should not have $C_{max}$ or $C_{min}$ values significantly higher than the immediate-release product, whereas efficacy could relate to the controlled-release product not having a $C_{max}$ or $C_{min}$ significantly lower than the immediate-release product. AUC could relate to both safety and efficacy, and the controlled- and immediate-release products should show comparable AUCs during a dosing interval at steady state. In addition, $C_{min}$ values should be comparable related to efficacy at the end of the dosing interval for the controlled-release product. Lower $C_{min}$ values for the controlled-release product would raise issues in this regard, and this is illustrated in Figure 5. The lower values

of $C_{min}$ for this controlled-release product, which was the subject of an NDA that contained no clinical data, caused concerns regarding its approvability. To allow the marketing of this controlled-release product without prior testing in patients to establish efficacy was considered to be unacceptable.

**Case III: Oral Controlled-Release Dosage Form of a Marketed Immediate-Release Drug for Which Extensive Pharmacodynamic/Pharmacokinetic Data Exist.** The same bioavailability/bioequivalence requirements as discussed in Case I apply to the controlled-release dosage form of a marketed immediate-release drug for which extensive PK/PD data exist. In certain cases the PK/PD information can serve in lieu of clinical trials to establish the safety and efficacy profile of the controlled-release drug product.

**Case IV: Nonoral Controlled-Release Dosage Form of a Marketed Immediate-Release Drug for Which Extensive Pharmacodynamic/Pharmacokinetic Data Exist.** The bioavailability studies described earlier (except for the food effect study) would be appropriate for the evaluation of a controlled-release formulation designed for nonoral route of administration. If an altered biotransformation pattern of active metabolites is observed, the need for clinical data becomes more apparent. One or more clinical efficacy studies would be required. In addition to bioavailability studies, special studies should be concerned with specific risk factors, for example, irritation and/or sensitization at the site of application or injection.

**Case V: Generic Equivalent of an Approved Controlled-Release Product.** The Drug Price Competition and Patent Term Restoration Act amendments of 1984 to the Food, Drug and Cosmetic Act of 1938 gave the FDA statutory authority to accept and approve for marketing ANDAs for generic substitutes of pioneer products, including those approved after 1962. To gain approval according to the law, ANDAs for a generic controlled-release drug product must, among other things, be both pharmaceutically equivalent and bioequivalent to the innovator controlled-release product, which is termed the *reference-listed drug product* as identified in FDA's *Approved Drug Products with Therapeutic Equivalence Ratings* (the "Orange Book").

***Pharmaceutical Equivalence.*** As defined in the Orange Book, to be pharmaceutically equivalent, the generic and pioneer formulations must (*1*) contain the same active ingredient, (*2*) contain the same strength of the active ingredient in the same dosage form, (*3*) be intended for the same route of administration, and (*4*) be labeled for the same conditions of use. The FDA does not require that the generic and reference-listed controlled-release products contain the same excipients or that the mechanism by which the release of the active drug substance from the formulation be the same (18). For substitution purposes, the two products have to be pharmaceutical equivalents.

***Bioequivalence Requirements.*** The same bioequivalence requirements apply to the establishment of the equivalence of the formulation used in efficacy trials if it is different from the formulation intended for marketing and generic controlled-release product approval. For development of a generic equivalent of an approved controlled-release product, the new generic formulation must be comparable with respect to rate and extent of bioavailability (usually using AUC, $C_{max}$, $C_{min}$, and the degree of fluctuation as criteria) in a crossover steady-state study compared to the reference controlled-release product.

In 1993, the Division of Bioequivalence of the Office of Generic Drugs issued a Guidance entitled "Oral Extended (Controlled) Release Dosage Form In Vivo Bioequivalence and In Vitro Dissolution Testing," which outlines the required studies for the approval of a controlled-release product (19). The following is a summary of the required studies as outlined in the 1993 guidance.

***Single-Dose Fasting Two-Way Crossover Bioequivalence Study.*** The objective of this study is to compare the rate (as measured by $C_{max}$) and extent of absorption (as measured by AUC) of a generic formulation with that of a reference-listed formulation when administered in equal labeled doses. The FDA-designated reference product is identified by the symbol + in the Orange Book.

*Design.* The study design is a single-dose, two-treatment, two-period, two-sequence crossover with an adequate washout period between the two phases of the study. An equal number of subjects should be randomly assigned to the two possible dosing sequences. An institutional review board should approve the proposed protocol for the study prior to its initiation.

*Selection of Subjects.* The applicant should enroll a number of subjects sufficient to ensure adequate statistical results. It is recommended that a minimum of 24 subjects be used in this study. More subjects may be required for a drug that exhibits high intrasubject variability in metrics of rate and extent of absorption. Subjects should be healthy volunteers, 18 to 50 years of age, and within 10% of ideal body weight for height and build. Written, informed consent must be obtained from all subjects before their acceptance into the study.

*Procedure.* Following an overnight fast of at least 10 hours, subjects should be administered a single dose of the test or reference product with 240 mL water. They should

continue fasting for 4 h after administration of the test or reference treatment.

*Blood Sampling.* In addition to the predose (0 h) sample, venous blood samples should be collected postdose so that there are at least four sampling time points on the ascending part and six or more on the descending part of the concentration-time curve. The biological matrix (plasma, serum, or whole blood) should be immediately frozen after collection and, as appropriate, centrifugation and kept frozen until assayed.

*Analysis of Blood Samples.* The active moieties should be assayed using a suitable analytical method validated with regard to specificity, accuracy, precision (both within and between days), limit of quantitation, linearity, and recovery. Stability of the samples under frozen conditions, at room temperature, and during freeze–thaw cycles, if appropriate, should be determined. If the analytical method is a chromatographic method, chromatograms of unknown samples, including all associated standard curve and quality control chromatograms, should be submitted for one-fifth of subjects, chosen at random (20).

*Pharmacokinetic Analysis of Data.* Calculation of area under the plasma concentration-time curve to the last quantifiable concentration (AUC 0—t) and to infinity. Calculation of $C_{max}$ and $T_{max}$ should be performed according to standard techniques.

*Statistical Analysis of Pharmacokinetic Data.* The log-transformed AUC and $C_{max}$ data should be analyzed statistically using analysis of variance. These two parameters for the test product should be shown to be within 80–125% of the reference product using the 90% confidence interval (21–23).

*Clinical Report and Adverse Reactions.* Subject medical histories, physical examination reports, and all incidents of adverse reactions to the study formulations should be reported.

***Multiple-Dose, Steady-State, Two-Way Crossover Bioequivalence Study.*** The objective of this study is to document that the steady-state rate and extent of absorption of the test controlled-release product is equivalent to the rate and extent of absorption of the reference-listed drug containing the same amount of the active ingredient in the same dosage form.

*Design.* The study design is a multiple-dose, two-treatment, two-period, two-sequence crossover with adequate washout period between the two phases of the study. An equal number of subjects should be randomly assigned to the two possible dosing sequences. Before initiation of the multiple-dose steady-state study, an institutional review board should approve the study protocol. Moreover, the steady-state study should be done after the completion of the single-dose study.

*Selection of Subjects.* Same as for the single-dose study.

*Procedures.* Controlled-release products that are administered once a day should be dosed following an overnight fast of at least 10 h; subjects should continue fasting for 4 h postdose. For controlled-release products that are dosed every 12 h, the morning dose should be given following an overnight fast of about 10 h, and subjects should continue fasting for 4 h postdose; the evening dose should be administered after a fast of at least 2 h, and subjects should continue fasting for 2 h postdose. Each dose should be administered with 240 mL water.

*Blood Sampling.* At least three trough concentrations ($C_0$) on three consecutive days should be determined to ascertain that the subjects are at steady state prior to measurement of rate and extent of absorption after a single-dose administration in a dosing interval at steady state. For controlled-release drug products administered more often than every 24 h, assessment of trough levels just prior to two consecutive doses is not recommended because a difference in the consecutive trough values may occur due to circadian rhythm irrespective of whether or not steady state has been attained. Adequate blood samples should be collected at appropriate times during a dosing interval at steady state to permit estimation of the total area under the concentration-time curve, peak concentration ($C_{max}$), the absolute lowest plasma concentration ($C_{min}$), and time to peak concentration ($T_{max}$).

*Pharmacokinetic Data.* The following pharmacokinetic data are to be reported for the evaluation of bioequivalence of the generic controlled-release product to the reference-listed product:

- Individual and mean blood drug concentration levels.
- Individual and mean trough levels ($C_0$) as well as the lowest plasma concentration ($C_{min}$).
- Individual and mean peak level ($C_{max}$).
- Calculation of individual and mean steady-state AUC.
- Individual and mean percent fluctuation; fluctuation for the test product should be evaluated for comparability with that for the reference product.
- Individual and mean time to peak concentration ($T_{max}$)

*Statistical Analysis of Pharmacokinetic Data.* Same as for the single-dose study.

*Clinical Report and Adverse Reactions.* Same as for the single-dose study.

For controlled-release capsule formulation marketed in multiple strengths, a single-dose bioequivalence study under fasting conditions is required only for the highest strength, provided that the compositions of the lower strengths are proportional to that of the highest strength and that all strengths contain identical beads or pellets. Single-dose bioequivalence studies may be waived for the lower strengths on the basis of acceptable dissolution profiles. Multiple-dose steady-state and single-dose food/fasting studies are to be conducted on the highest strength of the capsule formulation. For controlled-release products that are not beaded and are not compositionally proportional, the single-dose bioequivalence study is required for each strength, and the multiple-dose bioequivalence study is required only to the highest strength. This requirement can also be waived in the presence of an in vivo/in vitro correlation whose predictability has been established. For this waiver to be granted the following conditions have to be met:

- The lower strengths must be compositionally proportional or qualitatively the same

- The lower strengths must have the same release mechanism
- The lower strengths must have similar in vitro dissolution profiles
- The lower strengths must be manufactured using the same type of equipment and the same process, and at the same site as other strengths that have bioavailability data

In addition one of the following situations should exist:

- Bioequivalence has been established for all strengths of the reference-listed product.
- Dose proportionality has been established for the reference-listed product, and all reference product strengths are compositionally proportional or qualitatively the same; they must have the same release mechanism, and the in vitro dissolution profiles for all strengths must be similar.
- Bioequivalence has been established between the generic product and the reference-listed product at the highest and lowest strengths; for the reference-listed product, all strengths are compositionally proportional or qualitatively the same and have the same release mechanism, and the in vitro dissolution profiles are similar.

The criterion for granting such waivers is that the difference in predicted means of $C_{max}$ and AUC is no more than 10% based on dissolution profiles of the highest-strength and the lower-strength product.

***Single-Dose, Food/Fasting Study.*** The objective of this study is to document that the rate and extent of absorption of the generic controlled-release product is equivalent to the rate and extent of absorption of the reference-listed drug when both products are administered immediately after a high-fat meal and to assess the effect of a high-fat meal on the bioavailability of the generic controlled-release product.

*Statistical Analysis.* Absence of food effect will be concluded when the 90% confidence interval for the ratio of means (population geometric means based on log-transformed data) of fed and fasted treatments fall within 80–125% for AUC and 70–143% for $C_{max}$.

**Case VI: Controlled Release Dosage Form as a New Molecular Entity.** Independent of whether a controlled-release dosage form is evaluated by clinical studies, this dosage form should be characterized as described previously; that is, per dose proportionality, food effects, absorption characteristics (rate, pattern, and extent), and fluctuation. Moreover, the additional requirements for an immediate-release new molecular entity will also be applicable to controlled-release applications.

***Retention of Samples.*** The clinical laboratory conducting any in vivo study should retain an appropriately identified reserve sample of the test and reference products for a period of 5 years. Each reserve sample should consist of at least 200 dosage units. For more information on retention of bioequivalence samples, please refer to 21 CFR 320.63.V (24).

## Dissolution Testing for Controlled-Release Drug Products

In vitro dissolution testing is important for (1) providing process control and quality assurance, (2) determining stable release characteristics of the product over time, and (3) facilitating certain regulatory determinations (e.g., absence of effect of minor formulation changes or of change in manufacturing site on performance). In certain cases, especially for oral controlled-release formulations, the dissolution test, related to an in vivo/in vitro correlation (IVIVC), can serve not only as a quality control for the manufacturing process but also as an indicator of how the formulation will perform in vivo. A main objective of developing and evaluating an IVIVC is to establish the dissolution test as a surrogate for human bioequivalence studies.

An IVIVC describes a meaningful relationship between in vitro dissolution and expected in vivo bioavailability results. The availability of an IVIVC provides for excellent control of product quality, particularly for marketed batches of a drug product, that will not be tested in humans. The concept of IVIVC, particularly for oral controlled-release drug products, has been studied considerably over the years. Several workshops and publications have provided useful information regarding IVIVCs.

The *Report of the Workshop on CR Dosage Forms: Issues and Controversies* (1987) indicated that the state of science and technology at that time did not permit consistently meaningful IVIVC for oral controlled-release dosage forms and encouraged IVIVC as a future objective. Dissolution testing was considered useful only for process control, stability, minor formulation changes, and manufacturing site changes (7).

A U.S. Pharmacopeial Convention (USP) Pharmacopeial Forum stimuli article in 1988 established the classification of IVIVC into levels A, B, and C. In addition, USP Chapter 1088 describes techniques appropriate for level A, B, and C correlations and methods for establishing dissolution specifications (25).

A level A correlation is usually estimated by a two-stage procedure: deconvolution followed by comparison of the fraction of drug absorbed with the fraction of drug dissolved. A correlation of this type (shown in Figure 6) is generally linear and represents a point-to-point relationship between in vitro dissolution and the in vivo input rate (e.g., the in vivo dissolution of the drug from the dosage form). In a linear correlation, the in vitro dissolution and in vivo input curves may be directly superimposable or may be made to be superimposable by the use of a scaling factor. Nonlinear correlations, although uncommon, may also be appropriate (26).

A level B correlation uses the principles of statistical moment analysis. The mean in vitro dissolution time is compared either to the mean residence time or to the mean in vivo dissolution time. A level B correlation does not uniquely reflect the actual in vivo plasma level curve because a number of different in vivo curves will produce similar mean residence time values. An example of such a correlation is shown in Figure 7.

A level C correlation establishes a single-point relationship between a dissolution and a pharmacokinetic parameter. A level C correlation does not reflect the complete shape of the plasma concentration-time curve, which is important to the definition of the performance of oral controlled-release products. A multiple level C correlation relates one or several pharmacokinetic parameters of interest to the amount of drug dissolved at several time points of the dissolution profile. This is illustrated in Figure 8, where the relationship between the maximum plasma concentrations and percent dissolved at several time points is given (27).

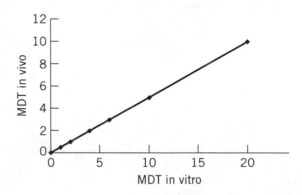

**Figure 7.** Level B correlation.

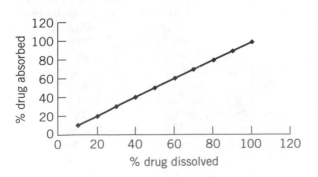

**Figure 6.** Level A correlation.

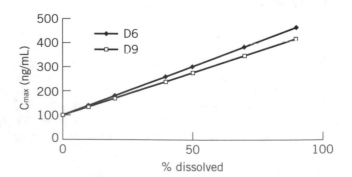

**Figure 8.** Multiple level C correlation.

The report entitled *In Vitro / In Vivo Testing and Correlation for Oral Controlled / Modified Release Dosage Forms* (8) concludes that although science and technology may not always permit meaningful IVIVC, the development of an IVIVC is an important objective on a product-by-product basis. Procedures for development, evaluation, and application of an IVIVC are described. Validation of dissolution specifications by a bioequivalence study involving two batches of product with dissolution profiles at the upper and lower dissolution specifications is suggested.

Further information related to IVIVCs was developed in a USP/American Association of Pharmaceutical Scientists/FDA-sponsored workshop that resulted in a report entitled *Workshop II Report: Scale-up of Oral Extended Release Dosage Forms* (20). This report identified the objectives of an IVIVC to be the use of dissolution as a surrogate for bioequivalence testing as well as an aid in setting dissolution specifications. The report concluded that dissolution may be used as a sensitive, reliable, and reproducible surrogate for bioequivalence testing. The report gave support to the concepts of USP Chapter 1088 and further found that an IVIVC may be useful for changes other than minor changes in formulation, equipment, process, manufacturing site, and batch size (28,29).

An FDA Guidance for Industry, "Extended Release Solid Oral Dosage Forms: Development, Evaluation and Application of In Vitro/In Vivo Correlations," provides recommendations for establishing useful IVIVCs (30). Human data should be utilized for regulatory consideration of an IVIVC. Bioavailability studies for IVIVC development should be performed with enough subjects to adequately characterize the performance of the drug product under study. Although crossover studies are preferred, parallel studies or cross-study analyses may be acceptable. The reference product in developing an IVIVC may be an intravenous solution, an aqueous oral solution, or an immediate-release product. IVIVCs are usually developed in the fasted state. When a drug is not tolerated in the fasted state, studies may be conducted in the fed state. Any in vitro dissolution method may be used to obtain the dissolution characteristics of the oral controlled-release dosage form, but the same system should be used for all formulations tested. The most commonly used dissolution apparatus is USP apparatus (basket) or II (paddle), used at compendially recognized rotation speeds (e.g., 100 rpm for the basket and 50–75 rpm for the paddle). In other cases, the dissolution properties of some oral controlled-release formulations may be determined with USP apparatus III (reciprocating cylinder) or IV (flow through cell). An aqueous medium, preferably not exceeding pH 6.8, is recommended as the initial medium for development of an IVIVC. For poorly soluble drugs, addition of surfactant may be appropriate. In general, nonaqueous and hydroalcoholic systems are discouraged unless all attempts with aqueous media are unsuccessful. The dissolution profiles of at least 12 individual dosage units from each lot should be determined. A suitable distribution of sampling points should be selected to define the profiles adequately (31–33).

IVIVCs are established in two stages. First, the relationship between dissolution characteristics and bioavailability characteristics needs to be determined. Second, the reliability of this relationship must be tested. The first stage may be thought of as developing an IVIVC, whereas the second stage may involve evaluation of predictability. The most commonly seen process for developing a level A IVIVC is to (1) develop formulations with different release rates, such as slow, medium, fast, or a single-release rate if dissolution is condition independent; (2) obtain in vitro dissolution profiles and in vivo plasma concentration profiles for these formulations; and (3) estimate the in vivo absorption or dissolution time course using an appropriate deconvolution technique for each formulation and subject [e.g., Wagner–Nelson, numerical deconvolution (34)]. These three steps establish the IVIVC model. Alternative approaches to developing level A IVIVCs are possible. The IVIVC relationship should be demonstrated consistently with two or more formulations with different release rates to result in corresponding differences in absorption profiles. Exceptions to this approach (i.e., use of only one formulation) may be considered for formulations for which in vitro dissolution is independent of the dissolution test conditions (e.g., medium, agitation, pH).

The in vitro dissolution methodology should adequately discriminate among formulations. Dissolution testing can be carried out during the formulation screening stage using several methods. Once a discriminating system is developed, dissolution conditions should be the same for all formulations tested in the bioavailability study for development of the correlation and should be fixed before further steps toward correlation evaluation are undertaken. During the early stages of correlation development, dissolution conditions may be altered to attempt to develop a one-to-one correlation between the in vitro dissolution profile and the in vivo dissolution profile.

Time scaling may be used as long as the time-scaling factor is the same for all formulations (35). An IVIVC that has been developed should be evaluated to demonstrate that predictability of in vivo performance of a drug product from its in vitro dissolution characteristics is maintained over a range of in vitro dissolution release rates and manufacturing changes. Because the objective of developing an IVIVC is to establish a predictive mathematical model describing the relationship between an in vitro property and a relevant in vivo response, a logical evaluation approach focuses on the estimation of predictive performance or, conversely, prediction error. Depending on the intended application of an IVIVC and the therapeutic index of the drug, evaluation of prediction error internally and/or externally may be appropriate. Evaluation of internal predictability is based on the initial data used to develop the IVIVC model. Evaluation of external predictability is based on additional test data sets that may reduce the number of bioequivalence studies performed during the initial approval process as well as with certain scale-up and postapproval changes. However, for the applications outlined in the following sections, the adequacy of the in vitro dissolution method to act as a surrogate for in vivo testing should be shown through an IVIVC for which predictability has been established (36). The criterion for granting in vivo bioavailability/bioequivalence waivers with an IVIVC is that the predicted $C_{max}$ and AUC from the in vitro dissolution data for the test and reference formulation should not differ by more than 20%. This is illustrated in Figure 9.

Waivers of bioequivalency studies are possible without an IVIVC for small changes in oral controlled-release drug products. For formulations consisting of beads in capsules, with the only difference between strengths being the number of beads, approval of lower strengths without an IVIVC is possible, provided bioavailability data are available for the highest strength and the dissolution profiles in several media are the same across strengths. Where the FDA Guidance for Industry Scale Up and Post-Approval Changes for Modified Release oral dosage forms (SUPAC-MR) *Modified Release Solid Oral Dosage Forms; Scale-Up and Postapproval Changes: Chemistry, Manufacturing, and Controls, In Vitro Dissolution Testing, and In Vivo Bioequivalence Documentation* (37) recommends a biostudy, biowaivers for the same changes made on lower strengths are possible without an IVIVC if (1) all strengths are compositionally proportional or qualitatively the same, (2) in vitro dissolution profiles of all strengths are similar, (3) all strengths have the same release mechanism, (4) bioequivalence has been demonstrated at the highest strength (comparing changed and unchanged drug product), and (5) dose proportionality has been demonstrated

for this oral controlled-release drug product. In the last circumstance, documentation of dose proportionality may not be necessary if bioequivalence has been demonstrated on the highest and lowest strengths of the drug product, comparing changed and unchanged drug product for both strengths as recommended in SUPAC-MR.

If an IVIVC has been established, waivers for more significant changes are possible. A biowaiver will likely be granted for an oral controlled-release drug product using an IVIVC for (1) level 3 process changes as defined in SUPAC-MR, (2) complete removal of or replacement of nonrelease-controlling excipients as defined in SUPAC-MR, and (3) level 3 changes in the release-controlling excipients as defined in SUPAC-MR.

If an IVIVC is developed with the highest strength, waivers for changes made on the highest strength and any lower strengths may be granted if these strengths are compositionally proportional or qualitatively the same, the in vitro dissolution profiles of all the strengths are similar, and all strengths have the same release mechanism.

This biowaiver is applicable to strengths lower than the highest strength, within the dosing range that has been established to be safe and effective, provided that the new strengths are compositionally proportional or qualitatively the same, have the same release mechanism, have similar in vitro dissolution profiles, and are manufactured using the same type of equipment and the same process at the same site as other strengths that have bioavailability data available.

Certain changes almost always necessitate in vivo bioavailability testing and in some cases might necessitate clinical trials, even in the presence of an IVIVC. These include the approval of a new formulation of an approved oral controlled-release drug product when the new formulation has a different release mechanism, approval of a dosage strength higher or lower than the doses that have been shown to be safe and effective in clinical trials, approval of another sponsor's oral controlled-release product even with the same release-controlling mechanism, and approval of a formulation change involving a non-release-controlling excipient in the drug product that may significantly affect drug absorption (38).

### Setting Dissolution Specifications

In vitro dissolution specifications should generally be based on the performance of the clinical/bioavailability lots. These specifications may sometimes be widened within the range allowed in the guidance so that scale-up lots, as well as stability lots, meet the specifications associated with the clinical/bioavailability lots. This approach is based on the use of the in vitro dissolution test as a quality control test without any in vivo significance, even though in certain cases (e.g., oral controlled-release formulations), the rate-limiting step in the absorption of the drug is the dissolution of the drug from the formulation. An IVIVC adds in vivo relevance to in vitro dissolution specifications, beyond batch-to-batch quality control. In this approach, the in vitro dissolution test becomes a meaningful predictor of in vivo performance of the formulation, and dissolution specifications may be used to minimize the possibility of releasing lots that would exhibit different in vivo performance.

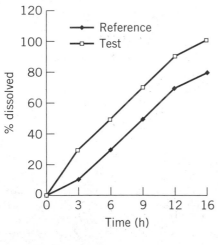

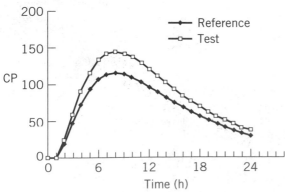

**Figure 9.** Criterion for granting an in vivo bioavailability waiver based on predicted $C_{max}$ and AUC from in vitro dissolution data using IVIVC.

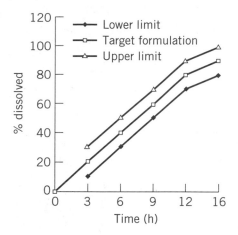

**Figure 10.** Upper and lower dissolution specifications in a case where no IVIVC exists.

A minimum of three time points is recommended to set the specifications. These time points should cover the early, middle, and late stages of the dissolution profile. The last time point should be the time point where at least 80% of drug has dissolved. If the maximum amount dissolved is less than 80%, the last time point should be the time when the plateau of the dissolution profile has been reached.

Specifications should be established based on average dissolution data for each lot under study, equivalent to USP stage 2 testing. Specifications that allow lots to pass at stage 1 of testing may result in lots with less than optimal in vivo performance passing these specifications at USP stage 2 or stage 3.

**Setting Dissolution Specifications without an IVIVC.** The recommended range at any dissolution time point specification is ± 10% deviation of the labeled claim from the mean dissolution profile obtained from the clinical/bioavailability lots as shown in Figure 10. In certain cases, reasonable deviations from the ± 10% range can be accepted provided that the range at any time point does not exceed 25%. Specifications greater than 25% may be acceptable based on evidence that lots (side batches) with mean dissolution profiles that are allowed by the upper and lower limit of the specifications are bioequivalent.

**Setting Dissolution Specifications Where an IVIVC Has Been Established.** Optimally, specifications should be established such that all lots that have dissolution profiles within the upper and lower limits of the specifications are bioequivalent. Less optimally but still possible, lots exhibiting dissolution profiles at the upper and lower dissolution limits should be bioequivalent to the clinical/bioavailability lots or to an appropriate reference standard. In the case where an IVIVC whose predictability has been established, dissolution specifications should be set in such a way that the average predicted $C_{max}$ and AUC that correspond to the lower and upper limit of the dissolution specifications should differ by no more than 20% from each other.

**Importance of Proper Dissolution Conditions.** Dissolution testing of controlled-release products is a very important test that can show a relationship to the expected bioavailability for the lots of controlled-release product being tested. However, the dissolution conditions during the test are very important and if not properly chosen can give results that are not predictive of the actual in bioavailability.

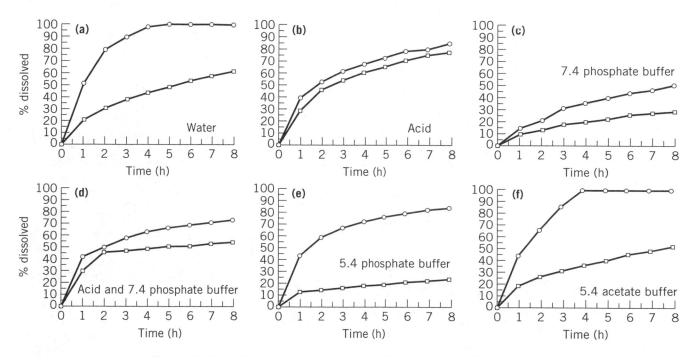

**Figure 11.** Quinidine gluconate dissolution profiles in several different media.

In Figure 11, dissolution profiles for two products tested under varying dissolution conditions are shown. Condition (b), which is dissolution testing in acid, shows similar dissolution profiles for the two products. If these two products were comparable in bioavailability, these conditions might be appropriate for testing these products; however, looking at other dissolution conditions, for example, in water, as can be seen in Figure 11(a) the data show significant differences in the dissolution profiles for these two products. If the products have different bioavailability characteristics, testing in, for example, water, might be more appropriate. For these two products it was established that the unapproved product had only approximately 40% of bioavailability of the reference product. Therefore, dissolution conditions shown in Figures 11(c)–(f) that distinguish between these two products—that is, conditions that show differing dissolution profiles—would be appropriate, whereas dissolution conditions that do not distinguish these two products would be inappropriate. In this particular example, the product with poor bioavailability was marketed without FDA approval. It is possible that only dissolution testing in acid was utilized by the manufacturer to show comparability between the two products. As it turns out, this was a very risky choice. The unapproved product, which was marketed, had very poor bioavailability that could not be determined by dissolution testing in acid. Only by conducting the bioequivalence study prior to marketing, establishing bioequivalence, then determining appropriate dissolution conditions, can bioequivalence between two controlled-release products be ensured (39).

## BIBLIOGRAPHY

1. A.T. Florence and P.U. Jani, *Drug Saf.* **10**(3), 233–266 (1994).

2. P.J. Marroum, *5th Int. Symp. Drug Dev.*, East Brunswick, N.J., May 15–17, 1997.

3. *Guidance for Providing Clinical Evidence of Effectiveness for Human Drug and Biological Products*, Center for Drug Evaluation and Research, Food and Drug Administration, Washington, D.C., 1998.

4. *Code of Federal Regulations*, Title 21, Food and Drug Administration, Washington, D.C., 21 CFR 314.126, 1998.

5. *Code of Federal Regulations*, Title 21, Food and Drug Administration, Washington, D.C., 21 CFR 320.25(f), 1998.

6. *Code of Federal Regulations*, Title 21, Food and Drug Administration, Washington, D.C., 21 CFR 314.54, 1998.

7. J.P. Skelly et al., *Pharm. Res.* **7**, 975–982 (1990).

8. J.P. Skelly et al., *Pharm. Res.* **4**, 75–77 (1987).

9. J.P. Skelly, *Division Guidelines for the Evaluation of Controlled Release Drug Products*, Center for Drug Evaluation and Research, Food and Drug Administration, Washington, D.C., 1984.

10. J.P. Skelly and W.H. Barr. in J.R. Robinson and V.H.L. Lee, eds., *Controlled Drug Delivery: Fundamentals and Application*, Dekker, New York, 1987, pp. 293–333.

11. *Draft Guidance on the Food-Effect Bioavailability and Bioequivalence Studies*, Center for Drug Evaluation and Research, Food and Drug Administration, Washington, D.C., 1997.

12. A. Karim, in A. Yacobi, and E. Holperin-Wolega, eds., *Oral Sustained Release Formulations, Design and Evaluation*, Pergamon, New York, 1987, pp. 293–333.

13. P.J. Marroum, *Nisoldipine Coat Core Clinical Pharmacology and Biopharmaceutics*, Center for Drug Evaluation and Research, Food and Drug Administration, Washington, D.C., 1994.

14. H.G. Schaefer et al., *Eur. J. Clin. Pharm.* **51**, 473–480 (1997).

15. S.T. Chen, *Nisoldipine C.C.*, Secondary Clinical Review, Division of Cardiorenal Drug Products, Center for Drug Evaluation and Research, Food and Drug Administration, Washington, D.C., 1994.

16. R. Kimball, *Medical Review of Viazem*, Division of Cardiorenal Drug Products, Center for Drug Evaluation and Research, Food and Drug Administration, Washington, D.C., 1994.

17. P.J. Marroum, *Core Clinical Pharmacology and Biopharmaceutics*, Center for Drug Evaluation and Research, Food and Drug Administration, Washington, D.C., 1994.

18. *Approved Drug Products with Therapeutic Equivalence Evaluations*, 18th ed., Center for Drug Evaluation and Research, Food and Drug Administration, Washington, D.C., 1998, pp. vii–xxi.

19. *Guidance for Oral Extended (Controlled) Release Dosage Forms: In Vivo Bioequivalence and In Vitro Dissolution Testing*, Office of Generic Drugs, Center for Drug Evaluation and Research, Food and Drug Administration, Washington, D.C., 1993.

20. V.P. Shah et al., *Pharm. Res.* **9**(4), 588–592 (1992).

21. *Guidance for Statistical Procedures for Bioequivalence Studies Using a Standard Two Treatment Crossover Design*, Office of Generic Drugs, Center for Drug Evaluation and Research, Food and Drug Administration, Washington, D.C., 1992.

22. D.J. Schuirman, *Drug Inf. J.* **24**, 315–323 (1990).

23. D.J. Schirman, *J. Biopharm. Pharmacokinet.* **15**, 657–680 (1987).

24. *Code of Federal Regulations*, Title 21, Food and Drug Administration, Washington, D.C., 21 CFR 320.63(v), 1998.

25. *The United States Pharmacopeia 23*, U.S. Pharmacopeial Convention, Rockville, Md., 1995, Chapter 1088, pp. 1924–1929.

26. H. Malinowski, in G. Amidon, J.R. Robinson, and R.L. Williams, eds., *Scientific Foundations for Regulating Drug Product Quality*, AAPS Press, Alexandria, Va., 1997, pp. 259–273.

27. P. Marroum, *Development of In Vivo In Vitro Correlations*, SUPAC-MR/IVIVC Guidance FDA Training Program Manual, Center for Drug Evaluation and Research, Food and Drug Administration, Washington, D.C., 1997, pp. 62–76.

28. J.P. Skelly et al., *Eur. J. Pharm. Biopharm.* **39**, 162–167 (1993).

29. H. Malinowski and J.D. Henderson, in P. Welling and F.L.S. Tse, eds., *Pharmacokinetics Regulatory-Industrial-Academic Perspectives*, 2nd ed., Dekker, New York, 1995, pp. 452–477.

30. H. Malinowski et al., *Dissolut. Technol.* **4**(4), 23–32 (1997).

31. M. Sicwert, *Eur. J. Drug Metab. Pharmacokinet.* **18**(1), 7–18 (1993).

32. J.P. Skelly and G.F. Shiu, *Eur. J. Drug Metab. Pharmacokinet.* **18**(1), 121–129 (1993).

33. H.H. Blume, I. McGilvery, and K.K. Midha, *Eur. J. Pharm. Sci.* **3**, 113–124 (1995).

34. J. Wagner, in J. Wagner, ed., *Pharmacokinetics for the Pharmaceutical Scientist*, Technomic Publishing, Lancaster, Pa., 1993, pp. 159–205.

35. D. Brockmeier, D. Voegele, and H.M. Hattingberg, *Drug Res.* **33**(1), 598–601 (1983).

36. P.J. Marroum, in G. Amidon, J.R. Robinson, and R.L. Williams, eds., *Scientific Foundations for Regulating Drug Product Quality*, AAPS Press, Alexandria, Va., 1997, pp. 305–319.

37. *Guidance for Modified Release Solid Oral Dosage Forms; Scale-Up and Post Approval Changes: Chemistry, Manufacturing, and Controls: In Vitro Dissolution Testing and In Vivo Bioequivalence Documentation*, Center for Drug Evaluation and Research, Food and Drug Administration, Washington, D.C., 1997.

38. H. Malinowski, *Ba, BE and Pharmacokinetics Studies*, FIP Bio-International 96, Business Center for Academic Societies of Japan, Tokyo, 1996.

39. V.K. Prasad et al., *Int. J. Pharm.* **13**, 1–8 (1982).

# H

# HYDROGELS

Anthony M. Lowman
Drexel University
Philadelphia, Pennsylvania

Nicholas A. Peppas
Purdue University
West Lafayette, Indiana

## KEY WORDS

Cross-linking

Diffusion

Erosion

Modeling

Permeability

Pore size

Responsive hydrogels

Swelling

## OUTLINE

Hydrogels are three-dimensional, water-swollen structures composed of mainly hydrophilic homopolymers or copolymers (1). They are rendered insoluble due to the presence of chemical or physical cross-links. The physical cross-links can be entanglements, crystallites, or weak associations such as van der Waals forces or hydrogen bonds. The cross-links provide the network structure and physical integrity.

Hydrogels are classified in a number of ways (1,2). They can be neutral or ionic based on the nature of the side groups. They can also be classified based on the network morphology as amorphous, semicrystalline, hydrogen-bonded structures, supermolecular structures, and hydrocolloidal aggregates. Additionally, in terms of their network structures, hydrogels can be classified as macroporous, microporous, or nonporous (1–3).

Since the development of poly(2-hydroxethyl methacrylate) gels in the early 1960s, hydrogels have been considered for use in a wide range of biomedical and pharmaceutical applications, mainly due to their high water content and rubbery nature. Because of these properties, hydrogel materials resemble natural living tissue more than any other class of synthetic biomaterials (1,4). Furthermore, the high water content allows these materials to exhibit excellent biocompatibility. Some applications of hydrogels include contact lenses, biosensors, sutures, dental materials, and controlled drug delivery (1,4–8). Park and Park (9) present a very useful commentary in relation to the biocompatibility of implantable hydrogel-based delivery systems.

The use of hydrogel-based controlled release systems that can swell in the presence of a biological fluid has been studied for over 30 years. A goal of many researchers is to develop hydrogel devices to provide time-independent, sustained release of bioactive agents (8). Researchers have attempted to achieve these goals by manipulation of the polymer composition and the device geometry. Hydrogel devices have been proposed for delivery by almost all of the conventional routes.

Two of the most important characteristics in evaluating the ability of a polymeric gel to function in a particular controlled release application are the network permeability and the swelling behavior. The permeability and swelling behavior of hydrogels are strongly dependent on the chemical nature of the polymer(s) composing the gel as well as the structure and morphology of the network. As a result, there are different mechanisms that control the release of drugs from hydrogel-based delivery devices. Such systems are classified by their drug release mechanism as diffusional-controlled release systems, swelling-controlled release systems, chemically controlled release systems, and environmentally responsive systems (2).

Hydrogels may exhibit swelling behavior dependent on the external environment. Thus, in the last 30 years there has been a major interest in the development and analysis of environmentally or physiologically responsive hydrogels (10). These hydrogels show drastic changes in their swelling ratio due to changes in their external pH, temperature, ionic strength, nature of the swelling agent, and electromagnetic radiation. Some advantages to environmentally or physiologically responsive hydrogels are relevant in drug delivery applications. In an exceptional new review in the field, am Ende and Mikos (11) offer a thorough physicochemical and mathematical interpretation of the conditions of diffusional release of various bioactive agents in hydrogels and other polymeric carriers, with emphasis on the conditions of stability of peptides and proteins during delivery.

An additional advantage of hydrogels, which has received considerable attention recently, is that they may provide desirable protection of drugs, peptides, and especially proteins from the potentially harsh environment in the vicinity of the release site (12). In particular, such carriers may be able to function as carriers for the oral delivery of sensitive proteins or peptides (13). Finally, hydrogels may be excellent candidates as biorecognizable biomaterials (14). As such, they can be used as targetable carriers of bioactive agents, as bioadhesive systems or as conjugates with desirable biological properties.

In this article, we present an overview of the structure and properties of hydrogels, the diffusional behavior of hydrogels, and the mechanisms for release from polymeric controlled release systems. Additionally, we review important contributions in this area with specific emphasis on new synthetic methods that promise to provide imaginative solutions of drug delivery problems.

## STRUCTURE AND PROPERTIES OF HYDROGELS

In order to evaluate the feasibility of using a particular hydrogel as a drug delivery device, it is important to know the structure and properties of the polymer network. The structure of an idealized hydrogel is shown in Figure 1. The most important parameters that define the structure and properties of swollen hydrogels are the polymer volume fraction in the swollen state, $v_{2,s}$, effective molecular

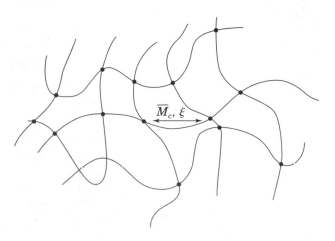

**Figure 1.** Schematic representation of the cross-linked structure of a hydrogel. $\overline{M}_c$ is the molecular weight of the polymer chains between cross-links (●) and $\xi$ is the network mesh size.

weight of the polymer chain between cross-linking points, $\overline{M}_c$, and network mesh or pore size, $\xi$ (15).

The polymer fraction of the polymer in the swollen gel is a measure of the amount of fluid that a hydrogel can incorporate into its structure.

$$v_{2,s} = \frac{\text{volume of polymer}}{\text{volume of swollen gel}} = \frac{V_p}{V_{\text{gel}}} = 1/Q \qquad (1)$$

This parameter can be determined using equilibrium swelling experiments (16). The molecular weight between cross-links is the average molecular weight of the polymer chains between junction points, both chemical and physical. This parameter provides a measure of the degree of cross-linking in the gel. This value is related to the degree of cross-linking in the gel ($X$) as:

$$X = \frac{M_o}{2M_c} \qquad (2)$$

Here, $M_o$ is the molecular weight of the repeating units making up the polymer chains.

The network mesh size represents the distance between consecutive cross-linking points and provides a measure of the porosity of the network. These parameters, which are not independent, can be determined theoretically or through a variety of experimental techniques.

### Equilibrium Swelling Theories

**Neutral Hydrogels.** Flory and Rehner (17) developed the initial depiction of the swelling of cross-linked polymer gels using a Gaussian distribution of the polymer chains. They developed a model to describe the equilibrium degree of cross-linked polymers that postulated that the degree to which a polymer network swelled was governed by the elastic retractive forces of the polymer chains and the thermodynamic compatibility of the polymer and the solvent molecules. In terms of the free energy of the system, the total free energy change upon swelling is written as:

$$\Delta G = \Delta G_{\text{elastic}} + \Delta G_{\text{mix}} \qquad (3)$$

Here, $\Delta G_{\text{elastic}}$ is the contribution due to the elastic retractive forces and $\Delta G_{\text{elastic}}$ represents the thermodynamic compatibility of the polymer and the swelling agent.

Upon differentiation of equation 3 with respect to the number of solvent molecules in the system (at constant T and P), an expression can be derived for the chemical potential change ($\Delta \mu$) in the solvent in terms of the contributions due to swelling.

$$\mu_1 - \mu_{1,0} = (\Delta \mu)_{\text{elastic}} + (\Delta \mu)_{\text{mix}} \qquad (4)$$

Here, $\mu_1$ is the chemical potential of the swelling agent within the gel and $\mu_{1,0}$ is the chemical potential of the pure fluid. At equilibrium, the chemical potential of the swelling agent inside and outside of the gel must be equal, therefore the elastic and mixing contributions to the chemical potential balance one another at equilibrium.

The chemical potential change upon mixing can be determined from the heat of mixing and the entropy of mixing. Using appropriate thermodynamic relationships, the chemical potential of mixing can be expressed as:

$$(\Delta\mu)_{mix} = RT(ln(1 - v_{2,s}) + \chi_1 v_{2,s}^2) \qquad (5)$$

where $\chi_1$ is the polymer solvent interaction parameter defined by Flory (17,18).

The elastic contribution to the chemical potential is determined from the statistical theory of rubber elasticity (18). The elastic free-energy is dependent on the number of polymer chains in the network, $v_e$, and the linear expansion factor, $\alpha$. For gels that were crosslinked in the absence of a solvent, the elastic contribution to the chemical potential is written as:

$$(\Delta\mu)_{elastic} = RT\left(\frac{V_1}{v\overline{M}_c}\right)\left(1 - \frac{2\overline{M}_c}{\overline{M}_n}\right)\left(v_{2,s}^{1/3} - \frac{v_{2,s}}{2}\right) \qquad (6)$$

where $v$ is the specific volume of the polymer, $V_1$ is the molar volume of the swelling agent, and $\overline{M}_n$ is the molecular weight of linear polymer chains prepared using the same conditions in the absence of a cross-linking agent. By combining equations 5 and 6, the swelling behavior of neutral hydrogels cross-linked in the absence of a solvent can be described by the following equation.

$$\frac{1}{\overline{M}_c} = \frac{2}{\overline{M}_n} - \frac{(v/V_1)\lfloor ln(1 - v_{2,s}) + v_{2,s} + \chi_1 v_{2,s}^2\rfloor}{\left(v_{2,s}^{1/3} - \frac{v_{2,s}}{2}\right)} \qquad (7)$$

In many cases, it is desirable to prepare hydrogels in the presence of a solvent. If the polymers were cross-linked in the presence of a solvent, the elastic contributions must account for the volume fraction density of the chains during cross-linking. Peppas and Merrill (19) modified the original Flory-Rehner to account for the changes in the elastic contributions to swelling. For polymer gels cross-linked in the presence of a solvent, the elastic contribution to the chemical potential is:

$$(\Delta\mu)_{elastic} = RT\left(\frac{V_1}{v\overline{M}_c}\right)\left(1 - \frac{2\overline{M}_c}{\overline{M}_n}\right)v_{2,r}\left(\left(\frac{v_{2,s}}{v_{2,r}}\right)^{1/3} - \frac{v_{2,s}}{2v_{2,r}}\right) \qquad (8)$$

Here, $v_{2,r}$ is the volume fraction of the polymer in the relaxed state. The relaxed state of the polymer is defined as the state of the polymer immediately after cross-linking of the polymer but prior to swelling or deswelling. For the case of gels prepared by cross-linking in the presence of a solvent, the equation for the swelling of the polymer gel can be obtained by combining equations 5 and 8 as the mixing contributions for both cases are the same. The swelling of hydrogels cross-linked in the presence of a solvent can then be written as:

$$\frac{1}{\overline{M}_c} = \frac{2}{\overline{M}_n} - \frac{(v/V_1)\lfloor ln(1 - v_{2,s}) + v_{2,s} + \chi_1 v_{2,s}^2\rfloor}{v_{2,r}((v_{2,s}/v_{2,r})^{1/3} - (v_{2,s}/2v_{2,r}))} \qquad (9)$$

By performing swelling experiments to determine $v_{2,r}$, the molecular weight between cross-links can be calculated for a particular gel using this equation (16).

**Ionic Hydrogels.** Ionic hydrogels contain pendent groups that are either cationic or anionic in nature. For anionic gels, the side groups of the gel are unionized below the $pK_a$ and the swelling of the gel is governed by thermodynamic compatibility of the polymer and the swelling agent. However, above the $pK_a$ of the network, the pendent groups are ionized and the gels swell to a large degree due to the development of a large osmotic swelling force due to the presence of the ions. In cationic gels, the pendent groups are unionized above the $pK_b$ of the network. When the gel is placed in fluid of pH less than this value, the basic groups are ionized and the gels swell to a large degree (Figure 2). The pH-dependent swelling behavior of ionic gels, which is completely reversible in nature, is depicted in Figure 3.

The theoretical description of the swelling of ionic hydrogels is much more complex than that of neutral hydrogels. Aside from the elastic and mixing contributions to swelling, the swelling of an ionic hydrogel is affected by the degree of ionization in the gel, ionization equilibrium between the gel and the swelling agent, and the nature of the counterions in the fluid. As the ionic content of a hydrogel is increased, in response to an environmental stimulus, increased repulsive forces develop and the network becomes more hydrophilic. The result is a more highly swollen network. Because of Donnan equilibrium, the

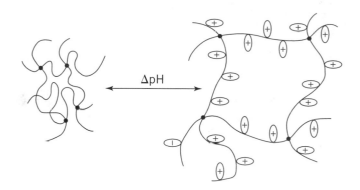

**Figure 2.** Expansion (swelling) of a cationic hydrogel due to ionization of pendent groups, at specific pH values.

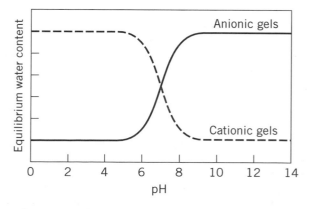

**Figure 3.** Equilibrium degree of swelling of anionic and cationic hydrogels as a function of the swelling solution pH.

chemical potential of the ions inside the gel must be equal to the chemical potential of the ions in the solvent outside of the gel (20). An ionization equilibrium is established in the form of a double layer of fixed charges on the pendent groups and counterions in the gel. Finally, the nature of counterions in the solvent affects the swelling of the gel. As the valence of the counterions increase, they are more strongly attracted to the gel and reduce the concentration of ions needed in the gel to satisfy Donnan equilibrium conditions.

The swelling behavior of polyelectrolyte gels was initially described as being a result of a balance between the elastic energy of the network and the osmotic pressure developed as a result of the ions (20–27). In electrolytic solutions, the osmotic pressure is associated with the development of a Donnan equilibrium. This pressure term is also affected by the fixed charges developed on the pendant chains. The elastic term is described by the Flory expression derived from assumptions of Gaussian chain distributions.

Models for the swelling of ionic hydrogels were developed by equating the three major contributions to the swelling of the networks. These contributions are due to mixing of the polymer and solvent, network elasticity, and ionic contributions. The general equation is given as

$$\Delta G = \Delta G_{mix} + \Delta G_{el} + \Delta G_{ion} \tag{10}$$

In terms of the chemical potential, the difference between the chemical potential of the swelling agent in the gel and outside of the gel is:

$$\mu_1 - \mu_{1,0} = (\Delta\mu)_{elastic} + (\Delta\mu)_{mix} + (\Delta\mu)_{ion} \tag{11}$$

For weakly charged polyelectrolytes, the elastic contribution and mixing contributions do not differ from the case of nonionic gels. However, for highly ionizable materials there are significant ionization effects and $\Delta\mu_{ion}$ is important. At equilibrium, the elastic, mixing, and ionic contributions must sum to 0.

The ionic contribution to the chemical potential is strongly dependent on the ionic strength and the nature of the ions. Brannon-Peppas and Peppas (26,27) and Ricka and Tanaka (20) both developed expressions to describe the ionic contributions to the swelling of polyelectrolytes. Assuming that the polymer networks under conditions of swelling behave similarly to dilute polymer solutions, the activity coefficients can be approximated as 1 and activities can be replaced with concentrations. Under these conditions, the ionic contribution to the chemical potential is described by the following:

$$(\Delta\mu)_{ion} = RTV_1\Delta c_{tot} \tag{12}$$

Here, $\Delta c_{tot}$ is the difference in the total concentration of mobile ions within the gel. The difference in the concentration of mobile ions is due to the fact that the charged polymer requires the same number of counterions to remain in the gel to achieve electroneutrality. The difference in the total ion concentration could then be calculated from the equilibrium condition for the salt.

Brannon-Peppas and Peppas (26,27) developed expressions for the ionic contributions to the swelling of polyelectrolytes for anionic and cationic materials. The ionic contribution for anionic network is:

$$(\Delta\mu)_{ion} = \frac{RTV_1}{4I}\left(\frac{v_{2,s}^2}{v}\right)\left(\frac{K_a}{10^{-pH} + K_a}\right)^2 \tag{13}$$

For the cationic network, the ionic contribution is:

$$(\Delta\mu)_{ion} = \frac{RTV_1}{4I}\left(\frac{v_{2,s}^2}{v}\right)\left(\frac{K_b}{10^{pH-14} + K_a}\right)^2 \tag{14}$$

In these expressions, $I$ is the ionic strength, $K_a$ and $K_b$ are the dissociation constants for the acid and base, respectively. It is significant to note that this expression has related the ionic contribution to the chemical potential to characteristics about the polymer/swelling agent that can readily determinable (e.g., $pH$, $K_a$, and $K_b$).

For the case of anionic polymer gels that were crosslinked in the presence of a solvent, the equilibrium swelling can be described by:

$$\frac{V_1}{4I}\left(\frac{v_{2,s}^2}{v}\right)\left(\frac{K_a}{10^{-pH} + K_a}\right)^2 = (ln(1 - v_{2,s}) + v_{2,s} + \chi_1 v_{2,s}^2)$$
$$+ \left(\frac{V_1}{v\overline{M}_c}\right)\left(1 - \frac{2\overline{M}_c}{\overline{M}_n}\right)v_{2,r}$$
$$\cdot \left(\left(\frac{v_{2,s}}{v_{2,r}}\right)^{1/3} - \frac{v_{2,s}}{2v_{2,r}}\right) \tag{15}$$

For the cationic hydrogels prepared in the presence of a solvent, the equilibrium swelling is described by the following expression:

$$\frac{V_1}{4I}\left(\frac{v_{2,s}^2}{v}\right)\left(\frac{K_b}{10^{pH-14} + K_a}\right)^2 = (ln(1 - v_{2,s}) + v_{2,s} + \chi_1 v_{2,s}^2)$$
$$+ \left(\frac{V_1}{v\overline{M}_c}\right)\left(1 - \frac{2\overline{M}_c}{\overline{M}_n}\right)v_{2,r}$$
$$\cdot \left(\left(\frac{v_{2,s}}{v_{2,r}}\right)^{1/3} - \frac{v_{2,s}}{2v_{2,r}}\right) \tag{16}$$

For the case of ionic hydrogels, the molecular weight between cross-links can be calculated by performing swelling experiments and applying equations 15 (anionic gels) or 16 (cationic gels).

### Rubber Elasticity Theory

Hydrogels are similar to natural rubbers in that they have the ability to respond to applied stresses in a nearly instantaneously and almost irreversible manner (18,28). These polymer networks have the ability to deform readily under low stresses. Also, following small deformations (less than 20%) most gels can fully recover from the deformation in a rapid fashion. Under these conditions, the behavior of the gels can be approximated to be elastic. This property can be exploited to calculate the cross-linking density or molecular weight between cross-links for a particular gel.

The elastic behavior of cross-linked polymers has been analyzed using classical thermodynamics, statistical thermodynamics, and phenomenological models. Based on classical thermodynamics, the equation of state for rubbers can be expressed as (18,28):

$$f = \left(\frac{\partial U}{\partial L}\right)_{T,V} + T\left(\frac{\partial f}{\partial T}\right)_{L,V} \tag{17}$$

where $f$ is the retractive force of the elastic polymer in response to an applied load, $U$ is the internal energy and $L$ is the length of the sample. For an ideal elastomer, $(\partial U/\partial L)_{T,V}$ is zero as deformation of the sample does not result in changes in the internal energy as the bonds are not stretched.

The connection between the classical and statistical thermodynamics can be made after applying the following Maxwell equation:

$$\left(\frac{\partial S}{\partial L}\right)_{T,V} = -\left(\frac{\partial f}{\partial T}\right)_{L,V} \tag{18}$$

By combining equations 17 and 18, the retractive force of an ideal elastomer can be written as:

$$f = -T\left(\frac{\partial S}{\partial L}\right)_{T,V} = -kT\left(\frac{\partial\, ln\Omega(r,\,T)}{\partial r}\right)_{T,V} \tag{19}$$

Here, $r$ is the end-to-end distance of the polymer chain and $\Omega$ is the probability that a polymer chain of length $r$ will have a specific conformation at a given temperature ($T$).

Upon evaluation of this equation for a system of $n$ chains, the change in the Helmholtz free energy upon elastic deformation can be written as:

$$(\varDelta A)_{elastic} = \frac{3nKT}{\bar{r}_f^2} \int_{\bar{r}_f}^{\bar{r}} r\, dr \tag{20}$$

Assuming that elastic deformation is not accompanied by a volume change, integration yields the classical equation for rubber elasticity.

$$\frac{f}{a} = \tau = \frac{\rho RT}{\overline{M}_c} \frac{\bar{r}_o^2}{\bar{r}_f^2}\left(\alpha - \frac{1}{\alpha^2}\right) \tag{21}$$

Here, $a$ is the cross-sectional area, $\tau$ is the stress, $\rho$ is the density of the polymer, $\alpha$ is the elongation ratio and $\bar{r}_o^2/\bar{r}_f^2$ is the front factor. The front factor is the ratio of the end-to-end distance of polymer chains in a real network and dilute chains in solution. Typically, this factor can be assumed less than or equal to one.

The previous expression was derived assuming an ideal network without any defects. In the ideal networks, all of the chains would participate equally to the elastic stress. However, in real networks, defects exist such as entanglements or dangling chain ends. After correcting for the case of real networks and assuming a front factor of one, the equation of state for rubber elasticity becomes:

$$\tau = \frac{\rho RT}{\overline{M}_c}\left(1 - \frac{2\overline{M}_c}{\overline{M}_n}\right)\left(\alpha - \frac{1}{\alpha^2}\right) \tag{22}$$

A similar analysis can be applied to swollen hydrogels where the polymer volume fraction in the network is less than unity. For the case of swollen gels, the number of cross-links per unit volume is different than for the case of unswollen polymer networks. The equation of state for real, swollen polymer gels can be expressed as:

$$\tau = \frac{\rho RT}{\overline{M}_c}\left(1 - \frac{2\overline{M}_c}{\overline{M}_n}\right)\left(\alpha - \frac{1}{\alpha^2}\right)v_{2,s}^{1/3} \tag{23}$$

This equation holds for networks cross-linked in the absence of any diluents. For the case of a swollen network cross-linked in the presence of a solvent, the equation of state can be written as (29):

$$\tau = \frac{\rho RT}{\overline{M}_c}\left(1 - \frac{2\overline{M}_c}{\overline{M}_n}\right)\left(\alpha - \frac{1}{\alpha^2}\right)\left(\frac{v_{2,s}}{v_{2,r}}\right)^{1/3} \tag{24}$$

For elongation of a polymer network along a single axis, the stress is inversely proportional to the molecular weight between cross-links in the polymer network. From mechanical analysis under short deformations, important structural information about the polymer networks can be obtained (30–32).

### Network Pore Size Calculation

One of the most important parameters in controlling the rate release of a drug from a hydrogel is the network pore size, $\xi$. The pore size can be determined theoretically or using a number of experimental techniques. A direct technique for measuring this parameter is quasielastic laser-light scattering (33) or electron microscopy. Some indirect experimental techniques for determination of the hydrogel pore size include mercury porosimetry (34,35), rubber elasticity measurements (30), or equilibrium swelling experiments (16,36). However, the indirect experiments allow for calculation of the porous volume (mercury porosimetry) or the molecular weight between cross-links (rubber elasticity analysis or swelling experiments).

Based on values for the cross-linking density or molecular weight between cross-links, the network pore size can be determined by calculating end-to-end distance of the swollen polymer chains between cross-linking points (16,18,36).

$$\xi = \alpha(\bar{r}_o^2)^{1/2} \tag{25}$$

In this expression, $\alpha$ is the elongation of the polymer chains in any direction and $(\bar{r}_o^2)^{1/2}$ is the unperturbed end-to-end distance of the polymer chains between cross-linking points. Assuming isotropic swelling of the gels, the elongation is related to the swollen polymer volume fraction.

$$\alpha = v_{2,s}^{-1/3} \tag{26}$$

The unperturbed chain distance is calculated through the Flory characteristic ratio, $C_n$.

$$(\bar{r}_o^2)^{1/2} = l(C_nN)^{1/2} \qquad (27)$$

Here, $l$ is the length of the bond along the backbone chain (1.54 Å for vinyl polymers). The number of links per chain, $N$, is related to the cross-link density as:

$$N = \frac{2\overline{M}_c}{M_o} \qquad (28)$$

Here, $M_o$ is the molecular weight of the monomeric repeating units. Upon combination of these equations, the pore size of a swollen polymeric network can be calculated using the following equation:

$$\xi = \left(\frac{2C_n\overline{M}_c}{M_o}\right)^{1/2} l v_{2,s}^{-1/3} \qquad (29)$$

## DIFFUSION IN HYDROGELS

The release of drugs and other solutes from hydrogels results from combination of classical diffusion in the polymer network and mass transfer limitations (2). To optimize a hydrogel system for a particular application, the fundamental mechanism of solute transport in the membranes must be understood completely. In this section, we focus on the mechanism of drug diffusion in hydrogels as well as the importance of network morphology in controlling the transport of drugs in hydrogels.

### Macroscopic Analysis

The transport or release of a drug through a polymeric controlled release device can be described by classical Fickian diffusion theory (2,37). This theory assumes that the governing factor for drug transport in the gels is ordinary diffusion. Drug delivery devices can be designed so that other mechanisms control the release rate such as gel swelling or polymer erosion.

For the case of one-dimensional transport, Fick's law can be expressed as (38):

$$J_i = -D_{ip}\frac{dC_i}{dX} \qquad (30)$$

Here, $J_i$ is the molar flux of the drug (mol/cm$^2$ s), $C_i$ is the concentration of drug and $D_{ip}$ is the diffusion coefficient of the drug in the polymer. For the case of a steady-state diffusion process, i.e., constant molar flux, and constant diffusion coefficient, equation 30 can be integrated to give the following expression:

$$J_i = K\frac{D_{ip}\Delta C_i}{\delta} \qquad (31)$$

Here, $\delta$ is the thickness of the hydrogel and $K$ is the partition coefficient, defined as

$$K = \frac{\text{drug concentration in gel}}{\text{drug concentration in solution}} \qquad (32)$$

For many drug delivery devices, the release rate will be time dependent. For non–steady-state diffusion problems,

Fick's second law is used to analyze the release behavior. Fick's second law is written as:

$$\frac{\partial C_i}{\partial t} = \frac{\partial}{\partial X}\left(D_{ip}\frac{\partial C_i}{\partial X}\right) \qquad (33)$$

This form of the equation is for one-dimensional transport with nonmoving boundaries. Equation 33 can be evaluated for the case of constant diffusion coefficients and concentration-dependent diffusion coefficients.

**Constant Diffusion Coefficients.** Equation 33 can be solved by application of the appropriate boundary conditions. Most commonly, perfect-sink conditions are assumed. Under these conditions, the following boundary and initial conditions are applicable (Fig. 4):

$$t = 0, \quad X < \pm\frac{\delta}{2}, \quad C_i = C_o \qquad (34a)$$

$$t > 0, \quad X = 0, \quad \frac{\partial C_i}{\partial X} = 0 \qquad (34b)$$

$$t > 0, \quad X = \pm\frac{\delta}{2}, \quad C_i = C_s \qquad (34c)$$

Here, $C_o$ is the initial drug concentration inside the gel and $C_s$ is the equilibrium bulk concentration. Upon application of the boundary conditions, Fick's second law for planar geometries can be integrated and the solution can be written in terms of the amount of drug released at a given time, $M_t$, normalized to the amount released at infinite times, $M_\infty$ (38).

$$\frac{M_t}{M_\infty} = 1 - \sum_{n=0}^{\infty} \frac{8}{(2n+1)^2\pi^2} exp\left[\frac{-D_{ip}(2n+1)^2\pi^2t}{\delta^2}\right] \qquad (35)$$

Alternatively, the solution to equation 33 can be written in terms of the integrated complimentary error function, $ierfcx$ (39).

$$\frac{M_t}{M_\infty} = 4\left(\frac{D_{ip}t}{\delta}\right)^{1/2}\left[\frac{1}{\pi^{1/2}} + 2\sum_{n=0}^{\infty}(-1)^n ierfc\frac{n\delta}{2(D_{ip}t)^{1/2}}\right] \qquad (36)$$

At short times, equation 36 can be approximated as:

$$\frac{M_t}{M_\infty} = 4\left(\frac{D_{ip}t}{\pi\delta}\right)^{1/2} \qquad (37)$$

**Concentration-Dependent Diffusion Coefficients.** In most systems, the drug diffusion coefficient is dependent on the

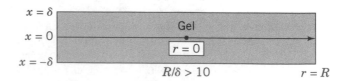

**Figure 4.** Depiction of the slab geometry used for one-dimensional analysis of Fick's second law.

drug concentration as well as the concentration of the swelling agent. In order to analyze the diffusive behavior of drug delivery systems when this is the case, one must choose an appropriate relationship between the diffusion coefficient and the drug concentration. Based on free-volume theory, Fujita (40) proposed the following relationship between the diffusion coefficient and the drug concentration in the gel.

$$D_{ip} = D_{iw} \, exp[-\beta(C_i - C_o)] \qquad (38)$$

Here, $D_{iw}$ is the diffusion coefficient in the pure solution, $\beta$ is a constant dependent on the system, and $C_o$ is the concentration of drug in solution. Additionally, a similar equation was written to relate the diffusion coefficient to the concentration of the swelling agent and the drug in the gel.

$$D_{ip} = D_{iw} \, exp[-\beta(C_s - C_i)] \qquad (39)$$

Here, $C_s$ is the swelling agent concentration.

### Effects of Network Morphology

The structure and morphology of a polymer network significantly affects the ability of a drug to diffuse through a hydrogel. For all types of release systems, the diffusion coefficient (or effective diffusion coefficient) of solutes in the polymer is dependent on a number of factors such as the structure and pore size of the network, the polymer composition, the water content, and the nature and size of the solute. Perhaps the most important parameter in evaluating a particular device for a specific application is the ratio of the hydrodynamic radius of the drug, $d_h$, to the network pore size, $\xi$ (Figure 5). Accordingly, hydrogels for controlled release applications are classified according to their pore size (3). The transport properties of drugs in each type of gel vary according to the structure and morphology of the network.

**Macroporous Hydrogels.** Macroporous hydrogels have large pores, usually between 0.1 and 1 $\mu$m. Typically, the pores of these gels are much larger than the diffusing species. In the case of these membranes, the pores are sufficiently large so that the solute diffusion coefficient can be described as the diffusion coefficient of the drug in the water-filled pores. The process of solute transport is hin-

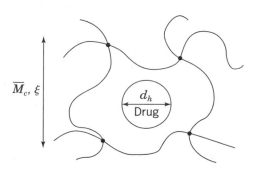

**Figure 5.** The effects of molecular size ($r_s$) on the diffusion of a solute in a network of pore size $\xi$.

dered by the presence of the macromolecular mesh. The solute diffusion coefficient can be characterized in terms of the diffusion coefficient of the solute in the pure solvent ($D_{iw}$) as well as the network porosity ($\epsilon$) and tortuosity ($\tau$). Additionally, the manner in which the solute partitions itself within the pore structure of the network affects the diffusion of the drug. This phenomenon is described in terms of the partition coefficient, $K_p$. These parameters can be incorporated to describe the transport of the drug in the membranes in terms of an effective diffusion coefficient ($D_{eff}$).

$$D_{eff} = D_{iw} \frac{K_p \epsilon}{\tau} \qquad (40)$$

**Microporous Hydrogels.** These membranes have pore sizes between 100 and 1000 Å. In these gels, the pores are water filled and drug transport occurs due to a combination of molecular diffusion and convection in the water-filled pores. In these gels, significant partitioning of the solute within the pore walls may occur for systems in which the drug and polymer are thermodynamically compatible. The effective diffusion coefficient can be expressed in a form similar to that for macroporous membranes.

Transport in microporous membranes is different from macroporous membranes because the pore size begins to approach the size of the diffusing solutes. Numerous researchers have attempted to describe transport for the case of the solute pore size being approximately equal to the network pore size. The rate of the diffusion coefficient in the membrane and pure solvent can be related to the ratio of the solute and pore radius.

$$\frac{D_{ip}}{D_{iw}} = (1 - \lambda^2)(1 - 2.104\lambda + 2.09\lambda^3 - 0.95\lambda^5) \qquad (41)$$

In this expression, the diffusion coefficient in the membrane, $D_{ip}$, is related to $\lambda$, the ratio of the solute radius ($r_s$) to the pore radius ($r_p$).

$$\lambda = \frac{r_s}{r_p} \qquad (42)$$

This theory was initially proposed by H. Faxen in 1923; however, other researchers have provided more recent corrections to this theory (41–43). An excellent review of all of these diffusional theories has been provided by Deen (44).

**Nonporous Hydrogels.** Nonporous gels have molecular sized pores equal to the macromolecular correlation length, $\xi$ (between 10 and 100 Å). Gels of this type are typically formed by the chemical or physical cross-linking of the polymer chains. In these gels, the polymer chains are densely packed and serve to severely limit solute transport. Additionally, the cross-links serve as barriers to diffusion. This distance between the physical obstructions is known as the mesh size. Transport of solutes in these membranes occurs only by diffusion.

The macromolecular mesh of nonporous membranes is not comparable to the pore structure microporous gels.

Therefore, the theories developed for the microporous membranes are nonapplicable to nonporous gels. The diffusional theories developed for nonporous membranes are based on the concept of free-volume. The free-volume is the area within the gel not occupied by the polymer chains (45). Diffusion of solutes in nonporous membranes will occur within this free volume.

The first theory describing transport in nonporous gels was presented by Yasuda et al. (46). This theory relates the normalized diffusion coefficient, the ratio of the diffusion coefficient of the solute in the membrane ($D_{2,13}$) to the diffusion coefficient of the solute in the pure solvent ($D_{2,1}$), to degree of hydration of the membrane, $H$ (g water/g swollen gel). The subscripts 1, 2, and 3 represent the solvent or water, solute, and the polymer. The normalized diffusion coefficient can be written as

$$\frac{D_{2,13}}{D_{2,1}} = \varphi(q_s)exp\left[-B\left(\frac{q_s}{V_{f,1}}\right)\left(\frac{1}{H} - 1\right)\right] \quad (43)$$

where $V_{f,1}$ is the free-volume occupied by the water, $\varphi$ is a sieving factor which provides a limiting mesh size below which solutes of cross-sectional area $q_s$ cannot pass, and $B$ is a parameter characteristic of the polymer. Based on this theory, a permeability coefficient for the drug in the swollen membrane, $P_{2,13}$ is given by

$$P_{2,13} = \frac{D_{2,13}K}{\delta} \quad (44)$$

Yasuda and Lamaze (47) confirmed this theory experimentally using urea, creatinin and secobarbital.

Peppas and Reinhart (48) also developed a theoretical model to describe solute transport in highly swollen, nonporous hydrogels. In this description, the diffusional jump lengths of the solute were assumed the same in the gel and the pure solvent. Additionally, the free-volume of the hydrogel was taken to be the same as the free-volume of the pure solvent. Using this approach, the normalized diffusion coefficient can be described in terms of the degree of swelling and the molecular weight of the polymer chains as

$$\frac{D_{2,13}}{D_{2,1}} = k_1\left(\frac{\overline{M}_c - \overline{M}_c^*}{\overline{M}_n - \overline{M}_c^*}\right)exp\left(-\frac{k_2 r_s^2}{Q - 1}\right) \quad (45)$$

where $k_1$ and $k_2$ are parameters based on the polymer structure, $Q$ is the degree of swelling (g swollen polymer/g dry polymer), $r_s$ is the solute radius, $\overline{M}_c$ is the molecular weight of the polymer chains between cross-links, $\overline{M}_n$ is the molecular weight of linear polymer chains prepared using the same conditions in the absence of a crosslinking agent and $\overline{M}_c^*$ is the critical molecular weight between cross-links below which a solute of size $r_s$ could not diffuse. In this depiction, the term ($\overline{M}_c - \overline{M}_c^*/\overline{M}_n - \overline{M}_c^*$) is comparable to the sieving factor ($\varphi$) presented by Yasuda. These researchers experimentally verified the dependence of the diffusion coefficient on the solute size and cross-linking density (49). Recently, the group of Prausnitz has studied the size exclusion effects in highly swollen networks (50).

Peppas and Moynihan (51) developed a theory for the case of moderately or poorly swollen, nonporous networks. In this case, the diffusional jump length of the solute in the membrane ($\lambda_{2,13}$) does not equal the diffusional jump length of the solute in the pure solvent ($\lambda_{2,13}$) and the free-volume of the gel does not equal the free-volume of the pure solvent. In this model, the normalized diffusion coefficient is written as

$$\frac{D_{2,13}}{D_{2,1}} = f(v_{2,s}^{-3/4})exp[k_3(\overline{M}_c - \overline{M}_n) - \pi r_s^2 l_s\Phi(V)] \quad (46)$$

where $f(v_{2,s}^{-3/4})$ is a parameter representing the characteristic size of the diffusional area, $k_3$ is a structural parameter and $l_s$ is the length of the solute. The free-volume contributions, $\Phi(V)$, are expressed as

$$\Phi(V) = \frac{V_1 - V_3}{(Q - 1)V_1^2 + V_1 V_3} \quad (47)$$

where $V_1$ and $V_3$ are the free-volumes of the polymer and the pure solvent.

Harland and Peppas (52) developed a theory for diffusion in a semicrystalline polymer. In this type of gel, the diffusion occurs only in the pores between the impermeable crystalline regions. The diffusion coefficient of the solute in the crystalline phase ($D_c$) can be expressed in terms of the diffusion coefficient in the amorphous phase ($D_a$) as

$$D_c = \frac{D_a(1 - v_c)}{\tau} \quad (48)$$

where $\tau$ is the tortuosity and $v_c$ is the volume fraction of the crystalline region.

For the case of semicrystalline hydrogels, the crystallites present the greatest obstacle to diffusion. Therefore, the diffusion coefficient for the gel in the swollen membrane is assumed to be equal to the diffusion coefficient of the solute in the crystalline regions. For the case of moderately or poorly swollen networks, the normalized diffusion coefficient can be written as

$$\frac{D_{2,13}}{D_{2,1}} = \frac{(1 - v_c)}{\tau} f(v_{2,s}^{-3/4})exp[k_3(\overline{M}_c - \overline{M}_n) - \pi r_s^2 l_s\Phi(V)] \quad (49)$$

For highly swollen networks, the normalized diffusion coefficient can be expressed as

$$\frac{D_{2,13}}{D_{2,1}} = \frac{(1 - v_c)}{\tau}\left(\frac{\overline{M}_c - \overline{M}_c^*}{\overline{M}_n - \overline{M}_c^*}\right)exp\left(-\frac{\pi r_s^2 l v_a}{V_w v_s}\right) \quad (50)$$

where $v_a$ is the volume fraction of the amorphous region, $v_s$ is the volume fraction of the solvent and $V_w$ is the volume fraction of the water.

### Experimental Determination of Diffusion Coefficients

**Membrane Permeability Experiments.** The membrane permeation method is used to study the diffusion coefficients of solutes through thin membranes. The drug permeates through an equilibrium-swollen membrane from a reservoir containing a high concentration of drug (donor cell) to a reservoir containing a lower concentration (receptor).

In these experiments, the drug concentration should be monitored over time in the receptor cell. The solute permeability, $P$, can be determined from the following expression.

$$ln\left(\frac{2C_o}{C_t} - 1\right) = \frac{2A}{Vl}Pt \qquad (51)$$

In this expression, $C_o$ is the donor cell concentration (initially), $C_t$ is the time-dependent receptor cell concentration, $A$ is the cross-sectional area of the membrane, $V$ is the volume of the cells and $l$ is the membrane thickness. A plot of $(Vl/A)ln(2C_o/C_t - 1)$ versus time yields a straight line of slope $P$ if the gel remains at equilibrium.

The diffusion coefficient is related to the partition coefficient by equation 44. The partition coefficient can be determined experimentally through equilibrium partitioning studies. In this type of experiment, hydrogels are swollen to equilibrium in drug solutions of concentration $C_o$. Once equilibrium has been reached, the partition coefficient is calculated as

$$K = -\frac{C_{m,eq}}{C_{s,eq}} \qquad (52)$$

where $C_{m,eq}$ is the equilibrium concentration of drug in the membrane and $C_{s,eq}$ is the equilibrium concentration of the drug in the solution.

**Controlled Release Experiments.** Another relatively easy technique for determination of diffusion coefficients is to measure the release of drug release from thin hydrogels (aspect ratio >10). In release experiments, the membranes containing the dispersed drug are placed in drug-free solutions and the concentration of the drug in the solutions is monitored over time. It is recommended that nearly perfect sink conditions be maintained throughout the experiment.

The diffusion coefficient can be found by fitting the release data to equation 37. This equation is valid for the initial 60% of the total amount released. This technique is most accurate for systems in which diffusion is the dominant mechanism for drug release. In order to determine whether a particular device is diffusion controlled, the early time release data can be fit to the following empirical relationship proposed by Ritger and Peppas (39).

$$\frac{M_t}{M_\infty} = kt^n \qquad (53)$$

The constants, $k$ and $n$, are characteristics of the drug–polymer system. The diffusional exponent, $n$, is dependent on the geometry of the device as well as the physical mechanism for release. For classical Fickian diffusion, Ritger and Peppas (39) determined that value for the diffusional

exponent was 0.5 for slab geometries, 0.45 for cylindrical devices, and 0.43 for spherical devices.

By determining the diffusional exponent, $n$, one can gain information about the physical mechanism controlling drug release from a particular device. Based on the diffusional exponent (53), the drug transport in a slab geometry is classified as Fickian diffusion ($n = 0.5$), Case II transport ($n = 1$), non-Fickian or anomalous transport ($0.5 < n < 1$) and Super Case II transport (Table 1). Representative release curves for each case are shown in Figure 6. For systems exhibiting Case II transport, the dominant mechanism for drug transport is due to polymer relaxation as the gels swells. These types of devices, known as swelling-controlled release systems, are described in more detail later. Anomalous transport occurs due to a coupling of Fickian diffusion and polymer relaxation.

Berens and Hofenberg (54) proposed the following model to describe the release behavior of dynamically swelling hydrogels.

$$\frac{M_t}{M_\infty} = k_1t + k_2t^{1/2} \qquad (54)$$

This expression describes the release rates in terms of relaxation-controlled transport process, $k_1t$, and the diffusion-controlled process, $k_2t^{1/2}$.

**Other Techniques.** There are several other methods for experimental determination of diffusion coefficients in polymeric systems. These techniques involve more complex instrumentation techniques that cannot be described in depth in this work. These techniques include NMR spec-

**Table 1. Drug Transport Mechanisms in Hydrogel Slabs**

| Diffusional exponent, $n$ | Type of transport | Time dependence |
|---|---|---|
| 0.5 | Fickian diffusion | $t^{1/2}$ |
| $0.5 < n < 1$ | Anomolous transport | $t^{n-1}$ |
| 1 | Case II transport | Time independent |
| $n > 1$ | Super Case II transport | $t^{n-1}$ |

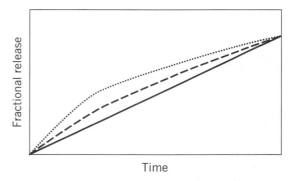

**Figure 6.** Comparison of the release behavior of systems exhibiting ($\cdots$) classical Fickian diffusion behavior, (— —) anamolous release behavior, and (——) zero-order release or Case II transport.

troscopy (55), scanning electron microscopy (56), Fourier transform infrared spectroscopy (57), and quasielastic laser-light scattering (33).

## CLASSIFICATIONS

Because of their nature, hydrogels can be used in many different types of controlled release systems. These systems are classified according to the mechanism controlling the release of the drug from the device. Hydrogel-based drug delivery systems are classified as diffusion-controlled systems, swelling-controlled systems, chemically controlled systems, and environmentally responsive systems (2). In this section, the mechanism of drug release in each type of system is described.

### Diffusion-Controlled Release Systems

Diffusion is the most common mechanism controlling release in hydrogel-based drug delivery system. There are two major types of diffusion-controlled systems; reservoir devices and membrane devices. Drug release from each type of system occurs by diffusion through the macromolecular mesh or through the water-filled pores.

**Reservoir Systems.** Reservoir systems consist of a polymeric membrane surrounding a core containing the drug (Fig. 7). Typically, reservoir devices are capsules, cylinders, slabs, or spheres. The rate-limiting step for drug release is diffusion through the outer membrane of the device. For a device with a membrane thickness of $\delta$, the molar flux of drug leaving the device is described by equation 31.

To maintain a constant release rate or flux of drug from the reservoir, the concentration difference must remain constant. This can be achieved by designing a device with excess solid drug in the core. Under these conditions, the internal solution in the core remains saturated. This type of device is an extremely useful device as it allows for time-independent or zero-order release.

The major drawback of this type of drug delivery system is the potential for catastrophic failure of the device. In the event that the outer membrane ruptures, the entire content of the device are delivered nearly instantaneously. When preparing these devices, care must be taken to ensure that the device does not contain pinholes or other defects that may lead to rupture.

**Matrix Systems.** In matrix devices, the drug is dispersed throughout the three-dimensional structure of the hydrogel (Fig. 8). Release occurs due to diffusion of the drug throughout the macromolecular mesh or water-filled pores. The fractional release from a one-dimensional device can be modeled using equation 33. In these systems, the release rate is proportional to time to the one-half power. This is significant in that it is impossible to obtain time independent or zero-order release in this type of system with simple geometries.

Drug can be incorporated into the gels by equilibrium partitioning, where the gel is swollen to equilibrium in concentrated drug solution, or during the polymerization reaction. Equilibrium partitioning is the favorable loading method for drug–polymer systems with large partition coefficients or for sensitive macromolecular drugs such as peptides or proteins that could be degraded during the polymerization.

### Swelling-Controlled Release Systems

In swelling-controlled release systems, the drug is dispersed within a glassy polymer. Upon contact with biological fluid, the polymer begins to swell. No drug diffusion occurs through the polymer phase. As the penetrant enters the glassy polymer, the glass transition temperature of the polymer is lowered allowing for relaxations of the macromolecular chains. The drug is able to diffuse out of the swollen, rubbery area of the polymers. This type of system is characterized by two moving fronts: the front separating the swollen (rubbery) portion and the glassy regions which moves with velocity, $v$, and the polymer–fluid interface (Fig. 9). The rate of drug release is controlled by the velocity and position of the front dividing the glassy and rubbery portions of the polymer.

For true swelling-controlled release systems, the diffusional exponent, $n$, is 1. This type of transport is known as Case II transport and results in zero-order release kinetics. However, in some cases, drug release occurs due to a combination of macromolecular relaxations and Fickian diffusion. In this case, the diffusional exponent is between 0.5 and 1. This type of transport is known as anomalous or non-Fickian transport. A complete mathematical treatment of this type of release behavior has been provided in two excellent reviews (37,53).

### Chemically Controlled Release Systems

There are two major types of chemically controlled release systems: erodible drug delivery systems and pendent chain systems (58,59). In erodible systems, drug release occurs due to degradation or dissolution of the hydrogel. In pendent chain systems, the drug is affixed to the polymer back-

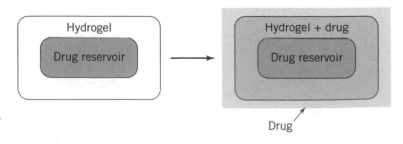

**Figure 7.** Schematic depiction of drug-release from a hydrogel-based reservoir delivery system.

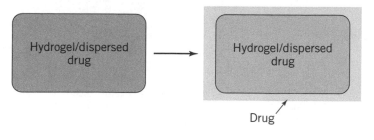

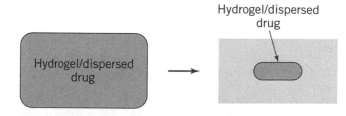

**Figure 8.** Schematic depiction of drug-release from a hydrogel-based matrix delivery system.

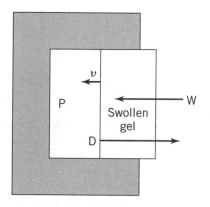

**Figure 9.** Schematic representation of the behavior of a one-dimensional swelling controlled release system. The water (W) penetrates the glassy polymer (P) to form a gel. The drug (D) is released through the swollen layer.

**Figure 10.** Schematic diagram of drug release from a hydrogel-based erodible delivery system.

bone through degradable linkages. As these linkages degrade, the drug is released.

**Erodible Drug Delivery Systems.** Erodible drug delivery systems, also known as degradable or absorbable release systems, can be either matrix or reservoir delivery systems. In reservoir devices, an erodible membrane surrounds the drug core. If the membrane erodes significantly after the drug release is complete, the dominant mechanism for release would be diffusion. Predictable, zero-order release could be obtained with these systems. In some cases, the erosion of the membrane occurs simultaneously with drug release. As the membrane thickness decreased due to erosion, the drug delivery rate would also change. Finally, in some erodible reservoir devices, the drug diffusion in the outer membrane does not occur. Under these conditions, drug release does not occur until the outer membrane erodes completely. In this type of device, the entire contents are released in a single, rapid burst. These types of systems have been investigated for use as enteric coatings.

For erodible matrix devices, the drug is dispersed within the three-dimensional structure of the hydrogel. Drug release is controlled by drug diffusion through the gel or erosion of the polymer. In true erosion-controlled devices, the rate of drug diffusion is significantly slower than the rate of polymer erosion, and the drug is released as the polymer erodes (Fig. 10).

In erodible system, there are three major mechanisms for erosion of the polymer. The first mechanism for erosion is the degradation of the cross-links. This degradation can

occur by hydrolysis of water-labile linkages, enzymatic degradation of the junctions, or dissolution of physical cross-links such as entanglements or crystallites in semi-crystalline polymers. The second mechanism for erosion is solubilization of insoluble or hydrophobic polymers. This could occur as a result of hydrolysis, ionization, or protonation of pendent groups along the polymer chains. The final mechanism of erosion is the degradation of backbone bonds to produce small molecular weight molecules. Typically, the degradation products are water soluble. This type of erosion can occur by hydrolysis of water-labile backbone linkages or by enzymatic degradation of backbone linkages. The most commonly studied erodible polymer systems are poly(lactic acid) (PLA), poly(glycolic acid) (PGA), and copolymers of PLA and PGA.

**Pendent Chain Systems.** Pendent chain systems consist of linear homo- or copolymers with the drug attached to the backbone chains. The drug is released from the polymer by hydrolysis or enzymatic degradation of the linkages (Fig. 11). Zero-order release can be obtained with these systems, provided that the cleavage of the drug is the rate-controlling mechanism.

### Environmentally Responsive Systems

Hydrogels may exhibit swelling behavior dependent on the external environment. Over the last 30 years there has been a significant interest in the development and analysis of environmentally or physiologically responsive hydrogels (10). Environmentally responsive materials show drastic changes in their swelling ratio due to changes in their external pH, temperature, ionic strength, nature and composition of the swelling agent, enzymatic or chemical reaction, and electrical or magnetic stimulus (60). In most responsive networks, a critical point exists at which this transition occurs.

Responsive hydrogels are unique in that there are many different mechanisms for drug release and many different

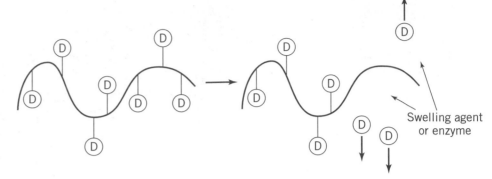

**Figure 11.** Schematic diagram of the release of drug from a pendent chain system due to scission of the bonds connecting the drug to the polymer backbone.

types of release systems based on these materials. For instance, in most cases drug release occurs when the gel is highly swollen or swelling and is typically controlled by gel swelling, drug diffusion, or a coupling of swelling and diffusion. However, in a few instances, drug release occurs during gel syneresis by a squeezing mechanism. Also, drug release can occur due to erosion of the polymer caused by environmentally responsive swelling.

Another interesting characteristic about many responsive gels is that the mechanism causing the network structural changes can be entirely reversible in nature. This behavior is depicted in Figure 12 for a pH- or temperature-responsive gel. The ability of these materials to exhibit rapid changes in their swelling behavior and pore structure in response to changes in environmental conditions lends these materials favorable characteristics as carriers for bioactive agents, including peptides and proteins. This type of behavior may allow these materials to serve as self-regulated, pulsatile drug delivery systems. This type of behavior is shown in Figure 13 for pH- or temperature-responsive gels. Initially, the gel is in an environment in which no swelling occurs. As a result, very little drug release occurs. However, when the environment changes and the gel swells, rapid drug release occurs (either by Fickian diffusion, anomalous transport, or Case II transport). When the gel collapses as the environment changes, the release can be turned off again. This can be repeated over numerous cycles. Such systems could be of extreme importance in the treatment of chronic diseases such as diabetes. Peppas (60) and Siegel (61) have presented detailed analyses of this type of behavior.

**pH-Sensitive Hydrogels.** One of the most widely studied types of physiologically responsive hydrogels is pH-responsive hydrogels. These hydrogels are swollen from ionic networks. These ionic networks contain either acidic or basic pendent groups. In aqueous media of appropriate pH and ionic strength, the pendent groups can ionize developing fixed charges on the gel as shown in Figure 2 for an ionic gel. The swelling behavior of these materials has been analyzed in a previous section.

There are many advantages to using ionic materials over neutral networks. All ionic materials exhibit a pH and ionic strength sensitivity. The swelling forces developed in these systems are increased over the nonionic materials. This increase in swelling force is due to the localization of fixed charges on the pendent groups. As a result, the mesh size of the polymeric networks can change significantly with small pH changes. In these materials, the drug diffusion coefficients and release rates vary greatly with environmental pH.

**Temperature-Sensitive Hydrogels.** Another class of environmentally sensitive materials that are being targeted for use in drug delivery applications is thermally sensitive polymers. This type of hydrogel exhibits temperature-sensitive swelling behavior due to a change in the polymer–swelling agent compatibility over the temperature range of interest. Temperature-sensitive polymers typically exhibit a lower critical solution temperature (LCST), below which the polymer is soluble. Above this temperature, the polymers are typically hydrophobic and do not swell significantly in water (62). However, below the LCST, the cross-linked gel swell to significantly higher degrees because of the increased compatibility with water. For polymers that exhibit this sort of swelling behavior, the rate of drug release would be dependent on the temperature. The highest release rates would occur when the temperature of the environment is below the LCST of the gel.

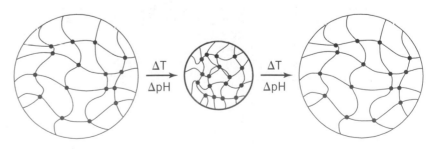

**Figure 12.** Swollen temperature- and pH-sensitive hydrogels may exhibit an abrupt change from the expanded (left) to the collapsed (syneresed) state (center) and then back to the expanded state (right) as temperature and pH change.

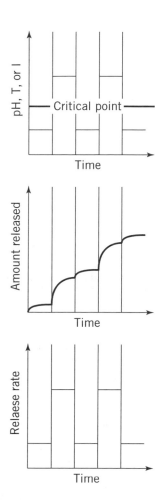

**Figure 13.** Cyclic change of pH, T, or ionic strength (I) leads to abrupt changes in the drug release rates at certain time intervals in some environmentally responsive polymers.

**Complexing Hydrogels.** Some hydrogels may exhibit environmental sensitivity due to the formation of polymer complexes. Polymer complexes are insoluble, macromolecular structures formed by the noncovalent association of polymers with the affinity for one another. The complexes form due to association of repeating units on different chains (interpolymer complexes) or on separate regions of the same chain (intrapolymer complexes). Polymer complexes are classified by the nature of the association as stereocomplexes, polyelectrolyte complexes, and hydrogen-bonded complexes (63). The stability of the associations is dependent on such factors as the nature of the swelling agent, temperature, type of dissolution medium, pH and ionic strength, network composition and structure, and length of the interacting polymer chains.

In this type of gel, complex formation results in the formation of physical cross-links in the gel. As the degree of effective cross-linking is increased, the network mesh size and degree of swelling is significantly reduced. As a result, the rate of drug release in these gels decreases dramatically upon the formation of interpolymer complexes.

**Materials Sensitive to Chemical or Enzymatic Reaction.** More recently, researchers have proposed controlled re-

lease devices that can respond to specific chemicals or enzymes in the body (64,65). Under normal conditions, the structure of the hydrogel would be sufficient to prevent drug release from occurring. However, in the presence of specific substances, chemical or enzymatic reactions could occur that may hydrolyze specific groups of the polymer chain resulting in an increased pore structure and rate of drug release from the gel. If degradation of the gel occurs, the systems cannot be fully reversible.

In some instances, enzymes can be incorporated into the structure of the hydrogel. In the presence of specific chemicals, the enzyme could trigger a reaction which would change the microenvironment of the hydrogel. Changes in the local microenvironment (such as pH or temperature) could lead to gel swelling or collapse. In these situations, the release rates would be altered significantly. This type of system could be completely reversible in nature.

**Magnetically Responsive Systems.** Magnetically responsive systems consist of polymers or copolymers containing magnetic microbeads (66). These systems can be prepared from most polymers; however, the most commonly used copolymer for these systems is the hydrophobic polymer, poly(ethylene-*co*-vinyl acetate) (PEVAc). Such systems are not typically classified as hydrogels because they do not swell to any appreciable degree. The drug release mechanism from a typical magnetic system is shown in Figure 14. The three-dimensional structure of these systems is such that no drug release can occur when no magnetic field is applied. However, when a magnetic field is applied, the microbeads pulsate allowing for the formation of micropores. Additionally, the pulsation of the beads "squeezes" the drug out of the gel through these pores. When the field is removed, drug release is halted rapidly. Such systems have been targeted for use as pulsatile drug delivery vehicles and have shown to exhibit extremely reproducible behavior.

## APPLICATIONS

### Neutral Hydrogels

A major goal in drug delivery is to develop systems that deliver drugs at a constant rate over an extended period. This can be achieved by using release systems in which gel swelling is the controlling mechanism for drug release. Researchers have also attempted to develop constant-release systems by alteration of device geometry and polymer composition. One of the first researchers to use hydrogels for swelling-controlled release was Good (67). In this work, glassy poly(2-hydroxyethyl methacrylate) (PHEMA) containing tripelennamine hydrochloride was swollen in water. The release rate of the solute was non-Fickian, but zero-order release was not obtained. The first such system in which zero-order release was observed was developed by Hopfenberg and Hsu (68). In these systems, cross-linked polystyrene was used to release red dyes into hexane. Other polymers that have been used extensively in controlled release systems include poly(vinyl alcohol) (PVA), poly(*N*-vinyl-2-pyrollidone) (PNVP), poly(ethylene glycol) (PEG), poly(ethylene oxide) (PEO), and PEVAc or

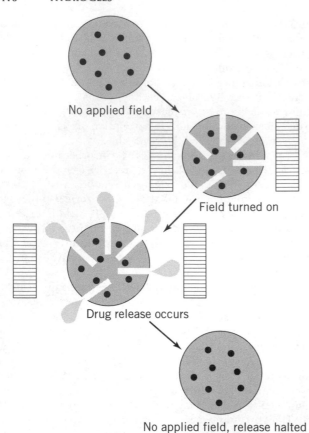

No applied field

Field turned on

Drug release occurs

No applied field, release halted

**Figure 14.** Schematic representation of drug release from a magnetically controlled system. Initially, when the field is off, no drug release occurs. When the field is turned on, the oscillations of the beads creates macropores (●) through which drug release occurs. Drug release is halted when the field is switched off.

copolymers thereof. In this section, we discuss applications of some of these materials.

**Poly(2-hydroxyethyl methacrylate).** PHEMA has been the most widely used polymer in drug delivery applications. It is an extremely hydrophilic polymer and is highly stable. The permeability of these membranes is easily controlled based on the degree of cross-linking used. Researchers have studied the swelling behavior, morphology, and diffusional behavior of PHEMA gels and copolymers thereof. The release of a wide range of drugs from these gels has also been studied. For example, Anderson et al. (69) studied the release behavior of hydrocortisone from PHEMA gels. The release behavior was non-Fickian, but true zero-order release was not achieved with these gels. Another significant study was performed by Sefton and Nishimura (70). They investigated the diffusional behavior of insulin in PHEMA-based hydrogels. Song et al. (71) developed one of the first pharmaceutically relevant zero-order release systems. They used a reservoir device consisting of a cross-linked PHEMA cylinder encapsulating a solution of silicon oil and progesterone. Zero-order release was obtained with these devices for up to 10 days. Lee (72,73) was able to use PHEMA to achieve zero-order re-

lease for short periods of time. In his work, oxprenolol was released from highly cross-linked PHEMA gels at constant rates for up to 3 hours.

The group of Peppas contributed greatly to the understanding of the underlying phenomena of macromolecular relaxations in swelling-controlled release systems. In particular, they studied hydrogels prepared from PHEMA and hydrophobic poly(methyl methacrylate) (PMMA). Franson and Peppas (74) prepared cross-linked copolymer gels of PHEMA-*co*-MAA of varying compositions. Theophylline release was studied and it was found that near zero-order release could be achieved using copolymers containing 90% PHEMA. Further studies by Davidson and Peppas (75) studied the effects of hydrophobicity and cross-link density on the release kinetics and diffusional properties of PHEMA-*co*-MMA membranes. Additionally, Korsmeyer and Peppas (76) examined the behavior of copolymer gels consisting of PHEMA and hydrophilic PNVP (PHEMA-*co*-NVP). In this work, zero-order release of theophylline was observed for up to 5 hours.

**Poly(vinyl alcohol).** Another hydrophilic polymer that has received attention is PVA. This material holds tremendous promise as a biological drug delivery device because it is nontoxic, hydrophilic, and exhibits good mucoadhesive properties (77). In one of the first applications of this material, Langer and Folkman (78) investigated the use of copolymers of PHEMA (Hydron®) and PVA as delivery vehicles for polypeptide drugs.

Two methods exist for the preparation of PVA gels. In the first method, linear PVA chains are cross-linked using glyoxal, gluteraldehyde, or borate. In the second method, pioneered by Peppas (79), semicrystalline gels were prepared by exposing aqueous solutions of PVA to repeated freezing and thawing. The freezing and thawing induced crystal formation in the materials and allowed for the formation of a network structure crosslinked with the quasipermanent crystallites. The latter method is the preferred method for preparation as it allows for the formation of an "ultrapure" network without the use of toxic cross-linking agents.

Korsmeyer and Peppas (80) prepared PVA gels by cross-linking with borate. They studied the swelling behavior and mechanical properties of these gels. In this work, the release rate of theophylline was dependent on the degree of cross-linking. Future studies by Moriomoto et al. (81) examined the release behavior of a wide range of drugs. In this work, they were able to release indomethacin, glucose, insulin, heparin, and albumin from chemically cross-linked PVA gels.

Since the development of the semicrystalline PVA gels by Peppas, significant work has been done in characterizing of these systems. Peppas and Hansen (82) studied the kinetics of crystal formation during the freezing and thawing process. Subsequent work by Urushizaki et al. (83) evaluated the effects of the number of freezing and thawing cycles on the networks properties. The gels became more rigid with increasing number of cycles. Peppas and Stauffer (84) have also investigated the effects of crystallization conditions such as freezing temperature, number of cycles, and freezing time on the structure and properties of the PVA networks.

Studies on the uses of PVA prepared by the freezing and thawing technique as controlled release devices have recently been reported. Peppas and Scott (85) and Ficek and Peppas (86) used PVA gels for the release of bovine serum albumin. In the work of Peppas and Scott, drug release occurred by classical Fickian diffusion. Ficek and Peppas (86) developed a method to prepare novel PVA microparticles containing bovine serum albumin by a freezing and thawing technique. In this work, they were able to release bovine serum albumin using these microparticles.

Other researchers have investigated the use of PVA gels as mucoadhesive delivery devices. The Tsutsumi et al. reported novel buccal delivery systems for ergotamine tartrate (87). Peppas and Mongia (88) have also considered PVA for mucoadhesive drug delivery applications. In this work, they investigated the mucoadhesive behavior of PVA gels prepared by the freezing and thawing technique. Additionally, they studied the release behavior of theophylline and oxprenolol from these materials. Additionally, the group of Peppas reported on the release behavior of ketanserin (89) and metronidazole (90) from these systems.

New phase erosion controlled release systems based on the semicrystalline phase have been reported by Mallapragada and Peppas (90,91). These systems exhibited an unusual molecular control of the drug or protein delivery by a simple dissolution of the carrier. Hydrophilic carriers pass through a process of chain unfolding from the semicrystalline phase to the amorphous one, eventually leading to complete chain disentanglement. It has been shown that PVA and PEG are the best systems for such release behavior, and that such devices have the potential to be used for a wide range of drug delivery applications release. A detailed mathematical analysis has been developed to analyze such swellable systems (91–93).

**Poly(ethylene oxide) and Poly(ethylene glycol).** Hydrogels of PEO and PEG have received significant attention in the last few years in biological drug delivery applications, especially because of their associated stealth characteristics and their protein resistance (94). Three major preparation techniques exist for the preparation of cross-linked PEG networks: (1) chemical crosslinking between PEG chains, (2) radiation cross-linking of PEG chains, and (3) chemical reaction of mono- and difunctional PEGs.

Some of the first chemically cross-linked PEG networks were prepared by McNeill and Graham (95). The cross-linked linear PEG chains used diisocyanates. These gels were used as reservoir devices for the controlled delivery of smaller molecular weight drugs. More recently, McNeill and Graham (96) have investigated the release behavior of small molecular weight solutes from PEG cross-linked with 1,2,6 hexanetriol. The release of proxyphylline from PEG spheres, slabs, and cylinders was studied. For each of the matrix devices, they observed non-Fickian release kinetics. Bromberg (97) also studied the release of chemically cross-linked PEG networks. In this work, PEG networks were cross-linked using tris(6-isocyanatohexyl)-isocyanurate. The kinetics of insulin release from PEG gels obeyed non-Fickian release kinetics.

The advantage of using radiation cross-linked PEO networks is that no toxic cross-linking agents are required. However, it is difficult to control the network structure of these materials. Stringer and Peppas (98) have recently prepared PEO hydrogels by radiation cross-linking. In this work, they analyzed the network structure in detail. Additionally, they investigated the diffusional behavior of smaller molecular weight drugs, such as theophylline, in these gels. Kofinas et al. (99) have prepared PEO hydrogels by a similar technique. In this work, they studied the diffusional behavior of two macromolecules, cytochrome C and hemoglobin, in these gels. They noted an interesting, yet previously unreported dependence between the cross-link density and protein diffusion coefficient and the initial molecular weight of the linear PEGs.

Peppas and coworkers (100–104) have presented exciting new methods for the preparation of PEG gels with controllable structures. In this work, highly cross-linked and tethered PEG gels were prepared from PEG dimethacrylates and monomethacrylates. The diffusional behavior of diltiazem and theophylline in these networks was studied. The technique presented in this work is promising for the development of a new class of functionalized PEG-containing gels that may be of use in a wide variety of drug delivery applications.

### pH-Sensitive Hydrogels

Hydrogels that have the ability to respond to pH changes have been studied extensively over the years. These gels typically contain side ionizable side groups such as carboxylic acids or amine groups (105,106). The most commonly studied ionic polymers include polyacrylamide (PAAm), poly(acrylic acid) (PAA), poly(methacrylic acid) (PMAA), poly(diethylaminoethyl methacrylate) (PDEAEMA), and poly(dimethylaminoethyl methacrylate) (PDMAEMA).

Cationic copolymers based on PDEAEMA and PDMAEMA have been studied by the groups of Peppas and Siegel. The group of Siegel has focused on the swelling and transport behavior of hydrophobic cationic gels. Siegel and Firestone (107,108) studied the swelling behavior of hydrophobic hydrogels of PDMAEMA and PMMA. Such systems were collapsed in solutions of pH greater than 6.6. However, in solutions of pH less than 6.6, such systems swelled due to protonation of the tertiary amine groups. The release of caffeine from these gels was studied (109). No caffeine was released in basic solutions; however, in neutral or slightly acidic solutions steady release of caffeine was observed for 10 days. More recently, Cornejo-Bravo and Siegel (110) have investigated the swelling behavior of hydrophobic copolymers of PDEAEMA and PMMA. Additionally, Siegel (111) has presented an excellent model of the dynamic behavior of ionic gels.

The groups of Peppas has studied the swelling behavior of more hydrophilic, cationic copolymers of P(DEAEMA-co-HEMA) and P(DMAEMA-co-HEMA) (112,113). These gels swelled in solutions of pH less than 7 and were in the collapsed state in basic solutions. These materials swelled to a greater degree than those prepared by Siegel. These materials were used to modulate the release behavior of protein and peptide drugs (114). Schwarte and Peppas (115) studied the swelling behavior of copolymers of PDEAEMA grafted with PEG. The permeability of dextrans of molecular weight 4,400 and 9,400 was studied. The membrane permeabilities in the swollen membranes (pH = 4.6) were two orders of magnitude greater than permeabilities of the collapsed membranes.

Anionic copolymers have received significant attention as well. The swelling and release characteristics of anionic copolymers of PMAA and PHEMA P(HEMA-co-MAA) have been investigated. In acidic media, the gels did not swell significantly; however, in neutral or basic media, the gels swelled to a high degree due to ionization of the pendent acid group (116,117). Brannon-Peppas and Peppas have also studied the oscillatory swelling behavior of these gels (118). Copolymer gels were transferred between acidic and basic solutions at specified time intervals. In acidic solutions, the polymer swelled due to the ionization of the pendent groups. In basic solutions, rapid gel syneresis occurred. Brannon-Peppas and Peppas (119) modeled the time-dependent swelling response to pH changes using a Boltzman superposition-based model. The pH-dependent release behavior of theophylline and proxyphylline from these anionic gels was also studied (120,121). Khare and Peppas (122) studied the pH-modulated release behavior of oxprenolol and theophylline from copolymers of P(HEMA-co-MAA) and P(HEMA-co-AA). In neutral or basic media, the drug release occurred rapidly by a non-Fickian mechanism. The release rate was slowed significantly in acidic media. In another study, am Ende and Peppas (123) examined the transport of ionic drugs of varying molecular weight in P(HEMA-co-AA). They compared experimental results to a free-volume–based theory and found that deviations occurred due to interactions between the ionized backbone chains and pendent acid groups. The swelling and release behavior of interpenetrating polymer networks of PVA and PAA was also investigated (124,125). These materials also exhibit strong pH-responsive swelling behavior. The permeability of these membranes was strongly dependent on the environmental pH and the size and ionic nature of the solute. Recent new studies have used ATR-FTIR spectroscopy to characterize the interactions between polyelectrolytes and solutes (125,126).

Heller et al. (127) studied the behavior of another type of pH-responsive hydrogel. In this work, they evaluated the pH-dependent release of insulin from degradable poly(ortho esters). Other researchers have used chitosan (CS) membranes for drug delivery applications. These materials exhibited pH-dependent swelling behavior due to gelation of CS upon contact with anions (128). Interpenetrating networks of CS and PEO have been proposed as drug delivery devices due to their pH-dependent swelling behavior (129,130). More recently, Calvo et al. (131) prepared novel CS-PEO microspheres. These systems were to provide a continuous release of entrapped bovine serum albumin for 1 week. In another study, methotrexate, an anticancer drug, was encapsulated in microspheres of pH-sensitive CS and alginate (132). Zero-order release of the drug was observed from the microspheres in pH = 1.2 buffer for greater than 1 week. The Kikuchi et al. (133) studied pH-responsive calcium-alginate gel beads. In such systems, modulated release of dextran was achieved by varying the pH and ionic strength of the environmental solution. Such systems may be promising for use in protein and peptide delivery applications.

**Temperature-Sensitive Hydrogels**

Some of the earliest work with temperature-sensitive hydrogels was done by the group of Tanaka (134). They synthesized with cross-linked poly(N-isopropylacrylamide) (PNIPAAm) and determined that the LCST of the PNIPAAm gels was 34.3°C. Below this temperature, significant gel swelling occurred. The transition about this point was reversible. They discovered that the transition temperature was raised by copolymerizing PNIPAAm with small amounts of ionic monomers. Beltran et al. (135) also worked with PNIPAAM gels containing ionic comonomers. They observed results similar to those achieved by Tanaka.

The earliest investigators studying PNIPAAm gels discovered that the response time of the materials in response to temperature changes was rather slow. Future studies focused on developing newer materials that had the ability to collapse and expand in a more rapid fashion. Dong and Hoffman (136) prepared heterogeneous gels containing PNIPAAm that collapsed at significantly faster rates than homopolymers of PNIPAAm. Kabra and Gehrke (137) developed new method to prepare PNIPAAm gels that resulted in significant increases in the swelling kinetics of the gels. They prepared gels below the LCST to produce a permanent phase-separated microstructure in the gels. These gels expanded at rates 120 faster and collapsed at rates 3,000 times faster than homogeneous PNIPAAm gels. The group of Okano (138,139) developed an ingenious method to prepare comb-type graft hydrogels of PNIPAAm. The main chain of the cross-linked PNIPAAm contained small molecular weight grafts of PNIPAAm. Under conditions of gel collapse (above the LCST), hydrophobic regions were developed in the pores of the gel resulting in a rapid collapse. These materials had the ability to collapse from a fully swollen conformation in less than 20 minutes, whereas comparable gels that did not contain graft chains required up to a month to fully collapse. Such systems show major promise for rapid and abrupt or oscillatory release of drugs, peptides, or proteins.

Thermoresponsive polymers may be particularly useful in a wide variety of drug delivery applications (62,140). Okano et al. (141) studied the temperature-dependent permeability of PNIPAAm gels. They were able to use these gels as "on-off" delivery devices in response to temperature fluctuations. This type of behavior was useful for controlling the release of insulin (142) and heparin (114). These gels were also used as "squeezing" systems (143). More recently, conjugates of PNIPAAm with trypsin have been developed (144). Such systems could be extremely targeted delivery of enzymes. Another promising application of these systems was explored by the Vernon et al. (145). In this work, islets of Langerhans were entrapped by thermal gelation of PNIPAAm for use as a rechargeable artificial pancreas.

**pH- and Temperature-Sensitive Hydrogels**

Over the last 10 years, researchers have developed a novel class of hydrogels that exhibit both pH- and temperature-sensitive swelling behavior. These materials may prove to be extremely useful in enzymatic or protein drug delivery applications. Hydrogels were prepared from PNIPAAm and PAA that exhibited dual sensitivities (146,147). These gels were able to respond rapidly to both temperature and pH changes. The Kim et al. investigated the use of such systems for carriers for insulin (148) and calcitonin (149). In general, these hydrogels only exhibited strong temperature-sensitive swelling behavior with large amounts of PNIPAAm in the gel. Cationic pH- and temperature-sensitive gels were prepared using polyamines and PNIPAAm (150). These systems were evaluated for local delivery of heparin.

Chen and Hoffman prepared new graft copolymers of PAA and PNIPAAm that responded more rapidly to external stimulus than previously studied materials (151). These materials exhibited increased temperature sensitivity due to the presence of the PNIPAAm grafts. Such systems were evaluated for use in prolonged mucosal delivery of bioactive agents, specifically peptide drugs (152).

Brazel and Peppas (153) studied the pH- and temperature-responsive swelling behavior of gels containing PNIPAAm and PMAA. These materials were used to modulate the release behavior of streptokinase and heparin in response to pH and temperature changes (154). Baker and Siegel (155) used similar hydrogels to modulate the glucose permeability. However, only large amounts of PNIPAAm were needed to observe large temperature sensitivities. The group of Peppas developed novel pH- and temperature-sensitive terpolymers of PHEMA, PMAA, and PNIPAAm (156). These systems were prepared to contain PNIPAAm-rich blocks, and as a result, these materials were able to exhibit strong temperature sensitivity with only 10% PNIPAAm in the gel. Using these materials, they were effectively able to modulate the release kinetics of streptokinase (157).

## Complexing Hydrogels

Another promising class of hydrogels that exhibit responsive behavior is complexing hydrogels. Osada studied complex formation in PMAA hydrogels (158). In acidic media, the PMAA membranes collapsed in the presence of linear PEG chains due to the formation of interpolymer complexes between the PMAA and PEG. The gels swelled when placed in neutral or basic media. The permeability of these membranes was strongly dependent on the environmental pH and PEG concentration (159). Similar results were observed with hydrogels of PAA and linear PEG (160). Interpenetrating polymer networks of PVA and PAA that exhibit pH- and weak temperature-sensitive behavior due to complexation between the polymers were prepared and the release behavior of indomethacin was studied (161,162).

The group of Peppas has developed a class graft copolymer gels of PMAA grafted with PEG, P(MAA-g-EG) (163–170; A.M. Lowman and N.A. Peppas, unpublished data). These gels exhibited pH-dependent swelling behavior due to the presence of acidic pendent groups and the formation of interpolymer complexes between the ether groups on the graft chains and protonated pendent groups. In these co-

valently cross-linked, complexing P(MAA-g-EG) hydrogels, complexation resulted in the formation of temporary physical cross-links due to hydrogen bonding between the PEG grafts and the PMAA pendent groups. The physical cross-links were reversible in nature and dependent on the pH and ionic strength of the environment. As a result, complexing hydrogels exhibit drastic changes in their mesh size over small changes of pH as shown in Figure 15. One particularly promising application for these systems is the oral delivery of protein and peptide drugs (170). As shown in Figure 16, these copolymers severely limited the release of insulin in acidic environments such as those found in the stomach. However, in conditions similar to those found in the intestines, insulin release occurred rapidly.

## Glucose-Sensitive Systems

Major developments have been reported in the use of environmentally responsive hydrogels as glucose-sensitive systems that could serve as self-regulated delivery devices for the treatment of diabetes. Typically, these systems have have been prepared by incorporating glucose oxidase into the hydrogel structure during the polymerization. In the presence of glucose, the glucose oxidase catalyzed the reaction between water and glucose to form gluconic acid. The gluconic acid lowered the pH of the microenvironment of the gel.

The first such systems developed by Kost et al. (171) consisted of glucose oxidase immobilized in hydrogels based on PHEMA and PDMAEMA. These systems exhibited glucose-sensitive swelling behavior. In the presence of glucose, gluconic acid was formed resulting in a decrease in the local pH. As a result, the cationic-based hydrogels swelled to larger degrees in the presence of glucose due to the production of gluconic acid. The glucose-responsive swelling behavior allowed for control over insulin permeation in these membranes by adjusting the environmental glucose concentrations (172,173). The kinetics of gel swelling and insulin release from cationic, glucose-sensitive hydrogels was also studied (174).

Glucose responsive systems were proposed that were based on anionic hydrogels (175,176). Ito et al. (175) prepared systems of porous cellulose membranes containing an insulin reservoir. The pores of these devices were grafted with PAA chains functionalized with glucose oxidase. In the presence of glucose, the decrease in environmental pH caused the PAA chains to collapse, opening the pores and allowing for insulin release. More recently, glucose-sensitive complexation gels of P(MAA-g-EG) were developed by the group of Peppas (176). In these gels, as the pH decreased in response to elevated glucose concentrations, interpolymer complexes formed resulting in rapid gel syneresis. The rapid collapse resulted in insulin release due to a "squeezing" phenomenon.

Other glucose responsive systems have been developed that take advantage of the formation of complexes between glucose molecules and polymeric pendent groups. Lee and Park (177) prepared erodible hydrogels containing allyl glucose and poly(vinyl pyrrolidone). These systems were cross-linked by the noncovalent associations between concanavalin-A (Con-A) and the glucose pendent groups.

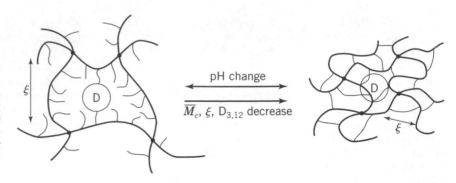

**Figure 15.** The effect of interpolymer complexation on the correlation length, $\zeta$, and the effective molecular weight between cross-links, $\overline{M}_c$, in P(MAA-*g*-EG) graft copolymer networks with permanent, chemical cross-links (●).

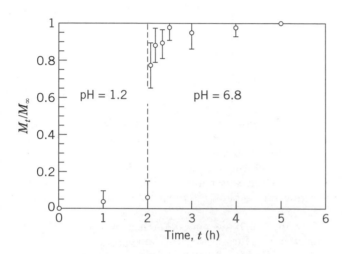

**Figure 16.** Controlled release of insulin in vitro from P(MAA-*g*-EG) microparticles simulated gastric fluid (pH = 1.2) for the first two hours and phosphate-buffered saline solutions (pH = 6.8) for the remaining three hours at 37°C (170).

In the presence of free glucose, the Con-A was bound to the free glucose, and the gels dissolved due to disruption of the physical cross-links. Newer materials developed by the group of Okano exhibited glucose responsive swelling behavior and insulin release (178–180). These gels were based on phenylboronic acid (PBA) and acrylamides. Another class of glucose-sensitive gels was prepared containing PBA, PNIPAAm, and PVA (181). These gels were designed to allow for the release of insulin at physiological pH and temperature.

## Oral Insulin Delivery Systems

One of the major objectives of researchers working in the controlled release field is to design an effective, oral insulin delivery system. However, this is a difficult task due to the degradation of the drug in the upper gastrointestinal (GI) system barrier and the slow transport of insulin across the lining of the colon into the bloodstream (182–184). Numerous attempts have been made by researchers to use hydrogels as carriers for oral delivery of insulin in order to protect the drug in the stomach and release it into more favorable regions of the GI tract.

Touitou and Rubinstein (185) designed a reservoir system consisting of insulin encapsulated by a polyacrylate gel. The coating was designed to dissolve only in the colon.

In this work, weak hypoglycemic effects were observed only with very high insulin doses and the addition of absorption enhancers. Saffran (186) developed a biodegradable hydrogel containing insulin. The device consisted of insulin dispersed in a terpolymer of styrene and PHEMA cross-linked with a difunctional azo-containing compound. The azo bond was cleaved by microflora present in the colon, and the polymer degraded, allowing for release of insulin into the colon. In this work, a hypoglycemic effect was obtained only with addition of absorption enhancers and protease inhibitors. However, the hypoglycemic effect obtained was not affected by the initial dosing.

Morishita et al. (187) administered insulin contained within Eudragit 100 gels. In these systems, the pH-responsive Eudragit® degraded in the upper small intestine, allowing for insulin release. They observed strong hypoglycemic effects in healthy and diabetic rats after the addition of absorption enhancers. Platé et al. (188) developed a hydrogel system containing immobilized insulin and protease inhibitors that was effective in lowering the blood glucose levels in rabbits. More recently, Mathiowitz et al. (189) have developed insulin containing poly(anhydride) microspheres. These materials adhered to the walls of the small intestine and released insulin based on degradation of the polymeric carrier. They observed a 30% decrease in the blood glucose levels of healthy rats. Lowman et al. (190) have developed a bioadhesive, complexation hydrogel system for oral delivery of insulin. This delivery system consisted of insulin-containing microparticles of crosslinked copolymers of P(MAA-*g*-EG). The P(MAA-*g*-EG) were more effective in delivering biologically active insulin than traditional enteric coating-type carriers because of the presence of the PEG grafts. The addition of PEG to the gels was critical because the PEG chains participate in the macromolecular complexes, function as a peptide stabilizer, and enhance the mucoadhesive characteristics of the gels. In this work, strong dose-dependent hypoglycemic effects were observed in healthy and diabetic rats following oral administration of these gels.

## Other Promising Applications

Promising new methods for the delivery of chemotherapeutic agents using hydrogels have been recently reported. Novel bioreconizable sugar-containing copolymers have been investigated for the use in targeted delivery of anticancer drugs (191–193). Peterson et al. have used poly(*N*-

2-hydroxypropyl methacrylamide) carriers for the treatment of ovarian cancer (193).

In the last few years there have been new creative methods of preparation of novel hydrophilic polymers and hydrogels that may represent the future in drug delivery applications. The focus in these studies has been the development of polymeric structures with precise molecular architectures. Stupp et al. (194) synthesized self-assembled triblock copolymer, nanostructures that may have very promising applications in controlled drug delivery. Novel biodegradable polymers, such as polyrotaxanes, have been developed that have particularly exciting molecular assemblies for drug delivery (195). Dendrimers and star polymers (196) are exciting new materials because of the large number of functional groups available in a very small volume. Such systems could have tremendous promise in drug-targeting applications. Merrill (197) has offered an exceptional review of PEO star polymers and applications of such systems in the biomedical and pharmaceutical fields. Griffith and Lopina (198) have prepared gels of controlled structure and large biological functionality by irradiation of PEO star polymers. The new structures discussed in this section could have particularly promising delivery applications when combined with emerging new technologies such as molecular imprinting (199,200).

A number of investigators have concentrated on the development of environmentally responsive gels that exhibit biodegradability. This can be achieved by a number of synthetic methods. Kopecek and associates (201) have developed biodegradable hydrogels by incorporating azo compounds. Bae and associates (202) have synthesized very promising biodegradable carriers by preparing 8-arm, star-shaped, block copolymers containing PLA and PEO. Another potentially useful biodegradable system is a photo-cross-linked polymer based on poly(L-lactic acid-co-L-aspartic acid) (203), which could be prepared in situ for delivery of antiinflammatory drugs following surgery.

## BIBLIOGRAPHY

1. N.A. Peppas, *Hydrogels in Medicine*, CRS Press, Boca Raton, Fla., 1986.

2. R. Langer and N.A. Peppas, *J. Mater. Sci.: Rev. Macromol. Chem. Phys.* **C23**(1), 61–126 (1983).

3. N.A. Peppas and D.L. Meadows, *J. Membr. Sci.* **16**, 361–377 (1983).

4. B.D. Ratner and A.S. Hoffman, in J.D. Andrade, ed., *Hydrogels for Medical and Related Applications*, American Chemical Society, Washington, D.C., 1976, pp. 1–36.

5. K. Park, *Controlled Release: Challenges and Strategies*, American Chemical Society, Washington, D.C., 1997.

6. N.A. Peppas, *Curr. Opin. Colloids Int. Sci.* **2**, 531–537 (1997).

7. N.A. Peppas and R. Langer, *Science* **263**, 1715–1720 (1994).

8. R. Baker, *Controlled Release of Biologically Active Agents*, Wiley, New York, 1987.

9. H. Park and K. Park, *Pharm. Res.* **13**, 1770–1776 (1996).

10. N.A. Peppas, *J. Bioact. Compat. Polym.* **6**, 241–246 (1991).

11. M.T. am Ende and A.G. Mikos, in L.M. Sanders and R.W. Hendren, eds., *Protein Delivery: Physical Systems*, Plenum, Tokyo, 1997, pp. 139–165.

12. V.H.L. Lee, M. Hashida, and Y. Mizushima, *Trends and Future Perspectives in Peptide and Protein Drug Delivery*, Harwood Academic Publishers, Chur, Switzerland, 1995.

13. J.A. Fix, *Pharm. Res.* **13**, 1760–1764 (1996).

14. J. Kopecek, P. Kopeckova, and V. Omelyanenko, in N. Ogata, S.W. Kim, J. Feijen, and T. Okano, eds., *Biomedical and Drug Delivery Systems*, Springer, Tokyo, 1996, pp. 91–95.

15. N.A. Peppas and A.G. Mikos, in N.A. Peppas, ed., *Hydrogels in Medicine and Pharmacy*, vol. 1, CRC Press, Boca Raton, Fla., 1986, pp. 1–25.

16. N.A. Peppas and B.D. Barr-Howell, in N.A. Peppas, ed., *Hydrogels in Medicine and Pharmacy*, vol. 1, CRC Press, Boca Raton, Fla., 1986, pp. 27–55.

17. P.J. Flory and J. Rehner, *J. Chem. Phys.* **11**, 521–526 (1943).

18. P.J. Flory, *Principles of Polymer Chemistry*, Cornell University Press, Ithaca, N.Y., 1953.

19. N.A. Peppas and E.W. Merrill, *J. Polym. Sci., Polym. Chem. Ed.*, **14**, 441–457 (1976).

20. J. Ricka and T. Tanaka, *Macromolecules* **17**, 2916–2921 (1984).

21. A. Katchalsky, *Experientia* **5**, 319–320 (1949).

22. T. Tanaka, *Polymer* **20**, 1404–1412 (1979).

23. M. Rubinstein, R.H. Colby, A.V. Dobrynin, and J.F. Joanny, *Macromolecules* **29**, 398–426 (1996).

24. U.P. Schroder and W. Opperman, in J.P. Cohen-Addad, ed., *The Physical Properties of Polymeric Gels*, Wiley, New York, 1996, pp. 19–38.

25. R. Skouri, F. Schoesseler, J.P. Munch, and S.J. Candau, *Macromolecules* **28**, 197–210 (1995).

26. L. Brannon-Peppas and N.A. Peppas, *Chem. Eng. Sci.* **46**, 715–722 (1991).

27. L. Brannon-Peppas and N.A. Peppas, in L. Brannon-Peppas and R.S. Harland, eds., *Absorbent Polymer Technology*, Elsevier, Amsterdam, 1990, pp. 67–75.

28. R.G. Treloar, *The Physics of Rubber Elasticity*, 2nd ed., Oxford University Press, Oxford, 1958.

29. N.A. Peppas and E.W. Merrill, *J. Appl. Polym. Sci.* **21**, 1763–1770 (1977).

30. A.M. Lowman and N.A. Peppas, *Macromolecules* **30**, 4959–4965 (1997).

31. K.S. Anseth, C.N. Bowman, and L. Brannon-Peppas, *Biomaterials* **17**, 1647–1657 (1996).

32. J.E. Mark, *Adv. Polym Sci.* **44**, 1–26 (1982).

33. R.S. Stock and W.H. Ray, *J. Polym. Sci., Polym. Phys. Ed.* **23**, 1393–1447 (1985).

34. D.N. Winslow, in E. Matijevic and R.J. Good, eds., *Surface and Colloid Science*, Plenum, New York, 1984, pp. 259–282.

35. A.G. Mikos et al., *J. Biomed. Mater. Res.* **27**, 183–189 (1993).

36. T. Canal and N.A. Peppas, *J. Biomed. Mater. Res.* **23**, 1183–1193 (1989).

37. B. Narasimhan and N.A. Peppas, in K. Park, ed., *Controlled Drug Delivery*, American Chemical Society, Washington, D.C., 1997, pp. 529–557.

38. J. Crank, *The Mathematics of Diffusion*, Oxford University Press, Oxford, 1956.

39. P.L. Ritger and N.A. Peppas, *J. Controlled Release* **5**, 23–26 (1987).

40. H. Fujita, *Fortschr. Hochpolym.-Forsch.* **3**, 1–14 (1961).

41. F.G. Smith, III and W.M. Deen, *J. Colloid Interface Sci.* **78**, 444–465 (1980).

42. D.M. Malone and J.L. Quinn, *Chem. Eng. Sci.* **33**, 1429–1440 (1978).

43. J.L. Anderson and J.A. Quinn, *Biophys. J.* **14**, 130–150 (1974).

44. W.M. Deen, *AIChE J.* **33**, 1409–1425 (1987).

45. J.D. Ferry, *Viscoelastic Properties of Polymers*, Wiley, New York, 1980.

46. H. Yasuda et al., *Makromol. Chem.* **126**, 177–186 (1969).

47. H. Yasuda and C.E. Lamaze, *J. Macromol. Sci., Phys.* **B5**, 111–134 (1971).

48. N.A. Peppas and C.T. Reinhart, *J. Membr. Sci.* **15**, 275–287 (1983).

49. C.T. Reinhart and N.A. Peppas, *J. Membr. Sci.* **18**, 227–239 (1984).

50. A.P. Sassi, H.W. Blanch, and J.M. Prausnitz, *J. Appl. Polym. Sci.* **59**, 1337–1346 (1996).

51. N.A. Peppas and H.J. Moynihan, *J. Appl. Polym. Sci.* **30**, 2589–2606 (1985).

52. R.S. Harland and N.A. Peppas, *Colloid Polym. Sci.* **267**, 218–225 (1989).

53. N.A. Peppas and R.W. Korsmeyer, in N.A. Peppas, ed., *Hydrogels in Medicine and Pharmacy*, vol. 3, CRC Press, Boca Raton, Fla., 1987, pp. 103–135.

54. A.R. Berens and H.B. Hofenberg, *Polymer* **19**, 490–496 (1978).

55. P. Stilbs, *Prog. NMR Spectrosc.* **19**, 1–45 (1987).

56. F.P. Price, P.T. Gilmore, E.L. Thomas, and R.L. Laurence, *J. Polym. Sci., Polym. Symp.* **63**, 33–44 (1978).

57. J.J. Sahlin and N.A. Peppas, *J. Biomater. Sci., Polym. Ed.* **8**, 421–436 (1997).

58. J. Heller and R.W. Baker, in R.W. Baker, ed., *Controlled Release of Bioactive Materials*, Academic Press, New York, 1980, pp. 1–37.

59. J. Heller, in J.M. Anderson and S.W. Kim, eds., *Advances in Drug Delivery Systems*, Plenum, New York, 1984, pp. 101–154.

60. N.A. Peppas, in R. Gurny, H.E. Juninger, and N.A. Peppas, eds., *Pulsatile Drug Delivery*, Wiss. Verlagsgese., Stuttgart, 1993, pp. 41–56.

61. R.A. Siegel, in K. Park, ed., *Controlled Drug Delivery*, American Chemical Society, Washington, D.C., 1997, pp. 1–27.

62. S.W. Kim, in N. Ogata, S.W. Kim, J. Feijen, and T. Okano, eds., *Advanced Biomaterials in Biomedical Engineering and Drug Delivery Systems*, Springer, Tokyo, 1996, pp. 125–133.

63. E.A. Bekturov and L.A. Bimendina, *Adv. Polym. Sci.* **43**, 100–147 (1981).

64. J. Heller, in J. Kost, ed., *Pulsed and Self-Regulated Drug Delivery Systems*, CRC Press, Boca Raton, Fla., 1990, pp. 93–122.

65. R. Langer, *Science* **249**, 1527–1533 (1990).

66. E.R. Edelman, J. Kost, H. Bobech, and R. Langer, *J. Biomed. Mater. Res.* **19**, 67–74 (1985).

67. W.R. Good, in R. Kostelnik, ed., *Polymeric Delivery Systems*, Gordon & Breach, New York, 1976, pp. 139–155.

68. H.B. Hopfenberg and K.C. Hsu, *Polym. Eng. Sci.* **18**, 1186 (1978).

69. J.M. Anderson et al., in J.D. Andrade, ed., *Hydrogels for Medical and Related Applications*, American Chemical Society, Washington, D.C., 1976, pp. 167–178.

70. M. Sefton and E. Nishimura, *J. Pharm. Sci.* **69**, 208–213 (1980).

71. S.Z. Song, J.R. Cardinal, S.H. Kim, and S.W. Kim, *J. Pharm. Sci.* **67**, 1352 (1981).

72. P.I. Lee, *Polymer* **25**, 973 (1984).

73. P.I. Lee, *J. Controlled Release* **73**, 1344 (1986).

74. N.M. Franson and N.A. Peppas, *J. Appl. Polym. Sci.* **28**, 1299 (1983).

75. G.W.R. Davidson and N.A. Peppas, *J. Controlled Release* **3**, 243 (1986).

76. R.W. Korsmeyer and N.A. Peppas, *J. Controlled Release* **89**, 1–9 (1984).

77. N.A. Peppas, in N.A. Peppas, ed., *Hydrogels in Medicine and Pharmacy*, vol. 2, CRC Press, Boca Raton, Fla., 1987, pp. 1–48.

78. R. Langer and J. Folkman, *Nature (London)* **263**, 970–974 (1976).

79. N.A. Peppas, *Makromol. Chem.* **176**, 3433–3440 (1975).

80. R. Korsmeyer and N.A. Peppas, *J. Membr. Sci.* **9**, 211–227 (1981).

81. K. Morimoto et al., *Pharm. Res.* **6**, 338–344 (1989).

82. N.A. Peppas and P.J. Hansen, *J. Appl. Polym. Sci.* **27**, 4787–4797 (1982).

83. F. Urushizaki, H. Yamaguchi, K. Nakamura, and S. Numajiri, *Int. J. Pharm.* **58**, 135–142 (1990).

84. N.A. Peppas and R.S. Stauffer, *J. Controlled Release* **16**, 305–310 (1991).

85. N.A. Peppas and J.E. Scott, *J. Controlled Release* **18**, 95–100 (1992).

86. B.J. Ficek and N.A. Peppas, *J. Controlled Release* **27**, 259–264 (1993).

87. K. Tsutsumi et al., *S.T.P. Pharm. Sci.* **4**, 230–236 (1994).

88. N.A. Peppas and N.K. Mongia, *Eur. J. Pharm. Biopharm.* **43**, 51–58 (1997).

89. N.K. Mongia, K.S. Anseth, and N.A. Peppas, *J. Biomater. Sci., Polym. Ed.* **7**, 1055–1064 (1996).

90. S.K. Mallapragada and N.A. Peppas, *J. Controlled Release* **45**, 87–94 (1997).

91. S.K. Mallapragada and N.A. Peppas, *AIChE J.* **43**, 870–876 (1997).

92. N.A. Peppas and P. Colombo, *J. Controlled Release* **45**, 35–40 (1997).

93. B. Narasimhan and N.A. Peppas, *J. Pharm. Sci.* **86**, 297–304 (1997).

94. N.B. Graham, in J.M. Harris, ed., *Poly(Ethylene Glycol) Chemistry, Biotechnical and Biomedical Applications*, Plenum, New York, 1992, pp. 263–281.

95. M.E. McNeill and N.B. Graham, *J. Controlled Release* **1**, 99–107 (1984).

96. M.E. McNeill and N.B. Graham, *J. Biomater. Sci., Polym. Ed.* **7**, 937–951 (1996).

97. L. Bromberg, *J. Appl. Polym. Sci.* **59**, 459–466 (1996).

98. J.L. Stringer and N.A. Peppas, *J. Controlled Release* **42**, 195–202 (1996).

99. P. Kofinas, V. Athanassiou, and E.W. Merrill, *Biomaterials* **17**, 1547–1550 (1996).

100. A.M. Lowman, T.D. Dziubla, and N.A. Peppas, *Polym. Prepri.* **38**, 622–623 (1997).

101. K.B. Keys, F.M. Andreopoulos, and N.A. Peppas, *Macromolecules* **31**, 8149–8156 (1998).

102. N.A. Peppas, *J. Biomater. Sci., Polym. Ed.* **9**, 535–542 (1998).

103. L.M. Schwarte and N.A. Peppas, *Polymer* **39**, 6057–6066 (1998).

104. H. Ichikawa and N.A. Peppas, *Polym. Prepri.* **40**, 363–364 (1999).

105. A.B. Scranton, B. Rangarajan, and J. Klier, *Adv. Polym. Sci.* **120**, 1–54 (1995).

106. W. Oppermann, in R.S. Harland and R.K. Prud'homme, eds., *Polyelectrolyte Gels: Properties, Preparation, and Applications*, American Chemical Society, Washington, D.C., 1992, pp. 159–170.

107. R.A. Siegel and B.A. Firestone, *Macromolecules* **21**, 3254–3259 (1988).

108. B.A. Firestone and R.A. Siegel, *Polym. Commun.* **29**, 204–208 (1988).

109. R.A. Siegel, M. Falamarzian, B.A. Firestone, and B.C. Moxley, *J. Controlled Release* **8**, 179–182 (1988).

110. J.M. Cornejo-Bravo and R.A. Siegel, *Biomaterials* **17**, 1187–1193 (1996).

111. R.A. Siegel, in J. Kost, ed., *Pulsed and Self-Regulated Drug Delivery*, CRC Press, Boca Raton, Fla., 1990, pp. 129–155.

112. D. Hariharan and N.A. Peppas, *Polymer* **37**, 149–161 (1996).

113. N.A. Peppas and D. Hariharan, *Bull. Gattefosse Rep.* **84**, 29–36 (1991).

114. A. Gutowska, Y.H. Bae, J. Feijen, and S.W. Kim, *J. Controlled Release* **22**, 95–104 (1992).

115. L.M. Schwarte and N.A. Peppas, *Polym. Prepr.* **38**(2), 596–597 (1997).

116. L. Brannon-Peppas and N.A. Peppas, *Biomaterials* **11**, 635–640 (1990).

117. J.H. Kou, G.L. Almidon, and P.I. Lee, *Pharm. Res.* **5**, 592–597 (1988).

118. L. Brannon-Peppas and N.A. Peppas, *Int. J. Pharm.* **70**, 53–57 (1991).

119. L. Brannon-Peppas and N.A. Peppas, *J. Controlled Release* **16**, 319–330 (1991).

120. R. Bettini, P. Colombo, and N.A. Peppas, *J. Controlled Release* **37**, 105–111 (1995).

121. L. Brannon-Peppas and N.A. Peppas, *J. Controlled Release* **8**, 267–274 (1989).

122. A.R. Khare and N.A. Peppas, *J. Biomater. Sci., Polym. Ed.* **4**(3), 275–289 (1993).

123. M.T. am Ende and N.A. Peppas, *J. Controlled Release* **48**, 47–56 (1997).

124. L.F. Gudeman and N.A. Peppas, *J. Membr. Sci.* **107**, 239–248 (1995).

125. N.A. Peppas and S.L. Wright, *Macromolecules* **29**, 8798–8804 (1996).

126. M.T. am Ende and N.A. Peppas, *Pharm. Res.* **12**, 2030–2035 (1995).

127. J. Heller, A.C. Chang, G. Rodd, and G.M. Grodsky, *J. Controlled Release* **11**, 193–201 (1990).

128. R. Bodmeier and O. Paeratakul, *J. Pharm. Sci.* **78**, 964–969 (1989).

129. S. Shiraishi, T. Imai, and M. Otagiri, *J. Controlled Release* **25**, 217–223 (1993).

130. K. Yao et al., *J. Appl. Polym. Sci.* **48**, 343–348 (1993).

131. P. Calvo, C. Remuñán-López, J.L. Vila-Jato, and M.J. Alonso, *J. Appl. Polym. Sci.* **63**, 125–132 (1997).

132. R. Narayani and K. Panduranga Rao, *J. Appl. Polym. Sci.* **58**, 1761–1769 (1995).

133. A. Kikuchi et al., *J. Controlled Release* **47**, 21–29 (1997).

134. S. Hirotsu, Y. Hirokawa, and T. Tanaka, *J. Chem. Phys.* **87**, 1392–1395 (1987).

135. S. Beltran et al., *Macromolecules* **24**, 549–551 (1991).

136. L.C. Dong and A.S. Hoffman, *J. Controlled Release* **13**, 21–31 (1990).

137. B.G. Kabra and S.H. Gehrke, *Polym. Commun.* **32**, 322–323 (1991).

138. R. Yoshida et al., *Nature (London)* **374**, 240–242 (1995).

139. Y. Kaneko et al., *Macromol. Symp.* **109**, 41–53 (1996).

140. A.S. Hoffman, *J. Controlled Release* **6**, 297–305 (1987).

141. T. Okano, Y.H. Bae, H. Jacobs, and S.W. Kim, *J. Controlled Release* **11**, 255–265 (1990).

142. Y.H. Bae, T. Okano, and S.W. Kim, *J. Controlled Release* **9**, 271–276 (1989).

143. R. Yoshida et al., *J. Controlled Release* **32**, 97–102 (1994).

144. M. Matsukata et al., *Bioconjugate Chem.* **7**, 96–101 (1996).

145. B. Vernon, A. Gutowska, S.W. Kim, and Y.H. Bae, *Macromol. Symp.* **109**, 155–167 (1996).

146. L.C. Dong, Q. Yan, and A.S. Hoffman, *J. Controlled Release* **19**, 171–178 (1992).

147. H. Feil, Y.H. Bae, and S.W. Kim, *Macromolecules* **25**, 5528–5530 (1992).

148. Y.H. Kim, Y.H. Bae, and S.W. Kim, *J. Controlled Release* **28**, 143–152 (1994).

149. A. Serres, M. Baudyš, and S.W. Kim, *Pharm. Res.* **13**(2), 196–201 (1996).

150. Y. Nabeshima et al., in N. Ogata, S.W. Kim, J. Feijen, and T. Okano, eds., *Advanced Biomaterials in Biomedical Engineering and Drug Delivery Systems*, Springer, Tokyo, 1996, pp. 315–316.

151. G.H. Chen and A.S. Hoffman, *Nature (London)* **373**, 49–52 (1995).

152. A.S. Hoffman et al., in N. Ogata, S.W. Kim, J. Feijen, and T. Okano, eds., *Advanced Biomaterials in Biomedical Engineering and Drug Delivery Systems*, Springer, Tokyo, 1996, pp. 62–66.

153. C.S. Brazel and N.A. Peppas, *Macromolecules* **28**, 8016–8020 (1995).

154. C.S. Brazel and N.A. Peppas, *J. Controlled Release* **39**, 57–64 (1996).

155. J.P. Baker and R.A. Siegel, *Macromol. Rapid Commun.* **17**, 409–415 (1996).

156. S.K. Vakkalanka and N.A. Peppas, *Polym. Bull.* **36**, 221–225 (1996).

157. S.K. Vakkalanka, C.S. Brazel, and N.A. Peppas, *J. Biomater. Sci. Polym. Ed.* **8**, 119–129 (1996).

158. Y. Osada, *J. Polym. Sci., Polym. Lett. Ed.* **18**, 281–286 (1980).

159. Y. Osada, K. Honda, and M. Ohta, *J. Membr. Sci.* **27**, 339–347 (1986).

160. S. Nishi and T. Kotaka, *Macromolecules* **19**, 978–984 (1986).

161. J. Byun, Y.M. Lee, and C.S. Cho, *J. Appl. Polym. Sci.* **61**, 697–702 (1996).

162. H.S. Shin, S.Y. Kim, and Y.M. Lee, *J. Appl. Polym. Sci.* **65**, 685–693 (1997).

163. J. Klier, A.B. Scranton, and N.A. Peppas, *Macromolecules* **23**, 4944–4949 (1990).

164. N.A. Peppas and J. Klier, *J. Controlled Release* **16**, 203–214 (1991).

165. C.L. Bell and N.A. Peppas, *J. Biomater. Sci., Polym. Ed.* **7**, 671–683 (1996).

166. C.L. Bell and N.A. Peppas, *Biomaterials* **17**, 1203–1218 (1996).

167. C.L. Bell and N.A. Peppas, *J. Controlled Release* **39**, 201–207 (1997).

168. C.L. Bell and N.A. Peppas, *Adv. Polym. Sci.* **122**, 125–175 (1995).

169. N.A. Peppas and A.M. Lowman, in S. Frøkjaer, L. Christup, and P. Krogsgaard-Larsen, eds., *Peptide and Protein Delivery*, Alfred Benzon Symposium, Munksgaard, Copenhagen, 1998, pp. 206–216.

170. A.M. Lowman, N.A. Peppas, M. Morishita, and T. Nagai, in I. McCullough and S. Shalaby, eds., *Tailored Polymeric Materials for Controlled Delivery Systems*, American Chemical Society, Washington, D.C., 1998, pp. 156–164.

171. J. Kost, T.A. Horbett, B.D. Ratner, and M. Singh, *J. Biomed. Mater. Res.* **19**, 1177–1133 (1984).

172. K. Ishihara, M. Kobayashi, and I. Shinohara, *Polym. J.* **16**, 625–631 (1984).

173. G. Albin, T.A. Horbett, and B.D. Ratner, *J. Controlled Release* **2**, 153–164 (1985).

174. M. Goldraich and J. Kost, *Clin. Mater.* **13**, 135–142 (1993).

175. Y. Ito, M. Casolaro, K. Kono, and Y. Imanishi, *J. Controlled Release* **10**, 195–203 (1989).

176. C.M. Hassan, F.J. Doyle, and N.A. Peppas, *Macromolecules* **30**, 6166–6173 (1997).

177. S.J. Lee and K. Park, *Polym. Prepr.* **35**(2), 391–392 (1994).

178. D. Shiino et al., *J. Controlled Release* **37**, 269–276 (1995).

179. D. Shiino et al., *J. Biomater. Sci. Polym. Ed.* **7**(8), 697–705 (1996).

180. T. Aoki et al., *Polym. J.* **28**(4), 371–374 (1996).

181. Hisamitsu, K. Kataoka, T. Okano, and Y. Sakurai, *Pharm. Res.* **14**(3), 289–293 (1997).

182. M. Saffran, B. Pansky, G.C. Budd, and F.E. Williams, *J. Controlled Release* **46**, 89–98 (1997).

183. V.H.L. Lee, S. Dodd-Kashi, G.M. Grass, and W. Rubas, in V.H.L. Lee, ed., *Protein and Peptide Drug Delivery*, Dekker, New York, 1991, pp. 691–740.

184. M. Saffran, in D.R. Friend, ed., *Oral Colon-Specific Drug Delivery*, CRC Press, Boca Raton, Fla., 1992, pp. 115–142.

185. E. Touitou and A. Rubinstein, *Int. J. Pharm.* **30**, 93–99 (1986).

186. M. Saffran et al., *Science* **233**, 1081–1084 (1986).

187. I. Morishita et al., *Int. J. Pharm.* **78**, 9–16 (1992).

188. N.A. Platé et al., *Vysokomol. Soedin.* **36**, 1876–1879 (1994).

189. E. Mathiowitz et al., *Nature (London)* **386**, 410–414 (1997).

190. M. Morishita et al., *Yakuzaigaku* **57**, 96–97 (1997).

191. R.C. Rathi, P. Kopecková, and J. Kopecek, *Macromol. Chem. Phys.* **198**, 1–16 (1997).

192. D.A. Putnam, J.G. Shiah, and J. Kopecek, *Biochem. Pharmacol.* **52**, 957–962 (1996).

193. C.M. Peterson et al., *Cancer Res.* **56**, 3980–3985 (1996).

194. S.I. Stupp et al., *Science* **276**, 384–389 (1997).

195. T. Ooya and N. Yui, *J. Biomater. Sci., Polym. Ed.* **8**, 437–445 (1997).

196. P.R. Dvornik and D.A. Tomalia, *Curr. Opin. Colloid Interface Sci.* **1**, 221–235 (1996).

197. E.W. Merrill, *J. Biomater. Sci., Polym. Ed.* **5**, 1–11 (1993).

198. L. Griffith and S.T. Lopina, *Macromolecules* **28**, 6787–6794 (1995).

199. K. Mosbach and O. Ramström, *Bio/Technology* **14**, 163–170 (1996).

200. S.H. Cheong et al., *Macromolecules* **30**, 1317–1322 (1997).

201. H. Ghandehari, P. Kopecková, P.Y. Yeh, and J. Kopecek, *Macromol. Chem. Phys.* **197**, 965–980 (1996).

202. Y.K. Choi, S.W. Kim, and Y.H. Bae, *Polym. Prepr.* **37**, 109–110 (1996).

203. J. Elisseeff, K. Anseth, R. Langer, and J.S. Hrkach, *Macromolecules* **30**, 2182–2184 (1996).

# IMMUNOISOLATED CELL THERAPY

MICHAEL J. LYSAGHT
Brown University
Providence, Rhode Island

PATRICK AEBISCHER
University of Lausanne Medical School
Lausanne, Switzerland

## KEY WORDS

Artificial liver
Artificial pancreas
Bioartificial xenograft
Encapsulation
Gene therapy
Macrocapsule
Membrane
Matrix
Microcapsule

## OUTLINE

## INTRODUCTION

In the late twentieth century, biomedical engineers and clinicians have increasingly adopted what may be termed the fundamental tenet of substitutive medicine: it is frequently the wiser choice to simply replace a failing organ than to struggle with attempts at repair and restoration. Such replacement can sometimes be accomplished with organ transplantation but more frequently relies upon man-made devices such as hemodialyzers, blood oxygenators, mechanical heart valves, pacemakers, and artificial hips or knees.

*Transplanted organs* more closely replicate natural function but are constrained in supply and require phar-macological immunosuppression. *Artificial organs* can be manufactured in large quantities and require no immu-nosupression but rarely offer the full functional capability of their natural counterparts. In the early seventies Dr. William Chick, who passed away in 1998, proposed a third approach to organ replacement: biohybrid organs, which combined living tissue and biomaterials into a single therapeutic device. Chick's primary interest was in the development of an artificial pancreas in which living cells were separated from the host by a selective membrane barrier. A satisfactory artificial pancreas still remains an elusive target, but immunoisolation, as this approach is now called, has provided the enabling technology for the development of a diverse collection of novel medical therapeutics. It has become a promising route to the delivery of gene therapy, and it offers the largest successful clinical experience thus far with the xenotransplantation of living animal-derived tissue into human hosts.

The concept of immunoisolation is illustrated in Figure 1. Living cells are ensconced in a supporting matrix and surrounded by a semipermeable membrane, which supports the transport of small molecules (oxygen, nutrients, and electrolytes) while restricting the passage of large molecules (immunoglobulins) and cells. The membrane is a mechanical analogue of pharmacological immunosupression. Both thwart the normally lethal response of a host immune system to foreign living tissue, albeit by different mechanisms. One important difference is that immunoisolatory membranes can often protect transplanted xenogeneic grafts (i.e., those originating in a different species than the recipient) whereas presently available immuno-supressive drugs cannot. Table 1 lists some of the more important classes of membranes, matrices, and cells used in immunoisolation over the past two decades. As the therapy developed, membranes and matrices have steadily evolved, but the types and classes of cells have changed dramatically. These advances in exploitation of cell and molecular biology have transformed immunoisolation from a specialized approach to a bioartificial pancreas to a multifaceted technology of potential utility whenever long-term site-specific protein delivery is required.

## BIOMATERIALS AND CELLS FOR IMMUNOISOLATION

### Membranes

Membranes are simply thin barriers capable of providing selective transport between adjacent phases. Synthetic biocompatible membranes are widely used in today's hemodialyzers, blood oxygenators, and plasma filters. With over 100 million hemodialyzers produced each year, the technology of medical-grade membranes has become highly advanced. The membranes most often used for immunoisolation are open-cell foams cast in the form of thin capillaries or "hollow fibers." These structures are made from thermoplastics such as polysulfone or acrylic copolymers. Fiber diameter is typically 500 to 1000 $\mu$m and a wall

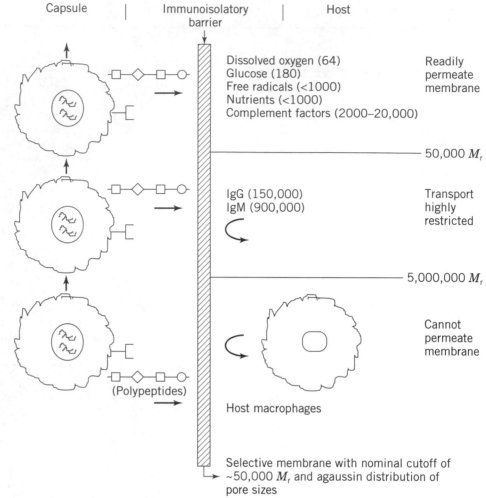

Capsule | Immunoisolatory barrier | Host

Dissolved oxygen (64)
Glucose (180)
Free radicals (<1000)
Nutrients (<1000)
Complement factors (2000–20,000)

Readily permeate membrane

— 50,000 $M_r$

IgG (150,000)
IgM (900,000)

Transport highly restricted

— 5,000,000 $M_r$

Cannot permeate membrane

(Polypeptides)

Host macrophages

Selective membrane with nominal cutoff of ~50,000 $M_r$ and agaussin distribution of pore sizes

**Figure 1.** Immunoisolated cell therapy relies upon a membrane to separate foreign secretory cells from a host. Molecules secreted by the cells, usually proteins, can pass through the membrane if they are smaller than 50,000 $M_r$. In similar fashion, oxygen glucose and molecules such as insulin can diffuse from the host to the cells. However, the membrane blocks the passage of host cells and restricts transport of host immunoglobulins. Membranes allow passage of complement factors and free radicals that appear to be generated primarily when an implant containing xenogeneic cells is directly implanted into host tissue.

thickness of about 15% of the diameter. In recent years, the physical strength of membranes for immunoisolation has been increased by adding metallic or polymeric supports. The composite structures are several thousand times as strong as hemodialysis membranes with comparable dimensions. Some investigators prefer to use a different kind of membrane prepared by the interfacial polymerization of weak polyelectrolytes. Typically, cells are suspended in a droplet containing an aqueous solution of alginate. Alginate is a naturally occurring polymer derived from seaweed, which contains a slight positive charge. The droplet is then bathed in a solution containing polylysine, a synthetic, negatively charged polypeptide. The two polymers react and the resultant film encapsulates the cell in a droplet with a diameter of about 500 $\mu$m. A third approach, still in development, applies a conformal coating of photopolymerizable acrylic film on the surface of cells or clusters of cells.

Membranes formulated to allow permeation of molecules as large as proteins while preventing passage of cells are termed microporous. Membranes capable of discriminating between small and large molecules are called semipermeable. As a general rule, microporous membranes are used with allogeneic cells and semipermeable membranes

with xenogeneic membranes. It is important to recognize that molecularly separative membranes contain a distribution of pore sizes and function as relative, rather than absolute, barriers to the transport of higher molecular species. A membrane with a nominal molecular weight cutoff of 50,000 daltons rejects well over 99% of IgG (60,000 daltons) or IgM (950,000 daltons) in a challenge solution but still permits passage of a very small fraction of the presenting immunoglobulins.

## Matrix

Cells in a capsule are usually supported on a matrix. First-generation matrices were prepared from gelled versions of the same polymers used to coat tissue culture dishes: alginate, agarose, or collagen. Such matrices were weak with a jellylike morphology. More recently, matrices have been prepared not as gels but as scaffolds or macroreticulated foams made from plastic resins, which are sometimes coated with water-soluble polymers. These highly tailored materials are more stable and lead to mechanically stronger implants than earlier gels.

## Cells

The first generation of cell technology for immunoisolation involved primary cells, which had been harvested from a

**Table 1. Elements of an Immunoisolatory Implant**

| | |
|---|---|
| *Cells* | |
| Primary | Islets, hepatocytes, adrenal chromaffin cells |
| Dividing | Cell lines, such as PC-12 cells which secrete dopamine |
| Engineered | Cell lines which have received the gene for proteins like factor VIII or EPO |
| Tailored | Engineered cells which also contain complement restriction factors, antiapoptotic genes, and which may be conditionally immortalized to reduce immunogenicity. |
| *Matrices* | |
| Hydrogels | Natural polymers such as alginate or collagen |
| Scaffolds | Urethane foams |
| *Membranes* | |
| Hemodialysis membranes | Polysulfone, PAN-VC with cutoff around 50,000 daltons |
| Weak polyelectrolysis | Films of polylysine and alginate |
| Conformal | Adherent surface coating |
| Microporous | Freely permeable to proteins (allograft only) |
| Reinforced | Membranes cast on a reinforcing structure |

donor, isolated, and purified. Since transplant-quality human cells, like human organs, are exceedingly scarce, investigators turned to porcine or bovine sources. Tissue was usually broken down into single-cell suspensions. For the pancreas, cells were maintained as islets, clusterlike organoids containing several thousand insulin-producing $\beta$ cells. Primary cells were true transplants; the graft originated in a living donor. The second class of cells to be developed for encapsulation was dividing cells from immortalized cell lines. Here, grafted cells originated in tissue culture flasks or in bioreactors. For example, the well-established PC-12 cell line was used to form capsules that secreted dopamine for the treatment of experimental Parkinson's disease. Of course, dividing cells did not continue to multiply indefinitely inside a capsule. Rather they reached a numerically stable population set by the carrying capacity of the capsule. Dividing cell lines vastly simplified the procurement of uniform stable populations of cells and fit in well with cell-banking techniques in widespread use in the biopharmaceutical industry. But only a few cell lines producing therapeutically useful products were available. So the next step was to engineer dividing cells so by adding the genes necessary to produce a desired protein. Engineered cells could be selected for high output (upwards of 1,000 ng/24h-million cells) and could be engineered to produce virtually any protein whose gene was available. More recently, genetic engineering has been employed for a different purpose: cells are being fine-tuned at the molecular level for their future life inside an immu-

noisolatory capsule. Using strategies borrowed from whole organ xenotransplantation, restriction factors have been added to the surface of grafts reducing their susceptibility to host complement. And antiapoptotic genes have been incorporated into the cell genome to enhance both performance and survival under the low-oxygen and nutrient-scarce environment that awaits cells after encapsulation. Investigators are also attempting to develop cell lines with a simple chemical switch capable of turning on and off proliferation to allow large-scale expansion of a cell line, which would not subsequently divide, inside a capsule.

## Implant Design

A wide variety of designs has been advanced to combine cell, matrix, and membrane into a device suitable for implantation. Designs vary widely but in all cases the maximum distance of a grafted cell from host is 200 to 500 $\mu$m. Cells further away from their ultimate source of nutrients and electrolytes simply fail to thrive. Several additional considerations apply to the selection of device design. Small beads or "microcapsules" have the optimal ratio of surface area to volume and require no redesign when scaling from small animal models to larger species. But once implanted they cannot be retrieved, and microcapsules do not conform readily to conventional medical-device or pharmaceutical quality assurance protocols. Larger implants or "macrocapsules" contain the entire therapeutic dose in a single device. These may be in the form of a sandwich of flat sheet membranes sealed at the periphery or a capillary sealed at the ends. These devices have the look and the feel of medical grade implants. They can be tested for membrane and seal integrity and some designs can be even be filled, or refilled, after implantation. All can be retrieved, and this feature facilitates the regulatory approval process. However, these designs are limited in the number of cells which can be contained in a practical design to around 5 million for a capillary device and around 100 million for a flat sheet device. Moreover, species scaling is complex. Consider the fact that a typical human weighs about 3,000 times more than a laboratory mouse, and it is problematic to make an implantable device 3,000 times larger, especially while not exceeding a thickness limitation of 500–1,000 $\mu$m. Flow-through designs, whether implantable or extracorporeal, draw heavily upon the technology already employed in available membrane devices such as hemodialyzers, blood oxygenators, and plasma filters. They have very large capacity and can hold several grams of foreign tissue if desired. Some designs incorporate a second set of fibers to provide oxygen to the mass of cells. Flow-through devices are well suited for use in extracorporeal circuits. But current flow-through implants are far more invasive than passive diffusion designs and require both surgical breaches of the host vasculature and lifelong, low-level anticoagulation.

## THERAPEUTIC APPLICATIONS OF IMMUNOISOLATION

Serious preclinical and clinical evaluation of immunoisolation began early in the 1990s. Funding availability, investigative resourcefulness, and simple happenstance have all contributed the relative rates of progress of the different applications, but the pacing item has usually been the type of cell required. Pain control and liver support both use primary cells; they have little else in common but both have moved into advanced clinical trials. In contrast, applications involving engineered cells and the delivery of gene therapy continue to offer enormous promise but remain earlier in their development. Engineered cells releasing ciliary derived neurotrophic factor (CNTF) have been studied in small-scale clinical trials for neurodegenerative diseases such as amyotrophic lateral sclerosis (ALS) and Huntington disease, but most gene therapy applications are still at the preclinical stage of development. And despite its formative role in the field, a clinically useful bioartificial pancreas remains an unfulfilled goal.

## Alleviation of Chronic Pain

The implant used for chronic pain comprises a short length of hollow fiber is filled with chromaffin cells harvested from calf adrenal glands. Adrenal chromaffin cells, so called because of their affinity for certain dyes, secrete the body's natural analgesics. These natural biochemicals are known to moderate pain by a variety of mechanisms including a partial blockage of the presynaptic release of the neurotransmitters required to send pain impulses from one neuron to the next. Prior to human evaluation, encapsulated chromaffin cell implants were shown to be highly effective in animal models of both acute and chronic pain. Device design and implant technique, which is analogous to a spinal tap, were finalized in sheep. The spinal column was chosen as the implant site because it provides an ideal anatomical location to intercept nerve impulses travelling from the peripheral to the central nervous system. The site is immunopriviliged and further allows the implant to be bathed in cerebrospinal fluid. The safety of the implant and its ability to maintain cell viability and secretory function has been confirmed in detailed animal models, a preliminary clinical investigation, and a Phase I clinical trial. The more than 100 patients in these series represent the most extensive clinical experience to date with cross-species transplantation of living tissue. The effectiveness of this therapy is now being evaluated in pivotal clinical trial to evaluate efficacy in a randomized, placebo-controlled, double-blind, multicenter trial. An empty implant was used as the placebo. Results of this trial are expected around the year 2000.

## Support of the Failing Liver

The liver is among the most biochemically complex organ of the human body and one of the few with a capacity for self-regeneration. Serious liver failure is a discouraging prognosis with the option of transplantation numerically limited to the fortunate few. The bioartificial liver is intended to maintain critical liver failure patients until a transplantable organ becomes available. The approach is simple in concept and more governed by the dictates of surgical pragmatism than the principles of pharmacological elegance. Hepatocytes are harvested from a porcine source and placed in the extraluminal chamber of a membrane exchange device designed along the principles of a hollow fiber hemodialyzer. Blood flows on one side of the membrane, the cells are on the other. In contrast to the pain implant, which contains a few milligrams of cells and is expected to last function months or years, the liver device contains between 20 and 200 g of purified hepatocytes, which are used for a very short period of 6 to 24 hours. Here, the mission of the membrane barrier is to prevent cell–cell contact while allowing protein traffic; hence, open microporous membranes are favored. Donor hepatocytes are believed to help clear substances normally removed by the liver and may even be active in synthesis. The procedure has been evaluated clinically in a Phase I trial involving just under 40 patients with diagnosed terminal liver failure. The device functioned as expected with no untoward safety problems; this has cleared the way for a larger multicenter randomized trial, which began with U.S. FDA approval late in 1998. Clinical response in the Phase I trial were very impressive indeed. However, the path to liver support systems is littered with a history of interventions that raised hopes in initial evaluation only to confound early enthusiasm in larger-scale application. The results of the Phase III trial, which will be conducted at several centers in the United States and Europe, are expected in 2002. Although the study design and stated goal is bridging patients to transplant, investigators will look carefully for signs of liver regeneration; the ultimate goal of liver support is recovery of liver function without transplantation.

The question naturally arises whether it is reasonable and prudent to obtain living components of implants and other biological devices from bovine (chromaffin cells) and porcine (hepatocytes) sources. FDA regulations for good manufacturing procedures (cGMPs) are scrupulously observed in cell harvesting to assure purity, identity, and freedom from foreign microorganisms. Concern about endogenous retroviruses from porcine sources or other adventitious zoonotic agents is another matter. No infectious pathogens have been identified in extensive testing and monitoring to date. An intact membrane would provide a formidable barrier for viral or retroviral transmission from graft to host. The clinical course of all patients is being closely monitored. Investigators are neither cavalier or complacent and recognize that experience in xenogeneic cell therapy may well shape the equally as important area of whole organ xenotransplantation.

## Parkinson's Disease

Parkinson's disease results from a deficiency of dopamine in the striatal region of the brain. Studies conducted in the early nineties used encapsulated PC-12 cell implants as a local source of dopamine in animal models of this condition. Although a naturally occurring mutant, PC-12 cells displayed most of the features of engineered cell lines including capacity for unlimited division and high constitutive release of a therapeutically useful product. Despite generally successful and widely published results in rodents and primates, the approach was never evaluated clinically. Numerous other avenues to treating Parkinson's disease are being investigated, including stem cell therapy, use of encapsulation to deliver growth factors to regenerate the brain's own dopamine producing circuitry, and human fetal transplantation.

### Delivery of Gene Therapy

Experience with PC-12 cells persuaded workers in the field of both the feasibility and practical benefits of dividing cell lines for encapsulation. Molecular biologists set about with characteristic enthusiasm to create multiple families of implantable cell lines producing medically useful proteins. In the broader sense, and as illustrated in Figure 2, this strategy amounted to use of immunoisolation as a pathway for delivery of gene therapy. Genes of interest were isolated and inserted into a carrier cell line using the conventional techniques of molecular biology. Cells were expanded, standardized, encapsulated, and implanted. The gene product was thereafter produced in situ. Because cells producing the gene were enclosed in a membrane capsule, investigators could reverse the experiment and remove the gene should this prove necessary or desirable. Some investigators preferred human-derived cell lines because allogeneic grafts do not normally secrete antigenic proteins and are thus easier to immunoisolate. Others advocated xenogeneic cell lines because they would be immediately destroyed in the event of escape from the capsule. Both approaches have merit and it is too early to tell which will prevail. A middle ground is also a possibility: xenogeneic cells can be humanized with complement restriction factors and other immune down-regulators. Alternatively, a variety of markers could be added to human cell lines to aid with their identification and destruction by the host immune system outside the capsule.

The first successful marriages of gene therapy and immunoisolation were reported in 1992 with cells transfected with the genes for neurotrophic factors and of lysosomal enzymes. Since that time a score of different protocols have been described in the literature (Table 2), and the number is likely to proliferate. In the first reported clinical trial, 6 patients diagnosed with ALS received implants containing a rodent cell line producing CNTF. Numerous preclinical studies and clinical trials not involving encapsulation had suggested that CNTF might be able to retard the deterioration of neurons that characterizes ALS and causes the eventual loss of peripheral motor activity. The device and implant protocol were very similar to that used for chronic pain; only the cells differed significantly from previous experience. In this trial, implanted cells survived and continued to release CNTF over the life of the experiment. The concentration of CNTF in monthly samples of spinal fluid averaged around 500 pg/mL. Progression of the disease was not markedly reduced relative to standard care although a much larger and longer trial would be needed to draw any firm conclusions about efficacy. Nevertheless, the intervention demonstrated production and delivery of the desired gene product at clinically useful concentrations and time frames. In this context the trial was at least as successful as the vast majority of clinical studies based upon ex vivo gene therapy. A second protocol has been initiated in Huntington patients, delivering the same product but this time with capsules implanted into the ventricles rather than into the cerebrospinal fluid. No results are yet available. Several additional applications are being evaluated in small and large animal models, as summarized in Table 2.

**(a)** Gene is added to viral vector.

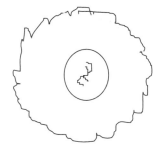

**(b)** Vector implants gene into nucleus of carrier cells.

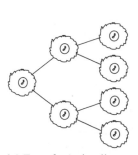

**(c)** Transfected cells are selected and proliferated.

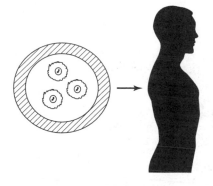

**(d)** Cells are encapsulated and implanted in host.

**Figure 2.** Encapsulation has been used as a method of delivering gene therapy. (**a,b**) Conventional molecular biological techniques are employed to add a gene coding for a therapeutic protein to a dividing cell line. (**c**) Modified cells are grown divided and grown up in vitro. (**d**) A population of the modified cells are then encapsulated and implanted in a host. This approach has been successfully applied in animal models for delivery of erythropoietin, factor IX, several neurotrophic factors, and other proteins implicated in monogenetic diseases. It has also been used clinically to deliver the neurotrophic factor CNTF to patients with amyotrophic lateral sclerosis and Huntington disease.

**Table 2. Encapsulation for Delivery of Gene Therapy**

| Target disease | Gene | Cell type | Status |
|---|---|---|---|
| ALS | *CNTF* | Hamster-BHK | Clinical, Phase I |
| Huntington | *CNTF* | Hamster-BHK | Clinical, Phase I |
| Hemophilia A | Factor VII | Murine | Preclinical, Rodent |
| Dwarfism | HGH | Murine myoblasts | Preclinical, Rodent |
| Anemia | EPO | Murine | Preclinical, Rodent |
| Type II diabetes | Glucagon-like peptide I | Murine | Preclinical, Rodent |
| Parkinson's | GDNF | Murine | Preclinical, Primate |
| Cancer | Cytochrome P450 2B1 | Murine | Preclinical, Rodent |

With its technical feasibility validated by the ALS trial, gene therapy is viewed by many as the most compelling application of immunoisolation. One practical problem is that biopharmaceutical firms owning rights to genes have usually invested massively in production facilities. Even a modest GMP level facility that produces a pharmaceutical product based upon recombinant protein costs upwards of $200 million. Such firms are understandably restrained in their enthusiasm for approaches that would render such facilities redundant. At the same time, the biotech start-ups with the technology for immunoisolation rarely have the resources to develop or acquire proprietary genes. For this reason commercial interest in encapsulation has tended to focus on therapeutic targets, such as the central nervous system, which remain beyond the reach of traditional drug development. Also favored are applications where cells release products in response to physiologic stimuli (e.g., the bioartificial pancreas) or metabolize circulating blood components (e.g., the bioartificial liver). Encapsulation may not fulfill its clinical promise of in situ drug delivery until the first round of gene products come off-patent.

## The Artificial Pancreas

The bioartificial liver and cell therapy for chronic pain are in Phase III clinical trials. Immunoisolated gene therapy has reached the clinic and is fecund with promise. What then of the bioartificial pancreas? The path from first inspiration to clinical success takes many strange twists, and this is well illustrated by the history of the bioartificial pancreas. Successful demonstration of the reversal of induced diabetes in rodents in the mid-seventies by immunoisolated cells was the original defining moments in the origin of tissue engineering. By now similar results have been routinely obtained in dozens of laboratories around the world. In fact, so compelling is the treatment of induced diabetes in rodents that no fewer than five original reports on this subject have appeared in *Science* magazine. But attempts to obtain similar results in larger animals, canines and primates, have generally proven disappointing. There is simply no widely accepted set of large animal experiments demonstrating long-term satisfactory reversal of diabetes and normalization of glucose kinetics. Even marginal results in larger animals have often required supplementary use of immunosuppression or some level of injected insulin.

Much of the complexity with the bioartificial pancreas comes down to the fact that about 700,000 islets, containing about 2 billion $\beta$ cells, would be required for transplantation. (This is two orders of magnitude more cells than have been successfully encapsulated in clinical implants to date.) Harvesting so many cells is one problem and housing them in a practical implant is another. Most successful rodent studies involve transplantation of about 500 rat islets to a mouse recipient. The 500 islets are usually handpicked from a partially digested rat pancreas. It is simply not practical to handpick the 700,000 islets required for a human graft, so investigators use semiautomated cell harvesting devices. Techniques for mass isolation have never reached the stage of consistently producing healthy batches of large islets given the natural variability in tissue sources and digestive enzymes. Furthermore, an islet is a not just a cluster of cells but a mini-organoid that enjoys its own vascular supply to distribute oxygen and nutrients to all of its constituent cells. In contrast, implants rely upon diffusion to provide adequate oxygenation and nutrition. The diffusion approach is manageable in small devices but very challenging in implants of the size required to house upwards of $10^9$ cells. Perhaps the most satisfactory results to date have been achieved with flow-through designs but these are highly invasive and their use would be justified only in the most threatening of patient circumstances. Yet another complexity is the dose of soluble antigens released by the large quantity of cells needed to replace pancreatic function. The issue exists with all implants but is far more formidable when grams rather than milligrams of tissue are implanted. Over the past two decades, these and other difficulties have frustrated the efforts of several brilliant and dedicated investigative teams to reproduce in large hosts the tantalizing results obtained early on in rodents.

Fortunately, the resourcefulness of molecular biology has opened a second path to the bioartificial pancreas. At least three groups have set about to develop glucose-response, insulin-secreting cell lines. Because of the mass of tissue involved, cell lines of human origin, or at least highly humanized animal cells, seem preferred. Cells must not lose integrity of function over 25–50 passages. To make device size manageable, these cells should produce insulin at least on the order of 1 ng/$10^6$ cells-24 hours. They cannot be overly sensitive to hypoxia or other insults encountered by encapsulated cells. At the same time, such cells cannot threaten the patient should they escape the immunoisolatory implant. Ten years ago, development of such a cell line would have been considered a quixotic undertaking. Today, emboldened by increasing knowledge of glucose regulation in the normal β cells and the emerging portfolio of antiapoptotic molecular technology, many investigators believe it to be within reach. We expect to see impressive large-animal results with manmade β cell lines within the next five years and hopefully, this time, a more rapid progression to clinical success.

## CONCLUDING PERSPECTIVES

In 1955, the term *artificial organ* provoked skepticism, incredulity, and even hostility. Twenty years later in 1975 both transplantation and organ replacement were medically accepted and rapidly growing. Today, two score years later, multiple manifestations of high technology substitutive medicine accounts for about 7% of the total global health care enterprise. In like fashion, we fully expect over the next two decades that tissue engineering, including immunoisolated cell therapy, will emerge from its investigative origins and play a fulcral role in twenty-first-century medicine.

## ACKNOWLEDGMENTS

Ms. Jacqueline Sutherland assisted the author with the preparation of the manuscript for this article.

## ADDITIONAL READING

P. Aebischer et al., *Nat. Med.* **2**, 1041 (1996).
P. Aebischer and M.I. Lysaght, *Xeno* **3**, 43–48 (1995).
P. Aebischer et al., *Exp. Neurol.* **126**, 151–158 (1994).
J. Brauker et al., *Transplant Proc.* **24**, 2924 (1992).
J. Brauker et al., *Hum. Gene Ther.* **9**, 879–888 (1998).
E. Buscher et al., *Anesthsiology* **85**, 1005–1012 (1996).
S.C. Chen et al., *Ann. NY Acad. Sci.* **831**, 350–360 (1997).
C.K. Colton, *Trends Biotechnol.* **14**, 158–162 (1996).
K.E. Dionne, B.M. Cain, and R.H. Li, *Biomaterials* **17**, 257–266 (1996).
N. Deglon et al., *Hum. Gene Ther.* **7**, 2135–2146 (1996).
C. Ezzell, *J. NIH Res.* **7**, 47–51 (1995).
P.E. Lacy et al., *Science* **254**, 1782–1784 (1991).
R. Lanza and W. Chick, *Ann. NY Acad. Sci.* **831**, 323–331 (1997).
P.J. Morris, *Trends Biotechnol.* **14**, 163–167 (1996).
C. Rinsch et al., *Hum. Gene Ther.* **8**, 1881–1889 (1997).
M.E. Sugamori and M.V. Sefton, *Trans. Am. Soc. Artif. Intern. Organs* **35**, 791–799 (1989).
S.J. Sullivan et al., *Science* **252**, 718–721 (1991).
B.A. Zielinski and P. Aebischer, *Biomat.* **15**, 1049–1056 (1994).

# IN VITRO–IN VIVO CORRELATION

STEPHEN S. HWANG
SUNEEL K. GUPTA
ALZA Corporation
Palo Alto, California

## KEY WORDS

Bioavailability
Compartmental analysis
Correlation
Deconvolution
Dissolution
Extended-release product
GI absorption
In vitro evaluation
In vivo evaluation
Linear system analysis
Manufacturing control
Oral dosage forms
Pharmacokinetics

## OUTLINE

## INTRODUCTION

Pharmaceutical companies rely heavily on the in vitro dissolution or release test to develop extended-release products and to ensure their performance in vivo. In 1971, Wagner (1) had already stated that "future research in dissolution rates should be directed mainly towards establishing correlations of the in vitro data with in vivo data." A strong correlation between the in vitro and the in vivo results means that the in vitro test results can predict the in vivo performance accurately and therefore indicates the test's usefulness as a tool for development and production control. To reach a valid correlation, one has to have valid methods to yield meaningful measurements, both in vitro and in vivo. It is also important that the scientific community reach a consensus about what it means by in vitro/in vivo correlation. The completion of these criteria was marked by the publication of "Stimuli" by the U.S. Pharmacopeial Convention's (USP's) subcommittee on Biopharmaceutics in *Pharmacopeial Forum* in 1988 (2) and the workshop on in vitro and in vivo testing and correlation for oral controlled/modified-release dosage forms in December 1988 cosponsored by the American Association of Pharmaceutical Scientists, the U.S. Food and Drug Administration (FDA), the Federation Internationale Pharmaceutique, and the USP. The outcome of this workshop is documented in the workshop report (3).

This article discusses the accepted definition of correlation, the available methods for evaluating in vivo data, the commonly used in vitro testing methods, the currently accepted methods of establishing the correlation, and the possible applications of such a correlation. This article also includes many of the published works on the correlation of the in vivo and the in vitro results. The examples include the correlation between the in vivo human data and the in vitro data, the in vivo dog data and the in vitro data, and the human data and the dog data.

## USP DEFINITION OF CORRELATION

The USP (4) defines three levels of in vitro and in vivo correlation: level A, level B, and level C. Level A is the highest category of correlation. It represents a point-to-point relationship between in vitro dissolution and in vivo input rate of the drug from the dosage form. Level B correlation examines the relationship between the mean in vitro dissolution time and the mean in vivo residence time or mean in vivo dissolution time. Like level A, level B utilizes all the in vitro and in vivo data but is not considered to be a point-to-point correlation because many different plasma drug concentration curves can give identical mean residence time values. Level C correlation examines the relationship between one dissolution time point (e.g., $t_{50\%}$) and one in vivo pharmacokinetic parameter (e.g., AUC, $C_{max}$, $T_{max}$). It represents a single-point correlation. It does not reflect the complete shape of the plasma drug concentration curve, the critical factor that defines the performance of modified-release products.

## IN VIVO EVALUATION

### Pharmacokinetic Study Design

To define the in vivo release profile of an extended-release product, one must include all necessary treatments in the study. Before discussing these treatments in the pivotal study, one should understand the drug kinetics over the dose and concentration range of interest, the mode of dosing (fed versus fasted), the time of dosing, the effect of body posture (5) or the influence of exercise (6), and many other details. It is beyond the scope of this article to enumerate what factors could affect the drug kinetics. In the simplest case when the drug kinetics is linear, time invariant, and not affected by any usual physiological activities or food, then the study needs minimally to include two treatments: the extended-release product and a reference product. Based on the concept of gastrointestinal (GI) bioavailability (7), the reference is ideally an oral solution. This reference can be modified without compromising the stated objective by using an oral suspension or an immediate-release dosage form, such as a tablet or a capsule. The basic idea is to have the reference readily available in the

GI tract for absorption. For in vitro/in vivo correlation, one could use an intravenous dose as the reference as well. This reference defines the drug disposition function and hence a clear absorption profile. However, this absorption profile differs from the in vitro release profile by a bioavailability factor. Administered as a bolus, the reference gives a direct observation of the characteristic function. From a purely theoretical point of view, the reference product can be administered with any known input function (e.g., constant infusion). In such a case, the characteristic function is not directly observed but derived.

In most cases, not everything is known about the drug kinetics. In such cases, one should be conservative in the study design to guard against many aforementioned experimental factors that may affect the drug kinetics. Instead of a single administration of the reference at any dose, one would use the actual therapeutic dose/regimen of the immediate-release product that the extended-release product is trying to match. For example, a 240-mg pseudoephedrine extended-release product designed for once-daily administration can be compared to four 60-mg doses of an oral solution given at 6-h intervals (8). This approach gives a useful direct pharmacokinetic comparison between the extended-release product and a product given at a therapeutically effective regimen. It guards against any complications such as nonlinear kinetics, food effects, and diurnal effects, which can invalidate a simulation based on the data from the simpler design. This design can be further expanded to study both the extended-release product and the immediate-release product at several dose levels to account fully for possible dose dependence in the kinetics. As usual, the study is limited by the number of blood samples that can be withdrawn from healthy volunteers. This limitation unavoidably results in realistic compromises in study design and objectives. The overall goals can be reached only through many studies that individually address a piece of the puzzle.

**Traditional Crossover Design.** Intersubject variability in pharmacokinetic parameters is typically much larger than intrasubject variability. Therefore, it is more efficient to use a crossover study design to discover the subtle differences between formulations. In a typical randomized two-way crossover study, half the subjects receive extended-release product followed by the reference product; the other half receive these two treatments in the reverse order.

**Stable Isotope-Labeled Drug As Reference.** Although the intrasubject variability in pharmacokinetic parameters is small when compared with the intersubject variability, it may still be large and demand a large sample size, and it makes the results of a small pilot study unreliable. Intrasubject variability can be reduced by giving both treatments concurrently as long as the analytical method can differentiate between the two different drug sources. This differentiation can be accomplished by labeling both sources of drug with different isotopes or labeling one source of drug with an isotope. For example, the reversible metabolism of sulindac has been investigated using different radioactive isotopes to label sulfoxide and sulfide (9).

The use of stable isotope-labeled drug has been reviewed by Browne (10) and is an excellent approach to increase the reliability of the results when the basic assumptions are met. A middle-of-the-road approach is to give two treatments sufficiently apart such that the carryover drug from the first treatment can be properly modeled from the data (11).

**Scintigraphic Monitoring.** Scintigraphic monitoring technology has blossomed over the past 10 years. Today, a trace amount of lanthanides can be placed into a commercial formulation. These metals, when neutron activated, become radioactive, providing the means for following the dosage form in vivo. During earlier studies, researchers used this method to account for the bioavailability variations potentially caused by the dosage's short transit time (12). As we begin to realize that both the rate and the extent of drug absorption could be site dependent, the appropriate derivation of the in vivo release profile becomes almost impossible without the location information.

## Mechanistic Considerations

After ingestion, the drug must be released from the dosage form and then dissolved. For some dosages, the release and dissolution can occur in one step rather than by two separate processes. The dissolved drug is simultaneously being degraded, emptied from any location, and absorbed through the GI membrane. The degradation, either chemically or microbially, can differ significantly from stomach to large intestine. The absorbed drug will then pass through the liver before appearing in the systemic circulation. The drug could be metabolized by the GI membrane and the liver, a process that is collectively known as first-pass metabolism because it happens before the drug's appearance in the systemic circulation. First-pass metabolism occurs for all drugs that undergo liver metabolism. The first-pass extraction ratio, E, accounts for part of the bioavailability loss. When the first-pass metabolism follows linear kinetics, then the extent of first-pass metabolism should remain unchanged regardless of the release rate from the dosage form. The impact of a nonlinear first-pass process is much more complex and has been discussed elsewhere (13).

## Theory

This section defines the mathematics in the derivation of the in vivo release profile of an extended-release product from the pharmacokinetic data generated from a study as mentioned earlier. The methods generally derive from compartment models or linear system properties. Mathematically, the two approaches yield the same answer. The linear-system–based deconvolution method makes fewer assumptions about the system, giving a more robust derived answer; the compartment approach gives a conceptual picture of the process and allows scientists to derive the necessary equations with the comfort of a physical process instead of mathematical abstraction.

**History.** It is not surprising, given the long history of applying compartmental analysis techniques in endocri-

nology, that the early development of pharmacokinetics is mostly, if not completely, based upon the compartment analysis. In 1963, Wagner and Nelson (14) published their method of deriving the absorption profile from pharmacokinetic data for a drug with a one-compartment disposition model. When applied to an oral data set alone, the method implies that the function describing the entire plasma drug concentration profile can be described by the convolution of two functions: a staircase-step function for the absorption process and a single-exponential disposition function. The method also requires that the absorption is completed before the last plasma drug sample in order to have a clean elimination phase. Loo and Riegelman (15) reported a mathematical equation to calculate the amount of drug absorbed when the drug disposition follows a two-compartment model. The application of a linear-system–based deconvolution approach to the pharmacokinetic data seems to have originated with the publication by Benet and Chiang (16). However, the publication lacks an explicit mathematical expression and numerical algorithm. The application of a linear-system approach to pharmacokinetics was broadly addressed by Cutler (17) in 1978. That same year, the linear-system–based deconvolution methods were reported by Vaughan and Dennis (18) and Cutler (19) using staircase function and polynomials as input functions.

**Compartment Models.** The cumulative amount of drug absorbed at any time $t$, $A_a(t)$, is the sum of the amount present in the body and the cumulative amount eliminated. The cumulative amount eliminated from the body is:

$$Cl \int_0^t C(\tau)d\tau$$

where $Cl$ is the body clearance for the drug and $C$ is the plasma or serum drug concentration. The clearance always derives from the dose, $D_{iv}$, and the area under the curve, $AUC_{iv}$, after a dose of the drug is given intravenously. The equation defining the amount of drug present in the body at time $t$ is disposition-model dependent. For a one-compartment disposition model, the amount of drug present in the body at time $t$ equals $C(t) V$, where $V$ is the distribution volume for the drug. Therefore, the cumulative amount of drug absorbed at any time t, $A_a(t)$, can be calculated from the following equation:

$$A_a(t) = VC(t) + Cl \int_0^t C(\tau)d\tau$$

If the drug-disposition model includes a peripheral compartment, compartment 2, then the equation defining the total amount of drug present in the body needs to include the amount of drug present in compartment 2. According to Wagner (20), the amount of drug in compartment 2 is given as $k_{12}e^{-k_{21}t}\int_0^t C(\tau)e^{k_{21}\tau}d\tau$. Thus, the cumulative amount of drug absorbed at any time t for a drug following a two-compartment disposition model is defined by the following equation:

$$A_a(t) = VC(t) + Cl \int_0^t C(\tau)d\tau + k_{12}e^{-k_{21}t} \int_0^t C(\tau)e^{k_{21}\tau}d\tau$$

The amount of drug absorbed is not a direct measurement of the in vivo performance of a dosage form. A theoretical purist would insist that the more appropriate measurement for the correlation to an in vitro release/dissolution measurement should be the corresponding in vivo measurement, such as the in vivo release/dissolution rate. This theoretical concern can be addressed by incorporating an absorption phase in the in vitro measurement to reflect the in vivo absorption process. The difference between the cumulative amount of drug released/dissolved, $A_r(t)$, and the cumulative amount absorbed is simply the amount of drug present in the GI tract (21), $G(t)$. When the GI absorption process is governed by a first-order process with a rate constant, $k_a$, and a bioavailability factor, $F$, then the amount of drug present in the GI tract is mathematically defined by the absorption rate.

$$G(t) = \frac{dA/dt}{Fk_a}$$

In this scenario, the cumulative amount of drug released/dissolved from an extended-release product is expressed as

$$A_r(t) = \frac{V}{k_aF}\left[\frac{dC}{dt} + (k + k_a)C + k_ak\int_0^t C(\tau)d\tau\right]$$

for a drug following the one-compartment disposition model, and

$$A_r(t) = \frac{V}{k_aF}\left[\frac{dC}{dt} + (k_{12} + k_{10} + k_a)C \right.$$
$$\left. + k_{10}k_a\int_0^t C(\tau)d\tau + (k_a - k_{21})k_{12}e^{-k_{21}t}\int_0^t C(\tau)e^{k_{21}\tau}d\tau\right]$$

for a drug following the two-compartment disposition model.

**Linear System.** In a linear-system approach, one considers the body as a system that yields responses based on the input. One of the necessary and sufficient conditions for a linear system is a response proportional to the input. When the input function is an impulse that is equivalent to a bolus dose, the response of the system is a characteristic function, $C^*$. For any input function, $in(\tau)$, the resulting plasma drug concentration, $C(t)$, is given as

$$C(t) = \int_0^t in(\tau)C^*(t - \tau)d\tau$$

The convolution integral is a simple multiplicative relationship in the Laplace domain, (i.e., $c(s) = in(s)c^*(s)$ where $c(s) = L\{C(t)\}$, and so on) and the derivation of the input function is a straightforward division, $in(s) = c(s)/c^*(s)$. When the impulse input is given intravenously, the characteristic function defines the disposition process, and the input function describes the input to the site where the

impulse dose is given. This input to the impulse-administering site is commonly understood as the amount absorbed, although it is not an exact measurement of absorption across the GI-tract membrane unless the impulse input is given at the absorption site of interest (e.g., the portal vein). Although the mathematics are deceptively simple, numerical assumptions are necessary to make the convolution equation useful. Here we follow the work of Vaughan and Dennis to illustrate the application of the deconvolution method to pharmacokinetic data. First, assume that the input is a staircase function (i.e., it consists of a finite set of rectangular pulses of intensity, $I_i$, from time $t_{i-1}$ to $t_i$, with $t_0 = 0$ and $t_i$ corresponding to the plasma drug sampling time). The input function can be explicitly given as follows:

$$I_1 = \frac{C(t_1)}{\int_0^{t_1} C^*(\tau)d\tau}$$

$$I_2 = \frac{C(t_2) - I_1\int_{t_1}^{t_2} C^*(\tau)d\tau}{\int_0^{t_2-t_1} C^*(\tau)d\tau}$$

$$\vdots$$

$$I_n = \frac{C(t_n) - \sum_j I_j\int_{t_j}^{t_n} C^*(\tau)d\tau}{\int_0^{t_n-t_{n-1}} C^*(\tau)d\tau}$$

This approach uses all the data from the impulse input without imposing a model and simply assumes the input as constant between sampling intervals. It is simple and easy to understand. Unfortunately, the inaccuracy of the deconvolution output tends to be larger than desirable because this method accepts every data point as exact and passes all noise to the derived input function.

For drugs undergoing reversible metabolism, the input can be a combination of the drug and its metabolite(s). Mathematically, the input is a vector, and the deconvolution process discussed above has to be expanded into an $n$-dimensional problem. This case has been reported elsewhere (22).

**Linear-System Basis in Compartment Model–Derived Equations.** Any linear compartment model meets the definition of a linear system. Therefore, it should not be surprising that any equation that derives from linear-system properties alone should also be applicable to any linear compartment model. However, the reverse is not true because the equations may have been based on assumptions specific to the particular compartment model. A compartment-model–based absorption equation may be invalid if one doubts whether the compartment model truly describes the drug disposition regardless of how closely the model can describe the data. A good example is the presence of three two-compartment models to describe the same plasma drug-disposition profile (23). A more general discussion of the model identification issues is found in other references (24). Interesting points related to the influence of the sampling site and lung metabolism on the derived pharmacokinetic parameters have also come up. Therefore, it is important to understand the deconvolution from the linear system perspective and determine how much of the derived information is independent of the model selection.

Any linear compartment model is a linear system with a multiexponential characteristic function

$$\sum_{i=1}^{n} A_i e^{-k_i t}$$

When the impulse dose is given intravenously, the multiexponential function is continuously declining and

$$\sum_{i=1}^{n} A_i > 0$$

The input function was derived by Veng-Pedersen (25). Wagner made the comparison between this approach and the compartment-model-derived absorption equations in 1983 (20). Specifically, the Laplace transformation of a monoexponential function, $a_1 e^{-k_1 t}$, is $a_1/(s + k_1)$, and the Laplace transformation of a biexponential function, $a_1 e^{-k_1 t} + a_2 e^{-k_2 t}$, is $a_1/(s + k_1) + a_2/(s + k_2)$. In the Laplace domain, the input function for a monoexponential characteristic function can be solved:

$$in(s) = \frac{c(s)}{a_1/(s + k_1)}$$

The solution for the cumulative input function is

$$A_a(t) = \frac{1}{a_1} \left[ C(t) + k_1 \int_0^t C(\tau) d\tau \right]$$

When a one-compartment disposition model is imposed, then $1/a_1$ becomes the distribution volume of the compartment. However, one can derive this equation without imposing the compartment model. Similarly, the input function for a biexponential system can be derived:

$$in(s) = \frac{c(s)}{a_1/(s + k_1) + a_2/(s + k_2)}$$

$$\frac{in(s)}{s} = \frac{c(s)}{a_1 + a_2} \left( 1 + \frac{\lambda_1 \lambda_2}{Rs} \right.$$
$$\left. + \left( \lambda_1 + \lambda_2 - R - \frac{\lambda_1 \lambda_2}{R} \right) \frac{1}{s + R} \right)$$

where $R = a_1 \lambda_2 + a_2 \lambda_1 / a_1 + a_2$.

The explicit cumulative-amount-absorbed equation can be solved by imposing inverse Laplace transformations on the previous equation:

$$A_a(t) = \frac{1}{a_1 + a_2} \left[ C(t) + \frac{\lambda_1 \lambda_2}{R} \int_0^t C(\tau) d\tau \right.$$
$$\left. + \left( \lambda_1 + \lambda_2 - R - \frac{\lambda_1 \lambda_2}{R} \right) Re^{-Rt} \int_0^t C(\tau) e^{R\tau} d\tau \right]$$

For the two-compartment model with elimination from the central compartment (1) only, the rate constants for the intercompartmental processes are commonly known as the following:

$$k_{21} = \frac{a_1 \lambda_2 + a_2 \lambda_1}{a_1 + a_2}$$

$$k_{10} = \frac{\lambda_1 \lambda_2}{k_{21}}$$

$$k_{12} = \lambda_1 + \lambda_2 - k_{21} - k_{10}$$

Thus, the equation for calculating the amount absorbed based on a biexponential characteristic function is identical to that derived from a two-compartment model. Therefore, the equation is valid even when the particular disposition model is not consistent with the actual situation such as elimination from the peripheral compartment.

### Numerical Algorithm

As mentioned previously, one way to approximate the input is to assume it is a staircase function. A polynomial function has been proposed by Cutler to approximate the input. Veng-Pedersen reported the use of a spline function (26) and a polyexponential function (27) to approximate the input function. One alternate approach is to assume the input function to be the solution of an $m$th-order ordinary differential equation (28), which is mathematically equivalent to a polyexponential function (29–32). The presence of physically impossible negative input caused by the data noise has been addressed by using a B-spline function (33) or a Fritsch-Butland nondecreasing cubic function to approximate the cumulative input function (34).

### Deviation from Ideal

Despite the mathematical simplicity and beauty, the real world is full of complications that, in some cases, invalidate the mathematics. Here are some of the complexities.

**In Vivo Dissolution.** It has been stated that one should use a readily available dosage as the reference to define the characteristic function, which is relatively simple for drugs with reasonable solubilities—that is, the therapeutic dose can be dissolved in a reasonable amount of water. When the drug is not so soluble, the situation becomes much more complicated. One must choose among a suspension, a nonaqueous solution, and an immediate-release dosage form, such as a tablet or a capsule. Each choice has its drawbacks. The most difficult issue is the uncertainty related to the in vivo dissolution for each choice.

**Absorption/Metabolism at Different GI Regions.** Studies have used various techniques, such as intubation and high frequency (HF) capsules, to investigate the extent of absorption of drugs introduced directly to the colon. For colonically introduced glibenclamide (35), acetaminophen (36), ondansetron (37), and nifedipine (38), the extent of absorption appears to be similar to that administered orally and supports the assumption of a uniform absorption process throughout the GI tract. Contrary to these uniform-absorption examples, the colonically introduced gepirone (39) has a higher bioavailability than an oral solution, and numerous drugs introduced into the lower GI tract have been reported to have a lower bioavailability than an oral solution—for example, ciprofloxacin (40), cimetidine, furosemide, hydrochlorothiazide (41), danazo (42), benazepril (43), isosorbide-5-mononitrate (44), amoxicillin (45), and sumatriptan (46). It seems prudent to investigate the absorption of the drug solution introduced into the lower GI tract experimentally. The usefulness of such information is demonstrated by an example—namely, the diminished bioavailability of an extended-release amoxicillin product that was published (47) after the publication of the amoxicillin intubation results. It is clear from these examples that useful information can be gained through independent experiment and that a simulation that takes account of location-dependent absorption, such as a model of $n$ absorption sites (48), will be more realistic than the uniform absorption assumption. The diminished absorption of metoprolol in the lower GI tract has been demonstrated by comparing the plasma drug concentration profiles after the intragastric infusion and an oral administration of an OROS® metoprolol tablet (49). The presystemic metabolism of a drug could be site dependent as well. Researchers have measured selegiline and its metabolites—$N$-desmethylselegiline, L-methamphetamine, and L-amphetamine—in the plasma samples collected after giving selegiline to eight healthy young males by the oral route and direct infusion to the duodenum, jejunum, and terminal ileum through a nasoenteric tube (50). The plasma selegiline concentration profile was the highest after oral dosing and the lowest after dosing to the terminal ileum. The plasma $N$-desmethylselegiline concentration profiles followed a pattern similar to that of selegiline; plasma $L$-methamphetamine and $L$-amphetamine concentration profiles were essentially the same regardless of the site of administration. This example clearly demonstrates the complexity of the problem.

## IN VITRO EVALUATION

Many methods exist for testing a dosage form for release/dissolution rate; more scientists are relying on the compendial methods. Typically, one uses USP apparatus 1, the basket method, and apparatus 2, the paddle method. The release/dissolution medium is typically aqueous, with surfactant if required. According to FDA guidance (51), the basket is run at 100 rpm, and the paddle is evaluated at 50 and 75 rpm. The dissolution medium (900 mL) of pH 1–1.5, 4–4.5, 6–6.5, and 7–7.5 is maintained at 37°C. The use of compendial equipment facilitates the across-laboratory comparison. However, this fact should not discourage scientists from exploring other methods to find a better-correlated test. Devane and Butler (52) gave a few examples of establishing in vitro/in vivo correlation where the adoption of initial dissolution method relies on the biopharmaceutical classification of the drug (53). The effect of food on the drug release from an extended-release product is a concern for the formulation scientists not just because of the regulatory agencies' requirement but also because of

the realistic expectation that many patients may take the product with food. It is therefore of interest that one can evaluate the food effect in vitro. One study investigated the in vitro dissolution behavior of four theophylline extended-release products utilizing the rotating-dialysis-cell method (54). The effects of pH, oil, and enzymes on the dissolution profiles were studied. The in vitro observations of the oil effect were related to the in vivo food effect reported in the literature, and the authors concluded that the rotating dialysis cell can be a useful tool in studying factors that may be responsible for dissolution-related food effects on the absorption of controlled-release products.

## CORRELATION BETWEEN THE IN VIVO AND THE IN VIVO PROFILES

In many examples, the correlation is established between the cumulative profiles. This practice may reflect

1. The way we measure the drug release/dissolution
2. The way Wagner and Nelson derived their equation
3. The fact that the cumulative profile is inherently stabler than the rate profile

A statistical concern exists with this approach: The observations at each time point are not truly independent. It is clear that the cumulative amount dissolved at time $t_i$ equals the cumulative amount dissolved at time $t_{i-1}$ plus the amount dissolved between $t_{i-1}$ and $t_i$. Therefore, the first observation is used $n$ times in the linear regression. It is statistically cleaner to use the incremental data rather than the cumulative data. Liu et al. (55) have demonstrated that the correlation of any two sets of cumulative fractions is inherently high. Even using randomly generated $x$–$y$ pairs, the average correlation coefficient was 0.935. Therefore, a $\chi$-square test using the incremental fraction data has been proposed to evaluate the correlation. It is more appropriate to show the correlation between the in vivo and the in vitro rates rather than the cumulative amounts.

In an example (56) of metoprolol tartrate extended-release tablet formulation, the authors evaluated several correlations' ability to predict the in vivo performance. This is an important criterion in selecting an in vitro method to control the in vivo performance of the extended-release product.

## EXAMPLES

The examples are collected to show how the correlation between the in vitro and the in vivo data is done in actual practice. The first group are examples using human in vivo data. The second group are examples using dog pharmacokinetic data. The third group are examples trying to correlate the human and the dog pharmacokinetic data.

The interest in correlation between the in vitro and the in vivo performance is not limited to the extended-release product. Actually, it seemed to have started with the work by Levy and Hollister (57) with the immediate-release aspirin tablet. In the case of immediate-release product, the correlation is theoretically more difficult because it is more difficult to separate the dissolution and the absorption processes when both are rapid.

Two commercial cinnarizine capsules have been studied in a group of 12 healthy male volunteers (58). The formulations have similar in vitro dissolution rates as measured by amount dissolved at 30 min at pH 1 and at pH 6 but very different dissolution rates at pH between 1 and 6. These two capsules had no distinguishable AUC and $C_{max}$ values in the subgroup of volunteers with high gastric acidity, whereas in the low-gastric-acidity group, they had statistically different AUC and $C_{max}$ values with a twofold difference in mean. This finding clearly demonstrated the importance of measuring the dissolution rate over a wide range of pH and understanding the important physiology parameters. In a study of five sugar-coated metronidazole tablets in 10 healthy male volunteers (59), there was no significant correlation between the in vitro dissolution parameters and the pharmacokinetic parameters. The authors stated that "although the in vivo findings seem to agree well with the dissolution rate/pH relationship, the relative values and the ranking between the tablets tested did not agree with those of dissolution rates." Therefore, the correlation between the in vitro and the in vivo performance seems to be more drug specific and difficult to generalize.

The relationship between in vitro and in vivo drug release from OROS® oxprenolol systems have been examined by analyzing the plasma drug concentration data from two pharmacokinetic studies (60). An intravenous dose was used as the reference in the first study, and an oral bolus was used as the reference in the second study. The in vivo release drug-release rate was derived using a numerical deconvolution technique (61). The in vivo drug release from the product in most volunteers followed the same pattern as measured in vitro; only after 6–8 h was the decline in the in vivo release rate somewhat greater than expected. The same product was also investigated by Bradbrook et al. (62). The percentage of dose remained to be absorbed as derived from the Loo-Riegelman method was almost identical to the percentage of dose remained to be released over the first few hours. The difference between the in vivo and the in vitro measurements seemed to increase with time but disappeared at the end of sampling period.

Two extended-release formulations of etodolac have been studied (63). In the pilot pharmacokinetic study, these two formulations were studied in 14 healthy male volunteers with an oral solution of equal dose. The in vitro dissolution data were used to simulate the expected plasma drug concentration for each formulation. There appeared to be a good agreement between the simulation and the observed plasma drug concentrations. The simulation was further carried out for formulations of other strengths used in a steady-state study. The in vivo performance was accurately predicted by the in vitro dissolution data.

Four cellulosic polymer-based matrix tablet formulations of chlorpheniramine maleate were studied in 24 healthy male volunteers (64). The amount of drug absorbed at various times was calculated according to the Wagner-Nelson method. The dosages were also evaluated

in vitro using USP apparatus 2 (50 rpm) and 1,000 mL of water. The plot of the amount released versus amount absorbed indicated a strong linear relationship for each formulation. The lines were approximately parallel with a different positive x-intercept for each formulation. After correcting for a lag time according to Levy and Hollister (57), a one-to-one correlation was established.

Three extended-release formulations of bromocriptine have been evaluated against an immediate-release capsule under fasting condition in a group of 14 healthy male volunteers (65). All extended-release formulations were based on a swelling hydrocolloid principle and contained hydroxypropylmethylcellulose (HPMC), microcrystalline cellulose, and cetyl palmitate but had quite different in vitro release profiles. The in vitro dissolution rate was measured in 500 mL of 0.1 HCl using USP apparatus 2 at 50 rpm. The dissolution profiles for all three extended-release formulations were essentially overlapping each other after normalizing to each formulation's mean dissolution time. For the three extended-release formulations, the in vitro and in vivo data are qualitatively in agreement with the bioavailability decreases with the prolonged in vitro release. However, this relationship breaks down when one considers the fastest dissolving extended-release formulation has a 123% bioavailability relative to the even faster-dissolving immediate-release capsule reference. Thus, this in vitro test seems to have only limited value in predicting the formulations' in vivo performance.

Six extended-release formulations of diltiazem HCl have been evaluated in 12 healthy volunteers (66). These included three multiple-unit dosage forms and one matrix tablet and two commercially available nondisintegrating tablets. The in vitro dissolution test used USP apparatus 2 and a distilled water medium with pH progression. The work demonstrated a linear relationship between the cumulative percentage released and the cumulative percentage absorbed for all three experimental multiple-unit dosages. There was also a statistically significant correlation between the pharmacokinetic parameter, mean residence time, and the in vitro dissolution parameter, mean dissolution time for all formulations studied.

Two theophylline extended-release formulations were compared with a solution under fasting condition in 12 healthy male volunteers (67). Both formulations consisted of the same spherical pellets but with different amounts of ethylcellulose/methylcellulose coating. Both formulations were demonstrated to have pH independent in vitro dissolution rates. The in vitro dissolution data were scaled using the Levy and Hollister technique to achieve a one-to-one correlation for each formulation. It is not clear from the publication whether one correlation equation can be established for both formulations. If the correlation is formulation specific, then it is of very little value for guiding further formulation work.

Five HPMC based extended-release formulations of diclofenac sodium have been evaluated in six healthy male volunteers (68). The release rate was controlled by the viscosity grade of the polymer and the amount of the polymer used. The authors reported a significant correlation between AUC, $C_{max}$, and the time for 50 or 80% drug to be released in vitro.

A marketed extended-release pseudoephedrine capsule and an experimental extended-release formulation of pseudoephedrine have been studied in six healthy volunteers (69). There was an evident linear relationship in the plot of cumulative fraction dissolved versus fraction absorbed. An extended-release OROS® formulation of pseudoephedrine HCl has been evaluated both in vitro and in 27 healthy male volunteers (8). The in vivo absorption profile was in good agreement with the in vitro release profile. There is a strong one-to-one relationship between the drug release rate and the absorption rate. Four cellulosic polymer-based matrix tablet formulations of pseudoephedrine sulfate were evaluated in 20 healthy male volunteers (70). The formulations were also evaluated in vitro using USP apparatus 2 at 50 rpm and 1,000 mL 0.1 N HCl. The results were essentially the same as what was observed for chlorpheniramine formulations, and a one-to-one correlation was established after correcting for the lag time and the difference in the first-order dissolution rate constant and the first-order absorption rate constant. The OROS nifedipine product, PROCARDIA XL, has been dosed to four mongrel dogs (71). The amount of drug released in the dog is in good agreement with the amount released in vitro.

A HPMC formulation of oxprenolol HCl has been evaluated in vitro and in five healthy mongrel dogs (72). A straight-line relationship was evident between the cumulative percent absorbed in vivo and the cumulative percent released in vitro. These results were somewhat disturbing because the more than twofold difference in the total AUC value reported between the conventional and the extended release formulations given at equivalent dose.

A series of seven 2-hydroxypropylcellulose (HPC)–based extended-release nifedipine formulations and one fast-release formulation were evaluated in gastric-acidity–controlled beagle dogs (73). The authors plotted the pharmacokinetic parameters, the absolute retarding parameter, mean residence time, and area under the plasma drug concentration curve against the mean in vitro dissolution time for each formulation. Although there is not a mathematical equation to describe the relationship, there are enough data points for interpolation. This approach clearly demonstrated the benefit of a well-designed experiment.

Three extended-release salbutamol products, Volmax®, and two HPMC-based matrix tablet formulations have been studied both in vitro and in vivo (74). An intravenous dose of salbutamol and an oral immediate-release product, Ventolin®, were also included in the in vivo study as the reference. The in vivo study used five mongrel dogs, and the in vitro study used USP apparatus 1 at 100 rpm and 500 mL of water as dissolution medium. The in vivo percentage of drug absorbed was calculated by numerical deconvolution using the KINBES® computer program. The in vitro dissolution profiles were generally faster than the corresponding in vivo absorption profiles for the three oral products; a good agreement between in vivo and in vitro results was evident for one of the HPMC formulations. In general, a higher amount dissolved in vitro than in vivo. This difference is at least partially due to the less than 100% bioavailability for oral salbutamol in dogs. A good

linear relationship seems to exist between the amount absorbed in vivo and the amount dissolved in vitro, with a positive intercept on the axis for the amount dissolved. Both the incomplete bioavailability and the relatively slow absorption process ($t_{1/2} = 2$ h) for the immediate-release oral product suggested that a better correlation might be achieved with the deconvoluted in vivo dissolution rate using the oral immediate-release product as the reference instead of the intravenous dose. Two ethylcellulose microsphere formulations of zidovudine were compared with zidovudine powder in four beagle dogs (75). The plasma drug concentration data were fitted to a first-order absorption/first-order elimination model. The observed in vitro release data were compared with the in vivo release profile generated from the best-fit absorption rate constant that, despite its name, actually reflected the drug-release process rather than the rapid-absorption process. Among three dissolution media, one was recommended on the basis of the agreement between the in vitro and in vivo results.

Two HPMC (Methocel® K4M) matrix tablets of zileuton were compared to an oral solution in nine beagle dogs using a single-dose design (76). The oral solution and immediate-release product plasma drug concentration profiles were fitted to polyexponential functions, and the matrix tablets' plasma drug concentration profiles were fitted to a smoothing cubic-spline function and then deconvoluted using the PCDON® software. The in vitro dissolution testing used USP apparatus 1 (100 rpm), 2 (100 rpm), and 3 (25 oscillations per min); and the dissolution medium was 10 mM sodium doecyl sulfate solution (900, 900, and 200 mL, respectively). The linear relationship was good between the in vitro and the in vivo data though the slope was approximately 2, indicating a much faster in vivo release rate. Despite the lack of a one-to-one correlation, a third HPMC (Methocel® K100LV) matrix tablet formulation was selected on the basis of the in vitro results and evaluated against an immediate-release tablet in eight beagle dogs using a multiple-dose design. The in vivo performance of the third formulation was consistent with that predicted from the in vitro dissolution data. Thus, the in vitro test method was a useful tool for selecting formulations despite the fact that it did not have a one-to-one correlation to the in vivo results.

Four theophylline extended-release formulations were evaluated in eight male beagle dogs under fasting condition (77). These formulations include theophylline/microcrystalline cellulose (Avicel® RC-581) beads, theophylline/microcrystalline cellulose (Avicel® PH-101) beads, Avicel® PH-101 beads coated with Surelease/HPMC, and a HPC/lactose/magnesium stearate matrix tablet. An intravenous dose of 80 mg theophylline, an oral theophylline solution, and an oral dose of theophylline syrup were included as references. The ratio of percentage cumulative released in vivo versus in vitro over time was evaluated as an indicator of the correlation. The in vivo release profiles for all four formulations are similar to the in vitro release profile in a pH 7.2 medium.

The GI physiology is quite different between dog and human. One notable difficulty is the relatively short GI transit time in dogs [reported to be between 6 and 12 h for four beagle dogs (78)]. The average gastric pH under fasting condition was 1.8 ($n = 4$) for beagle dogs and 1.1 ($n = 10$) for human ($P < .05$); the average gastric-emptying time was 100 min for dogs and 60 min for human ($P > .05$). It was also reported that the average intestinal pH for dogs was higher and more variable than that for humans (79). The importance of such a difference is not exactly clear and is most likely to be drug and formulation specific. In a study of four griseofulvin tablet formulations, the ultra-microsize tablet showed a lower bioavailability in humans (80) but was the most bioavailable formulation in beagle dogs (81). On the other hand, the microsized formulations showed similar in vivo results to those in humans. Four brands of uncoated diazepam tablets have been studied in humans (82) and in beagle dogs (83). The correlation between the in vivo pharmacokinetic parameters ($T_{max}$, AUC, and the plasma metabolite concentration at 0.25 and 0.5 h) and the in vitro dissolution parameters ($T_{30}$, $T_{50}$, and $T_{70}$, where $T_x$ is the time for $X\%$ of drug dissolved) were evaluated. Both the rate and the extent of the absorption in beagle dogs appeared to be correlated to the in vitro dissolution rate at pH 4.6 but not to the dissolution rate at pH 1.2. No statistically significant correlation existed in pharmacokinetic parameters between beagle dogs and humans. A consistent rank order was evident only between the serum drug concentration at 1 h in humans and the plasma drug concentration at 0.25 h in dogs; the rank order of other pharmacokinetic parameters was quite different between humans and dogs. In the study of fulfenamic acid (84) capsules, no significant correlation was evident between the dog and human results and the dissolution results, mainly because of the similar in vivo performance of all the products studied. Five tablet formulations of nalidixic acid have been studied in humans and beagle dogs (85). The rank order of tablets according to $C_{max}$ was exactly the same between humans and beagles. The area under the plasma drug concentration curve also showed the same rank order between humans and beagles except for one formulation with a poor disintegrating ability. The bioavailability of five brands of sugar-coated metronidazole tablets has been evaluated in humans, as discussed earlier, and in beagle dogs (86). The in vitro dissolution rate of these five tablets showed a wide range of pH dependence. The gastric acidity was measured for each volunteer and each dog. The relative values of $C_{max}$ and $AUC_{0-24}$ ranked similarly between corresponding gastric acidity groups of beagle dogs and humans, and a statistically significant correlation was evident between $AUC_{0-24}$ for beagle dogs and humans having low gastric acidity. It is also interesting to note that the tablet with a pH-independent dissolution rate had almost identical plasma drug concentration profiles between the low- and the high-gastric-acidity groups; the four formulations with pH-dependent dissolution rates had disparate plasma drug concentration profiles between the high-gastric-acidity and the low-gastric-acidity groups of dogs. Two commercial cinnarizine capsules, representing the extreme dissolution test results from 32 commercial capsules, were evaluated in 12 healthy beagle dogs (87). The gastric pH was quite variable from treatment to treatment and the between-dog variability in bioavailability cannot be statistically attributed to the gastric pH fac-

tor despite the statistically significant gastric pH effect on the bioavailability found in humans (88).

The food effect on the pharmacokinetics of two theophylline extended-release products has been evaluated in four male beagle dogs (89). The overall trends in relative bioavailability of these two products with and without food appeared to be similar to those reported in humans. A study of the food effect on a commercial theophylline extended-release product, Theo-24, in Hormel-Hanford miniswine (90) failed to reproduce the food effect observed in humans. Three extended-release theophylline matrix tablet formulations were evaluated with two commercial theophylline products in five mongrel dogs and six healthy male volunteers (91). Four formulations showed extended-release performance in both humans and dogs; one formulation showed incomplete absorption in both humans and dogs. For both human and dog data, the cumulative percent of dose absorbed appeared to be linearly related to the cumulative percent of dose released with a slope value close to one (92). Three extended-release theophylline capsule formulations with different in vitro dissolution rates were evaluated under fasting condition in 30 healthy male volunteers and six female beagle dogs (93). Although the agreement is not quantitative, the in vivo rate and the extent of absorption in both humans and dogs are in the same rank order as the in vitro dissolution rate. Qualitatively, the food exerted the same effect on the absorption profile for both humans and dogs. Four novel extended-release formulations of valproic acid have been studied in six healthy volunteers (94) and in five dogs (95,96). An apparent linear relationship existed between each formulation's absorption half-life for the human and that for the dogs (97). However, the relationship is not one-to-one implying that one cannot apply the dog absorption data directly to man and one needs to study a few formulations in both human and dog in order to establish the useful relationship. In addition, three extended-release formulations with diminished bioavailability in dogs were equally bioavailable as the immediate-release product in humans. For valproic acid, a wrong decision apparently could have been made if the formulation selection had been based on the dog pharmacokinetic data.

## APPLICATION OF IN VITRO/IN VIVO CORRELATIONS

The ultimate objective of in vitro dissolution/release rate testing is to ensure the product's in vivo performance. In too many cases, the correlation between the in vitro and the in vivo performance has not been explicitly established. Even in these cases, the dissolution tests are still performed because it is unthinkable to release a lot of product without the dissolution test. Because we have established the in vivo meaning of an in vitro dissolution test, it is important for us to apply the correlation and rely on the in vitro results to make decisions that we used to make on the basis of the pharmacokinetic data. The resulted savings in the resources must be good for all, although the extent of benefit may be quite different for each societal subgroup. The following are just some possible applications of the in vitro/in vivo correlations assuming a level A correlation.

## Dissolution/Release Rate Specifications

Without a correlation, the specifications of an in vitro test can be established only empirically. This approach is data driven but is valid only if all the batches have been extensively evaluated in clinical trials; furthermore, it probably can detect only relatively large differences between different batches. It is therefore more precise to set up the specification using the correlation to evaluate the in vivo consequences of the range. Clearly, the pharmacokinetic consequences alone are not sufficient to set up the specification. The pharmacodynamic knowledge is the key to making the specification clinically meaningful. In the absence of the information, some scientists may be willing to rely on the empirical bioequivalence range of $\pm 20\%$ as the first guidance. In the case of a one-to-one correlation, this automatically translates into a dissolution rate range of $\pm 20\%$. If the empirically derived dissolution range is much wider than $\pm 20\%$, then the companies invariably believe that the products have been punished by the presence of a one-to-one correlation. It is important for the regulatory agencies to recognize the scientific progress the industry has made in establishing the in vivo meaning of an in vitro test and reward the effort. Even though the batches may not be bioequivalent to each other, a product is still better controlled by an in vitro test with meaningful in vivo predictability than one without.

## Manufacturing Control

The extended-release products are distinguished through their input rate to the absorption site. Therefore, the rate of the drug release from these products is an important feature and should be carefully controlled and evaluated. The in vitro dissolution/release test is meaningful only when the test results are correlated to the products' in vivo performance.

## Process Change Assurance

The manufacturing processes of approved products are regulated by the regulatory agencies. The manufacturers are required to demonstrate that any kind of change, even an engineering improvement, does not cause changes in the finished product's in vivo performance. Consequently, many changes have to be supported by a bioequivalence study. With the availability of an in vitro test with a one-to-one correlation to the product's in vivo performance, a bioequivalence study should no longer be necessary. The dissolution/release tests measure only the rate process and cannot ensure the product performance for an excipient change. In such cases, the scientists and the regulatory agencies may consider a pilot pharmacokinetic study as an assurance that the new excipient does not inadvertently affect the absorption.

## CONCLUSIONS

The methods for derivation of the in vivo release and absorption profile have been well established. There are many methods for measuring the drug products' in vitro release/dissolution rate. The scientific community has reached the consensus to define what we mean by in vitro/in vivo correlation. Therefore, all the necessary tools are available to make the in vitro dissolution test have in vivo meaning.

## BIBLIOGRAPHY

1. J.G. Wagner, *Biopharmaceutics and Relevant Pharmacokinetics*, Drug Intelligence Publications, Hamilton, Ill., 1971, p. 114.
2. U.S. Pharmacopoeia Subcommittee on Biopharmaceutics, *Pharmacopeial Forum*, pp. 4160–4161 (1988).
3. J.P. Skelly et al., *Pharm. Res.* **7**, 975–982 (1990).
4. *The United States Pharmacopeia 23*, U.S. Pharmacopeial Convention, Rockville, Md., 1995, pp. 1924–1929.
5. M.S. Roberts and M.J. Denton, *Eur. J. Clin. Pharmacol.* **18**, 175–183 (1980).
6. M.A. van Baak, *Clin. Pharmacokinet.* **19**, 32–43 (1990).
7. W.R. Gillespie and P. Veng-Pedersen, *Biopharm. Drug Dispos.* **6**, 351–355 (1985).
8. S.S. Hwang et al., *J. Clin. Pharm.* **35**, 259–267 (1995).
9. D.E. Duggan, K.F. Hooke, and S.S. Hwang, *Drug Metab. Dispos.* **8**, 241–246 (1980).
10. T.R. Browne, *Clin. Pharmacokinet.* **18**, 423–433 (1990).
11. M.O. Karlsson and U. Bredberg, *Pharm. Res.* **6**, 817–821 (1989).
12. L.C. Feely and S.S. Davis, *Pharm. Res.* **6**, 274–278 (1989).
13. J.G. Wagner, in A. Yacobi and E. Halperin-Walega, eds., *Oral Sustained Release Formulations: Design and Evaluation*, Pergamon, New York, 1988, pp. 95–124.
14. J.G. Wagner and E.J. Nelson, *J. Pharm. Sci.* **52**, 610–611 (1963).
15. J.C.K. Loo and S. Riegelman, *J. Pharm. Sci.* **57**, 918–928 (1968).
16. L.Z. Benet and C.W.N. Chiang, *13th Nat. Meet. APhA Acad. Pharm. Sci.*, November 5–9, 1972, Abstr. 2, p. 169.
17. D.J. Cutler, *J. Pharmacokinet. Biopharm.* **6**, 265–282 (1978).
18. D.P. Vaughan and M. Dennis, *J. Pharm. Sci.* **67**, 663–665 (1978).
19. D.J. Cutler, *J. Pharmacokinet. Biopharm.* **6**, 243–263 (1978).
20. J.G. Wagner, *J. Pharm. Sci.* **72**, 838–842 (1983).
21. S.S. Hwang, W. Bayne, and F. Theeuwes, *J. Pharm. Sci.* **82**, 1145–1150 (1993).
22. S. Hwang and M. Knowles, *J. Pharm. Sci.* **83**, 629–631 (1994).
23. J.G. Wagner, *J. Pharmacokinet. Biopharm.* **3**, 457–478 (1975).
24. K.R. Godfrey, R.P. Jones, and R.F. Brown, *J. Pharmacokinet. Biopharm.* **8**, 633–648 (1980).
25. P. Veng-Pedersen, *J. Pharm. Sci.* **69**, 298–304 (1980).
26. P. Veng-Pedersen, *J. Pharm. Sci.* **69**, 305–311 (1980).
27. P. Veng-Pedersen, *J. Pharmacokinet. Biopharm.* **8**, 463–481 (1980).
28. S. Vajda, K.R. Godfrey, and P. Valko, *J. Pharmacokinet. Biopharm.* **16**, 85–107 (1988).
29. D. Verotta, *J. Pharmacokinet. Biopharm.* **18**, 483–489 (1990).
30. S. Vajda, K.R. Godfrey, and P. Valko, *J. Pharmacokinet. Biopharm.* **18**, 489–491 (1990).
31. P. Veng-Pedersen, *J. Pharmacokinet. Biopharm.* **18**, 491–495 (1990).
32. D. Verotta, *J. Pharmacokinet. Biopharm.* **18**, 495–499 (1990).

33. D. Verotta, *J. Pharmacokinet. Biopharm.* **21**, 609–636 (1993).

34. Z. Yu, S.S. Hwang, and S.K. Gupta, *Biopharm. Drug Dispos.* **18**, 475–488 (1997).

35. D. Brockmeier, H.-G. Grigoleit, and H. Leonhardt, *Eur. J. Clin. Pharmacol.* **29**, 193–197 (1985).

36. T. Gramatte and K. Richter, *Br. J. Clin Pharmacol.* **37**, 608–611 (1993).

37. P-H. Hsyu et al., *Pharm. Res.* **11**, 156–159 (1994).

38. H. Bode et al., *Eur. J. Clin. Pharmacol.* **50**, 195–201 (1996).

39. L.K. Tay et al., *J. Clin. Pharmacol.* **32**, 827–832 (1992).

40. S. Harder, U. Fuhr, D. Beermann, and A.H. Staib, *Br. J. Clin. Pharmacol.* **30**, 35–39 (1990).

41. S.A. Riley et al., *Aliment. Pharmacol. Ther.* **6**, 710–706 (1992).

42. W.N. Charman et al., *J. Clin. Pharmacol.* **33**, 1207–1213 (1993).

43. K.H. Chan et al., *Pharm. Res.* **11**, 432–437 (1994).

44. W.G. Kramer, *J. Clin. Pharmacol.* **34**, 1218–1221 (1994).

45. W.H. Barr et al., *Clin. Pharmacol. Ther.* **56**, 279–285 (1994).

46. P.E. Warner et al., *Pharm. Res.* **12**, 138–143 (1995).

47. J. Gottfries et al., *Scand. J. Gastroenterol.* **31**, 49–53 (1996).

48. Y. Plusquellec and G. Houin, *Arzneim. Forsch. / Drug Res.* **44**(I), 679–682 (1994).

49. S.J. Warrington et al., *Br. J. Clin. Pharmacol.* **19**, 19S–224S (1985).

50. J S. Barrett et al., *Pharm. Res.* **13**, 1535–1540 (1996).

51. Food and Drug Administration, *Guidance Oral Extended Release Dosage Forms: In Vivo Bioequivalence and In Vitro Dissolution Testing*, FDA, Washington, D.C.

52. J. Devane and J. Butler, *Pharm. Tech.* **21**(9), 146–159 (1997).

53. G.L. Amidon, H. Lennernäs, V.P. Shah, and J.R. Crison, *Pharm. Res.* **12**, 413–420 (1995).

54. S.K. El-Arini, G.K. Shiu, and J.P. Skelly, *Pharm. Res.* **7**, 1134–1140 (1990).

55. F-Y. Liu, N.C. Sambol, R.P. Giannini, and C.Y. Liu, *Pharm. Res.* **13**, 1501–1506 (1996).

56. N.D. Eddington et al., *Pharm. Res.* **15**, 466–473 (1998).

57. G. Levy and L. Hollister, *J. Pharm. Sci.* **53**, 1446–1452 (1964).

58. H. Ogata et al., *Int. J. Pharm.* **29**, 113–120 (1986).

59. H. Ogata et al., *Int. J. Pharm.* **23**, 277–288 (1985).

60. F. Langenbucher and J. Mysicka, *Br. J. Clin. Pharmacol.* **19**, 151S–162S (1985).

61. F. Langenbucher, *Pharm. Ind.* **44**, 1166–1172 (1982).

62. I.D. Bradbrook et al., *Br. J. Clin. Pharmacol.* **19**, 163S–169S (1985).

63. M. Dey et al., *Int. J. Pharm.* **49**, 121–128 (1989).

64. P. Mojaverian et al., *Pharm. Res.* **9**, 450–456 (1992).

65. J. Drewe and P. Guitard, *J. Pharm. Sci.* **82**, 132–137 (1993).

66. C. Caramella et al., *Biopharm. Drug Dispos.* **14**, 143–160 (1993).

67. K.-H. Yuen, A.A. Desmukh, and J.M. Newton, *Pharm. Res.* **10**, 588–592 (1993).

68. C.-H. Liu et al., *J. Pharm. Pharmacol.* **47**, 360–364 (1995).

69. V. Pade, J. Aluri, L. Manning, and S. Stavchansky, *Biopharm. Drug Dispos.* **16**, 381–391 (1995).

70. P. Mojaverian et al., *J. Pharm. Biomed. Analy.* **15**, 439–445 (1997).

71. D.R. Swanson, B.L. Barclay, P.S.L. Wong, and F. Theeuwes, *Am. J. Med.* **83**(Suppl. 6B), 3–9 (1987).

72. K.P. Devi et al., *Pharm. Res.* **6**, 313–317 (1989).

73. Z. Wang, F. Hirayama, and K. Uekama, *J. Pharm. Pharmacol.* **46**, 505–507 (1994).

74. R.M. Hernandes et al., *Int. J. Pharm.* **139**, 45–52 (1996).

75. K. Abu-Izze, L. Tambrallo, and D.R. Lu, *J. Pharm. Sci.* **86**, 554–559 (1997).

76. Y. Qiu, H. Cheskin, J. Briskin, and K. Engh, *J. Controlled Release* **45**, 249–256 (1997).

77. Z. Yu, J.B. Schwartz, and E.T. Sugita, *Biopharm. Drug Dispos.* **17**, 259–272 (1996).

78. W.A. Cressman and D. Sumner, *J. Pharm. Sci.* **60**, 132–134 (1971).

79. C.Y. Lui et al., *J. Pharm. Sci.* **75**, 271–274 (1986).

80. N. Aoyagi et al., *J. Pharm. Sci.* **71**, 1165–1169 (1982).

81. N. Aoyagi et al., *J. Pharm. Sci.* **71**, 1169–1172 (1982).

82. H. Ogata et al., *Int. J. Clin. Pharmacol. Ther. Toxicol.* **20**, 159–165 (1982).

83. H. Ogata et al., *Int. J. Clin. Pharmacol. Ther. Toxicol.* **20**, 576–581 (1982).

84. N. Kaniwa et al., *Int. J. Clin. Pharmacol. Ther. Toxicol.* **21**, 576–581 (1983).

85. H. Ogata et al., *Int. J. Clin Pharmacol. Ther. Toxicol.* **22**, 240–245 (1984).

86. H. Ogata et al., *Int. J. Pharm.* **23**, 289–298 (1985).

87. H. Ogata et al., *Int. J. Pharm.* **29**, 121–126 (1986).

88. H. Ogata et al., *Int. J. Pharm.* **29**, 113–120 (1986).

89. G.K. Shiu et al., *Pharm. Res.* **6**, 1039–1042 (1989).

90. G.K. Shiu et al., *Pharm. Res.* **5**, 48–52 (1988).

91. Z. Hussein, M. Bialer, M. Friedman, and I. Raz, *Int. J. Pharm.* **37**, 97–102 (1987).

92. Z. Hussein and M. Friedman, *Pharm. Res.* **7**, 1167–1171 (1990).

93. C.S. Cook et al., *Int. J. Pharm.* **60**, 125–132 (1990).

94. M. Bialer et al., *Biopharm. Drug Dispos.* **6**, 401–411 (1985).

95. M. Bialer, M. Friedman, and J. Dubrovsky, *Biopharm. Drug Dispos.* **5**, 1–10 (1984).

96. M. Bialer, M. Friedman, and J. Dubrovsky, *Int. J. Pharm.* **20**, 53–63 (1984).

97. M. Bialer, M. Friedman, and J. Dubrovsky, *Biopharm. Drug Dispos.* **7**, 495–500 (1986).

# INFECTIOUS DISEASE, DRUG DELIVERY TO TREAT

SHIGEFUMI MAESAKI
MOHAMMAD ASHLAF HOSSAIN
SHIGERU KOHNO
Nagasaki University School of Medicine
Nagasaki, Japan

## KEY WORDS

Antibiotics

Antifungal agents

Antimicrobial agents

Antimycobacterial agents

Antiparasitic agents

Antiviral agents

## OUTLINE

## INTRODUCTION

Infectious diseases are caused by pathogenic organisms
that enter the host, continue multiplying, and cause damage to the host tissue in many ways, resulting in symptoms
and signs that may range from trivial to life threatening
in nature depending on various host, agent, and environmental factors that facilitate growth and transmission.
Many antiinfective drugs have been developed since
1910, when Paul Ehrlich discovered salvarsan, and 1929,
when Alexander Fleming discovered penicillin. Various antimicrobial drugs are available for the treatment of infectious diseases. β-Lactam antibiotics, such as penicillin and
cephalosporin, are most commonly used because of their
wide antimicrobial spectrum, increased selective toxicity,
and potency. Third-generation cephems have wider spectra
and more potent antimicrobial activities than second-generation cephems. These are still less active against
methicillin-resistant *Staphylococcus aureus* (MRSA) and
*Pseudomonas aeruginosa*.

Antimicrobial agents are the substances produced by
some living organisms or prepared synthetically from naturally occurring substances that act on other living microorganisms and cause cell death or growth inhibition, resulting in inactivation of the pathogen and recovery from
the disease. Thus, they are either microbiocidal or microbiostatic in action. The antimicrobial agents obtained from
living organisms or semisynthetically prepared from naturally occurring substances are known as antibiotics. The
principle of antimicrobial chemotherapy is based on the
administration of the most strongly antimicrobial agents
against the causative microorganisms. An ideal drug for
treating infectious disease needs to be administered at a
concentration above the minimum inhibitory concentration (MIC) of the microorganisms. Moreover, it is necessary
for the drug to be safe and have minimum adverse effects.
In the past few decades, many kinds of antimicrobial
agents with broad efficacy against various kinds of microorganisms have been developed. Because antibiotic treatment of severe infections is not always successful, intensification of the treatment is needed. Targeting antibiotics
to infected tissues or cells by encapsulation in liposomes is
under investigation and may be of importance in the treatment of infections that prove refractory to conventional
forms of antibiotic therapy. For these reasons, drug delivery systems can be utilized to enhance the efficacy of antimicrobial agents against infectious disease.

Liposomes are microvesicles composed of continuous bilayers of phospholipids surrounding an aqueous phase.
Phospholipid, a component of the cell membrane, is amphipathic, consisting of hydrophilic heads and hydrophobic
tails. Liposomes have been extensively used to deliver antibiotics, antifungal agents, and antiviral compounds for
the treatment of infections in animal models. Three rationales have been invoked to justify the use of liposomes
as carriers of antiinfective agents: (*1*) passive targeting of
compounds to the mononuclear phagocytic system (MPS)
for enhanced effects at this site; (*2*) targeted drug delivery;
and (*3*) site-avoidance delivery, wherein the rate of transfer
of drug to the toxic site is reduced but not the drug-transfer
rate to the active site. All strategies result in an increase
in the therapeutic index of the drug; however, to obtain
optimal drug therapy, the characteristics for a particular
drug for site-specific delivery differ from those of site-avoidance delivery.

Future developments that will permit a larger variety
of agents to be delivered in liposomes include the development of fusogenic liposomes and of liposomes that avoid
the reticuloendothelial system. By modifying the lipid composition of the liposomes, it is possible to manipulate their
intracellular degradation and thereby the intracellular release and therapeutic availability of the antibiotic. Efficacy
of liposome-encapsulated antibiotics in the treatment of
infectious diseases outside the mononuclear phagocyte
system may be realized by manipulation of the liposome
composition. Evidence for this is found in the treatment of
systemic fungal infections, in which liposomes appear to
be very effective as a carrier of amphotericin B (AmB). The
most advanced application of liposome-based therapy is in
this field, and clinical studies with liposome-encapsulated
AmB have been in progress for several years.

Advanced liposomal therapeutic action has been attained by liposome surface modification, initially with specific glycolipids and subsequently with surface-grafted
polyethelenglycol (PEG), reducing in vivo rapid recognition
and uptake, and resulting in prolonged blood circulation
and providing selective localization in pathological sites.
The result is improved efficacy of encapsulated agents. The
surface PEG may produce a barrier, as described for colloids. Reduced in vivo uptake may result from inhibition
of plasma–protein adsorption, or opsonization, by the coating. Several physical studies support this mechanism, including those involving electrophoretic mobility. The dependence of blood circulation and tissue distribution on

PEG molecular weight correlates with $\zeta$-potential estimates of PEG-coating thickness. Effects on tissue distribution are reported for liver and spleen, the major phagocytic organs. The biological properties of these liposomes depend on the surface polymer rather than the lipid bilayer, yielding important advantages for lipid-mediated control of drug interaction and release without affecting the biodistribution.

The recent developments of the drug delivery systems and their application in treating different types of infectious diseases are described in the following paragraphs.

## LIPOSOMAL ANTIMICROBIAL AGENTS AGAINST THE CYTOZOIC MICROORGANISMS SALMONELLA TYPHIMURIUM, LISTERIA MONOCYTOGENES

Phagocytosis plays an important role in limiting invasion by pathogens. Several microorganisms can withstand the host cellular events after phagocytosis; thus, they can survive and multiply intracellularly. They are known as cytozoic microorganisms. The commonly believed mechanism is the altered production or inactivation of reactive oxygen metabolites induced by the toxins or other substances from the pathogen. Intracellular infections involving the MPS (parasitic, fungal, bacterial, and viral infections), an improved therapeutic index and reduced toxicity resulting from encapsulation of the antibiotics in liposomes in animal models have been demonstrated. By manipulating the liposomal composition, rates of uptake and intracellular degradation can be influenced, and thereby the rates at which liposome-encapsulated agents are released and become available to exert their therapeutic action are also influenced. With respect to the targeting of macrophage modulators at the MPS by means of liposomes for maximal stimulation of the nonspecific antimicrobial resistance, experimental evidence is now available of the potential usefulness of liposomes as carriers of these agents. This approach may also be of importance for the potentiation of treatment of severe infections.

There are three varieties of drug transportation into cells: passive transport, active transport, and phagocytosis. Macrolides are actively transported into the polymorphonuclear leukocytes via the nucleoside transport system or glycolytic pathway. Aminoglycosides are not the choice for intracytoplasmic infections such as salmonellosis. However, liposome-encapsulated cephalothin or streptomycin was effective in intraphagcytic killing of S. typhimurium and experimental salmonellosis. Desiderio and Campbell used multilamellar liposomes, composed of phosphatidylcholine, cholesterol, and phosohatidylserine, that entrapped an aqueous solution of cephalothin for in vitro experiments (1). Resident murine macrophages interiorized the liposome–antibiotics complex, and this event resulted in relatively high intracellular concentration of cephalothin. Treatment of infected macrophages with liposome-encapsulated cephalothin enhanced the intracellular killing of S. typhimurium compared with that of macrophages treated with free cephalothin. Tadakuma et al. reported its superior efficacy against experimental salmonellosis in mice; liposomal streptomycin was selectively delivered to the spleen and liver with concentrations in these organs about 100 times higher than those in mice receiving the free drugs (2).

Listeriosis is a severe disseminated infection in immunocompromised patients. Liposome-encapsulated ampicillin demonstrated efficacy for the treatment of murine disseminated listeriosis (3). All mice treated with liposome-encapsulated ampicillin survived. The in vitro susceptibility of C. trachomatis to liposome-encapsulated tetracycline was compared with that of free tetracycline. A chlamydia-infected mouse fibroblast monolayer was continuously exposed to varying concentrations of antibiotics. The MIC for anionic, cationic, and neutral liposomes containing tetracycline was approximately 2, 10, and 20 times smaller than that of free tetracycline (4).

Liposomal aminoglycosides showed excellent efficacy compared with free aminoglycosides or empty liposome. This therapeutic strategy should be developed for clinical application. Although the human defense mechanism against microbial infection is very complicated, biological response modifiers (BRMs) such as vaccination or cytokine therapy have been investigated. One of the most useful and protective vaccines for prevention of tuberculosis is the Mycobacterium vaccae vaccine, developed by Stanford et al. (5). As for the cytokines, interleukin 2, granulocyte macrophage-colony–stimulating factor, and tumor necrosis factor have very strong antimicrobial activity. Interferon alone, however, has weak efficacy and should be combined with other effective cytokines. These BRMs constitute an excellent strategy for antimicrobial chemotherapy.

## LIPOSOMAL ANTIMICROBIAL (PIPERACILLIN OR TOBRAMYCIN) AGENTS AGAINST BACTERIA

Penicillin proved to be miraculous in its action against gram-positive and gram-negative bacteria, diminishing their roles as nosocomial pathogens, and this resulted in its promiscuous use in the following years. Subsequently, semisynthetic penicillins, such as methicillin, ampicillin, amoxycillin, and then cephalosporins, were developed. Unfortunately, several bacteria capable of producing $\beta$-lactamase (and thus resistant to penicillin) were isolated. Penicillinase-producing organisms such as S. aureus have caused the virtual elimination of penicillin from therapy against this organism. S. aureus has become resistant to methicillin and is known as a very refractile organism that causes opportunistic infections. Piperacillin itself is known to be inactive against $\beta$-lactamase–producing S. aureus. However, Nacucchio et al. reported the enhanced antibacterial activity of liposome-encapsulated piperacillin against S. aureus (6). The results, expressed as the percentage of bacterial growth inhibition at 50% MIC of piperacillin, demonstrated that growth inhibition was the highest when piperacillin was liposome encapsulated. Even adsorption of piperacillin to liposomes rendered significant enhancement of the antistreptococcal capacity when compared with the effect of piperacillin alone. Exogenous $\beta$-lactamase hydrolyzed piperacillin and eradicated its antistreptococcal activity. Encapsulation of piperacillin within liposomes conferred a high degree of protection against hydrolysis. The enhanced antibacterial activity of liposome-encapsulated piperacillin was probably due to protection of the drug from hydrolysis by staphylococcal β-lactamase or stearic hindrance to the action of the β-lactamase by the lipid surface of liposomes.

Lower-respiratory-tract infections consisting of *P. aeruginosa* occur almost exclusively in patients with compromised local respiratory or systemic host defense mechanisms. Exposure to the hospital environment, particularly given in an intensive care setting, use of respiratory inhalation equipment, and prior antibiotic therapy, increases the likelihood of such infections. Lower-respiratory-tract colonization with mucoid strains of *P. aeruginosa* is a function of age in patients with cystic fibrosis. *P. aeruginosa* clearly plays a critical role in the progressive lung lesions and resulting disability observed in most patients with this disease. Under certain conditions *P. aeruginosa* produces a polysaccharide capsule, referred to as the *glycocalyx*. This material appears to form a matrix around the bacterium, anchoring it to its environment and protecting it from phagocytic cells and antibiotics.

Liposome-encapsulated anti-*Pseudomonas* agents were first developed in 1990 by Kotsifaki (7). The growth inhibition of four *P. aeruginosa* strains by liposome-trapped penicillin G indicated an association of its efficacy with the nature of the *O*-antigenic polymeric side chain. For instance, the P28-800 and PCF-95 strains, characterized by a rough polysaccharide chain, were the most susceptible, whereas strain P28-0, possessing an intact lipopolysaccharide, resisted the activity of the entrapped drug. Among the rough strains, P642, a β-lactamase producer, was not affected by the encapsulated drug. Lagace reported that ticarcillin- and tobramycin-resistant strains of *P. aeruginosa* were shown to have a markedly increased sensitivity to antibiotics enclosed in liposomes (8). The liposome-enclosed antibiotic was as effective against the β-lactamase–producing strain as against the non-β-lactamase–producing strain. The in vitro activities of encapsulated and free tobramycin were evaluated against *P. aeruginosa* by Poyner (9). The effects of free and liposome-encapsulated tobramycin on siderophore production by *P. aeruginosa* were examined and involved using MIC and sub-MIC quantities of tobramycin. Compared with free tobramycin, siderophore production was more effectively retarded in the presence of sub-MIC quantities of liposomal tobramycin, particularly at 0.5 MIC. Siderophore production was particularly reduced in the presence of positively charged phosphatidylcholine liposomes. Beaulac studied the efficacy of fluid liposome-encapsulated tobramycin against *P. aeruginosa* in an animal model of chronic pulmonary infection (10). Chronic infection in the lungs was established by intratracheal administration of a mucoid variant of *P. aeruginosa* prepared in agar beads, and antibiotic treatments were given intratracheally. Animals treated with encapsulated tobramycin in fluid liposomes had a number of colony-forming units (CFUs) less than the minimal CFUs of lungs for animals treated with encapsulated tobramycin in rigid liposomes, free antibiotic, or liposomes without tobramycin. Tobramycin measured in the lungs at 16 after the last treatment following the administration of encapsulated antibiotic was still ac-

tive. Low levels of tobramycin were detected in the kidneys after the administration of encapsulated antibiotic, while a higher concentration was detected in the kidneys following the administration of free antibiotic. These results suggest that local administration of fluid liposomes with encapsulated tobramycin may offer advantages over free antibiotics, including sustained concentration of the antibiotic, minimal systemic absorption, reduced toxicity, and increased efficacy, and that it could greatly improve the management of chronic pulmonary infection in cystic fibrosis patients.

## LIPOSOME-ENCAPSULATED ANTIMYCOBACTERIAL AGENTS

Mycobacteriosis, or mycobacterial infections, are caused by various species of the genus *Mycobacterium*. The most commonly encountered diseases are tuberculosis, caused by *Mycobacterium tuberculosis*, mycobacterial infections other than tuberculosis (MOTT), and leprosy, caused by *Mycobacterium leprae*. Antimycobacterial agents are those aimed at elimination or inactivation of infections caused by mycobacteria. The active drugs are expected to act on the agent within the cells or a caseous lesion. In recent years, the resurgence of multidrug-resistant (MDR) tuberculosis and the increased prevalence of atypical mycobacteriosis have been observed in immunocompromised patients, including those with AIDS or organ transplantation, and occasionally in immunocompetent patients as well. In the recent past, Rifampin-resistant strains have been on the increase. Strains of *Mycobacterium avium-intracellulare* complex (MAC) are common environmental organisms of low virulence. MACs have been frequently identified as the pathogen causing atypical mycobacteriosis in people with poor immune status. The disease differs in severity but is often debilitating and can be fatal. In addition to infections of the lung, fever, lymphadenopathy, weight loss, and so on, the disease may cause remote symptoms such as arthritis and osteomyelitis as a result of dissemination. A considerable number of patients with MAC infection do not respond to the conventional antimycobacterial agents. Thus, the quests for novel antimycobacterial drugs and modification of their cytotoxic potency by encapsulation in liposomes have been considered worthwhile.

Perhaps the most important recent advance in the field of infections due to MACs is the identification and development of more effective agents for the treatment and prevention of disseminated disease. These agents include clarithromycin, azithromycin, rifabutin and other rifamycins, ethambutol, clofazimine, fluoroquinolones, amikacin, and liposome-encapsulated gentamicin. Most clinicians currently use multidrug therapy to maximize efficacy and to minimize the emergence of resistance. Prospective clinical trials of multidrug regimens suggest that MAC colony counts in blood decline during therapy, usually accompanied by alleviation of clinical symptoms. The small size and short duration of these trials have not permitted an evaluation of survival or quality of life. Because the contribution of any single agent to multidrug trials is difficult to assess, short-term trials of monotherapy have been conducted recently; clarithromycin, azithromycin, ethambutol, and liposome-encapsulated gentamicin have proved most active.

The earliest report of a successful treatment for mycobacterial infection involved the use of liposome-encapsulated streptomycin. Vladimirsky reported that intravenously injecting mice infected with *M. tuberculosis* H37 RV with liposome-encapsulated streptomycin led to the amount of mycobacteria decreasing significantly in the spleen but not in the lungs, as compared with an injection of a buffered solution of streptomycin (12). Several kinds of aminoglycosides encapsulated in liposomes were investigated for therapeutic efficacy against MAC infection. Duzgunes studied the therapeutic effects of free and liposome-encapsulated amikacin on MAC infection by using the beige-mouse model of the disease (13). The results showed reduction of the CFU counts in the spleen and kidneys treated with liposome-encapsulated amikacin compared with those of both untreated controls and free-drug-treated mice (14). Liposomal kanamycin injections led to a greater reduction in the degree of gross pulmonary lesions and the growth of the organisms in the visceral organs (lungs, liver, spleen, and kidneys) of MAC-infected mice than did either free kanamycin alone or free kanamycin mixed with empty liposomal vesicles (15). Nightingale reported that a liposome-encapsulated gentamicin was given intravenously twice weekly for 4 weeks to AIDS patients with MAC bacteremia, with MAC colony counts in blood subsequently falling by 75% or more in all treated groups (16).

Other antimycobacterial agents were encapsulated in liposomes and investigated for their efficacy against MAC. Saito estimated the in vitro efficacy of liposome-entrapped rifampin against MAC infection induced in mice. Intraperitoneal injections of liposome-entrapped rifampin led to a greater reduction in bacterial growth in the lungs and spleen of infected mice than did free rifampin alone. However, liposome-entrapped rifampin given to mice via the intramuscular or subcutaneous route failed to show such an increased therapeutic efficacy. It was concluded that entrapment of rifampin in liposomal vesicles increased incorporation of the drug into host peritoneal macrophages and the activity of the agent against MAC phagocytosed into the macrophages (17). Majumdar investigated the therapeutic efficacies of liposome-encapsulated ciprofloxacin against growth of the MAC inside human peripheral blood monocyte/macrophages. Liposome-encapsulated ciprofloxacin was at least 50 times more effective against the intracellular bacteria than was the free drug: At a concentration of 0.1 μg/mL, liposome-encapsulated ciprofloxacin had greater antimycobacterial activity than the free drug at 5 μg/mL (18). The therapeutic efficacies of liposome-encapsulated ciprofloxacin and clarithromycin against MAC in a model of intramacrophage infection were evaluated by Onyeji. Liposome entrapment of clarithromycin significantly enhanced the activities of the drugs when compared with the antimycobacterial effects of equivalent concentrations of the free drugs, and liposome-encapsulated clarithromycin plus ethambutol was more effective in organism eradication than each agent used

singly (19). Mehta reported therapeutic efficacy of liposome encapsulation of clofazimine in the beige-mouse model of disseminated MAC infection. An equivalent dose of liposome-encapsulated clofazimine was more effective in eliminating the bacteria from the various organs studied, particularly from the liver. Moreover, because of the reduced toxicity of liposomal encapsulation of clofazimine, higher doses could be administered, resulting in a significant reduction in the numbers of CFU in the liver, spleen, and kidneys (20,21). In recent study, Deol investigated the therapeutic efficacies of isoniazid and rifampin encapsulated in lung-specific stealth liposomes against *M. tuberculosis* infection induced in mice. The efficacies of isoniazid and rifampin encapsulated in lung-specific stealth liposomes were higher than those of free drugs against tuberculosis, as evaluated on the basis of CFUs detected, organomegaly, and histopathology. Furthermore, liposomal drugs had marginal hepatotoxicities as determined from the levels of total bilirubin and hepatic enzymes in serum (22).

Aminoglycosides have been used for the treatment of mycobacterial infection (Fig. 1). Liposome-encapsulated amikacin showed significantly greater inhibitory activity against the survival of MAC inside mouse peritoneal macrophages than free drug (23). Liposome-encapsulated gentamicin significantly reduced the number of organisms in the spleen and liver of the infected beige mice, but it did not sterilize these organs (24). Liposome-encapsulated kanamycin was injected into mice infected with MAC once every week for up to 8 weeks, which led to a greater reduction in the degree of gross pulmonary lesions and in the growth of the organisms in the lung, liver, spleen, and kid-

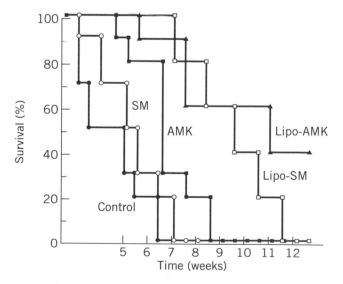

**Figure 1.** Survival rate over time in liposomal streptomycin and amikacin treatment for systemic *M. tuberculosis* infection in mice. *M. tuberculosis* (H37 Rv) ($2 \times 10^6$ CFU/ml) was injected intravenously to BALB/c mice (5 weeks old, male). Treatment was started 24 h after infection. Streptomycin (SM), amikacin (AMK), liposomal amikacin (Lipo-AMK), and liposomal streptomycin (Lipo-SM) were injected intravenously on days 3, 5, 7, and 9. *Source:* Based on data from Ref. 11.

neys (15). The effect of liposome-encapsulated streptomycin against MAC infection in beige mice regarding the liposome was studied (25). At four weeks, liposome-encapsulated streptomycin reduced the colony-forming units in the liver and spleen to about the same extent as a 50- to 100-fold higher dose of free drug. These observations suggest that liposomes were taken up by the infected macrophages of the liver and spleen and that the encapsulated streptomycin was delivered to the intracellular compartments containing mycobacteria. The effects of unilamellar and multilamellar liposomes were similar in this model. This suggests that similar quantities of streptomycin may have been delivered to the intracellular sites of infection, although higher tissue levels were achieved with multilamellar liposomes. It is likely that not all of the streptomycin in multilamellar vesicles was released into the phagosome internalizing the liposome.

## LIPOSOME-ENCAPSULATED ANTIFUNGAL AGENTS

Fungi are eukaryotic organisms that grow in a filamentous or yeastlike form or both. Fungi pathogenic to humans occur worldwide, some of which are common in defined geographic areas; in general, they can be divided into two basic groups, yeasts and molds. They are recognized as a significant and frequent cause of morbidity and mortality. Fungi may cause varying degrees of illnesses, starting from superficial to life-threatening, deep-seated mycosis. They are found to cause serious opportunistic infections in immunocompromised patients with, for example, leukemia, AIDS, and bone marrow transplants and in those treated for a long time with antineoplastic or immunosuppressive agents in particular. Frequent use of broad-spectrum antibacterial agents, intravenous catheters, and immunosuppressive therapies that are aimed at improving support to prolong survival, especially in critically ill patients, predisposes one to fungal infections.

Systemic fungal infections may arbitrarily be divided into two broad categories: endemic diseases such as blastomycosis, coccidioidomyosis, and histoplasmosis and opportunistic diseases such as aspergillosis, cryptococcosis, and candidosis, which occur almost exclusively in patient with impaired host defenses. Antifungal therapy is implemented according to several factors, such as the causative agent, the progression or invasion of the disease, and so on. Antifungal therapy may have to be administered empirically in febrile neutropenic patients who do not respond to treatment with an antibacterial drug (26).

Although newer pathogenic fungi causing deep-seated mycosis have emerged, antifungal agents have not substantially increased in number. There are few classes of agents that, in turn, have few mechanisms of action and limited response rates. The major groups of antifungal agents in clinical use are polyene antibiotics, azole derivatives, allylamine–thiocarbamates, morpholines, and miscellaneous compounds such as 5-fluorocytosin and griseofulvin. The polyenes and azoles are most widely used. Amphotericin B (AmB), nystatin and, natamycin are the polyene antifungal agents used for treatment of human diseases, with AmB being the only parenteral preparation with a broad range of antifungal activity. Over the past several years, increased efforts in both basic and clinical antifungal pharmacology have resulted in a number of entirely new reengineered or reconsidered compounds, that are at various stages of preclinical and early clinical development (27–29).

AmB is used to treat severe fungal infections. Its usefulness is compromised by a high incidence of adverse reactions, including fever, chills, nausea, vomiting, headache, and renal dysfunction with associated anemia, hypokalemia, and hypomagnesemia. Infrequently anaphylactic shock, thrombocytopenia, acute liver failure, flushing, vertigo, generalized seizures, and dose-related cardiac arrest or ventricular fibrillation have been reported. In children, overdose of AmB may cause cardiac arrhythmias and death. The damaging action of AmB to cells originates from its binding to sterols incorporated in cellular membranes (ergosterol in the case of fungal cells and cholesterol in mammalian cells); greater avidity for ergosterol-containing membranes than for cholesterol-containing membranes favors the clinical usefulness (30). Infusion-related side effects such as fever, rigors, and chills are common but not so alarming that AmB therapy should be discontinued, as would be recommended in the event of renal dysfunction. Renal insufficiency in association with azotemia, renal tubular acidosis, and impaired urinary concentrating ability resulting in electrolyte imbalance may be severe enough to lead to dose reduction or premature discontinuation of the drug (31).

Three ways have been suggested by which the therapeutic index of AmB might be improved: increasing the selectivity of drug-induced damage to fungal, as opposed to mammalian cells; decreasing toxicity to host cells bearing low-density lipoproteins (LDL) receptors; and decreasing toxicity for cells of the immune system. The lipid formulations of AmB are prepared with either phospholipids or detergents, and the resultant associations fall into two categories, liposomes and mixed micelles. Because only unbound AmB is active against cells, the damaging action of an AmB formulation depends on the ability of AmB to dissociate from complexes. At least three factors that can modify the distribution of AmB in vivo and affect the efficacy are suggested: its ability to bind to selected molecules found in plasma, its ability to bind to tissues and organs, and the ability of vesicles of differing sizes to traverse the epithelium. The lipid-based formulations are clearly less toxic, even at considerably higher doses; their pharmacokinetic properties and effectiveness differ in long-term assays because of different physicochemical characteristics (32). Adverse reactions to intravenous AmB are to some extent dependent on its mode of delivery. One potential problem with some of the new lipid formulations is their high manufacturing cost.

Three types of lipid-based formulations of AmB incorporated into liposomes, sheets, or discs are under intensive clinical investigation. The lipid-based formulations differ in size, structure, shape, lipid composition, and molar AmB content. Their physicochemical differences determine their thermodynamic stability, the distribution of AmB between the lipid formulation and lipoproteins, and their tissue distribution, levels in blood, uptake by macrophages, and penetration to the site of infection. Despite all these differences, they are all less toxic than AmB to mammalian cells, animals, and humans.

Many of the newer lipid formulations including AmB–lipid complex (ABLC, Abelcet), AmB colloidal dispersion (ABCD, Amphocil), and liposomal AmB (L-AmB, AmBisome) have been found to be relatively less toxic in animals and in vitro studies as well, but lowered toxicity in vitro does not always predict lower toxicity in vivo. These formulations are described in Table 1. At equivalent concentrations of AmB, deoxycolate AmB (D-AmB) has been found more effective in fungal infections than the lipid-based formulations. The small amount of free AmB that dissociates from the complexes impairs their antifungal effectiveness.

Liposomes are biocompatible, biodegradable, microvesicular systems for drug delivery owing to their amenability to controlled release and site-directed delivery. The recent development of lipid formulations with longer half-lives opens new therapeutic avenues in treating infections, including those in non-MPS tissues. The passive targeting of liposomes to the sites of infection is of great value with respect to clinical application. Liposomal entrapment can exchange the pharmacokinetics and hence reduce the toxicity of AmB. Liposomal AmB is AmB incorporated into lipid bilayer; AmBisome is a formulation of AmB in unilamellar liposomes.

Studies with mice showed that liposome encapsulation of AmB reduced its toxicity and allowed higher doses to be administered, thus increasing the therapeutic efficacy of the compound in animal models of systemic mycosis, including candidiasis (33). The priorities of liposomes differ depending on the composition of the lipids constituting the membrane. AmBisome's lack of effect on immune system cells at high concentrations may explain the decrease in toxic phenomena observed in vivo when this form of the drug is tested. The decrease in toxicity might then allow for the demonstration of the more positive effects of AmB on the immune system. Macrophages may function as a reservoir of AmB for intracellular and extracellular antimicrobial action.

The lipophilic AmB is located within the lipid layers of the liposome. Because the affinity of AmB to ergosterol in the fungus membrane is the highest among the antifungal agents, and its affinity to the lipid carrier and cholesterol in human cell membranes is the lowest, a selective transfer of AmB from the lipid carrier to its target, the fungus membrane, can be expected. The result is a great reduction in toxicity, which allows the administration of much higher doses of AmB. Although the antifungal activity of most lipid formulations is reduced to some extent, the use of much higher dosages might lead to an increased therapeutic index compared with conventional AmB. AmBisome, Amphocil, and ABLC share the advantage of low nephrotoxicity (34).

One trial involved sequestration of AmB from some host tissues while delivering the drug to the rethiculoendotherial cells where various intracellular pathogens, including *Cryptococcus neoformans*, reside. Mice were challenged intraperitoneally, intratracheally, or intracerebrally, and the course of infection in untreated control mice was compared with that of mice given D-AmB or L-AmB. Over six times more AmB in L-AmB than D-AmB could be delivered in mice without any acute toxicity. In murine pulmonary cryptococcosis, L-AmB–treated mice survived significantly longer than control mice; thus, L-AmB was found to be effective against pulmonary cryptococcosis. AmB from liposomes is readily available to the cryptococci. L-AmB may permit efficient delivery of much larger doses than are feasible with D-AmB. However, it is possible that the adjuvancy of either the AmB or the liposomes activates macrophages to kill more cryptococci and thus prolongs survival. Intracranially challenged mice treated with L-AmB had significantly reduced cell counts in brain and spleen; thus, it was suggested that effective intracranial concentrations of AmB were delivered in the liposomes and that liposomes may offer a way to protect at least some organs against the toxic effects of AmB (35).

In a controlled, randomized trial, a short antifungal prophylaxis course of AmBisome was found to reduce the incidence of proven invasive fungal infections significantly during the first month following liver transplantation surgery. AmBisome was well tolerated, although backache, thrombocytopenia, and renal function impairment were reported in a few patients (36). Aerosolized liposomal AmB deposited aerosolized form may be advantageous. This would provide delivery of the AmB to the lung tissue, but the drug would probably also get into systemic circulation. Because liposomal AmB is rapidly cleared by the reticuloendothelial system (RES), administration of higher doses is often required to obtain sufficient levels of drug in lung lead to side effects, in a murine model of pulmonary and systemic cryptococcosis, aerosolized liposomal AmB could be delivered to target and was found effective (37).

**Table 1. Characteristics of Some Lipid Formulation of AmB**

| Lipid formulation | Composition | Shape | Size |
|---|---|---|---|
| AmB–lipid complex (ABLC) | Dimyristoyl phosphatidylcholine Dimyristoyl phosphatidyl glycerol | Sheets | 1.6–11 $\mu$m |
| AmBisome | Hydrogenated phosphatidylcholine Cholesterol Distearoyl phosphatidylglycerol | Unilamellar Vesicles | 0.06 $\mu$m |
| AmB colloidal dispersion (ABCD) | Cholesteryl sulfate | Discs | 0.12 $\mu$m |
| Lipid nanosphere AmB | Soybean oil Egg lecithin Maltose | Vesicles | 25–50 nm |

Micelles are colloid particles formed by an aggregation of detergent molecules. Special drug career systems and dosage forms, such as nanoparticles and liposomes, hold the promise of overcoming the pharmacokinetic limitations. Nanoparticles are stable, solid colloidal particles of various sizes that consist of macromolecular material. Nanoparticles represent an interesting carrier system for specific enrichment in macrophage-containing organs like the liver and the spleen. Injectable nanoparticle carriers have important potential applications, as in site-specific drug delivery. Conventional carriers are generally eliminated by the RES. Although in deep mycoses, efforts to lower the toxicity of AmBisome, Abelcet, and Amphocil showed comparable efficacies, studies on the efficacies of NS-718, AmB encapsulated in lipid nanospheres are in progress in Japan. Lipid nanospheres are composed of equal amounts of egg lecithin and soybean oil. The carrier potentials of lipid nanosphere are characterized by lower uptake by the RES and good distribution to the sites of inflammation (38).

NS-718 was found to more effective than D-AmB or L-AmB against clinical isolates of *Candida albicans* and *Aspergillus fumigatus*. NS-718 was well tolerated and improved survival markedly at equivalent doses in treating pulmonary aspergillosis in rats. Increased activity was also supported by pharmacokinetic study (Fig. 2) (39). In another study, NS-718 was found to have better in vitro efficacy against clinical isolates of *C. neoformans* than other AmB formulations; it was well tolerated, and its efficacy was much higher than that of D-AmB or L-AmB in treating pulmonary cryptococcosis in mice (40) (Fig. 3).

The development of immunoliposomes is an important step in the pursuit for an effective cellular targeting system. Upon binding to the specific cellular receptor via an antibody–antigen link, the immunoliposomes cluster into the coated pits and are subsequently internalized into cellular vacuoles. The immunoliposomes are then processed

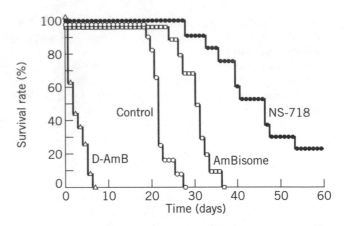

**Figure 3.** Survival rate over time of mice with pulmonary cryptococcosis, treated with NS-718. NS-718 (2mg/kg), AmB sodium deoxycholate (D-AmB) (2mg/kg), and AmBisome (2mg/kg) were injected intravenously once a day for 5 days after infection. *Source:* Based on data from Ref. 40.

through the endosome–lysosome system. Trials of their in vitro and in vivo antifungal activity, toxicity, and efficacy against systemic candidiasis in murine models showed prolonged circulation, reduced toxicity, and improved efficacy by PEG conjugated to L-AmB (41). In murine models of invasive pulmonary aspergillosis, PEG- and 34A monoclonal antibodies conjugated to L-AmB were found to have higher efficacy than those of L-AmB and PEG conjugated to L-AmB (Fig. 4). Conjugation of PEG could reduce uptake by the RES and prolong circulation time, and conjugation of 34A-monoclonal antibody to L-AmB could deliver the drug to the target tissue, that is, the endothelia of mouse pulmonary capillaries. Thus, it was suggested that functional liposomes encapsulating AmB could have great potential against refractory fungal infections because

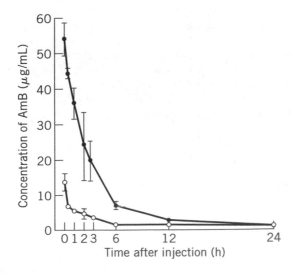

**Figure 2.** The concentrations of AmB in plasma after a single administration of 3mg/kg NS-718 and AmB sodium deoxycholate (D-AmB) in rats with invasive pulmonary aspergillosis. *Source:* Based on data from Ref. 43.

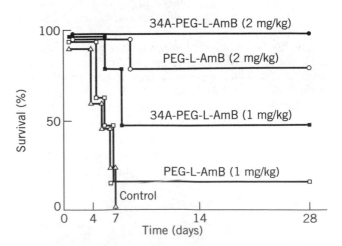

**Figure 4.** Effects of an immunoliposome, 34A-polyethylene glycol-liposomal-AmB (34A-PEG-L-AmB), on the survival rate over time in mice experimentally infected with invasive pulmonary aspergillosis. Liposomal AmB was injected intravenously in once a day for 5 days after infection. *Source:* Based on data from Ref. 43.

extended-circulation immunoliposomes are well tolerated and show comparable efficacy to other AmB formulations in murine invasive pulmonary aspergillosis (42,43).

ABLC is a formulation of lipid aggregates of dimyristoyl phosphatidylcholine (DMPC) and dimyristoyl phosphatidylglycerol (DMPG), and it has an unusual ribbonlike lipid structure that may attenuate the toxicity of AmB (44). Biological characteristics of ABLC were investigated in murine model of a variety of systemic mycoses such as candidiasis, aspergillosis, cryptococcosis, and histoplasmosis; ABLC was effective against all model infections studied, though comparable or a little lower in efficacies than D-AmB, but improved therapeutic index was noted. $LD_{50}$ of ABLC was more than 40 mg/kg compared to that of D-AmB, 3 mg/kg. Putatively, AmB is released gradually only in the infection site, most likely because yeast lipase disrupted the drug lipid interaction. Therefore, ABLC can be given at much higher doses, and therapeutic benefits can be achieved (45). In vitro antifungal activity and in vivo efficacy and acute toxicity in murine model of candidiasis were studied; ABLC was well tolerated and was found as effective as AmB (46).

In a multicenter (although not well-established) study, ABLC was found to have apparently better clinical and microbiological activity in patients with AIDS-associated cryptococcal meningitis and was significantly better tolerated than a standardized regimen of AmB but was not free from toxicities (47). Internalization of low-density-protein (LDL)-bound AmB into cells may result in toxicity, which may be minimized by a career like ABLC. In a patient with disseminated zygomycosis including renal lesions, the combined use of ABLC and granulocyte-colony-stimulating factor was successful; the patient's condition remained stable, and no relapse occurred after 1 year of follow-up (48).

AmB colloidal dispersion (ABCD) consists of a cholesteryl sulfate complex of AmB and is a disklike stable complex of the constituents in a 1:1 molar ratio. There is no free drug in this type of arrangement and low lipoprotein binding in serum or other milieu, suggesting that the drug moiety lipid-binding sites are minimally exposed and the preparation is in lyophilized form. In animal models, the half-life of ABCD is longer than that of D-AmB, which suggests that once lipid binding of ABCD occurs, dissociation occurs much slowly. AmB in ABCD reaches the antifungal target, but activity is variable. ABCD has been found effective in different animal models of systemic mycosis including coccidioidomycosis, candidiasis, and aspergillosis, and much higher doses could be administered safely (49). Efficacy and safety of ABCD was compared with that of a D-AmB suspension in a murine model of disseminated cryptococcosis, and improved therapeutic index was suggested (50). In a murine model of systemic cryptococcosis, the efficacies of commercially available lipid-based formulations and D-AmB were evaluated. The lipid-based formulations showed efficacy in prolongation of survival. ABCD and L-AmB showed equivalent potencies and were superior to ABLC; ABCD was better in clearing infections other than in the brain (51). Measurement of plasma concentration of AmB for 4 weeks after a single intravenous infusion of ABCD administered to healthy volunteers indicated that the pharmacokinetics of AmB following infusions do not differ significantly from those of D-AmB. Acute side effects reported were similar to those of D-AmB and were dose dependent. But no clinically significant changes in biochemical parameters in renal or liver function tests were noticed (52).

Four patients with cryptococcal meningitis were assessed clinically and were treated with ABCD, and it was suggested that despite having undetectable levels in the cerebrospinal fluid (CSF), ABCD is an efficacious alternative form of therapy for cryptococcal meningitis for patients intolerant to D-AmB (53). In an open-label multicenter study in hospitalized patients with invasive fungal infections and impaired renal function, ABCD was found to be effective and well tolerated in immunocompromised patients with invasive fungal infections and was safe for those with underlying renal impairment (54). In another multicenter study, ABCD was administered in patients with invasive mycoses, and clinical response and adverse effects were evaluated. The conditions of patients were also monitored by biochemical parameters, blood counts, routine urinalysis, culture, microscopy, radiography, and serologic findings. ABCD was found active and less nephrotoxic than AmB, particularly in patients nonresponsive or intolerant to D-AmB. It was also suggested that the colloidal particles of ABCD remain intact after intravenous injection and are rapidly cleared by hepatic reticuloendothelial cells, resulting in low intrarenal concentrations of AmB, which appeared to be the cause of reduction of nephrotoxicity (55). There are some noncommercially developed lipid-based formulations, such as multilamellar liposomal AmB, small unilamellar liposomal AmB, and AmB-intralipid.

Nystatin, another polyene derivative antibiotic, is to some extent similar to AmB in action, but parenteral administration being toxic, it should be used as a topical preparation. Incorporation of nystatin into multilamellar liposome was found to retain the in vitro antifungal activity and protected human erythrocytes from toxicity (56). Liposomal formulation of nystatin was well tolerated and improved survival at equivalent doses as compared to free nystatin; it showed markedly increased activity at higher doses in treating systemic candidiasis in mice. Clinical trials are under way to evaluate its efficacy and rational application (57). Fluconazole, a bis-triazole antifungal agent encapsulated in a biodegradable scleral implant, was studied in vitro and also in pigmented rabbits. Periodic measurement of released fluconazole and effects on ocular tissue by ophthalmoscopy, histology, and electrophysiological studies were done. Concentration of fluconazole in rabbit vitreous remained within 99% of MIC for C. albicans for 3 weeks, and no toxicity was noticed. Thus, usefulness of biodegradable, polymeric scleral implants containing fluconazole was suggested (58).

## LIPOSOME-ENCAPSULATED ANTIVIRAL AGENTS

Viruses are very small, "filterable" submicroscopic organisms, characterized by only a core of DNA or RNA in their genomes; they are surrounded by a protein coat and can replicate only in living cells. In some of the well-known viral infections, viral receptors are detected on host cells, which has been proposed to be important in causing disease. Many viral infections are self-limiting in nature and produce low-grade illness for short periods; others may cause prolonged suffering and at times life-threatening infections. Antiviral agents are aimed at suppression of viral replication; many of the currently available antiviral agents have therapeutic and prophylactic effects. However, viral containment or elimination is extremely difficult in individuals with impaired immune status.

In a murine model of respiratory Influenza A virus infection, improved therapeutic and prophylactic efficacies of the antiviral antibody were noted when liposomes were used as carriers for it. Thus, it was suggested that liposomes facilitated specific delivery and that targeting the antibodies to the infection site could enhance the in vivo protective efficacies of the antibody (59). In experimental Influenza A virus infection in mice, intranasal administration of 5,7,4'-trihydroxy-8-methoxyflavone solution in hydroxypropyl cellulose showed significant inhibition of virus proliferation in both nasal and bronchoalveolar cavities, which suggested potential for drug delivery by hydroxypropyl cellulose (60).

Human serum albumin (HSA) and negatively charged cis-aconitic anhydride–reacted HSA were found to covalently couple to liposome. After conjugation with polyethyleneglycol prolonged activity and dual attack on human immunodeficiency virus type 1 (HIV-1) life cycle in vitro and in vivo as a result of an effective drug delivery system it was suggested that the attachment of an anti-CD4 monoclonal antibody to the outer surface of liposomes resulted in their interaction with both monocytes and lymphocytes (61). Thus, it was suggested that immunoliposome targeting to CD4+ cells in human blood might provide a means of targeting antiviral agents in cells at risk of HIV infection (62). Nanoparticles prepared by emulsion polymerization from polyhexylcyanoacrylate were loaded with nucleoside analog zalcitabine or the HIV protease inhibitor saquinavir and then tested for antiviral activity in primary human monocytes/macrophage cells in vitro, showing comparable results to those of free drugs. The delivery of antiviral agents to the monocyte phagocytic system was enhanced by nanoparticulate formation of saquinavir and could also overcome pharmacokinetic problems; thus, improved activities of the drugs by enhancing cellular uptake and in HIV target cells and tissues were suggested (63).

A single intravitreal injection of liposome-encapsulated (S)-1-(3-hydroxy-2-phosphonylmethoxypropyl) cytosine (cidofovir) in a rabbit model of focal nonlethal Herpes simplex virus (HSV) type 1 viral retinitis was found to have remarkably protective and prolonged antiviral effect. The slow rate of release of cidofovir was considered responsible for the long-term effect (64). Intravitreal injection of acyclovir diphosphate dimystoylglycerol in a rabbit model of focal nonlethal HSV type 1 viral retinitis resulted in reduction of severity of infection and rapid clearing of vitreous and optical media as compared to the effects of gancyclovir, acyclovir, and buffer. The drug delivery system modified the antiviral effects and the duration of activity

as well, and treatment for cytomegalovirus retinitis and other retinal diseases was suggested (65).

## LIPOSOME-ENCAPSULATED ANTIPARASITIC AGENTS

Parasites are living organisms that grow, feed, and are sheltered on or in different organisms while contributing nothing to the survival of the host. Parasitic diseases comprise a large portion of health problems in underdeveloped or developing countries in tropical and subtropical areas. In developed countries, parasitic diseases occur infrequently, though contamination of water supplies, improperly cooked meat, or worldwide travel are important factors to consider. Some parasitic infections have been found in association with impaired immune status, such as results from AIDS.

The use of drug delivery systems in parasitic diseases other than lipid formulations of AmB in leishmaniasis are not well known and lag behind use against other infectious diseases. In experimental leishmaniasis, pentamidine isethionate and its methoxy derivative, encapsulated in mannose-grafted liposomes, showed improved therapeutic effects over pentamidine isethionate, though the methoxy derivative was less toxic. Sugar-grafted liposomes were suggested to have better carrier potential than liposomes or free drug (66). Colloidal carrier potential and the targeted and controlled delivery of albendazole encapsulated in polyhexylcyanoacrylate nanoparticle was evaluated in vitro and also in murine models of hepatic alveolar echinococcosis. Cytotoxicity study showed no toxicity for peritoneal macrophages but reduced viability of the parasites. In vivo, no acute toxicity of albendazole-loaded nanoparticles was observed, and the injectable form showed equivalent parasitostatic effect (67).

## BIBLIOGRAPHY

1. J.V. Desiderio and S.D. Campbell, *J. Infect. Dis.* **148**, 563–570 (1983).
2. T. Tadakuma et al., *Antimicrob. Agents Chemother.* **28**, 28–32 (1985).
3. E. Fattal et al., *Antimicrob. Agents Chemother.* **35**, 770–772 (1991).
4. H. Al-Awadhi, G.V. Stoke, and M. Reich, *J. Antimicrob. Chemother.* **30**, 303–311 (1992).
5. J.M. Grange et al., *Lancet* **3**, 1350–1352 (1995).
6. M.C. Nacucchio, M.J. Gatti Bellora, D.O. Sordelli, and M. D'Aquino, *J. Microencapsul.* **5**, 303–309 (1988).
7. H. Kotsifaki et al., *J. Chemother.* **2**, 82–86 (1990).
8. J. Lagace, M. Dubreuil, and S. Montplaisir, *J. Microencapsul.* **8**, 53–61 (1991).
9. E.A. Poyner, H.O. Alpar, and M.R. Brown, *J. Antimicrob. Chemother.* **34**, 43–52 (1994).
10. C. Beaulac et al., *Antimicrob. Agents Chemother.* **40**, 665–669 (1996).
11. H. Koga et al., *Tuberculosis* **69**, 55–60 (1993).
12. M.A. Vladimirsky and G.A. Ladigina, *Biomed. Pharmacother.* **36**, 8–9 (1982).
13. N. Duzgunes et al., *Antimicrob. Agents Chemother.* **40**, 2618–2621 (1996).

14. N. Duzgunes et al., *Antimicrob. Agents Chemother.* **32**, 1404–1411 (1988).

15. H. Tomioka, H. Saito, K. Sato, and T. Yoneyama, *Am. Rev. Respir. Dis.* **43**, 1421–1428 (1991).

16. S.D. Nightingale et al., *Antimicrob. Agents Chemother.* **37**, 1869–1872 (1993).

17. H. Saito and H. Tomioka, *Antimicrob. Agents Chemother.* **33**, 429–433 (1989).

18. S. Majumdar et al., *Antimicrob. Agents Chemother.* **36**, 2808–2815 (1992).

19. C.O. Onyeji, C.H. Nightingale, D.P. Nicolau, and R. Quintiliani, *Antimicrob. Agents Chemother.* **38**, 523–527 (1994).

20. R.T. Mehta et al., *Antimicrob. Agents Chemother.* **37**, 2584–2587 (1993).

21. R.T. Mehta, *Antimicrob. Agents Chemother.* **40**, 1893–1902 (1996).

22. P. Deol, G.K. Khuller, and K. Joshi, *Antimicrob. Agents Chemother.* **41**, 1211–1214 (1997).

23. L. Kesavalu et al., *Tubercle* **71**, 215–218 (1990).

24. S.P. Klemens, M.H. Cynamon, C.E. Swenson, and R.S. Ginsberg, *Antimicrob. Agents Chemother.* **34**, 967–970 (1990).

25. N. Duzgunes et al., *J. Infect. Dis.* **164**, 143–151 (1991).

26. G. Medoff et al., *Clin. Infect. Dis.* **15**(S1), S274–S281 (1992).

27. N.H. Georgopapadakou and T.J. Walsh, *Antimicrob. Agents Chemother.* **40**, 279–291 (1996).

28. R.J. Hay, *J. Am. Acad. Dermatol.* **31**, S82–S85 (1994).

29. A.H. Groll and T.J. Walsh, *Curr. Opin. Infect. Dis.* **10**, 449–458 (1997).

30. J.D. Cleary et al., *Ann. Pharmacother.* **27**, 715–719 (1993).

31. J.W. Hiemenz and T.J. Walsh, *Clin. Infect. Dis.* **22**(S2), S133–S144 (1996).

32. J. Brajtburg and J. Bolard, *Clin. Microbiol. Rev.* **9**, 512–531 (1996).

33. G. Lopez-Berestein et al., *J. Infect. Dis.* **147**, 939–945 (1983).

34. S. de Marie, *Leukemia* **10**(S2), S93–S96 (1996).

35. J.R. Graybill et al., *J. Infect. Dis.* **145**, 748–752 (1982).

36. J. Tollemar et al., *Transplantation* **59**, 45–50 (1995).

37. B.E. Gilbert, P.R. Wyde, and S.Z. Wilson, *Antimicrob. Agents Chemother.* **36**, 1466–1471 (1992).

38. J. Seki et al., *J. Controlled Release* **28**, 352–353 (1994).

39. T. Otsubo et al., *Antimicrob. Agents Chemother.* **43**, 471–475 (1999).

40. M.A. Hossain et al., *Antimicrob. Agents Chemother.* **42**, 1722–1725 (1998).

41. E.W.M. van Etten, M.T. ten Kate, L.E.T. Stearne, and I.A.J.M. Bakker-Woudenberg, *Antimicrob. Agents Chemother.* **39**, 1954–1958 (1995).

42. S. Kohno et al., *Adv. Drug Delivery Rev.* **24**, 325–329 (1997).

43. T. Otsubo et al., *Antimicrob. Agents Chemother.* **42**, 40–44 (1998).

44. A.S. Janoff et al., *Proc. Natl. Acad. Sci. U.S.A.* **85**, 6122–6126 (1988).

45. J.M. Clark et al., *Antimicrob. Agents Chemother.* **35**, 615–621 (1991).

46. K. Mitsutake et al., *Mycopathologia* **128**, 13–17 (1994).

47. P.K. Sharkey et al., *Clin. Infect. Dis.* **22**, 315–321 (1996).

48. C.E. Gonzalez, D.R. Couriel, and T.J. Walsh, *Clin. Infect. Dis.* **24**, 192–196 (1997).

49. D.A. Stevens, *J. Infect.* **28**(S1), S45–S49 (1994).

50. J.S. Hostetler, K.V. Clemons, L.H. Hanson, and D.A. Stevens, *Antimicrob. Agents Chemother.* **36**, 2656–2660 (1992).

51. K.V. Clemons and D.A. Stevens, *Antimicrob. Agents Chemother.* **42**, 899–902 (1998).

52. S.W. Sanders et al., *Antimicrob. Agents Chemother.* **35**, 1029–1034 (1991).

53. G. Valero and J.R. Graybill, *Antimicrob. Agents Chemother.* **39**, 2588–2590 (1995).

54. E.J. Anaissie et al., *Antimicrob. Agents Chemother.* **42**, 606–611 (1998).

55. A. Oppenheim, R. Herbrecht, and S. Kusne, *Clin. Infect. Dis.* **21**, 1145–1153 (1995).

56. R.T. Mehta et al., *Antimicrob. Agents Chemother.* **31**, 1897–1900 (1987).

57. R.T. Mehta et al., *Antimicrob. Agents Chemother.* **31**, 1891–1903 (1987).

58. H. Miyamoto et al., *Curr. Eye Res.* **16**, 939–935 (1997).

59. J.P. Wong, L.L. Stadnyk, and E.G. Saravolac, *Immunology* **81**, 280–284 (1994).

60. T. Nagai et al., *Biol. Pharm. Bull.* **20**, 1082–1085 (1997).

61. J.A.A.M. Kamps et al., *Biochim. Biophys. Acta* **1278**, 183–190 (1996).

62. N.C. Philips and C. Tsuoukas, *Cancer Detect. Prev.* **14**, 383–390 (1990).

63. R. Bender et al., *Antimicrob. Agents Chemother.* **40**, 1467–1471 (1996).

64. D. Kuppermann et al., *J. Infect. Dis.* **173**, 18–23 (1996).

65. I. Taskintuna et al., *Retina* **17**(1), 57–64 (1997).

66. G. Banerjee et al., *J. Antimicrob. Chemother.* **38**(1), 145–150 (1996).

67. J.M. Rodrigues, Jr. et al., *Int. J. Parasitol.* **25**, 1437-1441 (1995).

# INTELLIGENT DRUG DELIVERY SYSTEMS

JOSEPH KOST
Ben-Gurion University
Beer-Sheva, Israel

## KEY WORDS

Antibody interaction

Chelation

Electrically modulated systems

Glucose-responsive systems

Inflammation-sensitive systems

Intelligent polymers

Magnetically modulated systems

Modulated systems

Morphine-triggered delivery

pH sensitive systems

Photo-responsive systems

Responsive systems

Smart polymers

Temperature-sensitive systems

Ultrasonically modulated systems

Urea responsive systems

## OUTLINE

The ideal drug delivery system should provide therapeutics in response to physiological requirements, having the capacity to "sense" changes and alter the drug-release process accordingly. The basic approach that drug concentration–effect relationships are significantly invariant as a function of time in humans has led to the development of constant-rate drug delivery systems (1). Nevertheless, there are a number of clinical situations where such an approach may not be sufficient. These include the delivery of insulin for patients with diabetes mellitus, antiarrhythmics for patients with heart rhythm disorders, gastric acid inhibitors for ulcer control, and nitrates for patients with angina pectoris as well as selective $\beta$-blockade, birth control, general hormone replacement, immunization, and cancer chemotherapy. Recent studies in the field of chronopharmacology indicate that the onset of certain diseases exhibits strong circadian temporal dependency. Thus, drug delivery patterns can be further optimized by pulsed or self-regulated delivery, adjusted to the staging of biological rhythms (2).

Modulated drug delivery systems are devices that are implanted or injected into the body and are capable of releasing drugs in response to external stimuli such as magnetic fields, heat, ultrasound, pH, or concentration of a specific molecule. These "intelligent" delivery systems can be classified as open- or closed-loop systems. Open-loop control systems are those in which information about the controlled variable is not automatically used to adjust the system inputs to compensate for the change in the process variables, whereas in closed-loop control systems, the controlled variable is detected, and as a result the system output is adjusted accordingly. In the controlled drug delivery field, open-loop systems are known as *pulsatile* or *externally regulated*, and the closed-loop systems as *self-regulated*. The externally controlled devices apply external triggers for pulsatile delivery such as magnetic, ultrasonic,

thermal, electric, and electromagnetic irradiation, whereas in the self-regulated devices, the release rate is controlled by feedback information without any external intervention. The self-regulated systems utilize several approaches as rate-control mechanisms: pH-sensitive polymers, enzyme–substrate reactions, pH-sensitive drug solubility, competitive binding, antibody interactions, and metal concentration–dependent hydrolysis.

This article outlines the fundamental principles of both pulsatile and self-regulated systems for drug delivery.

## PULSATILE SYSTEMS

### Magnetically Modulated Systems

An early approach to pulsatile delivery involved incorporating magnetic beads in elastic polymers. When an oscillating magnetic field was applied, more drug was released (Fig. 1). Studies demonstrated that insulin and other molecules could be continuously released by embedding the hormone in a carrier such as ethylene-vinyl acetate copolymer (EVAc) (3). In vitro studies were then conducted characterizing the critical parameters affecting release rates (4). This information was utilized to design subcutaneous implants of EVAc–insulin, which decreased the glucose levels of diabetic rats for 105 days (5). The next step in designing a delivery system for use in diabetes was to develop a system that would be capable of releasing insulin at a higher rate upon demand. In vitro studies were conducted showing that EVAc–protein matrices containing magnetic beads exhibit enhanced release rates when placed in an oscillating external magnetic field (6–8). In vivo studies (9) showed that when polymeric matrices containing insulin and magnetic beads were implanted in diabetic rats, glucose levels could be repeatedly decreased on demand by application of an oscillating magnetic field.

The systems consists of drug powder dispersed within a polymeric matrix together with magnetic beads. One method of formulating this system is to add approximately 50% of the drug–polymer mixture to a glass mold that has been cooled to $-80°C$ using dry ice. The magnetic particles are added, followed by the remaining drug–polymer mixture (7). In addition to the experimental polymer matrices containing both magnets and insulin, controls, including polymer matrices that contained a magnet but no insulin and that contained insulin but no magnet, were made.

When a polymer matrix containing insulin and a magnet was implanted into a diabetic rat (9), the blood glucose

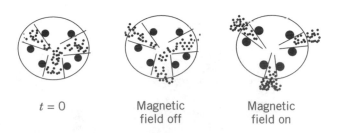

$t = 0$      Magnetic field off      Magnetic field on

**Figure 1.** Schematic presentation of a magnetic-triggered controlled-release system. Large dots are magnetic beads; small dots represent drug (1).

level fell from over 400 mg % to nearly 200 mg %. In the absence of the magnetic field, the blood glucose level remained near this value over the 51-day implantation period. However, every time the magnetic field was applied, the blood glucose level decreased. The average decrease for the three rats in the experimental group, which were triggered a total of 26 times, was 29.4%. The difference in glucose changes between the experimental group and the controls (14 rats, 97 triggers) was highly significant. These results were confirmed by an insulin radioimmunoassay (RIA).

The factors that are critical in controlling the release rates in these systems can be characterized by two main groups: (1) magnetic field characteristics and (2) mechanical properties of the polymer matrix. It was found that the extent of release enhancement increases as the magnetic field amplitude rises. When the frequency of the applied field was increased from 5 to 11 Hz, the release rate of bovine serum albumin (BSA) from EVAc copolymer matrices rose in a linear fashion (7). Saslavski et al. (10) investigated the effect of magnetic field frequency and repeated field application on insulin release from alginate matrices and found that with repeated applications, inverse effects can occur: High frequencies gave a significant release enhancement for the second magnetic field application, after which the enhancement level decreased owing to faster depletion at these frequencies.

The mechanical properties of the polymeric matrix also affect the extent of magnetic enhancement (6). For example, the modulus of elasticity of the EVAc copolymer can be easily altered by changing the vinyl acetate content of the copolymer. The release-rate enhancement induced by the magnetic field increase as the modulus of elasticity of EVAc decreases. A similar phenomenon was observed for the cross-linked alginate matrices: higher release-rate enhancement for less rigid matrices. Edelman et al. (11) also showed that enhanced release rates observed in response to an electromagnetic field (50 G, 60 Hz) applied for 4 min were independent of the duration of the interval between repeated pulses.

## Ultrasonically Modulated Systems

Kost et al. (12) suggested the feasibility of ultrasonic-controlled polymeric delivery systems in which release rates of substances can be repeatedly modulated at will from a position external to the delivery system. Both bioerodible and nonerodible polymers were used as drug carrier matrices. The bioerodible polymers evaluated were polyglycolide, polylactide, poly(bis(p-carboxyphenoxy) alkane anhydrides and their copolymers with sebacic acid. The releasing agents were p-nitroaniline, p-aminohippurate, bovine serum albumin, and insulin.

Enhanced polymer erosion and drug release were observed when the bioerodible samples were exposed to ultrasound. The systems response to the ultrasonic triggering was rapid (within 2 min) and reversible. The enhanced release was also observed in nonerodible systems exposed to ultrasound where the release is diffusion dependent. Release rates of zinc bovine insulin from EVAc matrices were 15 times higher when exposed to ultrasound compared

with the unexposed periods. It has also been demonstrated that the extent of enhancement can be regulated by the intensity, frequency, or duty cycle of the ultrasound (12).

To assess the effect of ultrasonic energy on the integrity of the releasing molecules, insulin samples were evaluated by high-pressure liquid chromatography (HPLC). No significant difference was detected between insulin samples exposed to ultrasound and unexposed samples, suggesting that the ultrasound is not degrading the releasing molecules (12).

In vivo studies have suggested the feasibility of ultrasound-mediated drug release enhancement (12). Implants composed of polyanhydride polymers loaded with 10% para-aminohippuric acid (PAH) were implanted subcutaneously in the back of catheterized rats. When exposed to ultrasound, a significant increase in the PAH concentration in urine was detected (400%) (Fig. 2). The Rats' skin histopathology of the ultrasound-treated area after an exposure of 1 h at 5 W/cm$^2$ did not reveal any differences between treated and untreated skin.

It was proposed (12) that cavitation and acoustic streaming are responsible for this augmented degradation and release. In experiments conducted in a degassed buffer, where cavitation was minimized, the observed enhancement in degradation and release rates was much smaller. It was also considered that several other parameters (temperature and mixing effects) might be responsible for the augmented release due to ultrasound. However, experiments were conducted that suggested that these parameters were not significant. A temperature rise of only 2.5°C was recorded in the samples during the triggering

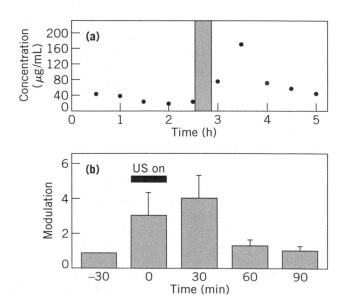

**Figure 2.** (a) Para-amino hippuric acid (PAH) concentration in the urine of Sprague-Dawley rats as a function of time before, during, and after a 20-min exposure to ultrasound (shaded area). (b) Modulation versus time expressed as a mean and standard deviation of four experimental rats. (Modulation was defined as the ratio of PAH concentration during and after the ultrasound [US] exposure to the mean of the PAH concentration before the exposure) (12).

period. A separate release experiment done at 40°C instead of 37°C showed that the rate increase was below 20%. To evaluate the ultrasound effect on the diffusion boundary layer, release experiments were performed under vigorous shaking. The increase of the release rates due to shaking were always below 20%. Therefore, it was concluded that the effect of the ultrasound on the augmented release cannot be due to mixing or temperature only.

Similar phenomena were observed by Miyazaki et al. (13) who evaluated the effect of ultrasound (1 MHz) on the release rates of insulin from ethylene-vinyl alcohol copolymer matrices and reservoir-type drug delivery systems. When diabetic rats receiving implants containing insulin were exposed to ultrasound (1 W/cm$^2$ for 30 min), a sharp drop in blood glucose levels was observed after the irradiation, indicating a rapid rate of release of insulin in the implanted site. The authors speculate that the ultrasound caused increased temperature in the delivery system, which may facilitate diffusion.

Over the past 40 years, numerous clinical reports have been published concerning phonophoresis (14). The technique involves placing the topical preparation on the skin over the area to be treated and massaging the area with an ultrasound probe. In the last decade, with the development of transdermal delivery as an important means of systemic drug administration, researchers investigated the possible application of ultrasound into transdermal delivery systems. Kost et al. (14–16) studied in rats and guinea pigs the effect of therapeutic ultrasound (1 MHz) on skin permeability of D-mannitol, a highly polar sugar alcohol; insulin, a high-molecular-weight polysaccharide; and physostigmine, a lipophilic anticholinesterase drug. Ultrasound nearly completely eliminated the lag time usually associated with transdermal delivery of drugs. Three to five minutes of ultrasound irradiation (1.5 W/cm$^2$ continuous wave or 3 W/cm$^2$ pulsed wave) increased the transdermal permeation of inulin and mannitol in rats by 5–20-fold within 1–2 h following ultrasound application. Ultrasound treatment also significantly increased the inhibition of cholinesterase during the first hour after application in both physostigmine-treated rats and guinea pigs.

Miyazaki et al. (17) performed similar studies evaluating the effect of ultrasound (1 MHz) on indomethacin permeation in rats. Pronounced effect of ultrasound on transdermal absorption for all three ranges of intensities (0.25, 0.5, and 0.75 W/cm$^2$) was observed. Bommannan et al. (18) examined the effects of ultrasound on the transdermal permeation of lanthanum nitrate, an electron-dense tracer. The results demonstrate that exposure of the skin to ultrasound can induce considerable and rapid tracer transport through an intercellular route. Prolonged exposure of the skin to high-frequency ultrasound (20 min, 16 MHz), however, resulted in structural alterations of epidermal morphology. Tachibana and Tachibana (19–21) have reported use of low-frequency ultrasound (48 KHz) to enhance transdermal transport of lidocaine and insulin across hairless mice skin. Low-frequency ultrasound has also been used by Mitragotri et al. (22,23) to enhance transport of various low-molecular-weight drugs including salicylic acid and corticosterone as well as high-molecular-

weight proteins including insulin, γ-interferon, and erythropoeitin across human skin in vitro and in vivo.

Mitragotri et al. (24) also evaluated the role played by various ultrasound-related phenomena, including cavitation, thermal effects, generation of convective velocities, and mechanical effects. The authors' experimental findings suggest that among all the ultrasound-related phenomena evaluated, cavitation plays the dominant role in sonophoresis using therapeutic ultrasound (frequency, 1–3 MHz; intensity, 0–2 W/cm$^2$). Confocal microscopy results indicate that cavitation occurs in the keratinocytes of the stratum corneum upon ultrasound exposure. The authors hypothesized that oscillations of the cavitation bubbles induce disorder in the stratum corneum lipid bilayers, thereby enhancing transdermal transport. The theoretical model developed to describe the effect of ultrasound on transdermal transport predicts that the sonophoretic enhancement depends most directly on the passive permeant diffusion coefficient in water and not on the permeant diffusion coefficient through the skin.

## Electrically Regulated Systems

Electrically controlled systems provide drug release by the action of an applied electric field on a rate-limiting membrane and/or directly on the solute, thus controlling its transport across the membrane. The electrophoretic migration of a charged macrosolute within a hydrated membrane results from the combined response to the electrical forces on the solute and its associated counterions in the adjacent electrolyte solution (25). Grimshaw et al. (26) reported four different mechanisms for the transport of proteins and neutral solutes across hydrogel membranes (1) electrically and chemically induced swelling of a membrane to alter the effective pore size and permeability, (2) electrophoretic augmentation of solute flux within a membrane, (3) electrosmotic augmentation of solute flux within a membrane, and (4) electrostatic partitioning of charged solutes into charged membranes.

Electrically controlled membrane permeability has also been of interest in the field of electrically controlled or enhanced transdermal drug delivery (e.g., iontophoresis, electroporation) (27,28).

Kwon et al. (29) studied the effect of electric current on solute release from cross-linked poly(2-acrylamido-2-methylpropane sulfonic acid-co-n-butylmethacrylate). Edrophonium chloride, a positively charged solute, was released in an on–off pattern from a matrix (monolithic) device with electric field. The mechanism was explained as an ion exchange between positive solute and hydroxonium ion, followed by fast release of the charged solute from the hydrogel. The fast release was attributed to the electrostatic force, squeezing effect, and the electroosmosis of the gel. However, the release of neutral solute was controlled by diffusion effected by swelling and deswelling of the gel.

Anionic gels as vehicles for electrically modulated drug delivery were studied by Hsu and Block (30). Agarose and combination of agarose with anionic polymers (polyacrylic acid, xanthan gum) were evaluated. The authors conclude the use of carbomer (polyacrylic acid) in conjunction with agarose enables the formulator to achieve zero-order release with electrical application. Increased anisotropicity of a gel system due to the application of electrical current could alter the effectiveness of a drug delivery system.

D'Emanuele and Staniforth (31) proposed a drug delivery device that consists of a polymer reservoir with a pair of electrodes placed across the rate-limiting membrane. By altering the magnitude of the electric field between the electrodes, the authors proposed to modulate the drug release rates in a controlled and predictable manner. A linear relationship was found between current and propanolol hydrochloride permeability through poly(2-hydroxyethyl methacrylate) (PHEMA) membranes cross-linked with ethylene glycol dimethacrylate (1% v/v). Buffer ionic strength and drug reservoir concentration as well as electrode polarity were found to have significant effects on drug permeability (32).

Labhasetwar et al. (33) propose similar approach for a cardiac drug delivery modulation. The authors studied a cardiac drug implant in dogs that is capable of electric current modulation. Cation-exchange membrane was used as an electrically sensitive rate-limiting barrier on the cardiac-contacting surface of the implant. The cardiac implant demonstrated in vitro drug-release rates that were responsive to current modulation. In vivo results in dogs have confirmed that electrical modulation resulted in regional coronary enhancement of the drug levels with current responsive increase in drug concentration.

A different approach for electrochemical-controlled release is based on polymers that bind and release bioactive compounds in response to an electric signal (34). The polymer has two redox states, only one of which is suitable for ion binding. Drug ions are bound in one redox state and released from the other. The attached electrodes serve to switch the redox states, and the amount of current passed can control the amount of ions released.

### Photoresponsive Systems

Photoresponsive gels change their physical or chemical properties reversibly upon photoradiation. A photoresponsive polymer consists of a photoreceptor, usually a photochromic chromophore, and a functional part. The optical signal is captured by the photochromic molecules, and then the isomerization of the chromophores in the photoreceptor converts it to a chemical signal. Photoinduced phase transition of gels was reported by Mamada et al. (35). Copolymer gels of $N$-isopropylacrylamide and a photosensitive molecule, bis(4-dimethylamino)phenyl)(4-vinylphenyl)methyl leucocyanide, showed a discontinuous volume-phase transition upon ultraviolet irradiation, caused by osmotic pressure of cyanide ions created by the ultraviolet irradiation. Suzuki and Tanaka (36) reported on phase transition in polymer gels induced by visible light, where the transition mechanism is due only to the direct heating of the network polymer by light. Yui et al. (37) proposed photoresponsive degradation of heterogeneous hydrogels comprising cross-linked hyaluronic acid and lipid microspheres for temporal drug delivery. A visible light–induced degradation of cross-linked hyaluronic acid gels by photochemical oxidation using methylene blue as a photosensitizer. (The hyaluronic acid gels were also proposed by the authors to be inflammation responsive [38].)

## RESPONSIVE SYSTEMS

Polymers that alter their characteristics in response to changes in their environment have been of great recent interest. Several research groups have been developing drug delivery systems based on these responsive polymers that more closely resemble the normal physiological process. In these devices, drug delivery is regulated by means of an interaction with the surrounding environment (feedback information) without any external intervention. The most commonly studied polymers having environmental sensitivity are either pH or temperature sensitive (Fig. 3).

### Temperature-Sensitive Systems

Temperature-sensitive polymers can be classified into two groups based on the origin of the thermosensitivity in aqueous media. The first is based on polymer–water interactions, especially specific hydrophobic/hydrophilic balancing effects and the configuration of side groups. The other is based on polymer–polymer interactions in addition to polymer–water interactions. When polymer networks swell in a solvent, there is usually a negligible or small positive enthalpy of mixing or dilution. Although a positive enthalpy change opposes the process, the large gain in the entropy drives it. In aqueous polymer solutions, the opposite is often observed. This unusual behavior is associated with a phenomenon of polymer-phase separation as the temperature is raised to a critical value, known as the lower critical solution temperature (LCST). $N$-alkyl acrylamide homopolymers and their copolymers, including acidic or basic comonomers show this LCST (39,40). Polymers characterized by LCST usually shrink, as the temperature is increased through the LCST. Lowering the temperature below LCST results in the swelling of the polymer. Bioactive agents such as drugs, enzymes, and antibodies may be immobilized on or within the temperature-sensitive polymers. Responsive drug-release patterns regulated by temperature changes have been recently demonstrated by several groups (39,41–52).

### Systems Sensitive to pH

The pH range of fluids in various segments of the gastrointestinal tract may provide environmental stimuli for responsive drug release. Studies by several research groups (53–65) have been performed on polymers containing weakly acidic or basic groups in the polymeric backbone. The charge density of the polymers depends on pH and

**Figure 3.** Schematic presentation of pH- or temperature-controlled drug delivery systems. Small dots represent drugs; lines represent polymer chains (1). ($\Delta$ can be positive or negative.)

ionic composition of the outer solution (the solution into which the polymer is exposed). Altering the pH of the solution will cause swelling or deswelling of the polymer. Thus, drug release from devices made from these polymers will display release rates that are pH dependent. Polyacidic polymers will be unswollen at low pH, because the acidic groups will be protonated and hence unionized. With increasing pH, polyacid polymers will swell. The opposite holds for polybasic polymers, because the ionization of the basic groups will increase with decreasing pH. Siegel et al. (66) found the swelling properties of the polybasic gels are influenced also by buffer composition (concentration and $pK_a$). A practical consequence proposed is that these gels may not reliably mediate pH-sensitive swelling-controlled release in oral applications, because the levels of buffer acids in the stomach (where swelling and release are expected to occur) generally cannot be controlled. However, the gels may be useful as mediators of pH-triggered release when precise rate control is of secondary importance.

Annaka and Tanaka (54) reported that more than two phases (swollen and collapsed) can be found in gels consisting of copolymers of randomly distributed positively and negatively charged groups. In these gels, polymer segments interact with each other through attractive or repulsive electrostatic interactions and through hydrogen bonding. The combination of these forces seems to result in the existence of several phases, each characterized by a distinct degree of swelling, with abrupt jumps between them. The existence of these phases presumably reflects the ability of macromolecular systems to adopt different stable conformations in response to changes in environmental conditions. For copolymer gels prepared from acrylic acid (the anionic constituent) and methacryl-amido-propyl-trimethyl ammonium chloride (460 mmol/240 mmol), the largest number of phases was seven. A similar approach was proposed by Bell and Peppas (64); membranes made from grafted poly(methacrylic acid-*g*-ethylene glycol) copolymer showed pH sensitivity due to complex formation and dissociation. Uncomplexed equilibrium swelling ratios were 40 to 90 times higher than those of complexed states and varied according to copolymer composition and polyethylene glycol graft length.

Giannos et al. (67) proposed temporally controlled drug delivery systems, coupling pH oscillators with membrane-diffusion properties. By changing the pH of a solution relative to the $pK_a$, a drug may be rendered charged or uncharged. Because only the uncharged form of a drug can permeate across lipophilic membranes, a temporally modulated delivery profile may be obtained with a pH oscillator in the donor solution.

Heller and Trescony (68) were the first to propose the use of pH-sensitive bioerodible polymers. In their approach, described in the section on systems utilizing enzymes, an enzyme–substrate reaction produces a pH change that is used to modulate the erosion of a pH-sensitive polymer containing a dispersed therapeutic agent.

Bioerodible hydrogels containing azoaromatic moieties were synthesized by Ghandehari et al. (69). Hydrogels with lower cross-linking density underwent a surface erosion process and degraded at a faster rate. Hydrogels with higher cross-linking densities degraded at a slower rate by a process where the degradation front moved inward to the center of the polymer.

Recently recombinant DNA methods were used to create artificial proteins that undergo reversible gelation in response to changes in pH or temperature (70). The proteins consist of terminal leucine zipper domains flanking a central, flexible, water-soluble polyelectrolyte segment. Formation of coiled-coil aggregates of the terminal domains in near-neutral aqueous solutions triggers formation of a three-dimensional polymer network, with the polyelectrolyte segment retaining solvent and preventing precipitation of the chain. Dissociation of the coiled-coil aggregates through elevation of pH or temperature causes dissolution of the gel and a return to the viscous behavior that is characteristic of polymer solution. The authors suggest these hydrogels have potential in bioengineering applications requiring encapsulation or controlled release of molecules and cellular species.

### Inflammation-Responsive Systems

Yui et al. (38) proposed an inflammation-responsive drug delivery system based on biodegradable hydrogels of cross-linked hyaluronic acid. Hyaluronic acid is specifically degraded by hydroxyl radicals, which are produced by phagocytic cells such as leukocytes and macrophages, locally at inflammatory sites. In their approach, drug-loaded lipid microspheres were dispersed into degradable matrices of cross-linked hyaluronic acid.

### Glucose and Other Saccharide-Sensitive Polymers

The basic principle of competitive binding and its application to controlled drug delivery was first presented by Brownlee and Cerami (71), who suggested the preparation of glycosylated insulins, which are complementary to the major combining site of carbohydrate-binding proteins such as Concavalin A (Con A). Con A is immobilized on sepharose beads. The glycosylated insulin, which is biologically active, is displaced from the Con A by glucose in response to, and proportional to, the amount of glucose present, which competes for the same binding sites. Kim et al. (72–79) found that the release rate of insulin also depends on the binding affinity of an insulin derivative to the Con A and can be influenced by the choice of saccharide group in glycosylated insulin. By encapsulating the glycosylated insulin-bound Con A with a suitable polymer that is permeable to both glucose and insulin, the glucose influx and insulin efflux would be controlled by the encapsulation membrane (Fig. 4).

It was found (73) that the glycosylated insulins are more stable against aggregation than commercial insulin and are also biologically active. The functionality of the intraperitoneally implanted device was tested in pancreatectomized dogs by an intravenous glucose tolerance test (IVGTT). The effect of an administered 500 mg/kg dextrose bolus on blood glucose level was compared with normal and pancreatectomized dogs without an implant. Figure 5 shows the results of this study (78). In addition, the blood glucose profile for a period of 2 days demonstrated that a diabetic dog, implanted with the self-regulating insulin

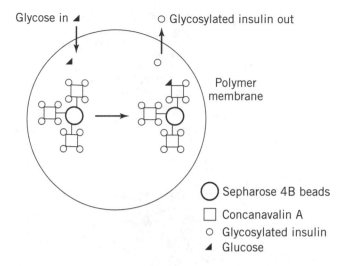

**Figure 4.** Schematic presentation of a self-regulated insulin delivery system based on competitive binding (79).

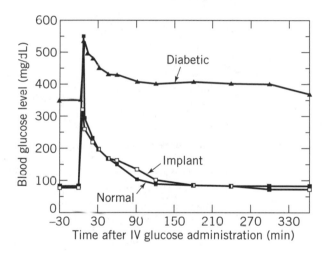

**Figure 5.** Peripheral blood glucose profiles of dogs administered bolus dextrose (500 mg/kg) during an intravenous glucose tolerance test. Blood glucose levels at $t = -30$ min show the overnight fasting level 30 min prior to bolus injection of dextrose (79).

delivery system, was capable of maintaining acceptable glucose levels (50–180 mg/dL) for the majority of the experiment (40 h) (75–77). Makino et al. (72) proposed a modification based on hydrophilic nylon microcapsules containing Con A and succinil-amidophenyl-glucopyranoside insulin.

Kokufata et al. (80) reported on a gel system that swells and shrinks in response to specific saccharides. The gel consists of a covalently cross-linked polymer network of $N$-isopropylacrylamide in which the lectin, Con A, is immobilized. Con A displays selective binding affinities for certain saccharides. For example, when the saccharide dextran sulphate is added to the gel, it swells to a volume up to 5 times greater. Replacing dextran sulphate with nonionic saccharide $\alpha$-methyl-D-mannopyranoside brings about collapse of the gel back to almost its native volume. The process is reversible and repeatable.

Taylor et al. (81) proposed similar approach for the delivery of insulin. The self-regulating delivery device, responsive to glucose, has been shown to operate in vitro. The device comprises an reservoir of insulin and a gel membrane that determines the delivery rates of insulin. The gel consists of a synthetic polysucrose and a lectin, Con A. The mechanism is one of displacement of the branched polysaccharide from the lectin receptors by incoming glucose. The gel loses its high viscosity as a result but reforms on removal of glucose, thus providing the rate-controlling barrier for the diffusion of insulin or any other antihyperglycemic drugs.

A similar approach was also presented by Park et al. (82–84), who synthesized glucose-sensitive membranes based on the interaction between polymer-bound glucose and Con A.

Kitano et al. (85) proposed a glucose-sensitive insulin-release system based on a sol–gel transition. A phenylboronic acid (PBA) moiety was incorporated in poly($N$-vinyl-2-pyrrolidone) by the radical copolymerization of $N$-vinyl-2-pyrrolidone with $m$-acrylamidophenylboronic acid (poly [NVP-$co$-PBA]). Insulin was incorporated into a polymer gel formed by a complex of poly(vinyl alcohol) with poly(NVP-$co$-PBA) (Fig. 6). PBA can form reversible covalent complexes with molecules having diol units, such as

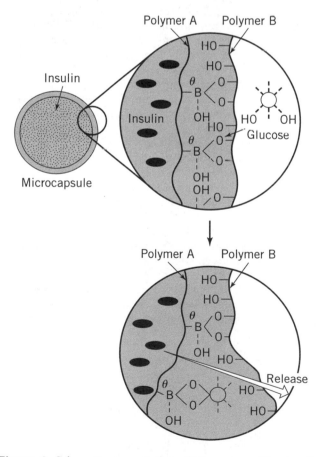

**Figure 6.** Schematic presentation of glucose-sensitive insulin-release system using PVA–poly(NVP-$co$-PBA) complex system (85).

glucose or PVA. With the addition of glucose, PVA in the PVA–boronate complex is replaced by glucose. This leads to a transformation of the system from gel to sol state, which facilitates the release of insulin from the polymeric complex. The same group (86) modified the approach suggesting glucose-responsive gels based on the complexation between polymers having PBA groups and PVA. The introduction of an amino group into PBA polymers was effective for increasing the complexation ability and the glucose responsivity at physiological pH.

## SYSTEMS UTILIZING ENZYMES

In this approach, the mechanism is based on an enzymatic reaction. One possible approach studied is an enzyme reaction that results in a pH change and a polymer system that can respond to that change.

### Urea Responsive Delivery

Heller and Trescony (68) were the first to attempt using immobilized enzymes to alter local pH and thus cause changes in polymer erosion rates. The proposed system is based on the conversion of urea to $NH_4HCO_3$ and $NH_4OH$ by the action of urease. As this reaction causes a pH increase, a polymer that is subjected to increased erosion at high pH is required.

The authors suggested a partially esterified copolymer of methyl vinyl ether and maleic anhydride (Scheme 1). This polymer displays release rates that are pH dependent. The polymer dissolves by ionization of the carboxylic acid group (87).

A schematic presentation of the experimental device is shown in Figure 7. The pH-sensitive polymer shown in Scheme 1 containing dispersed hydrocortisone is surrounded with urease immobilized in a hydrogel prepared by cross-linking a mixture of urease and bovine serum albumin with glutaraldehyde. When urea diffuses into the hydrogel, its interaction with the enzyme leads to a pH increase, therefore enhancing erosion of the pH-sensitive polymer with concomitant changes in the release of hydrocortisone.

Figure 8 shows release of hydrocortisone from the described device. Although the device has no therapeutic relevance, it established the feasibility of creating self-responsive delivery system.

Ishihara et al. (89,90) suggested a nonerodible system based on a similar idea. The system is comprised of a pH-sensitive membrane prepared by copolymerizing 4-

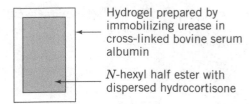

**Figure 7.** Schematic presentation of urea-sensitive drug delivery system (88).

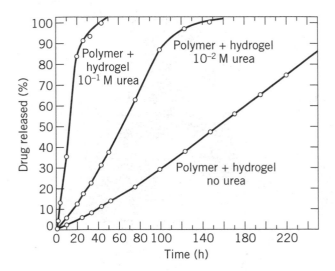

**Figure 8.** Hydrocortisone release from the N-hexyl ester of methylvinylether and maleic anhydride disks coated with immobilized urease in the presence and absence of external urea, 35°C, pH 6.25, hydrocortisone loading 10 wt % (68).

carboxyacrylanilide with methacrylate, sandwiched within a membrane containing urease immobilized in free radically crosslinked N,N-methylenebisacrylamide. The permeation of a model substance (1,4-bis(2-hydroxyethoxy) benzene varied with the urea concentration in the external solution.

### Glucose-Responsive Insulin Delivery

Glucose-responsive delivery systems utilizing enzymes are based on the enzymatic reaction of glucose oxidase with glucose shown in Scheme 2. Several approaches have been devised.

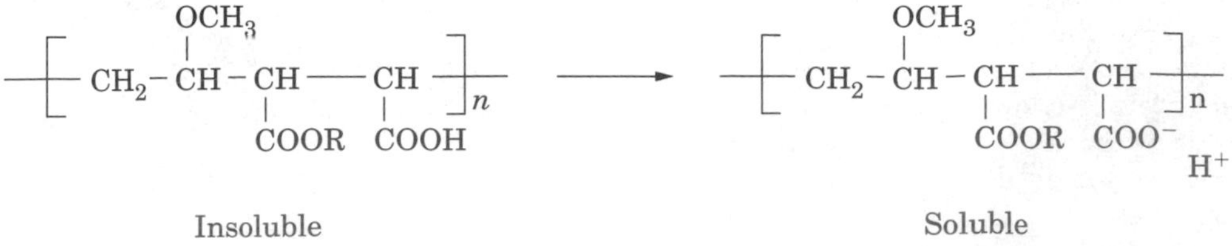

**Scheme 1.** Solubilization of a partially esterified copolymer of methyl vinyl ether and maleic anhydride by ionization of pendent carboxylic acid functions (87).

Insoluble

Soluble

$$\text{Glucose} \ + \ O_2 \ \xrightarrow{\begin{array}{c}\text{Glucose}\\\text{oxidase}\end{array}} \ \text{Gluconic acid} \ + \ H_2O_2$$

**Scheme 2.** Glucose oxidase catalyzes the conversion of glucose to gluconic acid and hydrogen peroxide.

Systems based on pH-sensitive polymers consist of immobilized glucose oxidase in a pH-responsive hydrogel, enclosing a saturated insulin solution or incorporated with insulin (91–99). As glucose diffuses into the hydrogel, glucose oxidase catalyzes its conversion to gluconic acid, thereby lowering the pH in the microenvironment of the hydrogel and causing swelling (Fig. 9). Because insulin should permeate the swelled hydrogel more rapidly, faster delivery of insulin in the presence of glucose is anticipated. As the glucose concentration decreases in response to the released insulin, the hydrogel should contract and decrease the rate of insulin delivery.

Horbett and coworkers (91–98) immobilized glucose oxidase in a cross-linked hydrogel made from *N,N*-dimethylaminoethyl methacrylate (DMA), hydroxyethyl methacrylate (HEMA), and tetraethylene glycol dimethacrylate (TEGDMA). Membranes were prepared at −70°C by radiation polymerization, previously shown to retain the enzymatic activity (100). To obtain sufficient insulin permeability through the gels, porous HEMA/DMA gels were prepared by polymerization under conditions that induce a separation into two phases during polymerization: one phase rich in polymer and the other rich in solvent plus unreacted monomer. When gelation occurs after the phase separation, the areas where the solvent/monomer phase

existed become fixed in place as pores in the polymer matrix. The authors used a dilute monomer solution to obtain porous gel, typically 1–10 μm in diameter (92).

The rate of insulin permeation through the membranes was measured in the absence of glucose in a standard transport cell; then glucose was added to one side of the cell to a concentration of 400 mg/dL while the permeation measurement was continued. The results indicated that the insulin transport rate is enhanced significantly by the addition of glucose. The average permeability after addition of 400 mg/dL glucose was 2.4 to 5.5 times higher than before glucose was added. When insulin permeabilities through the porous gels were measured in a flowing system in which permeabilities were measured with fluid flowing continuously past one side of the membrane, no effect of glucose concentration on insulin permeabilities could be detected. The authors propose that inappropriate design of the membranes used in the experiments is the explanation for their lack of response to glucose concentration (91).

A mathematical model describing these glucose-responsive hydrogels demonstrates two important points (91,92). (*1*) Progressive response to glucose concentration over a range of glucose concentrations can be achieved only with a sufficiently low glucose oxidase loading; otherwise, depletion of oxygen causes the system to become insensitive to glucose; and (*2*) a significant pH decrease in the membrane, with resultant swelling, can be achieved only if the amine concentration is sufficiently low that pH changes are not prevented by the buffering of the amines.

Although the great advantage of reservoir systems is the ease with which they can be designed to produce constant release-rate kinetics, their main disadvantage are leaks, which are dangerous because all the incorporated

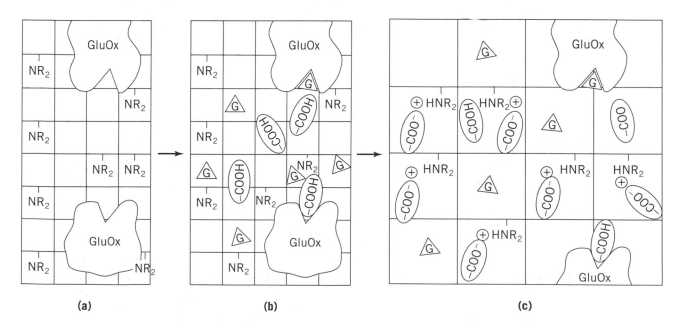

(a)                    (b)                    (c)

**Figure 9.** Mechanism of action of glucose-sensitive polymeric hydrogel. (**a**) In the absence of glucose, at physiologic pH, few of the amine groups are protonated. (**b**) In the presence of glucose, the glucose oxidase produces gluconic acid that can (**c**) protonate the amine groups. The fixed positive charge on the polymeric network led to electrostatic repulsion and membrane swelling and therefore enhanced insulin release (98).

drug could be rapidly released. In attempt to overcome this problem, Goldraich and Kost (99) proposed to incorporate the drug (insulin) and the enzyme (glucose oxidase) into the pH-responsive polymeric matrices.

Ishihara et al. (89,90,101–104) investigated two approaches for glucose-responsive insulin delivery systems; one approach is similar to that investigated by Horbett et al. (97). The polymers were prepared from 2-hydroxyethyl acrylate (HEA)-DMA and 4-trimethylsilystyrene (TMS) by radical polymerization of the corresponding monomers in dimethylformamide (DMF). The mole fractions of HEA, DMA, and TMS in the copolymer were 0.6, 0.2, and 0.2, respectively. Membranes were prepared by solvent casting. Capsules containing insulin and glucose oxidase were prepared by an interfacial precipitation method using gelatin as an emulsion stabilizer. The average diameter of the polymer capsules obtained was 1.5 mm (101,104). The water content of HEA–DMA–TMS copolymer membranes increased with a decrease in the pH of the medium. An especially drastic change was observed in the pH range of 6.3 to 6.15. The permeation of insulin through the copolymer membrane increased in response to pH decreases. The permeation rate of insulin at pH 6.1 was greater than that at pH 6.4 by about 42 times. The permeation of insulin through the copolymer membranes was very low in buffer solution without glucose. Addition of 0.2 M glucose to the upstream compartment induced an increase in the permeation rate of insulin. When glucose was removed, the permeation rates of insulin gradually returned to their original levels (101) (Fig. 10).

A similar approach for glucose-responsive insulin release based on hydrogels of poly(diethyl aminoethyl methacrylate-g-ethylene glycol) containing glucose oxidase and catalase was also proposed by Podual et al. (105).

The other approach proposed by Ishihara et al. (103) is based on an immobilized glucose oxidase membrane and a redox polymer having a nicotinamide moiety. The device consists of two membranes. One membrane, with the immobilized glucose oxidase, acts as a sensor for glucose and forms hydrogen peroxide by an enzymatic reaction; the other membrane is a redox polymer having a nicotinamide moiety that controls the permeation of insulin by an oxidation reaction with the formed hydrogen peroxide. The oxidation of the nicotinamide group increases hydrophilicity and therefore should enhance the permeability to water-soluble molecules such as insulin. The results showed relatively small increases in insulin permeability.

Iwata et al. (106,107) pretreated porous poly(vinylidene fluoride) membranes (average pore size of 0.22 $\mu$m) by air plasma, and subsequently acrylamide was graft-polymerized on the treated surface. The polyacrylamide was then hydrolyzed to poly(acrylic acid). In the pH range of 5 to 7, grafted poly(acrylic acid) chains are solvated and dissolved but cannot diffuse into the solution phase because they are grafted to the porous membrane. Thus, they effectively close the membrane pores. In the pH range of 1 to 5, the chains collapse and the permeability increases. To achieve the sensitivity of the system toward glucose, glucose oxidase was immobilized onto a poly(2-hydroxyethyl methacrylate) gel.

Ito et al. (108) adopted the approach proposed by Iwata et al. (107) using a porous cellulose membrane with surface-grafted poly(acrylic acid) as a pH-sensitive membrane. By immobilization of glucose oxidase onto the poly(acrylic acid)-grafted cellulose membrane, it became responsive to glucose concentrations (Fig. 11). Figure 12 shows the change in permeability of insulin through poly(acrylic acid)-grafted cellulose membranes following the addition of 0.2 M glucose to the buffered solution. The permeation coefficient after glucose addition was about 1.7 times that before the addition of glucose. The authors suggest to improve the proposed system (sensitivity of insulin

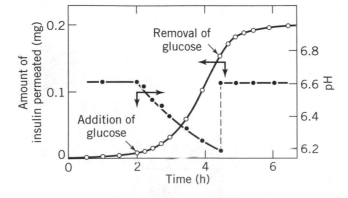

**Figure 10.** Effect of the addition and removal of glucose on insulin permeation through the HEA–DEA–TMS copolymer membrane and on the pH of the feed-side solution at 30°C. (HEA is 2-hydroxyethyl acrylate; DEA is N,N-diethylaminoethyl methacrylate; and TMS is 4-trimethylsilystyrenel; [●] pH; [○] insulin permeated [104]).

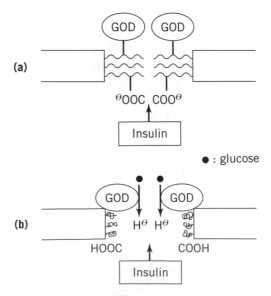

**Figure 11.** Principle of the controlled-release system of insulin. (a) In the absence of glucose, the chains of poly(acrylic acid) grafts are rodlike, lowering the porosity of the membrane and suppressing insulin permeation. (b) In the presence of glucose, gluconic acid (COOH) produced by glucose oxidase (GOD) protonates the poly(acrylic acid), making the graft chains coillike and opening the pores to enhance insulin permeation (108).

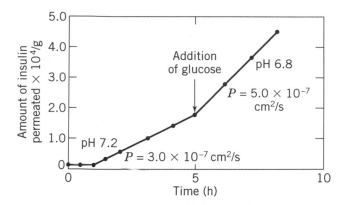

**Figure 12.** Permeation of insulin through poly(acrylic acid)-grafted membrane in 0.1 M tris-HCl–buffered solution. The concentration of added glucose is 0.2 M (108).

permeability to glucose concentrations) by modification of the graft chain: density, length, and size or density of pores.

Siegel and Firestone (109,110) proposed an implantable "mechanochemical" pump that functions by converting changes in blood glucose activity into a mechanical force, generated by the swelling polymer that pumps insulin out of the device.

More recently Siegel (111) proposed self-regulating oscillatory drug delivery based on a polymeric membrane whose permeability to the substrate of an enzyme-catalyzed reaction is inhibited by the product of that reaction. This negative feedback system can, under certain conditions, lead to oscillations in membrane permeability and in the levels of substrate and product in the device. Any one of these oscillating variables can then be used to drive a cyclic delivery process. Figure 13 illustrates a possible scheme for such drug delivery systems. The device consists of a chamber containing the drug to be delivered (D), plus an enzyme. The chamber is partially bounded by a membrane that controls solute influx and efflux. The substrate (S) for the enzyme is present externally at constant concentration and diffuses through the membrane into the chamber and is converted by the enzyme to a product (P). The product in turn diffuses out of the membrane. However, the product concentration in the chamber inhibitorily affects the permeabilities of the membrane to the sub-

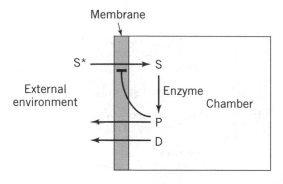

**Figure 13.** General scheme for proposed drug delivery oscillator (111).

strate. That is, increasing product concentration causes decreasing flux of substrate into the device. Siegel propose several means to control drug delivery based on the presented scheme. Drug solubility could be affected by substrate or product concentration, which will be seen to oscillate. Alternatively, the permeability of the membrane to drug can oscillate with time along with the substrate permeability.

Heller et al. (88,112,113) suggested a system in which insulin is immobilized in a pH-sensitive bioerodible polymer prepared from 3,9 bis (ethylidene 2,4,8,10-tetraoxaspirol(5,5)undecane and $N$-methyldiethanolamine (Scheme 3), which is surrounded by a hydrogel containing immobilized glucose oxidase. When glucose diffuses into the hydrogel and is oxidized to gluconic acid, the resultant lowered pH triggers enhanced polymer degradation and release of insulin from the polymer in proportion to the concentration of glucose. The response of the pH-sensitive polymers containing insulin to pH pulses was rapid. Insulin was rapidly released when the pH decreased from 7.4 to 5.0. Insulin release was shut off when the pH increased (Fig. 14). The amount of insulin released showed dependents on the pH change. However, when the in vitro studies were repeated in physiologic buffer, response of the device was only minimal. The authors found the synthesized amine containing polymer (Scheme 3) undergoes general acid catalysis where the catalyzing species is not hydronium ion but rather the specific buffer molecules used. Therefore, further development of this system will require the development of a bioerodible polymer that not only has adequate pH sensitivity but also undergoes specific ion catalysis.

Glucose-dependent insulin release was proposed by Langer and coworkers (114–116) based on the fact that insulin solubility is pH dependent. Insulin was incorporated into EVAc matrices in solid form. Thus, the release was governed by its dissolution and diffusion rates. Glucose oxidase was immobilized to sepharose beads that were incorporated along with insulin into EVAc matrices. When glucose entered the matrix, the produced gluconic acid caused a rise in insulin solubility and consequently enhanced release. To establish this mechanism in the physiological pH of 7.4, the insulin was modified by three additional lysine groups so that the resultant isoelectric point was 7.4. In vitro and in vivo studies demonstrated the response of the system to changes in glucose concentration. In the in vivo experiments, a catheter was inserted into the left jugular vein, and polymer matrices containing insulin and immobilized enzyme were implanted subcutaneously in the lower back of diabetic rats. Serum insulin concentrations were measured for different insulin matrix implants. A 2 M glucose solution was infused, 15 min into the experiments, through the catheter. Rats that received trilysine insulin/glucose oxidase matrices showed a 180% rise in serum insulin concentration that peaked at 45 min into the experiment. Control rats that received matrices containing no insulin or insulin but no glucose oxidase or diabetic rats without implants showed no change in serum insulin.

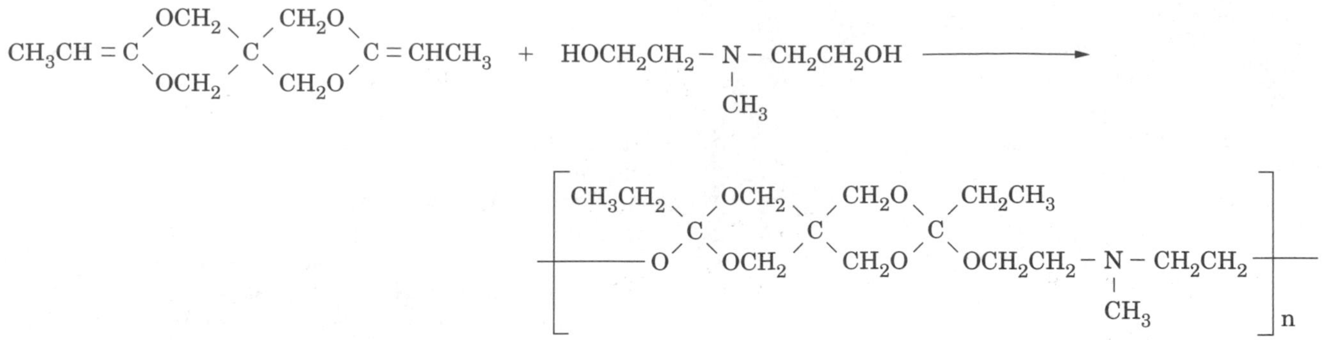

$$CH_3CH = C \overset{OCH_2}{\underset{OCH_2}{\diagdown}} C \overset{CH_2O}{\underset{CH_2O}{\diagup}} C = CHCH_3 \quad + \quad HOCH_2CH_2 - \underset{\underset{CH_3}{|}}{N} - CH_2CH_2OH \longrightarrow$$

$$\left[ \overset{CH_3CH_2}{\underset{O}{\diagdown}} C \overset{OCH_2}{\underset{OCH_2}{\diagup}} \overset{CH_2O}{\underset{CH_2O}{\diagdown}} C \overset{CH_2CH_3}{\underset{OCH_2CH_2 - \underset{\underset{CH_3}{|}}{N} - CH_2CH_2}{\diagup}} \right]_n$$

**Scheme 3.** Preparation of poly(orthoester)s containing tertiary amine groups in polymer backbone (122).

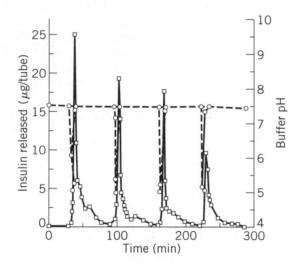

**Figure 14.** Release of insulin from a linear polymer prepared from 3,9-bis(ethylidene)-2,48,10-tetraoxaspiro [5,5] undecane and *N*-methyldiethanolamine as a function of external pH variations between pH 7.4 and 5.0 at 37°C: disks, $3 \times 0.55$ mm; insulin loading, 10 wt %; nonrecycling media was continuously perfused at a flow rate of 2 mL/min; total efflux was collected at 1–10 min intervals; (O) buffer pH; (□) insulin release (122).

### Morphine-Triggered Naltrexone Delivery Systems

Heller and coworkers (88,112,117–122) have been developing a naltrexone drug delivery system that would be passive until drug release is initiated by the appearance of morphine external to the device. Naltrexone is a long-acting opiate antagonist that blocks opiate-induced euphoria, and thus the intended use of this device is in the treatment of heroin addiction. Activation is based on the reversible inactivation of enzymes achieved by the covalent attachment of hapten close to the active site of the enzyme–hapten conjugate with the hapten antibody. Because the antibodies are large molecules, access of the substrate to the enzyme's active site is sterically inhibited, thus effectively rendering the enzyme inactive. Triggering of drug release is initiated by the appearance of morphine (hapten) in the tissue and dissociation of the enzyme–heptan–antibody complex rendering the enzyme active. This approach is being developed by incorporating the nal-

trexone in an bioerodible polymer (Fig. 15). This polymer matrix is then covered by a lipid layer that prevents water entry into the polymer matrix and therefore prevents its degradation and the release of naltroxane. The system is placed in a dialysis bag. The bag contains lipaze (enzyme) that is covalently attached to morphine and reversibly inactivated by antimorphine complexation. Thus, when morphine is present in the tissues surrounding the device, morphine diffuses into the dialysis bag and displaces the lipase–morphine conjugate from the antibody, allowing the now-activated enzyme to degrade the protective lipid layer, which in turn permits the polymeric core degradation and release into the body of the narcotic antagonist, naltrexone.

A key component of this morphine-responsive device is the ability to reversibly and completely inactivate an enzyme and to be able to rapidly disassociate the complex with concentrations as low as $10^{-8}$ to $10^{-9}$ M. To achieve this sensitivity, lipase was conjugated with several morphine analogs and complexed with affinity chromatography–purified polyclonal antimorphine antibodies. In vivo studies (118) suggest that the concentration of morphine in a device implanted in a typical heroin-addicted patient is estimated to be about $10^{-7}$ to $10^{-8}$ M. Recent studies have shown that reaching such sensitivity is possible (88).

### SYSTEMS UTILIZING ANTIBODY INTERACTIONS

Pitt et al. (123) proposed utilizing hapten–antibody interactions to suppress enzymatic degradation and permeabil-

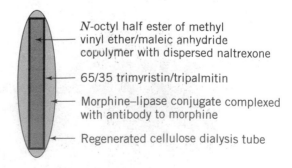

*N*-octyl half ester of methyl vinyl ether/maleic anhydride copolymer with dispersed naltrexone

65/35 trimyristin/tripalmitin

Morphine–lipase conjugate complexed with antibody to morphine

Regenerated cellulose dialysis tube

**Figure 15.** Schematic representation of triggered naltrexone delivery device (88).

ity of polymeric reservoirs or matrix drug delivery systems. The delivery device consists of naltrexone contained in a polymeric reservoir or dispersed in a polymeric matrix configuration. The device is coated by covalently grafting morphine to the surface. Exposure of the grafted surface to antibodies to morphine results in coating of the surface by the antibodies, a process that can be reversed by exposure to exogenous morphine. The presence of the antibodies on the surface or in the pores of the delivery device will block or impede the permeability of naltrexone in a reservoir configuration or enzyme-catalyzed surface degradation and concomitant release of the drug from a matrix device. A similar approach was proposed for responsive release of a contraceptive agent. The $\beta$-subunit of human chorionic gonadotropin (HCG) is grafted to the surface of the polymer, which is then exposed to antibodies to $\beta$-HCG. The appearance of HCG in the circulatory system (indicative of pregnancy) will cause release of a contraceptive drug. (HCG competes for the polymer-bound antibodies to HCG and initiates release of the contraceptive drug.)

Pitt et al. (123,124) also proposed a hypothetical reversible antibody system for controlled release of ethinyl estradiol (EE). EE stimulates biosynthesis of sex-hormone-binding globulin (SHBG). High serum levels of EE stimulate the production of SHBG, which increases the concentration of SHBG bound to the polymer surface and reduces the EE release rate. When the EE serum level falls, the SHBG level falls, as does binding of the SHBG to the polymer surface, producing an automatic increase in the EE release rate.

The polymers evaluated by Pitt and coworkers for covalent attachment of haptens were poly(ethylene vinyl acetate) copolymers for reservoir devices and hydroxylated polyester urethanes where the release is enzymatic degradation dependent. The polymers were prepared from $\epsilon$-caprolactone and $\delta$-valerolactone, lightly cross-linked with 1,6-hexanediisocyanate. The residual hydroxyl groups of the elastomer were shown to be suitable sites for hapten derivatization by treatment with chloroacetic anhydride. Biodegradability studies of the cross-linked polymers were conducted by implanting films subdermally in rabbits and measuring weight loss and dimensional changes with time. No weight loss was observed after 4 weeks of implantation. This results were in contrast with the immediate and rapid surface erosion of polymers prepared from $\epsilon$-caprolactone and $\delta$-valerolactone cross-linked with bis-2-2-($\epsilon$-caprolactone-4-yl)propane in which weight loss was detected within 2 weeks and increased thereafter in an approximately linear fashion (125).

## SYSTEMS UTILIZING CHELATION

Self-regulated delivery of drugs that function by chelation was also suggested (126). These include certain antibiotics and drugs for the treatment of arthritis as well as chelators used for the treatment of metal poisoning. The concept is based on the ability of metals to accelerate the hydrolysis of carboxylate or phosphate esters and amides by several orders of magnitude. Attachment of the chelator to a polymer chain by a covalent ester or amide link serves to prevent its premature loss by excretion and reduces its tox-

icity. In the presence of the specific ion, a complexing with the bound chelating agent will take place, followed by metal-accelerated hydrolysis and subsequent elimination of the chelated metal. Measurement of the rates of hydrolysis of polyvinyl alcohol coupled with quinaldic acid chelator (PVA-QA) in the presence of CO(II), Zn(II), Cu(II), and Ni(II) confirmed that it is possible to retain the susceptibility of the esters to metal-promoted hydrolysis in a polymer environment.

## CONCLUDING REMARKS

During the last three decades, polymeric-controlled drug delivery has become an important area of research and development. In this short time, a number of systems displaying constant or decreasing release rates have progressed from the laboratory to the clinic and clinical products. Although these polymeric-controlled delivery systems are advantageous compared to the conventional methods of drug administration, they are insensitive to the changing metabolic state. To more closely control the physiological requirements of the specific drugs, responsive mechanisms must be provided. The approaches discussed represent attempts conducted over the past two decades to achieve pulsatile release. It should be pointed out that these drug delivery systems are still in the developmental stage, and much research will have to be conducted for such systems to become practical clinical alternatives. Critical considerations are the biocompatibility and toxicology of these multicomponent polymer-based systems, the response times of these systems to stimuli, the ability to provide practical levels of the desired drug, and addressing necessary formulation issues in dosage or design (e.g., shelf life, sterilization, and reproducibility). A key issue in the practical utilization of the pulsatile, externally triggered systems (i.e., those that are magnetic, ultrasound, electrically regulated, and photoresponsive) will be the design of small portable trigger units that the patient can easily use. Ideally, such systems could be worn by the patient, as in a wristwatch-like system, and could be either preprogrammed to go on and off at specific times or turned on by the patient when needed. A critical issue in the development of responsive, self-regulated systems such as those containing enzymes or antibodies is the stability and/or potential leakage and possible immunogenicity of these bioactive agents. The successful development of responsive polymer delivery systems will be a significant challenge. Nevertheless, the considerable pharmacological benefit these systems could potentially provide—particularly given ongoing research in biotechnology, pharmacology, and medicine, which may provide new insights on the desirability and requirements for pulsatile release—should make this an important and fruitful area for future research.

## BIBLIOGRAPHY

1. R. Langer, *Nature (London)* **392**, 5–10 (1998).
2. J. Kost and R. Langer, *Adv. Drug Delivery Rev.* **6**, 19–50 (1991).

3. H. Creque, R. Langer, and J. Folkman, *Diabetes* **29**, 37–40 (1980).

4. L. Brown, L. Siemer, C. Munoz, and R. Langer, *Diabetes* **35**, 684–691 (1986).

5. L. Brown et al., *Diabetes* **35**, 692–697 (1986).

6. J. Kost, R. Noecker, E. Kunica, and R. Langer, *J. Biomed. Mater. Res.* **19**, 935–940 (1985).

7. E. Edelman, J. Kost, H. Bobeck, and R. Langer, *J. Biomed. Mater. Res.* **19**, 67–83 (1985).

8. D. Hsieh, R. Langer, and J. Folkman, *Proc. Natl. Acad. Sci. U.S.A.* **78**, 1863–1867 (1981).

9. J. Kost, J. Wolfrum, and R. Langer, *J. Biomed. Mater. Res.* **21**, 1367–1373 (1987).

10. O. Saslavski, P. Couvrer, and N. Peppas, in J. Heller et al., eds., *Controlled Release of Bioactive Materials*, Controlled Release Society, Basel, 1988, pp. 26–27.

11. E. Edelman, L. Brown, J. Taylor, and R. Langer, *J. Biomed. Mater. Res.* **21**, 339–353 (1987).

12. J. Kost, K. Leong, and R. Langer, *Proc. Natl. Acad. Sci. U.S.A.* **86**, 7663–7666 (1989).

13. S. Miyazaki, C. Yokouchi, and M. Takada, *J. Pharm. Pharmacol.* **40**, 716–717 (1988).

14. J. Kost, in B. Berner and S. Dinh, eds., *Electronically-Controlled Drug Delivery*, CRC Press, Boca Raton, Fla., 1998, pp. 215–228.

15. J. Kost and R. Sanger, in V. Shah and H. Maibach, eds., *Cutaneous Bioavailability, Bioequivalence and Penetration*, Plenum Press, 1994, pp. 91–104.

16. D. Levy, J. Kost, Y. Meshulam, and R. Langer, *J. Clin. Invest.* **83**, 2074–2078 (1989).

17. S. Miyazaki, O. Mizuoka, and M. Takada, *J. Pharm. Pharmacol.* **43**, 115–116 (1990).

18. D. Bommannan et al., *Pharm. Res.* **9**, 1043–1047 (1992).

19. K. Tachibana, *Pharm. Res.* **9**, 952–954 (1992).

20. K. Tachibana and S. Tachibana, *Anesthesiology* **78**, 1091–1096 (1993).

21. K. Tachibana and S. Tachibana, *J. Pharm. Pharmacol.* **43**, 270–271 (1991).

22. S. Mitragotri, D. Blankschtein, and R. Langer, *Pharm. Res.* **13**, 411–420 (1996).

23. S. Mitragotri, D. Blankschtein, and R. Langer, *Science* **269**, 850–853 (1995).

24. S. Mitragotri, D. Edwards, D. Blankschtein, and R. Langer, *J. Pharm. Sci.* **84**, 697–706 (1995).

25. A.J. Grodzinsky and P.E. Grimshaw, in J. Kost, ed., *Pulsed and Self-Regulated Drug Delivery*, CRC Press, Boca Raton, Fla., 1990, pp. 47–64.

26. P.E. Grimshaw, A.J. Grodzinsky, M.L. Yarmush, and D.M. Yarmush, *Chem. Eng. Sci.* **104**, 827–840 (1989).

27. M. Prausnitz, R. Bose, R. Langer, and J. Weaver, *Proc. Natl. Acad. Sci. U.S.A.* **90**, 10504–10508 (1993).

28. D. Rolf, *Pharm. Technol.* **12**, 130–140 (1988).

29. I.C. Kwon, Y.H. Bae, T. Okano, and S.W. Kim, *J. Controlled Release* **17**, 149–156 (1991).

30. C. Hsu and L. Block, *Pharm. Res.* **13**, 1865–1870 (1996).

31. A. D'Emanuele and J.N. Staniforth, *Pharm. Res.* **8**, 913–918 (1991).

32. A. D'Emanuele and J.N. Staniforth, *Pharm. Res.* **9**, 215–219 (1992).

33. V. Labhasetwar, T. Underwood, S. Schwendemann, and R. Levy, *Proc. Natl. Acad. Sci. U.S.A.* **92**, 2612–2616 (1995).

34. L.L. Miller, G.A. Smith, A. Chang, and Q. Zhou, *J. Controlled Release* **6**, 293–296 (1987).

35. A. Mamada, T. Tanaka, D. Kugwatchkakun, and M. Irie, *Macromolecules* **23**, 1517–1519 (1990).

36. A. Suzuki and T. Tanaka, *Nature (London)* **346**, 345–347 (1990).

37. N. Yui, T. Okano, and Y. Sakurai, *J. Controlled Release* **26**, 141–145 (1993).

38. N. Yui, T. Okano, and Y. Sakurai, *J. Controlled Release* **22**, 105–116 (1992).

39. A.S. Hoffman, *J. Controlled Release* **6**, 297–305 (1987).

40. T. Tanaka, in H.F. Mark and J.I. Kroschwitz, eds., *Encyclopedia of Polymer Science and Technology*, vol. 7, Wiley, New York, 1985, pp. 514–531.

41. T. Ueda, K. Ishihara, and N. Nakabayashi, *Macromol. Chem., Rapid Commun.* **11**, 345 (1990).

42. D.W. Urry et al., *Proc. Natl. Acad. Sci. U.S.A.* **85**, 3407–3411 (1988).

43. M. Yoshida et al., *Drug Des. Delivery* **7**, 159–174 (1991).

44. M. Palasis and H. Gehrke, *J. Controlled Release* **18**, 1–12 (1992).

45. T. Okano, Y.H. Bae, and S.W. Kim, in J. Kost, ed., *Pulsed and Self-Regulated Drug Delivery*, CRC Press, Boca Raton, Fla., 1990, pp. 17–46.

46. Y. Okahata, H. Noguchi, and T. Seki, *Macromolecules* **19**, 493–494 (1986).

47. H. Katano et al., *J. Controlled Release* **16**, 215–228 (1991).

48. S. Vakalanka, C. Brazel, and N. Peppas, *J. Biomater. Sci., Polym. Ed.* **8**, 119–129 (1996).

49. A. Serres, M. Baudys, and S. Kim, *Pharm. Res.* **13**, 196–201 (1996).

50. N. Ogata, *J. Controlled Release* **48**, 149–155 (1997).

51. J.E. Chung et al., *J. Controlled Release* **53**, 119–130 (1998).

52. M. Baundys, A. Serres, C. Ramkisoon, and S.W. Kim, *J. Controlled Release* **48**, 304–305 (1997).

53. L. Brannon-Peppas and N.A. Peppas, *J. Controlled Release* **8**, 267–274 (1989).

54. M. Annaka and T. Tanaka, *Nature (London)* **355**, 430–432 (1992).

55. B.A. Firestone and R.A. Siegel, *Polym. Commun.* **29**, 204–208 (1988).

56. L.-C. Dong and A.S. Hoffman, *Proc. Int. Symp. Controlled Release Bioact. Mater.* **17**, 325–326 (1990).

57. J.H. Kou, D. Fleisher, and G. Amidon, *J. Controlled Release* **12**, 241–250 (1990).

58. M. Pradny and J. Kopecek, *Makromol. Chem.* **191**, 1887–1897 (1990).

59. R.A. Siegel, M. Falmarzian, B.A. Firestone, and B.C. Moxley, *J. Controlled Release* **8**, 179–182 (1988).

60. K. Kono, F. Tabata, and T. Takagishi, *J. Membr. Sci.* **76**, 233–243 (1993).

61. D. Hariharan and N.A. Peppas, *J. Controlled Release* **23**, 123–136 (1993).

62. R.A. Siegel and B.A. Firestone, *Macromolecules* **21**, 3254–3259 (1988).

63. H. Allcock and A. Ambrosio, *Biomaterials* **17**, 2295–2302 (1996).

64. C. Bell and N. Peppas, *Biomaterials* **17**, 1201–1218 (1996).

65. K. Jarvinen et al., *Pharm. Res.* **15**, 802–805 (1998).

66. R.A. Siegel, I. Johannes, A. Hunt, and B.A. Firestone, *Pharm. Res.* **9**, 76–81 (1992).

67. S. Giannos, S. Dinh, and B. Berner, *J. Pharm. Sci.* **84**, 539–543 (1995).

68. J. Heller and P.V. Trescony, *J. Pharm. Sci.* **68**, 919–921 (1979).

69. H. Ghandehari, P. Kopeckova, and J. Kopecek, *Biomaterials* **18**, 861–872 (1997).

70. W. Petka et al., *Science* **281**, 389–392 (1998).

71. M. Brownlee and A. Cerami, *Science* **26**, 1190–1191 (1979).

72. K. Makino, E.J. Mack, T. Okano, and S.W. Kim, *J. Controlled Release* **12**, 235–239 (1990).

73. S. Sato, S.Y. Jeong, J.C. McRea, and S.W. Kim, *J. Controlled Release*, 67–77 (1984).

74. S. Sato, S.Y. Jeong, J.C. McRea, and S.W. Kim, *Pure Appl. Chem.* **56**, 1323–1328 (1984).

75. L. Seminoff et al., *Proc. Int. Symp. Controlled Release Bioact. Mater.* **15**, 161–161 (1988).

76. L. Seminoff and S.W. Kim, in J. Kost, ed., *Pulsed and Self-Regulated Drug Delivery*, CRC Press, Boca Raton, Fla., 1990.

77. S.W. Kim, et al., *J. Controlled Release* **11**, 193–201 (1990).

78. S.Y. Jeong, S.W. Kim, D. Holemberg, and J.C. McRea, *J. Controlled Release* **2**, 143–152 (1985).

79. S.Y. Jeong, S.W. Kim, M.J.D. Eenink, and J. Feijen, *J. Controlled Release* **1**, 57–66 (1984).

80. E. Kokufata, Y. Q. Zhang, and T. Tanaka, *Nature (London)* **351**, 302–304 (1991).

81. M. Taylor, S. Tanna, P. Taylor, and G. Adams, *J. Drug Target* **3**, 209–216 (1995).

82. A. Obaidat and K. Park, *Biomaterials* **11**, 801–806 (1997).

83. A. Obaidat and K. Park, *Pharm. Res.* **13**, 998–995 (1996).

84. S. Lee and K. Park, *J. Mol. Recognition* **9**, 549–557 (1996).

85. S. Kitano et al., *J. Controlled Release* **19**, 161–170 (1992).

86. I. Hisamitsu, K. Kataoka, T. Okano, and Y. Sakurai, *Pharm. Res.* **14**, 289–293 (1997).

87. J. Heller, R. Baker, R. Gale, and J. Rodin, *J. Appl. Polym. Sci.* **2**, 1991–2009 (1978).

88. J. Heller, in K. Park, ed., *Controlled Drug Delivery: Challenges and Strategies*, American Chemical Society, Washington, D.C., 1997, pp. 127–146.

89. K. Ishihara, N. Muramoto, H. Fuji, and I. Shinohara, *J. Polym. Sci., Polym. Chem. Ed.* **23**, 2841–2850 (1985).

90. K. Ishihara, N. Muramoto, H. Fuji, and I. Shinohara, *J. Polym. Sci., Polym. Lett. Ed.* **23**, 531–535 (1985).

91. G. Albin, T. Horbett, and B. Ratner, in J. Kost, ed., *Pulsed and Self-Regulated Drug Delivery*, CRC Press, Boca Raton, Fla., 1990, pp. 159–185.

92. G. Albin, T.A. Horbett, S.R. Miller, and N.L. Ricker, *J. Controlled Release* **6**, 267–291 (1987).

93. G. Albin, T.A. Horbett, and B.D. Ratner, *J. Controlled Release* **3**, 153–164 (1985).

94. L.A. Klumb and T.A. Horbett, *J. Controlled Release* **18**, 59–80 (1992).

95. L.A. Klumb and T.A. Horbett, *J. Controlled Release* **27**, 95–114 (1993).

96. J. Kost, T.A. Horbett, B.D. Ratner, and M. Singh, *J. Biomed. Mater. Res.* **19**, 1117–1133 (1985).

97. T. Horbett, J. Kost, and B. Ratner, *Am. Chem. Soc., Div. Polym. Chem.* **24**, 34–35 (1983).

98. T. Horbett, J. Kost, and B. Ratner, in S. Shalaby, A. Hoffman, T. Horbett, and B. Ratner, eds., *Polymers as Biomaterials*, Plenum, New York, 1984, pp. 193–207.

99. M. Goldraich and J. Kost, *Clin. Mater.* **13**, 135–142 (1993).

100. I. Kaetsu, M. Kumakura, and M. Yoshida, *Biotechnol. Bioeng.* **21**, 847–861 (1979).

101. K. Ishihara, *Proc. Int. Symp. Controlled Release Bioact. Mater.* **15**, 168–169 (1988).

102. K. Ishihara, M. Kobayashi, and I. Shonohara, *Macromol. Chem., Rapid Commun.* **4**, 327–331 (1983).

103. K. Ishihara, M. Kobayashi, N. Ishimaru, and I. Shinohara, *Polym. J.* **16**, 625–631 (1984).

104. K. Ishihara and K. Matsui, *J. Polym. Sci. Polym. Lett. Ed.* **24**, 413–417 (1986).

105. K. Podual, N. Peppas, and F. Doyle, *Proc. Int. Symp. Controlled Release Bioact. Mater.* **25**, 135–136 (1998).

106. H. Iwata and T. Matsuda, *J. Membr. Sci.* **38**, 185–199 (1988).

107. H. Iwata et al., *Proc. Int. Symp. Controlled Release Bioact. Mater.* **15**, 170–171 (1988).

108. Y. Ito, M. Casolaro, K. Kono, and Y. Imanishi, *J. Controlled Release* **10**, 195–203 (1989).

109. R. Siegel, in J. Kost, ed., *Pulsed and Self-Regulated Drug Delivery*, CRC Press, Boca Raton, Fla., 1990, pp. 129–157.

110. R.A. Siegel and B.A. Firestone, *J. Controlled Release* **11**, 181–192 (1990).

111. R. Siegel, in K. Park, ed., *Controlled Release: Challenges and Strategies*, American Chemical Society, Washington, D.C., 1997.

112. J. Heller, R.V. Sparer, and G.M. Zenter, in M. Chasin and R. Langer, eds., *Biodegradable Polymers as Drug Delivery Systems*, Dekker, New York, 1990, pp. 121–161.

113. J. Heller, *Adv. Drug Delivery Rev.* **10**, 163–204 (1993).

114. L. Brown, F. Ghodsian, and R. Langer, *Proc. Int. Symp. Controlled Release Bioact. Mater.* **15**, 116–167 (1988).

115. L. Brown, E. Edelman, F. Fishel-Ghodsian, and R. Langer, *J. Pharm. Sci.* **85**, 1341–1345 (1996).

116. F. Fischel-Ghodsian et al., *Proc. Natl. Acad. Sci. U.S.A.* **85**, 2403–2406 (1988).

117. K. Roskos, V. Tefft, and J. Heller, *J. Clin. Mater.* **13**, 109–119 (1993).

118. K.V. Roskos, B.K. Fritzinger, J.A. Tefft, and J. Heller, *Biomaterials* **16**, 1235–1239 (1995).

119. K.V. Roskos, J.A. Tefft, B.K. Fritzinger, and J. Heller, *J. Controlled Release* **19**, 145–160 (1992).

120. J.A. Tefft, K.V. Roskos, and J. Heller, *J. Biomed. Mater. Res.* **26**, 713–724 (1992).

121. G.R. Nakayama, K.V. Roskos, B.K. Fritzinger, and J. Heller, *J. Biomed. Mater. Res.* **29**, 1389–1396 (1995).

122. J. Heller, A.C. Chang, G. Rodd, and G.M. Grodsky, *J. Controlled Release* **13**, 295–304 (1990).

123. C.G. Pitt et al., *J. Controlled Release* **2**, 363–374 (1985).

124. C.G. Pitt, *Pharm. Int.*, 88–91 (1986).

125. C.G. Pitt, R.W. Hendren, and A. Schindler, *J. Controlled Release* **1**, 3–14 (1985).

126. C.G. Pitt et al., in J. Kost, ed., *Pulsed and Self-Regulated Drug Delivery*, CRC Press, Boca Raton, Fla., 1985, pp. 117–127.

# L

## LIPOSOMES

CHRISTOPHER J. KIRBY
Cortecs Research Laboratory
London, United Kingdom

GREGORY GREGORIADIS
University of London School of Pharmacy
London, United Kingdom

### KEY WORDS

Anticancer drugs

Antimicrobial agents

Cationic liposomes

Controlled release

Drug delivery

Drug targeting

Endocytosis

Gene delivery

Gene vaccines

Immunological adjuvants

Lamellae

Lipid polymorphism

Liposomes

Microencapsulation

Phospholipids

Steric stabilization

Tumor extravasation

### OUTLINE

### INTRODUCTION

The evolution of the science and technology of liposomes as a drug carrier has passed through a number of distinct phases. Lamellar structures in lipid–water systems had previously been studied as models of biomembranes and interpreted in terms of dynamic properties, solute sequestration, and release characteristics (1). The latter prompted the development of the liposome drug carrier concept (2), and in 1970 animals were, for the first time, injected with liposomes containing a variety of potentially therapeutic agents (3–6). This has now culminated in several injectable liposome-based drug formulations licensed for clinical use (5).

It has often been asked why it has taken so much time and effort to bring a liposomal drug onto the market. On the surface, this appears to be a legitimate question. However, the issue here is not the development of a new drug (which itself can take many years) but rather, the development of a whole approach (i.e., the use of a delivery system) to improve the efficacy of a great number of existing drugs. The liposome was adopted as a promising delivery system because its organized structure could accommodate drugs, depending on their solubility characteristics, in both the aqueous and lipid phases. In retrospect, it is not surprising that following initial work in vivo with liposomes containing therapeutic agents, several potential problems became apparent. First, although it was encouraging to note that entrapped enzymes, proteins, and lipid-soluble agents by and large acquired the pharmacokinetics of the carrier, water-soluble drugs of low molecular weight leaked extensively into the circulating blood. Secondly, there was rapid and quantitative interception of liposomes and their contents by the cells of the reticuloendothelial system (RES) through endocytosis, an event that appeared to limit the use of the system to, at best, the treatment of intralysosomal diseases or metal storage conditions. Equally disconcerting were the low levels of drug entrapment, vesicle size heterogeneity, and poor reproducibility and instability of formulations. Yet interest in liposomes, especially among academic workers, spread rapidly. We attribute this to the remarkable structural versatility of the system, which enables the design of countless liposome versions to satisfy particular needs in terms of both technology and optimal function in vivo.

In the ensuing years, great strides (5) were made toward understanding the way in which liposomes interact with the biological milieu at the molecular and cellular level. As a result, ways were found to improve liposomal stability in biological fluids so that drugs would not leak significantly, vesicle presence in the circulation would be extended, and targeting to alternative sites would be made possible. At the same time, important advances in liposome technology provided indications that many of the difficulties with entrapment and stability during storage could be overcome. Indeed, optimism generated by such progress led to the foundation of at least three liposome-based companies in the early 1980s, forcefully encouraging ideas and enthusiasm into the path of the realities of industrial research and development, and leading eventually to useful products. However, the contribution that liposome research has made (and will make) to targeted drug delivery is not just the happy prospect of additional liposome-based drugs. Of greater importance is the wealth of information that has become available on the ways that the body interacts with drug carriers and the harnessing of such information to circumvent anatomical and physiological barriers to optimal performance of the carrier. This cannot but prove instrumental in our efforts to ensure that old and new drugs act effectively.

## MOLECULAR ASPECTS OF LIPOSOME FORMATION

### Self Assembly of Lipids

Liposomes are microscopic aggregates of highly ordered lipid molecules which are normally dispersed in a hydrophilic solvent, typically water. This description applies to various types of microparticles, but the distinguishing feature of liposomes is that their lipid components are in a lamellar arrangement. When any molecule is dispersed in a solvent, its fate is dependent on the polarity of the molecule, which reflects the distribution of the electron clouds surrounding it. Polar molecules tend to dissolve in polar solvents such as water, whereas nonpolar ones are generally hydrophobic and only dissolve in nonpolar organic solvents. However, it is common to have both polar and nonpolar residues on the same molecule, in which case it is described as amphiphilic and its solvent dispersion characteristics are more complicated. Transient structuring of

the hydrogen bonds between water molecules allows very small amounts of amphiphile to dissolve in the free form, but at more appreciable concentrations, the hydrophobic residues must be masked from the polar environment. This is achieved by reorientation of the amphiphile molecules into ordered structures in which the hydrophobic regions are brought into proximity with each other while the polar groups are juxtaposed between them and the water continuum. How this is accomplished depends on the way in which adjacent amphiphile molecules are able to align with each other, and this is in turn dependent on their spatial geometry (1).

When the cross-sectional areas of the polar and nonpolar regions are approximately equal, the molecule is effectively cylindrical in shape and adjacent cylinders then align in a parallel arrangement to form a two-dimensional monolayer, with the polar head groups forming one surface of this monolayer and the fatty acid acyl chains the other. (Fig. 1). Two of these monolayers are organized back-to-back to form a bilayered sheet or lamella in which the hydrophobic regions are sandwiched between the polar groups and segregated from the bulk water. Such an arrangement is typical of naturally occurring phospholipids such as the phosphatidylcholines (lecithins), which form the major component of cell membranes.

On the other hand, if there is a significant difference between the cross-sectional areas of the polar and nonpolar regions, the amphiphile molecules approximate to the shape of a wedge rather than a cylinder and are unable to align in a planar arrangement (7). These differences may be due to steric factors such as head group size, or variation in the number of fatty acyl chains, or to electrostatic effects arising from attractive or repulsive forces between adjacent head groups. In these situations, the aligning amphiphiles are forced to form curved structures, the most common of which include hexagonal phases (cylinders) and micelles (ranging from spheres to elongated rods). Depending on which end of the molecule is widest, each of these structures adopts a normal or reverse (inside out) configuration. Thus, if the polar head group is broader than the nonpolar region, curvature will be "positive," with the polar group forming the outer surface of the curve (e.g., micelles [$L_1$ phase] and normal hexagonal [$H_I$] phase). Conversely, if the nonpolar tail is broader than the head group, curvature will be "negative," with the fatty acyl chains forming the outer surface (reverse micelles [$L_2$] or reverse hexagonal [$H_{II}$] phase). Another important nonbilayer phase is the cubic phase, alternative forms of which can resemble close-packed micelles or bicontinuous structures. It is quite normal to accommodate non-bilayer-forming amphiphiles into lipid bilayers by including amphiphiles of opposite shape tendency to compensate. Thus mixtures of wedges, inverse wedges, and cylinders can be formulated which still favors a planar arrangement.

### Lipid Polymorphism

This capability of amphiphiles to form a variety of structures and phases is known as polymorphism, and is of the utmost importance in normal biological processes such as cell trafficking, digestion, absorption, and fusion. An im-

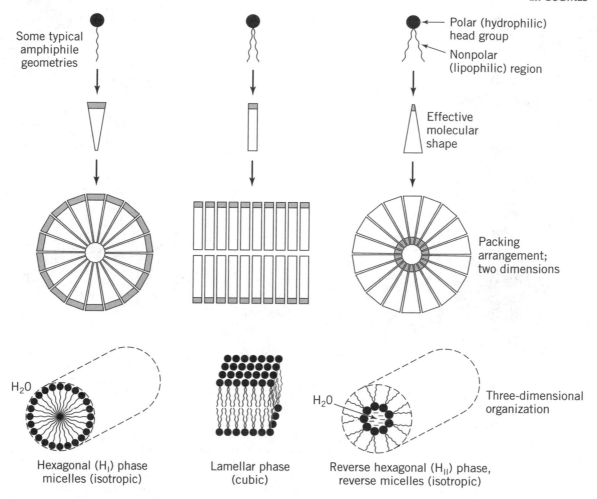

Some typical amphiphile geometries

Polar (hydrophilic) head group

Nonpolar (lipophilic) region

Effective molecular shape

Packing arrangement; two dimensions

Three-dimensional organization

$H_2O$

$H_2O$

Hexagonal ($H_I$) phase micelles (isotropic)

Lamellar phase (cubic)

Reverse hexagonal ($H_{II}$) phase, reverse micelles (isotropic)

**Figure 1.** Dependence of phase behavior on the molecular geometries of the individual amphiphile molecules. *Source:* Adapted from Ref. 7, with permission.

portant aspect is that different phases can interconvert, either by inward or outward migration of particular amphiphiles, or by individual amphiphiles changing their shape. For example, increased temperature can increase the dynamic motion of the hydrocarbon chains, effectively broadening the nonpolar region. Changes in pH can affect ionization of polar head groups and so alter their diameter, as can changes in their level of hydration. Thus, bilayer to nonbilayer transitions can take place, resulting in loss of membrane barrier function or increased fusogenicity. These transitions can be exploited to design lipid-based carrier systems with predetermined release properties which are triggered by changes in their microenvironment or in the local ambient conditions.

### Aqueous Dispersions of Lamellar Phase Lipid: Liposomes

Normal micelles, which show only short-range order, form spontaneously when the appropriate amphiphiles are added to water and dilute freely in excess water. For liquid crystalline phases, which also exhibit *long-range* order, the situation is different. Cubic and normal hexagonal phases do not disperse spontaneously, but procedures have been developed by which they can be converted into a colloidal

form without loss of short-range order, producing "cubosomes" and "hexasomes," respectively. When lamellar phase lipid is diluted with excess aqueous phase, areas of lamellae are able to detach from the bulk lipid and round off to form sealed spherical liposomes (Fig. 2), thereby ensuring that hydrophobic chains at the bilayer "edges" are not exposed to the aqueous phase. In practice, detachment of lamellae from uncharged bilayer lipid is extremely slow unless the mixture is agitated, in which case all of the lipid eventually becomes dispersed in the form of liposomes. Such structures were first described in the early 1960s (1) and within a short time led to intense activity by a large number of research groups throughout the world.

## LIPOSOMES AS DRUG DELIVERY SYSTEMS

### The Drug Delivery Concept

The common structural features shared by liposome bilayers and biological membranes led to their widespread early use as membrane models for investigating the structural and functional relationships of natural cell membranes and their components. For example, integral membrane

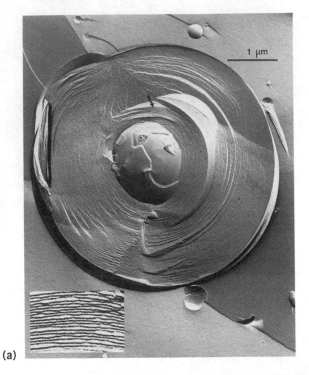

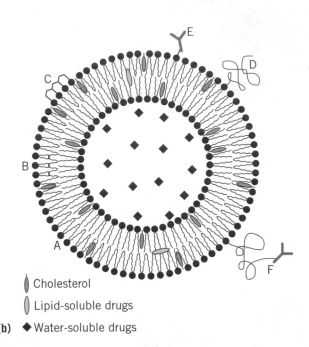

◖ Cholesterol

◖ Lipid-soluble drugs

**(b)** ◆ Water-soluble drugs

**Figure 2.** (**a**) Freeze-fracture electron micrograph of a multilamellar liposome prepared in water. The inset shows the area indicated by the arrow, magnified a further 6.7 times, to show the detail of the lamellae. *Source:* Micrograph courtesy of Mr. B. E. Brooker. (**b**) A diagrammatic representation of a unilamellar liposome showing the molecular structure of the lipid bilayer. The diagram also shows some of the modifications that can be made in order to tailor liposomes for specific applications, as discussed later in this article. Thus, A, B, and C are amphiphiles polymerized at the terminal and central regions of the fatty acyl chain, and at the polar head group, respectively. D denotes a poly(ethylene glycol) (PEG) modified phospholipid for preparing sterically stabilized liposomes, whereas E and F indicate ligand (antibody)-targeted amphiphiles with the targeting moiety attached to the head group and to the terminal region of a PEG chain, respectively.

proteins could be reconstituted into lipid bilayers to allow characterization in their natural state or to study their vectorial transport properties. However, their potential use as carriers for drugs, cosmetics, and many other types of active agent has excited a much greater volume of interest and led to their eventual commercialization (5). Such applications are dependent on the fact that liposomes are sealed structures, so that water-soluble substances dissolved in the internal aqueous phase are entrapped and isolated from the external environment. Many nonpolar compounds, as well as other amphiphiles, can also be included. For example, smaller alkanes partition into the central hydrocarbon region of the bilayer, whereas larger ones and amphiphilic molecules align themselves between and in parallel with the phospholipid chains. Insoluble substances such as colloidal metal particles have also been incorporated. Once inside, labile substances are protected from external conditions that might otherwise lead to their degradation. The liposomes themselves can be constructed with widely different physical structures, lipid composition, and surface properties, thus enabling a great deal of control over the fate of their contents. For these reasons, they have been explored very widely as delivery vehicles, particularly of drugs and other biologically active materi-

als (5,8). The basic aspects of the drug delivery concept are briefly summarized as follows.

- *Protection.* Both the active material and the host are protected by virtue of the membrane barrier function and the possibility of coencapsulation of protective agents.
- *Sustained release.* Such release is dependent on the ability to vary the permeability characteristics of the membrane by control of bilayer composition and lamellarity.
- *Controlled release.* Drug release is enabled by exploiting lipid phase transitions in response to external triggers such as changes in temperature or pH.
- *Targeted delivery.* Such delivery can be achieved by relying on natural attributes such as liposome size and surface charge to effect passive delivery to body organs or by incorporating antibodies or other ligands to aid delivery to specific cell types. At the same time, drugs can be directed away from vulnerable or unintended targets.
- *Internalization.* This occurs by encouraging cellular uptake via endocytosis or fusion mechanisms; for example, to deliver genetic materials into cells.

Examples of these control mechanisms are discussed later in this text.

Because the basic unit of liposome structure, the lipid bilayer, is itself formed spontaneously, this makes incorporation of additional levels of sophistication much more straightforward, thereby enabling the tailoring of liposomes for many different situations and applications. Their tremendous versatility, ease of construction, and similarity to natural membranes has given them many advantages over other systems that have been developed for use in drug delivery.

## Incorporation of Solutes

The amount of solute incorporated into the liposomes is defined both in terms of the encapsulation efficiency, which is the percentage of the starting material that becomes encapsulated, and the encapsulation capacity, which is the amount incorporated per unit amount of bilayer lipid. When material becomes encapsulated during liposome manufacture it is said to be passively entrapped, whereas if it is incorporated at a subsequent stage, it is described as actively loaded, or alternatively remotely loaded to indicate that incorporation has taken place at a site other than where manufacture took place. As already indicated, both hydrophilic and lipophilic materials may be incorporated, the former in the internal aqueous spaces and the latter within the lipid bilayer. The usual method of incorporating lipophilic substances is to dissolve them together with the amphiphile(s) in an organic solvent and then to remove the solvent so that the different bilayer components are left behind in a homogeneous, molecularly dispersed form. If an appropriate amount of the substance is added, then encapsulation efficiencies of 100% normally are achieved. However, the encapsulation *capacity* is likely to be much lower, and if added at greater than 5 mol % of the total lipid, it may well have adverse effects on the normal bilayer properties. On the other hand, water-soluble materials are able to partition between the internal and external water phases of the liposome, and encapsulation efficiencies generally are much lower than 100%, and dependent on the nature of the solute, the liposome components and the preparation method employed. These variables are discussed in the forthcoming sections.

## Morphology and Nomenclature of Liposomes

Lamellar-forming lipids, like other amphiphiles, normally exist in a densely packed crystalline form when in the dry state. Below a certain characteristic temperature, the nonpolar hydrocarbon chains are fully extended and rigid, but above this temperature thermal excitation is increased and the chains enter a more disordered, liquid-like state. If water is added to the lipid above this *transition temperature* ($T_c$), the polar head groups at the surface of the exposed amphiphile become hydrated and start to reorganize into the lamellar form. The water diffuses through this surface bilayer causing the underlying lipid to undergo a similar rearrangement, and the process is repeated until all of the lipid is organized into a series of parallel lamellae, each separated from the next by a layer of water. Once this liquid crystalline phase is fully swollen and in equilib-

rium with excess water, mild agitation allows portions of close-packed, multilamellar lipid to break away (1). The resulting large spherical liposomes, each consisting of numerous concentric bilayers in close apposition, alternating with layers of water, are known as multilamellar vesicles (MLV). These are heterogeneous in size, varying from a few hundreds of nanometers in diameter to around ten $\mu$m, in the latter case representing some hundreds of lamellae thick.

The main advantage of MLV is that they are simple to make and have a relatively rugged construction. However, because the interior is largely occupied by the internal lamellae, the volume available for solute incorporation is very limited, and under normal condition only a small percentage of the available material is taken up. Their large size range is generally considered to be a drawback for many medical applications requiring parenteral administration, because it leads to rapid clearance from the bloodstream by the cells of the RES. On the other hand, this effect has been exploited for passive targeting of substances to the fixed macrophages of the liver and spleen; e.g., for delivery of antimicrobial agents to microorganisms which are otherwise refractive to treatment because of their localization within these cells (9), chelating agents to remove accumulated metal ions (10), and exogenous enzymes to treat inherited enzyme deficiency conditions such as Gaucher's disease (5).

If the lipid contains a component of negatively charged amphiphile, then at low ionic strengths of the hydration media, the interlamellar distance is increased substantially due to charge-charge repulsive forces between adjacent lamellae. When the lipid is allowed to hydrate undisturbed followed by gentle swirling, individual lamellae are able to detach to form large unilamellar vesicles (LUV) (11). In practice, it is generally more convenient to use one of a variety of methods developed specifically for manufacturing this type of liposome. LUV vary in size from around 100 nm up to tens of micrometers in diameter, and because the interior is not occupied by internal lamellae, there is ample space for incorporation of "drug." This also means that any cells taking up LUV are less subject to lipid overload. However, the absence of additional lamellae means that they are more fragile than MLV and have increased permeability to small solutes.

The lower limit of liposome size is fixed by the maximum curvature that the lipid bilayer is able to achieve, around 21 nm in the case of unsaturated, natural phospholipids such as egg and soy lecithin. These single-layered liposomes are known as small unilamellar vesicles (SUV), and the upper limit of size is designated as 100 nm. Because of their small size, clearance from the systemic circulation is significantly reduced, so they remain circulating for longer and thus have a better chance of exerting the desired therapeutic effect in tissues other than those of the RES. However, the small size also decrees that they have a much lower capacity for drug entrapment, typically less than 1% of the material available.

SUV, together with MLV and LUV are the three major morphological categories of liposomes, and these may be subdivided further. For example, several types of MLV are known. Oligolamellar (or paucilamellar) vesicles have intermediate lamellarity between LUV and MLV, and their formation is favoured under certain conditions. Multivesicular liposomes (MVL) can also be prepared, in which several separate compartments are present within a single MLV (12). On the other hand, stable plurilamellar vesicles (SPLV) have a similar appearance to classical MLV under the microscope, but have different physical and biological properties owing to the fact that the latter appear to be in a state of osmotic compression (13). LUV and SUV are often subdivided according to their size. It is also common to classify liposomes according to the various manufacturing procedures, as described in the following section.

## PREPARATION OF LIPOSOMES IN THE LABORATORY

Early developments in liposome methodology were aimed particularly at achieving good solute entrapment (8). The move toward commercialization has placed increasing emphasis on such needs as stability, scale-up, reproducibility, and process validation (8). Numerous methods have been developed to meet widely different requirements. These can be broadly divided into two categories, those involving physical modification of existing bilayers and those involving generation of new bilayers by removal of a lipid solubilizing agent. The various methods available are discussed here from the perspective of the different structural types of liposome that they produce. Due to lack of space, this article can offer no more than a brief qualitative description of the more widely known procedures. For more in-depth details, the reader is referred to the various reviews, original papers, and manuals that are available (8,14,15).

### Oligo- and Multilamellar Vesicles

**Physical Methods.** *Simple "Hand-Shaken" MLV.* MLV may be prepared from single-source natural or synthetic lipids, by suspending the latter (1), preferably in a finely divided form, in an aqueous solution maintained at a temperature greater than the $T_c$ of the lipid. For unsaturated phospholipids such as egg and soy phosphatidylcholine (PC), which have $T_c$ values below 0°C, this is conveniently done at room temperature. Stirring or swirling considerably speeds up lipid hydration and liposome formation, and the possibility of lipid oxidation can be minimized by working in an inert atmosphere of nitrogen or argon. As the liposomes form, a small proportion of the solution and its associated solute becomes entrapped within the interlamellar spaces. Uptake may continue over several hours, though in practice two hours of gentle stirring is normally adequate to achieve near-maximal incorporation. At the end of this period, the loaded liposomes can be separated from nonencapsulated solute using a process such as centrifugation or dialysis.

It is often desirable to prepare liposomes from mixtures of amphiphile, for example to improve their stability or to impart functional properties such as charge. In this case it is essential that the different lipids be thoroughly mixed at the molecular level. This has traditionally been achieved in the laboratory by codissolving them in a common solvent such as a 2:1 (v/v) mixture of chloroform and methanol and

then removing the solvent. If done using a rotary evaporator, the lipid can be deposited as a thin film, which aids solvent removal and subsequent dispersion of the lipid. Hydration, separation, and harvesting can then be carried out as already described.

A major drawback of conventional MLV is their low efficiency for incorporation of water-soluble solutes, which is partly due to the fact that much of the volume is occupied by the internal lamellae. Thus, in neutral liposomes, only a few percent of the starting material may become entrapped. The encapsulation efficiency can be increased in several ways. Inclusion of a charged amphiphile, such as phosphatidyl glycerol or phosphatidic acid at a molar ratio of 10–20%, causes electrostatic repulsion between adjacent bilayers, leading to increased interlamellar separation, thus allowing more solute to be accommodated. However, if the solute itself is charged, entrapment may be increased or decreased depending on the relative sign, and may lead to solute association with the external surface. Freeze-drying of the lipid from solution in an organic solvent tends to form an expanded foam with an increased surface area, which increases the amount of aqueous phase that can be incorporated (16). A solvent with a high freezing point such as tertiary butanol should be used, so that it becomes deposited in the condenser rather than dissolving in the pump oil. Drying of the lipid onto an inert support, whether water-insoluble such as glass beads, or water-soluble such as sorbitol, also increases lipid surface area and may increase the level of solute entrapment.

It has been found that solute concentration in the interior of hand-shaken MLV is less that that in the external aqueous phase (13). This is because, as the phospholipid is hydrating, water passes across the semipermeable lamellae more rapidly than does the solute, and so the liposomes are correspondingly depleted. This can be alleviated by subjecting the suspension to cycles of freezing and thawing, which causes transient defects in the lamellae through which solute redistribution can take place.

*Dehydration/Rehydration Vesicles (DRV).* The DRV method was designed to achieve high levels of entrapment, particularly of sensitive biological molecules such as proteins and nucleic acids, using simple, mild conditions (17,18). The conventional MLV method described at the beginning of this section satisfies the latter two criteria, but since the multilayers form and seal off with the vast majority of the lipid never having come into contact with the solute, encapsulation efficiencies are extremely low. The intention of the DRV method is to maximize exposure of solute to the lipid before its final lamellar configuration has been fixed, so that the liposomes ultimately form around the solute. This is achieved by first preparing MLV in distilled water and then converting these to SUV using established procedures (to be described later), so that the phospholipid achieves the highest possible level of dispersion within an aqueous phase. Lipid packing constraints resulting from the high curvature of SUV mean that up to 70% of the total phospholipid will be present in the outer leaflet of the lamella. Thus when SUV are mixed with a solution of the material to be entrapped (again in distilled water), the majority of the amphiphile is directly exposed to the solute. At this stage the water is removed, most preferably by freeze-drying, to produce an intimate mixture of the dry components. The original vesicles are now in a metastable state, with the phospholipid head groups in close proximity and oriented toward the active component (Fig. 3). A small, defined amount of water is added (17–19), initiating fusion of the vesicles surrounding the active, and formation of larger liposomes, which now encapsulate a large proportion of the solute. The amount of water added at this stage is critical to achieving the optimal level of entrapment, a point that has been missed by some workers. Following the hydration step, the liposomes are diluted with an isotonic buffer such as phosphate-buffered saline (17), and washed to remove nonencapsulated material using a process such as centrifugation or dialysis. At the rehydration stage, there is an increase in the overall solute concentration (e.g., 10-fold) because of the reduced amount of water added (e.g., 10 times less than the original amount). Subsequent dilution with isotonic buffer produces a large osmotic gradient between the internal and external phases, which leads to a redistribution of solute, both within the forming liposome and with the external aqueous phase. This may be aided by increased permeability to solute as a result of bilayer stretching caused by hyperosmotic inflation. At this stage, liposome formation is completed and equilibration achieved.

A wide variety of active materials have been incorporated into DRV, ranging from small ionic compounds and cyclodextrin–drug complexes to proteins and DNA (17–21). Encapsulation efficiencies are high, typically 40–50% under the standardized conditions described, and these levels can be increased substantially by using additional lipid. For solutes that are used in small amounts, much higher proportions can be encapsulated, e.g., values greater than 70% are normally achieved for DNA. Retention of encapsulated carboxyfluorescein dye during incubation with mouse plasma at 37°C was found to be substantially higher than for SUV or reverse phase evaporation vesicles (REV). DRV may be oligo- or multilamellar depending on the conditions used to prepare them. Thus, when ascorbic acid was encapsulated at an initial concentration of 23 mM, freeze-fracture microscopy showed a normal multilamellar appearance. However, when a concentration of 115 mM was used, followed by equilibration in isotonic buffer, many were oligolamellar, with regions of multilayer alternating with broad expanses of aqueous phase (20) (Fig. 4). Liposomes retained 40% of the original solute, which in view of the high concentration at the rehydration stage (more than 1 M), represents a particularly high loading of ascorbic acid. This is presumably due to the increased internal aqueous volume available for solute entrapment. Increas-

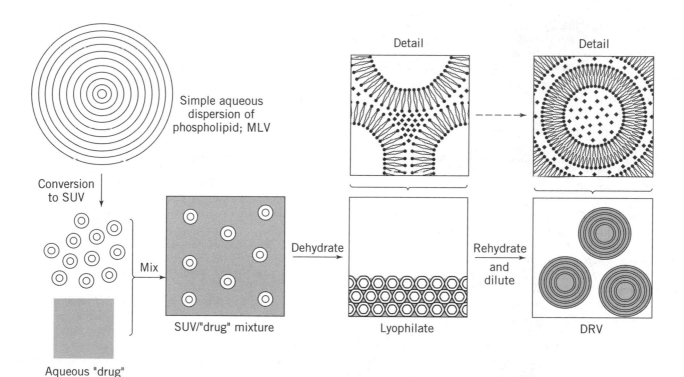

**Figure 3.** Manufacture of liposomes by the dehydration-rehydration method. A more detailed description of the individual steps is given in the text. The diagram is purely descriptive and no attempt has been made to draw any of the vesicles and other components to scale. During the freezing process, some deformation of the SUV is likely to occur as a result of osmotic shrinkage, caused by increasing external solute concentration arising from progressive formation of ice crystals. Further destabilization of the bilayer during dehydration facilitates rearrangement/fusion to form DRV during the subsequent rehydration step. The relative alignment of the individual drug and phospholipid molecules in the lyophilate (detail) is essentially similar to that in the final product.

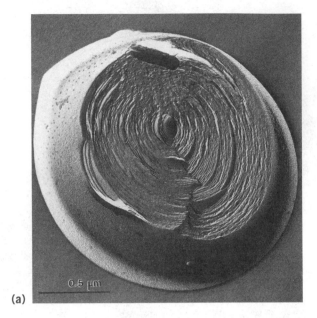

(a)

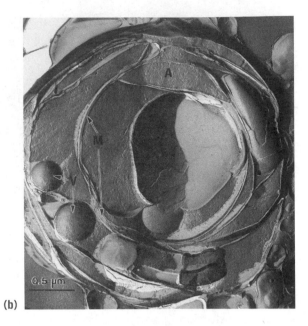

(b)

**Figure 4.** Freeze-fracture electron micrographs of liposomes containing ascorbic acid. DRV prepared in the presence of (**a**) 23 mM and (**b**) 115 mM ascorbic acid. M, regions of multilamellar lipid; V, internal vesicular structure; A, aqueous spaces containing ascorbic acid. *Source:* Courtesy of Mr. B. E. Brooker (see Ref. 20).

ing levels of entrapment at increased solute concentration have also been seen with sodium and potassium chloride, and it may be possible to harness controlled osmotic inflation to achieve still higher loading of other coencapsulated solutes such as macromolecules. These liposomes contained a number of internal vesicular structures, perhaps related to a heterogeneous solute distribution during the dehydration stage, and the particularly high osmotic gradient during rehydration.

The DRV procedure remains one of the most efficient methods for direct loading of drugs into liposomes. Unlike other high entrapment methods, it avoids the use of potentially damaging conditions such as sonication and the use of organic solvents or detergents, which might cause damage to sensitive biological materials. Its simplicity means that scale-up is relatively straightforward.

*Resizing of Liposomes.* For some applications, the large size and size heterogeneity of multilamellar liposomes is a disadvantage. Both parameters can be reduced by various physical processes that, when used to their full capability, result in the formation of unilamellar liposomes, and so are described more fully in the relevant section. Many of these procedures can be used under less extreme conditions to reduce size, lamellarity, and polydispersity while maintaining the essential oligo- or multilamellar nature of the liposomes. Such processes inevitably lead to reduced entrapment due to redistribution of solute molecules between the liposomes and the external aqueous phase, though losses can be reduced if processing is carried out in the presence of the mother liquor. Bath sonication, microfluidisation and membrane extrusion have been widely used in this respect. Of these, microfluidization (22) and membrane extrusion (C.J. Kirby, unpublished results) have been used to reduce the size range of DRV while still

retaining substantial proportions of the encapsulated solutes.

**Removal of Lipid-Solubilizing Agents.** In practice, the only commonly used agents during preparation of MLV are organic solvents (Fig. 5).

*Simple Hydration of Solvent-Solubilized Lipids.* Single- or mixed-source amphiphiles dissolved in organic solvents can be hydrated in solution without prior solvent removal. If the solvent is water-miscible such as ethanol or propylene glycol, it can be removed at the end of the procedure (e.g., by dialysis or filtration) or else allowed to remain. The *ethanol injection* procedure discussed later (23) was originally developed for producing SUV, but when the lipid is dissolved at increased concentration and added to the aqueous phase by simple admixture, the resulting liposomes are largely multilamellar. These might then be used to produce feedstock for further processing into other types of liposomes such as DRV. Alternatively, water-immiscible solvents such as ether, chloroform, or methylene chloride may be employed, removing them during processing by evaporation. As with hand-shaken MLV, solute encapsulation is generally very poor. More complex variations of this approach, designed to give improved encapsulation efficiencies, are be discussed later.

*Stable Plurilamellar Vesicles.* This is a variant of the REV method, which had been developed previously for preparing LUV. As with the DRV procedure, it aimed to increase the level of encapsulation of drugs in MLV by including them before the final lamellar configuration had been fixed. An aqueous phase containing the drug is emulsified together with excess of an ethereal solution of lipids, using bath sonication to form an oil-in-water type emulsion, while evaporating the solvent under a stream of ni-

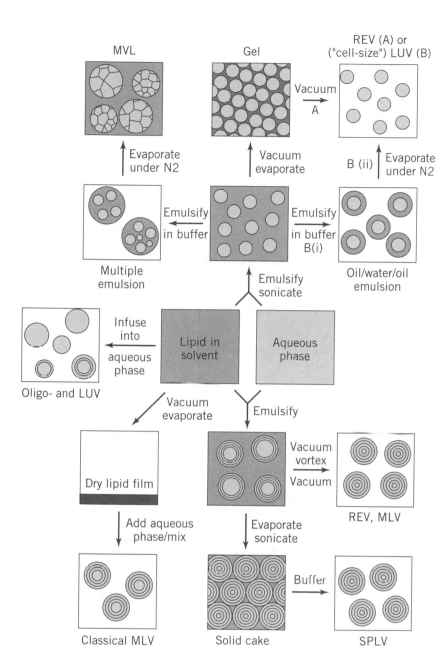

**Figure 5.** Solvent evaporation processes for liposome preparation. This scheme shows a considerably simplified summary of the different methodologies that are dependent on solvent evaporation, i.e., entailing the use of volatile solvents. No attempt is made at drawing to scale and various steps have been omitted. For convenience, steps A and B (i) and (ii) are shown as producing a similar product, whereas the "cell-sized" LUV derived from B (i) and (ii) may be considerably larger than the REV from step A. In addition, different processes require different solvents, but no such distinctions are made here.

trogen. The resulting cake is then suspended in buffer and washed. The important differences from the parent method are that the amount of lipid is proportionately higher and sonication is used throughout. Whereas the continuous water phase of REV is derived from the original disperse phase of the emulsion, in SPLV it is added later. It has been suggested that toward the end of the process, the aqueous cores collapse and are occupied by lamellae, with the solute load becoming distributed throughout the interlamellar spaces. Maximum levels of entrapment were reported to be 37%, but in terms of the amount of drug incorporated per unit amount of lipid, that level is about 10 times less than the typical value for DRV. In comparison with conventional MLV, SPLV were found to have improved physical stability and biological efficacy. This was claimed to be related to the fact that, unlike MLV, they are

not solute-depleted and are in a state of osmotic equilibrium with the surrounding media. The main drawbacks of the SPLV procedure are the need to use volatile organic solvents and sonication, both of which are serious obstacles to potential scale-up and incorporation of sensitive materials such as proteins.

*Multilamellar Reverse Phase Evaporation Vesicles (MLV-REV).* This is another derivation of the REV method, again differing from the latter in that more lipid and less water are used in order to encourage the formation of multilamellar liposomes (24). The main difference from the SPLV method is that the solvent is removed under vacuum in two stages, and the aqueous continuous phase for the liposomes is derived from the disperse phase of the emulsion (as with the parent REV method). The initial phase of evaporation leads to formation of a gel consisting of aqueous cores surrounded by multilayers of lipid. The gel is removed from the vacuum, vortexed, and then reevacuated. During the final stage, phase inversion occurs, wherein a proportion of the cores release their aqueous content, and this is used to form a continuous phase for the newly formed liposomes. Up to 91% of the solute was encapsulated within these MLV-REV, but again from the point of view of mass loading, this is only 50% of that which is typically achieved with DRV. As with the SPLV approach, a major drawback is the need for volatile organic solvents and the implications for safety and scale-up potential.

*Multivesicular Liposomes.* This approach involves the use of multiple, water-in-oil-in-water type emulsions (12). These are created by first dispersing the aqueous drug solution into a chloroform–ether organic phase containing dissolved amphiphile, thus forming a fine water-in-oil type emulsion, and then dispersing this in turn into a second aqueous phase to form the multiple emulsion. Compositions and conditions are chosen such that each droplet of organic disperse phase contains multiple droplets of the initial aqueous solution. The solvents are removed under a stream of nitrogen, wherein the amphiphile is deposited around the internal aqueous, drug-containing microdroplets, which end up as separate aqueous compartments within a single liposome. Therefore, these are actually multicompartment vesicles in which adjacent, nonconcentric compartments are separated by a single, common unilamellar boundary. The method can achieve very high encapsulation efficiencies of up to 89%, but requires specialized lipids and rather complex conditions. Again, limitations are imposed by the need to use volatile organic solvents, but it is noteworthy that the method has been successfully scaled-up for commercial use.

## Small Unilamellar Vesicles

Most of the commonly used methods for preparing SUV involve size-reduction of preexisting bilayers using procedures that have already been referred to in the previous section. The earliest, and probably still the most widely used approach at the bench scale, is by ultrasonic irradiation (25). SUV can be formed rapidly and conveniently by high-power probe sonication but are vulnerable to oxidative damage of unsaturated lipids. This can be reduced by maintaining them in an inert atmosphere during processing and by using a cooling bath to dissipate the large amounts of heat produced. Ideally the equipment should be fitted with an electronic controller to allow the sonic energy to be delivered in short pulses interspersed with longer cooling intervals (for example, 1 s sonication and 3–4 s cooling). Small particles of titanium shed from the probe are removed at the end of the process by centrifugation. A more gentle approach is to use bath sonication, where it is easier to control temperature and to maintain an inert atmosphere, and metal contamination is avoided. Unfortunately, processing time is substantially longer than when a probe sonicator is used.

Many of the literature reports describing preparation of SUV by probe sonication entail the lipid being processed at relatively low concentration. Interestingly, we have noted that when lipid concentration is increased progressively, up to a maximum of around 250 mg/mL, there is a substantial reduction in the time required to form clear SUV using probe sonication (C.J. Kirby, unpublished results, 1992). By this means, gram quantities can be processed in a relatively short time. This suggests that there is less chance of damage to the lipid when working at higher concentrations, a surprising observation because one would instinctively expect the opposite to be true. The reason may be that the high lipid concentration results in an increased likelihood of bilayer fragments being able to reanneal after high power sonication to form SUV. Whatever the reason, the phenomenon allows a certain level of increased throughput in what is otherwise a convenient but notoriously difficult procedure to scale up.

Alternative ways of preparing SUV by sizing make use of high pressure homogenization or extrusion. High-pressure homogenization devices include the French Pressure Cell (26), the Microfluidiser™ (27), and the Gaulin homogenizers (28). The latter two can have particularly high throughputs and employ high shear and cavitation effects to ultimately convert MLV to SUV. The Gaulin range can be used to forcibly hydrate a suspension of dry lipid and convert this to SUV in a single-step process. More impressively, differential scanning calorimetry (DSC) studies indicated that separate amphiphiles could be combined at the molecular level to form SUV having a homogeneous lipid distribution without the involvement of organic solvents (28). To our knowledge, no other mechanical procedure has been reported to be able to achieve this.

High-pressure extrusion involves forcing multilamellar liposomes at high pressure through membranes having "straight-through," defined size pores (29). The liposomes have to deform to pass through the small pores, as a result of which lamellar fragments break away and reseal to form small vesicles of similar diameter to that of the pore. Polycarbonate membranes are manufactured with pore diameters ranging from 8 $\mu$m down to 0.03 $\mu$m, and ceramic or stainless steel membranes are also available. Repeated cycling through small-diameter pores at temperatures greater than the $T_c$ of the lipid produces a homogeneous population of SUV. A particular advantage of the method is that the disruptive effects of sonication are avoided.

Rapid injection of an ethanolic solution of lipid into an aqueous solution produces a dispersion of vesicles of 30 to 110 nm in diameter depending on the concentration of lipid used (23). Compared with sonicated SUV, these have a relatively heterogeneous size range and have added disadvantages in that a dilute suspension of liposomes is formed, the method is restricted to using ethanol-soluble lipids, and the ethanol may need to be removed afterwards. However, a modification of this method, used to incorporate a lipid-soluble antifungal drug into liposomes, has been successfully scaled up to a batch size of hundreds of kilograms and regulatory approval obtained (30).

Dispersions of the acidic phospholipids, phosphatidic acid, or phosphatidylglycerol, either alone or mixed with other phospholipids, can be induced to form unilamellar liposomes simply by transiently increasing the pH. SUV can be favored by adjusting the experimental conditions (31). However, despite its apparent simplicity, the method is very restrictive in terms of the lipid compositions that can be used and the conditions of preparation. The vesicles tend to be of a heterogeneous size range, which is influenced by factors such as the nature of the other lipids present, ionic strength, cholesterol content, and the rate and extent of titration.

### Large Unilamellar Vesicles

LUV occupy the size range following on from SUV, but unlike the smaller SUV are not subject to lipid packing constraints. Their single bilayer membrane makes them well suited as model membrane systems whereas the large internal aqueous volume:lipid mass ratio means that the efficiency of drug encapsulation for both small and large molecules can be maximized. As with the liposome types already discussed, methods for preparing LUV fall into two categories, the first involving generation of new bilayers by removal of a lipid solubilizing agent, whereas the second involves physical modification of preformed bilayers. For LUV preparation, the solubilizing agents include detergents and chaotropic ions as well as organic solvents.

**Removal of Detergents and Chaotropic Agents.** For procedures involving detergent removal (32), the lipid is initially dissolved by an aqueous solution of the detergent to form mixed lipid–detergent micelles, and the detergent is then removed by a diffusion-based process such as dialysis, diafiltration, or gel chromatography. The method produces almost exclusively uni- or oligolamellar vesicles ranging in diameter from SUV size up to several micrometers depending on the conditions used. Ionic detergents, such as cholate and deoxycholate or nonionic detergents such as Triton × 100 and octylglucoside, have been used. Detergent removal methods have been especially useful for functional reconstitution of membrane proteins. Many of these proteins retain their activity better in the presence of the nonionic species but these often have a relatively low critical micellar concentration making them difficult to remove by the aforementioned procedures. Use of hydrophobic adsorbent beads has proved particularly useful in these cases. The principle of detergent removal was applied in the first commercially available apparatus for laboratory production of liposomes, the Lipoprep™, which relied on a fast flow dialysis cell to remove detergent (33). The approach has the considerable advantage of allowing close control of the final size of the liposomes, which have a homogeneous size range. Disadvantages are that encapsulation efficiency is low compared with most other methods of LUV preparation, and detergent removal by ordinary dialysis techniques is a time-consuming process that can take several days to reduce residual detergent to a negligible level. Even traces of detergent can have pronounced effects on liposome permeability and can greatly increase the transmembrane movement of phospholipids. Detergents may also have deleterious effects on the material being encapsulated. Giant liposomes have been prepared by solubilizing phospholipids in a micellar form, in solutions of chaotropic ions such as trichloroacetate, and then removing the latter by dialysis or dilution (34). The liposomes, which are uni- or oligolamellar have diameters in the region of 10–20 μm.

**Removal of Organic Solvents.** Solvent vaporization liposomes tend to be of a larger size range than those prepared by detergent removal. Three distinct types of process have been described, each involving addition of a solution of lipid in organic solvent, to an aqueous solution of the material to be encapsulated (see Fig. 5).

*Solvent Infusion.* Solvent such as diethyl ether, petroleum ether, ethylmethyl ether, or dichlorofluoromethane-containing dissolved lipid(s), is infused slowly into the aqueous phase, which is maintained at a temperature above the boiling point of the solvent so that bubbles are formed (35). The lipid is deposited as a multilayer around the vapor–water interphase, and as the solvent evaporates off, uni- and oligolamellar liposomes remain in solution. High encapsulation efficiencies (up to 46%) were reported when high concentrations of lipid (15 mM final concentration) were used, and most of the vesicles were in the size range 0.1–0.4 μm. The major disadvantage is the need for exposure of the active to organic solvents, with the possibility of damage to labile materials such as proteins.

*Reverse Phase Evaporation.* Modifications of this method aimed at preparing MLV have already been described (36). Those approaches proceed via the formation of a water-in-oil (diethyl ether) emulsion containing excess lipid. In the parent case, the amount of amphiphile used is greatly reduced such that, when all of the solvent has been removed (by rotary evaporation), there is just enough lipid to form a monolayer around each of the microdroplets of aqueous phase. At this stage emulsion inversion takes place. This involves collapse of a proportion of these inverted micelles so that their aqueous contents now form the new continuous phase, while their lipid components go to form an outer leaflet around the remaining structures, thus converting them from inverted micelles to a vesicular form. In the absence of cholesterol, these uni- and oligolamellar vesicles have diameters in the range of 0.05–0.5 μm, while with 50 mol % cholesterol, mean diameters are about 0.5 μm. High encapsulation efficiencies of up 65% can be achieved using hydrophilic solutes. Particular drawbacks of the method include the need to work with volatile organic solvents and to expose potentially labile solutes (e.g., proteins) to these, and the need to use sonication to form a stable emulsion.

*Double Emulsion Evaporation.* A variant of this procedure has already been described for the preparation of MVL (12). In fact, the original method (37) was used to prepare cell-sized liposomes of average diameter about 9 μm. There are again formed by preparing a multiple, water-in-oil-in-water type emulsion followed by solvent evaporation as described for MVL. However, the conditions differ in terms of lipid concentration, solvent composition, and duration of shaking, such that each droplet of organic phase contains only a single microdroplet of drug solution, and the final product is an aqueous suspension of LUV. Although only a small proportion (about 30%) of the lipid present actually becomes incorporated into liposomes, the captured volume is very high at around 52 liters per mole of lipid used. The main disadvantages are that the procedure is quite complex and that specialized lipid mixtures are required. As in the two preceding methods, the presence of organic solvents may prohibit its use for encapsulation of labile materials.

**Physical Modification of Existing Bilayers.** Extrusion procedures have already been referred to in the context of downsizing of MLV and ultimately of forming SUV. If MLV are repeatedly extruded under moderate pressure (<500 lb/in$^2$) through membranes containing straight-through, defined size pores, smallish LUV are formed, with diameters between 60 and 100 nm, depending on the pore size. These have trapped volumes of up to 3 $\mu$L/$\mu$mol lipid. Where appropriate, drugs can be actively loaded after liposome formation via transmembrane gradients. Alternatively, solute incorporation can be facilitated by prior intermixing with the lipid using freeze-thaw cycles or by using the DRV procedure. This may be particularly useful for entrapping macromolecules.

Exposure of sonicated phospholipid mixtures to alternate cycles of freezing and thawing followed by a brief sonication results in fusion of the SUV and can be used to prepare large, mainly unilamellar liposomes with a relatively large trapping capacity (38). This method was based on a procedure that had previously been used for reconstitution of membrane proteins and is mild in that it avoids the use of detergents and organic solvents, and the period of sonication is quite brief. Disadvantages are that sugars, divalent cations, and high concentrations of ions and proteins all inhibit the membrane fusion process, and the liposomes are heterogeneous in size.

Another procedure employing fusion to produce LUV makes use of electrostatic effects. SUV composed of negatively charged phospholipids are mixed with calcium ions which cause the vesicle to aggregate and then fuse (39). This results in the formation of "cochleate cylinders," which are rolled-up portions of lipid bilayer. Chelation of calcium by adding ethylene diamine tetraacetic acid (EDTA) results in conversion of the cochleates to large uni- and oligolamellar vesicles having a relatively large encapsulation capacity. The main limitation is that the method is only applicable to negatively charged lipids.

## SCALE-UP AND INDUSTRIAL MANUFACTURE OF LIPOSOMES

As in many areas of technology, the transition from an optimized research laboratory process to liposome manufacture at the production level is likely to be a daunting prospect, particularly because commercialization of liposome-based products is a relatively recent occurrence. In this context it is reassuring that many of the procedures discussed already have been successfully scaled up, and some are in commercial production. The preparation method chosen at the research phase is likely to have depended on the nature of the material being encapsulated and on the liposome type perceived to be optimal for its intended application. The latter is defined by parameters such as liposome size, lamellarity, permeability, fluidity, hydrophilicity, and other surface properties. Some methodologies are certainly more amenable to scale-up than others, and hopefully some thought will have been given

at that stage toward potential future requirements. The subject has been reviewed by various authors (40–42) and so is dealt with only briefly here.

When discussing the implications of scale-up, it is convenient to break down liposome manufacture into the different components of the process which are directly affected. The most important of these are the achievement of a molecular dispersion of different lipid species, hydration of the amphiphiles (wherein actual formation of the liposomes takes place), solute entrapment, and sizing. The first two are interdependent and so are discussed together.

### Lipid Intermixing and Hydration

Liposomes are often composed of lipid mixtures (e.g., to increase stability and to provide functionality) and organic solvents are commonly used to enable different lipids to be combined as a homogeneous molecular dispersion. Volatile solvents such as chloroform, methylene chloride, and diethylether have been widely used in this respect because they have high solubilizing capacity for lipids and are easy to remove by evaporation. On the other hand, for large-scale use they are difficult to work with and rigorous safety measures must be in place to guard against fire and toxicity hazards. Moreover, stringent regulations govern the levels of any trace residues that may be present in pharmaceutical products. Some of the methodologies that use them tend to be complex and so pose additional problems for large-scale production. However, volatile solvents have additional roles such as promoting lipid hydration via thin film formation and, in the case of the emulsion methods, in helping to configure liposome structure. In this context it is noteworthy that the MVL method (12), which involves the preparation of multiple, chloroform-containing emulsions, has been used for commercial scale production of Depotfoam™, which has been proposed as a sustained release depot system for localized injection via various routes. In this case however, the liposomal product has a unique configuration which has not been achieved using other methodologies, and as a general rule it is advisable to avoid large-scale use of procedures based on volatile solvents where a similar product can be prepared by other means.

Intermixing of lipid components is also achieved using water-miscible solvents with relatively low volatility such as ethanol, methanol, and propylene glycol, and these are easier to work with than volatile solvents. They also enable rapid hydration of the dissolved amphiphiles. An ethanol-based procedure has been applied for large-scale commercial production of a lipid-soluble antifungal drug econazole into liposomes, enabling 450-kg batch sizes to be produced (30). Ethanol is a pharmaceutically acceptable solvent and, in this case, was left in place at the end of the procedure, helping to maintain the sterility of the formulation. However in other situations its presence may be undesirable, necessitating its removal from the final product. Another disadvantage is the limited capacity of these solvents to dissolve certain liposome components such as certain other phospholipids and cholesterol, thus increasing the amount of solvent which has to be added. If necessary, removal can often be achieved using procedures such evaporation or dialysis and, in the case of larger liposomes, by centrifugation or ultrafiltration. Evaporation of smaller amounts of methanol or ethanol can be carried out rapidly and conveniently using a freeze-drier, but a cold trap of liquid nitrogen is needed to prevent contamination of the pump oil. A good compromise is to use a high boiling point solvent such as tertiary butanol which is readily deposited in the freeze-drier condenser.

Detergent removal methods have also been scaled up to produce larger volumes of liposomes, particularly containing bilayer-associated drugs. A range of lipophilic cytostatic drugs were incorporated in volumes of 0.1 to 5 liters for use in clinical trials (43), achieving incorporation efficiencies ranging from 85 to 100% compared with only 1.6% for a hydrophilic drug. Further scale-up to batches exceeding 10 liters was carried out using a topical steroid β-methasone (44), and a production unit capable of producing batches of between 100 and 200 L was reported. In this case, the methodology was chosen in order to produce monodisperse liposomes of defined size distribution. In all cases the initial association between lipid and drug was achieved by dissolution in organic solvents followed by evaporation to form a thin film, and where mixed lipid components are required, the same procedure would have to be followed. Apart from the likelihood of small amounts of residual detergent remaining in the product, a further drawback of this method for large-scale use is the relatively long time periods involved.

### Entrapment

For successful commercial usage of liposomes it is vital that sufficient material can be incorporated for the intended application and that it is retained by the vesicles until their subsequent use. This is particularly important for pharmaceutical applications where a shelf life of two years is often required. Both aspects are readily achievable for bilayer-associated substances but are much more of a problem for water-soluble ones and largely dictated by the properties of the latter. If low encapsulation efficiencies are acceptable, then it may be sufficient to use one of the mechanical procedures already described. On the other hand, various approaches have been used to increase the level of encapsulation and retention of hydrophilic substances. Interaction with the bilayers can be increased by including lipids of opposite charge, by preparing lipophilic derivatives (45), or in certain cases, by incorporating lipophilic chelating agents for the solute into the bilayer (46). Otherwise, manufacturing procedures that enable high encapsulation efficiencies can be used. If it is deemed appropriate to use volatile organic solvents, REV, or one of its derivative methods might be considered feasible. An alternative possibility is to use the DRV method, if necessary using extrusion or microfluidization to reduce the size of the resulting liposomes. This method has added advantages in that sensitive biological molecules can be incorporated with high efficiency, and also the lyophilized intermediate can, in theory, be stored indefinitely before being reconstituted as and when needed. This second point could help to alleviate the problems of leakage of smaller molecules during long-term storage.

Another approach to achieving high encapsulation efficiencies while eliminating the consequences of leakage during storage is to use the active loading principle (47,48) by means of which a drug can be remotely loaded into liposomes at the time it is needed, for example at the patients' bedside. This approach is dependent on the fact that small charged molecules have a very low ability to diffuse across lipid membranes, whereas if their charge is neutralized, many become lipophilic and able to cross much more readily. By passively incorporating ionizable substances into liposomes, and then adjusting the external pH, transmembrane pH or concentration gradients can be established and used to load certain drugs into the liposome. For example, a weakly basic drug such as doxorubicin added to liposomes with a relatively high external pH becomes deprotonated and thence lipophilic, enabling it to partition across the membrane. Once inside, it becomes protonated due to the low internal pH, preventing its exit and enabling more of the uncharged drug to enter. Other approaches can result in precipitation of the loaded drug by interaction with encapsulated cations, thus promoting further drug entry.

### Sizing

The suitability of high-pressure homogenization systems for scale-up of the sizing stage has already been discussed. In particular, the claimed ability of the Gaulin Micron range to be able to convert highly concentrated particulate suspensions of lipids directly to SUV by forced hydration, to achieve homogeneous lipid intermixing, and to be operated continuously (28) suggests that it is well fitted for effective scale-up of the earlier stages. The possibility of enabling incorporation of proteins and peptides with good retention of activity is also worthy of further investigation. One reported drawback is that the cavitational effects can result in mechanical erosion and contamination of the product with metal particles, but these could presumably be removed using appropriate filters. Also, because SUV are formed, the entrapped volumes are rather low. The Microfluidiser has also been widely used for preparing large volumes of small vesicles and has proved useful for producing empty SUV for scaling up of the DRV procedure.

Sizing by extrusion through polycarbonate membranes can be conveniently carried out in the laboratory using small, compact apparatus, resulting in liposomes with well-defined size characteristics (49). In earlier studies, the technique was limited due to clogging of the membrane resulting in reduced throughput, but it has now been successfully scaled up to enable processing of multiliter volumes. Initial improvements were achieved in batch mode by increasing the surface area of the membranes and using stirring to prevent clogging (43). More recently a continuous flow device was introduced which employed higher pressures and achieved flow rates of up to 500 mL/min. High lipid concentrations of up to 400 mg/mL could be processed, and no signs of membrane clogging were seen after using a range of lipid compositions (50).

### Other Issues

The major drive to develop liposomes has come from the pharmaceutical industry, and so particular emphasis has been placed on factors such as sterility, apyrogenicity, stability, and reproducibility. Sterility is normally achieved either by ultrafiltration of the final product through 0.2 $\mu m$ filters in the case of smaller oligo- and unilamellar liposomes, or by carrying out the whole manufacturing process in a sterile environment. Both approaches have been successful for a range of liposome-based products and so should pose no problems with forthcoming ones. Other approaches have been discussed (51). The adoption of good quality control procedures and the use of high-quality raw materials from approved suppliers should help to minimize problems of poor stability and pyrogenicity. If a solvent solubilization step is present, dissolved lipids can be ultrafiltered to remove pyrogenic substances. Antioxidants can be included to minimize oxidative degradation, particularly where unsaturated lipids are used. In some cases it might be feasible to freeze-dry the product in order to improve stability during storage, but this is generally most appropriate in the case of bilayer-soluble drugs.

## CHARACTERIZATION

There are numerous methods for the characterization of liposomes, depending on the aspect of the system under investigation (52). Such aspects include those that are related to the overall structure of liposomes, for instance vesicle size, lamellarity (i.e., number of bilayers), surface potential, and morphology. Others relate to the stability of liposomes in terms of either changes in vesicle size and other general characteristics on storage, or ability of vesicles to retain entrapped drugs in a given medium (e.g., buffer, biological fluids). Finally, it is often necessary to determine the extent, if any, of component modification during the production of liposomes or storage, especially in the case of unsaturated phospholipids that can be easily peroxidised or degraded.

### Overall Structure of Liposomes

Vesicle size can be determined either by hydrodynamic methods (53) (e.g., gel exclusion chromatography using Sepharose 4B or 2B, Sephacryl S500, and S1000) or by photon correlation spectroscopy (54–56) using instrumentation that is appropriate for the size range of the vesicles. Values are expressed as z average mean in nanometers or micrometers. A polydispersity index provides information on the range of vesicle size distribution. Size measurements can also be made in a z sizer which, in addition, also determines the z-potential (zP) of vesicles. zP values range from negative to positive and reflect the surface charge of the vesicles. The latter is particularly useful in formulations where the drug incorporated in liposomes is charged and thus able to modify the overall surface charge. The lamellarity of liposomes is determined by small-angle X-ray scattering (57,58) which can reveal whether bilayers are arranged nonconcentrically (multivesicular structures). Morphological observations of liposomes can be carried out by light microscopy (53) and electron microscopy (EM) (58) under a variety of conditions. For instance, bright field and phase contrast microscopy are used extensively, especially in conjunction with fluorescent dyes. However,

light microscopy is only suitable for large (e.g., LUV and MLV) and giant liposomes (53). With transmission electron microscopy, liposome samples in suspension are firstly negatively stained (e.g., with $OsO_4$, $UO^{2+}$) and then directly examined (58). The technique can reveal unilamellar, oligolamellar, or multilamellar structures. Alternatively, a three-dimensional image of liposomes is obtained by freeze-fracture electron microscopy (FFEM) where samples of liposome suspensions are quickly frozen, fractured at very low temperature (e.g., $-150°C$) and immediately replicated by appropriate means (59). Replicas are then examined by EM. In some cases, to facilitate identification of liposomes (e.g., in tissues where local structures can be mistaken for liposomes) these are labelled with gold particles which can be either entrapped within the vesicles or attached onto the vesicle surface using immunogold labeling (58). This, however, requires the presence of a protein on the liposomal surface. In cryoelectron microscopy a thin aqueous film on a specimen grid is formed by dipping the grid in and withdrawing it from the liposomal suspension (58). The film is rapidly vitrified, cooled to its melting point, and then examined by EM. As with FFEM, confirmation of vesicle images with entrapped gold particles or immunogold labeling is often necessary when biological samples are examined.

Other methods for the characterization of liposomes include spectroscopy, which can be used to measure the turbidity of a liposomal suspension (52). Sets of calibration curves are then employed to devise semiquantitative estimates of vesicle size or concentration. On the other hand, the structure and dynamics of the acyl chains of liposomal phospholipids can be determined by infrared or Raman spectroscopy, and the hydration of phospholipid polar heads (and their interaction with counterions) by Fourier transformation methods (60). Liposomal parameters such a liquid-crystalline transition temperature ($T_c$), or diffusion constants within bilayers can be measured by NMR and electron paramagnetic resonance (EPR) (60). In addition, electron resonance techniques can provide information on vesicle size and number of lamellae (60), and EPR spin labeling coupled with the addition of paramagnetic ions is a particularly useful method for the estimation of vesicle internal volumes. Finally, thermodynamic methods are useful in the study of $T_c$ energetics (DSC), or liposome dissolution heat (isothermal calorimetry) (60).

### Liposomal Stability

Changes in the structural characteristics of liposomes on storage can be monitored by one or more of the techniques already mentioned. These techniques, however, cannot detect subtle structural changes (e.g., vesicle porosity) that lead to the leakage of drugs entrapped in the aqueous phase of liposomes. Drug retention by liposomes on storage or in the presence of a biological fluid can be determined by a variety of techniques (60,61).

It is always appropriate that in designing the lipid composition of liposomes for the entrapment of a drug within their aqueous phase, the stability of the bilayers within a given milieu is monitored. A typical example is the effect of blood components on the ability of a liposomal formulation to retain entrapped solutes (61). To that end, one of the easiest and most effective methods is the use of calcein or carboxyfluorescein as marker solutes. Calcein and carboxyfluorescen self-quench at high concentrations (e.g., above 0.05 M) but acquire a green fluorescent colour on excessive dilution. Suspended liposomes containing quenched dye appear as brown or orange. However, destabilization of the vesicles leads to the leakage of some of the dye and its excessive dilution into the medium. Measurement of the resulting fluorescence by spectrofluorimetry reveals the extent of leakage. For liposomal formulations containing drugs, drug leakage on storage or in the biological milieu can be monitored by techniques that separate entrapped from leaked drug. This can be achieved by ultracentrifugation, dialysis, molecular size chromatography, or ion exchange chromatography.

## FATE OF LIPOSOMES IN VIVO AND ITS CONTROL

Following work in the early 1970s (2–6,62–67) with drug-containing MLV, several basic aspects of their behavior in vivo became apparent. For instance, rate of clearance of vesicles from the blood of intravenously injected rats was rapid, dose-dependent, and biphasic. It was subsequently observed that neutral MLV (68) and SUV (69) exhibit a slower rate of clearance than negatively and (more recently) positively charged MLV, and that SUV had a longer residence time than MLV (69). Another observation was that small, water-soluble drugs (e.g., 5-fluorouracil and penicillin G) leaked considerably from circulating MLV (6). In view of what we know today, this is hardly surprising as the MLV used contained little cholesterol. On the other hand, larger entrapped solutes such as albumin and amyloglycosidase did not appear to leak (3,4,68). Finally and perhaps more importantly, liposomes with entrapped materials were shown to end up in the fixed macrophages of the RES, mainly in the liver and spleen, through the lysosomotropic pathway (4,66,67). Within lysosomes, vesicles were seen (67) to lose their organized structure (presumably through the action of phospholipases) and to release their contents. Depending on their molecular size and ability to withstand the hostile milieu of the organelles, drugs could then act either locally (e.g., hydrolysis of stored sucrose by liposomal fructofuranosidase) (62) or, after diffusion through the lysosomal membrane, in other cell compartments (e.g., inhibition of DNA-directed RNA synthesis by liposomal actinomycin D in partially hepatectomized rats) (66). Such understanding of liposomal fate and behaviour led to a number of proposed applications, including the treatment of certain inherited metabolic disorders (4,62,70), intracellular infections (6,9), cancer (6,71), gene therapy (4), and because of antigen-presenting cell involvement in vesicle uptake in vivo, immunopotentiation and vaccine delivery (72).

The demonstration in the early 1970s that liposomes coated with cell-specific ligands (e.g., antibodies and asialoglycoproteins), can interact with, and be internalized by, alternative (other than macrophages) cells expressing appropriate receptors, both in vitro (73) and in vivo (73,74) culminated in the concept (73) of vesicle targeting and to novel avenues of liposome research and potential use. Perhaps more encouraging were indications that even ligand-free liposomes could enhance the delivery of drugs to solid tumors in vivo (74). It was soon realized (63), however, that success in this area would require not only quantitative retention of entrapped drugs by the vesicles (also needed for "passive" targeting to fixed macrophages), but also a sufficient time period for circulating (targeted, ligand-bearing) vesicles to encounter and associate with the target. Because vesicle behavior in vivo, resulting from a given structural feature of the system, must be closely related to the particular environment in which that system exists, knowledge of the influence of the biological milieu on both drug leakage and vesicle clearance rates appeared essential.

### Retention of Drugs by Liposomes

Initial indications as to why liposomes might become permeable to entrapped drugs, or destabilized in the presence of blood, thereby allowing drugs to leak out, came from the finding (75,76) that plasma high density lipoproteins (HDL) remove phospholipid molecules from the vesicle bilayer. It appeared logical to assume (77) that prevention of such HDL action, for instance by increasing the packing of the lipid components in the bilayer, would enable the vesicles to remain intact. Experiments to test this assumption were thus carried out, initially using MLV (77) and later with SUV (78). In both cases, additional cholesterol (e.g., a phospholipid: cholesterol molar ratio of 1:1) increased the bilayer stability in the presence of blood components (77–80) with very little loss of phospholipid or of an entrapped water-soluble marker, either in vitro (78) or after intravenous (78,79), intraperitoneal (78), or subcutaneous (81) injection. A role for HDL in the destabilization of liposomes was further substantiated by (1) the demonstration (82) that cholesterol-free SUV exposed to blood plasma from lipoprotein-deficient mice or a patient with congenital lecithin-cholesterol acyltransferase deficiency (characterized by low HDL levels) were more stable than when incubated with plasma from normal mice or humans; (2) the identification (82) of HDL as the only lipoprotein, among a number of lipoprotein species added to lipoprotein-deficient plasma, that could restore vesicle-destabilizing activity.

It was thus of interest to see whether reduction of vesicle leakage to solutes by cholesterol-induced bilayer packing could also be achieved with bilayers made rigid by substituting egg PC with high-melting phospholipids. As anticipated (77,80), SUV or MLV made of hydrogenated lecithin, distearyl phosphatidylcholine (DSPC) or sphingomyelin (SM) were less permeable to entrapped solutes in the presence of blood plasma, with bilayer stability becoming even greater when additional cholesterol was present (81–86). Thus, liposomes composed of equimolar DSPC and cholesterol were able to completely retain their marker solute (carboxyfluorescein) content, even after 48 hours of exposure to plasma at 37°C (84).

### Vesicle Clearance Rates: The Role of Lipid Composition

In the course of work on liposomal stability it was observed (83) that solute retention by cholesterol-rich SUV, al-

range of MLV, which predisposes them to removal from the RES, has been exploited to deliver various therapeutic agents to these cells, e.g., to replace genetically deficient enzymes (70) and to target antimicrobial agents to intracellular pathogens (9). By the same token, the small-size of SUV makes them more able to extravasate from leaky vasculature (95). However, passive targeting is limited in its scope, and tremendous effort has been directed toward the development of more universal levels of control by associating molecules with recognition properties with the liposome surface. The essential requirements (95) of this *active* targeting approach are summarized as follows:

- Characteristics of the target cell population must be identified which are qualitatively and/or quantitatively different from other cells that might be encountered by the delivery vehicle, and which are appropriate as recognition features for a targeting mechanism. Typical examples are antigenic determinants, including various types of membrane receptor, which are unique or overexpressed (by virtue of the aggressive growth rate of the cell).

- It must be possible to generate targeting ligands that are able to interact with some degree of exclusivity with these epitopes. Examples include antibodies (either intact or as active fragments), other proteins and peptides, lipoproteins, glycoproteins, and growth factors. Specific examples will be given later.

- The ligand must be capable of being attached stably and accessibly to the drug carrier so that interaction with the target epitope brings the liposome into close proximity with the cell surface.

- The liposome–ligand construct must be stable within the biological milieu and present for sufficient time to maximize the chances of it interacting with the target cell population.

- Binding should increase the chances of the drug reaching the cell interior, for example, by receptor-mediated endocytosis or membrane fusion events or by local diffusion across the cell membrane. Also, once internalized, the drug must retain its stability (e.g., against damage from lysosomal processing) and be available to exert its intended action.

**Ligand Coupling Strategies.** The general strategy of this approach is exemplified by the developments achieved with antibodies and what are now known as immunoliposomes. The earliest report of antibody-mediated liposome targeting appeared 24 years ago (73). In that first report, liposomes containing a radiolabeled antitumor drug, bleomycin, were associated with a series of polyclonal antibodies and then incubated with a range of cells against which the various antibodies had been raised. In all cases, radioactivity was higher in those cells incubated with the corresponding antibody than in the same cells incubated with the other, nonspecific antibodies, thus confirming an in vitro targeting effect. In the same study, asialofetuin, which is known to bind to galactose receptors on the liver parenchymal cells, was associated with doubly labeled liposomes and then administered to rats, some of which had

first been injected with free asialofetuin. The latter group of animals showed reduced association of the two radiolabels with the liver, thereby providing the first evidence of targeting of liposomes in vivo. A little later (74), in vivo targeting of liposomes was also demonstrated by the use of tumor-specific antibodies.

Many different coupling strategies have been used to associate the targeting ligands and the liposomes. The aforementioned study used a relatively crude approach whereby the two components were sonicated together, presumably enabling hydrophobic portions of the proteins to insert into the perturbed bilayers, and a similar approach by another group confirmed the targeting ability of the resulting vesicles (99). Others have used targeting moieties loosely bound to the liposome surface. Thus a hapten-specific antibody was used to bring about interaction between liposomes and cells, both of which were coated with the hapten (100), whereas in a separate study, liposomes opsonized with antibody were targeted to tumor cells resulting in their internalization by receptor-mediated endocytosis (101). The majority of the subsequent studies have used chemical coupling methods to link a targeting moiety to an amphiphilic molecule inserted into the liposome bilayer. Early approaches included immunoglobulin G (IgG) cross-linked to liposomal PE using glutaraldehyde (102), or to aldehyde groups formed by periodate-oxidation of liposomal glycosphingolipids (103). Another early method which became very popular involved covalent coupling of fatty acid chains directly to IgG using cross-linking agents, and then incorporation of the resultant molecules into liposomes by detergent dialysis (104). However, by far the most widely used approach to date involves the use of heterobifunctional cross-linking reagents for introducing thiol-based linkages between the lipid anchor and the ligand, thereby minimizing the chances of formation of homopolymers.

In the original method, thiol-reactive derivatives were prepared, both of the antibody (via free amino groups) and the PE anchor; the latter was incorporated into liposomes, and then free thiol groups were formed by reduction. Interaction of the two species resulted in coupling of the immunoglobulin molecule to the liposome surface via a thiol linkage (86,105). An alternative approach is to use the thiol groups generated when a F(ab')$_2$ fragment of IgG is reduced to form two Fab' fragments (106). A particular drawback with the thiol linkage is its limited stability in blood, and an important improvement was to replace this with the much more stable thiol–ether linkage (107). This is achieved by using maleimido-based thiol reagents to derivatize either the lipid anchor or the ligand so that subsequent interaction produces a stable thiol–ether bond. The same approach can of course also use Fab' fragments as already discussed (106). Over the years, many modifications and improvements have been made to this technique and are now widely used in many different targeting applications.

An alternative to the strategy of coupling the ligand directly to the liposome surface has been to make use of the high affinity avidin–biotin interaction (108). Avidin and streptavidin are glycoproteins with molecular weights of 67 and 60 kDa, respectively, either of which can tightly bind four molecules of low molecular weight biotin, and the components can be configured in different ways to achieve ligand binding. Biotinylation of the bilayer anchoring molecule and the ligand can be readily achieved via free amino groups and then bridged together by addition of (strept)avidin to form a tightly bound complex. Alternatively, (strept)avidin can be covalently bound to the bilayer using one of the aforementioned procedures and then complexed to a biotinylated ligand. Either approach provides a flexible way of enabling the separate components to be prepared and stored in advance and then put together on a "mix and match" basis as required.

Progress in the development of ligand-based liposome targeting has been held up by numerous obstacles that can affect various stages of the targeting process. During construction of the carrier, the ligand may be damaged or lose activity as a result of the conjugation process. Formation of IgG fragments may result in reduced affinity. On the other hand, binding of intact antibody has been observed to result in rapid clearance of the construct, probably as a result of the affinity of the Fc portion of the antibody for the Fc receptors in the liver. Moreover, chronic use of antibody-coated liposomes may lead to immune responses against the (foreign) proteins. This can be approached by preparing chimeric or else wholly humanized antibodies. Once in the bloodstream, exposure to blood components may lead to cleavage of the ligand from the carrier or else to drug leakage. On the other hand, if the liposome is too stable, it might not release its contents once its destination has been reached. Early attempts to combine the immunoliposome and Stealth approaches were problematic because the PEG chains obstructed access of antibody to the liposome surface during the conjugation procedure and to the cell surface epitope during the targeting procedure. This was eventually resolved both by using an avidin–biotin approach similar to that already described, or by conjugating the antibody to the terminal region of the PEG moiety.

At the level of the target cell, it has proved difficult to identify truly tumor-specific markers, in addition to which there may be considerable surface heterogeneity between different cells in a given tumor cell population. Moreover, it was demonstrated quite early that the simple fact of a carrier binding to its target epitope does not in any way guarantee its internalization (109), and other mechanisms for drug entry into the cell (e.g., passive diffusion or fusion) may well be ineffective. Different antibodies may themselves vary in their ability to be internalized. Reliance on overexpression of ubiquitous antigens known to internalize (e.g., transferrin, folate, and low density lipoproteins) is still accompanied by nonspecific binding to normal cells expressing these receptors. However, despite the gloomy picture that this may present, sound progress has been made such that positive results are now being achieved in vivo. These is discussed in more detail in the section "Applications for Liposomes".

## Polymerized Liposomes

The concept of liposome polymerization has been around for about 20 years but may still be some way from becoming a viable proposition. For a recent review, see Bonte et al. (110). The basic premise is that polymerizable amphiphiles can be used to form lipid vesicles and then made to polymerize using conventional triggers. The ideal outcome is a liposome of high but controllable stability with fluid, liquid crystalline properties. A variety of polymerizable phospholipids have been synthesised, with the reactive groups in different regions of the molecule as shown in Table 2. Thus, the polymer backbone can be in the head group region, through the midchain hydrocarbon region, or through the end of the hydrocarbon chains.

Polymerized liposomes are very stable in the presence of blood components and have been widely studied as potential drug delivery vehicles, where they have been seen as a possible solution to the instability of natural phospholipid liposomes. Their use in this respect would require that they be only partially polymerized to permit eventual drug release when required. To achieve this goal, windows of normal or trigger-responsive lipids would need to be included. In fact, by using mixtures of normal and polymerizable lipids, different domains can be formed, the size and distribution of which are dependent on the preparation conditions. This was well demonstrated in a study using a mixture of polymerizable octadecadienoyl phosphatidylcholine and nonpolymerizable dipalmitoyl phosphatidylcholine, where treatment of the resultant liposomes with solvents removed the normal lipids to leave "skeletonized" porous liposomes which could be observed by scanning electron microscopy. However, various technical hurdles must be overcome before polymerized liposomes are ready for use as drug delivery vehicles, and concerns remain about the possibility of toxic residues. Inert fragments of polymer backbone are known to accumulate in the body and biodegradable amphiphiles have been proposed as a way of overcoming this. Of the limited in vivo studies that have been carried out, one showed rapid clearance of methacrylate-derived liposomes from the blood of injected mice, by the liver and spleen.

A related potential application is the use of polymerized liposomes as an artificial blood substitute for transporting hemoglobin or synthetic oxygen carriers around the body. For example, synthetic hemoglobin embedded by copolymerization into the bilayer of imadazole dienoylphospholipid liposomes formed oxygen carriers which were stable for several months and were similar to human blood in terms of their oxygen-carrying capacity and binding affinity. Polymerized liposomes have also been used for studying membrane dynamics during various processes of cell biology. These include molecular recognition phenomena such as occur between membrane receptors and lectins, between cells and the extracellular matrix, and between phospholipase and other enzymes with the lipid bilayer. In addition, they have provided valuable information regarding the localized membrane changes that occur during processes such as endo- and exocytosis and membrane budding.

## pH- and Temperature-Sensitive Liposomes

These modifications were developed in order to increase the selectivity of drug action in the body via controlled recoated pits and be taken up by endocytosis. In addition their greater buffering capacity can give increased protection against low intralysosomal pH. In a related approach, protamines, which are natural polyamine components of sperm, have been included together with DC-Chol/DOPE, providing effective transfection of Chinese hamster ovary (CHO) cells (121). These natural cationic peptides may also have nuclear localization signals that might help them to gain entry into the nucleus. A similar system using 1,2-dioleoyloxy-3-(trimethylammonium) propane (DOTAP) liposomes together with protamine was injected intravenously into mice, showing expression of the luciferase gene in all organs, particularly in the lung (122).

## APPLICATIONS FOR LIPOSOMES

### Cancer

This application has been widely reviewed in recent publications (5,95,123–126). Possibilities for using liposomes for the treatment of cancer have been approached from several different directions. Chemotherapeutic agents, including conventional low molecular weight cytotoxic drugs as well as peptides and proteins, have been administered in undirected and in ligand-targeted liposomes. Cytokines and other immunomodulators have been encapsulated in MLV and used to activate macrophages and make them tumoricidal. Genetic material, including DNA, antisense molecules, and ribosomes, has been presented in liposomes to bring about different antitumor effects. Many of these approaches have given promising results and two formulations based on cytotoxic anthracyclines are now in commercial use.

A variety of cytotoxic drugs have been administered in liposomes via various routes to a range of animal tumor models (95,123–126). Much of this work has focused on the anthracyclines. Although for certain cytotoxic drugs encapsulation into liposomes was found to cause increased toxicity, early studies indicated that for others, including the anthracyclines, toxicity was reduced. This appears to be due to the fact that rapid uptake and sequestration of liposomes by the macrophages of the RES reduces the peak plasma levels of the drug and their associated toxicity, particularly to the myocardium. The macrophage-associated drug is ultimately released back into the circulation to produce an effect akin to continuous infusion. However, although the therapeutic benefits of conventional infusion are well known, the significance of such an effect in the context of liposomal delivery remains controversial, and certainly in some cases reduced toxicity is accompanied by reduced efficacy.

One effect that is clearly beneficial is the fact that many rapidly growing tumors have poorly formed, leaky vasculature through which liposomes of appropriate diameter can pass, and because of ineffective lymphatic drainage, will then accumulate. The longer the liposomes are able to circulate in the blood, the greater the level of accumulation. This effect is the basis of three different liposomal anthracycline products (Table 3) which have given successful results in clinical trials, two of which are on the market. Of these, TLC℠ D-99 (The Liposome Company),

which is still in Phase III, comprises doxorubicin actively loaded into egg PC–cholesterol liposomes of around 150 nm in diameter using a pH gradient (5,126). Comparative studies of dogs treated with TLC D-99 or with the same amount of free drug showed retention and occasionally increased antitumor activity in the former group, without the histological evidence of myocardial toxicity that was seen with the free drug. Several Phase III trials are ongoing in U.S. and Europe to evaluate TLC D-99 alone and in combination therapy in metastatic breast cancer.

Daunosome℠ (NeXstar, Inc.) consists of daunorubicin incorporated by a remote loading technique into 40–50 nm-diameter SUV composed of DSPC and cholesterol and is supplied as a ready to use aqueous liposome suspension (5,127) (see Table 3). This composition forms a particularly rigid bilayer which inhibits opsonization by the plasma lipoproteins, and this together with their small size results in an extended circulation time of 24 hours in the blood. It is also claimed that a substantial portion of the extravasated liposomes are taken up intact by the tumor, where they then release their contents over a prolonged period of time to produce sustained drug action within the tumor. The product has so far been approved for first-line therapy of AIDS-related Kaposi's sarcoma in the U.S. and other countries. Promising results against a variety of other tumors have led to Phase II clinical trials, which are currently underway to look at its utility for treating lymphoma, leukemia, and myeloma, as well as solid tumors in breast, ovary, and brain (127).

The third product, Doxil℠ (Sequus Pharmaceuticals) (see Table 3) is based on doxorubicin, which is again incorporated by active loading, this time into sterically stabilized liposomes comprising hydrogenated soy PC, PEG-DSPE (distearyl phosphatidylethanolamine) and cholesterol, and having an approximate diameter of 85 nm (5,95,128). The drug is in the form of a gel-like precipitate which enables high concentrations to be encapsulated and retained within the liposomes. The slightly increased size compared with DaunoXome (and hence increased loading), is achieved by virtue of the fact that sterically stabilized liposomes are less size dependent with respect to their clearance rate. Thus, circulation times of 15 to 30 hours in rodents and dogs and 40 to 60 hours in humans can be achieved with consequently increased accumulation in the vicinity of the tumor. The steric stabilization also results in decreased contact with other tissues such as the myocardium while circulating in the blood, with the drug remaining sequestered inside the liposomes during this time. Once localized in the tumor tissue, various mechanisms have been proposed to account for liposome breakdown and drug release, including low pH, the action of lipases and other released enzymes, and the consequences of engulfment by tumor-resident macrophages. Once released, doxorubicin is sufficiently lipid soluble to diffuse into the tumor cells, leading to their destruction (128). Doxil was the first liposomal drug approved by the FDA and has been on the market for treatment of AIDS-related Kaposi's sarcoma since 1995. It has also shown very promising results in a range of solid tumors including ovarian and breast carcinoma for which advanced clinical trials are underway (128). Sequus Pharmaceuticals is also developing cisplatin

**Table 3. Liposome-Based Products Developed or Under Development (UD)**

| Product | Drug | Target disease |
|---|---|---|
| *Aronex Pharmaceuticals, The Woodlands, TX* | | |
| Liposomal nystatin (trademark not yet given; intravenous) (UD) | Nystatin | Systemic fungal infections |
| Liposomal tretinoin (trademark not yet given; intravenous) (UD) | All-trans retinoic acid | Leukemia |
| Liposomal annamycin (trademark not yet given; intravenous) (UD) | Annamycin | Kaposi's sarcoma Retractory breast cancer |
| *IGI, Vineland Laboratories, Vineland, NJ* | | |
| Newcastle disease vaccine (intramuscular) (Novasomes) | Newcastle disease virus (killed) | Newcastle disease (chicken) |
| Avian Rheovirus vaccine (intramuscular) (Novasomes) | Avian rheovirus (killed) | For vaccination of breeder chickens; for passive protection of chicks against rheovirus infection |
| *NeXstar Pharmaceuticals (formerly Vestar), Boulder, CO* | | |
| AmBisome (intravenous) | Amphotericin B | Systemic fungal infections; visceral leishmaniasis |
| DaunoXome (intravenous) | Daunorubicin | First line treatment for advanced Kaposi's sarcoma. Breast cancer and other solid tumors |
| MiKasome (intravenous) (UD) | Amikacin | Serious bacterial infections |
| VincaXome (intravenous) (UD) | Vincristine | Solid tumors |
| *Novavax, Rockville, MD* | | |
| *E. coli* 0157:H7 vaccine (oral) (Novasomes) (UD) | *E. coli* 0157:H7 (killed) | *E. coli* 0157 infection |
| Shigella flexneri 2A vaccine (oral) (UD) | S. flexneri 2A (killed) | S. flexneri 2A infection |
| *SEQUUS Pharmaceuticals (formerly Liposome Technology), Menlo Park, CA* | | |
| Doxil (intravenous) | Doxorubicin | Kaposi's sarcoma. Refractory ovarian, recurrent breast, prostate and primary liver cancers |
| Amphotec™ (intravenous) | Amphotericin B | Systemic fungal infections |
| *The Liposome Company, Princeton, NJ* | | |
| D99 (intravenous) (UD) | Doxorubicin | Metastatic breast cancer |
| ABELCET™ (intravenous) (Ribbons) | Amphotericin B | Systemic fungal infections |
| C53 (intravenous) (UD) | Prostaglandin $E_1$ | Systemic inflammatory diseases |
| *Swiss Serum and Vaccine Institute, Berne, Switzerland* | | |
| Epaxal-Berna vaccine (intramuscular) (IRIV[a] liposomes) | Inactivated hepatitis A virions (HAV) (antigen; RG-SB strain) | Hepatitis A |
| Trivalent influenza vaccine (intramuscular) (UD) | Hemagglutinin and neuraminidase from $H_1N_1$, $H_3N_2$, and B strains according to recommendations by WHO | Influenza |
| HAV/HB_s-IRIV combined[b] vaccine (intramuscular) (UD) | HAV, genetically engineered hepatitis B antigens (HBs) | Hepatitis A and B |
| Diphtheria/tetanus/hepatitis A combined vaccine (intramuscular) (UD) | Diphtheria and $\alpha$ and $\beta$ tetanus toxoids; inactivated HAV virions | Diphtheria, tetanus, hepatitis A |
| Hepatitis A and B/diphtheria/tetanus/influenza supercombined vaccine[a] (intramuscular) (UD) | Inactivated HAV virons, HB_s, diphtheria, and $\alpha$ and $\beta$ tetanus toxoids, hemagglutinin and neuraminidase from influenza strains as in trivalent influenza vaccine | Hepatitis A and B, diphtheria, tetanus, and influenza |

[a]IRIV, immunopotentiating reconstituted influenza virosomes.
[b]Combined vaccines denote the presence of up to three antigens in the formulation.
[c]Supercombined vaccines denote the presence of more than three antigens.

in sterically stabilized liposomes, which may be effective in combination therapy with Doxil. Other companies are incorporating different cytotoxic drugs into liposomes, some of which are currently in Phase II clinical trials (5).

The beneficial effects of steric stabilization of liposomes using PEG are not completely clear cut. The accumulation efficiencies of long circulating (by virtue of their rigid bilayers) liposomes by a series of mouse tumors were investigated in the absence and presence of a surface coating of PEG. By comparing the area under the curve (AUC) for plasma drug levels with that of the tumor, it appeared that PEGylation actually reduced the proportion of drug that accumulated in the tumor (129). This occurred despite the fact that the sterically stabilized liposomes circulated for longer in the blood than did the corresponding conventional ones. Steric stabilization also tends to mediate against uptake of liposomes by the tumor cells in the same way that it prevents interaction with other cells. Given these facts, the conflict between the need to design liposomes which fully retain their drug content while circulating in the blood and the need to make this drug available to tumor cells after extravasation remains a difficult issue (129). Its resolution may ultimately depend on controlled release strategies such as the use of pH-sensitive liposomes that respond to reduced pH within the tumor microenvironment or of the combination of temperature-sensitive liposomes and localized hypothermia. Exciting possibilities have also been raised by the recent development of liposomes that can be made highly fusogenic by loss of PEG groups, which potentially enable them to fuse with adjacent cells (130).

The use of ligand-mediated targeting to increase the specificity of drug transfer to tumor cells is still at an early stage, but since the emergence of long-circulating liposomes, promising results in vivo are starting to appear. Early studies (131) reported substantial therapeutic effects when PEG-stabilized or $GM_1$-stabilized immunoliposomes, containing doxorubicin and a lipophilic prodrug, respectively, were used to treat newly established lung cancers in mice. In the former case, attempts to treat more advanced tumors were unsuccessful. Effective treatment involving receptor-mediated internalization was also achieved using immunoliposomes targeted against malignant cells within the vasculature (132). More recently a comprehensive series of studies has been carried out (133) using sterically stabilized, doxorubicin-containing liposomes targeted against cancers overexpressing the HER2/neu proto-oncogene. HER2 is an internalizing receptor, highly expressed in a wide range of tumors where it plays an important role in aggressive growth and so is unlikely to be down-regulated. It also has other important advantages as a target in that it is a readily accessible cell surface protein with infrequent expression in a limited range of normal cells. A highly reactive antibody is available which has been fully humanized to reduce the chances of immunogenicity, and in these studies was generally conjugated as the Fab' fragment to the terminal region of liposome-grafted PEG chains. After in vitro optimization, the system was tested against subcutaneous xenografts of different tumors showing various levels of HER2 expression. Interestingly, similar levels of gross uptake were seen in tumors with high and very low expression levels, whereas the antitumor effect was substantially higher in the former (134). This correlated with electron microscopic evidence using colloidal gold-labeled liposomes that in nontargeted systems (low HER2 expression or antibody-free liposomes), the label was concentrated in tumor-resident macrophages and in intercellular spaces, whereas in positively targeted systems it was present both between and *within* the tumor cells (133). This latter distribution presumably accounted for the superior antitumor effects of the targeted systems, whereas the observed gross similarities in tumor distribution using targeted and nontargeted systems agrees with earlier indications that extravasation is the rate-determining step (133).

A completely different strategy for anticancer therapy is to use liposomes to deliver activating factors, such as cytokines to macrophages, so that they become tumoricidal (135). Because the total number of available macrophages in the body is limited, the major indication is for the treatment of metastatic disease after surgical removal of the primary tumor. Activated macrophages ignore normal cells but selectively kill tumor cells, regardless of the numerous levels of biological heterogeneity that normally make metastases so difficult to treat. A widely used method of activation is to expose the macrophages to muramyl dipeptide (MDP), the minimal structural unit of Mycobacterium cell wall. By incorporating lipid soluble derivatives of MDP into MLV enriched with phosphatidylserine, these are preferentially taken up after intravenous injection by macrophages residing in the lung, which is one of the major sites of metastatic disease. Promising studies in mice were followed by successful treatment of dogs after surgical removal of osteosarcoma (135). The approach has progressed through Phase I and Phase II trials, where it was evaluated in combination with chemotherapy and is currently in Phase III for newly diagnosed osteosarcoma patients, where it is hoped that the treatment will destroy residual tumor cells that were not eliminated by chemotherapy. Adults and children have been treated with minimal side effects, and an added attraction is that activation of alveolar macrophages can be achieved by oral administration of liposomal MDP.

## Antimicrobial Agents

Liposomes have been widely investigated as delivery systems for treatment of bacterial, fungal, viral, and parasitic diseases, and the topic has been reviewed in recent publications (5,135–137). Therapeutic benefits can be brought about in various different ways. The use of liposomes for passive delivery of immunomodulators to macrophages has been discussed already with respect to cancer treatment, but this approach can also be used to bring about nonspecific resistance to infections in general (135). Thus a wide range of infectious diseases have been shown to respond to treatment by macrophages activated by MDP or cytokines such as interferon $\gamma$ (IFN-$\gamma$) associated with liposomes. Phagocytosed microorganisms often take up residence and multiply in the host cells, making them refractory to treatment by antimicrobial agents due to poor penetration or reduced intracellular activity of the drug.

However, this same phagocytic activity can be used to passively deliver particulate drug carriers such as liposomes to the interior of the cell while at the same time masking the toxicity of the drug to the host. Early studies (9) involved treatment of protozoal Leishmania infections using liposome-encapsulated antimonial drugs and produced substantial increases in therapeutic index due both to improved delivery and reduced toxicity. Similar effects have subsequently been shown in a wide range of bacterial, viral, and fungal infections (123,127).

In recent years there has been a dramatic increase in the incidence of potentially fatal, systemic fungal infections, particularly in patients who are immunocompromised as a result of treatment for diseases such as cancer, diabetes, and AIDS; as a result of organ transplantation; or due to various other factors. Amphotericin B is the drug of choice for treating such infections because its spectrum of activity is much larger than any other clinically acceptable drug (137). It is a lipophilic drug whose mode of action is to interact with ergosterol in plant and fungal membranes (the equivalent of cholesterol in mammalian cells), thereby causing membrane disruption and cell death. The drug does interact with cholesterol but to a lesser extent, and this partial specificity can be amplified by administering it in a particulate form so that it is taken up by the macrophages, which may themselves be hosts to the fungi but can also carry the drug to other sites of infection. Appropriate dosage forms can also reduce the rate of clearance from the bloodstream, enabling increased chance of extravasation through leaky vasculature due to inflammation of diseased tissue. While complexed within these formulations, amphotericin B is unavailable to interact with cell membranes in the body. Thus incorporation into liposomes and other lipidic complexes resulted in substantial reductions in toxicity compared with free drug or with the established formulation Fungizone™ (a deoxycholate-containing micellar dispersion), enabling doses up to five times larger than the latter to be given (127). Three such lipid formulations are now available (see Table 3). AmBisome™ (NeXstar Pharmaceuticals, Inc.) (5,127) is the only truly liposomal form, consisting of SUV composed of hydrogenated soy PC, distearoyl phosphatidylglycerol and cholesterol, stored as a lyophilized powder. Molecules of drug are believed to align in parallel to form cylindrical structures, two of which fit end to end to form a bilayer-spanning pore across which ions and other solutes can pass. These pores can be inserted into the fungal cell membrane, allowing rapid loss of cell contents. In vitro studies have shown binding of AmBisome liposomes to fungal cells, followed by liposome disintegration and subsequent cell disruption (127).

Amphotec™ (Sequus Pharmaceuticals) (5,138) is a non-liposomal form of amphotericin B. Instead it is a stable, disk-shaped complex of drug together with cholesterol sulphate, 115 nm in diameter and 4 nm thick. As with the previous formulation, it is stored in a lyophilized form and reconstituted for use. Comparative studies have indicated that its efficacy is approximately similar to that of AmBisome (138). The third formulation, known as ABELCET™ (The Liposome Company), is also not a true liposome, but is an interdigitated complex of lipid and drug, the lipid components comprising dimyristoyl phosphatidylcholine and dimyristoyl phosphatidylglycerol (5,126). Numerous studies at the research level have looked at the use of liposomal drugs to treat bacterial and viral infections and again the main focus has been on intracellular pathogens, particularly in immunocompromised patients. These have included targeting of the HIV virus itself, and of *Mycobacterium* species such as *M. avium*, which causes serious opportunistic infections in AIDS sufferers, and *M. tuberculosis*, which reflects the recent resurgence of multidrug resistant forms of TB. The MiKasome™ system (NeXstar) where the antibiotic amikacin is encapsulated in SUV composed of HSPC, DSPG and cholesterol is currently in Phase I clinical trials (127). Genetic approaches to cancer treatment are discussed in the following section.

## Gene Therapy

Liposome-based gene transfection systems have been promoted as a means of achieving the transfection efficacy of viral constructs without the associated risks. The earliest attempts at liposome-based gene transfer in vivo involved conventional entrapment in negatively charged vesicles, and transfection efficiencies were accordingly low. A plasmid containing the preproinsulin gene was injected intravenously into rats, a transient hypoglycaemic response was noted, and increased levels of insulin were detected in the blood, liver, and spleen (139). Improvements arose when increased insights into the mechanisms of uptake prompted the development of pH-sensitive liposomes (111), which could exploit the cellular endocytic mechanisms, and these were rapidly followed by the concept of cationic lipid delivery systems. However, the reactivity of cationic liposomes within the blood, and their rapid removal by the RES (after opsonization) (88), has meant that their application in vivo has generally been limited to localized delivery by various routes. For example, plasmids complexed with Lipofectin™ (DOTMA/DOPE) were injected into specific areas of the brain of Xenopus embryos to study gene function in the developing embryo, whereas expression in mammals was achieved in later studies after injection into mouse brain (140).

DNA–liposome complexes have been used to stimulate immunogenic responses against malignant tumors in vivo. Initial studies carried out in mice (141) involved direct injection into the tumor of the genes for foreign major histocompatibility complex proteins associated with DC-Chol cationic liposomes. The treatment resulted in complete tumor regression in many of the animals, and tumors remote from the injection site and not expressing the transfected gene product also underwent rejection. Based on these positive results, a human clinical trial was carried out whereby the human *HLA-B*7 gene in the same type of complex was injected into cutaneous tumors in human melanoma patients (142). Antigen expression, with generation of cytotoxic *T*-lymphocytes specific for *HLA-B*7 was seen in the tumors of all 5 patients, and in 1 case, complete regression of the primary tumor occurred.

Site-specific delivery within the vasculature was demonstrated by using a double balloon catheter to isolate a segment of artery, then injecting into this segment a plasmid associated with either DOTMA/DOPE liposomes or a retroviral vector. Transfection was confirmed histochemically and shown to be limited to the treated segment (143). Such selective transfection in vivo might be applicable for treatment of diseases such as atherosclerosis or cancer. A similar approach involving the use of antisense oligonucleotides was carried out for treatment of neointimal hyperplasia, which frequently occurs after balloon angioplasty, with neointima formation being completely inhibited for 2 weeks after treatment.

The lung is an ideal target for in vivo gene therapy because it is a discrete compartment that is directly accessible via the respiratory airways. Though it has been successfully targeted both by the intravenous and intratracheal routes, the latter is undoubtedly the most convenient, and expression of different reporter genes was readily demonstrated, initially by instillation of DOTMA/DOPE-gene complexes, and subsequently using aerosols. A major impetus for these model studies has been the need for a treatment for cystic fibrosis, which is one of the most common lethal hereditary diseases. After initial studies in mice, a placebo-controlled, human clinical trial was carried out using a cystic fibrosis transmembrane conductance regulator (CFTR) plasmid complexed with DC-Chol/DOPE in the form of an aerosol, administered intranasally to CF patients (144). This resulted in a partial restoration of 20% of normal CFTR activity persisting for up to 7 days, and with no evidence of adverse clinical or histological effects. It was concluded that such a result might be an acceptable clinical outcome, though the duration of effect would need to be extended. Similar studies carried out recently, using DOTAP and DC-Chol liposomes, have shown similar clinical results.

Despite the significant progress so far, there is still a long way to go before the ideal vector becomes a reality. Some headway has been made in improving stability within the bloodstream, and Stealth technology might eventually enable ligand-conjugated complexes to be delivered to specific cell types within the body (130). The rate-limiting step of gene transfer from the cell cytoplasm to the nucleus needs to be solved in order to obtain substantially increased expression levels. Other major goals include the achievement of long-term expression and the ability to regulate transcription of newly introduced genes to permit external control over their expression. Mechanisms for achieving these goals are being developed in the field of viral genetics and may eventually be applied in the liposome field.

### Liposomes as Immunological Adjuvants

Immunization against microbial infections and cancer is an attractive alternative to chemotherapy. However, vaccines consisting of attenuated organisms, although efficacious in producing diverse and persistent immune responses by mimicking natural infections usually without the disease, can be potentially unsafe. On the other hand, the extracellular localization of killed virus vaccines and their subsequent phagocytosis by professional antigen presenting cells (APC) or antigen-specific B cells, lead to MHC-II class restricted presentation and to T helper cell and humoral immunity but not to significant cytotoxic T cell (CTL) responses. Moreover, subunit vaccines produced from biological fluids may not be entirely free of infectious agents (145).

New-generation vaccines that are based on recombinantly made subunit and synthetic peptide antigens are usually nonimmunogenic and the need for immunopotentiation is well recognized (146). Although a great variety of structurally unrelated agents (immunological adjuvants) are capable of inducing immune responses to vaccine antigens, most of them are toxic (146). Surprisingly, for over 70 years the only immunological adjuvant licenced for use in humans was, until recently, aluminium salts or hydroxide (alum) (146). Alum, however, is far from ideal, fails to induce cell-mediated immunity and is often unsuitable for a variety of antigens. Twenty years after the discovery of the immunological adjuvant properties of liposomes (72) and a multitude of related animal immunization studies carried out in between (147,148), liposomes as adjuvants have come of age, with the first liposome-based vaccine (Epaxal-Berna; against hepatitis A) (5,149) licensed for use in humans (see Table 3). Epaxal-Berna is based on virosomes produced from the influenza virus (IRIV; immunopotentiating reconstituted influenza virosomes) (149). Vaccines based on novasomes (nonphospholipid biodegradable paucilamellar vesicles formed from single chain amphiphiles with or without other lipids) have also been licensed for the immunization of fowl against Newcastle virus disease and avian rheovirus (5). Other novasome-based (oral) vaccines against bacterial infections are under development (see Table 3) (5).

The way in which liposomes induce immune responses to antigens associated with them is not clear but has been attributed to a depot (slow antigen release) mechanism and the ability of vesicles and antigen content to migrate to regional lymph nodes after local injection (or, when given orally to be endocytozed by M cells which then provide antigen to the lymphoid cells in the Peyer's patches). In the case of liposomes, further improvement of adjuvanticity has been achieved by the use of coadjuvants such as lipopolysaccharide, positively charged lipids, and interleukin-2, and also by ligand-mediated targeting to antigen-presenting cells (147).

The approach adopted for the IRIV vaccines is of particular interest as it has been claimed (149) to combine several components that are known to contribute to immunostimulation and are, at the same time, harmless (e.g., egg PC). Thus, PE serves as a binding site for the hepatitis A virions (HAV). HAV attachment to host cells occurs by binding to the PE regions of their membrane, an appropriate feature of PE for Epaxal-Berna. It is also known to directly stimulate (in the form of liposomes) B cells and to generate antibodies without T-cell help. The inclusion of the influenza virus envelope phospholipids contributes to vesicle stability. Moreover, a key role is played by the haemagglutinin (HA) component: HA (the major antigen of the influenza virus) through its $HA_1$ epitope containing sialic acid, is expected to mediate IRIV binding to its receptors on macrophages and other antigen-presenting cells and subsequent receptor-mediated endocytosis, thus initiating a successful immune response. On the other hand, the HA$_2$ subunit of HA causes fusion of viral and endosomal membranes. The essential feature of IRIVs is that, in addition to the antigen they carry, they are also endowed with the fusion-inducing HA component that, presumably, mediates rapid release of the antigen into the target cell membranes. This proposed (149) mechanism for IRIV adjuvanticity is thought to account for the results obtained with the hepatitis A vaccine and the promising outlook of the trivalent influenza and "combined" or "supercombined" vaccines.

**Genetic Immunization**

A novel and exciting concept (150) developed recently, namely de novo production of the required vaccine antigen by the host's cells in vivo, promises to revolutionize vaccination, especially where vaccines are either ineffective or unavailable. The concept entails the direct injection of antigen-encoding plasmid DNA which, following its uptake by cells, finds its way to the nucleus where it transfects the cells episomally. Antigen so produced is recognized by the host as foreign and is then subjected to pathways similar to those undergone by the antigens of internalized viruses (but without their disadvantages) leading to protective humoral and cell-mediated immunity.

It was initially observed (151) that intramuscular injection of DNA vaccines leads to such types of immunity as CTL response. This was surprising because antigen presentation requires the function of professional APCs. However, myocytes which were shown to take up the plasmid, albeit only to a small extent and with only a fraction of cells participating in the uptake, are not professional APCs. Although myocytes carry MHC class I molecules and can present endogenously produced viral peptides to the CD8$^+$ cells to induce CTLs, they do so inefficiently as they lack vital costimulatory molecules such as the B7-1 molecule (151). It has thus been difficult to accept that antigen presentation leading to a CTL response occurs via myocytes. Instead, it was reported that CTL responses are, at least in part, the result of transfer of antigenic material between the muscle cells and professional APC. It is also likely that plasmid secreted by the myocytes or as such is taken up directly by APC (including dendritic cells) infiltrating the injection site. Moreover, the extent of DNA degradation by extracellular deoxyribonucleases is unknown but, depending on the time of its residence interstitially, degradation could be considerable. It follows that approaches to protect DNA from the extracellular biological milieu, introduce it into cells more efficiently, or target it to immunologically relevant cells, should contribute to optimal DNA vaccine design (152). One such approach is the use of liposomes.

The interaction of cationic SUV with DNA to form a condensed lipoplex has been discussed in an earlier section ("Cationic Liposomes"). The pathway to transfection is thought to commence with the binding of the positively charged complex to the negatively charged cell membranes followed by endocytosis. This leads to destabilization of the endosomal membrane whereupon, through lateral diffusion of anionic lipids from the cytoplasm-facing endosomal

monolayer, DNA is thought to be displaced from the complex and released into the cytosol (153). Intravenous application of the complexes has met with only modest success. Contributing causes include the formation of large aggregates between complexes and anionic plasma proteins leading to the neutralization of the cationic charge and rapid removal of the aggregates by the RES, and hence to reduction or even abolition of transfection activity. So far, results from a few reported studies on genetic immunization with SUV–plasmid DNA complexes have been rather disappointing in that immune responses were either similar to or modestly higher than those achieved with naked DNA (154). Therefore, conventional liposome entrapment procedures may offer an alternative approach.

**Liposome-Entrapped DNA.** It was recently proposed (155,156) that, as APC are a preferred alternative to muscle cells as targets for DNA vaccine uptake and expression, liposomes would be a suitable means of delivery of entrapped DNA to such cells. Locally injected liposomes are known (147) to be taken up avidly by APC, either after these have infiltrated into the site of injection or else within the lymphatics, and this event has been implicated in their immunoadjuvant activity. Liposomes would also protect their DNA content from deoxyribonuclease attack (152). Because of the structural versatility of the system, its tranfection efficiency could be further improved by the judicious choice of vesicle surface charge, size and lipid composition, or by the coentrapment of cytokine genes and other adjuvants (e.g., immunostimulatory sequences), together with the plasmid vaccine (152).

It has been shown that a variety of plasmid DNAs can be quantitatively entrapped into the aqueous phase of multilamellar liposomes by the mild dehydration-rehydration procedure. Incorporation values were, as expected, higher (57–90% of the amount used; about 10–500 $\mu$g) when a cationic lipid was present in the bilayers (152). The possibility that DNA was not entrapped within the bilayers of cationic liposomes but was rather complexed with their surface was examined by gel electrophoresis of cationic liposomes entrapping the DNA and of similar liposomes complexed with externally added DNA, in the presence of sodium dodecylsulphate (SDS). Whereas most of the complexed DNA was dissociated from the vesicles, presumably because of its displacement by the anionic SDS, there was very little dissociation of the entrapped DNA (157).

Plasmid-containing liposomes were tested in immunization experiments (155) using a plasmid (pRc/CMV HBS) encoding the S region of the hepatitis B surface antigen (HBsAg; subtype ayw). Figure 10 illustrates how mice (Balb/c) injected repeatedly by the intramuscular route with 5 or 10 $\mu$g plasmid entrapped in cationic liposomes elicited, at all times tested, much greater (up to 100-fold) antibody (IgG$_1$) responses against the encoded antigen than animals immunized with the naked DNA. This was also true for IFN-$\gamma$ and interleukin 4 (IL-4) levels in the spleens of immunized mice (155). In a more recent study, the role of the route of injection of the pRc/CMV HBS plasmid was examined for both humoral and cell-mediated immunity, using Balb/c mice and an outbred strain (T.O.) of the same species. Results (152) comparing responses for

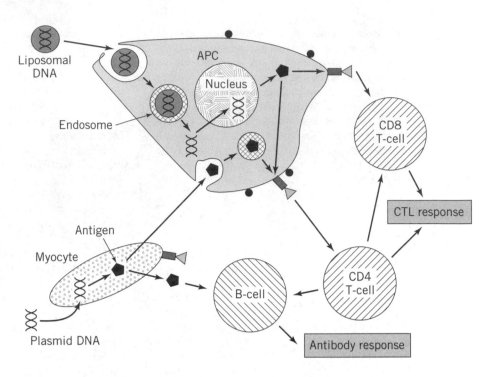

**Figure 10.** Schematic representation of proposed mechanisms of DNA immunization. Naked plasmid DNA is taken up by a small number of myocytes after intramuscular injection which are then transfected episomally. The produced antigen is released from the cells to interact with APC and thus induce immunity. In contrast, liposomal plasmid DNA interacts with APC directly. The way by which naked DNA enters the muscle cells to localize in the nucleus is unknown at present. For a proposed intracellular pathway of the liposomal DNA, see text. *Source:* Scheme designed by Dr. Yvonne Perrie, Centre for Drug Delivery Research.

liposome-entrapped and naked DNA indicate greater antibody (IgG$_1$) responses for the former not only by the intramuscular route but also after subcutaneous and intravenous injection. Interestingly, there was not much difference in the titers between the two strains, suggesting that immunization with liposomal pRc/CMV HBS is not MHC restricted. A similar pattern of results was obtained with IFN-$\gamma$ and IL-4 in the spleen (152). The ability of liposomes to mediate immune responses to entrapped, plasmid encoded antigens has recently been demonstrated with another plasmid encoding the *Mycobacterium leprae* heat shock protein 65 (158).

The way by which liposomal plasmid promotes immunity to the encoded antigen is not likely to involve muscle cells. Although cationic liposomes could in theory bind to the negatively charged myocytes and be taken up by them, protein in the interstitial fluid would neutralize the liposomal surface and thus be expected to interfere with such binding (152). Moreover, vesicle size (about 800 nm average diameter) would render access to the cells difficult if not impossible. It is more likely then that cationic liposomes are endocytosed by APC (see Fig. 10), possibly including dendritic cells. As discussed earlier, it appears that the key ingredient of the DNA liposomal formulations in enhancing immune responses is the cationic lipid. It is conceivable that some of the endocytosed DNA escapes the endocytic vacuoles prior to their fusion with lysosomes to enter the cytosol for eventual episomal transfection and presentation of the encoded antigen. It is perhaps at this stage of intracellular trafficking of DNA, spanning its putative escape from endosomes and access to their nucleus, that the cationic lipid (possibly together within the fusogenic phosphatidylethanolamine component), plays a significant but yet unravelled role (152). However, the question still remains as to why liposome-entrapped DNA is more efficient in inducing immunity to the encoded antigen than complexed DNA.

## Other Applications

Various other applications for liposomes have been explored or actively pursued but due to lack of space have not been discussed earlier. Among these, their use as carriers for radioisotopes or contrast agents for use in diagnostic imaging has received considerable attention over many years (159–161). A range of different imaging modalities have been used, the major ones being gamma scintigraphy ($\gamma$ S), computed tomography (CT), and magnetic resonance (MR), the basis behind all of these being to measure a signal difference between the target (e.g., a tumor) and the surrounding tissues. Carriers such as liposomes are used to amplify the signal differential, i.e., to apply contrast, with the contrast agents used for $\gamma$ S, CT, and MR being $\gamma$-emitting radionuclides, radiopaque materials, and paramagnetic metal ions, respectively. Imaging can be approached either by applying the contrast to the target or to the background, and the best results to date using liposomal systems have used the second approach, based on the fact that tumor cells generally have reduced phagocytic function compared with surrounding tissues. Thus the latter has an increased inclination to accumulate particle (liposome)-associated contrast and tumors register as holes within this background. Caution must be exercised due to the fact that areas of inflammation will also give a positive signal, and in fact good results have been obtained for imaging of arthritic areas. Because liposomes have a natural tendency to be cleared by the RES, they have been widely used for imaging hepatosplenic tumors and metastases and may be the systems of choice for CT and MR imaging in that application (159). However for $\gamma$ S, they are not significantly better than conventional colloidal systems, and their main promise is for imaging of non-RES tumors. Application of long-circulating, sterically stabilized liposomes may broaden the scope for tumor imaging due to their increased ability to extravasate from the tumor vasculature, and further developments in targeting will give additional potential. They are also useful for imaging the blood-pool, in order to detect vascular abnormalities such as atherosclerotic lesions, thrombi, tumors, and intravascular shunts (159). Another imaging technique that is giving promising results is ultrasound.

The first, large-scale commercial application for liposomes was in the field of cosmetics, and that industry remains by far the most prolific user, with over 200 different liposome-based products (161). These include a multitude of product types, ranging from humectants and tanning agents to hair restorers and antiaging products. Active components may be in the aqueous phase, including moisturizers, vitamins, plant extracts, and hydrolyzed proteins, or in the membrane bilayer, including lecithin itself, $\alpha$-tocopherol, retinoids, and carotenoids. In general, intact liposomes do not penetrate beyond the outermost layers of the stratum corneum, though some of their constituents may reach more deeply, by mechanisms which are as yet unclear. Notable exceptions are *transfersomes*, which seem to allow enhanced penetration and have been shown to permit entry of encapsulated peptide drugs into the systemic circulation (162).

It has been suggested that the various aspects of the drug delivery approach, in other words, increased stability of the encapsulant, passive (and active) targeting to different phases, and delayed and/or controlled release, can be used to enhance the activity of functional food ingredients in complex foods (163,164). For example, food enzymes could be used in a more selective and efficient manner, antioxidant vitamins could be used more effectively and various synthetic food additives (e.g., antimicrobial agents) could be replaced by safer naturally occurring ingredients (e.g., nisin or egg white lysozyme). As an illustration, hard cheeses, which can take over a year to ripen, can be produced in a fraction of the normal time using liposome-encapsulated enzymes, with retention of their normal properties and without the damaging consequences of the alternative approaches (165). This could potentially lead to considerable savings in terms of the costs of prolonged, controlled temperature storage and various other risk factors. Unfortunately, raw material and process costs currently outweigh these savings. In addition to the applications already discussed, many others have been developed and undoubtedly many more will emerge in the future. Experience to date suggests that some at least will have a major impact.

## BIBLIOGRAPHY

1. A.D. Bangham, M.M. Standish, and J.C. Watkins, *J. Mol. Biol.* **13**, 238–252 (1965).
2. G. Gregoriadis, P.D. Leathwood, and B.E. Ryman, *FEBS Lett.* **14**, 95–99 (1971).
3. G. Gregoriadis and B.E. Ryman, *Eur. J. Biochem.* **24**, 485–491 (1972).
4. G. Gregoriadis and B.E. Ryman, *Biochem. J.* **129**, 123–133 (1972).
5. G. Gregoriadis, *Trends Biotechnol.* **13**, 527–537 (1995).
6. G. Gregoriadis, *FEBS Lett.* **36**, 292–296 (1973).
7. J.N. Israelachvili, D.J. Mitchell, and B.W. Ninham, *Biochim. Biophys. Acta* **470**, 185–201 (1977).
8. G. Gregoriadis, ed. *Liposome Technology*, Vols. I, II, III, CRC Press, Boca Raton, Fla., 1984 (1st ed.), 1993 (2nd ed.).
9. C.R. Alving et al., *Proc. Natl. Acad. Sci. U.S.A.* **75**, 2959–2963 (1978).
10. Y.E. Rahman, in G. Gregoriadis, ed., *Liposomes as Drug Carriers*, Wiley, Winchester, 1988, pp. 485–495.
11. J.P. Reeves and R.M. Dowben, *J. Cell Physiol.* **73**, 49–60 (1969).
12. S. Kim et al., *Biochim. Biophys. Acta* **728**, 338–348 (1983).
13. S.M. Gruner, R.P. Lenk, A.S. Janoff, and M.J. Ostro, *Biochemistry* **24**, 2833–2843 (1985).
14. M.C. Woodle and D. Papahadjopoulos, *Methods Enzymol.*, **171**, 193–217 (1989).
15. R.R.C. New, *Liposomes, a Practical Approach*, Oxford IRL Press, New York, 1990.
16. U.S. Pat. 4,311,712 (Jan. 19, 1982), J.R. Evans, F.J.T. Fildes and J.E. Oliver (to Imperial Chemical Industries, Ltd.).
17. C.J. Kirby and G. Gregoriadis, *Bio/Technology* **2**, 979–984 (1984).
18. C.J. Kirby and G. Gregoriadis, in G. Gregoriadis, ed., *Liposome Technology*, 2nd ed., vol. 1, CRC Press, Boca Raton, Fla., 1984, pp. 19–27.
19. G. Gregoriadis et al., in J.E. Celis, ed., *Cell Biology: A Laboratory Handbook*, 2nd ed., vol. 4, Academic Press, Orlando, Fla., 1998, pp. 131–140.
20. C.J. Kirby et al., *Int. J. Food Sci. Technol.*, **26**, 437–449 (1991).
21. B. McCormack and G. Gregoriadis, *Biochim. Biophys. Acta* **1291**, 237–244 (1996).
22. G. Gregoriadis, H. da Silva, and A.T. Florence, *Int. J. Pharm.* **65**, 235–242 (1990).
23. S. Batzi and E.D. Korn, *Biochim. Biophys. Acta* **298**, 1015–1019 (1973).
24. C. Pidgeon, A.H. Hunt, and K. Dittrich, *Pharm. Res.* **3**, 23–34 (1986).
25. C.H. Huang, *Biochemistry* **8**, 344–352 (1969).
26. R.L. Hamilton et al., *J. Lipid Res.* **21**, 981–992 (1980).
27. E. Mayhew et al., *Biochim. Biophys. Acta* **775**, 169–174 (1984).
28. M.M. Brandl, D. Bachmann, M. Drechsler, and K.H. Bauer, in G. Gregoriadis, ed., *Liposome Technology*, 2nd ed., vol. 1, CRC Press, Boca Raton, Fla., 1992, pp. 49–65.
29. F. Olson et al., *Biochim. Biophys. Acta* **601**, 559–571 (1980).
30. R.W. Kriftner, in O. Braun-Falco, H.C. Korting, and H.I. Maibach, eds., *Liposome Dermatics*, Springer-Verlag, Berlin, 1992, pp. 91–100.
31. H. Hauser and N. Gains, *Proc. Natl. Acad. Sci. U.S.A.* **79**, 1683–1687 (1982).
32. Y. Kagawa and E. Racker, *J. Biol. Chem.* **246**, 5477–5487 (1971).
33. H.G. Weder and O. Zumbuehl, in G. Gregoriadis, ed., *Liposome Technology*, 1st ed., vol. 1, CRC Press, Boca Raton, Fla., 1983, pp. 79–107.
34. N. Oku and R.C. MacDonald, *Biochemistry* **22**, 855–863 (1983).

35. D.W. Deamer and A.D. Bangham, *Biochim. Biophys. Acta* **443**, 629–634 (1976).

36. F. Szoka and D. Papahadjopoulos, *Proc. Natl. Acad. Sci. U.S.A.* **75**, 4194 (1978).

37. S. Kim and G.M. Martin, *Biochim. Biophys. Acta* **646**, 1–9 (1981).

38. U. Pick, *Arch. Biochem. Biophys.* **212**, 186–194 (1981).

39. D. Papahadjopoulos, W.J. Vail, K. Jacobson, and G. Poste, *Biochim. Biophys. Acta* **394**, 483–491 (1975).

40. F.W. Martin, in P. Tyle, ed., *Specialized Drug Delivery Systems, Manufacturing and Production Technology*, Dekker, New York, 1990, pp. 267–316.

41. S. Amselem, A. Gabizon, and Y. Barenholz, in G. Gregoriadis ed., *Liposome Technology*, 2nd ed., vol. 1, CRC Press, Boca Raton, Fla., 1992, pp. 501–525.

42. E.C.A. van Winden, N.Z. Zuidam, and D.J.A. Crommelin, in D.D. Lasic and D. Papahadjopoulos, eds., *Medical Applications of Liposomes*, Elsevier, Amsterdam, 1998, pp. 567–604.

43. R.A. Schwendener, in G. Gregoriadis, ed., *Liposome Technology*, 2nd ed., vol. 1, CRC Press, Boca Raton, Fla., 1992, pp. 487–500.

44. H.G. Weder, in O. Braun-Falco, H.C. Korting, and H.I. Maibach, eds., *Liposome Dermatics*, Springer-Verlag, Berlin, 1992, pp. 101–109.

45. P. Van Hoogevest and P. Fankhauser, in G. Lopez-Berestein and I.J. Fidler, eds., *Liposomes in the Therapy of Infectious Diseases and Cancer*, Liss, New York, 1989, pp. 453–466.

46. European Pat. EP 198765-A (April 3, 1986), A. Rahman (to Georgetown University).

47. D.W. Deamer, R.C. Prince, and A.R. Crofts, *Biochim. Biophys. Acta* **274**, 323–335 (1972).

48. T.D. Madden et al., *Chem. Phys. Lipids* **53**, 37–46 (1990).

49. M.J. Hope, M.B. Bally, G. Webb, and P.R. Cullis, *Biochim. Biophys. Acta* **812**, 55–65 (1985).

50. T. Schneider, A. Sachse, G. Rossling, and M. Brandl, *Drug Dev. Ind. Pharm.* **20**, 2787–2807 (1994).

51. N.J. Zuidam, H. Talsma, and D.J.A. Crommelin, in Y. Barenholz and D.D. Lasic, eds., *Handbook of Non-medical Applications of Liposomes*, vol. 3, CRC Press, Boca Raton, Fla., 1995, pp. 71–80.

52. Y. Barenholz and S. Anselem, in G. Gregoriadis, ed., *Liposome Technology*, 2nd ed., vol. 1, CRC Press, Boca Raton, Fla., 1993, pp. 527–616.

53. D. Lichtenberg and Y. Barenholz, *Methods Biochem. Anal.*, **33**, 337–462 (1988).

54. H. Ruf, Y. Georgolis, and E. Grell, *Methods Enzymol.* **172**, 364–390 (1989).

55. S.W. Provencher, *Macromol. Chem.* **180**, 201–207 (1979).

56. Coulter ®Counter—*Medical and Biological Bibliography*, Coulter Electronics Ltd., Luton, England, 1990.

57. J. Boustra, G.S. Gooris, J.A. Van der Spek, and W. Bras, *Invest. Dermatol.* **97**, 1005–1012 (1991).

58. N. Skalko et al., *Biochim. Biophys. Acta* **1370**, 151–160 (1998).

59. G. Gregoriadis, N. Garcon, H. da Silva, and S. Sternberg, *Biochim. Biophys. Acta* **1147**, 185–193 (1993).

60. D.D. Lasic, *Liposomes: From Physics to Applications*, Elsevier, Amsterdam, 1973.

61. J. Senior and G. Gregoriadis, in G. Gregoriadis, ed., *Liposome Technology*, 1st ed., vol. 3, CRC Press, Boca Raton, Fla., 1984, pp. 262–282.

62. G. Gregoriadis and R. Buckland, *Nature (London)*, **244**, 170–172 (1973).

63. G. Gregoriadis, *N. Engl. J. Med.* **295**, 704–710, 765–770 (1976).

64. D. Papahadjopoulos, ed., *Liposomes and Their Uses in Biology and Medicine*, vol. 308, N.Y. Acad. Sci., New York, 1978.

65. G. Gregoriadis, ed., *Liposomes as Drug Carriers: Recent Trends and Progress*, Wiley, Chichester, U.K., 1988.

66. C.D.V. Black and G. Gregoriadis, *Biochem. Soc. Trans.* **2**, 869–871 (1974).

67. A.W. Segal et al., *Br. J. Exp. Pathol.* **55**, 320–327 (1974).

68. G. Gregoriadis and D. Neerunjun, *Eur. J. Biochem.* **47**, 179–185 (1974).

69. R.L. Juliano and B. Stamp, *Biochem. Biophys. Res. Commun.* **63**, 651–658 (1975).

70. P.E. Belchetz, I.P. Braidman, J.C.W. Crawley, and G. Gregoriadis, *Lancet* **2**, 116–117 (1977).

71. G. Gregoriadis and D. Neerunjun, *Res. Commun. Chem. Pathol.* **10**, 351–362 (1975).

72. A.C. Allison and G. Gregoriadis, *Nature (London)* **252**, 252 (1974).

73. G. Gregoriadis and D. Neerunjun, *Biochem. Biophys. Res. Commun.* **65**, 537–544 (1975).

74. D. Neerunjun, G. Gregoriadis, and R. Hunt, *Life Sci.* **21**, 357–370 (1977).

75. L. Krupp, A.V. Chobanian, and J.P. Brecher, *Biochem. Soc. Trans.* **17**, 590 (1989).

76. G. Scherphof, G. Roerdink, M. Waite, and J. Parks, *Biochim. Biophys. Acta* **542**, 296–307 (1978).

77. G. Gregoriadis and C. Davis, *Biochem. Biophys. Res. Commun.* **89**, 1287–1293 (1979).

78. C.J. Kirby, J. Clarke, and G. Gregoriadis, *Biochem. J.* **186**, 591–598 (1980).

79. C.J. Kirby and G. Gregoriadis, *Life Sci.* **27**, 2223–2230 (1980).

80. C.J. Kirby, J. Clarke, and G. Gregoriadis, *FEBS. Lett.* **111**, 324–328 (1980).

81. A. Tumer, C.J. Kirby, J. Senior, and G. Gregoriadis, *Biochim. Biophys. Acta* **760**, 119–125 (1983).

82. J. Senior, G. Gregoriadis, and K. Mitropoulos, *Biochim. Biophys. Acta* **760**, 111–118 (1983).

83. G. Gregoriadis and J. Senior, *FEBS Lett.* **119**, 43–46 (1980).

84. J. Senior and G. Gregoriadis, *FEBS Lett.* **145**, 109–114 (1982).

85. J. Senior, J.C.W. Crawley, and G. Gregoriadis, *Biochim. Biophys. Acta* **839**, 1–8 (1985).

86. B. Wolff and G. Gregoriadis, *Biochim. Biophys. Acta* **802**, 259–273 (1984).

87. G. Gregoriadis, *News Physiol. Sci.* **4**, 146–151 (1989).

88. D. Absolom, *Methods Enzymol.* **132**, 281–318 (1986).

89. C.D.V. Black and G. Gregoriadis, *Biochem. Soc. Trans.* **4**, 253–256 (1976).

90. F. Bonte and R. Juliano, *Chem. Phys. Lipids* **40**, 359 (1986).

91. H.C. Loughrey, M.B. Bally, L.W. Reinish, and P.R. Cullis, *Thromb. Haemostasis* **64**, 172–176 (1990).

92. R.L. Richards, H. Gewurtz, J. Siegel, and C.R. Alving, *J. Immunol.* **112**, 1185–1189 (1979).

93. J.D. Rossi and B.A. Wallace, *J. Biol. Chem.* **258**, 3327–3331 (1983).

94. S.M. Moghimi and H.M. Patel, in G. Gregoriadis, ed., *Liposome Technology*, 2nd ed., vol. 3, CRC Press, Boca Raton, Fla., 1993, pp. 44–58.

95. G. Gregoriadis and B. McCormack, eds., *Targeting of Drugs: Strategies for Stealth Therapeutic Systems*, Plenum, New York, 1998.

96. A.L. Klibanov, K. Maruyama, V.P. Torchilin, and L. Huang, *FEBS Lett.* **268**, 235–237 (1990).

97. R. Juliano and B. Stamp, *Biochem. Pharmacol.* **27**, 21–27 (1978).

98. C.J. Kirby and G. Gregoriadis, *Biochem. Pharmacol.* **32**, 609–615 (1983).

99. L. Huang and S.J. Kennel, *Biochemistry* **18**, 1702–1707 (1979).

100. J.N. Weinstein et al., *Science* **195**, 489–491 (1977).

101. L.D. Leserman, J.N. Weinstein, R. Blumenthal, and W.D. Terry, *Proc. Natl. Acad. Sci. U.S.A.* **77**, 4089–4093 (1980).

102. V.P. Torchilin, B.A. Khaw, V.N. Smirnov, and E. Haber, *Biochem. Biophys. Res. Commun.* **89**, 114–119 (1979).

103. T.D. Heath, R.T. Fraley, and D. Papahadjopoulos, *Science* **210**, 539–541 (1980).

104. A. Huang, L. Huang, and S.J. Kennel, *J. Biol. Chem.* **255**, 8015–8018 (1980).

105. L.D. Leserman, J. Barbet, F. Kourilsky, and J.N. Weinstein, *Nature (London)* **288**, 602–604 (1980).

106. F.J. Martin, W.L. Hubbell, and D. Papahadjopoulos, *Biochemistry* **20**, 4229–4438 (1981).

107. F.J. Martin and D. Papahadjopoulos, *J. Biol. Chem.* **257**, 286–288 (1982).

108. H.C. Loughrey, M.B. Bally, and P.R. Cullis, *Biochim. Biophys. Acta* **901**, 157–160 (1987).

109. J.N. Weinstein, R. Blumenthal, S.O. Sharrow, and P.A. Henkart, *Biochim. Biophys. Acta* **509**, 272–288 (1978).

110. F. Bonte, in F. Puisieux, P. Couvreur, J. Delattre, and J.-P. Devissaguet, eds., *Liposomes, New Systems and New Trends in Their Applications*, Editions de Santé, Paris, 1995, pp. 75–97.

111. M.B. Yatvin, W. Kreutz, B.A. Horwitz, and M. Shinitzky, *Science* **210**, 1253–1254 (1980).

112. R.M. Straubinger, K. Hong, D.S. Friend, and D. Papahadjopoulos, *Cell (Cambridge, Mass.)* **32**, 1069–1079 (1983).

113. D. Collins, D. Litzinger, and L. Huang, *Biochim. Biophys. Acta* **1025**, 234–242 (1990).

114. J.N. Weinstein, R.L. Magin, R.L. Cysyk, and D.S. Zaharko, *Cancer Res.* **40**, 1388–1395 (1980).

115. P.L. Felgner and G.M. Ringold, *Nature (London)* **337**, 387–388 (1989).

116. B. Sternberg, F.L. Sorgi, and L. Huang, *FEBS Lett.* **356**, 361–366 (1994).

117. S. Bhattacharya and L. Huang, in D.D. Lasic and D. Papahadjopoulos, eds., *Medical Applications of Liposomes*, Elsevier, Amsterdam, 1998, pp. 371–394.

118. C.J. Kirby, in R.A. Meyers, ed., *Encyclopaedia of Molecular Biology and Molecular Medicine*, VCH, Weinheim, 1996, pp. 415–425.

119. X. Gao and L. Huang, *Biochem. Biophys. Res. Commun.* **179**, 280–285 (1991).

120. J.P. Behr, *Tetrahedron Lett.* **27**, 5861–5864 (1986).

121. F.L. Sorgi, S. Bhattacharya, and L. Huang, *Gene Ther.* **4**, 961–968 (1997).

122. S. Li and L. Huang, *Gene Ther.* **4**, 891–900 (1997).

123. A.A. Gabizon, in F. Puisieux, P. Couvreur, J. Delattre, and J.-P. Devissaguet, eds., *Liposomes, New Systems and New Trends in their Applications*, Editions de Santé, Paris, 1995, pp. 507–522.

124. D.D. Lasic and F. Martin, eds., *Stealth Liposomes*, CRC Press, Boca Raton, Fla., 1995.

125. D.D. Lasic and D. Papahadjopoulos, eds., *Medical Applications of Liposomes*, Elsevier, Amsterdam, 1998.

126. C.E. Swenson, J. Freitag, and A.S. Janoff, in D.D. Lasic and D. Papahadjopoulos, eds., *Medical Applications of Liposomes*, Elsevier, Amsterdam, 1998, pp. 689–702.

127. P.G. Schmidt, J.P. Adler-Moore, E.A. Forssen, and R. Profitt, in D.D. Lasic and D. Papahadjopoulos, eds., *Medical Applications of Liposomes*, Elsevier, Amsterdam, 1998, pp. 703–732.

128. F.J. Martin, in D.D. Lasic and D. Papahadjopoulos, eds., *Medical Applications of Liposomes*, Elsevier, Amsterdam, 1998, pp. 635–688.

129. L.D. Mayer, P.R. Cullis, and M.B. Bally, in D.D. Lasic and D. Papahadjopoulos, eds., *Medical Applications of Liposomes*, Elsevier, Amsterdam, 1998, pp. 231–258.

130. D. Kirpotin et al., *FEBS Lett.* **388**, 115–118 (1996).

131. T.M. Allen, C.B. Hansen, and D.D. Stuart, in D.D. Lasic and D. Papahadjopoulos, eds., *Medical Applications of Liposomes*, Elsevier, Amsterdam, 1998, pp. 297–323.

132. T.M. Allen, I. Ahmad, D.E. Lopes de Menezes, and E.H. Moase, *Biochem. Soc. Trans.* **23**, 1073–1079 (1995).

133. D.B. Kirpotin et al., in D.D. Lasic and D. Papahadjopoulos, eds., *Medical Applications of Liposomes*, Elsevier, Amsterdam, 1998, pp. 325–345.

134. J.W. Park et al., *Cancer Lett.*, pp. 153–160 (1997).

135. L.L. Worth, I.J. Fidler, and E.S. Kleinermann, in D.D. Lasic and D. Papahadjopoulos, eds., *Medical Applications of Liposomes*, Elsevier, Amsterdam, 1998, pp. 47–60.

136. I.A.J.M. Bakker-Woudenberg, in F. Puisieux, P. Couvreur, J. Delattre, and J-P. Devissaguet, eds., *Liposomes, New Systems and New Trends in their Applications*, Editions de Santé, Paris, 1995, pp. 371–396

137. K.M. Wassan and G. Lopez-Berestein, in D.D. Lasic and D. Papahadjopoulos, eds., *Medical Applications of Liposomes*, Elsevier, Amsterdam, 1998, pp. 181–188.

138. P.K. Working, in D.D. Lasic and D. Papahadjopoulos, eds., *Medical Applications of Liposomes*, Elsevier, Amsterdam, 1998, pp. 605–624.

139. C. Nicolau et al., *Proc. Natl. Acad. Sci. U.S.A.* **80**, 1068–1072 (1983).

140. C.E. Holt, N. Garlick, and E. Cornel, *Neuron* **4**, 203–207 (1990).

141. G.E. Plautz et al., *Proc. Natl. Acad. Sci. U.S.A.* **90**, 4645 (1993).

142. G.J. Nabel et al., *Proc. Natl. Acad. Sci. U.S.A.* **90**, 11307–11311 (1993).

143. E. Nabel, G. Plautz, and G. Nabel, *Science* **249**, 285–1288 (1990).

144. N.J. Caplen et al., *Nat. Med.* **1**(1), 39–46 (1995).

145. G. Gregoriadis, B. McCormack, and A.C. Allison, eds., *New Generation Vaccines: The Role of Basic Immunology*, Plenum, New York, 1993.

146. R. Bomford, in R. Spier and B. Griffiths, eds., *Animal Cell Biotechnology*, vol. 2, Academic Press, London, p. 235.

147. G. Gregoriadis, *Immunol. Today* **11**, 89–97 (1990).

148. C.R. Alving, *J. Immunol. Methods* **140**, 1–13 (1991).

149. R. Gluck, in M.F. Powell and M.J. Newman, eds., *Vaccine Design: The Subunit and Adjuvant Approach*, Plenum, New York, p. 325.

150. M. Chattergoon, J. Boyer, and D.B. Weiner, *FASEB J.* **11**, 754–763 (1997).

151. H.L. Davis, R.G. Whalen, and B.A. Demeneix, *Hum. Gene Ther.* **4**, 151–159 (1993).

152. G. Gregoriadis, *Pharm. Res.* **15**, 661–670 (1998).

153. F.C. Szoka, Y. Xu, and O. Zelpati, *J. Liposome Res.* **6**, 567–587 (1996).

154. M. Sedegah, R. Hedstrom, P. Hobart, and and S.L. Hoffman, *Proc. Natl. Acad. Sci. U.S.A.* **91**, 9866 (1984).

155. G. Gregoriadis, R. Saffie, and B. De Souza, *FEBS. Lett.* **402**, 107–110 (1997).

156. G. Gregoriadis, B. McCormack, Y. Perrie, and R. Saffie, in D.D. Lasic and D. Papahadjopoulos, eds., *Medical Applications of Liposomes*, Elsevier, Amsterdam, 1998, pp. 61–73.

157. Y. Perrie and G. Gregoriadis, *J. Pharm. Pharmacol.* **50**, 155 (1998).

158. G. Gregoriadis and B. McCormack, *J. Liposome Res.* **8**, 48 (1998).

159. C. Tilcock, in J.R. Phillipot and F. Schuber, eds., *Liposomes as Tools in Basic Research and Industry*, CRC Press, Boca Raton, Fla., 1995, pp. 225–240.

160. V. Torchilin, in D.D. Lasic and D. Papahadjopoulos, eds., *Medical Applications of Liposomes*, Elsevier, Amsterdam, 1998, pp. 515–543.

161. D. Lasic, in D.D. Lasic, ed., *Liposomes, from Physics to Applications*, Elsevier, Amsterdam, 1993, pp. 477–488.

162. G. Cevc and G. Blume, *Biochim. Biophys. Acta* **1104**, 226–232 (1992).

163. C.J. Kirby, *Chem. Br.* **26**, 847–850 (1990).

164. C.J. Kirby, in G. Gregoriadis, *Liposome Technology*, 2nd ed., vol. 2, CRC Press, Boca Raton, Fla., 1993, pp. 216–231.

165. C.J. Kirby, B.E. Brooker, and B.A. Law, *Int. J. Food Sci. Technol.* **22**, 355–363 (1987).